Vol. 30 **Yearbook** **1976**

Synthetic Methods
of Organic Chemistry

With Reaction Titles and Cumulative Index
of Volumes 26–30

Synthetische Methoden
der Organischen Chemie

Jahrbuch mit deutschem Registerschlüssel
Mit Titeln und Generalregister der Bände 26–30

Editor	**William Theilheimer,** Ph. D., Nutley
Technical Editor	Bernhard Prijs, Ph. D., Basel
Editorial Consultant	John T. Plati, Ph. D., Nutley
	Karoly Kocsis, Ph. D., Basel
	Helmut Sigel, Ph. D., Basel
	John W. Scott, Ph. D., Nutley
Editorial Secretary	Helen Makus, Nutley
Production Manager	Denise Greder, Basel

S. Karger
Basel · München · Paris · London
New York · Sydney

Deutsche Ausgaben

Vol. 1	1946	1. Auflage
	1948	2. unveränderte Auflage
	1950	3. unveränderte Auflage
Vol. 2	1948	
Vol. 3	1949	with English Index Key
	1953	2. unveränderte Auflage
	1966	3. unveränderte Auflage
	1975	4. unveränderte Auflage
Vol. 4	1950	with English Index Key
	1966	2. unveränderte Auflage

English Editions

(ab Vol. 5: deutscher Registerschlüssel)

Vol. 1	1948	Interscience Publishers
	1975	(Karger) Second Edition
Vol. 2	1949	Interscience Publishers
	1975	(Karger) Second Edition
Vol. 5	1951	with Reaction Titles Vol. 1–5 and Cumulative Index
	1966	Second Edition
Vol. 6	1952	
	1975	Second Edition
Vol. 7	1953	
	1975	Second Edition
Vol. 8	1954	
	1975	Second Edition
Vol. 9	1955	
Vol. 10	1956	with Reaction Titles Vol. 6–10 and Cumulative Index
	1975	Second Edition
Vol. 11	1957	
	1975	Second Edition
Vol. 12	1958	
	1975	Second Edition
Vol. 13	1959	
	1975	Second Edition
Vol. 14	1960	
	1975	Second Edition
Vol. 15	1961	with Reaction Titles Vol. 11–15 and Cumulative Index
Vol. 16	1962	
Vol. 17	1963	
Vol. 18	1964	
Vol. 19	1965	
Vol. 20	1966	with Reaction Titles Vol. 16–20 and Cumulative Index
Vol. 21	1967	
Vol. 22	1968	
Vol. 23	1969	
Vol. 24	1970	
Vol. 25	1971	with Reaction Titles Vol. 21–25 and Cumulative Index
Vol. 26	1972	
Vol. 27	1973	
Vol. 28	1974	
Vol. 29	1975	

Cataloging in Publication
Synthetic methods of organic chemistry; yearbook v. 30, 1976. Basel, New York, Karger
Editor: 1948, W. Theilheimer
Every fifth volume includes a 5-year index, 1951
1. Chemistry, Organic-yearbooks
I. Theilheimer, Wilhelm, ed.
W1 SY66K

ISBN 3-8055-2256-8

All rights, including that of translation into other languages, reserved.
Photomechanic reproduction (photocopy, microcopy) of this book or parts thereof without special permission of the publishers is prohibited.
©
Copyright 1976 by S. Karger AG, Basel (Switzerland), Arnold-Böcklin-Strasse 25
Printed in Switzerland by Friedrich Reinhardt AG, Basel
ISBN 3-8055-2256-8

Contents
Inhalt

	Vol. 29	Vol. 30
Preface to Volume 30		V
From the Prefaces to the Preceding Volumes		V
Method of Classification		VII
High-Coverage Searches		IX
Trends in Synthetic Organic Chemistry, 1976		X
Reactions		1
Subject Index		544
Formula Index of Functional Combinations	526	
Abbreviations	538	
Journal Abbreviations	539	
Symbols	539	
Systematic Survey		747
Reagents	544	
Index of Supplementary References		750
Vorwort zu Band 30		V
Systematik	IX	
Entwicklungstendenzen der synthetischen organischen Chemie, 1976		X
Reaktionen		1
Alphabetisches Register		545
Formelregister komplexer Funktionen	526	
Deutscher Schlüssel zum Register		745
Abkürzungen	538	
Zeitschriften-Abkürzungen	539	
Symbole	539	
Systematische Übersicht		747
Hilfsstoffe	544	
Tabelle der Ergänzungszitate		750

Preface

This is the final volume of the sixth series. It contains a cumulative subject index and all reaction titles for Volumes 26–30, including recent supplementary references. This arrangement again reduces a five-volume search to a search through a single volume. Most of the references in this volume are to papers published between 1973 and 1975. The recommendations for high-coverage searches have been revised (s. page IX).

To prevent an unwieldy expansion of this volume, repetitive parts have been held to a minimum. However, the omitted parts can be found in Vol. 29, and the respective page numbers are listed on the contents page of this volume. Supplementary references to preceding series will be included in Vol. 31.

I again wish to thank my collaborators listed on the title page for their valuable advice and assistance, and other members of Hoffmann-La Roche, Inc., Nutley, for their kind cooperation.

Nutley, New Jersey, U.S.A., May 1976.   W. Th.

From the Prefaces to the Preceding Volumes

New methods for the synthesis of organic compounds and improvements of known methods are being recorded continuously in this series.

Reactions are classified on a simple though purely formal basis by symbols, which can be arranged systematically. Thus searches can be performed without knowledge of the current trivial or author names (e.g., "Oxidation" and "Friedel-Crafts reaction").

Users accustomed to the common notations will find these in the subject index. By consulting this index, use of the classification system may be avoided. It is thought that the volumes should be kept close at hand. The books should provide a quick survey, and obviate the immediate need for an elaborate library search. Syntheses are therefore recorded in the index by starting materials and end products, along with the systematic arrangement for the methods. This makes possible a sub-classification within the reaction symbols by reagents, a further methodical criterion. Complex compounds are indexed with cross reference under the related simpler compounds. General terms, such as synthesis, replacement, heterocyclics, may also be brought to the attention of the reader.

A table that indicates the sequence of the reagents may help the reader to locate reactions in the body of the text. This table also contains such frequently used reagents as NaOH and HCl, not included in the subject index.

A brief review, *Trends in Synthetic Organic Chemistry,* stresses highlights of general interest and calls attention to developments too recent to be included in the body of the text.

The abstracts are limited to the information needed for an appraisal of the applicability of a desired synthesis. In order to carry out a particular synthesis it is therefore advisable to have recourse to the original papers or, at least, to an abstract journal. In order to avoid repetition, selections are made on the basis of most detailed description and best yields, whenever the same method is used in similar cases. Continuations of papers already included will not be abstracted, unless they contain essentially new information. They may, however, be quoted at the place corresponding to the abstracted papers. These supplementary references (see page 750) make it possible to keep abstracts of previous volumes up-to-date.

Syntheses that are divided into their various steps and recorded in different places can be followed with the help of the notations *startg. m. f.* (starting material for the preparation of . . .) and *prepn. s.* (preparation, see).

Method of Classification

The following directions serve to explain the system of Classification.

1. Reaction Symbols

The first part of the symbol refers to the chemical bonds formed during the reaction. These bonds appear in the reaction symbols as the symbols for the two elements that have been linked together (e.g., the bond between hydrogen and nitrogen, as HN). The order of the elements is as follows: H, O, N, Hal (Halogen), S, and Rem (the remaining elements). C is always placed last.

The "principle of the latest position" is used whenever possible.

The methods of obtaining a particular chemical bond are subdivided according to types of formation. Four types are distinguished: addition ($\Downarrow$), rearrangement ($\cap$), exchange ($\downarrow\uparrow$), and elimination ($\Uparrow$). The last part of the symbol refers to the bonds which are destroyed in the reaction or to a characteristic element which is eliminated.

The following simplifying stipulations facilitate the use of the reaction symbols: *(1)* The chemical bond is rigidly classified according to the structure formula without taking the reaction mechanism into consideration. *(2)* Double or triple bonds are treated as being equivalent to two or three single bonds, respectively. *(3)* Generally speaking, only stable organic compounds are taken into consideration. Intermediary compounds, such as Grignard compounds and sodiomalonic esters, and inorganic reactants, such as nitric acid, are therefore not expressed in the reaction symbols.

Examples: see volume II, page VIII.
Systematic Survey: see page 747.

2. Reagents

A further subdivision, not included in the reaction symbols, is made on the basis of the reagents characteristic of the reaction. A table indicating the sequence of the reagents may be found on page 544 of vol. 29.

3. The material between the listings of the reagents is arranged with the simple examples first and the more complicated ones following.

4. When changes in more than one chemical bond occur during a reaction, as, for example, in the formation of a new ring, or if the reaction can be carried out in different ways, these reactions are introduced in several places when necessary. The main entry in such cases is placed usually according to the "principle of the latest position"; the other entries are cross-referenced back to it.

High-Coverage Searches

A search through Synthetic Methods provides a selection of key references from the journal literature. For greater coverage, as for bibliographies, a supplementary search through the following publications is suggested.

Chemical Reactions Documentation Service[1]
which also includes abstracts from patents and provides the data coded on magnetic tape and punched cards as additional retrieval tools.

Science Citation Index[2]
for which Synthetic Methods serves as a source of starting references.

Chemical Abstract Service[3]

References may not be included in Synthetic Methods

1) to reactions which are routinely performed by well known procedures,
2) to subjects which can be easily located in handbooks and indexes of abstract journals, such as the ring system of heterocyclics or the metal in case of organometallic compounds, and
3) to inadequately described procedures, especially where yields are not indicated.

References to less accessible publications such as those in the Russian or Japanese language are, as a rule, only included if the method in question is not described elsewhere.

---

[1] Derwent Publications Ltd., 128 Theobalds Road, London WCIX 8RP, England.
[2] Institute for Scientific Information, Philadelphia, Pa., USA.
[3] Chemical Abstracts Service, Columbus, Ohio, USA.

Trends in Synthetic Organic Chemistry
1976

In contrast to reactions at high temperature those at very low temperature are more difficult to perform and have therefore been much less investigated. Recently, the utility of cryochemical processes has been demonstrated by the smooth preparation of cyclopropanone from ketene and diazomethane at −145°[1].

Epoxides and episulfides can be deoxygenated and desulfurated respectively with retention of stereochemistry by a new reagent, 3-methylbenzothiazole-2-selone[2]. Dimethylphenylsilyllithium has been used for the *trans*-stereospecific deoxygenation of epoxides[3].

Bridgehead functionalization can be conveniently achieved with lead tetracetate[4]. Regio- and stereo-specific vicinal oxyamination of olefins by alkyl imido osmium compounds[5] has been reported as a novel reaction. A new method for the oxidation of alcohols with peracids depends upon catalysis by N-oxide radicals[6]. Optically active labile sec. bromides can be conveniently prepared from chiral alcohols by cautious reaction with PBr_3 and pyridine[7].

gem-Dialkylations of carbanions can be easily performed in the presence of 1,8-diazabicyclo[5.4.0]undec-7-ene[8]. α-Cyanoenamines are new intermediates which can be used for various syntheses such as the preparation of α-diketones from amides[9]. The latter can be conveniently obtained from acids with BF_3 in the presence of triethyl-

---

[1] E. F. Rothgery, R. J. Holt, and H. A. McGee, Jr., Am. Soc. *97*, 4971 (1975).
[2] V. Calò et al., Synthesis *1976*, 200.
[3] M. T. Reetz and M. Plachky, Synthesis *1976*, 199.
[4] S. R. Jones and J. M. Mellor, Synthesis *1976*, 32.
[5] K. B. Sharpless et al., Am. Soc. *97*, 2305 (1975); cf. J. Org. Chem. *41*, 177 (1976).
[6] J. A. Cella, J. A. Kelley, and E. F. Kenehan, J. Org. Chem. *40*, 1860 (1975); cf. B. Ganem, ibid. *40*, 1998.
[7] R. O. Hutchins, D. Masilamani, and C. A. Maryanoff, J. Org. Chem. *41*, 1071 (1976); cf. D. Landini, S. Quici, and F. Rolla, Synthesis *1975*, 430.
[8] H. Oediger and F. Möller, A. *1976*, 348.
[9] J. Toye and L. Ghosez, Am. Soc. *97*, 2276 (1975).

amine[10]. α,β-Ethylene-β-chloroketones can be easily prepared by treating β-diketones or β-ketoaldehydes with oxalyl chloride in an inert solvent[11]. Of several new methods for the preparation of nitriles from aldehydes[12], a particularly simple one merely uses hydroxylamine hydrochloride in refluxing dimethylformamide[13]. Ar. thioamides can be obtained in one step by Friedel-Crafts thioacylation with ethoxycarbonyl isothiocyanate[14].

A convenient preparation of isothiocyanates from amines, carbon disulfide, and Grignard reagents, such as ethylmagnesium bromide, has been published[15]. Phenyl N-phenylphosphoramidochloridate is a new phosphorylation agent[16].

Acid chlorides have been found to add easily to carbon-nitrogen double bonds of heterocyclics, e.g. of Δ^1-azirines, to form N-acyl-2-chloraziridines[17].

Hydroxyl groups, even if tertiary[18], can be protected as methylthiomethyl ethers. The protective group can be electively removed under neutral conditions in the presence of mercuric ion[19]. β-Methoxyethoxymethyl, a new protective group for hydroxyl, can be removed with Lewis acids such as $TiCl_4$ or $ZnBr_2$ under mild conditions[20]. Protection of carboxyl groups as 5-methylene-1,3-dioxanes has been recommended. This protective group avoids the introduction of new asymmetrical centers, while amenable to selective removal under mild conditions[21]. Removal of certain O-protective groups by radical anions can be performed under a series of controlled conditions to make it suitably preferential and selective[22].

The preparation of ketones by ozonization of sec. alcohols is recommended with certain limitations[23]. A convenient preparation of

[10] J. Tani, T. Oine, and I. Inoue, Synthesis *1975*, 714.
[11] R. D. Clark and C. H. Heathcock, J. Org. Chem. *41*, 636 (1976).
[12] Synth. Meth. *30*, 277.
[13] J. Liebscher and H. Hartmann, Z. Chem. *15*, 302 (1975).
[14] E. P. Papadopoulos, J. Org. Chem. *41*, 962 (1976).
[15] S. Sakai, T. Fujinami, and T. Aizawa, Bull. Chem. Soc. Japan *48*, 2981 (1975).
[16] W. S. Zielinski and Z. Leśnikowski, Synthesis *1976*, 185.
[17] A. Hassner, S. S. Burke, and J. Cheng-fan I, Am. Soc. *97*, 4692 (1975); cf. B. T. Golding and D. R. Hall, Soc. Perkin I *1975*, 1302.
[18] K. Yamada et al., Tetrah. Let. *1976*, 65.
[19] E. J. Corey and M. G. Bock, Tetrah. Let. *1975*, 2643, 3269.
[20] E. J. Corey, J.-L. Gras, and P. Ulrich, Tetrah. Let. *1976*, 809.
[21] E. J. Corey and J. W. Suggs, Tetrah. Let. *1975*, 3775.
[22] R. L. Letsinger and J. L. Finnan, Am. Soc. *97*, 7197 (1975).
[23] W. L. Waters et al., J. Org. Chem. *41*, 889 (1976); by dry ozonization cf. ref. 52.

3-arylated aldehydes and ketones from allylic alcohols has been published[24]. Tris(triphenylsilyl)vanadate is an excellent catalyst for the Meyer-Schuster rearrangement; α,β-unsadt. steroidal aldehydes have thus been produced from the corresponding ethynylcarbinols in high yield[25]. Novel and mild procedures for synthetically useful interchanges of acetal, thioacetal, and hemithioacetal groups have been described[26]. The preparation of glycosides, including di- and pseudo-saccharides, under mild conditions with amide acetals has been reported[27].

A regiospecific Baeyer-Villiger oxidation with ceric ion converts cyclic ketones into lactones, which may be different from those obtained by conventional peracid oxidation[28].

A simple total synthesis of prostaglandins uses monomeric formaldehyde as trapping agent for kinetic enolates[29]. α-Lithiated N,N-dimethylhydrazones have been successfully used as enolate equivalents[30]. Reactions via organocesium compounds have been published[31]. Protonation of lithium alkynyltrialkylborates with acid can be directed to achieve a double migration of alkyl groups from boron to carbon[32].

High yields of linear esters can be obtained by carbonylation of α-olefins in the presence of homogeneous platinum complexes and a Group IVB metal halide, such as $SnCl_2$, as co-catalyst[33]. Cuprous methyltrialkylborates are convenient intermediates for the synthesis of saturated nitriles from acrylonitrile, esters from acrylates, as well as trans-γ,δ-ethyleneketones from 1-acyl-2-vinylcyclopropanes[34]. Cyclohexyl- and β-phenethyl-trihalogenomethylmercury compounds are carbene transfer agents effective at room temperature[35]. β-Stannylalkylidenephosphoranes have been introduced as promising novel intermediates[36]. Carbonylchromium complexes may be used to increase reactivity, enhance selectivity, or protect substituents of arene rings. The activating power of the carbonylchromium group can be modified

---

[24] J. B. Melpolder and R. F. Heck, J. Org. Chem. *41*, 265 (1976); A. J. Chalk and S. A. Magennis, ibid. *41*, 273.
[25] G. L. Olson, K. D. Morgan, and G. Saucy, Synthesis *1976*, 25.
[26] E. J. Corey and T. Hase, Tetrah. Let. *1975*, 3267.
[27] S. Hanessian and J. Banoub, Tetrah. Let. *1976*, 657, 661.
[28] G. Mehta, P. N. Pandey, and T.-L. Ho, J. Org. Chem. *41*, 953 (1976).
[29] G. Stork and M. Isobe, Am. Soc. *97*, 6260 (1975); cf. ibid. *98*, 1583 (1976).
[30] E. J. Corey and D. Enders, Tetrah. Let. *1976*, 3, 11.
[31] N. Collignon, J. Organometal. Chem. *96*, 139 (1975); Bl. *1975*, 1821.
[32] M. M. Midland and H. C. Brown, J. Org. Chem. *40*, 2845 (1975).
[33] J. F. Knifton, J. Org. Chem. *41*, 793 (1976).
[34] N. Miyaura, M. Itoh, and A. Suzuki, Tetrah. Let. *1976*, 255.
[35] D. Seyferth, C. K. Haas, and D. Dagani, J. Organometal. Chem. *104*, 9 (1976).
[36] S. J. Hannon and T. G. Traylor, Chem. Commun. *1975*, 630.

by replacing one of the CO-groups by other ligands[37]. The complexes are also useful for the preparation of optically active compounds[38].

Both the increased yield and lower reaction temperature of anionic oxy-Cope processes should improve the synthetic utility of these and related rearrangements[39].

An aldol-type ring closure of steroid intermediates with high asymmetrical selectivity has been achieved in the presence of L-phenylalanine[40]. A highly stereoselective cyclopentene ring annelation has been published[41]. An efficient double cycloisomerization of dienic keto esters in the presence of stannic chloride has been reported, affording a direct entry into functionalized decalins[42].

A facile chroman ring closure is the starting point of a new synthesis of vitamine E and related compounds[43]. An efficient synthesis of *exo-* and *endo-*brevicomin, bicyclic ketals, has been reported[44]. A one-step 6,7-benzomorphan ring synthesis has been achieved by *m*-bridging of electron-deficient aromatics[45]. Uracils substituted in 5-position by a carbon chain can be obtained from 5-fluorouracil via a novel 1,4-fragmentation of a regioselectively 4,5-fused cyclobutane ring[46].

d-Biotin has been obtained by a stereospecific total synthesis from L-(+)-cysteine without a chemical resolution series characteristic of all previous syntheses. Part of this process is a remarkable rearrangement of a 4-vinylthiazolidine ring by bromination resulting in a 3-amino-4-bromotetrahydrothiophene ring[47].

The formation of undesirable stable emulsions in phase-transfer catalyzed reactions can be avoided by triphase catalysis with a polymer-based quaternary ammonium salt as catalyst, which can be removed by a simple filtration[48].

[37] G. Jaouen, A. Meyer, and G. Simonneaux, Chem. Commun. *1975*, 813.
[38] Synth. Meth. *30*, 540.
[39] D. A. Evans and A. M. Golob, Am. Soc. *97*, 4765 (1975).
[40] S. Danishefsky and P. Cain, Am. Soc. *97*, 5282 (1975).
[41] B. M. Trost and D. E. Keeley, Am. Soc. *98*, 248 (1976).
[42] R. W. Skeean, G. L. Trammell, and J. D. White, Tetrah. Let. *1976*, 525.
[43] J. W. Scott et al., Helv. *59*, 290 (1976).
[44] P. J. Kocienski and R. W. Ostrow, J. Org. Chem. *41*, 398 (1976).
[45] R. R. Bard and M. J. Strauss, Am. Soc. *97*, 3789 (1975).
[46] A. Wexler and J. S. Swenton, Am. Soc. *98*, 1602 (1976); 5-C-subst. pyrimidine nucleosides via organopalladium intermediates cf. D. E. Bergstrom and J. L. Ruth, ibid. *98*, 1587.
[47] P. N. Confalone et al., Am. Soc. *97*, 5936 (1975).
[48] S. L. Regen, Am. Soc. *97*, 5956 (1975).

Useful reactions on alumina surface have been reported, including displacements under mild conditions[49] and a preferential reduction of aldehydes to prim. alcohols with isopropanol[50]. Dehydrated chromatographic alumina has been recommended for the preparation of olefins from tosylates[51]. A novel oxidation method is the dry ozonization of saturated compounds absorbed on silica gel. As one application, tertiary alcohols can be conveniently obtained with retention of configuration[52].

Reductions with tetrabutylammonium borohydride can be performed in the absence of protic solvents[53]. A mixture of $TiCl_3$ and $LiAlH_4$ is a convenient reducing agent, which has been used for the reductive dimerization of oxo compounds to sym. ethylene derivatives[54], and of alcohols[55], also for the preparation of ethylene derivatives from oxido compounds[56]. Magnesium amalgam-titanium tetrachloride is a useful reagent for inter- and intra-molecular pinacolic coupling of oxo compounds[57]. Recently, active titanium metal powder has been found to be a superior reagent for the production of olefins by coupling of carbonyls or by reduction of glycols[58]. Phosphine complexes of transition metals, such as rhodium, have been chemically bonded to the surface of silica. Most of these catalysts retain substantial hydrogenation activity in the presence of mercaptans[59].

Anhydrous H_2O_2 can be conveniently stored and handled as the solid triethylenediamine $\cdot 2H_2O_2$ complex[60].

Pyridinium chlorochromate, a readily available stable reagent oxidizes a wide variety of alcohols to oxo compounds with high efficiency[61]. Also, a convenient synthesis of (–)-pulegone from (–)-citronellol through an oxidative ring closure with this reagent has been published[62].

[49] G. H. Posner et al., Tetrah. Let. *1975*, 3597.
[50] G. H. Posner and A. W. Runquist, Tetrah. Let. *1975*, 3601.
[51] G. H. Posner and G. M. Gurria, J. Org. Chem. *41*, 578 (1976).
[52] Z. Cohen et al., J. Org. Chem. *40*, 2141 (1975).
[53] D. J. Raber and W. C. Guida, J. Org. Chem. *41*, 690 (1976).
[54] Synth. Meth. *30*, 561.
[55] J. E. McMurry and M. Silvestri, J. Org. Chem. *40*, 2687 (1975).
[56] Synth. Meth. *30*, 662 suppl. *30*.
[57] E. J. Corey, R. L. Danheiser, and S. Chandrasekaran, J. Org. Chem. *41*, 260 (1976).
[58] J. E. McMurry and M. P. Fleming, J. Org. Chem. *41*, 896 (1976).
[59] K. G. Allum et al., J. Organometal. Chem. *107*, 390 (1976).
[60] P. G. Cookson, A. G. Davies, and N. Fazal, J. Organometal. Chem. *99*, C31 (1975).
[61] E. J. Corey and J. W. Suggs, Tetrah. Let. *1975*, 2647.
[62] E. J. Corey, H. E. Ensley, and J. W. Suggs, J. Org. Chem. *41*, 380 (1976).

The following references in Vol. 29 under Trends have been entered in this volume [63]:

4/366; 5/33; 6/581; 8/25; 11/591; 15/603; 16/499; 17/506; 18/144; 19/462; 20/619; 21/674; 23/487; 24/118; 30/135; 27/414; 28/414; 29/540; 30/266; 31/255; 32/98; 33/195; 34/361; 37/574; 38/168; 40/178; 41/336; 42/672; 43/338; 44/406; 50/40; 51/322; 52/365; 53/66; 54/80.

---

[63] The first figure refers to the footnote in Trends, Vol. 29, the second figure to the entry number to this volume.

Formation of H—O Bond

Uptake ⇓

Addition to Oxygen   HO⇓OO

Triethylamine	Et_3N
5-Hydroxytropolones	←
s. *26*, 1	
Thiourea	$(H_2N)_2CS$
Diols from cyclic peroxides	C

1.

3,6-Peroxy-α-damascone stirred 8 hrs. with thiourea in methanol → 3,6-dihydroxy-α-damascone. Y: 75%. K. H. Schulte-Elte, M. Gadola, and G. Ohloff, Helv. *56*, 2028 (1973); s. a. Y. Itō, M. Oda, and Y. Kitahara, Tetrah. Let. *1975*, 239.

Addition to Oxygen and Carbon   HO⇓OC

Sodium acetate	CH_3COONa
Glycols from oxido compds.	→ C(OH)C(OH)
s. *26*, 2	
Benzopinacol	
Quinols from quinones	←
s. *28*, 31	
Hydrazobenzene	PhNHNHPh

s. *26*, 263; with N,N-diethylhydroxylamine cf. S. Fujita and K. Sano, Tetrah. Let. *1975*, 1695

N,N-Diethylhydroxylamine	Et_2NOH
s. *26*, 263 suppl. 30	
Palladium-carbon	Pd-C
Intramolecular quinhydrones	←

2.

Startg. bisquinone hydrogenated with Pd-C in dioxane until the calculated amount of H_2 has been absorbed → product. Y: 70%. W. Rebafka and H. A. Staab, Ang. Ch. *85*, 831 (1973); *86*, 234 (1974).

Addition to Carbon   HO⇓CC

Electrolysis   ⚡
Cyclic alcohols from ethyleneketones
     Electrochemical reductive ring closure s. *28*, 1

Hydrazine/cupric sulfate/hydrogen peroxide $H_2N \cdot NH_2/CuSO_4/H_2O_2$
Cyclic alcohols from ethyleneketones
     Reductive transannular ring closure s. *28*, 2

Hydrogen peroxide s. under $H_2N \cdot NH_2$ H_2O_2

Rearrangement   ⟳

Oxygen/Oxygen Type   HO⟳OO

Triethylamine Et_3N
Lactols from cyclic peroxides

3.

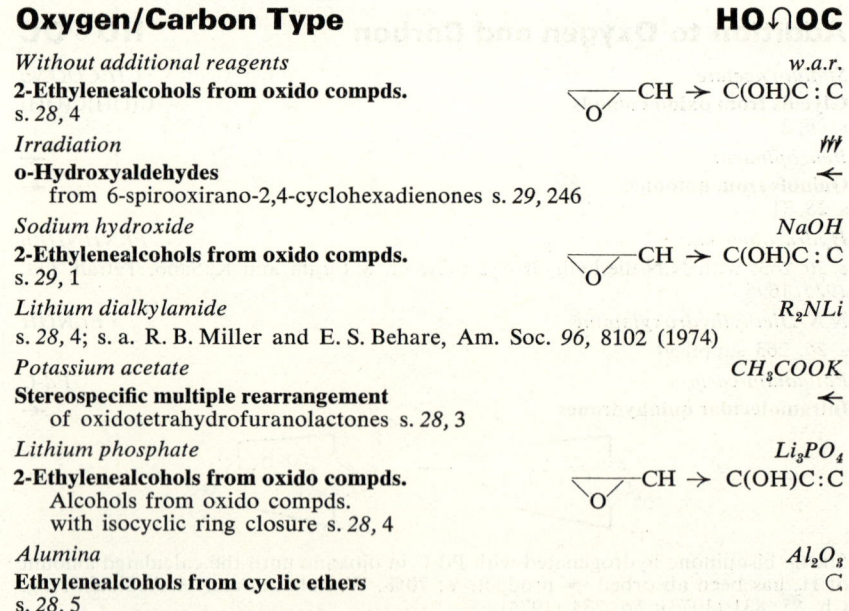

9,10-Dioxabicyclo[4.2.2]dec-7-ene refluxed 10 hrs. with triethylamine in methylene chloride → 1-hydroxy-9-oxabicyclo[4.2.1]non-7-ene. Y: 80%. Y. Kayama, M. Oda, and Y. Kitahara, Chem. Lett. *1974*, 345.

Oxygen/Carbon Type   HO⟳OC

Without additional reagents   w.a.r.
2-Ethylenealcohols from oxido compds.
s. *28*, 4

Irradiation   ⋕
o-Hydroxyaldehydes
     from 6-spirooxirano-2,4-cyclohexadienones s. *29*, 246

Sodium hydroxide   NaOH
2-Ethylenealcohols from oxido compds.
s. *29*, 1

Lithium dialkylamide R_2NLi
s. *28*, 4; s. a. R. B. Miller and E. S. Behare, Am. Soc. *96*, 8102 (1974)

Potassium acetate CH_3COOK
Stereospecific multiple rearrangement
     of oxidotetrahydrofuranolactones s. *28*, 3

Lithium phosphate Li_3PO_4
2-Ethylenealcohols from oxido compds.
     Alcohols from oxido compds.
     with isocyclic ring closure s. *28*, 4

Alumina Al_2O_3
Ethylenealcohols from cyclic ethers
s. *28*, 5

Boron fluoride BF_3
Quinols from cyclic 2-ene-1,4-diones ←
s. 7, 5 suppl. 28

Antimony trichloride $SbCl_3$
2-Ethylenealcohols from oxido compds. CH → C(OH)C:C
s. 29, 1

Sulfuric acid H_2SO_4
Ethylenealcohols from oxido compds. C
  Abnormal ring opening s. 28, 6

Hydrogen chloride HCl
2-Ethylenealcohols from oxido compds. CH → C(OH)C:CH
s. 28, 471; with diphenyl diselenide s. a. K. Mori,
Agri. Biol. Chem. (Tokyo) 38, 2045 (1974)

Exchange ↓↑

Nitrogen ↑   HO ↕ N

Sodium iodide/triethylamine NaI/Et_3N
Protection of carboxyl groups as acyloximes COON:C → COOH
  Removal of the protective group s. 27, 1

Halogen ↑   HO ↕ Hal

Irradiation ⚡
Photoreactions of 2,5-cyclohexadienones ←
  Phenols from 4-bromo-2,5-cyclohexadienones s. 27, 2

Sodium/potassium Na, K
3-Acetylenealcohols C
  from 3-chloro-4,5-dihydrofurans s. 29, 797

Hydrogen sulfide H_2S
Sulfinic acids from sulfonic acid chlorides $SO_2Cl \to SO_2H$
s. 6, 7 suppl. 29

Sulfur ↑   HO ↕ S

Without additional reagents w.a.r.
Selective O-detosylation OTs → OH
s. 28, 119

Potassium tert-butoxide $Me_3C \cdot OK$
Ring expansion of N-heterocyclic 2-aminomesylates Ö:
s. 30, 459

Sodium iodide NaI
Cleavage of sulfuric acid chloride esters $OSO_2Cl \to OH$
s. 27, 579

Lithium tetrahydridoaluminate $Li[AlH_4]$
O-Demesylation $OSO_2Me \to OH$
s. 30, 7

Dimethyl sulfoxide — Me_2SO
Solvolytic desulfation — $OSO_3^- \to OH$
of carbohydrates s. *27*, 3

2,4,6-Trinitrobenzenesulfonate — ←
Glycols — $O\overset{+}{S}R_2 \to OH$
from 1-hydroxy-2-(alkoxysulfonium) salts s. *27*, 129

2,6-Dichloroquinonechlorimine — ←
Reactions of 2,6-dichloroquinonechlorimine — ←
with phosphoric acid derivs. – Nucleotides from nucleoside pyrophosphates
s. *27*, 4

Hydrogen chloride — HCl
Alcohols from sulfuric acid monoesters — $OSO_3H \to OH$
s. *30*, 656

Remaining Elements ↑   HO ↓↑ Rem

Without additional reagents — w.a.r.
Cyclobutane-*cis*-1,2-diols — $OSi\leqslant \to OH$
from 1,2-di(siloxy)cyclobutanes s. *27*, 60

Carboxylic acids — $COOSi\leqslant \to COOH$
from carboxylic acid silyl esters s. *28*, 733

Sodium/alcohol — $NaOR$
Cyclic *trans*-glycols — $OSi\leqslant \to OH$
from cyclic *trans*-2-siloxyalcohols
s. *30*, 81

Methyl alcohol — $MeOH$
Cyclopropanol from silyloxycyclopropane ring
s. *29*, 829; with *n*-butyl alcohol, deltic acid, cf. D. Eggerding and R. West, Am. Soc. *97*, 207 (1975)

O-Deuteriomethanol — $MeOD$
Replacement of silyl groups by deuterium — $OSi\leqslant \to OD$
s. *27*, 20

Pyridinium trifluoroacetate — $C_5H_5\overset{+}{N}H\ CF_3COO^-$
Cleavage of alkoxysilanes
s. *29*, 881

Tetra-n-butylammonium fluoride — $n\text{-}Bu_4N^+\ F^-$
Protection of hydroxyl groups — $OSi\leqslant \to OH$
as *tert*-butyldimethylsilyl ethers
Selective removal of the protective group

4.

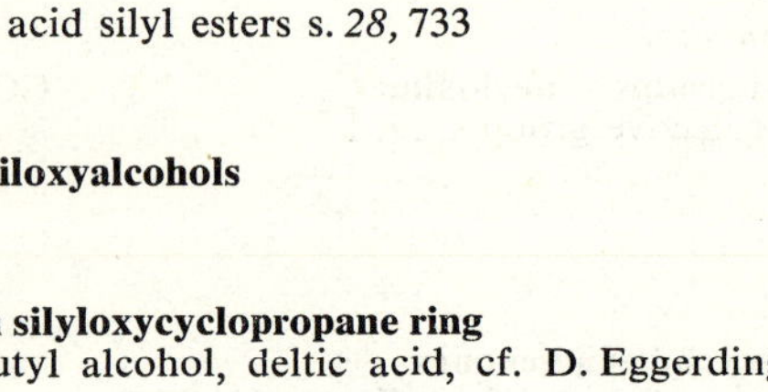

Startg. m. added to tetra-*n*-butylammonium fluoride in tetrahydrofuran, and the resulting soln. stirred 0.5 hr. → 0³′,N-dibenzoyldeoxycytidine. Y: 92%. F. e. s. K. K. Ogilvie, Can. J. Chem. *51*, 3799 (1973).

Hydrogen chloride   HCl
Protection of alcohol groups as silyl ethers
    Removal of the protective group s. *29*, 415

Phenols from aryloxysilanes   ArOSi≤ → ArOH
s. *29*, 602

Carbon ↑   HO ↕ C

Irradiation (s. a. under Et_3Al)
Removal of photo-sensitive protective groups
Protection of alcohol groups   OR → OH
as o-nitrobenzyl ethers

5.

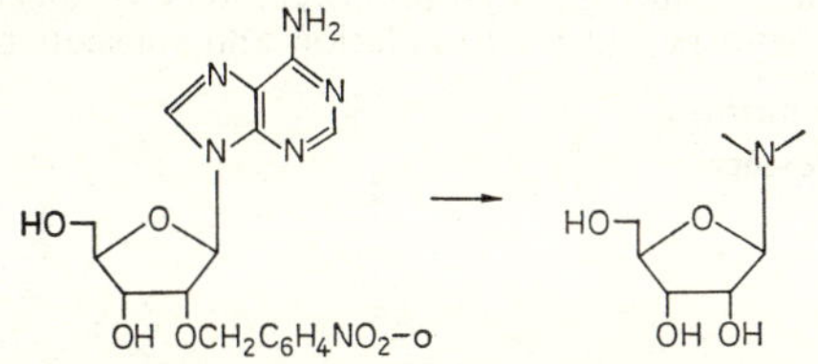

2′-O-(o-Nitrobenzyl)adenosine in methanol-dimethylformamide (2:1) irradiated 1 hr. with 350 nm light of a Rayonet photoreactor in a quartz vessel → adenosine. Y: ca. 100%. F. e. s. D. G. Bartholomew and A. D. Broom, Chem. Commun. *1975*, 38; **removal of a photo-sensitive o-nitrobenzyl ester polymer support** after solid-phase peptide synthesis cf. D. H. Rich and S. K. Gurwara, Am. Soc. *97*, 1575 (1975); **protection of phosphoryl hydroxy groups** (cf. Synth. Meth. *22*, 6) **as o-nitrobenzyl esters** s. M. Rubinstein, B. Amit, and A. Patchornik, Tetrah. Let. *1975*, 1445.

Protection of carboxyl groups   COO·C·COR → COOH
    as phenacyl esters – Selective removal of the protective group s. *29*, 2

Protection of carboxyl groups
    as benzoin esters – Selective removal of the photo-sensitive protective groups **as benzofurans** s. *27*, 5

Removal of photo-sensitive protective groups
    Azidoarylalkoxyl as O-protective group s. *27*, 6

Electrolysis
Removal of protective groups
    of 2-polyhalogenethyl-containing groups by electrochemical reduction at controlled potential s. *28*, 7

Sodium   Na
Diols from chloral acetals
s. *29*, 142

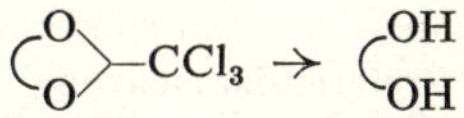

Lithium hydroxide   LiOH
Selective O-deacetylation   OAc → OH
    of carbohydrates s. *27*, 7

Alcohols from urethans   OCON< → OH
    Remote steric control – Removal of the p-phenylphenylcarbamyl directing group s. *28*, 8; alcohols from urethans with KOH as part of a process for the prepn. of stereoisomers from *meso*-compds. cf. A. Fischli et al., Helv. *58*, 564 (1975)

Sodium hydroxide   NaOH
Selective hydrolysis   COOR → COOH
of carboxylic acid esters
s. *30*, 10

Hydroxyl- from acoxyl-amines   >N-OAc → >N-OH
s. *26*, 88

Boronic acids from boronic acid esters   B(OR)$_2$ → B(OH)$_2$
s. *29*, 497

α,β-Ethylenecarboxylic acids   C
from dioxolanium salts s. *29*, 753

Potassium hydroxide (s. a. under Na/NH$_3$)   KOH
Protective of carboxyl groups   COOR → COOH
as carboxylic acid 2-(p-nitrophenylthio)ethyl esters – Selective removal of the protective group from peptides by oxidation and subsequent hydrolysis s. *26*, 3

Potassium hydroxide/amines   ←
Phenols from phenolesters   ArOAc → ArOH
s. *28*, 325

Sodium/alcohol   NaOR
Cleavage of ethers   OR → OH
s. *28*, 9

Sodium/liq. ammonia   Na/NH$_3$
Reductive fragmentation   ←
2-Ethylenealcohols from β,γ-oxidonitriles s. *27*, 8; s. a. J. Org. Chem. *40*, 1162 (1975)

Sodium/liq. ammonia-potassium hydroxide   Na/NH$_3$-KOH
Cleavage of benzyl ethers   OCH$_2$Ph → OH
s. *27*, 9

Sodium hydrogen carbonate   NaHCO$_3$
O-De(trifluoroacetylation)   OCOCF$_3$ → OH
s. *29*, 601

Preferential hydrolysis of phenolesters   ArOAc → ArOH
s. *28*, 86

Azulene ring   ◯
by stereospecific transannular ring closure s. *26*, 749

Sodium cyanide   NaCN
Phosphoric acid monoesters   C
from 2-alkoxy-2-oxo-1,3,2-dioxaphospholans s. *29*, 110

Sodium cyanide/hexamethylphosphoramide   NaCN/(Me$_2$N)$_3$PO
Preferential cleavage   COOMe → COOH
of carboxylic acid methyl esters

6.   (mesityl)—COOMe → (mesityl)—COOH

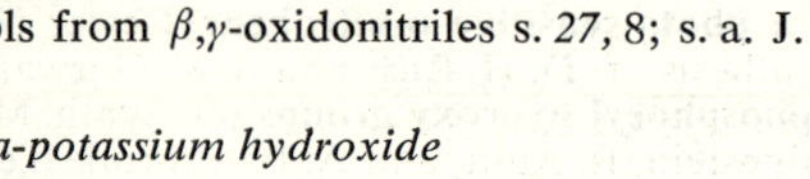

A 1:1 mixture of methyl mesitoate and ethyl benzoate allowed to react 24 hrs. at 75° with NaCN in hexamethylphosphoramide → mesitoic acid. Y: 85%. – No benzoic acid is obtained and 93% of the ethyl benzoate is recovered. F. e. s. P. Müller and B. Siegfried, Helv. *57*, 987 (1974).

Sodium methylmercaptide   MeSNa
9-Anthrylmethyl as protective group   COOCH$_2$Ar → COOH
    Removal from carboxyl – also protection of phenols and mercaptans s. 29, 3

Sodium ethylmercaptide   EtSNa
Cleavage of phenolethers   ArOR → ArOH
s. 27, 10; cf. G. L. Carlson, I. H. Hall, and C. Piantadosi, J. Med. Chem. *18*, 432 (1975)

Aq. ammonia   NH$_4$OH
Thionophosphoric acid diesters   ·O)$_2$P(O)SR → ·O)$_2$P(S)OH
    from monothiolphosphoric acid esters – P-Thionucleosides s. 28, 10

Ammonia   NH$_3$
O-Deacetylation   OAc → OH
s. 26, 138; 28, 483; s. a. Y. V. Tumanov et al., Ж. *44*, 1216 (1974); C. A. *81*, 49958

Amines s. under KOH   ←

Methylamine   MeNH$_2$
Phenols by O-demethylation   ArOMe → ArOH
s. 26, 4

Diisopropylamine   i-Pr$_2$NH
O-Deacylation   OAc → OH
s. 28, 11

1,5-Diazabicyclo[4.3.0]non-5-ene   ←
Cleavage   COOMe → COOH
    of hindered carboxylic acid methyl esters s. 29, 963

Copper s. under Zn   Cu

Cupric sulfate   CuSO$_4$
Preferential hydrolysis   COOR → COOH
    of pyridine-2-carboxylic acid esters s. 26, 5

Zinc   Zn
Protection of alcohol groups   OCOOCH$_2$CCl$_3$ → OH
    as β,β,β-trichlorethyl carbonates – Reductive removal of the protective group s. 28, 966

Zinc/ammonium chloride   Zn/NH$_4$Cl
Protection of alcohol groups   OCOAr → OH
    as o-nitrobenzoates – Reductive removal of the protective groups s. 29, 4

Zinc,copper/zinc chloride   Zn,Cu/ZnCl$_2$
Protection of alcohol groups   OCH$_2$CCl$_3$ → OH
    as β,β,β-trichlorethyl ethers – Reductive removal of the protective group s. 29, 5

Mercuric chloride   HgCl$_2$
Preferential deacetylation   OAc → OH
    of tert. alcohol groups s. 29, 6

Sodium tetrahydridoborate s. under I   Na[BH$_4$]

Lithium tetrahydroaluminate   Li[AlH$_4$]
Reductive O-deacetylation   OAc → OH
s. 26, 760; 27, 959

3-Allenealcohols from 1-acoxy-2-en-4-ynes   ←
s. 30, 24

Selective removal of O-protective groups OCON< → OH
Simultaneous formation of alcohols
from urethans and mesylates

7.

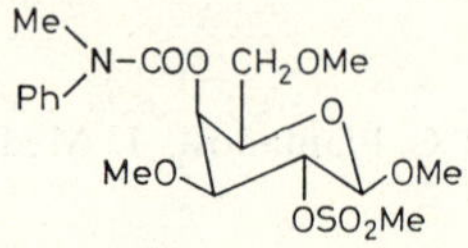

Methyl 3,6-di-O-methyl-4-O-[methyl(phenyl)carbamyl]-2-methanesulfonyl-β-D-galactopyranoside refluxed 3 hrs. with LiAlH$_4$ in tetrahydrofuran → methyl 3,6-di-O-methyl-β-D-galactopyranoside. Y: 90.7%. A. Penman and D. A. Rees, Soc. Perkin I *1973*, 2188.

Triethylaluminum/irradiation Et_3Al/⚡
Cleavage of ethers OR → OH
s. *29, 7*

Sodium dihydridobis-(2-methoxyethoxo)aluminate $Na[AlH_2(OCH_2CH_2OMe)_2]$
O-Deacetylation   OAc → OH
with formation of diols from lactones s. *30, 17*

Boron chloride BCl_3
Ether cleavage   OR → OH
with formation of o-γ-chlorophenols from chromans s. *30, 345*

Cleavage of sterically hindered carboxylic acid esters   COOR → COOH
with retention of phenolether groups s. *27, 11*

Boron bromide BBr_3
Cleavage of acetals
s. *29, 8*   CH(O-O) → CHO

Carboxylic acids from carboxylic acid esters   COOR → COOH
s. *28, 12*

Diethyl azodicarboxylate $EtOOC \cdot N : N \cdot COOEt$
Cleavage of allyl ethers   OCH$_2$CH : CH$_2$ → OH

8.
PhCH$_2$OCH$_2$CH=CH$_2$ + N=N (with COOEt groups)

↓

[PhCH$_2$OCH=CHCH$_2$N—NH (with COOEt groups)] → PhCH$_2$OH

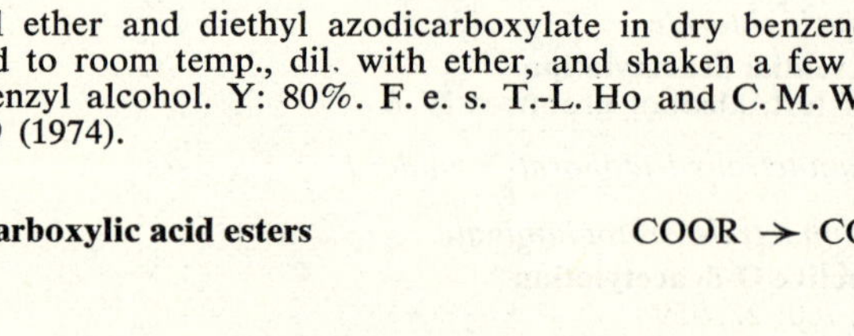

A soln. of allyl benzyl ether and diethyl azodicarboxylate in dry benzene refluxed overnight, cooled to room temp., dil. with ether, and shaken a few min. with dil. aq. HCl → benzyl alcohol. Y: 80%. F. e. s. T.-L. Ho and C. M. Wong, Synth. Commun. *4*, 109 (1974).

Enzymes   ←
Carboxylic acids from carboxylic acid esters   COOR → COOH
s. *28, 13*

α-Chymotrypsin   ←
Enzymatic resolution
L-α-Aminocarboxylic acids from their rac. esters s. *28, 13*

Microorganisms ←
Biochemical O-deacylation OAc → OH
 Alcohols from pivalates s. 26, 6

Phenol s. under HHal PhOH

Acetic acid CH_3COOH
Selective cleavage of O,O-isopropylidene derivs.
with retention of O,N-isopropylidene derivs. $\begin{matrix}C\cdot O\\C\cdot O\end{matrix}CMe_2 \rightarrow \begin{matrix}C\cdot OH\\C\cdot OH\end{matrix}$

A soln. of N-acetyl-6-O-acetyl-3-deoxy-1,2-N,O-isopropylidene-4,5-O-isopropylidene-*epi*-inosamine-1 in aq. 50%-acetic acid warmed 1 hr. at 60–70° → N-acetyl-6-O-acetyl-3-deoxy-1,2-N,O-isopropylidene-*epi*-inosamine-1. Y: 89%. A. Hasegawa and M. Nakajima, Carbohyd. Res. *29*, 239 (1973).

Trifluoroacetic acid CF_3COOH
Protection of phosphoryl hydroxyl groups $P\cdot OCMe_3$ → $P\cdot OH$
 as *tert*-butyl esters – Removal of the protective group s. 28, 170

Trichloroacetic acid CCl_3COOH
Cleavage of acetals $C(OR)_2$ → CO
s. 26, 7

2-Mercaptoethylamine $HSCH_2CH_2NH_2$
Protection of nucleoside hydroxyl groups as chloroacetate $OCOCH_2Cl$ → OH
 Preferential removal of the protective group s. 26, 8

Methyl iodide MeI
Protection of carboxyl groups $COOCH_2SMe$ → COOH
 as methylthiomethyl esters – Removal of the protective group s. 28, 14

Triphenyl phosphite/ozone-palladium/carbon $(PhO)_3P/O_3$-Pd/C
Oxidoreductive removal of O-protective anthryl groups ←
s. 28, 194

Ozone s. under $(PhO)_3P$ O_3

N-Hydroxybenzenesulfonamide $PhSO_2NHOH$
Cleavage of acetals $C(OR)_2$ → CO
s. 26, 176

Sulfuric acid H_2SO_4
Aldehydes from cyclic acetals $CH\begin{matrix}O\\O\end{matrix}$ → CHO
s. 27, 81

Selenium dioxide/acetic acid SeO_2/CH_3COOH
Oxidative cleavage OR → OH
 of β,γ-unsatd. ethers s. 26, 116

Chlorine Cl_2
1,3-Diols from 1,3-dioxanes $\begin{matrix}O\\O\end{matrix}$ → $\begin{matrix}OH\\OH\end{matrix}$
s. 29, 9

Iodine/sodium tetrahydridoborate $I/Na[BH_4]$
Hydroxy- from alkoxy-carboxylic acid esters OR → OH
 Retention of optical activity s. 26, 9

Perchloric acid   HClO
Cleavage of ketals
   preferential – with formation of ethyleneketals from   C(O-O) → CO
   2,3-oxidoalcohols s. *29*, 10

Ammonium chloride s. under Zn NH_4Cl

Hydrogen halide/phenol   HHal/PhOH
Cleavage of ethers and methylenedioxyarenes   ArOR → ArOH
   with N-deacetylation s. *28*, 15

Hydrogen chloride   HCl
Selective cleavage of mixed acetals   O·C·OR → OH
s. *26*, 10

Glycosides from acylglycosides   OAc → OH
s. *19*, 201 suppl. *29*

Catechols from catechol monoesters   ArOAc → ArOH
s. *29*, 123

Aldehydes from acetals $CH(OR)_2$ → CHO
s. *27*, 59
   from cyclic acetals s. *29*, 867

Prepn. of O-labeled compds. $CH(OR)_2$ → $CH^{18}O$
s. *29*, 11

Cleavage of O,O-isopropylidene derivs.
   Resolution of ketones by gas-liq. chromatography/   (CO,CO)CMe₂ → (C·OH, C·OH)
   after their conversion into diastereomeric ketals-
   Methods of resolution, review s. *26*, 11

Lactols from lactolides
s. *16*, 11 suppl. *29*   (O–C·OR) → (O–C·OH)

α-Ketocarboxylic acids   C:C(OR)COOR′ → CHCOCOOH
   from α-alkoxy-α,β-ethylenecarboxylic acid esters s. *29*, 200

α-Hydroxymethyl-α-aminocarboxylic acids   C
from 5-acylamino-4-oxo-1,3-dioxanes
s. *30*, 451

Nickel carbonyl $Ni(CO)_4$
O-Decarballyloxylation $OCOOCH_2CH:CH_2$ → OH
s. *29*, 28

Palladium-carbon (s. a. under $(PhO)_3P$)   Pd-C
O-Debenzylation OCH_2Ph → OH
s. *27*, 456; *29*, 739

Hydrogenolysis of carboxylic acid benzyl esters $COOCH_2Ph$ → COOH
s. *27*, 768

Palladium-carbon/hydrogen chloride   Pd-C/HCl
Protection of alcohol groups as benzyl ethers OCH_2Ph → OH
   Selective removal of the protective group s. *26*, 13

Platinum oxide PtO_2
O-Deacetylation   ←
   with lactone rearrangement s. *27*, 281

O-Decarbobenzoxylation $OCOOCH_2Ph$ → OH
s. *28*, 118

Via intermediates v.i.
Cleavage of phenolethers ArOR → ArOH
via aryloxysilanes s. *29*, 602

Elimination ⇑

Oxygen ↑ HO⇑O

Stannous chloride $SnCl_2$
Cyanohydrins from α-hydroperoxynitriles C(CN)OOH → C(CN)OH
s. *30*, 155

Triphenylphosphine Ph_3P
1-Azoalcohols from 1,1-azohydroperoxides OOH → OH

10.

A soln. of triphenylphosphine in *n*-pentane added with ice-cooling during 5 min. to a soln. of 1-(ethylazo)cyclohexyl hydroperoxide in the same solvent, and stirred 1.5 hrs. with ice-cooling → 1-(ethylazo)cyclohexanol. Y: 81%. F. e., **also from hydrazones** without isolation of the intermediate 1,1-azohydroperoxides (s. Synth. Meth. *22*, 122), s. M. Schulz, U. Missol, and H. Bohm, J. pr. *316*, 47 (1974); Z. Chem. *14*, 265 (1974).

Nitrogen ↑ HO⇑N

n-Butyllithium *n-BuLi*
Allenealcohols C
from cyclic 2-tosylhydrazonoethers – Fragmentation-type ring opening s. *27*, 12

Carbon ↑ HO⇑C

Sodium hydride NaH
Alcohols from ethers OR → OH
Optically active compds. – Resolution of alcohols s. *26*, 14

Sodium hydroxide NaOH
Hydroxylamines from N-oxides (NH→) N(→O)R → N·OH
s. *7*, 10 suppl. 29

p-Toluenesulfonic acid TsOH
Preferential hydrolysis $COOCMe_3$ → COOH
of carboxylic acid *tert*-butyl esters s. *27*, 235

α-Acetoxycarboxylic acids $C(:N_2)COO·CMe_3$ → CH(OAc)COOH
from α-diazocarboxylic acid *tert*-butyl esters s. *27*, 201

Formation of H—N Bond

Uptake ⇓

Addition to Nitrogen Atoms HN⇓N

Stannous chloride — $SnCl_2$
Sec. amines from nitrogen radicals N· → NH
s. *29*, 12

Addition to Nitrogen-Nitrogen Bond HN⇓NN

Potassium hydroxide — KOH
Hydrazo from azo compds. N:N → NHNH
    with simultaneous conversion of N-subst. carboxylic acid amides to oxo compds. s. *27*, 199

tert-Butyllithium — $Me_3C·Li$
Ureas from diaziridinones
    Hydrogen atom transfer – Electron transfer – Hydride transfer s. *29*, 13 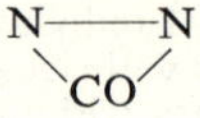
    N—N\CO → NHCONH

Sodium/liq. ammonia — Na/NH_3
Reductive ring opening C
    of N-condensed hydrazinium salts s. *26*, 460

Hydrazine s. under Cobalt boride — $H_2N·NH_2$

Cobalt boride-hydrazine ←
Hydrazo from azo compds. N:N → NHNH
s. *26*, 15

Addition to Nitrogen and Carbon HN⇓NC

Phenylhydroxylamine — PhNHOH
Azo compds. from azines ←
s. *26*, 16

Rearrangement ↺

Hydrogen/Oxygen Type HN↺HO

Basic alumina — Al_2O_3
Iminolactones from cyanohydrins
    C(OH)(CN) → C(O)(C:NH)
s. *21*, 33 suppl. *29*

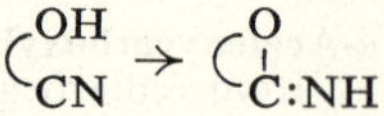

Hydrogen/Carbon Type HN∩HC

Sodium hydride *NaH*
Carboxylic acid amides from oxaziridines C
s. *26*, 17

Oxygen/Nitrogen Type HN∩ON

Without additional reagents w.a.r.
N-Subst. o-aminooxo comdps. C
  from 2,1-benzisoxazoles via 2,1-benzisoxazolium salts and 2,1-benzisoxazolines
  s. *26*, 18

Nitrogen/Nitrogen Type HN∩NN

Irradiation/air ☀/O_2
β-Aminonitriles from Δ^2-pyrazolines C
s. *28*, 889

Exchange ↓↑

Oxygen ↑ HN⇅O

Irradiation s. under Na_2S ☀
Lithium/liq. ammonia Li/NH_3
Sec. amines from N-oxyls N→O ⇸ NH
s. *27*, 13
Sodium/liq. ammonia Na/NH_3
Amines from nitro compds. NO_2 ⇸ NH_2
s. *1*, 23 suppl. *28*
Ethanolamine/pyridine $H_2NCH_2CH_2OH/C_5H_5N$
Replacement of chlorine by amino groups NO_2 ⇸ NH_2
  with and without reduction of nitro to amino groups – Aminoalcohols as reducing agents s. *28*, 16
1,8-Diazabicyclo[5.4.0]undec-7-ene s. under Se ←
Pyridine s. under $H_2NCH_2CH_2OH$ C_5H_5N
Zinc/acetic acid Zn/CH_3COOH
Semicarbazides from N-nitrosoureas N·CO·N·NO ⇸ N·CO·N·NH_2
s. *29*, 274
Zinc/ammonium chloride Zn/NH_4Cl
Hydroxylamines from nitro compds. NO_2 ⇸ NHOH
s. *26*, 28
Lithium tetrahydridoaluminate $Li[AlH_4]$
Aziridine ring from α,β-ethyleneoximes
s. *29*, 14

Lithium tetrahydridoaluminate/sulfuric acid	$Li[AlH_4]/H_2SO_4$
Diamines	$\begin{array}{c}NO_2\\CON<\end{array} \rightarrow \begin{array}{c}NH_2\\CH_2N<\end{array}$
from nitrocarboxylic acid amides s. 26, 64	
Chloroform s. under PtO_2	$CHCl_3$
Titanium trichloride	$TiCl_3$
Amines from nitro compds.	$NO_2 \rightarrow NH_2$
s. 17, 345 suppl. 29	
Hydrazine s. under Ni and Pd-C	$H_2N \cdot NH_2$
Sodium sulfide/irradiation	$Na_2S/\!/\!/\!/$
Amines from N-oxide radicals	$>NO \cdot \rightarrow >NH$
s. 28, 17	
Sodium dithionite	$Na_2S_2O_4$
Amines from nitro compds.	$NO_2 \rightarrow NH_2$
Nitramines from dinitro compds. s. 30, 330	
Selenium/1,8-diazabicyclo[5.4.0]undec-7-ene	←

10a. $PhNO_2 \rightarrow PhNH_2$

A mixture of 3 mmol nitrobenzene, 10 mg atoms Se, and 10 mmol 1,8-diazabicyclo[5.4.0]undec-7-ene in toluene stirred vigorously at 110° for 25 hrs. in a H_2-stream → aniline. Y: 93%. F. e. and reaction conditions s. K. Kondo, N. Sonoda, and H. Sakurai, Chem. Commun. 1975, 42.

Iron/acetic acid	Fe/CH_3COOH
Reduction of nitro groups	$NO_2 \rightarrow NHCOOR$
with O→N-carbalkoxyl migration s. 29, 15	
Triiron dodecacarbonyl	$Fe_3(CO)_{12}$
Amines from nitro compds.	$NO_2 \rightarrow NH_2$
Selective reduction s. 27, 14	
Nickel	Ni
Amines from alkoxylamines	$N \cdot OR \rightarrow NH$
s. 7, 45 suppl. 29	
Carboxylic acid amides	$CONHOCH_2Ph \rightarrow CONH_2$
from hydroxamic acid benzyl esters s. 29, 20	
Nickel/hydrazine	$Ni/H_2N \cdot NH_2$
Amines from nitro compds.	$NO_2 \rightarrow NH_2$
with and without reductive triazinone ring opening to o-aminocarboxylic acid amides s. 27, 15; amines from nitro compds. s. a. B. Stec and J. Szadowski, Przem. Chem. 53, 347 (1974) (Pol); C. A. 81, 152101	
Reductions with rhodium carbonyl	$Rh_6(CO)_{16}$
Amines from nitro compds.	$NO_2 \rightarrow NH_2$
s. 28, 18	
Palladium-carbon	$Pd-C$
1,2-Oximinohydroxylamines from 1,2-nitroximes	$NO_2 \rightarrow NHOH$
s. 26, 19	
Palladium-carbon/hydrazine	$Pd-C/H_2N \cdot NH_2$
Amines from nitroso compds.	$NO \rightarrow NH_2$
3-Aminoindolizines s. 29, 16	
Palladium-carbon/sulfuric acid	$Pd-C/H_2SO_4$
Hydroxylamines from nitroalkanes	$NO_2 \rightarrow NHOH$
s. 27, 16	

Platinum oxide	PtO_2
Amines from nitro compds. s. *27,* 15	$NO_2 \rightarrow NH_2$
Platinum oxide/chloroform	$PtO_2/CHCl_3$
Amine hydrochlorides by hydrogenation in the presence of chloroform s. *27,* 17	$C:NOH \rightarrow CHNH_2,HCl$

Nitrogen ↑   HN ⇊ N

Lithium tetrahydridoaluminate	$Li[AlH_4]$
Amines from azides s. *27,* 401	$N_3 \rightarrow NH_2$
Ureas s. under HCl	$(H_2N)_2CO$
Azodiisobutyronitrile s. under n-Bu$_3$SnH	$[Me_2C(CN)N:]_2$
Tri-n-butyltin hydride/azodiisobutyronitrile	$n\text{-}Bu_3SnH/[Me_2C(CN)N:]_2$
Cleavage of cyclic N-ylids s. *30,* 697	←
O,O-Diethyldithiophosphoric acid	$(EtO)_2PSSH$
Reductive cleavage of nitrogen-nitrogen bonds Amines from azo compds. s. *29,* 17	$ArN:NAr \rightarrow ArNH_2$
Hydrogen chloride	HCl
Protection of amino groups as N-nitrosamines Removal of the protective group s. *28,* 19	$N \cdot NO \rightarrow NH$
Hydrogen chloride/urea	$HCl/(H_2N)_2CO$
Amines from N-nitrosamines s. *28,* 590	
Nickel/hydrazine	$Ni/H_2N \cdot NH_2$
o-Aminocarboxylic acid amides from triazinone ring s. *27,* 15	C
Palladium-carbon	$Pd\text{-}C$
Amines from azides Aminolactones s. *29,* 324; 1,1,1-tris(aminomethyl)ethane s. L. J. Zompa and J.-P. Anselme, Org. Prep. Proced. Int. *6,* 103 (1974)	$N_3 \rightarrow NH_2$

Halogen ↑   HN ⇊ Hal

Sodium metabisulfite	$Na_2S_2O_5$
Replacement of N-halogen by N-hydrogen s. *27,* 548	$N \cdot Hal \rightarrow NH$
Hydrogen chloride	HCl
Selective N-dechlorination s. *23,* 30 suppl. *29*	$NCl \rightarrow NH$

Sulfur ↑   HN ⇊ S

Electrolysis	⚡
Selective N-detosylation by electrochemical reduction s. *27,* 18	$NTs \rightarrow NH$

Ammonium thiocyanate/2-methylindole ←
Removal of o-nitrophenylsulfenyl N-protective groups   N·SAr → NH
s. *27*, 19

Aluminum chloride $AlCl_3$
Imidazole ring rearrangements ←
Imidazoles from 5-alkylthio-4H-imidazoles s. *26*, 20

Anisole s. under CF_3SO_3H   PhOMe

p-Chlorobenzenethiol $p\text{-}ClC_6H_4SH$
Mono(carbobenzoxy)amines $N(SAr)COOCH_2Ph$ → $NHCOOCH_2Ph$
from N-carbobenzoxysulfenamides s. *29*, 412

Thioacetamide CH_3CSNH_2
Removal of o-nitrophenylsulfenyl N-protective groups   N·SAr → NH
s. *27*, 19

Trifluoromethanesulfonic acid/anisole $CF_3SO_3H/PhOMe$
Removal of protective groups   N·SAr → NH
s. *29*, 17a

Sulfuric acid H_2SO_4
N-Detosylation   NTs → NH
s. *29*, 359

Hydrogen fluoride   HF
N-Detosylation   NTs → NH
2-Amino- from 2 tosylamino-1,3,N,N-heterocyclics s. *26*, 438

Hydrogen chloride   HCl
Prim. amines from disulfenylamines $N(SR)_2$ → NH_2
s. *26*, 357

2-Aminoalcohols from 1,2,3-oxathiazolidine S-oxide ring   C
Stereospecific ring opening – Separation of diastereoisomers s. *29*, 18

Remaining Elements ↑   HN ⇋ Rem

Potassium hydroxide   KOH
Sec. amines from P-aminophosphonium salts $N \cdot P^+R_3$ → NH
s. *26*, 291

Lithium tetrahydridoaluminate $Li[AlH_4]$
N-Unsubst. aziridine ring $N \cdot P(O)(OR)_2$ → NH
from N-(aziridinyl)phosphonates s. *26*, 405

O-Deuteriomethanol   MeOD
Replacement of silyl groups by deuterium   Si≤ → D
s. *27*, 20

Hydrogen chloride   HCl
Amines from phosphoramidates $N \cdot PO(OR)_2$ → NH
s. *27*, 548

1,2-Halogenamines from 2-halogenophosphoramidates $N \cdot PO(OR)_2$ → NH
s. *27*, 548

Carbon ↑   HN ⇊ C

Irradiation s. under O_2

Electrolysis
Removal of 2-polyhalogenethyl-containing protective groups
s. *28,* 7

Sodium hydroxide (s. a. under Ni)   *NaOH*
Preferential and total N-deacetylation   NAc → NH
s. *28,* 20

Removal of alkali-labile N-protective carbalkoxy groups   NCOOR → NH
by 1,6-elimination s. *29,* 19; carbomethylsulfonylethoxyl as alkali-labile N-protective group cf. J. W. van Nispen and G. I. Tesser, Int. J. Peptide Protein Res. 7, 57 (1975)

Thioureas from carbalkoxythioureas   N·CS·N·COOR → N·CS·NH
s. *29,* 26

Pyrimidine ring opening   C
s. *29,* 273

Imidazoles from purines
o-Aminocarboxylic acid amides from pyrimidine ring – Carboxylic acid amides from hydroxamic acid benzyl esters s. *29,* 20

2-Aminoalcohols from 2-oxazolidone ring
s. *28,* 309

o-Aminophenols from benzoxazole-2-thiones
s. *26,* 21

Amidoximes from 1,2,4-oxadiazol-5-ones
s. *26,* 22

6-Amino-1,2,4-triazines
from s-triazolo[3,4-f][1,2,4]triazinium salts s. *27,* 21

Potassium hydroxide (s. a. under $H_2N \cdot NH_2$)   *KOH*
Amines from urethans   NCOOR → NH
s. *30,* 273

Amines from 1,2,4-triazolidine-3,5-diones   C
via hydrazines – Allylic functionalization s. *29,* 21

o-Hydroxycarboxylic acid amides
from 1,3-benzoxazine-2,4-diones s. *28,* 898

Potassium hydroxide/alcohol
Aziridine ring rearrangement
with N-decarbalkoxylation s. *29,* 319

Sodium/methanol   *Na/MeOH*
2,5-Epoxy-1,2,4,5-tetrahydro-3H-1,4-benzodiazepin-3-ones   O
from 3-acoxy-1,3-dihydro-2H-1,4-benzodiazepin-2-ones s. *27,* 22

Sodium carbonate s. under $R_3O^+BF_4^-$ Na_2CO_3

Potassium carbonate K_2CO_3
Protection of amino groups   NCOOR → NH
as carbocyano-*tert*-butoxyamines – Removal of the protective group – Peptides from peptide salts s. *28,* 21

Sodium nitrite/hydrogen chloride $NaNO_2/HCl$
Amines from cyanamides via ureas   N·CN → NH
s. 26, 23

Ammonia (s. a. under Ac_2O) NH_3
Preferential N-deacylation   NAc → NH
and selective O-deacylation s. 29, 357

Carbo-9-fluorenylmethoxyl as N-protective group   NCOOR → NH
Removal by β-elimination s. 26, 24

Prim. amines from phthalimides   (CO)(CO)NR → H_2NR
s. 26, 25

1-Benzoylpyrimidinium ring opening   C
2-Azapurines from purine nucleosides s. 27, 23

Triton B   ←
N-Dehydroxymethylation NCH_2OH → NH
s. 27, 456

Methylamine $MeNH_2$
Prim. amines from phthalimides   (CO)(CO)NR → H_2NR
s. 26, 25

Triethylamine Et_3N
Removal of N-substituents from 2-azetidinones   NH
s. 29, 22

Protection of amino groups   NCOOR → NH
as carbocyano-*tert*-butoxyamines – Removal of the protective group s. 28, 21

Cupric acetate s. under HCl $(CH_3COO)_2Cu$

Zinc   Zn
Carbotrichlorethoxyl as N-protective group   N·$COOCH_2CCl_3$ → NH
Reductive removal – 3,6-Dihydro-2H-1,2-thiazine 1-oxides s. 27, 24

Zinc/acetic acid Zn/CH_3COOH
Transamidation   NAc → NAc′
via N-acylcarboxylic acid amides s. 27, 476

Aluminum amalgam   Al,Hg
Stereospecific reduction of sulfoximines   ←
Sulfinic acid amides s. 28, 22

Sodium tetrahydridoborate $Na[BH_4]$
Selective reductive removal $NCOCF_3$ → NH
of N-protective trifluoroacetyl groups
s. 26, 26

Lithium tetrahydridoaluminate $Li[AlH_4]$
Pyrazolidines from N,N-Diels-Alder adducts   C
s. 27, 25

Boron fluoride BF_3
Selective N-deacylation   NAc → NH
s. 30, 120

Ethanol   EtOH
Selective removal of N-carbobromo-*tert*-butoxyl groups   NCOOR → NH
s. 28, 23

Propanol   *PrOH*
Protection of amino groups as tosylureas   NCONHTs → NH
**Removal of the protective group
under neutral conditions**

11.   p—CH$_3$C$_6$H$_4$SO$_2$NHCONHCH$_2$CONHCHCONHCH$_2$COOH
                                                    | CH$_2$Ph
↓
                      H$_2$NCH$_2$CONHCHCONHCH$_2$COOH
                                                      | CH$_2$Ph

A soln. of the startg. m. in propanol containing 5% water refluxed 2 hrs. → product. Y: 86%. F. e. s. H. Künzi and R. O. Studer, Helv. *58*, 139 (1975).

Dimethylformamide   *Me$_2$NCHO*
Preferential N-demethylation   ←
s. *29*, 23

Triethyloxonium fluoroborate/sodium carbonate   *Et$_3$O$^+$ BF$_4$$^-$/Na$_2CO_3$*
Selective N-deacetylation   NAc → NH
s. *29*, 24

Acetic anhydride/ammonia   *Ac$_2$O/NH$_3$*
**Removal of 1-oxidopyridin-2-ylmethyl protective groups
Protection of amino groups**   NR → NH

12.

A soln. of N^4-(1-oxido-2-pyridylmethyl)cytidine in acetic anhydride stirred 51 hrs. at 30°, concd. to dryness, dissolved in methanolic NH$_3$ in a stoppered vessel, and kept at room temp. overnight → cytidine. Y: 86.7%. F. e., **also protection of phosphoryl hydroxy groups**, s. Y. Mizuno et al., J. Org. Chem. *39*, 1250 (1974); **protection of alcohol groups as 1-oxido-2-picolyl ethers** s. ibid. *40*, 1385 (1975); **protection of sulfhydryl groups as 1-oxido-2-picolyl thioethers** s. Chem. Pharm. Bull. *22*, 2889 (1974); related protective groups cf. J. Org. Chem. *40*, 1391 (1975).

Phenol s. under Hydrogen halides   *PhOH*

Trifluoroacetic acid   *CF$_3$COOH*
Sec. from tert. amines   >NR → >NH
s. *27*, 26

N-De-*tert*-butylation   NHSO$_2$NHCMe$_3$ → NHSO$_2$NH$_2$
Monosubst. sulfamides s. *29*, 25

Sulfonyloxylamines   SO$_2$O·N·COOR → SO$_2$O·NH
from N-carbalkoxysulfonyloxylamines – Selective N-decarbo-*tert*-butoxylation
s. *27*, 110

Bromoform s. under Ph$_3$P   *CHBr$_3$*

Rose Bengal s. under O$_2$ ←

Hydrazine/potassium hydroxide   *H$_2$N·NH$_2$/KOH*
N-Deacylation   NAc → NH
s. *28*, 24

| *Methoxyamine* | $H_2N \cdot OMe$ |

Pteridin-4(3H)-one ring opening   C
   to 4,5-diamino-6-(1H)-pyrimidinones s. *26*, 27

| *Triphenylphosphine/bromoform* | $Ph_3P/CHBr_3$ |

Cleavage of nucleosides   ←
s. *28*, 25

| *Air s. a. under Ni and Pt* | O_2 |

| *Oxygen/ℎν/rose Bengal* | ← |

N-Demethylation   NMe → NH
s. *28*, 26

| *p-Toluenesulfonic acid* | TsOH |

Selective removal of protective groups from peptides   NTri → NH
   N-Detritylation s. *26*, 27a

| *Pyridine hydrochloride* | C_5H_5N,HCl |

Removal of acid-labile N-protective groups
s. *26*, 27a suppl. 29

| *Hydrogen halides/phenol* | HHal/PhOH |

N-Deacetylation   NAc → NH
s. *28*, 15

| *Hydrogen chloride* | HCl |

Amines from isonitriles   N : C → NH_2
s. *29*, 791

α-Amino-α,β-ethylenecarboxylic acid esters   C : C(NH_2)COOR
s. *29*, 739

3-Aminoalcohols NH_2
   from 3-isocyanoalcohols or 5,6-dihydro-4H-1,3-oxazines s. *29*, 617

N-Demethylation via formamides   NCHO → NH
s. *28*, 136

N-Deacetylation   NAc → NH
s. *28*, 20

α-Aminocarboxylic acids   C(NAc)CON< → C(NH)COOH
   from α-acylaminocarboxylic acid amides s. *28*, 288

N-Decarbalkoxylation   NCOOR → NH
s. *28*, 434

N-Decarbo-*tert*-butoxylation $NCOOCR_3$ → NH
s. *28*, 330; 3,4-dihydrobenz[d]-1,2-oxazine-1,4-diones s. *27*, 27

Acoxylamines from N-tritylacoxylamines   C(O)·O·NTri → C(O)·O·NH
s. *26*, 172

Monosubst. hydrazines from semicarbazides   N·N·CON< → N·NH
s. *29*, 274

**N,O-Bis(carbamyl)-
hydroxylamines**   NCON(OCON<)COOR → N·CO·NH(OCON<)
   from N-carbo-*tert*-butoxy-N,O-bis(carbamyl)hydroxylamines s. *26*, 292

Thioureas   N·CS·N·CO· → N·CS·NH
   from carbalkoxy- and acyl-thioureas – 1,1-Disubst. thioureas – Prepn. of thioureas, review, s. *29*, 26

Reductive removal   O
of o-nitrophenoxydimethylacetyl N-protective groups
s. *26*, 28

Prim. amines from phthalimides
s. 27, 429, 480

$\begin{array}{c}CO\\CO\end{array}\!\!>\!\!NR \rightarrow H_2NR$

α-Aminoketones from oxazoles
s. 28, 811

C

**α-Hydroxymethyl-α-aminocarboxylic acids
from 5-acylamino-4-oxo-1,3-dioxanes**
s. 30, 451

2-Arylpyrroles
from spiro-3H-indoline-3,2'-2'H-pyrroles s. 26, 961

Hydrogen chloride/cupric acetate $HCl/(CH_3COO)_2Cu$

Protection of amino groups as α-picolinoylamines   NAc → NH
Removal of the protective group s. 29, 27

Hydrogen bromide/formic acid $HBr/HCOOH$

Removal of N-diphenylmethyl groups $N \cdot CHPh_2$ → NH
Monosubst. guanidines s. 27, 28

Nickel/sodium hydroxide/air $Ni/NaOH/O_2$

Asym. synthesis of α-aminocarboxylic acids   C
via tetrahydro-1,4-oxazin-2-ones s. 29, 750

Nickel carbonyl $Ni(CO)_4$

N- and O-Decarballyloxylation $NCOOCH_2CH:CH_2$ → NH
Protection of hydroxyl and amino groups as carballyloxy derivs. s. 29, 28

Platinum/air Pt/O_2

N-Dealkylation   NR → NH
s. 27, 29

Platinum oxide/acetic acid PtO_2/CH_3COOH

**Protection of amino groups
as o-nitrocinnamoylamides
Reductive removal of the protective group** ←

13.

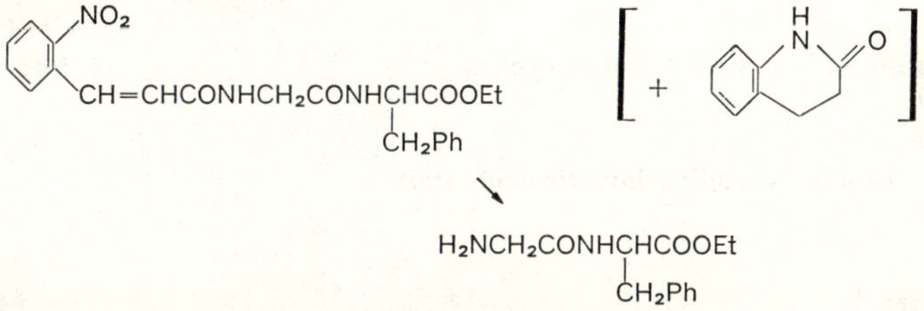

o-NO_2-C$_6$H$_4$-CH=CHCONHCH$_2$CONHCHCOOEt
                                            |
                                            CH$_2$Ph

[+ benzazepinone]

↓

H$_2$NCH$_2$CONHCHCOOEt
                  |
                  CH$_2$Ph

Ethyl N-o-nitrocinnamoylglycyl-L-phenylalaninate hydrogenated 1 hr. at 55° and 60 lb/in² with Pt-oxide in glacial acetic acid, and the product isolated as the hydrochloride → ethyl glycyl-L-phenylalaninate hydrochloride. Y: 82%. F. e. s. G. Just and G. Rosebery, Synth. Commun. 3, 447 (1973).

Via intermediates   v.i.

N-Demethylation via formamides   NMe → NH
s. 28, 136

N-Dealkylation via urethans   NR → NH
s. 23, 479

Cleavage of carboxylic acid amides   NCO → NH
with neighboring group participation s. 26, 28

Reductive removal   NAc → NH
of o-nitrophenoxydimethylacetyl-N-protective groups
s. 26, 28

Elimination ⇑

Oxygen ↑   HN⇑O

Lithium tetrahydridoaluminate $Li[AlH_4]$
Sulfonic acid amides from sulfhydroxamic acids $SO_2NHOH \rightarrow SO_2NH_2$
s. *29*, 29

Nickel   Ni
Lactams from cyclic hydroxamic acids $\underset{CO}{N(OH)} \rightarrow \underset{CO}{NH}$
s. *26*, 351

Cyclic 1-hydroxyamidines   ←
   from cyclic 1-hydroxyamidoximes s. *26*, 297
Sulfonic acid amides from sulfhydroxamic acids $SO_2NHOH \rightarrow SO_2NH_2$
s. *29*, 29

Nitrogen ↑   HN⇑N

Irradiation   ⥼
Carboxylic acid amides from hydrazides $CONHN< \rightarrow CONH_2$

14. $CH_3CONHN\begin{matrix}Me\\Ph\end{matrix} \longrightarrow CH_3CONH_2$

N-Methyl-N-phenylacethydrazide irradiated 4 hrs. in isopropanol under N_2 in a quartz tube with lamps having maximal emission at 254 nm → acetamide. Y: 76%. F. e., also in acetonitrile, and limitations s. R. S. Davidson and A. Lewis, Tetrah. Let. *1973*, 4679.

Lead tetraacetate $(CH_3COO)_4Pb$
Deamination of N-amino-N-heterocyclics $N \cdot NH_2 \rightarrow NH$
s. *26*, 480

Nickel   Ni
Urethans from hydrazodicarboxylic acid esters   ←
s. *30*, 208

Carbon ↑   HN⇑C

Without additional reagents   w.a.r.
1,2,4-Triazolidine-3,5-diones   ←
   from 1,2,4-triazolidine-3,5-dione 1,2-ylids s. *27*, 30

Irradiation   ⥼
Removal of photosensitive protective groups $NCHO \rightarrow NH$
   Anilines from formanilides s. *28*, 27

Potassium hydroxide   KOH
Mixed sec. amines   RNHR′
   from tert. amines and quaternary ammonium salts s. *27*, 31

Trifluoroacetic acid CF_3COOH
Sec. from *tert*-amines $>NR \rightarrow >NH$
s. *27*, 26

Formation of H—S Bond

Uptake ⇓

Addition to Sulfur   HS⇓SS

Methyl mercaptan   *MeSH*
Dithiols from cyclic disulfides   C
  s. *12*, 56 suppl. *29*

Exchange ⇅

Sulfur ↑   HS⇅S

Lithium tetrahydridoaluminate   *LiAlH$_4$*
1,2-Hydroxymercaptans   RSSR → 2 RSH
  from 2,2'-dihydroxydisulfides s. *28*, 525
Triphenylphosphine   *Ph$_3$P*
Mercaptans from disulfides
s. *17*, 45 suppl. *29*

Remaining Elements ↑   HS⇅Rem

O-Deuteriomethanol   *MeOD*
Replacement of silyl groups by deuterium   Si≤ → D
s. *27*, 20

Carbon ↑   HS⇅C

Electrolysis   ⚡
Removal of 2-polyhalogenethyl-containing protective groups
s. *28*, 7
Sodium hydroxide   *NaOH*
Mercaptans from isothiouronium salts $SC{<}^{NH_2}_{NH_2}]^+$ → SH
s. *26*, 593
Potassium hydroxide/2-methoxyethanol   *KOH/HOCH$_2$CH$_2$OMe*
o-Dithiols from 1,3-benzodithiole-2-thiones   C
s. *27*, 623
Zinc/hydrogen chloride   *Zn/HCl*
Mercaptans from thiocyanates   SCN → SH
s. *28*, 28

Cation exchanger	←
Mercaptans	SCHal$_3$ → SH
from methyl via trichloromethyl thioethers – Arylmercaptans s. *29*, 30	
2-Mercaptoethylamine	HSCH$_2$CH$_2$NHEt
Mercaptans from thiolic acid esters	SCOR → SH

15.   Me$_2$N—⟨ ⟩—NHCONHCOCH$_2$SAc

　　　　　↓ H$_2$NCH$_2$CH$_2$SH

　　　Me$_2$N—⟨ ⟩—NHCONHCOCH$_2$SH [+ H$_2$NCH$_2$CH$_2$SAc]

Startg. m. mixed at room temp. under argon with 1.1 moles cysteamine in O$_2$-free acetonitrile, and stirred 10 min. at 65° → product. Y: 97%. – Functional groups susceptible to hydrolysis and reduction are not affected by the above method. F. e. s. T. Endo, K. Oda, and T. Mukaiyama, Chem. Lett. *1974*, 443.

Hydrogen chloride	HCl
S-Deacetylation	SAc → SH
with esterification s. *28*, 147	

Formation of H—Rem Bond

Exchange ↓↑

Halogen ↑ 　　　　　　　　　　　　　　　　　　　　HRem ↓↑ Hal

Tributyltin hydride　　　　　　　　　　　　　　　　　　　　Bu$_3$SnH
1,3,2-Benzodithiaboroles
from 2-chloro-1,3,2-benzothiaboroles　　　　$\begin{smallmatrix}S\\S\end{smallmatrix}$⟩BHal → $\begin{smallmatrix}S\\S\end{smallmatrix}$⟩BH
s. *30*, 379

Formation of H—C Bond

Uptake ⇓

Addition to Oxygen and Nitrogen　　　　　　　　　HC⇓ON

Sodium/liq. ammonia/tert-butanol　　　　　　　　　　Na/NH$_3$/Me$_3$C·OH
β-Aminoketones from isoxazoles　　　　　　　　　　　　　　　　　C
s. *28*, 204

Addition to Oxygen and Carbon HC⇓OC

Reductions with *potassium-graphite* C_8K
s. *28*, 29

Alkali metal/hexamethylphosphoramide/tert-butanol ←
Alcohols from oxo compds. CO → CHOH
s. *26*, 47

Lithium/liq. ammonia Li/NH_3
Alcohols from oxido compds. ←
with skeletal rearrangement – Reduction of cyclopropyloxido compds. – Synthesis of functionalized strained ring compds. s. *29*, 31
Carboxylic acids from lactones
with preferential hydrogenation of carbon-carbon double bonds s. *27*, 32
Cyclopropanol ring from ethyleneketones ○
and reverse reaction s. *26*, 29

Lithium/liq. ammonia/ammonium chloride $Li/NH_3/NH_4Cl$
3-Ethylenealcohols from 2,4-dienones → CHC(OH)
with formation of alcohols from oxido compds. – Stereospecific reduction – 1α,3β-Dihydroxy-Δ^5-steroids s. *29*, 32

Lithium N-benzyl-tert-butylamide ←
Reductions with lithium amides CO → CHOH
Sec. alcohols from ketones s. *27*, 33

Sodium/liq. ammonia Na/NH_3
Alcohols from oxido compds. → CHC(OH)
s. *26*, 55
Alcohols from cyclic ethers C
s. *27*, 37 suppl. *28*

Alkali metal/hexamethylphosphoramide-tert-butanol $Na/(Me_2N)_3PO/Me_3C·OH$
Alcohols from oxo compds. CO → CHOH
s. *26*, 47

Cuprous iodide s. under K[BHsec-Bu₃] and Li[AlH(OMe)₃] CuI

Calcium/liq. ammonia Ca/NH_3
Sec. alcohols from ketones
Stereospecific reduction s. *17*, 57 suppl. *28*

Zinc Zn
α-Hydroxyketones from α-diketones COCO → CH(OH)CO
Ethylene derivs. from non-enolizable α-diketones s. *27*, 34

Zinc/acetic acid Zn/CH_3COOH
3-Hydroxy-4-ketotetrahydro- CO → CHOH
from 3,4-diketotetrahydro-thiophene 1,1-dioxides
s. *28*, 724
α-Hydroxyketones from β-diketones
with skeletal rearrangement s. *28*, 608

Zinc amalgam Zn,Hg
Sec. alcohols from ketones
s. *28*, 48

Diborane B_2H_6
Diols from ethyleneketones
s. *28*, 103

Borane/boron fluoride BH_3/BF_3
Hydroboration of O-heterocyclics C
　Ethylenealcohols s. *26,* 30

Sodium tetrahydridoborate $Na[BH_4]$
Sec. alcohols from ketones CO → CHOH
　s. *29,* 933

2-Ethylenealcohols from α,β**-ethyleneketones** C : C·CO → C : C·CHOH
　s. *26,* 809

Diols from cyclic glycols via dialdehydes CHO → CH_2OH
　s. *29,* 155

1,2-Bromohydrins from α**-bromoaldehydes** CBrCHO → $CBrCH_2OH$

16.

Excess $NaBH_4$ added to a soln. of the startg. m. (prepn. s. 484) in tetrahydrofuran containing a little water, and stirred 2–4 hrs. at 25° → product. Y: 93%. – The above conditions prevent epoxide formation and reductive dehalogenation. F. e. s. B. B. Snider, J. Org. Chem. *38,* 3961 (1973).

Sec. alcohols from ketones CO → CHOH
　with formation of lactams from aminocarboxylic acid esters – Stereospecific reduction s. *28,* 30

Reductive ring opening of bicyclic isoureas C
s. *29,* 33

Potassium tetrahydridoborate $K[BH_4]$
Degradation of carbohydrates CHO → CH_2OH
　to peracylglycitols s. *27,* 540

Lithium tetrahydridoaluminate $Li[AlH_4]$
Indanols from indanones CO → CHOH
　s. *29,* 840

α,β**-Ethyleneketones** ←
　from 3-keto-1,2-en-1-olethers s. *29,* 805

Alcohols from oxido compds. ⟨O⟩ → HC(OH)
　Preferential and stereospecific reduction s. *27,* 35;
　with preferential replacement of methoxy groups by hydrogen s. *29,* 59

Lithium tetrahydridoaluminate/aluminum chloride $Li[AlH_4]/AlCl_3$
Reductive ring opening of spiroketals C
s. *29,* 34

Lithium tetrahydridoaluminate/
di-tert-butyl ketone $Li[AlH_4]/Me_3C\cdot CO\cdot CMe_3$
Sec. alcohols from ketones CO → CHOH
　Stereospecific reduction s. *29,* 35

Reduction with magnesium bis(tetrahydridoaluminate) $Mg[AlH_4]_2$
2-Ethylenealcohols from α,β**-ethylenealdehydes** C : C·CHO → C : C·CH_2OH
　s. *27,* 36

Lithium dihydridodimesitylborate $Li[BH_2Ar_2]$
Stereospecific and preferential reduction of ketones CO → CHOH
s. *17,* 61 suppl. *29*

Lithium hydridotriethylborate	*Li[BHEt₃]*
Alcohols from oxido compds.	◯ → CHC(OH)
Regio- and stereo-specific reduction s. 27, 35 suppl. 29	
Potassium hydridotri-sec-butylborate	*K[BHsec-Bu₃]*
Hydroxy- from keto-carboxylic acid esters	CO → CHOH
Stereospecific reduction	
s. *12*, 64 suppl. *29*	
Potassium hydridotri-sec-butylborate/cuprous iodide	*K[BHsec-Bu₃]/CuI*
Sec. alcohols from ketones	
s. *30*, 40	
Diisobutylaluminum hydride	*i-Bu₂AlH*
Stereospecific reduction s. *24*, 61, suppl. *29;* tropine s. Y. Hayakawa and R. Noyori, Bull. Chem. Soc. Japan *47*, 2617 (1974)	
Triethylborane s. under Lithium hydridotri-tert-butoxoaluminate	*Et₃B*
(Dialkylamino)alanes	*RR'NAlH₂*
Asym. reduction	
s. *23*, 48 suppl. *29*	
Sodium dihydridobis-(2-methoxyethoxo)aluminate	*Na[AlH₂(OCH₂CH₂OMe)₂]*
Alcohols from oxido compds.	◯ → CHC(OH)
s. *27*, 35	
Diols from lactones	C

17.

with **O-deacetylation.** Startg. m. allowed to react 18 hrs. at 25° with Na-dihydridobis-(2-methoxyethoxo)aluminate in benzene → triol. Y: ca. 100%. K. Miyano et al., Tetrah. Let. *1974*, 1545.

Lithium hydridotrimethoxoaluminate/cuprous iodide	*Li[AlH(OMe)₃]/CuI*
Alcohols from oxido compds.	◯ → CHC(OH)
s. *29*, 78	
Lithium hydridotri-tert-butoxoaluminate/triethylborane	←
Alcohols from cyclic ethers	C
s. *27*, 37	
Boron fluoride s. under BH₃	*BF₃*
Aluminum chloride s. under Li[AlH₄]	*AlCl₃*
Reductions with *benzopinacol*	←
α-Hydroxyketones from α-diketones – Quinols from quinones s. *28*, 31	
Di-tert-butyl ketone s. under Li[AlH₄]	*Me₃C·CO·CMe₃*
Microorganism	←
Stereospecific reduction	CO → CHOH
Sec. alcohols from ketones s. *27*, 38; asym. reduction s. C. J. Sih et al., Am. Soc. *97*, 865 (1975).	
Yeast	←
Asym. reduction	C(OH)CO → C(OH)CHOH
Glycols from α-hydroxyketones s. *29*, 36	

Thiourea dioxide/sodium hydroxide $\underset{H_2N}{\overset{H_2N}{>}}C-SO_2^-/NaOH$

Sec. alcohols from ketones   CO → CHOH
s. 27, 39; stereospecific reduction s. N. Chatterjie et al., J. Med. Chem. 18, 490 (1975)

Hydrazine s. under Ni $H_2N \cdot NH_2$

Sodium tetracarbonylferrate(II) $Na_2Fe(CO)_4$
Aldehydocarboxylic acids
from dicarboxylic acid anhydrides s. 29, 64 $\underset{CO}{\overset{CO}{>}}O \rightarrow \underset{COOH}{\overset{CHO}{<}}$

Bis(dimethylglyoximato)cobalt(II)/quinine hydrochloride   ←
Asym. hydrogenation   CO → CHOH

18.   PhCOCOPh → PhCH(OH)COPh

A soln. of benzil in benzene hydrogenated at 10° in the presence of bis(dimethylglyoximato)cobalt(II) and equimolar amounts of quinine and its hydrochloride → (+)-benzoin. Y: 95%; optical Y: 71%. – The optical yield increases at lower temp. Also (−)-isomer with other chiral aminoalcohols s. Y. Ohgo et al., Chem. Lett. *1974*, 33.

Nickel/hydrazine Ni/H_2NNH_2
Aminofluoren-9-ols from nitrofluorenones
s. 27, 15

Rhodium-alumina $Rh-Al_2O_3$
Sec. alcohols from ketones
with benzene ring hydrogenation s. 26, 52

Palladium-barium sulfate/cyclohexene   ←
Alcohols from oxido compds. $\overset{O}{\triangle}$ → CHC(OH)
Preferential and selective transfer-hydrogenation
s. 28, 32; review of catalytic transfer-hydrogenation s. G. Brieger and T. J. Nestrick, Chem. Rev. 74, 567 (1974)

Palladium-carbon   Pd-C
Isoflavan-4-ols from isoflavones   ←
s. 29, 52
Partial and selective hydrogenation   CO → CHOH
of dicarboxylic acid imides s. 26, 31
Reductive Δ^2-oxazoline ring opening   C
α-Acylaminocarboxylic acid esters s. 26, 32

Iridium-carbon   Ir-C
2-Ethylenealcohols from α,β-ethylenealdehydes   C : C·CHO → C : C·CH$_2$OH
s. 27, 40

Iridium tetrachloride/trimethyl phosphite $IrCl_4/MeO)_3P$
Sec. alcohols from ketones   CO → CHOH
Stereospecific reduction s. 26, 33

Addition to Nitrogen and Carbon   HC⇓NC

Electrolysis   ↯
2-Ethyleneamines from α,β-ethylenenitriles   C : C·CN → C : C·CH$_2$NH$_2$
s. 27, 41

Sodium amalgam   Na,Hg
Acoxylamines from acyloximes   COO·N : C → COO·NHCH
s. 27, 42

Lithium/liq. ammonia   *Li/NH$_3$*
Cyclic tert. amines   C
    from N-condensed cyclimmonium salts – Selective reduction s. 29, 37

Sodium/liq. ammonia   *Na/NH$_3$*
Amines from aziridines   ▽N → CHC(NH·)
s. 26, 55

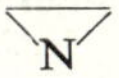

Sodium/liq. ammonia/ethanol   *Na/NH$_3$/EtOH*
1,4-Dihydropyridines from pyridines   Ⓗ
s. 27, 816

Aluminum amalgam   *Al,Hg*
Thiazolidines from Δ^2-thiazolines
s. 28, 176

Diborane   *B$_2$H$_6$*
2-Ethylene-*sec*-amines and sec. amines   C : N → CHNH
    from α,β-ethyleneazomethines s. 27, 43

2-Aminoalcohols from cyanohydrins   C(OH)CN → C(OH)CH$_2$NH$_2$
s. 16, 54 suppl. 29

2,3,4,5-Tetrahydro-1H-1,4-benzodiazepines   ←
    from 1,3-dihydro-2H-1,4-benzodiazepin-2-ones s. 29, 58

Sodium tetrahydridoborate   *NaBH$_4$*
Amines from imines   C : NH → CHNH$_2$
s. 26, 34

Hydrogenation   C : N → CHNH
of carbon-nitrogen with retention of nitrogen-phosphorus double bonds

19. 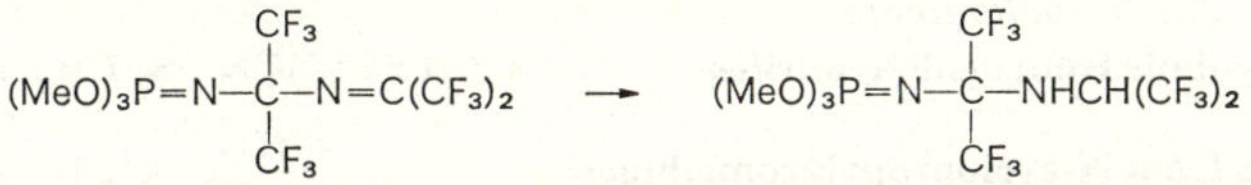   (MeO)$_3$P=N—C(CF$_3$)—N=C(CF$_3$)$_2$ →   (MeO)$_3$P=N—C(CF$_3$)—NHCH(CF$_3$)$_2$
                              (CF$_3$)   (CF$_3$)

Trimethoxy-[5,5,5-trifluoro-2,2,4-tris(trifluoromethyl)-1,3-diaza-3-pentenylidene]-phosphorane refluxed ca. 1 hr. with NaBH$_4$ in anhydrous ethanol → trimethoxy-[5,5,5-trifluoro-2,2,4-tris(trifluoromethyl)-1,3-diazapentylidene]phosphorane. Y: 85%. F. e. s. K. Burger et al., B. *107*, 1526 (1974).

N-Heteroaromatic prim. amines from nitriles   CN → CH$_2$NH$_2$

20.    N◯—CN   →   N◯—CH$_2$NH$_2$

4-Cyanopyridine refluxed 5 hrs. with 5 moles NaBH$_4$ in ethanol → 4-pyridylmethylamine. Y: 53%. – Functional groups usually resistant to reduction with NaBH$_4$ may become reducible by the electronic influence of the heteroaromatic ring. F. e. s. Y. Kikugawa et al., Chem. Pharm. Bull. *21*, 1927 (1973).

Aziridine from Δ^1-azirines   ▽N → ▽N
s. 29, 38   H

Preferential hydrogenation of N-heterocycles   Ⓗ
s. 26, 366
Dihydropteridinones
    3,4-Dihydro-7(8H)-pteridinones s. 28, 33
2,1-Benzisoxazolines
    from 2,1-benzisoxazolium salts s. 26, 18

Alcohols from carboxylic acid esters C
  with and without reductive lactam ring opening s. 29, 39
1,2,3,4,5,6-Hexahydro-1,6-benzodiazocines
  by nucleophilic ring opening s. 27, 44
Cyclobutanes from uracil dimers
  s. 27, 45
Epimerization of N-condensed heterocyclics ←
via N-condensed cyclimmonium salts
  s. 30, 167

Sodium tetrahydridoborate/boron fluoride $Na[BH_4]/BF_3$
Prim. amines from nitriles $CN \rightarrow CH_2NH_2$
  with retention of nitro groups s. 27, 46
2,3,4,5-Tetrahydro-1H-1,4-benzodiazepines ←
  from 1,3-dihydro-2H-1,4-benzodiazepin-2-ones s. 29, 58

Sodium tetrahydridoborate/cobaltous chloride $Na[BH_4]/CoCl_2$
Reductions with sodium tetrahydridoborate-transition metal salts $CN \rightarrow CH_2NH_2$
  Prim. amines from nitriles s. 26, 35

Sodium tetrahydridoborate/nickel $Na[BH_4]/Ni$
Prim. amines from nitriles
  s. 26, 35

Potassium tetrahydridoborate $K[BH_4]$
Sec. amines from azomethines $C:N \rightarrow CHNH$
  s. 29, 390

Tetrabutylammonium tetrahydridoborate/methyl iodide $Bu_4N^+[BH_4]^-/MeI$
Reactions with diborane in methylene chloride $CN \rightarrow CH_2NH_2$
  Prim. amines from nitriles – F. reductions, also Brown hydration s. 28, 34

Lithium tetrahydridoaluminate $Li[AlH_4]$
2-Aminoalcohols from α-siloxynitriles $C(OSi\leqslant)CN \rightarrow C(OH)CH_2NH_2$
  s. 27, 720
Sec. amines from N-cyclopropylazomethines $C:N \rightarrow CHNH$
  s. 27, 52
Reductive ring opening C
  of N-condensed cyclimmonium salts s. 26, 36
  of N-condensed cyclic amidines s. 26, 37
Reductive 4-quinazolone ring opening
  s. 29, 40
Ring opening of cyclimmonium salts
  with formation of allenes from acetylene derivs. s. 29, 41

Boron fluoride s. under $Na[BH_4]$ BF_3

Methyl iodide s. under $Bu_4N^+[BH_4]^-$ CH_3I

Organosilicon hydrides s. under $RhCl(Ph_3P)_3$ ←

Cobaltous chloride s. under $Na[BH_4]$ $CoCl_2$

Nickel s. a. under $Na[BH_4]$ and $Pt-Ni-Al_2O_3$ Ni

Nickel/ammonia Ni/NH_3
Diamines from ketonitriles $CN \rightarrow CH_2NH_2$
  s. 28, 362

Chlorotris(triphenylphosphine)rhodium(I)/organosilicon hydrides ←
Sec. amines from azomethines $C:N \rightarrow CHNH$
  s. 29, 42

Palladium-carbon	*Pd-C*
1,2-Dihydropyridines from pyridines s. 26, 38	Ⓗ
Platinum-nickel-alumina	*Pt-Ni-Al$_2$O$_3$*
Prim. amines from nitriles Partial and total reduction s. 26, 39	CN → CH$_2$NH$_2$
Platinum oxide	*PtO$_2$*
Sec. amines from azomethines s. 29, 371	C:N → CHNH

Addition to Sulfur and Carbon   HC↓SC

Irradiation s. under Ethanol	⚡
Sodium/liq. ammonia	*Na/NH$_3$*
Reductive 6H-1,3-thiazine ring opening	C

21.

Small pieces of Na added at –33° to a soln. of 2-amino-4,6,6-trimethyl-6H-1,3-thiazine in anhydrous ammonia → N-(1,3-dimethyl-1-butenyl)thiourea. Y: 96%. F. e. s. S. Hoff and A. P. Blok, R. *93*, 78 (1974).

Tri-n-butylamine s. under H$_2$S	*n-Bu$_3$N*
Sodium tetrahydridoborate	*Na[BH$_4$]*
Mercaptans from thioketones 2-Adamantanethiols s. 27, 47	CS → CHSH
Ethanol/irradiation	*EtOH/⚡*
Sulfhydryl from thiocarbonyl groups Selective reduction s. 26, 40	CS → CHSH
Hydrogen sulfide/tri-n-butylamine	*H$_2$S/n-Bu$_3$N*
Reductive ring contraction Thiazolidines from 2,3-dihydro-4H-1,4-thiazines – also reverse oxidative ring expansion s. 27, 48	Ö

Addition to Carbon and Carbon   HC↓CC

Electrolysis	⚡
Electrocatalytic hydrogenation	Ⓗ

22.

3-Methoxy-1,3,5(10),8-estratetraen-17β-ol in ethanol containing 10%-H$_2$SO$_4$ electrolyzed 2 hrs. at 20° and 0.4 amp. with a Pt-cathode catalytically activated by a

Pd-deposit prepared by electrolysis of a 2%-palladium chloride soln. in 1 N HCl
→ 3-methoxy-8α-estra-1,3,5(10)-trien-17β-ol. Y: ca. 90%. – The 8β-isomer can be
obtained at higher temp. F. e. s. K. Junghans, B. *107*, 3191 (1974).

Lithium   Li
Ring opening by reductive cleavage   Ⓒ
    of 1,3-diene central bonds s. 27, 49

Cesium,potassium,sodium   Cs,K,Na
Reactions with cesium alloys   Ⓞ
    Reductive ring closure to isocyclics s. 28, 35

Lithium-methanol   Li-MeOH
cis-**Stilbenes from tolans**   C ≡ C → CH : CH
    cis-Stibene-α,α'-D₂ s. 26, 41

Sodium hydride s. under Li/NH₃   NaH

n-Butyllithium s. under Cobalt 2-ethylhexanoate   n-BuLi

Pentyn-1-yllithium s. under CuH $C_3H_7C \equiv CLi$

Lithium/liq. ammonia   Li/NH₃
Preferential hydrogenation   C : C → CHCH
    of carbon-carbon double bonds with formation of carboxylic acids from
    lactones s. 27, 32
Carboxylic acids from α,β-ethylenecarboxylic acids   C:C·COOH → CHCHCOOH
s. 26, 54
Preferential hydrogenation   Ⓗ
    of isocycles or N-heterocycles s. 26, 42

Lithium/liq. ammonia-sodium hydride   Li/NH₃-NaH
Partial hydrogenation of isocycles
    with phenol ring retention s. 26, 43

Lithium/liq. ammonia/water   Li/NH₃/H₂O
Partial ring hydrogenation
    of arylcarboxylic acids s. 26, 44

Lithium/liq. ammonia/methanol   Li/NH₃/MeOH
Birch reduction
    Preferential hydrogenation of isocycles s. 26, 42

Lithium/liq. ammonia/ethanol   Li/NH₃/EtOH
3-Ethylenealcohols from 2,4-dienones   C:C·C:C·CO → CH·C:C·CH·CHOH
    Stereospecific reduction s. 29, 928

Lithium/liq. ammonia/isopropyl alcohol   Li/NH₃/Me₂CHOH
Birch reduction of cyclic tert. amines   Ⓗ
with and without subsequent conjugation

23.

N-Methylindoline in tetrahydrofuran added to liq. NH₃ along with isopropanol,
stirred, Li-ribbon added in small pieces, and the product distilled

rapidly in an apparatus washed with aq. NH_3 and dried → 4,7-dihydro-
Y: 86%.

from a small amount of Dowex 50W-X2 cation exchanger, H-form → 4,5-dihydro-N-methylindoline.
Y: 81%.

P. Radlick, R. Klem, and H. T. Crawford, J. Org. Chem. 39, 1587 (1974).

Lithium/liq. ammonia/tert-butyl alcohol $Li/NH_3/Me_3C \cdot OH$
Partial hydrogenation of isocycles
s. 28, 989

Lithium/liq. ammonia/ammonium chloride $Li/NH_3/NH_4Cl$
3-Ethylenealcohols from 2,4-dienones $C:C \cdot C:C \cdot CO$ → $CH \cdot C:C \cdot CHOH$
s. 29, 32

Lithium ammonia complex/tert-amyl alcohol   ←
Reduction with lithium ammonia complex $C:C \cdot CO$ → $CHCHCO$
    Ketones from α,β-ethyleneketones – Stereospecific hydrogenation s. 27, 50

Lithium/ethylamine
and lithium/ethylenediamine $Li/EtNH_2$ and $Li/H_2NCH_2CH_2NH_2$
Cyclobutane from bicyclo[1.1.0]butane ring   C
s. 26, 45

Sodium/liq. ammonia Na/NH_3
Reductive ring opening of isocyclics   C
s. 29, 43

Sodium/rubidium carbonate-butylamine Na/Rb_2CO_3-$BuNH_2$
Ring hydrogenation with alkali metal catalysts   Ⓗ
    Partial hydrogenation s. 26, 46

Potassium/liq. ammonia K/NH_3
Partial hydrogenation of isocycles
    with phenol ring retention s. 26, 43

Reduction with *alkali metal/*
hexamethylphosphoramide-tert-butanol $Na/(Me_2N)_3PO$-$Me_3C \cdot OH$
Hydrogenation of unactivated carbon-carbon multiple bonds   ←
    Alcohols from oxo compds. – Steric effect of metal and medium – Reduction of anthracenes s. 26, 47

Rubidium carbonate s. under Na Rb_2CO_3

Sodium cyanide s. under Na[BH₄] $NaCN$

Butylamine s. under Na/Rb₂CO₃ $BuNH_2$

Cuprous hydride/pentyn-1-yllithium $CuH/C_3H_7C \equiv CLi$
Oxo from α,β-ethyleneoxo compds. $C:C \cdot CO$ → $CHCHCO$
s. 29, 44

Calcium/liq. ammonia Ca/NH_3
cis- and trans-Ethylene from acetylene derivs. $C \equiv C$ → $CH:CH$
s. 28, 36

Zinc/n-propyl alcohol Zn/n-$PrOH$
cis-Ethylene from acetylene derivs.
    Selective hydrogenation s. 27, 51

Zinc/acetic acid Zn/CH_3COOH
γ,δ-Ethyleneketones from 2,4-dienones $C:C \cdot C:C \cdot CO$ → $C:C \cdot CH \cdot CH \cdot CO$
s. 29, 928

Zinc-amalgam   *Zn,Hg*
Preferential hydrogenation   C : C·N< → CHCH·N<
of carbon-carbon double bonds − Amines from enamines s. 29, 45

Mercuric acetate s. under Na[BH₄]   *(CH₃COO)₂Hg*

Diborane   *B₂H₆*
Sec. amines from α,β-ethyleneazomethines   C : C·C : N → CHCHCHNH
s. 27, 43

Sodium tetrahydridoborate (s. a. under (CH₃COO)₂Ni and PdCl₂)   *Na[BH₄]*
Cyclic 1,1-aminoethers   C : C → CHCH
from cyclic 1-alkoxyenamines s. 28, 793

Nitro compds. and oximes   C : C·NO₂ → CHCHNO₂
from 1,1-nitroethylene derivs. s. 28, 37

Selective ring hydrogenation of nitroarenes   Ⓗ
1,5-Dinitro-3-azabicyclo[3.3.1]non-6-enes from 1,3-dinitrobenzenes via 3,5-bis-*aci*-nitrocyclohexenes s. 26, 48

Sodium tetrahydridoborate/sodium cyanide   *Na[BH₄]/NaCN*
Reductive ring closure of cyclimmonium salts   O
s. 26, 761

Sodium tetrahydridoborate/pyridine   *Na[BH₄]/C₅H₅N*
Selective reduction of carbon-carbon double bonds   C : C → CHCH
Arylmethyl- from arylmethylene-1,3-indandiones s. 26, 49

Sodium tetrahydridoborate/mercuric acetate   *Na[BH₄]/(CH₃COO)₂Hg*
Tert. amines from enamines   C : C·N< → CHCH·N<
s. 28, 38

Sodium tetrahydridoborate-carbon s. under H₂PtCl₆   *Na[BH₄]-C*

Sodium tetrahydridoborate/cobaltous chloride   *Na[BH₄]/CoCl₂*
Carboxylic from ethylenecarboxylic acid esters   C : C → CHCH
s. 26, 35

Lithium tetrahydridoaluminate   *Li[AlH₄]*
3-Allenealcohols from 1-acoxy-2-en-4-ynes   ←

24.   CH₃—CHCH=CHC≡CH → CH₃—CHCH₂CH=C=CH₂
                | |
                OAc   OH

Startg. m. refluxed 16 hrs. with a suspension of LiAlH₄ in anhydrous ether → hexa-1,2-dien-5-ol. Y: ca. 100%. − Reduction of the acetates occurs more easily than reduction of the free alcohols. F. e. s. M. Santelli and M. Bertrand, Bl. *1973*, 2331; deuteriated products and stereochemistry s. ibid. *1973*, 2335.

Sec. amines from N-cyclopropylazomethines   C
s. 27, 52

Lithium tetrahydroaluminate/aluminum chloride   *Li[AlH₄]/AlCl₃*
1,2,3,6-Tetrahydropyridine from 2-pyridone ring   ←
s. 26, 63

Sodium dihydridobis-(2-methoxyethoxo)aluminate   *Na[AlH₂(OCH₂CH₂OMe)₂]*
β-Aminoketones   CO·C : C·N< → COCHCHN<
from vinylogous carboxylic acid amides s. 29, 46

Aluminum chloride s. under Li[AlH₄]   *AlCl₃*

n-Propyl alcohol s. under Zn and Fe   *n-PrOH*

Formamides s. under Formic acid   *RNHCHO*

Formic acid s. a. under RuHCl(Ph₃P)₃   HCOOH

Formic acid/formamides   HCOOH/RNHCHO
Stereospecific hydrogenation   C : C → CHCH
  of enaminolactams to lactams with formation of formamides from ketones s. *27*, 53

Polymethylhydrosiloxane s. under Pd-C   ←

Triphenyltin hydride   Ph₃SnH
Preferential and selective hydrogenation
  of carbon-carbon double bonds – Reduction with organotin hydrides, review s. *26*, 50

Triphenyltin deuteride   Ph₃SnD
β-Deuterio- from α,β-ethylene- via α,β-dideuterio-ketones C : C·CO → CDCDCO
  Preferential and selective deuteriation of carbon-carbon double bonds s. *29*, 47

Potassium azodicarboxylate/acetic acid   KOOC·N : N·COOK/CH₃COOH
Preferential hydrogenation   C : C·COOH → CHCHCOOH
  of aliphatic carbon-carbon double bonds – Carboxylic from α,β-ethylene-carboxylic acids under mild conditions – In situ generation of diimide s. *28*, 39

Phosphorus s. under HI   P

Copper chromite   CuCr₂O₄
Total hydrogenation   Ⓗ
  of oxo-N-heterocyclics s. *26*, 67

Molybdenocene hydride   (C₅H₅)₂MoH₂
Ethylene derivs. from dienes   C : C → CHCH
s. *29*, 54

Polystyrenetricarbonylchromium   ←
Z-β,γ-Ethylene- from 2,4-diene-carboxylic acid esters C:C·C:C → CH·C:C·CH
Geospecific selective hydrogenation

25.

Methyl sorbate in cyclohexane hydrogenated 24 hrs. at 160° under 500 psi initial H₂-pressure with polystyrenetricarbonylchromium, a polymer-based homogeneous catalyst, prepared by reaction of a swollen 1%-divinylbenzene cross-linked polystyrene with hexacarbonylchromium → methyl Z-3-hexenoate. Y: 97.4%. C. U. Pittman, Jr., B. T. Kim, and W. M. Douglas, J. Org. Chem. *40*, 590 (1975); **review of polymer-based homogeneous catalysts** s. Z. M. Michalska and D. E. Webster, Chem. Tech. *5*, 117 (1975).

Hydrogen iodide/phosphorus   HI/P
Preferential hydrogenation   C : C → CHCH
  of aliphatic carbon-carbon double bonds with replacement of bromine by hydrogen s. *26*, 73

Rhenium sulfide   Re₂S₇
Hydrogenation of N-heterocycles   Ⓗ
s. *29*, 389

Iron/n-propyl alcohol   Fe/n-PrOH
cis-Ethylene from acetylene derivs.   C ≡ C → CH : CH
s. *28*, 36

Iron carbonyl/sodium hydroxide $\qquad Fe(CO)_5/NaOH$
Selective hydrogenation $\qquad$ C : C → CHCH
of carbon-carbon double bonds s. 28, 40

Cobalt 2-ethylhexanoate/n-butyllithium ←
Hydrogenation of carbon-carbon double bonds
s. 27, 54

Cobaltous chloride s. a. under Na[BH$_4$] $\qquad CoCl_2$

Cobaltous chloride/dimethylglyoxime $\qquad CoCl_2/[CH_3C(:NOH)]_2$
Reduction with bis(dimethylglyoximato)cobalt(II)
Selective hydrogenation of activated carbon-carbon double bonds s. 27, 55

Nickel $\qquad Ni$
Lactones from ethylenelactones $\qquad$ C : C → CHCH
Retention of allylic oxygen – Hydroxycarboxylic acids from lactones and reverse reaction s. 27, 56

5-Alkyl- from 5-alkylidene-4-imidazolones
s. 26, 193

α-**Acylaminocarboxylic acid amides** $\qquad$ C
from 4-alkylidene-5-oxazolones s. 28, 288

Nickel,aluminum $\qquad Ni,Al$
Carboxylic acids from α,β-ethylenecarboxylic acids C : C · COOH → CHCHCOOH
s. 26, 51

Potassium hexacyanodiniccolate(I) $\qquad K_4[Ni_2(CN)_6]$
Hydrogenation of carbon-carbon double bonds $\qquad$ C : C → CHCH
s. 24, 66

Nickel acetate/sodium tetrahydridoborate $\qquad (CH_3COO)_2Ni/Na[BH_4]$
Reductions with P-2 nickel
Preferential hydrogenation of carbon-carbon double bonds s. 29, 48; preferential and selective hydrogenation with P-1 nickel, β-dihydroionone from ionone, s. P. Lombardi, G. *104*, 867 (1974)

Ruthenium complex ←
Partial homogeneous hydrogenation
of carbon-carbon double bonds s. 28, 869

Chlorohydridotris(triphenylphosphine)ruthenium/
formic acid $\qquad RuHCl(Ph_3P)_3/HCOOH$
Homogeneous hydrogenation $\qquad$ C : CH$_2$ → CHCH$_3$
of terminal ethylene derivs. s. 29, 49

Dichlorotris(triphenylphosphine)ruthenium(II)/
ethylene glycol $\qquad RuCl_2(Ph_3P)_3/HOCH_2CH_2OH$
Ketones from α,β-ethyleneketones $\qquad$ C : C · CO → CHCHCO
s. 29, 50

Rhodium-alumina $\qquad Rh\text{-}Al_2O_3$
Benzene ring hydrogenation $\qquad$ Ⓗ
with formation of sec. alcohols from ketones s. 26, 52

Rhodium-carbon $\qquad Rh\text{-}C$
Ring hydrogenation
s. 23, 66; s. a. K. Chebaane, M. Guyot, and D. Molho, Bl. *1975*, 244

Rhodium carbonyl $\qquad Rh_6(CO)_{16}$
Selective hydrogenation $\qquad$ C : C → CHCH
of conjugated carbon-carbon double bonds s. 28, 18

Diphosphine-rhodium(I) catalyst ←
Asym. homogeneous hydrogenation
s. *27*, 57; s. a. A. P. Stoll and R. Süess, Helv. *57*, 2487 (1974); cf. A. Levi, G. Modena, and G. Scorrano, Chem. Commun. *1975*, 6

Palladium-calcium carbonate $Pd\text{-}CaCO_3$
Partial and stereospecific hydrogenation
of carbon-carbon double bonds s. *28*, 42

Palladium-calcium carbonate/pyridine $Pd\text{-}CaCO_3/C_5H_5N$
Preferential and selective hydrogenation
of carbon-carbon double bonds – Retention of oxido groups s. *27*, 58

Palladium-alumina $Pd\text{-}Al_2O_3$
Partial ring hydrogenation Ⓗ
s. *29*, 51

Palladium-carbon $Pd\text{-}C$

Aldehydes from α,β-ethyleneacetals
s. *27*, 59
$C:C\cdot CH\underset{O}{\overset{O}{\diagdown}} \rightarrow CHCHCHO$

α-Methyl- from α-methylene-ketones $CO\cdot C:CH_2 \rightarrow CO\cdot CHCH_3$
Stereospecific hydrogenation s. *27*, 89

**α-Amino- from α-amino-α,β-ethylene-carboxylic
acid esters** $C:C(NH_2)COOR \rightarrow CHCH(NH_2)COOR$
s. *29*, 739

Direction of stereospecificity $C:C \rightarrow CHCH$
by functional groups – Hydrogenation of carbon-carbon double bonds s. *28*, 43

Cycloocta-1,5-diene from -1,3,5-triene ring Ⓗ
s. *29*, 56

Cyclobutane-*cis*-1,2-diols
from 1,2-di(siloxy)cyclobut-1-enes via 1,2-di(siloxy)cyclobutanes s. *27*, 60

Isoflavonones and isoflavan-4-ols
from isoflavones s. *29*, 52

Pyrrolizidines from dihydropyrrolizines
Stereospecific ring hydrogenation s. *30*, 631

Preferential hydrogenation $C:C \rightarrow CHCH$
of carbon-carbon double bonds after steric blocking of a neighboring hydroxyl group as dimethylisopropylsilyl ether – Protection of alcohol groups as *tert*-butyldimethylsilyl ethers s. *28*, 44; limitations of this protection method cf. G. H. Dodd, B. T. Golding, and P. V. Ioannou, Chem. Commun. *1975*, 249

gem-**Hydrogenolysis** C
Double bicyclo[1.1.0]butane ring opening s. *26*, 53

Palladium-carbon/benzyl alcohol $Pd\text{-}C/PhCH_2OH$
Transfer hydrogenation ←
Preferential and selective hydrogenation of carbon-carbon multiple bonds
s. *29*, 53

Palladium-carbon/acetic acid $Pd\text{-}C/CH_3COOH$
Hydrogenation of carbon-carbon double bonds $C:C \rightarrow CHCH$
with reduction of formyl to methyl groups s. *26*, 68

Palladium-carbon/polymethylhydrosiloxane ←
Reduction with polymethylhydrosiloxane $CO\cdot C:C \rightarrow CO\cdot CHCH$
Ketones from α,β-ethyleneketones s. *28*, 45

Palladium-lead-calcium carbonate $Pd\text{-}Pb\text{-}CaCO_3$
Preferential hydrogenation   C : C → CHCH
of carbon-carbon double bonds s. *17*, 94

Palladium-thorium dioxide $Pd\text{-}ThO_2$
Ethylene derivs. from dienes   C : C·C : C → CH·C : C·CH
s. *29*, 54; from 1,3-dienes by an iridium phosphine complex with irradiation cf. W. Strohmeier and L. Weigelt, J. Organometal. Chem. *82*, 417 (1974)

Palladium oxide   *PdO*
Asym. hydrogenation of proline derivs.   C : C → CHCH
Asym. synthesis of α-aminocarboxylic acids and peptides – Review of asym. synthesis s. *29*, 55

Palladous chloride/sodium tetrahydridoborate $PdCl_2/Na[BH_4]$
Preferential stereospecific ring hydrogenation   Ⓗ
with catalyst prepared in situ s. *27*, 61; reductions with a non-pyrophoric Pd-catalyst prepared by reduction of PdCl$_2$ with NaBH$_4$ cf. T. W. Russell and D. M. Duncan, J. Org. Chem. *39*, 3050 (1974)

Hexachloroplatinic acid/sodium tetrahydridoborate-carbon/-
hydrogen chloride $H_2PtCl_6/Na[BH_4]\text{-}C/HCl$
Cycloalkane ring   ←
from cyclic β,γ-ethyleneketones s. *29*, 933

Via intermediates   *v.i.*
Cycloocta-1,5-diene from -1,3,5-triene ring via rhodium complexes   Ⓗ
s. *29*, 56

Exchange   ⇵

Oxygen ↑   HC ⇵ O

Irradiation (s. a. under Cl_3SiH)   ⚡
Replacement of acoxy groups by hydrogen   OAc → H
s. *26*, 984

Replacement of acetoxy groups by hydrogen   ←
Migration of carbon-carbon double bonds s. *28*, 46

Sodium (s. a. under $(C_5H_5)_2TiCl_2$)   *Na*
Replacement of alkoxy groups by hydrogen   OR → H
preferential s. *30*, 47

Potassium hydroxide s. under $H_2N\cdot NH_2$ *KOH*

Lithium/liq. ammonia Li/NH_3
Hydrocarbons from oxo compds.   CO → CH_2
s. *28*, 47

Carboxylic from β-ketocarboxylic acid esters   C : C·OR → $CHCH_2$
via β-alkoxy-α,β-ethylenecarboxylic acid esters – Stereospecific reduction of enolethers – Carboxylic acids, also from α,β-ethylenecarboxylic acids s. *26*, 54

Lithium/liq. ammonia/1-methoxy-2-propanol $Li/NH_3/CH_3CH(OH)CH_2OMe$
Sec. amines from tert. N-oxides   C

26.

Reductive ring opening. 2 g. startg. crude N-oxide added to liq. NH$_3$ containing 1-methoxy-2-propanol, then Li added portionwise with vigorous stirring, which is continued for 6 min., and the product isolated as the hydrochloride → 1.2 g. 1,2,3,4,5,6,7,8-octahydro-10,11-dimethoxy-3-benzazecine hydrochloride. F. e. s. J. P. Yardley, Synthesis *1973*, 543.

Lithium/liq. ammonia/cobalt $Li/NH_3/Co$
Hydrocarbons from oxo compds. $CO → CH_2$
s. *28*, 47

Lithium/ethylamine/tert-butyl alcohol $Li/EtNH_2/Me_3C \cdot OH$
Hydrocarbons from phosphorodiamidates $O \cdot PO(N<)_2 → H$
s. *28*, 65

Sodium/liq. ammonia Na/NH_3
Reductive opening of heterocycles   C
    by alkali metals in liq. ammonia – Hydrocarbons from ketals – also alcohols from oxido compds., and amines from aziridines s. *26*, 55

Tri-n-propylamine s. under Cl$_3$SiH $n\text{-}Pr_3N$

Pyridine s. under Na[BH$_4$] C_5H_5N

Calcium chloride s. under Na[BH$_4$] $CaCl_2$

Lithium dimethylcuprate $Li[CuMe_2]$
Ketones from α-acetoxy- and α-bromo-ketones   COC(OAc) → COCH

27.

2α-Acetoxy-5α-cholestan-3-one added at 0° under N$_2$ to Li-dimethylcuprate in ether, and the product isolated after 4 min. → 5α-cholestan-3-one. Y: 82%.
F. e. s. J. R. Bull and A. Tuinman, Tetrah. Let. *1973*, 4349.

3β-Acetoxy-5α-bromocholestan-6-one after 2 min. → 3β-acetoxy-5α-cholestan-6-one. Y: 95%.

Cuprous iodide s. under Li[AlH(OMe)$_3$]   CuI

Zinc   Zn
Allenes from acoxy-2-acetylenes $C(OAc)C \equiv C → C:C:CH$
s. *28*, 115

Zinc/formic acid   *Zn/HCOOH*
Methylene and *sec*-alcohol from keto groups $CO \to CH_2$
s. *28*, 48

Zinc/acetic acid   *Zn/CH₃COOH*
Ketones from α,β-oxidoketones   CO—△—O → CO·CHCH

28.

Startg. m. heated at 100° in acetic acid containing Zn-powder → 3-hydrocinnamoylindole. Y: 95%. V. E. Zhigachev, Y. I. Smushkevich, and N. N. Suvorov, Tr. Mosk. Khim.-Tekhnol. Inst. *74*, 59 (1973); C. A. *82*, 31201.

Oximes from 1,1-nitroethylene derivs. $C:C\cdot NO_2 \to CHC:NOH$
s. *28*, 37

Zinc/silica/hydrogen $Zn/SiO_2/H_2$
Hydrocarbons from ketones $CO \to CH_2$
s. *28*, 49

Borane BH_3
Prim. alcohols from carboxylic acids $COOH \to CH_2OH$
    Selective reduction – Reduction of sterically hindered acids – 2-Aminoalcohols from α-aminocarboxylic acids – also reduction via enolethers s. *29*, 57 – Labeled compds. s. *29*, 89

Hexahydropyridazines $CO \to CH_2$
    from hexahydropyridazine-3,6-diones – Reductions of cyclic and acyclic hydrazides with borane s. *26*, 56

Borane/trimethyl borate/dimethyl sulfide $BH_3/(MeO)_3B/Me_2S$
Prim. alcohols from carboxylic acids $COOH \to CH_2OH$

29.   p—O₂NC₆H₄COOH → p—O₂NC₆H₄CH₂OH

Selective reduction. Borane-dimethyl sulfide added dropwise under N₂ during 1 hr. to a refluxing soln. of p-nitrobenzoic acid and trimethyl borate in tetrahydrofuran, stirred and refluxed an additional 3 hrs. → p-nitrobenzyl alcohol. Y: 97.6%. F. e. s. C. F. Lane et al., J. Org. Chem. *39*, 3052 (1974).

Diborane B_2H_6
Chromans from 4-chromanones $CO \to CH_2$
s. *24*, 80 suppl. 29

2,3,4,5-Tetrahydro- and 2,3-dihydro-1H-1,4-benzodiazepines   ←
    from 1,3-dihydro-2H-1,4-benzodiazepin-2-ones – Retention of nitro groups s. *29*, 58

Sodium tetrahydridoborate $Na[BH_4]$
Methyl from carboxyl groups $COOH \to CH_3$
    via alcohols and tosylates s. *28*, 52

Prim. alcohols from carboxylic acid esters $COOR \to CH_2OH$
    Selective reduction s. *26*, 57; *29*, 39

Sec. amines from 1-alkoxyazomethines $C(OR)\cdot N:C \to CH\cdot NH\cdot CH$
s. *27*, 62

Sec. amines from iminoesters $N:C\cdot OR \to NH\cdot CH_2\cdot$
s. *29*, 369

Aliphatic nitro compds. $ONO_2 \to H$
    from nitronitric acid esters s. *27*, 63

Lactones from dicarboxylic acid monoesters
s. *18*, 92 suppl. *28*

$$\begin{matrix} COOH \\ COOR \end{matrix} \rightarrow \begin{matrix} CO \\ \diagdown \\ CH_2O \end{matrix}$$

Sodium tetrahydridoborate/pyridine — $Na[BH_4]/C_5H_5N$

Cyclic acetals from 1,2-acoxymesylates
s. *27*, 81

Sodium tetrahydridoborate/calcium chloride — $Na[BH_4]/CaCl_2$

Alcohols from carboxylic acid esters — $COOR \rightarrow CH_2OH$
s. *27*, 64

Sodium tetrahydridoborate/boron fluoride — $Na[BH_4]/BF_3$

Alcohol and methyl from carboxyl groups
s. *27*, 65

$$COOH \diagup^{CH_3}_{CH_2OH}$$

2,3,4,5-Tetrahydro- and 2,3-dihydro-1H-1,4-benzodiazepines
from 1,3-dihydro-2H-1,4-benzodiazepin-2-ones – Retention of nitro groups
s. *29*, 58

Sodium tetrahydridoborate/trialkyloxonium fluoroborate — $Na[BH_4]/R_3O^+BF_4^-$

Cyclic imines from lactams
s. *23*, 80 suppl. *27*

$$\begin{matrix} NH \\ | \\ CO \end{matrix} \rightarrow \begin{matrix} NH \\ | \\ CH_2 \end{matrix}$$

Sodium tetrahydridoborate-carbon s. under H_2PtCl_4 — $Na[BH_4]$-C

Lithium tetrahydridoaluminate — $Li[AlH_4]$

Preferential replacement of methoxy groups — $OMe \rightarrow H$
by hydrogen with formation of alcohols from oxido compds. s. *29*, 59

Hydrocarbons from sulfuric acid monoesters — $OSO_2O^- \rightarrow H$
s. *26*, 58

Alcohols from carboxylic acid esters — $COOR \rightarrow CH_2OH$
s. *28*, 53

Selective reduction
Alcohols from carboxylic acid esters with replacement of tosyloxy groups by hydrogen s. *28*, 50

Ethers from acetals — $C(OR)(OR') \rightarrow CH(OR)$
s. *26*, 59

2-Allenealcohols — $C(OR) \cdot C \equiv C \rightarrow C : C : CH$
from 2-acetylene-1,4-diol monoethers s. *25*, 52

Isochromenes from isocoumarins
s. *27*, 66

$$\begin{matrix} C \\ | \\ CO \end{matrix} \rightarrow \begin{matrix} O \\ | \\ CH_2 \end{matrix}$$

N-Methyl from N-carbalkoxyl — $NCOOR \rightarrow NMe$
s. *21*, 401

Amines from carboxylic acid amides — $CON< \rightarrow CH_2N<$
Labeled compds. s. *26*, 389

Aziridines and 2-aminoalcohols
from 2-chloracylamines – Stereospecific ring closure s. *26*, 60

Cyclic enamines from cyclic enacylamines — $CON \cdot C:C \rightarrow CH_2N \cdot C:C$
s. *29*, 60

Pyrazolidines from N,N-Diels-Alder adducts
s. *27*, 25

Reductive 4-pyrimidinone ring opening
s. *27*, 67

Reductive Beckmann rearrangement
Ring expansion s. *26*, 61

Spirodiamines by reduction of lactams
with Stevens rearrangement of ammonium ylids s. 30, 531

$\underset{CO}{\overset{N\cdot}{C}} \rightarrow \underset{CH_2}{\overset{N\cdot}{C}}$

2,3,3a,4,9,9a-Hexahydro-1H-pyrrolo[2,3-b]quinoxalines ←
from 1-benzimidazolylsuccinimides s. 29, 329

Lithium tetrahydridoaluminate/aluminum chloride $Li[AlH_4]/AlCl_3$
Methylamines from urethans   NCOOR → NMe
s. 21, 401 suppl. 29

Alcohols from carboxylic acid esters   COOR → CH_2OH
s. 26, 62

1,2,3,6-Tetrahydropyridine from 2-pyridone ring ←
N-Condensed rings s. 26, 63

Lithium tetrahydridoaluminate/sulfuric acid $Li[AlH_4]/H_2SO_4$
Diamines from nitrocarboxylic acid amides   CON< → CH_2N<
s. 26, 64

Bis-(3-methyl-2-butyl)borane/acetic acid R_2BH/CH_3COOH
cis-2-Ethyleneethers or α-alkoxyketones $C \equiv C \cdot C(OR)_2 \rightarrow CH:CH \cdot CH(OR)$
from α-acetyleneacetals s. 29, 61

Diisobutylaluminum hydride R_2AlH
Alcohols from carboxylic acid esters   COOR → CH_2OH
s. 28, 53

Acetoxy compds. from carboxylic acid esters   COOR → CH_2OAc
s. 27, 68

Lactols from dicarboxylic acid esters $\underset{COOR}{\overset{COOR}{\diagup}} \rightarrow \underset{C \cdot OH}{\overset{CH_2O}{\diagup}}$
s. 29, 62

Sodium trihydridocyanoborate $Na[BH_3CN]$
Hydrocarbons from tosylates   OTs → H
s. 27, 74

Sodium trihydridocyanoborate/tosylhydrazine $Na[BH_3CN]/TsNHNH_2$
Hydrocarbons from oxo compds.   CO → CH_2
Selective reduction s. 26, 65

Lithium hydridotrimethoxoaluminate/cuprous iodide $Li[AlH(OMe)_3]/CuI$
Replacement of sulfonyloxy groups OSO_2R → H
by hydrogen s. 29, 78

Trimethyl borate s. under BH_3 $(MeO)_3B$
Boron fluoride s. under $Na[BH_4]$ BF_3
Aluminum chloride s. under $LiAlH_4$ $AlCl_3$
Trialkyloxonium fluoroborate s. under $Na[BH_4]$ $R_3O^+ BF_4^-$
Di-tert-butyl peroxide s. under Cl_3SiH $(Me_3C \cdot O \cdot)_2$
Formic acid s. under Zn   HCOOH
Trifluoroacetic acid s. under Et_3SiH CF_3COOH
Dimethyl sulfide s. under BH_3 Me_2S
Acetyl chloride s. under Ph_3SnH CH_3COCl

Triethylsilane/trifluoroacetic acid Et_3SiH/CF_3COOH
Hydrocarbons from oxo compds.   CO → CH_2
Selective reduction s. 29, 63

Trichlorosilane/tri-n-propylamine $Cl_3SiH/n\text{-}Pr_3N$
Methyl from carboxyl groups   COOH → CH_3
Reactions of trichlorosilane in the presence of tert amines, review s. 26, 66

Trichlorosilane/di-tert-butyl peroxide/irradiation $Cl_3SiH/(Me_3C\cdot O\cdot)_2$/↟↟↟
Cyclic ethers from lactones 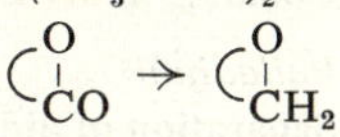
s. 27, 69   C(–O–CO) → C(–O–CH_2)

Titanocene dichloride/sodium $(C_5H_5)_2TiCl_2/Na$
Hydrocarbons   ←
from aldehydes, oxido compds., and carboxylic acid esters

30.   n—$C_{10}H_{21}CH_2CHO$ n—$C_{10}H_{21}CH_2COOEt$ n—$C_8H_{17}CH$—CH_2 (epoxide)

                        ↘   ↙   ↓

                    n—$C_{10}H_{21}CH_2CH_3$ n—$C_8H_{17}CH_2CH_3$

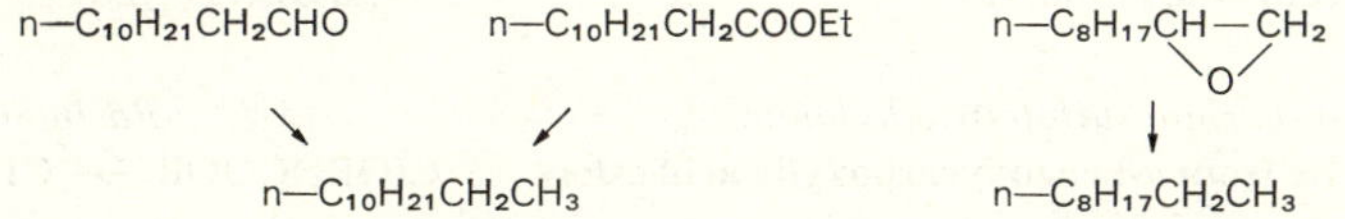

Dodecanal added to a dark greenish mixture obtained by stirring titanocene dichloride and Na-sand in benzene under an inert atmosphere (no O_2 or N_2), which is maintained throughout the reaction, and stirred 72 hrs. at room temp. → dodecane. Y: 71%; from ethyl dodecanoate, Y: 64–69%. Similarly: Decane from 1-decene oxide. Y: 68–81%. F. e. s. E. E. van Tamelen and J. A. Gladysz, Am. Soc. *96*, 5290 (1974); *polymer-based titanocene dichloride* cf. W. D. Bonds et al., ibid. *97*, 2128 (1975).

Triphenyltin hydride/acetyl chloride Ph_3SnH/CH_3COCl
Hydrocarbones from ketones   CO → CH_2

31.   Fc—$COCH_3$ →   Fc—CH_2CH_3

Acetylferrocene allowed to react 4.5 hrs. at ca. 25° under N_2 with triphenyltin hydride and acetyl chloride in stirred anhydrous benzene → ethylferrocene. Y: 90%. F. e. s. H. Patin and R. Dabard, Bl. *1973*, 2756, 2760.

Hydrazine/potassium hydroxide $H_2N\cdot NH_2/KOH$
Wolff-Kishner-Huang Minlon reduction of ozonides CH_2
s. *30*, 46

Tosylhydrazine s. under $Na[BH_3CN]$ $TsNHNH_2$

Phosphorus s. under HI   P

Copper chromite $CuCr_2O_4$
Total hydrogenation of oxo-N-heterocyclics   CO → CH_2
s. 26, 67

Iodine s. under HI   I

Hydrogen iodide   HI
Ketones from α,β-ethylene-α-hydroxyketones $CO\cdot C(OH):C$ → $CH_2CO\cdot CH$
Preferential reduction s. 28, 51

Hydrogen iodide/phosphorus/iodine   HI/P/I
9,10-Dihydroanthracenes from anthraquinones   CO → CH_2

32.

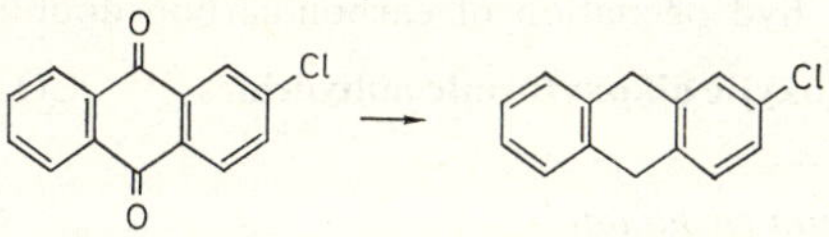

         2-chloroanthraquinone → 2-chloro-9,10-dihydroanthracene

A mixture of 2-chloroanthraquinone, red P, iodine, and 47%-hydriodic acid heated 24 hrs. at 140° in an oil bath → 2-chloro-9,10-dihydroanthracene. Y: 90%. F. e. s. R. N. Renaud and J. C. Stephens, Can. J. Chem. *52*, 1229 (1974).

Sodium tetracarbonylferrate(II)   *$Na_2Fe(CO)_4$*
Aldehydes from carboxylic acid anhydrides $(RCO)_2O \rightarrow RCHO$
   Aldehydrocarboxylic acids from dicarboxylic acid anhydrides s. *29*, 64

Cobalt s. under Li/NH₃   *Co*

Palladium   *Pd*
Separation of aldehydes $ArCHO \rightarrow ArCH_3$
   via benzylic hydrogenolysis s. *30*, 149

Palladium-calcium carbonate   *$Pd\text{-}CaCO_3$*
Arenes from O-arylisoureas $ArOC(:N\cdot)NH\cdot \rightarrow ArH$
s. *30*, 50

Palladium-barium sulfate/triethylamine   *$Pd\text{-}BaSO_4/Et_3N$*
Carboxylic from α-benzoxycarboxylic acid esters $C(OBz)COOR \rightarrow CHCOOR$
s. *27*, 92

Palladium-carbon   *Pd-C*
Selective benzylic hydrogenolysis $C(OR)N< \rightarrow CHN<$
   Amines from 1,1-aminoethers s. *28*, 57

Selective hydrogenation and oxidation   C
of O-bridged isocycles

33.

1,4-Dihydro-1,4-dimethyl-1,4-epoxynaphthalene hydrogenated with 10%-Pd-C in methanol at 40 psi until H₂-uptake ceases after 20 min. → 1,4-dimethyl-1,2,3,4-tetrahydronaphthalene. Y: 97%.

mixed with benzene, water, and *Aliquat 336 as phase transfer catalyst*, treated 0.5–1 hr. with a mixture of KMnO₄ and MgSO₄ at such a rate that gentle reflux occurs, and refluxing continued 0.5 hr. → o-diacetylbenzene. Y: 69%.

M. S. Newman, H. M. Dali, and W. M. Hung, J. Org. Chem. **40**, 262 (1975).

Palladium-carbon/triethylamine   *$Pd\text{-}C/Et_3N$*
Arenes from phenols $ArOSO_2R \rightarrow ArH$
   via aryl sulfonates, also via iminocarbonic acid esters s. *29*, 65; via sulfuric acid monoaryl esters cf. W. Lonsky, H. Traitler, and K. Kratzl, Soc. Perkin I 1975, 169

Palladium-carbon/acetic acid   *$Pd\text{-}C/CH_3COOH$*
Methyl from formyl groups $CHO \rightarrow CH_3$
   with and without hydrogenation of carbon-carbon double bonds s. *26*, 68

Aldehydes from carboxylic alkoxyformic anhydrides $CO\cdot O\cdot COOR \rightarrow CHO$
s. *28*, 66

Palladium-lead-calcium carbonate/-
lithium tetrahydridoaluminate   *$Pd\text{-}Pb\text{-}CaCO_3/LiAlH_4$*
3-Ethylenealcohols $C(OR)C \equiv C \rightarrow C:CH\cdot CH_2$
   from 2-acetylene-1,4-diol monoethers s. *29*, 66

*Hexachloroplatinic acid/-
sodium tetrahydridoborate-carbon/hydrogen chloride* $H_2PtCl_6/Na[BH_4]$-C/HCl
Cycloalkane ring ←
from cyclic β,γ-ethyleneketones s. 29, 933

Via intermediates v.i.
Methyl from carboxyl groups COOH → CH_3
via alcohols and tosylates s. 28, 52
Hydrocarbons CO → CH_2
from oxo compds. via tosylhydrazones s. 29, 67
Cycloalkane ring ←
from cyclic β,γ-ethyleneketones s. 29, 933
Prim. alcohols from carboxylic acids COOH → CH_2OH
via enolesters from isoxazolium salts s. 29, 57
Aldehydes from carboxylic acid esters COOR → CHO
via alcohols s. 28, 53; s. a. T. M. Cresp et al., Soc. Perkin I *1974*, 2435

Nitrogen ↑ HC ⇊ N

Electrolysis ↯
Cathodic deamination NH_2 → H
β-Keto- from α-amino-β-keto-carboxylic acid esters s. 29, 68
Replacement of diazo groups by hydrogen $C:N_2$ → CH_2
s. 29, 69

Potassium/hexamethylphosphoramide/tert-butanol $K/(Me_2N)_3PO/Me_3C\cdot OH$
Alcohols from N,N-disubst. carboxylic acid amides $CONR_2$ → CH_2OH
s. 28, 54

Sodium/alcohol NaOR
Nitriles, also α-alkylated, and α-cyanocarboxylic acid esters ←
from 1,1-cyano(azocarboxylic acid esters) – Ketones from nitriles s. 28, 55

Zinc Zn
α-Aminocarboxylic acids $C(:N\cdot N<)COOH$ → $CH(NH_2)COOH$
from α-ketocarboxylic acid hydrazones s. 7, 77

Zinc/acetic acid Zn/CH_3COOH
Replacement of diazo groups by hydrogen ←
with migration of carbon-carbon double bonds – 1,3-Cyclohexadiene from cyclohexenedione ring s. 27, 70

Sodium tetrahydridoborate $Na[BH_4]$
Hydrocarbons from tosylhydrazones $C:N\cdot NHTs$ → CH_2
s. 29, 67; also formation of diazo compds. (cf. Synth. Meth. *12*, 571), phosphonic and α-diazophosphonic esters, s. G. Rosini and G. Baccolini, Synthesis *1975*, 44

34.

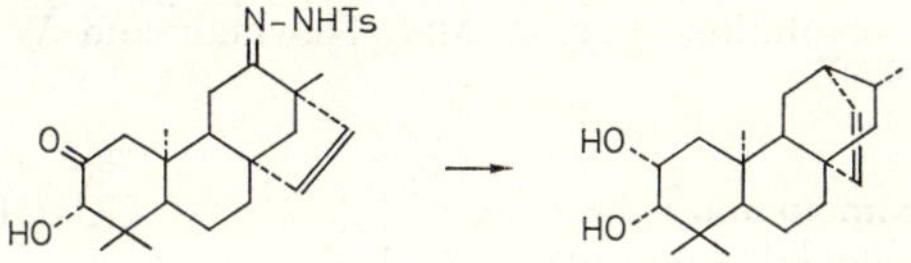

with skeletal rearrangement. *ent*-3β-Hydroxybeyer-15-en-2-one 12-tosylhydrazone allowed to react with $NaBH_4$ in ethanol-dioxane → *ent*-16S-2β,3β-dihydroxyatis-13-ene. Y: > 73%. K. H. Pegel, L. Phillips et al., Chem. Commun. *1973*, 552.

Ar. N-methylamines ArN·Me
from ar. N-(succinimidomethyl)amines s. *28*, 56

Sodium tetrahydridoaluminate $Na[AlH_4]$
Aldehydes from N,N-disubst. carboxylic acid amides $CONR_2 \rightarrow CHO$
s. *26*, 69

Lithium tetrahydridoaluminate $Li[AlH_4]$
N-Hydroxy-N-heterocyclics N< → H
from isocyclic 2-aminonitroso compds. s. *27*, 331

Sodium trihydridocyanoborate $Na[BH_3CN]$
Hydrocarbons from tosylhydrazones C:N·NHTs → CH_2
s. *29*, 67

Sodium dihydridobis(2-methoxyethoxo)aluminate $NaAlH_2(OCH_2CH_2OMe)_2$
Allenes C≡C·C·N< → CH:C:C
from quaternary 2-acetyleneammonium salts

35.

A soln. of Na-bis-(2-methoxyethoxy)aluminum hydride in dry tetrahydrofuran added at 20–30° during 0.5 hr. to a stirred suspension of 4-hydroxy-4-(6-methoxy-2-naphthyl)-1-trimethylammonio-2-pentyne iodide in the same solvent, and stirring continued 2 hrs. at 25° → 4-hydroxy-4-(6-methoxy-2-naphthyl)-1,2-pentadiene. Y: 84%. F. e. with $LiAlH_4$ in pyridine s. E. Galantay, I. Basco, and R. V. Coombs, Synthesis *1974*, 344.

Glass powder s. under $(Me_2N)_3PO$ ←

Amyl nitrite/tetrahydrofuran ←
Simple deamination NH_2 → H
s. *29*, 70

Hexamethylphosphoramide/glass powder ←
Ar. deamination $ArNH_2$ → ArH
via diazonium mercuric salt complexes

36.

7-Chloronaphthalene-1-diazonium bromide mercuric bromide complex (prepared by diazotization of 7-chloro-1-naphthylamine followed by treatment with mercuric bromide; Y: 92–99%) added to a stirred mixture of hexamethylphosphoramide and powdered soft glass, and the product isolated after ca. 10 min. → 7-chloronaphthalene. Y: 93%. – Finely divided solids such as glass powder may facilitate reactions with gas evolution. F. e. s. M. S. Newman and W. M. Hung, J. Org. Chem. *39*, 1317 (1974).

Palladium-carbon Pd-C
Alcohols from α-bromoketones CO·CBr → CH(OH)CH
via 3-acetoxyquinuclidinium salts s. *29*, 71

Partial and selective benzylic hydrogenolysis ←
Tert. amines from 1,1-diamines, 1,1-aminoethers and monoacyl-1,1-diamines s. *28*, 57

Halogen ↑   HC ↕ Hal

Irradiation (s. a. under Et₃N and i-PrOH)	⇋
Partial replacement of chlorine by hydrogen s. *29, 73*	Cl → H
Aldehydes from carboxylic acid bromides s. *27, 76*	COBr → CHO
Electrolysis	↯
Acetylene from polyhalogenoethylene derivs. s. *27, 961*	C ≡ C
Lithium-tert-butanol	Li-Me₃COH
Hydrocarbons from 1,1-dichlorides s. *29, 72*	CCl₂ → CH₂
Sodium hydride	NaH
Replacement of iodine by hydrogen	I → H

37.

α-Iodonaphthalene added under N₂ to NaH in tetrahydrofuran, and refluxed 48 hrs. → naphthalene. Y: 88%. F. e. and limitations s. R. B. Nelson and G. W. Gribble, J. Org. Chem. **39**, 1425 (1974).

Sodium amalgam	Na,Hg
Partial and preferential replacement of chlorine by hydrogen s. *29, 73*	Cl → H
Sodium/alcohol s. under RSH	NaOR
n-Butyllithium/hydrogen bromide	n-BuLi/HBr
Halogen-metal interconversion Reaction via 1,1-halogenoorganolithium compds. – Bromides from 1,1-dibromides – α-Bromo-germanes and -silanes s. *26, 70*	CBr₂ → CHBr
Lithium iodide s. under BF₃	LiI
Triethylamine/irradiation	Et₃N/⇋
Replacement of chlorine by hydrogen s. *29, 74*	Cl → H
Tri-n-butylamine s. under HSiCl₃	n-Bu₃N
Pyridine s. under S₂O₄²⁻	C₅H₅N
Copper s. under Zn	Cu
Silver s. under Zn	Ag
Lithium dimethylcuprate	Li[CuMe₂]
Ketones from α-bromoketones s. *30, 27*	CO·CBr → CO·CH
Cuprous iodide s. under K[BHsec-Bu₃] and Li[AlH(OMe)₃]	CuI
Zinc	Zn
α,β-Ethylene- from β-chlor-α,β-ethylene-aldehydes s. *28, 58*	CCl:C·CHO → CH:C·CHO

Zinc/dimethylformamide Zn/Me_2NCHO
Replacement of halogen by hydrogen   Hal → H

38. $[CH_2]_{11}$ CHBr → $[CH_2]_{11}$ CH$_2$

Bromocyclododecane, Zn-dust, and dimethylformamide stirred 4 hrs. at 130–140° preferably under N_2, then poured into water → cyclododecane. Y: 65%. F. e. s. G. Mehta and S. K. Kapoor, J. Organometal. Chem. **66**, C33 (1974).

Zinc/acetic acid Zn/CH_3COOH
Methyl groups from 1,1,1-trichlorides CCl_3 → CH_3
s. 27, 71

Zinc/ammonium chloride Zn/NH_4Cl
Cyclobutanone from 2,2-dichlorocyclobutanone ring CCl_2 → CH_2
s. 29, 75

Zinc/copper Zn/Cu
Replacement of halogen by deuterium   Hal → D
s. 30, 39

α,β-Ethyleneketones   CO·C : C·Hal → CO·C : CH
from α,β-ethylene-β-halogenoketones
s. 30, 39

Zinc/silver Zn/Ag

39. (4,4-dimethyl-2-chlorocyclohex-2-enone) → (4,4-dimethylcyclohex-2-enone)

A methanolic soln. of the startg. m. added to methanol-moist Zn,Ag-couple (prepared from Zn-dust and anhydrous Ag-acetate in hot acetic acid), and vigorously stirred 15–30 min. at room temp. → product. Y: 81%. F. e. and limitations s. R. D. Clark and C. H. Heathcock, J. Org. Chem. **38**, 3658 (1973); with Zn,Cu-couple prepared from Zn-powder and cupric acetate, also **replacement of halogen by deuterium**, s. R. M. Blankenship, K. A. Burdett, and J. S. Swenton, J. Org. Chem. **39**, 2300 (1974).

Sodium tetrahydridoborate $Na[BH_4]$
Alcohols from 1,2-nitrosohalides   CHalC(NO) → CHC(OH)
s. 27, 202

Oxido compds. from α-halogenoketones   COCHal → (epoxide)
s. 28, 59

Amines from 2-chloramines   ←
   with rearrangement s. 29, 76

Sodium tetrahydridoborate/valeric acid $Na[BH_4]/C_4H_9COOH$
Replacement of *tert*-halogen by hydrogen   ≥C·Hal → ≥C·H
s. 27, 72; with Li-9-borabicyclo[3.3.1]nonane ate complexes, also replacement of allylic and benzylic halogen, cf. Y. Yamamoto et al., Am. Soc. **97**, 2559 (1975)

Lithium tetrahydridoaluminate $Li[AlH_4]$
Methyl from trifluoromethyl groups CF_3 → CH_3
s. 27, 73

Reductions with
lithium tetrahydridoaluminate-titanium tetrachloride $Li[AlH_4]/TiCl_4$
Replacement of ar. and vinylic halogen by hydrogen Hal → H
s. 29, 77; replacement of ar. halogen with Raney-Ni in ethanolic KOH s. A. J. de Koning, Org. Prep. Proc. Intern. 7, 31 (1975)

Lithium hydridotriethylborate $Li[BHEt_3]$
Replacement of halogen by hydrogen
s. 28, 60

Reductions with
potassium hydrido-tri-sec-butylborate/cuprous iodide $K[BHsec-Bu_3]/CuI$
Replacement of halogen by hydrogen
Sec. alcohols from ketones

40. $C_5H_{11}-C\equiv CI$ → $C_5H_{11}-C\equiv CH$

   Br—⟨⟩—COOH → ⟨⟩—COOH

   ⟨⟩=O → ⟨⟩—OH

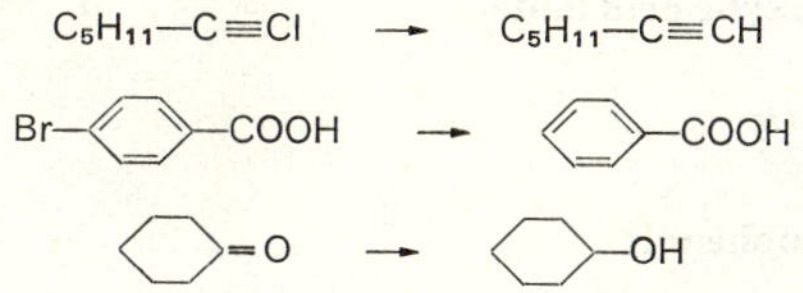

1-Iodohept-1-yne allowed to react 15 min. at 0°with a reagent prepared from 3 equivalents K-tri-*sec*-butylboron hydride and 1.5 equivalents CuI in tetrahydrofuran. → hept-1-yne. Y: 94%. – Similarly: 4-Bromobenzoic acid at 25° for 12 hrs. → benzoic acid. Y: 86%. – Cyclohexanone at 0° for 0.5 hr. → cyclohexanol. Y: 84%. F. e., also ethylene from acetylene derivs., s. T. Yoshida and E. Negishi, Chem. Commun. *1974*, 762.

Sodium trihydridocyanoborate $Na[BH_3CN]$
Selective reductions ←
   with sodium trihydridocyanoborate in hexamethylphosphoramide – Hydrocarbons from alcohols, halides, and tosylates s. 27, 74; Na-trihydridocyanoborate as selective reducing agent, review, s. C. F. Lane, Synthesis *1975*, 135

Tetrabutylammonium trihydridocyanoborate $Bu_4N^+[BH_3CN]^-$
Selective reductions I → H
s. 27, 74 suppl. 29

Sodium dihydridobis-(2-methoxyethoxo)aluminate/
palladium chloride complexes ←
Replacement of halogen by hydrogen Hal → H
s. 28, 60

Lithium hydridotrimethoxoaluminate/cuprous iodide $Li[AlH(OMe)_3]/CuI$
Replacement of halogen Hal → H
   and sulfonyloxy groups by hydrogen – Alcohols from oxido compds. s. 29, 78

Boron fluoride/lithium iodide BF_3/LiI
Ketones from α-halogenoketones COCHal → COCH
s. 26, 71

Isopropyl alcohol s. a. under Fe(CO)$_5$ Me_2CHOH

Isopropyl alcohol/irradiation $Me_2CHOH/⚡$
Preferential replacement of chlorine by hydrogen Cl → H
s. 27, 77

Valeric acid s. under Na[BH$_4$] C_4H_9COOH

Mercaptan-sodium/alcohol $RSH/NaOR$
Ketones from α-subst. ketones COCH
   from α-halogeno- and α-alkylthio-ketones s. 27, 75

Thiophenol	*PhSH*
Replacement of halogen by hydrogen s. *20*, 79	Hal → H
Trichlorosilane/tri-n-butylamine	*HSiCl$_3$/n-Bu$_3$N*
2-Ketothioethers from 1-chloro-2-ketothioethers s. *28*, 61	CO·CCl(S·) → CO·CH(S·)
Titanium trichloride	*TiCl$_3$*
Replacement of halogen by hydrogen Ketones from α-halogenoketones s. *29*, 79	CO·CHal → CO·CH
Titanium tetrachloride s. under Li[AlH$_4$]	*TiCl$_4$*
Tributyltin hydride	*Bu$_3$SnH*
Aldehydes from carboxylic acid halides s. *27*, 76	COHal → CHO
Phosphorus s. under HI	*P*
Triphenylphosphine	*Ph$_3$P*
Phenols from o-bromophenols s. *26*, 550	Br → H
Dithionite/pyridine	*S$_2$O$_4^{2-}$/C$_5$H$_5$N*
Dehalogenation via pyridinium salts Ketones from α-halogenoketones s. *29*, 80	CO·CHal → CO·CH
Hydrogen iodide/phosphorus	*HI/P*
Replacement of bromine by hydrogen with preferential hydrogenation of aliphatic carbon-carbon double bonds s. *26*, 72	Br → H
Iron carbonyl/isopropyl alcohol	*Fe(CO)$_5$/Me$_2$CHOH*
1,1-Di- from 1,1,1-tri-halides Preferential replacement of halogen by hydrogen s. *27*, 77	CHal$_3$ → CHHal$_2$
Sodium tetracarbonylferrate(II)	*Na$_2$Fe(CO)$_4$*
Aldehydes from carboxylic acid chlorides s. *27*, 76	COCl → CHO
Nickel/potassium hydroxide	*Ni/KOH*
Replacement of ar. halogen by hydrogen s. *29*, 77 suppl. *30*	ArHal → ArH
Nickel/triethylamine	*Ni/Et$_3$N*
Preferential replacement of sec. chlorine by hydrogen s. *27*, 78	Cl → H
Nickel carbonyl	*Ni(CO)$_4$*
Preferential replacement of halogen by hydrogen – 1,1-Di- from 1,1,1-tri-halides s. *27*, 77	CHal$_3$ → CHHal$_2$
Palladium-barium sulfate/sodium acetate	*Pd-BaSO$_4$/CH$_3$COONa*
Hydrocarbons from α-halogenomesylates s. *26*, 73	Hal → H
Palladium-carbon	*Pd-C*
Replacement of chlorine by hydrogen s. *28*, 903	Cl → H
Palladium-carbon/sodium acetate	*Pd-C/CH$_3$COONa*
Aldehydes from carboxylic acid chlorides Modified Rosenmund-Zetzsche reduction s. *27*, 79	COCl → CHO

Palladium-carbon/diethylamine Pd-C/Et_2NH
Replacement of bromine by hydrogen   Br → H
s. 27, 80

Palladium-carbon/triethylamine Pd-C/Et_3N
Hydrocarbons from 1,2-ethylene-1,2-dichlorides   CCl : CCl → CH_2CH_2
Preferential replacement of chlorine by hydrogen s. 28, 62

Palladium-carbon/ethyldiisopropylamine Pd-C/i-Pr_2NEt
Aldehydes from carboxylic acid chlorides   COCl → CHO
Modified Rosenmund-Zetzsche reduction s. 27, 79

Palladium chloride complexes s. under $Na[AlH_2(OCH_2CH_2OMe)_2]$ ←

Via intermediates   v.i.
Aldehydes from carboxylic acid chlorides   COCl → CHO
via 1,2-acoxymesylates and cyclic acetals s. 27, 81

Sulfur ↑   HC ↟ S

Irradiation s. under $Na[BH_4]$ ⇈

Electrolysis
Hydrocarbons from sulfonium salts $C \cdot \overset{+}{S}R_2$ → CH
s. 29, 260

Sodium/alcohol s. under RSH   NaOR

Sodium/liq. ammonia Na/NH_3
Ethylene derivs. from 2-ethylenealcohols $C(OH) \cdot C : C$ → $C : C \cdot CH$
via 2-ethylenisothiouronium salts
with double bond migration

41.

(±)-17β-Hydroxy-3-methoxy-17α-vinylestra-1,3,5(10)-triene added to a soln. of thiourea and concd. HCl in methanol, stirred and refluxed 4 hrs. → (±)-S-{[17(20)E]-3-methoxy-19-norpregna-1,3,5(10),17(20)-tetraen-21-yl} thiouronium chloride (Y: 80%) stirred 5 hrs. at −45 to −35° with 10 moles Na in liq. NH_3-anhydrous tetrahydrofuran → (±)-[17(20)E]-3-methoxypregna-1,3,5(10),17(20)-tetraene (Y: ca. 100%). F. e. s. H. Schick, Pharmazie 30, 30 (1975).

Sulfinic acids from benzyl sulfones SO_2CH_2Ph → SO_2H
s. 26, 74

Cupric chloride s. under $Li[AlH_4]$ $CuCl_2$

Zinc/acetic acid Zn/CH_3COOH
Replacement of sulfonyl groups by hydrogen SO_2R → H
γ-Diketones from α-sulfonyl-γ-diketones s. 29, 792

Zinc chloride s. under $Li[AlH_4]$ $ZnCl_2$

Sodium tetrahydridoborate $Na[BH_4]$
Elimination of sulfur dioxide   C
Reductive elimination, also oxidative elimination s. 26, 75

Sodium tetrahydridoborate/irradiation $\quad Na[BH_4]/$ ⚡
Hydrocarbons from sulfoxonium ylids $\quad C:S(O)R_2 \rightarrow CH_2$
s. *27*, 82

Lithium tetrahydridoaluminate/cupric chloride $\quad Li[AlH_4]/CuCl_2$
Hydrocarbons from thioethers and mercaptals $\quad \leftarrow$
s. *28*, 63

*Lithium tetrahydridoaluminate/cupric chloride/-
zinc chloride* $\quad Li[AlH_4]/CuCl_2/ZnCl_2$
Hydrocarbons from mercaptals
s. *28*, 63 $\quad C\overset{S}{\underset{S}{\diagdown}} \rightarrow CH_2$

Lithium tetrahydridoaluminate/titanium tetrachloride $\quad Li[AlH_4]/TiCl_4$
Hydrocarbons $\quad \leftarrow$
from thioethers and mercaptals s. *28*, 63

Mercaptan-sodium/alcohol $\quad RSH/NaOR$
Ketones from α-alkylthioketones $\quad CO \cdot C(SR) \rightarrow CO \cdot CH$
s. *27*, 75

Titanium tetrachloride s. under Li[AlH₄] $\quad TiCl_4$

Phenyl hydrazine $\quad PhNHNH_2$
Benzoxazole from 2H-1,4-benzoxazine ring $\quad $ O:
s. *28*, 408

Chromous sulfate $\quad CrSO_4$
Replacement of sulfonyl groups by hydrogen $\quad SO_2R \rightarrow H$
γ-Diketones from α-sulfonyl-γ-diketones s. *29*, 792

Nickel $\quad Ni$
Methyl from hydroxymethyl groups $\quad CH_2OH \rightarrow CH_3$
via 2-pyrimidinyl thioethers s. *27*, 91
Hydrocarbons from sulfoxonium ylids $\quad C:S(O)R_2 \rightarrow CH_2$
s. *27*, 82
Alcohols from 2-hydroxythioethers $\quad SR \rightarrow H$
s. *28*, 524
Prim. alcohols and methyl groups $\quad COSR \overset{\nearrow CH_2OH}{\searrow CH_3}$
from thiolic acid esters s. *26*, 76
Ring opening by desulfuration $\quad C$
s. *26*, 691
Reductive isothiazolid-3-one ring opening

42.

Ph—[ring: Ph, S, CONHCMe₃, N-CMe₃, O] → Ph—CH—CHPh / CO CONHCMe₃ / NH-CMe₃

An alc. soln. of 2-*tert*-butyl-4,5-diphenyl-5-*tert*-butylcarbamoyl-1,2-thiazolidin-3-one refluxed 3 hrs. with Raney-Ni → N,N'-di-*tert*-butyl-2,3-diphenylbutanediamide. Y: 85%. F. e. s. T. Minami and T. Agawa, J. Org. Chem. *39*, 1210 (1974).

Nickel/sodium hydroxide $\quad Ni/NaOH$
Hydrodesulfination
s. *29*, 90

Nickel boride $\quad Ni_2B$
Hydrocarbons from sulfido compds. $\quad \underset{S}{\diagdown\diagup} \rightarrow CHCH$
s. *27*, 976

Remaining Elements ↑   HC ↯ Rem

Without additional reagents   *w.a.r.*
Ketones   COC : P ≦ → COCH$_2$
   from (2-acylalkylidene)phosphoranes s. *30,* 582

Lithium   *Li*
Demercuration   Hg· → H
s. *29,* 172

Sodium hydroxide   *NaOH*
β-Deuterio- from α,β-dideuterio-ketones   CDCDCO → CDCHCO
s. *29,* 47

Potassium carbonate   *K$_2$CO$_3$*
Preferential dedeuteriation   SO$_2$CD$_2$ → SO$_2$CH$_2$
   Preferential deuteriation s. *29,* 81

Sodium tetrahydridoborate   *Na[BH$_4$]*
Demercuration   Hg· → H
s. *29,* 172

3-Ethylenealcohols   C : C·C : C(OSi ≦) → C : C·CH·CHOH
   from α,β-ethyleneketones via (1,3-dien)oxysilanes – Stereospecific reduction
s. *27,* 83

Lithium tetrahydridoaluminate   *Li[AlH$_4$]*
***trans*-Ethylene derivs.**   ←
from β,γ-ethylenephosphonic acid esters and halides

43.   CH$_3$CH=CHCH$_2$PO(OEt)$_2$ → CH$_3$CH=CHCHPO(OEt)$_2$
       + BrCH$_2$Ph   CH$_2$Ph

           CH$_3$CH$_2$ \\ ═ / CH$_2$Ph

n-Butyllithium added under argon pressure at –60° to a stirred soln. of a mixture of diethyl *cis*- and *trans*-2-butenyl phosphonates in tetrahydrofuran, stirring continued 1 hr. at the same temp., benzyl bromide added dropwise, stirred 0.5 hr. at –60°, allowed to warm to room temp., the crude intermediate (Y: 83% if isolated) dissolved in dry ether, ice-cooled, treated with LiAlH$_4$, and stirred 1 hr. at 0° → *trans*-1-phenyl-2-pentene (Y: 89%). F. e. s. K. Kondo, A. Negishi, and D. Tunemoto, Ang. Ch. *86,* 415 (1974).

Capric acid   *C$_9$H$_{19}$COOH*
Hydrocarbons from boronic acid esters   B(OR)$_2$ → H
s. *27,* 84

Hydrogen sulfide   *H$_2$S*
Hydrocarbons from monoorganomercury compds.   C : C·HgOAc → C : CH
s. *30,* 156

Hydrogen chloride   *HCl*
Selective replacement of phosphonic acid ester groups   PO(OR)$_2$ → H
   by hydrogen s. *29,* 82

Nickel   *Ni*
Replacement of alkylseleno groups   SeR → H
   by hydrogen s. *29,* 83

Carbon ↑   HC ⇅ C

Without additional reagents   *w.a.r.*
Ketones from α-aminomethyleneketones   COC : C·N< → COCH$_2$
s. *30*, 589

Sodium s. under Fe(acac)$_3$ *Na*

Potassium/hexamethylphosphoramide   *K/(Me$_2$N)$_3$PO*
Reductive replacement of cyano groups   CN → H
  by hydrogen s. *27*, 87; cf. C. Fabre, M. H. A. Salem, and Z. Welvart, Bl. *1975*, 178

Sodium hydroxide   *NaOH*
C-Deacylation   COR → H
s. *30*, 593

S-Chloromethyl from S-1,2-dichlorovinyl groups   CCl : CHCl → CH$_2$Cl
s. *26*, 77

Ketones   CO
  from α,β-ethylene-γ-hydroperoxy-γ-lactones – Retroaldol reaction s. *26*, 227

Sodium carbonate   *Na$_2$CO$_3$*
4-Alkylpyridines from 5-(4-pyridyl)barbituric acids   C
s. *28*, 746

Potassium cyanide   *KCN*
Deuterioaldehydes from benzils   COCO → CDO
s. *28*, 64

Sodium chloride/dimethyl sulfoxide   *NaCl/Me$_2$SO*
Decarbalkoxylation   COOR → H
  Carboxylic from malonic acid esters s. *29*, 84

Isopropylamine   *Me$_2$CHNH$_2$*
Retro-Mannich reaction   CH$_2$NR$_2$ → H
s. *27*, 85

Piperidine   *C$_5$H$_{11}$N*
Decarbalkoxylation   COOR → H
  Carboxylic from malonic acid esters s. *29*, 84 suppl. *30*

1,5-Diazabicyclo[4.3.0]non-5-ene or triethylenediamine   ←
  with dehydrobromination s. *29*, 963

Barium hydroxide   *Ba(OH)$_2$*
Carboxylic from β-ketocarboxylic acid esters   C(COR)COOR → CHCOOR
s. *29*, 499

Methylmagnesium iodide   *MeMgI*
Anthrones from anthranol esters   ←
s. *29*, 921

Zinc acetate   *(CH$_3$COO)$_2$Zn*
C-Deacylation   COR → H
  β,γ-Ethyleneketones from β,γ-ethylene-β'-diketones s. *29*, 85

Boric acid   *H$_3$BO$_3$*
γ-Ketocarboxylic acid esters   COOR → H
  from γ-keto-β-dicarboxylic acid esters s. *29*, 86

Isopropyl alcohol s. under HCl   *Me$_2$CHOH*

Formic acid s. under HCl   *HCOOH*

Dimethyl sulfoxide (s. a. under NaCl)   *Me$_2$SO*

Decarbalkoxylation   COOR → H
Carboxylic from malonic acid esters s. *29*, 84; with piperidine cf. F. Texier, E. Marchand, and R. Carrié, Tetrahedron *30*, 3185 (1974)

Sulfuric acid   *H$_2$SO$_4$*

C-Deacetylation   C·Ac → CH
Heterofulvenes s. *30*, 248

Replacement of cyano groups by hydrogen   CN → H
with formation of pyrazolo[3,4-b]pyridine from pyrazolo[1,5-a]pyrimidine ring s. *26*, 756

Triphenylcarbonium perchlorate   *Ph$_3$C$^+$ ClO$_4^-$*

4-Methylpyrylium ring by C-cleavage ←
s. *27*, 86

Hydrogen chloride   *HCl*

3-(1-Isoquinolyl)phthalides ←
from 2-acyl-1,2-dihydroisoquinoline-1-spiro-2'-(1',3'-indandiones)

44.

A soln. of 2-acetyl-6,7-dimethoxy-1,2-dihydroisoquinoline-1-spiro-2'- (1',3'-indandione) in 12 M HCl and ethanol refluxed 12 hrs. → 3-(6,7-dimethoxy-1-isoquinolyl)phthalide. Y: 83%. F. e. s. V. Smula et al., Can. J. Chem. *51*, 3287 (1973).

Hydrogen chloride/isopropyl alcohol   *HCl/Me$_2$CHOH*

Reductive C-cleavage of octahydroxanthene-1,8-diones   C

45.

Startg. m. heated 6 hrs. at 80° in isopropanol containing HCl → 2,2'-methylenebis(cyclohexane-1,3-dione). Y: 71–80%. F. e. s. O. S. Wolfbeis and H. Junek, Tetrah. Let. *1973*, 4905.

Hydrogen chloride/formic acid   *HCl/HCOOH*

C-Deacylation   COR → H
s. *29*, 236

Ferric acetoacetonate/sodium   *Fe(acac)$_3$/Na*

Reductive replacement of cyano groups   CN → H
by hydrogen s. *27*, 87

Nickel   *Ni*

Replacement of thiolcarbalkoxyl groups   COSEt → H
by hydrogen s. *26*, 78

C-Debenzylation ←
Partial C-debenzylation s. *26*, 12

Via intermediates   v.i.

Hydrocarbons from ethylene derivs.   C : C → CH_2
Degradation via ozonides

46.

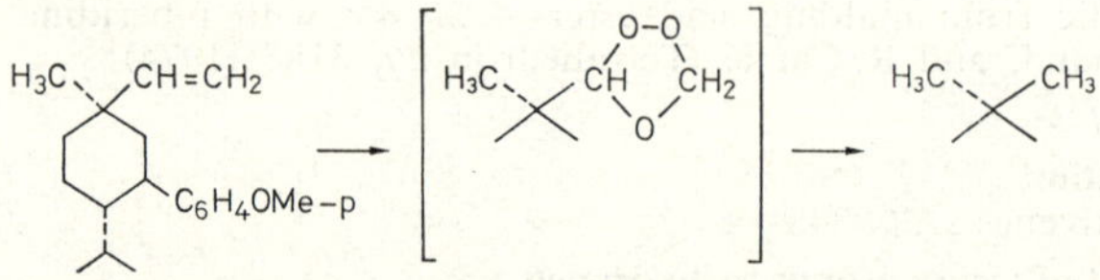

An O_3-O_2-stream passed at ca. –20° for 45 min. through a soln. of 0.173 g. startg. m. in ethyl acetate, the solvent removed at ca. 40°/50 mm, the resulting crude ozonide treated with 98%-hydrazine hydrate, KOH-pellets, and diethylene glycol, heated in an oil bath under N_2 at 150° for 1.5 hrs., then at 200° to remove water and excess hydrazine, finally refluxed 3 hrs. → 0.088 g. product. Also isomer s. A. S. C. Prakasa Rao, U. R. Nayak, and S. Dev, Tetrahedron *30*, 1107 (1974).

C-Decarbobenzoxylation $COOCH_2Ph$ → H
s. *27*, 768

Alcohols from cyanocyclopropanes   ▷—CN → C(OH)C·CH
via β-lactones s. *29*, 87

Elimination   ⇑

Oxygen ↑   HC⇑O

Irradiation s. under MeOH   ⫙

Sodium   Na
Hydrocarbons from alcohols   OH → H
with preferential replacement of alkoxy groups by hydrogen

47.

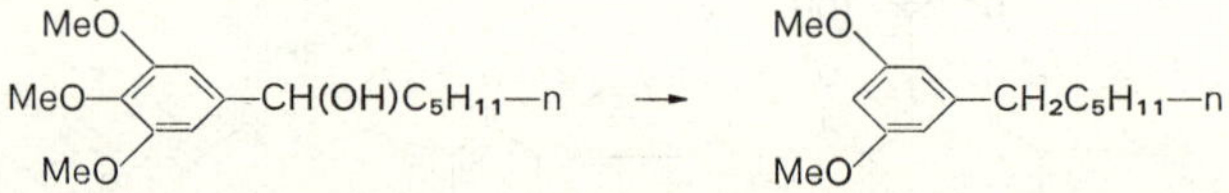

tert-Butanol added dropwise to a stirred mixture of the startg. m., Na-sand, and benzene heated to 80–85° oil bath temp., and stirring continued 3–4 hrs. → 1-hexyl-3,5-dimethoxybenzene. Y: 83–92%. F. e. s. K. Bailey, Can. J. Chem. *52*, 2136 (1974).

Lithium/liq. ammonia Li/NH_3
Removal of benzylic oxygen   ArC(OH) → ArCH
     Hydrocarbons from benzylic alcohols – also retention of benzylic hydroxyl
     s. *27*, 88

Potassium carbonate s. under H_2CO K_2CO_3

Lithium tetrahydridoborate $Li[BH_4]$
Hydroxycarboxylic acids from ketolactones   ←
     Stereospecific reduction s. *27*, 505

Sodium tetrahydridoborate-hydrogen chloride $Na[BH_4]$-HCl
Replacement of hydroxyl by hydrogen   OH → H
s. *26*, 79

Sodium tetrahydridoborate-carbon s. under H_2PtCl_6 $Na[BH_4]$-C

Lithium tetrahydridoaluminate $Li[AlH_4]$

1,3-Dienes from 2-allenealcohols   C(OH)C : C : C → C : C·CH : C
s. *29*, 88; reduction of allenealcohols s. R. Baudouy and J. Gore, Tetrahedron *31*, 383 (1975)

Arenes by high-temp. reduction   ArOH → ArH
with lithium tetrahydridoaluminate

48.

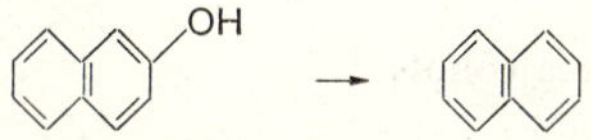

from phenols. A soln. of β-naphthol and 10 moles $LiAlH_4$ in anhydrous tetrahydrofuran stirred 15 min., most of the solvent removed by distillation under reduced pressure, and the residue distilled at 0.1 mm → naphthalene. Y: 43%. F. e., also from ar. compds. with amino groups and other functions, s. T. Severin and I. Ipach, Synthesis *1973*, 796.

Sodium trihydridocyanoborate/triphenyl
phosphite methiodide $Na[BH_3CN]/(PhO)_3\overset{+}{P}Me\ I^-$

Hydrocarbons from alcohols   OH → H
s. *27*, 74

Methanol/irradiation   MeOH/⫽

Replacement of hydroxyl by hydrogen
    Methyl from hydroxymethyl groups s. *26*, 80

Formaldehyde/potassium carbonate H_2CO/K_2CO_3

α-Methylketones   C : CHOH → C : CH_2 (→ $CHCH_3$)
    from α-hydroxymethyleneketones via α-methyleneketones — Stereospecific hydrogenation s. *27*, 89; α-methyleneketones s. a. H. Hashimoto et al., Tetrah. Let. *1974*, 3745

Titanium trichloride $TiCl_3$

Cyclopent-2-enones   OH → H
    from cyclopent-2-en-4-ol-1-ones s. *27*, 772

Stannous chloride $SnCl_2$

Replacement of hydroxyl by hydrogen   OH → H
s. *26*, 665

Triphenyl phosphite methiodide s. under $Na[BH_3CN]$ $(PhO)_3\overset{+}{P}Me\ I^-$

Hydrogen chloride s. under $Na[BH_4]$ HCl

Nickel   Ni

Carboxylic acid amides from hydroxamic acids   CONHOH → $CONH_2$
s. *27*, 90

Palladium   Pd

Simultaneous removal   OH → H
of benzylic oxygen, O-*tert*-butyl and O,O-isopropylidene groups

49.

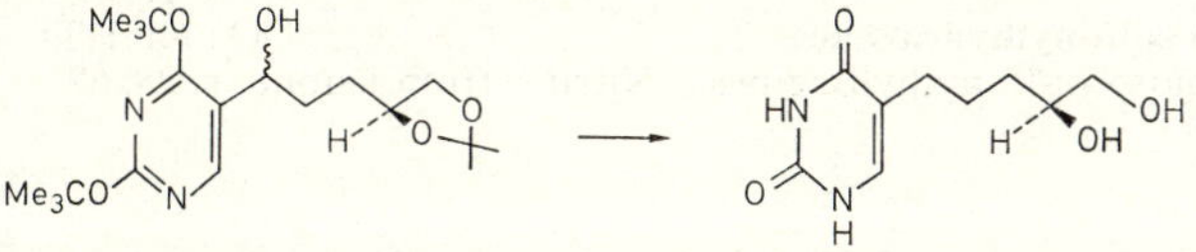

(S)-5-(O-Isopropylidene-1′,4′,5′-trihydroxypentyl)-2,4-di-(*tert*-butoxy)pyrimidine in moist methanol hydrogenated 18 hrs. at room temp. with Pd-black → (S)-5-(4′,5′-dihydroxypentyl)uracil. Y: ca. 100%. H. Hayashi et al., Am. Soc. *95*, 8749 (1973).

Palladium-carbon/sulfuric acid $Pd\text{-}C/H_2SO_4$
Removal of benzylic hydroxyl   ArC(OH) → ArCH
s. *27*, 872

Dichlorobis(triphenylphosphine)palladium(II) $[PdCl_2(PPh_3)_2]$
Alcohols from glycols   C(OH)C(OH) → CHC(OH)
s. *26*, 81

Hexachloroplatinic acid/sodium tetrahydridoborate-
carbon/hydrogen chloride $H_2PtCl_6/Na[BH_4]\text{-}C/HCl$
Hydrocarbons from ethylenealcohols   ←
s. *29*, 933

Via intermediates   v.i.
Methyl from hydroxymethyl groups CH_2OH → CH_3
    via 2-pyrimidinyl thioethers s. *27*, 91

Deoxygenation via phosphoric acid derivs.   OH → H
    Hydrocarbons from alcohols via phosphorodiamidates – also deoxygenation of ketones and conversion of keto groups into carbon-carbon double bonds s. *28*, 65

Ethylene derivs. from 2-ethylenealcohols   C:C·C(OH) → C:C·CH
    via 2-ethylenisothiouronium salts with double bond migration s. *30*, 41

Replacement of hydroxyl by hydrogen via isoureas   ArOH → ArH
Arenes from phenols

50.

β-Naphthol stirred 3 days at 45° with 2 equivalents dicyclohexylcarbodiimide, then hydrogenated 6 hrs. with Pd-CaCO₃ in ethyl acetate → naphthalene. Y: 87%. F. e., also with isolation of the intermediate O-arylisoureas, s. E. Vowinkel, C. Wolff, and H.-J. Baese, B. *107*, 907, 1213 (1974); **hydrocarbons from alcohols** cf. ibid. *107*, 1353.

Arenes from phenols
    via aryl sulfonates s. *29*, 65

Aldehydes from carboxylic acids   COOH → CHO
    via prim. alcohols – Labeled compds. s. *29*, 89
    via carboxylic alkoxyformic anhydrides – α-Acylaminoaldehydes s. *28*, 66; reduction with Na-tetracarbonylferrate(II) (cf. Synth. Meth. *29*, 64) s. Y. Watanabe et al., Tetrah. Let. *1975*, 1063

Carboxylic acid esters   C(OH)COOR → CHCOOR
    from α-hydroxy- via α-benzoxy-carboxylic acid esters s. *27*, 92

Nitrogen ↑   HC⇑N

Without additional reagents   w.a.r.
Nitriles from sulfonylhydrazones   C:NNHTs → CHCN
    via α-cyano-N-sulfonylhydrazines – Nitriles from ketones s. *28*, 67

Sulfur ↑   HC⇑S

Irradiation   ⚡
Photochemical desulfuration   ←
    1,2,4-Triazoles from 1,2,4-triazoline-3-thiones s. *26*, 82

Trifluoromethanesulfonic acid CF_3SO_3H
Replacement of sulfonic acid groups by hydrogen $SO_3H \to H$
s. *17*, 559 suppl. *28*

Nickel Ni
Hydrodesulfination ←
s. *29*, 90

Chlorotris(triphenylphosphine)rhodium(I) $RhCl(Ph_3P)_3$
Hydrogenolysis of sulfur compds.
Hydrocarbons from mercaptans $SH \to H$

51. PhCH₂SH → PhCH₃

Benzyl mercaptan hydrogenated with chlorotris(triphenylphosphine)rhodium(I) at 100°/30 atm. ⇢ toluene. Y: ca. 100%. F. e. s. A. S. Berenblyum et al., Izvest. *1973*, 2650; C. A. *80*, 95402.

Carbon ↑ HC⇑C

Without additional reagents w.a.r.
Decarboxylation $COOH \to H$
    Ketones from β-ketocarboxylic acids s. *28*, 733
1,3-Dienes from alkylideneglutaconic acids
s. *29*, 91
Amines from α-aminocarboxylic acids $C(NH_2)COOH \to CHNH_2$
s. *28*, 68
Aldehydes
    from glycidic acid *tert*-butyl esters s. *26*, 859

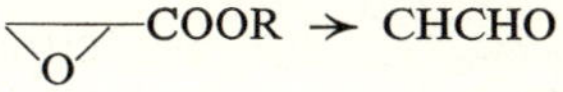

Irradiation (s. a. under PhNMe₂ and Ph₂CO) ⚡
C-De-*tert*-butylation $CMe_3 \to H$
    Partial de-*tert*-butylation – Phenols s. *26*, 83
Photodecarboxylation $COOH \to H$
s. *27*, 93 suppl. *29*

n-Butyllithium n-BuLi
Decarboxamidation $CONH_2 \to H$
s. *26*, 84

Diphenylamine Ph_2NH
Amines from α-aminocarboxylic acids $C(NH_2)COOH \to CHNH_2$
s. *28*, 68

Dimethylaniline/irradiation PhNMe₂/⚡
C-Decarbinolation $C(OH){<} \to H$
s. *26*, 85

Pyridine C_5H_5N
Decarboxylation $COOH \to H$
s. *26*, 86

2,4-Dimethylpyridine-sodium/alcohol ←
2,4-Dienecarboxylic from alkylideneglutaconic acids ←

52.

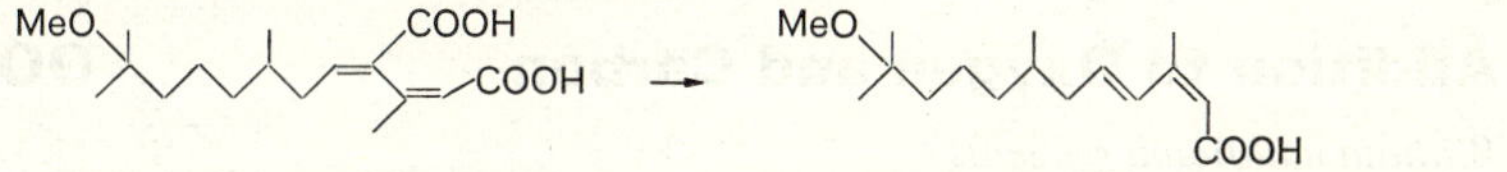

A soln. of (2Z,4E)-4-carboxy-11-methoxy-3,7,11-trimethyl-2,4-dodecadienoic acid and 2,4-dimethylpyridine in dry toluene heated at 100° under N₂ until decarboxylation is complete after 3 hrs., cooled to 70°, purged with N_2, methanolic 25%-Na-methoxide added, and held ca. 1 hr. at 70° under N_2 → (2Z,4E)-11-methoxy-3,7,11-trimethyl-2,4-dodecadienoic acid. Y: 93%. F. e. s. C. A. Henrick et al., J. Org. Chem. *40*, 1 (1975).

Cuprous oxide/quinoline/2,2′-dipyridyl ←
Double ring contraction Ö:
of highly-condensed polyisocyclics with decarboxylation s. *29*, 92

Cupric acetate $(CH_3COO)_2Cu$
Amines from α-aminocarboxylic acids $C(NH_2)COOH → CHNH_2$
s. *28*, 68 suppl. 29

Magnesium methoxide/liq. ammonia $(MeO)_2Mg/NH_3$
Decarbonylation ←
with migration of carbon-carbon double bonds s. *26*, 87

Montmorillonite ←
Reactions in sheet-silicate intercalates ←
Anilines from diaminostilbenes s. *29*, 93

Benzophenone/irradiation $Ph_2CO/⚡$
Photodecarboxylation $COOH → H$
s. *27*, 93

Acetic anhydride Ac_2O
α,β-Ethylenecarboxylic acids
from α,β-ethylene-α-dicarboxylic acids

53. ⟦S⟧–CH=C(COOH)$_2$ → ⟦S⟧–CH=CHCOOH

A suspension of (2-thienyl)methylenemalonic acid in acetic anhydride refluxed 15 min. until gas evolution has ceased → *trans*-3-(2-thienyl)acrylic acid. Y: 85.7% containing 5% *cis*-isomer. F. e. s. Y. Taniguchi and H. Kato, Chem. Pharm. Bull. *21*, 2070 (1973).

Trifluoroacetic acid CF_3COOH
C-De-*tert*-butylation $CMe_3 → H$
Phenols s. *26*, 83 suppl. 29

Hydrogen chloride HCl
Decarboxylation $COOH → H$
s. *27*, 768

Chlorotris(triphenylphosphine)rhodium(I) $RhCl(Ph_3P)_3$
Decarbonylation of 2-ethylenealcohols $C:C·CH_2OH → CHCH_2$
s. *27*, 94

Formation of O—O Bond

Uptake ⇓

Addition to Oxygen and Carbon   OO⇓OC

Without additional reagents w.a.r.
1,2-Hydroxyhydroperoxides ▽O → $C(OOH)C(OH)$
from oxido compds. s. *28*, 69

Exchange

Hydrogen ↑ OO ↕ H
Fluorine/potassium salt F/K^+
Acyl peroxides from carboxylic acids $2\ COOH \rightarrow CO \cdot OO \cdot OC$
s. *27*, 95

Elimination

Hydrogen ↑ OO ↑ H
Sodium peroxide/methanesulfonic acid $Na_2O_2/MeSO_3H$
Dicarboxylic acid peroxides from dicarboxylic acids

$$\begin{matrix} \text{COOH} \\ \text{COOH} \end{matrix} \rightarrow \begin{matrix} \text{CO} \cdot \text{O} \\ | \\ \text{CO} \cdot \text{O} \end{matrix}$$

54.

Na_2O_2 added cautiously with stirring at ca. 5° to methanesulfonic acid, after 15 min. cyclohexane-1,1-dicarboxylic acid added, and stirred overnight at 5° → product. Y: 75%. F. e. s. A. H. Alberts, H. Wynberg, and J. Strating, Synth. Commun. *3*, 297 (1973).

Formation of O—N Bond

Uptake

Addition to Hydrogen and Nitrogen ON ↓ HN
Via intermediates v.i.
Hydroxylamines $NH \rightarrow N \cdot OH$
from amines via benzoxylamines s. *26*, 88

Addition to Nitrogen ON ↓ N
Irradiation ⧸⧸⧸
Nitro from nitroso compds. $NO \rightarrow NO_2$
s. *12*, 151 suppl. 28
Nitric from nitrous acid esters $ONO \rightarrow ONO_2$
s. *26*, 89
Diol mononitrates $\begin{matrix}\text{CH}\\\text{ONO}\end{matrix} \rightarrow \begin{matrix}\text{C} \cdot ONO_2\\\text{OH}\end{matrix}$
from nitrous acid esters s. *30*, 95

Sodium nitrite/sulfuric acid $NaNO_2/H_2SO_4$
Nitro from nitroso compds. $NO \rightarrow NO_2$
s. *27*, 356

Trifluoroacetic anhydride s. under H_2O_2	$(CF_3CO)_2O$
Dichloromaleic anhydride s. under H_2O_2	←
Hydroperoxides s. under $MoCl_5$	$R \cdot OOH$
Formic acid s. under H_2O_2	$HCOOH$
Peroxytrifluoroacetic acid s. $H_2O_2/(CF_3CO)_2O$	CF_3COO_2H
m-Chloroperoxybenzoic acid	$m\text{-}ClC_6H_4COO_2H$
N-Oxides from amines	$\geqslant N \rightarrow \geqslant N\rightarrow O$
3-Oxazoline N-oxides s. 26, 90	
N-Oxides	
s. 26, 90	
Hydrogen peroxide/trifluoroacetic anhydride	$H_2O_2/(CF_3CO)_2O$
Aliphatic nitro compds. from oximes	C : NOH → $CHNO_2$
Carbohydrates s. 27, 96	
Azoxy from azo compds.	N : N → N(O) : N
s. 27, 97	
Hydrogen peroxide/dichloromaleic anhydride	←
Cyclic N-oxides	N→O
Pyridazine 1-oxides from pyridazines s. 29, 94	
Hydrogen peroxide/formic acid	$H_2O_2/HCOOH$
Ring expansion	☉
by Meisenheimer rearrangement of cyclic N-oxides s. 26, 151	
Hydrogen peroxide/sulfuric acid/acetic acid	$H_2O_2/H_2SO_4/CH_3COOH$
N-Oxides	⌒N → ⌒N→O
of weakly basic N-heterocyclics s. 27, 98	
Oxidations with *molybdenum pentachloride/hydroperoxides*	$MoCl_5/R \cdot OOH$
s. 27, 184	
Potassium permanganate	$KMnO_4$
Nitro from nitroso compds.	NO → NO_2
s. 30, 315	

Rearrangement

Hydrogen/Nitrogen Type

Phosphate buffer ←
N-N′-Alkoxy group migration

55.

An aq. soln. of 1-ethoxy-2′,3′-O-isopropylideneadenosine hydriodide · 1/2 ethanol passed through a column of Amberlite IRA-402 (HCO_3^-), the resulting free base eluted with water, the eluate concd. in vacuo, phosphate buffer of pH 7 added, and heated 5 hrs. in a boiling water bath ⇢ product. Y: 84%. F. e., also isolation of the intermediate, s. T. Fujii et al., Chem. Pharm. Bull. *21*, 1676 (1973).

Oxygen/Nitrogen Type ON⋂ON

Without additional reagents   w.a.r.
4-Aminofuroxans
    from glyoximes via 3-aminofuroxans s. *27,* 99

Nitrogen/Carbon Type ON⋂NC

Without additional reagents   w.a.r.
O-Carbamylhydroxylamines
from N-hydroxyureas   N·CO·N(OH)· ⇢ N·CO·ONH·
s. *29,* 95

Exchange ↓↑

Hydrogen ↑ ON↕H

Without additional reagents   w.a.r.
Nitric acid esters from alcohols   OH ⇢ ONO_2
s. *26,* 361
Acetic anhydride Ac_2O
s. *17,* 149 suppl. *29*
Peroxybenzoic acid $PhCOO_2H$
Nitroso compds. from amines NH_2 ⇢ NO
s. *26,* 91
m-Chloroperoxybenzoic acid $m\text{-}ClC_6H_4COO_2H$
s. *26,* 91
Chromium trioxide/sulfuric acid CrO_3/H_2SO_4
Nitro compds. from amines NH_2 ⇢ NO_2
    Chemistry of carboranes, review, s. *28,* 70; review of carboranes s. a. W. Stumpf, Chem. Ztg. *99,* 1 (1975)

Oxygen ↑ ON↕O

Without additional reagents   w.a.r.
Acoxylamines from amines   NH ⇢ N·OAc
s. *26,* 88

Carbon ↑   ON ↕ C

Potassium hydroxide/hydrogen peroxide   *KOH/H$_2$O$_2$*
Cyclic azoxy compds.   C
   from 1,2,4-triazoli-3,5-dione ring – from cyclic hydrazo-N,N'-dicarboxylic acid esters s. *26*, 92; from 1,2,4-triazoli-3,5-dione ring s. a. Am. Soc. *96*, 7839 (1974)

Elimination ⇑

Hydrogen ↑   ON ⇑ H

Oxidations with *silver carbonate-Celite*   *Ag$_2$CO$_3$-Celite*
Nitroso compds. from hydroxylamines   NHOH → NO
   Azo N,N'-dioxides from aliphatic hydroxylamines s. *27*, 100; f. oxidations of nitrogen compds. s. M. Fetizon et al., Tetrahedron *31*, 165 (1975); oxidations of diols, mostly to lactones, s. ibid. *31*, 171; cyclic azo-N,N'-dioxides cf. P. Singh, J. Org. Chem. *40*, 1405 (1975); F. D. Greene and K. E. Gilbert, ibid. *40*, 1409; J. P. Snyder, M. L. Heyman, and E. N. Suciu, ibid. *40*, 1395

Oxygen ↑   ON ⇑ O

Sodium azide   *NaN$_3$*
Benzofurazans   O
from o-dinitro compds. or benzofurazan oxides

56.

3,4-Dinitroanisole | 5-Methoxybenzofurazan oxide
   heated 1–2 hrs. until gas evolution has ceased in ethylene glycol
with ca. 2 moles NaN$_3$ at 150° → | with ca. 1 mole NaN$_3$ at 140° → 5-methoxybenzofurazan. Y: 70%.
F. e. s. L. Di Nunno, S. Florio, and P. E. Todesco, Soc. Perkin I *1973*, 1954.

Trimethyl phosphite   *(MeO)$_3$P*
4-Acylbenzofurazans   ←
   from 7-nitro-2,1-benzisoxazoles s. *27*, 101

Nitrogen ↑   ON ⇑ N

Without additional reagents   *w.a.r.*
[1,2,5]Oxadiazolo[3,4-b]pyridine 1-oxides   ←
   from 4-nitro-1,2,3,4-tetrazolopyridines s. *27*, 102

Formation of O–Hal Bond

Exchange	⇅

Hydrogen ↑   OHal ⇅ H

Without additional reagents   *w.a.r.*
O-Perchlorination   OH → $OClO_3$
s. *30*, 195

Halogen ↑   OHal ⇅ Hal

Pyridine C_5H_5N
Iodosoacetates from iododichlorides ICl_2 → $I(OAc)_2$
s. *27*, 543

Formation of O—S Bond

Uptake	⇓

Addition to Hydrogen and Oxygen   OS⇓HO

Pyridine s. under Sulfamic acids C_5H_5N
Sulfamic acids/pyridine   >N-SO_3H/C_5H_5N
Sulfuric acid monoesters   OH → OSO_2OH
s. *27*, 103

Addition to Hydrogen and Sulfur   OS⇓HS

m-Chloroperoxybenzoic acid m-$ClC_6H_4COO_2H$
Sulfinic acids from mercaptans   SH → SO_2H
s. *12*, 156 suppl. 29

Addition to Sulfur   OS⇓S

Irradiation s. under Methylene blue   ⚡
Tetrachloroauric(III) acid $HAuCl_4$
Sulfoxides from thioethers   >S → >SO
    Stereospecific oxidation s. *18*, 141 suppl. 29

Ammonium ceric nitrate	$(NH_4)_2Ce(NO_3)_6$
Sulfoxides from thioethers s. *28*, 70a	
Ethyl alcohol s. under SO_2Cl_2	*EtOH*
Peroxyacetic acid s. H_2O_2/Ac_2O and H_2O_2/CH_3COOH	CH_3COO_2H
m-Chloroperoxybenzoic acid	$m\text{-}ClC_6H_4COO_2H$
S-Oxides s. *27*, 734	$> SO$
Sulfoxides from thioethers s. *30*, 685	$> S \twoheadrightarrow SO$
Sulfinic from sulfenic acid esters	$S(OR) \twoheadrightarrow SO(OR)$

57. $Cl_3C\text{—}S\text{—}O\text{—}CH_2Ph \longrightarrow Cl_3C\text{—}S(=O)\text{—}O\text{—}CH_2Ph$

Benzyl trichloromethanesulfenate treated at 0° with m-chloroperoxybenzoic acid in methylene chloride $\twoheadrightarrow$ benzyl trichloromethanesulfinate. Y: almost 100%. F. e. s. S. Braverman and Y. Duar, Tetrah. Let. *1975*, 343.

Methylene blue/irradiation	←
Thiolsulfinic acid esters from dialkyl disulfides – Oxidation with singlet oxygen s. *29*, 96	$S \cdot S \twoheadrightarrow S \cdot SO$
N-Chloronylon-66	←
Sulfoxides from thioethers s. *28*, 70a	$> S \twoheadrightarrow > SO$
Hydrogen peroxide s. a. under $Mo(CO)_6$, NH_4MoO_4, and $MoCl_5$	H_2O_2
Hydrogen peroxide/potassium hydroxide	H_2O_2/KOH
Acoxysulfones from hydroxythioethers s. *26*, 93	$> S \twoheadrightarrow > SO_2$
Hydrogen peroxide/acetic anhydride	H_2O_2/Ac_2O
Sulfoxides from thioethers s. *28*, 70a	$> S \twoheadrightarrow > SO$
Hydrogen peroxide/acetic acid	H_2O_2/CH_3COOH
1,1-Sulfinylthioethers from mercaptals s. *29*, 97	$C(S\text{–})_2 \twoheadrightarrow C(SO\text{–})(S\text{–})$
Sulfuryl chloride/ethyl alcohol	$SO_2Cl_2/EtOH$
Sulfoxides from thioethers s. *29*, 98	$> S \twoheadrightarrow > SO$
Molybdenum hexacarbonyl/hydrogen peroxide	$Mo(CO)_6/H_2O_2$
Sulfoxides and sulfones from thioethers s. *17*, 153 suppl. 28	$> S \Bigg\langle \begin{array}{l} > SO \\ > SO_2 \end{array}$
Ammonium molybdate/hydrogen peroxide	NH_4MoO_4/H_2O_2
1,1-Disulfoxides from mercaptals s. *29*, 99	$C(S\text{–})_2 \twoheadrightarrow C(SO\text{–})_2$
Sulfones from thioethers s. *26*, 3	$> S \twoheadrightarrow > SO_2$
Molybdenum pentachloride/hydrogen peroxide	$MoCl_5/H_2O_2$
Sulfoxides and sulfones from thioethers s. *17*, 153 suppl. 28	$> S \Bigg\langle \begin{array}{l} > SO \\ > SO_2 \end{array}$

Potassium permanganate $KMnO_4$
Sulfuric from sulfurous acid esters $(RO)_2SO \rightarrow (RO)_2SO_2$
s. 27, 104

Potassium permanganate/pyridine $KMnO_4/C_5H_5N$
N-Sulfonylsulfoximines $>S:N \cdot SO_2 \cdot \rightarrow > S(O):N \cdot SO_2 \cdot$
from N-sulfonylsulfilimines s. 26, 94

Rearrangement   ↻

Hydrogen/Oxygen Type   OS∩HO

Sodium hydride NaH
O-Sulfonyl group migration   ←
s. 27, 105

Oxygen/Nitrogen Type   OS∩ON

Irradiation   ⇝
Nitrososulfones from nitrosulfoxides   ←
s. 27, 106

Exchange   ↓↑

Hydrogen ↑   OS↓↑H

Potassium hydroxide/carbon tetrachloride KOH/CCl_4
α,β-Ethylenesulfonic acids from sulfones   ←
Ramberg-Bäcklund-type rearrangement s. 29, 100

Oxygen ↑   OS↓↑O

Without additional reagents   w.a.r.
Skeletal rearrangement   ←
with O-tosylation s. 29, 699

Sodium carbonate Na_2CO_3
Sulfonic acid esters from anhydrides $OH \rightarrow OSO_2R$
s. 27, 107 suppl. 30

Pyridine C_5H_5N
Enesulfonates s. 27, 107; triflates with anhydrous Na-carbonate cf. M. Hanack et al., Org. Synth. 54, 84 (1974)

Dicyclohexylcarbodiimide $RN:C:NR$
Sulfinic acid esters from alcohols $OH \rightarrow OS(O)Ar$
s. 27, 108

Nitrogen ↑   OS ⇅ N

Sodium methoxide   *NaOMe*
O-Sulfination   OH → OS(O)R
    with N-sulfinylphthalimides s. *29*, 289

Reactions with N-tosylimidazole   OH → OTs
    Carbohydrate derivs. s. *29*, 101

Halogen ↑   OS ⇅ Hal

Sodium hydride   *NaH*
2-Aminomesylates from 2-aminoalcohols   OH → OSO$_2$Me
s. *30*, 459

Potassium hydroxide   *KOH*
Labeled aryl mesylates   ArOH → ArOSO$_2$Me
s. *29*, 102

Sodium hydrogen carbonate   *NaHCO$_3$*
Carbamyl chlorosulfines   NHCO·CCl$_2$SCl → NHCO·CCl:S:O
    from α-carbamyl-α,α-dichloromethylsulfenyl chlorides s. *29*, 103

Triethylamine   *Et$_3$N*
Mesylation of alcohols   OH → OSO$_2$Me
    Mesylates as intermediates s. *27*, 109; also mesylation of phenols and N-mesylation with quaternary methylsulfonylammonium salts, such as diethylmethyl-(methylsulfonyl)ammonium fluorosulfonate, MeSO$_2$N$^+$ Et$_2$Me FSO$_3^-$, cf. J. F. King and J. R. du Manoir, Am. Soc. *97*, 2566 (1975)

Chlorides   OH → OSO$_2$Me → Cl
from alcohols via mesylates
s. *30*, 366

Selective O- and N-tosylation   OH → OTs

58.

o-aminophenol + ClSO$_2$C$_6$H$_4$CH$_3$—p

↙   ↘

2-aminophenyl OSO$_2$C$_6$H$_4$CH$_3$—p   2-hydroxyphenyl NHSO$_2$C$_6$H$_4$CH$_3$—p

Triethylamine | Pyridine
added at 0° to a suspension of o-aminophenol in methylene chloride, then tosyl chloride added, and allowed to stand 1 hr. at 20° →
tosylate. Y: 94%. | tosylamide. Y: 95%.
The stronger base causes O-tosylation. K. Kurita, Chem. & Ind. *1974*, 345.

Sulfonyloxylamines from sulfonic acid chlorides   SO$_2$Cl → SO$_2$·O·NH$_2$
    via N-carbalkoxysulfonyloxylamines – Selective N-decarbo-*tert*-butoxylation
s. *27*, 110

N-Sulfonyloxysulfonamides   SO$_2$NHOH → SO$_2$NHOSO$_2$Ar
    from sulfhydroxamic acids s. *26*, 95

Pyridine   *C$_5$H$_5$N*
O-Mesylation   OH → OSO$_2$Me
s. *29*, 622

| Benzenesulfonates from alcohols | OH → OSO$_2$Ph |
s. 27, 401

Tosylates from alcohols OH → OTs
s. 28, 52, 119

2-Azido-tosylates from -alcohols
s. 30, 278

Silver salt Ag$^+$

Sulfonic acid esters from chlorides Cl → OSO$_2$R
1-Alkoxy-, 1-aryloxy-, and 1-acoxy-1-sulfonyloxy compds. s. 27, 111

Carbon ↑ OS ↕ C

Without additional reagents w.a.r.

Reactions with pyrosulfuryl fluoride OSO$_2$F
Fluorosulfonic acid esters from ethers s. 27, 112

Hydrogen peroxide/formic acid H$_2$O$_2$/HCOOH

Sulfonic acids ·S·CON< → ·SO$_3^-$
from thiolcarbamic acid esters – Ar. sulfonic acids from phenols s. 26, 96

Potassium permanganate KMnO$_4$

Sulfonic acids from sultones SO$_2$R → SO$_2$OH
s. 29, 104

Elimination ⇑

Carbon ↑ OS ⇑ C

Sodium hydroxide NaOH

Sulfoxides from alkoxysulfonium salts >S$^+$—OR → >SO
s. 27, 126

N-Chlorosuccinimide CH$_2$CO—NCl / CH$_2$CO

Sultines O
1,2-Oxathiolane 2-oxides – Ethylenesultines and their conversion to 1,3-dienes
s. 29, 105

Formation of O—Rem Bond

Uptake ⇓

Addition to Hydrogen and Oxygen ORem ⇓ HO

Pyridine C$_5$H$_5$N

Thiophosphorylation of nucleosides OH → OPS(OH)$_2$
s. 27, 113

2,6-Lutidine ←
Nucleotides from nucleosides OH → OPO(OH)$_2$
s. *8*, 146 suppl. *29*

Cyanogen (CN)$_2$
1-Phosphorylation of carbohydrates OH → OPO(OH)$_2$
s. *27*, 114

Phosphorus oxide chloride/trimethyl phosphate POCl$_3$/(MeO)$_3$PO
Preferential phosphorylation OH → OPO(OH)$_2$
of nucleoside hydroxyls s. *26*, 97

Addition to Hydrogen and Remaining Elements ORem⇓HRem

Sodium Na
Phosphoric acid esters CO → CH·O·PO(OR)$_2$
from dialkyl phosphites and ketones s. *28*, 523

Addition to Oxygen and Oxygen ORem⇓OO

Without additional reagents w.a.r.
Insertion of phosphines ☉
into cyclic oxygen-oxygen bonds – 1,3,2-Dioxaphospholanes from 1,2-dioxetanes s. *29*, 106

Addition to Oxygen and Carbon ORem⇓OC

Without additional reagents w.a.r.
α-Hydroxyphosphoric acid esters ▽O → C(OH)·C·OPO(OR)$_2$
from oxido compds. s. *26*, 98

Ring O,P-ylids ←
s. *27*, 115

Chlorotris(triphenylphosphine)rodium(I) RhCl(Ph$_3$P)$_3$
Alkoxysilanes from oxo compds. CO → CH·OSi ≦
s. *29*, 107; asym. synthesis cf. T. Hayashi, K. Yamamoto, and M. Kumada, Tetrah. Let. *1974*, 4405

Addition to Remaining Elements ORem⇓Rem

Without additional reagents w.a.r.
1,1-Acoxyselenides >Se → >Se(OAc)$_2$
from selenides and acyl peroxides via diacyloxyselenuranes s. *29*, 108

Sodium hydride NaH
Phosphinic acids from phosphinous acid esters >P·OR → P(O)OH
s. *30*, 159

Dimethyl sulfoxide Me$_2$SO
Phosphinic acid esters from chlorophosphines R$_2$PCl → R$_2$P(O)OR
via phosphinic acid chlorides s. *26*, 99

Iodobenzene dichloride PhICl$_2$
Selenoxides from selenides >Se → >SeO
s. *26*, 101

Hydrogen peroxide H_2O_2
Phosphoric from phosphorous acid triamides $>N)_3P \rightarrow >N)_3PO$
s. *26*, 100

Periodate IO_4^-
Selenoxides from selenides $>Se \rightarrow >SeO$
s. *26*, 101

Rearrangement ↻

Oxygen/Nitrogen Type   ORem↻ON

Without additional reagents w.a.r.
Silylaminosiloxanes ←
   from N,N-disilylsiloxylamines s. *28*, 301

Remaining Elements/Carbon Type   ORem↻RemC

Irradiation ⚡
3-Oxo-4-silaspiro[2.6]nonanes ○
   from silacyclohexan-2-ones s. *27*, 708
Sodium/alcohol NaOR
Phosphoric
from α-hydroxyphosphonic acid esters $C(OH) \cdot PO(OR)_2 \rightarrow CH \cdot O \cdot PO(OR)_2$
s. *28*, 71

Exchange ↓↑

Hydrogen ↑   ORem↮H

Without additional reagents w.a.r.
Phosphorous and phosphinic acid monoesters $OH \rightarrow OP(OH)_2$
   Azeotropic water entrainment s. *26*, 102
Triethylamine Et_3N
Phosphorodichloridates from alcohols $OH \rightarrow OPOCl_2$
s. *28*, 72
Pyridine s. under N,O-Bis(trimethylsilyl)acetamide C_5H_5N
Mercuric chloride/triethylamine $HgCl_2/Et_3N$
Phosphoric acid monoesters from alcohols $OH \rightarrow OPO(OH)_2$
s. *23*, 123
Mercuric chloride/N-methylimidazole ←
s. *23*, 123 suppl. *29*
N-Methylimidazole s. under HgCl_2 ←
N,O-Bis(trimethylsilyl)acetamide ←
Preferential protection of phenol groups $ArOH \rightarrow ArOSi\leqslant$
   as aryloxysilanes s. *28*, 74

N,O-Bis(trimethylsilyl)acetamide/pyridine ←
Protection of alcohol groups OH → OSi
  as silyl ethers s. 29, 964
Trimethyl phosphate $(MeO)_3PO$
Phosphorous acid monoesters OH → OP(OH)$_2$
  also phosphorylation s. 26, 103
Chlorotris(triphenylphosphine)rhodium(I) $RhCl(Ph_3P)_3$
Alkoxysilanes from alcohols and silanes OH → OSi
s. 29, 109
Via intermediates v.i.
Phosphorylation OH → OPO(OH)$_2$
  via 2-alkoxy- or 2-aryloxy-2-oxo-1,3,2-dioxaphospholanes – Phosphorylbetaines
  – Phosphoric acid monoamids s. 29, 110

Oxygen ↑   ORem ⇈ O

Without additional reagents w.a.r.
Activation of glycol groups O
  as cyclic dialkoxystannanes – Nucleoside derivs. s. 29, 111
Glycosyl phosphates C
from carbohydrate orthocarboxylic acid esters

59.

3,4,6-Tri-O-acetyl-α-D-glucopyranose 1,2-(*tert*-butyl orthoacetate) allowed to react 0.5 hr. at 19–20° with dibenzyl hydrogen phosphate in dry benzene → 2,3,4,6-tetra-O-acetyl-β-D-glucopyranosyl dibenzyl phosphate. Y: 85%. F. e. s. L. V. Volkova, L. L. Danilov, and R. P. Evstigneeva, Carbohyd. Res. *32*, 165 (1974).

Iodocarboxylates from iodoso compds. IO → I(OAc)$_2$
  (Dibenzoyldioxyiodo)benzenes as oxidants s. 27, 116

Sodium sulfate Na_2SO_4
Diphosphoric acid monoesters OH → O·P(O)(OH)·O·P(O)(OH)$_2$
s. 29, 112; without Na-sulfate cf. R. I. Ho, L. Corman, and W. O. Foye, J. Pharm. Sci. *63*, 1474 (1974).

Pyridine s. under (CH$_3$COO)$_2$Cu, CuCl$_2$, RN : C : NR, and ArSO$_2$Cl C_5H_5N

Cupric acetate/pyridine $(CH_3COO)_2Cu/C_5H_5N$
Phosphorylation with 8-(phosporyloxy)quinolines ←
s. 29, 113

Cupric chloride/pyridine $CuCl_2/C_5H_5N$
Phosphorylation with 8-(phosporyloxy)quinolines OH → OPO(OH)(OR)
  Mixed phosphoric acid diesters – Sym. diphosphoric acid diesters s. 29, 113

Diethyl azodicarboxylate s. under Ph$_3$P ROOC·N : N·COOR

Dimethylformamide s. under SOCl$_2$ and TsCl Me_2NCHO

Dicyclohexylcarbodiimide/pyridine $RN : C : NR/C_5H_5N$
Protection of phosphoryl hydroxyl groups P·OH → P·OR
  with 2-α-pyridylethanol s. 26, 104

Triphenyl phosphine/diethyl azodicarboxylate $Ph_3P/ROOC \cdot N : N \cdot COOR$
Preferential phosphorylation $OH \rightarrow OPO(OH)_2$
s. 26, 105

Thionyl chloride/dimethylformamide $SOCl_2/Me_2NCHO$
Diphosphoric acid diester monobetaines   ←
s. 28, 73

p-Toluenesulfonyl chloride/dimethylformamide $TsCl/Me_2NCHO$
s. 28, 73

Nitrogen ↑   ORem ↓↑ N

Without additional reagents   w.a.r.
Preferential protection of phenol groups $ArOH \rightarrow ArOSi \leqq$
  as aryloxysilane s. 28, 74

2-Germa-1,3-dioxaheterocycles   ○
  from diols and diaminogermanes s. 29, 114

Mixed phosphoric acid diesters $OPO(OH)(OR)$
from phosphoric acid monoamide monoesters

60.

A suspension of hydrogen phenyl N-(2-aminophenyl)phosphoramidate in *n*-propanol refluxed 24 hrs. at 97° → phenyl *n*-propyl phosphate. Y: 82% as the o-phenylenediammonium salt. F. e. s. T. Koizumi, Y. Arai, and E. Yoshii, Chem. Pharm. Bull. 22, 468 (1974).

Irradiation   ⫴
Phosphoric from nitrous and phosphorous acid esters $C \cdot ONO \rightarrow C \cdot OPO(OR)_2$
s. 28, 80

Sodium hydroxide $NaOH$
Selective hydrolysis of phosphoramidates $P \cdot N < \rightarrow P \cdot OH$
s. 27, 117

Pyridine C_5H_5N
(1,3-Dien)oxysilanes
from α,β-ethyleneketones $CHC : C \cdot CO \rightarrow C : C \cdot C : C \cdot OSi \leqq$
s. 27, 83

Ammonium ceric nitrate $(NH_4)_2Ce(NO_3)_6$
Oxidative phosphorylation $OH \rightarrow OPO(OR)_2$
  with 1-phosphoryl-1,4-dihydropyridines s. 29, 115

Bromine Br_2
Mixed phosphoric acid diesters $OPO(OH)(OR)$
s. 28, 75

Hydrogen chloride HCl
Selective hydrolysis of phosphoramidates $P \cdot N < \rightarrow P \cdot OH$
s. 27, 117

Halogen ↑ ORem ↓↑ Hal

Without additional reagents w.a.r.
Borinic acid esters from halogenoboranes >BHal → >B·OR
  s. 27, 649

Rearrangement of α-bromoboranes ←
  with nucleophiles − B→C-Alkyl migration − Borinic acids s. 27, 118

Carboxylic acid silyl esters COOH → COOSi≦
  from carboxylic acids s. 27, 512

Phosphinic acid esters >PO(O)Cl → >P(O)OR
  from phosphinic acid chlorides s. 26, 99

N-Halogenosulfonylaminophosphoryl dichlorides HalSO$_2$·N·POCl$_2$
  via O→N-alkyl migration s. 27, 119

Sodium hydride NaH
Enoxysilanes from ketones CHCO → C:C·OSi≦
  s. 27, 720

Phosphinous acid esters >PCl → >P·OR
  from chlorophosphines s. 30, 159

Phosphoroperoxidates (RO)$_2$POCl → (RO)$_2$P(O)OOR'
  from phosphorochloridates s. 28, 76

Potassium hydroxide KOH
Monoperoxyphosphonic acid esters PCl → P·OOR
  from phosphonochloridates s. 27, 120

n-Butyllithium n-BuLi
Phosphorodiamidates from alcohols OH → OPO(N<)$_2$
  s. 28, 65

Lithium N-isopropylcyclohexylamide RR'NLi
O-Silyl O-alkyl keteneacetals CHCOOR → C:C(OR)(OSi≦)
  from carboxylic acid esters s. 28, 77

Lithium bis(trimethylsilyl)amide (Me$_3$Si)$_2$NLi
(1,3-Dien)-2-oxysilanes CHCO → C:C·O·Si≦
  from α,β-ethyleneketones via dienolate ions s. 29, 116

Triethylamine Et$_3$N
Enoxysilanes from oxo compds.
  s. 29, 829

Mixed phosphoric acid esters PHal$_2$ → P(OR)(OR')
  from phosphorodichloridates s. 29, 117

2-Alkoxy-2-oxo-1,3,2-dioxaphospholanes PHal → P·OR
  s. 29, 110

Mixed thionophosphonic acid esters ·P(S)Cl(OR) → ·P(S)(OR)(OR')
  from thionophosphonochloridates s. 26, 106

Dimethylaniline PhNMe$_2$
Phosphorylation of alcohols OH → OPO$_3$H$_2$
  via dithiophosphorous acid esters s. 27, 121

Copper Cu
Mixed thionophosphonic acid esters ·P(S)Cl$_2$ → ·P(S)(OR)(OR')
  from thionophosphonic acid dichlorides via thionophosphonochloridates
  s. 26, 106

Calcium chloride $CaCl_2$
Mixed phosphoric acid esters   ←
s. *26*, 107

Magnesium   Mg
s. *26*, 107

Imidazole   ←
O-Silylation of alcohol groups   OH → OSi≦

61.

Startg. m. treated 22 hrs. at 35° with 1.5 moles dimethyl-*tert*-butylsilyl chloride and 2 equivalents imidazole in dimethylformamide → silyl ether. Y: 85%. E. J. Corey and J. Mann, Am. Soc. *95*, 6832 (1973); f. e. s. ibid. *97*, 2287 (1975); **protection of nucleoside alcohol groups as silyl ethers** s. a. K. K. Ogilvie et al., Tetrah. Let. *1974*, 2861, 2865.

Dimethyl sulfoxide Me_2SO
Sym. diphosphoric acid diesters   2 · P(O)Cl$_2$ → · P(O)(OH) · O · P(O)(OH) ·
from phosphorodichloridates – Diphosphonic acids from phosphonic acid dichlorides s. *29*, 118

Sulfur ↑   ORem ↓↑ S

Silver salt/pyridine Ag^+/C_5H_5N
Mixed diphosphoric acid P¹,P²-diesters   O · P(O)(OH) · O · P(O)(OR)OH
from thionophosphoric acid monoesters

62.

A soln. of bis(triethylammonium) isoamyl phosphorothioate in dry pyridine added gradually with vigorous stirring to a suspension of disilver p-nitrophenyl phosphate in dry pyridine, and the product isolated after 5 hrs. as the dicyclohexylammonium salt → dicyclohexylammonium P¹-isoamyl P²-p-nitrophenyl pyrophosphate. Y: 83%. F. e. s. I. Nakagawa and T. Hata, Bull. Chem. Soc. Japan *46*, 3275 (1973).

Dimethyl sulfoxide/sulfuric acid Me_2SO/H_2SO_4
Phosphine oxides from phosphine sulfides   PS → PO
    Inversion of configuration – Also arsine oxides from arsine sulfides s. *29*, 119

Thionyl chloride $SOCl_2$
Phosphonic from thionophosphonic acid dichlorides   P(S)Cl$_2$ → P(O)Cl$_2$
s. *28*, 78

Sulfuric acid s. under Me_2SO $\qquad\qquad H_2SO_4$
Iodine $\qquad\qquad I$
Phosphoric acid monoesters $\qquad\qquad O \cdot P(SR)_2 \rightarrow O \cdot PO(OH)_2$
  from dithiophosphorous acid esters s. 27, 121

Remaining Elements ↑ $\qquad\qquad$ ORem ↓↑ Rem

Dimethyl sulfoxide/sulfuric acid $\qquad\qquad Me_2SO/H_2SO_4$
Replacement of P-selenium by P-oxygen $\qquad\qquad \geqslant PSe \rightarrow \geqslant PO$
s. 29, 119

Carbon ↑ $\qquad\qquad$ ORem ↓↑ C

Without additional reagents $\qquad\qquad$ w.a.r.
α-Aminocarboxylic borinic anhydrides $\qquad\qquad\leftarrow$
  Stabilized borazanes s. 26, 108
Tetraalkyl- from pentaalkyl-antimonanes $\qquad\qquad R_5Sb \rightarrow R_4SbOAc$
s. 28, 79
Irradiation $\qquad\qquad$ ⚡
Phosphoric from nitrous and phosphorous acid esters $\qquad\qquad ONO \rightarrow OPO(OR)_2$
s. 28, 80
Trimethylamine oxide $\qquad\qquad Me_3N{\rightarrow}O$
N→P-Oxygen transfer
  Cyclic phosphoric acid diesters s. 29, 120 $\qquad\qquad \underset{O}{\overset{O}{\diagdown}}P\cdot OR \rightarrow \underset{O}{\overset{O}{\diagdown}}PO(OH)$

Boryl pivalate $\qquad\qquad Me_3C \cdot COOBR_2$
Boryl carboxylates $\qquad\qquad R \cdot COOBR'_2$
s. 29, 121
Diethylboryl pivalate $\qquad\qquad Me_3C \cdot COOBR_2$
Cleavage of boron-carbon bonds $\qquad\qquad R_3B \rightarrow R_2B \cdot O \cdot C : C$
  catalyzed by pivalic acid derivs. – Enoxyboranes from ketones s. 28, 81; borinic acid esters from hydroxy compds. s. A. 1975, 352
Gallium chloride $\qquad\qquad GaCl_3$
Aryloxysilanes from phenolethers $\qquad\qquad ArOR \rightarrow ArO \cdot Si \leqslant$
s. 29, 602

Formation of O—C Bond

Uptake $\qquad\qquad$ ⇓

Addition to Hydrogen and Oxygen $\qquad\qquad$ OC ⇓ HO

Calcium carbide/potassium hydroxide $\qquad\qquad CaC_2/KOH$
O-Vinyloximes from oximes $\qquad\qquad C:NOH \rightarrow C:NOCH:CH_2$
s. 27, 122

Addition to Hydrogen and Carbon   OC↓HC

Irradiation s. under H_3PO_3   ///

Sodium hydroxide/dimethyl sulfoxide $NaOH/Me_2SO$
Triarylcarbinols   H → OH
s. 29, 122

Lithium diisopropylamide (s. a. under
Molybdenum peroxide complex) i-Pr_2NLi
α-Hydroxycarboxylic from carboxylic acids   CHCOOH → C(OH)COOH
s. 28, 82; α-hydroxylation of carboxylic acid derivs. cf. H. H. Wasserman and
B. H. Lipshutz, Tetrah. Let. 1975, 1731

Triton B   ←
Autoxidation   H → OH
Alcohols from hydrocarbons s. 27, 917

Cupric chloride/dimethylamine $CuCl_2/Me_2NH$

63.

α-Hydroxylactones from lactones. O_2 passed 3 hrs. at room temp. into a soln. of 7-chlorodeoxycamptothecin in dimethylformamide containing $CuCl_2$ and a little aq. 40%-dimethylamine → 7-chlorocamptothecin. Y: 95%. F. e. s. M. Boch et al., B. 105, 2126 (1972); s. a. C. S. F. Tang, C. J. Morrow, and H. Rapoport, Am. Soc. 97, 159 (1975).

Acetic anhydride s. under CrO_3 Ac_2O

Peroxyacetic acid s. H_2O_2/CH_3COOH CH_3COO_2H

Lead tetraacetate $(CH_3COO)_4Pb$
α-Hydroxylactols from lactols

64.

A mixture of 18,20-oxido-20-hydroxy-4-pregnen-3-one, Pb-tetraacetate, and acetic acid stirred 6 hrs. at room temp., the resulting product mixture dissolved in methanol, 0.1 N KOH added, and stirred 2 hrs. at room temp. → 18,20-oxido-20,21-dihydroxy-4-pregnen-3-one. Y: 58%. – A longer oxidation time decreases the yield. W. D. Slaunwhite and A. J. Solo, J. Pharm. Sci. 64, 168 (1975).

Phosphoric acid/air/irradiation H_3PO_4/O_2///
1-Hydroxyphenazines from phenazines   ArH → ArOH
also 1-alkoxyphenazines s. 26, 109

Hydrogen peroxide/acetic acid $\quad H_2O_2/CH_3COOH$
1,2,4-Triazin-5-ones from 1,2,4-triazines $\quad$ HetH → HetOH
s. *28*, 83

Oxidations with *sulfur trioxide* $\quad SO_3$
Alcohols from hydrocarbons $\quad$ H → OH
Benzyl alcohols s. *27*, 123

Sulfamic acid s. under NaClO₂ $\quad H_2NSO_3H$

Chromium trioxide/acetic anhydride/acetic acid $\quad CrO_3/Ac_2O/CH_3COOH$
Alcohols from hydrocarbons $\quad$ H → OH
Bridgehead functionalization s. *27*, 124

Molybdenum peroxide complex/lithium diisopropylamide $\quad$ ←
C-Hydroxylation of enolates

65.

A soln. of ethyl bicyclo[2.2.2]octene-5-carboxylate in dry tetrahydrofuran added dropwise at –70° under N₂ to Li-diisopropylamide in tetrahydrofuran-hexane, after 30 min. at –70° powdered MoO₅-pyridine-hexamethylphosphoramide complex added in one portion with vigorous stirring, after 1 hr. at –70° the temp. allowed to rise to 0° → ethyl 5-hydroxybicyclo[2.2.2]octene-5-carboxylate. Y: 80–85%. – Similarly: 3β-Tetrahydro-2-pyranyloxyandrost-5-en-17-one → 16α-hydroxy-3β-tetrahydro-2-pyranyloxyandrost-5-en-17-one. Y: 75%. F. e. s. E. Vedejs, Am. Soc. *96*, 5944 (1944).

Sodium chlorite/sulfamic acid $\quad NaClO_2/H_2NSO_3H$
Carboxylic acids from aldehydes $\quad$ CHO → COOH

66.

An aq. soln. of Na-chlorite added to an aq. soln. of vanillin and sulfamic acid as chlorine scavenger, and the precipitated product isolated after 1 hr. → vanillic acid. Y: 84%. F. e., **also with resorcinol** as chlorine scavenger, s. B. O. Lindgren and T. Nilsson, Acta Chem. Scand. *27*, 888 (1973).

Sodium periodate s. under RuO₄ $\quad NaIO_4$

Ruthenium tetroxide/sodium periodate $\quad RuO_4/NaIO_4$
Double benzylic oxidation $\quad$ ←
9α-Hydroxy-6-ketosteroids s. *29*, 178

Via intermediates $\quad$ v.i.
Catechols from phenols $\quad$ ArH → ArOH
via catechol monoesters s. *29*, 123

Catechols from phenols   ArCOR → ArOH
via Dakin oxidation of o-hydroxyacetophenones

67.

HO—⟨⟩—CH₂CHCOOH   →   HO—⟨⟩—CH₂CHCOOH
CH₃COCl +   NH₂   CH₃CO   NH₂

HO—⟨⟩—CH₂CHCOOH
HO   NH₂

Ca. 4 moles AlCl₃ added portionwise during 5 min. to a suspension of L-tyrosine in nitrobenzene causing the temp. to rise to 52°, then acetyl chloride added all at once, the resulting soln. stirred 6 hrs. at 100°, and the product isolated as the hydrochloride → 3-acetyl-L-tyrosine hydrochloride (Y: 89%) dissolved in aq. NaOH in an inert atmosphere maintained throughout the reaction, 30%-H₂O₂ added in one portion at −10°, whereupon the temp. rises to ca. 44°, and stirring continued 0.5 hr. at 35° after the exothermic reaction has subsided → 3-(3,4-dihydroxyphenyl)-L-alanine (Y: 75%). H. Bretschneider et al., Helv. 56, 2857 (1973).

Addition to Oxygen   OC↓OO

Without additional reagents   w.a.r.
Stereospecific formation   CH → C·OOH
     of hydroperoxides from hydrocarbons with migration of carbon-carbon double bonds s. 26, 110

Lithium diisopropylamide i-Pr_2NLi
α-Hydroperoxynitriles from nitriles   C(CN)H → C(CN)OOH
s. 30, 155

Sodium carbonate Na_2CO_3
Hydroperoxides by autoxidation   H → OOH
s. 25, 114 suppl. 29

Addition to Oxygen and Nitrogen   OC↓ON

Chlorides Cl^-
Alkoxylamines from oxoazonium salts   >N⁺=O → >N·OR
s. 27, 125

Hydrogen chloride   HCl
Skeletal rearrangement of O,O,N-heterocyclics   ←
s. 26, 111

Addition to Oxygen and Sulfur   OC↓OS

Trialkyloxonium fluoroborates $R_3O^+BF_4^-$
Epimerization of cyclic sulfoxides   ←
     via alkoxysulfonium salts s. 27, 126; prepn. of alkoxy- and other sulfonium salts cf. Y. Hara and M. Matsuda, Chem. Commun. 1974, 919

Addition to Oxygen and Carbon   OC⇓OC

Without additional reagents   w.a.r.
O-Acylation of vinylogous carboxylic acid amides   ←
s. 26, 112

Acoxyperchlorates from acyl perchlorates   C

68.

$$CH_3-C\begin{matrix}O\\OClO_3\end{matrix}$$

$$\triangle_O \quad + \qquad \diagup\diagdown \qquad + \quad \square_O$$

CH₃COOCH₂CH₂OClO₃   CH₃COO[CH₂]₄OClO₃

Ethylene oxide | Tetrahydrofuran
    added at 0° to a stirred soln. of acetyl perchlorate in CCl₄, and stirred
                0.5 hr. → solns. of
2-perchloratoethyl acetate. | 4-perchloratobutyl acetate.
Y: 89% by spectroscopy. | Y: 78% by spectroscopy.
F. e. s. K. Baum and C. D. Beard, J. Org. Chem. *40*, 81 (1975).

3-Alkoxymethyl-4-hydroxy-3,4-dihydro-2(1H)-quinazolones   ←
    from 4,5-epoxy-1,3,4,5-tetrahydro-2H-1,4-benzodiazepin-2-ones s. 26, 113

Irradiation/magnesium sulfate   *hν*/MgSO₄
7-Hydroxy-4,5-dihydro-1,3-benzodioxepins   O
    from p-quinones and oxo compds. with *tert*-butyl rearrangement s. 28, 84

Sodium hydroxide   NaOH
Dihydroxyoxido from dioxido compds.   ←
    with rearrangement – Stereospecific reaction s. 26, 114

Sodium hydroxide/alcohol
Hydroxycarboxylic acids from lactones $\begin{matrix}O\\|\\CO\end{matrix} \rightarrow \begin{matrix}OH\\\\COOH\end{matrix}$
s. 27, 56

Hydroxycarboxylic acid esters from lactones
Partial conversion $\begin{matrix}O\\|\\CO\end{matrix} \rightarrow \begin{matrix}OH\\\\COOR\end{matrix}$
Vulpinic acids from pulvinic acid dilactones

69.
    MeOH + [3,3'-dichloropulvinic acid lactone] → [3,3'-dichlorovulpinic acid structure]

3,3′-Dichloropulvinic acid lactone stirred 5 min. with 18 N NaOH and methanol
→ 3,3′-dichlorovulpinic acid. Y: 86%. F. e. s. F. R. Foden, J. McCormick, and
D. M. O'Mant, J. Med. Chem. *18*, 199 (1975).

Potassium hydroxide s. under HgO   KOH

Potassium carbonate   K₂CO₃
Dicarboxylic acid esters $\begin{matrix}CO\\CO\end{matrix}\rangle O \rightarrow \begin{matrix}COOR\\COOR\end{matrix}$
    from dicarboxylic acid anhydrides s. 27, 127

Triethylamine Et_3N
Ketene acylals   ←

70.
$$Ph_2C=C=O + Ph_2CHCO\text{-}O\text{-}COCHPh_2 \rightarrow Ph_2C=C(O\text{-}COCHPh_2)(O\text{-}CHPh_2CO)$$

Diphenylacetic anhydride and diphenylketene stored 1 month at 25° in the presence of a trace of triethylamine → product. Y: 80%. Also formation of naphthalene-1,3-diol esters s. L. A. Feiler, R. Huisgen, and P. Koppitz, Chem. Commun. *1974*, 405.

2-Acylaminoalcohols from Δ^2-oxazolines   C
s. *28*, 85

Dodecyldimethylamine $C_{12}H_{25}NMe_2$
Glycol monoesters from oxido compds. $\triangle_O \rightarrow C(OH)C(OAc)$
s. *27*, 128

Pyridine C_5H_5N
Acylals from aldehydes $CHO \rightarrow CH(OCOR)_2$
s. *27*, 759

Preferential acylation of alcohol groups
   with retention of phenol groups – Dicarboxylic acid   (CO)(CO)O → (COOH)(COOR)
   monoesters from dicarboxylic acid anhydrides s. *28*, 86

Mercuric oxide/potassium hydroxide   *HgO/KOH*
Coumarilic acids from coumarins   Ọ
s. *28*, 87

Magnesium sulfate s. under ₩ $MgSO_4$
Sodium tetrahydridoborate/hydrogen chloride $NaBH_4/HCl$
Alkoxylactams from dicarboxylic acid imides   ←
s. *27*, 947

Boron fluoride BF_3
Total O-acetylation of carbohydrates   C
   Acetolysis of glycosides s. *28*, 88

Diacoxy compds. from cyclic ethers   ←
   with isocyclic skeletal rearrangement s. *28*, 89

Cation exchanger   ←
s-Trioxanes from aldehydes

71.
$$3\ n\text{-}C_9H_{19}CHO \rightarrow \text{1,3,5-tri(}n\text{-}C_9H_{19}\text{)-1,3,5-trioxane}$$

Cyclotrimerization. Amberlyst 15 cation exchanger added at 0° to *n*-decanal and pentane, stirred 1 hr. → product. Y: 95%. F. e. s. M. Cossu and E. Ucciani, C. r. *277* (C), 1145 (1973).

Stannic chloride $\qquad$ $SnCl_4$

1,3-Dioxolanes from oxido compds. and ketones $\qquad$ CO → C(−O−)(−O−)

72.

Acetylcyclopentadienylmanganese tricarbonyl allowed to react 1–3 hrs. at 40° with epichlorohydrin in CCl_4 containing $SnCl_4$ → product. Y: 96%. Y. M. Paushkin and L. P. Romanovskaya, Dokl. Akad. Nauk Beloruss. SSR *17*, 1025 (1973) (Russ); C. A. *80*, 48135; s. a. C. Fauran et al., Chim. Therap. *9*, 281 (1974).

Nitric acid/acetic acid $\qquad$ HNO_3/CH_3COOH
2-Ene-1,4-diones from furans $\qquad$ C
s. *28*, 90

Phosphorus pentoxide $\qquad$ P_2O_5
Mixed 1,1-diacoxy compds. $\qquad$ CO → C(OAc)(OAc′)
s. *29*, 124

Ozone $\qquad$ O_3
Stereospecific ozonization of glycosides $\qquad$ C
s. *28*, 235

Hydrogen peroxide s. under HCl $\qquad$ H_2O_2

p-Toluenesulfonic acid $\qquad$ TsOH
Δ^2-Oxazoline ring opening $\qquad$ C
   Glycosylation with carbohydrate Δ^2-oxazoline derivs. s. *14*, 174 suppl. *29*

γ-Diketals from furans $\qquad$ ←
s. *26*, 115

2,4,6-Trinitrobenzenesulfonic acid $\qquad$ $2,4,6\text{-}(NO_2)_3C_6H_2SO_3H$
Stereospecific formation of glycols $\qquad$ → C(OH)C(OH)
   from oxido compds. via 1-hydroxy-2-(alkoxysulfonium) salts s. *27*, 129

Selenium dioxide/acetic acid $\qquad$ SeO_2/CH_3COOH
Oxidative cleavage of β,γ-unsatd. ethers $\qquad$ ←
s. *26*, 116

Hydrogen chloride $\qquad$ HCl
5-Subst. β-(2-furyl)propionic acids $\qquad$ ←
   from α,β-ethylene-β-(2-furyl)ketones – Furan ring rearrangement s. *29*, 125

Hydrogen chloride/hydrogen peroxide $\qquad$ HCl/H_2O_2
Oxidative furan ring opening $\qquad$ C
s. *26*, 117; cf. Zh. Prikl. Khim. *47*, 1366 (1974); C. A. *81*, 151445

Addition to Nitrogen and Carbon $\qquad$ OC⇓NC

Without additional reagents $\qquad$ w.a.r.
N-Carbamyloxy from N-hydroxy compds. $\qquad$ NOH → NO·CO·NHR
s. *29*, 126

Reactions of N-fluoroimines $\qquad$ C:NF → C(NHF)OR
s. *29*, 127

O-Arylisoureas $\qquad$ ArOH → ArOC(:NR)NHR
   from carbodiimides and phenols s. *30*, 340

α-(O-Carbamylhydroxylamino)-　　　　　　　　　　>NOH → >NO·CONHR
from α-hydroxylamino-carboxylic acid amides
s. 30, 125

Allophanates　　　　　　　　　　　　　　　　OH → O·CO·NR·CONHR
from 2 isocyanate molecules s. 29, 661

N-Carbo-*tert*-butoxy-N,O-bis(carbamyl)-　　　　　　　　　　　⟋OCON<
hydroxylamines　　　　　　　　　　　　　　>NCO–N
s. 26, 292　　　　　　　　　　　　　　　　　　　　　　⟍COOCMe₃

Aziridine ring opening　　　　　　　　　　　　　　　　C
s. 26, 118

Irradiation (s. a. under Ph₂CO)　　　　　　　　　　　　　ttt
Diaziridines from 3-pyrazolidone betaines　　　　　　　　Ö:
s. 26, 320

Sodium hydroxide　　　　　　　　　　　　　　　　　NaOH
Azacyclols　　　　　　　　　　　　　　　　　　　　←
　3a-Hydroxy-3H-1,2,3a,4,9,10-hexahydropyrrolo[2,1-b]quinazol-9-ones s. 26, 119
γ-Aminocarboxylic acids from 2-pyrrolidones　　　　　　　C
　γ-Aryl-γ-aminocarboxylic acids s. 26, 120
Isoquinolinium ring opening

73.　MeO　　　　　　　Br⁻ + ClBz　　　MeO　　　　　　Bz
　　　　　　N⁺　　　　　　　　→　　　　　　　　N
　　MeO　　　　CH₂–C₆H₄COOEt–o　　MeO　　　CHO　CH₂–C₆H₄COOEt–o

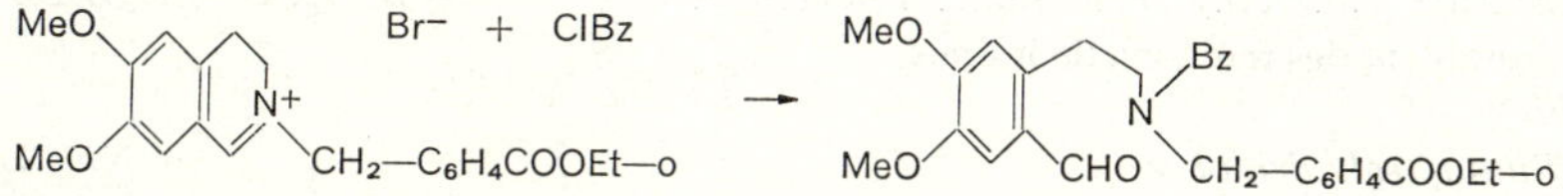

Startg. m. cooled and treated with excess aq. NaOH and benzoyl chloride →
product. Y: 98%. M. Shamma and L. Töke, Chem. Commun. *1973*, 740.

Potassium hydroxide　　　　　　　　　　　　　　　　KOH
2-Oxo-1,3-N,N-heterocyclics　　　　　　　　　　　　　←
　from 1,3-N,N-heterocyclics s. 29, 129
3,4-Dihydroisoquinolinium ring opening　　　　　　　　C
s. 28, 693

Sodium/alcohol　　　　　　　　　　　　　　　　　　NaOR
α-Cyaniminoesters from 1,1-dinitriles　　　　　CN → C(:NH)OR
s. 29, 130
o-Alkoxy-o′-amino-N-heterocyclics from dinitriles　　　　　O

74.　　　　　　　　　　　　　　　　　　　　　　　OEt
　　　　　　　CN　　　　　　　　　　　　　　　　N
　　　　　　　　+ HOEt　　→
　　　　　N　　CN　　　　　　　　　N
　　　　　　　　　　　　　　　　　　　　　NH₂

Naphthyridines. Startg. m. added to a soln. of Na in anhydrous ethanol, and
refluxed 3 hrs. → 8-amino-6-ethoxy-1,7-naphthyridine. Y: 70%. F. e. and limitations s. F. Alhaique, F. M. Riccieri, and E. Santucci, Tetrah. Let. *1975*, 173.

Sodium ethoxide　　　　　　　　　　　　　　　　　NaOEt
Amide acetals　　　　　　　　　　　　　　　　>N·C(OR)₂·
　from iminoester fluoroborates s. 26, 933

Sodium carbonate/sodium hydrogen carbonate　　　Na₂CO₃/NaHCO₃
γ-Hydroxy-γ-lactams from cyanoketones　　　　　　　　O
s. 26, 121

Sodium hydrogen carbonate	$NaHCO_3$
α-Aminocarboxylic acid esters	C
from aziridinones via their O-alkylation s. 27, 192	
Ring expansion by pseudobase isomerization	☉
s. 27, 130	
Sodium iodide/ozone	NaI/O_3
Oxidative rearrangement	$N(Ac)C : C \rightarrow N : C \cdot C(OAc)$
of 2,3,4,4a-tetrahydrocarbazoles s. 27, 131	
Ammonia	NH_3
Carboxylic acid amides from nitriles	$CN \rightarrow CONH_2$
Continuous process s. 29, 196	
Triethylamine	Et_3N
Iminocarbonic from cyanic acid esters	$O \cdot C \equiv N \rightarrow O \cdot C(:NH)OR$
s. 27, 214	
Hofmann degradation of cyclimmonium salts	C
with lactone ring opening s. 28, 91	
Pyridine	C_5H_5N
Acyl carbamyl peroxides	$N:C:O \rightarrow NHCO \cdot OO \cdot COR$
from peroxycarboxylic acids and isocyanates s. 26, 122	
Mercuric acetate/ethylenediaminetetraacetic acid	$(CH_3COO)_2Hg/EDTA$
Aminoaldehydes from cyclic imines	C
s. 28, 92	
Sodium tetrahydridoborate	$Na[BH_4]$
Iminoesters from nitriles	$CN \rightarrow C(:NH)OR$

75.

[Structure: 2,4-Quinolinedicarbonitrile + HOEt → ethyl 2-(4-cyanoquinoline)imidate]

Preferential conversion. 2,4-Quinolinedicarbonitrile, ethanol, 1/16 mole $NaBH_4$, and diphenyl ether refluxed 2 hrs. → ethyl 2-(4-cyanoquinoline)imidate. Y: 84%. F. e. and limitations s. H. Watanabe, Y. Kikugawa, and S. Yamada, Chem. Pharm. Bull. *21*, 465 (1973).

Thallic nitrate	$Tl(NO_3)_3$
Urethans from isonitriles	$N:C \rightarrow NHCOOR$
s. 28, 93	
Fluoroboric acid/tert-butyl alcohol	$HBF_4/Me_3C \cdot OH$
Reactions with carbodiimides	←
Guanylureas s. 29, 131	
tert-Butyl alcohol s. under HBF_4	$Me_3C \cdot OH$
Benzophenone/irradiation	$Ph_2CO/↯$
Carboxylic acid amides from nitriles	$CN \rightarrow CONH_2$
Neighboring group participation s. 27, 132	
Peroxyacetic acid	CH_3COO_2H
Insertion of oxygen	$C \cdot N \rightarrow C \cdot O \cdot N$
into nitrogen-carbon bonds s. 26, 123	
m-Chloroperoxybenzoic acid	$m\text{-}ClC_6H_4COO_2H$
3-Alkoxyoxaziridines from iminoesters	N—O
s. 29, 132; s. a. J. Org. Chem. *39*, 3855 (1974)	

Titanium tetrachloride	$TiCl_4$
Carboxylic acid amides from nitriles s. *11*, 194 suppl. 28	$CN \rightarrow CONH_2$
Polyphosphoric acid s. *26*, 989	$H(PO_3H)_xOH$
Ozone s. under NaI	O_3
Chromyl chloride	CrO_2Cl_2
Carboxylic acid amides from azomethines s. *26*, 124	$CH:N \rightarrow CONH$
Hydrogen chloride	HCl
Iminoesters from nitriles s. *30*, 220	$CN \rightarrow C(:NH)OR$
α-**Acoxycarboxylic acid amides** from α-acoxynitriles s. *27*, 709	$C(OAc)CN \rightarrow C(OAc)CONH_2$
α-**Ketocarboxylic acid amides** from acylcyanides s. *27*, 207	$COCN \rightarrow COCONH_2$
Ureas from cyanamides s. *26*, 23; s. a. J. Anatol and J. Berecoechea, Synthesis *1975*, 111	$N \cdot CN \rightarrow N \cdot CONH_2$
2-Carbamyl-2-cyanophosphonium salts from alkylidenemalononitriles s. *28*, 588	←
Palladous chloride	$PdCl_2$
Carboxylic acid amides from nitriles s. *21*, 167 suppl. 29	$CN \rightarrow CONH_2$

Addition to Sulfur and Carbon   OC⇓SC

Irradiation s. under Methylene blue	∰
Sodium hydroxide	$NaOH$
2′-Hydroxybenzophenone-2-sulfinic acids from thioxanthone 10,10-dioxides s. *28*, 224	C
Peroxyacetic acid s. H_2O_2/CH_3COOH	CH_3COO_2H
Hydrogen peroxide/acetic acid	H_2O_2/CH_3COOH
Hydroxysulfonic acids from cyclic thioethers **Oxidative 8,5′-S-cyclonucleoside ring opening**	C

76.

A soln. of the startg. m. in acetic acid treated overnight at 30° with 30%-H_2O_2 → 9-(β-D-xylofuranosyl)adenine-8-sulfonic acid. Y: 81%. Y. Mizuno, C. Kaneko, and Y. Oikawa, J. Org. Chem. *39*, 1440 (1974).

Methylene blue/irradiation	←
Reactions of S-heterocyclics with photogenerated singlet oxygen via thiozonides – Isocycles from S-heterocycles – Review of singlet oxygen s. *26*, 125	Ö:

Addition to Remaining Elements   OC ⇓ Rem

Sodium hydroxide   *NaOH*
4-Phosphabicyclo[3.1.0]hex-2-ene 4-oxides   ←
  from 1-iodomethylphospholium salts and 4-iodomethyl-2-pholene 1-oxides
  s. 27, 133

Addition to Carbon-Carbon Bonds   OC ⇓ CC

Without additional reagents   *w.a.r.*
Hydrocarbons from ethylene derivs. $C:C \rightarrow CH_2$
Degradation via ozonides
  s. 30, 46

Oxido compds. from ethylene derivs. $C:C \rightarrow \triangle\!\!\!\!{}_O$
  Oxidoboranes s. 29, 133

1,2-Alkoxyperoxides $C:C \rightarrow C(OR)C(\cdot OOR)$
from ethylene derivs. and trioxides

77.

Octafluorocyclopentene allowed to react 8 hrs. in vacuo at 67° with bis(trifluoromethyl) trioxide → 1,2,2,3,3,4,4,5-octafluoro-5-[(trifluoromethyl)dioxy]cyclopentyl trifluoromethyl ether. Y:80% (52% *cis*- and 48% *trans*-isomer). F. e. s. F. A. Hohorst, J. V. Paukstelis, and D. D. DesMarteau, J. Org. Chem. **39**, 1298 (1974).

Reactions with acylynamines   ←
  β-Acoxy-α,β-ethylenecarboxylic acid amides s. 29, 134

Fluorosulfonic acid esters $C:C \rightarrow CHC \cdot OSO_2F$

78.

from ethylene derivs. Fluorosulfonic acid allowed to react with propylene at –40° under dry N_2 → isopropyl fluorosulfonate. Y: 83%. F. e., **also from cyclopropanes,** s. G. A. Olah, J. Nishimura, and Y. K. Mo, Synthesis *1973*, 661.

Sulfocarboxylic acid anhydrides $>C:C:O \rightarrow >CHCO \cdot OSO_2R$
from ketenes and sulfonic acids

79.

Methanesulfonic acid added at –20° to a soln. of pentamethyleneketene in chloroform, and stirred 0.5 hr. at 0° → cyclohexanecarbonyl methanesulfonate. Y: 98.5%. F. e. s. E. Tempesti et al., Soc. Perkin I *1974*, 771.

Mixed orthocarboxylic acid esters   C
  from cyclopropanone ketals s. 29, 135

Oxidative ring contraction   ←
  to spiro-N-heterocyclics s. 29, 136

Irradiation (s. a. under Xylene, Nitroarenes, CH_3COOH,
Methylene blue, and HCl)   ⇈
Carboxylic acid esters $C:C \cdot CHO \rightarrow CHCHCOOR$
  from α,β-ethylenealdehydes s. 28, 94

Carboxylic acid esters and amides ←
  from α,β-ethyleneketones s. 27, 134

3,5-Dienecarboxylic acid derivs. C
  from hemiquinolesters – Geospecific ring opening s. 28, 95

Carboxylic acid esters
  from cyclopropylketones s. 28, 96

Bicyclic ethylenecarboxylic acid esters ←
  from bicyclic dienones – Skeletal rearrangement – Homoelectrocyclic reaction
  s. 26, 126

O-Heterocyclics from isocyclics ☉
  Ring expansion s. 26, 127

Lactolides from cyclic ketones
  s. 29, 137

Fused cyclopropane ring ←
  from tropone adducts s. 28, 97

Argon laser ←
Reactions by laser irradiation O
1,2,4-Trioxane ring

80.

The argon laser shows considerable promise as a light source for the rapid prepn. of both thermally and photochemically sensitive unusual products on the gram scale. – E: A soln. of cyclooctatetraene and p-benzoquinone in CCl_4 open to the air stirred 0.5 hr. and irradiated 2.75 hrs. with an argon laser → 4′a,10′a-dihydrospiro-(2,5-cyclohexadiene-1,3′-cycloocta-*as*-trioxin)-4-one. Y: 39%. R. M. Wilson et al., Am. Soc. *96*, 2955 (1974).

Electrolysis ↯
Anodic oxidations of hindered phenols ←
  Hemiquinols, hemiquinol-ethers and -esters s. 28, 98

1,3-Dialkoxy comps. from cyclopropanes ▽ → C(OR)·C·(OR)
  s. 27, 135

Sodium hydride/oxygen NaH/O_2
Spiro-N-heterocyclics ←
  from 2β-oxiranyl-1,2,3,4,6,7,12,12b-octahydroindolo[2,3-a]quinolizines –
  Oxidative skeletal rearrangement s. 28, 99

Sodium hydroxide (s. a. under MeSO₃H) NaOH
Unsym. peroxides from ethylene derivs. C:C → CHC·OOR
  s. 26, 128

Cyclobtanone ring opening C
  s. 30, 487

Potassium hydroxide KOH
Disaccharides C:C → CHC·O·
  s. 26, 129

Sodium/alcohol NaOR
Enolethers from allenes C:C:C → C:C(OR)CH
  6-Alkoxyfulvenes from fulveneallenes s. 29, 138

β-Alkoxy-β,γ-ethylene-   C : C : C · COOR → C : C(OR)CHCOOR
from α-allene-carboxylic acid esters
s. 29, 159

1,3-Benzodioxoles   O
from acetylenedicarboxylic acid esters s. 28, 100

γ,δ-Ethylene-ε-hydroxycarboxylic acid esters   C
from oxidocyclobutanones – Fragmentation-type ring opening s. 28, 101

Sodium n-propoxide   NaOPr-n
α-Nitroketals via 2-nitroenolethers   C : C · OR → CH · C(OR)(OR')
s. 26, 130

Potassium/tert-butanol/dimethyl sulfoxide-oxygen $K/Me_3C \cdot OH/Me_2SO\text{-}O_2$
Isocyclic ring opening   C
of fused thiophene derivs. to carboxylic acids by autoxidation s. 26, 131

Butyllithium s. under BH_3   BuLi

Lithium 2,2,6,6-tetramethylpiperidide/hydrogen peroxide $\complement NLi/H_2O_2$
Reactions via eneboranes   CH · C ≡ CH → C : CHCH₂OH
Ethylenealcohols from acetylene derivs. – 2-Ethylenealcohols –
Hydroallylation of ethylene derivs. s. 29, 139

Sodium hydrogen carbonate s. under HCl $NaHCO_3$

Sodium azide s. under $MeSO_3H$ NaN_3

Sodium nitrite/acetic acid $NaNO_2/CH_3COOH$
Partial cis-hydroxylation   C : C → C(OH)C(OH)
of carbon-carbon double bonds s. 27, 147

Pyridine s. under Peroxymaleic acid C_5H_5N

Silver carbonate Ag_2CO_3
α,β-Ethylenealdehydes   C(OAc)C ≡ CH → C : C · CHO
from 1-ethynylacoxy compds. s. 28, 191

Cupric nitrate s. under $PdCl_2$ $Cu(NO_3)_2$

Silver nitrate s. under Br $AgNO_3$

Zinc s. under CrO_2Cl_2   Zn

Mercuric oxide/sulfuric acid HgO/H_2SO_4
Ketones from acetylene derivs.   C ≡ C → CH₂CO
with formation of tetrahydropyrans from 5-aminoalcohols s. 29, 140; ketones from acetylene derivs. s. a. M. Newman and V. Lee, J. Org. Chem. **40**, 381 (1975)

Mercuric acetate (s. a. under BF_3 and $Li_2[PdCl_4]$) $(CH_3COO)_2Hg$
2-Ethylenealcohols from cyclopropane ring   C
s. 26, 132

Mercuric acetate/sodium tetrahydridoborate $(CH_3COO)_2Hg/Na[BH_4]$
Mercuration-demercuration   C : C → CHC · OOR
Unsym. peroxides from ethylene derivs. – β-Peroxycarboxylic acid esters
s. 29, 141

Mercuric trifluoroacetate/sodium tetrahydridoborate $(CF_3COO)_2Hg/Na[BH_4]$
Diols from ethylenealcohols   C : C · (OH) → CH · C(OH) · C(OH)
via chloral acetals – Regio- and stereo-specific oxymercuration-demercuration –
Cyclic cis-glycols s. 29, 142

Mercuric sulfate $\qquad HgSO_4$
**γ-Keto- from β,γ-acetylene-carboxylic
acid esters** $\qquad C\equiv C\cdot CH_2COOR \rightarrow COCH_2CH_2COOR$
s. *28*, 783

Mercuric sulfate/sulfuric acid $\qquad HgSO_4/H_2SO_4$
2-Ethylenealcohols from allenes $\qquad C:C:C \rightarrow C:CH\cdot C(OH)$
s. *27*, 136

α,β-Ethyleneketones from 1,2,3-trienes $\qquad C:C:C:C \rightarrow C:CHCOCH$
s. *26*, 133

Borane/butyllithium $\qquad BH_3/BuLi$
1,3-Diol esters from ethylene derivs. $\qquad CH\cdot C:C \rightarrow C(OAc)\cdot CH\cdot C(OAc)$
via 1,3-diols s. *28*, 102

Borane/hydrogen peroxide $\qquad BH_3/H_2O_2$
Diols from ethyleneketones $\qquad \leftarrow$
s. *28*, 103

Diol monoesters from ethylenealcohols $\qquad C:C \rightarrow CHC(OH)$
s. *27*, 137

Hydroperoxides from ethylene derivs. $\qquad C:C \rightarrow CHC(OOH)$
Autoxidation of organoboranes – Synthetic applications of organoboranes, review – Radical reactions with boranes, review s. *27*, 138; review of hydroperoxides and peroxides cf. W. M. Weigert and P. Kleinschmidt, Chem. Ztg. *98*, 583 (1974)

Borane-dimethyl sulfide/hydrogen peroxide $\qquad BH_3\text{-}Me_2S/H_2O_2$
Alcohols from ethylene derivs. $\qquad C:C \rightarrow CHC(OH)$
Brown hydration s. *21*, 174 suppl. 29

Diborane-hydrogen peroxide $\qquad B_2H_6\text{-}H_2O_2$
**Cyclic *trans*-glycols from cyclic enoxysilanes
via cyclic *trans*-2-siloxyalcohols**

81.

1-(Trimethylsilyloxy)cyclohexene allowed to react with diborane in tetrahydrofuran followed by H_2O_2/NaOH-treatment → *trans*-2-(trimethylsilyloxy)cyclohexanol (Y: 95%) allowed to react 2 hrs. with Na-methoxide in methanol → *trans*-cyclohexane-1,2-diol (Y: 96%). F. e. s. H. Kono and Y. Nagai, Org. Prep. Proced. Int. *6*, 19 (1974).

Diborane-hydrogen peroxide/sodium hydroxide $\qquad B_2H_6\text{-}H_2O_2/NaOH$
Glycols from α,β-ethyleneketones $\qquad C:C\cdot CO \rightarrow CHC(OH)CH(OH)$
Stereospecific reduction – Borane-dimethyl sulfide as storable hydroboration agent s. *26*, 134

Sodium tetrahydridoborate s. under $(CH_3COO)_2Hg$, $(CF_3COO)_2Hg$,
and O_3 $\qquad Na[BH_4]$

*Tetrabutylammonium tetrahydridoborate/
methyl iodide-hydrogen peroxide* $\qquad Bu_4N^+[BH_4]^-/MeI\text{-}H_2O_2$
Brown hydration $\qquad C:C \rightarrow CHC(OH)$
s. *28*, 34

Thallic trifluoroacetate $\qquad (CF_3COO)_3Tl$
Oxophlorins $\qquad CO$
s. *29*, 143

Oxidations with *thallic nitrate* $Tl(NO_3)_3$
Oxo compds. from ethylene derivs. with rearrangement ←
Oxidation of acetylene derivs. s. 27, 141
1,6-Addition to 1,3,5-trienes ←

82.

A methanolic soln. of the startg. m. added to a vigorously stirred soln. of thallic nitrate trihydrate in methanol, and the product isolated after 10 min. → 3-methoxy-1-(3-methoxy-1-buten-1-yl)-2,6,6-trimethyl-1-cyclohexene. Y: 97.8%. F. Kienzle and R. E. Minder, Helv. *58*, 27 (1975).

Thallic sulfate $Tl_2(SO_4)_3$
3-Chromanols ○
s. *28*, 653

Boron fluoride BF_3
Protection of alcohol groups C : C → CHC(OR)
as tetrahydro-2-pyranyl ethers s. *27*, 139
O,O-Heterocyclics ○
from 2 ethylenealcohol molecules s. *29*, 144

Boron fluoride/mercuric acetate $BF_3/(CH_3COO)_2Hg$
Mercuration of acetylene derivs. C ≡ C → CH : C(OAc)
Enol acetates s. *7*, 201 suppl. 29

Fluoroboric acid HBF_4
Reactions of acetylene derivs. C ≡ C → CO·CH(·O·N:)
with cationoid electrophiles – 2-Ketoalkoximes from aliphatic nitro compds. – β-Diketones from carboxylic aicd chlorides s. *28*, 104

Fluoroboric acid/acetic acid HBF_4/CH_3COOH
Protection of carbon-carbon triple bonds C : C → CHC(OH)
as cobalt carbonyl complex – Selective hydration of carbon-carbon double bonds s. *27*, 140

Diethylalumium chloride Et_2AlCl
Alcohols from ethylene derivs.
via hydroalumination s. *27*, 142

Ammonium ceric nitrate $(NH_4)_2Ce(NO_3)_6$
Oxidative cyclobutane ring opening C
s. *27*, 143

Xylene $C_6H_4Me_2$
Preferential epoxidation C : C → △O

83.

Conjugated dienones and diene esters are epoxidized at the γ,δ-double bond by *molecular oxygen* when heated in solvents which have readily abstractable H-atoms. – E: A soln. of the startg. m. in xylene heated 24 hrs. at 120–130° in the presence of air → epoxide. Y: 77%; startg. m. recovered 22%. F. e. and solvents s. H. Hart and P. B. Lavrik, J. Org. Chem. *39*, 1793 (1974).

Xylene/irradiation ←
Addition of protic solvents ←
to cyclic carbon-carbon double bonds
s. *26*, 135

Crown polyether s. under KMnO₄ ←

Nitroarenes/irradiation $ArNO_2$/𝑊
Photochemical reactions of ar. nitro compds. C

84.

Oxidative ring opening. A soln. of 1,4-dimethoxynaphthalene and *m*-chloronitrobenzene in benzene irradiated under N_2 with the Pyrex-filtered light of a high-pressure Hg-lamp → product. Y: 70%. F. e., also with α-nitronaphthalene, s. I. Saito, M. Takami, and T. Matsuura, Tetrah. Let. *1974*, 659; reaction with ar. amines cf. ibid. *1974*, 661.

Acetic anhydride s. under H_2O_2 Ac_2O

Cumene hydroperoxide ←
6-Acyl-1,6-dihydropyridazines ←
from 1-iminopyridinium ylids s. *27*, 144

Dowex 50W-4 ←
β,γ-Ethylenealdehyde derivs. C
from enecyclopropyl ethers − β,γ-Ethyleneacetals s. *29*, 145

Formic acid s. under H_2O_2 $HCOOH$

Acetic acid CH_3COOH
2,3-Dihydrofurans from α,β-ethyleneketones O
via tetrahydrofurylium salts s. *28*, 105

Acetic acid/irradiation CH_3COOH/𝑊
Skeletal isomerization of bicyclic dienones ←
s. *28*, 106

Trifluoroacetic acid s. under $(CH_3COO)_4Pb$ CF_3COOH

Peroxyacetic acid s. under H_2SO_4 CH_3COO_2H

m-Chloroperoxybenzoic acid m-$ClC_6H_4COO_2H$
Oxido compds. from ethylene derivs. C:C → △O
s. *26*, 584

Direction of ring closure to O-heterocycles O
s. *28*, 107

m-Chloroperoxybenzoic acid/hydrogen chloride m-$ClC_6H_4COO_2H/HCl$
s. *28*, 107

Peroxymaleic acid/pyridine ←
Oxido compds. from ethylene derivs. C:C → △O
s. *26*, 136

Peroxyphthalic acid ←
2-Hydroxyindoxyls from indoles ←

85.

2 equivalents of an ethereal soln. of monoperoxyphthalic acid added at 0° during 3 hrs. to 2-*tert*-butylindole in the same solvent, and the product isolated after 21 hrs. at 0° → 2-hydroxy-2-*tert*-butylindoxyl. Y: 50%. F. e. s. E. Braudeau, S. David, and J.-C. Fischer, Tetrahedron 30, 1445 (1974).

Methylene blue/irradiation ←
3,6-Dihydro-o-dioxins from 1,3-dienes O
s. 29, 146
Preferential oxidative ring opening C
  by singlet oxygen – Dialdehydes from cyclic ethylene derivs. – Tert. amino groups as singlet-oxygen quenchers s. 29, 147

Methyl iodide s. under $Bu_4N^+[BH_4]^-$ MeI

Rose Bengal/irradiation ←
1,2-Dioxetane ring $C:C \rightarrow$ ▭
from carbon-carbon double bonds $O-O$

86.

Oxidation with singlet oxygen. 3-Methyl-2-phenylindene irradiated at –78° with light > 450 nm in 1:1 methanol-acetone containing rose Bengal as sensitizer in the presence of oxygen → product. Y: 54%. F. e. s. P. A. Burns and C. S. Foote, Am. Soc. 96, 4339 (1974); **singlet oxygen, review,** s. W. Adam, Chem. Ztg. 99, 142 (1975); **1-phospha-2,8,9-trioxaadamantane ozonide as singlet oxygen source** cf. A. P. Schaap, K. Kees, and A. L. Thayer, J. Org. Chem. 40, 1185 (1975).

3,6-Dihydro-o-dioxins from 1,3-dienes O
s. 29, 146

Perfluoroacetone s. under H_2O_2 $(CF_3)_2CO$

N-Bromosuccinimide/dimethyl sulfoxide $\begin{matrix}CH_2CO\\|\\CH_2CO\end{matrix}NBr/Me_2SO$

α-Dioxo compds. from acetylene derivs. $C \equiv C \rightarrow COCO$
s. 27, 145

Lead tetraacetate $(CH_3COO)_4Pb$
Reaction of α-diketones with lead tetraacetate C
  Dicarboxylic acid esters from cyclic α-diketones s. 27, 146

Lead tetraacetate/trifluoroacetic acid $(CH_3COO)_4Pb/CF_3COOH$
Oxo compds. from ethylene derivs. ←
  with rearrangement s. 27, 141 suppl. 28

Cyanogen azide s. under $AgClO_4$ $N \equiv C \cdot N_3$

Dinitrogen trioxide/acetic acid N_2O_3/CH_3COOH
Partial *cis*-hydroxylation $C:C \rightarrow C(OH)C(OH)$
  of carbon-carbon double bonds s. 27, 147

Nitric acid HNO_3
2,5-Dihydrofuran ring from 1,3-dienes ○

87.

A soln. of 3,4-bis(diphenylmethylene)succinic anhydride and concd. HNO_3 in acetonitrile refluxed 10 min. → 6,6,8,8-tetraphenyl-3,7-dioxabicyclo[3.3.0]oct-1(5)-ene-2,4-dione. Y: 89%. F. e. s. F. Toda and E. Todo, Bull. Chem. Soc. Japan *47*, 348 (1974).

Triphenylphosphine/ozone Ph_3P/O_3
1,4-Dihydroxy-2,3-diketones $C \equiv C \rightarrow CO \cdot CO$
from 2-acetylene-1,4-diols s. *30*, 167

Trimethyl phosphite/ozone $(MeO)_3P/O_3$
Azulene from naphthalene ring ←
via oxidative ring opening – Transannular ring closure s. *28*, 108

Antimony trichloride $SbCl_3$
Ethers from ethylene derivs. $C:C \rightarrow CHC(OR)$
Preferential addition s. *27*, 148

Oxygen s. under NaH and K/Me₃C·OH/Me₂SO O_2

Ozone s. a. under Ph₃P and (MeO)₃P O_3

Ozone/sodium tetrahydridoborate $O_3/Na[BH_4]$
Preferential oxidative ring opening C
Hydroxycarboxylic acid esters from cyclic enolethers s. *28*, 109

Ozone/sulfur dioxide O_3/SO_2
4H-1,4-Thiazine 1,1-dioxides ☉
from 2,5-dihydrothiophene 1,1-dioxides via 2,6-dialkoxy-1,4-oxathiane 4,4-dioxides – Acetals from ethylene derivs. s. *29*, 148

Ozone/nickel O_3/Ni
Cleavage of ethylene derivs. C
to alcohols and amines – Diols and diamines from cyclic ethylene derivs. s. *27*, 149

Hydrogen peroxide (s. a. under Lithium 2,2,6,6-tetramethylpiperidide, BH₃, B₂H₆, Bu₄N⁺[BH₄]⁻, and Fe(acac)₃) H_2O_2
Glycidic from α,β-ethylenecarboxylic acid esters $C:C \cdot COOR \rightarrow$ ◁—COOR
s. *28*, 110; peroxide chemistry, reviews, s. W. M. Weigert et al., Chem. Ztg. *99*, 106 (1975); A. Blaschette and D. Brandes, ibid. *99*, 125

Hydrogen peroxide/sodium hydroxide $H_2O_2/NaOH$
Hydroxycarboxylic acids from cyclic ketones C
s. *29*, 149

Hydrogen peroxide/formic acid/sodium carbonate $H_2O_2/HCOOH/Na_2CO_3$
Glycols from ethylene derivs. $C:C \rightarrow C(OH)C(OH)$
s. *27*, 150

Hydrogen peroxide/acetic acid H_2O_2/CH_3COOH
4-Azadioxindoles from 4-azaoxindoles $C:C(OH) \rightarrow C(OH)CO$
s. *27*, 151

Oxidations with *hydrogen peroxide/perfluoroacetone* $H_2O_2/(CF_3)_2CO$
Baeyer-Villiger oxidation ←
s. *26*, 137

Hydrogen peroxide/sulfuric acid/acetic anhydride $H_2O_2/H_2SO_4/Ac_2O$
2,4-Cyclohexadienones from benzene ring ←
s. *28*, 111

Baeyer-Villiger oxidation ←
s. *26*, 137

Sulfur dioxide s. under O_3 SO_2

Sulfuryl chloride SO_2Cl_2
1-Acylamino-1-peroxides from enacyclamines CO·N·C:C → CO·N·C(OOR)CH
s. *28*, 112

Methanesulfonic acid-sodium azide-sodium hydroxide $MeSO_3H\text{-}NaN_3\text{-}NaOH$
Ethylenecarboxylic acids from cyclic ketones   C
Fragmentation-type ring opening

88.

NaN₃ added at 0° during 45 min. to a well-stirred soln. of diamantanone in methanesulfonic acid, stirring continued 2.5 hrs. at 20°, poured onto ice, treated with aq. NaOH and NaOH-pellets, then stirred 18 hrs. → tetracyclo[7.3.1.0²,⁷.0⁶,¹¹]-tridec-3-ene-12-carboxylic acid (startg. m. f. 523). Y: 41%. F. Blaney, D. Faulkner, and M. A. McKervey, Synth. Commun. *3*, 435 (1973).

p-Toluenesulfonic acid   *TsOH*
Protection of alcohol groups   C:C → CHC(OR)
    as 4-methoxytetrahydropyran-4-yl derivs. – Mixed O-heterocyclic ketals from enolethers – O-Deacetylation s. *26*, 138

1,3,3-Trialkoxy compds.   C:C·CO → C(OR)·CH·C(OR)₂
    from α,β-ethyleneketones s. *26*, 178

Oxidations with *o-sulfoperoxybenzoic acid* $o\text{-}C_6H_4(SO_3H)(COO_2H)$
***trans*-Glycols from ethylene derivs.**   C:C → C(OH)C(OH)
s. *27*, 152

Sulfuric acid (s. a. under H_2O_2) H_2SO_4
Alcohols from ethylene derivs.   C:C → CHC(OH)
    Protection of aldehyde groups as amine adducts – as enamines s. *29*, 150

2-Decalones   O
    Bicyclic ketones from dienynes s. *29*, 678

4-Hydroxy-2-pyrones from 2,4-diynoic acids
s. *29*, 151

Sulfuric acid/peroxyacetic acid H_2SO_4/CH_3COO_2H
Glycol esters from ethylene derivs.   C:C → C(OAc)C(OAc)
s. *29*, 152

Potassium persulfate/sulfuric acid $K_2S_2O_8/H_2SO_4$
Enolesters from α,β-ethyleneketones   C:C·COR → C:C·OCOR

89.   Ph—CH=CH—COC₆H₄F—p → Ph—CH=CH—OCOC₆H₄F—p

Oxidation. A mixture of 4′-fluorochalcone, K-peroxydisulfate, concd. H₂SO₄, and glacial acetic acid kept 50 hrs. at room temp. with intermittent shaking → *trans*-1-phenyl-2-(4-fluorobenzoyloxy)ethylene. Y: 37%. F. e. s. D. N. Dhar and R. C. Munjal, Synthesis *1973*, 542.

Chromium(III) sulfate — $Cr_2(SO_4)_3$
Alcohols from ethylene derivs. — C : C → CHC(OH)
s. 26, 139

Chromium trioxide/acetic acid — CrO_3/CH_3COOH
Oxido compds. from ethylene derivs. — C : C → △O
Preferential formation s. 27, 153
Oxidative ring opening — C
of cyclic ketones s. 27, 154

Chromyl chloride/zinc — CrO_2Cl_2/Zn
Ketones from ethylene derivs. — C : CH → CHCO
s. 28, 481

Molybdenum peroxide-hexamethylphosphoramide complex — $MoO_5 \cdot (Me_2N)_3PO$
Stereospecific addition of oxygen — C : C → △O
to carbon-carbon double bonds – 5α,6α-Oxidosteroids s. 28, 113

Bromine/silver nitrate — $Br/AgNO_3$
Ketones from ethylene derivs. — C : CH → CHCO
s. 27, 155

Iodine — I
Glycol ethers and 1,4-dioxanes — C : C → C(OR)C(OR)
from ethylene derivs. s. 28, 114

tert-Butyl hypochlorite — $Me_3C \cdot OCl$
Oxindoles ←
from indoles and 3-chloroindolenines with 2→3-position shift of substituents
s. 29, 153

Sodium hypochlorite s. a. under RuO_2 — $NaOCl$

Sodium hypochlorite/acetic acid — $NaOCl/CH_3COOH$
γ-Lactones from cyclobutanones — ⊙
s. 26, 140

Sodium hypochlorite/sulfuric acid — $NaOCl/H_2SO_4$
Oxido compds. from ethylene derivs. — C : C → △O
Stereospecific addition s. 26, 141

Silver perchlorate/cyanogen azide — $AgClO_4/N \equiv C \cdot N_3$
Cyanogen azide ring expansion — ⊙
Cyclic ketones from enisocycles s. 29, 154

Periodate — IO_4^-
6-Spirooxirano-2,4-cyclohexadienones ←
from hindered o-hydroxymethylphenols s. 29, 246
Oxidative ring opening of N-heterocycles — C
s. 28, 903
α-Carboxy- from α-keto-lactams — ⊙
with ring contraction
3-Carboxy-2-azetidinone ring

90.

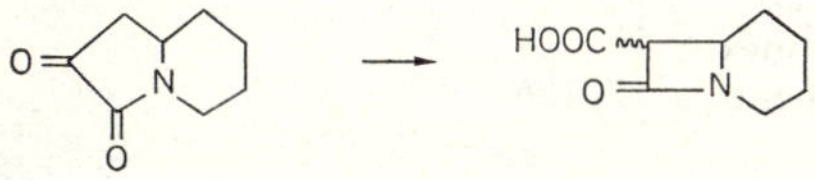

1-Azabicyclo[4.3.0]nonane-8,9-dione allowed to react 20 min. with excess periodate in Li-phosphate buffer of pH 6.3 → 7-carboxy-8-oxo-1-azabicyclo[4.2.0]octane. Y: 70%. F. e. s. D. R. Bender et al., J. Org. Chem. *38*, 3439 (1973); *40*, 1264 (1975).

Sodium periodate/sodium hydroxide $NaIO_4/NaOH$
Diols from cyclic glycols via dialdehydes C
with β-elimination of phosphoryl – Degradation of polyribonucleosides
s. 29, 155

Periodic acid HIO_4
Oxidative cleavage of oxido compds.
s. 21, 273 suppl. 29

Hydrogen chloride HCl
α-**Acylamino-α-alkoxy-** C : C·NAc → CH·C(OR)NAc
and α-**acylamino-α-hydroxy-** from α-acylamino-α,β-ethylene-carboxylic acid esters
s. 29, 156

o-**Ketoacetic acids** C
from 2,2-dialkoxy-1,3-dihydroxyindans s. 26, 243

Hydrogen chloride/irradiation HCl/hv
Skeletal rearrangement of tropolones ←
s. 26, 142

Hydrogen chloride/sodium hydrogen carbonate $HCl/NaHCO_3$
α-**Acylamino-α-hydroxy-** C : C·NAc → CH·C(OH)NAc
from α-acylamino-α,β-ethylene-carboxylic acid esters
s. 29, 156

Manganese dioxide MnO_2
Oxidative cleavage of glycols C
under mild conditions – Diketones from cyclic glycols s. 29, 157

Potassium permanganate/crown polyethers ←
Oxidative opening of isocycles
s. 28, 198

Ferric acetoacetonate/hydrogen peroxide $Fe(acac)_3/H_2O_2$
Oxido compds. from ethylene derivs. C : C → △O
Stereoselective addition

91.

5β,6β-**Oxidosteroids.** 30%-H_2O_2 added dropwise to a stirred soln. of cholesterol and ferric acetylacetonate in acetonitrile, and stirred 40 min. at 40° → cholesterol β-epoxide (Y: 68%) and α-epoxide (Y: 17%). F. e. s. M. Tohma, T. Tomita, and M. Kimura, Tetrah. Let. *1973*, 4359.

Ferric chloride $FeCl_3$
Allene-ketals and -ketones C
from enecyclopropenes s. 27, 156

Nickel s. under O_3 Ni

Ruthenium dioxide/sodium hypochlorite $RuO_2/NaOCl$
α-**Dioxo compds. from acetylene derivs.** C ≡ C → COCO
s. 27, 145

Palladous chloride/cupric nitrate $PdCl_2/Cu(NO_3)_2$
Acetals from ethylene derivs. $C:CH \rightarrow CHC(OR)_2$
   3,3-Dialkoxytetrahydro- from 2,5-dihydro-furans – also formation of ketones
   s. 26, 143

Lithium tetrachloropalladate/mercuric acetate $Li_2[PdCl_4]/(CH_3COO)_2Hg$
Ketones from ethylene derivs. $C:CH \rightarrow CHCO$
s. 27, 157

Osmium tetroxide/pyridine OsO_4/C_5H_5N
$α,α'$-**Dihydroxyketones** $C(OAc)C \equiv CH \rightarrow C(OH)COCH(OH)$
   from acoxy-2-acetylenes via allenes s. 28, 115

Tris(triphenylphosphine)platinum(0) $(Ph_3P)_3Pt(0)$
Homogeneous metal catalysis $C:C \rightarrow CHC(OAc)$
   Acoxy compds. from ethylene derivs. – Partial and stereospecific addition
   s. 29, 158

Via intermediates v.i.
$β$-**Keto- from** $α$-**allene-** $C:C:C \cdot COOR \rightarrow CHCOCHCOOR$
via $β$-**alkoxy-**$β,γ$-**ethylene-carboxylic acid esters**
s. 29, 159

$α$-**Cyanoketones from cyanoallenes** $C:C:C \cdot CN \rightarrow CHCOCHCN$
   via $β$-amino-$α,β$-ethylenenitriles s. 29, 160

Carbonyl groups from carbon-carbon bonds $C:C \rightarrow CHCO$
   via boronic acid esters – Cyclic thiolic acid esters – Substituent-directed
   metalation – Metalation of 2-acetylenethioethers s. 27, 158

Oxido compds. from ethylene derivs. $C:C \rightarrow \overset{\triangle}{O}$
via 1,2-bromohydrins
   via *trans*-1,2-bromohydrins – Optically active compds. s. 26, 144

$α$-**Aminoketones from enamines**
   via rearrangement of $α$-halogenazomethinium salts – Cyclic $α$-aminoketones
   s. 28, 116

Rearrangement

Hydrogen/Oxygen Type   OC⌒HO

Without additional reagents w.a.r.
Migration of cyclic carbonic acid ester groups
   1,3-Dioxol-2-ones from 2-oxo-1,3-dioxanes s. 28, 118

Irradiation (s. a. under Xylene)
Hydroxycarboxylic acid aryl esters
   from aryloxycarboxylic acids s. 29, 161
$$\begin{matrix} OAr \\ COOH \end{matrix} \rightarrow \begin{matrix} OH \\ COOAr \end{matrix}$$

N-Methylpyrrolidine s. under $Mg(NO_3)_2$ C_4H_8NMe

Pyridine C_5H_5N
Skeletal rearrangement ←
with ring expansion

92.

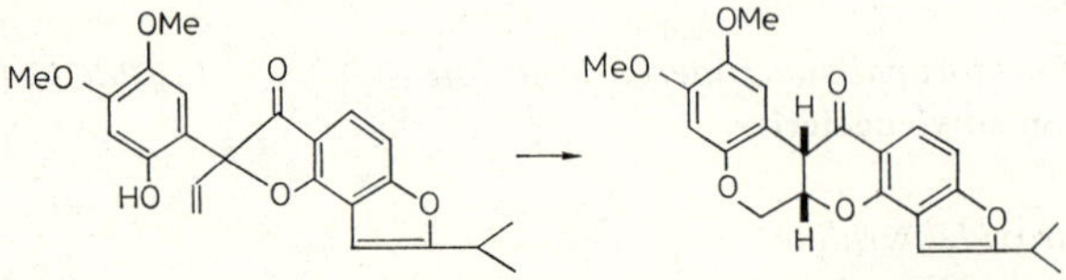

of O-heterocyclics. 7-Isopropyl-2-(2-hydroxy-4,5-dimethoxyphenyl)-2-vinylbenzo-[1,2-b:3,4-b]difuran-3(2H)-one in pyridine heated overnight at 100° in a sealed vessel → (±)-isorotenone. Y: 80%. F. e. s. L. Crombie, P. W. Freeman, and D. A. Whiting, Soc. Perkin I *1973*, 1277.

Cuprous oxide Cu_2O
5,6-Dihydro-4H-1,3-oxazines from 3-isocyanoalcohols O
s. *29*, 617

Magnesium nitrate/N-methylpyrrolidine $Mg(NO_3)_2/C_4H_8NMe$
α-Hydroxythiolic acid esters COCH(OH)(SR) → CH(OH)C(O)SR
from 1-hydroxy-2-ketothioethers s. *26*, 145

Mercuric acetate/sodium tetrahydridoborate $(CH_3COO)_2Hg/Na[BH_4]$
Lactones from ethylenecarboxylic acids O
by oxymercuration-demercuration s. *28*, 117

Boron fluoride BF_3
Lactones from ethylenecarboxylic acids
s. *21*, 184 suppl. *29*

Xylene/irradiation ←
Oxetane ring from 2-ethylenealcohols
with ring contraction s. *26*, 146

Acetic acid CH_3COOH
Spirolactones from ethylenecarboxylic acids
s. *27*, 505

Antimony pentafluoride s. under HF SbF_5

Sulfuric acid H_2SO_4
Lactones from ethylenecarboxylic acids
s. *30*, 523

Hydrogen fluoride/antimony pentafluoride HF/SbF_5
Reactions in superacidic medium ←
Dearomatization of phenols s. *27*, 159

Platinum oxide PtO_2
O-Decarbobenzoxylation Ö
with migration of cyclic carbonic acid ester groups – 1,3-Dioxol-2-ones from 2-oxo-1,3-dioxanes s. *28*, 118

Via intermediates v.i.
Quinones from 6-hydroxychromenes C
s. *27*, 160

Hydrogen/Carbon Type   OC∩HC

Without additional reagents   w.a.r.
Carboxylic acid esters from cyclic acetals   C
s. 26, 147

γ-Irradiation   ∰
Carboxylic acids from lactols   C
Preferential formation

93.

Carbohydrate derivs. α-Lactose monohydrate crystals irradiated at room temp. with $4 \cdot 10^{20}$ eV/g. ^{60}Co-γ-rays → 5-deoxylactobionic acid. Y: 60% based on startg. m. consumed. M. Dizdaroglu et al., A. *1973*, 1592.

Sodium hydride   NaH
Carboxylic acid esters from α-hydroxyketones $C(OH)COCH_3$ → $CHCH_2COOR$
   Favorskii-type rearrangement s. 29, 162

Potassium hydroxide   KOH
Epimerization of hydrobenzoins   ←
s. 26, 148 suppl. 29

Sodium/alcohol   NaOR
*erythro-threo-***Epimerization**   ←
   of acetals s. 29, 163

Butyllithium   BuLi
Epimerization of glycols   ←
s. 26, 148

Potassium cyanate   KOCN
Isomerization of oxido compds.   ←
s. 26, 750

Magnesium bromide $MgBr_2$
Ketones from oxido compds.   △O → CHCO
s. 27, 161

Thallic nitrate s. under $HClO_4$ $Tl(NO_3)_3$

Boron fluoride BF_3
   Directed stereospecific ring opening – Epimerization
   of aryl groups s. 26, 149

Enzyme   ←
Keto from glycol groups $C(OH)CH(OH)$ → $CHCO$
s. 27, 259

Di-tert-butyl peroxide (Me₃C·O·)₂
Acoxy compds. from acetals C
with rearrangement

94.

A mixture of methyl 2,3-di-O-acetyl-4,6-O-benzylidene-α-D-glucopyranoside and di-*tert*-butyl peroxide stirred and heated 7 hrs. at 140° → methyl 2,3-di-O-acetyl-4-O-benzoyl-6-deoxy-α-D-glucopyranoside. Y: 41%. F. e. s. L. M. Jeppesen, I. Lundt, and C. Pedersen, Acta Chem. Scand. *27*, 3579 (1973).

p-Toluenesulfonic acid TsOH
Glycolic acid esters OCH₂COOR
by intramolecular redoxidation
s. *30*, 115

Perchloric acid/thallic nitrate HClO₄/Tl(NO₃)₃
Arylacetic acid esters ArCOCH₃ → ArCH₂COOR
from aryl methyl ketones s. *27*, 162

Hydrogen bromide HBr
Ketones from oxido compds.
Directed stereospecific ring opening s. *26*, 149 → CHCO

Chlorotris(triphenylphosphine)rhodium(I) RhCl(Ph₃P)₃
s. *17*, 204 suppl. 29

Palladium-carbon Pd-C
Ring opening by disproportionation C
Acylamines from cyclic 1,1-aminoethers – Inversion of configuration – Stereospecific H-labeling at benzylic carbon s. *29*, 164

Via intermediates v.i.
Epimerization of sec. hydroxyl via tosylates ←
s. *28*, 119

Oxygen/Nitrogen Type OC∩ON

Without additional reagents w.a.r.
3-α-Acoxy-3-acyl- N : N · N(OH)CH → N : N · N(Ac)C(OAc)
from 3-hydroxy-triazenes
s. *29*, 346

Irradiation (s. a. under CH₃COOH) ⫯
Diol mononitrates from nitrous acid esters

95.

3β-Acetoxy-5α-lanostan-7α-yl nitrite in dry benzene irradiated 7 hrs. in a slow stream of dry oxygen with the Pyrex-filtered light of a 125 w. high-pressure Hg-vapor lamp → 3β-acetoxy-7α-hydroxy-5α-lanostan-32-yl nitrate. Y: 44%. J. Allen, D. H. R. Barton et al., Soc. Perkin I *1973*, 2402.

o-Hydroxyazo from azoxy compds. ←

96.

4'-Chloroazoxybenzene-2-carboxylic acid irradiated 24 hrs. with an unfiltered 100 w. medium-pressure Hg-arc in 95%-ethanol → 2-(4-chloro-2-hydroxyphenyl-azo)benzoic acid. Y: ca. 100%. F. e. s. B. C. Gunn and M. F. G. Stevens, Soc. Perkin I *1973*, 1682.

Cyclic oximinoethers from ethylenenitrites ◯

97.

A soln. of the startg. m. in benzene irradiated 8–10 hrs. with cooling at 30° under N₂ with the Pyrex-filtered light of a high-pressure Hg-lamp → 2,5-dimethyl-tetrahydrofurfuraldoxime. Y: 62.5%. F. e. s. M.-P. Bertrand and J.-M. Surzur, Bl. *1973*, 2393; cf. ibid. *1973*, 2399.

6-Hydroxy-1,2-oxazine ring
  from unsatd. nitro compds. s. *28*, 120

Sodium/alcohol	*NaOR*
2-Alkoxy-1,3-N,N-heterocyclics	◉
from N-heterocyclic oxime tosylates s. *28*, 121	
Sodium carbonate	*Na₂CO₃*
Unsubst. carboxylic acid amides from oximes s. *29*, 165	CH : NOH → CONH₂
Pyridine s. under TsCl	*C₅H₅N*
Aluminum chloride	*AlCl₃*
Phthalimides	
from 2,3-benzoxazin-1-ones s. *30*, 268	
Acetic anhydride	*Ac₂O*
5,6-Dihydro-4H-furo[3,2-b]pyrid-2-ones	←
from 4,5,6,7-tetrahydro-4-oxoisoxazolo[2,3-a]pyridinium salts s. *29*, 166	
Trifluoroacetic anhydride (s. a. under CF₃COOH)	*(CF₃CO)₂O*
N-Heterocyclic carbinols from N-oxides s. *9*, 243 suppl. *29*	←
Acetic acid/irradiation	*CH₃COOH/⚡*
Photochemical Beckmann rearrangement	◉
Lactams from cyclic oximes – Ring expansion s. *26*, 150	
Trifluoroacetic acid/trifluoroacetic anhydride	*CF₃COOH/(CF₃CO)₂O*
***β*-Rearrangement of N-oxides** s. *29*, 167	←
Silica	*SiO₂*
Carboxylic acid amides from aldoximes	CH : NOH → CONH₂

98. PhCH=NOH → PhCONH₂

A mixture of benzaldoxime and silica gel, which has been activated at 130–140°, refluxed 69 hrs. in anhydrous xylene → benzamide. Y: 92%. – Similarly: Acet-aldoxime → acetamide. Y: 89%. F. e. s. J. B. Chattopadhyaya and A. V. R. Rao, Tetrahedron *30*, 2899 (1974).

Hexamethylphosphoramide $\qquad$ $(Me_2N)_3PO$
Beckmann rearrangement $\qquad$ C(:NOH)R → CONHR
s. 26, 308 suppl. 28

Polyphosphoric acid $\qquad$ $H(PO_3H)_xOH$
Isomeric carboxylic acid amides $\qquad$ CONHR
  by Beckmann rearrangement and Schmidt reaction s. 26, 308

β-Ketolactams from cyclic β-diketones $\qquad$ ⊙
  via cyclic β-alkoxy-α,β-ethyleneoximes s. 27, 163

Phosphorus oxide chloride $\qquad$ $POCl_3$
N→C-Sulfonyloxy group migration $\qquad$ N·OTs → C·OTs
s. 27, 164

Beckmann rearrangment $\qquad$ ⊙
4-Hydroxypyrimidine from 3-nitrosopyrrole ring

99.

POCl₃ added to a soln. of 1,3-dimethyl-5-nitroso-6-phenylpyrrolo[2,3-d]pyrimidine-2,4(1H,3H)-dione in dimethylformamide, and refluxed 1 hr. → 1,3-dimethyl-5-hydroxy-7-phenylpyrimido[4,5-d]pyrimidine-2,4(1H,3H)-dione. Y: 60%. F. e. s. F. Yoneda et al., Chem. Pharm. Bull. 21, 473 (1973).

p-Toluenesulfonyl chloride/pyridine $\qquad$ $TsCl/C_5H_5N$
Beckmann rearrangement of nitrones
  Lactams from cyclic nitrones – Ring expansion s. 28, 122

Sulfuric acid $\qquad$ H_2SO_4
Oxazolidine from 1,2,4-oxadiazoline ring $\qquad$ ←
s. 29, 168

Carbon/Sulfur Type $\qquad$ OC↷OS

Sodium acetate $\qquad$ CH_3COONa
α-Acoxythiolic acid esters $\qquad$ COCH₂SO· → CH(OAc)COS·
from β-ketosulfoxides
Pummerer rearrangement

100.

  n—C₃H₇COCH₂—S—C₆H₄CH₃—p → n—C₃H₇CH—C(=O)S—C₆H₄CH₃—p
                ‖ |
                O   OAc
  + OAc₂

A stirred mixture of 1-(p-tolylsulfinyl)-2-pentanone, Na-acetate, and acetic anhydride heated during 1 hr. to reflux, which is maintained 2 additional hrs. → S-(p-tolyl) 2-acetoxypentanethioate. Y: 77%. F. e. s. S. Iriuchijima, K. Maniwa, and G. Tsuchihashi, Am. Soc. 97, 596 (1975).

Oxygen/Carbon Type $\qquad$ OC↷OC

Without additional reagents $\qquad$ w.a.r.
2-Alkoxy-1,3-oxazin-6-ones $\qquad$ ○
  from α,β-ethylene-β-isocyanatocarboxylic acid esters and reverse reaction
  s. 27, 165

Nitrogen/Carbon Type OC⌒NC

Without additional reagents w.a.r.
Aminoethers from aminoalcohols ←
  by rearrangement s. 28, 123
Meisenheimer rearrangement
2-Ethylenealkoxylamines $C:C\cdot C\cdot N(O)R_2 \rightarrow C(O\cdot NR_2)C:C$
from 2-ethyleneamine N-oxides

101.

Crude N-geranyl-N,N-dimethylamine oxide thermolyzed and distilled during 1 hr. at 75–80°/0.05 mm → product. Y: 81%. – This is part of a sequence for the **isomerization of 2-ethylenealcohols.** F. e. s. V. Rautenstrauch, Helv. 56, 2492 (1973).

Ring expansion ⊙
  by Meisenheimer rearrangement – Dihydro-1,2-oxazine ring s. 26, 151
4,4a,5,7-Tetrahydrocyclopenta-1,3,4-oxadiazines ←
  from 2,3-diazabicyclo[2.2.1]hept-5-enes s. 27, 166
Irradiation ⚡
Coumarins and carbostyrils ←
  from 3-o-hydroxy- and 3-o-amino-arylidenesuccinimides respectively s. 26, 152
Lithium tetrahydridoaluminate $Li[AlH_4]$
O- from N-Heterocyclic cage compds. ←
s. 27, 167
Pyridine hydrochloride C_5H_5N,HCl
8-Oxa-2,3-diazabicyclo[3.2.1]oct-6-enes ←
  from 1,2-diazabicyclo[3.2.0]-3-hepten-6-ols s. 29, 169
Hydrogen chloride HCl
3-Aminosydnone imine salts ○
s. 26, 828

Sulfur/Carbon Type OC⌒SC

Acetic acid CH_3COOH
S→O-Acyl group migration SAc → OAc
s. 9, 248
Silica SiO_2
s. 9, 248 suppl. 29

Carbon/Carbon Type OC⌒CC

Without additional reagents w.a.r.
γ,δ-Ethyleneketones C.
  from (vinylcyclopropyl)carbinols – Thermal sigmatropic 1,5-rearrangement
  s. 28, 124

Difuranylacetylenes from dienediynedials ○

102.

A soln. of 3,8-diphenyl-2,8-decadiene-4,6-diynedial in benzene kept 8 hrs. at 50° → bis(3-phenyl-2-furanyl)acetylene. Y: 86%. F. e. s. R. Muneyuki et al., Bull. Chem. Soc. Japan *46*, 2565 (1973).

3-Subst. isochromenes ☉
   from 1-acylbenzocyclobutenes s. *27*, 169

Enolethers from oxo compds. ←
by retro-Claisen rearrangement

103.

Pin-2-en-9-al heated at 210° in a sealed tube → product. Y: ca. 100%. – Also with 9-methylpin-2-en-9-one s. Y. Bessiere-Chretien and C. Grison, Chem. Commun. *1973*, 549.

Irradiation ⊬

O-Heterocyclics from oxido compds. ←

104.

A soln. of 4,5-epoxy-4,5-dihydropyrene in dry methylene chloride irradiated 15 hrs. under N_2 with 254 nm UV-light → phenanthro[4,5-bcd]oxepin. Y: 36.4%. B. L. Van Duuren, G. Witz, and S. C. Agarwal, J. Org. Chem. *39*, 1032 (1974).

O-Heterocyclics ○
   by double ring closure of ethyleneoxo compds. – Intramolecular Paterno-Büchi reaction s. *26*, 153

Sodium hydride   NaH
1,3-Oxathiolanes
s. *27*, 170

Sodium hydroxide   NaOH
Δ⁴-Isoxazol-3-ones
   from α,β-acetylenehydroxamic acids s. *28*, 125

Ketones from alcohols with ring opening   C
s. *19*, 211

Cleavage of β-dioxo compds. ←
   Skeletal 1,3-cyclohexadione ring rearrangement s. *29*, 170

Sodium/alcohol   NaOR
4H-1,4-Oxaphosphorin 4-oxides ○
s. *28*, 126

Sodium acetate CH_3COONa
4-Chromanone ring
   with triethylamine cf. *28*, 127; with Na-acetate s. H. Omokawa, S. Kouya, and K. Yamashita, Agri. Biol. Chem. (Tokyo) *39*, 393 (1975)

Triethylamine Et_3N
4-Chromanone ring
  via Fries rearrangement s. 28, 127

2,3-Dihydrofurans from cyclopropyl ketones ☉
s. 26, 154; with NaH cf. S. Danishefsky and J. Dynak, Tetrah. Let. *1975*, 79

Cuprous stearate $C_{17}H_{35}COOCu$
γ-Enollactones
  from cyclopropenecarboxylic acids – 2-Alkoxyfurans from cyclopropenecarboxylic acid esters s. 26, 155

Aluminum oxide Al_2O_3
Oxido compds. from 2-ethylenealcohols C(OH)C : C → ⟨O⟩—CH
s. 29, 1

Coumarans ○
s. 11, 224 suppl. 28

Triphenylphosphine Ph_3P
2,3-Dihydrofurans from cyclopropyl ketones ☉
s. 26, 154

Phosphorus tribromide PBr_3
Cyclic ethers from ethylenealcohols ○
  14β-Hydroxysteroids s. 27, 171

p- from o-Quinones ←
s. 27, 172

p-Toluenesulfonic acid $TsOH$
Dithiolortholactones ○
  from hydroxyketene mercaptals s. 29, 552

Sulfuric acid H_2SO_4
α-Diketones from β,γ-ethylene-α-hydroxyketones
  Rearrangement s. 29, 171

2-Benzopyr-3-ones ☉
  from 1,3-dihydroxy-2-indanones s. 26, 243

Lactones ←
from cyclopropylcarboxylic acids

105.

Startg. acid heated at 150° with aq. 50%-sulfuric acid for 1 hr. or with perchloric acid for 10 hrs. → lactone. Y: 85%. J. S. Bindra et al., Am. Soc. *95*, 7522 (1973).

Perchloric acid $HClO_4$
s. 30, 105

Sodium periodate $NaIO_4$
1,3-Benzodioxoles from o-hydroxycarbinols ○
s. 28, 207

Tetramethylammonium chloride $Me_4N^+Cl^-$
Skeletal rearrangement of spirocyclics ⇐
s. 27, 173

Palladium-carbon/hydrogen $Pd\text{-}C/H_2$
Oxo compds. from ethylenealcohols ⇐
s. 28, 128

Via intermediates v.i.
Cyclic ethers from ethylenealcohols ○
by oxymercuration-demercuration s. 29, 172; s. a. R. O. Klaus, H. Tobler, and C. Ganter, Helv. 57, 2517 (1974)

Exchange ↓↑

Hydrogen ↑ OC ↓↑ H

Without additional reagents w.a.r.
o-Alkoxy-p-quinones from p-quinones H → OR
s. 27, 174

Ketolactones from lactones CH_2 → CO
s. 27, 505

Catechol monoesters from phenols ArH → ArOAc
s. 29, 123

Oxidative cyclopropane ring opening C
s. 26, 156

Irradiation (s. a. under $HgBr_2$ and Hematoporphyrin) ///
Oxo compds. from hydrocarbons CH_2 → CO
Aldehyde from methyl groups s. 26, 157

Electrolysis (s. a. under $HON(SO_3Na)_2$) ⚡
Electrooxidative rearrangement ⇐
with alkoxylation – Electrochemical oxidation and reduction, review s. 26, 158

Hemiquinol-ethers and -esters ⇐
from phenols s. 28, 98

Anodic allylic oxidation H → OAc
Acoxy-2-ethylenes from ethylene derivs. s. 29, 173

Sodium hydride NaH
Oxidative skeletal rearrangement ⇐
of N-heterocyclics s. 28, 129

Sodium hydroxide/H_2O_2 $NaOH/H_2O_2$
(Hydroxythiol)thionocarbonic acid esters OH → O·C(:S)S·OH
of carbohydrates s. 27, 175

Sodium methoxide s. under Se NaOR

Sodium acetate s. under I CH_3COONa

Sodium fluoride   NaF
Trifluoromethyl ethers   OH → OCF$_3$
from alcohols via fluoroformates

106.
$$CH_3-\underset{NO_2}{\overset{NO_2}{C}}-CH_2OH + COF_2 \longrightarrow CH_3-\underset{NO_2}{\overset{NO_2}{C}}-CH_2OCOF \longrightarrow CH_3-\underset{NO_2}{\overset{NO_2}{C}}-CH_2OCF_3$$

Excess *carbonyl fluoride* distilled into an evacuated autoclave containing the startg. m. and NaF precooled at −78°, then gently rocked 64 hrs. at room temp. under 800–1100 psig. autogenous pressure → fluoroformate (Y: 70%) cooled to −78° in an autoclave, sulfur tetrafluoride and *anhydrous HF* condensed into it, rocked and heated 48 hrs. at 85° under 400–700 psig. autogenous pressure → product (Y: 73%). F. e. s. H. M. Peters et al., J. Chem. Eng. Data *20*, 118, 113 (1975).

Triethylamine   Et$_3$N
Mixed carbonic acid esters   OH → O·CO·OR
s. *27*, 404

Triethylamine/cyanogen bromide   Et$_3$N/BrCN
Cyclic iminocarbonic acid esters   ◯
s. *27*, 176

Pyridine s. under CuCl and (CH$_3$COO)$_4$Pb   C$_5$H$_5$N

Cuprous chloride/pyridine   CuCl/C$_5$H$_5$N
Copper salt-amine catalyzed autoxidation   ←
of phenols − Hemiquinolethers from phenols s. *28*, 130

Silver nitrate/persulfate   AgNO$_3$/S$_2$O$_8^{2-}$
Keto from methylene groups   CH$_2$ → CO
s. *27*, 177

Mercuric acetate/lithium tetrachloropalladate   (CH$_3$COO)$_2$Hg/Li$_2$PdCl$_4$
1,3-Dioxolanes
from ethylene derivs. and glycols s. *29*, 174   C:CH → CH·C⟨^O_O⟩

Mercuric bromide/irradiation   HgBr$_2$/hν
Allylic oxidation   C:C·CH$_2$ → C:C·CO
α,β-Ethyleneketones from ethylene derivs. s. *28*, 131

Borax s. under Me$_3$C·OCl   Na$_2$B$_4$O$_7$

Thallic acetate   (CH$_3$COO)$_3$Tl
α-Acoxycarboxylic acids   2 R·CH·COOH → R·C(OCO·CH·R)·COOH
from 2-carboxylic acid molecules s. *27*, 178

Ammonium ceric nitrate   (NH$_4$)$_2$Ce(NO$_3$)$_6$
Glycol mononitrates from hydrocarbons   CHCH → C(OH)C(ONO$_2$)
s. *28*, 132

Methanol s. under Dichlorodicyanoquinone   MeOH

Hematoporphyrin/irradiation   ←
Quinones from phenols   ←
s. *27*, 179

Nitrobenzene   PhNO$_2$
1,3-Oxazin-4-ones from Δ^3-pyrrol-2-ones   ⊙
s. *29*, 175

Acetic anhydride s. *under KMnO₄* $\quad\quad Ac_2O$

2,2,6,6-Tetramethyl-1-oxopiperidinium chloride $\quad\quad\leftarrow$
α-Dioxo compds. from ketones $\quad\quad COCH_2 \rightarrow COCO$
s. *29*, 176

Hydroperoxides s. *under MoCl₅* $\quad\quad R \cdot OOH$

Benzoyl peroxide $\quad\quad (PhCOO)_2$
Oxo compds. from hydrocarbons $\quad\quad CH_2 \rightarrow CO$
s. *26*, 157

Benzoyloxylation of indoles $\quad\quad H \rightarrow OBz$
s. *26*, 159

2,3-Dichloro-5,6-dicyanoquinone/methanol $\quad\quad\leftarrow$
Phenolketones from phenols $\quad\quad CH_2 \rightarrow CO$
s. *26*, 160

Cyanogen bromide s. *under Et₃N* $\quad\quad BrCN$

N-Bromosuccinimide $\quad\quad \begin{matrix}CH_2CO\\ \end{matrix}\!\!\!\!\!\!\!\!\!\!\!\!NBr\\ CH_2CO$

Ketones $\quad\quad CO$
  via acoxy compds. and sec. alcohols s. *26*, 161

Lead tetraacetate $\quad\quad (CH_3COO)_4Pb$
4-Hydroxypyrimidine from 3-aminopyrrole ring
s. *27*, 363

Lead tetraacetate/pyridine $\quad\quad (CH_3COO)_4Pb/C_5H_5N$
1,1-Diacoxy compds. from alcohols $\quad\quad CHOH \rightarrow C(OAc)_2$
  with N-heterocyclic ring contraction s. *28*, 133

Hexafluorophosphate $\quad\quad PF_6^-$
Carbatricobalt decacarbonyl derivs. $\quad\quad\leftarrow$
  Acylation of alcohols, amines, and mercaptans s. *27*, 180

Ozone $\quad\quad O_3$
Keto from methylene groups $\quad\quad CH_2 \rightarrow CO$
s. *26*, 162

Hydrogen peroxide s. *under NaOH and Na₂MoO₄* $\quad\quad H_2O_2$

N,N-Dichlorobenzenesulfonamide/
sodiobenzenesulfonamide $\quad\quad PhSO_2NCl_2/PhSO_2NHNa$
Benzylic carbonyl from methylene groups $\quad\quad ArCH_2 \rightarrow ArCO$
s. *27*, 181

Sodium hydroxylaminedisulfonate/electrolysis $\quad\quad HON(SO_3Na)_2/\updownarrow$
Quinones from phenols $\quad\quad\leftarrow$
s. *28*, 721

Sulfuric acid $\quad\quad H_2SO_4$
Ketones from hydrocarbons $\quad\quad CH_2 \rightarrow CO$
s. *28*, 134

Persulfate s. *under AgNO₃* $\quad\quad S_2O_8^{2-}$

Selenium/sodium methoxide   *Se/NaOR*
Urethans from acylamines   >NCHO → >NCOOR

107.   Me$_2$N—CHO + NaOMe → Me$_2$N—COOMe

N,N-Dimethylformamide, Na-methoxide, and metallic selenium allowed to stand 20 hrs. at room temp. in tetrahydrofuran under N$_2$ → methyl N,N-dimethylcarbamate. Y: 36%. F. e. s. K. Kondo, N. Sonoda, and H. Sakurai, Chem. Commun. *1974*, 160.

Selenium/oxygen   *Se/O$_2$*
Sym. carbonic acid esters from alcohols   2 ROH → (RO)$_2$CO
s. *28*, 322; s. a. Bull. Chem. Soc. Japan *48*, 108 (1975)

Selenium dioxide   *SeO$_2$*
α-Alkoximinooxo compds. from alkoximes   C(:NOR)CH$_2$ → C(:NOR)CO
s. *27*, 182
2-Formylbenzo[b]selenophenes   ←
     from selenochromenes – 2-Formylbenzo[b]thiophene from thiochromene s. *26*, 163

Sodium dichromate/sulfuric acid   *Na$_2$Cr$_2$O$_7$/H$_2$SO$_4$*
Saccharins   O
s. *26*, 164

Chromium trioxide/pyridine   *CrO$_3$/C$_5$H$_5$N*
α,β-Acetyleneketones from acetylene derivs.   C≡C·CH$_2$ → C≡C·CO
s. *28*, 135

Chromium trioxide/acetic acid   *CrO$_3$/CH$_3$COOH*
Benzylic oxidation   ArCH$_2$ → ArCO
     Ketones from hydrocarbons s. *26*, 165

Chromium trioxide/sulfuric acid   *CrO$_3$/H$_2$SO$_4$*
Phthalimides from phthalimidines   –N(CH$_2$/CO–) → –N(CO/CO)
s. *27*, 183

Sodium molybdate/hydrogen peroxide   *Na$_2$MoO$_4$/H$_2$O$_2$*
2,5-Cyclohexadien-4-onyl hydroperoxides   ←
     from phenols s. *22*, 170 suppl. 29

Oxidation with *molybdenum pentachloride/hydroperoxides*   *MoCl$_5$/ROOH*
α-Hydroxyketones from ethylene derivs.   C:C → C(OH)CO
     Hydroxydiketones from ethylenealcohols – Ketones from sec. alcohols –Oxido compds. from ethylene derivs. – Ketosugars – N-Oxidations s. *27*, 184

Bromine/sodium hydroxide   *Br/NaOH*
Lactams from cyclic tert. amines   C(N–)(CH$_2$) → C(N–)(CO)
s. *30*, 152

Iodine   *I*
Phenolesters   ArH → ArOCOR
     from arenes and acyl peroxides s. *26*, 166

Iodine/sodium acetate   *I/CH$_3$COONa*
1-Azoethers from hydrazones   NHC:C → N:N·C(OR)
s. *27*, 185

tert-Butyl hypochlorite/borax   *Me$_3$C·OCl/Na$_2$B$_4$O$_7$*
Replacement of hydrogen by alkoxy groups   H → OR
s. *29*, 177

Potassium permanganate/acetic anhydride	$KMnO_4/Ac_2O$
α-**Diketones from ethylene derivs.**	C : C → COCO
s. *27*, 186	
Potassium permanganate/acetic acid	$KMnO_4/CH_3COOH$
N-Demethylation via formamides	NCH_3 → NCHO

Formamides from methylamines with and without simultaneous oxidation of methylene to keto groups s. *28*, 136

Nickel peroxide/sodium hydroxide	$NiO_2/NaOH$
Carboxylic acids from alcohols	CH_2OH → COOH

Selective oxidation s. *27*, 187

Oxidations with *sodium ruthenate*	Na_2RuO_4
Carboxylic acids from alcohols	

s. *28*, 137

Ruthenium tetroxide/sodium periodate	$RuO_4/NaIO_4$
Double benzylic oxidation	←

9α-Hydroxy-6-ketosteroids s. *29*, 178

Palladium-carbon/p-cymene	←
Dehydrogenative oxidation	

Ketones from hydrocarbons with aromatization s. *29*, 179

Lithium tetrachloropalladate(II) s. under *($CH_3COO)_2Hg$*	Li_2PdCl_4
Via intermediates	v.i.
Remote functionalization	CH_2 → CO

of non-activated methylene groups – Direction by complex formation – Keto from methylene groups – Carbon-carbon double bonds s. *27*, 188

Reactions via selenides	CH·C : C → C : C·C(OAc)

Acoxy-2-ethylenes from ethylene derivs. with double-bond shift via 2-acoxyselenides *29*, 180

Oxygen ↑   OC ⇈ O

Without additional reagents	w.a.r.
Chloromethyl ethers from alcohols	OH → OCH_2Cl
s. *26*, 167	
1,1-Hydroxyperoxides	C(OH)(COOR)
s. *28*, 138	
3β-Acoxy-Δ⁵-steroids	←
from 6β-alkoxy-3α,5α-cyclosteroids	

108.

A soln. of 25-hydroxy-6β-methoxy-3α,5-cyclo-5α-cholestane in glacial acetic acid stirred 24 hrs. at 70° → 25-hydroxycholesteryl 3-acetate. Y: 94%. J. J. Partridge, S. Faber, and M. R. Uskoković, Helv. *57*, 764 (1974); with acetic anhydride in the presence of BF_3 at low temp. s. C. R. Narayanan, S. R. Prakash, and B. A. Nagasampagi, Chem. & Ind. *1974*, 966.

Carboxylic acid esters from carboxylic acids   COOH → COOR
s. *29*, 181
   and *tert*-butyl ethers
   s. *26*, 168
   and isoureas
   s. *29*, 547; s. a. W. Friedrichsen, E. Büldt, and R. Schmidt, Tetrah. Let. *1975*, 1137
Iminoesters   CONH → C(OR) : N
   from carboxylic acid amides and fluorosulfonic acid esters s. *28*, 139
Sulfonic acid esters from sulfonic acids SO_2OH → SO_2OR
   and chloroformic acid esters s. *26*, 169

109.
$$2\ p\text{—}CH_3C_6H_4\text{—}SO_2OH + C_2H_5CO\text{—}\overset{O}{\overset{\|}{P}}(OEt)_2 \longrightarrow 2\ p\text{—}CH_3C_6H_4\text{—}SO_2OC_2H_5$$
$$\left[+\ C_2H_5CO\text{—}\overset{O}{\overset{\|}{P}}(OH)_2\right]$$

and acylphosphonic acid esters. p-Toluenesulfonic acid refluxed 8 hrs. with diethyl propionylphosphonate in benzene or toluene → ethyl p-toluenesulfonate. Y: 92.6%. F. e. s. P. Golborn, Synth. Commun. *1973*, 273.

Irradiation s. under HCl   ⇤

Sodium hydride   *NaH*
Enolesters from ketones   CHCO → C : C(OAc)
s. *28*, 140 suppl. *30*

Sodium hydroxide   *NaOH*
Ethers from alcohols   OH → OR
   with anion-induced skeletal rearrangement s. *27*, 189

n-Butyllithium   *n-BuLi*
Enolesters from ketones   CHCO → C : C(OAc)
s. *28*, 140; with NaH cf. R. A. Auerbach et al., Org. Synth. *54*, 49 (1974)

Sodium carbonate Na_2CO_3
Glycosidation   OH → OR
   with N-carbalkoxylation s. *30*, 297

Potassium salt K^+
Phenolethers   ArOH → ArOR
   from phenols and carboxylic acid esters s. *29*, 182

Triethylamine s. under ClCOOEt Et_3N

N-Methylmorpholine s. under Isobutyl chloroformate   ⇤

Pyridine (s. a. under HBO₂ and BrCN) C_5H_5N
O-Acetylation   OH → OAc
s. *26*, 939

2,6-Di-tert-butylpyridine   ⇤
1,4-Dioxanes   O
   from glycols and 1,2-ditosylates s. *29*, 183

4-Dialkylaminopyridines   ⇤
4-Dialkylaminopyridines as acylation catalysts   OH → OAc
   Acylation of hindered hydroxy compds. s. *29*, 184

Cupric acetate s. under HCl $(CH_3COO)_2Cu$

Cupric chloride $\qquad CuCl_2$
Aryl isopropyl carbonates from arenes $\qquad$ ArH → ArOCOOCHMe$_2$
s. *26*, 170

Calcium sulfate s. under Cation exchanger $\qquad CaSO_4$

Zinc chloride $\qquad ZnCl_2$
Dienone-phenol rearrangement $\qquad$ ←
Phenol acetates from dienones s. *27*, 190

Metaboric acid/pyridine $\qquad HBO_2/C_5H_5N$
Preferential O-acylation $\qquad$ OH → OAc
s. *27*, 191

Alumina $\qquad Al_2O_3$
3β-Acoxy-Δ⁵-steroids $\qquad$ ←
from 6β-hydroxy-3α,5α-cyclosteroids s. *29*, 185

Molecular sieve $\qquad$ ←
Acetals from oxo compds. $\qquad$ CO → C(OR)$_2$
s. *28*, 141

Boron fluoride $\qquad BF_3$
O,O-Propylidene derivs. of carbohydrates
s. *26*, 171

Total O-acylation of carbohydrates $\qquad$ OH → OAc
s. *28*, 88

Lactolides from lactols
s. *28*, 146 suppl. *29*

3β-Acoxy-Δ⁵-steroids $\qquad$ ←
from 6β-alkoxy-3α,5α-cyclosteroids
s. *30*, 108

Aluminum chloride-polymer $\qquad$ ←
Reactions with polymer-protected reagents $\qquad$ COOH → COOR
Carboxylic acid esters from carboxylic acids s. *29*, 186

Diethyl azodicarboxylate (s. a. under Ph₃P) $\qquad$ ROOC·N:N·COOR
Carboxylic acid esters
from carboxylic acids and phosphorodiamidites s. *28*, 142

Dimethylformamide s. under (RO)₂SO₂ $\qquad HCONMe_2$

Dicyclohexylcarbodiimide $\qquad$ RN:C:NR
Acoxylamines $\qquad$ COOH → COONH$_2$
from carboxylic acids via N-tritylacoxylamines s. *26*, 172

Trialkyloxonium fluoroborates $\qquad R_3O^+ BF_4^-$
Iminoester fluoroborates $\qquad$ CON< → C(OR):N⁺< BF$_4^-$
from carboxylic acid amides s. *26*, 933

Isouronium salts from ureas $\qquad$ ←
Cyclic isouronium salts s. *26*, 173

O-Alkylation of mesoionic heterocyclics $\qquad$ ←
s. *26*, 174

α-Aminocarboxylic acid esters
from aziridinones via their O-alkylation s. *27*, 192 $\qquad$ → C(NH)COOR

6-Alkoxy-2-azapyrylium salts
from 1,2-oxazin-6-ones

110.

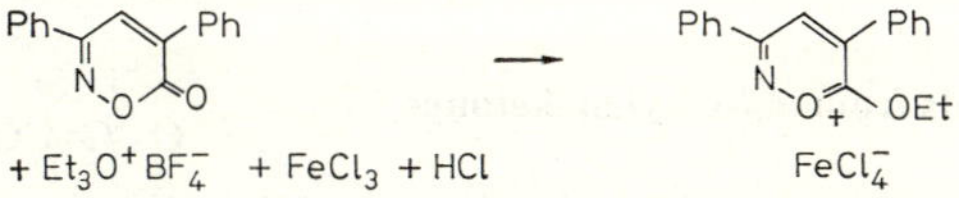

Ph–[N,O,C(=O),CH=C(Ph)] + Et$_3$O$^+$BF$_4^-$ + FeCl$_3$ + HCl ⟶ Ph–[N,O$^+$,C(OEt)=CH–C(Ph)] FeCl$_4^-$

A soln. of 3,5-diphenyl-1,2-oxazin-6-one in methylene chloride mixed with a soln. of triethyloxonium fluoroborate in the same solvent, which is removed after 3 days, the residue dissolved in acetic acid, and treated with ferric chloride in concd. HCl → 6-ethoxy-3,5-diphenyl-2-azapyrylium tetrachloroferrate. Y: 50%. – Also 6-methoxy analog with dimethyl sulfate s. O. P. Shelyapin, I. V. Samartseva, and L. A. Pavolova, Zh. Org. Khim. *9*, 1987 (1973); C. A. *79*, 137062.

Ethyl orthoformate (s. a. under HCl)   HC(OEt)$_3$

Ethyl chloroformate/triethylamine   ClCOOEt/Et$_3$N

Carboxylic acid aryl esters   COOH → COOAr
    from carboxylic acids and phenols – Sequential polypeptides s. *27*, 193

Isobutyl chloroformate/N-methylmorpholine ←

N-Acoxy- from N-hydroxy-ureas   N·CO·N·OH → N·CO·N·OAc
    via carboxylic alkoxyformic anhydrides s. *27*, 194

Cation exchanger/calcium sulfate ←

Cyclic acetals   O
s. *11*, 233 suppl. 29

Amberlyst 15 ←

Acetals and enolethers   CO → C(OR)$_2$
from oxo compds. and orthoformic acid esters

111.

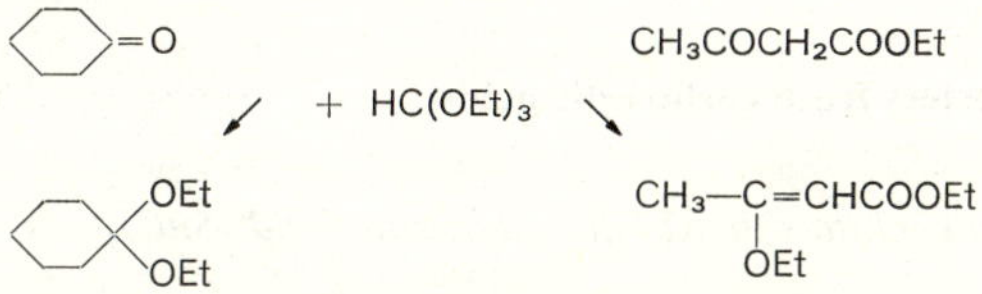

cyclohexanone =O + HC(OEt)$_3$ ⟶ cyclohexanone diethyl acetal (OEt, OEt)

CH$_3$COCH$_2$COOEt ⟶ CH$_3$–C(OEt)=CHCOOEt

Amberlyst 15 cation exchanger, a macroreticular sulfonic acid resin based on styrene-divinylbenzene, is a superior catalyst for the prepn. of enolethers and acetals. – E:
Cyclohexanone, triethyl orthoformate, and Amberlyst 15, shaken 3–4 hrs. at 0–5° under N$_2$ → cyclohexanone diethyl acetal. Y: ca. 100%. | Ethyl acetoacetate, triethyl orthoformate, and Amberlyst 15, shaken 3–4 hrs. at 0–5° under N$_2$ → enol ether. Y: 84%.
F. e. s. S. A. Patwardhan and S. Dev, Synthesis *1974*, 348.

Polycyclic acetallactones   O
from cyclic triols and oxocarboxylic acids

112.

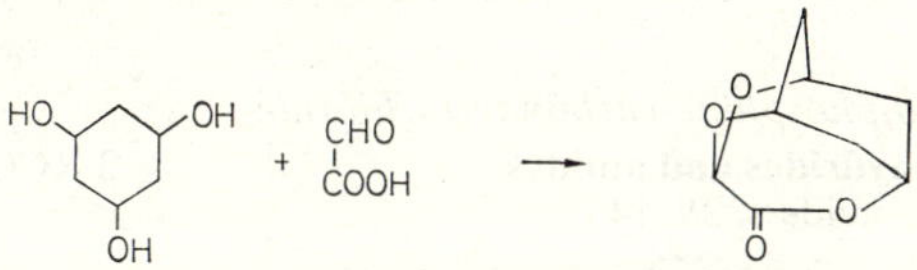

cis-cyclohexane-1,3,5-triol (HO, OH, OH) + CHO–COOH ⟶ [bicyclic acetallactone]

A soln. of *cis*-cyclohexane-1,3,5-triol in 1,2-dimethoxyethane refluxed 0.5 hr. with an equal weight of glyoxylic acid monohydrate and excess Amberlyst 15 cation exchanger → product. Y: 85%. R. B. Woodward et al., Am. Soc. *95*, 6853 (1973).

Trifluoroacetic acid s. under Et₃SiH $\qquad$ CF_3COOH

Oxalic acid s. under SnCl₄ $\qquad$ $(COOH)_2$

Fumaric acid $\qquad$ ←
α,β-Ethyleneketals from α,β-ethyleneketones
s. *30*, 114 $\qquad$ C : C·CO → C : C·C⟨O,O⟩

Dialkyl sulfate-dimethylformamide $\qquad$ $(RO)_2SO_2$-$HCONMe_2$
Acetals from aldehydes $\qquad$ CHO → CH(OR)₂

113. $C_2H_5CHO + HC\langle^{OEt}_{+NMe_2}\ EtOSO_2O^- + HOEt \rightarrow C_2H_5CH(OEt)_2\ [+EtOSO_2OH, +HCONMe_2]$

Propionaldehyde added dropwise under anhydrous conditions to a vigorously stirred mixture of dimethylformamide-diethyl sulfate adduct and ethanol, and allowed to stand 4 hrs. → propionaldehyde diethyl acetal. Y: 76%. F. e. s. W. Kantlehner, H.-D. Gutbrod, and P. Gross, A. *1974*, 690.

α,α'-Dipyridyl disulfide s. under Ph₃P $\qquad$ ←

Carbon tetrachloride s. under (Me₂N)₃P $\qquad$ CCl_4

Cyanogen bromide/pyridine $\qquad$ $BrCN/C_5H_5N$
Carboxylic acid anhydrides $\qquad$ 2 RCOOH → (RCO)₂O
from carboxylic acids s. *28*, 143

Triethylsilane/trifluoroacetic acid $\qquad$ Et_3SiH/CF_3COOH
Ethers from oxo compds. $\qquad$ CO → CH·O·R
s. *24*, 126

Silicon tetrachloride $\qquad$ $SiCl_4$
Carboxylic acid esters from carboxylic acids $\qquad$ COOH → COOR
s. *12*, 238

Stannic chloride/dicyclohexylcarbodiimide/oxalic acid $\quad$ $SnCl_4/RN:C:NR/(COOH)_2$
Acetals $\qquad$ CO → C⟨O,O⟩
s. *29*, 187

Triphenylphosphine/diethyl azodicarboxylate $\qquad$ $Ph_3P/ROOC·N:N·COOR$
O-Acylation of hydroxy compds. $\qquad$ OH → OAc
with epimerization of steroids – Preferential O-acylation – Carbohydrate derivs. s. *29*, 188; glycerides, prevention of O-acyl group migration s. R. Aneja et al., Chem. Commun. *1974*, 963

Triphenylphosphine/α,α'-dipyridyl disulfide $\qquad$ ←
Active carboxylic acid esters $\qquad$ COOR
by redoxesterification – Optically active compds. – Redoxphosphorylation
s. *26*, 175

Tris(dimethylamino)phosphine/carbon tetrachloride $\qquad$ $(Me_2N)_3P/CCl_4$
Carboxylic acid anhydrides and amides $\qquad$ 2 RCOOH → (RCO)₂O
from carboxylic acids s. *28*, 144

Chlorotris(dimethylamino)phosphonium perchlorate $\qquad$ $\overset{+}{ClP}(NMe_2)_3\ ClO_4^-$
Carboxylic acid anhydrides
from carboxylic acids s. *28*, 144

Hydrogen peroxide/potassium hydroxide H_2O_2/KOH
Acoxysulfones from hydroxythioethers   OH → OAc
s. *26*, 93

Thionyl chloride $SOCl_2$
Sulfonic acid esters SO_2OH → SO_2OR
from sulfonic acids and alcohols s. *29*, 189

N-Hydroxybenzenesulfonamide $PhSO_2NHOH$
Acetals   CO → $C(OR)_2$
and their cleavage s. *26*, 176

Methanesulfonic acid $MeSO_3H$
Glycuronides   OH → OR
s. *28*, 145

p-Toluenesulfonic acid   TsOH
Ethers   OH → OR
from alcohols and phosphorous acid diesters s. *27*, 195

Protection of alcohols
as 9-phenyl-9-anthronyl ethers – Removal of the protective group by Wolff-Kishner reduction s. *26*, 177

Enolethers   CHCO → C:C·OR
from ketones and orthoformic acid esters s. *29*, 190

α,β- **and** β,γ-**Ethyleneketals from** α,β-**ethyleneketones**   ←

114.

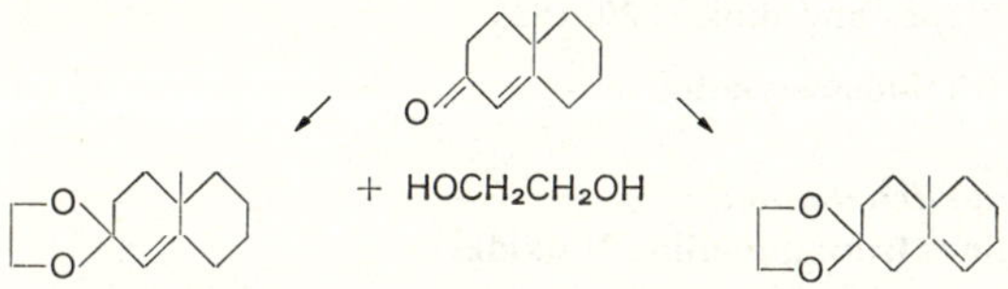

Startg. enone refluxed 1–3 days with ethylene glycol and benzene with azeotropic water entrainment in the presence of

fumaric acid → α,β-ethyleneketal. | p-toluenesulfonic acid → β,γ-ethylene-
Conversion: 90%. | ketal. Conversion: ca. 100%.
F. e. s. J. W. De Leeuw et al., R. *92*, 1047 (1973).

Acetoxy-2-ethylenes from 2-ethylenealcohols   C(OH)·C:C → C:C·C(OAc)
with allyl rearrangement – Bishomologization of ketones s. *29*, 191; allyl rearrangement in acetic acid-anhydride, steroids, cf. D. O. Olsen and J. H. Babler, J. Org. Chem. *40*, 255 (1975).

1,3,3-Trialkoxy compds.   C:C·CO → C(OR)·CH·C(OR)$_2$
from α,β-ethyleneketones s. *26*, 178

O-Alkylglycolic acid esters   COCHO → CH(OR)COOR
by intramolecular redoxidation

115.

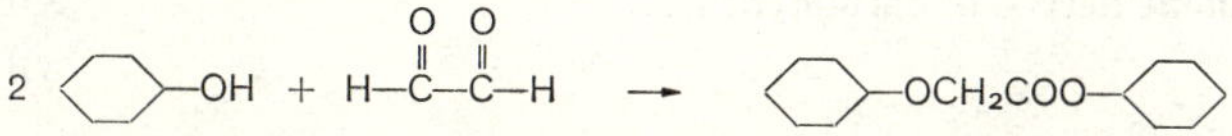

A mixture of cyclohexanol, 40%-glyoxal, and some p-toluenesulfonic acid refluxed at 121° with azeotropic removal of a water-cyclohexanol mixture whereby the temp. is maintained at 121° by reducing the pressure as distillation proceeds → cyclohexyl cyclohexoxyacetate. Y: 90.3%. F. e., also from the glyoxal diacetals, and limitations s. J. M. Kliegman and R. K. Barnes, J. Org. Chem. *39*, 1772 (1974).

Cyclic orthocarboxylic acid esters ←
  from lactones s. *26*, 179

4-Oxo-1,3-dioxanes ◯
  from β-hydroxycarboxylic acids s. *27*, 196

m-Benzenedisulfonic acid $m\text{-}C_6H_4(SO_3H)_2$
Partial protection of dialdehydes
  on polymer support s. *12*, 205 suppl. 29   CHO → CH(O-O)

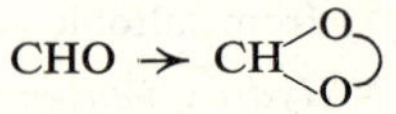

Sulfuric acid H_2SO_4
Lactolides from lactols
s. *28*, 146   C·OH → C·OR

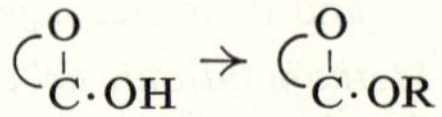

Carboxylic acid esters   COOH → COOR
  from carboxylic acids and *tert*-butyl ethers s. *26*, 168

Perchloric acid $HClO_4$
Carboxylic acid aryl esters   COOH → COOAr
  from carboxylic acids and phenols – Azeotropic water entrainment s. *26*, 180

4-Alkoxybenzopyrylium salts ◯
  4-Alkoxyflavylium salts s. *29*, 192

Hydrogen chloride HCl
Acetals from aldehydes   CHO → CH(OR)$_2$
s. *27*, 220

Ketals   (OH,OH) → (O,O)
  from enoxysilanes and diols s. *29*, 193

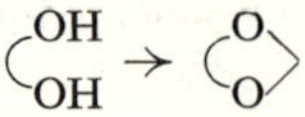

Esterification with S-deacetylation ←
s. *28*, 147

Hydrogen chloride/irradiation $HCl/∭$
6-Alkoxyquinolines from quinoline N-oxides ←

116.   MeOH + [2-cyano-4-methylquinoline 1-oxide] → [2-cyano-6-methoxy-4-methylquinoline]

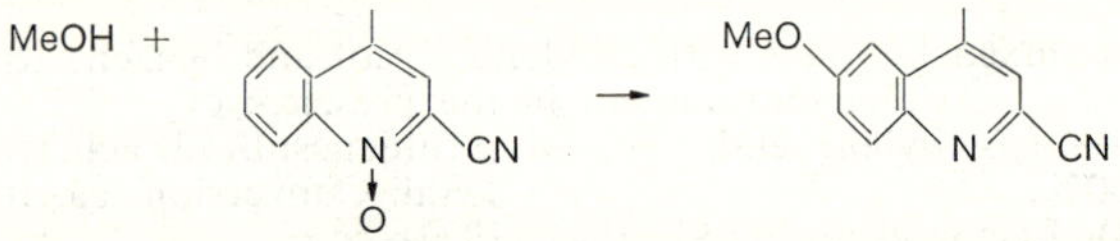

2-Cyano-4-methylquinoline 1-oxide irradiated 3 hrs. in methanol containing concd. HCl with the Pyrex-filtered light of a 450 w. high-pressure Hg-lamp → 2-cyano-6-methoxy-4-methylquinoline. Y: 45%. F. e., also 6-chloroquinolines, s. C. Kaneko et al., Chem. Lett. **1974**, 133.

Hydrogen chloride/cupric acetate $HCl/(CH_3COO)_2Cu$
Carboxylic acid esters from carboxylic acids   COOH → COOR
  at room temp. s. *10*, 171 suppl. 29

Hydrogen chloride/ethyl orthoformate $HCl/HC(OEt)_3$
O,O-Alkylidene derivs. of carbohydrates   (OH,OH) → (O,O)
s. *19*, 239

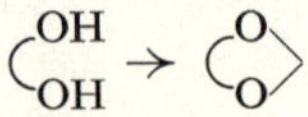

Ferric sulfate $Fe_2(SO_4)_3$
Carboxylic acid esters from carboxylic acids   COOH → COOR
  Continuous reactor s. *28*, 148

Nickel-silica $Ni\text{-}SiO_2$
Sym. ethers from alcohols   2 ROH → ROR
s. *26*, 181

Tetracarbonyldichlorodirhodium $[Rh(CO)_2Cl]_2$
α,β-**Ethylene-acetals from -aldehydes**   C : C·CHO → C : C·CH(OR)$_2$
s. *9*, 265 suppl. 29

Nitrogen ↑   OC ↟ N

Without additional reagents   *w.a.r.*
2-Step Vilsmeier aldehyde synthesis   H → CHO
s. *27*, 797

O-Acetylation with N-acetylsaccharin   OH → OAc
s. *29*, 394

Carboxylic acid esters   COOH → COOR
     from non-reactive carboxylic acids via quaternary ammonium carboxylates –
     Methyl esters – Ethers from quaternary ammonium hydroxides and acetates
     s. *26*, 182; benzyl esters s. K. Williams and B. Halpern, Synthesis *1974*, 727

Peroxides   OOH → OOR
     from hydroperoxides and diazo compds. s. *27*, 197

Eneurethans   C : C·CON$_3$ → C : C·NHCOOR
     from α,β-ethylenecarboxylic acid azides s. *29*, 237

Irradiation (s. a. under KOH)   ⚡
Phenols from diazonium salts   ArN$_2^+$ → ArOH

117.

     o-(N$_2^+$ BF$_4^-$)(SMe)C$_6$H$_4$ → o-(OH)(SMe)C$_6$H$_4$

2-Phase medium. An aq. soln. of o-(methylthio)benzenediazonium fluoroborate under a benzene layer irradiated with the quartz-filtered light of a 125 w. medium-pressure Hg-lamp whereby the product is driven into the benzene layer by a N$_2$-stream → o-(methylthio)phenol. Y: 91%. F. e. s. H. Böttcher and H. G. O. Becker, Z. Chem. *14*, 100 (1974).

Aldehydes from azides   CH$_2$N$_3$ → CHO
s. *28*, 149

Carboxylic acids from carboxylic acid amides   CON< → COOH
     Removal of photo-sensitive carboxyl-protective groups s. *29*, 194; review of photo-sensitive protective groups s. Israel J. Chem. *12*, 103 (1974)

Sodium hydride s. a. under NO$_2$ *NaH*

Sodium hydride/dimethyl sulfoxide-benzaldoxime   *NaH/Me$_2$SO-PhCH : NOH*
Phenols from ar. nitro compds.   ArNO$_2$ → ArOH

118.   NC–C$_6$H$_4$–NO$_2$ + HON=CHPh → [NC–C$_6$H$_4$–O–N=CHPh] → NC–C$_6$H$_4$–OH

     2 equivalents benzaldoxime in dimethyl sulfoxide added slowly to a soln. of 2 equivalents Na-methylsulfinylmethide prepared by heating NaH in dimethyl sulfoxide at 70° until a clear soln. results, stirred 0.5 hr. at 70°, cooled, treated with p-nitrobenzonitrile in dimethyl sulfoxide, and stirred 46 hrs. at 25° → p-cyanophenol. Y: 94%. F. e. s. R. D. Knudsen and H. R. Snyder, J. Org. Chem. *39*, 3343 (1974).

Sodium hydroxide   *NaOH*
1,3-Dialdehydes from β-amino-α,β-ethylenealdehydes   C(CHO)$_2$
s. *13*, 786

Chromenes O
from o-hydroxyaldehydes via coumarins and subsequent enaminelactone rearrangement s. 27, 198

Dicarboxylic acids from ditosylates C
via cyclic β-amino-α,β-ethylenenitriles s. 29, 195

2-Cyclohexenones from 1,4-dihydropyridines ←
s. 27, 816

Potassium hydroxide *KOH*
Oxo compds. from acylamines CHNHAc → CO
with formation of hydrazo from azo compds. s. 27, 199

β-Aminocarboxylic acids from Δ^2-pyrazolines C
s. 27, 200

**α-Ketocarboxylic acids
from α-ketocarboxylic acid amides** COCONH$_2$ → COCOOH
s. 27, 207

Potassium hydroxide/potassium iodide/irradiation *KOH/KI/⚡*
Aldehydes from nitriles CN → CHO
s. 26, 183

Sodium/alcohol (s. a. under O_3 and H_2SO_4) *NaOR*
Carboxylic acid esters CHCH$_2$COOR
from 1-ethynyl-N-nitroso-urethans and -ureas s. 28, 150

Amide acetals from α,α-diaminonitriles ($>$N)$_2$C·CN → $>$N·C(OR)$_2$
s. 28, 187

Sodium ethoxide *NaOEt*
O-Acetylation OH → OAc
with N-acetylsaccharin s. 29, 394

Sodium hydrogen carbonate *NaHCO$_3$*
Vinylogous acyl compds. OH → OAc
Sym. 3-keto-1,2-en-1-olethers from vinylogous carboxamidium salts s. 26, 184

Ketones from cyclopropylamines C

119.   Ph\/\NH$_2$,HCl → Ph\/\C(=O)
       Ph/   Ph/

Satd. NaHCO$_3$-soln. added to an aq. soln. of 1-methyl-2,2-diphenylcyclopropylamine hydrochloride, and stirred 24 hrs. at room temp. → 4,4-diphenyl-2-butanone. Y: 76% as the 2,4-dinitrophenylhydrazone. H. M. Walborsky and P. E. Ronman, J. Org. Chem. *38*, 4213 (1973).

Sodium acetate *CH$_3$COONa*
2-Cyclohexenenone ring ←
from cyclic 1,3-dienamines s. 27, 950

Sodium nitrate (s. a. under n-PrONO) *NaNO$_2$*
Replacement of amino groups by oxygen ←
s. 28, 152 suppl. 29

Potassium iodide s. under KOH *KI*

Ammonia *NH$_3$*
Carboxylic acids and carboxylic acid amides from nitriles CN → COOH
Continuous process s. 29, 196

Triethylamine	*Et₃N*
Carbamic carboxylic anhydrides ←	
from N-carbamylpyridinium salts s. 28, 151	
Pyridine s. under (CH₃COO)₄Pb	*C₅H₅N*
Copper	*Cu*
α-**Acetoxycarboxylic acids**	C(:N)COOCMe₃ → CH(OAc)COOH
from α-diazocarboxylic acid *tert*-butyl esters s. 27, 201	
Silver oxide	*Ag₂O*
Wolff rearrangement	COCHN₂ → CH₂COOR

Dicarboxylic acid esters from carbalkoxy-α-diazoketones s. 30, 456; review s. H. Meier and K.-P. Zeller, Ang. Ch. 87, 52 (1975)

Cuprous iodide	*CuI*
α-**Ketocarboxylic esters from** α-**diazoketones**	COCHN₂ → COCOOR
s. 27, 204	
Barium nitrite	*Ba(NO₂)₂*
Replacement of amino groups by oxygen ←	
s. 28, 152	
Mercuric oxide	*HgO*
α-**Dioxo compds. from** α-**aminoketones**	COCH(N<) → COCO
s. 28, 153	
Mercuric acetate	*(CH₃COO)₂Hg*
s. 28, 153	
Sodium tetrahydridoborate/sodium hydroxide	*Na[BH₄]/NaOH*
Alcohols from oximes	CO·C:NOH → CH(OH)CH(OH)
Glycols from α-isonitrosoketones s. 26, 185	
Alcohols from 1,2-nitrosohalides	CClC(NO) → CHC(OH)
Preferential *anti*-Markownikoff hydration s. 27, 202	
Aluminum isopropoxide	*Al(OR)₃*
Oxo compds. from oximes	C:NOH → CO
s. 28, 154	
Thallic nitrate	*Tl(NO₃)₃*
α,β-**Acetylenecarboxylic acid esters**	C

from 5-pyrazolones – α,β-Acetylene- from β-keto-carboxylic acid esters – α-Allene- from α-alkyl-β-keto-carboxylic acid esters s. 27, 203

Boron fluoride	*BF₃*
2-Oxazolidone from 3-acyl-2-iminooxazolidine ring ←	
Selective N-deacylation	
Heterocyclic carbohydrate derivs.	

120.

A mixture of 2-imino-N',O³',⁵'-tribenzoyl-β-D-arabinofuro[1',2':4,5]oxazoline, BF₃-etherate, and methanol refluxed 2 hrs. under anhydrous conditions → 3',5'-di-O-benzoyl-β-D-arabinofuro[1',2':4,5]oxazolin-2-one. Y: 99%. A. Holý, Coll. *38*, 3912 (1973).

Formaldehyde s. under CH_3COOH CH_2O

Benzaldoxime s. under NaH/Me_2SO $PhCH:NOH$

N,N'-Carbonyldiimidazole   ←

Cyclic carbonic acid esters from diols
s. 29, 197

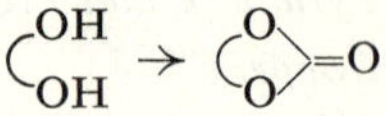

$$\begin{matrix}OH\\OH\end{matrix} \rightarrow \begin{matrix}O\\O\end{matrix}\!\!=\!O$$

Triethyloxonium fluoroborate $R_3O^+\,BF_4^-$

Ring closures of ethylenenitriles   ○

121.

α,β-**Ethyleneketones.** 2-(β-Cyanoethyl)methylenecycloheptane and triethyloxonium fluoroborate heated 2 hrs. at 85°, aq. *HCl* added, and boiled 1 hr. → bicyclo-[5.4.0]undec-7-en-9-one. Y: 29%. F. e. s. F. Johnson et al., J. Org. Chem. *39*, 1434 (1974).

Dowex 50W-X2   ←

α-**Alkoxyketones and α-ketocarboxylic acid esters** $COC(:N_2) \rightarrow COCH(OR)$
from α-diazoketones s. 27, 204

Acetic acid CH_3COOH

2-Alkoxy- from 2-amino-1,3-dithioles
s. 28, 155

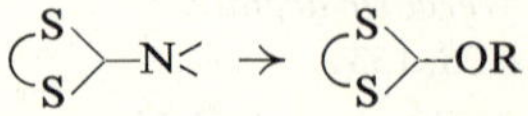

$\begin{matrix}S\\S\end{matrix}\!\!-\!\!N\!\!< \;\rightarrow\; \begin{matrix}S\\S\end{matrix}\!\!-\!\!OR$

5-Halogenopyrrole-2-carboxaldehydes   ←
from 2-aminomethylene-5-halogeno-2H-pyrroles s. 28, 754

Ring closure by intramolecular nucleophilic substitution   ○
Ferrocenes s. 26, 186

Stereospecific 1-aminocyclobutene ring opening   C
s. 28, 156

Acetic acid/formaldehyde-hydrogen chloride $CH_3COOH/CH_2O\text{-}HCl$

Ketones from prim. amines $CHNH_2 \rightarrow CO$
s. 26, 187

Oxalic acid $(COOH)_2$

Ketones from azomethines $C:N \rightarrow CO$
s. 27, 890

Carbon oxide sulfide COS

Alcohols from amines $C\cdot N< \;\rightarrow\; C\cdot OH$
s. 29, 531

Silica SiO_2

Oxo from *aci*-nitro compds. $C:NO_2^- \rightarrow CO$
s. 30, 572

Titanium trichloride $TiCl_3$

Oxo compds. from oximes $C:NOH \rightarrow CO$
also from other oxo compd. N-derivs. s. 26, 188; reductive cleavage of 2,4-dinitrophenylhydrazones with $TiCl_3$ s. J. E. Mc Murry and M. Silvestri, J. Org. Chem. *40*, 1502 (1975).

Oxo from aliphatic nitro compds. $CHNO_2 \rightarrow CO$
s. 28, 157

Stannous chloride	SnCl$_2$
Partial O-alkylation	OH → OR
s. 26, 189	
Lead tetraacetate	(CH$_3$COO)$_4$Pb
Reaction of hydrazones with lead tetraacetate	C : N·NH$_2$ → CH(OAc)
Acoxy compds. from hydrazones s. 26, 190	
Lead tetraacetate/pyridine	(CH$_3$COO)$_4$Pb/C$_5$H$_5$N
Protection of carboxyl groups	CON(R)NHR → COOH
as N,N'-diisopropylhydrazides – Oxidative removal of the protective group s. 28, 158	
Nitrogen dioxide/sodium hydride	NO$_2$/NaH
Ketones from acylamines	CHNHAc → CO
s. 29, 198	
n-Propyl nitrite/sodium nitrite	n-PrONO/NaNO$_2$
Ketones from aliphatic nitro compds.	CHNO$_2$ → CO
s. 28, 157	
Isoamyl nitrite	i-C$_5$H$_{11}$ONO
Replacement of imino groups by oxygen	C : NH → CO
Thiolcarbamic acid esters from isothiouronium salts s. 28, 159	
Phosphate buffer	←
Aldehydes-1-d from 1,1-di(pyridinium salts)	←
s. 29, 211	
Oxygen s. under (Ph$_3$P)$_4$Pd	O$_2$
Ozone/dimethyl sulfide-sodium/alcohol	O$_3$/Me$_2$S-NaOR
Oxo from aliphatic nitro compds.	CHNO$_2$ → CO
s. 29, 199	
p-Toluenesulfonic acid	TsOH
Ketones from azomethines	C : N → CO
s. 29, 647	
p-Toluenesulfonic acid hydrate	TsOH·H$_2$O
Cleavage of 2,4-dinitrophenylhydrazones	C : NNHAr → CO
via hydration s. 27, 205	
Sulfuric acid	H$_2$SO$_4$
Replacement of amino by alkoxy groups	NH$_2$ → OR
s. 26, 191	
Ketones from eneurethans	C : C·N·COOR → CHCO
s. 29, 237	
α-Aminocarboxylic acids from hydantoins	C
s. 30, 568	
Sulfuric acid-sodium/alcohol	H$_2$SO$_4$/NaOR
Acetals from aliphatic nitro compds.	CH(NO$_2$) → C(OR)$_2$
Modified Nef reaction	

122.

A soln. of the startg. m. in 0.5 N methanolic Na-methoxide added dropwise at −35° to a soln. of sulfuric acid in methanol → product. Y: 78%. F. e. s. R. M. Jacobson, Tetrah. Let. *1974*, 3215.

Selenic acid H_2SeO_4
Sym. ethers from diazo compds. $2 >CN_2 \rightarrow >CHOCH<$

123. $2\ Ph_2CN_2 + H_2O \longrightarrow Ph_2CH-O-CHPh_2$

80%-Selenic acid added at –5 to –3° to diphenyldiazomethane in CCl_4 until the mixture is decolorized $\rightarrow$ dibenzhydryl ether. Y: 98%. A. L. Fridman, Y. S. Andreichikov, and V. L. Gein, Zh. Org. Khim. *9*, 1082 (1973); C. A. *79*, 52917.

Chromous acetate $(CH_3COO)_2Cr$
Ketones from oxime acetates $C:N \cdot OAc \rightarrow CO$
under mild conditions s. *26*, 192

Perchloric acid $HClO_4$
Alcohols from diazo compds. $>CN_2 \rightarrow >CHOH$

124.

0.2 N perchloric acid added with swirling to a soln. of β,β,β-trichloroethyl 6-diazopenicillanate in acetone, and stored overnight at 5° $\rightarrow$ β,β,β-trichloroethyl 6β-hydroxypenicillanate. Y: 60%. F. e. s. J. C. Sheehan et al., J. Org. Chem. *39*, 1444 (1974).

α-Hydroxy- from α-diazo-ketones $COCHN_2 \rightarrow COCH_2OH$
s. *9*, 296 suppl. 29

Hydrochlorides ←
Thiolic acid esters from thioiminoesters $C(:NH)SR \rightarrow COSR$
s. *27*, 624

Hydriodides ←
N-Acylthiolcarbamic acid esters $CON \cdot C(:NH)SR \rightarrow CON \cdot COSR$
from N-acylisothiouronium salts s. *28*, 160

Hydrogen chloride HCl
Ketones from azomethines $C:N \rightarrow CO$
s. *29*, 504

Tropolone ring ←
from cyclic α-isonitrosoketones s. *27*, 206

α-Nitroketones $CH(NO)C \cdot NO_2 \rightarrow COC \cdot NO_2$
from 1,2-nitronitroso compds. s. *29*, 318

α-Cyanoketones $C(N<):C \cdot CN \rightarrow CO \cdot CH \cdot CN$
from β-amino-α,β-ethylenenitriles s. *29*, 160

α-Ketocarboxylic acids $COCONHCMe_3 \rightarrow COCOOH$
from α-ketocarboxylic acid *tert*-butylamides
s. *27*, 326

α-Ketocarboxylic acids $C:C(NAc) \cdot COOR \rightarrow CHCOCOOH$
from α-acylamino- or α-alkoxy-α,β-ethylenecarboxylic acid esters
s. *29*, 200

α-Aminocarboxylic acids $CON< \rightarrow COOH$
from α-aminocarboxylic acid amides
s. *30*, 290

α-**Amino- and α-acylamino-carboxylic acids** C(NAc)CON< → C(NH)COOH
from α-acylaminocarboxylic acid amides s. 28, 288   C(NAc)COOH

α-**Keto- from α-hydroxylamino-** CH(N·OH)CON< → COCON<
via α-(O-carbamylhydroxylamino)-carboxylic acid amides

125. PhCHCONH—⟨⟩ → PhCHCONH—⟨⟩ → PhCOCONH—⟨⟩
      | |
      MeN—OH   MeN—O—CONHPh
      + O=C=N—Ph

α-(N-Methylhydroxylamino)phenylacetic cyclohexylamide refluxed 10–15 min. with an equimolar amount of phenyl isocyanate in a small amount of methylene chloride → 2-[N-methyl-O-(phenylcarbamoyl)hydroxylamino]phenylacetic cyclohexylamide (Y: 95%) refluxed 15–20 min. in 2 N HCl-methanol with swirling → phenylglyoxylic cyclohexylamide (Y: 94%). F. e. s. D. Moderhack and G. Zinner, Chem. Ztg. 98, 110 (1974).

Chroman-2-carboxylic acids   O
from acoxycyanohydrins s. 30, 183

α-**Aminocarboxylic acids**   C
from 5-alkylidene-4-imidazolones s. 26, 193

γ-**Lactones**   ☉
from cyanocyclopropanes s. 29, 87

Oxidations with *potassium ferrate* K_2FeO_4
Aldehydes from amines CH_2NH_2 → CHO
α,β-Ethylenealdehydes from 2-ethylenealcohols s. 26, 194

Palladium-carbon   Pd-C
Oxo compds. from aminocyclopropane ring   C
s. 28, 161

Palladium-carbon/hydrogen chloride   Pd-C/HCl
Alcohols from nitriles   CN → CH_2OH
s. 26, 195

Tetrakis(triphenylphosphine)palladium/oxygen $(Ph_3P)_4Pd$
Ketones from oximes   C:NOH → CO
under mild conditions s. 29, 201

Via intermediates   v.i.
1,3-Dialdehydes from β-**amino-**α,β-**ethylenealdehydes** $C(CHO)_2$
s. 13, 786 suppl. 29

α-**Ketocarboxylic acids**   COCN → COCOOH
from acylcyanides via α-ketocarboxylic acid amides s. 27, 207

Halogen ↑   OC ⇄ Hal

Without additional reagents   w.a.r.
Aldehydes from 1,1-dichlorides $CHCl_2$ → CHO
s. 29, 201a

Hydroxycyclopropenones

126.

Phenyltrichlorocyclopropene dissolved at 0° in acetone, ice added, after 4 hrs. at 0° allowed to warm for 1 hr., then most of the acetone evaporated under reduced pressure → phenylhydroxycyclopropenone. Y: 66%. F. e. s. J. S. Chickos, E. Patton, and R. West, J. Org. Chem. *39*, 1647 (1974).

α-Acoxy from α-hydroxy-carboxylic acids   OH → OAc

127.   n—$C_{17}H_{35}$COCl + HOCHCOOH   →   n—$C_{17}H_{35}$COOCHCOOH
                               | |
                               CH_3 CH_3

Anhydrous lactic acid heated with 1 equivalent stearoyl chloride at 120° under N_2 → stearoyllactic acid. Y: 93–97%. A. E. Baskaeva et al., Maslo-Zir. Prom. *1974* (2), 40 (Russ); C. A. *80*, 107951.

Acoxycarboxylic acid chlorides
from hydroxycarboxylic acids s. *29*, 514

Carboxylic acid anhydrides   2 RCOCl → $(RCO)_2O$
from 2 carboxylic acid chloride molecules s. *28*, 162

1,4-Dioxa-2,5-diazines
from hydroximinochlorides and amidoximes s. *27*, 208

Irradiation
Carboxylic acid esters from α-chloroketones
with stereospecific skeletal rearrangement s. *27*, 209

Sodium hydride   *NaH*
1-(Alkoxy)enolethers from ketones   CHCO → C : C · O · C · OR
s. *28*, 803

**β-Alkoxy-α,β-ethylenecarboxylic
acid esters**   CO · CH · COOR → C(OR') : C · COOR
from β-ketocarboxylic acid esters s. *26*, 54

Methylenedioxy compds. from diols

128.

9-Hydroxymethylprotoberberine heated 15 min. at 80° under N_2 with excess dry methylene chloride and NaH in dry dimethylformamide → product. Y: ca. 100%. T. Kametani et al., Heterocycles *3*, 143 (1975).

Sodium hydride/sodium hydroxide   *NaH/NaOH*
Carbamic acid anhydrides from carbamyl chlorides   2 N · COCl → N · $CO)_2O$
    Carbanilic anhydrides s. *26*, 196

Sodium hydride/dimethyl sulfoxide   *NaH/Me_2SO*
Total O- and N-methylation   OH → OMe
s. *29*, 410

Lithium oxide — Li_2O
Dicarboxylic acid anhydrides
from dicarboxylic acid chlorides s. 26, 411

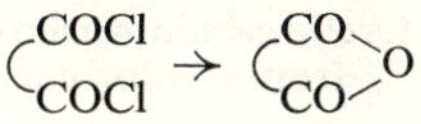

$\begin{array}{c}COCl\\COCl\end{array} \rightarrow \begin{array}{c}CO\\CO\end{array}\!\!\!>\!\!O$

Sodium hydroxide — $NaOH$
Lactams from o-α-halogeno-N-heterocyclics ☉
Ring expansion s. 17, 263

Sodium hydroxide/hexamethylphosphoramide — $NaOH/(Me_2N)_3PO$
Carboxylic acid esters from halides — $COOH \rightarrow COOR$
s. 28, 165

Potassium hydroxide — KOH
Diaminocyclopropenones ←
s. 29, 574

Potassium hydroxide/calcium carbide — KOH/CaC_2
Vinyl ethers from bromides — $Br \rightarrow OCH:CH_2$
s. 28, 163

Potassium hydroxide/hexamethylphosphoramide — $KOH/(Me_2N)_3PO$
Carboxylic acid esters from halides — $COOH \rightarrow COOR$
s. 28, 165

Sodium/alcohol — $NaOR$
Phenolethers from bromonium salts — $Ar_2Br^+ \rightarrow ArOR$
s. 29, 202

1-Alkoxy- from 1-halogen-1-alkoximes — $C(:NOR)Hal \rightarrow C(:NOR)OR'$
s. 26, 197

α-Alkoxyglycidic acid esters — $CHalCO \rightarrow \triangle\!\!-OR$
from β-halogeno-α-ketocarboxylic acid esters

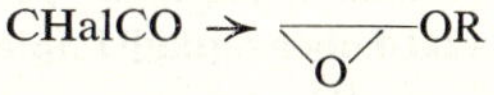

129. $C_2H_5\text{—}CH\text{—}CO\text{—}COOEt \xrightarrow{+\ NaOEt} C_2H_5\underset{O}{\triangle}\overset{OEt}{\underset{COOEt}{}}$
      $\qquad\quad |$
      $\qquad\ \ Cl$

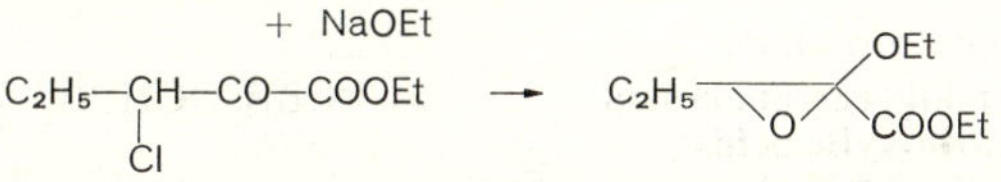

Ethyl β-chloro-α-ketopentanoate allowed to react at room temp. with Na-ethoxide in ethanol → product. Y: 58%. F. e. s. B. Moraud and J.-C. Combret, C. r. *277* (C), 523 (1973).

4,5-Dihydro-3H-1,2,5-benzothia(IV)diazepines ☉
from 3,4-dihydro-1,2,4-benzothia(IV)diazines

130.

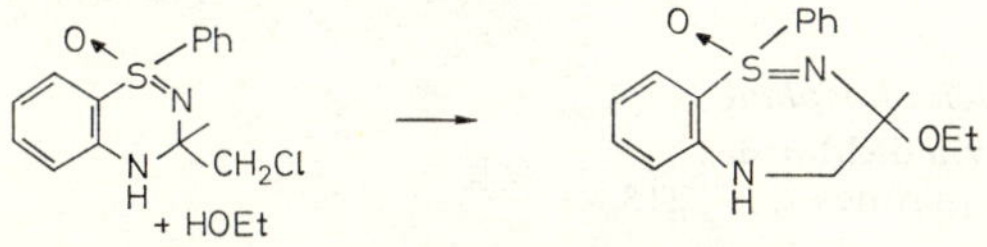

Startg. m. stirred 2 hrs. at 25° in ethanolic 1 N Na-ethoxide → 3-ethoxy-3-methyl-1-phenyl-4,5-dihydro-3H-1,2,5-benzothia(IV)diazepine 1-oxide. Y: 97%. F. e. s. E. Cohnen and J. Mahnke, B. *106*, 3368 (1973).

Sodium/alcohol-dimethyl sulfoxide — $NaOR/Me_2SO$
Selective O-alkylation — $OH \rightarrow OR$
Prevention of N-quaternization – 3-Alkoxypyridines s. 26, 198

Sodium ethoxide — $NaOR$
Nucleophilic substitution at acetylenic C-atoms — $C\equiv C\cdot Cl \rightarrow C\equiv C\cdot OR$
Alkoxyacetylenes from acetylenehalides s. 27, 210

Potassium/tert-butanol $K/Me_3C \cdot OH$

Cyclic α-iminocarboxylic acids Ö:
  from α-chlorolactams s. 27, 211

3-Chloro-2-*tert*-butoxy-2,5,6,7-tetrahydrooxepins C
  from 7,7-dichloro-2-oxabicyclo[4.1.0]heptanes s. 27, 972

Replacement of chlorine by alkoxyl ←
  with position shift s. 27, 212

Methyllithium s. under CuI MeLi

Sodium hydrogen carbonate $NaHCO_3$

α,β-Ethylene-β-halogenocarboxylic acid amides $COCH_2 \to C(Hal):C \cdot CON<$
  from ketones via immonium salts s. 28, 717

N,O-Diacylquinone iminoximes ←
  from quinone iminoximes s. 27, 461

N-Acoxyurethans $2\ ROCOCl \to ROCO \cdot N \cdot OCOR$
  s. 26, 412

Benzoxazole from 2H-1,4-benzoxazine ring Ö:
  s. 26, 199

Potassium carbonate K_2CO_3

O-Alkylation of phenols $ArOH \to ArOR$
  s. 26, 200

α-Hydroxyaldehydes $C(OH)CHCl_2 \to C(OH)CHO$
  from 2-hydroxy-1,1-dichlorides s. 28, 164

Potassium carbonate/potassium iodide K_2CO_3/KI

12H-Dibenzo[d,g][1,3]dioxocin-6-carboxylic acids O
  s. 29, 203

Potassium hydrogen carbonate $KHCO_3$

Geospecific Favorskii rearrangement $CBrCOCH_2Br \to C:CHCOOH$
***cis*-α,β-Ethylenecarboxylic acids**
  from methyl α,α'-dibromoketones

131.  $CH_3CHBrCOCH_2Br \longrightarrow$ $\underset{CH_3}{H}\!\!>\!\!C\!=\!C\!\!<\!\!\underset{COOH}{H}$

1,3-Dibromo-2-butanone added during 5 min. to aq. $KHCO_3$, and stirred 2–3 hrs. → crude isocrotonic acid. Y: 69–77%. F. e. s. C. Rappe, Org. Synth. 53, 123 (1973).

Sodium dihydrogen phosphate NaH_2PO_4

O-Heterocyclics via dichlorides O
  Tetrahydro-4-pyrones s. 27, 213

Sodium fluoride NaF

Reactions with acetylenedicarboxylic acid fluoride ←
Acetylenedicarboxylic acid esters from alcohols

132. 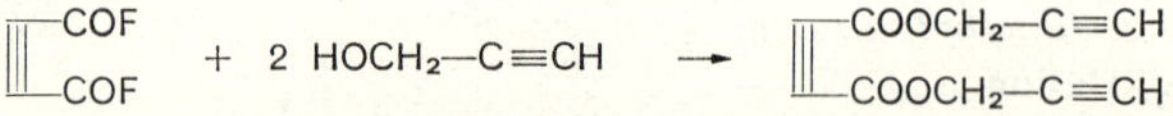 $\begin{matrix}-COF\\-COF\end{matrix}$ $+\ 2\ HOCH_2-C\equiv CH \longrightarrow$ $\begin{matrix}-COOCH_2-C\equiv CH\\-COOCH_2-C\equiv CH\end{matrix}$

A mixture of propargyl alcohol and NaF in dry benzene treated dropwise at 15° with a soln. of acetylenedicarbonyl fluoride in the same solvent, and stirred 18 hrs. at 25° → dipropargyl acetylenedicarboxylate. Y: 80%. F. e. and reactions s. F. E. Herkes and H. E. Simmons, J. Org. Chem. 40, 420, 423 (1975).

Potassium iodide(s. a. under K_2CO_3)   KI
Potassium iodide/ozone KI/O_3
α-**Ketocarboxylic acid esters**   C ≡ CBr → COCOOR
from acetylenebromides

133.
$$C_6H_{13}-C\equiv C-Br + HOMe \rightarrow \left[C_6H_{13}-\underset{OMe}{\overset{OOH}{C}}-COOMe \right] \xrightarrow{KI} C_6H_{13}COCOOMe$$

An O_3–O_2-stream passed at –30° through a soln. of 1-bromo-1-octyne in dry methanol until 90–95% of the startg. m. has reacted, dry N_2 passed through to remove excess O_3, KI added portionwise at ca. –30° with stirring, allowed to stand 0.5 hr. at room temp., then aq. 0.1 N Na-thiosulfate added to neutralize the liberated iodine → product. Y: 50%. F. e. s. S. Cacchi, L. Caglioti, and P. Zappelli, J. Org. Chem. *38*, 3653 (1973).

Sodium salt Na^+
Iminocarbonic acid esters $(ArO)_2C:NH$
    Sym. iminocarbonic acid esters – Mixed iminocarbonic acid esters from cyanic acid esters s. *27*, 214

Sodium salt/dimethyl sulfoxide Na^+/Me_2SO
Carboxylic acid esters from halides   COOH → COOR
s. *28*, 165

Ammonium hydroxide NH_4OH
α-**Aminoketones**   ⊙
    from α-halogenazomethinium salts – Rearrangement s. *28*, 116

Triethylamine Et_3N
Partial benzoylation of phenol groups   ArOH → ArOBz
s. *27*, 215

Diazoacetylation of alcohols   OH → $OCOCHN_2$
s. *26*, 201

2-Amino-1,3-dioxole ring   ◯
    from o-diols and chloroformiminium chlorides s. *29*, 204

Pyridine C_5H_5N
O,O-Benzylidene derivs.
    of carbohydrates s. *28*, 166 $\begin{matrix}C\cdot OH\\C\cdot OH\end{matrix} \rightarrow \begin{matrix}C\cdot O\\C\cdot O\end{matrix}CHPh$

α-**Benzoxy- from α-hydroxy-carboxylic acid esters**   C(OH)COOR → C(OBz)COOR
s. *27*, 92

Chlorocarboxylic acid esters   COCl → COOR
    from chlorocarboxylic acid chlorides s. *29*, 516

Carbalkoxydicarboxylic acid anhydrides
s. *27*, 216

1,2-Acoxymesylates
    from carboxylic acid chlorides s. *27*, 81

Acyl peroxides   2 COCl → C(O)·OO·(O)C
    from carboxylic acid chlorides – Acetyleneacyl peroxides – Sym. and mixed acyl peroxides s. *26*, 202

2,6-Lutidine   ←
Modified Königs-Knorr reaction   Cl → OR
    Oligosaccharides s. *23*, 242 suppl. *29*

Copper s. under Zn   Cu

Cuprous oxide s. under RN : C Cu_2O

Silver oxide Ag_2O
Carboxylic acid esters from carboxylic acids   COOH → COOR
s. 26, 421

Silver γ-hydroxyvalerate $CH_3CH(OH)CH_2CH_2COOAg$
Acylglycosides from acylhalogenosugars   OH → OR
s. 26, 203

Silver trifluoroacetate CF_3COOAg
4-Alkoxy-2,5-cyclohexadienone imines   ←
from anilines s. 26, 204

Silver salicylate $o\text{-}HOC_6H_4COOAg$
Acylglycosides from acylhalogenosugars   OH → OR
s. 26, 203 suppl. 28

Cyclic orthocarboxylic acid esters of carbohydrates   ○
s. 27, 217

Silver nitrate $AgNO_3$
Ring expansion of isocyclics   ⊙
by 1,1-dihalogenocyclopropane ring opening s. 26, 205

Cuprous iodide/methyllithium   CuI/MeLi
Reactions with cuprous alkoxides
Phenolethers from aryl halides   ArHal → ArOR

134.
```
              MeLi + HOBu—n
                    ↓
         ⎡  LiOBu—n  ⎤
         ⎢     ↓     ⎥  + CuCl
  PhI +  ⎣  CuOBu—n  ⎦            →    PhOBu—n
```

A soln. of Li-*n*-butoxide prepared from *n*-butanol and methyllithium in dimethoxyethane added to 0.1–1 mole cuprous chloride under dry N_2, stirred 0.5 hr., *pyridine* added, and the resulting mixture containing cuprous *n*-butoxide refluxed 12 hrs. with iodobenzene → *n*-butyl phenyl ether. Y: 70–98%. – This is the best available procedure for the conversion of aryl halides into alkyl aryl ethers. F. e. and reactions s. G. M. Whitesides, J. S. Sadowski, and J. Lilburn, Am. Soc. 96, 2829 (1974).

Silver salt Ag^+
O-Substitution of maleimides   ←
s. 29, 419

Selective O-alkylation of uracils   OR
s. 26, 206

Calcium oxide/dimethyl sulfoxide CaO/Me_2SO
Carboxylic acid esters from halides   COOH → COOR
s. 28, 165

Calcium carbide s. under KOH CaC_2

Magnesium-diborane/hydrogen peroxide $Mg\text{-}B_2H_6/H_2O_2$
Replacement of halogen by hydroxyl   Hal → OH
s. 27, 232

Zinc,copper	Zn,Cu
4-Alkylidene-2-amino-1,3-dioxolanes	O
from α,α'-dibromoketones and carboxylic acid amides s. 28, 167	
Cadmium carbonate	$CdCO_3$
Aryl glycosides	Br → OAr
s. 27, 218	
Mercuric carboxylates	$(RCOO)_2Hg$
Reaction of alkyl halides with mercuric carboxylates	Hal → OCOR
s. 30, 135	
Mercurous nitrate	$HgNO_3$
Reaction of alkyl halides with mercury salts	Br → ONO_2

135. $Cl[CH_2]_4Br + HgNO_3 \longrightarrow Cl[CH_2]_4ONO_2$

High yields of **nitrates, carboxylates, alcohols, and ethers** can be obtained **from alkyl bromides** by reaction with mercury salts. – Preferential replacement: 1-Bromo-4-chlorobutane refluxed 3 hrs. at 85° with mercurous nitrate in dry 1,2-dimethoxyethane → 4-chlorobutyl nitrate. Y: 98%. F. e. s. A. McKillop and M. E. Ford, Tetrahedron 30, 2467 (1974); with mercuric carboxylates, catalysis by *triacyl borates*, comparison with Ag- and Na-carboxylates, s. R. C. Larock, J. Org. Chem. 39, 3721 (1974).

Diborane s. under Mg	B_2H_6
Lithium tetrahydridoaluminate	$Li[AlH_4]$
2-Aminoalcohols from 2-chloracylamines $CCl·C·N·COR$ → $C(OR)·C·N·CH_2R$	
s. 26, 60	
Silver fluoroborate s. under Me_2SO	$AgBF_4$
Tritylpyridinium fluoroborate	$Ph_3C·\overset{+}{N}C_5H_5\ BF_4^-$
Preferential O-tritylation	OH → OTri
s. 8, 275 suppl. 29	
Aluminum halide	$AlHal_3$
Skeletal rearrangement	←
of N-halogenoadamantan-1-amines s. 28, 168	
Triethylammonium acetate	$Et_3NH^+\ CH_3COO^-$
Partial replacement of chlorine by acetoxy groups	←
with position shift s. 28, 169	
Tetramethylammonium salts	Me_4N^+
Mixed phosphorous acid diesters	P·OH → P·OR
Phosphorous acid monoesters s. 29, 205	
Phosphoric acid monoesters	Hal → OPO_3H_2
from halides – Protection of phosphoryl hydroxyl groups as *tert*-butyl esters s. 28, 170; mixed phosphoric acid diesters from monoesters and halides cf. R. A. Bauman, Synthesis 1974, 870	
Imidazole	←
Preferential O-acylation	OH → OAc
of carbohydrates s. 26, 207; s. a. J.-C. Jacquinet, J.-M. Petit, and P. Sinay, Carbohyd. Res. 38, 305 (1974)	
Isonitriles/cuprous oxide	$RN:C/Cu_2O$
Carboxylic acid esters	COOH → COOR
from carboxylic acids and halides s. 27, 219	
Triethyloxonium fluoroborate s. under Et_3SiH	$Et_3O^+\ BF_4^-$

Formic acid   *HCOOH*
α,β-**Ethylenecarboxylic from carboxylic acids**   C : C·COOH
s. 27, 918

Trifluoroacetic acid CF_3COOH
Selective acylation of phenol groups   ArOH → ArOAc
s. 29, 206; prevention of aporphine ring opening (cf. Synth. Meth. 28, 336)
s. Synthesis 1975, 249

Dimethyl sulfoxide (s. a. under Na^+ and CaO) Me_2SO
Replacement of halogen by hydroxyl   Hal → OH
α-Oxo-N-heterocyclics s. 29, 207

Carboxylic acids from hydroximinohalides   C(Hal) : NOH → COOH
s. 28, 171

α-**Ketocarboxylic acids from** α,α-**dibromoketones** $COCHBr_2$ → COCOOH
s. 29, 208

Lactams from o-iodocyclimmonium salts   ←

136.

A soln. of the startg. m. in dimethyl sulfoxide stored 2 weeks at room temp. → product. Y: ca. 100%. R. E. Lyle et al., J. Org. Chem. 38, 3268 (1973).

Dimethyl sulfoxide/silver fluoroborate $Me_2SO/AgBF_4$
Oxo compds. from halides   CHHal → CO
     Formoxy-aldehydes from -bromides s. 30, 358

Dimethyl sulfoxide/epichlorhydrin Me_2SO/epichlorhydrin
α,β-**Ethylene-**α-**hydroxyketones**   CH·CBr·CO → C : C(OH)·CO
     from α-halogenoketones s. 26, 208

Triethylsilane/triethyloxonium fluoroborate $Et_3SiH/Et_3O^+ BF_4^-$
Aldehydes   CN → CHO
     from nitriles via nitrilium salts s. 29, 209

Titanium tetrachloride $TiCl_4$
Ketones from α,β-**ethylenechlorides**   CCl : C → COCH
s. 29, 210

Nitric acid HNO_3
Aldehyde from chloromethyl groups CH_2Cl → CHO
     Carboxylic acids s. 26, 209

Hexamethylphosphoramide s. under KOH $(Me_2N)_3PO$

Air s. under Pd-C O_2

Ozone s. under KI O_3

Hydrogen peroxide/sodium hydroxide $H_2O_2/NaOH$
Replacement of chlorine by oxygen   ←
     at room temp. – Preferential replacement – Strongly nucleophilic HOO^- ions
s. 26, 210

Sulfur trioxide s. under Sulfuric acid SO_3

Methanesulfonic acid $MeSO_3H$
Acylisocyanates   C(Cl) : NCOCl → CONCO
from N-(chloroformyl)iminochlorides

PhS—C(Cl)(N—COCl) → PhS—CO—N=C=O

Ph—C(Cl)(N—COCl) → Ph—CO—N=C=O

Methanesulfonic acid added dropwise at 50° during 1 hr. to the startg. m., and heated 2 hrs. at 100° → (phenylthio)carbonylisocyanate. Y: 92%. – Similarly: Benzoylisocyanate. Y: 90%. F. e. s. H. Hagemann, Ang. Ch. *85*, 1058 (1973).

Trifluoromethanesulfonic acid CF_3SO_3H
Phenolesters   COCl → COOAr
    from phenols and carboxylic acid chlorides s. *16*, 279 suppl. 29

Sulfuric acid H_2SO_4
α,β-Ethyleneketones   C(SR)CHC(Hal) : C → C : C·COCH
    from 3-ethylene-3-halogenothioethers s. *28*, 961

Sulfuric acid/sulfur trioxide H_2SO_4/SO_3
Aldehydes from 1,1-dihalides $CH(Hal)_2$ → CHO
s. *29*, 847

Perchlorates ClO_4^-
Nucleophilic substitution   Cl → OR
    of chlorotropenium salts s. *26*, 540

Silver perchlorate $AgClO_4$
Ring expansion of isocyclics   ⊙
    by 1,1-dihalogenocyclopropane ring opening – Cyclic 2′-alkoxy-α,β-ethylene-halides s. *26*, 205

Hydrochlorides   ←
Preferential and selective O-acylation   OH → OAc
    Protection of amino groups by protonation s. *28*, 172

β-Alkoxycarboxylic acid esters from β-chloriminoesters   ←

ClCH₂CH₂C(NH·HCl)(OEt) + 2 HOEt → EtOCH₂CH₂COOEt [+ EtCl + NH₄Cl]

Ethyl 3-chloropropionimidate hydrochloride dissolved in abs. ethanol, stirred and warmed slowly to 40° during 5 hrs. → ethyl 3-ethoxypropionate. Y: 92%. F. e. s. C. L. Schilling, Jr., J. Org. Chem. *39*, 1770 (1974).

Hydrogen chloride   HCl
α,β-Dibrom- from α,β-ethylene-aldehydes   Br → OR
    with subsequent preferential replacement of bromine by alkoxyl and acetalation s. *27*, 220

γ-Lactones from 1-dicyanomethyl-2-bromides   ○
s. *26*, 707

Palladium-carbon/sodium hydroxide/air $Pd\text{-}C/NaOH/O_2$
Replacement of bromine by hydroxyl   ←
    with rearrangement s. *27*, 221

Via intermediates   v.i.
Aldehydes-1-d $CH(Hal)_2$ → CDO
    from 1,1-dihalides via 1,1-di(pyridinium salts) s. *29*, 211

Sulfur ↑   OC ⇈ S

Without additional reagents   w.a.r.
Nucleophilic substitution   ←
  of sulfonium ylid groups s. 27, 222

Irradiation/oxygen   ⇈⇈/O_2
Ketones from thioketones   CS → CO
  by singlet oxygen s. 29, 220

Sodium hydride (s. a. under Imidazole)   NaH
O-Methylation   CHCO → C : C(OMe)
s. 27, 223

Sodium hydroxide (s. a. under FSO_3Me)   NaOH
2-Aminoalcohols $SO_2N \cdot C \cdot C \cdot Cl$ → $N \cdot C \cdot C \cdot OH$
  from 2-chlorosulfonic acid amides s. 29, 213

Sodium hydroxide/tetrabutylammonium iodide $NaOH/Bu_4N^+I^-$
O-Methylation of alcohols   OH → OMe

139. $PhCH_2CH_2OH + (MeO)_2SO_2$ → $PhCH_2CH_2OMe$ [+ $MeOSO_2ONa$]

A soln. of phenethyl alcohol and tetrabutylammonium iodide as *phase transfer catalyst* in petroleum ether stirred 15–30 min. with 50%-NaOH, dimethyl sulfate added dropwise with cooling below 45°, and stirred 2–3 hrs. → methyl phenethyl ether. Y: 90%. – Sec. alcohols show less reactivity and tert. alcohols do not react at all. F. e. s. A. Merz, Ang. Ch. **85**, 868 (1973).

Potassium hydroxide   KOH
Ethers from sulfonium salts $S^+<$ → OR
s. 27, 224

Sodium/alcohol   NaOR
Thionocarboxylic acid esters   ←
  from thioacyl disulfides s. 27, 225

Potassium tert-butoxide $KOCMe_3$
Enolethers from fluorosulfonic acid esters   CHCO → C : C(OR)
s. 28, 173

Sodium hydrogen carbonate/dimethyl sulfoxide $NaHCO_3/Me_2SO$
Cyclic carbonic acid esters
from diol monosulfonates

140.

c-3-Hydroxy-2-methyl-4-p-tolylnorbornan-r-2-ylmethyl tosylate in dimethyl sulfoxide containing $NaHCO_3$ heated 0.5 hr. at 100–110° → 7-methyl-1-p-tolyl-3,5-dioxatricyclo[6.2.1.0^{2,7}]undecan-4-one. Y: 78%. – Hydrogen carbonate anion is a more effective nucleophile than dimethyl sulfoxide. The reaction is favored by high temp. and brief exposure. F. e. s. N. Bosworth, P. Magnus, and R. Moore, Soc. Perkin I *1973*, 2694.

Potassium carbonate K_2CO_3
Acoxy compds. from mesylates   ⊙
  with indan-to-azulene ring expansion s. 27, 226

Sodium acetate CH_3COONa
Acetolysis of brosylates $OSO_2Ar \twoheadrightarrow OAc$
s. 26, 760

Potassium acetate CH_3COOK
Acoxy compds. from tosylates $OTs \twoheadrightarrow OAc$
with epimerization s. 27, 227

Sodium salt Na^+
s. 26, 944

Ammonium hydroxide s. under H_2O_2 NH_4OH

Triethylamine s. under PhC(: NOH)Cl and Cl Et_3N

Cupric oxide s. under $CuCl_2$ CuO

Silver oxide Ag_2O
Cleavage of mercaptals $C(S-)_2 \twoheadrightarrow CO$
s. 28, 182
Ethyleneacetals
from cyclic 1,1-chlorothioethers s. 29, 972
Oxo compds. from s-trithianes
s. 29, 219

Cupric chloride $CuCl_2$
Nucleophilic substitution $SR \twoheadrightarrow OR$
of 2-thiopyridine oxide groups – Ethers from thioethers s. 28, 174
Cleavage of mercaptals $C(S-)_2 \twoheadrightarrow CO$
s. 28, 182

Cupric chloride/cupric oxide $CuCl_2/CuO$
γ-**Diketones** $C(SR)_2 \twoheadrightarrow CO$
from α,β-ethyleneketones and mercaptals via γ-ketomercaptals s. 28, 175

Carboxylic from dithiocarboxylic acid esters $CSSR \twoheadrightarrow COOR'$
s. 29, 214

Barium oxide/barium hydroxide $BaO/Ba(OH)_2$
Total methylation of phenolic hydroxyls $ArOH \twoheadrightarrow ArOMe$
with retention of halogen s. 27, 228

Calcium carbonate s. under $HgCl_2$ $CaCO_3$

Mercuric oxide s. a. under $HgCl_2$ HgO

Mercuric oxide/boron fluoride HgO/BF_3
Carbohydrates from thioglycosides $SR \twoheadrightarrow OH$

Startg. m. allowed to react 4 hrs. at 20° with 2.5 equivalents of yellow HgO and 1 equivalent BF_3-etherate in aq. tetrahydrofuran $\twoheadrightarrow$ α-D-galactopyranose deriv. Y: 91%. S. David and J.-C. Fischer, C. r. 277 (C), 179 (1973).

Oxo compds. from cyclic mercaptals
s. 30, 569

Aldehydes from 1,3-dithianes
s. 26, 211
$CH\genfrac{}{}{0pt}{}{\diagup S}{\diagdown S}\Big)$ → CHO

Mercuric carboxylates $(RCOO)_2Hg$
Carboxylic from thionocarboxylic acid esters CSOR → COOR
s. 28, 213

Mercuric acetate $(CH_3COO)_2Hg$
Glycosides from thioglycosides SR → OR'
with anomerization s. 29, 215

Mercuric acetate/boron fluoride $(CH_3COO)_2Hg/BF_3$
Acylals from 1,3-dithianes C
s. 26, 211

Mercuric chloride $HgCl_2$
Ketones from thioenolethers C : C·SR → CHCO
s. 29, 802

α,β-**Ethylenealdehydes** $C(SR)·C : CH(SR)$ → C : C·CHO
from 1,3-bis(alkylthio)ethylenes s. 26, 862
Synthesis of aldehydes CO → $C(OH)CH_2CHO$
via Δ^2-thiazolines and thiazolidines – β-Hydroxyaldehydes s. 28, 176
β-**Ketocarboxylic acid esters** $C(SR) : C·COOR'$ → COCHCOOR''
from β-alkylthio-α,β-ethylenecarboxylic acid esters
s. 30, 584

Mercuric chloride/calcium carbonate $HgCl_2/CaCO_3$
Ketones from cyclic mercaptals
s. 28, 656
$C\genfrac{}{}{0pt}{}{\diagup S}{\diagdown S}\Big)$ → CO

Mercuric chloride/mercuric oxide $HgCl_2·HgO$
Aldehydes-1-d from s-trithianes-d₃ C CHO
s. 30, 688
Carboxylic acid esters from 1,3-dithianes
via 2-methylthio-1,3-dithianes – also carboxylic acids
s. 28, 177
$CH\genfrac{}{}{0pt}{}{\diagup S}{\diagdown S}\Big)$ → COOR

Thallic trifluoroacetate $(CH_3COO)_3Tl$
Cleavage of mercaptals $C(S-)_2$ → CO
s. 28, 182

Boron fluoride (s. a. under HgO and $(CH_3COO)_2Hg$) BF_3
Benzils and indoles ←
from 2-aminoenol sulfonates s. 28, 178

Silver fluoroborate $AgBF_4$
Carboxylic acid esters COSR → COOR'
from thiolic acid 2-pyridyl esters
s. 30, 142

Ammonium ceric nitrate $(NH_4)_2Ce(NO_3)_6$
Oxo compds. from cyclic mercaptals
s. 27, 230
$C\genfrac{}{}{0pt}{}{\diagup S}{\diagdown S}\Big)$ → CO

Dimethylformamide-dialkyl sulfate $Me_2NCHO·(RO)_2SO_2$
Carboxylic acid esters from carboxylic acids COOH → COOR
s. 28, 179

Imidazole/sodium hydride ←
O-Thiobenzoylation OH → OCSAr
s. 29, 947

Trialkyloxonium fluoroborate $\qquad R_3O^+ BF_4^-$
Replacement of sulfur by oxygen $\qquad CSN< \rightarrow CON<$
 Carboxylic acid amides from thioamides s. *28*, 180
Anion exchanger $\qquad \leftarrow$
Acoxy compds. from tosylates $\qquad OTs \rightarrow OAc$
 with O-acyl group migration s. *26*, 212
Formic acid $\qquad HCOOH$
Monocyclic α-hydroxyketones $\qquad $ C ☉
 from bicyclic mercaptals – Ring expansion of cyclic ethylene derivs. s. *27*, 229
Acetic acid $\qquad CH_3COOH$
Hydrolysis of alkylthio-N-heterocyclics $\qquad \leftarrow$
 s. *26*, 229
Lactams
 via cyclic 1,1-iminosulfones s. *29*, 648 $\qquad$ C(=N)·SO_2R $\rightarrow$ C(NH)CO

Trifluoroacetic acid (s. a. under H_2O_2) $\qquad CF_3COOH$
Ketones $\qquad \leftarrow$
 from cyclopropane mercaptals s. *29*, 216
Dimethyl sulfoxide s. under $NaHCO_3$, H_2SO_4, and I $\qquad Me_2SO$
Dialkyl sulfate s. under Me_2NCHO $\qquad (RO)_2SO_2$
Methyl iodide $\qquad CH_3I$
Oxo compds. from mercaptals $\qquad C(S-)_2 \rightarrow CO$
 s. *27*, 230
Benzhydroximic chloride/triethylamine $\qquad PhC(:NOH)Cl/Et_3N$
Replacement of sulfur by oxygen $\qquad CS \rightarrow CO$
 s. *28*, 181
N-Bromosuccinimide $\qquad \begin{smallmatrix}CH_2CO\\ \ \ |\ \ \ \ \ \ \ \ \rangle NBr \\ CH_2CO\end{smallmatrix}$

Oxo compds. from cyclic mercaptals
 from 1,3-dithianes – α-Ketocarboxylic acid esters – Retention $\qquad$ C(S)(S) $\rightarrow$ CO
 of carbon-carbon double bonds s. *27*, 230
Chloramine-T $\qquad \leftarrow$
Protection of oxo compds.
 as cyclic mercaptals – Removal of the protective group s. *27*, 271
Trimethyl phosphite $\qquad (MeO)_3P$
2-Ethylenealcohols from β,γ-ethylenesulfoxides $\ $ C(SOR)·C:C $\rightarrow$ C:C·C(OH)
 with rearrangement s. *30*, 406
Synthesis of 2-ethylenealcohols $\qquad \leftarrow$
 via α-alkylation of β,γ-ethylenesulfoxides – Allyl rearrangement – β,γ-Ethylene-
 sulfoxides as intermediates, review – Stereospecific synthesis of 2-ethylene-
 alcohols s. *29*, 217

Oxygen s. under ⇈ $\qquad O_2$
Hydrogen peroxide s. a. under NaOCl $\qquad H_2O_2$
Hydrogen peroxide/ammonium hydroxide $\qquad H_2O_2/NH_4OH$
Replacement of sulfhydryl by oxygen $\qquad \leftarrow$
 s. *29*, 218
Hydrogen peroxide/trifluoroacetic acid $\qquad H_2O_2/CF_3COOH$
Replacement of alkylthio groups by oxygen $\qquad \leftarrow$
 N-Monosubst. barbituric acids s. *29*, 408

Methyl fluorosulfonate/sodium hydroxide   *FSO₃Me/NaOH*
Cleavage of mercaptals   C(S-)₂ → CO
 via S,S'-dimethylation s. *28*, 182

p-Toluenesulfonic acid   *TsOH*
β-Ketocarboxylic acid esters   COC : C(SR)₂ → COCHCOOR'
 from α-ketoketene mercaptals s. *30*, 466

Sulfuric acid   *H₂SO₄*
Thiolic acid esters from ketene mercaptals   C : C(SR)₂ → CHC(O)SR
s. *26*, 213

Sulfuric acid/dimethyl sulfoxide   *H₂SO₄/Me₂SO*
Carbonyl from thio- and seleno-carbonyl groups   CS → CO
s. *27*, 231

Chlorine/triethylamine   *Cl/Et₃N*
Replacement of alkylthio by alkoxy groups   SR → OR'
s. *28*, 183

Iodine/sodium hydrogen carbonate   *I/NaHCO₃*
Replacement of alkylthio groups   ←
 Aldehydes from mercaptals – α-Hydroxyaldehydes – Thiolic from trithio-orthocarboxylic acid esters – α-Ketothiolic acid esters s. *26*, 214

Iodine/dimethyl sulfoxide   *I/Me₂SO*
Oxo compds.   C
 from s-trithianes and mercaptals s. *29*, 219

Sodium hypochlorite/hydrogen peroxide   *NaOCl/H₂O₂*
Ketones from thioketones   CS → CO
 by singlet oxygen s. *29*, 220

Silver perchlorate   *AgClO₄*
Carboxylic acid esters   COSR → COOR'
from thiolic acid 2-pyridyl esters

142.

 PhCH₂CH₂COS—(2-pyridyl) + HOCHMe₂ → PhCH₂CH₂COOCHMe₂

An equimolar amount of AgClO₄ or AgBF₄ added to a soln. of 2-pyridyl 3-phenyl-propionothiolate and 2-propanol in benzene, and the product isolated after 10 min. at room temp. → 2-propyl 3-phenylpropionate. Y: 70–85%. Also lactones s. H. Gerlach and A. Thalmann, Helv. *57*, 2661 (1974).

Perchloric acid   *HClO₄*
Oxo compds. from 1,1-sulfinylthioethers   C(SR)(SOR) → CO
s. *29*, 667

Tetrabutylammonium iodide s. under NaOH   *Bu₄N⁺ I⁻*

Hydrogen chloride   *HCl*
Acetals from 1,1-hydroxysulfonic acids   C(OH)SO₃H → C(OR)₂
s. *26*, 215

Anhydrosugar acetals   CH(OR)₂
s. *26*, 216

Arylacetic acid esters   C : C(SR)SOR → CHCOOR
 from 1-sulfinylthioenolethers s. *28*, 735

Carboxylic acids   C : C(SO₂R)NHCHO → CHCOOH
 from 1-(sulfonyl)eneformylamines s. *28*, 729

Remaining Elements ↑ OC ↟ Rem

Without additional reagents	*w.a.r.*
Hydroperoxides from dichloroboranes s. 29, 221	$R \cdot BCl_2 \rightarrow R \cdot OOH$
O,O-Heterocyclics s. 28, 184	◯
Sodium hydroxide (s. a. under H_2O_2)	*NaOH*
Oxo compds. from silyloxycyclopropane ring s. 29, 829	▽—OSi≤ → CH·C·CO
Carboxylic acid esters from carboxylic acids and phosphates – Hindered compds. s. 29, 222	COOH → COOR
Potassium carbonate	K_2CO_3
Phenolethers from aryloxysilanes s. 26, 217	ArOSi≤ → ArOR
Pyridine	C_5H_5N
Replacement of siloxy by acoxy groups Preferential replacement of prim. groups s. 29, 223	OSi≤ → OAc
Diborane/hydrogen peroxide	B_2H_6/H_2O_2
Boranes as intermediates Phenols from arylmercury compds. – Replacement of halogen by hydroxyl s. 27, 232; phenols and arylboronic acids from arylthallium bis(trifluoroacetates) cf. S. W. Breuer et al., Chem. Commun. *1975*, 36	ArHg· → ArOH
Glucosyltransferase	←
Glycosides s. 28, 185	←
m-Chloroperoxybenzoic acid	$m\text{-}ClC_6H_4COO_2H$
Ketones from enesilanes	C:C·Si≤ → CHCO

143.

Startg. m. kept 4 hrs. at room temp. with a slight excess of *m*-chloroperoxybenzoic acid in methylene chloride → product. Y: 90%. F. e., also with final formic acid treatment, s. G. Stork and M. E. Jung, Am. Soc. *96*, 3683 (1974).

α-Hydroxyketones from enoxysilanes   C:C·OSi≤ → C(OH)CO

144.

A soln. of *m*-chloroperoxybenzoic acid in dry hexane cooled in an ice-methanol bath, treated during ca. 5 min. with a soln. of the startg. m. in the same solvent, stirred 45 min., the resulting *m*-chlorobenzoic acid removed by filtration, concd., and partitioned 3 hrs. between ether and 10%-NaOH-soln. to effect hydrolysis → product. Y: 77%. F. e., also acidic hydrolysis, s. G. M. Rubottom, M. A. Vazquez, and D. R. Pelegrina, Tetrah. Let. *1974*, 4319.

Dimethyl sulfoxide s. under H_2SO_4	Me_2SO
Ozone	O_3
Ozonization of organomercury compds. Carboxylic acids from dialkylmercury compds. – Ketones from monoorganomercury halides s. 26, 218	←

Ozone/sodium tetrahydridoborate-hydrogen chloride $O_3/Na[BH_4]$-HCl
Lactones from cyclic enoxysilanes
Oxidative ring expansion

145.

Startg. silyl enolether in methanol treated at −78° with the equivalent amount of ozone, then with excess $NaBH_4$, warmed to room temp., the solvent evaporated, and the residue stirred briefly with aq. 10%-HCl → lactone. Y: 93%. F. e. s. R. D. Clark and C. H. Heathcock, Tetrah. Let. *1974,* 1713.

Hydrogen peroxide s. a. under B_2H_6 H_2O_2

Hydrogen peroxide/sodium hydroxide $H_2O_2/NaOH$
2-Ene-1,4-diols from dihydro-1,2-oxaborines   C
s. *28,* 583

Sulfuric acid/dimethyl sulfoxide H_2SO_4/Me_2SO
Carbonyl from selenocarbonyl groups   CSe → CO
s. *27,* 231

Bromine   Br
α-Diketones from 1,2-di(siloxy)alkenes   C(OSi≦) : C(OSi≦) → COCO
Spiro[adamantane-2,1′-cyclobutane-2′,3′-diones] s. *27,* 233

Carbon ↑   OC↕C

Without additional reagents   w.a.r.
Nucleophilic substitution
    of carbamyloxy by alkoxy groups   OCON< → OR
    s. *27,* 374

    of aryloxycarbonyloxy groups   O·CO·OAr → OR
    Glycosyl compds. s. *29,* 224

Protection of glycol groups   O
    as 2-amino-1,3-dioxolanes s. *27,* 234

Irradiation s. under Iodine   ⧚

Sodium hydride   NaH
Mixed from active carbonic acid esters   OCOOR → OCOOR′
    Transesterification s. *26,* 219

Sodium hydroxide   NaOH
Hydroxycarboxylic acids   C
    from α-hydroxylactols – Degradation with loss of 1 C-atom s. *29,* 225

Potassium hydroxide   KOH
Ketones from α-halogenonitriles   C(CN)Hal → CO
s. *26,* 718

Carboxylic acids from lactams   C
    Retro-Mannich reaction s. *28,* 186

Carboxylic acids
    from 1-bromo-1-trifluoroacetoxycyclopropane ring – Stereospecific introduction of angular carboxyl groups s. *29,* 226

Sodium/alcohol   *NaOR*
Carboxylic acid esters $COCCl_3 \rightarrow COOR$
   from trichloromethyl ketones s. 29, 816

Amide acetals $(>N)_2C \cdot CN \rightarrow >NC(OR)_2$
   from α,α-diaminonitriles s. 28, 187

Sodium alkoxide
Amide acetals and 1,1,1-alkoxydiamines $(>N)_2C \cdot CN \rightarrow (>N)_2C \cdot OR$
   from α,α-diaminonitriles s. 28, 187

Potassium/alcohol   *KOR*
Transesterification $COOAr \rightarrow COOR$
of carboxylic acid phenyl esters

146. $Me_3C \cdot OH + PhOCOPh \rightarrow Me_3C \cdot OCOPh$

Tert. benzoates. *tert*-Butanol added dropwise under N_2 to K in *liq. NH_3*, a small amount of $FeCl_3$ added to catalyze the formation of K-*tert*-butoxide, then solid phenyl benzoate added with stirring, and the product isolated after 50 min. → *tert*-butyl benzoate. Y: 77%. F. e. and limitation s. R. A. Rossi and R. Hayos de Rossi, J. Org. Chem. *39*, 855 (1974).

Potassium/tert-butanol-molecular sieve   ←
Dicarboxylic acid monoesters $COOR \rightarrow COOR'$
   via mixed dicarboxylic acid esters – Preferential transesterification s. 27, 235

Potassium tert-butoxide   *KOCMe_3*
Cleavage of diaryl ketones $ArCOAr' \rightarrow ArCOOH$
s. 28, 600

Hydrolytic anthraquinone ring cleavage   C
s. 26, 220

Potassium tert-butoxide/oxygen   *KOCMe_3/O_2*
α-Keto- from α-cyano-carboxylic acid esters $CH(CN)COOR \rightarrow COCOOR$
   Autoxidation s. 27, 236

Potassium carbonate K_2CO_3
Carboxylic acid esters $COCCl_3 \rightarrow COOR$
   from trichloromethyl ketones s. 29, 227

Sodium hydrogen sulfite   *NaHSO_3*
2,5-Cyclohexadienone ring   C
   with O-heteroring opening s. 28, 188

Sodium salt/dimethylformamide Na^+/Me_2NCHO
O-Acylbenzoins $2\,ArCOCOAr \rightarrow ArCOCH(OCOAr)Ar$
   from 2 benzil molecules s. 26, 221

Triethylamine   *Et_3N*
Carboxylic acid esters $COOH \rightarrow COOR$
   from carboxylic acids and enolethers s. 28, 189
   from trichloromethyl ketones $COCCl_3 \rightarrow COOR$
   s. 29, 227

N-Acylcarboxylic acid amides   C
   from 4-alkylidene-5-oxazolones s. 29, 228

Pyridine   *C_5H_5H*
Phosphorylation of alcohols $OH \rightarrow OPO_3H_2$
   with p-nitrophenyl phosphate, also with P¹,P²-di-p-nitrophenyl pyrophosphate s. 28, 190

Mixed phosphoric acid diesters $\quad \cdot O \cdot P(O)(OR)OH \rightarrow \cdot O \cdot P(O)(OR')OH$
  via pyridinium phosphate betaines s. 26, 222

Silver oxide $\hfill Ag_2O$

Transetherification $\hfill OR \rightarrow OR'$
s. 27, 237

Silver carbonate $\hfill Ag_2CO_3$

α,β-**Ethylenealdehydes** $\quad C(OAc)C \equiv CH \rightarrow C:C \cdot CHO$
  from 1-ethynylacoxy compds. s. 28, 191

Calcium carbonate s. under $(CH_3COO)_4Pb$ $\hfill CaCO_3$

Mercuric oxide s. under Iodine $\hfill HgO$

Magnesium sulfate s. under $KMnO_4$ $\hfill MgSO_4$

Zinc chloride $\hfill ZnCl_2$

Transacetalation
  of carbohydrate derivs. s. 26, 223 $\quad \langle_O^O\rangle CHR \rightarrow \langle_O^O\rangle CHR'$

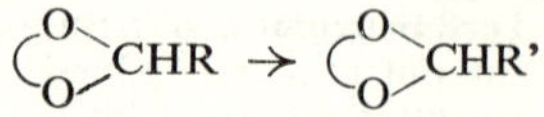

Bis-(3-methyl-2-butyl)borane $\hfill R_2BH$

α-**Alkoxyketones from** α-**acetyleneacetals** $\quad C \equiv C \cdot (OR)_2 \rightarrow CH_2COCH(OR)$
s. 29, 61

Thallous ethoxide $\hfill TlOEt$

Oxo compounds from glycols $\quad C(OH)C(OH) \rightarrow 2\,CO$
s. 29, 229

Molecular sieve s. under $K/Me_3C \cdot OH$ $\hfill \leftarrow$

Thallic nitrate $\hfill Tl(NO_3)_3$

Oxo compds. from glycols $\quad C(OH)C(OH) \rightarrow 2\,CO$
s. 29, 229

Thallic nitrate/perchloric acid $\hfill Tl(NO_3)_3/HClO_4$

Benzils from chalcones $\quad ArCOCH:CHAr \rightarrow ArCOCOAr$
s. 27, 238

Boron bromide $\hfill BBr_3$

Transesterification $\hfill COOR \rightarrow COOR'$
s. 30, 262

Triphenylcarbonium fluoroborate $\hfill Ph_3C^+\ BF_4^-$

Cleavage of acetals $\hfill \leftarrow$
  α-Hydroxyketones from acetals – Removal of protective groups by hydride transfer s. 27, 239

Aluminum chloride/hydrogen chloride $\hfill AlCl_3/HCl$

Ketones $\hfill R_2CO$
  from N,N-dichloro-(tert-carbin)amines with C-cleavage – C→N-1,2-Alkyl migration s. 27, 240

Crown polyether s. under $KMnO_4$ $\hfill \leftarrow$

Dimethylformamide s. under Na^+ $\hfill Me_2NCHO$

Dowex 50W-4 $\hfill \leftarrow$

β,γ-**Ethyleneacetals** $\hfill C$
  from enecyclopropyl ethers s. 29, 145

Dowex 50-Xl(H^+) $\hfill \leftarrow$

Transesterification $\hfill COOH \rightarrow COOR$
  Acylaminocarboxylic acid esters from acylaminocarboxylic acids – Comparison of cation exchangers – Ion exchangers as catalysts, review s. 26, 224

o-Nitrophenol ←
O-Glycosides from carbophenoxy glycosides ←
s. *30*, 150
Formic acid s. under O_3 and HCl   HCOOH
Acetic acid (s. a. under O_3) CH_3COOH
Nitrous acid esters   OH → ONO
   by transesterification s. *26*, 225
β-Cyanoketones   C(N<)CN → CO
   from 1-amino-1,3-dinitriles s. *29*, 663
Trifluoroacetic acid CF_3COOH
Protection of *vic*-hydroxyl groups   O
   as 2-alkoxy-1,3-dioxolanes s. *28*, 268
Oxalic acid $(COOH)_2$
Aldehydes from α-aminonitriles   CH(N<)CN → CHO
   β,γ-Ethylenealdehydes s. *29*, 901
Peroxyacetic acid s. H_2O_2/CH_3COOH CH_3COO_2H
Dimethyl sulfide s. under O_3 Me_2S
Methyl iodide   MeI
Carboxylic acid esters and amides $COOCH_2SMe$ → COOR
   from carboxylic acid methylthiomethyl esters s. *29*, 230
Acetyl chloride CH_3COCl
Acetoxy compds.   OAc
   from tetrahydro-2-pyranyl ethers s. *27*, 241
Lead tetraacetate/calcium carbonate $(CH_3COO)_4Pb/CaCO_3$
Ketones from α-hydroxyketones   CH(OH)COR → CO
   Oxidative degradation of side chains s. *26*, 226
Acoxy-2-ethylenes from 3-ethylenealcohols ←
s. *27*, 243
Lead tetraacetate/acetic acid $(CH_3COO)_4Pb/CH_3COOH$
Ketones from α-hydroxyhydrazones   C(OH)C:N·N< → CO
   Oxidative degradation s. *27*, 244
Nitric acid/sulfuric acid HNO_3/H_2SO_4
Oxidative degradation ←
of isocycles in N-heterocyclics
Pyridine from quinoline ring

147.   [quinoline-CBr₃] → [HOOC-pyridine-COOH] → [MeOOC-pyridine-COOMe]

Labeled compds. A soln. of 2-tribromomethylquinoline-2-^{13}C in concd. H_2SO_4 stirred and heated to 150°, the resulting bromine vapors displaced by argon, the temp. raised to 260°, concd. HNO_3 added cautiously during 2 hrs., then excess HNO_3 removed by a slow stream of argon while the temp. rises to 275°, cooled to room temp., added slowly to cold methanol, refluxed 18 hrs. with slow volume reduction by periodic distillation of the methanol → dimethyl pyridine-2,5-dicarboxylate-2-^{13}C. Y: 71%. T. A. Bryson et al., J. Org. Chem. *39*, 1158 (1974).

Oxygen s. under $KOCMe_3$ and Cobaltocene O_2
Ozone (s. a. under $(CH_3COO)_2Co$) O_3
Carboxylic acids from acetylene derivs.   C≡C → COOH
s. *27*, 245

Ketones from phenol ring
via α,β-ethylene-γ-hydroperoxy-γ-lactones with retroaldol reaction s. *26*, 227

Ozone/sodium hydrogen carbonate $O_3/NaHCO_3$
**Ketocarboxylic acids
from cyclic 2-ethylenealcohols**

148.

2%-O$_3$ in O$_2$ bubbled at –70° through a soln. of 2,5-dimethyl-3-isopropylidene-bicyclo[3.2.0]heptan-2-ol in ethyl acetate until an excess of ozone is present, the solvent evaporated below 20°, satd. aq. NaHCO$_3$ added, and stirred overnight → keto acid. Y: 71%. R. L. Cargill and B. W. Wright, J. Org. Chem. *40*, 120 (1975).

Ozone/acetic acid O_3/CH_3COOH
Ketocarboxylic acids
by abnormal ozonolysis s. *28*, 192

Ozone/dimethyl sulfide O_3/Me_2S
Aldehydes from ethylene derivs. CH : C → CHO
Oxidative cleavage
Removal of benzaldehyde by benzylic hydrogenolysis

149.

Ozone bubbled for 2 min. at –70° through a soln. of (E)-(S)-O-isopropylidene-1-phenyl-1-hexene-5,6-diol in dry methanol, then excess dimethyl sulfide added, tightly stoppered, kept overnight at –70° then 0.5 hr at room temp., the solvent evaporated in vacuo, dimethyl sulfide removed completely by 3-fold evaporation with dry methanol, then hydrogenated with Pd-black in dry methanol to convert the resulting benzaldehyde into toluene → (S)-O-isopropylidenepentane-4,5-diol-1-al. Y: 95%. – A large excess of Pd-black is used to counteract traces of dimethyl sulfide, a catalyst poison. H. Hayashi et al., Am. Soc. *95*, 8749 (1973); ozone, review, s. R. Criegee, Chem. Ztg. *99*, 138 (1975).

Ozone/hydrogen peroxide/formic acid $O_3/H_2O_2/HCOOH$
Carboxylic acids from ethylene derivs. CH : C → COOH
s. *27*, 246

Ozone/sulfur dioxide O_3/SO_2
Acetals from ethylene derivs. C : C → C(OR)$_2$
s. *29*, 148

Ozone/hydrogen chloride O_3/HCl
Oxindoles
from 3-cyano-1,4-dihydroquinolines s. *29*, 231

Ozone/nickel O_3/Ni
Cleavage of ethylene derivs. to alcohols C : C → CHOH
s. *27*, 149

Hydrogen peroxide s. a. under O_3 H_2O_2

Hydrogen peroxide/sodium hydroxide $H_2O_2/NaOH$
Oxidative cleavage of α,β-ethyleneketones
Ring opening of isocycles s. *26*, 228

Hydrogen peroxide/acetic acid H_2O_2/CH_3COOH
Oxidative cleavage of α-diketones $RCOCOR' \rightarrow R\cdot COOH + HOOC\cdot R'$
s. *28*, 193

p-Toluenesulfonamide $TsNH_2$
O- and N-Glycosides
from carbophenoxy glycosides
N- from O-Glycosides

150.

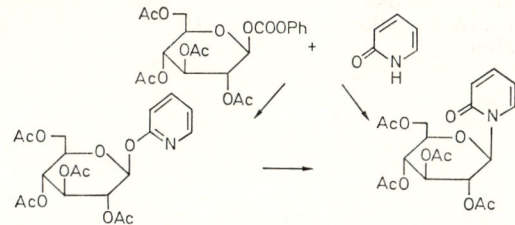

2,3,4,6-Tetra-O-acetyl-1-O-phenoxycarbonyl-β-D-glucopyranose added at 135–140°
to a homogeneous prefused mixture of 2(1H)-pyridone and
1 equivalent | 5 equivalents
p-toluenesulfonamide, and stirred 1 hr. at 135–140°/20 mm to remove the phenol
liberated →

2-(2,3,4,6-tetra-O-acetyl-β-D-glucopyrano- | 1-(2,3,4,6-tetra-O-acetyl-β-D-gluco-
syloxy)pyridine. Y: 78.5%; 84.5% with | pyranosyl-2(1H)-pyridone.
o-nitrophenol instead of tosylamide. | Y: 71.4%.

The O-glycoside can be converted into the N-isomer in 78% yield by heating
with 5 equivalents tosylamide at 150°. M. Yamada et al., Carbohyd. Res. *31*, 151
(1973).

Acetyl tosylate CH_3COOTs
Acetoxy compds. from ethers   OR → OAc
s. *29*, 232

p-Toluenesulfonic acid   TsOH
Protection of alcohol groups as 9-anthryl ethers   OR → OR'
Transetherification – Oxidoreductive removal of the protective group s. *28*, 194

Preferential transacetalation

151.

Cyclic acetals. 8a-Methyl-1,6-dioxo-1,2,3,4,6,7,8,8a-octahydronaphthalene stirred
30 hrs. at 25° in 2-ethyl-2-methyl-1,3-dioxolane containing 2% ethylene glycol
and p-toluenesulfonic acid → 1,1-ethylenedioxy-8a-methyl-6-oxo-1,2,3,4,6,7,8,8a-
octahydronaphthalene. Y: 80%. F. e. s. G. Bauduin and Y. Pietrasanta, Tetrahedron *29*, 4225 (1973).

Cyclic mixed acetals   O
from acoxyaldehydes and orthoformic acid esters s. *28*, 195

| *Sulfuric acid* | H_2SO_4 |

Carboxylic acid esters   COOH → COOR
    from carboxylic acids and enol ethers s. *28,* 189

| *Persulfate* | $S_2O_8^{2-}$ |

Baeyer-Villiger oxidation   ←
    s. *26,* 137

Chromium trioxide/acetic acid CrO_3/CH_3COOH

Hemiquinols from phenolethers   ←
    s. *29,* 233

Bromine/sodium hydroxide   *Br/NaOH*

**Lactams from cyclic tert. amines
with and without C-demethylation
Preferential oxidation**

$$\underset{CHCH_3}{\overset{N-}{\diagup}} \rightarrow \underset{CO}{\overset{N-}{\diagup}}$$

152.

4 moles | 2.0 moles
    bromine dissolved in methylene chloride, added dropwise to a soln. of
5α-conanine | 21-nor-5α-conanine
in methylene chloride, stirred with aq. NaOH-soln., and the product isolated
                    after stirring at room temp.
for 1 hr. | for 3 hrs.
                    → 21-nor-5α-conanin-20-one. Y: 80%.
F. e., also with N-bromosuccinimide, and prepn. of **cyclic azomethinium salts**
(cf. Synth. Meth. *18,* 533), s. A. Picot and X. Lusinchi, Synthesis *1975,* 109.

Iodine/mercuric oxide/irradiation   *I/HgO/⚡*

Aldehydes from 1-acyl-3-hydroxypiperidines   CHO
    s. *27,* 247

Sodium hypochlorite s. under RuO_2 and RuO_4   *NaOCl*

Perchloric acid (s. a. under $Tl(NO_3)_3$) $HClO_4$

O,O-Isopropylidene derivs. of carbohydrates
    with acetone dimethyl ketal s. *21,* 276 suppl. 28

$$\underset{C \cdot OH}{\overset{C \cdot OH}{\diagup}} \rightarrow \underset{C \cdot O}{\overset{C \cdot O}{\diagup}}\!\!\times$$

Acoxy compds. from ethers   OR → OCOR′
    s. *27,* 248

Hydroxylactones   ☉
    from oxidocarboxylic acid esters s. *29,* 234

Pyridine hydrochloride C_5H_5N,HCl

Benzofurans from o-alkyloxyacetylenes   O
    s. *28,* 196

Quaternary ammonium halide s. under $KMnO_4$ $R_4N^+\ Hal^-$

Tetraethylammonium iodide $Et_4N^+\ I^-$

β-Hydroxyethylation of phenols   ArOH → $ArOCH_2CH_2OH$
    s. *29,* 235

Hydrogen chloride (s. a. under AlCl₃)　　　　　　　　　　　　　　　　　　　HCl
Hydrolysis of alkylthio-N-heterocyclics　　　　　　　　　　　　　　　　　　C
with and without pyrimidine ring opening s. 26, 229; o-aminoaldehydes by pyrimidine ring opening s. G. Evens and P. Caluwe, J. Org. Chem. 40, 1438 (1975)

Asym. synthesis of α-aminocarboxylic acids
via tetrahydro-1,4-oxazin-2-ones s. 29, 750

1,5-Dioxocin-2,6-diones and 4-pyrones　　　　　　　　　　　　　　　　　　⊙
from 2 2,3-dihydrofuran-2,3-dione molecules s. 28, 197

Hydrogen chloride/formic acid　　　　　　　　　　　　　　　　　　　HCl/HCOOH
Carboxylic acids from ketones　　　　　　　　　　　　　　　　　　　COR → H
C-Deacylation s. 29, 236

Hydrogen chloride/acetic acid　　　　　　　　　　　　　　　　　　HCl/CH₃COOH
γ-Diketones from enaminoenollactones　　　　　　　　　　　　　　　　　　C
s. 27, 249

Manganese dioxide　　　　　　　　　　　　　　　　　　　　　　　　　　MnO₂
Oxidative cleavage of glycols　　　　　　　　　　　　　　　C(OH)C(OH) → CO
s. 29, 157

2(5H)-Furanone from 2H-pyran ring　　　　　　　　　　　　　　　　　　⊙

153.

A soln. of 1,3,7,7-tetramethyl-2-oxabicyclo[4.4.0]-3,5-decadiene in benzene refluxed 36 hrs. with activated MnO₂ → 1,5,5-trimethyl-9-oxa-8-oxobicyclo[4.3.0]-6-nonene. Y: 91%. F. startg. m. s. B. R. von Wartburg and H. R. Wolf, Helv. 57, 916 (1974).

Potassium permanganate/magnesium sulfate　　　　　　　　　　　　　KMnO₄/MgSO₄
Selective oxidation of O-bridged isocycles　　　　　　　　　　　　　　　　C
Phase transfer catalysis s. 30, 33

Potassium permanganate/crown polyether　　　　　　　　　　　　　　　　　←
Crown polyethers as complexing agents　　　　　　　　　　　　　　　　　C
Oxidations with potassium permanganate in benzene s. 28, 198

Potassium permanganate/quaternary ammonium halide　　　　　　　KMnO₄/R₄N⁺ Hal⁻
Oxidations with potassium permanganate in benzene　　　　　　　　　　　　←
s. 28, 198 suppl. 29

Cobaltocene-oxygen complex　　　　　　　　　　　　　　　　　　[(C₅H₅)₂Co]₂O₂
Oxidative cleavage of α-diketones　　　　　　　　　　　　　　COCO → 2 COOH
s. 28, 193

Cobaltous acetate/acetic acid/ozone　　　　　　　　　　　(CH₃COO)₂Co/CH₃COOH/O₃
Carboxylic acids from isocycles　　　　　　　　　　　　　　　　　　　　C
s. 26, 230

Nickel s. under O₃　　　　　　　　　　　　　　　　　　　　　　　　　　　Ni

Ruthenium dioxide/sodium hypochlorite　　　　　　　　　　　　　　　RuO₂/NaOCl
Carboxylic acids from acetylene derivs.　　　　　　　　　　　　　C ≡ C → COOH
s. 27, 245

Ruthenium tetroxide	RuO_4
Aldehydes from ethylene derivs.	CH : C → CHO

154.

$3\alpha,7\alpha,12\alpha$-Triacetoxy-24,24-diphenyl-5β-chol-23-ene in acetone-water titrated with a soln. of RuO_4 in CCl_4 → $3\alpha,7\alpha,12\alpha$-triacetoxy-24-nor-5β-cholan-23-al. Y: 73%. Y. Shalon and W. H. Elliott, Synth. Commun. **3**, 287 (1973).

Ruthenium tetroxide/sodium hypochlorite	$RuO_4/NaOCl$
Oxidative ring opening	C

s. **24**, 251 suppl. **29**; cf. D. C. Ayres and A. M. M. Hossain, Soc. Perkin I *1975*, 707

Dichlorobis(benzonitrile)palladium(II)	$PdCl_2(PhCN)_2$
Transetherification of enolethers geospecific s. **27**, 250	C : C(OR) → C : C(OR')

Via intermediates	v.i.
Ketones from nitriles via α-hydroperoxynitriles and cyanohydrins	CHCN → CO

155. PhCH$_2$CH(CH$_3$)CN → [PhCH$_2$-C(CH$_3$)(OOH)CN] $\xrightarrow{SnCl_2}$ [PhCH$_2$-C(CH$_3$)(OH)CN] $\xrightarrow{NaOH}$ PhCH$_2$COCH$_3$

Startg. m. in tetrahydrofuran added at –78° under N_2 to a soln. of Li-diisopropylamide in the same solvent, dry O_2 introduced during 0.5 hr. at –78°, treated with $SnCl_2$ in 2 M HCl, stirred 0.5 hr. at 0°, and washed with 1 M NaOH during ether-water work up → phenylpropanone. Y: 82%. F. e., also isolation of the intermediates, s. S. J. Selikson and D. S. Watt, J. Org. Chem. **40**, 267 (1975).

Ketones from α,β-ethylenecarboxylic acid azides via eneurethans – 3-Chromanones s. **29**, 237	C : C·CON$_3$ → CHCO

β,γ-Ethyleneoxo compds. from 2-alkoxyenecyclopropanes	←

156.

An aq. soln. of mercuric acetate added to a stirred soln. of 2-methoxy-2-methyl-cyclohexylidenecyclopropane in ethanol-water, and after 5 min. H_2S passed in for 5 min. → 4-cyclohexylidene-2-butanone. Y: 85%. F. e. s. M. S. Newman and M. C. Vander Zwan, J. Org. Chem. **39**, 1186 (1974).

Elimination ⇑

Hydrogen ↑   OC⇑H

Irradiation (s. a. under Rose Bengal and Ph_2C : NBr)   ∰

5-Hydroxybenzoxazoles   O
   from o-amino-p-quinones s. **29**, 238

Electrolysis ↯
Heterocyclics by anodic oxidation
s. *29*, 239

Lithium methoxide s. under Me₃C·OCl   *LiOMe*

Sodium/alcohol s. under Rose Bengal   *NaOR*

Sodium hydrogen carbonate s. under I   *NaHCO₃*

Potassium iodide s. under I   *KI*

Triethylamine s. under N-Chlorosuccinimide   *Et₃N*

Pyridine s. under PhICl₂, N-Halogenosaccharin, and SO₃   *C₅H₅N*

Cupric oxide   *CuO*
Oxo compds. from alcohols   CHOH → CO
s. *27*, 251

Silver oxide/magnesium sulfate   *Ag₂O/MgSO₄*
Quinones from quinols   ←
s. *28*, 611

Silver carbonate-Celite   ←
Aldonolactones from carbohydrates
s. *19*, 308 suppl. *28;* lactones from lactols s. a. H. J. J. Loozen, E. F. Godefroi, and J. S. M. M. Besters, J. Org. Chem. *40*, 892 (1975)

$$\underset{CHOH}{C=O} \rightarrow \underset{CO}{C=O}$$

Aluminum isopropoxide   *Al(OR)₃*
Ketones from sec. alcohols   CHOH → CO
s. *26*, 161

Ammonium ceric nitrate   *(NH₄)₂Ce(NO₃)₆*
Aldehydes from prim. alcohols   CH₂OH → CHO
   Labeled compds. s. *29*, 89

Benzalacetone s. under RuCl₂(Ph₃P)₃   *PhCH : CHCOCH₃*

Carbodiimides s. under Me₂SO   *RN : C : NR*

Acetic anhydride s. under Me₂SO and HClO₄   *Ac₂O*

Hydroperoxides s. under MoCl₅   *R·OOH*

Dimethyl sulfide s. under N-Chlorosuccinimide   *Me₂S*

Thioanisole-chlorine complex   ←
Oxo compds. from alcohols   CHOH → CO
s. *28*, 200; with polymer-based thioanisole, also monooxidation of diols, cf. G. A. Crosby, N. M. Weinshenker, and H.-S. Uh, Am. Soc. *97*, 2232 (1975)

Dimethyl sulfoxide s. a. under SO₃–C₅H₅N   *Me₂SO*

Dimethyl sulfoxide/carbodiimides   *Me₂SO/RN : C : NR*
Oxo compds. from alcohols   CHOH → CO
   Pfitzner-Moffat oxidation s. *19*, 307

Dimethyl sulfoxide/acetic anhydride   *Me₂SO/Ac₂O*
Pulvinic acid dilactones   ←
   from 2,5-dihydroxy-p-quinones s. *28*, 199

Dimethyl sulfoxide-chlorine complex   *Me₂SO·Cl₂*
Oxo compds. from alcohols   CHOH → CO
s. *28*, 200

Phenol iodosoacetate $\qquad$ $PhI(OAc)_2$
3,4-Dihydrocoumarins by oxidative ring closure $\qquad$ ○
s. 27, 252

Rose Bengal/irradiation-sodium/alcohol $\qquad$ ←
Oxo compds. from alcohols $\qquad$ CHOH → CO
s. 29, 240

2,3-Dichloro-5,6-dicyanoquinone $\qquad$ ←
1,3-Benzodioxole-2-spirocyclohexadien-4′-ones $\qquad$ ○
s. 26, 231

Iodobenzene dichloride/pyridine $\qquad$ $PhICl_2/C_5H_5N$
Ketones from sec. alcohols $\qquad$ CHOH → CO
s. 29, 244

Phenyl-N-bromoketimine/irradiation $\qquad$ $Ph_2C:NBr$/⧸⧸⧸
Oxo compds. from alcohols
s. 27, 253

N-Chlorosuccinimide-triethylamine $\qquad$ $\begin{matrix}CH_2CO\\|\\CH_2CO\end{matrix}\!\!\!\!\diagdown\!\!\!\diagup NCl/Et_3N$
Quinones from quinols $\qquad$ ←
s. 28, 512 suppl. 30

N-Chlorosuccinimide/dimethyl sulfide-triethylamine $\qquad$ $\begin{matrix}CH_2CO\\|\\CH_2CO\end{matrix}\!\!\!\!\diagdown\!\!\!\diagup NCl/Me_2S-Et_3N$
Oxo compds. from alcohols $\qquad$ CHOH → CO
under mild conditions s. 28, 201

Lead dioxide $\qquad$ PbO_2
4-Oxo- from 4-hydroxypyrazoles $\qquad$ ←
s. 27, 254

Lead tetraacetate $\qquad$ $(CH_3COO)_4Pb$
1,3,4-Oxadiazoles from acylhydrazones $\qquad$ ○
Oxidative ring closure – Nitrilimines as intermediates – s-Triazolo[4,3-b]benzothiazoles – Oxidative ring closures of oxo compd. N-derivs., review s. 26, 232

Nitrogen dioxide $\qquad$ NO_2
Quinones from quinols $\qquad$ ←
s. 28, 512; with N-chlorosuccinimide-triethylamine complex, a mild oxidant, also bianthronyls from anthrones (cf. Synth. Meth. *11*, 807), cf. H. D. Durst, M. P. Mack, and F. Wudl, J. Org. Chem. *40*, 268 (1975)

Nitric acid/perchloric acid $\qquad$ $HNO_3/HClO_4$
Oxo compds. from alcohols $\qquad$ CHOH → CO
Oxidation of mono- and di-arylcarbinols s. 29, 241

Triphenylphosphine dibromide $\qquad$ Ph_3PBr_2
Benzils from benzoins $\qquad$ ArCH(OH)COAr → ArCOCOAr
s. 9, 357 suppl. 28

Bismuth trioxide $\qquad$ Bi_2O_3
α-Diketones $\qquad$ COCHOH → COCO
from glycols via α-hydroxyketones s. 26, 233

Oxygen s. under Cobalt oxide $\qquad$ O_2

N-Chlorosaccharin/pyridine $\qquad$ ←
Reactions with N-chlorosaccharin $\qquad$ ←
s. 28, 202 suppl. 29

N-Bromosaccharin/pyridine ←
Oxidations with N-bromosaccharin CHOH → CO
s. 28, 202

Sulfur trioxide-pyridine/dimethyl sulfide SO_3-C_5H_5N/Me_2SO
α,β-**Ethyleneoxo compds.** C(OAc)CH·CHOH → C:C·CO
from 1,3-diol monoesters s. 29, 242

Sodium dichromate $Na_2Cr_2O_7$
Oxo compds. from alcohols CHOH → CO
s. 26, 234; in dimethyl sulfoxide as solvent in the presence of concd. H_2SO_4 cf.
Y. S. Rao, J. Org. Chem. *39*, 3304 (1974)

Sodium dichromate/sulfuric acid $Na_2Cr_2O_7$/H_2SO_4
s. 26, 234 suppl. 29

Chromiun trioxide CrO_3
α-**Keto- from α-hydroxy-carboxylic**
acid *tert*-butylamides CH(OH)CONHCMe$_3$ → COCONHCMe$_3$
s. 27, 326

Chromiun trioxide/pyridine CrO_3/C_5H_5N
Oxo compds. from alcohols CHOH → CO
Improved Collins reagent s. 26, 235

Chromium trioxide-graphite CrO_3-C
Aldehydes from prim. alcohols CH$_2$OH → CHO
s. 28, 203

Chromium trioxide/3,5-dimethylpyrazole ←
Oxo compds. from alcohols CHOH → CO
s. 26, 235 suppl. 29

Chromium trioxide/sulfuric acid CrO_3/H_2SO_4
Keto- from hydroxy-alkoximes
s. 28, 896

Molybdenum pentachloride/hydroperoxides $MoCl_5$/R·OOH
Ketones from sec. alcohols CHOH → CO
s. 27, 184

Chlorine s. under Thioanisole and Me_2SO Cl

Iodine/potassium iodide/sodium hydrogen carbonate I/KI/$NaHCO_3$
Isomerization of α,β-ethyleneketones O
via α,β-ethyleneoximes, isoxazoles, and β-aminoketones s. 28, 204

Reactions with *iodine/pyridine* I/C_5H_5N
Lactones from carboxylic acids
s. 28, 205 $\underset{COOH}{\overset{H}{C}} \rightarrow \underset{CO}{\overset{O}{C}}$

tert-Butyl hypochlorite/lithium methoxide Me_3C·OCl/LiOMe
Stereospecific oxidative ring closure O
Spirooxazolid-4-one ring s. 29, 243

Sodium hypochlorite s. under $RuCl_3$ NaOCl

Sodium chlorite/sulfuric acid $NaClO_2$/H_2SO_4
Ketones from sec. alcohols CHOH → CO
s. 29, 244

Perchloric acid (s. a. under HNO_3) $HClO_4$

Perchloric acid/acetic anhydride $\qquad HClO_4/Ac_2O$
N-(4-En-Δ^2-oxazolin-5-yliden)ammonium salts $\qquad$ O
s. 28, 206

Sodium periodate (s. a. under RuO_2) $\qquad NaIO_4$
1,3-Benzodioxoles
from o-hydroxycarbinols – Oxidative rearrangement s. 28, 207

Manganese dioxide $\qquad MnO_2$
Aldehydes from prim. alcohols $\qquad CH_2OH \rightarrow CHO$
s. 28, 53

Quinoline-3,4-dione 1-oxides $\qquad \leftarrow$
from 1,3-dihydroxy-4(1H)-quinolones s. 28, 447

1,3-Benzodioxole-2-spirocyclohexadien-4'-ones $\qquad$ O
s. 26, 231

δ-(2-Furyl)ketones $\qquad \leftarrow$

157.

cis-3-(1,2-Epoxy-2,6,6-trimethylcyclohexyl)prop-2-en-1-ol refluxed 18 hrs. with MnO_2 in benzene → 6-methyl-6-(2-furyl)heptan-2-one. Y: 92%. F. e. s. B. R. von Wartburg and H. R. Wolf, Helv. 57, 916 (1974).

Potassium ferrate(VI) $\qquad K_2FeO_4$
α,β-Ethylenealdehydes from 2-ethylenealcohols $\qquad C:C\cdot CH_2OH \rightarrow C:C\cdot CHO$
s. 26, 194

Ferric chloride $\qquad FeCl_3$
Ethyleneketones from cyclopropanol ring $\qquad$ C
s. 26, 29

Cobalt oxide/oxygen $\qquad \leftarrow$
Autoxidation $\qquad CHOH \rightarrow CO$
Oxo compds. from alcohols s. 26, 463

Nickel $\qquad Ni$
γ-Lactones from γ-hydroxyketones $\qquad$ O
s. 29, 245

Ruthenium dioxide/sodium periodate $\qquad RuO_2/NaIO_4$
Ketones from sec. alcohols $\qquad CHOH \rightarrow CO$
s. 28, 208; carbohydrate derivs. s. J. S. Brimacombe, J. Minshall, and C. W. Smith, Soc. Perkin I *1975*, 682

Keto- from hydroxy-lactones – Ketocarboxylic acids from lactones – Comparison of oxidants s. 28, 208

Dichlorotris(triphenylphosphine)ruthenium(II)/-
benzalacetone $\qquad RuCl_2(Ph_3P)_3/PhCH:CHCOCH_3$
Transfer-dehydrogenation $\qquad CH(OH)CH(OH) \rightarrow COCO$
α-Diketones from glycols s. 28, 209; α,β-ethyleneketones from 2-ethylenealcohols cf. Y. Sasson and G. L. Rempel, Can. J. Chem. 52, 3825 (1974)

Ruthenium trichloride/sodium hypochlorite $\qquad RuCl_3/NaOCl$
Ketones from sec. alcohols $\qquad CHOH \rightarrow CO$
s. 27, 280

Palladium-carbon   *Pd-C*
Oxidative furan ring closure   O

158.

Air passed slowly for 10 hrs. into a refluxing soln. of 4-hydroxy-3-(4-methoxyphenyl)-7-methoxycoumarin in diphenyl ether → coumestrol dimethyl ether. Y: 25%. T. Kappe and A. Brandner, Z. Naturf. *29b*, 292 (1974).

Platinum oxide PtO_2
Lactones from diols
s. *13*, 314 suppl. 29

Via intermediates   *v.i.*
Ketones from sec. alcohols   CHOH → CO
**Phosphinic acids from chlorophosphines
via phosphinous acid esters**

159.

1-Decalol and chlorodiphenylphosphine added successively to a NaH-mineral oil dispersion in toluene, after the vigorous reaction has ceased refluxed and stirred 3 hrs. under an air stream → 1-decalone (Y: 88% as the 2,4-dinitrophenylhydrazone) and diphenylphosphinic acid (Y: 85%). F. e., also isolation of intermediates, s. I. Shahak and Y. Sasson, Synthesis *1974*, 358.

o-Hydroxyaldehydes CH_2OH → CHO
from hindered o-hydroxymethylphenols via 6-spirooxirano-2,4-cyclohexadienones s. *29*, 246

Furo[2,3-b]quinolines   O
s. *29*, 247

Oxygen ↑   OC⇑O

Without additional reagents   *w.a.r.*
Ketones from phosphoroperoxidates $CH \cdot O \cdot O \cdot PO(OR)_2$ → CO
s. *29*, 248

1,1-Acoxyselenides   ←
from diacyloxyselenuranes s. *29*, 108

Enollactones from ketocarboxylic acids   O
s. *27*, 919

Chromenes by cyclodehydration
s. *28*, 210

**Benzofuro[2,3-c]pyrrole ring
from cyclic phenolnitrones**

159a.

A soln. of 2-ethyl-4-(2-hydroxy-3-methoxyphenyl)-3-methyl-1-pyrroline 1-oxide in xylene refluxed 45 min. under N_2, and the product isolated as the hydrochloride

→ 3-ethyl-3a,8b-dihydro-5-methoxy-3a-methyl-1H-benzofuro[2,3-c]pyrrole hydrochloride. Y: 76.7%. F. e. s. S. Klutchko et al., J. Org. Chem. *38*, 3012 (1973); prepn. of the startg. m. cf. ibid. *38*, 3049.

Oxazoles from 2-ketonitrones
s. *28*, 219

Irradiation (s. a. under Hematoporphyrin) ⟊
Protection of carboxyl groups COOR → COOH
as benzoin esters – Selective removal of the photo-sensitive protective group
s. *27*, 5

Sodium hydride (s. a. under N-Tosylimidazole) NaH
Cyclic ethers from diol monotosylates O
Steroidal spiroethers s. *26*, 236

Oxido compds. C(OH)C(OSO₂R) → ▽
from glycol monosulfonates s. *26*, 237

Cyclic ethers from dimesylates
Stereospecific ring closure s. *7*, 355 suppl. *28*

C·OSO₂Me   C
| →   \\O
C·OSO₂Me   C/

Sodium hydroxide NaOH
Cyclic phosphoramidates O

160.

Stereoisomers. 1 N NaOH added to N-(5'-deoxy-5'-adenosyl)-N-octylphosphoric acid bis-(p-nitrophenyl ester) amide in acetone, and allowed to stand 2 hrs. at room temp. → 5'-deoxy-5'-(octylamino)adenosine-3'-phosphoric acid p-nitrophenyl ester 3',5'-cycloamide. Y: 90%. Also isolation of the P-diastereomer s. A. Murayama et al., B. *106*, 3127 (1973).

Potassium hydroxide KOH
5,6-Dihydro-2-pyrones and 4-hydroxytetrahydro-2-pyrones
from β,δ-dihydroxycarboxylic acid *tert*-butyl esters s. *27*, 255

2,1-Benzisoxazoles from nitro compds.
s. *27*, 256

2H-Pyrano[3,2-c]quinolines ☉
s. *29*, 249

Potassium hydroxide/alcohol
Redoxidation of cyclic N-oxides ←

161.

2-Phenyl-3-(α-hydroxybenzyl)quinoxaline 1-oxide refluxed 1 hr. in methanolic 10%-KOH → 2-phenyl-3-benzoylquinoxaline. Y: 85%. F. e. s. M. J. Haddadin et al., Tetrahedron *30*, 659 (1974); cf. U. Eholzer et al., Synthesis *1974*, 296.

Sodium/alcohol   *NaOR*
Oxido compds. from glycol monosulfonates $C(OH)C(OSO_2R) \to$ 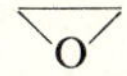
    Carbohydrate derivs. s. 26, 237

Δ^2-**Oxazoline ring from 1-acoxy-2-acylamines**
    with double inversion – Carbohydrate derivs. s. 27, 257

Amines   ←
Carboxylic acid esters $CH(OR) \cdot OOH \to COOR$
    from 1,1-alkoxyhydroperoxides s. 28, 215

Pyrrolidine s. under CH_3COOH   ←

Pyridine s. under $PhSO_2Cl$ C_5H_5N

Mercuric salt s. under Thionocarboxylic acid esters Hg^{2+}

Sodium tetrahydridoborate $Na[BH_4]$
Lactones from ketocarboxylic acids $\overset{\text{CO}}{\underset{\text{COOH}}{\diagup}} \to \overset{\text{CHO}}{\underset{\text{CO}}{\diagup}}$
    Steric control s. 20, 53

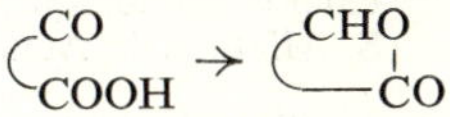

Sodium tetrahydridoborate/sodium hydroxide $Na[BH_4]/NaOH$
α,β-**Ethylene-γ-lactones**
    from α,β-ethylene-γ-ketocarboxylic acids s. 27, 258

Diisobutylaluminum hydride/sulfuric acid $i\text{-}Bu_2AlH/H_2SO_4$
Lactones from ketocarboxylic acids
    Stereospecific ring closure s. 28, 211

Clay   ←
Cyclic ethers from diols
s. 26, 238

Dicyclohexylcarbodiimide $RN:C:NR$
Pummerer rearrangement
    3,1-Benzoxathian-4-ones – Chirality S→C-transfer s. 29, 250 suppl. 30

Hematoporphyrin/irradiation   ←
δ-**(2-Furyl)ketones**   ←

162.

A stirred soln. of *trans*-5,6-dihydroxy-5,6-dihydro-β-ionone in methylene chloride containing hematoporphyrin as sensitizer irradiated 40 hrs. under N_2 with the Pyrex-filtered light of a high-pressure Hg-lamp → 6-methyl-6-(5-methylfur-2-yl)-heptan-2-one. Y: 80%. Also from epoxy instead of glycol compds., and f. dyes as sensitizers, s. W. Skorianetz and G. Ohloff, Helv. 56, 2151 (1973).

Acetic anhydride (s. a. under H_2SO_4) Ac_2O
Lactones from hydroxycarboxylic acids $\overset{\text{OH}}{\underset{\text{COOH}}{\diagup}} \to \overset{\text{O}}{\underset{\text{CO}}{\diagup}}$
s. 27, 505

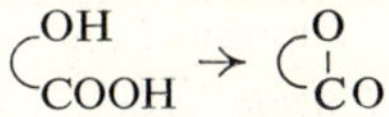

5-Oxazolones
    from α-acylaminocarboxylic acids s. 30, 451

4-Alkylidene-2-vinyl-5-oxazolones

163.

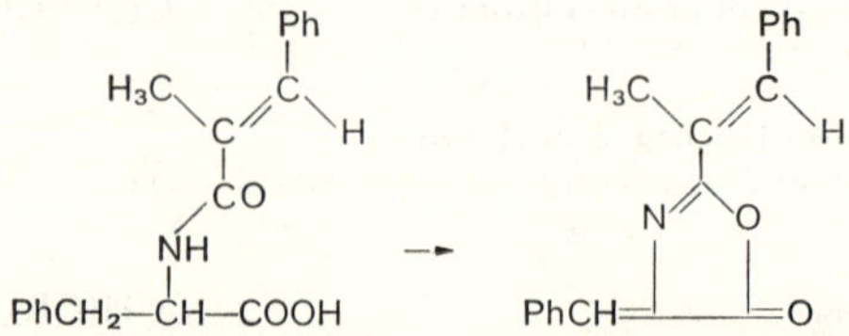

Pyridine hydrobromide perbromide added to a soln. of N-[(E)-α-methylcinnamoyl]-DL-phenylalanine in acetic anhydride containing a little pyridine, warmed to 80°, cooled to room temp., more pyridine added, and stirred for 10 min. at room temp. → azlactone deriv. Y: 62%. F. e. s. J. M. Riordan and C. H. Stammer, J. Org. Chem. *39*, 654 (1974).

Ring closures with sulfoxides
Pummerer rearrangement
3,1-Benzoxathian-4-ones from o-carboxysulfoxides s. *29*, 250; with dicyclohexylcarbodiimide, intramolecular S→C-transfer of chirality, cf. B. Stridsberg and S. Allenmark, Acta Chem. Scand. *B28*, 591 (1974)

Propionic anhydride $(C_2H_5CO)_2O$

2-Aminofurans from β-ketocarboxylic acid amides via 2,3-dihydrofurfurylideneammonium salts

164.

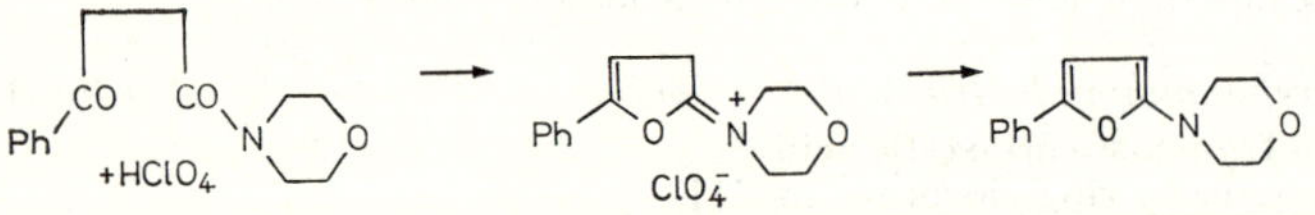

A stirred suspension of β-benzoylpropionomorpholide in propionic anhydride cooled below 5°, treated slowly with perchloric acid, and stirring continued 0.5 hr. → intermediate salt (Y: 97%) suspended in ether, triethylamine added dropwise with stirring, and the product isolated after 0.5 hr. → 2-morpholino-5-phenylfuran (Y: 96%). F. e., also in acetic anhydride, s. G. V. Boyd and K. Heatherington, Soc. Perkin I *1973*, 2523.

Trifluoroacetic anhydride $(CF_3CO)_2O$

Carboxylic acid anhydrides from carboxylic acids   2 COOH → CO·O·OC
Dicarboxylic acid anhydrides s. *27*, 260

Oxazoles from α-acylaminocarboxylic acid amides   O

165.

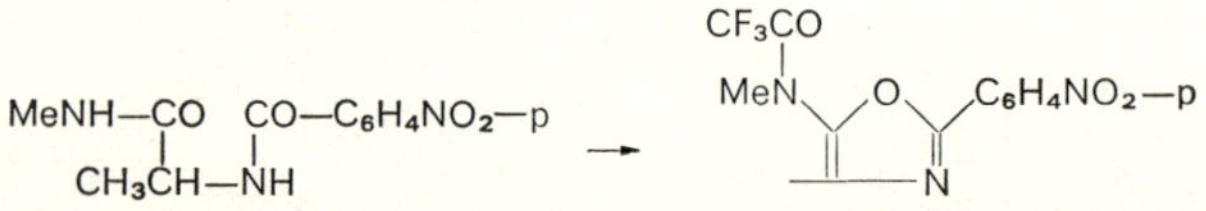

Trifluoroacetic anhydride added to a soln. of the startg. m. in anhydrous chloroform, and stirred 3 hrs. at room temp. → 4-methyl-5-(N-methyl-N-trifluoroacetylamino)-2-p-nitrophenyloxazole. Y: 91%. F. e. s. D. Clerin and J.-P. Fleury, Bl. *1973*, 3127, 3134.

Amberlyst 15   ←

O-Heterocyclics by cyclodehydration
Furans from γ-diketones s. *29*, 251

Formic acid   HCOOH

1,2,4-Dioxazol-3-ones
from N-acylperoxycarbamic acids s. *28*, 212

Acetic acid/pyrrolidine ←
O-Heterocyclics from diols
    with cleavage of ketals and migration of a carbon-carbon double bond s. 27, 261

Trifluoroacetic acid CF_3COOH
Furoic from 3-deoxy-2-ketoaldonic acids
    s. 29, 252

**Heterocyclics from β-ketosulfoxides
via Pummerer rearrangement**

166.

A soln. of the startg. β-keto sulfoxide and trifluoroacetic acid in benzene refluxed 40–90 min. until tlc indicates the absence of startg. m. → product. Y: 81%. F. e. s. D. T. Connor and M. von Strandtmann, J. Org. Chem. *39*, 1594 (1974).

Oxalic acid $(COOH)_2$
Tetrahydrofurylideneacetic acid esters
    from ε-hydroxy-β-ketocarboxylic acid esters s. 30, 448

**4-Hydroxy-3(2H)-furanones
from 2-acetylene-1,4-diols via 1,4-dihydroxy-2,3-diketones**

167.

3-Hexyne-2,5-diol ozonized 45 min. at –15 to –13° with an O_3–O_2-stream in methanol, then treated at –15° during 0.5 hr. with an ethereal soln. of triphenylphosphine, allowed to come to room temp., and stirred overnight → crude 2,5-dihydroxyhexane-3,4-dione (Y: 52% if isolated) extracted with water, treated with oxalic acid, and refluxed 4 hrs. under argon → 4-hydroxy-2,5-dimethyl-3(2H)-furanone. Overall Y: 58.3%. F. e. s. L. Re, B. Maurer, and G. Ohloff, Helv. *56*, 1882 (1973).

Thionocarboxylic acid esters/mercuric salt $RCSOR/Hg^{2+}$
Carboxylic acid anhydrides and esters $(RCOO)_2Hg \to (RCO)_2O$
    from mercuric carboxylates and thionocarboxylic acid esters s. 28, 213

Dimethyl sulfoxide Me_2SO
Aldehydes and carboxylic acids   C
    from ozonides s. 28, 214

Carboxylic acid esters $CH(OR) \cdot OOH \to COOR$
    from 1,1-alkoxyhydroperoxides s. 28, 215

2,2′-Dipyridyl disulfide s. under Ph_3P   ←

Triphenylmethyl chloride $Ph_3C \cdot Cl$
Bicyclic enolethers
    from cyclic hydroxyketones s. 21, 294 suppl. 29

Triphenylphosphine/2,2'-dipyridyl disulfide
Macrolactonization by double activation

168.

Prostaglandin $F_{2\alpha}$ 11,15-bis(tetrahydro-2-pyranyl) ether treated 15 hrs. at 25° with 1.5 equivalents each of 2,2'-dipyridyl disulfide and triphenylphosphine in dry xylene soln., dil. with more xylene, and refluxed 5 hrs. under N_2 → product. Y: 90%.– Both carboxyl and hydroxyl functions are activated by internal proton transfer resulting in an electrostatically driven ring closure without the need for basic or acidic catalysts. F. e. s. E. J. Corey, K. C. Nicolaou, and L. S. Melvin, Jr., Am. Soc. 97, 653/4 (1975); cf. ibid. 97, 2287.

Triphenyldiethoxyphosphorane $\qquad$ $Ph_3P(OEt)_2$
Heterocycles by cyclodehydration
  with triphenyldiethoxyphosphorane s. 28, 216

Polyphosphoric acid $\qquad$ $H(PO_3H)_xOH$
α,β-Ethylene-γ-lactones
  from γ-ketocarboxylic acids with isocyclic skeletal rearrangement s. 28, 217

N-Tosylimidazole/sodium hydride
Oxido compds. from glycols $\qquad$ C(OH)C(OH) → C–O–C
  Anhydrosugars s. 29, 101

Thionyl chloride $\qquad$ $SOCl_2$
Ketones from 1,1-diols $\qquad$ $C(OH)_2$ → CO
s. 26, 239

2-(γ,δ-Ethylene)-5,6-dihydro-4H-1,3-oxazines
  via Claisen-type rearrangement of 2-ene-1-(β,γ-ethylene)tetrahydro-1,3-oxazines
  s. 29, 704

4H-3,1-Benzoxazin-4-ones
  from o-acylaminocarboxylic acids s. 29, 391

Thionyl bromide $\qquad$ $SOBr_2$
Dicarboxylic acid anhydrides $\qquad$ (COOH)(COOH) → (CO)(CO)O
  from dicarboxylic acids s. 30, 364

Benzenesulfonyl chloride/pyridine $\qquad$ $PhSO_2Cl/C_5H_5N$
β-Lactones from β-hydroxycarboxylic acids $\qquad$ (OH)(COOH) → (O)(CO)
s. 27, 981

p-Toluenesulfonic acid $\qquad$ TsOH
α,β-Ethylene-γ-lactones
  from γ-ketocarboxylic acids s. 29, 253

3,4-Dihydro-2-pyrones
s. 26, 240

1,6-Anhydrosugars from glycosides
s. 26, 241

Flav-2-ene ring closure
  with subsequent Vilsmeier aldehyde synthesis s. 26, 242

Sulfuric acid (s. a. under i-Bu$_2$AlH) H_2SO_4

γ-Lactones from β-hydroxycarboxylic acids
s. *27*, 262

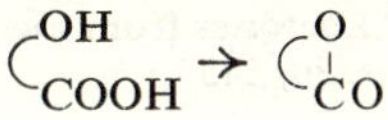

Lactones from hydroxycarboxylic acids
with skeletal rearrangement s. *28*, 218

2-Benzopyr-3-ones
from 2,2-dialkoxy-1,3-dihydroxyindans via o-ketoacetic acids, and from 1,3-dihydroxy-2-indanones s. *26*, 243

Sulfuric acid/acetic anhydride H_2SO_4/Ac_2O

Oxazoles from 2-ketonitrones
s. *28*, 219

Hydrogen chloride   *HCl*

Lactones from hydroxycarboxylic acids
s. *27*, 56, 505

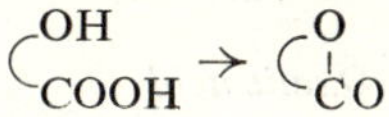

Hydrogen bromide   *HBr*

Ring closure with cleavage of ethers
Pyranol[2,3-b]pyridines s. *26*, 244

Ferric chloride $FeCl_3$

Δ²-Oxazoline ring from 1-acoxy-2-acylamines
s. *27*, 257

Via intermediates   *v.i.*

**Lactones from hydroxycarboxylic acids
via hydroxythiolic acid esters**

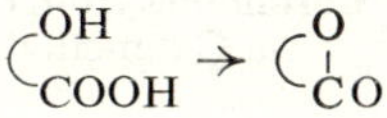

169.

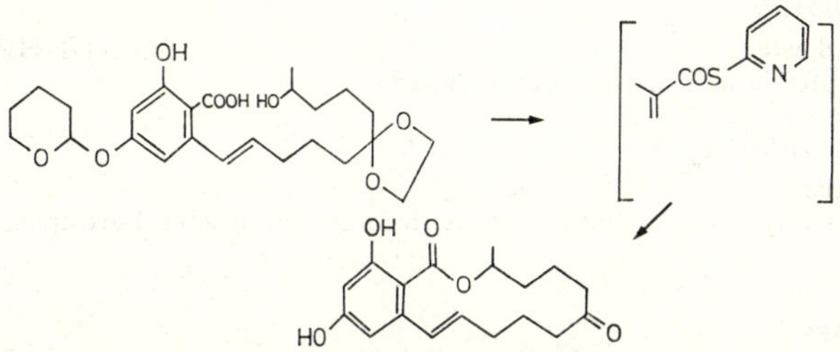

Macrocyclic lactones. Startg. m. converted into the 2-pyridinethiol ester according to Synth. Meth. *26,* 621, the resulting mixture containing the crude intermediate dil. with dry benzene, added slowly from a mechanically driven syringe under argon during ca. 15 hrs. to refluxing benzene, then the protective groups removed by warming 5 hrs. at 60° in acetic acid-water-tetrahydrofuran → (±)-zearalenone. Y: 75%. F. e. s. E. J. Corey and K. C. Nicolaou, Am. Soc. *96,* 5614 (1974).

Nitrogen ↑   OC⇑N

Irradiation   *hv*

2,1-Benzisoxazoles
from 1,2,3-benzotriazine 3-oxides s. *29*, 254

Sodium/tert-butanol $Na/Me_3C \cdot OH$

α,β-Ethyleneoximes and Δ²-isoxazolines
from 1,3-aminooximes s. *27*, 263

Pyridine C_5H_5N
Lactones from mesyloxycarboxylic acid amides
s. *26*, 245

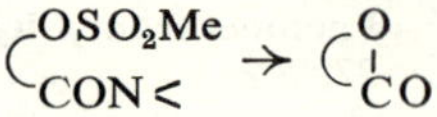

$\begin{matrix} \text{OSO}_2\text{Me} \\ \text{CON}< \end{matrix} \rightarrow \begin{matrix} \text{O} \\ | \\ \text{CO} \end{matrix}$

Intramolecular nucleophilic substitution ○
 of tert. amino groups – Anthr[2,1-d]oxazole-6,11-diones s. *26*, 246

Mercuric oxide/sulfuric acid HgO/H_2SO_4
Tetrahydropyrans from 5-aminoalcohols
s. *29*, 140

Cation exchanger ←
[1]Benzopyrano[4,3-b]-1,5-benzodiazepines
s. *29*, 255

Oxalic acid $(COOH)_2$
Aldehydes from cyclic 1,1-aminoethers
s. *28*, 793

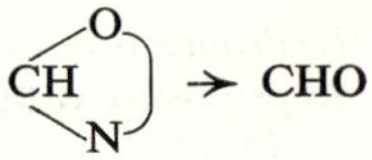

$\overset{O}{\underset{N}{CH}} \rightarrow CHO$

Sulfuric acid (s. a. under HgO) H_2SO_4
O-Heterocyclics from acylaminoalcohols ○
s. *27*, 264

Chromones from o-hydroxyketones
 via C-dimethylaminomethylenation s. *26*, 247

Selenium dioxide SeO_2
Sulfinic acid esters $SO_2NHNHR \rightarrow S(O)OR$
 from sulfonic acid hydrazides s. *29*, 256

Ammonium chloride NH_4Cl
Oxo compds. C
 from tetrahydro-1,3-oxazine ring with retention of acetal groups s. *29*, 257

Halogen ↑   OC⇑Hal

Without additional reagents w.a.r.
1,3-Dioxolan-2-ones ○
s. *29*, 258

2-Amino-Δ^2-oxazolines
 from halogenureas s. *30*, 291

Furan from cyclopropane ring ☉
s. *26*, 248

Sodium hydride NaH
Benzofurans ○
s. *28*, 220

4(5H)-Oxazolones
 from α-halogenodicarboxylic acid imides s. *27*, 265

Potassium hydroxide　　　　　　　　　　　　　　　　　　　　　　　　KOH
Flavones from o'-hydroxychalcone dibromides
Emilewicz-Kostanecki ring closure

170.

An ethanolic suspension of 3'-bromo-2'-hydroxy-4',6'-dimethoxychalcone dibromide treated 2 hrs. *at room temp.* with aq. *0.2 M KOH*, and dil. with water → 8-bromo-5,7-dimethoxyflavone. Y: ca. 100%. J. A. Donnelly and H. J. Doran, Tetrah. Let. *1974*, 4083.

1,4-Oxazin-3-ones
s. *26*, 250

Sodium methoxide　　　　　　　　　　　　　　　　　　　　　　　　NaOMe
Arene oxides
s. *28*, 221; cf. Am. Soc. *97*, 3185 (1975)

Potassium/tert-butanol　　　　　　　　　　　　　　　　　　　$K/Me_3C \cdot OH$
Oxido compds. from *trans*-1,2-bromohydrins　　　　CBrC(OH) →
　　Optically active compds. s. *26*, 144

Potassium tert-butoxide　　　　　　　　　　　　　　　　　　　　$KOCMe_3$
α,β-Oxidosulfonic acid amides　　C(OH)CHal(SO$_2$N<) → ─SO$_2$N<
s. *28*, 222

n-Butyllithium　　　　　　　　　　　　　　　　　　　　　　　　　　*n-BuLi*
α-Chloroketones from 2,2-dichloroalcohols　　　　　　　　　　　　CClCO
s. *28*, 945

Sodium hydrogen carbonate　　　　　　　　　　　　　　　　　　　$NaHCO_3$
Quinone methids　　　　　　　　　　　　　　　　　　　　　　　　←
s. *27*, 266

Potassium carbonate s. under CuO　　　　　　　　　　　　　　　　K_2CO_3

Sodium salt　　　　　　　　　　　　　　　　　　　　　　　　　　　Na^+
1H,3H-4,2,1,5-Benzoxathiadiazocine 2,2-dioxides　　　　　　　　　O
s. *26*, 249

Triethylamine　　　　　　　　　　　　　　　　　　　　　　　　　　Et_3N
1-Subst. isochromene from 1,2,3,5-tetrahydroisoquinoline ring　　CN → CO
s. *27*, 267

Furo[2,3-b]quinolines　　　　　　　　　　　　　　　　　　　　　　O
s. *29*, 247

1,4-Oxazin-3-ones
s. *26*, 250

Pyridine (s. a. under CuO)　　　　　　　　　　　　　　　　　　　　C_5H_5N
Oxido compds. from 1,2-bromohydrins　　　　　　　　　CBrC(OH) →
s. *12*, 206 suppl. 28

2,7-Dioxatwistans　　　　　　　　　　　　　　　　　　　　　　　　O
s. *28*, 223

Cupric oxide/potassium carbonate/pyridine　　　　　　　$CuO/K_2CO_3/C_5H_5N$
O-Heterocycles
　　by intramolecular Ullmann reaction s. *29*, 259

Alumina Al_2O_3
Cyclic ethers from oxoiodides ←

171.

A soln. of 3,4-seco-4-iodocholest-5-en-3-al in *n*-hexane absorbed on neutral alumina, and eluted with benzene and *n*-hexane → 4-oxa-A-homocholest-5-ene. Y: 92%. H. Suginome and K. Kato, Tetrah. Let. *1973*, 4143.

Stannic chloride $SnCl_4$
Reactions with 5-chloro-2,4-dienals ⊙

172.

Pyrylium salts. *cis,trans*-2,3,4,5-Tetrachloro-5-phenyl-2,4-pentadienal in anhydrous CS_2 warmed 14 hrs. with $SnCl_4$ at 50° → 3,4,5-trichloro-2-phenylpyrylium pentachlorostannate. Y: 89%. Also with the *cis,cis*-isomer, and f. reactions s. A. Roedig et al., B. *107*, 1136 (1974).

Phosphoric acid H_3PO_4
Insertion of carbonyl into carbon-carbon bonds ⊙
via C-dichloromethylation s. *27*, 726

Sulfur ↑ OC⇑S

Without additional reagents w.a.r.
Lactones
from hydroxythiolic acid esters – Macrocyclic lactones
s. *30*, 169

O-Heterocyclics by pyrolysis ☉
s. *27*, 268

Irradiation s. under Methylene blue ☼

Electrolysis ⚡
Lactones from carboxysulfonium salts ⊙
Hydrocarbons from sulfonium salts s. *29*, 260

Sodium hydroxide (s. a. under $R_3O^+BF_4^-$ and Br) *NaOH*
Xanthones from thioxanthone 10,10-dioxides
via 2′-hydroxybenzophenone-2-sulfinic acids s. *28*, 224

Potassium hydroxide/methanol *KOH*
Δ^2-1,2,3-Oxadiazoline 2-oxides
from 2-bromo-N-sulfonyl-N-nitramines

173.

KOH-powder added in small portions at 20° to a suspension of N-(β-bromoethyl)-N,m-dinitro-p-toluenesulfonamide in methanol, and stirred 3.5 hrs. → 4,5-dihydro-1,2,3-oxadiazole 2-oxide. Y: 84.5%. F. e. s. O. A. Luk'yanov et al., Izvest. *1973*, 1294; C. A. *79*, 78697.

Sodium/alcohol   *NaOR*
Dihydropyrans from ketosulfonium salts
s. 27, 269

Cupric acetate   *$(CH_3COO)_2Cu$*
Δ^2-Oxazolines from 2-hydroxythioureas
s. 27, 406

Cycloalkenes   ←
[2.3]-Sigmatropic rearrangement   C(SOAr)·C : CCl → C : C·CO
α,β-Ethyleneketones from γ-chloro-β,γ-ethylenesulfoxides s. 29, 261

Trialkyloxonium fluoroborate/sodium hydroxide $R_3O^+BF_4^-/NaOH$
Oxido compds.   CO → △O
from ketones via 2-hydroxythioethers s. 29, 262;
cf. A. Anciaux et al., Tetrah. Let. *1975*, 1617

Methylene blue/irradiation   ←
Ketones from sulfines   C : S : O → CO
Carboxylic acid chlorides from 1-chlorosulfines s. 27, 270

Chloramine-T   ←
Protection of oxo compds.   C
as cyclic monothioacetals or mercaptals – Removal of the protective group under mild conditions s. 27, 271

Tri-n-butylphosphine n-Bu_3P
Ethers from sulfenic acid esters   ·S·OR → OR
s. 27, 272

Bromine/sodium hydroxide   *Br/NaOH*
1,3,4-Oxadiazol-2(3H)-ones   O
from thionocarbazic acid esters

174.
    (CH₃)₂C=CH-CO   SC-OMe   (CH₃)₂C=CH⎤O⎡O
             | | →   N — NH
            NH — NH

Bromine added dropwise at 0–5° to a methanolic soln. of O-methyl 3-(3,3-dimethylacryloyl)thiocarbazate, decanted from the resulting sulfur, evaporated to dryness, and the residue boiled briefly with 2 N NaOH → 5-(isobuten-1-yl)-1,3,4-oxadiazol-2(3H)-one. Y: 43%. F. e. s. K. Rüfenacht, Helv. *57*, 487 (1974).

Via intermediates   *v.i.*
Lactones from alkylthiocarboxylic acids
via carboxysulfonium salts s. 29, 263

Remaining Elements ↑   **OC⇑Rem**

Without additional reagents   *w.a.r.*
Oxetanes
from (3-halogenalkoxy)stannanes s. 28, 225

Irradiation/bromotrichloromethane   *⇝/BrCCl₃*
Oxo compds. via alkoxystannanes   CHOSn≤ → CO
from alcohols s. 28, 226; oxidation with bromine cf. K. Saigo, A. Morikawa, and T. Mukaiyama, Chem. Lett. *1975*, 145

Sodium hydride   *NaH*
Ketones from phosphinous acid esters   CHOP< → CO
s. 30, 159

Potassium tert-butoxide   *KOCMe₃*
Oxido compds. and ketones   C(HgCl)CH(OH) → CHCO\O (epoxide)
from 2-hydroxymonoorganomercury chlorides s. *29*, 264

p-Toluenesulfonic acid   *TsOH*
Oxo compds. from enol phosphates   C:C·OP(O)(OR)₂ → CH·CO
Mono-α-dechlorination of oxo compds. s. *26*, 251

Sulfuric acid   *H₂SO₄*
s. *26*, 251

Chromium trioxide/sulfuric acid   *CrO₃/H₂SO₄*
α,β-**Ethyleneketones**   C
from 2-oxa-3-borolenes s. *28*, 866

Carbon ↑   OC⇑C

Without additional reagents   *w.a.r.*
Ethers from carbonic acid esters   ·OCOOR → OR
s. *27*, 273; s. a. R. Taylor, Tetrah. Let. *1975*, 593

Ketenes from O-silyl O-alkyl keteneacetals   C:C(OR)Si≤ → C:C:O
s. *28*, 227

Acyloximes from 1,1-acoxynitrones   C:N(O)C·OAc → C:NOAc

175.   Ph₂C=N—CHPh → Ph₂C=N—OAc
            | | [+ OHC—Ph]
            O OAc

C,C-Diphenyl-N-(α-acetoxybenzyl)nitrone heated in water or benzene → O-acetylbenzophenone oxime. Y: 75%. L. A. Neiman and S. V. Zhukova, Zh. Org. Khim. *9*, 1986 (1973); C. A. *79*, 136703.

Lactones from quaternary carbalkoxyammonium salts
s. *28*, 228   N⁺≤\COOR → C\CO (O)

Ethylenelactones from carboxyethyleneurethans   ←

176.

A soln. of the startg. m. in 4:1 water-dimethoxyethane refluxed 18 hrs. at pH 7 → product. Y: 99%. – The heterolysis of the p-arylurethano unit provides the driving force for lactonization. E. J. Corey and G. Moinet, Am. Soc. *95*, 6831 (1973).

*Irradiation (s. a. under 2,3-Diphenylquinoxaline, Ph₂CO,
and N-Iodosuccinimide)*   *hν*
Ketones from α-hydroxyketones   C(OH)COR → CO
s. *27*, 674

Electrolysis
Alcohols from carboxylic acids   C·COOH → C·OH
Degradation with loss of 1 C-atom

177.   n—C₈H₁₇—CH=CH—C₆H₁₃CH₂COOH → n—C₈H₁₇—CH=CH—C₆H₁₃CH₂OH

A soln. of oleic acid and NaOH in methanol-water electrolyzed 7 hrs. at ca. 20 v. and 0.8–0.1 amp. between a Pt-anode and a Hg-cathode → heptadec-*cis*-8-enol. Y: 44%. F. e. s. F. Gunstone, C. Scrimgeour, and S. Vedanayagam, Chem. Commun. *1974*, 916.

α,β-Ethyleneketones from enolesters   CH·C : C(OAc) → C : C·CO
s. *28*, 229 suppl. *29*

Sodium hydride   NaH
Heterocyclics from enephosphonium salts   ○

178.

Startg. m. refluxed in acetonitrile in the presence of a catalytic amount of NaH → 2,5-dimethylbenzoxazole. Y: 80%. – Similarly: 2-Methylquinazol-4-one. Y: 75%. F. e. s. E. Schweizer and S. V. DeVoe, J. Org. Chem. *40*, 144 (1975).

Sodium hydroxide (s. a. under H_2O_2)   NaOH
Ketones from cyanohydrins   C(OH)CN → CO
s. *30*, 155

Ketones from α-acoxynitriles   C(OAc)CN → CO
s. *30*, 557

3-Acyl-4-hydroxy-5,6-dihydro-2-pyrones   ←
  from 5-carbalkoxy-2,3-dihydro-2-pyrones s. *29*, 265

Potassium hydroxide/alcohol   KOH
Ring closure with α-amino-β,γ-ethylenenitriles   ○
  Benzoxazoles s. *27*, 535

Sodium nitrite/hydrogen chloride $NaNO_2/HCl$
Quinones from quinol monoethers   ←
s. *27*, 274

Lithium chloride   LiCl
Oxetanes from 2-oxo-1,3-dioxanes   ○
s. *29*, 666

Dimethylaniline $PhNMe_2$
C-Cleavage of α-ketoaldehyde aldoxime side chains   ←
s. *29*, 267

Pyridine C_5H_5N
α,β-Ethyleneketones   CH·C : C(OAc) → C : C·CO
  from enolesters via γ-halogenolesters s. *28*, 229; cf. U. Ravid and R. Ikan, Synth. Commun. *5*, 137 (1975)

2,3-Diphenylquinoxaline/irradiation ←
Aldehydes from 2-aminoalcohols $CH(OH)C \cdot NH_2 \rightarrow CHO$

179. $PhCH(OH)CH_2NHPh \longrightarrow PhCHO$

C-Cleavage. A soln. of 2-anilino-1-phenylethanol in benzene irradiated under N_2 with light of maximum emission at 350 nm in the presence of 2,3-diphenylquinoxaline as sensitizer → benzaldehyde. Y: 92%. F. e. and sensitizers s. R. S. Davidson and S. P. Orton, Chem. Commun. *1974*, 209.

Silver carbonate-Celite ←
Ketones from 2-acetylenealcohols $C(OH)C\equiv CH \rightarrow CO$
Cleavage s. *28*, 230

Zinc Zn
Reductive removal of protective groups
Protection of carbonyl groups as bromoethylethylene acetals, $C{<}^O_O \rightarrow CO$
also as mono- or di-2,2,2-trichloroethyl acetals s. *28*, 231

Mercuric oxide s. under Iodine HgO

Alumina Al_2O_3
Acetylenebenzofurans from cumulenehalides O
Activation of unsatd. halides by adsorption – Ring closure with ether cleavage s. *26*, 252

Triphenylcarbonium salt Ph_3C^+
Oxo compds. from trityl ethers $CHOTri \rightarrow CO$
s. *28*, 232

Benzophenone/irradiation Ph_2CO/⧫⧫⧫
Lactones from lactolides $C{<}^O_{CH(OR)} \rightarrow C{<}^O_{CO}$
s. *28*, 233

Acetic anhydride s. under HCl Ac_2O

N-Iodosuccinimide/irradiation $CH_2CO{>}NI$/⧫⧫⧫
  CH_2CO
Oxo compds. from alcohols with C-cleavage $>C(OH)R \rightarrow >CO$
Ketones from tert. alcohols s. *29*, 268

Stannous chloride/hydrogen chloride $SnCl_2/HCl$
6-Chromanols O

180.

SnCl$_2$ in concd. HCl added to a soln. of 2,5-diacetoxy-3-(3-methylbut-2-enyl)-toluene in dioxane, and refluxed 8 hrs. → 2,2,8-trimethylchroman-6-ol. Y: 96%. F. e. s. S. Inoue et al., Soc. Perkin I *1974*, 2097.

Lead tetraacetate $(CH_3COO)_4Pb$
α- from β-Hydroxycarboxylic acid esters C(OH)COOR
by oxidative cleavage s. *26*, 253

Nitrogen dioxide-nitric oxide NO_2/NO
Ketones from oxazolidine N-oxyls C
s. *28*, 234

Phosphorus oxide chloride $POCl_3$
Benzothiazol-2-ones ←
  from benzothiazole N-oxides s. *29*, 269

Ozone (s. a. under Iodine) O_3
Ozonization of acetals $CH(OR)_2 \rightarrow COOR$
  Carboxylic acid esters – Stereospecific ozonization of glycosides s. *28*, 235

Hydrogen peroxide/sodium hydroxide $H_2O_2/NaOH$
Dakin oxidation $ArCOR \rightarrow ArOH$
  Catechols from o-hydroxyacetophenones s. *30*, 67

Hydrogen peroxide/sulfuric acid H_2O_2/H_2SO_4
3,4-Benzocoumarins from diphenic acids ○
s. *26*, 254

Thionyl chloride $SOCl_2$
Ketones from α-hydroxycarboxylic acids $C(OH)COOH \rightarrow CO$
  Ring contraction of cyclic ketones s. *27*, 275

p-Toluenesulfonic acid $TsOH$
β-Alkylthio-α,β-ethylenecarboxylic acid esters $(C:C:C(OR)_2 \rightarrow C:CHCOOR$
  from 1,1-dialkoxy-3-alkylthioallenes s. *30*, 584

3,4-Dihydro-2-pyrones ○
s. *26*, 240

Sulfuric acid H_2SO_4
Ring expansion of O-heterocyclics ○
  Conversion of furan derivs. into monosaccharides – 6-Hydroxy-2,6-dihydro-3-pyrones from furylcarbinols by oxidative ring expansion s. *27*, 326

Iodine/mercuric oxide/ozone $I/HgO/O_3$
Carboxylic acid esters $R \cdot CO \cdot O \cdot OC \cdot R \rightarrow R \cdot COOR$
from carboxylic acid anhydrides
  by oxidative decarbonylation s. *26*, 255

Sodium hypochlorite s. under $RuCl_3$ $NaOCl$

Periodate IO_4^-
Pentoses from hexoses ←
s. *27*, 277

Hydrogen chloride HCl
Ketones from α-siloxynitriles $C(CN) \cdot OSi \leqq \rightarrow CO$
s. *29*, 808

α,β-Ethyleneoxo compds. from allenyl ethers $C:C:C(OR) \rightarrow C:CHCO$
s. *27*, 835

β-Keto- from β-alkoxy-β,γ-ethylene-carboxylic
acid esters $C:C(OR)C \cdot COOR \rightarrow CHCOC \cdot COOR$
s. *29*, 159

5-β-Ketocyclopent-2-enones ←
from enolesters via cyclobutane ring closure and opening

181.

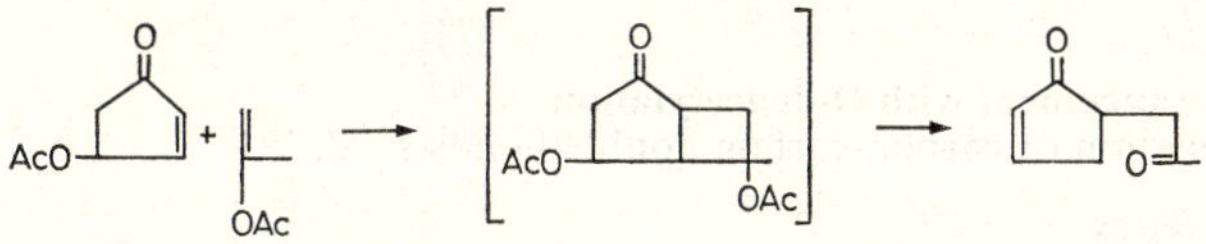

A soln. of 4-acetoxy-2-cyclopenten-1-one and isopropenyl acetate in benzene irradiated 2 hrs. at 0° and 14 hrs. at room temp. with the Pyrex-filtered light of a 450 w. high-pressure Hg-vapor lamp, the crude adduct dissolved in tetrahydrofuran, treated with N HCl, and stirred 96 hrs. at room temp. under N_2 → 5-(2'-oxopropyl)-2-cyclopenten-1-one. Overall Y: 63%. F. e. s. H. J. Liu, Synth. Commun. *3,* 441 (1973).

3(2H)-Furanones from α,β-acetylene-γ-alkoxyketones   O

182.

Dry HCl-gas passed with water-cooling into a soln. of 4-methoxy-1,4,4-triphenyl-2-butyn-1-one in methanol → 2,2,5-triphenyl-3(2H)-furanone. Y: 90%. F. e. s. W. Ried and A. Marhold, B. *107,* 1714 (1974).

Chroman-2-carboxylic acids from acoxycyanohydrins

183.

A suspension of crude 2-cyano-4-(2,5-diacetoxy-3,4,6-trimethylphenyl)butan-2-ol in concd. HCl heated 40 hrs. at 90° under N_2 → 6-hydroxy-2,5,7,8-tetramethylchroman-2-carboxylic acid. Y: 66%. J. W. Scott et al., J. Am. Oil Chemists' Soc. *51,* 200 (1974).

Cyclic α-diketones   C
    from 4,5-dichloro-1,3-dioxolan-2-one ring s. *30,* 468

(α-Diketo)phenols
s. *27,* 278

Hydrogen chloride/acetic anhydride HCl/Ac_2O
α-Oxo-N-heterocyclics from o-carboxy-N-oxides   ←
s. *26,* 256

Hydrogen bromide/acetic acid HBr/CH_3COOH
Ring closure with alkoxyhalides   O
s. *27,* 279

3,4-Disubst. 2(5H)-furanones   ←
s. *27,* 710

Nickel   Ni
Oxo from formoxy compds.   >CHOCHO → >CO

184.

Cyclohexyl formate heated 11 hrs. at 230–250° with Raney-Ni → cyclohexanone. Y: 62%. F. e. s. M. R. Sandner and D. J. Trecker, J. Org. Chem. *38,* 3954 (1973).

Oxidations with Ruthenium trichloride-hypochlorite $RuCl_3/ClO^-$
Ketones from sec. alcohols   C(OH)Ar → CO
    Oxidative elimination of phenyl groups s. *27,* 280

Platinum oxide PtO_2
Lactone rearrangement with O-deacetylation   ←
    Hydrogenation of carbon-carbon double bonds s. *27,* 281

Via intermediates   v.i.
Ketones from α-halogenocarboxylic acid halides   C(Hal)COHal → CO
    Curtius degradation with loss of 1 C-atom s. *28,* 630

Formation of N—N Bond

Uptake ⇩

Addition to Nitrogen NN⇩N

Without additional reagents w.a.r.
Azido- from diazo-N-heterocyclics ←

185.

5-Diazoimidazole-4-carboxamide added during 40 min. to a soln. of hydrazine (≥ 95%) and abs. ethanol, stirred 20 hrs. at room temp. in the dark → 5-azidoimidazole-4-carboxamide. Y: 80%. F. e. s. Y. F. Shealy and C. A. O'Dell, J. Heterocyclic Chem. 10, 839 (1973).

Potassium salt K^+
Reaction of potassium triorganocyanoborates ←
   N-Azocyanoboron betaines – N-Diazonitrilium group s. 27, 282

Pyridine C_5H_5N
N-Chlorodiimide N'-oxides NO → N(O) : NCl
   from nitroso compds. s. 29, 270

Mesitylenesulfonyloxylamine $ArSO_2ONH_2$
N-Aminocyclimmonium salts
   from N-heterocyclics s. 28, 243;
   s. a. J. Heterocyclic Chem. 12, 107 (1975)

Hydroxylamine-O-sulfonic acid $H_2NO \cdot SO_3H$
Triazanium salts from hydrazines $>N \cdot NH_2$ → $>\overset{+}{N}(NH_2)_2$
s. 29, 271

Hydroxylamine O-sulfonic acid/potassium
hydrogen carbonate $H_2NO \cdot SO_3H/KHCO_3$
1-Aminopyridazinium salts ←
   from pyridazines s. 28, 236

Rearrangement ↻

Hydrogen/Nitrogen Type NN↻HN

Without additional reagents w.a.r.
v-Triazolo[3,4-a]pyrimidin-5(4H)-ones ←
   from v-triazolo[3,4-a]pyrimidin-7(4H)-ones and reverse rearrangement s. 29, 272

Piperidine ←
v-Triazolo[3,4-a]pyrimidin-7(4H)-ones ←
from v-triazolo[3,4-a]pyrimidin-5(4H)-ones s. *29*, 272

Hydrogen chloride *HCl*
Migration of N-nitroso groups ←

186. ⟨⟩—N—CO—N—CH₂CH₂Cl → ⟨⟩—N—CO—N—CH₂CH₂Cl
         | | | |
         NO   H   H   NO

A suspension of crude startg. m. in concd. HCl-ethanol stirred 1 hr. at 0–5° → N-(2-chlorethyl)-N'-cyclohexyl-N-nitrosourea. Y: ca. 95%. T. P. Johnston, G. S. McCaleb, and J. A. Montgomery, J. Med. Chem. *18*, 104 (1975).

Oxygen/Nitrogen Type NN⌒ON

Without additional reagents w.a.r.
5,11-Dihydro-11-oxopyrido[1,2-b]cinnolin-6-ium ylids ←
from 3-(2-pyridyl)-2,1-benzisoxazoles

187.

A soln. of 3-(2-pyridyl)-2,1-benzisoxazole in 1,2,4-trichlorobenzene refluxed 18 hrs. at ca. 215° → 5,11-dihydro-11-oxopyrido[2,2-b]cinnolin-6-ium hydroxide inner salt. Y: 75–94%. F. e. s. R. Y. Ning, W. Y. Chen, and L. H. Sternbach, J. Heterocyclic Chem. *11*, 125 (1974).

Potassium hydroxide *KOH*
3-Acylamino-1,2,4-triazol-5-one ←
from 3-ureido-1,2,4-oxadiazole ring s. *26*, 257

Exchange ↕

Hydrogen ↑ NN↕H

Potassium hydroxide s. under K₃[Fe(CN)₆] *KOH*

Sodium acetate *CH₃COONa*
N-Nitrosoacylamines from acylamines NHAc → N(NO)Ac
s. *28*, 935

Sodium nitrite (s. a. under HClO₄) *NaNO₂*
N-Condensed 1,2,3-triazine from N-condensed pyrimidine ring ○
via pyrimidine ring opening s. *29*, 273

Sodium nitrite/fluoroboric acid $\quad NaNO_2/HBF_4$
o-Azo-N-oxides from o-amino-N-oxides
  via N-condensed 1,2,3,5-oxatriazolium salts s. *27*, 283

Sodium nitrite/hydrogen chloride $\quad NaNO_2/HCl$
Monosubst. hydrazines from ureas $\quad NHCON \rightarrow NHNH_2$
  via N-nitrosoureas and semicarbazides s. *29*, 274

N-Nitrosohydrazines from hydrazines $\quad >N\cdot NH \rightarrow\; >N\cdot N\cdot NO$
s. *26*, 828

Silver carbonate-Celite $\quad \leftarrow$
Azo N,N′-dioxides $\quad 2\,NHOH \rightarrow N(O):N(O)$
  from aliphatic hydroxylamines s. *27*, 100

Fluoroboric acid s. under $NaNO_2$ $\quad HBF_4$

Ammonium ceric nitrate $\quad (NH_4)_2Ce(NO_3)_6$
Azine N,N′-dioxides from oximes $\quad 2\,C:NOH \rightarrow C:N(O)N(O):C$
s. *26*, 260

Acetic anhydride s. under CH_3COOH $\quad Ac_2O$

Acetic acid/acetic anhydride $\quad CH_3COOH/Ac_2O$
Amine N-nitrimides $\quad N^+\cdot N^-\cdot NO_2$

188.

Nitric acid in acetic anhydride added dropwise at 0° during 1 hr. to 1-amino-2,6-dimethylpyridinium nitrate in acetic anhydride-acetic acid, and the product isolated after 2 more hrs. at 0° → 2,6-dimethylpyridine 1-nitrimide. Y: 90%. F. e. and methods s. J. Epsztajn, A. R. Katritzky et al., Soc. Perkin I *1973*, 2622, 2624.

Trichloroacetic acid s. under RONO $\quad CCl_3COOH$

Diphenyl selenoxide $\quad Ph_2SeO$
Oxidations with selenoxides $\quad \leftarrow$
  Oxidative dimerization s. *29*, 275

Isoamyl nitrite $\quad RONO$
N-Nitrosation $\quad NH \rightarrow N\cdot NO$
  N-Nitroso-1,2,3,4-tetrahydroisoquinolines s. *30*, 334

Isoamyl nitrite/trichloroacetic acid $\quad RONO/CCl_3COOH$
3,4-Dihydro-3,1,2-benzothiadiazin-4-ones $\quad \bigcirc$
  from o-aminothiolic acids s. *27*, 293

Nitrosyl chloride $\quad NOCl$
N-Nitrimines from oximes $\quad C:NOH \rightarrow C:N\cdot NO_2$
  also with preferential replacement of bromine by chlorine s. *26*, 258

Chloramine $\quad NH_2Cl$
Hydrazine from amines $\quad NH \rightarrow N\cdot NH_2$
s. *26*, 259

Sodium hypobromite $\quad NaOBr$
Azo N,N′-dioxides $\quad 2\,NHOH \rightarrow N(O):N(O)$
  from aliphatic hydroxylamines s. *27*, 100 suppl. *28;* cf. *14*, 146

Perchloric acid/sodium nitrite $\quad HClO_4/NaNO_2$
N-Nitrosocyanamides from cyanamides $\quad NHCN \rightarrow N(NO)CN$
s. *29*, 276

Potassium hexacyanoferrate(III)/potassium hydroxide $K_3[Fe(CN)_6]/KOH$
Sym. azo compds. from amines $2\,NH_2 \rightarrow N:N$
  Quinone monoimines s. 28, 237

Nickel peroxide NiO_2
Azine N,N'-dioxides from oximes $2\,C:NOH \rightarrow C:N(O)N(O):C$
s. 26, 260

Oxygen ↑   NN ⇅ O

Without additional reagents   w.a.r.
Azoxy compds. by coupling   ←
s. 29, 277

Thallium Tl
Azoxy compds. from 2 nitro compd. molecules $2\cdot NO_2 \rightarrow \cdot N(O):N\cdot$
s. 26, 261

Sodium tetrahydridoborate/cobaltous chloride $Na[BH_4]/CoCl_2$
s. 26, 35

Acetic acid CH_3COOH
Triazenes from nitroso compds. $NO \rightarrow N:N\cdot N<$

189.   PhNO + H₂N—NHCOOCMe₃   →   PhN=N—NHCOOCMe₃

3-Carbalkoxytriazenes. *tert*-Butyl carbazate and nitrosobenzene stirred 85 hrs. at room temp. in ethanol containing 96%-acetic acid → *tert*-butyl 3-(phenylimino)-carbazate. Y: 76%. F. e., **also 3-acyltriazenes from carboxylic acid hydrazides**, s. R. Kreher and K. Goth, A. *1973*, 1750.

Nitrogen monoxide NO
Diazonium nitrates from nitroso compds. $ArNO \rightarrow ArN_2^+$

190.   ⟨N⟩—⟨C₆H₄⟩—NO + 2 NO → ⟨ ⟩—N₂⁺NO₃⁻

Nitrogen monoxide passed 0.5 hr. into a soln. of p-pyrrolidinonitrosobenzene in dry CHCl₃/CCl₄ under N₂ until the green color of the startg. m. has faded → p-pyrrolidinobenzenediazonium nitrate. Y: 79%. F. e. s. F. Effenberger, W. Kurtz, and P. Fischer, B. *107*, 1285 (1974).

Mesitylenesulfonyloxylamine $ArSO_2ONH_2$
3,4-Dihydro-5H-2,3-benzodiazepines   ☉
s. 29, 278

Cobaltous chloride s. under Na[BH₄] $CoCl_2$

Nitrogen ↑   NN ⇅ N

Sodium nitrite/hydrogen chloride $NaNO_2/HCl$
N-Nitrosamines from hydrazines $N\cdot NH_2 \rightarrow N\cdot NO$
s. 28, 238

Triethylamine Et_3N
Carbalkoxyhydrazones $CHN_3 \rightarrow C:N\cdot N<$
from azides and N-carbalkoxysulfonyloxylamines

191.   n—C₅H₁₁CH₂N₃ + p—O₂NC₆H₄—SO₂ONHCOOEt → n—C₅H₁₁CH=N—NHCOOEt

Equimolar amounts of *n*-hexyl azide and N-p-nitrobenzenesulfonoxyurethan warmed 4 hrs. at 30° in nitromethane in the presence of triethylamine → *n*-hexaldehydecarbethoxyhydrazone. Y: 90%; conversion 20%. H. H. Gibson, Jr., C. H. Bundy, and H. R. Gaddy III, Tetrah. Let. *1973*, 3801.

Copper $\qquad Cu$
Sym. azo compds. and diaryls $\qquad 2\,ArN_2^+ \rightarrow ArN:NAr$
   from diazonium salts s. *29*, 279

Stannous chloride $\qquad SnCl_2$
7-Acyl-1,2,3-benzothiadiazoles $\qquad \leftarrow$
   from 7-amino-1,2-benzisothiazoles s. *28*, 239

Halogen ↑ $\qquad$ NN ⫤ Hal

Potassium hydroxide $\qquad KOH$
Unsym. azoxy compds. $\qquad RNO + Cl_2NR' \rightarrow RN(O):NR'$
   from nitroso compds. and N,N-dichloramines s. *29*, 280; without KOH, in the presence of CuCl cf. Chem. Commun. *1975*, 312

Potassium/tert-butanol $\qquad K/Me_3C \cdot OH$
Azo compds. from N-halogenamines $\qquad 2\,NHBr \rightarrow N:N$
s. *27*, 284

Cuprous chloride $\qquad CuCl$
Unsym. azo compds. $\qquad RNO + Cl_2NR' \rightarrow RN(O):NR'$
   from nitroso compds. and N,N-dichloramines s. *29*, 280 suppl. *30*

Sulfur ↑ $\qquad$ NN ⫤ S

Triethylamine $\qquad Et_3N$
1,2,3,4-Tetrahydro-1,2,4,5-tetrazine $\qquad \odot$
   from Δ^3-imidazoline-5-thione ring s. *29*, 281

Remaining Elements ↑ $\qquad$ NN ⫤ Rem

Without additional reagents $\qquad w.a.r.$
Azo compds. $\qquad ArNO \rightarrow ArN:NAr'$
   from nitroso compds. and arsine imines s. *28*, 240

Carbon ↑ $\qquad$ NN ⫤ C

Sodium nitrite/sulfuric acid $\qquad NaNO_2/H_2SO_4$
o-Azonitriles $\qquad C$
   from isatin-3-hydrazones s. *29*, 282

Silver nitrite $\qquad AgNO_2$
N-Nitrosamines from tert. amines $\qquad NR \rightarrow N \cdot NO$
s. *22*, 288 suppl. *29*

Nitrosyl fluoroborate $\qquad NO^+\,BF_4^-$
Reactions with nitrosyl salts $\qquad ArN:C< \rightarrow ArN_2^+$
   Diazonium salts from azomethines s. *26*, 262

Elimination ⇑

Hydrogen ↑   NN⇑H

Electrolysis   ↯
Tetrazolium salts from formazans   ○
   s. *10*, 255 suppl. 29

Sodium acetate s. under I CH_3COONa

Pyridine s. under *N-Bromosuccinimide* C_5N_5N

Cuprous chloride s. under O_2 $CuCl$

Calcium carbonate s. under $(CH_3COO)_4Pb$ $CaCO_3$

p-Benzoquinone   ⤺
Azo from hydrazo compds.   NHNH → N : N
   Quinols from quinones s. *26*, 263

Di-tert-butyl peroxide $Me_3C \cdot OO \cdot CMe_3$
Azo- from hydrazino-silanes   NHNHSi ⟋ → N : N · Si ⟋
   s. *27*, 285

N-Bromosuccinimide/pyridine $\begin{matrix}CH_2CO\\|\\CH_2CO\end{matrix}\!\!\!\!\!\diagdown\!\!\!\mathrm{NBr}/C_5H_5N$

N-Carbamylazoformic acid esters   CONHNHCOOR → CON : N · COOR
   from 1-carbalkoxysemicarbazides

192.   PhNHCONHNHCOOEt   →   PhNHCO—N=N—COOEt

4-Phenyl-1-ethoxycarbonylsemicarbazide allowed to react at room temp. with equimolar quantities of N-bromosuccinimide and pyridine in methylene chloride → ethyl N-phenylcarbamoylazoformate. Y: > 90%. F. e. s. G. T. Knight et al., Chem. Commun. *1974*, 193.

Lead tetraacetate $(CH_3COO)_4Pb$
Diazo compds. from hydrazones   C : NNH_2 → C : N_2
   s. *27*, 444

Δ¹-3-Pyrazolones from 3-pyrazolidones   NHNH → N : N
   at low temp. s. *26*, 264

v-Triazolo[4,5-d]pyrimidines   ○
   s. *28*, 241

Lead tetraacetate/calcium carbonate $(CH_3COO)_4Pb/CaCO_3$
Δ³-1,3,4-Thiadiazolines   NHNH → N : N
   from 1,3,4-thiadiazolidines s. *30*, 566

Oxygen s. a. under Pd-C O_2

Oxygen/cuprous chloride $O_2/CuCl$
Diazo compds. from hydrazones   C : NNH_2 → C : N_2
   s. *27*, 444 suppl. 29

Iodine/sodium acetate I/CH_3COONa
Oxidative ring closure of ar. nitro compds.   ○
   s. *27*, 286

Periodic acid HIO_4
Azo from hydrazo compds.   NHNH → N : N
   s. *26*, 265

Potassium permanganate — $KMnO_4$
Aliphatic-ar. azo compds. s. *29*, 283
Palladium-carbon/oxygen — Pd-C/O_2
Alicyclic azo compds. s. *29*, 283

Oxygen ↑ NN⇑O

Triethylamine — Et_3N
Δ^2-1,3,4-Oxadiazol-5-ones
from 1-acyl-3-sulfonyloxyureas

193.
PhCONHCONHOSO$_2$-(mesityl) → Ph-C(=N-NH)-O-C=O (cyclic)

Triethylamine added dropwise to an ice-cooled soln. of the startg. m. in chloroform → 2-phenyl-5-oxo-4,5-dihydro-1,3,4-oxadiazole. Y: 64%. Y. Tamura et al., Synthesis *1974*, 361.

Nickel/sodium hydroxide — $Ni/NaOH$
Benzo[c]cinnoline 5,6-dioxides
s. *26*, 266

Nitrogen ↑ NN⇑N

Irradiation
Diaziridines from Δ^2-tetrazolines

194.

A soln. of 1-methyl-4,5-diphenyl-Δ^2-tetrazoline in methylene dichloride irradiated 20 hrs. at room temp. under N_2 with a high-pressure Hg-arc → 1-methyl-2,3-diphenyldiaziridine (2 stereoisomers). Y: 62%. F. e. s. T. Akiyama et al., Chem. Lett. *1974*, 185.

Stannous chloride — $SnCl_2$
Indazole from 3,4-dihydro-1,2,3-benzotriazine ring
s. *28*, 242

Carbon ↑ NN⇑C

Potassium hydroxide/alcohol — KOH
5-Alkoxy-1,2,3-triazoles
s. *27*, 287

Potassium/tert-butanol s. under $Me_3C \cdot OCl$ — $K/Me_3C \cdot OH$

Silver nitrate/sodium acetate — $AgNO_3/CH_3COONa$
Oxidative retrocoupling of azo compds. ←
s. *29*, 284

Chloramine — NH_2Cl
1,2,3-Triazoles from 1,2,4-triazin-3(2H)-ones
s. *27*, 289

tert-Butyl hypochlorite-potassium/tert-butanol $Me_3C \cdot OCl\text{-}K/Me_3C \cdot OH$
Azo compds. from ureas   NHCONH $\rightarrow$ N:N
s. *27*, 288

Formation of N—Hal Bond

Exchange   ↓↑

Hydrogen ↑   NHal ↓↑ H

Without additional reagents   w.a.r.
N-Chloro-N-nitramines from N-nitramines $NHNO_2 \rightarrow NClNO_2$
s. *27*, 290

N-Monochlorourethans by comproportionation $OCONH_2 \rightarrow OCONHCl$
s. *29*, 485

Perchloration with dichlorine heptoxide $H \rightarrow ClO_3$

195.

Et₂>NH + O₃Cl—O—ClO₃   HOPr → Et₂>N—ClO₃   O₃ClOPr

N-Perchloration.	**O-Perchloration.**
2 moles of 2-ethylaziridine	1-Propanol

added at 0° with stirring to a soln. of dichlorine heptoxide in CCl_4, and the product isolated

after 5 min. $\rightarrow$ N-perchloryl-2-ethyl-aziridine. Y: 83%. (Am. Soc. *96*, 3237)	after 18 hrs. stirring at room temp. $\rightarrow$ propyl perchlorate. Y: 73%. (Am. Soc. *96*, 3233)

F. e. s. K. Baum and C. D. Beard, Am. Soc. *96*, 3233, 3237 (1974); cf. J. Org. Chem. *40*, 81 (1975).

Sodium hydroxide   NaOH
N-Chloro-N-nitramines from N-nitramines $NHNO_2 \rightarrow NClNO_2$
s. *27*, 290

Fluoroxytrifluoromethane CF_3OF
N-Fluorination   NH $\rightarrow$ NF

196.   p—F—$C_6H_4SO_2$NHMe → p—F—$C_6H_4SO_2$NFMe

A mixture of fluoroxytrifluoromethane and N_2 bubbled slowly at 0° through a soln. of N-methyl-p-fluorobenzenesulfonamide in methylene chloride and Freon $\rightarrow$ N,p-difluoro-N-methylbenzenesulfonamide. Y: 85%. F. e. s. D. H. R. Barton et al., Soc. Perkin I *1974*, 732.

N,N-Dihalogenobenzenesulfonamides $PhSO_2NHal_2$
N-Halogenation   NH $\rightarrow$ NHal
    of disilazanes s. *22*, 565 suppl. 29

Acetyl hypobromite CH_3COOBr
N-Bromination of amides and imides $CONH \rightarrow CONBr$
s. 26, 267

Calcium hypochlorite $Ca(OCl)_2$
N-Chloramines as intermediates $NH \rightarrow NCl$
s. 26, 204

Oxygen ↑   NHal ↕ O

Without additional reagents   *w.a.r.*
N-Halogenimines from oximes $C:NOH \rightarrow C:NHal$

197.

Partial conversion. A soln. of cyclohexane-1,2-dione dioxime in abs. ethanol satd. at ca. 0° with dry HCl and stored several days in a refrigerator → 2-chloriminocyclohexanone oxime hydrochloride. Y: 64%. Y. Kobayashi and S. Wakamatsu, Bull. Chem. Soc. Japan *46*, 1816 (1973).

Formation of N—S Bond

Uptake   ⇓

Addition to Hydrogen and Nitrogen   NS ⇓ HN

Without additional reagents   *w.a.r.*
N-Thiols $NH \rightarrow N \cdot SH$
s. 26, 268

Addition to Nitrogen   NS ⇓ NN

Without additional reagents   *w.a.r.*
Sulfonic acid hydrazides from azo compds. $N:N \rightarrow NHN \cdot SO_2R$
s. 30, 384

Addition to Nitrogen and Halogen   NS ⇓ NHal

Iodine   *I*
Iminothionyl chlorides from N,N-dichloramines $NCl_2 \rightarrow N:SCl_2$
s. 26, 269

Addition to Nitrogen and Carbon NS↓NC

Potassium metabisulfite $K_2S_2O_5$
Carbamylsulfonic acids from isocyanates $N:C:O \rightarrow NHCOSO_3H$
s. *29, 285*
Sulfuryl chloride/dimethylformamide SO_2Cl_2/Me_2NCHO
N-Thioiminohalides $CN \rightarrow CHal:NSR$
   from nitriles and disulfides s. *30, 347*

Addition to Sulfur NS↓S

Mesitylenesulfonyloxylamine $ArSO_2ONH_2$
Aminations ←
   N-Aminocyclimmonium salts from N-heterocyclics – Aminosulfenium salts from thioethers – S-Aminosulfoxonium salts from sulfoxides s. *28, 243*
Sulfoximines from sulfoxides $>S-O \rightarrow >S\overset{O}{\underset{NH}{<}}$
s. *8,* 370 suppl. *29*

Rearrangement ↻

Sulfur/Carbon Type NS∩SC

Without additional reagents w.a.r.
Thiooxime carbamates $C:N \cdot O \cdot CSN< \rightarrow C:N \cdot S \cdot CON<$
   from oxime thiocarbamates s. *27, 291*

Exchange ↕

Hydrogen ↑ NS↕H

Without additional reagents w.a.r.
N,N′-Thiobisphthalimide as sulfur-transfer agent ◯
   2,1,3-S,N,N-Heterocyclics s. *28, 244*
Sodium hydrogen carbonate $NaHCO_3$
Sulfamoyl azides from amines $NH \rightarrow NSO_2N_3$
s. *28, 245*
Sodium acetate CH_3COONa
Azo compds. $ArN_2^+ \rightarrow ArN:NSR$
from diazonium salts via diazo-N-sulfides

198. $p-O_2NC_6H_4\overset{+}{N}\equiv N\ Cl^- + HSPh \longrightarrow p-O_2NC_6H_4N=N-SPh +$ [naphthol structure with HO]

   $p-O_2NC_6H_4N=N-$[naphthol with HO]

A cold soln. of p-nitrobenzenediazonium chloride buffered at pH 5–6 by addition of Na-acetate added below 5° during 20 min. to a stirred ethanolic soln. of thiophenol, and stirring continued 0.5 hr. at the same temp. → p-nitrobenzenediazo phenyl sulfide (Y: ca. 85%) allowed to react with β-naphthol in 95%-ethanol → p-nitrophenylazo-1-naphth-2-ol (Y: ca. 100%). F. e. s. A. B. Sakla et al., Helv. 57, 481 (1974).

Triethylamine	Et_3N
Ring closures with amidrazones	O

1,2,3,5-Thiatriazole S-oxides – also 1,2,4-triazole ring s. 26, 270

Pyridine (s. a. under Iodine)	C_5H_5N

1,2,5-Thiadiazolidine 1-oxides
s. 26, 271

Mercury acetamide	$(CH_3CONH)_2Hg$
Sym. sulfinodiimines	$2\,NH_2 \rightarrow N:S:N$

s. 26, 589

Dicyclohexylcarbodiimide/sulfuric acid	$RN:C:NR/H_2SO_4$
5-Hydroxy-1,3,2-oxathiazolium betaines	O

from α-mercaptocarboxylic acids s. 27, 292

Acetic acid	CH_3COOH

1,2,3-Benzothiadiazoles from o-aminomercaptans
s. 27, 623

Trichloroacetic acid s. under RONO	CCl_3COOH
Bis(dimethylamino)sulfane s. under Hal	$Me_2N \cdot S \cdot NMe_2$
Lead tetraacetate	$(CH_3COO)_4Pb$
N-Sulfonylsulfoximines	$SO_2NH_2 \rightarrow SO_2N:S(O)<$

from sulfonic acid amides and sulfoxides – also sulfonylsulfilimines from sulfonamides and thioethers s. 26, 272

N-Sulfamoylsulfilimines	$NH_2 \rightarrow N:S<$

from sulfamides and thioethers s. 28, 246

Isoamyl nitrite/trichloroacetic acid	$RONO/CCl_3COOH$
3,4-Dihydro-3,1,2-benzothiadiazin-4-ones	O

from 3,1-benzoxazine-2,4-diones via o-aminothiolic acids – Diazotization of aminothiolic acids s. 27, 293

Sodium p-toluenesulfonate s. under I	$NaOTs$
Sodium hydroxylamine-O-sulfonate	H_2NOSO_3Na
S-Acylhydrosulfamines	$COSH \rightarrow COSNH_2$

s. 29, 286

Sulfuric acid s. under RN : C : NR	H_2SO_4
Halogen/bis(dimethylamino)sulfane	$Hal/Me_2N \cdot S \cdot NMe_2$
Sulfinodiimines from amines	$2\,NH_2 \rightarrow N:S:N$

s. 26, 273

Bromine	Br
N-Monosubst. sulfone diimines	$>S:NH \rightarrow >S(:NH)(:NR)$

from sulfilimines s. 27, 294

Sulfonic acid bisiminoamides	$SH \rightarrow S(:NR)_2NHR$

from mercaptans s. 28, 247

Iodine/pyridine/sodium p-toluenesulfonate	$I/C_5H_5N/NaOTs$
Reactions in liq. sulfur dioxide	$2\,NH \rightarrow N \cdot SO_2 \cdot N$

with iodine and pyridine, of amines s. 29, 287

Sodium hypochlorite   *NaOCl*
1-Imino-1-sulfenamides $CSNH \rightarrow C(:N\cdot)SNH_2$
    from carboxylic acid thioamides s. 28, 248

Via intermediates   *v.i.*
N-Thionylphosphorimidates $(RO)_2P(O)NH_2 \rightarrow (RO)_2P(O)N:S:O$
    from phosphoramidates s. 28, 249

Oxygen ↑   **NS ⇅ O**

Without additional reagents   *w.a.r.*
Aminosulfenium from alkoxysulfonium salts $>S^+(OR) \rightarrow >S^+N<$
s. 28, 250

Sodium azide/phosphoric acid/phosphorus pentoxide $NaN_3/H_3PO_4/P_2O_5$
Benz[d]isothia(IV)zol-3-one 1-oxides   O
    from o-carbalkoxysulfoxides s. 26, 360

Triethylamine   *Et_3N*
Protection of amino groups $NH \rightarrow N\cdot SO_2CF_3$
    as trifluoromethanesulfonamides – Gabriel-type synthesis s. 29, 288

Phosphorus pentoxide s. under NaN_3   *P_2O_5*

Phosphoric acid s. under NaN_3   *H_3PO_4*

Nitrogen ↑   **NS ⇅ N**

Without additional reagents   *w.a.r.*
S-Transamination   ←
s. 28, 251

Sulfenamides from amines and sulfenimides $NH \rightarrow N\cdot SR$
s. 26, 274

N- and O-Sulfination $NH \rightarrow N\cdot SOR$
    with N-sulfinylphthalimides s. 29, 289

N-Sulfonation $NH \rightarrow N\cdot SO_2R$
    with N-sulfonyl-N'-methylimidazolium chlorides in homogeneous aq. medium
    s. 27, 295

Sulfonylguanidines $SO_2NH_2 \rightarrow SO_2N:C(NH_2)_2$
    from sulfonic acid amides s. 28, 252

Cuprous chloride   *CuCl*
Reactions of azides with mercaptans $SH \rightarrow S\cdot NH\cdot$
    Sulfenamides s. 27, 296

N-Methyl-N-tosylpyrrolidinium perchlorate   ←
N-Tosylation $NH \rightarrow NTs$
    also selective s. 28, 253

Halogen ↑   **NS ⇅ Hal**

Without additional reagents   *w.a.r.*
Amino- from chloro-sulfenium hexachloroantimonates $NH \rightarrow N\cdot S^+<$
s. 28, 254

Reactions with chlorothio(dichloroformimido)cyanamide   ←
    Aminothio(dichloroformimido)cyanamides s. 26, 275

Sym. bi-N-sulfonylsulfinodiimides $2\,SO_2NCl_2 \rightarrow SO_2N:S:NSO_2$
  from N,N-dichlorosulfonamides s. 28, 557

Sodium hydride NaH
1,4,3,5-Oxathiadiazine 4,4-dioxides O
  from carbalkoxysulfamoyl chlorides s. 29, 290

Sodium salt Na⁺
N-Sulfonyl-S-(sulfonylimino)sulfinic acid amides $SH \rightarrow S(:NSO_2R)NHSO_2R$
  from mercaptans and N,N-dichlorosulfonamides s. 27, 298

Triethylamine Et_3N
Cyclic N-sulfenylcarbamic carboxylic anhydrides $SCl \rightarrow S\cdot N{<}$
  also use as acylating agents s. 28, 255

Thiosulfenimides $SH \rightarrow S\cdot S\cdot N{<}^{CO}_{CO}$
  from mercaptans via thiosulfenyl chlorides s. 28, 256

Sulfhydroxamic acids $NHOH \rightarrow N(OH)SO_2R$
  from hydroxylamines and sulfonic acid chlorides s. 28, 257

Sulfur-hydrazine compds. $\substack{SCl \\ SCl} \rightarrow \substack{S\cdot N\cdot \\ | \\ S\cdot N\cdot}$
  s. 27, 299

Pyridine C_5H_5N
Selective N-tosylation $NH \rightarrow NTs$
  s. 30, 58

Sulfhydroxamic acids $NHOH \rightarrow N(OH)SO_2R$
  from hydroxylamines and sulfonic acid chlorides s. 28, 257

Sulfur ↑   NS ⇅ S

Potassium tert-butoxide $KOCMe_3$
Sulfinodiimines from thionylimines $2\,N:SO \rightarrow N:S:N$
  Diarylsulfinodiimines s. 26, 276

Silver nitrate $AgNO_3$
Sulfenamides from disulfides $RSSR \rightarrow RSN$
  s. 28, 258

Remaining Elements ↑   NS ⇅ Rem

Without additional reagents w.a.r.
Sulfenamides $S\cdot OR \rightarrow S\cdot N{<}$
  from sulfenic acid esters and aminoboranes s. 29, 291

Sulfur dichloride SCl_2
Amide-N-sulfenyl chlorides $CON\cdot Si{\leq} \rightarrow CON\cdot SCl$
  from N-silylcarboxylic acid amides s. 28, 259

Carbon ↑   NS ⇅ C

Butyllithium C_4H_9Li
Acetylenesulfenamides $C\equiv C\cdot SCN \rightarrow C\equiv C\cdot SN$
  from acetylenethiocyanates s. 28, 260

Elimination ⇑

Hydrogen ↑   NS⇑H

Sodium hydroxide   *NaOH*
N-Imidoylsulfilimines $NHS^+ < \rightarrow N^- \cdot S^+$

198 a.   Ph–C(NH)(NH–$\overset{+}{S}$Me$_2$) Cl⁻ → Ph–C(NH)(N⁻–$\overset{+}{S}$Me$_2$)

2 N NaOH added below 5° to a stirred soln. of the startg. m. in chloroform-water → N-benzimidoyl-S,S-dimethylsulfilimine. Y: ca. 100%. T. Fuchigami and K. Odo, Chem. Lett. *1974*, 247.

Triethylamine Et_3N
N-Acylsulfilimines $>S^+ \cdot N^- Ac$
   from acylaminosulfenium salts s. 27, 300

Formation of N—Rem Bond

Uptake ⇓

Addition to Nitrogen   NRem⇓NN

Irradiation/di-tert-butyl peroxide $h\nu/Me_3C \cdot OO \cdot CMe_3$
Addition of silanes $SiH + N:N \rightarrow Si \cdot N \cdot NH \cdot$
to nitrogen-nitrogen double bonds
s. 27, 301

Addition to Nitrogen and Carbon   NRem⇓NC

Without additional reagents   *w.a.r.*
1,1-Dichloro(trichlorophosphazo) compds.   ←
   from nitriles with chlorination s. 27, 302
Reaction of quinone diimines with phosphorous acid esters   ←
s. 27, 303
2-Phosphaquinazolinium salts   ○
   from o-cyanophosphine imines s. 26, 277

Addition to Remaining Elements   NRem⇓Rem

Without additional reagents   *w.a.r.*
Phosphazides   ←
s. 28, 261

Exchange ↓↑

Hydrogen ↑   NRem ↮ H

Without additional reagents	*w.a.r.*
Trihalogenophosphazocarbonyl compds. s. 26, 278	$CONH_2 \rightarrow CON:PCl_3$
Trichlorophosphazosulfonyl compds. from sulfonic acid amides s. 27, 304	$SO_2NH_2 \rightarrow SO_2N:PCl_3$
Sodium hydroxide	*NaOH*
o-Hydroxyphosphine imines from o-aminophenols s. 28, 262	$NH_2 \rightarrow N:P\leqslant$
Triethylamine s. under CCl_4	Et_3N
Carbon tetrachloride	CCl_4
P,N-Heterocyclics s. 29, 292	O
Carbon tetrachloride/triethylamine	CCl_4/Et_3N
Phosphorylation of diamines	$PH \rightarrow PN<$
Phosphoromonoamidates from diaryl phosphites – Thionophosphoramidates from thionophosphorochloridates s. 28, 263	
Lead tetraacetate	$(CH_3COO)_4Pb$
Arsine imines from arsines	$NH_2 \rightarrow N:As\leqslant$

199. $Ph_3As + H_2NSO_2C_6H_4CH_3-p \longrightarrow Ph_3As=N-SO_2C_6H_4CH_3-p$

Triphenylarsine allowed to react with p-toluenesulfonamide in methylene chloride in the presence of Pb-tetraacetate → N-p-toluenesulfonyltriphenylarsinimine. Y: 90%. F. e. and method s. J. I. G. Cadogan and I. Gosney, Chem. Commun. *1973*, 586; cf. Soc. Perkin I *1974*, 460, 466.

Sodium hypohalite	*NaOHal*
N-Halogeno-N-sodiosulfonic acid amides s. 27, 305	$SO_2NH_2 \rightarrow SO_2NClNa$

Oxygen ↑   NRem ↮ O

Without additional reagents	*w.a.r.*
7-Hydro-8-boraxanthines s. 27, 306	O
Ring closures with boronic acids B,N,S-Heterocyclics s. 27, 307	
Pyridine	C_5H_5N
Phosphorylation	$NH \rightarrow N \cdot PO(OR)OH$
with 8-(phosphoryloxy)quinolines, of amines and alcohols s. 29, 293	

Nitrogen ↑   NRem ↮ N

Without additional reagents	*w.a.r.*
Acylimidophosphorisocyanatidates from carboxylic acid azides and phosphorisocyanatidites s. 27, 308	←

1,3,2,4-Diazadiphosphetidine 2,4-dioxides ○
from prim. amines − 1,3,2,4-Diazadiphosphetidines s. 29, 294

n-Butyllithium   *n-BuLi*
Phosphoric acid diamide hydrazides $PO \cdot NR_2 \rightarrow PO \cdot N \cdot NH$
from phosphoric acid triamides and azo compds.

200. $(Me_2N)_3PO \rightarrow [(Me_2N)_2P\overset{O}{\underset{Li}{\diagdown}}] + \underset{N-Ph}{\overset{N-Ph}{\|}} \rightarrow (Me_2N)_2P\overset{O}{\underset{NHPh}{\diagdown N-Ph}}$

n-Butyllithium in hexane added at 0–25° to hexamethylphosphoramide in tetrahydrofuran, and, after 1 hr., treated with azobenzene → product. Y: ca. 70%.
E. M. Kaiser, J. D. Petty, and L. E. Solter, J. Organometal. Chem. *61*, C1 (1973).

Halogen ↑   NRem ⇵ Hal

Lithium   *Li*
Reductive N-silylation   ←
1,4-Disilyl-1,4-dihydropyrazines from pyrazines s. 26, 279

Sodium hydride   *NaH*
N-Silylketenimines and α-silylnitriles $CHCN \rightarrow C:C:N \cdot Si \leqslant$
from nitriles and chlorosilanes − Silylating agents s. 29, 295

Butyllithium   *BuLi*
Amino- from halogeno-stibines $SbHal \rightarrow Sb \cdot N <$
s. 26, 280

Lithium diisopropylamide   *i-Pr₂NLi*
C,N-Disilylketenimines $CH_2CN \rightarrow C(Si \leqslant):C:N \cdot Si \leqslant$
from nitriles and chlorosilanes s. 30, 438

Triethylamine   *Et₃N*
Thionophosphoramidates $(RO)_2P(S)Cl \rightarrow (RO)_2P(S)N<$
from thionophosphorochloridates s. 28, 263

P-Spiroheterocyclics   ○
from α-aminocarboxylic acids and 2-chloro-1,3,2-dioxaphospholanes s. 28, 264;
spirophosphoranes, review, s. R. Burgada, Bl. *1975*, 407

Sulfur ↑   NRem ⇵ S

Without additional reagents   *w.a.r.*
N-Sulfonylphosphine imines $SO_2N:S< \rightarrow SO_2N:P \leqslant$
from phosphines and sulfonylsulfilimines s. 30, 700

Partial replacement of cyclic P-sulfur   ←
by cyclic P-silylnitrogen in perthiophosphonic acid anhydrides s. 29, 296

1,3,4,2,5-Triazadiborolidines   ○
s. 26, 281

Sodium salt   *Na⁺*
Borylation with boryl sulfonates   ←
s. 27, 309

Carbon ↑   NRem ⇅ C

Without additional reagents   w.a.r.

Aminoboranes from trialkylboranes   NH → N·B<
s. 29, 297

Cleavage of plumbanes and stannanes   ←
   with 1-chlorobenzotriazole s. 27, 310

Phosphoramidates N_3 → $NHPO(OR)_2$
   from azides and phosphorous acid esters s. 27, 311; mixed phosphoramidates in the presence of excess LiCl cf. R. L. Letsinger and G. A. Heavner, Tetrah. Let. *1975*, 147

N-Phosphonyldicarboxylic acid imides from phosphites $(RO)_3P$ → $(RO)_2P(O)N<$
Arbuzov-Michaelis-type rearrangement

201.

Triethyl phosphite added during 10–15 min. with ice-water cooling under N_2 to a suspension of N-bromosuccinimide in dry CS_2, allowed to warm to room temp. and stirred for an additional hr. → N-(diethylphosphonyl)succinimide. Y: 92%. F. e., also inverse addition, and limitations s. D. J. Scharf, J. Org. Chem. **39**, 922 (1974).

Lithium chloride   LiCl
Mixed phosphoramidates   NHPO(OR)(OR′)
s. 27, 311 suppl. 30

Formation of N—C Bond

Uptake   ⇓

Addition to Hydrogen and Oxygen   NC⇓HO

Potassium hydroxide   KOH
3,4-Dihydro-2H-1,3-benzoxazine-2-thiones   O
   from o-hydroxyaldehydes and isothiocyanates s. 29, 550

Addition to Hydrogen and Oxygen   NC⇓HN

Triethylamine s. under H_2Se Et_3N

Cuprous chloride   CuCl
Formamides from amines   NH → NCHO
   with carbon monoxide s. 26, 282

Aluminum chloride $AlCl_3$
anhydro-**2-Hydroxy-4-oxopyrido[1,2-a]pyrimidinium hydroxides**   O
   from 2-aminopyridines s. 26, 283

Hydrogen selenide-triethylamine $\quad H_2Se\text{-}Et_3N$
Formamides from amines $\quad NH \to NCHO$
    with carbon monoxide s. *26,* 282 suppl. *29*
Hydrogen chloride $\quad HCl$
N-Hydroxyformamidines $\quad N:C \to N:CH\cdot N(OH)\cdot$
from hydroxylamines and isonitriles

202.

PhNHOH + C=N—⟨⟩ → Ph\N—C⟨N—⟨⟩ / HO   H

Cyclohexyl isocyanide added with cooling to phenylhydroxylamine hydrochloride (or equimolar amounts of phenylhydroxylamine and HCl) in methanol, and allowed to stand at room temp. until the isocyanide odor has disappeared → product. Y: 98%. F. e. s. G. Zinner, W. Heuer, and D. Moderhack, Chem. Ztg. *98,* 159 (1974).

Thioureas from amines $\quad NH \to NCSN<$
s. *1,* 363

Addition to Hydrogen and Carbon $\quad$ NC↓HC

Without additional reagents $\quad$ w.a.r.
Aliphatic-ar. azo compds. from diazo compds. $\quad C:N_2 \to CH\cdot N:NAr$
    with O-alkylation s. *26,* 284

Addition to Oxygen and Nitrogen $\quad$ NC↓ON

Without additional reagents $\quad$ w.a.r.
Hydroxylamines from nitroso compds. $\quad NO \to N(OH)R$
s. *27,* 313; enehydroxylamines cf. R. K. Howe and P. A. Berger, J. Org. Chem. *39,* 3498 (1974)

Addition to Oxygen and Carbon $\quad$ NC↓OC

Without additional reagents $\quad$ w.a.r.
Aminoalcohols $\quad \leftarrow$
    from dicarboxylic acid anhydrides s. *28,* 364
Reactions with o-quinone methids $\quad \leftarrow$

203.

Ph₂C⟨ ⟩=O + H₂NMe → Ph₂C—NHMe⟨ ⟩—OH

o-α-Aminophenols. 1,2-Naphthoquinone 1-diphenylmethid refluxed with methylamine → product. Y: 80%. F. e. and reactions s. M. A. F. Elkaschef, F. M. E. Abdel-Megeid, and S. M. M. Elzein, Acta Chim. (Budapest) *79,* 411 (1973) (Eng); C. A. *80,* 70576.

7-β-Hydroxypurinium salts from oxido compds. $\quad \leftarrow$

204.

Ethylene oxide bubbled at room temp. into a slurry of guanosine and glacial acetic acid → 7-(2-hydroxyethyl)guanosinium acetate. Y: 93%. F. e. s. R. Roe, Jr., J. S. Paul, and P. O'B. Montgomery, Jr., J. Heterocyclic Chem. *10,* 849 (1973).

Reactions of difluoroamine   CO → C(OH)NF$_2$
1-(Difluoroamino)alcohols from oxo compds. – Reactions of difluoroamine, review s. 26, 285

Reaction of aldehydes with thiourea   CHO → CH(OH)NHCSNH$_2$
1-Hydroxythioureas s. 26, 286

Addition of enolesters   ←
with carbon-oxygen cleavage – Addition to nitrogen-nitrogen double bonds s. 27, 314

1,4-Dihydropyridine ring   CO → CN
from 4H-pyran ring and amines s. 27, 315

Transannular ring closure   O
9-Azabicyclo[3.3.1]nonane-1,5-diol ring

205.

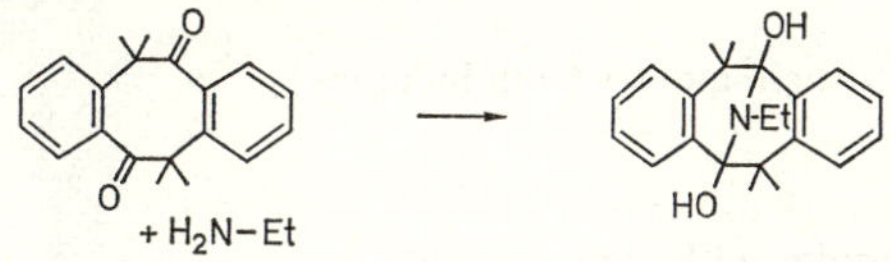

An ethanolic soln. of 3,3,7,7-tetramethyl-1,2,5,6-dibenzocyclooctane-4,8-dione treated with a satd. ethanolic soln. of ethylamine, and warmed 1.5 hrs. at 40° → N-ethyl-4,4,8,8-tetramethyl-2,3,6,7-dibenzo-9-azabicyclo[3.3.1]nonane-1,5-diol. Y: 85.3%. F.e. s. R. M. Lagidze et al., Khim. Prir. Soedin. *1973*, 188; C. A. *79*, 5241.

[2+2]-Cycloaddition of carbodiimides
1,3-Oxazetidines from isocyanates s. 28, 265

1,2,4-Dioxazolidine ring by aminoperoxidation

206.

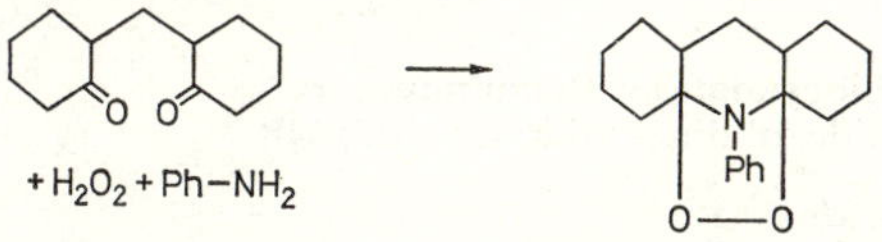

2,2′-Methylenedicyclohexanone allowed to react 10–12 hrs. at room temp. with H$_2$O$_2$ and aniline in aq. alcohol → product. Y: 78%. F. e. s. V. A. Kaminskii, V. I. Alekseev, and M. N. Tilichenko, Khim. Geterotsikl. Soedin. *1972*, 1708; C. A. *78*, 72083.

2-Imino-2H-1,3,5-oxadiazin-4(3H)-ones
from acylisocyanates and carbodiimides s. 28, 266

γ-Ketocarboxylic acid amides   C
from γ-enollactones s. 26, 287

β-Amino acid peptides
from 4,5-dihydro-1,3-oxazin-6-ones

207.

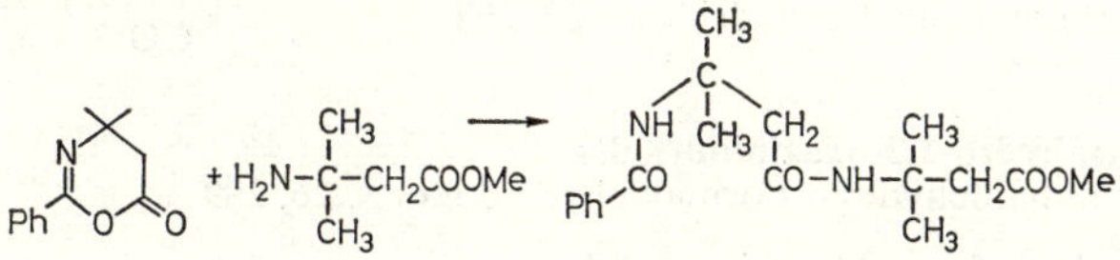

4,5-Dihydro-4,4-dimethyl-2-phenyl-1,3-oxazin-6-one refluxed 3 hrs. with methyl β-aminoisovalerate in acetonitrile → methyl β-benzoylaminoisovaleryl-β-aminoisovalerate. Y: 97.5%. F. e. s. C. N. C. Drey, J. Lowbridge, and R. J. Ridge, Soc. Perkin I *1973*, 2001.

o-Acylaminocarboxylic acid amides
   from 4H-3,1-benzoxazin-4-ones s. 29, 391

Oxazole ring opening
   s. 26, 332

4-Hydroxyoxazolid-2-ones   ←
   from 1,3-dioxol-2-ones s. 28, 365

Oxadiazabicycloalkanes   ◎
   from cyclic amidines and oxido compds. s. 27, 316

Sodium   Na
Self-condensation of β-cyanoketones   ←
   s. 29, 774

Potassium amide KNH_2
Unsubst. carboxylic acid amides from lactones   C
   s. 26, 288

Lithium chloride   LiCl
Reaction of hydantoins with oxido compds. $\triangle_O \rightarrow C(OH)C \cdot N <$
   s. 28, 267

Lithium bromide/tributylphosphine oxide $LiBr/Bu_3PO$
2-Oxazolidones   O
   from isocyanates and oxido compds. s. 27, 317

Potassium chloride   KCl
Reaction of hydantoins with oxido compounds $\triangle_O \rightarrow C(OH)C \cdot N <$
   s. 28, 267

Collidine   ←
Reaction of isothiocyanates with aminosugars   ←
   5-Hydroxyimidazolidine-2-thiones s. 27, 318

Magnesium chloride $MgCl_2$
2-Azidoalcohols from oxido compds. $\triangle_O \rightarrow C(OH)C \cdot N_3$
   s. 30, 278

Trifluoroacetic acid CF_3COOH
Simultaneous protection by different protective groups $NH \rightarrow N \cdot C(OR)Ar$
   Protection of *vic*-hydroxyl groups as 2-alkoxy-1,3-dioxolanes and of amino groups as 1,1-aminoethers s. 28, 268

Tributylphosphine oxide s. under LiBr Bu_3PO

Vanadium pentoxide-phosphorus pentoxide-alumina $V_2O_5\text{-}P_2O_5\text{-}Al_2O_3$
Dicarboxylic acid monoamides   (CO)₂O → COOH/CONH₂
   from dicarboxylic acid anhydrides s. 29, 298

Hydrochlorides   ←
6-Hydroxytetrahydro-1,3-oxazinium salts   O
   from β-aminoaldehydes – Formaldehyde donor s. 26, 289

Hydroxyalkoximes from dihydropyran ring   C
   s. 28, 896

2-Imino-1,3,4-oxadiazoline ring opening
   1-Acyl-5-carbamyldiaminoguanidines s. 28, 269

Addition to Nitrogen NC↓NN

Without additional reagents *w.a.r.*
C-Functionalization of pyrimidines ←
with diethyl azodicarboxylate
Purines from pyrimidines

208.

A stirred suspension of 1,3-dimethyl-6-aminouracil in a mixture of diethyl azo-dicarboxylate and chlorobenzene heated at 150–160° oil bath temp. until an exothermic reaction starts and continues for several min., then heating resumed for 20 min. → 1,3-dimethyl-5-(1,2-dicarbethoxyhydrazino)-6-aminouracil (Y: 70% as the hemihydrate) stirred and refluxed 0.5 hr. with Raney-Ni in abs. ethanol → 1,3-dimethyl-5-carbethoxyamino-6-aminouracil (Y: 75%) heated 45 min. at 175–235° under N_2 → 1,3-dimethyluric acid (Y: 93%). F. e. of the functionalization s. E. C. Taylor and F. Sowinski, J. Org. Chem. *39*, 907 (1974); 8-subst. xanthines by a 1-step process cf. F. Yoneda, S. Matsumoto, and M. Higuchi, Chem. Commun. *1975*, 146; **substituting addition** of diethyl azodicarboxylate, isocyclics, s. a. F. P. Colonna, G. Pitacco, and E. Valentin, Chem. Commun. *1975*, 71.

Cycloadditions O
  with 4-hydroxy-1,2,3-triazolium betaines s. *28*, 270

Lithium diisopropylamide *i-Pr₂NLi*
Trisubst. hydrazines N : N → NRNH
  from azo and CH-acidic compds. s. *27*, 319

Di-tert-butyl peroxide $Me_3C \cdot OO \cdot CMe_3$
Reactions with acyl radicals RN : NR → RN(Ac)NHR
  Acylhydrazines from aldehydes and azo compds. s. *28*, 271

Addition to Nitrogen and Sulfur NC↓NS

Without additional reagents *w.a.r.*
1,2,3-Oxathiazetidine 2-oxides O
  from thionylimines and aldehydes s. *28*, 272

3-Imino-1,2,4-thiadiazetidine 1-oxides
  from carbodiimides and N-thionylsulfonamides s. *29*, 473

Addition to Nitrogen and Remaining Elements NC↓NRem

Without additional reagents *w.a.r.*
Carbamic acid silyl esters from aminosilanes N·Si≦ → N·COOSi≦
s. *29*, 438

1-(α-Siloxy)imidazoles from aldehydes CHO → CH(OSi≦)N⊃
s. 26, 290

Sec. amines from phosphine imines and halides N : PR₃ → NHR'
via P-aminophosphonium salts s. 26, 291

Imidinophosphoranes P : NR → P : N·C : NR
from phosphine imines and nitriles s. 27, 684

Addition to Nitrogen and Carbon NC⇓NC

Without additional reagents w.a.r.
Preferential nucleophilic addition N : C → NHC·N<
to carbon-nitrogen double bonds – 1,1-Diamines from cyclic azomethines
s. 29, 299

Selective N-aminoethylation NH → NCH₂CH₂NH₂
s. 28, 273

Ureas from isocyanates ←
with migration of 1,3,5-triazin-2-yl groups s. 30, 229

Halogenoureas from halogenisocyanates N : C : O → NHCONH₂
s. 30, 291

α-Chloroenureas C : C(Cl)N : C : O → C : C(Cl)NHCON<
from α-chloroenisocyanates
s. 28, 472

Biurets from isocyanates N : C : O → NHCONHCONH₂
s. 29, 300

N,O-Bis(carbamyl)hydroxylamines NHCO·NH·O·CONH·
via N-carbo-*tert*-butoxy-N,O-bis(carbamyl)hydroxylamines s. 26, 292

Reaction of hydroxylamine with isocyanates ←
N-hydroxyureas, 3-hydroxybiurets and N,N,O-tris(carbamyl)hydroxylamines
s. 26, 293

Alkoxyureas and 3-alkoxybiurets ←
from alkoxylamines and isocyanates

209. PhN=C=O + H₂NOCH₂Ph

  ↙ 1   2 ↘

PhNHCONHOCH₂Ph   PhNHCO—N—CONHPh
                              |
                              OCH₂Ph

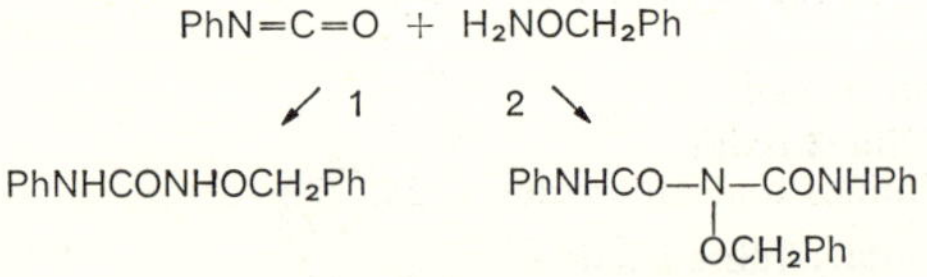

Phenyl isocyanate
dissolved in benzene, added dropwise to a soln. of the equimolar amount of benzyloxyamine in the same solvent, and stirred 3 hrs. at room temp. → 1-benzyloxy-3-phenylurea. Y: > 90%. F. e. s. | added to a soln. of 0.25 mole benzyloxyamine in benzene, and refluxed 1 hr. → 1,5-diphenyl-3-benzyloxybiuret. Y: 99%.

J. H. Cooley et al., J. Chem. Eng. Data *19*, 100 (1974).

Hydroxyguanidines from disubst. cyanamides N·C≡N → N·C(NH₂) : NOH
s. 26, 294

N-Carbamyl- and N-thiocarbamyl-amidines
from isocyanates and isothiocyanates respectively s. 26, 295

Phosphinylthiosemicarbazides   N : C : S → NHCSNHNHP(O)R$_2$
from phosphinic acid hydrazides and isothiocyanates – 1-Phosphinyl-4-acyl-thiosemicarbazides s. 27, 320

Reactions with thiophosphinylisothiocyanates   N : C : S → NHCSN<

210.
$$Ph_2P-N=C=S + H_2NPh \longrightarrow Ph_2P-NHCSNHPh$$
$$\overset{\|}{S} \qquad\qquad\qquad\qquad \overset{\|}{S}$$

Thiophosphinylthioureas. Aniline mixed with diphenylphosphinothioyl isothiocyanate and heated a few min. at 80–90° after the exothermic reaction has subsided → 3-diphenylphosphinothioyl-1-phenylthiourea. Y: almost 100%. F. e. and reactions s. I. Ojima, K. Akiba, and N. Inamoto, Bull. Chem. Soc. Japan 46, 2559 (1973).

4-Alkoxy-1,3-diazetidin-2-one ring
from O-alkyllactims and isocyanate s. 29, 301

Parabanic acids from cyanamides
s. 28, 274

Reactions of oxo compds. with halogenosulfonyl isocyanates
5,6-Dihydro-4H-1,3,5-dioxazin-4-one-5-sulfonylhalides

211.
2 PhCHO + O=C=N-SO$_2$Cl ⟶ [Ph, Ph-substituted 1,3,5-dioxazine with N-SO$_2$Cl]

Chlorosulfonyl isocyanate added dropwise at –50 to –40° under dry N$_2$ to a stirred soln. of benzaldehyde in abs. methyl chloride → 4-oxo-2,6-diphenyl-5,6-dihydro-4H-1,3,5-dioxazine-5-sulfonyl chloride. Y: 80%. F. e. and reactions s. K. Clauss, H.-J. Friedrich, and H. Jensen, A. 1974, 561.

3,4,6,7-Tetrahydro-2H-thiazolo[3,2-a]-s-triazin-4-ones
from 2,4-diaza-1,3-diene groups and isocyanates s. 27, 321

1,3,5-Oxadiazine-4,6-diones
from acylisocyanates and isocyanates s. 27, 322

P,N-Heterocyclics
from phosphonic acid diisocyanates s. 29, 302

Aziridine ring opening
with carboxylic acids and derivs. – 2-Acylaminoalcohols s. 26, 296

Cycloadditions with diaziridinones
4-Acyl-1,2,4-triazolidine-3,5-diones from acylisocyanates

212.
Bz–N=C=O + O=C–N–N⟨ ⟶ 1,2,4-triazolidine-3,5-dione product

Di-*tert*-butyldiaziridinone heated 6 hrs. at 80° with benzoyl isocyanate in benzene → 1,2-di-*tert*-butyl-4-benzoyl-1,2,4-triazolidine-3,5-dione. Y: 73%. F. e. s. Y. Ohshiro et al., Chem. Lett. 1974, 383.

Reactions of oxaziridines with heterocumulenes
1,2,4-Oxadiazolid-5-ones from isocyanates

213.

Ph−N=C=O + Ph−⟨N−CHMe$_2$⟩O ⟶ Ph−N−C(=O)−O−N(CHMe$_2$)−Ph (cyclic)

Phenyl isocyanate added dropwise with stirring to a soln. of 2-isopropyl-3-phenyloxaziridine in benzene, and refluxed 13 hrs. → 2-isopropyl-3,4-diphenyl-1,2,4-oxadiazolidin-5-one. Y: 95%. F. e. s. M. Komatsu et al., J. Org. Chem. *39*, 948, 957 (1974).

Reactions of 2,3-dihydrothiazolo[3,2-a]benzimidazol-3-ones C
4-Thiazolidone ring opening s. 27, 323

Ring contraction of N-heterocyclics Ö:
via cyclic quaternary ammonium salts s. 26, 297

Electrolysis ⅄
Acylamines H → NHAc
from hydrocarbons and nitriles s. 29, 334

Sodium hydroxide (s. a. under C_5H_5N) NaOH
Cyclic 1-hydroxyamidines from cyanoketones O
via cyclic 1-hydroxyamidoximes − Bridge amidines s. 26, 298
4-Imidazolidones
from α-aminonitriles and oxo compds. s. 28, 275

Lithium/liq. ammonia Li/NH$_3$
Reductive N-alkylation of N-heterocyclics ←
1-Subst. 1,4-dihydroquinolines from quinolines

214.

quinoline + BrBu-n ⟶ 1-n-butyl-1,4-dihydroquinoline (Bu-n on N)

A soln. of quinoline in dry ether followed by Li added to liq. NH$_3$, stirred 0.5 hr. at ca. −33° under N$_2$, cooled to ca. −70°, n-butyl bromide slowly injected from a syringe, and stirring continued 15 min. → 1-n-butyl-1,4-dihydroquinoline. Y: 84%. F. e. s. A. J. Birch and P. G. Lehman, Soc. Perkin I *1973*, 2754.

Sodium amide NaNH$_2$
Amidines Ar·C≡N → Ar·C(:NH)NHAr'
from ar. nitriles and ar. amines s. 26, 299

Sodium hydrogen carbonate NaHCO$_3$
Sulfonylformamidoximes SO$_2$CN → SO$_2$C(:NOH)NH$_2$
from sulfonylcyanides s. 29, 303

Potassium acetate CH$_3$COOK
3-Hydroxaminoaziridines from 2H-azirines N:C → NHC(NHOH)
s. 28, 276

Potassium thiocyanate s. under CH$_3$COOH KSCN
Sodium salt Na$^+$
Sulfonylguanidines SO$_2$NH$_2$ + C(:NR)$_2$ → SO$_2$N:C(NHR)$_2$
from carbodiimides s. 26, 410

Potassium salt s. under NH$_2$Cl K$^+$
Triethylamine Et$_3$N
1,2-Diaminopyrrole ring O
from α-diazoketones and α,β-ethylenenitriles s. 26, 466

2-Dimethylaminoethanol $\quad Me_2NCH_2CH_2OH$
Phthalocyanines from o-dinitriles
s. 27, 435

Pyridine $\quad C_5H_5N$
N,N,O-Tris(carbamyl)hydroxylamines $\quad 3\ RN:C:O \rightarrow (RNHCO)_2N \cdot OCONHR$
from isocyanates s. 26, 293

Pyridine-sodium hydroxide $\quad C_5H_5N\text{-}NaOH$
Ethyleneacylamines $\quad C$
from cyclic tert. amines s. 26, 300

Cuprous oxide $\quad Cu_2O$
Thioacylureas $\quad CSNH \rightarrow CS \cdot N \cdot CO \cdot NH \cdot$
from carboxylic acid thioamides and isocyanates

215.

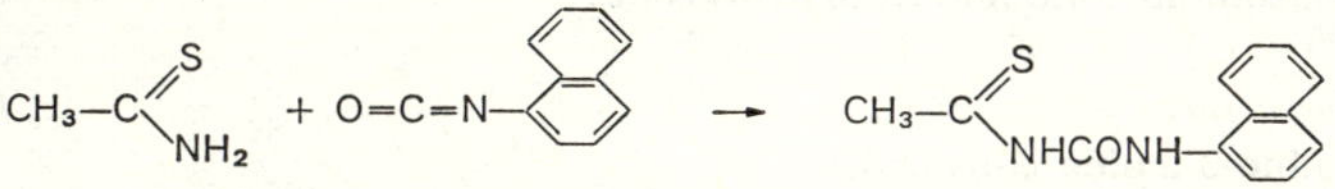

$CH_3-C(=S)NH_2$ + $O=C=N-\text{(naphthyl)}$ → $CH_3-C(=S)NHCONH-\text{(naphthyl)}$

A soln. of thioacetamide, α-naphthyl isocyanate, and Cu$_2$O in anhydrous xylene refluxed 1 hr. ↠ 1-thioacetyl-3-α-naphthylurea. Y: 60%. F. e. s. V. I. Cohen, J. Org. Chem. 39, 3043 (1974).

Silver carbonate $\quad Ag_2CO_3$
Selective N-carbamylation $\quad NH \rightarrow NCONHR$
s. 29, 304

Magnesium oxide/cyanogen bromide $\quad MgO/BrCN$
Hydroxycyanamides $\quad C$
from cyclic tert. amines s. 28, 277

Trialkylborane $\quad R_3B$
Acylamidines $\quad CONH \rightarrow CON \cdot C(:NH) \cdot$
from carboxylic acid amides and nitriles s. 29, 305

Boron fluoride $\quad BF_3$
Ureas from cyanamides $\quad N \cdot CN \rightarrow N \cdot CO \cdot NHR$

216.

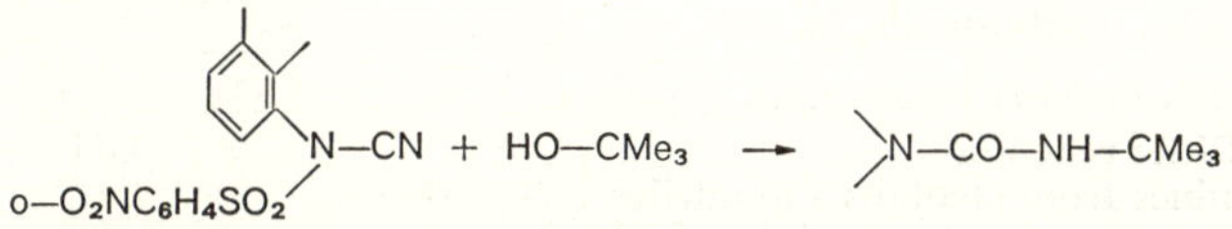

(2,3-dimethylphenyl)(o-O$_2$NC$_6$H$_4$SO$_2$)N—CN + HO—CMe$_3$ → ⟩N—CO—NH—CMe$_3$

BF$_3$-etherate followed by dioxane added to a stirred mixture of N-(2-nitrobenzenesulfonyl)-N-(2,3-dimethylphenyl)cyanamide and *tert*-butanol, allowed to stand overnight at room temp., then warmed 1 hr. at 60° ↠ N-(2-nitrobenzenesulfonyl)-N-(2,3-dimethylphenyl)-N'-*tert*-butylurea. Y: 77%. F. e. s. J. Anatol and J. Berecoechea, Synthesis *1975*, 111; Bl. *1975*, 395.

Δ²-Imidazolines from aziridines and nitriles ☉

217.

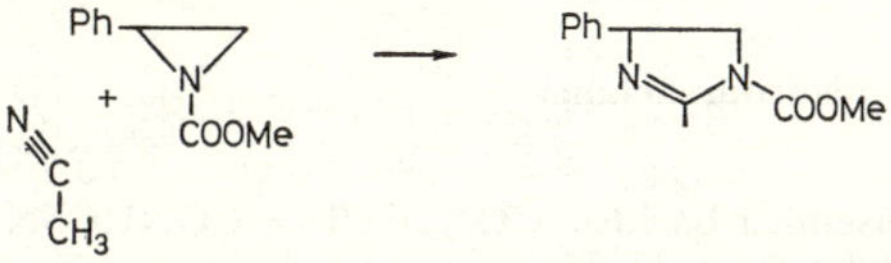

Ph-aziridine-COOMe + N≡C-CH$_3$ → Ph-imidazoline-COOMe

1-Methoxycarbonyl-2-phenylaziridine added dropwise under N$_2$ at 81° to a mixture of refluxing acetonitrile and a catalytic amount of BF$_3$-etherate, and refluxing continued 4 hrs. ↠ 1-methoxycarbonyl-2-methyl-4-phenyl-2-imidazoline. Y: 82%. F. e. s. T. Hiyama et al., Tetrahedron 29, 3137 (1973).

Fluoroboric acid/tert-butanol $HBF_4/Me_3C \cdot OH$
Reactions with carbodiimides ←
Guanylureas s. 29, 131

Aluminum chloride $AlCl_3$
Sulfonylureas from isocyanates $N:C:O \rightarrow NHCONHSO_2R$
s. 8, 394 suppl. 29

tert-Butanol s. under HBF$_4$ $Me_3C \cdot OH$

Formamide $HCONH_2$
1-Formaminoindoles from cinnolines ○:
s. 27, 324

Acetic acid/potassium thiocyanate $CH_3COOH/KSCN$
1,2,4-Triazolidine-3-thiones from hydrazones ○
s. 29, 306

Trifluoroacetic acid CF_3COOH
5-Acylimino-3-amino-ψ-oxazoles
s. 27, 325

Cyanogen bromide s. under MgO $BrCN$

Chloramine/potassium salt NH_2Cl/K^+
3-Aminoisothiazole ring

218.

A soln. of chloramine added with stirring to an aq. soln. of the startg. m., and the product isolated after 3 hrs. → ethyl 5-(3-amino-4-cyano-5-isothiazolyl)-4-ethoxy-2-methylthiophene-3-carboxylate. Y: 83%. F. e. s. K. Hartke and G. Gölz, A. *1973*, 1644.

Phosphorus pentachloride PCl_5
Trisaryloxy-1,3,5-triazines
from cyanic acid esters s. 26, 301

Carbonium hexachloroantimonates ←
Modified Ritter reaction $OH \rightarrow NHCOR$
Acylamines from alcohols and nitriles s. 29, 307

Sulfuric acid H_2SO_4
α-Ketocarboxylic acids from cyanohydrins $CH(OH)CN \rightarrow COCOOH$
via Ritter reaction and α-ketocarboxylic acid *tert*-butylamides s. 27, 326

Sulfuric acid/acetic acid H_2SO_4/CH_3COOH
1,1-Di(thiolurethans) $CHO \rightarrow CH(NHCOSR)_2$
from thiocyanates and aldehydes s. 29, 308

Hydrochlorides ←
Aminohydroxamic acids from lactams
s. 27, 492

$$\underset{CO}{\overset{NH}{C}} \rightarrow \underset{CONHOH}{\overset{NH_2}{C}}$$

4-Acyl-1-amidinothiosemicarbazides $CON:C:S \rightarrow CONHCSNHNHC(:NH)NH_2$
from acylisothiocyanates s. 27, 327

Hydrogen chloride HCl
Amidinothioureas $N \cdot C \equiv N \rightarrow N \cdot C(:NH) \cdot N \cdot CSNH_2$
from cyanamides and thioureas s. 26, 302

Triazines from nitriles
Trimerization s. 26, 303

Ferric chloride $FeCl_3$

α,β-**Ethyleneamidines from β-chloronitriles**   CN → C(: NR)NHR′

219.
$ClCH_2CH_2CN + ClCHMe_2 + FeCl_3 \longrightarrow [ClCH_2CH_2C\equiv \overset{+}{N}CHMe_2 \; FeCl_4^-]$

$\downarrow + H_2NPh$

$CH_2=CHC\begin{smallmatrix}NHPh\\HCl\\NCHMe_2\end{smallmatrix} \longleftarrow ClCH_2CH_2C\begin{smallmatrix}NHPh\\N-CHMe_2\end{smallmatrix} \overset{NaOH}{\longleftarrow} [ClCH_2CH_2C\begin{smallmatrix}NHPh\\+\\NHCHMe_2\end{smallmatrix} FeCl_4^-]$

β-Chloropropionitrile added to ice-cooled $FeCl_3$ in isopropyl chloride, stirred 3 hrs. at room temp., excess isopropyl chloride distilled off in vacuo, the residue dissolved in methylene chloride, cooled to ca. –10°, aniline in methylene chloride added dropwise, the cooling bath removed, stirring continued 1 hr. at room temp. under N_2, the solvent distilled off in vacuo, treated with ice-water and aq. 30%-NaOH, crude β-chloro-N-isopropyl-N′-phenylpropionamidine isolated, and allowed to stand 1 day at room temp. → N-isopropyl-N′-phenylacrylamidine. Y: 43%. R. Fuks, Bl. Soc. chim. Belg. *82*, 571 (1973).

Via intermediates   *v.i.*

Amidines from nitriles via iminoesters   CN → C(: NH)NH$_2$

220.
$Me_2CHOOC-\underset{NHBz}{C}=CH-\langle\!\!\!\bigcirc\!\!\!\rangle-CN + HOMe \longrightarrow \rangle-C\begin{smallmatrix}OMe\\NH, HCl\end{smallmatrix} \longrightarrow \rangle-C\begin{smallmatrix}NH_2\\NH, HCl\end{smallmatrix}$

Abs. methanol added to a soln. of isopropyl 4-cyano-α-benzoylaminocinnamate in abs. dioxane-ether, then dry HCl passed at 5–8° for 7–8 hrs. into the soln., and stored 3 days at 0–5° → crude iminoester hydrochloride (Y: 90–95%) treated with abs. ethanol, a 10%-soln. of NH$_3$ in abs. ethanol added with shaking, which is continued 12 hrs. at room temp., then warmed 3 hrs. at 60°, and the product isolated as the hydrochloride → isopropyl 4-amidino-α-benzoylaminocinnamate hydrochloride (Y: 90%). F. e. s. P. Richter et al., Pharmazie *28*, 585 (1973).

Imidazoles from nitriles
Ring closure via acetals s. 26, 304

Addition to Sulfur and Carbon   NC↓SC

Without additional reagents   *w.a.r.*
**5-Sulfonylimino-1,2,3,4-thiatriazolines
from azides and sulfonylisothiocyanates**

221.
$PhCH_2N_3 + \underset{N-SO_2C_6H_4CH_3-p}{\overset{S}{\underset{\|}{C}}\!\!\diagup} \longrightarrow PhCH_2-N\!\!\underset{N-SO_2C_6H_4CH_3-p}{\overset{N=\!\!=\!\!N}{\diagdown\!\!\!S}}$

Equimolar amounts of benzyl azide and tosyl isothiocyanate allowed to react 20 hrs. at room temp. in CCl_4 → 4-benzyl-5-tosylimino-1,2,3,4-thiatriazoline. Y: 70%. F. e. and reactions of the products s. G. L'abbé et al., Am. Soc. *96*, 3973 (1974); conversion of the product into other heterocyclics with heterocumulenes cf. J. Org. Chem. *40*, 1728 (1975).

Sulfinovinylbetaines   C
from thiirene 1,1-dioxides s. 30, 424

Sodium azide NaN_3
5-Amino-1,2,3,4-thiatriazoles
  from isothiocyanates s. *28,* 278

Silver nitrate $AgNO_3$
2-Aminomercaptans from sulfido compds. $\diagdown_S\diagup \twoheadrightarrow C(SH)C(N<)$
  s. *19,* 379 suppl. *28*

Addition to Carbon NC↓CC

Without additional reagents w.a.r.
Aminolactols from lactones $C:C \twoheadrightarrow CHC(N<)$
  s. *28,* 607
Enamines from acetylene derivs. $C\equiv C \twoheadrightarrow CH:C\cdot N<$
  3-Hydroxyquinolines via acetylenedicarboxylic acid ester adducts s. *26,* 939
β-Amino-α,β-ethylenenitriles $C:C:C\cdot CN \twoheadrightarrow CH\cdot C(N<):C\cdot CN$
  from cyanoallenes s. *29,* 160
Nucleophilic addition $C:C \twoheadrightarrow CHC\cdot N_3$
Azides from ethylene derivs.

222.

Thiochromone 1,1-dioxide in chloroform treated with a soln. of hydrazoic acid in the same solvent, and the product isolated after 15 min. at 25° ⇾ product. Y: 95%. Also addition of HCl s. I. W. J. Still, M. T. Thomas, and A. M. Clish, Can. J. Chem. *53,* 276 (1975).

α-Azomethinioamidinium salts ←
  from 1,1,2-triaminoethylenes s. *28,* 279
Nucleophilic reactions with aminosulfenium ylids $C\equiv C \twoheadrightarrow CH:C\cdot \overset{-}{N}\cdot \overset{+}{S}$
  s. *29,* 309
α-Diazo(sulfonylamidines) ←
  from sulfonic acid azides and ynamines s. *29,* 310
4-Piperidone ring ◯
  from cross-conjugated dienones s. *27,* 328
[2+2]-Cycloaddition to tetramethoxyethylene
  1,2-Oxazetidines from ethylene derivs. and nitroso compds. s. *27,* 329; reactions of tetraalkoxyethylenes cf. P. H. J. Ooms, J. W. Scheeren, and R. J. E. Nivard, Synthesis *1975,* 260, 263
5,6-Dihydro-4H-1,3,4-oxadiazine ring
  from diacylazo compds. s. *28,* 280
Reaction of sulfinodiimines with ketenes
  1,3-Cycloaddition – 1,2,5-Thiadiazolid-4-ones from sulfinodiimines and ketenes s. *26,* 305
β-Amino-α,β-ethylenecarboxylic acid aziridides ←
  from 2 aziridine ring molecules and cyclopropenones – Fragmentation of aziridines s. *26,* 306
Insertion of azimines ←
  into carbon-carbon triple bonds s. *28,* 281

Irradiation (s. a. under O_2) ↯
Carboxylic acid amides ←
  from α,β-ethyleneketones s. 27, 134

Potassium K
N-Vinylation with acetylene NH → NCH : CH_2
  s. 27, 330

Sodium/alcohol NaOR
2,3-Dihydro-4H-1,3-benzoxazin-4-ones ○
  from o-hydroxycarboxylic acid amides s. 26, 307

sec-Butyllithium sec-BuLi
Amines from ethylene derivs. C : C → CHC(N<)
  s. 24, 331 suppl. 28

Sodium azide (s. a. under ICl) NaN_3
1,2,3-Triazoles from acetylene derivs. ○
  s. 8, 404 suppl. 29

Sodium azide/cyanogen chloride NaN_3/ClCN
Cyanimines and N-cyanoaziridines ←
  from ethylene derivs. – Ketones from ethylene derivs. s. 28, 282

Sodium nitrite/hydrogen chloride $NaNO_2$/HCl
1-Alkoxy-2-oximino- from enamino-lactams CH : C → C(: NOH)C(OR)
  s. 29, 311

Sodium salt s. under NH_2Cl Na^+

Cupric nitrate $Cu(NO_3)_2$
C-Cleavage with electrophilic substitution C
  s. 29, 312

Cuprous chloride CuCl
N-Vinylation with acetylene NH → NCH : CH_2
  s. 27, 330

Lithium tetrahydridoaluminate $Li[AlH_4]$
N-Hydroxy-N-heterocyclics ⊙
  from isocyclic 2-nitrosamine dimers s. 27, 331

Thallic acetate $(CH_3COO)_3Tl$
1,2-Diamines from ethylene derivs. C : C → C(N<)C(N<)
  s. 10, 280 suppl. 29

Y-Zeolite(Zn^{2+}) ←
Azomethines C ≡ C → CH_2C : N ·
  from amines and acetylene derivs. s. 28, 283

Ammonium ceric nitrate $(NH_4)_2Ce(NO_3)_6$
1,2-Azidonitrates from ethylene derivs. C : C → C(ONO_2)C·N_3
  s. 28, 284

m-Chloroperoxybenzoic acid m-$ClC_6H_4COO_2H$
Cyclic acylaminoalcohols ○
  from ethyleneacylamines – 2-Azaadamantanes s. 29, 313; s. a. J. Org. Chem. 39, 3822 (1974)

Methylene blue s. under O_2 ←

Tribenzenesulfenamide $(PhS)_3N$
Reactions with trisulfenamides ⊙
  Pyrimidines from pyrroles – Prepn. of trisulfenamides s. 29, 314

Cyanogen chloride s. under NaN₃   ClCN

Trimethylsilyl azide s. under (CH₃COO)₄Pb   Me₃SiN₃

Titanium trichloride   TiCl₃
Radical amination with ring closure   ←
s. 27, 332

Lead tetraacetate/trimethylsilyl azide   (CH₃COO)₄Pb/Me₃SiN₃
Cyanoketones from cyclic ethylene derivs.   C
s. 26, 325

Hydrazoic acid   HN₃
Isomeric carboxylic acid amides   COR → CONHR
    by Schmidt reaction and Beckmann rearrangement s. 26, 308
Schmidt reaction of ketones
s. 26, 308

n-Butyl nitrite/acetic acid   *n-BuONO/CH₃COOH*
Hydrolytic nitrosation   C
    Oximinolactones from bicyclic enolethers − N-Acetyloximinolactams s. 29, 315

n-Butyl nitrite/hydrogen chloride   *n-BuONO/HCl*
s. 29, 315

Nitrosyl chloride   NOCl
α-Oximinocarboxylic acids and derivs. from ketenes   ←
    α-Chloro-α-oximinocarboxylic acid esters s. 28, 285

Nitryl fluoroborate $NO_2^+ \ BF_4^-$
1,2-Nitroacylamines   C : C → C(NO₂)C·NHAc
    from ethylene derivs. and nitriles − Extended Ritter reaction s. 28, 286

Chloramine/sodium salt NH_2Cl/Na^+
Acylamines from oxo compds.   CO → NHCO
s. 29, 316

Oxygen/methylene blue/irradition   ←
Quinazolines from indoles   ☉

223.

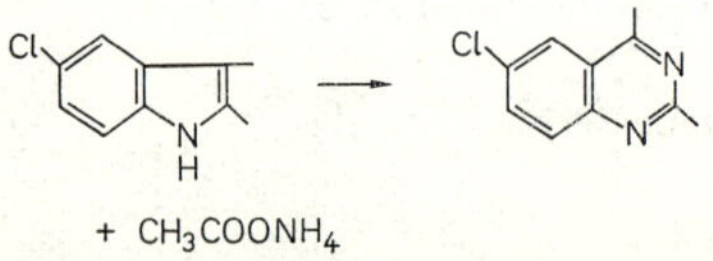

+ CH₃COONH₄

NH₄-acetate added to a soln. of 2,3-dimethyl-5-chloroindole in aq. methanol, and the resulting soln. containing methylene blue as sensitizer irradiated 7 hrs. at room temp. with 2 incandescent lamps in a stream of oxygen → 2,4-dimethyl-6-chloroquinazoline. Y: 72%. F. e. s. K. Maeda, T. Mishima, and T. Hayashi, Bull. Chem. Soc. Japan 47, 334 (1974).

Ozone s. under Ni   O₃

Hydroxylamine-O-sulfonic acid   H₂NOSO₃H
Dibenzo-1,4-diazepines from acridinium salts   ☉
s. 28, 287

Sulfuric acid   H₂SO₄
5,6-Dihydro-4H-1,3-oxazines   ○
    from 3-ethylenealcohols and nitriles, also from nitriles, ethylene derivs., and formaldehyde s. 26, 309

Iodine monochloride/sodium azide ICl/NaN_3
Azidotetrazoles ←
from nitriles and ethylene derivs. – Hassner-Ritter reaction s. *29*, 317

Hydrogen chloride HCl
α-**Nitroketones and 1,1-nitroethylene derivs.**   C : C → C(NO)C(NO$_2$)
from ethylene derivs. via 1,2-nitronitroso compds. s. *29*, 318

Nickel Ni
α-**Amino- and α-acylamino-carboxylic acids**   C
from 4-alkylidene-5-oxazolones via α-acylaminocarboxylic acid amides s. *28*, 288; asym. synthesis of α-acylaminocarboxylic acid amides with optically active amines on Pd-catalyst cf. E. I. Karpeiskaya et al., Izvest. *1974*, 1443; C. A. *81*, 152602

Nickel/ozone Ni/O_3
Diamines from cyclic ethylene derivs.
s. *27*, 149   C=C → CHN< / CHN<

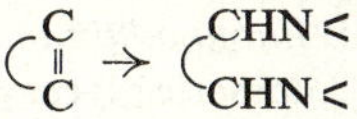

Via intermediates   *v.i.*
Pyrimidine nucleosides ←
from carbohydrate 2-amino-Δ²-oxazoline ring s. *28*, 341

Rearrangement   ↻

Hydrogen/Nitrogen Type   NC↻HN

Without additional reagents   *w.a.r.*
Nucleophilic ar. Smiles-type N-N-rearrangement ←
with preceding replacement of halogen by amino groups s. *26*, 310

Irradiation   ⫲
Carbostyrils ←
from 3-o-aminoarylidenesuccinimides s. *26*, 152

Potassium hydroxide/alcohol KOH
Aziridine ring rearrangement ←
with N-decarbalkoxylation s. *29*, 319

Pyridine C_5H_5N
4-Imino-2-oxazolidones from cyanourethans   O
s. *26*, 311

Acetonitrile   *MeCN*
Quinolizidones by rearrangement ←
s. *26*, 312

Hydrochlorides ←
4-Amino-1H-1,5-benzodiazepines   O
s. *29*, 320

Hydrogen chloride   HCl
Pyrazolo[1,5-a]pyrimidines

224.

Ethyl (4-ethoxycarbonyl-5-aminopyrazol-1-yl)methylenecyanoacetate refluxed 4 hrs. in ethanol containing a little concd. HCl → diethyl 7-aminopyrazolo[1,5-a]-pyrimidine-3,6-dicarboxylate. Y: 69–93%. F. e. and startg. m. s. K. Saito et al., Bull. Chem. Soc. Japan *47*, 476 (1974).

Hydrogen/Carbon Type   NC∩HC

Without additional reagents   w.a.r.
Anomeric glycosyl hydrazines
s. *28*, 289

Aziridine ring opening   C
s. *28*, 290

Sodium/alcohol   NaOR
α,β-Ethylenenitriles from pyridazine ring
s. *28*, 291

Apiezon
Indole from Δ¹-azirine ring
s. *27*, 335

Oxygen/Nitrogen Type   NC∩ON

Without additional reagents   w.a.r.
4-Nitroalcohols from nitric acid esters $CHC \cdot C \cdot C(ONO_2) \to C(NO_2)C \cdot C \cdot C(OH)$
    Functionalization of non-activated groups s. *28*, 292

    Δ¹-Azirine-3-carboxylic acid esters
       from 5-alkoxyisoxazoles s. *26*, 313

Irradiation
1-Hydroxy-3-oxospiro[indoline-2,4'-(4'H)pyrans]
       from 4-(o-nitrobenzylidene)-4H-pyrans s. *29*, 321

    Δ¹-Azirine-3-carboxamides
       from 5-aminoisoxazoles s. *29*, 322; isoxazole ring conversion into other heterocycles, review, s. Synthesis *1975*, 20

Sodium hydroxide   NaOH
3-Indazolones from 2-hydroxyindazoles
s. *26*, 324

Sodium methoxide   NaOMe
Isoxazole ring opening   C
    α,β-Ethylene-β-hydroxynitriles s. *28*, 293

Water H_2O
Dimroth rearrangement of N-alkoxy compds.
s. *27*, 168

Hydrogen chloride   *HCl*
1-(Azomethylimino)phthalans   ☉
   from 4-(alkylidenehydrazino)-1H-2,3-benzoxazines s. *28*, 294

3-(o-α-Hydroxyphenyl)-1,2,4-1H-triazoles   ←
   from 4-hydrazino-1H-2,3-benzoxazines s. *28*, 295

Oxygen/Carbon Type   NC↻OC

Without additional reagents   *w.a.r.*
[3.3]-Sigmatropic rearrangement   ←
   2-Ethyleneacylamines from 2-ethyleneiminoesters s. *29*, 325

α,ω-Rearrangement   ←
   α,ω-Di(acylamines) from α,ω-di(iminoesters)

225.

400 mg. ethylene bis-N-phenylbenzimidate refluxed 22.5 hrs. in chlorobenzene → 260 mg. bisamide. – A simple benzimidate, *n*-propyl N-phenylbenzimidate, was thermally stable at the temp. sufficient for the rearrangement of the bisimidates. F. e. s. D. H. R. Barton and S. Prabhakar, Soc. Perkin I *1974*, 781.

Acylhydrazines from 1,1-aryloxyhydrazones $C(OR):N \cdot NH \rightarrow CONHNR$
s. *27*, 336; s. a. Bull. Chem. Soc. Japan *48*, 365 (1975)

N-Halogenosulfonylaminophosphoryl dichlorides $N(SO_2Hal) \cdot POCl_2$
   via O→N-alkyl migration s. *27*, 119

1,4,6-Triazahepta-3,5-dien-7-ones   ←
   from 1,3,5-triazahepta-1,3-dien-7-ones s. *27*, 337

Mesoionic pyrimidines   ←
   by O→N-alkyl migration s. *29*, 323

*lin-ang-***Rearrangement**   ←
   of N-condensed imidazole ring s. *26*, 339

Hydroxylactams   ☉
   from azido- via amino-lactones s. *29*, 324

Sodium hydride s. under O_2 *NaH*

Sodium/alcohol   *NaOR*
Pyrrolo[1,2-d][1,2,4]triazine   ←
   from 2-(pyrrol-2-yl)-1,3,4-oxadiazole ring

226.

5-Methyl-2-(pyrrol-2-yl)-1,3,4-oxadiazole refluxed 48 hrs. in a soln. of Na-*n*-propoxide in *n*-propanol → product. Y: 80%. F. e. s. M. Robba, D. Maume, and J. C. Lancelot, Tetrah. Let. *1973*, 3239, 3235.

Triethylamine Et_3N
Acylhydrazines from 1,1-alkoxyhydrazones   C(OR) : N·NH → CONHNR
s. 27, 336 suppl. 29

1,3-Oxazin-4-one ring   O
    from o-cyanatocarboxylic acid esters s. 26, 314

Silver bromide $AgBr$
N- from O-Subst. maleimides ←
s. 29, 419

Mercuric trifluoroacetate $(CF_3COO)_2Hg$
[3.3]-Sigmatropic rearrangement ←
    2-Ethyleneacylamines from 2-ethyleniminoesters s. 29, 325

Dimethylformamide s. under SOCl$_2$ Me_2NCHO

Dimethyl sulfate $(MeO)_2SO_2$
N-Methyllactams from O-methyllactims   C(=N)–OMe → C(=N·Me)–CO
s. 28, 296

Oxygen/sodium hydride O_2/NaH
Rearrangement ←
    of 2β-oxiranyl-1,2,3,4,6,7,12,12b-octahydroindolo[2,3-a]quinolizines s. 28, 99

p-Toluenesulfonamide $TsNH_2$
N- from O-glycosides ←
s. 30, 150

Thionyl chloride/dimethylformamide $SOCl_2/Me_2NCHO$
O→N-Alkyl migration ←
with replacement of hydroxyl by chlorine

227.

Dry dimethylformamide added to a soln. of 1-(2-hydroxyethoxy)-3-methylphthalazin-4(3H)-one in dry chloroform, then SOCl$_2$ added dropwise with stirring, and refluxed 7 hrs. → 2-(2-chloroethyl)-3-methylphthalazine-1,4(2H,3H)-dione. Y: 82%. B. G. Pring and C.-G. Swahn, Acta Chem. Scand. 27, 1891 (1973).

Hydrogen chloride HCl
4-Hydroxyindoles from 4-aminobenzofurans ←
s. 28, 297

Nitrogen/Nitrogen Type   NC∩NN

Without additional reagents   w.a.r.
N→C-Nitro group migration ←
s. 29, 326

Pyrimidines from pyridazines ←
s. 29, 327

Irradiation   ⚡
Rearrangement of cyclic N-ylids   CH : N$^+$·N$^-$Ac → C(NHAc) : N
s. 28, 298

Cumene hydroperoxide $PhCMe_2 \cdot OOH$
6-Acyl-1,6-dihydropyridazines
from 1-iminopyridinium ylids s. 27, 144

Formic acid $HCOOH$
Rearrangement of triazolotriazines
s. 26, 315

Nitrogen/Carbon Type   NC∩NC

Without additional reagents   w.a.r.
Dimroth rearrangement
under anhydrous conditions s. 30, 252

N-Carbamyl and N-thiocarbamyl group migration
s. 27, 338

2,3,5,6-Tetrahydro-1H-pyrrolo[2,1-b][1,3]benzodiazepines
from 1-[2-(1-pyrrolinyl)]indolines s. 27, 339

Irradiation
Diazirines from diazo compds. and reverse reaction $N^-:N^+:C \rightarrow N:N$
Light specificity s. 29, 328

Sodium hydroxide $NaOH$
Dimroth rearrangement
4-Aminopyrimidines from 6-imino-1,6-dihydropyrimidines

228.

5-Ethyl-1,6-dihydro-6-imino-1-methylpyrimidine hydriodide and 2 N NaOH heated 3 hrs. at 80° → 5-ethyl-4-methylaminopyrimidine. Y: 94%. F. e. s. D. J. Brown and K. Ienega, Soc. Perkin I *1974*, 372.

Sodium iodide NaI
2,3,5,6-Tetrahydro-1H-imidazo[1,2-a]imidazoles
s. 27, 340

Mercuric bromide $HgBr_2$
N-Glycosyl migration
s. 26, 316

Lithium tetrahydridoaluminate $Li[AlH_4]$
2,3,3a,4,9,9a-Hexahydro-1H-pyrrolo[2,3-b]quinoxalines
from 1-benzimidazolylsuccinimides s. 29, 329

Acetic anhydride Ac_2O
2,3-Dihydro-4H-1,4-thiazin-3-one 1,1-dioxide
from 2,3,4,7-tetrahydro-1,4,5-thiadiazepin-3-one 1,1-dioxide ring s. 28, 299

Iodine I
Δ²-Imidazolines from N-imidoylethylenimines
s. 16, 410 suppl. 29

Sulfur/Carbon Type NC∩SC

Without additional reagents w.a.r.
Carboxylic acid thioamides from thioiminoesters C(:N·)S· → C(S)N<
s. *26*, 317
Migration of 1,3,5-triazin-2-yl groups ←
Ureas from isocyanates

229.

Phenyl isocyanate added to a stirred soln. of S-(4,6-dimethoxy-s-triazin-2-yl)-2-aminothiophenol in benzene, and stirred 3 hrs. → N-(4,6-dimethoxy-s-triazin-2-yl)-N-phenylcarbamoyl-2-aminothiophenol. Y: 95%. F. e., also O→N-migration, and limitations s. T. Shiojima et al., Bull. Chem. Soc. Japan *46*, 2549 (1973).

Potassium hydroxide KOH
o-Aminosulfinic acids ←
from o-aminosulfones – Smiles rearrangement s. *28*, 300

Remaining Elements/Carbon Type NC∩RemC

Without additional reagents w.a.r.
Silylaminosiloxanes ←
from N,N-disilylsiloxylamines s. *28*, 301

Carbon/Carbon Type NC∩CC

Without additional reagents w.a.r.
C,N-Positional interchange ←
of acyl groups s. *30*, 527
Isocyanates from carboxylic acid azides CON₃ → N:C:O
s. *30*, 460
2,3-Dihydro-4(1H)-quinazolones ○
s. *28*, 302
1,2,3-Triazines from cyclopropenyl azides ⊙
s. *29*, 330
Ring contraction by 2 C-atoms ○⃜
of N,N-heterocycles s. *28*, 303
Δ²-1,2,4-Triazoline from indone ring ←
s. *27*, 341
Irradiation ⁂
Carbon-nitrogen from carbon-carbon double bonds C:C·NH → CH·C:N
s. *28*, 304
Phenanthrid-6(5H)-ones ←
from phenanthridine N-oxides s. *28*, 305
Benzimidazol-2-ones ←
from benzimidazole 3-oxides s. *26*, 990
Oximinoketones C
from cyclic nitrous acid esters s. *27*, 342

Rearrangement of cyclic ethylenenitrites
s. 27, 343

Lactams from spirooxaziridines
s. 29, 331; 4-azahomoadamantanes s. Synthesis 1974, 812

1,3-Oxazepines
from 2,3-oxazabicyclo[3.2.0]hepta-3,6-dienes s. 29, 332

Sodium/alcohol — NaOR
3,4-Dihydro-2H-1,3-benzothiazin-4-ones
s. 27, 344

Ammonium hydroxide — NH_4OH
2,3-Dihydro-4H-1,4-thiazines
s. 26, 318

Diethylamine — Et_2NH
O→N-Acyl group migration
s. 28, 306

Triethylamine s. under $(PhO)_2PON_3$ — Et_3N

Pyridine s. under $(CH_3COO)_4Pb$ — C_5H_5N

Mercuric chloride-sodium tetrahydridoborate — $HgCl_2$-$Na[BH_4]$
Heterocyclics
by intramolecular solvomercuration-demercuration – N-Heterocyclics from ethyleneamines via intramolecular aminomercuration s. 27, 345

Lithium tetrahydridoaluminate — $Li[AlH_4]$
N-Heterocyclics by reductive ring closure
s. 28, 307

Aziridine ring from α,β-ethyleneoximes
s. 29, 14

Acetic acid — CH_3COOH
Pyrazole from pyrazolenine ring
van Alphen rearrangement s. 27, 346

Trifluoroacetic acid — CF_3COOH
2-Amino-5,6-dihydro-4H-1,3-thiazines
from 2-ethylenisothiouronium salts s. 28, 308

Lead tetraacetate — $(CH_3COO)_4Pb$
Pyrimido[4,5-d]pyrimidine from pyrrolo[2.3-d]pyrimidine ring

230.

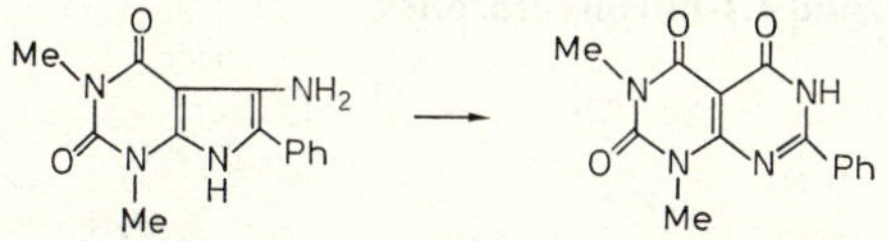

Oxidative ring expansion. Pb-tetraacetate added portionwise to a suspension of 1,3-dimethyl-5-amino-6-phenylpyrrolo[2,3-d]-2,4(1H,3H)-pyrimidinedione in acetic acid, stirred and heated 3 hrs. at 90° → 1,3-dimethyl-5-hydroxy-7-phenylpyrimido-[4,5-d]-2,4(1H,3H)-pyrimidinedione. Y: 90%. F. e., **also by reductive and nucleophile-induced ring expansion,** s. F. Yoneda and M. Higuchi, Bull. Chem. Soc. Japan 46, 3849 (1973).

Lead tetraacetate/pyridine $(CH_3COO)_4Pb/C_5H_5N$
2-Aminoalcohols from β-hydroxycarboxylic acid amides $CONH_2 \rightarrow NH_2$
via 2-oxazolidone ring – Hofmann-type degradation with retention of configuration s. 28, 309

Diphenyl phosphoryl azide/triethylamine $(PhO)_2PON_3/Et_3N$
Urethans from carboxylic acids $COOH \rightarrow NHCOOR$
Simplified Curtius degradation s. 28, 353

Phosphoric acid H_3PO_4
4,5-Dihydropyrrolo[1,2-c]quinazolines ☉
from spiro-3H-indoline-3,2′-2′H-pyrroles s. 26, 961

Phosphorus oxide chloride $POCl_3$
Benzothiazol-2-ones ←
from benzothiazole N-oxides s. 29, 269

p-Toluenesulfonyl chloride/sodium hydroxide $TsCl/NaOH$
Benzimidazol-2-ones ←
from benzimidazole-3-oxides s. 26, 990; s. a. Soc. Perkin I *1975*, 212

Sulfuric acid H_2SO_4
3H,5H-Thiazolo[4,3-a]isoindol-5-ones ←
from 2-isothioureidoindan-1,3-diones s. 26, 319

Hydrochlorides ←
N-Condensed cyclimmonium salts ○
by benzocyclobutene ring opening with dehydrogenation s. 28, 310

Heteropolar Bond Type NC∩Het

Irradiation ⁀⁀
Diaziridines from 3-pyrazolidone betaines ○
s. 26, 320

Exchange ↓↑

Hydrogen ↑ NC↕H

Without additional reagents w.a.r.
Direction of nitration $H \rightarrow NO_2$
s. 26, 328
1,1,1-Trinitro compds. from aldoximes $CH:NOH \rightarrow C(NO_2)_3$
s. 27, 347
1,1-Nitroazo compds. and 1,1-nitrohydrazones ←

231.

from hydrazones. A 3 M soln. of NO_2 in methylene chloride added slowly to a soln. of acetone 2,4-dinitrophenylhydrazone in the same solvent, and after 0.5 hr.

washed with satd. NaHCO₃-soln. → *gem*-nitroazo compd. Y: 80%. − Similarly: Benzaldehyde 2,4-dinitrophenylhydrazone → *gem*-nitrohydrazone. Y: 76%. F. e. s. T. A. B. M. Bolsman and T. J. de Boer, Tetrahedron *29*, 3929 (1973).

Reaction of alkyl nitrites with organic salts ←
with sulfonium and pyridinium salts − α-Ketohydroximinohalides from 2-ketosulfonium salts and by nitrosation of β-ketosulfoxides s. *27*, 348

Uracil ring O
from o-aminocarboxylic acid amides s. *29*, 333

Cyclic N-cyanoguanidines
from diamines and dimethyl cyaniminodithiocarbonate s. *28*, 311

2,4-Quinazolinediones
from 3,1-benzoxazine-2,4-diones and hindered amines via o-aminocarboxylic acid amides s. *27*, 349

Quinolizinium ring opening C
s. *27*, 350

Irradiation (s. a. under Me₃C·ONO and O₂) ⚡
Difluoroamination H → NF₂
s. *27*, 351

Electrolysis ⚡
Anodic coupling ←
Diarylamines from ar. amines s. *28*, 703 suppl. 29

Acylamines from hydrocarbons and nitriles H → NHCOR
ω-Acylamination of carboxylic acid esters s. *29*, 334; remote acylamination of ketones s. Am. Soc. *97*, 853 (1975)

Sodium hydride/tosyl azide *NaH/TsN₃*
Replacement of hydrogen by azido groups H → N₃

232.   PhCH(COOEt)₂ → PhC(COOEt)₂
        + TsN₃ |
                         N₃

A soln. of diethyl phenylmalonate added dropwise at room temp. under N₂ to a suspension of NaH in dry glyme, when gas evolution has ceased a soln. of tosyl azide in the above solvent added dropwise during 0.5 hr., stirred 1 hr. at 35–40° then an additional 2 hrs. → diethyl azidophenylmalonate. Y: 77%. F. e. s. S. J. Weininger et al., J. Org. Chem. *39*, 1591 (1974).

Sodium hydroxide *NaOH*
Cyclic o-azo-N-oxides C
from N-condensed 1,2,3,5-oxatriazolium salts s. *27*, 283

Sodium/alcohol *NaOR*
Tetraazamacroheterocyclics O
Cyclotetramerization of 3-carbamylpyridinium salts

233.

1-Benzyloxymethyl-3-carbamoylpyridinium chloride added to a soln. of Na in abs. alcohol, and worked up after 20 min. → product. Y: 97%. F. e. s. W.-H. Gündel, Z. Naturf. *28b*, 471 (1973); *29b*, 556 (1974).

Potassium/alcohol KOR
1,2,3-Triazole ring from enamines
Diazo group transfer s. 26, 321

n-Butyllithium n-BuLi
Reactions via metalated aliphatic-ar. azo compds. $N:N\cdot CH \to NR\cdot N:C$
Subst. hydrazones – Anionic [3+2]-cycloaddition – Metalated azomethines
s. 27, 352

Lithium diisopropylamide i-Pr$_2$NLi
α-Aminocarboxylic from carboxylic acids $CHCOOH \to C(NH_2)COOH$
s. 28, 312; s. a. Chem. Pharm. Bull. 23, 167 (1975)

Potassium carbonate K$_2$CO$_3$
Aminoacetamides from 2 amine molecules $2 >NH \to >NCH_2CON<$
s. 28, 313

2-Imidazolidone ring from 1,2-diamines O
s. 3, 375

10bH-Oxazolo[3,2-c][1,3]benzoxazine-2(3H),5-diones
by double ring closure s. 27, 353

Sodium acetate CH$_3$COONa
Oxidative acylation of prim. nitro compds. ←
s. 26, 322

2-Amino-Δ^2-oxazolines from 2-aminoalcohols O
Stereospecific ring closure with retention or inversion of configuration, interconversion of stereoisomeric 2-aminocohols s. 29, 335

Cyclic ar. azo compds. by coupling
11H-Dibenzo[c,f][1,2,5]triazepines s. 27, 354

5-Imino-Δ^2-1,3,4-thiadiazolines
from thiocyanates and diazonium salts

234.
$$\begin{array}{c} Bz-CH_2-S \\ + \quad | \\ N_2^+ \; Cl^- \; CN \\ | \\ Ph \end{array} \longrightarrow \begin{array}{c} Bz\underset{|}{\overset{}{\diagup}}S \\ N\diagdown \quad \diagup \\ \; N \quad NH \\ | \\ Ph \end{array}$$

Startg. thiocyanate allowed to react with benzenediazonium chloride in Na-acetate-buffered soln. → 5-imino-4-phenyl-2-benzoyl-Δ^2-1,3,4-thiadiazoline. Y: 75%. F. e. s. A. S. Shawali and A. O. Abdelhamid, Tetrah. Let. 1975, 163.

Sodium cyanide s. under MnO$_2$ NaCN

Ammonium thiocyanate NH$_4$SCN
Sym. thioureas from amines $2\,NH \to N\cdot CS\cdot N$
s. 26, 323

Sodium nitrite s. a. under ROOC·N : N·COOR and AgClO$_4$ NaNO$_2$

Sodium nitrite/sodium sulfite NaNO$_2$/Na$_2$SO$_3$
3-Indazolones O
from o-aminoketones via 2-hydroxyindazoles s. 26, 324

Sodium nitrite/acid NaNO$_2$/H$^+$
Furoxan ring from hydroxylamines
s. 29, 338

Sodium nitrite/acetic acid $NaNO_2/CH_3COOH$
α-**Oximinoazomethines from enamines**   ←

235.

Isobasic isoquinolines. An aq. soln. of $NaNO_2$ added dropwise with stirring at 5–10° during 1 hr. to a soln. of the startg. m. in acetic acid, and stirring continued 4 hrs. at the same temp. → 2-(6,7-dimethoxy-3,4-dihydro-1-isoquinolinyl)-2-(hydroximino)acetonitrile. Y: 77%. F. e. s. K. Harsányi, K. Takács, and É. Benedek, A. *1973*, 1606.

Isoalloxazine 5-oxides   ○
     from 6-chloro- via 6-anilino-uracils s. *28*, 314

Pyrimido[5,4-e]-*as*-triazine-5,7(6H,1H)-diones
     s. *27*, 355

Ring closure with nitrosation
     1-Oxo-6,6a-dithia-2-azapentalenes from 1,2-dithiolium salts – 1-Oxa-6,6a-diselena-2-azapentalenes from 1,2-diselenolium salts s. *28*, 315

Sodium nitrite/sulfuric acid $NaNO_2/H_2SO_4$
Nitration   H → NO_2
     Nitro from nitroso compds. s. *27*, 356

1-Acetoximino-2-nitrites   C : CH → C(ONO)C : NOAc
     from ethylene derivs. s. *29*, 336

Potassium nitrate/sulfuric acid KNO_3/H_2SO_4
Cyclic N-oxides via nitration   ○
     s. *29*, 337

Furoxan ring from hydroxylamines
     s. *29*, 338

Diethylamine s. under H_2O_2 Et_2NH

Triethylamine Et_3N
N,N-Dicyanamides from prim. amines   NH → N·CN
     s. *29*, 342

Cyclic N-chlorocarbonyl-1-alkoxyenamines   ←
     from cyclic iminoesters s. *27*, 466

2-Oxazolone ring as O,N-protective group   ○
     s. *27*, 357; s. a. Tetrahedron *30*, 3701 (1974)

Triethylamine/diazonium fluoroborate   ←
Diazo compds. from amines via triazenes $CHNH_2$ → CN_2

236.

Ethyl alanate allowed to react at –70° with 2 equivalents 2,4-dinitrobenzenediazonium fluoroborate and slightly more than 4 equivalents triethylamine in *dimethylformamide* → product. Y: 75%. – This amine diazotization occurs under very mild, essentially neutral conditions. F. e. s. J. F. McGarrity, Chem. Commun. *1974*, 558; diazo compds. from amines with isoamyl nitrite in the presence of a little acetic acid s. M. Takamura et al., Tetrahedron *31*, 227 (1975).

Pyridine (s. a. under ArN : C : S) C_5H_5N
4-Hydrazono-5-pyrazolones $CH_2 \rightarrow C:N \cdot NHAr$
s. *29*, 450

Reactions of N-heterocyclics with sulfonic acid azides   ←
3-Diazo-2-sulfonyliminoindolines from indoles s. *29*, 339

Ring closure with carbon disulfide   ○
Imidazo[1,5-a]pyrazine-3-thiols s. *27*, 358

Pyridine/nitrosyl chloride $C_5H_5N/NOCl$
Nitrile from methyl groups $CH_3 \rightarrow CN$

237.

2,4,6-Trinitrotoluene added at –10° to a soln. of nitrosyl chloride in pyridine, and stirred 6 hrs. at –10 to –5° → 2,4,6-trinitrobenzonitrile. Y: 60.9%. M. E. Sitzmann and J. C. Dacons, J. Org. Chem. *38*, 4363 (1973); f. method cf. 342, 590.

Silver acetylides $Ag \cdot C \equiv C \cdot R$
Ynazo compds. from diazonium salts $N_2^+ \rightarrow N:N \cdot C \equiv C \cdot$

238.

An ethanolic soln. of p-chlorophenyldiazonium chloride (or bromide) added dropwise at room temp. to a vigorously stirred suspension of Ag-phenylacetylide in chloroform, and stirring continued 0.5 hr. → product. Y: 70%. F. e. s. S. J. Huang et al., J. Org. Chem. *40*, 124 (1975).

Silver acetate CH_3COOAg
Carbonylation of amines   ←
s. *28*, 322

Silver cyanide/tert-butyl isocyanide $AgCN/Me_3C \cdot N:C$
1,3-Heterocyclics   ○
s. *29*, 340

Aluminum nitrate $Al(NO_3)_3$
Ar. mononitration $ArH \rightarrow ArNO_2$

239.

p-Hydroxybenzoic acid added in small portions with occasional cooling to a stirred suspension of $Al(NO_3)_3 \cdot 9 H_2O$ in acetic anhydride, and the product isolated after 1.5 hrs. → 3-nitro-4-hydroxybenzoic acid. Y: 80%. F. e. s. K. Rajamohan and N. V. Subba Rao, Indian J. Chem. *11*, 1076 (1974).

Diethyl azodicarboxylate $ROOC \cdot N:N \cdot COOR$
Alloxazines   ○
s. *28*, 316

Diethyl azodicarboxylate/sodium nitrite $ROOC \cdot N:N \cdot COOR/NaNO_2$
Cyclic N-oxides
by oxidative ring closure – Toxoflavin 4-oxides s. *28*, 317

Diazonium fluoroborate s. under Et_3N ←

Urea (s. a. under CF_3COOH) $(NH_2)_2CO$

2-Oxo-1,3-N,N-heterocyclics from diamines
s. *21*, 393 suppl. *29*

Dicyclohexylcarbodiimide/phosphoric acid $RN:C:NR/H_3PO_4$

N-Methylthiomethylnitrones from oximes $C:NOH \rightarrow C:N(O)CH_2SMe$
s. *27*, 359

tert-Butyl isocyanide s. under AgCN and $PdCl_2$ $Me_3C \cdot N:C$

Acetic acid CH_3COOH

3-Acyl-1,2,4-oxadiazoles O
from α-acylaminoketones, also f. 1,2,4-oxadiazoles s. *28*, 318

Trifluoroacetic acid/urea $CF_3COOH/(NH_2)_2CO$

Ar. nitration $ArH \rightarrow ArNO_2$

240.

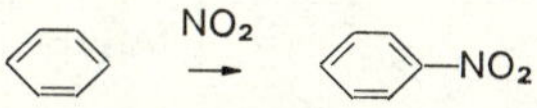

Nitrogen dioxide bubbled 1 hr. at 50° through a soln. of benzene in trifluoroacetic acid containing urea → nitrobenzene. Y :99%. F. e. s. R. O. C. Norman, W. J. E. Parr, and C. B. Thomas, Soc. Perkin I *1974*, 369; **nitration** of ar. and heteroar. compds. **by transition metal-nitrato complexes** cf. R. G. Coombes and L. W. Russell, Soc. Perkin I *1974*, 1751; **orientation control** via ar. thallation (cf. Synth. Meth. *25*, 389) cf. B. Davies and C. B. Thomas, Soc. Perkin I *1975*, 65.

Methylene blue s. under O_2 ←

Aryl isothiocyanates/pyridine $ArN:C:S/C_5H_5N$

2-Mercaptopyrimidine ring O
from o-aminocarboxylic acid amides s. *27*, 360

Phosgene $COCl_2$

2,5-Oxazolidiones
from α-aminocarboxylic acids s. *14*, 240

Trichloromethyl chloroformate $ClCOOCCl_3$

2,5-Oxazolidiones
from α-aminocarboxylic acids s. *14*, 240 suppl. *29*

Oxalyl chloride $(COCl)_2$

Isocyanates from amines $NH_2 \rightarrow N:C:O$
s. *27*, 361

**2,3,5,10,11,11a-Hexahydro-
1H-imidazo[5,1-c][1,4]benzodiazepine-1,3,11-triones
from 1,3-dihydro-2H-1,4-benzodiazepin-2-one-3-carboxamides**

241.

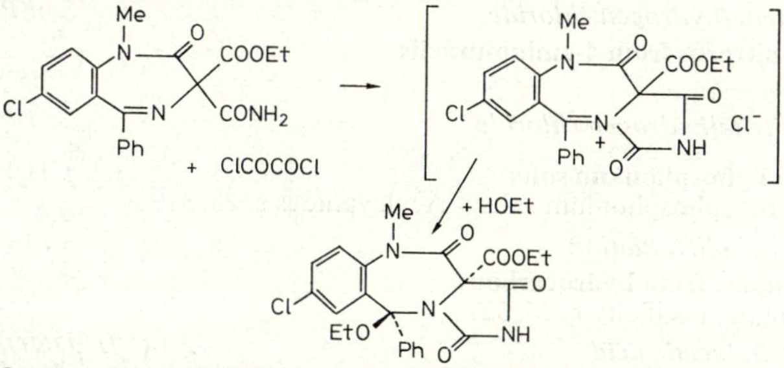

Stereospecific ring closure. Oxalyl chloride added all at once at 10° to a suspension of 3-carbamoyl-7-chloro-3-ethoxycarbonyl-2,3-dihydro-1-methyl-5-phenyl-1H-

1,4-benzodiazepin-2-one in dry ethylene chloride, refluxed 5 hrs., cooled to 10°, ethanol added, and stirred 15 min. at room temp. → 7-chloro-5β-ethoxy-11aα-ethoxycarbonyl-2,3,5,10,11,11a-hexahydro-10-methyl-5α-phenyl-1H-imidazo[5,1-c]-[1,4]benzodiazepine-1,3,11-trione. Y: 65%. F. e. and reactions of the products s. R. Jaunin and W. Arnold, Helv. 56, 2569 (1973).

Thiophosgene $\qquad CSCl_2$
1,1-Chloroisothiocyanates and enisothiocyanates $\qquad$ C : NH → CCIN : C : S
from imines

242.

Ph₂C=NH → Ph₂C(Cl)(N=C=S)

Ph(Me₂CH)C=NH → [Ph(Me₂CH)C(Cl)(N=C=S)] → Ph(Me₂C=)C-N=C=S

Benzophenone | Isopropyl phenyl ketone
imine in toluene added dropwise with stirring to an ice-cooled soln. of thiophosgene in dry toluene, and refluxed 2–3 hrs. →
diphenylchloromethyl | 2-methyl-1-phenylpropenyl
Y: 82%. | isothiocyanate. | Y: 73%.
F. e. s. W. I. Gorbatenko, W. A. Bondar, and L. I. Samaraj, Ang. Ch. 85, 866 (1973).

Trimethylsilyl azide s. under $(CH_3COO)_4Pb$ $\qquad Me_3SiN_3$

Lead tetraacetate $\qquad (CH_3COO)_4Pb$
α-Azidoketones from ethylene derivs. $\qquad$ C : CH → C(N₃)CO
Reactions with lead(IV) acetate azides, review, s. 27, 362

Hexahydro-1,2,4,5-tetrazines $\qquad$ ○
by oxidative dimerization of semicarbazides s. 26, 326

4-Hydroxypyrimidine from 3-aminopyrrole ring $\qquad$ ⊙
s. 27, 363

Reactions with *lead tetraacetate/trimethylsilyl azide* $\qquad (CH_3COO)_4Pb/Me_3SiN_3$
Ethyleneazides and cyanoketones from ethylene derivs. ←
s. 26, 325

n-Propyl nitrite/hydrogen chloride $\qquad n\text{-}PrONO/HCl$
4-Amino-5-nitroso- from 4-amino-uracils $\qquad$ H → NO
s. 29, 186

Isopropyl nitrite/hydrogen chloride $\qquad i\text{-}PrONO/HCl$
Nitriles from phosphonium salts $\qquad CH_2\overset{+}{P}{\leqslant} \to CN$
via α-nitrosophosphonium salts – Acylcyanides s. 28, 319

tert-Butyl nitrite/irradiation $\qquad Me_3C\cdot ONO/\cancel{h\nu}$
Nitroso compds. from hydrocarbons $\qquad$ H → NO
Wavelength specificity s. 26, 327

Isoamyl nitrite/acetic acid $\qquad i\text{-}C_5H_{11}ONO/CH_3COOH$
Diazo compds. from amines $\qquad$ CHNH₂ → CN₂
s. 30, 236

Nitrosyl chloride s. under C_5H_5N — $NOCl$

n-Propyl nitrate s. under $NaNH_2$ — $n\text{-}PrONO_2$

Acetyl nitrate — CH_3COONO_2
Direction of nitration — $H \to NO_2$
s. *26*, 328

Nitryl fluoroborate — $NO_2^+\ BF_4^-$
Nitration of phosphines via phosphonium salts
    Protection of phosphorus(III) compds. as phosphonium salts s. *27*, 364; also nitration with nitryl trifluoromethanesulfonate s. F. Effenberger and J. Geke, Synthesis *1975*, 40; L. M. Yagupol'skii, I. I. Maletina, and V. V. Orda, Zh. Org. Khim. *10*, 2226 (1974); C. A. *82*, 16443

Nitryl chloride — NO_2Cl
Nitration
s. *28*, 320

Diphenyl phosphite — $(PhO)_2P\cdot OH$
Sym. ureas and thioureas
from amines — $RNH_2 \begin{matrix} \nearrow (RNH)_2CS \\ \searrow (RNH)_2CO \end{matrix}$

243.
$$+\ CO_2 \quad \overset{2\ PhNH_2}{\swarrow \quad \searrow} \quad +\ CS_2$$
$$PhNHCONHPh \qquad PhNHCSNHPh$$

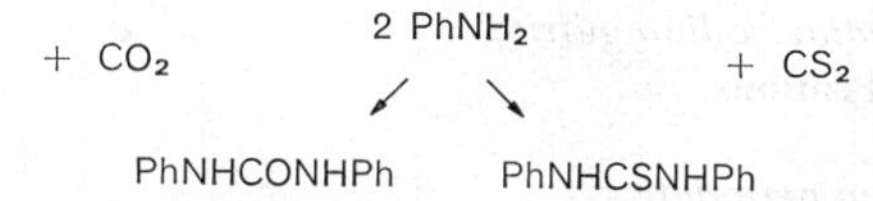

A mixture of equimolar amounts of aniline and diphenyl phosphite in pyridine kept 4 hrs. at 40°
with passage of $CO_2 \to$ N,N'-diphenylurea. Y: 85%. | with $CS_2 \to$ N,N'-diphenylthiourea.
Also in the presence of *imidazole* in dimethylformamide, and f. e. s. | Y: ca. 100%.
N. Yamazaki, F. Higashi, and T. Iguchi, Tetrah. Let. *1974*, 1191.

Phosphoric acid s. under $RN : C : NR$ — H_3PO_4

Oxygen s. a. under Se — O_2

Oxygen/methylene blue/irradiation — ←
1,3,5-Triazines from imidazoles — ☉
s. *27*, 365

Hydrogen peroxide/diethylamine — H_2O_2/Et_2NH
Isothiocyanates and sym. thioureas from amines — ←
s. *27*, 366

Sulfur — S
Ureas from amines and carbon monoxide $RNH + CO + NHR' \to RN\cdot CO\cdot NR'$
s. *26*, 329

N,N-Dichlorobenzenesulfonamide/
sodiobenzenesulfonamide — $PhSO_2NCl_2/PhSO_2NHNa$
N-Sulfonylimines from hydrocarbons — $CH_2 \to C:NSO_2R$
s. *27*, 181

Polymer-based sulfonic acid azide — ←
Diazo compds. by diazo group transfer — $CH_2 \to C:N_2$
s. *20*, 271 suppl. *29*

Benzenesulfonyl azide — $PhSO_2N_3$
Reactions of N-heterocyclics with sulfonic acid azides — ◯
    N-Condensed 1,2,3-triazole ring s. *28*, 321

Tosyl azide s. under NaH — TsN_3

Trifluoromethanesulfonic acid — CF_3SO_3H
Ar. nitration — H → NO_2
s. *10*, 292 suppl. 29

Sulfuric acid — H_2SO_4
Direction of nitration
s. *27*, 367

Sulfuric acid/sulfur trioxide — H_2SO_4/SO_3
s. *27*, 367

Selenium/oxygen — Se/O_2
Sym. ureas from amines and carbon monoxide — 2 >NH → >NCON<
   Sym. carbonic acid esters from alcohols – Urethans from alcohols, amines, and CO s. *28*, 322

Iodine s. under $AgClO_4$ — I

Sodium iodate — $NaIO_3$
2,5-Diamino-p-quinones from p-quinols ←
s. *27*, 368

Silver perchlorate/iodine/sodium nitrite — $AgClO_4/I/NaNO_2$
Nitration via radical cations — H → NO_2
s. *27*, 369

Tetraethylammonium periodate — $Et_4N^+IO_4^-$
N-Acyl-3,6-dihydro-2H-1,2-oxazine ring — O
from hydroxamic acids and 1,3-dienes

244.

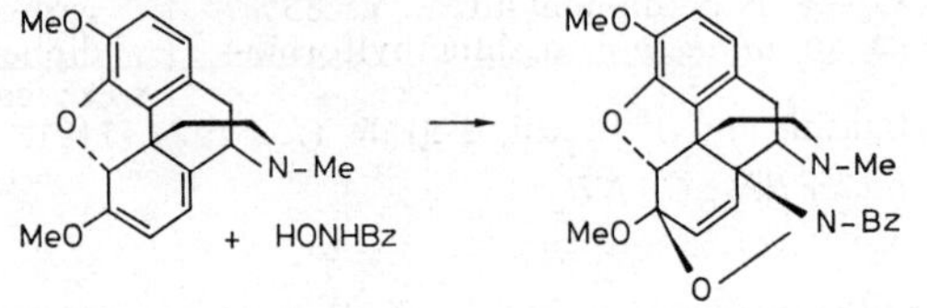

Benzohydroxamic acid added at 0° to a stirred mixture of thebaine in ethyl acetate and tetraethylammonium periodate in aq. acetic acid-Na-acetate of pH ca. 6, and the product isolated as the hydriodide → adduct. Y: 97%. F. e. s. G. W. Kirby and J. G. Sweeny, Chem. Commun. *1973*, 704.

Hydrochlorides ←
Diisocyanates — NH_2 → N:C:O
   from aminohydroxamic acids s. *27*, 492

4-Amino-1,2,4-triazolidine-3,5-diones — O
   from hydrazines s. *29*, 341

Hydrogen chloride (s. a. under i-PrONO) — HCl
2-Aminobenzimidazoles
from o-diamines and cyanamide

245.

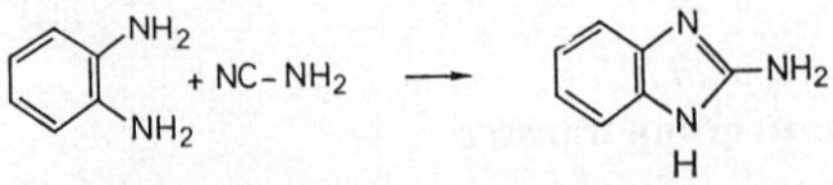

Aq. 50%-cyanamide added dropwise to a boiling soln. of o-phenylenediamine in concd. HCl, heated 1 hr. at 100°, aq. 50%-*NaOH*-soln. added, and refluxed until NH_3-evolution subsides → 2-aminobenzimidazole. Y: 92%. S. Weiss et al., Ang. Ch. *85*, 866 (1973).

Manganese dioxide MnO_2
β-Aminoketones from 2-ethylenealcohols $CH(OH)C:C \rightarrow CO \cdot CH \cdot C \cdot N <$
s. 27, 370

β-Amino-α,β-ethylenealdehydes $C \equiv C \cdot CH_2OH \rightarrow C(N<):CHCHO$
from 2-acetylenealcohols s. 28, 361

Oxidation with addition of nucleophiles $C \equiv C \cdot CH(OH) \rightarrow C(N<):CHCO$
β-Amino-α,β-ethyleneketones from 2-acetylenealcohols

246.   PhCH(OH)—C≡C—CH₂C(OH)Me₂   →   PhCO—CH=C—CH₂C(OH)Me₂
         + HN⌒O   N⌒O

γ-Manganese dioxide followed by 1-phenyl-5-methyl-2-hexyne-1,5-diol added with stirring to a soln. of morpholine in anhydrous ether, and kept 2.5 hrs. at room temp. → 5-hydroxy-1-phenyl-5-methyl-3-morpholino-2-hexen-1-one. Y: 86.4%. F. e. s. L. I. Vereshchagin et al., Zh. Org. Khim. 9, 1355 (1973); C. A. 79, 91899.

Manganese dioxide/sodium cyanide $MnO_2/NaCN$
Carboxylic acid amides from aldehydes $CHO \rightarrow CON<$
s. 27, 371

Potassium hexacyanoferrate(III) $K_3[Fe(CN)_6]$
Oxidative dimerization of N-heterocyclics   ←
     Biimidazoles s. 27, 372

3-Aminofuroxans from glyoximes $H \rightarrow NH_2$
s. 27, 99

2,5-Dihydropyrazines from 2 ketone molecules   O
     via 2 α-aminoketone molecules s. 28, 323

Palladous chloride/tert-butyl isocyanide $PdCl_2/Me_3C \cdot N:C$
1,3-Heterocyclics
s. 29, 340

Via intermediates   v.i.
α-Diazoketones from ketones $COCH_2 \rightarrow CO \cdot C:N_2$

247.   PhCH=CHCOCH₃ + EtOOC—COOEt   PhCH=CHCOCHN₂
                ↘ NaOEt   TsN₃ ↗
                     PhCH=CHCOC̄HCOCOOEt
                          Na⁺

α'-Diazo-α,β-ethyleneketones. A soln. of benzylideneacetone and diethyl oxalate in abs. ethanol added dropwise below 10° during 15–20 min. to a soln. of Na in ethanol, stirring continued 2–3 hrs., the resulting crude Na-compds. suspended in ethanol, ice-cooled, stirred, and treated dropwise with tosyl azide below 10°, then stirred 2–3 hrs. at room temp. → 4-diazo-3-oxo-1-phenyl-1-butene. Y: 73%. F. e. s. R. E. Harmon, V. K. Sood, and S. K. Gupta, Synthesis 1974, 577.

N,N-Dicyanamides from prim. amines $NH_2 \rightarrow N(CN)_2$
s. 29, 342

3-Hydroxybiurets $N \cdot CO \cdot N(OH) \cdot CO \cdot NH_2$
     from 3-hydroxyureas s. 26, 330

8-Subst. purines   O
     from 4-amino- via 4-amino-5-nitroso-uracils s. 29, 343

3-Aminoisoxazoles
     from α,β-ethylenenitriles via α,β-dibromonitriles s. 26, 409

Oxygen ↑   NC ⇵ O

Without additional reagents   w.a.r.

N-Alkylation with amide acetals   NH → NR
s. 27, 373

N-Alkylation with fluorosulfonic acid esters
4(5H)-Imino-1H-1,5-benzodiazepin-2(3H)-ones – Methyl fluorosulfonate as methylating agent, review s. 29, 344

Reactions of 3-hydroxyphthalides   OH → N<
with amines s. 28, 324

2,1-Benzisoxazolium salts   ←
s. 26, 18

Replacement of acoxy by amino and hydroxy groups   OAc → N<
s. 28, 325

1,2-Nitramines from 1,2-acoxynitro compds.
Carbohydrate derivs. s. 28, 342

Replacement of acoxy by amino groups   ←
with rearrangement s. 28, 326

Nucleophilic substitution of carbamyloxy groups   OCON< → N'<
by amino and alkoxy groups s. 27, 374

N-Aminomethylation   NH → NCH$_2$N<
s. 26, 871

β-Amino-α,β-ethyleneketones from β-diketones   CO·CH·CO → CO·C:C(N<)
s. 26, 331

Imines from ketones   CO → C:NH
with retention of halogen s. 26, 433

Azomethines from amines and oxo compds.   NH$_2$ + OC → N:C
from aldehydes s. 28, 356
Sec. from prim. amines and oxo compds. via azomethines s. 29, 390

Alkoximes from alkoxylamines   O·NH$_2$ → O·N:C
s. 27, 425 suppl. 29

Azines from ketones   2 >CO → >C:N·N:C<
s. 30, 566

Selective N-formylation   NH → NCHO
with isopropenyl formate s. 27, 375

2-Acylamino-2-hydroxyindan-1,3-diones   C(OH)$_2$ → C(OH)NAc
s. 29, 345

o-Acetamino- from o-amino-carboxylic acid amides   NH → NAc
s. 28, 437

α-Mercaptocarboxylic acid amides   COOH → CON<
from the acids s. 28, 327

o-Alkylthiocarboxylic acid amides   COOR → CON<
from o-mercaptocarboxylic acid esters – O→S-Alkyl migration s. 27, 376

Phthalimides from alcohols   OH → N<$^{OC}_{OC}$
s. 27, 429

o-Hydroxyamidines   NH$_2$ → N:C·N<
Amidines from amide acetals – Oxazole ring opening – N-Condensed oxazolidine ring opening by aminolysis s. 26, 332

Diamidides C(:N·)NH₂ → C(:N·)N:C·N<
from amidines and amide acetals s. 26, 333

Acylhydrazones COOR → CON·N:C
from carboxylic acid esters and oxo compds. s. 27, 377

3-α-Acoxy-3-acyl- N:N·N(OH)CH → N:N·N(Ac)C(OAc)
from 3-hydroxy-triazenes
s. 29, 346

N-Hydroxythioureas from hydroxylamines 2 RNHOH → RN(OH)·CS·NHR
s. 26, 334

Thiourethans from dithioldicarbonic acid esters NH → NCOSR<
Protection of amino groups s. 27, 378

Dithiourethans from xanthates SC(:S)OR → SC(:S)N<
s. 28, 328

2-Acylaminoquinoxalines ←
from 2-oximino-1,2,3,4-tetrahydroquinoxalines s. 28, 372

N-Condensed lactams ←
from ketocarboxylic acids or isocoumarins and 2-aminoacetals via isocarbostyrils s. 28, 329

N-Aminodicarboxylic acid imides ⌒O → ⌒N−
from dicarboxylic acid anhydrides − N-Decarbo-*tert*-butoxylation
s. 28, 330

Replacement of cyclic oxygen by cyclic nitrogen with subsequent C-deacetylation

248.

$$\underset{+\ H_2NBu}{\text{NC}\diagdown_C\diagup^{COCH_3}} \longrightarrow \underset{Bu}{\text{NC}\diagdown_C\diagup^{COCH_3}} \longrightarrow \underset{Bu}{\text{NC}\diagdown_C\diagup^{H}}$$

Heterofulvenes. A soln. of 2,6-dimethyl-4H-pyran-4-ylidenecyanoacetone in butylamine refluxed 1–2 hrs. → N-butyl-2,6-dimethyl-1,4-dihydropyridin-4-ylidenecyanoacetone (Y: 98%) dissolved in methanol, treated with some concd. H₂SO₄, and kept ca. 1 hr. at room temp. → N-butyl-2,6-dimethyl-1,4-dihydropyridin-4-ylideneacetonitrile (Y: 94%). F. e., also 4H-thiopyran derivs., s. I. Belsky, H. Dodiuk, and Y. Shvo, J. Org. Chem. *39*, 989 (1974); replacement of cyclic oxygen by nitrogen in *dimethylformamide* cf. H.-J. Timpe, A. V. El'cov, and N. I. Rtiscev, Chimia *28*, 657 (1974).

3-Aminopyrrole ring O
from γ-diketones and hydrazines with N,N-cleavage s. 28, 331

Isoindolenines
s. 29, 347

3-Hydroxyisoquinolines
s. 26, 335; cf. G. Simchen and M. Häfner, A. *1974*, 1802

6,7-Dihydroisoquinoline-1(2H),3(5H)-diones
by double ring closure s. 28, 332

1,6-Naphthyridin-5(6H)-ones
s. 28, 333

1,2,3,9b-Tetrahydro-5H-imidazo[2,1-a]isoindol-5-ones by double ring closure

249.

A mixture of 2-(3-fluorobenzoyl)benzoic acid, ethylenediamine, and toluene stirred and refluxed 5–24 hrs. in a Dean-Stark apparatus until no more water is separated → 9b-(3-fluorophenyl)-1,2,3,9b-tetrahydro-5H-imidazo[2,1-a]isoindol-5-one. Y: 85%. F. e. s. P. Aeberli et al., J. Med. Chem. *18*, 177 (1975); analogs cf. ibid. *18*, 182.

4-Imidazolidones
from α-alkylideneaminocarboxylic acid esters s. *27*, 379

1,3-Dihydroxyimidazolidines
from 1,2-di(hydroxylamines) s. *29*, 348

3-Hydroxy-4-imidazolidones
from α-aminohydroxamic acids and aldehydes s. *26*, 336

Hydantoins from cyanohydrins
s. *30*, 568

Δ³-Imidazoline 3-oxides
from *anti*-1,2-aminooximes s. *26*, 337

Δ⁴-Imidazoline-2-thiones
s. *18*, 461

1-Hydroxyimidazoles
from α-isonitrosoketones and aldehydes s. *26*, 338

N-Condensed imidazole ring
from o-α-cyanocarboxylic acids – Double ring closure with subsequent *lin-ang*-rearrangement s. *26*, 339

N-Condensed heterocyclics by double ring closure
with subsequent N-condensed 3,4-dihydro-2-pyridone ring synthesis s. *27*, 380

N-Condensed imidazolium ring
from α-chloraldehydes s. *29*, 349

Ring closures with β-alkoxy-α,β-ethylenecarboxylic acid esters

250.

4(3H)-Pyrimidinones from orthocarboxylic acid esters. | **2(1H)-Pyridone-5-carboxylic acid esters.**

Liq. NH₃ added with isopropanol/Dry Ice-cooling to ethyl β-ethoxyacrylate and ethyl orthoacetate in an autoclave, then heated 12 hrs. at 120° → 2-methyl-4(3H)-pyrimidinone. Y: 86%. F. e. and method for 2,6-disubst. analogs s. V. D. Adams and R. C. Anderson, Synthesis *1974*, 286. | ethyl 2(1H)-pyridone-5-carboxylate. Y: 90%.

5-Acyl-4(3H)-pyrimidinones
from β-amino-α,β-ethylenecarboxylic acid amides and 2 ketene molecules
s. 26, 340

Diazasteroids
Ring closure with O-alkyllactims – 8,13-Diazasteroids s. 29, 350

Pyrimidines from α-cyanoacetals
s. 27, 381

4-Acyloximino-1,2,3,4-tetrahydroquinazolines
from aldehydes s. 28, 334

3-Amino-4-hydroxy-3,4-dihydroquinazolines
from o-acylaminoketones s. 27, 382

**N-Condensed 4(1H)-pyrimidinone ring
from acetylenedicarboxylic acid esters**

251.

Dimethyl acetylenedicarboxylate added at 0° to a stirred ethereal soln. of 2-amino-4,6,6-trimethyl-1,3-thiazine, and stirring continued 2 hrs. ↛ product. Y: 72%.
M. N. Sharma, Current Sci. *43*, 147 (1974); f. e. s. ibid. *43*, 179.

**4-Aminopyrazolo[3,4-d]pyrimidines from carboxylic acid amides
via Dimroth rearrangement under anhydrous conditions**

252.

Anhydrous N-ethylformamide added to 5-amino-1-methylpyrazole-3,4-dicarbonitrile, and the resulting soln. refluxed 5.5 hrs. under N_2 ↛ 4-ethylamino-1-methyl-pyrazolo[3,4-d]pyrimidine-3-carbonitrile. Y: 81% based on startg. m. consumed; conversion: 75%. F. e. s. S. M. Hecht and D. Werner, Soc. Perkin I *1973*, 1903.

Ring closure with malonic acid aryl esters
Pyrimidonium betaine ring s. 27, 383

Pyrazine ring
s. 29, 351
from o-nitrosamines s. 26, 341

2,3-Dihydroxypyrazines from 2-aminoacetals
s. 27, 511

1,4-Diazepine-5,7-dione ring
s. 26, 342

Ring closures of dialdehydes with diamines
with and without hydride shift s. 26, 384

Oxazolidines from 2-aminoalcohols and aldehydes
Asym. synthesis s. 28, 335

5-Amino-Δ^2-1,2,3-triazolines
from azides and oxo compds. s. 26, 343

Δ^2-1,2,4-Triazolines
from amidrazones and aldehydes s. 27, 387

Reaction of aldehydes with ammonia

253.

$$3(CH_3)_2CHCHO + 3\,NH_3 \longrightarrow$$

Hexahydro-1,3,5-triazines. Isobutyraldehyde added during 15 min. at 5–10° to stirred aq. concd. NH₄OH, and kept 24 hrs. at 0° → 2,4,6-triisopropyl-1,3,5-hexahydrotriazine. Y: 78%. F. e. and products s. A. T. Nielsen et al., J. Org. Chem. *38*, 3288 (1973).

Imidazo[2,1-c]-*as*-triazines
  from oxazolium salts s. *26*, 344

s-Triazolo[4,3-d][1,3,4]benzotriazepines
  from 4-amino-3-(2-aminophenyl)-4H-1,2,4-triazoles s. *29*, 352

5,6-Dihydro-4H-1,2,5-oxadiazines
  from *syn*-1,2-aminooximes and aldehydes s. *27*, 386

Ring closures with squaric acid 1,2-dihydrazides
  1,2,4,5-Tetrazepine ring s. *29*, 353

1,5,10,14,18, 23-Hexaaza[26]annulenes
  s. *29*, 354

Double ring closures
  N-Condensed heterocyclics from 1,5-dialdehydes
    s. *26*, 345
  Pyrazolo[1,5-c]pyrimidines from 1,3,5-triketones
    s. *27*, 385
  Ring closures with o-hydroxylaminocarboxylic acid esters
    s. *26*, 346

Acylamines from cyclic tert. amines
  with aromatization s. *28*, 336

Cyclopent-2-enone from furan ring
  s. *27*, 388

Acyllactone rearrangement with amines
  s. *27*, 389

Fragmentation of azo compds.
  Fragmentation-acylation s. *27*, 390

N-Condensed 5,6-dihydro-4(3H)-pyrimidinone ring
  from O-alkyllactims and 2-azetidinones s. *26*, 347

2-Benzoxazolone ring opening
  1-(o-Hydroxyaryl)hydantoin ring s. *26*, 348

Pyrazole ring from α,β-oxidooxo compds.
  Anchimeric and frangomeric process s. *26*, 349

1-Aminophthalazine from 3-alkoxyisoisoindolenine ring
  s. *26*, 350

3-Nitrosopyrrole ring expansion
  to 4-aminopyrimidine ring s. *29*, 355

3-Hydroxymethyl-1,2,4-triazoles from Δ²-4-oxazolones
  s. *27*, 391

Electrolysis

5,7-Dihydroxy-6H-dibenzo[d,f][1,3]diazepines
  from oxo compds. s. *29*, 362 suppl. *30*

Lithium/palladium-carbon *Li/Pd-C*
Sec. amines from carboxylic acids COOH → CH$_2$NHR
s. 27, 392

Sodium *Na*
Imines from acetals C(OR)$_2$ → C:NH
s. 27, 393

Potassium/potassium amide *K/KNH$_2$*
Alkali metal-promoted replacement of ar. substituents ArNH$_2$
Ar. amines from ethers s. 27, 459

Sodium hydride *NaH*
Cyclic enamines from oxido compds. NH → N·C:C
α-**Carboxy-N-styrylheterocyclics**

254.

<chemical scheme: ethyl pyrrole-2-carboxylate + styrene oxide → N-styryl-pyrrole-2-carboxylic acid>

NaH-powder added to a stirred soln. of ethyl pyrrole-2-carboxylate in dry dimethylformamide, when H$_2$-evolution has ceased after 2–3 hrs. styrene oxide added, heated 1 hr. at 110–120°, cooled, and ice-water added → N-(*trans*-styryl)-pyrrole-2-carboxylic acid. Y: 82%. F. e. s. J. Rokach, Y. Girard, and J. G. Atkinson, Can. J. Chem. *51*, 3765 (1973).

N-Subst. carboxylic acid amides COOR → CON<
from carboxylic acid esters s. 27, 394

Sodium hydride/trialkyloxonium fluoroborates *NaH/R$_3$O$^+$ BF$_4^-$*
N-Alkylation of sulfoximines S(:O):NH → S(:O):NR
s. 28, 337

Sodium hydride/P-aminophosphonium salt
Reactions with P-aminophosphonium salts
Amines and azides from alcohols R·OH ⇄ R·N< , R·N$_3$

255.

<chemical scheme: cyclopropylcarbinol → cyclopropyl-CH$_2$ONa + HN(Me)(Bu-n) + Ph$_3$P-N(Me)(Ph) I$^-$ → cyclopropyl-CH$_2$N(Me)(Bu-n); PhCH$_2$OH → PhCH$_2$ONa + NaN$_3$ → PhCH$_2$N$_3$>

Cyclopropylcarbinol allowed to react with NaH in dimethylformamide, the resulting soln. treated in one portion with a soln. of N,N-methylphenylaminotriphenylphosphonium iodide and N-methyl-*n*-butylamine in dimethylformamide, stirred and heated 2 hrs. at 80° → N,N-methyl-*n*-butylcyclopropylcarbinylamine. Y: 92%. – Similarly with benzyl alcohol and NaN$_3$ → benzyl azide. Y: 90%. F. amines s. Y. Tanigawa, S. Murahashi, and I. Moritani, Tetrah. Let. *1975*, 471.

Sodium hydroxide *NaOH*
Quinone monoimines from quinones CO → C:NH
s. 27, 395

Lactams from lactones C(=O)-O → C(=O)-NH
via cyclic hydroxamic acids – Bridge lactams s. 26, 351

4-Aminopyrimidine ring from o-aminoamidines

Adenines. A soln. of 5-amino-1-β-D-ribofuranosylimidazole-4-carboxamidine cyclic 3′,5′-phosphate, o-chlorobenzaldehyde, and NaOH in dimethylformamide-water stirred 3 hrs., then a soln. of chloranil in dimethylformamide added, and stirring continued 1 hr. → 2-(2-chlorophenyl)adenosine cyclic 3′,5′-phosphate. Y: 81%. F. e., also with aliphatic aldehydes, dehydrogenation also with 10%-Pd-on-carbon, s. R. B. Meyer, Jr., D. A. Shuman, and R. K. Robins, Am. Soc. 96, 4962 (1974).

2,5-Dihydropyrazines
from 2 α-aminoketone molecules s. 28, 323

1,2,3,4-Tetrahydropyrimido[1,6-a]benzimidazoles
s. 26, 352

v-Triazolo[3,4-a]pyrimidines from β-diketones
s. 27, 396

Potassium hydroxide (s. a. under $PhNO_2$, Air and $Fe(CO)_5$)   KOH
Replacement of acoxy by amino and hydroxy groups   OAc → N<
s. 28, 325

Sodium/alcohol   NaOR
3-Subst. 1-acylguanidines from aldehydes   ←
Redox-condensation s. 29, 356

Pyrimidines with EMCE
s. 26, 353

Ring closures with sulfone diimines
4,5-Dihydro-3H-1,2,6-thia(IV)diazine-3,5-diones from malonic acid esters s. 26, 354

5-Pyrazolones from 5-isoxazolones   ←
s. 27, 397

Pyrimidine from coumarin ring   ←
s. 26, 355; f. method s. K. Takagi and M. Hubert-Habart, Chim. Ther. 9, 681 (1974)

Sodium alkoxide
Purines from ureidomalonic acid esters
via pyrimidines – 8-Hydroxyguanines s. 26, 356

Sodium methoxide   NaOMe
Pyrimidines from isocyanates
Uracils s. 28, 338

Sodium pyridinolate   NaOHet
Sodium 2- or 4-pyridinolates as catalysts
Pyrimidines from α-cyanocarboxylic acid esters – N-Subst. carboxylic acid amides from esters s. 27, 398

n-Butyllithium   n-BuLi
Prim. amines via (disulfen)amides NH_2
from tosylates, also from bromides and ethylene derivs. s. 26, 357

Carboxylic acid amides   COOR → CON<
 from carboxylic acid esters s. 26, 358

Potassium amide (s. a. under K) KNH_2
4(1H)-Pyridones   O
 from β-diketones and nitriles s. 26, 359

Sodium hydrogen carbonate $NaHCO_3$
N-Acylation with p-nitrophenyl esters   NH → NAc
 Protection of α-aminocarboxylic acid groups s. 21, 521

Potassium carbonate K_2CO_3
Hexahydro-1,5-diazocines   O
 from 1,3-diamines and α,β-ethyleneketones s. 27, 399

Sodium acetate CH_3COONa
Preferential and selective N-acylation   NH → NAc
 via total acylation – Selective O-deacylation and preferential N-deacylation s. 29, 357

4-Imidazolones from 5-oxazolones   ←

257.

A mixture of 4-benzylidene-2-m-tolyl-2-oxazolin-5-one and aniline refluxed 3 hrs. in acetic acid containing fused Na-acetate → 4-benzylidene-1-phenyl-2-m-tolyl-2-imidazolin-5-one. Y: 73–87%. F. e. s. A. M. Islam, A. M. Khalil, and M. S. El-Houseni, Australian J. Chem. 26, 1701 (1973).

1,2,4-Triazine ring from diazonium salts   O
 v-Triazolo[5,1-c]-as-triazines s. 29, 358

Potassium cyanide/hydrazine $KCN/H_2N \cdot NH_2$
Nitriles from aldehydes   CHO → CN
s. 28, 339

Sodium thiocyanate s. under HCl $NaSCN$

Ammonium thiocyanate NH_4SCN
3,4-Dihydro-2(1H)-quinazolinethiones   O

258.

NH_4-thiocyanate added to a stirred soln. of 2-isopropylaminobenzhydrol in aq. concd. HCl-dioxane, and heated 0.5 hr. at 100° → 3,4-dihydro-1-isopropyl-4-phenyl-2(1H)-quinazolinethione. Y: 85%. F. e. s. R. V. Coombs et al., J. Med. Chem. 16, 1237 (1973).

Lithium azide LiN_3
Azides from nitrates ONO_2 → N_3
 Tert. azides s. 27, 400

Sodium azide NaN_3
Prim. amines from alcohols   OH → NH_2
 via benzenesulfonates and azides – Polyamines from polyalcohols s. 27, 401

Sodium azide/phosphoric acid/phosphorus pentoxide $NaN_3/H_3PO_4/P_2O_5$
Benz[d]isothia(IV)zol-3-one 1-oxides   O
    from o-carbalkoxysulfoxides s. 26, 360

Sodium azide/sulfuric acid NaN_3/H_2SO_4
Ar. amines from cyclic ketones   ←
with C-alkyl migration

259.

7,7-Dimethyl-3-phenyl-5-oxo-5,6,7,8-tetrahydrocinnoline treated 2 hrs. at room temp. with NaN_3 in sulfuric acid → 5-amino-6,7-dimethyl-3-phenylcinnoline. Y: 75%. Also by Semmler aromatization of oximes (s. Synth. Meth. 2, 778) s. K. Nagarajan and R. K. Shah, Chem. Commun. *1973*, 926.

Benzo[e]-1,2,4-thiadiazine S-oxides   O
    s. 28, 340

Sodium nitrite $NaNO_2$
Aliphatic nitro compds.   ROH → RNO_2
    from alcohols via nitric acid esters s. 26, 361

Potassium fluoride   KF
β-Iminodithiocarboxylic acids   COCH → C(:N·)C·CSSH
    from ketones s. 30, 464

Sodium salt Na^+
Crown heterocyclics   O
    Crown polyamines via crown polytosylamines s. 29, 359; f. crown heterocyclics s. S. A. G. Högberg and D. J. Cram, J. Org. Chem. *40*, 151 (1975)

1,2,4-Triazin-5-one-3-thiones
    s. 26, 387

Ammonium hydroxide (s. a. under HBr) NH_4OH
Pyrimidine nucleosides
    from carbohydrates and cyanoacetylenes via carbohydrate 2-amino-Δ^2-oxazoline ring s. 28, 341

1,3,4,5-Tetrahydro-1,3,6-oxadiazepino[3,4-a]benzimidazoles
    s. 26, 362

Dicyclohexylamine $(C_6H_{11})_2NH$
Protection of amino groups with β-dicarbonyl compds.   NH → N·C:C·COR
    as (2-acyl-1-methylvinyl)amines s. 27, 402

Trimethylamine s. under $MeSO_2Cl$ Me_3N

Triethylamine (s. a. under $CuCl_2$, Dimethylpropynylsulfonium Et_3N
bromide, $(EtO)_2P(O)CN$, and $(PhO)_2PON_3$)
1,2-Nitramines from 1,2-acoxynitro compds.   OAc → N<
    Carbohydrate-amino acid compds. s. 28, 342

Selective N-acylation   NH → NAc
    with carboxylic acid p-nitrophenyl esters s. 26, 363

Protection of prim. amino groups
    as phthalimides s. 27, 403 NH_2 → N(OC)(OC)

Urethans via mixed carbonic acid esters   OH → OCON<
    s. 27, 404

Reactions with hydroxamic O-sulfates ←
Ureas s. 26, 364
Anomalous 2-imidazolone ring closure
from p-quinones s. 27, 405
Triethylamine/oxime carbonates ←
Peptide synthesis COOH → CON<

260.
$$\text{PhCH}_2\text{OOC—NHCH}_2\text{COOH} + \text{H}_2\text{N}\overset{\text{CH}_2\text{CHMe}_2}{\underset{|}{\text{CH}}}\text{COOEt}$$

$$\downarrow \text{Me}_2\text{CHCH}_2\text{OCOON}=\text{C}\underset{\text{COO}_2\text{Et}}{\overset{\text{CN}}{\diagup}}$$

$$\text{PhCH}_2\text{OOC—NHCH}_2\text{CON}\overset{\text{CH}_2\text{CHMe}_2}{\underset{|}{\text{H}\text{CH}}}\text{COOEt}$$

A soln. of ethyl 2-isobutyloxycarbonyloximino-2-cyanoacetate in ethyl acetate added dropwise at –15° to a soln. of N-benzyloxycarbonylglycine and triethylamine in ethyl acetate, stirred 0.5 hr. at the same temp., then a soln. of ethyl L-leucinate hydrochloride and triethylamine in methylene chloride added dropwise, and stirred 1 hr. → ethyl N-benzyloxycarbonylglycyl-L-leucinate. Y: 80%. F. e., mostly **esterifications with oxime carbonates**, s. M. Itoh, Bull. Chem. Soc. Japan *47*, 471 (1974); peptide synthesis with μ-**oxobis[tris(dimethylamino)phosphonium] bis(fluoroborate)**, $(\text{Me}_2\text{N})_3\text{P}^+\text{-O-P}^+(\text{NMe}_2)_3$, $2\ \text{BF}_4^-$ Bates reagent, and **N-methylmorpholine** s. A. J. Bates et al., Helv. *58*, 688 (1975).

Piperidine $C_5H_{10}N$
Oximes from ketones CO → C : NOH
with epimerization s. 28, 343
v-Triazolo[3,4-a]-pyrimidines from β-diketones ○
s. 27, 396
Pyridine (s. a. under $(PhO)_2POH$, Iodine, and $Rh_6(CO)_{16}$) C_5H_5N
N-Acylation with cyclic acylphosphoric acid esters NH → NAc
Preferential N-acylation – Steric control s. 28, 344
Nitriles from aldehydes CHO → CN
via aldoximes under mild conditions s. 30, 277
Pyridine N-oxide and Δ^2-isoxazoline from 4-enepyran ring ←
s. 29, 360
Cupric acetate $(CH_3COO)_2Cu$
Reaction of cupric acetate with thioureas NH → NAc
N-Acetylation – Δ^2-Oxazolines from 2-hydroxythioureas s. 27, 406
Cuprous chloride CuCl
2-Imino-1,3-O,N-heterocyclics ○
from carbodiimides and diols s. 29, 361
Cupric chloride/triethylamine $CuCl_2/Et_3N$
Peptides by di- and oligo-merization CON<
s. 26, 365
Calcium oxide CaO
Δ^2-Pyrroline ring from cyclopropylketones ⊙
s. 27, 407
Magnesium/hydrogen chloride Mg/HCl
5,7-Dihydroxy-6H-dibenzo[d,f][1,3]diazepines ○
from oxo compds. s. 29, 362; by electrolysis cf. Acta Chem. Scand. *B28*, 539 (1974)

Zinc/propionic acid   *Zn/C$_2$H$_5$COOH*
3-Benzazepine from isoquinoline ring ◐
s. *28*, 345

Mercuric sulfate   *HgSO$_4$*
N-Vinylation with vinyl esters   NH → N·CH : CH$_2$
s. *16*, 442

Mercuric chloride s. under (PhO)$_2$POH   *HgCl$_2$*

Sodium tetrahydridoborate   *Na[BH$_4$]*
Preferential reductive N-alkylation   NH → NR
with preceding reductive hydrogenation of N-heterocycles s. *26*, 366; reductive N-methylation s. a. I. Noguchi and D. B. MacLean, Can. J. Chem. *53*, 125 (1975)

Pyrrolo[3,4-b]quinolines   ○

261.

[Reaction scheme: 2,3-dibenzoylquinoline + H$_2$NMe → 2-methyl-1,3-diphenylpyrrolo[3,4-b]quinoline]

NaBH$_4$ added to a hot ethanolic soln. of 2,3-dibenzoylquinoline and 40%-methylamine → 2-methyl-1,3-diphenylpyrrolo[3,4-b]quinoline. Y: 75%. Also **isoindoles** s. M. J. Haddadin and N. C. Chelhot, Tetrah. Let. *1973*, 5185.

Lithium tetrahydridoaluminate   *Li[AlH$_4$]*
Reductive N-alkylation   CONH → CH$_2$NR
of carboxylic acid amides s. *29*, 363

Sodium trihydridocyanoborate   *Na[BH$_3$CN]*
Amines from oxo compds.   CO → CHN<
Retention of N-oxide radicals s. *19*, 448 suppl. 29

Tert. from sec. amines and oxo compds.   NH → NMe
Reductive N-methylation of amines – Tetraalkylhydrazines by reductive N-alkylation – α-Methylenelactones from lactones via reductive amination s. *28*, 346

Cyclic imines from dioxo compds.   ○
N-Labeled compds. s. *28*, 347

Boric acid esters/p-toluenesulfonic acid   *(RO)$_3$B/TsOH*
Carboxylic acid amides from carboxylic acids   COOH → CON<
s. *26*, 367

Alumina   *Al$_2$O$_3$*
Carboxylic acid amides from carboxylic acids   COOH → CONH$_2$
s. *26*, 368

1-Subst. benzimidazoles   ○
from o-nitramines and alcohols s. *26*, 369

Molecular sieve   ←
Azomethines   NH$_2$ + OC → N : C
from amines and aldehydes s. *30*, 463

N-Condensed 1,3-N,N-heterocyclic 1,1-iminoalcohols   ←
from diketones and o-diamines – Double ring closure s. *29*, 364

Boron fluoride   *BF$_3$*
Quinone monoimines from quinones   CO → C : NH
s. *27*, 395

Ring closure of 2-alkoxy-2,3-dihydro-4H-1,3-benzoxazin-4-ones ○
with 1,3-dienes s. *28,* 348

Boron bromide BBr_3
Reaction of carboxylic acid esters COOR → CON<
with boron bromide

262. PhCH₂COOCH₂Ph
      ↘
       PhCH₂COOMe
       ↘
        PhCH₂CONHPh

| **Transesterification.** | **Carboxylic acid amides from esters.** |
| A soln. of benzyl | A soln. of methyl |
| phenylacetate in methylene chloride or benzene treated with a soln. of BBr_3 in benzene, stirred |
| 1.5 hrs. at room temp., methanol added, and stirred 5 hrs. at room temp. → methyl phenylacetate. Y: 83%. | 5 hrs. at room temp., aniline added, and stirred 3 hrs. at room temp. → phenylacetanilide. Y: 81%. |

F. e. s. H. Yazawa, K. Tanaka, and K. Kariyone, Tetrah. Let. *1974,* 3995.

Fluoroborates ←
Amino- from alkoxy-carbonium salts OR → N<
s. *28,* 359

4H-1,2-Diazepines from pyrylium salts ⊙
s. *26,* 370

Quaternary N-carbalkoxyammonium fluoroborate ←
Peptide synthesis COOH → CON<

263. PhCH₂OOC—NHCH₂COOH + H₂NCH₂COOEt

              Me
              |
      ↓   ⟨ ⟩—N⁺—COOCH₂CHMe₂   BF₄⁻
              |
              Me

      PhCH₂OOC—NHCH₂CONHCH₂COOEt

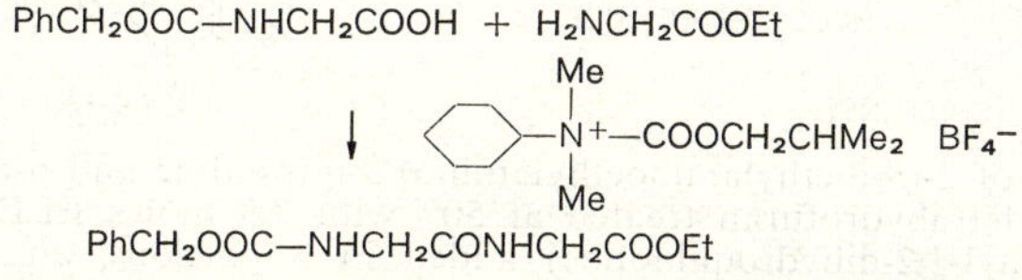

N-Isobutyloxycarbonyl-N,N-dimethylcyclohexylammonium fluoroborate added in small portions at –10° to a stirred soln. of N-benzyloxycarbonylglycine in dimethylformamide, stirred 15 min. at –10° and 15 min. at room temp., recooled to –10°, small portions of ethyl glycinate added with stirring, and allowed to stand 1 day at room temp. → ethyl N-benzyloxycarbonylglycylglycinate. Y: 90%.
F. e. s. J. V. Paukstelis and M. Kim, J. Org. Chem. *39,* 1499 (1974).

Carbon monoxide s. under $Rh_6(CO)_{16}$ CO
Ammonium acetate CH_3COONH_4
Sym. 1,1'-dienamines ←

264.

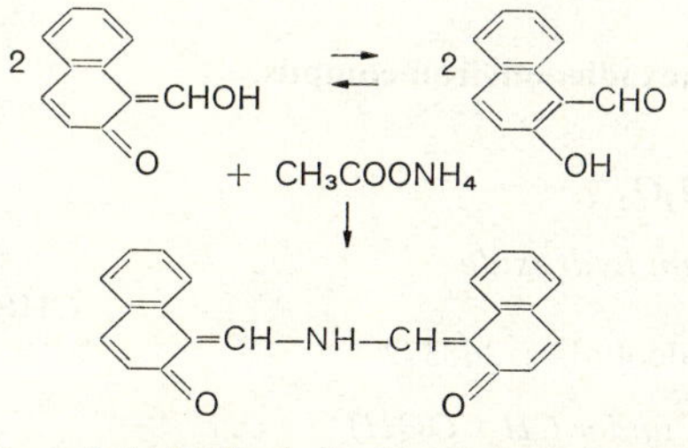

2 [naphthalenone]=CHOH ⇌ 2 [naphthalene]—CHO / OH
   + CH₃COONH₄
   ↓
[naphthalenone]=CH—NH—CH=[naphthalenone]

A soln. of 2-hydroxy-1-naphthaldehyde in 1 : 10 glacial acetic acid-benzene heated with NH₄-acetate until no more H₂O separates in the distillate → bis-(1,2-dihydro-

2-oxo-1-naphthylidenemethyl)amine. Y: 85%. F. e., also formation of 3-acyl-
pyridines (cf. Synth. Meth. *12*, 838) from non-cyclic startg. m., s. P. Ollinger,
W. Remp, and H. Junek, M. *105*, 346 (1974).

Isoquinol-3(4H)-ones from 2-arylacetic acid molecules   O
via self-acylation s. *26*, 371

Benzyltrimethylammonium salt s. under 1,2,4-Triazole $PhCH_2\overset{+}{N}Me_3$

Piperidinium acetate $C_5H_{10}\overset{+}{N}H_2, CH_3COO^-$
Pyrimidine ring
from β-amino-α,β-ethyleneoxo compds. s. *27*, 408

Diazomethane CH_2N_2
Bis(chlorosulfonyl)imides from ethers $N(SO_2Cl)_2$
s. *27*, 409

Diethyl azodicarboxylate s. under Ph_3P $ROOC \cdot N : N \cdot COOR$

Urea s. under H_2NSO_3H $(NH_2)_2CO$

2-Ethoxy-N-ethoxycarbonyl-1,2-dihydroquinoline   EEDQ
N-Subst. carboxylic acid amides   COOH → CON<
from carboxylic acids s. *23*, 415

Benzimidazoles   O
from o-diamines and carboxylic acids

265.

A stirred soln. of 2-(β-diethylaminoethylamino)-5-nitraniline and p-ethoxyphenyl-
acetic acid in tetrahydrofuran treated at 50° with 3,5 moles EEDQ (2-ethoxy-
N-ethoxycarbonyl-1,2-dihydroquinoline) added in 3 portions, and the product
isolated after 9 days → etonitazene. Y: ca. 100%. F. I. Carroll and M. C. Cole-
man, J. Med. Chem. *18*, 318 (1975).

1,2,4-Triazole/benzyltrimethylammonium salt   ←
Urethans from mixed carbonic acid esters   NH → NCOOCMe₃
N-Carbo-*tert*-butoxylation s. *28*, 375

Tetramethylguanidine $Me_2NC(:NH)NMe_2$
s. *28*, 375 suppl. *29*

Trialkyloxonium fluoroborate (s. a. under NaH) $R_3O^+ BF_4^-$
N-Arylation   NH → NAr
with tricarbonylcyclohexadienoneiron compds.
s. *30*, 266

Acetonitrile s. under H_2O_2   MeCN

Nitrobenzene/potassium hydroxide $PhNO_2/KOH$
Azomethines   CH(OH) + H_2N → C : N
from amines and alcohols s. *26*, 372

Acetic anhydride (s. a. under CH_3COOH) Ac_2O
Purines from 4-aminopyrimidines and nitroso compds.   O
7-Aryltheophyllines s. *29*, 365

Trifluoroacetic anhydride $(CF_3CO)_2O$
Carboxylic acid amides from carboxylic acids   COOH → CON<
s. 24, 408 suppl. 28

Cation exchanger   ←
Δ²-Imidazolines from carboxylic acids   O
   Benzimidazoles from o-diamines s. 27, 410

Ascorbic acid   ←
Enehydroxylamines   COCHCO → COC:C·NOH
   from hydroxylamines s. 30, 651

Formic acid   HCOOH
Formamides from ketones   CO → CH·N·CHO
s. 27, 53

N-Methylation of sulfoximines   S(:O):NH → S(:O):NR
s. 28, 337

Lactams from ketocarboxylic acids   O
   Stereospecific ring closure s. 28, 349

2,3-Dihydro-1,4-diazepine ring
   2,3-Dihydro-6H-indeno[1,2-e][1,4]diazepin-6-ones from 2-acylindan-1,3-diones s. 26, 373

Acetic acid (s. a. under O_2 and NH_4Cl) CH_3COOH
N-Arylation   NH → NAr
with tricarbonylcyclohexadienoneiron complexes

266.

            Tricarbonylcyclohexadienoneiron

| and aniline in glacial acetic acid heated overnight at 75° under N_2 → diphenylamine. Y: 90%. | and triethyloxonium fluoroborate in methylene chloride stirred overnight under N_2, then 2 moles cyclohexylamine added → N-cyclohexylaniline. Y: 70%. |

F. e. s. A. J. Birch and I. D. Jenkins, Tetrah. Let. *1975*, 119.

Hydrazones from ketones   CO → C:N·N<
s. 26, 940

N,N'-Dicarboxylic acid diimides   ←
   from N-aminodicarboxylic acid imides s. 29, 366

Enaminolactams from ketocarboxylic acids   O
s. 29, 367

Benzo[c]quinolizinium from pyrylium salts
s. 27, 411

Imidazoles from α-acylaminoketones

267.

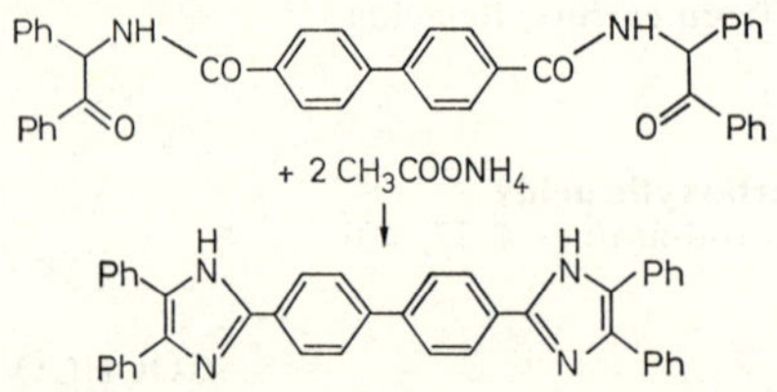

Startg. m. refluxed 5 hrs. with NH$_4$-acetate in glacial acetic acid → 4,4'-bis-(4,5-diphenyl-2-imidazolyl)biphenyl. Y: 87%. F. e. s. P. Schneiders, J. Heinze, and H. Baumgärtel, B. *106*, 2415 (1973).

Imidazole 3-oxides
from α-dioxo compd. monoximes and aldehydes – 2-Imidazolones – Oxazole N-oxides s. *27, 412*

7-Hydroxy-s-triazolo[1,5-a]pyrimidines
s. *29, 368*

3-Hydroxypyrrolo[1,2-a]benzimidazol-1-ones
from 3-hydroxycyclobut-3-ene-1,2-diones and o-diamines s. *27, 413*

v-Triazolo[3,4-a]pyrimidines from β-diketones
s. *27, 396*

Reactions of 2,3-benzoxazin-1-ones

268.

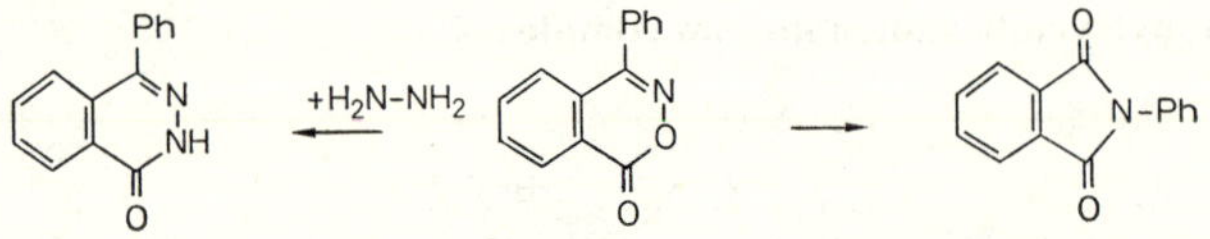

Phthalazones | **Phthalimides**
4-Phenyl-2,3-benzoxazin-1-one
dissolved in glacial acetic acid, treated dropwise with excess hydrazine hydrate, and allowed to stand 1 hr. with occasional shaking → 4-phenylphthalazin-1-one. Y: 80%. | mixed with AlCl$_3$ in nitrobenzene or chlorobenzene, and refluxed 15 min. → N-phenylphthalimide. Y: 90%.

F. e. and reactions s. F. G. Baddar, A. F. M. Fahmy, and N. F. Aly, Soc. Perkin I *1973*, 2448.

Tricycloquinazolines
from 3 2,1-benzisoxazole molecules

269.

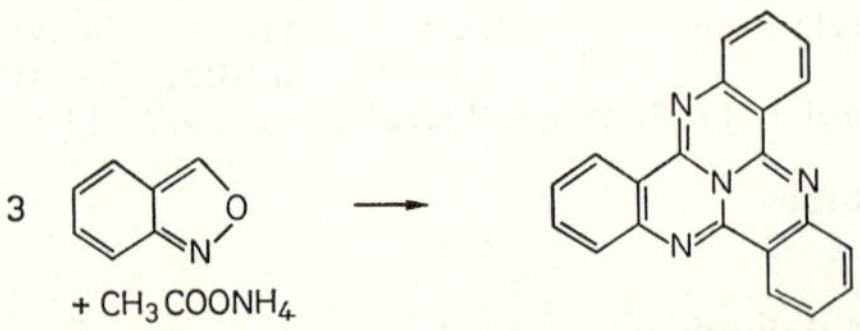

A mixture of anthranil, NH$_4$-acetate, acetic acid, and sulfolane stirred and heated 10 hrs. at 170° → tricycloquinazoline. Y: 51%. F. e. s. F. Yoneda and K. Mera, Chem. Pharm. Bull. *21*, 1610 (1973).

Acetic acid/acetic anhydride CH_3COOH/Ac_2O

Pyridazin-3,4-diol ring
from dicarboxylic acids s. *26, 374*

Propionic acid (s. a. under Zn) C_2H_5COOH

α-**Acylaminonitriles from cyanohydrins** $C(CN)OH \rightarrow C(CN)N \cdot Ac$
s. *26*, 375

dl-Isovaline $C_2H_5C(CH_3)(NH_2)COOH$

Prim. amines from ar. aldehydes $ArCHO \rightarrow ArCH_2NH_2$
s. *27*, 414

Trifluoroacetic acid CF_3COOH

Monoalkylation of amines $NH_2 \rightarrow NHR$
with orthocarboxylic acid esters via iminoesters s. *29*, 369

Oxime carbonates s. under Et_3N ←

Dimethylpropynylsulfonium bromide/triethylamine ←

Carboxylic acid amides from carboxylic acids $COOH \rightarrow CON<$
s. *26*, 376

Diphenyl sulfite $(PhO)_2SO$

Peptides

270.
$$PhCH_2OOC-NHCH_2COOH + H_2N-\overset{COOEt}{\underset{|}{C}}HCH_2-C_6H_4OH-p$$
$$\downarrow$$
$$Ph-CH_2OOC-NHCH_2CONH-\overset{COOEt}{\underset{|}{C}}HCH_2-C_6H_4OH-p$$

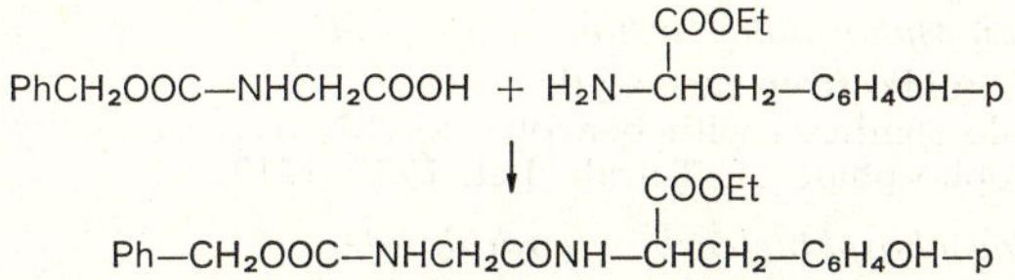

Diphenyl sulfite added to a mixture of equimolar amounts of benzyloxycarbonyl-glycine, ethyl L-tyrosinate hydrochloride, and *triethylamine* in *pyridine*, stirred 12 hrs. at room temp., and 6 hrs. at 40° → ethyl benzyloxycarbonylglycyl-L-tyrosinate. Y: 72%. F. e. s. N. Yamazaki, F. Higashi, and M. Niwano, Tetrahedron *30*, 1319 (1974).

Carbon tetrahalide s. under Ph_3P $CHal_4$

Carbon tetrachloride (s. a. under $(Et_2N)_3P$) CCl_4

sym-**Octahydroacridinium salts from *prim*-amines** O
s. *27*, 415

Phosgene $COCl_2$

N-Carbamylamidinium chlorides $N \cdot CO \cdot NH \rightarrow N \cdot CO \cdot \overset{+}{N} \cdot C : NR_2$
from ureas and carboxylic acid amides s. *29*, 370

Trimethylchlorosilane/hexamethyldisilazane $Me_3SiCl/Me_3SiNHSiMe_3$

Preferential replacement of oxygen by amino groups ←
s. *28*, 350

Titanium amides $Ti(NR_2)_4$

Trisubst. amidines $CONHR \rightarrow C(:NR)NR'_2$
from carboxylic acid amides s. *27*, 416

Titanium tetrachloride $TiCl_4$

Hindered sec. from prim. amines and oxo compds. $CO + H_2N \rightarrow CHNH$
via azomethines s. *29*, 371; azomethines s. a. D. A. Evans and L. A. Domeier, Org. Synth. *54*, 93 (1974).

Carboxylic acid amides from carboxylic acids $COOH \rightarrow CON<$
s. *26*, 377

Hydrazine s. under KCN $H_2N \cdot NH_2$

Benzalazine/sulfuric acid $PhCH:N \cdot N:CHPh/H_2SO_4$

Monosubst. hydrazines from oxonium salts $NHNH_2$
s. *28*, 351

Hydrazoic acid s. under NaN₃ HN_3

Hydroxylamines $RNHOH$
anti-Aldoximes from aldehydes CHO → CH : NOH
s. 27, 960

Triphenylphosphine s. a. under $(Ac_2CH)_2Pd$ Ph_3P

Triphenylphosphine/diethyl azodicarboxylate $Ph_3P/ROOC \cdot N : N \cdot COOR$
Alkylation with alcohols NH → NR
 N-Alkylation s. 28, 753

Phthalimides from alcohols OH → N(OC)(OC)
s. 27, 429

Triphenylphosphine/carbon tetrahalide $Ph_3P/CHal_4$
N-Subst. carboxylic acid amides COOH → CON<
 from carboxylic acids – Peptides s. 27, 417

P-Aminophosphonium salt s. under NaH ←

Tris(dimethylamino)phosphine/carbon tetrachloride $(Me_2N)_3P/CCl_4$
Carboxylic acid amides from carboxylic acids
s. 28, 144; peptide synthesis with benzotriazolyl-N-oxytrisdimethylaminophosphonium hexafluorophosphate cf. Tetrah. Let. *1975*, 1219

Tris(diethylamino)phosphine/carbon tetrachloride $(Et_2N)_3P/CCl_4$
Peptide synthesis
s. 27, 418

Phosphoric acid amides $(R_2N)_3PO$
Tert. amines OH → NR₂
s. 28, 352

Replacement of oxygen by amino groups ←
s. 26, 378

Phenyl phosphorodiamidate $PhOPO(NH_2)_2$
Direct replacement of hydroxyl by amino groups OH → NH₂
s. 29, 372

Diethyl phosphorocyanidate/triethylamine $(EtO)_2P(O)CN/Et_3N$
N-Subst. carboxylic acid amides from carboxylic acids COOH → CON<
 Peptides s. 28, 353; s. a. W.-Y. Chen and R. K. Olsen, J. Org. Chem. **40**, 350 (1975)

Azidotris(dimethylamino)phosphonium hexafluorophosphate $N_3\overset{+}{P}(NMe_2)_3PF_6^-$
Peptides
s. 28, 144 suppl. 29

Diphenyl phosphorazidate/triethylamine $(PhO)_2PON_3/Et_3N$
Reactions with diphenyl phosphoryl azide
 Peptide synthesis by fragment coupling – Urethans from carboxylic acids – Simplified Curtius degradation – Coupling reagents in peptide synthesis, review – Review of peptide synthesis s. 28, 353

Phosphorus pentoxide (s. a. under NaN₃) P_2O_5
N-Acetylation NH → NAc
s. 29, 373

Diethyl methylphosphonite $(EtO)_2PMe$
2-Amino-3H-azepines from nitrobenzenes ☉
s. 26, 379

Diphenyl phosphite/pyridine $\qquad$ $(PhO)_2POH/C_5H_5N$
Diphenyl phosphite-pyridine as acylating agent $\qquad$ ←
  Peptides and carboxylic acid esters s. 27, 419
Diphenyl phosphite/pyridine/mercuric chloride $\qquad$ $(PhO)_2POH/C_5H_5N/HgCl_2$
N-(Chlorophosphoryl)pyridinium betaines $\qquad$ COOH → CON<
as acylating agents
  Peptides and carboxylic acid esters s. 27, 419
Triethyl phosphite $\qquad$ $(EtO)_3P$
Nitrenoid insertion $\qquad$ ⊙
  1H-Azepines from benzenes s. 27, 420
Polyphosphoric acid esters $\qquad$ ←
2,3-Dihydro-4H-1,3-benzoxazin-4-ones $\qquad$ O
  from o-hydroxycarboxylic acid amides and oxo compds. s. 29, 374
Polyphosphoric acid $\qquad$ $H(PO_3H)_xOH$
Carboxylic acid amides $\qquad$ COOH → CON<
  from carboxylic and sulfamic acids s. 29, 375
Phthalimides from prim. amines $\qquad$ $NH_2 → N{<}^{OC}_{OC}$
s. 1, 332/3 suppl. 28
1,3,4-Oxadiazoles $\qquad$ O
  from 2 carboxylic acid molecules s. 26, 380
Phosphoric acid s. under NaN₃ $\qquad$ H_3PO_4
Chlorotris(dimethylamino)phosphonium perchlorate $\qquad$ $ClP^+(NMe_2)_3\ ClO_4^-$
Peptides $\qquad$ COOH → CON<
s. 28, 144
Phosphorus tribromide $\qquad$ PBr_3
Quinazolines $\qquad$ O
  from o-aminoketones and nitriles s. 27, 804
Phosphorus oxide chloride $\qquad$ $POCl_3$
Replacement of oxygen by amino groups $\qquad$ ←
s. 29, 376
8-*tert*-Aminopurines $\qquad$ O
  from 4-amino-5-nitrosopyrimidines and formamides s. 28, 354; from formamide diethyl acetals without POCl₃ cf. Synthesis 1975, 264
Quinazolines from 3-nitrosoindoles $\qquad$ ⊙
s. 28, 355
Phosphorus pentachloride $\qquad$ PCl_5
Amidines from carboxylic acid amides $\qquad$ C(:N·)N<
s. 10, 329
Phosphorus pentachloride/methanesulfonamide $\qquad$ $PCl_5/MeSO_2NH_2$
Nitriles from carboxylic acids $\qquad$ COOH → CN
s. 27, 422
Air s. a. under HBr $\qquad$ O_2
Air/potassium hydroxide $\qquad$ O_2/KOH
2-Hydroxy-7H-dibenzo[de,h]quinol-7-ones $\qquad$ O
s. 29, 377
Oxygen/acetic acid $\qquad$ O_2/CH_3COOH
2,6-Dihydro-6-benzoxazolones
s. 29, 378

Hydrogen peroxide/acetonitrile $H_2O_2/MeCN$
Azines from oxo compds. and ammonia $2\ CO \twoheadrightarrow C:N\cdot N:C$
s. 27, 421

p-Toluenesulfonamide $TsNH_2$
N-Glycosides   ←
from carbophenoxy glycosides s. 30, 150

o-Sulfobenzoic anhydride/potassium salt   ←
Peptide synthesis with o-sulfobenzoic anhydride $COOH \twoheadrightarrow CON<$

271.

[reaction scheme: potassium N-phthaloylglycinate + o-sulfobenzoic anhydride → mixed anhydride intermediate (phthalimido-CH₂CO-O-OC-C₆H₄-SO₃K) + H₂N-CH(CH₃)COOH → N-phthaloylglycylalanine (phthalimido-CH₂-CONH-CH(CH₃)-COOH)]

o-Sulfobenzoic anhydride added to potassium N-phthaloylglycinate in hot dry benzene, refluxed ca. 2 hrs., alanine added, and refluxing continued ca. 3 hrs. → N-phthaloylglycylalanine. Y: 85%. F. e. s. A. K. Koul et al., Indian J. Chem. *11*, 612 (1973); **peptide synthesis with N-sulfonyloxy compds.** cf. M. Itoh et al., Tetrah. Let. *1974*, 3089; **review of peptide bond formation, reviews of acylation** in general s. J. H. Jones, Chem. & Ind. *1974*, 723 and f. papers of this series.

Methanesulfonyl chloride/trimethylamine $MeSO_2Cl/Me_3N$
Amines from alcohols $OH \twoheadrightarrow N<$
s. 29, 379

p-Toluenesulfonyl chloride $TsCl$
Phenolbetaines   ←
s. 26, 381

Methyl fluorosulfonate FSO_2OMe
N-Subst. carboxylic acid amides $COOH \twoheadrightarrow CON<$
from carboxylic acids s. 29, 230

p-Toluenesulfonic acid (s. a. under (RO)₃B) $TsOH$
Azomethines from amines $CH + H_2N \twoheadrightarrow C:N$
and aldehydes – 4-Benzoylbenzylideneanilines s. 28, 356
and ketones s. 29, 504

α,β-Ethylenehydrazones $C:C\cdot CO \twoheadrightarrow C:C\cdot C:N\cdot N<$
from α,β-ethyleneketones s. 27, 832

Enethiolurethans $S\cdot CONH \twoheadrightarrow S\cdot CON\cdot C:C<$
from thiolcarbamic acid esters and aldehydes s. 26, 382

2,2-Disulfonylenamines $C:COR \twoheadrightarrow C:C\cdot N<$
from 2,2-disulfonylenolethers

272. $(MeSO_2)_2C=CH-OEt + H_2NPh \longrightarrow (MeSO_2)_2C=CH-NHPh$

A soln. of 2,2-bis(methylsulfonyl)vinyl ethyl ether, aniline, and a little p-toluene-sulfonic acid in chloroform allowed to stand 4.5 hrs. → N-[2,2-bis(methylsulfonyl)-vinyl]aniline. Y: 83%. F. e. and related compds. s. A. R. Friedman and D. R. Graber, J. Org. Chem. *39*, 1432 (1974).

Cyclic 1,1-iminoalcohols from lactols ←
s. *29*, 380

Pyrroles from ketones and hydrazines ○
via hydrazones s. *26*, 842

6,7-Dihydro-1H-1,2,5-triazepines
s. *29*, 381

Sulfamic acid/urea $H_2NSO_3H/(NH_2)_2CO$

Nitriles from carboxylic acids   COOH → CN
s. *27*, 422

Hydroxylamine-O-sulfonic acid $H_2NO \cdot SO_3H$

Nitriles from aldehydes   CHO → CN
s. *30*, 277

Sulfuric acid (s. a. under PhCH : N·N : CHPh) H_2SO_4

Monomethylation of amines NH_2 → NHMe
of ar. nitroamines s. *26*, 383

Amines from alcohols via urethans   OH → NHCOOR

273.   Ph₂CHOH + H₂NCOOCH₂Ph → Ph₂CHNHCOOCH₂Ph → Ph₂CHNH₂

Concd. H₂SO₄ added to a soln. of benzhydrol and benzyl carbamate in glacial acetic acid, and stirred 18 hrs. at room temp. → benzyl N-benzhydrylcarbamate (Y: 85%) added to a soln. of KOH in 95%-ethanol, and refluxed 4 hrs. → benzhydrylamine (Y: 94% as the acetate). F. e. s. J. E. Ollmann and D. T. Witiak, J. Org. Chem. *39,* 1589 (1974).

Quinolines from ketones ○
s. *27*, 423

2-Aminopyrimidines from β-diacetals
Labeled compds. s. *28*, 357

N-Condensed oxazolidine ring
from 2-aminoalcohols and ketones s. *19*, 464 suppl. 29

Iodine/pyridine I/C_5H_5N

N-Subst. carboxylic acid amides   COOH → CON<
from carboxylic acids s. *29*, 287

Perchlorates ClO_4^-

Monoalkylation of amines NH_2 → NHR
with alkoxytriaminophosphonium salts s. *28*, 358

Amino- from alkoxy-carbonium salts   OR → N<
Aminotropenium salts s. *28*, 359

Perchloric acid $HClO_4$

6-Azo-1H-1,4-diazepinium salts ○
s. *27*, 424

Bromides Br^-

Nucleophilic ring closure with cyclimmonium salts
1,9a-Dihydro-4H-pyrido[2,1-c][1,2,4]-triazines s. *29*, 382

Ammonium chloride/acetic acid NH_4Cl/CH_3COOH

4H-1,4-Thiazine 1,1-dioxides ←
from 2,6-dialkoxy-1,4-oxathiane 4,4-dioxides s. *29*, 148

Hydrochlorides ←

Alkoximes from alkoxylamines $O \cdot NH_2$ → $O \cdot N : C$
s. *27*, 425

β-Alkoxy-α,β-ethyleneoximes   CO·CH·CO → C(OR):C·C:NOH
from β-diketones
s. 27, 163

Acylamines   NH → NAc
from amine hydrochlorides and orthocarboxylic acid esters s. 26, 384

Amidines from iminoesters   C(OR):NH → C(NH$_2$):NH
s. 30, 220

Lactamimides from O-alkyllactims $\underset{C·OR}{\overset{N}{C}} \rightarrow \underset{C:N·}{\overset{NH}{C}}$
s. 29, 383

Imidazoles   O
from iminoesters and α-hydroxyketones s. 29, 384

5-Hydroxy-Δ^2-imidazoline ring
from O-alkyllactims and α-aminoketones s. 29, 385

Pyrazolo[1,5-a]quinazolin-5(4H)-ones
from o-hydrazinocarboxylic acids s. 26, 385

Pyridine hydrochloride C_5H_5N,HCl

Protection of carbonyl groups   CO → C:N·O·C·S·
as 1-(organothio)alkoximes s. 29, 386

Tetraethylammonium bromide $Et_4N^+Br^-$

N-1-Glycitylheterocyclics   C
1,3-Dioxolan-2-one ring opening

274.

A mixture of 1,2:3,4-di-O-isopropylidene-D-mannitol 5,6-carbonate, theophylline, and tetraethylammonium bromide in dimethylformamide heated 8 hrs. at 50–160° → 7-(1-deoxy-3,4:5,6-di-O-isopropylidene-D-mannitol-1-yl)theophylline. Y: 75%. – Similarly with succinimide → N-(1-deoxy-3,4:5,6-di-O-isopropylidene-D-mannitol-1-yl)succinimide. Y: 82%. F. e. s. H. Komura, T. Yoshino, and Y. Ishido, Carbohyd. Res. *31*, 154 (1973).

Hydrogen chloride   *HCl*

Polymethine cyanines   →
s. 27, 426

Tosylhydrazones from ketones   CO → C:NNHTs
s. 27, 949

Lactams from lactones $\underset{CO}{\overset{O}{C}} \rightarrow \underset{CO}{\overset{N-}{C}}$
s. 29, 387

Pyrazoles from α-acetyleneoxo compds.   O
s. 26, 386

1,2,4-Triazine-3-thiones
from α-ketoaldehyde aldoximes – 1,2,4-Triazin-5-one-3-thiones s. 26, 387

Pyrimido[5,4-e]-*as*-triazines
s. 28, 360

Hydrogen chloride/sodium thiocyanate   HCl/NaSCN
5-Mercapto-Δ^4-1,2,4-triazolines
from oxo compds. and hydrazines s. 29, 388

Hydrogen chloride/acetic acid HCl/CH_3COOH
Perhydro-1,3-N,N-heterocyclics
s. 27, 427

Hydrogen bromide/ammonium hydroxide/air $HBr/NH_4OH/O_2$
5-Hydroxyindoles

275.

A soln. of 1-(2,5-dimethoxy-4-methylphenyl)-2-aminopropane refluxed 12 hrs. in 48%-HBr, poured into ice-water, the pH adjusted to 9,5 by careful addition of concd. NH_4OH, and stirred overnight in air → 5-hydroxy-2,6-dimethylindole. Y: 68%. J. S. Zweig and N. Castagnoli, Jr., J. Med. Chem. *17*, 747 (1974).

Rhenium sulfide Re_2S_7
Ring hydrogenation of N-heterocycles   ←
with and without N-alkylation by alcohols s. 29, 389

Iron carbonyl/potassium hydroxide $Fe(CO)_5/KOH$
N-Alkylation with oxo compds.   CO → CHNHR
Sec. from prim. amines

276.   PhCHO + H₂NPh → PhCH₂NHPh

1 N alc. KOH and $Fe(CO)_5$ added to an alc. soln. of benzaldehyde and aniline, then stirred 5–10 hrs. at room temp. under CO → N-benzylaniline. Y: 98%. – This simple method proceeds under mild conditions and is applicable to low boiling oxo compds. F. e. s. Y. Watanabe et al., Tetrah. Let. *1974*, 1879; **also tert. amines,** f. process, s. G. P. Boldrini, M. Panunzio, and A. Umani-Ronchi, Synthesis *1974*, 733; f. catalysts cf. L. Markó and J. Bakos, J. Organometal. Chem. *81*, 411 (1974).

Nickel   *Ni*
N-Alkylation with alcohols   >NH → >NR
Azeotropic water entrainment s. 26, 388; N-alkylation of indoles cf. H. Plieninger, H. P. Kraemer, and C. Roth, B. *108*, 1776 (1975); with Pd-black cf. S.-I. Murahashi, T. Shimamura, and I. Moritani, Chem. Commun. *1974*, 931

Diamines from cyanoketones   CO → $CHNH_2$
s. 28, 362

Hexarhodium hexadecacarbonyl/carbon monoxide/pyridine $Rh_6(CO)_{16}/CO/C_5H_5N$
Azomethines $CHO + O_2N$ → CH:N
from nitro compds. and aldehydes s. 28, 363

Palladium   *Pd*
N-Alkylation with alcohols   >NH → >NR
s. 26, 388 suppl. 30

Palladium-carbon (s. a. under Li)   *Pd-C*
Quaternary ammonium salts $NO_2 \rightarrow \overset{+}{N}\leqslant$
from nitro compds. s. 27, 428

Palladium acetoacetonate/triphenylphosphine $(Ac_2CH)_2Pd/Ph_3P$
N-Allylation $NH \rightarrow NCH_2CH:CH_2$
s. 26, 827

Via intermediates v.i.
Prim. amines $OH \rightarrow NH_2$
from alcohols via phthalimides – Inverted amines from optically-active alcohols
s. 27, 429

Amines from carboxylic acids $COOH \rightarrow CH_2NH_2$
via carboxylic acid chlorides and amides – Labeled compds. s. 26, 389

Sec. from prim. amines and oxo compds. $NH_2 \rightarrow NHR$
via azomethines s. 29, 390

Hindered sec. amines s. 29, 371

o-Acylaminocarboxylic acid amides ←
from o-acylaminocarboxylic acids via 4H-3,1-benzoxazin-4-ones s. 29, 391

Nitriles from aldehydes via aldoximes $CHO \rightarrow CN$

277.

under mild conditions. 2-Thiophenealdehyde and pyridine added to an aq. soln. of hydroxylamine hydrochloride, stirred 1 hr. at room temp., cupric sulfate and a soln. of *triethylamine* in methylene chloride added followed by dicyclohexyl-carbodiimide in the same solvent, and the product isolated after 2 hrs. → 2-thiophenecarbonitrile. Y: 93%. – F. e., also compds. with unprotected amino and hydroxyl groups, and sterically hindered compds., also limitations, s. E. Vowinkel and J. Bartel, B. *107*, 1221 (1974); with hydroxylamine-O-sulfonic acid cf. C. Fizet and J. Streith, Tetrah. Let. *1974*, 3187; with O-2,4-dinitrophenylhydroxylamine cf. M. J. Miller and G. M. Loudon, J. Org. Chem. *40*, 126 (1975).

Reactions with *cis*-benzene trioxide
Aziridine ring from oxido compds. $O \rightarrow NH$
via 2-azido-alcohols and -tosylates

278.

cis-Benzene trioxide warmed 48 hrs. at 65° with 6 equivalents NaN_3 in $MgCl_2$-buffered methanol → 3,5,6-triazido-1,2,4-cyclohexanetriol (Y: 98%) kept 14 days at 20° with 4.5-equivalents tosyl chloride in pyridine → intermediate (Y: ca. 100%) treated 10 min. at 20° with 4.5 moles $LiAlH_4$ in tetrahydrofuran → *cis*-benzene triimine. Overall Y: 45%. R. Schwesinger and H. Prinzbach, Ang. Ch. *85*, 1107 (1973); f. reactions s. ibid. *85*, 1110, 1111.

Aminoalcohols C
from dicarboxylic acid anhydrides s. 28, 364

Protection of prim. amino groups $CO \rightarrow CN$
as Δ^4-oxazol-2-ones s. 28, 365

Nitrogen ↑

Without additional reagents w.a.r.
Amines from azoles ←
s. 29, 561

N-Dialkoxymethylation $NH \rightarrow N \cdot C \underset{O}{\overset{O}{\diagup}}$

279.

[structure: 5,5-dimethyl-1,3-dioxane-2-NEt₃ BF₄⁻ + HN-C(Ph)=N-Ph → 5,5-dimethyl-1,3-dioxane-2-N(Ph)-C(Ph)=N-Ph]

1,1,1-Dialkoxyamidines. 5,5-Dimethyl-2-triethylammonio-1,3-dioxane fluoroborate (prepn. s. Synth. Meth. 27, 995) stirred 2 hrs. at 0° with 1,3-diphenylbenzamidine in abs. methylene chloride → product. Y: 90%. F. e. s. W. Tritschler and S. Kabuss, Synthesis 1973, 423; s. a. Synth. Meth. 27, 430.

Quinone monoimines   ←
from phenols and diazocyanides s. 27, 442

Reactions with N-nitrimines $C : N \cdot NO_2 \rightarrow C : N(O) \cdot$
Nitrones from hydroxylamines s. 26, 390

Hydrazones from enamines $C : C \cdot NH_2 \rightarrow CH \cdot C : N \cdot N$
s. 29, 392

Trifluoromethanesulfonamides as acylating agents $NH \rightarrow NAc$
s. 29, 393

N- and O-Acetylation
under mild neutral conditions with N-acetylsaccharin – Selective N-acetylation
s. 29, 394

Acylamines $\overset{+}{N} \leqslant \rightarrow NHAc$
from quaternary ammonium salts s. 28, 463

Squaric acid diamides   ←
by exchange of amino groups s. 28, 366

Amide acetals   ←
from quaternary 1,1,1-dialkoxyammonium salts s. 27, 430; s. a. Synthesis 1975, 272

Carboxylic acid hydrazides $CON_3 \rightarrow CO \cdot N \cdot N <$
from carboxylic acid azides

280.   PhCH=CHCON₃ + H₂N—NHPh → PhCH=CHCONHNHPh

A soln. of cinnamoyl azide and phenylhydrazine in benzene stirred 0.5 hr. at room temp. → product. Y: 70%. F. e., also with hydrazine, s. A. F. M. Fahmy and H. A. Fadel, Indian J. Chem. 11, 735 (1973).

N-Acyliminoesters $C : C(OR) \rightarrow CHC(OR) : NAc$
from carboxylic acid azides and enolethers, also N-sulfonyliminoesters
s. 29, 395

Reactions with 1-(N-iminoformyl)imidazoles $>NH \rightarrow >N \cdot CH : N \cdot$

281.

[scheme: imidazole-CH=N—CMe₃ center;
+ PhNH₂ → PhNH—CH=N—CMe₃ (left);
+ (CH₃CO)₂CH₂ → (CH₃CO)₂C=CHNHCMe₃ (right)]

Formamidines from amines. | **Enamines from active methylene groups.**
An equimolar mixture of 1-(N-*tert*-butyliminoformyl)imidazole and aniline heated 5 hrs. at 100° → formamidine deriv. Y: 94%. | and acetylacetone warmed 6 hrs. at 50° → enamine deriv. Y: 80%.
F. e. and reactions s. Y. Ito, Y. Inubushi, and T. Saegusa, Tetrah. Let. 1974, 1283; formamidines s. a. Synth. Commun. 4, 289 (1974).

Reactions with azolide analogs
1,2,4-Triazol-1-yl as leaving group
Allophanates from alcohols

282.
$$2 \underset{N}{\overset{N-N-CONH_2}{\diagdown\diagup}} + HOPh \longrightarrow H_2N-CO-NH-COOPh$$

A mixture of phenol and 2 moles 1-carbamoyl-1,2,4-triazole heated 3 hrs. at 80° → phenyl allophanate. Y: 72%. F. e. and reactions s. H. G. O. Becker and V. Eisenschmidt, *J. pr. 315*, 640 (1973).

Cyclic azomethine imides from diazo compds.

283.

A soln. of 4-phenyl-1,2,4-triazoline-3,5-dione in abs. benzene added dropwise with ice-cooling and vigorous stirring to a soln. of diazofluorene in the same solvent → 3,5-dioxo-4-phenyl-1-(fluoren-9-yl)tetrahydro-1H-s-triazole. Y: 93%. Spiroring closures with the product s. W. Ried and S.-H. Lim, *A. 1973*, 1141.

N-Sulfonyliminoesters
  from enolethers and sulfonic acid azides with isocyclic ring contraction
  s. *29*, 396

N′-Substitution of sulfonylureas $SO_2N \cdot CO \cdot NH \rightarrow SO_2N \cdot CO \cdot NR$
s. *26*, 391

2(1H)-Pyridone ring   ○
s. *28*, 367

Isoquinol-3(2H)-ones
s. *28*, 368

Piperazines
  from 2-keteneacetal and 2 sulfonic acid azide molecules s. *27*, 431

2-Alkoxy-4-imino-Δ²-oxazolines
  from ketenimines and azidoformic acid esters s. *27*, 432

Heterocyclics from diazonium salts and nitriles
  4H-3,1-Benzoxazines s. *28*, 369

Purines
  from 4-amino-5-nitrosopyrimidines and hydrazones s. *27*, 433

Pteridine ring
  from enamines and 4-amino-5-nitrosopyrimidines – Subsequent aromatization to benzo[g]pteridines s. *26*, 392

Pteridines from enamines
s. *26*, 448

3-Amino-1,2,4-triazole ring from carbodiimides
s. *26*, 393

1,2,4-Triazine ring from o-nitrosamines
s. *27*, 434

Cyclic sulfinodiimines from diamines
s. *26*, 394

Conjugated N-macroheterocyclics from o-dinitriles
  Phthalocyanines from o-dinitriles and NH_3 or from 1,3-diiminoisoindolines – Metal halogenophthalocyanines s. *27*, 435

Hexaazamacroheterocyclics

284.

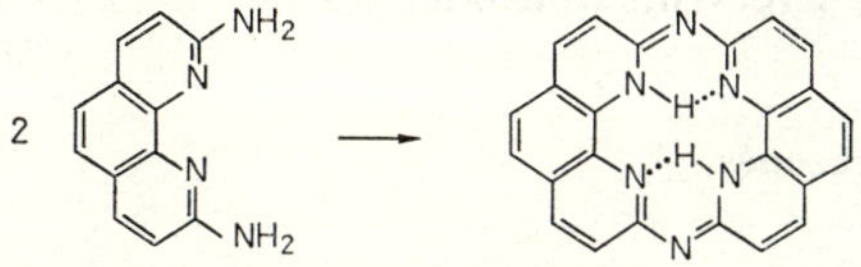

2,9-Diamino-1,10-phenanthroline heated under N_2 whereupon needles separated at 280° with NH_3-evolution, after complete solidification heating continued 1 hr. at 300° ⇥ product. Y: almost 100%. F. method s. S. Ogawa, T. Yamaguchi, and N. Gotoh, Soc. Perkin I *1974*, 976.

1,2,3-Triazine ring opening   C

285.

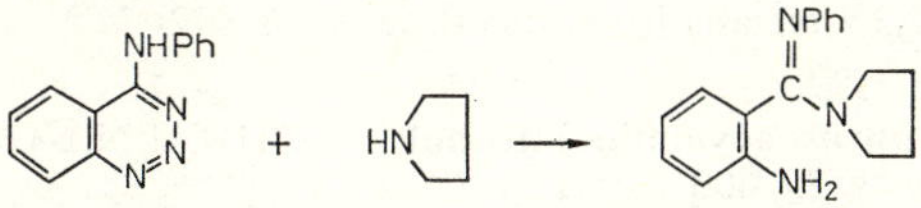

o-Aminoamidines. A soln. of 4-anilino-1,2,3-benzotriazine in pyrrolidine refluxed 7 hrs., then stored several days at 0° ⇥ 2-amino-N'-phenyl-N,N-tetramethylenebenzamidine. Y: 90%. F. e., products, and limitations s. M. F. G. Stevens, Soc. Perkin I *1974*, 615.

Reactions of silyl azides   ⊙
  Uracils s. *28*, 370

Tetrahydro-1,2,4-triazin-6-ones
  from 2-carbonyl-2H-azirines s. *28*, 371

Ring contraction with sulfonic acid azides   ⊙
  s. *29*, 434

6H-1,3-Oxazine-6-thiones   ←
  from pyridinium N-imides and cyclopropenethiones – Review of amin-N-imines s. *27*, 436

Irradiation   ₩

Pyridinium cyclopentadienylids   ←
  from diazocyclopentadienes s. *29*, 397

N-Acylaziridine ring
  from ethylene derivs. and carboxylic acid azides s. *29*, 398

Electrolysis   ↯

Peptide synthesis by in situ activation   CON<
  Anodic oxidation of hydrazides s. *26*, 398

Sodium hydride   NaH

Sec. amines from 2 prim. amine molecules $2\, RNH_2 \rightarrow RNHR$
  s. *28*, 378

Azides from amines $NH_2 \rightarrow N_3$
  s. *27*, 438 suppl. *30*

Sodium hydroxide   NaOH

Urethans from amines and azidoformic acid esters $NH \rightarrow NCOOCH_2Ar$
  Prepn. at constant pH – Carbo-p-methoxybenzoyl as N-protective group s. *27*, 437; N-carbobenzoxylation s. P. Hermann and E. Schreier, J. pr. *316*, 719 (1974)

Oxazole ring   O
  from o-nitrophenols and pyridinium salts via pyridinium ylids – Naphthoxazoles s. *26*, 395

Sodium/butanol Na/BuOH
Conjugated N-macroheterocyclics from o-dinitriles
s. 27, 435

Sodium ethoxide NaOEt
Sec. amines from hydrazones NHN : C → NHR
s. 29, 399

Methyllithium MeLi
Azides from amines via amine anions $NH_2 \rightarrow N_3$
s. 27, 438; with NaH instead of methyllithium cf. H. Quast and P. Eckert, A. *1974*, 1727

Sodium hydrogen carbonate $NaHCO_3$
2-Acylaminoquinoxalines ○
via 2-oximino-1,2,3,4-tetrahydroquinoxalines s. 28, 372

Sodium azide NaN_3
Reactions with 2-(N-nitrosoacylamino)alcohols $C(OH)CH_2N(NO)Ac \rightarrow C:CHN_3$
Terminal enazides s. 29, 400
Diazonium salt reactions ○
N-Condensed tetrazole ring – Thiazolo[3,2-d]tetrazoles s. 27, 439

Disodium hydrogen phosphate Na_2HPO_4
N-Cyaniminoesters from iminoesters C(OR) : NH → C(OR) : N·CN
s. 27, 440

Triethylamine Et_3N
Amide acetals ←
from quaternary 1,1,1-dialkoxyammonium salts s. 27, 430
N-Acylaminopyridinium betaines ←
also N-aminopyridinium salts s. 28, 373
Urethans NH → NCOOR
from amines and azidoformic acid esters – N-Carbo-*tert*-butoxylation s. 27, 437

Cadmium acetate $(CH_3COO)_2Cd$
1,3-O,N-Heterocyclics ○
from nitriles and aminoalcohols s. 28, 374

Zinc chloride s. under H_3PO_4 $ZnCl_2$

Mercuric bromide $HgBr_2$
Nucleosides by transglycosidation ←
s. 27, 441

Fluoroborates BF_4^-
Quaternary N-acylammonium fluoroborates NH → NAc
as acylating agents

286.

Ph—CHCH₃ | NH₂ + [cyclohexyl]—N⁺(Me)(Me)—Ac BF₄⁻ → PhCHCH₃ | NHAc

N-Acylation. dl-α-Phenethylamine added to a soln. of N-acetyl-N,N-dimethylcyclohexylammonium fluoroborate in acetonitrile, and allowed to stand 1 hr. → crude dl-α-phenethylacetamide. Y: 99%. F. e. s. J. V. Paukstelis and M. Kim, J. Org. Chem. *39*, 1503 (1974).

Aluminum chloride $AlCl_3$
Pyrimidines from β-iminoazomethines ○
and nitriles, also from β-diimines s. 26, 396; 1,2-dihydropyrimidines from β-iminoazomethines and azomethines s. Synthesis *1974*, 720

Aluminum chloride/phenol $AlCl_3/PhOH$
Sec. ar. amines from azides $N_3 \rightarrow NHAr$
s. 29, 402 suppl. 30

Benzyltrimethylammonium salt $PhCH_2\overset{+}{N}Me_3$
Urethans $NH \rightarrow NCOOR$
from 1-carbalkoxy-1,2,4-triazoles or mixed carbonic acid esters – N-Carbo-*tert*-butoxylation s. 28, 375

N-Thioacylimidazole ←
N-Thioacylation $NH \rightarrow N \cdot CSR$
Thioacylamines – Thioacylhydrazines s. 26, 397

Tetramethylguanidine ←
N-Carbo-*tert*-butoxylation $NH \rightarrow NCOOCMe_3$
s. 27, 437; s. a. J. Turk, G. T. Panse, and G. R. Marshall, J. Org. Chem. 40, 953 (1975)

Phenols (s. a. under $AlCl_3$) $ArOH$
Quinone monoimines and
N′-cyanoazoformamidines $2 N : N \cdot CN \rightarrow N : N \cdot C(:NH)NHCN$
from diazocyanides and phenols – N,N-Cleavage of azo groups s. 27, 442

Acetic acid CH_3COOH
Transamination ←
of aminomethylenemalonic acid esters s. 27, 443

Electrophilic substitution of azo groups
Replacement by nitro groups $N : N \cdot R \rightarrow NO_2$

287.

70%-HNO_3 added portionwise during 2 hrs. to a stirred soln. of 1-phenylazo-2-naphthol in acetic acid $\rightarrow$ 1,6-dinitro-2-naphthol. Y: 75%. Also replacement by other groups s. N. J. Bunce, Soc. Perkin I 1974, 942.

Trifluoroacetic acid CF_3COOH
Sec. ar. amines from azides $N_3 \rightarrow NHAr$
s. 29, 402; Friedel-Crafts-type synthesis with $AlCl_3$/PhOH
cf. K. Nakamura, A. Ohno, and S. Oka, Synthesis 1974, 882

N-Bromosuccinimide $\begin{matrix}CH_2CO \\ | \\ CH_2CO\end{matrix}\!\!\!>\!\!NBr$

Peptide synthesis by in situ activation $CON<$
by oxidation, also anodic, of hydrazides – Solid-phase synthesis – Activation by formation of enolesters – Peptide synthesis, review s. 26, 398

Silicic acid H_2SiO_3
Enurethans from ethylene derivs. $C:CH \rightarrow C:C \cdot NHCOOR$
and N-alkoxycarbonyliminopyridinium ylids s. 28, 376

Lead tetraacetate $(CH_3COO)_4Pb$
2,4,6,8(1H,3H,7H,9H)-Pyrimido[5,4-g]pteridinetetrone 5-oxides O
from 6-amino-5-nitrosouracils – Oxidative dimerization s. 28, 377

Nitryl fluoroborate $NO_2^+ BF_4^-$
Ring closures with enazides
Furoxans s. 29, 401

Triphenylphosphine Ph_3P
Pyrazines from α-azidoketones
s. 26, 399

Phosphorus pentoxide P_2O_5
Diazo compds. C : NH → C : N$_2$
from imines via hydrazones s. 27, 444

Phosphoric acid/zinc chloride $H_3PO_4/ZnCl_2$
Sec. amines from 2 prim. amine molecules 2 RNH$_2$ → RNHR
Phenoxazines from 2 o-aminophenol molecules s. 28, 378

Phosphorus oxide chloride $POCl_3$
Acylamidrazones C
from tetrazoles and carboxylic acid amides

288.

POCl$_3$ added at room temp. to a soln. of 6-ethyl-3-(1H-tetrazol-5-yl)chromone in dimethylformamide, then water added and neutralized with NaOH-soln. → N,N-dimethylformamide N²-(6-ethylchromone-3)-carbohydrazone. Y: 83%. F. e. s. A. Nohara, Tetrah. Let. *1974*, 1187.

Sulfuric acid H_2SO_4
1,1-Di(acylamines) from morpholinals C(NAc)$_2$
s. 29, 403

Sym. N,N′-diacylhydrazines 2 AcNHNH$_2$ → AcNHNHAc
from carboxylic acid hydrazides s. 27, 445

3-Pyrazolidones O
from α,β-ethylenenitriles and hydrazines s. 28, 379

Hydrochlorides ←
Azomethines C : NH → C : NR
from imines and amines – Aminoazomethines s. 29, 404

Transamidation of amidines ←

289.

N,N-Dimethyl-N′-1-anthraquinonylformamidine hydrochloride stirred 1 hr. in aniline → N-phenyl-N′-1-anthraquinonylformamidine. Y: 92%. – The 1-anthraquinonylamidines are highly reactive in the protonated form, but are passive in the form of the bases. F. e. and reactions s. L. B. Krasnowa, S. I. Popov, and N. S. Dokunikhin, Zh. Org. Khim. *9*, 1494 (1973); Eng. transl. 1523; C. A. *79*, 91826.

Amidrazones from amidines C(N<) : NH → C(N<) : N·NH$_2$
s. 28, 380

Monosubst. 1,2,4,5-tetrazines from iminoesters O
s. 28, 381

Hydrogen chloride HCl
Azacyanines ←
α,β-Diazatrimethinecyanines, also azamero-, azamethine-, α-azatrimethine-, and azapentamethine-cyanines s. 27, 446

Hydrogen bromide/acetic acid HBr/CH_3COOH
Dimerization to triple-N-condensed dicyclimmonium salts ○
s. 26, 400

Iron carbonyl $Fe(CO)_5$
Unsym. carbodiimides $N_3 \rightarrow N:C:NR$
from azides and isonitriles s. 27, 447

Nickel Ni
One-step purine synthesis ○
s. 28, 382

Palladium Pd
Deaminative condensation of amines ←
s. 29, 405

Halogen ↑   NC ⇈ Hal

Without additional reagents w.a.r.
Replacement of halogen by amino groups $Hal \rightarrow NH_2$
with nucleophilic, ar. N-N-rearrangement s. 26, 310

α-Aminocarboxylic acids $Cl \rightarrow N<$
from α-chlorocarboxylic acid amides
via α-aminocarboxylic acid amides

290. $Me_3SiCH_2NH_2 + ClCH_2CONH_2 \rightarrow Me_3SiCH_2NHCH_2CONH_2$
  $\downarrow HCl$
  $Me_3SiCH_2NHCH_2COOH \leftarrow Me_3SiCH_2NHCH_2COOH, HCl$

Chloracetamide and trimethylsilylmethylamine warmed to 40° with stirring, after subsiding of the reaction dissolved in 1 N HCl, refluxed 14 hrs., and the product isolated as the hydrochloride → N-trimethylsilylmethylglycine hydrochloride (Y: 98%) dissolved in abs. ethanol, treated with propylene oxide, and stirred 16 hrs. at 20–25° → N-trimethylsilylmethylglycine (Y: 95%). W. Fink, Helv. 57, 1042 (1974); **aminocarboxylic acids from their hydrohalides via silylation** cf. S. V. Rogozhin, Y. A. Davidovich, and A. I. Yurtanov, Synthesis 1975, 113; **synthesis of α-aminocarboxylic acids, review**, cf. E. N. Safonova and V. M. Belikov, Russ. Chem. Rev. 43, 745 (1974) (Eng. transl.).

Fluoroamines from chlorofluorides $Cl \rightarrow N<$
s. 29, 512

1-Subst. 1,4-dihydro-4-iminoquinolines ←
from 4-aminoquinolines and prim. halides s. 26, 401

Hydroxylamines from halides $Hal \rightarrow NHOH$
from fluorides – from chlorides with retention of fluorine s. 26, 402

α-Aminoketones from 1,2-oxido-1-halides $Hal \rightarrow C(N<)CO$
β-Amino-α-ketocarboxylic acid derivs. s. 26, 403;
f. reactions of 1,2-oxido-1-halides cf. P. Coutrot and C. Legris, Synthesis 1975, 118

Reactions of
N-(chloroformyl)iminochlorides $C(Hal):N \cdot COCl \rightarrow C(N<):N \cdot CON<$
N-Carbamylamidines s. 28, 383

6-Anilino- from 6-halogeno-uracils $Hal \rightarrow N<$
s. 28, 314

Reaction of chlorimidium salts with nucleophiles ←
Carboxylic acid amides and amidinium salts s. 26, 404

N-(Chloroglyoxyloyl)iminoesters C(OR) : NH → C(OR) : NCOCOCl
  from iminoesters s. 26, 458

N-Unsubst. aziridine ring
  from 1,2-azidoiodides via N-(aziridinyl)phosphonates – Stereospecific
  ring closure – Ring closure with other phosphorus(III) nucleophiles s. 26, 405

2-Cyanoaziridines
  from α,β-ethylene-α-halogenonitriles and prim. amines s. 27, 468

Cyclic imines from dihalides
s. 27, 448

Isoindolines from hydrazines
  via cyclic quaternary hydrazinium salts s. 27, 449

2-Amino-Δ¹-pyrrolines from γ-bromonitriles
s. 27, 450

Pyrroles from acetylenehalogenhydrins
s. 27, 451

Alkoxyoxindoles from alkoxylamines
  N→C-Alkoxy group migration s. 28, 384

2-Aminopyrrolo[2,3-b]quinoxalines
s. 27, 452

Ring closure with Vilsmeier salts
  Pyrazolo[4,3-b]-1,4-benzoxazines s. 27, 453

2-Amino-Δ²-oxazolines
from 2-halogenisocyanates via halogenoureas

291. $CH_2=CH-CH-N=C=O + NH_3 \rightarrow CH_2=CH-CH-NHCONH_2$
          | |
          CH_2Br CH_2Br

                                              NaOH ↓
                        ← $CH_2=CH-\underset{O}{\underset{|}{\diagdown}}\underset{NH_2,HBr}{\overset{N}{\diagup}}$
                      NH₂

Startg. m. added dropwise at 2–4° during 50–70 min. to vigorously stirred aq. NH₃, stirring continued 5–10 min. at 2–5°, the resulting bromourea deriv. heated 10–25 min. at 84–88° in water, and treated at 2–5° with 10 N NaOH → product. Y: 90%. F. e. and reactions of the products s. K.-D. Kampe, A. *1974*, 593.

5-Aminooxazoles from 1-(dichloromethylene)acylamines

292. $Cl_2C=C-PO(OEt)_2$ $Me_2N-\underset{O}{\diagdown}\underset{Ph}{\overset{N}{\diagup}}-PO(OEt)_2$
         Me₂NH + |
                    NH   →
                    |
                   OC
                    |
                   Ph

An soln. of dimethylamine in tetrahydrofuran added to a soln. of diethyl (1-benzamido-2,2-dichlorovinyl)phosphonate in the same solvent, and the product isolated after 1 week → diethyl (5-dimethylamino-2-phenyl-4-oxazolyl)phosphonate. Y: 82%. B. S. Drach and E. P. Sviridov, Ж. *43*, 1648 (1973); C. A. *80*, 3577; Zh. Org. Khim. *10*, 1271 (1974); C. A. *81*, 91402.

Cyclic hydroxamic acids from o-aminohydroxamic acids
s. 28, 385

4,5,6,7-Tetrahydro-1H-1,2-diazepines from hydrazines
s. 28, 386

3-Alkoxy-4-quinazolones
  from o-aminohydroxamic acid esters and carboxylic acid chlorides s. 29, 406

2,5-Piperazinediones
s. 27, 454

N-Condensed 4-quinazoline ring
from o-aminocarboxylic acids s. 27, 455

Pyrimidopteridine 10-oxides
s. 28, 387

2-Oxazolone ring
from isocyanates and cyclic N-oxides

293.

3,5-Dibromopyridine N-oxide heated 7 hrs. at 110° with phenyl isocyanate in dimethylformamide → 2,3-dihydro-2-oxo-3-phenyl-6-bromooxazolo[4,5-b]pyridine. Y: 70%. T. Hisano, T. Matsuoka, and M. Ichikawa, Heterocycles 2, 163 (1974); s. a. M. Hamana, H. Noda, and M. Aoyama, ibid. 2, 167.

1,5-Diazabicyclo[3.3.0]octane-3,4,6,8-tetraones
s. 26, 406

N-Condensed heterocyclics
from o-aminocarboxylic acid amides − Double ring closure s. 26, 407

1-Aza-4,6-dioxabicyclo[3.3.0]octanes
s. 29, 407

Imidazolinylhydrazonohalides
from trihalogenodiazabutadienes s. 27, 529

Amidines from 2,2-dichloroaziridines   C
s. 28, 388

Ring expansion by 1,1-dihalogenocyclopropane ring opening   ☉
s. 26, 408

Replacement of chlorine by amino groups
with ring expansion s. 28, 389

Pyrazoles from p-chloro-N-heterocyclics   ←
s. 28, 390

Potassium s. under KNH$_2$ K

Lithium hydride   LiH

N-Monosubst. barbituric acids   NH → NR
via N,S-dialkylation of 2-thiobarbituric acids s. 29, 408

Sodium hydride   NaH

Benzyloxymethyl as composite N-protective group   NH → NCH$_2$OCH$_2$Ph
Stepwise removal of the protective group s. 27, 456

Sodium hydride as reducing agent   C:N → CHNAc
2-Acyl-1,2-dihydroisoquinolines from isoquinolines s. 29, 409

Carboxylic acid amides from oxaziridines   N → CON<
s. 26, 17

Sodium hydride/dimethyl sulfoxide   NaH/Me$_2$SO

Total N- and O-methylation   NH → NR
s. 29, 410

Sodium hydroxide   NaOH

α-Acylamino- from α-amino-carboxylic acids   NH → NAc
s. 30, 451

3-Aminoisoxazoles   ○
from α,β-ethylenenitriles via α,β-dibromonitriles s. 26, 409

1,3,5-Triazines from biguanides

294.

A soln. of 2-furoyl chloride in acetone added dropwise with ice-cooling and vigorous stirring to a soln. of 1,1,5,5-tetramethylbiguanide in aq. 5%-NaOH, and allowed to warm to room temp. with continued stirring during 2 hrs. → product. Y: 88%. F. e. s. R. Vanderhoek, G. Allen, and J. A. Settepani, J. Med. Chem. 16, 1305 (1973).

Potassium hydroxide   KOH
N-Alkylation in dimethyl sulfoxide   NH → NR
of carboxylic acid amides s. 27, 457

N-Carbalkoxytetrahydro-1,2-oxazines   ○
from 1,4-dibromides and N-hydroxyurethans

295.

1,4-Dibromobutane refluxed 6 hrs. on a water bath with 3 moles N-hydroxyurethan and 3 moles KOH in abs. ethanol → N-carbethoxytetrahydro-1,2-oxazine. Y: 72%. F. e. s. F. G. Riddell and D. A. R. Williams, Tetrahedron 30, 1083 (1974).

Sodium/alcohol   NaOR
Isomeric nitroimidazoles   NH → NR
by N-alkylation s. 28, 391

N-Phosphorylaziridines   C : C → N◁
from ethylene derivs. and N,N-dibromophosphoramidates s. 29, 411

3-Amino-Δ²-1,2,4-thiadiazoline 1,1-dioxides   ○
and sulfonylguanidines
from carbodiimides s. 26, 410

Potassium tert-butoxide $KOCMe_3$
Mono(carbobenzoxy)amines   Hal → $NHCOOCH_2Ph$
from halides via N-carbobenzoxysulfenamides s. 29, 412

n-Butyllithium   n-BuLi
Ynamines from difluoromethylene compds. $CH : CF_2$ → C≡C·N<
s. 27, 458

n-Butyllithium/triethylenediamine   ←
Di- and tri-aroylamines   ←
s. 26, 420

Phenyllithium   PhLi
1,4-Diaryl-2,5-piperazinediones   ○
s. 28, 392

Sodium amide $NaNH_2$
Tert. from prim. amines and halides NH_2 → NR_2
s. 28, 393

Replacement of halogen by amino groups ←

296. [structure: 4-chloroethynyl-1,3,5-trimethylpyrazole] + NaNH₂ → [structure: 4-ethynyl-5-aminomethyl-1,3-dimethylpyrazole]

with rearrangement. 4-β-Chlorethynyl-1,3,5-trimethylpyrazole allowed to react with NaNH₂ in liq. NH₃ → 4-ethynyl-5-aminomethyl-1,3-dimethylpyrazole. Y: 85%. M. S. Shvartsberg and S. F. Vasilevskii, Izvest. *1973*, 2166; C. A. *80*, 27162.

Potassium amide/potassium KNH_2/K
Alkali metal-promoted replacement   ArHal → ArNH₂
   of ar. substituents – Ar. amines from halides and ethers s. *27*, 459

Lithium nitride Li_3N
Dicarboxylic acid imides and anhydrides
   from dicarboxylic acid chlorides s. *26*, 411   –COCl / –COCl → –CO–NH / –CO–

Sodium carbonate Na_2CO_3
N-Carbalkoxylation   NH → NCOOR

297. [structure: 2-amino-2-deoxy-D-glucose] + ClCOOMe → [structure: methyl 2-deoxy-2-methoxycarbonylamino-α-D-glucopyranoside]

with glycosidation. 2-Amino-2-deoxy-D-glucose hydrochloride dissolved in methanol containing sodium, powdered anhydrous Na-carbonate added followed by dropwise addition of excess methyl chlorocarbonate with stirring, and refluxed overnight → methyl 2-deoxy-2-methoxycarbonylamino-α-D-glucopyranoside. Y: 83%. F. e. s. S. Otani, Bull. Chem. Soc. Japan *47*, 781 (1974).

   selective, with retention of alcohol and guanidino groups s. *30*, 303

Sodium carbonate/sodium iodide Na_2CO_3/NaI
Cyclic enamines from 4-acetylenechlorides   O
s. *30*, 637

Sodium hydrogen carbonate $NaHCO_3$
N-Arylhydroxamic acids   COCl → CON(OH)Ar
   from carboxylic acid chlorides and arylhydroxylamines s. *27*, 460

N-Carbobenzoxylation   NH → NCOOR
s. *28*, 394

N-Hydroxy- and N-acoxy-urethans ←
s. *26*, 412

Quinone N,O-diacyliminoximes ←
   from quinone iminoximes s. *27*, 461

Potassium carbonate K_2CO_3
3-Aminopyrazoles from α,β-ethylene-α-halogenonitriles   O

298. CH₂=C(Cl)–CN + H₂N–NH₂ → [3(5)-aminopyrazole]

Hydrazine hydrate added to aq. K-carbonate, then 2-chloroacrylonitrile added dropwise at 5–10° under N₂ with vigorous stirring during 1 hr., stirring continued 1 hr. at room temp. and 1.5 hrs. at 40–50°, and allowed to stand overnight → 3(5)-aminopyrazole. Y: 70%. G. Ege and P. Arnold, Ang. Ch. *86*, 237 (1974).

Sodium acetate CH_3COONa
Pteridines from α,α-dichloroketones

299.

An aq. soln. of the startg. tetraaminopyrimidine dihydrochloride added dropwise at room temp. during 6 hrs. to a stirred aq. soln. of 1,1,3-trichloroacetone and Na-acetate while the pH is kept at 5.5–6.0 by additions of HCl and NaOH, stirring continued overnight at room temp. → 2,4-diamino-7-chloromethylpteridine. Y: 80%. D. C. Suster et al., J. Med. Chem. *17*, 758 (1974).

Potassium acetate CH_3COOK
6,7,8,9-Tetrahydro-7a,11a-ethanooxyindolo[1,2-h][1,7]-naphthyridin-10(11H)-ones from 3-quinuclidones via 6,7-dihydropyrido[1,2-a]indol-9(8H)-ones

300.

A mixture of 2-(o-fluorobenzylidene)-3-quinuclidinone and K-acetate in diglyme stirred and refluxed 20 hrs. under argon → 6,7-dihydro-8-(β-acetoxyethyl)pyrido-[1,2-a]indol-9(8H)-one (Y: 79%) dissolved with acrylamide and K-*tert*-butoxide in dioxane, stirred and refluxed 1 hr., ethanol added to the intermediate Michael adduct slowly followed by additional K-*tert*-butoxide, and refluxing continued 9 hrs. → 6,7,8,9-tetrahydro-7a,11a-ethanooxyindolo[1,2-h][1,7]naphthyridin-10-(11H)-one (Y: 61%). F. e. of the rearrangement s. D. L. Coffen, D. A. Katonak, and F. Wong, Am. Soc. *96*, 3966 (1974).

6H-Benz[b]indolo[3,2,1-d,e][1,5]naphthyridines from 10-aryl-3,4-dihydro-2H-1,4-ethanobenzo[b][1,5]naphthyridines

301.

A mixture of 8-chloro-10-(2-fluoro-5-nitrophenyl)-3,4-dihydro-2H-1,4-ethanobenzo-[b][1,5]naphthyridine and K-acetate in diglyme refluxed 3 hrs. with stirring under argon, cooled, water and concd. HCl added, then heated on a steam bath to hydrolyze the acetoxy group → 12-chloro-7,8-dihydro-8-(2-hydroxyethyl)-2-nitro-6H-benz[b]indolo[3,2,1-d,e][1,5]naphthyridine. Y: 42%. D. L. Coffen and F. Wong, J. Org. Chem. *39*, 1765 (1974).

Potassium thiocyanate   KSCN
4-Organothio-Δ^3-oxazoline-2-thiones
    by benzilic acid-type rearrangement s. *28*, 395

Ammonium thiocyanate	NH_4SCN
Formamidinoylisothiocyanates from chloroformamidines s. 26, 413	←
Thiophospinylthioureas from thiophosphinic acid chlorides and amines s. 28, 396	$P(:S)Cl \rightarrow P(:S)NHCSN<$
Sodium azide	NaN_3
Azides from halides s. 29, 424	$Hal \rightarrow N_3$
α-Azido-α,β-ethyleneketones from α,β-dibromoketones s. 26, 414	$COCHBrCBr \rightarrow COC(N_3):C$
Carboxylic acid azides from carboxylic acid chlorides s. 30, 460	$COCl \rightarrow CON_3$
2-Amino-Δ¹-azirines from 1-chlorenamines s. 27, 462	○
Sodium nitrite	$NaNO_2$
3-Nitroisoxazoles from 2-acetylenebromides s. 28, 397	
Potassium nitrite	KNO_2
Nucleophilic substitution on 1,2-halogenoximes 1,2-Nitroximes s. 27, 463	$C(NO_2)C:NOH$
Sodium pyrophosphate s. under Mg	$Na_4P_2O_7$
Potassium bromide s. under $Bu_4N^+ OH^-$	KBr
Sodium iodide s. under Na_2CO_3	NaI
Potassium iodide	KI
1,4-Diazabicyclo[4.n.0)alkan-5-ones by double ring closure s. 27, 464	○
Sodium salt	Na^+
Selective N-alkylation of nucleosides s. 26, 415	$NH \rightarrow NR$
1 Acyl-1-arylhydrazines s. 28, 409	$N(Ar)Ac \cdot N<$
Potassium salt	K^+
Phthalimides from chlorides s. 27, 480	$Cl \rightarrow N\begin{smallmatrix}OC \\ OC\end{smallmatrix}$
N-Sulfonylphthalimides s. 26, 416	$\begin{smallmatrix}CO \\ CO\end{smallmatrix}N-SO_2R$
Ammonia	NH_3
N-Alkylation s. 29, 413	$NH \rightarrow NR$
Tetrabutylammonium hydroxide	$Bu_4N^+ OH^-$
Preferential N-alkylation via dianions s. 26, 417	
Tetrabutylammonium hydroxide/potassium bromide	$Bu_4N^+OH^-/KBr$
N-Alkylation of hydrazines by ion pair extraction s. 29, 414	$NNH \rightarrow NNR$
Triethylamine (s. a. under Pd-C)	Et_3N
Enamines from α,β-ethylenehalides s. 27, 465	$C:C \cdot Hal \rightarrow C:C \cdot N<$

Regiospecific N-acetylation of imidazoles   NH → NAc

302.

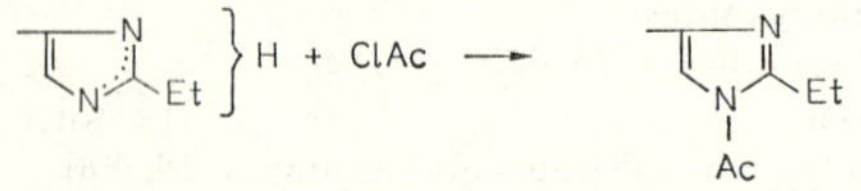

Acetyl chloride added dropwise at 5–10° to a soln. of the startg. m. in acetonitrile containing triethylamine, and allowed to stand 1 hr. at room temp. → crude 1-acetyl-2-ethyl-4-methylimidazole. Y: 93%. F. e. s. E. F. Godefroi and J. H. F. M. Mentjens, R. *93*, 56 (1974).

Selective N-acylation
  with in situ protection of alcohol groups as silyl ethers s. *29*, 415

N-Subst. carboxylic acid amides   COCl → CON<
  from carboxylic acid chlorides – N-*tert*-Butylbenzamides s. *29*, 416

Cyclic 1-alkoxyenacylamines from cyclic iminoesters   ←
  N-Chlorocarbonyl-1-alkoxyenamines s. *27*, 466

Phthalimides from prim. amines
  under mild conditions – Optically active compds. s. *29*, 417   NH₂ → N⟨OC / OC

Selective N-carbalkoxylation   NH → NCOOR
with retention of alcohol groups

303.

+ ClCOOEt

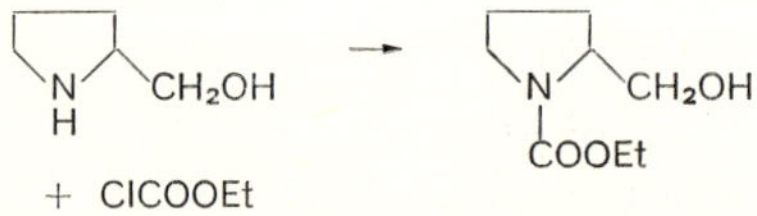

A cooled soln. of 4.5 g. L-prolinol and triethylamine in aq. 50%-acetone treated with ethyl chloroformate, and stirred 3 hrs. → 5.9 g. N-ethoxycarbonyl-L-prolinol. W. Wiegrebe et al., Helv. *57*, 301 (1974); selective N-carbobenzoxylation in the presence of Na-carbonate s. F. Schneider et al., ibid. *57*, 434; retention of alcohol and guanidino groups s. S. Umezawa et al., J. Antibiotics (Tokyo) *27*, 997 (1974).

N-Carbaryloxylation   NH → NCOOAr
s. *27*, 512

N,N-Diacylhydrazines   N·NH₂ → N·NAc₂
  from carboxylic acid chlorides s. *27*, 467

2-Cyanoaziridines 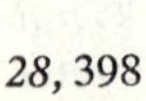
  from α,β-ethylene-α-halogenonitriles and prim. amines s. *27*, 468

2-Amino-1,3,2-diazaphospholi-4,5-dione 2-oxides
s. *26*, 418

s-Triazolo[3,4-b][1,3,4]oxadiazoles
  from tribromodiazabutadienes – Double ring closure s. *27*, 469

Ethyldiisopropylamine s. under CuCl   *i-Pr₂NEt*

N-Ethylmorpholine   ←

1-Aminooxindoles
  from α-halogenocarboxylic acid halides and 1,1-disubst. hydrazines s. *28*, 398

Pyridine (s. a. under HN₃)   *C₅H₅N*

Replacement of chlorine by amino groups   Cl → N<
s. *28*, 16

1-Acyl-1-arylhydrazines   N(Ar)Ac·N
s. *28*, 409

3-Acylaminoisocoumarins
  from o-carboxyphenylacetonitriles s. *27*, 470

2,6-Lutidine ⇐
Quaternary ammonium salts R_4N^+
  from prim. and sec. amines s. 26, 419
N,N,N-Tri(acylamines) $(RCO)_3N$
  from carboxylic acid chlorides and amides s. 26, 420

Quinoline ⇐
Sec. from prim. amines and halides $NH_2 \rightarrow NHR$
  s. 28, 16 suppl. 29

Copper Cu
Sulfonylamination with N,N-dichlorosulfonamides $H \rightarrow NHSO_2R$

304.

$$\text{(tetrahydropyran)} + Cl_2NSO_2Me \rightarrow \text{(N-(2-tetrahydropyranyl)methanesulfonamide)} \; NHSO_2Me$$

N,N-Dichloromethanesulfonamide added in small portions at 5–7° with stirring under N_2 to a suspension of Cu-powder in tetrahydropyran ⇸ N-(2-tetrahydropyranyl)methanesulfonamide. Y: 52% based on the dichlorosulfonamide. F. e. s. T. Shingaki et al., Chem. Lett. 1973, 1243.

Silver oxide Ag_2O
N-Methylation of α-aminocarboxylic acid derivs. $NH \rightarrow NMe$
  with formation of methyl esters s. 26, 421

Cuprous trifluoromethanesulfonate CF_3SO_3Cu
Ar. prim. amines from ar. halides $ArHal \rightarrow ArNH_2$
  s. 30, 592

Cupric sulfate/stannous chloride $CuSO_4/SnCl_2$
Diarylamines $ArNH_2 + HalAr' \rightarrow ArNHAr'$
  from ar. amines and halides – Modified Ullmann reaction s. 29, 418

Cuprous chloride/ethyldiisopropylamine $CuCl/i\text{-}Pr_2NEt$
N-Acyllactams from carboxylic acid chlorides
  s. 27, 471

$$\underset{NH}{\overset{CO}{\diagup}} \rightarrow \underset{N \cdot COR}{\overset{CO}{\diagup}}$$

Cuprous iodide CuI
Gabriel synthesis $Hal \rightarrow N(CO)(CO)$
  with ar. halides s. 6, 479 suppl. 24

Silver nitrate $AgNO_3$
o- and p-Nitroureas from chloroformamidines ⇐
  s. 28, 399

Silver sulfate Ag_2SO_4
N-Substitution of carboxylic acid amides $CONH \rightarrow CONR$
  s. 27, 472

Silver salt Ag^+
N- and O-Substitution of maleimides $NH \rightarrow NR$
  N- from O-Subst. maleimides s. 29, 419; N-glycosylmaleimides s. J. Org. Chem. 40, 24 (1975)

Magnesium Mg
Tert. amines from N-halogenamines $R_2N \cdot Hal \rightarrow R_2N \cdot R'$

305. $BuMgCl + ClNEt_2 \rightarrow BuNEt_2 \cdot MgCl_2 \rightarrow BuNEt_2$

N-Chlorodiethylamine allowed to react with butylmagnesium chloride in dry ether, and the resulting intermediate (Y: 93%) treated with 30%-NaOH ⇸ butyldiethylamine (Y: 84%). F. e. s. I. M. Bortovoi, Tr. Tomsk. Gos. Univ. 237, 64 (1973) (Russ); C. A. 80, 70221.

Disubst. hydroxylamines $\quad NO_2 \rightarrow N(OH)R$
from nitro compds. and halides s. 27, 473

Magnesium/sodium pyrophosphate $\quad Mg/Na_4P_2O_7$
Azides from halides $\quad Hal \rightarrow N_3$
via tosyltriazene salts s. 26, 422

Magnesium-nickel/aluminum $\quad Mg\text{-}Ni/Al$
Amines and azides $\quad Hal \rightarrow NH_2$
from halides via tosyltriazene salts s. 26, 422

Methylmagnesium chloride $\quad MeMgCl$
Tert. from prim. amines and halides $\quad NH_2 \rightarrow NR_2$
s. 26, 423

Thallous ethoxide $\quad TlOEt$
N-Alkylation $\quad NH \rightarrow NR$
s. 26, 424

Trialkyloxonium fluoroborates $\quad R_3O^+ BF_4^-$
Ring closure via O-alkyllactims $\quad \bigcirc$
s. 29, 425

Aluminum chloride $\quad AlCl_3$
1H-1,2,4-Triazoles
from 1,1-chlorohydrazones and nitriles s. 29, 420

Formaldehyde $\quad CH_2O$
α-Amino- from α-bromo-carboxylic acids $\quad CBrCOOH \rightarrow C(NH_2)COOH$
s. 26, 425

Hexamethylenetetramine $\quad C_6H_{12}N_4$
1,3-Dihydro-2H-1,4-benzodiazepin-2-ones $\quad \bigcirc$
s. 27, 474

2-Methylimidazole $\quad \leftarrow$
Selective N-acylation $\quad NH \rightarrow NAc$
s. 29, 421

Anion exchanger $\quad \leftarrow$
Protection of diequatorial *trans*-2-aminoalcohol groups $\quad \bigcirc$
as oxazolidones s. 26, 426

Phenol $\quad PhOH$
5H-Thiazolo[2,3-b]quinazol-5-ones
from 2-chlorothiazoles s. 27, 475

Acetic acid $\quad CH_3COOH$
4,5-Dichloro-3(2H)-pyridazones from hydrazines
s. 26, 427

1,3-Dihydro-2H-1,4-benzodiazepin-2-ones

306.

5-Ferrocenyl derivs. A soln. of 2-bromo-2′-ferrocenoylacetanilide in methylene chloride added at −78° during 20 min. to liq. NH$_3$ with stirring, which is continued

6 hrs. at reflux, then the cooling bath removed, NH_3 allowed to evaporate overnight, filtered, concd., dissolved in methanol containing glacial acetic acid, and refluxed 5 hrs. → 1,3-dihydro-5-ferrocenyl-2H-1,4-benzodiazepin-2-one. Y: 91%.
F. e. s. R. Kalish, T. V. Steppe, and A. Walser, J. Med. Chem. 18, 222 (1975).

N-Trimethylsilyltrifluoroacetamide $CF_3CONHSiMe_3$
Transamidation N·COR ↛ N·COR'
    via selective formation of N-acylcarboxylic acid amides s. 27, 476

Trimethylsilyl azide Me_3SiN_3
Isocyanates COCl ↛ N:C:O
    from carboxylic acid chlorides and anhydrides – Modified Curtius degradation s. 28, 400

Stannous chloride s. under $CuSO_4$ $SnCl_2$

Ammonium azide NH_4N_3
1,1-Diazides from 1,1-dichlorides CCl_2 ↛ $C(N_3)_2$
s. 26, 428

Tetrabutylammonium azide $Bu_4N^+ N_3^-$
Carboxylic acid azides COCl ↛ CON_3
    from carboxylic acid chlorides s. 30, 460

Hydrazoic acid/pyridine HN_3/C_5H_5N
Isocyanates from carboxylic acid chlorides COCl ↛ N:C:O
    via carboxylic acid azides s. 3, 378 suppl. 29

Triphenylphosphine Ph_3P
Oxazoles ◯
    from α-azidoketones and carboxylic acid chlorides s. 27, 477

Phosphonitrile chloride ↚
Nitriles from carboxylic acid chlorides COCl ↛ CN
s. 8, 517 suppl. 28

Sulfuric acid H_2SO_4
Difluoronitromethyl ketones $COCF_2NO_2$
s. 29, 422

Hydrochlorides ↚
2,4-Diaminopyrimidine ring from o-aminonitriles ◯
s. 28, 401
Ring closure with iminoesters
    3,4-Dihydro-2(1H)-oxopyrido[1,2-a]pyrimidinium chlorides from 2-aminopyridines s. 27, 478
1,4-Oxazine ring from α-bromoketones
    7H-Pyrimido[4,5-b][1,4]oxazines s. 26, 429

Hydrogen chloride HCl
Benzoxazole-2-aldoximes
    from o-aminophenols s. 26, 430

Ferric chloride $FeCl_3$
Subst. amidines CN ↛ C(:N·)N<
    from nitriles, halides, and amines via nitrilium salts s. 29, 423; 30, 219

Nickel/aluminum s. under Mg Ni/Al

Palladium-carbon/triethylamine $Pd-C/Et_3N$
N-Arylhydroxamic acids ArNO ↛ ArN(OH)COR
    from carboxylic acid chlorides and nitroso compds. s. 27, 479

Via intermediates *v.i.*

Amines from halides via azides Hal → NH₂
Preferential replacement of halogen by azido groups – Halogenoglycosamines s. *29*, 424

Prim. amines from chlorides Cl → NH₂
via phthalimides – Gabriel synthesis s. *27*, 480

Ring closure via O-alkyllactims ◯
2,3,3a,4-Tetrahydro-1H-pyrrolo[2,3-b]quinolines s. *29*, 425

Imidazo[1,2-a]pyrazine ring
s. *26*, 431

Sulfur ↑ NC ↕ S

Without additional reagents *w.a.r.*

N-Alkylation NH → NR
with trichloromethanesulfonic acid esters s. *24*, 620 suppl. *29*

Imines from thioketones C : S → C : N ·
s. *29*, 426

Azo compds. from diazo-N-sulfides N : N · SR → N : N · Ar
s. *30*, 198

Acylation with acyl dithiocarbamates NH → NAc
Selective N-acylation s. *28*, 402; also O- and S-acylation s. Indian J. Chem. *13*, 35 (1975)

Carboxylic acid amides COOH → CON<
from carboxylic acids and sulfurous acid diamides s. *30*, 654

Carboxylic acid amides from thiolic acids COSH → CON<
s. *28*, 403

Replacement of cyclic sulfur ⌒S → ⌒N-
by cyclic nitrogen with deuteriation s. *26*, 432

1,6a-Dithia-6-azapentalenes ←

307. S—S—S + H₂NMe → S—S—N—Me

from 6a-thiathiophthenes. Aq. methylamine added at room temp. to a soln. of 6a-thiathiophthene in acetonitrile, and the product isolated after 5 min. → 6-methyl-1,6a-dithia-6-azapentalene. Y: 90%. F. e. and method s. J. G. Dingwall et al., Soc. Perkin I *1973*, 2351.

Reaction of o-(chloromethylsulfonylamino)ketones ◯
with ammonia – 1,2-Dihydroquinazolines with SO₂-elimination – Imines from ketones with retention of halogen s. *26*, 433

2-Amino-1,3-N,N-heterocyclics
from diamines and isothiocyanates – 2-Aminoperimidines s. *26*, 434

1,2,4-Triazole ring
from carboxylic acid hydrazides and thioamides – 4H-s-Triazolo[4,3-a][1,4]-benzodiazepines s. *28*, 404

5-Amino-3-mercapto- and 3,5-diamino-1,2,4-triazoles
from isodithiobiurets s. *29*, 427

7,8,9,10,10a,11-Hexahydro-13H-pyrido[1',2' : 3,4]imidazo[2,1-b]quinazol-13-ones
s. *28*, 405

Hexahydro-1,3,5-triazine ring
by trimerization of isothioureas s. 26, 435; cf. S. Ishida et al., Bull. Chem. Soc. Japan 48, 956 (1975)

Sulfonylureas C
from 2-sulfonylimino-1,3-oxathiolanes s. 26, 436

1,2,4-Triazoles from 4-azo-Δ^2-5-thiazolones ←
s. 28, 406

Sodium hydride *NaH*
N-Methylation NH → NMe
with dimethylsulfoxonium methylid s. 26, 437

Tert. from prim. amines NH_2 → NR_2
s. 29, 428

Replacement of hydrogen by azido groups H → N_3
s. 30, 232

Sodium hydroxide *NaOH*
2-Amino-1,3-N,N-heterocyclics O
from diamines via 2-tosylamino-1,3-N,N-heterocyclics s. 26, 438

Tetrazoles
from sulfonylhydrazones and diazonium salts

308.
$$Ph-CH{\stackrel{N-NHSO_2Ph}{}} + Cl^- {}^+N_2Ph \longrightarrow Ph-C{\stackrel{N-NHSO_2Ph}{N=N-Ph}} \xrightarrow{NaOH} Ph-C{\stackrel{N=N}{N-N-Ph}} \;[+ HSO_2Ph]$$

A soln. of benzenediazonium chloride (prepared from aniline, $NaNO_2$, and HCl) in aq. ethanol added at 2-5° to a stirred soln. of benzaldehyde phenylsulfonylhydrazone and NaOH in ethanol → 2,5-diphenyltetrazole. Y: 51%. F. e. s. S. Ito et al., Chem. Lett. 1973, 1071.

Sodium/alcohol *NaOR*
Partial replacement of alkylthio by amino groups SR → N<

309.
$$N≡C-N=C(SMe)_2 + H_2NCH_2COOEt \longrightarrow N≡C-N=C{\stackrel{SMe}{NHCH_2COOEt}}$$

Abs. ethanolic Na-ethoxide soln. added to a stirred mixture of dimethyl cyanimidodithiocarbonate, ethyl glycinate hydrochloride, and abs. ethanol, stirred and refluxed 6 hrs. → ethyl N-[cyanimino(methylthio)methyl]glycinate. Y: 74%. F. e. and reactions of the products s. J. Gante and G. Mohr, B. 108, 174 (1975).

Pyrimidines from α-ketoketene mercaptals O

310.

p-$MeOC_6H_4$—CO—CH=C$(SMe)_2$ + HOEt + H_2N-C(=NH)-NH_2 → p-$MeOC_6H_4$-[4-ethoxy-2-amino-pyrimidine]

+ H_2N-C(=NH_2)-S (thiourea) → p-$MeOC_6H_4$-[4-ethoxy-2-mercapto-pyrimidine]

1-(4-Methoxyphenyl)-3-bis(methylthio)-2-propen-1-one (prepn. s. 466) added to a mixture of guanidine nitrate and ethanolic Na-ethoxide, stirred and refluxed ca. 8 hrs. → 2-amino-4-ethoxy-6-(4-methoxyphenyl)pyrimidine. Y: 92%. – Similarly with thiourea: 2-Mercapto-4-ethoxy-6-(4-methoxyphenyl)pyrimidine. Y: 85%. F. e. and process s. S. M. S. Chauhan and H. Junjappa, Synthesis 1974, 880.

Sodium/alcohol-acetyl chloride   NaOR/AcCl
Ring contraction of α-sulfinyllactams   Ö:

311.

Startg. lactam sulfoxide allowed to react 20 min. at 0° with 2.2 equivalents acetyl chloride followed by addition of 4 equivalents 0.1 M Na-methoxide in methanol, and stirred 14 hrs. at 22° → dl-vincamine. Y: 80%. Also dl-epivincamine s. J. L. Herrmann et al., Am. Soc. 96, 3702 (1974).

Potassium carbonate K_2CO_3
N-Allylation of sulfonic acid amides SO_2NH → $SO_2NCH_2CH:CH_2$
s. 27, 481

Sodium azide NaN_3
1,2,3-Triazoles from α,β-ethylenesulfones   O
s. 29, 429

Sodium salt Na^+
N-Carbobenzoxylation   NH → $NCOOCH_2Ph$
    in homogeneous aq. soln. – Selective N-carbobenzoxylation s. 26, 439

Alkoxythiocarbonylhydrazines   N·N·C(:S)OR
    also [(alkylthio)thiocarbonyl]hydrazines s. 26, 440

Diethylamine Et_2NH
Aziridine from 2-azetidinone ring   Ö:
s. 26, 441

Triethylamine Et_3N
N-Carbalkoxylation   NH → NCOOR
    with monothiolcarbonic acid esters s. 29, 430

Silver nitrate $AgNO_3$
Sulfenimines from disulfides   RSSR → RS·N:C<
s. 28, 407

Mercuric chloride $HgCl_2$
Replacement of alkylthio by ar. amino groups   SR → NAr
s. 27, 482

Fluoroborates BF_4^-
Thiopyrylium ring opening   C

312.

A soln. of aniline in acetonitrile added at 40–50° to a soln. of thiopyrylium fluoroborate in the same solvent, and stirring continued ca. 1 hr. → 5-(phenylamino)-N-phenyl-2,4-pentadienyliminium fluoroborate. Y: 90%. F. e. s. Z. Yoshida et al., J. Org. Chem. 38, 3990 (1973).

Water	H_2O
Enamines from sulfilimines s. *30*, 387	$N:S< \twoheadrightarrow NHC:CH$
Acetyl chloride s. under NaOR	*AcCl*
Polyphosphoric acid	$H(PO_3H)_xOH$
Sym. azines s. *27*, 483	$C:N \cdot N:C$
Perchlorates	ClO_4^-
Pyridinium from pyrylium perchlorates and thionylimines s. *29*, 431	$CO \rightarrow CN$
Reactions with dithiocarbamidium salts 2-Amino-1,3-heterocyclics, also 1-aminothioenolethers s. *27*, 484	O
Hydrochlorides	←
Replacement of alkylthio by hydrazine groups Benzoxazole from 2H-1,4-benzoxazine ring s. *28*, 408	$SR \twoheadrightarrow N \cdot N<$
Hydrogen chloride	*HCl*
Ring closure of ethylenetosylamines	O

313.

Startg. m. kept 40 hrs. at 34° in HCl-methanol → product. Y: 85%. F. e. s. W. N. Speckamp and J. Dijkink, Heterocycles *2*, 291 (1974).

Via intermediates	*v.i.*
1-Acyl-1-arylhydrazines from 1-arylhydrazine-2-sulfonic acids s. *28*, 409	$NHArN \cdot SO_3^- \twoheadrightarrow N(Ar)Ac \cdot NH$
Carbodiimides from thionylimines and hydroxyiminohalides s. *26*, 442	$N:C:N$

Remaining Elements ↑   NC ⇅ Rem

Without additional reagents	*w.a.r.*
Sec. amines from azides and boranes s. *27*, 485	$N_3 \twoheadrightarrow NHR$
Tert. amines from phosphoric acid amides s. *28*, 352	$P \cdot NR_2 \twoheadrightarrow C \cdot NR_2$
1-Substitution of tetrazoles s. *29*, 432	←
α-Aminoketones from 1,2-di(siloxy)-1-ethylene derivs. s. *28*, 410	$C(OSi\leqslant):C(OSi\leqslant) \rightarrow CO \cdot CH \cdot N<$
Hydrazones and trisubst. hydrazines from N-nitrosamines s. *29*, 433; reaction with Grignard reagents s. a. P. R. Farina and H. Tieckelmann, J. Org. Chem. *40*, 1070 (1975).	$N \cdot NO \rightleftarrows \begin{array}{l} N \cdot N:C \\ N \cdot NHR \end{array}$

**Reaction of N-nitrosamines
with organolithium compds.**   CHN·NO → C(OR)N·NHR'

314.
$$\text{EtOH} + \underset{CH_3}{\overset{CH_3}{>}}\!\!N\!-\!NO + \text{LiCMe}_3 \longrightarrow \underset{\text{EtOCH}_2}{\overset{CH_3}{>}}\!\!N\!-\!NHCMe_3$$

1,1-Alkoxyhydrazines. A soln. of *tert*-butyllithium in *n*-pentane added during 1 hr. at −70° under argon to a well-stirred soln. of dimethylnitrosamine in anhydrous ether, after an additional hr. quenched with abs. ethanol and brought to 0° → 1-methyl-1-ethoxymethyl-2-*tert*-butylhydrazine. Y: 84%. F. e. s. P. K. Farina and H. Tieckelmann, J. Org. Chem. *38*, 4259 (1973).

Sym. thioureas and isothiocyanates   2 NSi≤ → N·CS·N
  from silylamines s. 27, 486

Isocyanates from aminosilanes and disilazanes   N : C : O
  at low temp. s. *26*, 443; in a flow system under pressure s. Ж. *44*, 2156 (1974); C. A. *82*, 73081

Partial replacement of phosphonyl by amino groups   PO(OR)$_2$ → NH$_2$
  s. *27*, 487

1,2-Diamines   CBr·C·N·PO(OR)$_2$ → C(N<)·C(N<)
  from 2-bromophosphoramidates s. *28*, 411

Carboxylic from phosphoric acid amides   COOH → CON<
  s. *22*, 484

Reactions with phosphine imines   COCN → C(:N·)CN
  Tetrazoles from carboxylic acid chlorides − 1,1-Iminocyanides from acylcyanides s. *26*, 444

Acylaminosilanes from disilazanes   N(Si≤)$_2$ → N(Ac)Si≤
  s. *28*, 412

Heterocyclics from silyl derivs. and phosgene   O
  2,5-Oxazolidiones − Reactivity of silyloxy groups s. *27*, 488

Ring contraction with sulfonic acid azides   Ö:
  N-Acylsulfonic acid amides from enoxysilanes s. *29*, 434

Sodium azide   NaN$_3$
Tetrazoles from carboxylic acid chlorides   O
s. *26*, 444

Triethylamine s. under Me$_3$SiCl   Et$_3$N

Cadmium carbonate   CdCO$_3$

Acyclic-sugar nucleosides   ⊂N·HgCl → ⊂NR
  N-Thioglycosyl-N-heterocyclics s. *29*, 435; from 1-alkylthio-1-pyridinium salts s. *29*, 436

Mercuric acetate/sulfuric acid   (CH$_3$COO)$_2$Hg/H$_2$SO$_4$
(N-Vinyl)oxo- from silyloxy-N-heterocyclics   ←
s. *26*, 445

Zinc chloride   ZnCl$_2$
N-Glycosyl-N-heterocyclics   ⊂NR
s. *26*, 446 suppl. *29*

Trimethylchlorosilane/triethylamine   Me$_3$SiCl/Et$_3$N
Isothiocyanates from silylamines   NHSi≤ → N : C : S
s. *27*, 486

Stannic chloride SnCl₄

N-Glycosyl-N-heterocyclics OAc → N⟩
from acyl glycosides s. *26,* 446; s. a. J. Org. Chem. *39,* 3654–3672 (1974); P. D. Cook et al., Am. Soc. *97,* 2916 (1975); with trimethylsilyl perchlorate or triflate instead of SnCl₄ cf. H. Vorbrüggen and K. Krolikiewicz, Ang. Ch. *87,* 417 (1975)

Nitrosyl chloride NOCl
Nitroso compds. from stannanes Sn≤ → NO
also from silanes s. *27,* 489

Nitro compds. ArSn≤ → ArNO₂
from stannanes via nitroso compds.

315. Me₃Sn—⟨⟩—SnMe₃ → [Me₃Sn—⟨⟩—NO] → Me₃Sn—⟨⟩—NO₂

Partial conversion. A soln. of p-bis(trimethylstannyl)benzene in anhydrous ether treated at room temp. under N₂ with 0.2 mole nitrosyl chloride in ether, stirred 0.5 hr., evaporated to dryness, ethyl acetate added followed by an aq. soln. of K-permanganate and H₂SO₄, shaken 45 min. → trimethyl-p-nitrophenylstannane. Y: 64%. – The product can not be prepared by nitration of trimethylphenylstannane. F. e. s. C. Eaborn, I. D. Jenkins, and D. R. M. Walton, Soc. Perkin I *1974,* 870.

p-Toluenesulfonic acid TsOH
Enamines from oxo compds. CHCO → C:C·N≤
s. *29,* 437

Silver perchlorate AgClO₄
2-Thionucleosides ←

316.

A soln. of AgClO₄ in benzene added to 4-(trimethylsilyloxy)-2-(trimethylsilylthio)-6-azapyrimidine and 2,3,5-tribenzoyl-D-ribofuranosyl chloride in abs. benzene, and the product isolated after 0.5 hr. at 24° → 2′,3′,5′-tribenzoyl-2-thio-6-azauridine. Y: 84.3%. F. e. s. H. Vorbrüggen and P. Strehlke, B. *106,* 3039 (1973).

Hydrochlorides ←
Replacement of oxygen by amino groups ←
s. *26,* 447

Via intermediates v.i.
Nitriles CH₂P⁺≤ → CN
from phosphonium via α-nitrosophosphonium salts s. *28,* 319

Isocyanates from aminosilanes NHSi≤ → N:C:O
via carbamic acid silyl esters – Phosgene-free prepn. of isocyanates s. *29,* 438

N-Subst. aziridine ring O
from 1,2-azidoiodides and dichloroboranes – N-Subst. aziridines with well defined stereochemistry s. *29,* 439

Carbon ↑   NC ⇈ C

Without additional reagents   *w.a.r.*

Amines   Hal → N<
    from halides and acylamines s. *24*, 491

Tert. amines from halides
    N-Methyl-N-phenylglycine anilides s. *28*, 413

2- from 1-Alkyltetrazoles   ←
s. *28*, 414

Enamines from formamides   N·CHO → N·CH:C
s. *28*, 415

Sec. acylamines   >$\overset{+}{N}R_2$ → >NAc
    from quaternary ammonium salts s. *29*, 440

α,α-Dichloro-β-ketocarboxylic acid esters   NH → NAc
    as acylating agents s. *29*, 441

Carboxylic acid amides   COCH$_2$COR → CONH$_2$
    from β-diketones and N-chloramines s. *28*, 416

Ureas   N·COCH$_2$COR → N·CO·N<
    from amines and β-ketocarboxylic acid amides s. *28*, 417

3-Oxo-2-(1-pyridinio)-1-indenolates   ←

317.

Startg. m. allowed to react a few min. with pyridine in tetrahydrofuran → 3-oxo-2-(1-pyridinio)-1-indenolate. Y: 60%. H. Junek, A. Hermetter, and H. Fischer-Colbrie, Ang. Ch. *86*, 380 (1974).

Dicarboxylic acid imides   CO → CN
    from dicarboxylic acid anhydrides and formamides — Carboxydicarboxylic acid imides s. *29*, 442

Phthalimidines |CO / |CO → |CH / |CO ⟩N−
    from o-dioxo compds. and isocyanates s. *27*, 490;
    s. a. J. Org. Chem. *39*, 3924 (1974).

Indazoles   O
s. *27*, 491

Pteridines from enamines
s. *26*, 448

1,4,7-Triazatricyclo[5,2,1,0^{4,10}]decanes
s. *26*, 450

Diisocyanates from lactams   _NH / _CO → _N:C:O / _N:C:O
    via aminohydroxamic acids s. *27*, 492

Fragmentation of aziridines   C
s. *26*, 306

o-Aminocarboxylic acid amides
    from 3,1-benzoxazine-2,4-diones and hindered amines s. *27*, 349

4-Quinazolones   ←
    from 3,1-benzoxazine-2,4-diones and carboxylic acid amides s. *28*, 418

Synthesis of heterocyclics
with cyclic carbon dioxide eliminators s. 26, 449

8-Arylpurines from 4-aminopyrimidines

6-Amino-4-hydroxy-2-phenyl-5-(4-phenylurazol-1-yl)pyrimidine (from 6-amino-4-hydroxy-2-phenylpyrimidine and 4-phenyl-1,2,4-triazoline-3,5-dione in dioxane at room temp.) fused 1 hr. at 250° with ca. 2 equivalents p-chlorobenzaldehyde → product. Y: 70%. – *4-Phenyl-1,2,4-triazoline-3,5-dione* acts as nitrogen source for N-7 of the purine ring. F. e. s. F. Yoneda, S. Matsumoto, and M. Higuchi, Chem. Commun. *1974*, 551.

Lithium hydride   *LiH*
N-Alkylation with chloroformic acid esters   NH → NR
Preferential N-alkylation s. 28, 419

Sodium hydride   *NaH*
Cyclic enamines from oxido compds.   ⊂NH → ⊂N·C:C
α-Carboxy-N-styrylheterocyclics s. 30, 254

Potassium hydroxide   *KOH*
Exchange of substituents of tert. amines   NR → NR'
s. 27, 493

1-Hydroxybenzimidazole 3-oxides
from benzofurazan oxides and β-ketosulfones or nitroalkanes – f. N-heterocyclics – 2H-Benzimidazole 1,3-dioxides s. 28, 420

Potassium hydroxide-sodium nitrite/hydrogen chloride   *KOH-NaNO₂/HCl*
α-Ketoaldehyde aldoximes   COCHCOOR → COCH:NOH
from β-ketocarboxylic acid esters

Ethyl 3-(indol-3-yl)-3-oxopropionate stirred 42.5 hrs. at room temp. with aq. KOH added in 2 portions, adjusted to pH 7 with 3%-HCl, filtered, and the water-cooled filtrate treated portionwise and alternately with aq. NaNO₂ and 3%-HCl → indol-3-ylglyoxal 1-oxime. Y: 92%. T. P. Karpetsky and E. H. White, Tetrahedron *29*, 3761 (1973).

Sodium/alcohol   *NaOR*
1,3,5-Triazine ring
from 4H-1,3-benzoxazin-4-one ring – Prepn. of o-hydroxyphenyl-s-triazines s. 28, 421

Sodium compound s. under *TsN₃*

Sodium amide $NaNH_2$
Replacement of trifluoromethyl by amino groups $CF_3 \twoheadrightarrow NH_2$
   in heterocyclics s. 27, 494

Lithium diisopropylamide $i\text{-}Pr_2NLi$
Aliphatic nitro compds. $COOH \twoheadrightarrow NO_2$
   from carboxylic acids s. 26, 923

Sodium nitrite/hydrogen chloride s. under KOH $NaNO_2/HCl$

Sodium benzylmercaptide $PhCH_2SNa$
Carboxylic acid amides $CHO \twoheadrightarrow CH_2CON{<}$
   from aldehydes via α-alkoxy-α,β-ethylenenitriles – Synthesis with addition of
   1 C-atom s. 29, 443

Sodium salt Na^+
N-Aroylsulfonic acid amides $ArCOCH_3 \twoheadrightarrow ArCONHSO_2Ar'$
   from aryl methyl ketones s. 26, 451

Ammonium hydroxide s. under O_2 NH_4OH

Morpholine   ←
Dihalogenomethylenehydrazines $NH\cdot N : C(Ac)_2 \twoheadrightarrow NH\cdot N : C(Hal)_2$
   from 1,2,3-triketone 2-monohydrazones via Japp-Klingemann cleavage
   of α-azo-α,α-dihalogenoketones s. 28, 422

Pyridine C_5H_5N
Ring closure with removal of substituents   ◯
   Isoxazolo[4,5-b]quinoxalines – Flavazoles – Leaving groups s. 26, 452

Cupric acetoacetonate   ←
Fragmentation of N-heterocyclics   C
   with diazo compds. s. 29, 779

Cuprous iodide CuI
Sec. acylamines from unsatd. tert. amines R_2NAc
s. 29, 444

Fluoroborates   ←
Vinylogous amidinium salts   ←
   from vinylogous carboxylic acid amides and alkoxyaminomethinium salts
   s. 26, 453

Dimethylformamide ethylene acetal   ←
N-Transalkylation with position shift   ←

320.

A mixture of 3-phenyltoxoflavin and dimethylformamide ethylene acetal refluxed
2 hrs. → 1-(2-hydroxyethyl)-3-methyl-6-phenyl-7-azalumazine. Y: 72%. F. e. s.
F. Yoneda and T. Nagamatsu, J. Heterocyclic Chem. *11*, 271 (1974).

Oxonium fluoroborate $R_3O^+ BF_4^-$
Preferential N-alkylation of N-heterocyclics   ←
   Acyl as N-protective group – also alkylation with carboxonium salts s. 26, 454

Trifluoroacetic acid $\qquad CF_3COOH$
Thiazolidine ring $\qquad$ ○
s. *28*, 423

Dimethyl sulfoxide/1,1,1-trichloro-3,3,3-trifluoroacetone $\qquad Me_2SO/CF_3COCCl_3$
Neutral N-trifluoroacetylation $\qquad NH \to NCOCF_3$
s. *26*, 455

Thiophosgene $\qquad CSCl_2$
Isothiocyanates and nitriles $\qquad \twoheadleftarrow$
from silylamidines s. *27*, 495

Dibutyltin oxide $\qquad Bu_2SnO$
Azomethines $\qquad NHCOOR + OC \to N:C$
from oxo compds. and urethans s. *26*, 456

Reactions with *bismuth triacetate* $\qquad (CH_3COO)_3Bi$
Transamidation of carboxylic acid amides $\qquad NAc \to NAc'$
s. *28*, 424

Air/hydrogen bromide/ammonium hydroxide $\qquad O_2/HBr/NH_4OH$
5-Hydroxyindoles $\qquad$ ○
s. *30*, 275

Ozone/nickel $\qquad O_3/Ni$
Cleavage of ethylene derivs. to amines $\qquad C:C \to CHN<$
s. *27*, 149

Sulfur $\qquad S$
Carboxylic acid thiomorpholides $\qquad CSCOOH \to CSN<$
from α-thioketocarboxylic acids s. *27*, 496

Tosyl azide/sodium compound $\qquad \twoheadleftarrow$
Diazo group transfer $\qquad CO \cdot CHCO \cdot COOR \to CO \cdot CN_2$
α'-Diazo-α,β-ethyleneketones s. *30*, 247

p-Toluenesulfonic acid $\qquad TsOH$
Perimidines from 2-acylindan-1,3-diones $\qquad \twoheadleftarrow$
s. *27*, 497

Hydrogen chloride $\qquad HCl$
Chloro-1,3,5-triazines $\qquad$ ○
from 3 nitrile molecules s. *26*, 457

Hydrogen bromide s. under O_2 $\qquad HBr$

Cobalt carbonyl $\qquad Co(CO)_4$
Azomethines from isocyanates $\qquad N:C:O \to N:C<$
s. *28*, 425

Cobaltous chloride $\qquad CoCl_2$
1,1-Dipyrazolyl compds. from ketones $\qquad \twoheadleftarrow$
s. *28*, 426

Nickel s. under O_3 $\qquad Ni$

Via intermediates $\qquad v.i.$
Acylisocyanates from iminoesters $\qquad C(:NH)OR \to CON:C:O$
via N-(chloroglyoxyloyl)iminoesters – Review of acylisocyanates s. *26*, 458

Elimination ⇑

Hydrogen ↑ NC⇑H

Without additional reagents	w.a.r.
8-Subst. purines	○
from 4-amino-5-nitrosopyrimidines s. *29*, 343	
Irradiation s. under NaOH and ArN : NAr	⚡
Electrolysis	⚡
Heterocyclics	
by anodic oxidation s. *29*, 239	
Sodium	Na
Intramolecular nucleophilic ring closure	
at pyridine o-position s. *28*, 427	
Sodium hydride	NaH
4-Aminoquinazolines from 3-aminoindazoles	☉
s. *27*, 498	
Sodium hydroxide/irradiation	NaOH/⚡
1-Hydroxy-3H-indol-2-ones	○
from ar. nitro compds. s. *27*, 499	
Potassium hydroxide	KOH
1,1,2-Trisubst. hydrazines	←
by rearrangement of 1,1,1-trisubst. hydrazinium salts s. *26*, 459	
Sodium/alcohol	NaOR
Formation of N-methylene bridges	←
from and reductive ring opening of N-condensed hydrazinium salts s. *26*, 460	
Sodium hydrogen carbonate s. under Me₃C·OCl	NaHCO₃
Sodium nitrite/acetic acid	NaNO₂/CH₃COOH
1,2,4,5-Tetrazines	CHNH → C:N
from 1,6-dihydro-1,2,4,5-tetrazines s. *28*, 428	
Triethylamine s. under Me₃C·OCl	Et₃N
Pyridine s. under (CH₃COO)₄Pb	C₅H₅N
Cuprous chloride/oxygen	CuCl/O₂
1,4-Dicyano-1,3-dienes from o-diamines	C
s. *28*, 429	
Mercuric acetate	(CH₃COO)₂Hg
4-Hydroxypyridazinium betaines	←
s. *26*, 940	
Lactam rearrangement	○
with aromatization s. *28*, 748	
Azo compds./irradiation	ArN : NAr/⚡
2(1H)-Quinoxalones	CHNH → C:N
from 3,4-dihydro-2(1H)-quinoxalones s. *26*, 461	
Peroxidase/hydrogen peroxide	←
Oxidative ring closure	○
N-Condensed cyclimmonium salts from cyclic tert. amines s. *29*, 904	

Chloranil ←
Pyrimidines from 1,2-dihydropyrimidines   CHNH → C:N
s. *30*, 256

4-Hydroxyisoquinolinium salts ←
  from 2,3-dihydroisoquinol-4(1H)-ones s. *26*, 462

Lead tetraacetate $(CH_3COO)_4Pb$
Azomethines from amines   C:NR
Stieglitz-type rearrangement
s. *24*, 504; s. a. J. Org. Chem. *39*, 3932 (1974)

Nitrilimines as intermediates   ○
s. *26*, 232

Oxidative and reductive   C
pyrrolo[1,2-c]pyrimidine ring opening
s. *28*, 430

1,2,3-Benzotriazines from 1-aminoindazoles   ⊙
s. *28*, 431

Lead tetraacetate/potassium carbonate $(CH_3COO)_4Pb/K_2CO_3$
N-Condensed aziridine ring   ○
  from ethyleneamines s. *27*, 500

Lead tetraacetate/pyridine $(CH_3COO)_4Pb/C_5H_5N$
2-Oxazolidone ring
  from β-hydroxycarboxylic acid amides s. *28*, 309

Vanadium pentoxide/sulfuric acid V_2O_5/H_2SO_4
Epimerization of N-condensed heterocyclics ←
via N-condensed cyclimmonium salts

321.

Reserpine allowed to react 3 hrs. at room temp. with V_2O_5–H_2SO_4-reagent in water containing tartaric acid → 3,4-dehydroreserpine sulfate (Y: ca. 100%) dissolved in methanol, stirred and treated during ca. 1 hr. with small portions of $NaBH_4$ → 3α-isomer (Y: 85%). F. e. s. R. Stainier, H. P. Husson, and C. L. Lapière, J. Pharm. Belg. *28*, 307 (1973); *vanadium pentoxide-titanium dioxide catalyst* cf. B. V. Suvorov et al., Vestn. Akad. Nauk Kaz. SSR *1974*, 16 (Russ); C. A. *82*, 30432.

Oxygen s. a. under CuCl, Cobalt oxide, and Pd-C O_2

Air

1,2-Di(azomethines) from ene-1,2-diamines   NH·C:C·NH → N:C·C:N
s. *26*, 683

Hydrogen peroxide s. under Peroxidase H_2O_2

Potassium nitrosodisulfonate $ON(SO_3K)_2$
3,4-Dihydro- from 1,2,3,4-tetrahydroisoquinolines   CHNH → C:N
s. *30*, 642

Oxidations with *persulfate* $K_2S_2O_8$
Phenanthridones from biphenyl-2-carboxamides ○
s. *29*, 445

Bitosylsulfinodiimide $TsN:S:NTs$
Dehydrogenations with bitosylsulfinodiimide CHNH → C:N

322.   PhCH₂NHTs → PhCH=NTs

N-Benzyl-p-toluenesulfonamide heated 4 hrs. at 120° with bitosylsulfinodiimide → N-benzylidene-p-toluenesulfonamide. Y: 70%. F. e. s. G. Kresze and N. Schönberger, A. *1974*, 847.

Bromine *Br*
1,2,4-Triazol-3-ones from semicarbazones ○
s. *27*, 501

Iodine *I*
4-Cyanoquinolinium salts ←
from 4-cyano-1,4-dihydroquinolines s. *27*, 679

tert-Butyl hypochlorite/sodium hydrogen carbonate $Me_3C \cdot OCl/NaHCO_3$
1,1-Iminocyanides from α-aminonitriles NHCH(CN) → N:C(CN)
s. *29*, 645

tert-Butyl hypochlorite/triethylamine $Me_3C \cdot OCl/Et_3N$
Cyclic azomethines from cyclic sec. amines CHNH → C:N
s. *20*, 373 suppl. 28

Potassium permanganate $KMnO_4$
4(1H)-Quinazolones s. *13*, 383

Potassium hexacyanoferrate(III) $K_3[Fe(CN)_6]$
Quinone diimines from p-diamines ←

323.

A soln. of tetrachloro-p-phenylenediamine in benzene stirred 6 hrs. at room temp. with an aq. soln. of K-hexacyanoferrate(III) and *KOH* → tetrachloro-p-benzoquinone diimine. Y: 86%. M. Hedayatullah, J.-C. Raoult, and L. Denivelle, Bl. *1973*, 2702.

Cobalt oxide/oxygen ←
Autoxidation CH₂NH₂ → CN
Nitriles from prim. amines – Oxo compds. from alcohols s. *26*, 463

Palladium-carbon *Pd-C*
Pyrimidines from 1,2-dihydropyrimidines CHNH → C:N
s. *30*, 256

Palladium-carbon/oxygen $Pd-C/O_2$
Indolizine ring opening C
s. *29*, 446

Platinum oxide PtO_2
4-Hydroxypyridazinium betaines ←
s. *26*, 940

Via intermediates *v.i.*
3,4-Dihydro- from 1,2,3,4-tetrahydro- CHNH → C:N
via N-nitroso-1,2,3,4-tetrahydro-isoquinolines
s. *30*, 334

Oxygen ↑   NC↑O

Without additional reagents   *w.a.r.*

Cyclic acylimines   C(OH)NH → C : N
Isoindolenones s. *29*, 447

Alkoxysulfonylisocyanates   OSO$_2$NHCOOR → OSO$_2$N : C : O
from alkoxysulfonylcarbamic acid esters – Aryloxysulfonylisocyanates s. *29*, 448

Ring closure with alcohols   O
7,8,9,10-Tetrahydropyrido[2,1-c][1,2,4]benzothiadiazine 5,5-dioxides s. *29*, 449

N-Condensed 4-hydroxy-2-pyridone ring
s. *27*, 503

Pyridine ring from 2,4-dieneoximes
s. *22*, 507

1-Subst. 5-pyrazolones
from β-ketocarboxylic acid hydrazides with rearrangement s. *26*, 464

Imidazole-4-carboxylic acid esters
by Claisen-type rearrangement s. *27*, 502

Purines from pyrimidines
s. *30*, 208

Pyrimido[1,2-a][1,8]naphthyrid-10-ones
and 6-hydroxyanthyridines s. *26*, 465

Pyrimido[4,5-d]pyrimidine-2,4,5,7-tetrones
s. *26*, 678

1,2-Benzisothiazole 1,1-dioxides
from o-acylsulfonic acid amides s. *30*, 330

4H-s-Triazolo[4,3-a][1,4]benzodiazepines

324.

Labeled compds. Acetic acid-1-^{14}C 2-(7-chloro-5-phenyl-3H-1,4-benzodiazepin-2-yl)hydrazide heated 15 min. at 205°/ca. 90 mm in an oil bath → 8-chloro-1-methyl-6-phenyl-4H-s-triazolo[4,3-a][1,4]benzodiazepine-1-^{14}C. Y: 85.7%. F. e. s. R. S. P. Hsi, J. Labelled Compds. *9*, 435 (1973); non-labeled compds., also f. methods s. K. Meguro et al., Chem. Pharm. Bull. *21*, 2382, 2375 (1973).

Tetrahydropyrazolo[3,4-e][1,2,4]triazine-3,7-diones
via 3-ethoxycarbonylamino-4-hydrazono-5-pyrazolones s. *29*, 450

4,5-Dihydro-1,3,2,4,6-thia(VI)thiatriazin-5-one 3,3-dioxides
from N-carbalkoxysulfamoylsulfone diimines s. *27*, 504

Irradiation (s. a. under (EtO)$_3$P)   ⋕

11H-Indeno[1,2-b]quinolines

325.

A soln. of *trans*-2-(2-aminobenzylidene)-1-indanone in anhydrous methanol irradiated 4⅓ hrs. with the Pyrex-filtered light of a 450 w. Hg-arc lamp → 11Hindeno[1,2-b]quinoline. Y: 97.5%. F. e. s. D. C. Lankin and H. Zimmer, J. Heterocyclic Chem. *10*, 1035 (1973).

Electrolysis ↯
Cyclic hydroxamic acids
   from nitrocarboxylic acids s. *28*, 432

Sodium hydride/dimethyl sulfoxide NaH/Me_2SO
Aziridines $C(NH\cdot)\cdot C(OTs) \rightarrow \overline{N}$
   from 2-tosyloxy- or 2-halogen-amines – N-Arylaziridines s. *28*, 433

Sodium hydroxide (s. a. under $CHCl_3$) $NaOH$
o-Carbalkoxy-N-heterocyclics O
   from N-carbalkoxyaminocyanohydrin tosylates via aminocyanohydrin tosylates
   – 2- and 4-Quinuclidinecarboxylic acid esters s. *28*, 434
Rheadan ring skeleton
   via dihydrobenzazepine ring and spirolactones – Lactone ring expansion via
   ketolactones and hydroxycarboxylic acids s. *27*, 505
4-Hydroxy-1,2,4-triazoles
   from N^3-hydroxy-N^1-acylamidrazones s. *27*, 506

Potassium hydroxide s. a. under $SOCl_2$ KOH

Potassium hydroxide/ethanol
Pyrrolo[1,2-b][1,2,4]triazine ring
   from α-diazoketones and α,β-ethylenenitriles via 1,2-diaminopyrrole ring
   s. *26*, 466

Sodium/alcohol (s. a. under $Na[BH_4]$) $NaOR$
Pyrido[1,2-a]-1,3,5-triazines
s. *28*, 435

Potassium/tert-butanol $K/Me_3C\cdot OH$
N-Ring closure with tosylates
s. *27*, 507
   Azetidines s. *16*, 926

Sodium nitrite/sulfuric acid $NaNO_2/H_2SO_4$
N-Condensed N-heterocyclics
s. *28*, 436

Sodium salt s. under $PhCH_2Cl$ Na^+

Ammonia (s. a. under Ni) NH_3
Dicarboxylic acid imides COOR → CO
   from dicarboxylic acid amide esters s. *28*, 622   CONH   CO N–

4-Pyrimidinone ring from o-aminocarboxylic acid amides O
   via o-acylaminocarboxylic acid amides – Pyrimido[4,5-b][1,8]naphthyrid-4(3H)-
   ones s. *28*, 437

1,3-Propanediamine s. under CO $H_2N[CH_2]_3NH_2$

Triethylamine (s. a. under Me_3SiCl, $TiCl_4$ and $POCl_3$) Et_3N
Aziridines from 2-(tosyloxy)amines $C(NH\cdot)\cdot C(OTs) \rightarrow \overline{N}$
s. *28*, 433

2-Pyridone ring via 5-hydroxy-5,6-dihydro-2-pyridone ring O
   4,5-Dihydrofuro[3,2-c]pyrid-4-ones s. *27*, 508

2-Pyridone ←
Lactams from aminocarboxylic acid esters   NH–   N–
s. *29*, 451   COOR → CO

Pyridine (s. a. under CH_3COOH, Cyanuric chloride, $TiCl_4$, and $POCl_3$) C_5H_5N
Azetidinium ring from 3-aminoalcohols
s. 27, 509

Cuprous acetate s. under CO CH_3COOCu

Cupric sulfate s. under RN : C : NR $CuSO_4$

Zinc/acetic acid Zn/CH_3COOH
Adenines from 7-acylaminofurazano[3,4-d]pyrimidines ←
s. 27, 521

Sodium tetrahydridoborate $Na[BH_4]$
Lactams from aminocarboxylic acid esters $\overset{NH-}{\underset{COOR}{}} \rightarrow \overset{N-}{\underset{CO}{}}$
s. 28, 30

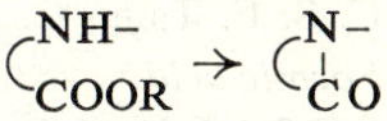

Sodium tetrahydridoborate-sodium/alcohol $Na[BH_4]/NaOR$
Reductive ring closure of nitro compds. ○
Phenazines s. 27, 510

Alumina Al_2O_3
Piperidines from aminoalcohols
s. 26, 467

1,2,3-Benzotriazin-4(3H)-ones
from o-carbalkoxytriazenes s. 29, 452

Basic alumina
2-Azanaphthoquinones $\overset{NH}{\underset{C(OAc)}{}} \rightarrow \overset{N}{\underset{C}{}}$
by elimination of acetic acid s. 26, 468

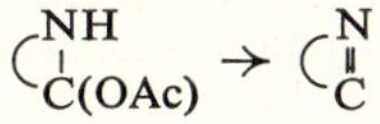

Molecular sieve ←
3H-1,3-Benzodiazepines ○
s. 29, 453

Carbon monoxide/1,3-propanediamine/-
cuprous acetate $CO/H_2N[CH_2]_3NH_2/CH_3COOCu$
Oximes from nitro compds. $CHNO_2 \rightarrow C:NOH$

326. ⬡—NO₂ → ⬡=NOH

Cuprous acetate dissolved at 85° in stirred degassed 1,3-propanediamine under 1 atm. CO, nitrocyclohexane added slowly, and stirred rapidly 2–4 hrs. at 85° under CO → cyclohexanone oxime. Y: 81%. F. e. s. J. F. Knifton, J. Org. Chem. *38*, 3296 (1973).

Dicyclohexylcarbodiimide $RN:C:NR$
Nitriles from aldoximes $CH:NOH \rightarrow CN$
s. 28, 440

Quinazolino[2,3-c][1,4]benzothiazin-12-ones ○
by double ring closure s. 26, 469

Dicyclohexylcarbodiimide/cupric sulfate $RN:C:NR/CuSO_4$
Nitriles from aldoximes $CH:NOH \rightarrow CN$
s. 30, 277

N,N'-Carbonyldiimidazole ←
s. 28, 440

Trichloroacetonitrile CCl_3CN
s. 28, 440

Acetic anhydride Ac_2O

Quinoxaline ring from N-nitrosanilines
s. *29*, 454

α-Methylenelactams by rearrangement ←

327.

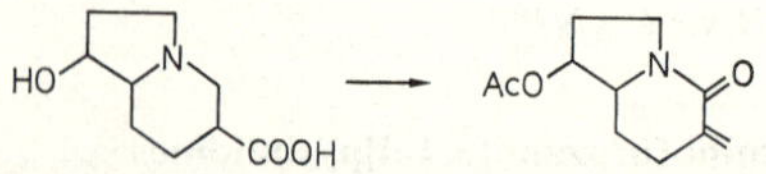

A soln. of 1-hydroxyoctahydroindolizine-6-carboxylic acid in acetic anhydride refluxed 2.5 hrs. → 1-acetoxy-6-methylene-5-oxooctahydroindolizine. Y: 84%. C. S. F. Tang, C. J. Morrow, and H. Rapoport, Am. Soc. *97*, 159 (1975).

Formic acid $HCOOH$

o-(1,2,4-Triazol-4-yl)ketones ←
from 3-acylamino-4-hydroxy-3,4-dihydroquinazolines s. *29*, 455

Acetic acid CH_3COOH

Dicarboxylic acid imides
from dicarboxylic acid amide esters s. *28*, 622

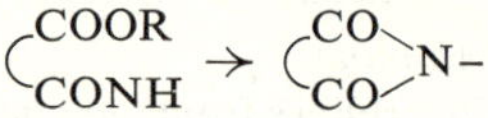

2,3-Dihydroxypyrazines from 2-aminoacetals
s. *27*, 511

1,3-Dihydro-2H-1,4-benzodiazepin-2-ones
from 3-amino-4-hydroxy-3,4-dihydrocarbostyrils s. *28*, 437

Acetic acid/pyridine CH_3COOH/C_5H_5N

Cyclic azomethines from aminoketones
1,3-Dihydrothieno[2,3-e][1,4]diazepin-2-ones s. *28*, 438

Trifluoroacetic acid CF_3COOH

Dihydropyrrolo[2,1-c][1,3,4H]oxazines and pyrroles ←
from 5,6-dihydro-4H-1,3-oxazines s. *28*, 439

Thiourea $(NH_2)_2CS$

Azomethines from oxaziridine ring

328.

An ethanolic soln. of *cis*-2,2-dimethyl-3-phenyl-6-oxa-1-azabicyclo[3.1.0]hexane and thiourea allowed to stand 8 hrs. at room temp. → 5,5-dimethyl-4-phenyl-1-pyrroline. Y: 82%. F. e. and reagents s. D. St. C. Black and K. G. Watson, Australian J. Chem. *26*, 2159 (1973).

Benzyl chloride/sodium salt $PhCH_2Cl/Na^+$

Oximes from nitro compds. $CHNO_2 \to C:NOH$
s. *28*, 644

Chloroform-sodium hydroxide/-
benzyltriethylammonium chloride $CHCl_3\text{-}NaOH/PhCH_2\overset{+}{N}Et_3Cl^-$

Prepn. of nitriles in alkaline medium $CONH_2 \to CN$
with dichloromethylene – Nitriles from aldoximes, carboxylic acid amides and thioamides – Cyanamides from ureas – Phase-transfer catalyzed reactions, review, s. *29*, 456; review s. a. Chem. Technol. *5*, 210 (1975)

Carbon tetrachloride s. under Ph_3P CCl_4

Cyanuric chloride/pyridine ←

Nitriles from aldoximes $CH:NOH \to CN$
s. *28*, 440

Hexamethylcyclotrisilazane ←
Nitriles from carboxylic acid amides and aldoximes CN
s. *26*, 501

Trimethylchlorosilane/triethylamine Me_3SiCl/Et_3N
Isocyanatocarboxylic acid silyl esters NHCOOPh → N:C:O
  from aminocarboxylic acids via aminocarboxylic acid silyl ester hydrochlorides and carbophenoxyaminocarboxylic acid silyl esters s. *27*, 512

Phenyltrichlorosilane $PhSiCl_3$
Isocyanates NHCOOSi≤ → N:C:O
  from carbamic acid silyl esters s. *29*, 438

Titanium tetrachloride/triethylamine $TiCl_4/Et_3N$
Nitriles from carboxylic acid amides $CONH_2$ → CN
s. *27*, 514

Titanium tetrachloride/pyridine $TiCl_4/C_5H_5N$
Nitriles from aldoximes CH:NOH → CN
  also via aldoxime carbonates s. *27*, 513

Triphenylphosphine/carbon tetrachloride Ph_3P/CCl_4
Nitriles from carboxylic acid amides $CONH_2$ → CN
s. *27*, 514

Triphenylphosphine/carbon tetrachloride/
triethylamine $Ph_3P/CCl_4/Et_3N$
Isonitriles from formamides NHCHO → N:C
  Nitriles from aldoximes s. *26*, 470

Tris(diethylamino)phosphine $(Et_2N)_3P$
Nitriles from carboxylic acid amides $CONH_2$ → CN
s. *28*, 442

Hexamethylphosphoramide $(Me_2N)_3PO$
Nitriles from aldoximes CH:NOH → CN
s. *28*, 440 suppl. 29
Nitriles from carboxylic acid amides $CONH_2$ → CN
s. *26*, 501

Phosphorus pentoxide s. under H_3PO_4 P_2O_5

Triphenyl diethoxyphosphorane $Ph_3P(OEt)_2$
Heterocycles by dehydrative ring closure O
s. *28*, 216

Triethyl phosphite/irradiation $(EtO)_3P/\text{hv}$
Dinitriles from furazans C
s. *27*, 515

Polyphosphoric acid $H(PO_3H)_xOH$
Pyrazole ring O
  from β-diketone monohydrazones s. *28*, 441

Phosphoric acid/phosphorus pentoxide H_3PO_4/P_2O_5
Purines from pyrimidines
  8-Hydroxyguanines s. *26*, 356

Phosphonitrile chloride ←
Nitriles from carboxylic acid amides $CONH_2$ → CN
s. *28*, 442

Triphenylphosphine dibromide Ph_3PBr_2
Aziridines from 2-aminoalcohols 
  Ring closure with Walden inversion s. 26, 471; azetidines from
  3-aminoalcohols cf. J. P. Freeman and P. J. Mondron, Synthesis 1974, 894

Phosphorus oxide chloride/triethylamine $POCl_3/Et_3N$
Nitriles from carboxylic acid amides $CONH_2 \rightarrow CN$
s. 28, 442

Phosphorus oxide chloride/pyridine $POCl_3/C_5H_5N$
Carbodiimides from amidoximes $C(:NOH)NH \cdot \rightarrow \cdot N:C:N \cdot$
s. 27, 516

Phosphorus pentachloride PCl_5
3,4-Dihydro-1,2-benzothiazine-3(2H)-one 1,1-dioxides
s. 28, 443

Thionyl chloride $SOCl_2$
Ring contraction of N-heterocyclics
  via cyclic quaternary ammonium salts s. 26, 298

Thionyl chloride/potassium hydroxide $SOCl_2/KOH$
Cyanoaldehydes
  from cyclic 2-oximinoalcohols – Beckmann fragmentation s. 27, 517

p-Toluenesulfonic acid $TsOH$
Cyclic tert. amines from aminoalcohols
s. 26, 472

Enaminolactams from ketocarboxylic acid amides
s. 27, 518

3,4-Dihydro-1,2-benzothiazin-3(2H)-one 1,1-dioxides
s. 28, 443

N-Condensed 1,3,5-triazine ring

329.

Double ring closure. Some p-toluenesulfonic acid added to a soln. of the startg. m. in refluxing diphenyl ether, and refluxing continued 15 min. → 2-methyl-9-carbethoxy-1,3,4,7-tetraazacycl[3.3.3.]azine. Y: 45%. O. Ceder and K. Rosén, Acta Chem. Scand. 27, 2421 (1973).

Sulfuric acid H_2SO_4
Phthalimidines
s. 26, 473

Imidazoles from acetals
s. 26, 304

Iodine/hydrogen chloride I/HCl
Isoquinolines from aminoketals
s. 28, 444

Perchlorates ClO_4^-
Cyclic azomethinium salts from *sec*-aminoketones
s. 28, 445

Perchloric acid HClO$_4$
Phthalimidines
s. 26, 473

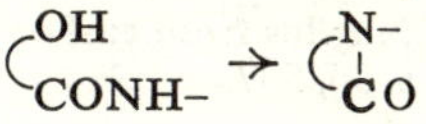

$\underset{\text{CONH-}}{\text{OH}} \rightarrow \underset{\text{CO}}{\text{N-}}$

Ethyldiisopropylamine hydrobromide i-Pr$_2$EtN,HBr
Ring closure with N-oxides O
Benzo[1,2-b]indolizines s. 29, 457

Benzyltriethylammonium chloride s. under CHCl$_3$ PhCH$_2$N$^+$ Et$_3$ Cl$^-$

Hydrogen chloride HCl
Lactams from aimnocarboxylic acids
s. 30, 331

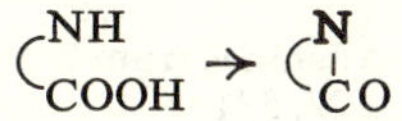

$\underset{\text{COOH}}{\text{NH}} \rightarrow \underset{\text{CO}}{\text{N}}$

Ring closure with enolethers O
Pyrazol-3-ones s. 26, 474
3-Hydroxyhydantoins
from α-isocyanato- via α-N-hydroxyureido-carboxylic acid esters s. 27, 519
1,2-Benzisothiazole 1,1-dioxides
from o-acylsulfonic acid amides
Amines from nitro compds.

330.

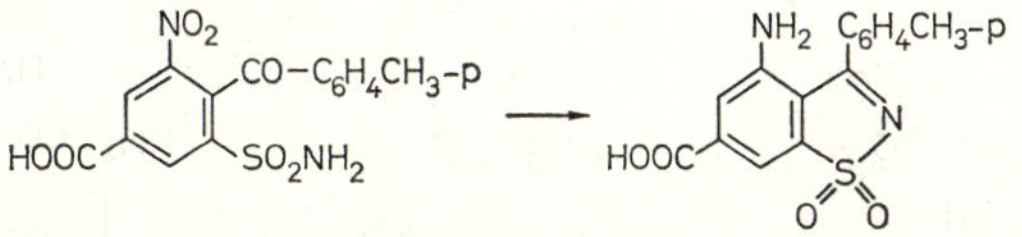

Solid Na-dithionite added during ca. 0.5 hr. to a stirred soln. of 4-p-toluoyl-3-nitro-5-sulfamylbenzoic acid in water-pyridine, stirring continued ca. 1 hr., evaporated in vacuo, the residue dissolved in water, acidified with excess concd. HCl, and heated 0.5 hr. on a steam bath → 4-amino-3-p-tolyl-1,2-benzisothiazole-6-carboxylic acid 1,1-dioxide. Y: 48%. – Ring closure occurs easily after the reduction. F. e., **also partial reduction of dinitro compds. to nitramines**, s. O. B. T. and C. K. Nielsen, and P. W. Feit, J. Med. Chem. 16, 1170 (1973); ring closure by melting without HCl s. ibid. 18, 41 (1975).

Heterocyclics from carbohydrates
Ring closure with sulfonic acid esters – Dihydro-1,4-thiazine ring s. 28, 446

1,3-Dihydro-4H-furo[3,4-d][1,3]benzodiazepin-3-ones
s. 26, 475

3,5-Dioxo-6,7-benzopyrrolo[1,2-c]thiazoles
by double ring closure s. 27, 520

Quinoline-3,4-dione 1-oxides ←
via 1,3-dihydroxy-4(1H)-quinolones s. 28, 447

Hydrogen chloride/acetic acid HCl/CH$_3$COOH
3,4,5,6-Tetrahydro-2,6-methano-2H-1,3-benzoxazocines O
s. 29, 458

Iron/hydrogen chloride Fe/HCl
Reductive ring closure
3,4-Dihydro-5H-1,4-benzodiazepin-5-ones – 1,4-Benzodiazepines, reviews
s. 26, 476

Nickel/ammonia Ni/NH$_3$
2,3-Dihydro-1H-1,4-benzodiazepines
s. 29, 749

Chlorotris(triphenylphosphine)rhodium(I) $RhCl(Ph_3P)_3$
Nitriles from carboxylic acid amides $CONH_2 \rightarrow CN$
s. *26*, 507

Palladium-carbon/cyclohexene   ←
Reductive pyrrolo[1,2-c]pyrimidine ring opening   C
s. *28*, 430

Palladium-carbon/acetic acid $Pd\text{-}C/CH_3COOH$
Syntheses of N-heterocycles   ←
via reductive furazan ring opening s. *27*, 521

Adenines from 7-acylaminofurazano[3,4-d]pyrimidines   ←
s. *27*, 521

Platinum-carbon/hydrogen chloride $Pt\text{-}C/HCl$
Oxindoles and 1-hydroxyoxindoles   O
from nitrocarboxylic acids

331.

A soln. of o-nitrophenylglycine hydrochloride in aq. methanol containing HCl hydrogenated 0.5 hr. with 5%-Pt-on-carbon under ca. 3 atm. H_2 → 3-amino-1-hydroxy-2-indolinone hydrochloride. Y: 62%.

in water hydrogenated 12 hrs. with Pd-black under ca. 3 atm. H_2 → o-aminophenylglycine (Y: 61%) in aq. methanol treated with concd. HCl, and, after 1 hr., evaporated to dryness in vacuo → 3-amino-2-indolinone hydrochloride (Y: 97%).

A. L. Davis, D. R. Smith, and T. J. McCord, J. Med. Chem. *16*, 1043 (1973).

Via intermediates   v.i.
Acylimines $C(OH)NHAc \rightarrow C:NAc$
from 1,1-acylaminoalcohols via 1,1-acylaminochlorides

332. $ClCH_2-CCl_2-CH(OH)-NHCOC_3H_7-n$ → $ClCH_2-CCl_2-CHCl-NHCOC_3H_7-n$

$ClCH_2-CCl_2-CH=NCOC_3H_7-n$

N-(1-Hydroxy-2,2,3-trichloropropyl)-*n*-butyramide refluxed 3 hrs. with $SOCl_2$ in benzene, evaporated in vacuo, the crude intermediate dissolved in benzene, stirred and treated dropwise at 30° under anhydrous conditions during 15 min. with a soln. of triethylamine in benzene → N-*n*-butyryl-2,2,3-trichloropropionaldimine. Y: 69%. F. e. s. H. Zinner, W.-E. Siems, and G. Erfurt, J. pr. *316*, 63 (1974).

Oxindoles from nitrocarboxylic acids   O
s. *30*, 331

Nitrogen ↑   NC↑N

Without additional reagents   w.a.r.

2-Imidazolidones from β-aminoaminimides   ○
s. 28, 448

**Benzimidazoline-2-thiones and amines
from o-aminothioureas**

333.

N-o-Aminophenyl-N'-phenylthiourea heated 15 min. above its m. p. at 155°
→ benzimidazoline-2-thione (Y: 95%) and aniline (Y: almost 100%). F. e. s.
A.-M. M. E. Omar, S. A. S. El-Dine, and A. A. B. Hazzaa, Pharmazie 28, 682
(1973).

N-Condensed N-acylpyrazole ring
s. 29, 459

Dicarboxylic acid imides
from dicarboxylic acid amides s. 28, 449

1-Cyanoimidazoles by thermolysis   C
s. 28, 450

Azetines from cyclopropyl azides   ⊙
Regiospecific ring expansion s. 27, 524

Δ³-2-Imidazolones from 3-azido-2-azetidinones   Ọ:
s. 28, 451

o-Cyano-N-hydroxy-N-heterocyclics
from o-azido-N-oxides s. 28, 452

Ring contraction of 3-oxo-1,2-N,N-heterocyclics
s. 27, 522

Cyclic cyanamides   ←
from N-condensed tetrazole ring s. 27, 523

Irradiation (s. a. under CF_3COOH and O_2)   ⚡

Azomethines from azides CRN_3 → $C:NR$
with rearrangement s. 29, 460

3-Indazolones   Ọ:
from 3-amino-1,2,3-benzotriazin-4(3H)-ones s. 29, 461

2H-Benzo[c]-1,2-thiazete 1,1-dioxides
and dibenzo[c,e]-1,2-thiazine 1,1-dioxides from 2H-benzo[e]-1,2,3,4-thiatriazine
1,1-dioxides – Stable β-sultam s. 29, 462

Sodium hydride   NaH
Azomethines from N-nitrosamines $CN \cdot N(NO)$ → $C:N$
s. 28, 453

Sodium hydroxide   NaOH
3,4-Dihydro- from 1,2,3,4-tetrahydro- $CH \cdot N \cdot NO$ → $C:N$
via 1,2,3,4-tetrahydro-N-nitroso-isoquinolines

334.

A mixture of 1,2,3,4-tetrahydroisoquinoline and 4 moles isoamyl nitrite allowed
to stand 1 hr. at room temp. → crude N-nitroso-1,2,3,4-tetrahydroisoquinoline

(Y: 90%) in ethanol refluxed 10 hrs. with aq. 20%-NaOH → 3,4-dihydroisoquinoline (Y: 70%). F. e. and limitations s. K. Sakane et al., Bull. Chem. Soc. Japan 47, 1297 (1974).

4-Pyridones   O
s. 29, 468

5-Cyanouracils
s. 28, 454

Triethylenediamine

Δ^1-**Azirines from α,β-ethyleneazides** 
s. 18, 331 suppl. 29

Copper/quinoline Cu/C_9H_7N
4-Alkylidene-5-imidazolones
from α,β-ethylene-α-(1-tetrazolyl)carboxylic acid amides s. 29, 463

Mercuric oxide/potassium hydroxide HgO/KOH
Nitriles from hydrazones $CH:NNH_2 \rightarrow CN$
s. 27, 525

Sodium tetrahydridoborate $Na[BH_4]$
O,N-Heterocyclics from azidohydroxynitriles   O
Reductive double ring closure s. 29, 464

Lithium tetrahydroaluminate $Li[AlH_4]$
Aziridine ring from 2-azidotosylates $C(OTs)C(N_3) \rightarrow$ 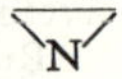
s. 30, 278

Levulinic acid/hydrogen chloride $CH_3COCH_2CH_2COOH/HCl$
Ring closure with oximes   O
5H-Pyrrolo[2,1-c][1,4]benzodiazepines s. 26, 478

Formic acid $HCOOH$
Ring closure with diazo comdps.
2,3-Dihydro-4H-1,2-benzothiazin-4-one 1,1-dioxides s. 29, 465

Trifluoroacetic acid/irradiation $CF_3COOH/h\nu$
9H-Pyrimido[4,5-b]indoles
Tetrazole-azidoazomethine tautomerism – 1,2,3-Triazole ring-chain tautomerism – 1,2,3-Triazoline-2-diazoamine tautomerism s. 29, 466

Ring closure with *cyanogen bromide* $BrCN$
Cyclimmonium salts from di-*tert*-amines   O
s. 26, 479

Lead tetraacetate $(CH_3COO)_4Pb$
Cyanimines from guanylhydrazones $C:N\cdot NHC(:NH)NH_2 \rightarrow C:N\cdot CN$
s. 28, 455

Reactions of N-amino-N-heterocyclics with lead tetraacetate   C
2 Nitrile molecules from 4-amino-1,2,4-triazoles – Deamination of N-amino-N-heterocyclics s. 26, 480

Triethyl phosphite $(EtO)_3P$
Benzoxazoles from o-acoxynitriles   O
s. 28, 456

Polyphosphoric acid $H(PO_3H)_xOH$
Ring closure with hydrazones
Imidazo[1,5-a]quinoxal-4-ones s. 29, 467

Oxygen/irradiation	$O_2/✦$
Nitriles from hydrazones s. 26, 481	$CH:N\cdot N< \twoheadrightarrow CN$
Sulfuric acid/acetic acid	H_2SO_4/CH_3COOH
Ring closure with azo compds. 2-Deoxyalloxazines s. 26, 482	O
Hydrochlorides	←
Lactams from aminocarboxylic acid amides s. 26, 483	$\underset{CO}{\overset{N-}{C}}$
Hydrogen chloride (s. a. under $CH_3COCH_2CH_2COOH$)	HCl
4-Pyridones s. 29, 468	O
Nickel	Ni
Reductive ring closure of nitrooximes	

335.

A suspension of the startg. m. in ethanol hydrogenated 48 hrs. at 49° with Raney-Ni, evaporated to dryness in vacuo, triturated with ethanol under N_2, mixed with N,N-dimethylacetamide, and heated 3 min. at 85° under N_2 → ethyl 1,2-dihydro-3-[(p-methoxycarbonyl-N-methylanilino)methyl]-5(6H)-oxopyrido[3,4-b]pyrazine-7-carbamate. Y: 60%. R. D. Elliott, C. Temple, Jr., and J. A. Montgomery, J. Med. Chem. 17, 553 (1974).

Palladium-carbon	Pd-C
Heteroring closure with dimethylformamide dimethyl acetal Indoles via trans-β-amino-2-nitrostyrenes – Enamines from amide acetals s. 29, 469	

Halogen ↑   NC⇑Hal

Without additional reagents	w.a.r.
Indoles from halogenamines s. 27, 526	O
Polyhalogenopyrazolo[1,5-a]pyridinium salts by double ring closure s. 26, 484	
2-Pyrimido[1,2-a]pyrimid-2-ones s. 26, 485	
3-Azido-1,2,4-triazoles from 1-(N-tetrazol-5-ylhydrazono)-1-halides s. 28, 457	←
Irradiation	✦
N-Condensed cyclimmonium salts by cyclo-N-quaternization s. 27, 527	O

Functionalization of non-activated groups ←
   by rearrangement of bromonitrones s. 27, 528

Sodium hydride   *NaH*
N-Tosylazetidines O
   Optically active compds. s. 29, 470
2-Ketolactams from 2-bromolactamols Ọ
   Ring contraction s. 29, 471

Sodium hydride/dimethyl sulfoxide   *NaH/Me$_2$SO*
Aziridines from 2-halogenamines   CHalC(NH) → △N
   s. 28, 433

Sodium hydroxide   *NaOH*
N-Cyanoaziridine ring from 2-halogenocyanamides △N
   s. 26, 521
2,3-Dihydro-1H-pyrrolo[2,1-c][1,2,4]benzothiadiazine 5,5-dioxides O
   by double ring closure s. 28, 457

Potassium hydroxide   *KOH*
Aziridines from ethylene derivs.   ClC(NCOOR) → △N
   via 1,2-iodourethans s. 28, 458

Sodium acetate   *CH$_3$COONa*
4-Hydroxypyrazoles O
   via bromination of α-diketone monohydrazones s. 26, 486

Ammonia   *NH$_3$*
1-Bromo-N-fluoroimines   CBr$_2$NF$_2$ → CBr : NF
   from 1,1-dibromo-N,N-difluoroamines s. 26, 487

Triethylamine   *Et$_3$N*
α,β-Ethylene-N-fluoroimines   CHNF$_2$ → C : NF
   from 3-chloro-N,N-difluoroamines s. 28, 459
Acylimines from 1,1-acylaminohalides   CHalNHAc → C : NAc
   s. 30, 332

1,8-Bis(dimethylamino)naphthalene ←
Si,S,N-Heterocyclics O

336.   [structure: N–S–Si(Me)Me / NH–BrCH$_2$ → N–S–Si(Me)Me ring-closed imidazo[2,1-b]thiazole]

A suspension of the startg. m. in tetrahydrofuran containing 1,8-bis(dimethylamino)naphthalene (s. Synth. Meth. 29, 428), a strong, non-nucleophilic base, stirred 24 hrs. at room temp. → 2-dimethylsila-3H-imidazo[2,1-b]thiazole. Y: 80–85%. F. e. s. H. Alper and M. S. Wolin, J. Org. Chem. 40, 437 (1975).

Pyridine   *C$_5$N$_5$N*
1,3-Diazetidine-2,4-diones N◇N
   from 2 isocyanate molecules via allophanoyl chlorides s. 26, 488

Silver oxide   *Ag$_2$O*
N-Heterocyclics from N-chloramines O
   Transannular ring closure s. 28, 460

Sodium tetrahydridoborate   *Na[BH$_4$]*
N-Condensed heterocyclics
   by double ring closure s. 27, 946

Lithium tetrahydridoaluminate $Li[AlH_4]$
Aziridines
from 2-chloracylamines s. 26, 60

Water H_2O
Reactions with polyhalogenodiazabutadienes   O
Dihydroimidazotriazoles via imidazolinylhydrazonohalides s. 27, 529

Titanium tetrachloride/titanium trichloride $TiCl_4/TiCl_3$
N-Heterocyclics from N-halogenamines
Regio- and stereo-specific intramolecular radical addition of amines –
Redoxidative ring closure s. 27, 530

Triiron dodecacarbonyl/methanol $Fe_3(CO)_{12}/MeOH$
Nitriles from hydroximinohalides   C(Hal) : NOH → CN

337.

$$p\text{-}ClC_6H_4C(Cl)=NOH \xrightarrow[]{Fe_3(CO)_{12} + HOCH_3 \;\;[HFe_3(CO)_{11}^-]} p\text{-}ClC_6H_4C\equiv N$$

A mixture of triiron dodecacarbonyl and methanol stirred and refluxed 8 hrs. in benzene, cooled, 4-chlorobenzohydroxamoyl chloride added, and refluxed 17–22 hrs. → 4-chlorobenzonitrile. Y: 86%. F. e., also with $Fe(CO)_5$, s. N. A. Genco, R. A. Partis, and H. Alper, J. Org. Chem. *38*, 4365 (1973).

Sulfur ↑   NC⇑S

Without additional reagents   w.a.r.
Dihydropyrindines from β-arylsulfonic acid azides   O

338.

Flash vacuum pyrolysis. β-(p-Tolyl)ethylsulfonyl azide flash-pyrolyzed in vacuo at 650° → product. Y: 70.2%. F. e. s. R. A. Abramovitch and W. D. Holcomb, Am. Soc. *97*, 676 (1975); flash vacuum pyrolysis s. a. 665.

Ar. nitriles   C CN
by 1,2,5-thiadiazole 1,1-dioxide ring opening – Dinitriles s. 29, 472

Isocyanates from hydroxamic acids   C
via 1,3,2,4-dioxathiazole 2-oxides – 1,3,2,4-Dioxathiazole 2-oxides from nitrile oxides s. 26, 489

Monosulfonylcarbodiimides and thionylimines
from carbodiimides and N-thionylsulfonamides via 3-imino-1,2,4-thiadiazetidine 1-oxides s. 29, 473

N-Condensed 1,2,4-triazole ring   ←
by elimination of sulfur dioxide – s-Triazolo[3,4-a]pyridines s. 26, 490

Sodium hydroxide (s. a. under $CHCl_3$)   NaOH
Elimination of sulfur dioxide $SO_2N< $ → $N<$
from sulfonic acid amides s. 29, 213

Sodium salt Na^+
1,7-Electrocyclic ring closure   O
2,3-Benzodiazepines s. 27, 531; 1,2-benzodiazepine ring s. Soc. Perkin I *1975*, 102

Triethylamine (s. a. under Cyanuric chloride) Et_3N
Elimination of sulfur dioxide $SO_2N< \twoheadrightarrow N<$
from sulfonic acid amides

339.

$$\text{Cl}_2\text{C}_5\text{N(Cl)}_2\text{-SO}_2\text{NHCH}_2\text{CH}_2\text{OH} \rightarrow \text{(pyridyl)-NHCH}_2\text{CH}_2\text{OH}$$

A soln. of triethylamine in ethanol added at room temp. to a stirred soln. of tetrachloro-4-pyridyl-(N-2-hydroxyethyl)sulfonamide in the same solvent, and kept at room temp. overnight $\twoheadrightarrow$ N-(tetrachloro-4-pyridyl)ethanolamine. Y: 85%. B. Iddon, H. Suschitzky, and A. W. Thompson, Soc. Perkin I *1973*, 2971.

4-Hydroxypyrazoles from sulfonium salts ○
s. *29*, 474

Zinc/acetic acid Zn/CH_3COOH
N-Condensed tricyclic guanidines
s. *29*, 475

Quinones ⇐
Carbodiimides from thioureas $NHCSNH \twoheadrightarrow N:C:N$
s. *27*, 532

Triethylammonium salt ⇐
Reactions of N-carbalkoxysulfamic acid esters ⇐
Urethans – Stereospecific β-elimination s. *26*, 491

Phenylpropiolamidines ⇐
Isothiocyanates from dithiocarbamic acids $NHCSSH \twoheadrightarrow N:C:S$

340.

$$Ph-C\equiv C-C(NPh)(NHPh)$$

$$p-CH_3C_6H_4NHCSSH, Et_3N \longrightarrow p-CH_3C_6H_4N=C=S$$

$$\left[+ Ph-C(SH)=CH-C(NPh)(NHPh) \right]$$

A mixture of triethylammonium N-p-tolyldithiocarbamate and N,N'-diphenylphenylpropiolamidine in ethanol stirred 3.5 hrs. at room temp. $\twoheadrightarrow$ p-tolyl isothiocyanate. Y: 85.9%. F. e. s. H. Fujita, R. Endo, and K. Murayama, Chem. Lett. *1973*, 883.

Methyl iodide MeI
2-Aminobenzimidazoles from o-aminothioureas ○
s. *29*, 476; s. a. Pharmazie *30*, 83 (1975); cf. ibid. *30*, 85

Chloroform-sodium hydroxide/
benzyltriethylammonium chloride $CHCl_3-NaOH/PhCH_2N^+Et_3\ Cl^-$
Nitriles from carboxylic acid thioamides $CSNH_2 \twoheadrightarrow CN$
s. *29*, 456

Cyanuric chloride/triethylamine ⇐
Carbodiimides from thioureas $NHCSNH \twoheadrightarrow N:C:N$
s. *27*, 532

Triphenylphosphine Ph_3P
Heterocumulenes from heterocyclics C
s. *29*, 477

Tris(dimethylamino)phosphine $\qquad$ $(Me_2N)_3P$
Dicarboxylic acid imides from sulfenimides
s. 26, 492

$\begin{matrix} CO \\ CO \end{matrix}\!\!\rangle N\!-\!SR \rightarrow \begin{matrix} CO \\ CO \end{matrix}\!\!\rangle N\!-\!R$

Chlorine/sodium hydrogen carbonate $\qquad$ $Cl/NaHCO_3$
1,1-Sulfonylisocyanates $\qquad$ $SO_2 \cdot C \cdot NHCOSR \rightarrow SO_2 \cdot C \cdot N : C : O$
from 1-sulfonyl-1-thiolcarbamic acid esters s. 26, 493

Remaining Elements ↑ $\qquad$ NC⇑Rem

Without additional reagents $\qquad$ *w.a.r.*
3H-Pyrrolenines $\qquad$ O
s. 27, 533
Irradiation $\qquad$ ⚡
Quinolines from phosphorimidates
s. 28, 461
Sodium/alcohol $\qquad$ *NaOR*
Nitriles from α-nitrosophosphonium salts $\qquad$ $CH(NO)\overset{+}{P}\!\!\leq\; \rightarrow\; CN$
s. 28, 319
Acetyl chloride $\qquad$ *AcCl*
Nitriles from N-silylcarboxylic acid amides $\qquad$ $CONHSi\!\leq\; \rightarrow\; CN$
s. 28, 462
Benzoyl chloride $\qquad$ *BzCl*
s. 28, 462

Carbon ↑ $\qquad$ NC⇑C

Without additional reagents $\qquad$ *w.a.r.*
Nitro compds. from acyl nitrates $\qquad$ $COONO_2 \rightarrow NO_2$
Decarboxylative nitration s. 26, 494
Acylamines $\qquad$ $AcN \cdot COOR \rightarrow AcN \cdot R$
from acylurethans and from quaternary ammonium salts s. 28, 463
Nitriles from ketenimines $\qquad$ ←
Synthesis and reactions of ketenimines, review s. 26, 495; cf. L. A. Singer and K. W. Lee, Commun. *1974*, 962
Nitriles by 1,2,5-thiadiazole 1,1-dioxide ring opening $\qquad$ C CN
s. 29, 472
Acylisocyanates $\qquad$ $C(OR):NCOCOCl \rightarrow CON:C:O$
from N-(chloroglyoxyloyl)iminoesters s. 26, 458
2 Cyanoketene molecules from 2,5-diazido-p-quinones $\qquad$ C
s. 26, 496
Isothiocyanates from 1,4,2-oxathiazoles
s. 28, 527
Irradiation (s. a. under Ph_2CO) $\qquad$ ⚡
Cyclic N-oxides from nitro compds. $\qquad$ O
s. 28, 464
Photodecarboxylation of heterocyclics $\qquad$ Ö:
s. 27, 534

Benzimidazole from quinoxaline 1-oxide ring

341.

A methanolic soln. of 300 mg. methyl quinoxalin-2-ylcarbamate 1-oxide irradiated 5 hrs. with Pyrex-filtered u. v. light → 220 mg. methyl benzimidazol-2-ylcarbamate. F. e. s. R. A. Burrell, J. M. Cox, and E. G. Savins, Soc. Perkin I *1973*, 2707.

Sodium hydride	*NaH*
Heterocyclics from enephosphonium salts s. *30*, 178	○
Sodium hydroxide	*NaOH*
Naphthostyrils from naphthalimides s. *29*, 478	○
2,3-Dihydro-1H-1,4-benzodiazepines from 2,3-piperazinediones s. *29*, 479	←
Potassium hydroxide s. *29*, 479	*KOH*
Potassium hydroxide/alcohol	*KOH*
Heterocyclics from α-amino-β,γ-ethylenenitriles Benzimidazoles and benzoxazoles s. *27*, 535	○
Sodium/alcohol	*NaOR*
Nitriles from oxo compds. via 1-(sulfonyl)eneformylamines – Synthesis with addition of 1 C-atom s. *29*, 480	CO → CHCN
Sodium acetate s. under *Ac₂O*	*CH₃COONa*
Dicyclohexylethylamine	*(C₆H₁₁)₂NEt*
Azomethines from 5-oxazolidones s. *27*, 536	C
Calcium oxide	*CaO*
1-Subst. cyclic azomethines from N-acyllactams s. *28*, 465	$\underset{CO}{\overset{N \cdot COR}{C}} \rightarrow \underset{C \cdot R}{\overset{N}{C}}$
Mercuric acetate	*(CH₃COO)₂Hg*
Lactam rearrangement with aromatization s. *28*, 748	←
Water	*H₂O*
2-Quinazolones from indole-1,2-dicarboximide ozonides s. *29*, 481	C
Benzophenone/irradiation	*Ph₂CO/ʜʜ*
Nitriles from azomethines s. *28*, 466; cf. Chem. Lett. *1974*, 1403	CH : NR → CN
Ethyl chloroformate	*ClCOOEt*
3-Subst. 1,2,4-triazoles from N-1,2,4-triazol-4-ylamidines – 1,3,4-Oxadiazoles s. *26*, 497	←
Acetic anhydride/sodium acetate	*Ac₂O/CH₃COONa*
4-Hydroxyquinoline-2-carboxylic acids by aromatization and N-deacetylation s. *26*, 498	←

Diisopropyl peroxydicarbonate | $i\text{-}Pr \cdot OCO \cdot OO \cdot COO \cdot Pr\text{-}i$
Nitriles from aldimines | $CH:NCMe_3 \rightarrow CN$
s. 27, 537

Acetic acid | CH_3COOH
Ethylene derivs. from pyrimidines | C
s. 26, 499

Trifluoroacetic acid | CF_3COOH
Δ^2-Thiazolines from penicillin ring skeleton | O
s. 26, 500

Peroxyacetic acid s. H_2O_2/CH_3COOH | CH_3COO_2H

Hexamethylcyclotrisilazane | ←
Nitriles from N-subst. and unsubst. carboxylic acid amides | $CON< \rightarrow CN$
s. 26, 501

Triphenyl phosphite | $(PhO)_3P$
Nitriles from furazans and furoxans | C
s. 28, 467

Polyphosphoric acid | $H(PO_3H)_xOH$
Lactams from dicarboxylic acid amide esters
via dicarboxylic acid imides by decarbonylation s. 26, 502

Ring closure with removal of N-protective *tert*-butyl groups | O
Saccharins s. 27, 538

Oxygen/potassium hydroxide | O_2/KOH
Carboxylic acid amides from 1-acylsemicarbazides $CON \cdot N \cdot CON< \rightarrow CON<$
s. 26, 503

Hydrogen peroxide/acetic acid | H_2O_2/CH_3COOH
Oxindoles from 3-hydroxyquinolines | Ö:
s. 26, 504

Thionyl chloride | $SOCl_2$
Nitriles from α-isonitrosoketones | $C(:NOH)COR \rightarrow CN$

342.

A mixture of α-oximino-α-(1-ethyl-2-benzimidazolyl)acetophenone, $SOCl_2$, and methylene chloride refluxed 15 min. → 2-cyano-1-ethylbenzimidazole. Y: 73%. – This is the final step of sequence for the conversion of methyl to cyano groups (cf. 237). F. e. s. J. D. Albright and R. G. Shepherd, J. Heterocyclic Chem. *10*, 899 (1973).

1,2,3,4-Tetrahydroquinoxalines | Ö:
from 4-hydroxy-2,3,4,5-tetrahydro-1H-1,4-benzodiazepines s. 27, 539

Sodium hypochlorite | $NaOCl$
Degradation of carbohydrates to peracetyl glycitols | $CONH_2 \rightarrow NH_2$
via Hofmann degradation of carboxylic acid amides s. 27, 540

Pyridine hydrochloride | C_5H_5N,HCl
Benzoxazoles from o-alkoxyoximes | O
Beckmann rearrangement s. 26, 505

Hydrogen chloride   HCl

Isoindolines
via partial formation of hexamethylenetetraminium salts s. 27, 541

Ring closures with iminoesters
Pyrrolo[2,3-d]pyridazines s. 29, 482

1H-Imidazo[1,2-b]-s-triazoles
s. 27, 542

1H-Indazoles from 1H-1,2,4-benzotriazepines
s. 29, 483

4-Alkylidene-2-quinazolones ←
from 1-carbamyl-1,2-dihydroquinolines – 4-Methylene-2-quinazolones
s. 26, 506

Thieno[3,2-b]indoles ←
from 3-iminothiophene-2-spiro-3-indolenines

343.

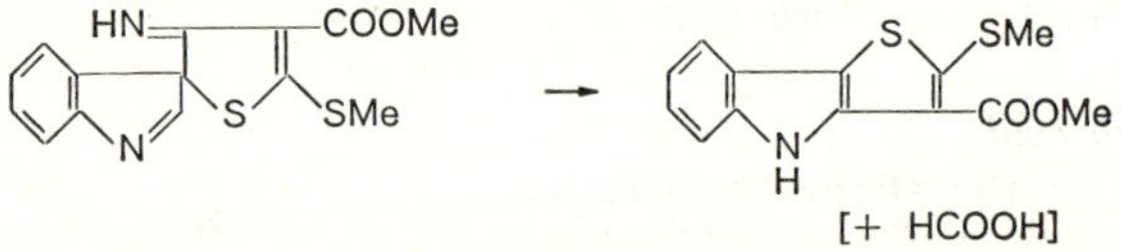

3-Imino-4-methoxycarbonyl-5-(methylthio)thiophene-2-spiro-3(3H)-indole hydrochloride added to a soln. of 10% HCl in methanol, and refluxed 5 hrs. on a steam bath → 2-methylthio-3-(methoxycarbonyl)thieno[3,2-b]indole. Y: 60–80%. F. startg. m. s. G. Kobayashi et al., Chem. Pharm. Bull. *21*, 2344 (1973).

Hydrogen iodide   HI

Δ¹-Pyrrolines from 2-alkylidenepyrrolidines ←
s. 26, 851

Chlorotris(triphenylphosphine)rhodium(I) $RhCl(Ph_3P)_3$
Nitriles   CON< → CN
from N-subst. and unsubst. carboxylic acid amides s. 26, 507

Via intermediates   v.i.
Pteridines from purinium salts
s. 29, 484

Formation of Hal—Hal Bond

Uptake   ⇓

Addition to Halogen   HalHal⇓Hal

Sulfuryl chloride SO_2Cl_2
Iodosoacetates   I → I(OAc)₂
from iodides via iododichlorides s. 27, 543

Formation of Hal—S Bond

Exchange ↓↑

Hydrogen ↑   HalS ↕ H

Without additional reagents   *w.a.r.*
Iminosulfonic acid chlorides   S(O)NH· → S(O)(:N·)Cl
   from sulfinic acid amides s. 28, 468
Sulfuryl chloride SO_2Cl_2
α-Carbamyl-α,α-dichloromethylsulfenyl chlorides CH_2SH → $C(SCl)Cl_2$
   from α-mercaptoacetamides s. 28, 492

Halogen ↑   HalS ↕ Hal

Without additional reagents   *w.a.r.*
Sulfenyl-bromides from -chlorides   SCl → SBr
s. 28, 516

Formation of Hal—Rem Bond

Uptake ⇓

Addition to Remaining Elements   HalRem ⇓ Rem

Phosphorus pentachloride PCl_5
Protection of amino groups   ←
   as platinum complexes s. 30, 361

Exchange ↓↑

Oxygen ↑   HalRem ↕ O

Thionyl halide $SOHal_2$
C,As-Dihalogenarsines from arsinic acids   ←
5,10-Dihalogeno-5,10-dihydroacridarsines

344. Startg. m. treated at room temp. with excess $SOCl_2$ → 5,10-dichloro-5,10-dihydroacridarsine. Y: 63%. – Also Br-analog with $SOBr_2$ s. R. J. M. Weustink, C. Jongsma, and F. Bickelhaupt, Tetrah. Let. *1975*, 199.

Remaining Elements ↑   HalRem ↓↑ Rem

Without additional reagents   w.a.r.
N-Sulfonyliminoseleninyl chlorides   SeSi≤ → Se(Cl) : NSO$_2$R
    from silyl selenides and N,N-dichlorosulfonamides – N-Sulfonyliminosulfinyl chlorides from silyl sulfides s. *28*, 469

Formation of Hal—C Bond

Uptake   ⇊

Addition to Oxygen and Nitrogen   HalC⇊ON

Irradiation   ⫯⫯⫯
5-Subst. o-aminoketones from 2,1-benzisoxazoles   C
    Photochemical opening of heterocycles s. *28*, 470

Addition to Oxygen and Carbon   HalC⇊OC

Without additional reagents   w.a.r.
Simultaneous formation   ▽$_O$ → CHalC(OH)
    of 1,2-halogenhydrins and 2-ethylenealcohols from oxido compds. s. *28*, 471

1,2-Chlorohydrin chlorosulfites   ▽$_O$ → C(OSOCl)CCl
    from oxido compds. s. *29*, 486

Zinc chloride   ZnCl$_2$
β-Chlorethyl ethers from chlorides   Cl → OCH$_2$CH$_2$Cl
s. *26*, 508

Boron chloride   BCl$_3$
2-γ-Chlorophenols from chromans   C

345.

with ether cleavage. 6-Acetyl-7,8-dimethoxy-2,2-dimethylchroman allowed to react 1 hr. at 0° with excess BCl$_3$ in methylene chloride → product. Y: 92%. F. e. and limitations s. R. J. Molyneux, Chem. Commun. *1974*, 318.

Aluminum chloride   AlCl$_3$
1,2-Acoxyhalides from oxido compds.   ▽$_O$ → C(OAc)CHal
s. *20*, 175 suppl. *29*

Cation exchanger   ←
Acoxyhalides   CO → C(Hal)(OAc)
    from cyclic ethers and carboxylic acids s. *29*, 487

Acetic acid   CH$_3$COOH
Isochromene ring opening   C
s. *27*, 544

Acetyl halide   *AcHal*
Cyclonucleoside ring opening
    s. *29*, 488

Iron carbonyl   *Fe(CO)$_5$*
Cleavage of ethers with carboxylic acid halides
    Acoxyhalides from cyclic ethers s. *26*, 509

Via intermediates   *v.i.*
Chlorocarboxylic acid esters
    from lactones s. *29*, 516 $\overset{O}{\underset{CO}{C}} \rightarrow \overset{Cl}{\underset{COOR}{C}}$

Addition to Nitrogen and Carbon   HalC↓NC

Without additional reagents   *w.a.r.*
2-Halogenamines from aziridines   ⟨N⟩ → N(R)C·CHal

346.

    O$_2$N—C$_6$H$_2$(NO$_2$)$_2$—Cl + PrN⟨ ⟶ O$_2$N—C$_6$H$_2$(NO$_2$)$_2$—N(Pr)—CH$_2$CH$_2$Cl

A soln. of 1-propylaziridine in benzene added dropwise at 70° during 1 hr. to a soln. of 1-chloro-2,4,6-trinitrobenzene in the same solvent, and stirring continued 2 hrs. at 70° → N-propyl-N-(2-chlorethyl)-2,4,6-trinitroaniline. Y: ca. 100%. F. e. s. H. J. Nestler and H. Bestian, A. *1974*, 460.

1-Chloroenamidine hydrochlorides   ←
    from 2 nitrile molecules – 4,6-Dichloropyrimidines s. *29*, 489
1,1'-Dichloroazo compds. from ketazines   C:N·N:C → C(Cl)N:N·C(Cl)
    s. *29*, 832
α-Chloroeneureas from nitriles   CHC≡N → C:C(Cl)N:C:O
    via α-chloroenisocyanates s. *28*, 472
Reaction of phosgene with azomethines   CH·C:C·C:N → C:C·C:C·N·COCl
    1,3-Dienecarbamyl chlorides from α,β-ethyleneazomethines via 3-chlor-
    1-ethylenecarbamyl chlorides s. *27*, 545
Reactions with N-(chlorocarbonyl)isocyanate   C:N → CCl·N·CO·N:C:O
    α-Chloro-N-carbonylureas from azomethines – Heterocyclics s. *28*, 473
N-Chlorosulfonylformiminochlorides   SO$_2$CN → SO$_2$C(:NCl)Cl
    from sulfonylcyanides s. *26*, 510
Ring closures with carbodiimides and carboxylic acid chlorides   O
    s. *26*, 511
4-Chlorimidazoles
    s. *27*, 546
4-Halogeno-2-quinazolones from o-cyanoisocyanates
    also 4-halogenoquinazoline-2-thiones from o-cyanoisothiocyanates s. *26*, 512
2-Phosphaquinazolinium salts
    from o-cyanophosphine imines s. *26*, 277

Magnesium bromide   *MgBr$_2$*
α-Bromocarboxylic acid amides   △=O → CBrCON<
    from aziridinones s. *28*, 474

Boron chloride   *BCl$_3$*
Allophanoyl chlorides   2 RN:C:O → RNHCONRCOCl
    from 2 isocyanate molecules s. *26*, 488

Dimethylformamide s. under SO_2Cl_2   *Me_2NCHO*

Sulfuryl chloride/dimethylformamide   *SO_2Cl_2/Me_2NCHO*

N-Thioiminochlorides   CN ⇢ CCl : N · SR
from nitriles and disulfides

347. $CCl_3-C\equiv N$ + MeS—SMe ⟶ $\begin{array}{c}Cl\\ \diagdown\\ CCl_3 \diagup\end{array} C=N-SMe$

Equivalent amounts of trichloroacetonitrile and dimethyl disulfide, and a little dimethylformamide as catalyst mixed at room temp., sulfuryl chloride added dropwise, and allowed to stand overnight at room temp. ⇢ product. Y: 82%. F. e. and method s. J. Geevers and W. P. Trompen, Tetrah. Let. *1974*, 1691.

Tetraethylammonium chloride   *$Et_4N^+\ Cl^-$*

Addition of sulfenylchlorides to active cyano groups
N-(Alkylthio)sulfonylformiminochlorides

348. $Ph-SO_2-C\equiv N$ + ClSMe ⟶ $Ph-SO_2-\underset{\underset{Cl}{|}}{C}=N-SMe$

Benzenesulfonyl cyanide allowed to react at 0° with methanesulfenyl chloride in tetrahydrofuran in the presence of a catalytic amount of tetraethylammonium chloride ⇢ product. Y: 93%. F. e. and reactions s. H. Kristinsson, Tetrah. Let. *1973*, 4489.

Tetrabutylammonium chloride   *$Bu_4N^+\ Cl^-$*

Thiobis(sulfonylformimidoyl chlorides)   2 CN ⇢ CCl : N · N : CCl
from sulfonylcyanides

349. $2\ MeSO_2C\equiv N$ + SCl_2 ⟶ $MeSO_2-\underset{\underset{Cl}{|}}{C}=N-S-N=\underset{\underset{Cl}{|}}{C}-SO_2Me$

Methanesulfonyl cyanide allowed to react at 0° with sulfur dichloride in methylene chloride containing tetrabutylammonium chloride as catalyst ⇢ thiobis(methanesulfonylformimidoyl chloride). Y: 91%. F. e. s. M. S. A. Vrijland, Tetrah. Let. *1974*, 837.

Addition to Sulfur and Carbon   HalC⇊SC

Without additional reagents   *w.a.r.*
2-Halogenosulfenic acid alkoxymethyl esters   C
 from 1,1-alkoxyhalides and thiirane 1-oxides s. *28*, 475

Addition to Carbon   HalC⇊CC

Without additional reagents   *w.a.r.*
1,2-Alkoxyhalides   C : C ⇢ C(OR)CHal
 from ethylene derivs. and hypochlorites s. *26*, 513
1,2-Chloroiodides from ethylene derivatives   C : C ⇢ CClCl
 s. *26*, 514
2,3-Ethylene-1,4-dibromides from 1,3-dienes   C : C·C : C ⇢ CBr·C : C·CBr
 s. *29*, 490; from isoprene s. M. Ohsugi, I. Ichimoto, and H. Ueda, Agri. Biol. Chem. (Tokyo) *38*, 1925 (1974)
α,β-Dibrom- from α,β-ethylene-aldehydes   C : C·CHO ⇢ CBrCBrCHO
 s. *27*, 220

α-**Iodoacetals from enolethers** C : C(OR) → ClC(OR)$_2$
s. 27, 547

1,2-Chloroperoxides C : C → CHalC(OOR)
from ethylene derivs. and O-chloroperoxides

350.   CF$_3$—OO—Cl + CFH=CHCl   →   CF$_3$—OO—CFH—CHCl$_2$

1-Chloro-2-fluoroethylene added portionwise at –111° during several hrs. to chloroperoxytrifluoromethane until an excess is present, allowed to warm slowly to –78°, kept ca. 33 hrs. at this temp., and if necessary allowed to warm until the yellow color disappears or the temp. reaches 0° → product. Y: 75%. F. e. and limitations s. N. Walker and D. D. Des Marteau, Am. Soc. 97, 13 (1975); **1,2-halogenoperoxides via peroxymercuration** (s. Synth. Meth. 26, 637) cf. A. J. Bloodworth and I. M. Griffin, Soc. Perkin I 1975, 695.

1,2-Halogenamines C : C → CHalC(NH$_2$)
from ethylene derivs. and N,N-dihalogenophosphoramidates s. 27, 548

α-**Halogenazomethinium salts** C : C·N< → CHalC : N$^+$<
from enamines s. 28, 116

Addition of N-chlorocyanamides ←
to carbon-carbon double bonds s. 26, 515

Perhalogenoperchloric acid esters C : C → CHalC · OClO$_3$
s. 29, 491

Heterotwistanoids O
s. 29, 596

Bicyclic iodotetrazolylisocyclics
from cyclic dienes and nitriles
Transannular ring closure

351.

$$\begin{array}{c} CH_3 \\ | \\ CN \\ | \\ IN_3 \end{array} + \bigcirc \longrightarrow$$

cis,cis-1,5-Cyclononadiene in acetonitrile added slowly at 0° during 0.5 hr. to iodine azide in acetonitrile (prepn. in situ s. Synth. Meth. 29, 317), and stirred overnight at room temp. → product. Y: 80%. F. e. s. S. N. Moorthy, D. Devaprabhakara, and K. G. Das, Tetrah. Let. 1975, 257.

Stereospecific benzvalene ring opening C
Deuterio compds. s. 27, 549

Dibromides from dienes ←
with cyclopropane ring closure s. 26, 516

1,3-Dibromides from cyclopropane ring C
with subsequent migration of bromide – 3,5-Cyclosteroid ring opening
s. 29, 492

Spirocyclopropane ring opening
s. 28, 476

Carboxylic acid esters from 4,7-dioxaspiro[2,4]heptanes
s. 29, 493

Spiroisocyclics from cyclopropane ring ←
s. 28, 477

Irradiation ₩
2-Halogenacylamines C : C → CHalC(NAc)
from ethylene derivs. and N-halogenocarboxylic acid amides s. 28, 478

Pyridine (s. a. under $AgNO_3$) $\qquad C_5H_5N$
Reactions with poly(hydrogen fluoride)-pyridine ▽ → CH·C·CF
Hydrofluorination of ethylene derivs., acetylene derivs., and cyclopropanes – Halogenofluorination of ethylene and acetylene derivs. – Nitrofluorination s. *29*, 494

Dihalides from ethylene derivs. ←
with rearrangement – *exo-endo*-Rearrangement s. *26*, 517

Silver nitrite $\qquad AgNO_2$
1,2-Nitriodides from ethylene derivs. C : C → ClC(NO_2)
s. *27*, 550

Silver nitrate/pyridine $\qquad AgNO_3/C_5H_5N$
Halogenofluorination C : C → CFCHal
s. *29*, 494

Silver nitrate/pyridine/iodine monochloride $\qquad AgNO_3/C_5H_5N/ICl$
Addition of iodonium nitrate to carbon-carbon double bonds ←
1,2-Iodhydrin nitrates s. *26*, 518

Cupric chloride $\qquad CuCl_2$
1,2-Chloroiodides from ethylene derivs. C : C → CClCI
s. *26*, 514

Calcium carbonate/N-bromosuccinimide $\qquad CaCO_3 | \begin{matrix} CH_2CO \\ CH_2CO \end{matrix} \rangle NBr$
***trans*-1,2-Bromohydrins from ethylene derivatives** C : C → C(OH)CBr
Optically active compds. s. *26*, 144

9-Borabicyclo[3.3.1]nonane ←
Sec. bromides C : CHR → CHCHBrR
by *anti*-Markownikoff hydrobromination s. *26*, 519; s. a. C. P. Pinazzi, D. Derouet, and J. C. Brosse, Bl. *1975*, 201

Thallous carboxylates $\qquad RCOOTl$
1,2-Acoxyiodides from ethylene derivs. C : C → ClC(OAc)
Modified Prévost reaction s. *13*, 209 suppl. 28

Dimethylformamide (s. a. under N-Chlorosuccinimide) $\qquad Me_2NCHO$
1,2-Dichlorides from ethylene derivs. C : C → CClCCl
s. *27*, 551

Nitrobenzene $\qquad PhNO_2$
1,1-Chlorofluorides from α,β-ethylenechlorides C : CCl → CHCClF
s. *29*, 496

2,4,4,6-Tetrabromo-2,5-cyclohexadienone ←
Addition to carbon-carbon double bonds C : C → CBrC(OR)
1,2-Alkoxybromides from ethylene derivs. s. *28*, 479

N-Chlorosuccinimide/dimethylformamide $\qquad \begin{matrix} CH_2CO \\ CH_2CO \end{matrix} \rangle NCl / Me_2NCHO$
1,2-Formoxychlorides from ethylene derivs. C : C → CClC(OCHO)
s. *28*, 480

N-Bromosuccinimide (s. a. under $CaCO_3$) $\qquad \begin{matrix} CH_2CO \\ CH_2CO \end{matrix} \rangle NBr$
Cyclic bromohydrins from dienes ○
s. *26*, 520

N-Cyanoaziridine ring ▽N
from ethylene derivs. via 2-halogenocyanamides s. *26*, 521

Iodobenzene dibromide $PhIBr_2$
Preferential formation of *trans*-1,2-dibromides   C : C → CBrCBr
from ethylene derivs. s. *27*, 552

Trichloramine NCl_3
1,2-Dichlorides from ethylene derivs.   C : C → CClCCl
s. *26*, 522

Antimony chlorides   ←
1,2-Chlorohalides from ethylene derivs.   C : C → CClCHal
s. *23*, 544 suppl. *29*

Thionyl bromide $SOBr_2$
1,2-Dibromides from ethylene derivs.   C : C → CBrCBr
s. *30*, 364

Sulfuryl chloride SO_2Cl_2
2-(ω-Halogenalkoxy)halides   ←
from cyclic ethers and ethylene derivs. s. *27*, 553

Chromous chloride $CrCl_2$
1,2-Halogenurethans from ethylene derivs.   C : C → CHalC(N · COOR)
s. *27*, 554

Chromyl chloride CrO_2Cl_2
α-Chloroketones and ketones   C : CH → CCl · CO
from ethylene derivs. s. *28*, 481

Iodine monochloride s. under $AgNO_3$ ICl

Calcium hypochlorite $Ca(OCl)_2$
Ethylenehalogenhydrins from allenes   C : C : C → C : CHalC(OH)
s. *27*, 555

Via intermediates   v.i.
α,β-Ethylenehalides   C ≡ C → CH : CHal
from acetylene derivs. and boronic acid esters via α,β-ethyleneboronic acids – Geospecific replacement s. *29*, 497

Rearrangement   ↻

Hydrogen/Carbon Type   HalC↻HC

Aluminum halide $AlHal_3$
1,3- from 1,2-Dihalides   CHCHalCHal → CHalCHCHal
s. *29*, 498

Hydrogen bromide/acetic acid HBr/CH_3COOH
Migration of bromine   ←
s. *29*, 492

Nitrogen/Carbon Type   HalC↻NC

Without additional reagents   w.a.r.
Chlorocarbodiimides   C
from 1-aziridinecarboximidoyl chlorides s. *27*, 556

Triethylamine Et_3N
Cyclic chlorourethans ←
    from cyclic chloroformoxyimines s. 27, 557
Hydrogen chloride HCl
Chlorocarbodiimides C
    from 1-aziridinecarboximidoyl chlorides s. 27, 556

Halogen/Remaining Elements Type   HalC∩HalRem

Without additional reagents w.a.r.
Rearrangement of (polyhalogeno)silanes ←
s. 26, 523

Carbon/Carbon Type   HalC∩CC

Irradiation/olefin ⚡/C : C
2-Halogenisocyanates C
    from N-halogeno-2-azetidinones s. 28, 482
Irradiation/acetic acid ⚡/CH_3COOH
o-α-Chloro-N-heterocyclics O
    from ethylene-N-chloramines – 2-α-Chloropyrrolidines s. 26, 524; with Ag_2O
    cf. R. Tadayoni et al., Tetrah. Let. *1975*, 735

Exchange   ↓↑

Hydrogen ↑   HalC↕H

Without additional reagents w.a.r.
Trichloromethyl from methyl thioethers $SCH_3 \rightarrow SCCl_3$
s. 29, 30
1,1-Chlorazo compds. from hydrazones C:N·NH· → CCl·N:N·
s. 26, 558
Thiocarbamyl halides from thioformamides N·C(:S)H → N·C(:S)Hal
s. 26, 525
Irradiation (s. a. under $CBrCl_3$ and N-Bromosuccinimide) ⚡
Bromination H → Br
s. 26, 486
Irradiation/fluoroxytrifluoromethane ⚡/CF_3OF
Photofluorination H → F
    with fluoroxytrifluoromethane s. 27, 559
Irradiation/iodobenzene dichloride ⚡/$PhICl_2$
1-Monochlorination of ethers H → Cl
s. 26, 526
Irradiation/N-chlorodiisopropylamine/sulfuric acid ⚡/$i\text{-}Pr_2NCl/H_2SO_4$
Chlorination
    Directed (ω–1)-monochlorination s. 26, 527

Irradiation/thionyl chloride   *hν/SOCl₂*
α-**Chlorocarboxylic acid esters**   CHCOOH → CClCOOR
    from carboxylic acids s. *10*, 429 suppl. *29*

Sodium hydride   *NaH*
α-**Bromocarboxylic acid esters**   H → Br
    from halides via β-ketocarboxylic acid esters – Synthesis with addition of 2 C-atoms s. *29*, 499

Sodium hydroxide   *NaOH*
α,α-**Dibromo-α-sulfonylsulfonic** CH_2 → CBr_2
from α-sulfonylsulfonic acid amides
s. *30*, 355

Sodium methoxide   *NaOMe*
1,1,1-Ethylenenitroiodides   C : CH(NO)$_2$ → C : CI(NO$_2$)
    from 1,1-nitroethylene derivs. s. *26*, 528

Triphenylmethyllithium/
tetramethylethylenediamine-ethylene dibromide $Ph_3C \cdot Li/TMEDA\text{-}BrCH_2CH_2Br$
α-**Bromination of γ-lactones**   H → Br
s. *28*, 895

Lithium N-isopropylcyclohexylamide   ←
α-**Halogenocarboxylic from carboxylic acid esters**   CHCOOR → CHalCOOR
s. *27*, 560

Sodium fluoride s. under CF₃OF   *NaF*

Sodium salt   *Na⁺*
γ-**Halogeno-β-lactones**   O
    from β,γ-ethylenecarboxylic acids – Additions with intramolecular ring closure, review, s. *27*, 561; preferential formation cf. E. J. Corey and H. S. Sachdev, J. Org. Chem. *40*, 579 (1975)

Triethylamine   *Et₃N*
4-β-Chloro-2-oxazolidones   →
from 3-hydroxypyrrolidines

352.

1-Methyl-3-pyrrolidinol in chloroform added with ice-cooling below 10° to a soln. of phosgene in the same solvent, stirring in the cold continued 45 min. and during addition of triethylamine, then allowed to warm to room temp. → 5-(2-chloroethyl)-3-methyl-2-oxazolidinone. Y: 52%. F. e. s. M. L. Fielden et al., J. Med. Chem. *16*, 1124 (1973).

Pyridine (s. a. under PhICl₂ and N-Bromosuccinimide) C_5H_5N
2-Bromomethyltetrahydrofurans   O
    from 4-ethylenealcohols s. *29*, 495

Pyridine/thionyl chloride $C_5H_5N/SOCl_2$
Chlorination   H → Cl
s. *16*, 590 suppl. *28*

Cupric chloride/hydrogen chloride $CuCl_2/HCl$
α,β-**Ethylene-γ-halogenoxo compds.**   CH·C : C·CO → CHal·C : C·CO
    from α,β-ethyleneoxo compds. s. *29*, 500

Mercuric oxide	*HgO*
Bromination s. *29*, 501	H → Br
Aluminum chloride s. under S_2Cl_2	$AlCl_3$
Ethyl alcohol	*EtOH*
Reaction of aldoximes with chlorine Hydroximinochlorides – 1,1-Dichlorides s. *27*, 562	CH : NOH → C(: NOH)Cl
tert-Butyl alcohol s. under H_2SO_4	$Me_3C \cdot OH$
Azodiisobutyronitrile s. under N-Bromosuccinimide	$[Me_2C(CN)N:]_2$
Acetic acid	CH_3COOH
Fluorination under mild conditions s. *28*, 483	H → F
1,1-Bromohydrazones from hydrazones s. *29*, 502	CH : N·N< → C(Br) : N·N<
Propionic acid	C_2H_5COOH
Chlorination 4,6-Dichloro-4,6-dien-3-onesteroids s. *27*, 563	H → Cl
Peroxycarboxylic acids/titanium tetrachloride	$RCOO_2H/TiCl_4$
Ar. chlorination under mild conditions s. *26*, 529	ArH → ArCl
Peroxytrifluoroacetic acid s. under $TiCl_4$	CF_3COO_2H
Ethylene dibromide s. under $Ph_3C \cdot Li$	$BrCH_2CH_2Br$
Bromotrichloromethane/irradiation	$CBrCl_3$/⫶⫶⫶
α-Monobromination of boranes s. *28*, 484	B·CH → B·CBr
2,4,4,6-Tetrabromo-2,5-cyclohexadienone	←
Bromination s. *28*, 491	H → Br
Monobromination of ar. amines s. *28*, 485	ArH → ArBr
Fluoroxytrifluoromethane/sodium fluoride	CF_3OF/NaF
1,1,N-Trifluoroamines from azomethines s. *29*, 503	CH : N· → CF_2NF
Iodobenzene dichloride s. a. under Irradiation	$PhICl_2$
Iodobenzene dichloride/pyridine	$PhICl_2/C_5H_5N$
Monohalogenation of ar. amines Monochlorination – α′-Monobromination of α,β-ethyleneketones s. *28*, 485	ArH → ArHal
N-Chlorodiisopropylamine s. under Irradiation	$i\text{-}Pr_2NCl$
2-Bromo-2-cyano-N,N-dimethylacetamide	$Me_2NCOCHBrCN$
α-Monobromination of ketones	$COCH_2$ → COCHBr

353.

A soln. of 2-benzylidenecyclohexanone and 2-bromo-2-cyano-N,N-dimethylacetamide in benzene stirred and refluxed 2 hrs. → 2-bromo-6-benzylidenecyclohexanone. Y: 88%. F. e., also with 2-bromomalononitrile, s. M. Sekiya, K. Ito, and K. Suzuki, Tetrahedron *31*, 231 (1975).

2-Bromomalononitrile $\qquad$ $BrCH(CN)_2$
s. *30*, 353

N,N-Dichlorourea/hydrogen chloride $\qquad$ NH_2CONCl_2/HCl
Chlorination under mild conditions $\qquad$ H → Cl
s. *28*, 486

N-Halogenosuccinimide $\qquad$ $\begin{matrix} CH_2CO \\ | \\ CH_2CO \end{matrix} \rangle NHal$

2-Halogenenamines $\qquad$ CHal : C·N<
s. *28*, 487

N-Halogenosuccinimide/hydrogen halide $\qquad$ $\begin{matrix} CH_2CO \\ | \\ CH_2CO \end{matrix} \rangle NHal/HHal$

α-Halogenocarboxylic acid halides $\qquad$ CHCOHal → CHal'COHal
 from carboxylic acid halides − α-Bromocarboxylic acid chlorides s. *27*, 565;
 α-chlorination, and α-iodination with iodine s. Tetrah. Let. *1974*, 3235

N-Chlorosuccinimide $\qquad$ $\begin{matrix} CH_2CO \\ | \\ CH_2CO \end{matrix} \rangle NCl$

Acetylenechlorides $\qquad$ C ≡ CH → C ≡ CCl
s. *13*, 573 suppl. 29

Dichloromethyl ketones $\qquad$ $CHCH_3$ → $COCHCl_2$
 via preferential chlorination of azomethines s. *29*, 504

N-Bromosuccinimide/irradiation $\qquad$ $\begin{matrix} CH_2CO \\ | \\ CH_2CO \end{matrix} \rangle NBr/\hspace{-2mm}/\hspace{-2mm}/\hspace{-2mm}/$

Bromination with cyclopropane ring opening $\qquad$ C
 Cyclopropylcarbinyl-allylcarbinyl radical rearrangement s. *28*, 488

N-Bromosuccinimide/pyridine $\qquad$ $\begin{matrix} CH_2CO \\ | \\ CH_2CO \end{matrix} \rangle NBr/C_5H_5N$

α-Bromosulfoxides from sulfoxides $\qquad$ SOCH → SOCBr
s. *27*, 564

N-Bromosuccinimide/azodiisobutyronitrile $\qquad$ $\begin{matrix} CH_2CO \\ | \\ CH_2CO \end{matrix} \rangle NBr/[Me_2C(CN)N:]_2$

γ-Halogenenolesters from enolesters $\qquad$ CHC : C(OAc) → C(Hal)C : C(OAc)
s. *28*, 229

Trichloroisocyanuric acid/sulfuric acid $\qquad$ ←
Chlorination $\qquad$ H → Cl
s. *26*, 530

1-Chlorobenzotriazole $\qquad$ ←
Chlorination of carbazoles $\qquad$ ArH → ArCl
s. *28*, 489

α-Chlorosulfoxides from sulfoxides $\qquad$ SOCH → SOCCl
s. *27*, 564

3-Chloroindolenine from indole ring $\qquad$ ←
s. *28*, 490

Thiophosgene $\qquad$ $CSCl_2$
1,1-Chloroisothiocyanates from imines $\qquad$ C : NH → C(Cl)N : C : S
s. *30*, 242

Titanium tetrachloride s. a. under $RCOO_2H$ $\qquad$ $TiCl_4$

Titanium tetrachloride/peroxytrifluoroacetic acid $\qquad$ $TiCl_4/CF_3COO_2H$
Halogenation $\qquad$ ArH → ArHal
with titanium tetrachloride – Bromination with 2,4,4,6-tetrabromo-2,5-cyclohexadienone s. *28*, 491

Titanium tetrachloride/hydrogen peroxide $\qquad$ $TiCl_4/H_2O_2$
Chlorination $\qquad$ ArH → ArCl
s. *28*, 491

Phosphorus tribromide $\qquad$ PBr_3
Halogenation with skeletal rearrangement $\qquad$ H → Br
Bicyclo[3.3.1]nonanes with bridgehead functions s. *27*, 566

Hydrogen peroxide s. under $TiCl_4$ $\qquad$ H_2O_2

Thionyl chloride s. under ⇈ *and* C_5H_5N $\qquad$ $SOCl_2$

Chlorosulfonyl isocyanate $\qquad$ $O:C:N \cdot SO_2Cl$
Chlorosulfonyl isocyanate as chlorinating agent $\qquad$ H → Cl
s. *29*, 505

Sulfuryl chloride (s. a. under S_2Cl_2) $\qquad$ SO_2Cl_2
1,1-Chlorothioethers from thioethers $\qquad$ CHSR → C(Cl)SR
s. *30*, 687

α-Carbamyl-α,α-dichloromethylsulfenyl chlorides $\qquad$ CH_2SH → $C(Cl_2)SCl$
from α-mercaptoacetamides s. *28*, 492

Halogenosulfones from hydroxysulfoxides $\qquad$ ←
s. *27*, 567

Sulfuric acid (s. a. under Irradiation $\qquad$ H_2SO_4
and Trichloroisocyanuric acid)
Directed chlorination of carboxylic acids $\qquad$ H → Cl
ω-Chlorination s. *26*, 531

Sulfuric acid/tert-butyl alcohol $\qquad$ $H_2SO_4/Me_3C \cdot OH$
Preferential halogenation $\qquad$ H → Hal
s. *29*, 506

Chlorosulfonic acid (s. a. under Iodine) $\qquad$ $ClSO_3H$
Chlorination $\qquad$ H → Cl
s. *28*, 493

Sulfur monochloride/aluminum chloride/sulfuryl chloride $\qquad$ $S_2Cl_2/AlCl_3/SO_2Cl_2$

354.

Sulfuryl chloride added dropwise with stirring at 20° during 80 min. to 2-ethylphenol containing catalytic amounts of $AlCl_3$ and S_2Cl_2, and the product isolated after 5 hrs. → 4-chloro-2-ethylphenol. Y: 82.3%. E. Schrötter and W. Krüger, Pharmazie *29*, 269 (1974).

Bromine-graphite $\qquad$ Br-C
Monobromination $\qquad$ H → Br
s. *28*, 485

Iodine $\qquad$ I
Ar. perchlorination $\qquad$ ←
s. *27*, 568

Iodine/chlorosulfonic acid $\qquad$ $I/ClSO_3H$
Ar. polyhalides $\qquad$ ←
s. *30*, 376

Bromine chloride $\qquad$ $BrCl$
Bromination with bromine chloride $\qquad$ H → Br
s. *20*, 417 suppl. 29

tert-Butyl hypochlorite $\qquad$ $Me_3C \cdot OCl$
Chlorination $\qquad$ H → Cl
s. *29*, 984; of hydroxypyridines s. *29*, 507

Allylic chlorination $\qquad$ C : C·CH → C : C·CCl
s. *27*, 569

Preferential chlorination $\qquad$ H → Cl
    2-Amino-3-chloro-1,4-quinones s. *28*, 494

Sodium hypochlorite $\qquad$ $NaOCl$
α,α-Dihalogeno-α-sulfonylsulfonic $\qquad$ CH_2 → CCl_2
from α-sulfonylsulfonic acid amides

355.
$MeSO_2CH_2SO_2N\!\!\diagup\!\!O$ $\qquad$ $MeSO_2CH_2SO_2NMe_2$

↓ $\qquad\qquad\qquad\qquad\qquad\qquad$ ↓

$MeSO_2CCl_2SO_2N\!\!\diagup\!\!O$ $\qquad$ $MeSO_2CBr_2SO_2NMe_2$

Chlorination. 5%-Na-hypochlorite soln. added to a soln. of 1-(methylsulfonyl)methanesulfonyl morpholide in dioxane, and stirred 1 hr. at room temp. → 1,1-dichloro-1-(methylsulfonyl)methanesulfonyl morpholide. Y: 98%. | **Bromination.** N,N-Dimethyl-1-(methylsulfonyl)methanesulfonamide added to aq. NaOH, cooled to 5°, bromine added dropwise during 15 min. with stirring, which is continued for 15 min. → 1,1-dibromo-N,N-dimethyl-1-(methylsulfonyl)methanesulfonamide. Y: 90%.

F. e. s. C. T. Goralski and T. C. Klingler, J. Chem. Eng. Data *19*, 189 (1974).

Potassium hypobromite $\qquad$ $KOBr$

Halogenoquinolines $\qquad$ H → Hal
s. *26*, 532

Hydrochlorides $\qquad$ ←
Bromination of N-heterocyclics $\qquad$ H → Br
s. *29*, 508

Hydrogen halide s. under N-Bromosuccinimide $\qquad$ $HHal$
Hydrogen chloride s. under $CuCl_2$ $\qquad$ HCl
Hydrogen bromide $\qquad$ HBr
Bromination of 2,1,3-benzothiadiazoles $\qquad$ H → Br
s. *26*, 533

Oxygen ↑ $\qquad\qquad\qquad\qquad$ HalC ↕ O

Without additional reagents $\qquad$ *w.a.r.*
1,3-Halogenamines from 2-aminoalcohols $\qquad$ ←
    Rearrangement s. *26*, 534

1,2-Halogenoximes from 1,1-nitroethylene derivs. $\qquad$ C : C(NO$_2$) → CHalC : NOH
s. *27*, 570

1,2-Acoxyhalides from glycols $\qquad$ C(OH)C(OH) → CBrC(OAc)
    Regiospecific reaction s. *29*, 509

Heterocyclics from tetraacetylethylene ○
    2-α-Halogenofurans from 2-ene-1,4-diones s. 27, 571

Sodium hydroxide/chloroform $NaOH/CHCl_3$
Chlorides from hydroxy compds. OH → Cl
    from alcohols s. 26, 535

Methyllithium/lithium chloride/tosyl chloride $MeLi/LiCl/TsCl$
β,γ-Ethylenehalides from 2-ethylenealcohols C : C·C(OH) → C : C·CHal
s. 24, 840; s. a. Org. Synth. 54, 68 (1974); with triphenylphosphine – CCl_4 (s. Synth. Meth. 21, 607) cf. ibid. 54, 63

Lithium chloride s. under MeLi and $MeSO_2Cl$ $LiCl$

Potassium chloride/Aliquat 336 $KCl/(n-C_8H_{17})_3MeN^+Cl^-$
Chlorides from mesylates OSO_2R → Cl
    Phase transfer catalysis s. 30, 366

Sodium salt (s. a. under $(COCl)_2$) Na^+
Nucleophilic substitution at reactive C-atoms OSO_2CF_3 → Cl
    Trifluoromethanesulfonyloxyl as reactive leaving group – α-Subst. sulfones s. 28, 495

Pyridine s. a. under $COCl_2$, TsCl, and SO_2Cl_2 C_5H_5N

Pyridine/N-halogenosuccinimide $C_5H_5N \mid \begin{matrix} CH_2CO \\ | \\ CH_2CO \end{matrix} NHal$
Halides from carbazic acid esters $OCONHNH_2$ → Hal
    Replacement of hydroxyl by halogen s. 28, 496

Pyridine-poly(hydrogen fluoride) $C_5H_5N \cdot (HF)_x$
Fluorides from alcohols OH → F
s. 24, 582, suppl. 29

Cupric chloride $CuCl_2$
Replacement of hydroxyl by chlorine OH → Cl
s. 28, 980

Magnesium iodide MgI_2
Iodides from tosylates OTs → I
s. 17, 621 suppl. 29

Zinc chloride s. a. under $SOCl_2$ $ZnCl_2$

Zinc chloride/α,α-dichloromethyl ether $ZnCl_2/Cl_2CHOMe$
Acylochlorosugars ←
    from carbohydrate orthocarboxylic acid esters s. 27, 572

Molecular sieve ←
1,1-Halogenothioethers from sulfoxides SO·CH → S·CHal
s. 28, 497

Boron fluoride s. under MoF_6 BF_3

Formaldehyde CH_2O
o-Halogeno-N-heterocyclics from cyclic N-oxides ←
s. 26, 536

Dimethylformamide s. under Ph_3P Me_2NCHO

2,4-Dimethylphenyl cyanate ←
Replacement of hydroxyl by chlorine OH → Cl
    Carbohydrates s. 28, 498

N-Methyl-N,N'-dicyclohexylcarbodiimidium iodide ←
Iodides from alcohols ROH → RI
  Inversion of configuration s. *27*, 573

Acetic acid CH_3COOH
Iodonium salts from iodosohydroxy tosylates ←
s. *27*, 574

Acetic acid/phosphorus tribromide CH_3COOH/PBr_3
Anionotropic rearrangement ←
  2-Bromo-1,3-diones from 3-acetylenealcohols s. *26*, 537

Acetic acid/phosphorus oxide chloride $CH_3COOH/POCl_3$
2,3-Dihydrofuran ring opening C
  with replacement of oxygen by chlorine s. *26*, 538

Dimethyl sulfide/N-halogenosuccinimide $Me_2S / \genfrac{}{}{0pt}{}{CH_2CO}{CH_2CO}\!\!>\!\!NHal$

Preferential replacement of allylic and benzylic hydroxyl OH → Hal
  by halogen via alkoxysulfonium salts under mild conditions s. *28*, 499

Organic halides RHal
Halides from phosphorous acid derivs. $OP(NR_2)_2$ → Hal

356.

from phosphorodiamidites. A mixture of 1,4 : 3,6-dianhydro-D-mannitol 2,5-bis-(tetraethylphosphorodiamidite) and benzyl chloride heated 6 hrs. at 130° → 1,4 : 3,6-dianhydro-2,5-dichloro-2,5-dideoxy-L-iditol. Y: 79%. F. e. s. E. E. Nifant'ev, M. P. Koroteev, and N. S. Rabovskaya, Ж. *43*, 1806 (1973); C. A. *79*, 137414.

Trityl chloride $Ph_3C \cdot Cl$
Acoxyhalides from cyclic orthocarboxylic acid esters ←
  1,2-Acoxyhalides from 2-alkoxy-1,3-dioxolanes s. *28*, 500

Chloroform s. under NaOH $CHCl_3$

Carbon tetrahalide s. under Ph_3P $CHal_4$

Carbon tetrachloride s. under Ph_3P CCl_4

p-Bis(trichloromethyl)benzene s. under $FeCl_3$ ←

α,α-Dichloromethyl ethers s. under $ZnCl_2$ Cl_2CHOMe

2-Chloro-1,1,2-trifluorotriethylamine $CHFClCF_2NEt_2$
Allenecarboxylic acid fluorides ←
  from cyclopropenonecarbinols s. *28*, 501

Trichloromethylisocyanide dichloride s. under $FeCl_3$ $Cl_3C \cdot N : CCl_2$

N-Halogenosuccinimide s. under C_5H_5N and Me_2S $\genfrac{}{}{0pt}{}{CH_2CO}{CH_2CO}\!\!>\!\!NHal$

Cyanuric fluoride ←
Carboxylic acid fluorides from carboxylic acids COOH → COF
s. *3*, 463 suppl. *28*

Cyanuric chloride ←
Chlorides from alcohols ROH → RCl
s. *26*, 539

Phosgene $COCl_2$
Tricarboxylic acid trichlorides C COCl
from carboxydicarboxylic acid anhydrides s. *29*, 513

Reactions of ar. nitrones ←
o-Chloroazomethines s. *30*, 363

Chlorotropenium salts from tropones ←
Nucleophilic substitution s. *26*, 540

Phosgene/pyridine $COCl_2/C_5H_5N$
Iminochlorides from carboxylic acid amides CONH· → C(Cl):N·
s. *27*, 575

Acetyl chloride s. under $SOCl_2$ AcCl

Trichloroacetyl halide CCl_3COHal
α,β-Ethylenehalides from ketones CHCO → C:CHal
s. *26*, 541

Oxalyl chloride $(COCl)_2$
β-Chloro-α,β-ethyleneketones COCHCO → COC:CCl
from β-diketones s. *14*, 107 suppl. 29

Chlorotropenium salts from tropones ←
Nucleophilic substitution s. *26*, 540

Oxalyl chloride/sodium salt $(COCl)_2/Na^+$
3-Chloro-2-azetidinones ☉
from 2-aziridinecarboxylic acids

357.

Na-*cis*-1-*tert*-butyl-3-methyl-2-aziridinecarboxylate added slowly to a soln. of oxalyl chloride in benzene, and the resulting slurry stirred 1 hr. at room temp. → *cis*-1-*tert*-butyl-3-chloro-4-methyl-2-azetidinone. Y: 79%. F. e. s. J. A. Deyrup and S. C. Clough, J. Org. Chem. *39*, 902 (1974).

Acetylsalicyloyl chloride ←
1,2-Acetoxychlorides from glycols C(OH)C(OH) → CClC(OAc)
s. *26*, 547

Trimethylchlorosilane Me_3SiCl
1,2-Acoxychlorides
from 2-alkoxy-1,3-dioxolanes s. *28*, 500 suppl. 29

Hexadecyltributylphosphonium bromide $C_{16}H_{33}\overset{+}{P}Bu_3Br^-$
Replacement of hydroxyl by chlorine OH → Cl
s. *16*, 596 suppl. 29

Triphenylphosphine/carbon tetrahalide $Ph_3P/CHal_4$
***cis*-1,2-Dihalides from oxido compds.** → CHalCHal
s. *29*, 510

Triphenylphosphine/carbon tetrachloride Ph_3P/CCl_4
Chloroformamidines from ureas >N·CO·NH· → >N·C(Cl):N·
s. *29*, 511

Triphenylphosphine/bromine/dimethylformamide $Ph_3P/Br/Me_2NCHO$

Formoxyaldehydes from diols via formoxybromides

$\begin{matrix} OH \\ OH \end{matrix} \rightarrow \begin{matrix} Br \\ OCHO \end{matrix}$

358. $CH_3\overset{OH}{\underset{|}{C}}H[CH_2]_8CH_2OH \rightarrow CH_3\overset{OCHO}{\underset{|}{C}}H[CH_2]_8CH_2Br \rightarrow CH_3\overset{OCHO}{\underset{|}{C}}H[CH_2]_8CHO$

A soln. of 4 equivalents triphenylphosphine in dry dimethylformamide treated dropwise at 0° under N_2 with 4 equivalents bromine followed by 1 equivalent 1,10-undecanediol in a small amount of dimethylformamide, and kept 120 hrs. at –20° → 10-formoxyundecyl bromide (Y: 64%) allowed to react at room temp. with Ag-fluoroborate in dimethyl sulfoxide, then *triethylamine* added → product (Y: 60%). F. e. s. R. K. Boeckman, Jr., and B. Ganem, Tetrah. Let. *1974*, 913; f. oxidations of **halides to oxo compds.** s. ibid. *1974*, 917.

Hexamethylphosphoramide s. under SOHal$_2$ $(Me_2N)_3PO$

Reactions with *polymer-bound phosphine dichlorides* ←

Carboxylic acid chlorides from carboxylic acids COOH → COCl

359. $\underset{Ph}{\overset{Ph}{\diagdown}}$polym—PCl$_2$

PhCH$_2$COOH ⟶ PhCH$_2$COCl

Beads of cross-linked styrene polymer-bound diphenylphosphine dichloride suspended in methylene chloride containing phenylacetic acid, and after 0.5 hr. the beads removed by filtration → phenylacetyl chloride. Y: ca. 100%. – The reagent can be regenerated from the resulting phosphine oxide with phosgene. F. e. and reactions s. H. M. Relles and R. W. Schluenz, Am. Soc. *96*, 6469 (1974).

Triphenylphosphine dibromide Ph_3PBr_2

Bromides from ethers $R_2O \rightarrow 2\,RBr$

Cleavage under neutral conditions s. 27, 576

Phenyltetrafluorophosphorane $PhPF_4$

Fluoroamines $C(OSi\mathrel{\unicode{x2264}})CCl \rightarrow CFCCl$

from chlor(alkoxysilanes) via chlorofluorides – Fluorides from alkoxysilanes s. *29*, 512; s. a. D. J. Costa, N. E. Boutin, and J. C. Riess, Tetrahedron *30*, 3793 (1974); steroids s. Bl. *1974*, 2861

Phosphorus trichloride PCl_3

Combined prepn. of carboxylic acid chlorides COOH → COCl
and 1-hydroxy-1,1-diphosphonic acid
s. *29*, 593

Phosphorus oxide chloride (s. a. under CH$_3$COOH and PCl$_5$) $POCl_3$

6-Chloroimidazo[2,1-b]thiazoles O
s. *26*, 542

Phosphorus pentachloride PCl_5

Alkyl chlorides from alkyl salicylates OCOR → Cl

360. CH$_3$CH$_2$CH$_2$OOC—⟨HO-C$_6$H$_4$⟩ ⟶ CH$_3$CH$_2$CH$_2$Cl

A soln. of *n*-propyl salicylate in dry methylene chloride added dropwise during 0.5 hr. to a stirred slurry of PCl$_5$ in methylene chloride with gentle suction, stirring and stronger suction continued 0.5 hr. to remove solvent and HCl → *n*-propyl chloride. Y: ca. 100%. F. e. s. A. G. Pinkus and W. H. Lin, Synthesis *1974*, 279.

Protection of amino groups as platinum complexes
Carboxylic acid chlorides from carboxylic acids   COOH → COCl

361.

HOOC−CH$_2$NH$_2$ \\ Pt / Cl \\ Cl / \\ NH$_2$CH$_2$−COOH   ⟶   ClOC−CH$_2$NH$_2$ \\ Pt / Cl \\ Cl / Cl \\ NH$_2$CH$_2$−COCl
trans

A heterogeneous mixture of the startg. m. and PCl$_5$ in dry chloroform stirred vigorously at 25° for 3 days → tetrachlorobis(glycyl chloride)platinum(IV). Y: 96%. − Free α-amino acid chlorides are not stable. F. e. s. B. Purucker and W. Beck, B. *107*, 3476 (1974).

β-**Chloronitriles**   C : C·CONH$_2$ → CCl·CH·CN
from α,β-ethylenecarboxylic acid amides s. *28*, 502

1-Halogen-1-aroximes   CONHOAr → CCl(: NOAr)
from hydroxamic acid esters s. *26*, 543

1,3-Dichloroisoquinolines from 2-oximino-1-indanones   ☉
via ring opening by Beckmann rearrangement s. *26*, 544

Phosphorus pentachloride/phosphorus oxide chloride   *PCl$_5$/POCl$_3$*
Chloropyridine from pyridone ring   ←
s. *28*, 903

Phosphorus pentabromide   *PBr$_5$*
Replacement of hydroxyl by bromine   OH → Br

362.   BzNHCH$_2$OH   ⟶   BzNHCH$_2$Br

N-(Hydroxymethyl)benzamide added during 10 min. at 0° with rigorous exclusion of moisture to an ethereal soln. of PBr$_5$, stirred 2 hrs. at 0° and 3 hrs. at 25° → N-(bromomethyl)benzamide. Y: ca. 100%. B. M. Trost et al., Am. Soc. *95*, 7813 (1973).

Vanadium tetrachloride   *VCl$_4$*
Replacement of hydroxyl by chlorine   ←
with allyl rearrangement s. *28*, 503

Thionyl halide   *SOHal$_2$*
C,As-Dihalogenarsines from arsinic acids   ←
5,10-Dihalogeno-5,10-dihydroarsanthracenes
s. *30*, 344

Thionyl halide/hexamethylphosphoramide   *SOHal$_2$/(Me$_2$N)$_3$PO*
Preferential replacement of prim. hydroxyl by halogen   OH → Hal
in nucleosides s. *26*, 545

Thionyl chloride   *SOCl$_2$*
Replacement of hydroxyl by chlorine   OH → Cl
s. *29*, 798

1,1-Acylaminohalides   C(NAc)OH → C(NAc)Hal
from 1,1-acylaminoalcohols s. *30*, 332

Chlorides from acoxy compds.   OAc → Cl
s. *28*, 504

Carboxylic acid chlorides from carboxylic acids   COOH → COCl
s. *26*, 957
    with and without formation of dicarboxylic acid chlorides from dicarboxylic acid anhydrides − Chloroformyldicarboxylic acid anhydrides s. *29*, 513

Acoxycarboxylic acid chlorides
from hydroxycarboxylic acids s. 29, 514
Reactions of ar. nitrones ←

363.   PhCH=N–⟨⟩   →   PhCH=N–⟨⟩
              | |
              O   Cl

o-Chloroazomethines. 1 molar equivalent thionyl chloride or phosgene added at room temp. to a soln. of α,N-diphenylnitrone in a minimum amount of benzene → benzylidene-o-chloraniline hydrochloride. Y: 81%. F. e. s. D. Liotta et al., J. Org. Chem. 38, 3445 (1973); 39, 2718 (1974); **reaction with oxalyl chloride** s. ibid. 39, 1975 (1974).

Chlorazines from acylhydrazones   CONHN : C → C(Cl) : N·N : C
s. 27, 577

2-Chlorothioenolethers   CH : C·SOR → CCl : C·SR
from α,β-ethylenesulfoxides s. 26, 546

Beckmann fragmentation   C
Chloronitriles from cyclic oximes s. 29, 515

Thionyl chloride/zinc chloride $SOCl_2/ZnCl_2$
Chlorocarboxylic acid esters   O   Cl
from lactones via chlorocarboxylic acid chlorides s. 29,   C   →   C
516; chlorocarboxylic acid chlorides s. a. I. I. Grandberg | |
and N. I. Bobrova, Izv. Timiryazevsk. S.-kh. Akad. 1974, 198 (Russ); C. A. 82, 30905

Thionyl chloride/dimethylformamide $SOCl_2/Me_2NCHO$
Replacement of hydroxyl by chlorine   OH → Cl
with O→N-alkyl migration s. 30, 227

Thionyl chloride/acetyl chloride $SOCl_2/AcCl$
1,1-Alkoxychlorides from acetals $C(OR)_2$ → $C(OR)Cl$
s. 30, 567

Thionyl bromide $SOBr_2$
Reactions with thionyl bromide   COOH → COBr

364.   CH₃CH₂COOH   CH₂=CHCOOH
                    ↓   ↓
            CH₃CH₂COBr   CH₂BrCHBrCOOH

Carboxylic acid bromides from carboxylic acids. Propanoic acid mixed with $SOBr_2$, the resulting soln. warmed 0.5 hr. at 50–60° on a water bath, more $SOBr_2$ added, and warming continued 0.5 hr. → propionyl bromide. Y: 90%. | **1,2-Dibromides from ethylene derivs.** Acrylic acid stirred and mixed with $SOBr_2$, and the product isolated after 0.5 hr. at 20° → α,β-dibromopropionic acid. Y: 90%.

F. e., also perpn. of **dicarboxylic acid anhydrides from dicarboxylic acids**, s. S. D. Saraf and M. Zaki, Synthesis 1973, 612.

Methanesulfonyl chloride/lithium chloride $MeSO_2Cl/LiCl$
β,γ-Ethylenechlorides from 2-ethylenealcohols
s. 27, 578; via 2,4-dinitrophenyl ethers cf. S. Czernecki and C. Georgoulis, Bl. 1975, 405

Tosyl chloride s. a. under MeLi   *TsCl*

Tosyl chloride/pyridine $TsCl/C_5H_5N$
s. 28, 505

Sulfuryl chloride/pyridine $\qquad SO_2Cl_2/C_5H_5N$
Partial replacement of hydroxyl by chlorine $\qquad OH \rightarrow Cl$
via chloro(chlorosulfates) s. 27, 579

Dialkylaminosulfur trifluorides $\qquad R_2NSF_3$
Fluorides $\qquad COOH \rightarrow COF$
Carboxylic acid fluorides from carboxylic acids s. 19, 635 suppl. 29

Diethylaminosulfur trifluoride $\qquad Et_2NSF_3$
Replacement of oxygen by fluorine
1,1-Difluorides from oxo compds. $\qquad CO \rightarrow CF_2$

365. $\qquad (CH_3)_2CHCH_2CHO \rightarrow (CH_3)_2CHCH_2CHF_2$

Dialkylaminosulfur trifluorides and bis(dialkylamino)sulfur difluorides are easy to handle fluorinating agents useful for replacing hydroxyl and carbonyl oxygen by fluorine under very mild conditions. They are particularly useful in fluorinating sensitive alcohols and aldehydes. – E: Isovaleraldehyde added slowly at 25° to a soln. of diethylaminosulfur trifluoride in CCl_3F, and stirred 0.5 hr. $\rightarrow$ 1,1-difluoro-3-methylbutane. Y: 80%. F. e. s. W. J. Middleton, J. Org. Chem. 40, 574 (1975).

Sulfur tetrafluoride $\qquad SF_4$
Trifluoromethyl ethers $\qquad OCOF \rightarrow OCF_3$
from fluoroformates s. 30, 106

Selenium tetrafluoride $\qquad SeF_4$
Fluorides $\qquad F$
s. 14, 620 suppl. 29

Molybdenum hexafluoride/boron fluoride $\qquad MoF_6/BF_3$
1,1-Difluorides from oxo compds. $\qquad CO \rightarrow CF_2$
s. 28, 506; s. a. Tetrahedron 31, 391 (1975)

Bromine s. under Ph_3P $\qquad Br$

Aliquat 336 s. under KCl $\qquad (n-C_8H_{17})_3MeN^+Cl^-$

Poly(hydrogen fluoride) s. under C_5H_5N $\qquad (HF)_x$

Ferric chloride/p-bis(trichloromethyl)benzene $\qquad \leftarrow$
Carboxylic acid halides $\qquad COOR \rightarrow COHal$
from carboxylic acid esters – Chlorocarboxylic acid chlorides from lactones s. 29, 517

Ferric chloride/trichloromethylisocyanide dichloride $\qquad FeCl_3/Cl_3C \cdot N : CCl_2$
1,1-Dichlorides from aldehydes $\qquad CHO \rightarrow CHCl_2$
s. 29, 518

Via intermediates $\qquad v.i.$
Chlorides from alcohols via mesylates $\qquad OH \rightarrow Cl$

366. ArCH₂OH + ClSO₂Me → [ArCH₂OSO₂Me] → ArCH₂Cl + KCl
(Ar = 2,4-dimethyl-5-methoxyphenyl, MeO-substituted)

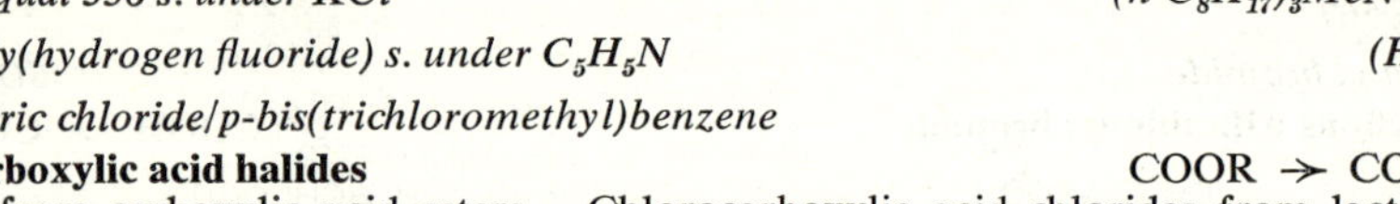

Phase transfer catalysis. Methanesulfonyl chloride added to a stirred soln. of 2,4-dimethyl-5-methoxybenzyl alcohol and triethylamine in benzene, stirred 4 hrs. at room temp., filtered, the filtrate washed with cold satd. brine (NaCl-soln.), Aliquat 336 (methyltrioctylammonium chloride), aq. and powdered KCl added, and stirred 20 hrs. $\rightarrow$ crude 2,4-dimethyl-5-methoxybenzyl chloride. Y: 93%. M. S. Newman and H. M. Chung, J. Org. Chem. 39, 1036 (1974); *crown polyethers as phase transfer catalysts* in anion promoted reactions such as nucleophilic substitution cf. D. Landini, F. Montanari, and F. M. Pirisi, Chem. Commun. 1974, 879.

Nitrogen ↑   HalC ⇵ N

Without additional reagents   w.a.r.
1,1-Dichlorides from aldoximes $CH:NOH \rightarrow CHCl_2$
s. *27*, 562

1,2-Acoxyhalides from glycols $C(OH)C(OH) \rightarrow CClC(OAc)$
and nitriles, also from glycols and acetylsalicyloyl chloride s. *26*, 547

α-Chloro- from α-diazo-sulfones $SO_2 \cdot C:N_2 \rightarrow SO_2 \cdot CHCl$

367.   p—CH₃C₆H₄—SO₂—C—C₆H₅ → p—CH₃C₆H₄—SO₂—CHClC₆H₅
                     ‖
                     N₂

α-Phenyltosyldiazomethane allowed to react with HCl in ether → α-chlorobenzyl p-tolyl sulfone. Y: ca. 100%. A. M. D. van Leusen, and J. Strating, Tetrah. Let. *1973*, 5207.

Irradiation   ⇝
Replacement of nitro groups by chlorine $NO_2 \rightarrow Cl$
s. *26*, 548

Sodium nitrite s. under CuHal $NaNO_2$

Potassium hydrogen fluoride KHF_2
Replacement of ammonium groups by nucleophiles $\overset{+}{N}\leqslant \rightarrow F$
   by fluorine s. *28*, 507

Lithium bromide $LiBr$
Reactions with 1-nitroxido compds.   [epoxide-NO₂] $\rightarrow$ CBrCO

368.

α-**Bromoketones.** Methyl 2,3-anhydro-4,6-O-benzylidene-3-C-nitro-β-D-allopyranoside allowed to react at room temp. with excess Li-bromide monohydrate in tetrahydrofuran → methyl 4,6-O-benzylidene-2-bromo-2-deoxy-β-D-*ribo*-3-hexulopyranoside. Y: 75%. F. e. and reactions s. S. Kumazawa et al., Ang. Ch. *85*, 992 (1973).

Cuprous halide/sodium nitrite $CuHal/NaNO_2$
Replacement of amino groups by halogen $NH_2 \rightarrow Hal$
s. *11*, 628; *14*, 623

Cupric bromide $CuBr_2$
Sandmeyer reaction in dimethyl sulfoxide $N_2^+ BF_4^- \rightarrow Hal$
   Halides from diazonium fluoroborates s. *26*, 549

Ethyl chloroformate $ClCOOEt$
Chlorides from tert. amines $NR_2 \rightarrow Cl$
s. *28*, 508

Acetic acid CH_3COOH
Phenols from diazo oxides via o-bromophenols   ←
   Monodehydroxylation of o-diols s. *26*, 550

Nitrosyl chloride	*NOCl*
Replacement of amino groups by chlorine	$NH_2 \to Cl$

369.

4-Amino-2,2,5,5-tetrakis(trifluoromethyl)-3-imidazoline shaken 1 hr. at 60° with nitrosyl chloride in a Pt-lined autoclave → 4-chloro-2,2,5,5-tetrakis(trifluoromethyl)-3-imidazoline. Y: 85%. W. J. Middleton, D. Metzger, and E. A. Donald, J. Heterocyclic Chem. *10*, 997 (1973).

Phosphorus pentachloride/hydrogen chloride	PCl_5/HCl
*α,α-**Dichlorocarboxylic acids from nitriles***	$CH_3CN \to CCl_2COOH$
s. *29*, 519	
Sulfuryl chloride	SO_2Cl_2
5,5-Dichlorobarbituric acids	←
s. *28*, 509	
tert-Butyl hypochlorite	$Me_3C \cdot OCl$
1,1-Alkoxyhalides from diazo compds.	$C:N_2 \to CCl(OR)$
s. *27*, 580	
Via intermediates	*v.i.*
*α,α-**Dihalogenaldehydes***	$CHal:C \cdot N< \to C(Hal)(Hal')CO$
from 2-halogenenamines	

370.

An ethereal soln. of bromine added at –60° to a soln. of the startg. m. in abs. ether, then treated with water until the intermediate precipitate has dissolved → α-bromo-α-chloroaldehyde. Y: 66%. F. e. s. L. and P. Duhamel, and J.-M. Poirier, Tetrah. Let. *1973*, 4237.

Halogen ↑ HalC ↕ Hal

Potassium fluoride	*KF*
Trifluoromethyl from trichloromethyl groups	$CCl_3 \to CF_3$
s. *28*, 516	
Silver chlorodifluoroacetate	$CF_2ClCOOAg$
Replacement of bromine by chlorine	$Br \to Cl$
s. *27*, 581	
Cuprous chloride	*CuCl*
s. *29*, 821	
Silver fluoride	*AgF*
Replacement of chlorine by fluorine	$Cl \to F$

371.

7,7-Dichloro-2,5-diphenylbenzocyclopropene stirred 24 hrs. at room temp. in the dark with a suspension of a large excess of AgF in dry acetonitrile → 7,7-difluoro-2,5-diphenylbenzocyclopropene. Y: 85%. P. Müller, Chem. Commun. *1973*, 895.

Silver fluoroborate $AgBF_4$
    Preferential replacement – Glycosyl fluorides s. 26, 551
Aluminum bromide $AlBr_3$
Replacement of chlorine by bromine   Cl → Br
s. 28, 510

Anion exchanger   ←
Sulfonic acid fluorides SO_2Cl → SO_2F
    from sulfonic acid chlorides s. 29, 520

Nitrosyl chloride   NOCl
Replacement of bromine by chlorine   Br → Cl
s. 26, 258

Antimony pentachloride-graphite/sodium thiosulfate $SbCl_5 \cdot C_{24}/Na_2S_2O_3$
    also replacement of iodine s. 29, 521

Ferric chloride $FeCl_3$
Replacement of tert. and benzylic chlorine by iodine   Cl → I

372.   Me₃C—Cl + NaI → Me₃C—I

tert-Butyl chloride stirred 2 hrs. at room temp. with 2 moles NaI in the presence of 0.02 mole $FeCl_3$ in a non-polar solvent → tert-butyl iodide. Y: 99%. – Simple prim. and sec. alkyl chlorides do not react. F. e. s. J. A. Miller and M. J. Nunn, Tetrah. Let. *1974*, 2691.

Sulfur ↑   HalC ⇅ S

Without additional reagents   w.a.r.
Replacement of sulfinyl and sulfonyl groups   SOR ↘
    by halogen s. 29, 522 SO_2R ↗   Hal

α-Ketohydroximinohalides $COCH_2\overset{+}{S}$< → COC(:NOH)Hal
    from 2-ketosulfonium salts s. 27, 348

Bis(chlorocarbonyl)amines   C
    from 1,2,4-dithiazoli-3,5-diones s. 27, 582

Methyl iodide   MeI
Cleavage of 2-alkylthio-Δ^2-thiazolines   ←
s. 28, 802

Molybdenum hexafluoride MoF_6
Trifluoromethyl ethers   O·C(S)Cl → O·CF_3
    from chlorothionoformic acid esters s. 29, 523

Remaining Elements ↑   HalC ⇅ Rem

Without additional reagents   w.a.r.
Halides from diorganomercury compds. R_2Hg → 2 RHal

373.

Ferrocenes. Iodine in 1,2-dichloroethane added to a refluxing soln. of bis-(2-iodoferrocenyl)mercury in the same solvent, heated 10 min. on a steam bath, and stirred until the soln. turns yellow → 1,2-diiodoferrocene. Y: 98%. P. V. Roling and M. D. Rausch, J. Org. Chem. *39*, 1420 (1974).

α-Halogenocarbonyl compds. from enoxysilanes   C : C · OSi ≤ → CHalCO

374.

PhCH$_2$CH=CH—OSiMe$_3$ → PhCH$_2$CHClCHO

Bromine in CCl$_4$ added dropwise with stirring during 3 hrs. to the startg. m. in the same solvent → 2-bromobutyrolactone. Y: 84%. – Similarly with chlorine: 2-Chloro-3-phenylpropanal. Y: 95%. F. e. s. R. H. Reuss and A. Hassner, J. Org. Chem. *39*, 1785 (1974).

Halides from organocobalt compds.   Hal
  β-Halogenostyrenes by geospecific replacement s. *26*, 552

Sodium hydroxide   *NaOH*
trans-α,β-Ethyleneiodides   C : C · B(OH)$_2$ → C : C · I
  from α,β-ethyleneboronic acids s. *29*, 497

Oxalyl chloride/dimethylformamide   *(COCl)$_2$/Me$_2$NCHO*
Carboxylic acid chlorides   COOSi ≤ → COCl
  from carboxylic acid silyl esters – α-Ketocarboxylic acid chlorides s. *29*, 524

Iodine monochloride   *ICl*
Halides from stannanes   Sn ≤ → Hal
  3-Subst. benzocyclobutenes s. *28*, 511

Carbon ↑   HalC ↕ C

Without additional reagents   *w.a.r.*
α-Azo-α,α-dihalogenoketones   C(Ac) : NNH → C(N : N)Cl$_2$
  from 1,2,3-triketone 2-monohydrazones s. *28*, 422

Irradiation s. under HgO   ☼

Sodium hydroxide/potassium iodide   *NaOH/KI*
1,1-Diiodides from ketones   CHCOR → CI$_2$

375.

```
     C₆H₄NO₂—p                    C₆H₄NO₂—p
        |                            |
    HO—C—H                       HO—C—H
        |                            |
     H—C—NHCOCH₂COCH₃      →     H—C—NHCOCHI₂
        |                            |
      CH₂OH                        CH₂OH
```

Diiodomethyl compds. An aq. soln. of I$_2$ and KI added dropwise with stirring in the dark to a soln. of D-*threo*-N-[2-hydroxy-1-hydroxymethyl-2-(4-nitrophenyl)-ethyl]acetoacetamide in aq. NaOH → iodamphenicol. Y: 60%. – The product could not be obtained by transhalogenation of chloramphenicol. F. e. s. A. Stephen, G. Schulz, and H. Reinshagen, A. *1974*, 363.

Sodium hydrogen carbonate s. under BaCO$_3$ *NaHCO$_3$*
Potassium iodide s. under NaOH   *KI*
Barium carbonate-sodium hydrogen carbonate   *BaCO$_3$-NaHCO$_3$*
α,β-Ethylene-γ-halogenaldehydes   C : C · C : CH(OAc) → CCl · C : C · CHO
  from 1-acoxy-1,3-dienes s. *26*, 553

Mercuric oxide/irradiation	*HgO/////*
Bromides from carboxylic acids s. *16*, 620 suppl. *28*	COOH → Br
Benzoyl peroxide	*(PhCOO)₂*
Iodides from carboxylic acids Degradation with loss of 1 C-atom s. *26*, 554	COOH → I
Acetic acid	*CH₃COOH*
Replacement of *tert*-butyl by halogen in quinones via quinols s. *28*, 512	CMe₃ → Hal
Phosphorus pentachloride	*PCl₅*
Acoxyhalides from cyclic 1-carboxyacetals s. *27*, 583	C
Chlorosulfonic acid s. under I	*ClSO₃H*
Iodine/chlorosulfonic acid	*I/ClSO₃H*
Ar. polyhalides – Halogenolysis	←

376.

o-Cl-C₆H₄-COOH → C₆Br₅Cl (pentabromochlorobenzene)

A slight excess of bromine added slowly with stirring to o-chlorobenzoic acid and iodine as halogen carrier in chlorosulfonic acid, and heated 5 hrs. on a steam bath with continued stirring → pentabromochlorobenzene. Y: 67%. F. e. s. A. I. Hashem, J. Appl. Chem. Biotechnol. *23*, 621 (1973).

Sodium hypochlorite/acetic acid	*NaOCl/CH₃COOH*
Chlorides from aldehydes s. *29*, 525	CHO → Cl

Formation of S—S Bond

Uptake   ⇓

Addition to Sulfur and Sulfur   SS ⇓ SS

Sodium/alcohol	*NaOR*
1,2-Dithiane 1,1-dioxide ring opening s. *28*, 513	C

Addition to Sulfur and Carbon   SS ⇓ SC

Periodates	*IO₄⁻*
S-Sulfonylthiourethans from C-sulfonylthioformamides s. *28*, 514	SO₂CS·N< → SO₂SCO·N<

Exchange ↓↑

Hydrogen ↑   SS ↓↑ H

Without additional reagents   w.a.r.
Thiosulfenyl chlorides from mercaptans   SH → S·SCl
s. 28, 256
o-Benzotrithiol-2-ones from o-dithiols   ◯
s. 28, 515
Dicyclohexylcarbodiimide   RN : C : NR
Thiosulfuric acid S-monoesters   SH → S·SO$_3^-$
from mercaptans s. 26, 557
8-Cyano-3,10-dimethylisoalloxazine   ←
Oxidations with electron-deficient isoalloxazines
Sym. disulfides from mercaptans   2 RSH → RSSR

377.

2 PhSH ⟶ Ph–S–S–Ph

Thiophenol allowed to react with 3,10-dimethyl-8-cyanoisoalloxazine in aq. acetonitrile under argon → diphenyl disulfide. Y: 98%. – The oxidant is quantitatively regenerated in the presence of air. Also oxidation of nitroalkanes s. I. Yokoe and T. C. Bruice, Am. Soc. 97, 450 (1975).

Dimethyl dithiobis(thioformate)   [MeOC(S)S·]$_2$
Sym. disulfides from mercaptans
s. 26, 559 suppl. 29
Morpholinosulfenyl chloride   ←
Sym. bis(thiophosphinyl)trisulfides   2 >P(S)SH → >P(S)·S·S·S·P(S)<
from dithiophosphinic acids s. 26, 558
Nitric acid   HNO$_3$
Sym. disulfides from mercaptans   2 RSH → RSSR
s. 26, 559
Chloramine   NH$_2$Cl
     f. oxidations with chloramine s. 26, 560; sym. disulfides with 2,4,4,6-tetrabromo-2,5-cyclohexadienone cf. T. Ho, T. W. Hall, and C. M. Wong, Synthesis 1974, 872

Oxygen ↑   SS ↓↑ O

Sulfuric acid   H$_2$SO$_4$
Thiolsulfinic acid esters   C
from 2 episulfoxide molecules s. 26, 561

Nitrogen ↑   SS ↓↑ N

Without additional reagents   w.a.r.
Disulfides from mercaptans and sulfenimides   RSH → RSSR'
     also trisulfides from hydrodisulfides and sulfenimides s. 26, 562

Acetic acid CH_3COOH
Thiolsulfonic acid esters from sulfinic acids $SO_2H \rightarrow SO_2SR$
    and sulfenamides or sulfenic acid esters s. 26, 563

Halogen ↑   SS ↾↿ Hal

Without additional reagents   w.a.r.
Hydropolysulfides $[S]_xH$

378.   PhSSCl + H$_2$S   →   PhSSSH

A soln. of phenylchlorodisulfane in CS_2 added dropwise with vigorous stirring at $-70°$ to excess liquefied H$_2$S, and stirring continued 1 hr. at the same temp. → phenyltrisulfane. Y: 90%. F. e., also hydrodisulfides and hydrotetrasulfides, s. H. J. Langer and J. B. Hyne, Can. J. Chem. *51*, 3403 (1973).

Potassium fluoride   KF
Sym. disulfides from sulfenylchlorides   RSSR
    via sulfenylbromides – Trifluoromethyl from trichloromethyl groups s. 28, 516

Sodium salt Na^+
Sulfonylthioamines $RSO_2H \rightarrow RSO_2S \cdot N<$
    from sulfinic acids and aminosulfenyl chlorides s. 28, 517

Ethyldiisopropylamine s. under $H_2N \cdot NH_2$ $i\text{-}Pr_2NEt$

Tri-n-propylamine s. under Cl_3SiH $n\text{-}Pr_3N$

Trichlorosilane/tri-n-propylamine $Cl_3SiH/n\text{-}Pr_3N$
Sym. disulfides by reduction $2\ RSO_2Cl \rightarrow RSSR$
    s. 26, 564

Hydrazine/ethyldiisopropylamine $H_2N \cdot NH_2/i\text{-}Pr_2NEt$
o,o'-Dicyanodisulfides from 1,2-benzisothiazoles   C
    s. 27, 584

N-Hydrogenopyridinium salts $C_5H_5\overset{+}{N}H$
Sulfonyldisulfides $RSO_2SH \rightarrow RSO_2SSR'$
    from thiolsulfonic acids and sulfenylchlorides s. 28, 518

Molybdenum hexacarbonyl $Mo(CO)_6$
Sym. disulfides $2\ RSO_2Cl \rightarrow RSSR$
    from sulfonic acid chlorides s. 26, 564

Sulfur ↑   SS ↾↿ S

Sodium hydrogen carbonate $NaHCl_3$
Sym. from unsym. disulfides $2\ RSSR' \rightarrow RSSR + R'SSR'$
    Disproportionation s. 29, 526

Triethylamine Et_3N
Unsym. disulfides $RS \cdot S \cdot COOR \rightarrow RSSR'$
    from sulfenylthiocarbonic acid esters s. 26, 565; protection of sulfhydryl groups as sulfenylthiocarbonic acid esters cf. R. G. Hiskey, N. Muthukumaraswamy, and R. R. Vunnam, J. Org. Chem. *40*, 950 (1975)

Silver nitrate $AgNO_3$
Thiolsulfonic acid esters $SO_2Na \rightarrow SO_2SR$
    from sulfinic acids and disulfides s. 27, 585

Remaining Elements ↑   SS ↕ Rem

Without additional reagents   w.a.r.
Disulfides   S·OR → S·SR′
from sulfenic acid esters and thioboron compds. s. 27, 586

Carbon ↑   SS ↕ C

Without additional reagents   w.a.r.
Benzothiazoles via 1,2,4-benzodithiazines   ○
s. 26, 566

Elimination   ⇑

Hydrogen ↑   SS ⇑ H

Potassium iodide s. under H_2O_2   KI
Triethylamine s. under I Et_3N
Dimethyl sulfoxide Me_2SO
1,2-Dithiolanes from 1,3-dithiols   ○
s. 27, 642 suppl. 29
Hydrogen peroxide/potassium iodide H_2O_2/KI
1,2-Dithiolanes from 1,3-dithiols
s. 27, 642
Iodine/triethylamine I/Et_3N
s. 27, 642

Carbon ↑   SS ⇑ C

m-Chloroperoxybenzoic acid $m\text{-}ClC_6H_4COO_2H$
Disulfides from mercaptals   S·C·S → S·S
3,6-Epidithiopiperazine-2,5-diones s. 29, 527
Bromine/sodium hydrogen carbonate $Br/NaHCO_3$
Disulfides from mercaptals $S·CH_2·S → S·S$
s. 28, 519
Iodine/acetic acid I/CH_3COOH
Cyclic disulfides from dithioethers   ○
s. 12, 639 suppl. 28

Formation of S—Rem Bond

Uptake   ⇓

Addition to Sulfur and Carbon   SRem ⇓ SC

Trimethylsilylacetamide $CH_3CONHSiMe_3$
Sulfenic acid silyl esters   ○
by penicillin S-oxide ring opening s. 29, 528

Exchange

Hydrogen ↑   SRem ⇅ H

Without additional reagents   w.a.r.
**1,3,2-Benzodithiaboroles from o-dithiols
via 2-chloro-1,3,2-benzodithiaboroles**

379.

Toluene-3,4-dithiol allowed to react 5 hrs. with boron chloride, whereby HCl evolves slowly → 2-chloro-5-methyl-1,3,2-benzodithiaborole (Y: 85%) treated under N_2 with tributyltin hydride, and heated 2 hrs. at 90° → 5-methyl-1,3,2-benzodithiaborole (Y: 90%). G. Srivastava, Soc. Perkin I *1974*, 916.

Mercury bis(phenylacetylide) $(PhC \equiv C)_2Hg$
Mercuric mercaptides $\cdot S]_2Hg$
from thioureas and carboxylic acid thioamides s. *26*, 567

Morpholinosulfenyl chloride   ←
Trithiodiphosphonic acid esters   ←
from O-alkylthiophosphonic acids s. *29*, 529

Chloramine NH_2Cl
Selenosulfides and diselenides   SeH ⇄ SeSR / SeSeR
from selenols s. *27*, 587

Chlorotris(triphenylphosphine)rhodium(I) $RhCl(Ph_3P)_3$
Silyl sulfides   SH + HSi⟨ → S·Si⟨
from mercaptans and silanes s. *29*, 530

Oxygen ↑   SRem ⇅ O

Boron sulfide B_2S_3
Phosphine sulfides from phosphine oxides $R_3PO \rightarrow R_3PS$
s. *28*, 550

O,O-Diethyl dithiophosphoric acid $(EtO)_2P(S)SH$
Replacement of oxygen by sulfur   ⩾PO → ⩾PS
s. *28*, 550

Nitrogen ↑   SRem ⇅ N

Without additional reagents   w.a.r.
As-Heterocyclics from aminoarsines   ○
s. *26*, 568

Imidazole   ←
S-Silylation   SH → S·Si⟨

380. $Me_3Si-NH-SiMe_3$

PhSH ⎯⎯⎯⎯⎯⎯⎯→ PhS—SiMe₃

Thiophenol stirred and refluxed 3 hrs. with ca. 2 moles hexamethyldisilazane in the presence of a small amount of imidazole → product. Y: 93%. F. e. s. R. S. Glass, J. Organometal. Chem. *61*, 83 (1973).

Halogen ↑   SRem ⇅ Hal

Lithium tetrahydridoaluminate $Li[AlH_4]$
Replacement of group IV element halogen $\geqslant GeHal \rightarrow \geqslant GeSR$
    by alkylthio and alkylseleno groups s. *28*, 520

Formation of S—C Bond

Uptake ⇓

Addition to Hydrogen and Carbon   SC⇓HC

Without additional reagents   *w.a.r.*
Reactions of thioketones with allenes   ←
    1,3-Dienes – 2-Alkylthio-1,3-dienes s. *28*, 521
    Addition of thioketones to allenes s. *27*, 588
Irradiation   ⫲
Insertion of sulfur dioxide $CH \rightarrow C \cdot SO_2H$
    into hydrogen-carbon bonds – α-Subst. sulfinic acids s. *26*, 569

Addition to Oxygen and Carbon   SC⇓OC

Without additional reagents   *w.a.r.*
1,1-Hydroxythioethers from aldehydes $CHO \rightarrow CH(OH)SR$
    s. *28*, 522
Sodium   *Na*
Monothiolphosphoric acid esters $CO \rightarrow CHSPO(OR)_2$
    from thionophosphorous acid esters and ketones – Phosphoric acid esters
    from dialkyl phosphites and ketones s. *28*, 523
Sodium hydride   *NaH*
Carboxythioethers from lactones
    s. *27*, 618
Sodium salt   *Na⁺*
Alcohols from oxido compds.   ▽O → C(OH)C(SR)
    via 2-hydroxythioethers – Regiospecific ring opening
    s. *28*, 524
Triton B   ←
1,2-Hydroxymercaptans from oxido compds.   ▽O → C(OH)C(SH)
    s. *26*, 570
Sodium tetrahydridoborate/sulfur $Na[BH_4]/S$
1,2-Hydroxymercaptans   2 ▽O → $C(OH)C \cdot SS \cdot C(OH)$
    from oxido compds. via 2,2'-dihydroxy-
    disulfides s. *28*, 525
Sulfur s. under Na[BH₄]   *S*
Sulfuric acid H_2SO_4
Quinolsulfonium salts   ←
    from quinones and thioethers s. *27*, 589

Addition to Nitrogen and Carbon SC↓NC

Without additional reagents w.a.r.
Reaction of carbon oxide sulfide with amines NR → N·C(O)SR
Insertion into nitrogen-carbon bonds – Thiolurethans – Alcohols s. *29*, 531
N-Thioformylacylamines N : C → N(CHS)COR
from thiolic acids and isonitriles

381.

$$\text{Et–C}_6\text{H}_3(\text{Et})\text{–N=C} + \underset{\underset{\text{CH}_2\text{Cl}}{|}}{\text{C}}\overset{\text{O}}{\underset{\text{SH}}{}} \longrightarrow \text{Et–C}_6\text{H}_3(\text{Et})\text{–N}\begin{array}{c}\text{CH=S}\\ \text{COCH}_2\text{Cl}\end{array}$$

2,6-Diethylphenyl isocyanide and chlorothioacetic acid added to ether, and allowed to stand 16 hrs. → 2′,6′-diethyl-N-thioformyl-2-chloroacetanilide. Y: 84%. F. e. s. J. P. Chupp and K. L. Leschinsky, J. Org. Chem. *40*, 66 (1975).

Acyl carbamyl sulfides NCO → NHCOS·COR
from thiolic acids and isocyanates

382. PhN=C=O + HSCOPh → PhNHCO—S—COPh

Thiobenzoic acid allowed to react at room temp. with phenyl isocyanate in *n*-hexane → benzoyl phenylcarbamyl sulfide. Y: 86%. F. e., **also acyl acylcarbamyl sulfides from acyl isocyanates**, s. S. Motoki, T. Saito, and H. Kagami, Bull. Chem. Soc. Japan *47*, 775 (1974).

Isothioureas from carbodiimides RN : C : NR → RNHC(SR′) : NR

383.

$$p\text{–CH}_3\text{C}_6\text{H}_4\text{–SH} + \underset{\underset{\text{N–C}_6\text{H}_{11}}{\parallel}}{\overset{\overset{\text{N–C}_6\text{H}_{11}}{\parallel}}{\text{C}}} \longrightarrow p\text{–CH}_3\text{C}_6\text{H}_4\text{–S–}\underset{\underset{\text{NH–C}_6\text{H}_{11}}{|}}{\overset{\overset{\text{N–C}_6\text{H}_{11}}{\parallel}}{\text{C}}}$$

N,N-Dicyclohexylcarbodiimide in abs. acetone treated at 0° with thio-*p*-cresol, and kept 10 hrs. at the same temp. → N,N′-dicyclohexyl-S-(p-tolyl)isothiourea. Y: 94%. F. e. s. E. Vowinkel and G. Claussen, B. *107*, 898 (1974).

Dithiocarbamic acid esters N : C : S → NHC(S)SR
from isothiocyanates and mercaptans s. *21*, 635 suppl.
Carbamyl dithiocarboxylates N : C : O → NHC(O)SC(S)
from dithiocarboxylic acids and isocyanates s. *29*, 532
Dithiocarbazic acid esters N·N : C : S → N·NHCSSR
from N-isothiocyanatoamines s. *26*, 571
Addition of sulfinic acids SO₂N : C → SO₂NHC·SO₂R
to carbon-nitrogen and nitrogen-nitrogen double bonds

384.

PhSO₂N=CHPh   O⟨   ⟩N—CO—N=N—CO—N⟨   ⟩O
↓   + p—CH₃C₆H₄SO₂H   ↓
PhSO₂NHCHPh   O⟨   ⟩N—CO—N—NH—CO—N⟨   ⟩O
       | |
   SO₂C₆H₄CH₃—p   SO₂C₆H₄CH₃—p

α-(Sulfonylamino)sulfones from N-sulfonylimines.	Sulfonic acid hydrazides from azo compds.
A soln. of *p*-toluenesulfinic acid and the startg. N-sulfonylimine in anhydrous tetrahydrofuran allowed to stand several hrs. at room temp. → product. Y: 90%. F. e. s. P. Messinger, Arch. Pharm. *307*, 348 (1974).	and azodicarboxylic acid bismorpholide in dioxane allowed to stand 12 hrs. at room temp. → (N-tosyl)hydrazodicarboxylic acid bismorpholide. Y: 87%.

1,3-Thiazetidines from carbodiimides 
s. 27, 590

2-Amino-4,7-dihydro-1,3-thiazepines
from 2-vinylaziridine and isothiocyanates s. 28, 526

5,6-Dihydro-4H-1,3,5-thiadiazin-4-ones
from azomethines s. 27, 591

Carbamylisothiocyanates and their reactions
2-Amino-5,6-dihydro-4H-1,3,5-thiadiazin 4-ones from azomethines – Carbamylthiocyanates s. 26, 572

Ring closures with thiocarbamylisothiocyanates
2-Amino-5,6-dihydro-4H-1,3,5-thiadiazine-4-thiones from azomethines s. 27, 591

Isothiocyanates
from nitrile oxides via 1,4,2-oxathiazoles s. 28, 527

Irradiation

Thioacylimines $\qquad$ $C \equiv N + S:C \rightarrow C(S)N:C$
from thioketones and nitriles s. 29, 533

Sodium succinate buffer

Thiolurethans $\qquad$ $SH \rightarrow S \cdot CONH \cdot$
from mercaptans and isocyanates s. 15, 457 suppl. 28

Sodium mercaptide $\qquad$ *NaSR*

1-Alkoxy-1-alkylthio-1-amines $\qquad$ $N \cdot C(OR)SR'$
s. 28, 528

Diethylamine $\qquad$ Et_2NH

2-Aminothiophene ring
s. 29, 534

Thiourea $\qquad$ $(NH_2)_2CS$

Thioureas from carbodiimides $\qquad$ $N:C:N \rightarrow NHCSNH$
s. 27, 592

p-Toluenesulfonic acid $\qquad$ *TsOH*

1,3,4-Thiadiazolidines
from azines s. 30, 566

Hydrogen chloride $\qquad$ *HCl*

Thioiminoester hydrochlorides from nitriles $\qquad$ $CN \rightarrow C(:NH)SR$
s. 27, 624

Addition to Sulfur $\qquad$ SC⇓SS

Sulfuryl chloride/calcium carbonate $\qquad$ $SO_2Cl_2/CaCO_3$

2a,5a-Dichloro-1,4-dithianes
s. 26, 573

Addition to Sulfur and Remaining Elements $\qquad$ SC⇓SRem

Without additional reagents $\qquad$ *w.a.r.*

1,3,4-Thiaoxaborepanes from 1,2-thiaborolanes
s. 28, 529

Addition to Sulfur and Carbon   SC↓SC

Without additional reagents   *w.a.r.*
S-Alkylation
of vinylogous carboxylic acid thioamides s. 26, 574

Δ³-1,3,4-Thiadiazoline 1-oxides
from diazo compds. and sulfines s. 29, 535

Irradiation
1,3-Dithietanes
by dimerization of thioketones s. 29, 536

Lithium/alcohol   *LiOR*
**Reaction of thiirane 1,1-dioxides
with strong nucleophiles**

385.   PhSH + H–C(Me)(Me)–S(O)(O)–C–H → PhS–CH(Me)–C(Me)(H)–SO₂Li

Stereospecific ring opening. 2,3-Dimethylthiirane 1,1-dioxide and thiophenol added to an equimolar amount of Li-methoxide in methanol, and stirred until the mixture reacts neutral ⇢ product. Y: 85%. F. e. s. E. Vilsmaier and G. Becker, Synthesis *1975*, 55.

Organolithium compds.   *RLi*
Thiophilic addition   C:S:O ⇢ CH·S(O)R
Sulfoxides from sulfines s. 26, 575

Butyllithium/triethylenediamine
Cyclopropane from dihydrothiopyran ring
s. 26, 576

Sodium salt   *Na⁺*
Reactions of 2-amino-1,3-dithiolanium salts
Ambident electrophiles s. 26, 577

Magnesium   *Mg*
Sec. thioethers from thioketones   C:S ⇢ CH·SR
s. 29, 641

Sodium tetrahydridoborate/sulfur   *Na[BH₄]/S*
1,2-Dithiols from sulfido compds.   ⟨S⟩ ⇢ C(SH)C(SH)
s. 28, 525

Addition to Remaining Elements and Carbon   SC↓RemC

Without additional reagents   *w.a.r.*
Monoorganomercury sulfonates   RHgR ⇢ RSO₂OHgR
from diorganomercury compds. s. 27, 593

Silasultones
also germasultones s. 26, 578

Addition to Carbon Atoms SC↓C

Without additional reagents w.a.r.
Isothiocyanates from isonitriles N : C → N : C : S
s. *28*, 530
Irradiation ⊬
s. *28*, 530

Addition to Carbon-Carbon Bonds SC↓CC

Without additional reagents w.a.r.
Reactions of sulfur dichloride with acetylene derivs. ←
  Sym. 2-halogenothioenolethers – Thianaphthenes s. *26*, 579
**Thioiminoesters and ketene mercaptals
from ketenimines and mercaptans** $C:C:N \cdot \begin{smallmatrix} CHC(SR):N \cdot \\ C:C(SR)_2 \end{smallmatrix}$

386.
$$Ph_2C=C=N-C_6H_4Cl-p \qquad Ph_2C=C=N-Ph$$
$$\swarrow 1 \quad + \quad HSPh \quad 2 \searrow$$
$$Ph_2CH-\underset{\underset{SPh}{|}}{C}=N-C_6H_4Cl-p \qquad Ph_2C=C(SPh)_2$$

A soln. of diphenylketene
N-(p-chlorophenyl)imine and thiophenol in benzene refluxed 24 hrs. at 80° → phenyl N-(p-chlorophenyl)-diphenylthiacetimidate. Y: 73%. | N-phenylimine in excess thiophenol refluxed 24 hrs. *at 169°* with introduction of dry HCl → diphenylketene diphenylmercaptal. Y: 67.7%.
F. e. s. M. W. Barker, S. C. Lauderdale, and J. R. West, J. Org. Chem. *38*, 3951 (1973).

Reaction of sulfilimines with acetylene derivs. ←

387.
$$MeOOC-C\equiv C-COOMe$$
$$+$$
$$\underset{C_6H_4NO_2-p}{N=SMe_2}$$

$$MeOOC-\underset{\underset{C_6H_4NO_2-p}{N}}{\overset{\|}{C}}-\underset{SMe_2}{\overset{\|}{C}}-COOMe \qquad \underset{C_6H_4NO_2-p}{\overset{MeOOC}{\underset{N}{\diagdown}}}\hspace{-0.5em}\underset{H\cdots O}{\diagup}C-OMe$$

α-Iminosulfonium ylids. | **Enamines.**
N-p-Nitrophenyliminodimethylsulfurane allowed to react with dimethyl acetylenedicarboxylate
in dry benzene for several hrs. → ylid. Y: 56%. | *in moist chloroform* → dimethyl N-p-nitrophenylaminofumalate. Y: ca. 100%.
F. e. and products s. Y. Hayashi, D. Swern et al., Tetrah. Let. *1974*, 1071.

β-Halogenosulfones C : C → CHal·C·SO₂R
  from sulfonic acid halides s. *21*, 642 suppl. 29
α,β-Ethylene-β-halogenosulfones C ≡ C → CHal : C·SO₂R
  from ethylene derivs. and sulfonic acid halides – Geospecific addition
  s. *27*, 598 suppl. 30
α,β-Ethylene-β'-iodosulfones C : C : C → C : C(SO₂R)·Cl
  from allenes and sulfonic acid iodides s. *29*, 537

Dithiocarboxylic acid esters   C : C → CH·C·SC(S)R
from dithiocarboxylic acids and ethylene derivs. – Markownikoff-type and Michael-type additions s. 29, 538

Dithiourethans from ethylene derivs.   C : C → CHC·SC(S)N<
s. 27, 594

Cyclic thioethers from ethylenemercaptans   O
S-Bridges s. 27, 595

Benz[b]indeno[2,1-d]thiophenes
by double ring closure s. 28, 531

4-Thiazolidones from thiocarbohydrazones
s. 27, 596

1,2-Dithiolanes from 2-enethietanes   ⊙
s. 28, 532

Irradiation   ⋀⋀

α-Fluorosulfinylperfluorocarboxylic acid fluorides   C : CF$_2$ → C(SOF)COF
from terminal perfluoroalkenes s. 26, 580

α,β-Ethylene-β-iodosulfones   C ≡ C → CI : C·SO$_2$R
from acetylene derivs. and sulfonic acid iodides s. 27, 598

Sodium hydroxide   NaOH
1,4,2-Dithiazine 1,1-dioxides   O
from α,β-ethylenesulfonic acid amides s. 28, 533

Sodium methoxide   NaOMe
S-Cyanoethylation   SH → S·C·CH·CN
s. 30, 397

Sodium acetate   CH$_3$COONa
β-Ketomercaptans   CO·C : C → CO·CH·C(SH)
from α,β-ethyleneketones s. 29, 539

Potassium thiocyanate/acetic acid   KSCN/CH$_3$COOH
Thiolic acid esters from ethylene derivs.   C : C → CHC·SAc
Preferential addition s. 27, 597

Trimethylamine s. under HCOOH   Me$_3$N

Triethylamine   Et$_3$N
Alkylthioketene aminals   ←
by rearrangement of diaminomethylene sulfonium salts s. 26, 581

Δ^2-Thiazolines from enisonitriles   O

388.
$$H_2S + \underset{(CH_3)_2C=C-COOEt}{C\diagdown N} \longrightarrow \underset{COOEt}{S\diagup N}$$

H$_2$S passed at 20–40° into a stirred soln. of the startg. m. and triethylamine in chloroform until spectroscopy indicates completion of the reaction → ethyl 5,5-dimethyl-2-thiazoline-4-carboxylate. Y: 80%. F. e. s. U. Schöllkopf and D. Hoppe, Ang. Ch. 85, 1102 (1973).

Cupric chloride   CuCl$_2$
α,β-Ethylene-β-halogenosulfones   C ≡ C → CHal : C·SO$_2$R
from acetylene derivs. and sulfonic acid halides – Geospecific addition s. 27, 598; also thermal addition without catalyst s. J. Org. Chem. 39, 3867 (1974)

Calcium carbonate s. under (CH$_3$COO)$_4$Pb   CaCO$_3$

Azodiisobutyronitrile $[Me_2C(CN)N:]_2$
β-Alkylthio-β-lactones from mercaptans   C : C → CHC(SR)
s. 28, 534

Diisopropyl peroxydicarbonate $(Me_2CH \cdot OCO \cdot O \cdot)_2$
2-Chlorosulfonyl isocyanates   C : C → CClC·SO₂N : C : O
from ethylene derivs. s. 26, 613

Formic acid/trimethylamine $HCOOH/Me_3N$
Sym. sulfones from ethylene derivs.   2 C : C → CH·C)₂SO₂
s. 27, 599

Acetic acid CH_3COOH
2-Chlorothiocyanates from ethylene derivs.   C : C → CCl·C(SCN)
s. 29, 540

Thiirane oxide   ←
3,6-Dihydro-1,2-dithiin 1-oxides from 1,3 dienes   O
s. 28, 535

Titanium tetrachloride $TiCl_4$
Thioethers from ethylene derivs.   C : C → CHC(SR)
s. 4, 543 suppl. 28

Lead tetraacetate/calcium carbonate $(CH_3COO)_4Pb/CaCO_3$
Cyclic thioethers from ethylenemercaptans   O
S-Bridges s. 27, 595

Hexafluoroisobutenylidene sulfate   ←
β-Sultones from ethylene derivs.

389.
$$\underset{CH_2}{\overset{CF_2}{\|}} + O_2S\underset{O}{\overset{O}{\diamond}}=C(CF_3)_2 \longrightarrow \begin{matrix}CF_2-O\\|\;\;\;\;\;|\\CH_2-SO_2\end{matrix} \;\; [+\; O=C=C(CF_3)_2]$$

Vinylidene fluoride bubbled slowly at 10–20° into hexafluoroisobutenylidene sulfate until hexafluorodimethylketene evolution ceases → 2,2-difluoroethane β-sultone. Y: 98%. F. e. s. G. A. Sokol'skii et al., Khim. Geterotsikl. Soedin. *1973*, 178; C. A. **78**, 136135.

Iron carbonyl $Fe(CO)_5$
Markownikoff addition   C : C(SR) → CHC(SR)₂
Mercaptals from thioenolethers s. 27, 600

Rearrangement ↰

Hydrogen/Oxygen Type   SC∩HO

Potassium hydroxide   KOH
3-Acyl-2,3-dihydrobenzothiophenes   ←
s. 27, 755

Hydrogen/Carbon Type SC∩HC

Irradiation ⇝
S-Heterocyclics from thioketones ○
s. *29*, 541

Triethylamine Et_3N
α-(Alkylthio)thiopyrans ○
s. *26*, 582

p-Toluenesulfonic acid TsOH
α(Alkylthio)-thiophenes and -thiopyrans
also with preceding rearrangement of ketene mercaptals s. *26*, 582

Oxygen/Carbon Type SC∩OC

Without additional reagents w.a.r.
O→S-Alkyl migration OR → SR
s. *27*, 376

Diaxial-diequatorial rearrangement ←
with positional interchange of arylthio and sulfonyl groups s. *26*, 583

Aluminum chloride $AlCl_3$
Dithiolcarbonic acid esters from xanthates OCSSR → SCOSR
s. *18*, 649 suppl. *29*

Trialkyloxonium fluoroborate $R_3O^+ BF_4^-$
Thiolic from thionocarboxylic acid esters CSOR → COSR
s. *27*, 601

m-Chloroperoxybenzoic acid $m\text{-}ClC_6H_4COO_2H$
Ring expansion of cyclic α,β-ethylenesulfones ⊙
via rearrangement of α,β-oxidosulfones – also acid-catalyzed and thermal
rearrangement of α,β-oxidosulfoxides s. *26*, 584

Nitrogen/Sulfur Type SC∩NS

Sodium/alcohol NaOR
1,3-Dithietane from 3-isothiazolone ring ⊙
Dimerization s. *26*, 585

Methyllithium MeLi
Aminosulfones from sulfonic acid amides ←
s. *27*, 602 suppl. *30*

Sulfuric acid H_2SO_4
s. *27*, 602; with methyl- (or *n*-butyl-)lithium at 25° cf. S. J. Shafer and
W. D. Closson, J. Org. Chem. *40*, 889 (1975)

Halogen/Carbon Type SC∩HalC

Without additional reagents w.a.r.
Cephalosporin from penicillin ring skeleton ⊙
s. *29*, 542; s. a. S. Kukolja et al., Am. Soc. *97*, 3192 (1975)

Sulfur/Sulfur Type   SC∩SS

Irradiation   ⚡
Spiro-1,3-S,S-heterocyclics   ○
from isocyclic enedisulfides

390.

A soln. of the startg. m. in 96%-ethanol irradiated 18 hrs. with 2537 Å light of a low-pressure Hg-arc in a quartz vessel → product. Y: 79%. L. Dalgaard and S.-O. Lawesson, Tetrah. Let. *1973*, 4319.

Sulfur/Carbon Type   SC∩SC

Without additional reagents   w.a.r.
Rearrangement of sulfido compds.   C
s. 26, 586

Carbon/Carbon Type   SC∩CC

Irradiation   ⚡
Radical ring closure   ○
  Cyclic thioenolethers from acetylenemercaptans s. 26, 587
Thiazoles from isothiazoles   ←
  Photoisomerization s. 27, 603

Ethyldiisopropylamine   i-Pr₂NEt
2-Ene-1,3-dithioles   ○
from dithiocarboxylic acid propargyl esters

391.

A mixture of the startg. m. and ethyldiisopropylamine in xylene added all at once to xylene heated at 135°, and heating continued 20 min. at 135–140° → product. Y: 91%. F. e. s. J. Meijer et al., R. *92*, 1067 (1973).

Benzoyl peroxide   (PhCOO)₂
Ketosulfones from enol sulfonates   ←
  Radical rearrangement – β-Ketosulfones – 1,5-Rearrangement s. 26, 588
Antimony pentachloride   SbCl₅
2,3,4,7-Tetrahydro-1,4-thiazepin-7-one ring   C
  from penicillin ring skeleton s. 27, 604

Exchange   ↓↑

Hydrogen ↑   SC↕H

Without additional reagents   w.a.r.
Sym. diaryl disulfides and sym. sulfinodiimines   2 ArH → ArSSAr
s. 26, 589

Acyl aryl disulfides   ArSSAc
from acylsulfenylchlorides s. 28, 536

Thianaphthenes   O
s. 26, 579

2,7-Dithiaisotwistanes
s. 28, 537

Ring closure with enamines
Δ^4-2-Thiazolones s. 29, 543

Reactions of N-cyaniminodithiocarbonic acid esters
2,2′-Iminobis(benzothiazoles) from o-aminomercaptans – 2-Alkylthiobenzimidazoles from o-diamines s. 28, 538

5-Alkylthio-1,3,4-thiariazol-2(3H)-ones
from dithiocarbazic acid esters – The chemistry of phosgene, review s. 28, 539

Reactions with ketene and sulfur dioxide
2,1,5-Benzothiadiazepin-4-one 2-oxides s. 26, 590

Electrolysis   ⇃

Sulfonium salts from 2 thioether molecules   ←
Anodic synthesis s. 27, 605

Potassium oxide s. under Cr_2O_3 K_2O

Potassium iodide s. under I   KI

Pyridine s. under $SOCl_2$ C_5H_5N

Montmorillonite/sulfur monochloride   ←

(poly-S)-**Macroheterocyclics**   O
s. 29, 544

Aluminum chloride $AlCl_3$

Replacement of hydrogen by alkylthio groups   H → SR
s. 27, 638

Lanthanum oxide s. under Cr_2O_3 La_2O_3

Acetic anhydride Ac_2O

1,2-Oxathiin 2,2-dioxide ring   O
s. 28, 540

N-Chlorosuccinimide $\begin{matrix}CH_2CO\\CH_2CO\end{matrix}\!\!>\!\!NCl$

Thioethers from mercaptans   SH → SR
s. 26, 592

Lead thiocyanate s. under Br $Pb(SCN)_2$

Phosphorus oxide chloride $POCl_3$

α-Chlorosulfonylcarboxylic acid chlorides   CHCOOH → C(SO$_2$Cl)COCl
from carboxylic acids s. 28, 541

Rhodanamine H_2NSCN

Thiocyanation with rhodanamine   H → SCN
s. 28, 542

Thionyl chloride/pyridine $SOCl_2/C_5H_5N$

Thianaphthenes   O
s. 29, 545

Thiazole ring
4,5,6,7-Tetrahydrothiazolo[4,5-d]pyrimidine-5,7-diones s. 26, 591

Sulfuryl chloride SO_2Cl_2
1,3-Dithiolanes from 1,2-dithiols
s. *27*, 606

Sulfamic acids $RNHSO_3H$
Sulfonation H → SO_3H
s. *15*, 463 suppl. *28*

Sulfur monochloride (s. a. under Montmorillonite) S_2Cl_2
Benzothiazoles from ar. amines ○

392.

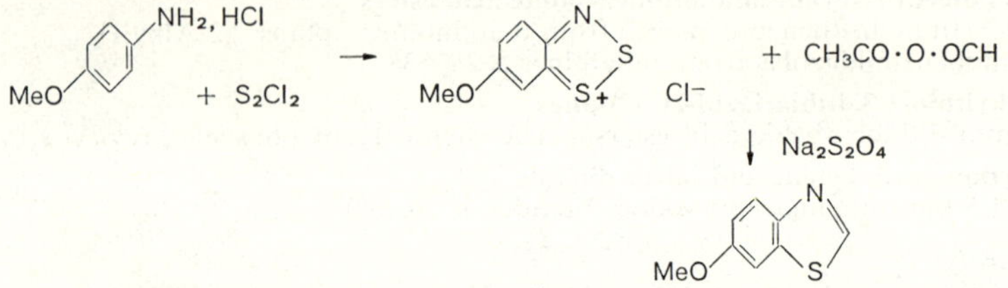

A mixture of 150 g. p-anisidine hydrochloride and sulfur monochloride stirred 20 hrs. at room temp. then 5 hrs. at 70°, the soln. of the resulting intermediate (215 g.) in ice-water treated with 4 N NaOH followed by Na-dithionite, heated to boiling, cooled, treated at 4° with acetic formic anhydride, and kept 24 hrs. at 4° → 6-methoxybenzothiazole. Y: 62%. M. D. Friedman et al., J. Med. Chem. *16*, 1314 (1973).

Sulfur dichloride SCl_2
Sym. thioethers RSR
s. *20*, 444 suppl. *28, 29*

Chromium oxide-lanthanum oxide aluminum
oxide-potassium oxide $Cr_2O_3\text{-}La_2O_3\text{-}Al_2O_3\text{-}K_2O$
Thiophenes from 1,3-dienes ○
s. *28*, 543

Bromine/lead thiocyanate $Br/Pb(SCN)_2$
Electrophilic substitution of ketene mercaptals H → SCN
s. *28*, 544

Iodine/potassium iodide I/KI
Thioethers from mercaptans SH → SR
s. *26*, 592

Manganese dioxide MnO_2
Oxidative sulfonation H → SO_3H
s. *29*, 546

Ferric chloride $FeCl_3$
Replacement of hydrogen by sulfhydryl H → SH
via quinonoid intermediates and isothiouronium salts s. *26*, 593

Oxygen ↑ SC ↕ O

Without additional reagents w.a.r.
Alkylation with N,N'-dicyclohexylisoureas SH → SR
Thioethers from mercaptans – Carboxylic acid esters from carboxylic acids
s. *29*, 547

Mannich reaction with prim. amines
s. *26*, 594

$NH_2 \rightarrow N(CH_2S\cdot)_2$

Amide mercaptals from amide acetals
s. *26*, 595

$C(N<)(OR)_2 \rightarrow C(N<)(SR')_2$

Tetrathiophosphoric acid esters
from alcohols s. *28*, 545

$3 \; ROH \rightarrow (RS)_3PS$

N-Condensed heterocyclics from 1,5-dialdehydes
s. *26*, 345

○

Alkylidene-S-heterocyclics
from α,β-ethylene-α-mercaptocarboxylic acids
  5-Alkylidene-4-thiazolidones – 2-Alkylidene-4,5-dihydrothiophen-3-ones – 4-Alkylidene-1,3-oxathiolan-5(4H)-ones s. *29*, 548

4-Thiazolidone ring
s. *30*, 399

2-Aminothiazoles from ketones
s. *27*, 607

1,4-Thiazin-3-one ring
from o-aminomercaptans and α,β-ethylenecarboxylic acid esters

393.

A mixture of N-benzyl-5-methoxycarbonylmethylidene-2-thionothiazolid-4-one and o-aminobenzenethiol in methanol refluxed 5 min. → product. Y: 80%. F. e. s. H. Nagase, Chem. Pharm. Bull. **22**, 42 (1974).

Tetrahydro-1,3,5-thiadiazine-2-thiones
s. *26*, 596

Irradiation s. under Cyclohexene

Sodium hydride   *NaH*
1,1-Disulfoxides
from sulfinic acid esters and sulfoxides

$CHSO\cdot R \rightarrow C(SO\cdot R)SO\cdot R'$

394.   p—CH₃C₆H₄—SOOMe + MeSOMe → p—CH₃C₆H₄—SOCH₂SOMe

A soln. of methyl p-toluenesulfinate in tetrahydrofuran added dropwise at −5 to 0° during 10 min. to methylsulfinylcarbanion (prepared from dimethyl sulfoxide and NaH) in tetrahydrofuran, and stirred 1 hr. at 0° → product (mixture of *meso*- and *dl*-isomers). Y: 85%. F. e. s. N. Kunieda, J. Nokami, and M. Kinoshita, Chem. Lett. *1973*, 871.

Sodium hydroxide   *NaOH*
Reaction of nucleophiles with oxido compds.   ←
  Thioenolethers – Reactions of oxido compds., review s. *26*, 597

4-Hydroxycephams

395.

4-Mercapto-4-methyl-2-pentanone dissolved at −3° in 2 N NaOH, and added dropwise at 0–5° during 1–1,5 hrs. to an soln. of 4-acetoxy-2-azetidinone → 4-hydroxy-2,2,4-trimethylcepham. Y: 94%. F.e. s. H. W. Schnabel, D. Grimm, and H. Jensen, A. *1974*, 477.

Potassium hydroxide   KOH
1,3-Dithiolane-2-thiones
from oxido compds. s. 28, 546

Sodium/alcohol   NaOR
Thioethers   SH → SR
from mercaptans and phosphates

396.   PhSH + OP(OEt)$_3$ → PhSEt

Thiophenol added under N$_2$ to methanolic Na-methoxide, stirred 15 min. at room temp., 0.9 mole triethyl phosphate added with stirring, and refluxed 3 hrs. → ethyl phenyl sulfide. Y: 94%. F. e., also with other phosphorus compds., s. P. Savignac and P. Coutrot, Synthesis *1974*, 818.

Sodium methoxide   NaOMe
S-Cyanoethylation with and without ring closure

397.

3-Cyanothiochroman-4-ones. Methyl thiosalicylate in dioxane treated with a few drops of Na-methoxide and acrylonitrile, and the product isolated after 8 hrs. at 30° → 3-cyano-2,3-dihydrobenzothiopyran-4-one. Y: 82%. A. P. Momsenko, Zh. Org. Khim. *9*, 775 (1973); C. A. *79*, 31797.

A soln. of thiosalicylic acid in dioxane treated with Na-methoxide to give an alkaline reaction, then acrylonitrile added, after 2 hrs. at 70° left to stand 12 hrs. → o-(β-cyanoethylthio)benzoic acid. Y: 70%.

Sodium hydrogen carbonate/phosphorus pentasulfide   NaHCO$_3$/P$_2$S$_5$
Replacement of carbonyl oxygen by sulfur   CO → CS
s. 22, 799 suppl. 28

Potassium thioacetate   CH$_3$COSK
Thiolactones from lactones
s. 26, 598

Sodium sulfide   Na$_2$S
Dithiohemiacetals from aldehydes   CHO → CH(SH)(SR)
s. 27, 608

Cyclic 2-hydroxythioethers ←
  from sulfonyloxyoxido compds. s. 29, 549

Lithium salt/p-toluenesulfonyl chloride $Li^+/TsCl$
Thioethers from alcohols $OH \rightarrow SR$
s. 28, 547

Sodium salt Na^+
Thioethers from acoxy compds. $OAc \rightarrow SR$
  with 2,2'-cyclonucleoside ring opening s. 26, 599

Triethylamine (s. a. under $TiCl_4$) Et_3N
3,4-Dihydro-2H-1,3-benzothiazin-2-ones ○
and 3,4-dihydro-2H-1,3-benzoxazine-2-thiones
  from o-hydroxyaldehydes and isothiocyanates s. 29, 550

Piperidine $C_5H_{11}N$
5-Arylthioimidazoles
  from 2-cyanonitrones and arylmercaptans s. 29, 551

Pyridine C_5H_5N
S-Formylation of thiophenols $SH \rightarrow SCHO$
  with acetic formic anhydride s. 26, 600

3-Chlorothianaphthene-2-carbonyl chlorides ○
  from cinnamic acids s. 28, 548

Trimethylaluminum Me_3Al
Protection against nucleophilic attack ←
  of carbalkoxyl groups as ketene mercaptals and lactone groups as cyclic dithiolortholactones via hydroxyketene mercaptals s. 29, 552

Thiolic from carboxylic acid esters $COOR \rightarrow COSR'$
s. 29, 553

Alumina Al_2O_3
Sym. thioethers from alcohols $2\ ROH \rightarrow R_2S$
s. 27, 609

Boron fluoride BF_3
Selective S-alkylation SR
s. 28, 549

Boron fluoride/acetic acid BF_3/CH_3COOH
Cyclic mercaptals from acetals ○
  1,3-Dithianes s. 26, 601

Cyclohexene/irradiation ←
Benzothiazolines from alcohols
s. 27, 610

Dimethylformamide dineopentyl acetal $Me_2NCH(OR)_2$
Methyl from hydroxymethyl groups $OH \rightarrow SHet$
  via 2-pyrimidinyl thioethers s. 27, 91

Imidazole/phosphorus pentasulfide ←
2,1-Benzisothiazoles from 2,1-benzisoxazoles $C{-}O \rightarrow C{-}S$
s. 29, 554

N,N'-Carbonyldiimidazole ←
Thiolic acid esters $SH \rightarrow SCOR$
  from carboxylic acids and mercaptans s. 27, 611; with diethyl phosphorocyanidate or diphenyl phosphorazidate instead of N,N'-carbonyldiimidazole cf. S. Yamada, Y. Yokoyama, and T. Shioiri, J. Org. Chem. 39, 3302 (1974)

Formic acid   HCOOH
Mannich-type condensation $SO_2H \rightarrow SO_2CH_2N(OH)\cdot$
with sulfinic acids and hydroxylamines s. 27, 612

Trifluoroacetic acid CF_3COOH
Relayed introduction of alkylthio groups   SR
into carbohydrates s. 29, 555

Carbon disulfide CS_2
Thiolactams from cyclic nitrones
s. 27, 613

$\underset{N \rightarrow O}{\overset{CH}{C}} \rightarrow \underset{NH}{\overset{CS}{C}}$

Titanium tetrachloride/triethylamine $TiCl_4/Et_3N$
Thioenolethers from oxo compds. $CHCO \rightarrow C:C\cdot SR$
s. 29, 556

Trimethyl phosphite $(MeO)_3P$
1-Chloro-2-ketothioethers $COCO \rightarrow COCCl(SR)$
from α-diketones and sulfenylchlorides via 1,3,2-dioxapholenes – Reactions of sulfenylhalides, review s. 26, 602

O,O-Diethyl dithiophosphoric acid $(EtO)_2P(S)SH$
Replacement of oxygen by sulfur   ←
s. 28, 550

Phosphorus pentasulfide (s. a. under Imidazole) P_2S_5
Thioureas from ureas $(>N)_2CO \rightarrow (>N)_2CS$
s. 27, 614

1,3-Dioxole-2-thiones from 1,3-dioxol-2-ones $CO \rightarrow CS$
s. 29, 557

Δ²-Thiazolines from N-acylaziridines   ⊙
s. 27, 615

Ring closures with α-acylaminoketones   ○
Thiazoles s. 26, 603

3H-1,2,4-Dithiazoles from acylimines

398.

$\underset{F_3C}{\overset{F_3C}{>}}C=N-COCHMe_2 \xrightarrow{P_2S_5} \underset{F_3C}{\overset{F_3C}{>}}\overset{S-S}{\underset{N}{C}}CHMe_2$

Equimolar amounts of N-(2,2,2-trifluoro-1-trifluoromethylethylidene)isobutyramide and phosphorus pentasulfide vigorously stirred and heated 48 hrs. at 140° → 5-isopropyl-3,3-bis(trifluoromethyl)-3H-1,2,4-dithiazole. Y: 70%. F. e. s. K. Burger, J. Albanbauer, and W. Strych, Synthesis *1975*, 57.

p-Toluenesulfonyl chloride s. under Li⁺   TsCl

p-Toluenesulfonic acid   TsOH
4-Thiazolidone ring

399.

[indolenine-COOEt + HS-CH₂-COOEt → thiazolidine-fused indoline structure]

A soln. of 2-ethoxycarbonyl-3,3-dimethylindolenine, ethyl thioglycolate, and a trace of p-toluenesulfonic acid in benzene refluxed 20 hrs. → ethyl 9,9-dimethyl-

3-oxo-2,3,9,9a-tetrahydrothiazolo[3,2-a]indole-9a-carboxylate. Y: 62%. Also with thioglycolic acid in the absence of p-toluenesulfonic acid, and f. e. s. H. C. J. Ottenheijm, N. P. E. Vermeulen, and L. F. J. M. Breuer, A. *1974*, 206.

Sulfuric acid H_2SO_4

2,3-Dihydro-4H-1,3-thiazin-4-ones
from oxo compds. and α,β-ethylene-β-mercaptocarboxylic acid amides s. *28*, 551

Dihydro-1,3-thiazine-6-thione ring
from oxo compds. and β-iminodithiocarboxylic acids s. *27*, 616

Hydrogen chloride HCl

2-Ethylenisothiouronium salts
from 2-ethylenealcohols with rearrangement s. *30*, 41
$C:C \cdot C \cdot S \begin{bmatrix} NH_2 \\ NH_2 \end{bmatrix}^+$

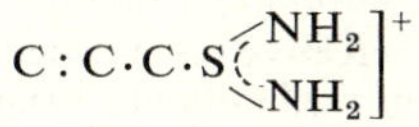

Benzothiazole ring from quinones ○
s. *27*, 617

2-Amino-1,3-thiazines
from α,β-ethyleneketones s. *28*, 552

1,3,4-Thiadiazolium salts
from carboxylic acid thiohydrazide hydrochlorides s. *28*, 553

Hydrogen chloride/acetic acid HCl/CH_3COOH

Replacement of cyclic oxygen by sulfur $\bigcirc O \rightarrow \bigcirc S$
s. *29*, 558

Via intermediates v.i.

Thioethers from alcohols $OH \rightarrow SR$
s. *28*, 547

Thiolactones from lactones
via carboxythioethers s. *27*, 618
$\underset{CO}{C}{-}O \rightarrow \underset{CO}{C}{-}S$

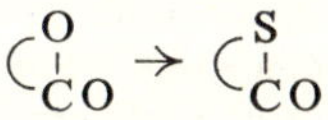

Nitrogen ↑ SC ↕ N

Without additional reagents w.a.r.

Thioethers and amines from azoles $SH \rightarrow SR$
s. *29*, 561

α-Chlorosulfoxides $SO \cdot Cl \rightarrow SO \cdot C \cdot Cl$
from sulfinic acid chlorides and diazo compds. s. *28*, 554

Sulfones $SO_2H \rightarrow SO_2R$
from tert. amines and sulfinic acids s. *29*, 559

Sulfones from azo compounds $N:NR \rightarrow SO_2R$
s. *26*, 604

Subst. sulfonic acid amides from sulfamides $H \rightarrow SO_2N<$
s. *29*, 560

Replacement of imino groups by sulfur $C:NR \rightarrow C:S$
s. *28*, 555

Ring closures with sulfur ○
2-Aminothiazole ring from enamines s. *27*, 619

1,3-Benzodithiole-2-thiones ←
from 1,2,3-benzothiadiazoles s. *27*, 623

Irradiation ⇝

Sulfonium and sulfoxonium ylids ←
from diazo compds. s. *27*, 622

Reactions of α-diazoketones with sulfur dioxide ○
2,3-Dihydro-1,4-oxathiin-2-one 4,4-dioxides s. 29, 562

Lithium   *Li*
Thioethers from halides and sulfenimides   Hal → SR
s. 30, 401

Sodium hydride/carbon disulfide NaH/CS_2
Thiolic acids   CON< → COSH
from prim. and sec. carboxylic acid amides s. 29, 563

Sodium hydroxide   *NaOH*
**Selective replacement
of acylamino by organothio groups**   >N·C·N(CO)(CO) → >N·C·SR
1,1-Aminothioethers s. 30, 576

6-Organothiopteridines from pyridinium salts   ○
s. 27, 620

Potassium/tert-butyl alcohol $K/Me_3C·OH$
Ring closures with azidoquinones

400.

<img: reaction scheme showing 2-azido-5-tert-butyl-1,4-benzoquinone + HSCH₂COOEt → bicyclic product>

2-Azido-5-*tert*-butyl-1,4-benzoquinone allowed to react with ethyl mercaptoacetate in *tert*-butanol in the presence of K-*tert*-butoxide → product. Y: 83%. F. e. s. G. Cajipe, D. Rutolo, and H. W. Moore, Tetrah. Let. *1973*, 4695; indole derivs. cf. P. Germeraad and H. W. Moore, J. Org. Chem. **39**, 774, 781 (1974); **chemistry of azidoquinones, review,** s. Chem. Soc. Rev. **2**, 415 (1974).

Sodium hydrogen sulfide/liq. ammonia $NaHS/NH_3$
Sym. disulfides from nitro compds.   2 RNO_2 → RSSR
s. 29, 564

Triethylamine Et_3N
Thioketones from imines   C:NH → CS
s. 26, 605

C-Sulfenylation with N-sulfenyl compds.   H → SR
C-Disulfenylation with sulfenimides s. 28, 556

Pyridine C_5H_5N
Replacement of amino groups by sulfur   ←
s. 27, 621

Copper   *Cu*
Sulfoxonium ylids from diazo compds. $C:N_2$ → C:S(O)<
s. 27, 622

Cuprous cyanide   *CuCN*
s. 27, 622

Cupric sulfate $CuSO_4$
Sulfonium and sulfoxonium ylids $C:N_2$ → C:S<
from diazo compds. s. 27, 622

Magnesium   *Mg*
Thioethers   Hal → SR
from halides and sulfenimides

401.   PhCH$_2$CH$_2$Br + (N-(phenylthio)phthalimide) → PhCH$_2$CH$_2$—S—Ph

A soln. of N-(phenylthio)phthalimide in dry benzene added under N$_2$ to a stirred ethereal soln. of phenethylmagnesium bromide prepared from phenethyl bromide and Mg in ether, and refluxed 2–3 hrs. → phenyl phenethyl sulfide. Y: 68%. F. e., also with Li, s. M. Furukawa, T. Suda, and S. Hayashi, Synthesis *1974*, 282.

Aluminum chloride   *AlCl$_3$*
Triarylsulfonium salts from tosylsulfilimines   S : NTs → S$^+$\\Ar
s. *26*, 606

Sulfones from sulfonic acid azides   SO$_2$N$_3$ → SO$_2$R
s. *29*, 782

Lithium isopropane nitronate   *[Me$_2$C(NO$_2$)]$^-$Li$^+$*
Reactive nucleophiles   ←
as activators of unreactive nucleophiles
    Preferential replacement of nitro groups by sulfonyl groups s. *26*, 607

Carbon disulfide s. under NaH   *CS$_2$*

Sulfuric acid   *H$_2$SO$_4$*
Mercaptals from morpholinals   C(N<)$_2$ → C(SR)$_2$

402.   BzCH(morpholino)$_2$ + 2 HSCH$_2$Ph → BzCH(SCH$_2$Ph)$_2$

α-**Ketoaldehyde mercaptals.** Startg. *gem*-dimorpholine compd. allowed to react with benzyl mercaptan in acetic acid containing concd. H$_2$SO$_4$ → product. Y: 90%. F. e. s. Y. Le Floc'h, A. Brault, and M. Kerfanto, Bl. *1973*, 3499.

Hydrogen chloride   *HCl*
Ketene mercaptals   C : C : NR → C : C(SR')$_2$
    from ketenimines and mercaptans s. *30*, 386

Via intermediates   *v.i.*
o-Dithiols from o-aminomercaptans   NH$_2$ → SH
    via 1,2,3-benzothiadiazoles and 1,3-benzodithiole-2-thiones s. *27*, 623

Thiolic acid esters from nitriles   CN → COSR
    via thioiminoester hydrochlorides – α-Hydroxythiolic acid esters from cyanohydrins s. *27*, 624

Halogen ↑   SC ↯ Hal

Without additional reagents   *w.a.r.*
Thioethers from sulfinic acid chlorides   ArSOCl → ArSAr'
s. *30*, 409

Reactions with hexamethyldisilthianes   Cl$_2$ → S
    Thioketones from 1,1-dichlorides – Isothiocyanates from iminocarbonylchlorides – Sym. bi-N-sulfonylsulfinodiimides from N,N-dichlorosulfonamides s. *28*, 557

β-Ketosulfoxides SOCl → SO·C·COR
from enolizable ketones and sulfinic acid chlorides

403.   o—$O_2NC_6H_4$—SOCl + CH_3COCH_3 →   o—$O_2NC_6H_4$—$SOCH_2COCH_3$

A mixture of o-nitrobenzenesulfinyl chloride and anhydrous acetone allowed to stand 10–12 hrs. with occasional shaking → o-nitrophenylsulfinylacetone. Y: 80%. F. e. s. N. K. Chapovskaya, L. K. Knyazeva, and N. S. Zefirov, Zh. Org. Khim. 9, 1014 (1973); C. A. 79, 52950.

1,1-Arylthiohydrazones C(Hal) : N·N< → C(SAr) : N·N<
from 1,1-halogenohydrazones s. 29, 565

Acylation of thioureas ←
S-Acylisothiouronium salts, also N-acylation s. 27, 625

Ring closures with di(sulfenylchlorides) ○
2-Chloro-1,4-dithienes s. 26, 608

Benzoxathianones from thioketones
s. 29, 566

1,2-Benzisothiazoles
s. 26, 609

4-Chloro-2-thiazolidones
from S-chloroisothiocarbamyl chlorides and aldehydes s. 26, 610

2-Acylbenzothiazoles
from o-azosulfenylbromides and methyl ketones s. 29, 567

Δ^2-Thiazolinium salts from carboxylic acid thioamides
Reactions of carboxylic acid thioamides, review s. 27, 626

Dihydrothiazolo[3,2-a]pyridinium salts
Betaines s. 26, 611

1,3-Thiazinium betaine ring
s. 28, 558

1-Imino-1H,3H-thiazolo[3,4-a]benzimidazoles
s. 28, 559

5,6,7,8-Tetrahydrothiazolo[3,2-a][1,3]diazepines
s. 26, 612

Ring closure with o-dichlorides
2,3-Dihydro-2-iminothiazolo[4,5-b]quinoxalines from thioureas s. 28, 560

2-Alkylthio-1,3,4-thiadiazoles
from dithiocarbazic acid esters and carboxylic acid chlorides s. 29, 568

Irradiation ⧊

Radical reactions of chlorosulfonyl isocyanates ←
with ethylene derivs. – Isothiazolid-3-one 1,1-dioxides – 2-Chlorosulfonyl isocyanates s. 26, 613

Lithium hydride LiH
N,S-Dialkylation of 2-thiobarbituric acids SR
s. 29, 408

Sodium hydride NaH
Thioiminoesters from carboxylic acid thioamides CS·NH → C(:N)SR
s. 5, 440 suppl. 28

Sodium hydroxide (s. a. under $Na_2S_2O_3$ and $(NH_2)_2CS$) NaOH
Dithiacyclophanes ○
High dilution technique – Dithiaparacyclophanes s. 26, 614; with KOH, dithia[3.3]cyclophanes cf. R. H. Mitchell, T. Otsubo, and V. Boekelheide, Tetrah. Let. 1975, 219

Potassium hydroxide   *KOH*
Bis(alkylthio)methylenehydrazones from hydrazones $C:NNH_2 \rightarrow C:NN:C(SR)_2$
s. 26, 615

Reactions of sulfinic acid with haloforms $SO_2H \rightarrow SO_2CHCl_2$
Dihalogenomethyl sulfones – α,β-Ethylenesulfonic acids – Ramberg-Bäcklund rearrangement s. 27, 627

Sodium/alcohol   *NaOR*
Cyclopropanone mercaptals $CCl_2 \rightarrow C(SR)_2$
from 1,1-dichlorocyclopropanes s. 29, 569

Ring closures with 3-hydroxy-1-mercaptoisoquinolines   O
s. 29, 570

Sodium ethoxide   *NaOEt*
Thioethers from halides   Hal $\rightarrow$ SR
s. 26, 616

Potassium tert-butoxide   *KOCMe$_3$*
Δ^2-Thiazoline ring opening   C
with subsequent transesterification of thioiminoesters s. 29, 571

Butyllithium   *BuLi*
(α-Sulfonylalkylidene)phosphoranes $SO_2F \rightarrow SO_2C:P\leqslant$
from phosphonium salts and sulfonic acid fluorides, also with rearrangement

404.
$$PhCH_2SO_2F \longrightarrow PhCH_2SO_2-\underset{Ph}{C}=PPh_3$$
$$+ \; CH_2\overset{+}{P}Ph_3Br^-$$
$$\underset{Ph}{|}$$
$$CH_3SO_2F \longrightarrow PhCH_2SO_2CH=PPh_3$$

A soln. of butyllithium in hexane added dropwise at room temp. under N$_2$ to a stirred suspension of benzyltriphenylphosphonium bromide in tetrahydrofuran, after 30 min. stirring benzylsulfonyl fluoride in tetrahydrofuran added dropwise during ca. 15 min., and stirring continued 0.5 hr. $\rightarrow$ α-benzylsulfonylbenzylidenetriphenylphosphorane. Y: 91%. – Similarly with methanesulfonyl fluoride $\rightarrow$ benzylsulfonylmethylenetriphenylphosphorane. Y: 95%. F. e., also with sulfonic acid anhydrides, s. B. A. Reith, J. Strating, and A. M. van Leusen, J. Org. Chem. 39, 2728 (1974).

Thiophene ring opening   C
s. 27, 628

Butyllithium/bis(phenylsulfonyl) sulfide   *BuLi/(PhSO$_2$)$_2$S*
Sym. thioethers from halides   2 RHal $\rightarrow$ R$_2$S
s. 27, 629

n-Butyllithium   *n-BuLi*
β,γ-Ethylenesulfoxides   SCl $\rightarrow$ S(O)R
from 2-ethylenealcohols and sulfenylchlorides s. 29, 572

Lithium diisopropylamide   *t-Pr$_2$NLi*
β,γ-Ethylenemercaptals   SR
s. 29, 709

Sodium carbonate   *Na$_2$CO$_3$*
1,3-Thiazepine ring   O
s. 28, 561

Sodium thiocyanate   *NaSCN*
Thioaroyl thiocyanates   Hal $\rightarrow$ SCN
s. 28, 562

Potassium thiocyanate   KSCN
4-Subst. $\Delta^{4,6}$-steroids
s. *29*, 573

Potassium methylxanthate   MeOCSSK

2,4-Diene-1,3-dithietanes from ketene dichlorides

405.

$$2\ Bz{-}CH{=}CCl_2 \xrightarrow{MeO{-}C(S)SK} Bz{-}CH{=}\underset{S}{\overset{S}{\diamond}}{=}CH{-}Bz$$

Desaurins. Benzoylketene dichloride stirred with K-methylxanthate in aq. 90%-methanol → product. Y: 93%. F. e. and methods s. W. Schroth, D. Schmiedl, and A. Hildebrandt, Z. Chem. *14*, 92 (1974); reactions cf. ibid. *14*, 186.

Potassium ethylxanthate   EtOCSSK
Sym. thioethers from halides   2 RHal → R$_2$S
s. *27*, 629

Sodium sulfide   Na$_2$S
Reactions with chlorocyclopropenium salts
    Diamino-cyclopropenones and -cyclopropenethiones s. *29*, 574

Sodium hydrogen sulfide   NaSH
S-Cyclonucleosides
s. *27*, 630

Sodium thiosulfate   Na$_2$S$_2$O$_3$
Replacement of halogen by sulfhydryl   Hal → SH
s. *26*, 617

Sodium thiosulfate/sulfuric acid-sodium hydroxide   Na$_2$S$_2$O$_3$/H$_2$SO$_4$-NaOH
Sulfonylthioureas   SO$_2$Cl → SO$_2$NHCSNH$_2$
    from sulfonic acid chlorides s. *27*, 631

Lithium salt   Li$^+$
Synthesis of thioethers from mercaptans   CHSH → CRSR'
s. *29*, 620

Potassium salt   K$^+$
Acyl alkoxycarbonyl sulfides   ROCOS$^-$ → ROCOSCOR
    also diacyl sulfides s. *26*, 618
2-Ketoxanthates from α-chloroketones   Cl → SC(S)OR
s. *29*, 588

Triethylamine   Et$_3$N
Epimerized *trans*- from *cis*-2-ethylenealcohols   C:C·C(OH)
    via rearranged β,γ-ethylenesulfoxides from sulfenyl chlorides

406.

Startg. *cis*-dl-ketol ester in ether containing ca. 3 equivalents triethylamine treated at room temp. with 1.4 equivalents p-toluenesulfenyl chloride → intermediate dl-sulfoxide (Y: 87%) treated at room temp. with excess trimethyl phosphite in methanol → *trans*-dl-ketol ester (Y: 92%). F. e. s. J. G. Miller et al., Am. Soc. *96*, 6774 (1974).

Ketene S,N-acylals and ketene N-acyl-S,N-acylals ←
  from N-acylcarboxylic acid thioamides s. 29, 575
Hydrazonyl sulfides ←
from 1,1-halogenohydrazones and thioacylhydrazines

407.

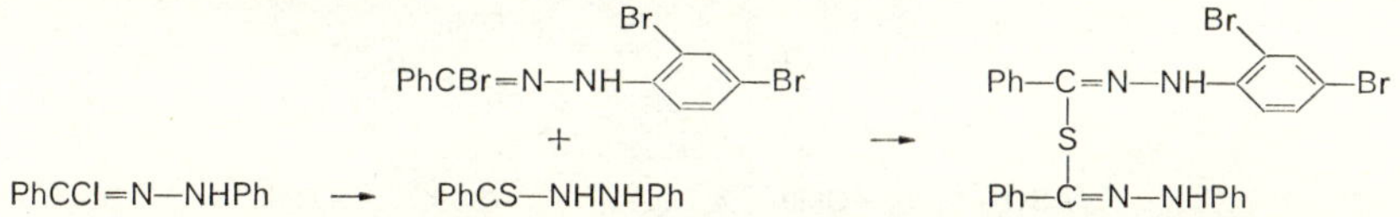

PhCBr=N—NH—C₆H₃Br—Br   Ph—C(Br)=N—NH—C₆H₃Br—Br
            + |
                                     S
PhCCl=N—NHPh → PhCS—NHNHPh |
                                 Ph—C=N—NHPh

A mixture of α-chlorobenzaldehyde phenylhydrazone, 2 moles triethylamine, and chloroform satd. at 0° with H_2S, after 0.5 hr. 4 moles triethylamine added dropwise, and allowed to stand 2 hrs. at room temp. ↠ N-phenyl-N'-thiobenzoylhydrazine (Y: 92%) dissolved with α-bromobenzaldehyde 2,4-dibromophenylhydrazone in chloroform, and treated with 1.1 moles triethylamine ↠ product (Y: 90%). F. e. and procedures s. P. Wolkoff et al., Can. J. Chem. 52, 879 (1974).

1,2-Thiazetidine 1,1-dioxides ○
  from sulfamoyl chlorides s. 28, 563
Benzothiazole N-oxides
s. 28, 564
1,3,5-Oxathiazolines
from thioketones and hydroximinochlorides

408.

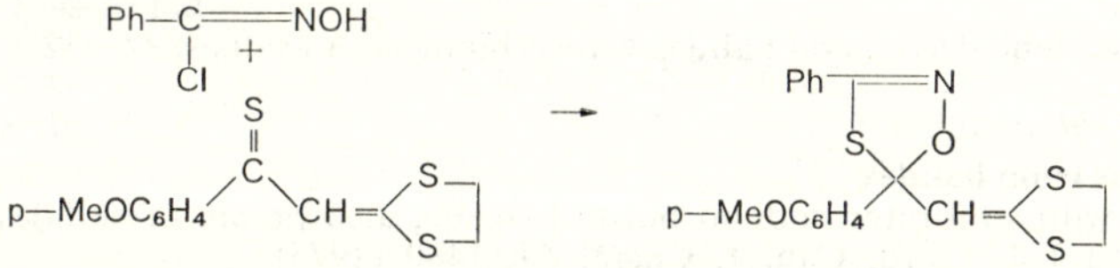

          Ph—C=NOH
          | +
          Cl
          S
          ‖
p—MeOC₆H₄—C   S—   Ph—=N
           \CH=⟨ | →   S O
              S—   p—MeOC₆H₄   CH=⟨S—
                                              S—

Excess triethylamine added slowly to an equimolar soln. of the startg. thioketone and benzhydroxamoyl chloride in ether ↠ product. Y: 85%. F. e. s. J. Maignan and J. Vialle, Bl. 1973, 2388.

Pyridine	C_5H_5N
Allenesulfoxides	C(OH)C≡C → C:C:C(SO·)
from 2-acetylenealcohols and sulfenylchlorides s. 28, 565	
2-Acyl-3-aminothianaphthenes	←
from 3-chloro-1,2-benzisothiazolium chlorides s. 29, 576	
Magnesium	*Mg*
Sulfinic acid esters	Hal → S(O)OR
from sulfurous acid esters and halides s. 29, 577	
Zinc	*Zn*
Mercaptals from 1,1-alkoxyhalides	C(OR)Hal → C(SR')₂
s. 28, 566	
n-Propylmagnesium bromide	*n-PrMgBr*
4H-1,3-Benzothiazines	○
s. 28, 567	
Sulfurated sodium boron hydride	$NaBH_2S_3$
Mercaptans from halides	Hal → SH
s. 28, 569 suppl. 30	
Dimethylthallium compds.	←
Replacement of halogen in iminocarbonylhalides	←
Iminodithiocarbonic acid esters s. 29, 578	

Aluminum chloride $AlCl_3$
Sulfoxides and thioethers   ArSOCl → ArSOAr′
from sulfinic acid chlorides

409.

p-Toluenesulfinyl chloride and anisole

kept 24 hrs. at room temp. in the presence of $AlCl_3$ → p-anisyl p-tolyl sulfoxide. Y: 81%.	refluxed 3 hrs. → p-anisyl p-tolyl sulfide. Y: 70% by glpc.

T. Fujisawa, M. Kakutani, and N. Kobayashi, Bull. Chem. Soc. Japan *46*, 3615 (1973).

Acetic anhydride Ac_2O
1,3-Oxathiolium salts   O
    from carboxylic acid thioamides and α-halogenoketones – Condensed 1,3-oxathiolium salts from thioamides and quinones s. *28*, 568

Thiourea/sodium hydroxide $(NH_2)_2CS/NaOH$
Thioethers   C:C → CH·C(SR)
    from ethylene derivs. and halides via isothiouronium salts s. *27*, 632

1-Acetyl-2-thiourea $AcNHCSNH_2$
Mercaptans from halides   Hal → SH
s. *28*, 569; with sulfurated sodium boron hydride, also prepn. of sulfides, cf. J.-R. Brindle and J.-L. Liard, Can. J. Chem. *53*, 1480 (1975)

Phosphorus pentasulfide P_2S_5
Carboxylic acid thioamides from iminochlorides   CCl:NR → CSNHR

410.

Startg. m. allowed to react with H_2S in the presence of P_4S_{10} as catalyst → N-methylpropionic thioamide. Y: 69–89%. F. e. s. K. Jakopcic and B. Karaman, Bull. Sci., Cons. Acad. Sci. Arts RSF Yougoslavie, Sect. A *18* (4–6), 65 (1973) (Eng); C. A. *79*, 78050.

Antimony pentahalide $SbHal_5$
Sulfones from sulfonic acid halides $SO_2Hal → SO_2Ar$

411.

A mixture of methanesulfonyl fluoride-antimony pentafluoride complex and benzene stirred 12 hrs. at 25° in a water bath → methyl phenyl sulfone. Y: 86.4%. F. e. s. G. A. Olah and H. C. Lin, Synthesis *1974*, 342.

Bis(phenylsulfonyl)sulfide s. under BuLi $(PhSO_2)_2S$

p-Toluenesulfonic acid   *TsOH*
2,3-Dihydro-1,4-dithiins from α-bromoketones   O
s. *26*, 619

Sulfur ↑ SC ↨ S

Without additional reagents — w.a.r.
Reactions of sulfonium salts with sulfenamides
s. 28, 830

Sulfenic from thiolsulfinic acid esters
and acetylene derivs. s. 27, 634

1,1-Sulfinyldisulfides SO·C·SS·
from thiolsulfinic and esters s. 29, 579

Self-thioalkylation of thiolsulfonic acid esters
s. 27, 633

Irradiation (s. a. under BH₃)
Thioethers from halides and disulfides Hal → SR
Preferential replacement of halogen s. 27, 635

Sodium hydroxide NaOH
Δ²-Thiazolines ○
from dithiocarboxylic acid carboxymethyl esters and 2-aminomercaptans

412.

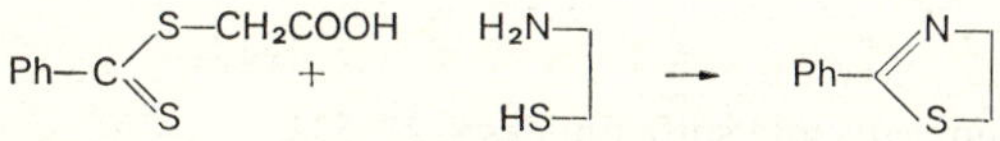

Ph—C(S—CH₂COOH)(=S) + H₂N—/HS— → Ph—C(N—/S—)

under mild conditions. A soln. of thiobenzoylthioacetic acid and cysteamine in methanol adjusted to pH 8 by additions of aq. NaOH stirred at room temp. under slightly reduced pressure to remove generated H₂S until the red color of the thiobenzoylthioacetic acid has disappeared → 2-phenyl-Δ²-thiazoline. Y: 87%. F. e. s. N. Suzuki and Y. Izawa, Tetrah. Let. *1974*, 1863.

2-Amino-Δ²-thiazolines
from thiosulfuric acid S-monoesters s. 28, 570

Sodium/alcohol NaOR
Thioethers from active methylene compds. SSR → SR'
and disulfides s. 29, 580

Thioethers from hindered phenols ArH → ArSAr'
and aryl disulfides s. 28, 571

n-Butyllithium n-BuLi
2-Alkylthio-1,3-dithianes H → SR
from 1,3-dithianes s. 28, 177

tert-Butyllithium Me₃C·Li
Reactions via 1,1-ethyleneorganolithium compds.
Thioenolethers C:CBr → C:C·SR
from α,β-ethylenebromides and disulfides

413.

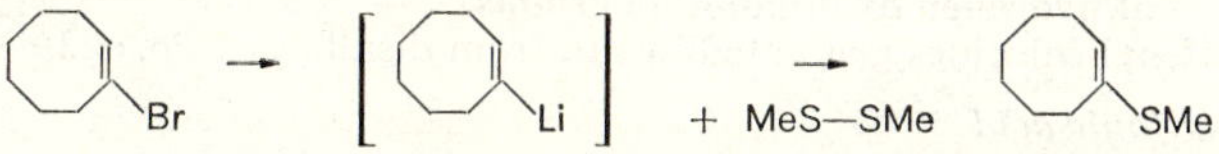

[cyclooctene-Br] → [cyclooctene-Li] + MeS—SMe → [cyclooctene-SMe]

tert-Butyllithium in pentane added at –78° under argon during 10 min. to a soln. of 1-bromo-1-cyclooctene in tetrahydrofuran, after 1.5 hrs. treated with dimethyl disulfide, and allowed to warm during 4 hrs. → 1-methylthio-1-cyclooctene. Y: 80%. F. e. and reactions s. D. Seebach and H. Neumann, B. *107*, 847 (1974).

Lithium N-isopropyl-N-cyclohexylamide $i\text{-}Pr(C_6H_{11})NLi$
C-Sulfenylation   H → SR

414.   n—$C_8H_{17}$$CH_2$COOEt → n—$C_8H_{17}$CHCOOEt
                                                                      |
                                                                      SPh

    + Ph—S—S—Ph   ↓

                                         [n—$C_8H_{17}$$COCH_3$]

of carboxylic acid esters. Ethyl decanoate in tetrahydrofuran treated at –78° with Li-N-isopropylcyclohexylamide, then at –25° to room temp. with phenyl disulfide → ethyl α-phenylthiodecanoate. Y: 91%. – This is the first stage of a 5-step process for the prepn. of **methyl ketones from carboxylic acid esters**, also for the **isomerization of ketones** (cf. Synth. Meth. *23*, 169) **via their α-sulfenylation.** F. e. s. B. M. Trost, K. Hiroi, and S. Kurozumi, Am. Soc. *97*, 438 (1975); prepn. of **α,β-ethylene-** and **α-keto-carboxylic acid esters via α,α-disulfenylation,** cf. J. Org. Chem. *40*, 148 (1975); **oxidative decarboxylation** cf. Am. Soc. *97*, 3528 (1975).

Potassium cyanate   KOCN
2H-1,3,5-Thiadiazin-2-ones   ☉
    from 1,2,4-dithiazolium salts s. *27*, 636

Triethylamine Et_3N
Thioenolethers   ←
    from sulfonium salts and sulfenimides s. *28*, 572

Magnesium   Mg
Dithiocarbamic acid esters $>N \cdot C(S)S \cdot]_2$ → $2\,N \cdot C(S)SR$
    from bisthiuram disulfides s. *26*, 620

Borane/irradiation $BH_3/\text{\textit{\Hbar}}$
Thioethers   C:C → CHC(SR)
    from ethylene derivs. and disulfides s. *27*, 637

Borane/Air BH_3/O_2
    via hydroboration s. *27*, 637

Aluminum chloride $AlCl_3$
Electrophilic substitution   ArH → ArSR
    Ar. thioethers from thiolsulfonic acid esters s. *27*, 638

Triphenylphosphine Ph_3P
Thiolic acid esters   COOH → COSR
    from carboxylic acids and disulfides s. *26*, 621

Air s. under BH_3 O_2

Sodium dithionite $Na_2S_2O_4$
Benzothiazoles from ar. amines   ←
s. *30*, 392

Sulfuryl chloride SO_2Cl_2
Replacement of hydrogen by organothio groups   H → SR
    with sulfenyl chlorides generated in situ from disulfides s. *26*, 622

p-Toluenesulfonic acid   TsOH
Transesterification   ○
    of tetrathioorthocarbonic acid esters – Tetrathiospirocyclics s. *26*, 623

1,4-Benzothiazine ring
    from o-aminodisulfides and ketones s. *26*, 624

Remaining Elements ↑   SC ↕ Rem

Without additional reagents   *w.a.r.*
N-(Alkylthiocarbonyl)ureas   ←
    from isocyanates and thioboric acid esters s. 27, 639
Replacement of cyclic tin by cyclic carbon   ←
    Carbonic acid derivs. – Spiroorthocarbonic acid esters s. 26, 625

Irradiation/potassium thiocyanate   *hν/KSCN*
Ar. thiocyanates $ArTl(OCOCF_3)_2 \rightarrow ArSCN$
    from arylthallium bis(trifluoroacetates) s. 27, 888

Disodium hydrogen phosphate Na_2HPO_4
Thiolic from selenolic acid esters
Selective S-acylation   SH → SCOR
Protection of sulfhydryl groups

415.
    $BzSeCH_2CH_2\overset{+}{N}Me_3$ Br^- + $HSCH_2CH(NH_2)COOH$ → $BzSCH_2CH(NH_2)COOH$

A mixture of L(–)-cysteine hydrochloride and benzoyl selenol choline bromide stirred 24 hrs. in water, and the product precipitated by adding disodium phosphate soln. to pH 6.2 → S-benzoylcysteine. Y: 89%. F. e. s. A. Makriyannis, W. H. H. Günther, and H. G. Mautner, Am. Soc. 95, 8403 (1973).

Pyridine C_5H_5N
N-Condensed 1,5,3-dithiazepine ring   ○
    from mercuric acetylides and isothiocyanates s. 28, 573

Aluminum chloride $AlCl_3$
Thiolic from carboxylic acid esters and silyl sulfides   COOR → COSR′
    s. 29, 581

Carbon ↑   SC ↕ C

Without additional reagents   *w.a.r.*
Transesterification of thioiminoesters   C(:N·)SR → C(:N·)SR′
    s. 29, 571

1,2,4-Thiadiazoles   ←
from 1,3,4-oxathiazol-2-ones and nitriles
Nitrile sulfides as intermediates

416.
    p-ClC₆H₄–⟨N,S,C=O,O⟩ → [p-ClC₆H₄–C(≡N)–S + N≡C–C₆H₄Cl-p] → p-ClC₆H₄–⟨N,S,N=C–C₆H₄Cl-p⟩

5-p-Chlorophenyl-1,3,4-oxathiazol-2-one added at 190° in 3 portions at 5-min. intervals to stirred p-chlorobenzonitrile, and stirring continued 10 min. at 190° → 3,5-bis-(p-chlorophenyl)-1,2,4-thiadiazole. Y: 62–71%. F. e. s. R. K. Howe and J. E. Franz, J. Org. Chem. 39, 962 (1974).

Lithium   *Li*
Alkyl from aryl sulfoxides   SOAr → SOR
    Optically active compds. s. 29, 582

Sodium hydroxide   *NaOH*
Dithiocarboxylic acid esters   CSSR → CSSR′
    by transesterification s. 26, 626

Sodium hydroxide/benzyltriethylammonium chloride $NaOH/PhCH_2N^+Et_3\ Cl^-$
Thioethers from thiocyanates   SCN → SR

417. $PhCH_2SCN + HCCl_3 \longrightarrow PhCH_2SCCl_3$

Aq. 50%-NaOH added portionwise to a vigorously stirred soln. of benzyl thiocyanate, chloroform, and benzyltriethylammonium chloride, and kept 3–4 hrs. at 40° → benzyl trichloromethyl sulfide. Y: 80%. F. e. s. M. Mąkosza and M. Fedoryński, Synthesis *1974*, 274.

Potassium hydroxide   KOH
Cyclic 1,1-hydroxythioethers   O
s. *28*, 574

o-Aminothiolic acids   C
from 3,1-benzoxazine-2,4-diones s. *27*, 293

Potassium tert-butoxide $Me_3C \cdot OK$
α,α-Disulfonylethers $C(OR)(SO_2R)_2$
from α-halogeno-α-sulfonylethers s. *29*, 583

Sodium cyanide   NaCN
Thioethers   SCN → SR
from thiocyanates and alcohols or from thioiminocarbonic acid esters s. *28*, 575

Sodium phenylmercaptide   PhSNa
Isothioureas from N-cyanourethans   N(CN)COOR → NHC(:NH)SR

418. $p-O_2NC_6H_4CH_2N(CN)(COOCH_2Ph) + HSPh \longrightarrow p-O_2NC_6H_4CH_2NH-C(=NH)(SPh) \cdot HCl$

Benzyl N-p-nitrobenzylcyanocarbamate allowed to react 16 hrs. at room temp. with 4 equivalents benzenethiol and a catalytic amount of Na-benzenethiolate in acetonitrile, then treated with ethereal HCl → N-p-nitrobenzyl-S-phenylisothiourea hydrochloride. Y: 80%. F. e. s. T. Taguchi, Y. Sato, and T. Mukaiyama, Chem. Lett. *1973*, 1225.

Sodium sulfide Na_2S
Thioethers from thiocyanates   SCN → SR

419.   (2-CF₃-C₆H₄)-SCN → [(2-CF₃-C₆H₄)-SH] + ClCH₂CH₂COOH → (2-CF₃-C₆H₄)-SCH₂CH₂COOH

A soln. of 2-trifluoromethylthiocyanatobenzene in ethanol heated with concd. aq. Na_2S (s. Synth. Meth. *2*, 54) on a water bath, treated with Na-β-chloropropionate, and heated 1.5 hrs. at 80° → γ-(2-trifluoromethylphenylthio)propionic acid. Y: 87%. F. e. s. P. Jacquignon et al., Experientia *30*, 452 (1974).

Triethylamine Et_3N
Mercaptans from α-aminonitriles   C(CN)N< → CHSH

420.   (3,4-(MeO)₂C₆H₃)-CH(CN)(NHCHMe₂) → (3,4-(MeO)₂C₆H₃)-CH₂SH

A soln. of α-(3,4-dimethoxyphenyl)-α-isopropylaminoacetonitrile in anhydrous pyridine-triethylamine treated 2 hrs. with a slow H_2S-stream, then kept 16 hrs. at room temp. in a sealed flask → 3,4-dimethoxytoluene-α-thiol. Y: 92%. F. e. s. R. Crossley and A. C. W. Curran, Soc. Perkin I *1974*, 2327.

Piperidine	$C_5H_{11}N$
Sym. thioethers from thiolic acid esters s. *29*, 584	2 RSCOR' → R$_2$S
Benzyltriethylammonium chloride s. under NaOH	$PhCH_2N^+Et_3$ Cl^-
Hydrogen bromide/acetic acid **1,1-Aminothioethers from Δ^3-5-oxazolones** s. *26*, 627	HBr/CH_3COOH C

Elimination ⇈

Oxygen ↑   SC⇈O

Sodium/alcohol   NaOR
S-Cyclonucleosides ○
s. *26*, 628

Sodium hydrogen carbonate s. under (PhO)$_2$CO $NaHCO_3$

Acylazo compds. ←
Cephalosporins from penicillin S-oxides ⊙
s. *29*, 585; f. method cf. J. J. de Koning et al., J. Org. Chem. *40*, 1346 (1975)

Acetic anhydride Ac_2O
S(IV)-Heterocyclics ←
s. *24*, 642

Diphenyl carbonate/sodium hydrogen carbonate $(PhO)_2CO/NaHCO_3$
8,2'-Cyclonucleosides ←
s. *27*, 640

Phosgene $COCl_2$
Thiocyanates from N-formylsulfenamides   S·NHCHO → S·CN
s. *28*, 576

Trimethyl phosphite $(MeO)_3P$
Rearrangement of penicillin ring S-oxides ←
s. *26*, 629

Polyphosphoric acid $H(PO_3H)_xOH$
5-Aminothiazoles ○
from α-acylaminocarboxylic acid thioamides s. *27*, 641

Thionyl chloride $SOCl_2$
**1,3-S,N-Heterocyclics
from hydroxycarboxylic acid thioamides**

421.

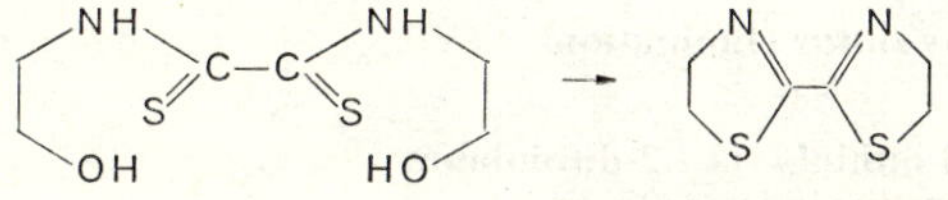

4 moles SOCl$_2$ added in 2 equal portions to a stirred suspension of N,N'-bis-(2-hydroxyethyl)dithiooxamide in toluene, and stirred 1.5 hrs. at 45° after the initial exothermic reaction has subsided → 2,2'-bi-2-thiazoline dihydrochloride. Y: 99%. F. e. s. D. A. Tomalia and J. N. Paige, J. Org. Chem. *38*, 3949 (1973).

Nitrogen ↑   SC⇑N

Without additional reagents   w.a.r.
Phenothiazine ring with rearrangement
    Review of nitrenes s. 26, 630

Potassium hydroxide   KOH
1,4-Dithiafulvenes
    from 2 1,2,3-thiadiazole molecules s. 28, 577

Halogen ↑   SC⇑Hal

Without additional reagents   w.a.r.
Δ^2-Thiazoline ring
    Powerful ambident neighboring groups s. 26, 631
2-Amino-1,3,4-thiadiazines
    from α-chlorothiosemicarbazones s. 12, 945

Sulfur ↑   SC⇑S

Without additional reagents   w.a.r.
Benzo[c]thiophenes from 2,3-benzodithianes
s. 29, 586

Sodium/alcohol   NaOR
s. 29, 586

Sodium thiosulfate/acetic acid $Na_2S_2O_3/CH_3COOH$
Benzothiazoles from 1,2,4-benzodithiazines
s. 26, 566

Sodium iodide   NaI
Chlorothionoformic acid esters
    from chloro(chlorosulfenyl)oxymethyl disulfides s. 26, 632

Pyridine s. under $(CF_3CO)_2O$ C_5H_5N

Copper   Cu
Thioethers from thiolsulfonic acid esters $SO_2SR \rightarrow SR$
    by SO_2-elimination s. 29, 587

Trifluoroacetic anhydride/pyridine $(CF_3CO)_2O/C_5H_5N$
Thietane-2,4-diones
    from α-dithiolic acids s. 28, 578

Tris(dimethylamino)phosphine $(Me_2N)_3P$
Ring contraction by sulfur elimination
s. 27, 642

Thietanes from 1,3-dithiols via 1,2-dithiolanes
    Thioethers from disulfides s. 27, 642

Perchloric acid $HClO_4$
1,3-Dithiol-2-one ring
    from α-chloroketones via 2-ketoxanthates s. 29, 588

Carbon ↑ SC⇑C

Without additional reagents   w.a.r.
Thiolic acid esters $CO \cdot S \cdot CO \cdot SR \to COSR$
    from acyl polythiocarbonates s. 28, 579

2-Amino-1,3-dithietan-2-ylium salts
    from 1,1-bis(dithiocarbamic acid esters)

422.

Startg. m. treated 5–30 min. at room temp. with 70%-$HClO_4$ → 2-dimethylamino-4-phenyl-1,3-dithietan-2-ylium perchlorate. Y: 90%. F. e. and method s. Y. Ueno and M. Okawara, Chem. Lett. *1973*, 863.

Thiazaphospholanes   O
s. 26, 633

Benzo[c]thiophenes from 2-benzothiopyr-3-ones   Ö:
    Decarbonylation s. 26, 634

Irradiation   ⇝
α-Dithiones from 1,3-dithiol-2-ones   C
    α-Dithione-1,2-dithiete equilibrium s. 29, 589

Decarbonylation of thiolactones $\overset{S}{\underset{CO}{C}} \to CS$
s. 27, 643

Sodium carbonate Na_2CO_3
2,3-Dihydro-6H-1,3-thiazine ring   O
s. 29, 590

Sodium cyanide   NaCN
Thioethers from thioiminocarbonic acid esters $SC(:NH)OR \to SR$
s. 28, 575

Aluminum chloride $AlCl_3$
Thiocoumarins   O
    also selenocoumarins s. 26, 635

Aluminum chloride/thionyl chloride $AlCl_3/SOCl_2$
Thiophene S-oxide ring from sulfinic acids
    with C-de-*tert*-butylation s. 29, 591

Trifluoroacetic anhydride $(CF_3CO)_2O$
Thiolactones from carboxythioethers $\overset{SR}{\underset{COOH}{C}} \to \overset{S}{\underset{CO}{C}}$
s. 27, 618

Thiourea $SC(NH_2)_2$
Thioindoxyls   O
    from o-alkylthio-ω-diazoacetophenones s. 27, 644

Thionyl chloride s. under $AlCl_3$ $SOCl_2$

Hydrogen bromide/acetic acid HBr/CH_3COOH
Heterocyclics by ether cleavage
    Thianaphthenes s. 26, 657

Formation of Rem—Rem Bond

Exchange ↓↑

Hydrogen ↑   RemRem ↓↑ H

Chloramine
Diselenides from selenols
s. *27*, 587

$$2 \cdot \text{SeH} \xrightarrow{NH_2Cl} \cdot \text{SeSe} \cdot$$

Halogen ↑   RemRem ↓↑ Hal

Sodium
Hypodiphosphorous acid esters
   from phosphorochloridites s. *28*, 580
Tetrathiohypodiphosphorous acid esters
   from dithiophosphorochloridites

$$>\text{PHal} \xrightarrow{Na} >\text{P} \cdot \text{P}<$$

423.

2 [benzodithia-P—Cl] → [benzodithia-P—P-dithiabenzo]

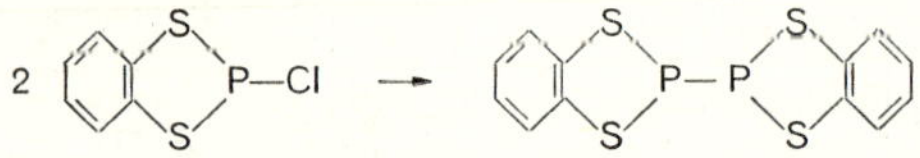

o-Phenylene dithiochlorophosphite refluxed 72 hrs. with a vigorously stirred Na-dispersion in o-xylene → bis-(o-phenylenedithio)hypodiphosphite. Y: 49%. M. Baudler et al, Z. Naturf. *28b*, 363 (1973).

Lithium tetrahydridoaluminate $Li[AlH_4]$
Replacement of group IV element halogen
   by alkylseleno groups s. *28*, 520   GeHal → GeSeR

Silyl selenide $(R_3Si)_2Se$
Boroselenines from dihalogenoboranes
s. *28*, 581

Elimination ⇑

Remaining Elements ↑   RemRem ⇑ Rem

Irradiation
1,2-Disilacyclohexanes
   from 1,2,3-trisilacycloheptanes s. *28*, 582

Formation of Rem—C Bond

Uptake

Addition to Oxygen and Remaining Elements   RemC↓ORem

Phenyl acetate CH_3COOPh
Distannoxanes from diorganotin oxides $3 >Sn:O \rightarrow \geqslant Sn \cdot O \cdot Sn \leqslant$
s. *27*, 645

Addition to Oxygen and Carbon   RemC↓OC

Without additional reagents   w.a.r.
2,3-Dihydroxybenzenephosphonic acid esters   ←
    from o-quinones and dialkyl phosphites s. *29*, 592
Butyllithium   BuLi
Synthesis of 2-ene-1,4-diols   ☺
    from boranes and furans via dihydro-1,2-oxaborins s. *28*, 583
Di-n-butylamine n-Bu_2NH
1-Hydroxy-1,1-diphosphonic acid esters $CO \rightarrow C(OH)PO(OR)_2$
    from acylphosphonic acid esters and dialkyl phosphites s. *28*, 584
Sodium tetrahydridoborate $Na[BH_4]$
Organic selenium compds. from elemental selenium   C
    Dicarboxydiselenides from lactones s. *28*, 585
Phosphorus trichloride PCl_3
Combined prepn. of 1-hydroxy-1,1-diphosphonic acids $COOH \rightarrow C(OH)(PO_3H_2)_2$
    and carboxylic acid chlorides from carboxylic acids s. *29*, 593

Addition to Nitrogen and Carbon   RemC↓NC

Triethylamine Et_3N
Thiocarbamylphosphine oxides $>P(O)H \rightarrow >P(O)C(S)NHR$
    from isothiocyanates s. *26*, 636
Zinc chloride/triethylamine $ZnCl_2/Et_3N$
N-Silylenamines from nitriles $CHCN \rightarrow C:CHNHSi \leqslant$
s. *27*, 646

Addition to Halogen and Carbon   RemC↓HalC

Without additional reagents   w.a.r.
Dibromoalanes from bromides $Br \rightarrow AlBr_2$
s. *29*, 594

Insertion of metals into halogen-carbon bonds ←
  Tribromogermanes s. 27, 647

Cuprous chloride CuCl
Halogenosilanes from halides ←
s. 28, 586

Addition to Sulfur and Carbon RemC⇓SC

Without additional reagents w.a.r.
Sulfinovinylbetaines from thiirene 1,1-dioxides C

424.

Hexamethylphosphorous triamide | 1,5-Diazabicyclo[4.3.0]non-5-ene
in benzene added in one portion at room temp. to a soln. of 2,3-diphenylthiirene 1,1-dioxide in dry O$_2$-free benzene → products. Y: ca. 100%.
F. **sulfinovinylphosphonium betaines**, also **sulfovinylbetaines** by oxidation, s. B. B. Jarvis and W. P. Tong, Synthesis *1975*, 102.

Sulfur-donor ligand o-metalated complexes ←
s. 29, 643

Addition to Remaining Elements RemC⇓RemRem

Dichlorobis(triphenylphosphine)palladium(II) [PdCl$_2$(Ph$_3$P)$_2$]
Reaction of disilanes with acetylene derivs. ⊙

425.

Si-Heterocyclics. A soln. of 1,1,2,2-tetramethyl-1,2-disilacyclopentane, dimethyl acetylenedicarboxylate, and dichlorobis(triphenylphosphine)palladium in benzene refluxed 3 hrs. under N$_2$ → dimethyl 1,1,4,4-tetramethyl-1,4-disilacyclohept-2-ene-2,3-dicarboxylate. Y: 83.4%. F. e. s. H. Sakurai, Y. Kamiyama, and Y. Nakadaira, Am. Soc. *97*, 931 (1975).

Addition to
Remaining Elements and Carbon RemC⇓RemC

Aluminum chloride AlCl$_3$
Organosiliconphosphorus compds. C
s. 28, 587

Addition to Carbon Atoms   RemC⇓C

Without additional reagents   w.a.r.
Protection of carbon-carbon triple bonds   ←
as cobalt carbonyl complexes s. 27, 140

Addition to Carbon-Carbon Bonds   RemC⇓CC

Without additional reagents   w.a.r.
2-Aminomercuration $C:C \rightarrow C(N<)C \cdot Hg^+$
s. 29, 595

Peroxymercuration of ethylene derivs. $C:C \rightarrow C(OOR)C \cdot Hg^+$
Markownikoff addition – Hydroperoxymercuration – α-Alkoxyketones from enolesters s. 26, 637; s. a. A. J. Bloodworth and I. M. Griffin, Soc. Perkin I 1975, 195

Heterotwistanoids   O
s. 29, 596

1,1-Diboro compds. from acetylene derivs. $C \equiv C \rightarrow CH_2C(B<)_2$
s. 30, 427

2-Aminoeneboranes   ←
from aminoboranes and acetylene derivs. s. 29, 597

Di(ene)chloroboranes $2 C \equiv C \rightarrow (CH:C)_2BCl$
from 2 acetylene deriv. molecules s. 28, 666

α,β-Ethyleneboronic acid esters $C \equiv C \rightarrow CH:C \cdot B(O \cdot)_2$
from acetylene derivs. and boronic acid esters s. 29, 497

B-Heterocyclics from dienes   O
s. 27, 648

1,3,2-Benzodioxaboroles   ←
from ethylene derivs. s. 26, 638

2-Acetoxyvinylthallic acetates $C:C \rightarrow C(OAc):C \cdot Tl(OAc)_2$
from acetylene derivs. s. 29, 598

α-Cyano-α,β-ethylenecarboxylic acid amides $C:C(CN)_2 \rightarrow C:C(CN)CONH_2$
from alkylidenemalononitriles via 2-carbamyl-2-cyanophosphonium salts s. 28, 588

α-Bromophosphonic acid dibromides $C:C \rightarrow CBr \cdot C \cdot P(O)Br_2$
from ethylene derivs. s. 29, 599

α-Phosphoranylideneketenes   C
from cyclopropenones s. 28, 589

γ-Irradiation   ⟋⟋⟋
Hydrostannylation
s. 19, 719 suppl. 29

Potassium tert-butoxide $KOCMe_3$
Poly-*tert*-phosphines   ←
from enephosphine sulfides s. 29, 600; s. a. Am. Soc. 97, 46, 53 (1975)

4,5-Dihydro-3-azaphospholes   O
Reactions of vinyl isocyanide with group V element(III) compds. s. 26, 639

Silver trifluoroacetate CF_3COOAg
2-Hydroxyselenides $C:C \rightarrow C(OH)C \cdot SeR$
from ethylene derivs. and selenylbromides s. 29, 601

Magnesium   Mg
1,2-Disilyl-1-ethylene from acetylene derivs. $C \equiv C \rightarrow C(Si \leqslant):C(Si \leqslant)$
with subsequent cleavage of phenolethers via aryloxysilanes s. 29, 602

β-Silyl- from α,β-ethylene-oxo compds. COC : C → COCHC·Si≦
1,4-Addition s. 26, 640

Borane BH_3
Monoorganomercury compds. C : C → CHC·Hg⁺
from ethylene derivs. – *anti* Markownikoff hydrohalogenation – Reduction to hydrocarbons s. 26, 641

Lithium tetrahydridoborate/boron chloride $Li[BH_4]/BCl_3$
Dialkylchloroboranes and borinic acid esters 2 C : C → CHC·)₂BCl
from ethylene derivs. s. 27, 649

Azodiisobutyronitrile $[Me_2C(CN)N :]_2$
1,4-Dihydrostannins from 1,4-diynes

426.

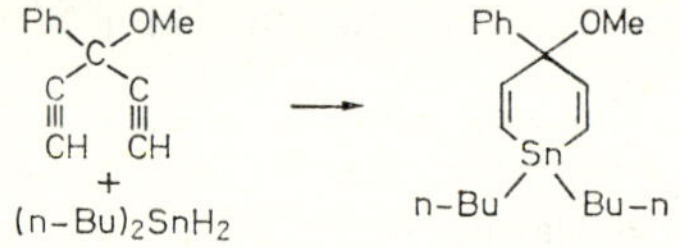

Ph_C_OMe
  C≡C
  CH CH
  +
(n-Bu)₂SnH₂

→

Ph_C_OMe (on Sn ring with n-Bu, Bu-n)

Startg. m. refluxed with di-*n*-butyltin dihydride in methylcyclohexane in the presence of azodiisobutyronitrile → 1,1-di-*n*-butyl-4-methoxy-4-phenyl-1,4-dihydrostannin. Y: 68%. F. e. s. G. Märkl and F. Kneidl, Ang. Ch. 85, 990 (1973); Tetrah. Let. 1975, 2411.

Di-tert-butyl peroxide $Me_3C·OO·CMe_3$
Phosphinic acid esters RR'P(O)OR''
from phosphonous acid monoesters and ethylene derivs. s. 11, 712 suppl. 29

Triphenylphosphine s. under Pd and H₂PtCl₆ Ph_3P

Bromine Br
2-Acoxyselenides C : C → C(OAc)C·SeR
from ethylene derivs. and diselenides s. 29, 180

Nickel carbonyl $Ni(CO)_4$
Hydrosilylation C : C → CHC(Si≦)
Regiospecific addition s. 29, 603

Palladium/triphenylphosphine Pd/Ph_3P
1,4-Addition and preferential addition
s. 26, 642; silylation with hydridotetrakis(triphenylphosphine)rhodium(I), RhH(PPh₃)₄, cf. H. Kono, N. Wakao, and I. Ojima, Chem. Lett. 1975, 189

Hexachloroplatinic acid/triphenylphosphine H_2PtCl_6/Ph_3P
Regiospecific addition
s. 29, 603

Via intermediates v.i.
1,1-Dimercury compds. C ≡ C → CH₂C(HgCl)₂
from acetylene derivs. via 1,1-diboro compds.

427. 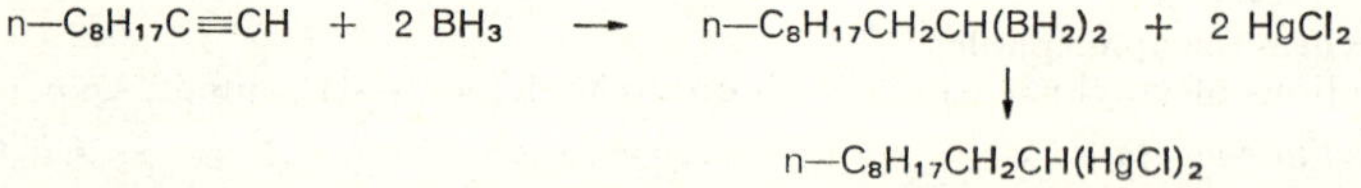 n—C₈H₁₇C≡CH + 2 BH₃ → n—C₈H₁₇CH₂CH(BH₂)₂ + 2 HgCl₂
↓
n—C₈H₁₇CH₂CH(HgCl)₂

1-Decyne allowed to react at 0° with borane in tetrahydrofuran under N₂, stirred 1 hr. at room temp., then treated slowly at 0° with methanol, after 1 hr. at room temp. HgCl₂ added followed by dropwise addition of aq. 5 M NaOH with vigorous stirring, which is continued 24 hrs. at room temp. → product. Y: 77%. F. e. s. R. C. Larock, J. Organometal. Chem. 61, 27 (1973).

Rearrangement ↻

Oxygen/Remaining Elements Type   RemC↻ORem

n-Butyllithium   n-BuLi
α-Hydroxyphosphonic from phosphoric acid derivs.   P·OCH → P·C(OH)<
s. *29*, 604

Exchange ↓↑

Hydrogen ↑   RemC↕H

Without additional reagents   w.a.r.
Deuteriation   H → D
    with replacement of cyclic sulfur by cyclic nitrogen s. *26*, 432
Mercuration of ar. amines   ArH → ArHgOCOR
Mono- and di-mercuration

428.

o-Toluidine stirred overnight with mercuric acetate, 0.25 mole in water → mono- | 2 moles in ethanol → di-mercury deriv. Y: 80%.
F. e. s. J. Arient and J. Podstata, Coll. *39*, 955 (1974).

β-(Acetoxymercury)ketones from cyclopropanols   C
s. *26*, 643

Telluride dichlorides   2 ArH → Ar$_2$TeCl$_2$
s. *28*, 967

Sodium hydride/dimethyl sulfoxide-D$_6$ NaH/(CD$_3$)$_2$SO
Replacement of benzylic hydrogen by deuterium   ArCH → ArCD
s. *27*, 650

Sodium hydroxide s. under LiBr   NaOH

Sodium deuterioxide   NaOD
α-Labeled amines from N-nitrosamines   ←
s. *28*, 590

Sodium/ethanol-O-d   Na/EtOD
Deuteriation   H → D
s. *19*, 722

Potassium tert-butoxide   KOCMe$_3$
   via tricarbonylchromium(0) complexes s. 28, 591

Potassium carbonate   K$_2$CO$_3$
Preferential deuteriation
s. 29, 81

Lithium bromide/sodium hydroxide   LiBr/NaOH
Polyfluoromercury compds.   2 ArH → Ar$_2$Hg
s. 26, 644

Pyridine (s. a. under Ph$_3$PBr$_2$)   C$_5$H$_5$N
Preferential deuteriation   H → D
s. 29, 81

Aldehydes-2-d$_2$ CH$_2$CHO → CD$_2$CHO
s. 27, 651

Cuprous tert-butoxide   Cu·OCMe$_3$
Metalation with copper *tert*-butoxides   ←
   Cuprous acetylides s. 27, 652

Mercuric acetate   (CH$_3$COO)$_2$Hg
Cyclic ethers from ethylenealcohols
   by oxymercuration-demercuration s. 29, 172

Aluminum chloride   AlCl$_3$
Phenoxaphosphine 10-oxides
   from diaryl ethers s. 30, 541

Acetic anhydride   Ac$_2$O
H-Labeled α-acylaminocarboxylic acids   H → T
s. 26, 645

Dimethyl sulfoxide-D$_6$ s. under NaH   (CD$_3$)$_2$SO

Triphenylphosphine dibromide/pyridine   Ph$_3$PBr$_2$/C$_5$H$_5$N
Deuterioaldehydes from aldehydes   CHO → CDO
   via 1,1-dibromides and 1,1-dicyclimmonium salts prepared in situ s. 27, 653

Selenium monochloride   Se$_2$Cl$_2$
Phenoselenazines
s. 26, 646

Benzoselenopheno[2,3-b]benzoselenophenes
   from 1,1-diarylethylenes s. 28, 592

Hydrogen chloride   HCl
Ar. deuteriation at high temp.   ArH → ArD

429. C$_6$H$_5$—COOH → C$_6$D$_5$—COOH

Benzoic acid heated 75 hrs. at 275° with 4%-HCl in D$_2$O → benzoic acid-d$_5$. Y: > 95%; % exchange 94%. – Similarly: Aniline → aniline-d$_5$. Y: 72%; % exchange 97%. Also phenol-d$_5$ s. N. H. Werstiuk and T. Kadai, Can. J. Chem. 52, 2169 (1974).

Oxygen ↑   RemC ↓↑ O

Without additional reagents   *w.a.r.*
P-Aminomethylation   PH → P·CH$_2$N<
s. *29*, 605

α-Cyanophosphonium salts   C : CBr(NO$_2$) → C(CN)$\overset{+}{P}$ ≦
from 1,1,1-ethylenenitrobromides and phosphines s. *29*, 606

Se,N-Heterocyclics   O
Perhydro-1,5,3,7-diselenadiazocines s. *28*, 593

n-Butyllithium   *n-BuLi*
N-Heterocyclic phosphonic acid esters   ←
from dialkyl phosphite and alkoxylcyclimmonium salts s. *27*, 654

Polyethyleneimine   ←
Di- from mono-organomercury compds.   2 RHg$^+$ → R$_2$Hg
s. *26*, 647

Zinc chloride/phosphorus trichloride   *ZnCl/PCl$_3$*
α-Alkoxyenephosphonic acid esters   CH·C(OR)$_3$ → C : C(OR)·PO(OR)$_2$
from phosphorous and orthocarboxylic acid esters via α,α-dialkoxyphosphonic acid esters – Review of orthocarboxylic acid esters s. *29*, 607

Borane   *BH$_3$*
Borinic acid esters from trialkylboranes   >B·OAr
s. *27*, 659

Lithium tetrahydridoaluminate   *Li[AlH$_4$]*
Unsym. trialkylboranes   >B·OAr → >BR
from borinic acid esters s. *27*, 659

Acetic anhydride   *Ac$_2$O*
Arsonium ylids   C : As ≦
from arsine oxides s. *29*, 608

Phosphorus trichloride s. under ZnCl$_2$ *PCl$_3$*

Nitrogen ↑   RemC ↓↑ N

Without additional reagents   *w.a.r.*
1-Halogenodiorganomercury compds.   HgHal → Hg·C·Hal
from monoorganomercury halides and diazo compds. s. *27*, 655

Phosphines and arsines from amines   N< ⇄ As< / P<
s. *29*, 609

β-Carbalkoxyenephosphonic acid esters   C(NO$_2$) : CH → CH : C·PO(OR)$_2$
from α,β-ethylene-α-nitrocarboxylic acid esters s. *28*, 594

Telluro- and seleno-acetals   ←
from diazo compds. and ditellurides or diselenides s. *26*, 648

Sodium   *Na*
α-Aminophosphonic acid esters   C(NH$_2$)P(O)(OR)$_2$
from dialkyl phosphites and azines s. *29*, 610

Copper   *Cu*
Silanes from diazo compds. $>CN_2 \rightarrow >CHSi\leqslant$

430.
$$\text{Ph—Si(Me)(Me)—H} + N_2CHCOOMe \rightarrow \text{Ph—Si(Me)(Me)—CH}_2COOMe$$

A mixture of phenyldimethylsilane, methyl diazoacetate, and commercial Cu-powder stirred and heated intermittently at 90° under N_2 for 4–15 min. → methyl dimethylphenylsilylacetate. Y: 90%. F. e. s. H. Watanabe et al., J. Organometal. Chem. *69*, 389 (1974).

Acetic acid CH_3COOH
Selenothiolcarbamic acid esters $\cdot S \cdot C(:Se)N<$
 from isothiouronium salts s. *26*, 652

Halogen ↑   **RemC ⇅ Hal**

Without additional reagents   w.a.r.
Diiodoboranes from iodides $I \rightarrow BI_2$
 s. *26*, 649

1-Phosphonioiminoesters from 1,1,1-ethylenenitrohalides   ←

431.
$$Ph_3P + Ph-CH=C(Br)(NO_2) + HOMe \rightarrow Ph_3\overset{+}{P}-CH_2-N=C(OMe)(Ph)\ Br^-$$

1-Bromo-1-nitro-2-phenylethylene in methanol added in one portion at room temp. to triphenylphosphine in the same solvent, and stirred 24 hrs. → (α-methoxybenzylideneaminomethyl)triphenylphosphonium bromide. Y: 55%. F. e. and products s. C. J. Devlin and B. J. Walker, Soc. Perkin I *1974*, 453.

Formamidiniumphosphonic acid anhydrides   ←
 from chloroformamidinium chlorides s. *28*, 595

Selenides from selenylhalides   SeHal → SeR
 s. *29*, 912

1-Phosphonio-1-selenoniomethylid salts   ←
 from alkylidenephosphoranes and selenide dichlorides

432.
$$\text{[ ]}SeCl_2 + 2\ CH=PPh_3(COOEt) \rightarrow [\text{[ ]}Se\cdots C\cdots PPh_3(COOEt)]^+ Cl^-$$
$$[\ + CH_2-\overset{+}{P}Ph_3(COOEt)\ Cl^-\ ]$$

A soln. of 0.41 g. 1,1-dichlorotetrahydroselenophene in abs. tetrahydrofuran added dropwise at 20° during 1 hr. to a soln. of 1.39 g. ethoxycarbonylmethylenetriphenylphosphorane in the same solvent, and stirring continued 1 hr. → (tetramethyleneselenonio)(triphenylphosphonio)(ethoxycarbonyl)methylid chloride. Y: 89%. N. N. Magdesieva and R. A. Kyandzhetsian, Zh. Org. Khim. *9*, 1755 (1973); C. A. *79*, 126223.

Electrolysis   ↯
Sym. dialkylmercury compds. from α,ω-dibromides R_2Hg
Controlled-potential electroreduction

433. $2\ BrCH_2CH_2CH_2CH_2Br \rightarrow (CH_3CH_2CH_2CH_2)_2Hg$

1,4-Dibromobutane electrolyzed with a stirred Hg-cathode at –2.5 v. vs. satd. calomel electrode in dimethylformamide containing tetrabutylammonium fluoroborate as supporting electrolyte ⇥ di-*n*-butylmercury. Y: 93%. F. e., also isolation of di-*n*-hexyldimercury, (n-C_6H_{13}Hg)$_2$, a mercury(I) compd., s. J. Casanova and H. R. Rogers, Am. Soc. **96**, 1942 (1974).

Sodium hydride   NaH
α-Silylnitriles   CHCN ⇥ C(Si ≦)CN
from nitriles and chlorosilanes s. 29, 295

Sodium hydroxide   NaOH
(α-Sulfonylalkylidene)phosphoranes $SO_2 \cdot C : P ≦$
Review of cyclic phosphonium ylids s. 27, 656

Sodium compound   ⇐
Selenonium ylids from selenide dichlorides   >CHNa ⇥ C : SeR$_2$

434.   Ph$_2$SeCl$_2$ + NaCHAc$_2$ ⟶ Ph$_2$Se=CAc$_2$

A methanolic soln. of diphenylselenium dichloride treated at room temp. with a soln. of Na-acetoacetonate in the same solvent, and stirred 3 hrs. ⇥ diacetylmethylenediphenylselenurane. Y: 60%. K. H. Wei et al., Am. Soc. **96**, 4099 (1974); **from selenoxides and selenonium imides** cf. S. Tamagaki and K. Sakaki, Chem. Lett. **1975**, 503.

Potassium/alcohol   KOR
1,1-Mercurydiazo compds.   CH(: N$_2$) ⇥ C(: N$_2$)HgR

435.   MeHgCl + KOEt ⟶ [MeHgOEt] + HC—COOMe ⟶ MeHg—C—COOMe
                                                 ‖   ‖
       [+ KCl]   N$_2$ N$_2$

Methylmercuric chloride stirred at 60° in abs. ethanol and treated with ethanolic K-ethoxide, stirring continued 4 hrs. at 60°, KCl removed by filtration, methyl diazoacetate added dropwise at 0°, and stirred 15 min. after completion of the addition ⇥ methyl methylmercuridiazoacetate. Y: 86%. F. e. s. S. J. Valenty and P. S. Skell, J. Org. Chem. **38**, 3937 (1973).

Ethyllithium   EtLi
1-Organoseleno-1-organothio-1-en-3-ynes   C
by selenophene ring opening

436. [reaction scheme: 3-bromo-2-methyl-5-(methylthio)selenophene + BrEt ⟶ Z-MeS/SeEt/CH$_3$ substituted enyne]

Ethereal ethyllithium added to a stirred soln. of 3-bromo-2-methyl-5-(methylthio)selenophene in anhydrous ether, stirred 15 min., ethyl bromide added, and stirred 4 hrs. ⇥ Z-1-ethylseleno-1-methylthiopent-1-en-3-yne. Y: 55%. S. Gronowitz and T. Frejd, Acta Chem. Scand. **27**, 2242 (1973).

Butyllithium   BuLi
1-Alkoxyallenylsilanes   C : C : C(OR)(Si ≦)
s. 27, 835

Lithium diisopropylamide   i-Pr$_2$NLi
Reactions of metalated N-nitrosamines   N(NO)·CH ⇥ N(NO)·C·Si ≦
with heteroelectrophiles

437.   Me$_3$C—N—Me ⟶ [Me$_3$C—N—CH$_2$Li] + ClSiMe$_3$ ⟶ Me$_3$C—N—CH$_2$SiMe$_3$
                 | | |
                 NO   NO   NO

N-Nitrosomethyl-*tert*-butylamine added at −78° to a stirred soln. of Li-diisopropylamide in anhydrous tetrahydrofuran, and after 10 min. treated with trimethylchlorosilane ⇥ N-nitrosotrimethylsilylmethyl-*tert*-butylamine. Y: 80%. Also alkylation and reaction with other heteroelectrophiles such as disulfides, diselenides, or chlorostannanes s. D. Seebach and D. Enders, J. Med. Chem. **17**,

1225 (1974); **reactions of metalated N-nitrosamines with electrophiles, review,** s. Ang. Ch. *87*, 1 (1975); **1-alkylation of N-nitrosamines** cf. D. H. R. Barton et al., Soc. Perkin I *1975*, 579.

C,N-Disilylketenimines $CH_2CN \to C(Si\lessdot) : C : N \cdot Si\lessdot$
from nitriles and chlorosilanes

438.

$$PhCH_2CH_2CN + 2\ ClSi\underset{Me}{\overset{Me}{\diagup}}\!\!-CMe_3 \longrightarrow \underset{PhCH_2}{\overset{Me_3C-Si-Me}{\diagup}}C=C=N-Si\underset{Me}{\overset{Me}{\diagup}}\!\!-CMe_3$$

3-Phenylpropionitrile stirred 5 min. under N_2 at $-78°$ with a soln. of 2 moles Li-diisopropylamide in tetrahydrofuran, and 4 moles *tert*-butyldimethylsilyl chloride in tetrahydrofuran added $\to$ product. Y: ca. 100%. F. e., also N-silylketenimines (cf. Synth. Meth. *29*, 295), s. D. S. Watt, Synth. Commun. *4*, 127 (1974).

Triethylamine   *Et₃N*
(α-Azoalkylidene)phosphoranes $C(Hal) : N \cdot NH \cdot \to C(: PR_3) \cdot N : N \cdot$
    from 1,1-hydrazonohalides and phosphines s. *29*, 611

Pyridine s. under *EtMgBr*   *C₅H₅N*

Magnesium   *Mg*
1,2-Disilanes $C : C \to C(Si\lessdot)C(Si\lessdot)$
    from chlorosilanes and ethylene derivatives – 1,2-Digermanes s. *26*, 650

3-Hydroxystannanes $Sn \cdot O \cdot C \cdot C \cdot CBr \to Sn \cdot C \cdot C \cdot C \cdot OH$
from 3-stannoxybromides
s. *27*, 657

Ethylmagnesium bromide   *EtMgBr*
1-Acetylenephosphonic acid esters $C \equiv CH \to C \equiv C \cdot PO(OR)_2$
    from acetylene derivs. and phosphorochloridates – 1-Acetylenethionophosphonic acid esters s. *28*, 596

Ethylmagnesium bromide/pyridine   *EtMgBr/C₅H₅N*
1-Acetylenethionophosphonic acid esters $C \equiv C \cdot PS(OR)_2$
s. *28*, 596

Sodium dihydridobis-(2-methoxyethoxo)aluminate $NaAlH_2(OCH_2CH_2OMe)_2$
Formation of phosphorus-carbon bonds $>P(O)OR \to >P(O)R'$
    Phosphine oxides from phosphinic acid esters s. *27*, 658

Aluminum bromide   *AlBr₃*
Phosphonic acid dichlorides from halides $Hal \to POCl_2$
s. *26*, 651

Sulfur ↑   RemC ↕ S

Without additional reagents   *w.a.r.*
α-Alkylthio- and α-mercapto-phosphonic acid esters $SH \to PO(OR)_2$
from phosphorous acid esters and 1,1-dithiols

439.

$$+ P(OMe)_3 \quad \underset{SH}{\overset{SH}{\diagdown}}\!\!\bigcirc\!\! \quad + P(OPr-i)_3$$

$$\underset{PO(OMe)_2}{\overset{SMe}{\diagdown}}\!\!\bigcirc \qquad\qquad \underset{PO(OPr-i)_2}{\overset{SH}{\diagdown}}\!\!\bigcirc$$

Excess trimethyl phosphite added dropwise with stirring under N_2 to a soln. of 1,1-cyclohexanedithiol in toluene, and refluxed several hrs. $\to$ dimethyl 1-methylthiocyclohexylphosphonate. Y: 70%. – Similarly with triisopropyl phosphite → diisopropyl 1-mercaptocyclohexylphosphonate. Y: 81%. F. e. s. Z. Yoshida, T. Kawase, and S. Yoneda, Tetrah. Let. *1975*, 235.

3,6-Dihydro-2H-1,2-thiaphosph(V)orin 2-sulfides ←
from perthiophosphonic acid anhydrides and 1,3-dienes s. *29*, 612

2,6-Diamino-1,4-thiaphosph(V)orin 4-sulfides ←
from perthiophosphonic acid anhydrides and 2 ynamine molecules s. *29*, 613

Sodium hydrogen carbonate $NaHCO_3$
Selenoureas and selenonothiocarbamic acid esters $N \cdot C(: Se) \cdot N$
from isothiouronium salts s. *26*, 652

Remaining Elements ↑   RemC ⇅ Rem

Without additional reagents w.a.r.
C-Transilylation ←
s. *26*, 653

P-Heterocyclics from dihalides and diphosphines ○
Märkl cyclic phosphonium salt synthesis s. *28*, 597

Arsenanes from arsine dichlorides $\geqslant AsCl_2 \rightarrow\ \geqslant AsR_2$

$$Me_3AsCl_2 + 2\ LiMe \rightarrow Me_5As$$

A soln. of methyllithium in dimethyl ether added slowly at $-60°$ under N_2 with exclusion of moisture to a suspension of trimethylarsine dichloride in the same solvent, and stirred 6 hrs. at $-60°$ → pentamethylarsenane. Y: 80%. K.-H. Mitschke and H. Schmidbaur, B. *106*, 3645 (1973).

Sodium hydroxide $NaOH$
1,1-Dimercury from 1,1-diboro compds. $C(B<)_2 \rightarrow C(HgCl)_2$
s. *30*, 427

Alane etherates ←
Tetraalkylstannanes R_4Sn
s. *26*, 654

Aluminum chloride $AlCl_3$
Partial replacement of silyl by dichlorophosphino groups $SiR_3 \rightarrow PCl_2$

$$Me_3Si\text{—}\langle\rangle\text{—}SiMe_3 + PCl_3 \rightarrow Me_3Si\text{—}\langle\rangle\text{—}PCl_2$$
$$[+\ ClSiMe_3]$$

A mixture of PCl_3 and $AlCl_3$ stirred 1 hr. at room temp., then a soln. of p-bis-(trimethylsilyl)benzene in PCl_3 added dropwise with stirring, which is continued 0.5 hr. at room temp., refluxed 3 hrs., $POCl_3$ added to the hot mixture, stirred 0.5 hr., and the resulting $AlCl_3 \cdot POCl_3$-complex removed by filtration → p-trimethylsilylphenyldichlorophosphine. Y: 80%. K. Dey, J. Indian Chem. Soc. *50*, 224 (1973).

Via intermediates v.i.
Mixed trialkylboranes $>BR \rightarrow\ >BR'$
via borinic acid esters s. *27*, 659; from dialkylbromoboranes cf. A. Pelter, K. Rowe, and K. Smith, Chem. Commun. *1975*, 532

Carbon ↑   RemC ⇅ C

Without additional reagents w.a.r.
α-Mercaptophosphonic acid esters $P(OR)_3 \rightarrow C(SH)PO(OR)_2$
from phosphorous acid esters and 1,1-dithiols
s. *30*, 439

Glow discharge $\quad\quad\quad\quad\quad\quad\quad\quad\quad\quad\quad\quad\quad\quad\quad\quad\quad\quad\quad\quad\quad\quad\quad\quad$ ⇃
Trifluoromethylmetal compds. $\quad\quad\quad\quad\quad\quad\quad\quad\quad\quad\quad\quad\quad\quad\quad$ $CF_3 \cdot M$

442.
$$CF_3\!-\!CF_3 \longrightarrow \begin{bmatrix} 2\ CF_3 \cdot \\ 4\ CF_3 \cdot \end{bmatrix} \begin{array}{l} +\ HgI_2 \longrightarrow (CF_3)_2Hg \\ +\ SnI_4 \longrightarrow (CF_3)_4Sn \end{array}$$
$$2\ CF_3\!-\!CF_3 \longrightarrow$$

Hexafluoroethane passed during 5 hrs. at ≤ 0.1 mm through a radiofrequency glow discharge field of ca. 23 w. at 4.8–10 MHz, and allowed to react with $HgI_2 \twoheadrightarrow$ bis(trifluoromethyl)mercury. Y: 95% based on HgI_2 consumed. – Similarly with $SnI_4 \twoheadrightarrow$ tetrakis(trifluoromethyl)tin. Y: > 90%. F. e. s. R. J. Lagow et al., Am. Soc. *97*, 518 (1975); **reactions by glow discharge, review,** s. K. Gorzny, Chem. Ztg. *99*, 257 (1975).

Sodium iodide/acetic acid $\quad\quad\quad\quad\quad\quad\quad\quad\quad\quad\quad\quad\quad\quad$ NaI/CH_3COOH
α-Hydroxyphosphine oxides $\quad\quad\quad\quad\quad\quad\quad\quad\quad$ $CO \twoheadrightarrow C(OH)P(O)R_2$
from oxo compds. and phosphinylformic acid benzyl esters s. *29*, 614

Mercuric acetate $\quad\quad\quad\quad\quad\quad\quad\quad\quad\quad\quad\quad\quad\quad\quad\quad\quad$ $(CH_3COO)_2Hg$
β,γ-Ethyleneoxo compds. $\quad\quad\quad\quad\quad\quad\quad\quad\quad\quad\quad\quad\quad\quad\quad\quad\quad\quad$ C
from 2-alkoxyenecyclopropanes s. *30*, 156

Selenium dioxide $\quad\quad\quad\quad\quad\quad\quad\quad\quad\quad\quad\quad\quad\quad\quad\quad\quad\quad\quad\quad$ SeO_2
1,2,3-Selenadiazoles $\quad\quad\quad\quad\quad\quad\quad\quad\quad\quad\quad\quad\quad\quad\quad\quad\quad\quad\quad$ O
s. *26*, 655

Hydrochlorides $\quad\quad\quad\quad\quad\quad\quad\quad\quad\quad\quad\quad\quad\quad\quad\quad\quad\quad\quad\quad\quad$ ←
Phosphonic acid esters $\quad\quad\quad\quad\quad\quad\quad\quad\quad\quad\quad\quad\quad\quad\quad\quad\quad\quad\quad$ ←
from phosphites and N-heterocyclics s. *27*, 660

Nickel chloride $\quad\quad\quad\quad\quad\quad\quad\quad\quad\quad\quad\quad\quad\quad\quad\quad\quad\quad\quad\quad\quad$ $NiCl_2$
Ar. phosphonic and phosphinic acid esters $\quad\quad\quad\quad\quad\quad\quad\quad$ ←
from halides and phosphorous or phosphonous acid esters respectively s. *26*, 656

Elimination ⇑

Hydrogen ↑ $\quad\quad\quad\quad\quad\quad\quad\quad\quad\quad\quad\quad\quad\quad\quad\quad\quad\quad\quad\quad\quad\quad\quad$ RemC⇑H

Sodium amide $\quad\quad\quad\quad\quad\quad\quad\quad\quad\quad\quad\quad\quad\quad\quad\quad\quad\quad\quad\quad\quad\quad$ $NaNH_2$
Cumulated ylids
Vinylidenephosphoranes $\quad\quad\quad\quad\quad\quad$ $\geqslant \overset{+}{P}\cdot CH:C \twoheadrightarrow \geqslant P:C:C$

443. $\quad\quad\quad\quad\quad\quad$ $Ph_3\overset{+}{P}\!-\!CH\!=\!C(OEt)_2\ BF_4^- \longrightarrow Ph_3P\!=\!C\!=\!C(OEt)_2$

Dry (2,2-diethoxyvinyl)triphenylphosphonium tetrafluoroborate added with stirring to a suspension of $NaNH_2$ in liq. NH_3, which is then allowed to evaporate, and the residue extracted 0.5 hr. with refluxing abs. benzene → (2,2-diethoxyvinylidene)triphenylphosphorane. Y: 48%. H.-J. Bestmann, R. W. Saalfrank, and J. P. Snyder, B. *106*, 2601 (1973).

Halogen ↑ $\quad\quad\quad\quad\quad\quad\quad\quad\quad\quad\quad\quad\quad\quad\quad\quad\quad\quad\quad\quad\quad\quad$ RemC⇑Hal

1,5-Diazabicyclo[5.4.0]undec-5-ene $\quad\quad\quad\quad\quad\quad\quad\quad\quad\quad\quad\quad\quad\quad$ ←
Acridarsines $\quad\quad\quad\quad\quad\quad\quad\quad\quad\quad\quad\quad\quad\quad\quad\quad\quad\quad\quad\quad\quad\quad\quad\quad$ ←
s. *27*, 661

Carbon ↑   RemC⇑C

n-Butyllithium   *n-BuLi*
1,1-Aminosilanes   ←
from quaternary 1,2-silylammonium salts

444.   Ph₃SiCH₂CH₂ṄMe₃ I⁻ → Ph₃SiCH₂NMe₂ [+ CH₂=CH₂]

A small excess of *n*-butyllithium added at –25 to –15° to a soln. of trimethyl-(2-triphenylsilylethyl)ammonium iodide in ether-hexamethylphosphoramide ⇢ (dimethylaminomethyl)triphenylsilane. Y: 56.5%. F. e. s. Y. Sato, Y. Ban, and H. Shirai, Chem. Commun. *1974*, 182.

Aluminum chloride $AlCl_3$
Selenocoumarins   ○
s. *26*, 635

Hydrogen chloride   *HCl*
P-Carbohydrates   ⊙

445.

A soln. of methyl 5-deoxy-5-(ethylphosphinyl)-2,3-O-isopropylidene-β-D-ribofuranoside refluxed 24 hrs. in 0.25 *M* HCl under N₂ ⇢ 5-deoxy-5-(ethylphosphinyl)-D-ribopyranose. Y: 97%. S. Inokawa et al., Carbohyd. Res. *30*, 127 (1973).

Hydrogen bromide/acetic acid HBr/CH_3COOH
Heterocyclics by ether cleavage   ○
     Selenanaphthenes – Thianaphthenes s. *26*, 657

Formation of C—C Bond

Uptake   ⇓

Addition to Hydrogen and Carbon   CC⇓HC

Potassium tert-butoxide s. under n-BuLi $Me_3C·OK$

n-Butyllithium   *n-BuLi*
α-Aminocarboxylic acids from isonitriles   CH·N : C ⇢ C(COOH)NH₂
s. *29*, 898

*n-Butyllithium/potassium tert-butoxide/
lithium bromide*   *n-BuLi/Me₃C·OK/LiBr*
Metalation   H ⇢ COOH
s. *28*, 599

Lithium diisopropylamide	*i-Pr$_2$NLi*
α-Dicarboxylic acid monoesters	
from carboxylic acid esters s. *28*, 598	
Lithium bromide s. under n-BuLi	*LiBr*
Diphenylcalcium	*Ph$_2$Ca*
C-Carboxylation of hydrocarbons	
Metalation s. *28*, 599	
Phenyltrihalogenomethylmercury	*PhHgCHal$_3$*
Insertion of dihalogenomethylene groups	H → CHHal$_2$
into hydrogen-carbon bonds s. *26*, 658	
Sulfuric acid/nitric acid	*H$_2$SO$_4$/HNO$_3$*
C-Carboxylation	H → COOH
s. *29*, 615	
Via intermediates	*v.i.*
Ar. carboxylation	ArH → ArCOOH
via diaryl ketones s. *28*, 600	

Addition to Oxygen and Carbon   CC⇓OC

Without additional reagents	*w.a.r.*
Synthesis of α-hydroxyketones	COCHO → COCH(OH)R
from α-ketoaldehydes	

446.

Pyruvaldehyde added to an aq. soln. of 4-hydroxy-6-methyl-2-pyridone, and stirred 2 days at room temp. → product. Y: ca. 100%. F. e. s. J. A. Findlay and F. Y. Shum, Synth. Commun. *1973*, 355.

Reaction of formaldehyde with ketones	←
β-Acoxyketones –5-Acyl-1,3-dioxanes s. *26*, 659	
Reactions of activated negatively-subst. quinones	←
s. *27*, 662	
Reactions with vinylogous carboxylic acid amides	H → CH(OH)CCl$_3$
Chloral condensation s. *26*, 660	
Carbalkoxy-α-diazoketones	CO⟩O → COOMe / COCHN$_2$
from dicarboxylic acid anhydrides	
s. *30*, 456	
Homologization of cyclic ketones	☉
via α-hydroxydiazo compds. s. *26*, 661	
Irradiation (s. a. under Li)	⚡
Functionalization of non-activated methylene groups	CH → C·C(OH)
s. *27*, 188	
3-Oxa-4-silaspiro[2,6]nonanes	←
from silacyclohexan-2-ones s. *27*, 708	
Lithium	*Li*
One-step Grignard-type reaction	C·OH
Alcohols from oxo compds. or carboxylic acid esters and halides s. *26*, 662	

Tert. alcohols from ketones CO → C(OH)R
Reversal of stereospecificity s. 30, 452a

Allylic syntheses with 2-ethylenemesitoates CO → C(OH)·C·C : C
via 2-ethyleneorganolithium compds. – 3-Ethylenealcohols from ketones –
1,5-Dienes – Allylic syntheses with ethylene derivs. s. 28, 601

Reactions with ethylene sulfate C
Synthesis of sulfuric acid monoesters s. 28, 773

Lithium/irradiation Li/⫲

Prim. alcohols from cyclic ethers
δ-Hydroxy-n-butylation of arenes with tetrahydrofuran s. 26, 663

Lithium/naphthalene $Li/C_{10}H_8$

β-Hydroxycarboxylic acids from oxo compds. CO → C(OH)·C·COOH
β-Hydroxy-δ-lactones from β-hydroxyketones s. 26, 664

Lithium/naphthalene-diethylamine $Li/C_{10}H_8\text{-}Et_2NH$

α,β-Ethylene-δ-hydroxycarboxylic acids CO → C(OH)·C·C : C·COOH
from ketones s. 27, 663

Sodium hydride s. a. under n-BuLi NaH

Sodium hydride/alcohol NaH/EtOH

1,2-Halogenhydrins CO → C(OH)CHal
from oxo compds. and halides s. 29, 616

Sodium hydride/dimethyl sulfoxide NaH/Me_2SO

Bismethylene transfer ←
s. 27, 664

Sodium hydroxide NaOH

C-Ethylation via α-hydroxyethylation H → C_2H_5
s. 26, 665

Potassium hydroxide/alcohol KOH

Sec. alcohols from aldehydes CHO → CH(OH)R
s. 29, 762

2-Diazoalcohols from oxo compds. CO → C(OH)C(: N_2)
s. 26, 668 suppl. 28

Potassium compd. ←

Reaction of Δ^2-5-oxazolones with CH-acidic compds. ←
1-Acyl-2-amino-Δ^2-pyrrol-4-ones s. 26, 666

Butyllithium BuLi

Metalation of isonitriles CH(N : C) → $C(NH_2)·C·C(OH)$
3-Aminoalcohols from isonitriles and oxido compds. via 3-isocyanoalcohols
and also via 5,6-dihydro-4H-1,3-oxazines s. 29, 617; syntheses via α-metalated
isonitriles, review, s. D. Hoppe, Ang. Ch. 86, 878 (1974)

2-Isocyanoalcohols from oxo compds. CO → C(OH)·C·N : C
s. 26, 667

2-Diazoalcohols from oxo compds. CO → C(OH)C(: N_2)
s. 26, 668

Metalation of diazo compds. at low temp. CO → C(OH)C(: N_2)COOR
α-Diazo-β-hydroxycarboxylic acid esters from oxo compds. s. 26, 668; s. a. A.
1974, 1767

n-Butyllithium n-BuLi

Tert. alcohols from ketones CO → C(OH)R
s. 28, 176

2-Hydroxythioethers from oxo compds. CO → C(OH)·C(SR)
s. 29, 262 suppl. 30

n-Butyllithium/sodium hydride *n-BuLi/NaH*
γ-Hydroxy-β- and keto-carboxylic acid esters CO → C(OH)·C·CO
Stepwise prepn. of dicarbanions

447. (CH₃)₂CO + CH₃COCH₂COOMe → (CH₃)₂C(OH)CH₂COCH₂COOMe

Methyl acetoacetate added dropwise at 0° under N₂ to 1.1 moles NaH-mineral oil dispersion in tetrahydrofuran, stirred 10 min., then a soln. of 1.05 moles *n*-butyllithium in hexane added dropwise, stirred 10 min. at 0°, acetone added in one portion, and again stirred 10 min. at 0° → methyl 5-hydroxy-5-methyl-3-oxohexanoate. Y: 72%. F. e. s. S. N. Huckin and L. Weiler, Can. J. Chem. **52**, 2157 (1974).

Synthesis via β-ketosulfoxide dicarbanions CO → C(OH)·C·CO·C·SO·
Stepwise prepn. of dicarbanions – δ-Hydroxy-β-ketosulfoxides s. 29, 618

Tetrahydrofurylideneacetic acid esters ▽O → C(OH)·C·C·CO
from β-ketocarboxylic acid esters and oxido compds.
via ε-hydroxy-β-ketocarboxylic acid esters

448.

Propylene oxide added at 0° to the dicarbanion of ethyl acetoacetate prepared according to Synth. Meth. **26**, 861, warmed to room temp., the crude intermediate isolated after 3 hrs. and refluxed 2 hrs. in an inert atmosphere with an equal weight of oxalic acid in methylene chloride → ethyl α-(tetrahydro-5-methyl-2-furylidene)acetate. Y: 62%. F. e., **also tetrahydrothiophene analogs from sulfido compds.**, s. T. A. Bryson, J. Org. Chem. **38**, 3428 (1973).

n-Butyllithium/tetramethylethylenediamine *n-BuLi/TMEDA*
Preferential allyclic metalation CO → C(OH)·CH·C:C
3-Ethylenealcohols from ethylene derivs. and oxo compds. – C-β-Hydroxy-ethylation – Synthesis of ethylenethioethers – Regiospecific effect of complexing agents – Ambident character of allylic metalation s. 29, 619

o-Metalation of ar. amines ←
s. 19, 736 suppl. 29

Syntheses via S,C-dianions CHSH → CRSR'
1-Metalation of mercaptans – Synthesis of thioethers s. 29, 620

n-Butyllithium/triethylenediamine ←
2-Hydroxythioethers from ketones CO → C(OH)·C(SR)
s. 29, 262; without triethylenediamine, also from aldehydes, cf. S. Song, M. Shiono, and T. Mukaiyama, Chem. Lett. **1974**, 1161; preferably metalation of tetrahydropyranyl enolethers cf. J. Hartmann, M. Stähle, and M. Schlosser, Synthesis **1974**, 888

tert-Butyllithium/tetramethylethylenediamine *Me₃C·Li/TMEDA*
Metalation of enolethers CHO → CH(OH)CO
α-Hydroxyketones from aldehydes s. 29, 621; prepn. of a soln. of *tert*-butyllithium in pentane cf. W. N. Smith, Jr., J. Organometal. Chem. **82**, 1 (1974).

Phenyllithium *PhLi*
Sec. alcohols from aldehydes CHO → CH(OH)R
s. 27, 665

Lithium/liq. ammonia *Li/NH₃*
Alcohols from oxido compds. ▽O → C(OH)CR
Reductive alkylation s. 27, 666

Lithium diisopropylamide   *i-Pr$_2$NLi*
C-Methylenation   CH$_2$ → C : CH$_2$
    via C-hydroxymethylation and O-mesylation – α-Methylenelactones s. *29, 622*;
    synthesis of α-methylenelactones, review, s. P. A. Grieco, Synthesis *1975*, 67

α-Hydroxycarboxylic acid thioamides   CO → C(OH)CSN<
    from oxo compds. and thioformamides s. *29*, 623

Lithium dicyclohexylamide   *(C$_6$H$_{11}$)$_2$NLi*
2,2-Di- and 2,2,2-tri-halogenalcohols   CO → C(OH)CHal$_3$
from oxo compds.

449.

      ⬡=O + HCBr$_3$ → cyclohexyl-C(OH)(CBr$_3$)

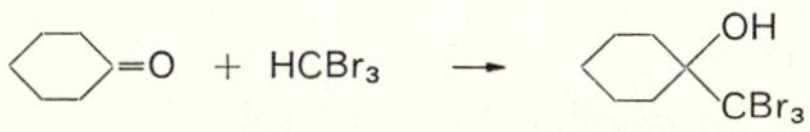

A well-stirred soln. of cyclohexanone and bromoform in dry tetrahydrofuran cooled to –78°, treated dropwise during 15 min. with Li-dicyclohexylamide prepared from *n*-butyllithium and dicyclohexylamine in dry tetrahydrofuran, and kept 1 hr. at the same temp. → 1-(tribromomethyl)cyclohexanol. Y: 91%. F. e. s. H. Taguchi, H. Yamamoto, and H. Nozaki, Am. Soc. *96*, 3010 (1974).

Sodium amide   *NaNH$_2$*
Low temp. Claisen condensation
Stereospecific synthesis   CHO → CH(OH)R
of sec. alcohols from aldehydes

450.

                                              OH
                                                |
        PhCHO + CH$_2$COOMe → PhCH—CH—COOMe
                            | |
                           Ph   Ph

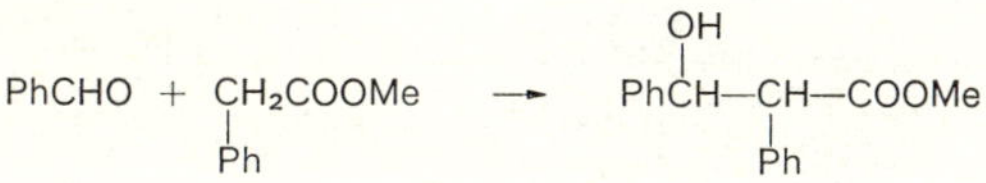

Powdered NaNH$_2$ followed by benzaldehyde added at –24° to a soln. of methyl phenylacetate in abs. *ether*, and stirred 2 hrs. at the above temp. → methyl *threo*-2,3-diphenyl-3-hydroxypropanoate. Y: 72%. F. e. s. B. Kurtev, C. Kratchanov, and N. Kirtchev, Synthesis *1975*, 106.

Sodium amide/lithium bromide   *NaNH$_2$/LiBr*
β-Hydroxyketones   C(OH)·C·CO
s. *5*, 467 suppl. 29

Potassium amide   *KNH$_2$*
β-Hydroxymercaptals from oxido compds.   △O → C(OH)·C·C(SR)$_2$
    C-β-Hydroxyethylation s. *29*, 624

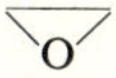

Sodium carbonate   *Na$_2$CO$_3$*
3,4-Dihydroxy-2-imino-2,5-dihydrofurans   O
    from aldehydes s. *9*, 823

Potassium cyanide   *KCN*
α-Hydroxyamidoximes from aldehydes   CHO → CH(OH)C(:NOH)NH$_2$
    Synthesis with addition of 1 C-atom s. *27*, 667

Lactamols from diketones
    Stereospecific double ring closure s. *28*, 603   O

Potassium cyanide · 18-crown-6 polyether   ←
α-Siloxynitriles from oxo compds.   CO → C(CN)OSi≤
s. *28*, 610 suppl. 29

Lithium bromide s. under NaNH$_2$ *LiBr*
Diethylamine (s. a. under Li/C$_{10}$H$_8$)   *Et$_2$NH*

Ring expansions with α-diazophosphorus compds.
via α-hydroxydiazo compds. – 3-Hydroxycarbostyrils from isatins s. 29, 625

2-Imino-2,3-dihydropyrans from flavanones
s. 26, 173

Triethylamine Et_3N

o-Amino-N-heterocyclics from cyanamide
s. 28, 604

Thiophenes from 1,3-oxathiol-2-ylidenimmonium salts
s. 28, 605; cf. Heterocycles 3, 217 (1975)

Pyridine C_5H_5N

Labeled cyanohydrins $CO \rightarrow C(OH)\overset{x}{C}N$
from oxo compds. s. 30, 568

**α-Hydroxymethyl-α-aminocarboxylic acids
from α-aminocarboxylic acids
via α-acylaminocarboxylic acids, 5-oxazolones,
and 5-acylamino-4-oxo-1,3-dioxanes**

451.

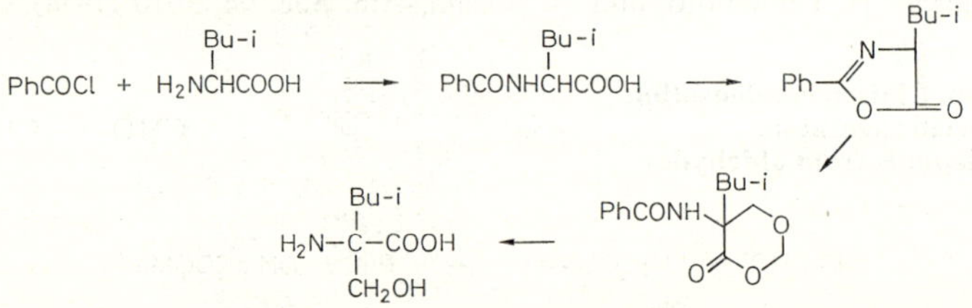

PhCOCl + H$_2$NCHCH(Bu-i)COOH ⟶ PhCONHCH(Bu-i)COOH ⟶ [oxazolone with Bu-i]

[dioxane with Bu-i, PhCONH] ⟶ H$_2$N–C(Bu-i)(CH$_2$OH)–COOH

DL-Leucine allowed to react with benzoyl chloride and 2 N NaOH, the resulting crude benzoyl-DL-leucine heated 0.5 hr. at 90–100° with acetic anhydride, the crude 4-isobutyl-5-oxo-2-phenyl-2-oxazoline obtained dissolved in pyridine, treated with aq. 35%-formaldehyde, shaken 8 hrs., dil. with water, and the precipitated 5-benzoylamino-5-isobutyl-4-oxo-1,3-dioxane boiled 5 hrs. in 5 N HCl → α-isobutylserine. Overall Y: 56%. F. e. s. Z. J. Kamiński, M. T. Leplawy, and J. Zabrocki, Synthesis 1973, 792.

Cupric oxide Cu_2O

Δ²-Oxazolines
from isonitriles and oxo compds. s. 28, 606; synthesis of heterocyclics via copper-isonitrile complexes, review, s. Synthesis 1975, 291

Cuprous oxide/sulfuric acid Cu_2O/H_2SO_4

Tert. carboxylic acids from alcohols

452.

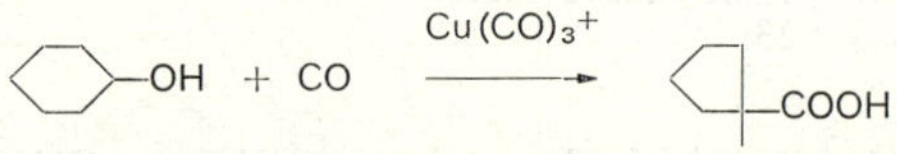

Cyclohexanol–OH + CO $\xrightarrow{Cu(CO)_3^+}$ 1-methylcyclopentane–COOH

Carbonylation. Cyclohexanol added dropwise by syringe at ca. 30° under 1 atm. CO during 20 min. to a suspension of *Cu(I)-carbonyl* prepared by vigorous stirring of a mixture of Cu$_2$O and 98%-H$_2$SO$_4$ under CO, and the product isolated when CO-absorption has ceased after 1–3 hrs. → 1-methylcyclopentanecarboxylic acid. Y: 80%. F. e. s. Y. Souma and H. Sano, Bull. Chem. Soc. Japan 46, 3237 (1973).

Lithium dialkylcuprates $Li[CuR_2]$

Synthesis of alcohols from oxido compds. $\overset{\triangle}{O} \rightarrow C(OH)CR$
with retention of satd. carbonyl moieties s. 26, 669; with Li- or Na-alanates, catalysis by Ni-halides, cf. G. Boireau et al., Tetrah. Let. 1975, 2521

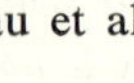

Cuprous iodide s. under Mg   *CuI*
Magnesium (s. a. under EtMgBr)   *Mg*
Tert. alcohols from ketones   CO → C(OH)R
s. *28*, 922

452a.

Reversal of stereospecificity. An ethereal soln. of methyl 2,3-di-O-methyl-6-O-trityl-α-D-*xylo*-hexopyranosid-4-ulose treated at −80° with an ethereal soln. of methylmagnesium iodide prepared from Mg | LiBr-free methyllithium, and and methyl iodide, and stirred 2.5 hrs. at −80° | stirred 1.5 hrs. at −80°
→ methyl 2,3-di-O-methyl-4-C-methyl-6-O-trityl-α-D-
galactopyranoside. Y: 94%. | glucopyranoside. Y: 70%.
Limitations s. M. Miljkovic et al., J. Org. Chem. *39*, 1379 (1974).

Grignard syntheses with amines
s. *1*, 681

Aminolactols from lactones   ←
Retention of cyano groups s. *28*, 607

Synthesis of acoxy compds.   ←
from 2 aldehyde molecules s. *27*, 668

2,4-Dienols from 2H-pyrans   C
s. *27*, 669

5,6-Dienols from 2-acetylenetetrahydropyrans
s. *29*, 626

Ring opening of 1,3-heterocycles
s. *26*, 670

Magnesium/cuprous iodide   *Mg/CuI*
Regiospecific electrophilic syntheses   ←
with α,β-ethyleneketones
via 1,4-addition and magnesium enolates

453.

Aldol condensation. Methylmagnesium iodide prepared from Mg and methyl iodide prepared from Mg and methyl iodide in abs. ether treated at −5° with powdered CuI, after stirring 5 min. 3-methyl-2-cyclohexenone in abs. ether added, stirring continued 0.5 hr. at −5°, a soln. of acetaldehyde in abs. ether added with efficient stirring at −15 to −10°, stirred 0.5 hr. at 0° and 0.5 hr. at 25° → product. Y: 75%. – Substitution takes place at the hindered 2-position. F. e., also regiospecific acylation and alkylation, s. F. Näf and R. Decorzant, Helv. *57*, 1317 (1974).

Magnesium/sugar deriv.   ←
Asym. synthesis   CO → C(OH)R
Tert. alcohols from ketones s. *26*, 671; 2-aminoalcohols cf. A. Gaset, P. Audoye, and A. Lattes, J. Appl. Chem. Biotechnol. *25*, 13, 19 (1975)

Zinc (s. a. under TiCl₄) Zn
Reformatskii synthesis ←
s. *26*, 938
   with trimethylsilyl α-bromocarboxylates – β-Hydroxycarboxylic acids from oxo compds. s. *27*, 670
   Stereospecific Reformatskii synthesis s. *21*, 305

β-Hydroxynitriles CO → C(OH)·C·CN
   from oxo compds. and α-bromonitriles, also 1,4-addition s. *26*, 672

Zinc/trimethyl borate Zn/(MeO)₃B
Reformatskii synthesis CO → C(OH)·C·COOR
   β-Hydroxycarboxylic acid esters from oxo compds. – Improved procedure s. *26*, 673; with activated indium cf. L.-C. Chao and R. D. Rieke, J. Org. Chem. *40*, 2253 (1975).

Zinc/acetic acid Zn/CH₃COOH
α-Hydroxyketones from β-diketones ←
   with skeletal rearrangement s. *28*, 608

Ethylmagnesium bromide EtMgBr
Ethynyldiols from lactols C
454.

Stereospecific formation. Acetylene bubbled 0.5 hr. at −78° through anhydrous tetrahydrofuran, ethylmagnesium bromide in ether added, allowed to warm slowly to room temp., a soln. of *cis*-hexahydro-4-(6-methoxy--2-naphthyl)-2,2,3a-trimethyl-2H-cyclopenta[b]furan-6a-ol in tetrahydrofuran added dropwise, and stirred 19 hrs. at room temp. under N₂ → (1S,2S,5S)-2-ethynyl-2-hydroxy-5-(6-methoxy-2-naphthyl)-α,α,1-trimethylcyclopentaneethanol. Y: 86%. J. H. Dygos and L. J. Chinn, J. Org. Chem. *38*, 4319 (1973).

Ethylmagnesium bromide/magnesium EtMgBr/Mg
1,4-Diols and tetrahydrofuran ring ←
   from ketones and 1,3-chlorohydrins – Spiro-O-heterocyclics s. *28*, 609

Ethylmagnesium iodide EtMgI
α-**Metalation of sulfoxides** CO → C(OH)R
β-**Hydroxy-α-sulfinylcarboxylic acid esters**
from α-sulfinylcarboxylic acid esters and oxo compds.
455.

An ethereal soln. of ethylmagnesium iodide added at 0° to a vigorously stirred soln. of ethyl α-phenylsulfinylacetate in the same solvent, after 3 min. stirring benzaldehyde added, and stirred 1.5 hrs. at room temp. → ethyl 2-phenylsulfinyl-3-hydroxy-3-phenylpropionate. Y: 84%. F. e. s. N. Kunieda, J. Nokami, and M. Kinoshita, Tetrah. Let. *1974*, 3997; **f. reactions of α-sulfinylcarboxylic acid esters** cf. J. J. A. van Asten and R. Louw, Tetrah. Let. *1975*, 671.

Zinc chloride $ZnCl_2$
1,4-Addition with aldol condensation ←
s. *30*, 489

Zinc iodide ZnI_2
α**-Siloxynitriles from oxo compds.** CO → C(CN)OSi≦
Protection of carbonyl groups s. *28*, 610; without catalyst cf. H. Neef, J. pr. *316*, 817 (1974)

Indium In
Reformatskii synthesis CO → C(OH)·C·COOR
β-Hydroxycarboxylic acid esters from oxo compds. s. *26*, 673 suppl. *30*

Trialkylalanes R_3Al
Tert. alcohols from ketones CO → C(OH)R
Stereospecific synthesis – Steric effect of solvents s. *27*, 671

Trimethyl borate s. under Zn $(MeO)_3B$

Boric acid/acetic acid HBO_2/CH_3COOH
C-Hydroxymethylation H → CH₂OH
s. *29*, 627

Aluminum isoproxide $(RO)_3Al$
β**-Hydroxycarboxylic acid esters from oxo compds.** CO → C(OH)·C·COOR
s. *29*, 629

Boron fluoride BF_3
2-Iminotetrahydrofurans ☉
from isonitriles and oxetanes s. *27*, 672

2,3-Diiminotetrahydrofurans
from oxido compds. and 2 isonitrile molecules s. *27*, 673

Aluminum chloride $AlCl_3$
4-Aryl-2-hydroxycyclobut-2-enones ←
from cyclobutenediones s. *26*, 674

α**-Siloxynitriles from oxo compds.** CO → C(CN)OSi≦
s. *28*, 610

γ**-Diketones from 2(3H)-furanones** C
s. *29*, 822

Aluminum chloride/hydrogen chloride $AlCl_3/HCl$
Synthesis of indan-1-one-3-carboxylic acids ○
with C-alkyl migration s. *28*, 659

Gallium chloride $GaCl_3$
Synthesis of alcohols CO → C(OH)R
from oxo compds. and silanes s. *30*, 624

Tetraalkylammonium cyanides $R_4N^+ CN^-$
as bases and nucleophiles
s. *29*, 628

Tetrabutylammonium cyanide $Bu_4N^+ CN^-$
Reactions with tetraalkylammonium salts 2 ArCHO → ArCH(OH)COAr
Benzoins from aldehydes – Organic-inorganic contact s. *26*, 675

Anion exchanger/liq. ammonia ⇽
1-Ethynylalcohols from ketones   CO → C(OH)C≡CH
s. *24*, 663 suppl. *28*

Formic acid   *HCOOH*
Ketones from α-diketones   H → COR
  via α-hydroxyketones by synthesis and cleavage s. *27*, 674

Trifluoroacetic acid   *CF₃COOH*
Arylation of quinones   H → Ar
  via arylquinols s. *28*, 611

Thiazolium salts ⇽
Acyloins from 2 aldehyde molecules   2 CHO → CH(OH)CO
s. *29*, 630; s. a. H. Stetter and H. Kuhlmann, Tetrah. Let. *1974*, 4505

Titanium isopropoxide   *(RO)₄Ti*
β-Hydroxycarboxylic acid esters from oxo compds.   CO → C(OH)·C·COOR
  and ketene s. *29*, 629

Titanium tetrachloride/zinc   *TiCl₄/Zn*
Pinacols from 2 oxo compd. molecules   2 CO → C(OH)C(OH)
s. *30*, 561

Antimony pentachloride-liq. sulfur dioxide   *SbCl₅/SO₂*
Carboxylic acids from alcohols   OH → COOH
  by carbonylation s. *28*, 612

Sulfuric acid (s. a. under Cu₂O)   *H₂SO₄*
s. *28*, 612

Cobalt carbonyl   *Co(CO)₄*
α-Acylaminocarboxylic acids from aldehydes   CHO → CH(NAc)COOH
  also from ethylene derivs. s. *28*, 613

Via intermediates   *v.i.*
1-(Aminomethyl)alcohols   CO → C(OH)CH₂NH₂
  from ketones via α-siloxynitriles s. *27*, 720

**Homologous dicarboxylic acid esters
from dicarboxylic acid anhydrides
via carbalkoxy-α-diazoketones**

456.

A dry ethereal soln. of diazomethane prepared from nitrosomethylurea added at 0° to a soln. of 1-benzylpyrrolidine-2,5-dicarboxylic anhydride in dry benzene, allowed to warm to room temp. overnight, evaporated to dryness, the resulting crude methyl 1-benzyl-5-diazoacetylpyrrolidine-2-carboxylate dissolved in dry methanol, treated with Ag₂O, and refluxed 2 hrs. → methyl 2-(1-benzyl-5-methoxycarbonylpyrrolidine-2-yl)acetate. Y: 53%. – This shortened process may be successful where the usual Arndt-Eistert synthesis via acid chlorides (s. Synth. Meth. *22*, 813) fails. F. e. s. E. W. Della and M. Kendall, Soc. Perkin I *1973*, 2729.

Addition to Nitrogen and Carbon CC↓NC

Without additional reagents w.a.r.
Nucleophilic addition of phenols N : C → NHCAr
to azomethines s. 26, 676

2,4-Dienecarboxylic acid amides ←
from cyclopropenones and enamines – Insertion of 3-C-atoms into nitrogen-carbon bonds s. 28, 614

Reactions of 2,4-diaminothiophenes N : C : O → NHCOR
Subst. carboxylic acid amides from isocyanates s. 26, 677

Reactions of trichloroacetyl isocyanate ←
with allenes – Reactions of tetramethylallene s. 28, 615

α,β-Ethylenecarboxylic acid thioamides CO → C : C·CSNHR
from alkylidenephosphoranes s. 26, 903

4-Carboxy-2-pyrrolidones ←
from azomethines and succinic acid anhydrides s. 28, 616

Heterocyclics from ketenes and heterocumulenes ○
2,4-Azetidinediones s. 27, 675

Uracil ring from amines
with isocyanatoformic acid esters – Pyrimido[4,5-d]pyrimidine-2,4,5,7-tetrones s. 26, 678

Enazides △N → C(R) : CN₃
from Δ¹-azirines and diazo compds.

457.

Ph—⟨N⟩—Ph + PhCHN₂ → Ph\C(N₃)/CH₂Ph·Ph

Phenyldiazomethane in cyclohexane added dropwise to an ethereal soln. of 2,3-diphenyl-2H-azirine, and refluxed 72 hrs. under N₂ → 1-azido-1,2,3-triphenylprop-1-ene. Y: 71%. J. H. Bowie, B. Nussey, and A. D. Ward, Australian J. Chem. 26, 2547 (1973).

Irradiation ⚡
Δ³-Oxazolines ☉
from Δ¹-azirines and oxo compds.

458.

Ph—⟨N⟩— + H\C(=O)—CHMe₂ → Ph—⟨N-O⟩—CHMe₂

A soln. of 2,2-dimethyl-3-phenyl-2H-azirine in benzene and isobutyraldehyde irradiated 1 hr. with the Pyrex-filtered light of a high-pressure Hg-lamp → 5-isopropyl-2,2-dimethyl-4-phenyl-3-oxazoline. Y: 80%. F. e. s. H. Giezendanner et al., Helv. 56, 2611 (1973); from ketones s. P. Claus et al., ibid. 57, 2173 (1974).

Ring expansion of N-heterocyclics via 2-aminomesylates
2-Aminoalcohols from azomethines
Stereospecific addition to carbon-nitrogen double bonds

459.

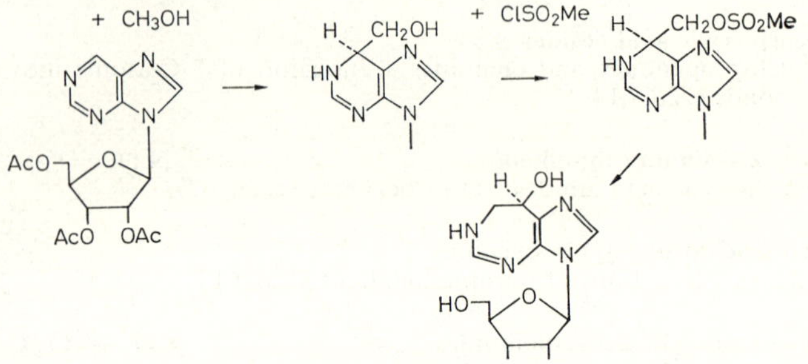

9-(2,3,5-Tri-O-acetyl-β-D-ribofuranosyl)purine in methanol irradiated 3.5 hrs. at 5° under argon with a 10 w. low-pressure Hg-lamp → 9-(2,3,5-tri-O-acetyl-β-D-ribofuranosyl)-6-hydroxymethyl-1,6-dihydropurine (Y: 96%) treated 17 hrs. at 0° with mesyl chloride in anhydrous dimethoxyethane in the presence of NaH → mesylate (Y: 91%) treated immediately at 0° with K-*tert*-butoxide in dimethoxyethane for 18 hrs. → coformycin (Y: 38%). M. Ohno et al., Am. Soc. **96**, 4326 (1974).

Lithium   Li
Acylamines from carboxylic acid chlorides   N : C : O → NHCOR
via carboxylic acid azides and isocyanates

460.

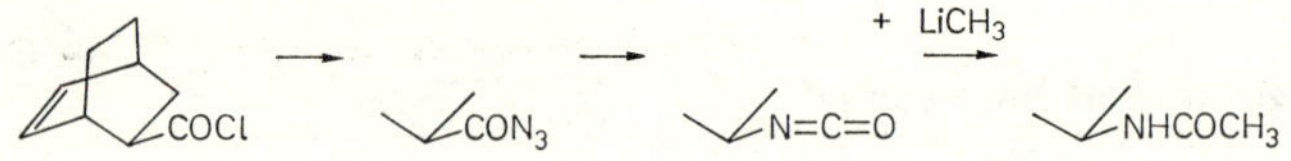

Curtius degradation. Aq. NaN₃ added at 0° to a soln. of *exo*-5-chlorocarbonyl-bicyclo[2.2.2]oct-2-ene in acetone, the resulting crude acid azide isolated after 20 min. stirring at 0°, dissolved in dry benzene, refluxed 2 hrs., the crude isocyanate obtained dissolved in anhydrous ether and added dropwise at 0° to a dry ethereal soln. of methyllithium → *exo*-5-acetamidobicyclo[2.2.2]oct-2-ene. Overall Y: 87%. F. e. s. N. A. LeBel, R. M. Cherluck, and E. A. Curtis, Synthesis *1973*, 678; with tetrabutylammonium azide instead of NaN₃ cf. A. Brändström, Bo Lamm, and I. Palmertz, Acta Chem. Scand. **B28**, 699 (1974).

Sodium   Na
1,2-Diamines from azomethines   2 C : N · → C(NH·)C(NH·)
Dimerization s. *24*, 681 suppl. 28

Sodium/dimethyl sulfoxide   Na/Me₂SO
Amidines from carbodiimides   N : C : N → N : CRNH
α,β-Acetyleneamidines s. *28*, 617; cf. J. Pornet and L. Miginiac, Bl. *1974*, 994

Sodium hydride   NaH
Subst. carboxylic acid amides from isocyanates   N : C : O → NHCOR
s. *27*, 676

C-Acylation with diacylamines   C
Ketocarboxylic acid amides from dicarboxylic acid imides s. *26*, 679

2-Amino-1-sulfinylthioleethers $C \equiv N \rightarrow C(NH_2) : C(SR)(SOR)$
from nitriles – Synthesis of α-aminocarboxylic acid derivs. from nitriles with addition of 1 C-atom s. 29, 631

2-Halogenothiolurethans $CHO \rightarrow CHClC \cdot SCONH_2$
from thiocyanates and aldehydes s. 28, 618

2-Oxazolidinethiones O
from isothiocyanates and oxo compds. s. 29, 632

Sodium/alcohol NaOR
β-Amino-α,β-ethylenenitriles $2\ CH_2CN \rightarrow C(CN) : C(NH_2)CH_2 \cdot$
from 2 nitrile molecules s. 27, 677

2,4-Diaminothiophenes O
from isothiocyanates, nitriles, and halides s. 26, 892

Potassium tert-butoxide $KOCMe_3$
2-Oxazolidinethiones
from isothiocyanates and oxo compds. s. 29, 632

Organolithium compds. RLi
4-Subst. 1,4-dihydropyridine from pyridine ring ←
s. 29, 633

Methyllithium/diisopropylamine $MeLi/i\text{-}Pr_2NH$
β-Amino-α,β-ethyleneazomethines $CN \rightarrow C(NH_2) : C \cdot C : NR$
from azomethines and nitriles
α-Metalation of azomethines

461.
$$CH_3CH=N-\langle\bigcirc\rangle$$
$$\downarrow$$
$$Ph-C\equiv N + [LiCH_2CH=N-\langle\bigcirc\rangle] \longrightarrow Ph-\underset{NH_2}{C}=CHCH=N-\langle\bigcirc\rangle$$

Diisopropylamine stirred and treated at 0° with an ethereal methyllithium soln., 10 min. after CH_4-evolution has ceased ethylidenecyclohexylamine in ether added dropwise at 0°, after 15 min. cooled to –78°, treated with benzonitrile in ether, allowed to warm to room temp. and to stand 12 hrs. → product. Y: 91%. F. e. s. G. Wittig, S. Fischer, and M. Tanaka, A. *1973*, 1075; α-metalation of azomethines s. a. T. Cuvigny, M. Larchevêque, and H. Normant, C. r. 277 (C), 1511 (1973); A. *1975*, 719; cf. Bl. *1973*, 2985, 2989; J. Organometal. Chem. 70, C5 (1974); K. Takabe et al., Synth. Commun. *5*, 227 (1975); synthesis of γ-diketones from azomethines cf. Tetrah. Let. *1974*, 1237; Synthesis *1975*, 256; **α- and γ-alkylation of α,β-ethyleneazomethines** cf. Tetrah. Let. *1975*, 1239.

Butyllithium BuLi
Eneformylamines from isonitriles $CO \rightarrow C : C \cdot NHCHO$
s. 26, 680

Lithium amide $LiNH_2$
Synthesis of acylamines from acylimines $C : NAc \rightarrow CRNAc$
s. 26, 681

2-Acylaminophosphonic acid esters ←
from acylimines and phosphonic acid esters s. 29, 880

Lithium diethylamide Et_2NLi
Asym. syntheses with sulfoxides $N : C \rightarrow NHCR$
Synthesis of sec. amines from azomethines s. 29, 634

Sodium amide   *NaNH$_2$*
Sec. amines from azomethines   N : C → NHCR
   2-Aminophosphinic acid esters – β-Aminocarboxylic acids s. 26, 682

Sodium cyanide   *NaCN*
1,2-Di(azomethines) from 2 azomethine molecules   ←
   via ene-1,2-diamines – Benzil dianils via α,α′-dianilinostilbenes s. 26, 683

Pseudocyanides   ←
   s. 26, 764

3H-Imidazo[1,5-b]-s-triazoles   O
   from 3 nitrile molecules s. 27, 678

1,2,3,4,5,6-Hexahydro-1,6-benzodiazocines   C
   by nucleophilic ring opening s. 27, 44

Sodium cyanide/acetic acid   *NaCN/CH$_3$COOH*
α-Aminonitriles from azomethines   C : N → C(CN)NH
   s. 27, 781

Potassium cyanide (s. a. under Me$_2$C(OH)CN)   *KCN*
4-Cyanoquinolines from quinolinium salts   ←
   via 4-cyano-1,4-dihydroquinolines and 4-cyanoquinolinium salts s. 27, 679

Triethylamine   *Et$_3$N*
Base-catalyzed reactions   N : C : O → NHCOR
of active methylene compds. with isocyanates
   Subst. carboxylic acid amides – Pyrrolidine-2,3,5-triones s. 29, 635

β-Ketocarboxylic acid amides   C(OSi≤) : C → CO·C·CONHR
   from enoxysilanes and isocyanates s. 30, 462

6-Amino-2-pyridinethione ring   O
   from isothiocyanates s. 27, 680

Thiophenes   ←
   from oxathiol-2-ylidenimmonium salts s. 28, 605

Pyridine   *C$_5$H$_5$N*
β-Ketocarboxylic acid amides   C(OSi≤) : C → CO·C·CONHR
from enoxysilanes and isocyanates

462.   [cyclopentene-OSiMe$_3$]   +   O=C=NPh   →   [cyclopentene-OSiMe$_3$, CONHPh]   →   [cyclopentanone =O, CONHPh]

Unlike ordinary enolethers, **enoxysilanes react like enamines** as nucleophiles toward isocyanates (cf. Synth. Meth. 20, 502). – E: 1-Trimethylsilyloxycyclopentene and phenyl isocyanate heated 12 hrs. at 130° in the presence of a catalytic amount of a tert. amine such as pyridine or triethylamine, then treated with methanol and water → 2-(N-phenylcarbamoyl)cyclopentanone. Y: 94%. F. e. s. I. Ojima and S. Inaba, Tetrah. Let. *1973*, 4271; **N-sulfonyl analogs**, without base, cf. Chem. Lett. *1974*, 1069.

Cupric acetoacetonate   *[(CH$_3$CO)$_2$CH]$_2$Cu*
Aminomethylene compds. from isonitriles   CH$_2$ → C : CHNH·
   s. 28, 619

Lithium diorganocuprates   *Li[CuR$_2$]*
4-Subst. 1-carbalkoxy-1,4-dihydropyridines   ←
   from pyridines s. 26, 684 suppl. 29

Magnesium   *Mg*
Synthesis of sec. from prim. amines,   CHO → CHRNHR'
aldehydes, and halides via azomethines

463. [3,4-dichlorobenzaldehyde] + H$_2$NEt → [3,4-dichlorobenzal=NEt] → [3,4-Cl$_2$C$_6$H$_3$CHNHEt·CH$_2$Ph]
+ ClCH$_2$Ph

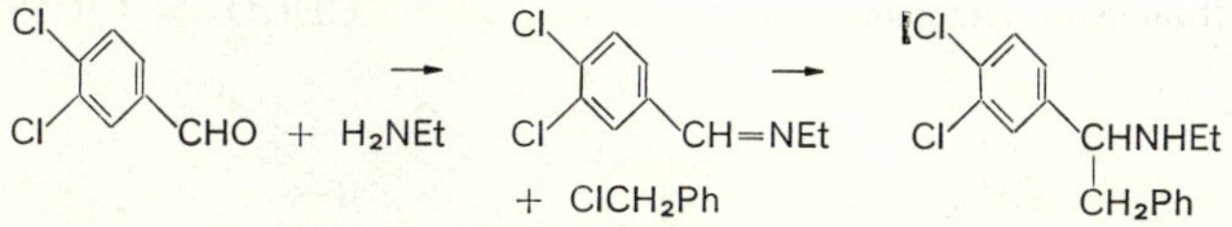

A mixture of 3,4-dichlorobenzaldehyde, anhydrous ethylamine, and molecular sieve Linde 4A in benzene stirred overnight, the resulting crude azomethine dissolved in anhydrous ether, added dropwise to a stirred soln. of benzylmagnesium chloride prepared from Mg and benzyl chloride in ether, stirring continued 3 hrs. at room temp., and treated cautiously with an aq. soln. of NH$_4$Cl → product. Y: 90%. F. e. s. D. D. Miller et al., J. Med. Chem. *18*, 99 (1975).

Prim. (*tert*-carbin)amines from nitriles   C≡N → CR$_2$NH$_2$
     s. *29*, 636

2-Subst. 1-carbethoxy-1,2-dihydropyridines   ←
     from pyridines s. *26*, 684

Δ^4-Isoxazolines from isoxazolium salts   ←
     s. *26*, 934

Ketocarboxylic acid amides   C
     from dicarboxylic acid imides s. *27*, 681

2-δ-Ketoindoles
     from 6,7,8,9-tetrahydropyrido[1,2-a]indol-6-ones s. *29*, 637

Methoxymagnesium methyl carbonate   *MeOMgOCOOMe*
β-Aminoketones from azomethines   N : C → NHC·C·COR
     s. *28*, 620; s. a. Synthesis *1974*, 284

Zinc   *Zn*
β-Aminocarboxylic acid esters   N : C → NHC·C·COOR
     from azomethines and α-bromocarboxylic acid esters – 1,2- and 1,4-addition of organometallic compds. to α,β-ethyleneazomethines s. *29*, 638

Di(isopinocampheyl)borane/acetone cyanohydrin   *R$_2$BH/Me$_2$C(OH)(CN)*
α-Aminocarboxylic acids from nitriles   CN → CH(NH$_2$)COOH
     Asym. synthesis with addition of 1 C-atom s. *28*, 621

Aluminum chloride   *AlCl$_3$*
β-Iminoazomethines and β-diketones   ←
     from azomethines and nitriles s. *26*, 685

Acetone cyanohydrin/potassium cyanide   *Me$_2$C(OH)CN/KCN*
1-Cyano-1,2-ene-1,2-diamines   N : C·CH : N → NH·C : C(CN)·NH
     from 1,2-di(azomethines) s. *29*, 639

Acetic acid   *CH$_3$COOH*
α-Cyano-N-sulfonylhydrazines   C : N·N·Ts → C(CN)NHN·Ts
     from sulfonylhydrazones s. *28*, 67

Hydrogen chloride   *HCl*
Quinoline-4-carboxylic acid amides   ☉
     from isatins and enamines – Dicarboxylic acid imides from dicarboxylic acid amide esters s. *28*, 622

Addition to Sulfur and Carbon CC⇓SC

Without additional reagents w.a.r.
β-Iminodithiocarboxylic acids from ketones   CHCO → C(CSSH)C : NR

464.

A mixture of cyclopentanone and cyclohexylamine refluxed 2–5 hrs. with anhydrous KF, which is then removed by filtration, and CS$_2$ added gradually below 0° to the stirred filtrate → product. Y: 61%. F. e. and limitations s. T. Takeshima et al., Soc. Perkin I *1974*, 914.

Reaction of 2,4-diaminothiophenes with dienophiles
1,3-Cyclohexadiene from thiophene ring s. *29*, 640

Irradiation
Photoreactions of 4-thiouracils with α,β-ethylenenitriles

465.

A soln. of 1,3,6-trimethyl-4-thiouracil and acrylonitrile in methylene chloride irradiated under N$_2$ in a Pyrex apparatus with a Hanau QF 81 lamp → product. Y: 70%. F. e. and reactions s. J. L. Fourrey, P. Jouin, and J. Moron, Tetrah. Let. *1973*, 3229.

Sodium hydride NaH
β-Ketocarboxylic acid esters   COCH$_2$ → COC : C(SR)$_2$
from ketones via α-ketoketene mercaptals

466. p—MeOC$_6$H$_4$—COCH$_3$ + CS$_2$ + IMe → p—MeOC$_6$H$_4$—CO—CH=C(SMe)$_2$
                              ↓ + 2 HOEt
        + H$_2$O
     p—MeOC$_6$H$_4$—COCH$_2$COOEt ← [p—MeOC$_6$H$_4$—CO—CH=C(OEt)$_2$]

4-Methoxyacetophenone mixed with 80%-NaH in mineral oil, CS$_2$, methyl iodide, and dry benzene, dimethylacetamide added portionwise, and stirred 1 hr. with occasional cooling → 1-(4-methoxyphenyl)-3-bis(methylthio)-2-propen-1-one (Y: 80%) (startg. m. f. 310) dissolved in benzene, ethanol and p-toluenesulfonic acid added, stirred 2 hrs. at room temp., refluxed 15 min., and washed with bicarbonate soln. → ethyl p-methoxybenzoylacetate (Y: 90%). – The reactions occur readily both with methyl and methylene groups; 2 such groups on both sides of a keto group may react simultaneously. F. e. s. I. Shahak and Y. Sasson, Tetrah. Let. *1973*, 4207.

Reactions with cyclic S-aminosulfoxonium salts
via their ylids – Oxido- and cyclopropyl-sulfinic acid amides s. *27*, 682

Vinyllithium $\qquad CH_2 : CHLi$
Ring expansion ◌
  by 2 C-atoms with vinyllithium s. 27, 683; prepn. of a soln. of vinyllithium in tetrahydrofuran s. W. N. Smith, Jr., J. Organometal. Chem. 82, 7 (1974); syntheses with trifluorovinyllithium cf. J. F. Normant et al., Synthesis 1975, 122

Magnesium $\qquad Mg$
Reaction of thioketones with Grignard compds. ←
  Tert. mercaptans and sec. thioethers s. 29, 641
Thiophene ring opening C
s. 29, 642

Tallium(I) compound ←
C-Alkylation with cyclic sulfonium salts
s. 28, 623

Via intermediates v.i.
3H-Benzo[c]thiophen-1-ones ○
  via sulfur-donor ligand o-metalated complexes s. 29, 643

Addition to
Remaining Elements and Carbon $\qquad$ CC↓RemC

Without additional reagents $\qquad$ w.a.r.
Insertion of C-atoms into carbon-silicon bonds $\qquad$ $C \cdot Si \leqslant \rightarrow C \cdot C \cdot Si \leqslant$
s. 27, 883
Syntheses with o-silylheterocyclics ←
s. 29, 881
Reactions of nitriles with phosphoranes $\qquad C : P \leqslant \rightarrow C : C(R) \cdot N : P \leqslant$
  Enimino- from alkylidene-phosphoranes – Imidinophosphoranes from phosphine imines s. 27, 684
Ring expansion of P-heterocyclics ◌
s. 27, 685

Butyllithium $\qquad$ BuLi
Wittig synthesis with cyclic phosphonium salts C
  Retention of phosphorus s. 26, 686

Addition to Carbon Atoms $\qquad$ CC↓C

Sodium cyanide $\qquad$ NaCN
α,α-Alkoxyaminonitriles and 1,1,1-aminodinitriles ←
  from alkoxyaminomethinium salts s. 28, 624

Addition to Carbon-Carbon Bond $\qquad$ CC↓CC

Without additional reagents $\qquad$ w.a.r.
Ene synthesis ←
s. 29, 676
Thermal C-alkylation of phenols $\qquad C : C \rightarrow CHCAr$
  with ethylene derivs. s. 27, 686
Synthesis of halides from ethylene derivs. $\qquad C : C \rightarrow CHalCR$
s. 30, 607

2-Diazoamines $\quad$ C : C(N<) → CHC(N<)·C(: N_2)
from diazo comdps. and enamines s. 28, 625

β-Acoxycarboxylic acid esters $\quad$ CHO → CH(OAc)·C·COOR
from aldehydes and α-alkoxyenolesters s. 26, 687

α,β-Ethylenecarboxylic acid amides $\quad$ C ≡ C → CH : C·CONHR
from isonitriles and acetylene derivs. – Multifunctional compds. s. 26, 205

Sym. allenedicarboxylic acid amides from ynamines $\quad$ ←
s. 27, 687

α,β-Ethyleniminoesters $\quad$ N : C → N : C(OR)·C : C
from acetylene derivs. and isonitriles s. 29, 644

1,1,1-Aminodinitriles from α-aminonitriles $\quad$ N·CHCN → N·C(CN)$_2$
via 1,1-iminocyanides s. 29, 645

Reactions of sulfonium cyclopentadienylids $\quad$ C ≡ C → CH : CR
α,β-Ethylene- from acetylene-dicarboxylic acid esters s. 28, 626

β-Amino-α-diazosulfones $\quad$ C : C → CHC·C(: N_2)SO_2·
from α-diazosulfones and enamines s. 27, 688

α-Cyano-1,3-dithiolane-$\Delta^{2,\alpha}$-thioacetyl bromides $\quad$ ←
from 1,3-dithiolane-2-thiones s. 26, 689

Synthesis of boranes from ethylene derivs. $\quad$ C : C → CRC·BR_2
s. 27, 689; prepn. of organoboranes, review, s. K. Smith, Chem. Soc. Rev. 3, 443 (1974)

m-Subst. N-heterocyclics from cyclic N-oxides $\quad$ ←
s. 29, 646

[2+1]-Cycloaddition $\quad$ ▽
Iminocyclopropenes from acetylene derivs. and isonitriles s. 27, 690

Reaction of electron-deficient with electron-rich ethylene derivs. $\quad$ □

467.

Cyclobutanes. An ethereal soln. of dimethyl 1-cyanoethylene-1,2-dicarboxylate cooled at 5° added dropwise to N,N-dimethylisobutenylamine cooled at –15°, and after 15 min. evaporated below 35° → cyclobutane deriv. Y: 78%.

Interchange of ethylene derivs. via cyclobutanes. An ethereal soln. of N,N-dimethylvinylamine added at –55° to a soln. of tricyanoethylene in toluene, after 1 day allowed to warm to room temp., and the solvents evaporated under high vacuum → 1,1-dicyano-2-N,N-dimethylaminoethylene. Y: 24%.

F. e. s. H. K. Hall, Jr., and P. Ykman, Am. Soc. 97, 800 (1975); **olefin metathesis, review,** s. R. J. Haines and G. J. Leigh, Chem. Soc. Rev. 4, 155 (1975).

2-Amino-3-halogenocyclobutenes from ynamines
with halogen migration s. 28, 627

Ring closures with bis(trifluoromethyl)ketene
Cyclobutane-1,3-diones from ketenes − Effect of solvent s. 27, 691

Direction of cycloaddition
[2+2]- versus [2+4]-Cycloaddition of enamines to 3-alkylideneoxindoles s. 27, 692

Cycloaddition with ynamines
4-Pyrones from 2 ynamine molecules s. 27, 693

1,2,3,4-Tetrahydro-1,4-methanonaphthalenes
from indenes s. 28, 628

Cyclopolyenes as dienophiles
s. 26, 758

Diene synthesis with anthracenes
Ar. p-dimethylation s. 27, 694; diene synthesis with anthracene s. a. D. N. Butler, A. Barrette, and R. A. Snow, Synth. Commun. 5, 101 (1975).

Reactions with halogeno-1,3-dioxol-2-ones
Cyclic α-diketones via diene synthesis

468.

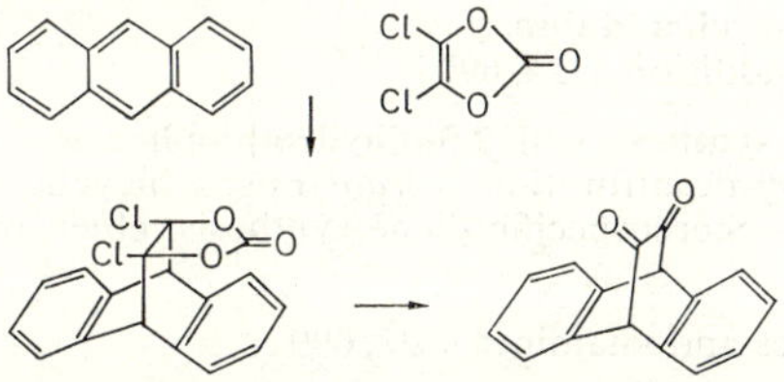

A soln. of dichlorovinylene carbonate and anthracene in abs. xylene refluxed 100 hrs. → intermediate (Y: 84%) refluxed 12 hrs. in dioxane-water containing HCl → 9,10-dihydro-9,10-dioxoethanoanthracene (Y: 78%). F. e. s. H.-D. Scharf and W. Küsters, B. 105, 564 (1972); review s. Ang. Ch. 86, 567 (1974).

**Diene synthesis
with maleic thioanhydride and with silicon and germanium compds.**
s. 27, 695

High-pressure diene synthesis
with azomethines s. 27, 696

Diene synthesis with cyclimmonium salts
and enolethers − Diene synthesis with acridizinium salts, and with norbornene derivs. s. 27, 697

Diene synthesis with acyl- and sulfonyl-oximes
s. 27, 698

Linear annellation
Diene synthesis with O-transacylation s. 28, 629

Methylenecarbonyl bridges
via diene synthesis and Curtius degradation − Ketones from α-chlorocarboxylic acid chlorides s. 28, 630

Protection of carbon-carbon double bonds
by diene synthesis s. 28, 631; s. a. M. Golfier and T. Prangé, Bl. 1974, 1158; as Fe-complex cf. K. M. Nicholas, Am. Soc. 97, 3254 (1975)

Diene syntheses with 1,2-dicyanocyclobutene

as dienophile. | **via 2,3-dicyano-1,3-butadiene.**

1,2-Dicyanocyclobutene added with vigorous stirring *below 30°* to an ethereal soln. of cyclopentadiene, and the product isolated after 0.5–1 hr. → *endo*-2,5-dicyanotricyclo[4.2.1.0^{2,5}]non-7-ene. Y: 90%. | and cyclopentene dissolved with some hydroquinone in benzene, and heated 16 hrs. *at 135°* in an autoclave → 3,4-dicyanobicyclo[4.3.0]non-2-ene. Y: 65%. — These reactions of 2,3-dicyano-1,3-butadiene are typical **diene syntheses with inverse electronic demand.**

F. e. s. D. Belluš et al., Helv. **56**, 3004 (1973).

Stereospecific [6+4]-cycloaddition
also [8+2]-cycloaddition s. *26,* 690

Stereospecific diene synthesis with 2,5-dihydrothiophenes
Ring opening by desulfuration – *trans*-Fused bicyclic isocyclics with angular methyl groups – Stereospecific diene synthesis, effect of substituents s. *26,* 691

4-Aminoquinolines
from ketenimines and ynamines s. *27,* 699

Diene syntheses with vinylketenimines
and electron-deficient as well as electron-rich dienophiles – 2-Cyclohexenones – 4-Aminoquinolines from ynamines – Diene syntheses with phosphonium salts s. *29,* 647

Diene synthesis
with ethylene-N-heterocyclics, and with α,β-ethyleneketones s. *27,* 700

Cyclic ketones by carbon monoxide insertion
Iron π-allyl complexes s. *28,* 632

O-Heterocyclics from enamines
s. *28,* 633

4-Imino-2-azetidinones
from ketenes and carbodiimides s. *26,* 692

3-Pyrrolidones
from allenes and nitrones s. *26,* 693

Stereospecific 1,3-dipolar cycloaddition
Δ^3-Pyrroline ring from isoquinolinium methylides s. *29,* 649

2,4,4',6'-Tetraoxospiro[pyrrolidin-1,2'-(pyrrolidino[1,3,4-c]pyrrolidines)]
from azomethines and 2 maleimide molecules s. *27,* 701

4(1H)-Pyridones
from 1-azirines and cyclopropenones s. *28,* 634

Pyridines from 1,3-diynes
s. *29,* 650

Reactions of acetylene derivs.
having both electron-donating and -accepting groups
  Ring closures – Carbostyrils from isocyanates s. 26, 694
1,3-Dipolar cycloaddition with pyridinium betaines
s. 26, 695
Azepine ring
  from 2 acetylenedicarboxylic acid ester molecules – Syntheses with acetylenecarboxylic acid esters, review s. 28, 635
Cycloadditions with thiete 1,1-dioxides
  and 1,3-dienamines s. 26, 696
S-Heterocyclics by [8+2]-cycloaddition
s. 28, 636
Reactions with thiophosgene S-oxide
s. 26, 697
Dihydrothiopyrans
  from 1,3-dienes – Dihydro-Δ^2-thiopyrans from vinylogous carboxylic acid thioamides s. 28, 637
Thiopyran ring

470.

Dimethyl acetylenedicarboxylate added at room temp. to a soln. of methyl 1,2-dimethylindole-3-dithiocarboxylate in dioxane or dimethylformamide, and heated 5–10 min. on a steam bath → 4a,5-dimethyl-1-methylthio-3,4-bis(methoxycarbonyl)-4aH-thiopyrano[4,3-b]indole. Y: 90–95%. F. e. s. Y. Tominaga et al., Chem. Pharm. Bull. *21*, 2770 (1973).

4-Hydroxy-Δ^2-dehydroquinuclidines from 4-piperidones

471.

Triacetoneamine treated 24–36 hrs. at 20° with dimethyl acetylenedicarboxylate in hexane or ether → dimethyl 4-hydroxy-6,6,7,7-tetramethyl-Δ^2-dehydroquinuclidine-2,3-dicarboxylate. Y: 58–70%. F. e. s. R. G. Kostyanovskii, Z. E. Samoilova, and M. Zarifova, Izvest. *1973*, 1681; C. A. *79*, 105054.

N-Condensed heterocyclics
  from benzocyclobutenes – Epimers s. 29, 651; s. a. Soc. Perkin I *1975*, 737
Uracil-1,3-disulfofluorides
  from acetylene derivatives s. 26, 698
5-Acyl-4(1H)-pyrimidinones
  from β-amino-α,β-ethylenecarboxylic acid amides and 2 ketene molecules s. 26, 340
Reverse-oriented 1,3-dipolar cycloaddition
  4-Aminoisoxazolidines from nitrones and enamines s. 26, 699; 1,3-dipolar cycloaddition with nitrones, review, s. D. St. Clair Black, R. F. Crozier, and V. C. Davis, Synthesis *1975*, 205
Δ^2-Isoxazoline ring from nitrile oxides
  Cycloaddition to benzvalenes s. 29, 652

2,1-Oxazabicyclo[4.1.0]heptanes
 from 4,5-dihydro-6H-1,2-oxazine 2-oxides and acetylene derivs. s. *29*, 653

6-Alkoxy-5,6-dihydro-4H-1,3-oxazines
 from acylimines and enolethers – [4+2]-Cycloaddition s. *29*, 654

Bicyclic N-condensed N,N-dialkoxylamines
 from 5,6-dihydro-4H-1,2-oxazine 2-oxides – N-Condensed isoxazolidine ring s. *27*, 702

5,6-Dihydro-4H-1,2-oxazine 2-oxide ring
 from 1,1-nitroethylene derivs. and enamines s. *26*, 700

Benz[b]indeno[2,1-b]thiophenes
 by double ring closure s. *28*, 531

1-Amino-2,10-dioxabicyclo[4.4.0]deca-3,8-dienes
 from α,β-ethyleneketones and ynamines s. *27*, 703

Cycloaddition to 4H-pyrazol-4-one N-oxides
 Pyrazolo[2,3-b]isoxazol-4-ones s. *26*, 701

1,2,4,5-Tetraazapentalenes
 from 2 acetyleneazo compd. molecules s. *26*, 702

Δ^4-Pyrrol-2-ones from enamines
s. *26*, 703

Cyclopentane from cyclopropane ring
 Thermal addition of electron-deficient carbon-carbon multiple bonds to strained carbocycles, review s. *26*, 704; s. a. A. A. P. Noorstrand, H. Steinberg, and T. J. de Boer, Tetrah. Let. *1975*, 2611

[$\sigma 2 + \pi 2 + \pi 2$]-Cycloaddition with enecyclopropanes
s. *29*, 655

3-Aminofulvenes
 from enecyclopropenes and ynamides s. *29*, 656

Ring expansion by 2 C-atoms with acetylene derivs.
 of cyclic sulfonium ylids s. *27*, 705; of N-heterocyclics s. R. M. Acheson, G. Paglietti, and P. A. Tasker, Soc. Perkin I *1974*, 2496

[8]Paracycloph-4-enes
by double spirocyclopropane ring opening

472.

trans-1-Cyano-1,3-butadiene heated 5 hrs. with dispiro[2.2.2.2]deca-4,9-diene in benzene at 160° in a sealed glass ampoule under argon → product. Y: 69%. F. e., also with *cis*-1-cyano-1,3-butadiene, s. T. Tsuji and S. Nishida, Am. Soc. *95*, 7519 (1973).

Δ^2-Pyrrol-4-ones
from cyclopropenones and azomethines

473.

Equimolar amounts of diphenylcyclopropenone and acetone methylimine heated 0.5 hr. at 80° in 2-methoxyethanol → 1,5,5-trimethyl-2,3-diphenyl-Δ^2-pyrrolin-4-one. Y: 70–95%. F. e. s. T. Eicher and J. L. Weber, Tetrah. Let. *1974*, 1381.

Heterocyclics from cyclobutanes

474.

A soln. of the startg. m. in acetonitrile kept 3 weeks at room temp. → tetrahydropyridine deriv. Y: 99%. – Similarly 1 week with benzaldehyde → oxane deriv. Y: 95%. F. e. s. R. Schug and R. Huisgen, Chem. Commun. 1975, 60.

1(1H)-Benzazepines from indoles
s. 26, 705

N-Condensed N-acylpyrazole ring
from cyclic α-diazoketones – Ring expansion with ring closure s. 29, 675; s. a. M. Martin and M. Regitz, A. 1974, 1702

Reaction of cyclopropenones with N-heterocyclics
N-Condensed heterocyclics – F. reactions of cyclopropenones s. 27, 706

4-Amino-Δ⁴-thiazolines
from aziridines and isothiocyanates – Aziridine ring expansion, review – Heterocyclics by aziridine ring expansion, review, s. 26, 706

Cycloaddition with skeletal rearrangement

475.

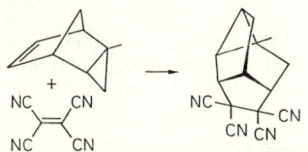

2-exo-Methyl-endo-tricyclo[3.2.1.0²,⁴]oct-6-ene allowed to react 1 week at room temp. with tetracyanoethylene in methylene chloride → 4,4,5,5-tetracyano-8-methyltetracyclo[4.2.2.0²,⁸.0³,¹⁰]decane. Y: >80%. J. M. Coxon, M. de Bruijn, and C. K. Lau, Tetrah. Let. 1975, 337.

Irradiation (s. a. under CuBr, R₃B,
Me₂CO, Ph₂CO, CH₃COOH, and HCl)

Synthesis of ketones from aldehydes   CHO → COR
s. 16, 737 suppl. 28

β-Cyanosulfones   C:C → C(CN)C·SO₂·
from ethylene derivs. and sulfonylcyanides s. 28, 638

γ-Lactones
from ethylene derivatives and 1,1,1-bromodinitriles via 1-dicyanomethyl-2-bromides, also stereospecific ring closure, and 1,1-dicyanocyclopropanes s. 26, 707

Cyclobutane ring   2 C:C → □
by dimerization of ethylene derivs. and reverse reaction – Light specificity – Stereospecific photodimerization s. 27, 704; phase effects s. F. H. Quina and D. G. Whitten, Am. Soc. 97, 1602 (1975); furo[3,2-g]coumarin derivs. s. J. Gervais and F. C. De Schryver, Photochem. Photobiol. 21, 71 (1975)
by cycloaddition s. 30, 181

Naphthalenes
from aryl ketones and acetylene derivs. via 1,4-dihydro-1-naphthols s. 28, 639

[4+4]-Cycloaddition
s. 29, 658

1,3-Addition to benzene ring
476.

A soln. of anisole and cyclopentene in cyclohexane irradiated at 254 nm → product. Y: 85%. – This is the first stage of a **stereospecific synthesis of hydroazulenes**. R. Srinivasan, V. Y. Merritt, and G. Subrahmanyam, Tetrah. Let. *1974*, 2715.

Isocyclic cage compds.
by multiple ring closure s. 28, 640

[8+2]-Cycloaddition
8-Oxabicyclo[5.3.0]decatrienes from tropones s. 27, 707

O-Heterocycles
by ring closure with ether rearrangement
477.

A soln. of 2-methoxy-1,4-naphthoquinone and 2 moles styrene in benzene irradiated in a glass tube by a 300 w. high-pressure Hg-arc lamp through a 5 cm. thick water layer → product. Y: 69%. K. Maruyama and T. Otsuki, Chem. Lett. *1974*, 129.

Azetidine ring from ethylene derivs.
s. 29, 659

Tetrahydrofuran ring
from cyclopropane ring and ketones s. 28, 641

1,5-Diazabicyclo[3.3.0]octanes from azines
s. 28, 642

Δ^1-Pyrrolines
from Δ^1-azirines and electron-deficient ethylene derivs. s. 26, 708

Ring expansion by 2 atoms of Δ^1-azirines
to Δ^3-5-oxazolones – Photochemistry of carbon-nitrogen double bonds and heterocyclics, reviews, s. 28, 643

Cyclopropylacetic acid esters
from cyclopent-2-en-1-ones s. 26, 709

3-Oxa-4-silaspiro[2.6]nonanes
from silacyclohexan-2-ones s. 27, 708

Electrolysis
Reactions with electrolytically generated bases   C : C → CHCR
Michael addition s. 25, 633 suppl. 29

Lithium (s. a. under CuI)   *Li*
Reaction of organolithium compds.   ←
    with unsatd. and satd. amines s. 27, 712
1,2,4,5-Tetrahydro-2,5-methano-3,1-benzoxazepines   ○
    from quinolines s. 28, 924

Lithium/tetramethylethylenediamine   *Li/TMEDA*
Synthesis of alcohols   C : C·C(OH) → CH·CR·C(OH)
    from 2-ethylenealcohols s. 29, 660

Sodium   *Na*
Enolesters from 2 ketene molecules   C : C(OAc)
s. 26, 710
α-Acoxycarboxylic acid amides   CO → C(CN)OAc
    from oxo compds. and acylcyanides via α-acoxynitriles s. 27, 709
3,4-Disubst. 2(5H)-furanones   ←
    via Michael addition s. 27, 710

Sodium hydride   *NaH*
β,γ-Ethylene-β-dicarboxylic acid esters   C≡C → CH$_2$C : C
    from acetylenedicarboxylic acid esters s. 26, 711
α-Ketocarboxylic acid ester mercaptals   C : C → CHC·(SR)$_2$COOR
    from ethylene derivs. s. 29, 661
Additions of CH-acidic compds. to 9-nitroanthracenes   ←
    9-Subst. anthracenes – Oximes from nitro compds. s. 28, 644

Sodium hydroxide   *NaOH*
Addition of active methylene groups   ←
    to acetylene-N-heterocyclics s. 29, 662
Bridge bicyclo[3.2.1]octane ring from quinones   ○
s. 27, 711

Potassium hydroxide/alcohol   *KOH*
β-Cyanoketones   C(N<)CN → CO·C·C·CN
    from α-aminonitriles via 1-amino-1,3-dinitriles s. 29, 663

Lithium triethylcarboxide s. under BH$_3$ *Et$_3$C·OLi*

Sodium/alcohol   *NaOR*
3-Cyclohexanolones from α,β-ethyleneketones   ○

478.

Startg. m. allowed to stand 24 hrs. at room temp. with ethyl acetoacetate in ethanol containing Na-ethoxide → product. Y: 76–86%. F. e. s. L. S. Stanishevskii, I. G. Tishchenko, and A. M. Zvonok, Vestsi Akad. Navuk Belarus. SSR, Ser. Khim. Navuk *1974* (1), 59 (Russ); C. A. *80,* 108265.

Prolines from α,β-ethyleneoxo compds.
s. 29, 664

Cyclohexanone ring from cross-conjugated dienones
    Double Michael addition – 4,5-Benz-2-azaspiro[5.5]-undecane-1,3,9-triones
s. 29, 665

8,17-Diazadibenzo[c,j]tetracyclo[7.3.1.0²,⁹.0⁷,¹⁶]tridecanes   ○
from 2 isoquinolinium salt molecules and active methylene compds. s. *28*, 645

Potassium tert-butoxide   *KOCMe₃*
Michael addition   C : C → CHCR
s. *28*, 646

Organolithium compds. s. a. under CuI   *RLi*

Organolithium compds./N,N,N',N'-tetramethylethylenediamine   *RLi/TMEDA*
Reaction of acetylene derivs.   C≡C → CH : CR
with organolithium compds.
Synthesis of ethylene from acetylene compds.

479.
$$Ph-C≡C-CH_2OH \longrightarrow Ph-\overset{H}{\underset{C_4H_9}{C}}=C-CH_2OH$$

Geospecific addition. 3-Phenylpropargyl alcohol allowed to react 3 hrs. at –30 to 20° with 2.5 equivalents butyllithium in ether containing 0.2 equivalent TMEDA → product. Y: ca. 90%. F. e. s. L.-I. Olsson and A. Claesson, Tetrah. Let. *1974*, 2161.

Methyllithium (s. a. under i-Bu₂AlH)   *MeLi*
2-Cyclohexenone ring   ○
from α,β-ethylene-α-silylketones – Annelation – Michael addition under aprotic conditions s. *29*, 666

n-Butyllithium (s. a. under CuI and Et₂AlCl)   *n-BuLi*
Syntheses with 1,1-sulfinylthioethers   C : C → CHCR
Michael addition s. *27*, 830 suppl. *28*

α-Lateral metalation of heterocyclics   ←
Dimerization and reverse reaction

480.

Ph—⟨thiazole⟩—CH₃ ⇌ [Ph—⟨thiazole⟩—CH₂Li] ⇌ Ph—⟨thiazole⟩—CH₂—⟨thiazole(CH₃)⟩—Ph

n-Butyllithium in hexane added dropwise at –78° under N₂ to a stirred soln. of 2-methyl-4-phenylthiazole in dry tetrahydrofuran, allowed to warm to room temp., stirring continued 8 hrs., and quenched with water → dimer (Y: 90%) heated 0.5 hr. at 150–160° →2-methyl-4-phenylthiazole (Y: 96%). F. e. s. G. Knaus and A. I. Meyers, J. Org. Chem. *39*, 1189, 1192 (1974).

tert-Butyllithium s. under CuI   *Me₃C·Li*

Lithium diisopropylamide   *i-Pr₂NLi*
Synthesis of iodides from ethylene derivs.   C : C → CRCl

481.

[structure: CH₃, COO-CMe₃ on CH, MeS, n-C₈H₁₇-furanone] + I₂ → [structure: CH₃, COO-CMe₃ on C, MeS, H, I, n-C₈H₁₇-lactone]

Startg. m. in tetrahydrofuran added at –78° during 20 min. to a soln. of *tert*-butyl α-lithio-α-thiomethylpropionate prepared from *tert*-butyl α-thiomethyl-propionate and Li-diisopropylamide (s. R. J. Cregge et al., Tetrah. Let. *1973*, 2425) in tetrahydrofuran, stirred 2 hrs. at the same temp., iodine added at –78°, and stirring continued 0.5 hr. at –78° → iodo lactone. Y: 93%. J. L. Herrmann, M. H. Berger, and R. H. Schlessinger, Am. Soc. *95*, 7923 (1973).

Michael addition C : C → CHCR
α-Alkylthiocarboxylic acid esters s. 28, 646

Syntheses with ketene mercaptal S-monoxides CH → C·C·COR
1,4-Dicarbonyl compds. – β-Ketocarboxylic from carboxylic acid esters s. 29, 667

Syntheses with protected cyanohydrins COC : C → COCHC(OR)CN
via masked acyl carbanions

482.

1,4-Addition takes place in high yield with protected cyanohydrins derived from α,β-ethylenealdehydes. – E: A soln. of the protected cyanohydrin of crotonaldehyde in hexamethylphosphoramide added dropwise at –78° under N_2 to a vigorously stirred soln. of Li-diisopropylamide, stirring continued 5 min., a soln. of 3-methylcyclohex-2-enone in tetrahydrofuran added, after 5–10 min. warmed to 0°, and quenched with water → product. Y: 70–85%. – This clean 1,4-addition is in striking contrast to the result with satd. alkoxynitriles, which give high yields of 1,2-addition products. F. e. s. G. Stork and L. Maldonado, Am. Soc. 96, 5272 (1974).

Stereospecific cyclopentane ring annelation O

483.

Startg. m. treated 1–2 hrs. at –78° then 3–4 hrs. at –30° with Li-diisopropylamide in tetrahydrofuran, then a soln. of 2 equivalents ethyl acrylate in the same solvent added slowly, kept several hrs. at –78°, and allowed to warm to room temp. overnight → product. Y: 95% based on startg. m. consumed; conversion 65%. – A single diastereomer is obtained. F. e. s. J. P. Marino and W. B. Mesbergen, Am. Soc. 96, 4050 (1974).

Anionic [3+2]-cycloaddition
Pyrrolidines s. 26, 712

Lithium diisobutylamide i-Bu_2NLi
Michael addition of carboxylic acid amides C : C → CHC·C·CON<
s. 28, 647

Potassium carbonate K_2CO_3
Diene synthesis with α-bromo-α,β-ethylenealdehydes O

484.

α-Bromoacrolein, 2,3-dimethylbutadiene, K-carbonate, and hydroquinone kept 18 hrs. at 55° in benzene under N_2 → product (startg. m. f. 16). Y: 95%. F. e. and **order of reactivity of dienes** s. B. B. Snider, J. Org. Chem. 38, 3961 (1973).

Pyrazole from N-aminoimidazolium ring

485.

3-Amino-1-methylbenzimidazolium mesitylenesulfonate warmed 5 hrs. at 40–50° with dibenzoylacetylene in dimethylformamide in the presence of K-carbonate → product. Y: 70%. F. e. s. Y. Tamura et al., Chem. & Ind. *1973*, 952.

Sodium cyanide   NaCN
Ketones from aldehydes and ethylene derivs.   CHO → CO·C·CH
s. *28*, 648; thiazolium salts as catalysts cf. Tetrah. Let. *1974*, 4505; Synthesis *1975*, 379

Mandelonitrile benzoates from benzils   COCOR → CH(CN)OCOR

486.   p—$O_2NC_6H_4$—COCOPh + NaCN → p—$O_2NC_6H_4$—CH(CN)(OCOPh)

4-Nitrobenzil added in one portion at room temp. under N_2 to a stirred mixture of NaCN and dimethyl sulfoxide, and after 1.5 min. poured into acidified ice-water slurry → 4-nitromandelonitrile benzoate. Y: 58%. W. C. Reardon, J. E. Wilson, and J. C. Trisler, J. Org. Chem. *39*, 1596 (1974).

Lithium thiophenoxide s. under CuI   PhSLi

Cyclohexylamine s. under CH_3COOH $C_6H_{11}NH_2$

Diethylamine Et_2NH
2-Imino-2,3-dihydropyrans   O
    from chalcones and flavones s. *26*, 713

Triethylamine Et_3N
C-Acetoacetylation with diketene   H → $COCH_2COCH_3$
s. *24*, 708 suppl. 29

1,2-Addition of 2 different C-functions   ☐
to carbon-carbon double bonds via cyclobutanone ring

487.

2-Chlorocarbonyl-1,3-dithiane allowed to react with a 2-fold excess of cyclopentadiene in ether in the presence of triethylamine → 1 : 1 adduct (Y: 70%) heated 3 hrs. at 70° with aq. NaOH → product (Y: 98%). – This method is regio- and stereo-specific. E. Cossement, R. Binamé, and L. Ghosez, Tetrah. Let. *1974*, 997; **reductive alkylation of carbonyl oxygen via cyclobutanone ring** cf. B. M. Trost, M. J. Bogdanowicz, and J. Kern, Am. Soc. *97*, 2218 (1975).

N,N,N',N'-Tetramethylethylenediamine s. under Li and RLi   *TMEDA*
Pyridine C_5H_5N
1,4-Addition with boranes   C : C → CHCR
s. *28,* 649
Double Michael addition with azomethines   N : C·CH$_2$ → N : C·CR$_2$
s. *29,* 668
Silver oxide Ag_2O
***anti*-Markownikoff addition**   C : C → CH·C·C·COR
to carbon-carbon double bonds

488.   n—C$_6$H$_{13}$CH=CH$_2$ + CH$_3$COCH$_3$ → n—C$_6$H$_{13}$CH$_2$CH$_2$CH$_2$COCH$_3$

Synthesis of ketones. 1-Octene added to a suspension of argentic oxide in *dry* acetone, stirred and refluxed 13 hrs. under N$_2$ → 2-undecanone. Y: 81%. F. additions of acetone s. M. Hájek, P. Šilhavý, and J. Málek, Tetrah. Let. *1974,* 3193.

2-Diazoamines   C : C(N<) → CHC(N<)·C(: N$_2$)
from diazo compds. and enamines s. *28,* 625

Cuprous acetylides $R-C\equiv C-Cu$
cis-**2-Azetidinones**   O
from acetylene derivs. and nitrones s. *28,* 650

1-Pentynylcopper s.under (Me$_2$N)$_3$P n-C$_3$H$_7$–C≡C–Cu

Lithium dialkylcuprates (s. a. under ZnCl$_2$) $Li[CuR_2]$
1,4-Addition with aldol condensation   ←
β-**Hydroxy- and α,β-ethylene-ketones**

489.

Cyclohexen-3-one added at 0° under N$_2$ to an ethereal soln. of Li-dimethylcuprate, stirred 0.5 hr. at 0°, a satd. ethereal soln. of anhydrous ZnCl$_2$ added followed by acetaldehyde, and stirring continued 5 min. → *β*-hydroxyketone (Y: 97%) refluxed in benzene in the presence of a catalytic amount of p-toluenesulfonic acid → product (Y: 87%). F. e. and limitations s. K. K. Heng and R. A. J. Smith, Tetrah. Let. *1975,* 589; α,β-ethylene- from *β*-hydroxy-ketones s. a. R. A. Micheli et al., J. Org. Chem. *40,* 675 (1975).

Silver carbonate Ag_2CO_3
2-Diazoamines   C : C(N<) → CHC(N<)·C(: N$_2$)
from diazo compds. and enamines s. *28,* 625

Cuprous cyanide s. under Iodine	*CuCN*
Cuprous halide s. under Mg	*CuHal*
Cuprous chloride s. under Mg	*CuCl*
Cuprous bromide s. a. under Mg	*CuBr*
Cuprous bromide/irradiation	*CuBr/*⚡

Cyclobutane ring   □
by stereospecific dimerization of ethylene derivs. s. *27,* 704

Cuprous iodide s. a. under Mg	*CuI*
Cuprous iodide/lithium	*CuI/Li*

1,4-Addition   O
with intramolecular aldol condensation, also with replacement of bromine by hydrogen – *β*-Hydroxyketones s. *29,* 669

Cuprous iodide/organolithium compds.   *CuI/RLi*
Geospecific 1,6-addition   C : C·C : C·COOR → CR·C : C·CHCOOR
    β,γ-Ethylene from 2,4-diene-carboxylic acid esters – Organocopper compds. as reactants s. 29, 670

Cuprous iodide/n-butyllithium   *CuI/n-BuLi*
γ-Ketomercaptals   COC : C → CO·CH·C·C(SR)$_2$
    from α,β-ethyleneketones and mercaptals s. 28, 175

Cuprous iodide/tert. butyllithium   *CuI/Me$_3$C·Li*
Synthesis of enesilanes by 1,4-addition   COC : C → COCH·C·C(Si≦) : C

490.

    to α,β-ethyleneketones. Startg. halogenosilane allowed to react 1–2 hrs. at –78° with 2 equivalents *tert*-butyllithium in ether, treated 15–30 min. at –50 to –20° with CuI in tetrahydrofuran then at –78° with 1 molar equivalent 2-cyclohexen-1-one, and warmed to –10° during ca. 1 hr. → product. Y: > 90%. F. e. s. R. K. Boeckman, Jr., and K. J. Bruza, Tetrah. Let. *1974*, 3365.

Cuprous iodide/lithium thiophenoxide   *CuI/PhSLi*
Syntheses with mixed lithium cuprates   C : C·CO → CR·CHCO
    1,4-Addition – α-Subst. ketones from α,α'-dibromoketones – Ketones from carboxylic acid chlorides – Convenient prepn. of Li-phenylthio(alkyl)cuprates s. 29, 671

Magnesium   *Mg*
4-Subst. 4,5-dihydropyridazinium from pyridazinium salts   C : C → CHCR
    s. 26, 714

Reaction of Grignard compds. with unsatd. amines   C≡C → CH : CR
    Addition to carbon-carbon multiple bonds s. 27, 712

Magnesium/cuprous halide   *Mg/CuHal*
Synthesis of ethylene from acetylene derivs.   C≡C → CH : CR
    s. 26, 715

Magnesium/cuprous chloride   *Mg/CuCl*
1,4-Addition to α-methyleneketones   CO·C : CH$_2$ → CO·CHCH$_2$R
Total synthesis of steroids

491.

    A soln. of *m*-methoxybenzyl chloride in anhydrous tetrahydrofuran added dropwise during 55 min. to a stirred refluxing suspension of Mg-turnings in the same solvent, stirring and refluxing continued 0.5 hr., cooled to room temp., more tetrahydrofuran and CuCl added, stirred 0.5 hr., then a soln. of (+)-(1S,3aR,7aS)-

1-*tert*-butoxy-7a-methyl-3a,6,7,7a-tetrahydro-4-methyleneindan-5(4H)-one (prepn. s. R. A. Micheli et al., J. Org. Chem. 40, 675) in tetrahydrofuran added with stirring during 15 min., poured into a stirred aq. 1 N H_2SO_4-ice mixture, and stirring continued 5 min. → (+)-17β-*tert*-butoxy-3-methoxyestra-1,3,5(10),9(11)-tetraene. Y: 71%. Also isolation of the intermediate, ring closure with HCl, and 13β-ethyl analog s. N. Cohen et al., J. Org. Chem. 40, 681 (1975).

Synthesis of enephosphine sulfides $\quad P(S) \cdot C \equiv C \rightarrow P(S) \cdot CH : CR$
from ynephosphine sulfides s. 26, 715

Magnesium/cuprous bromide $\qquad Mg/CuBr$
Syntheses via vinylcopper compds. $\qquad C \equiv CH \rightarrow CR : CH_2$
Terminal ethylene from acetylene derivs.
Regio- and geo-specific addition

$$C_2H_5MgBr \xrightarrow{CuBr} \left[C_2H_5Cu\right] \xrightarrow{+ n-C_4H_9-C \equiv CH} \left[\begin{matrix}C_2H_5\\n-C_4H_9\end{matrix}\!\!>\!\!C=CHCu\right]$$

$$\downarrow HCl$$

$$\begin{matrix}C_2H_5\\n-C_4H_9\end{matrix}\!\!>\!\!C=CH_2$$

Ethylmagnesium bromide added at –35° under an inert atmosphere to a suspension of CuBr in ether, after 0.5 hr. at –35° 1-hexyne added, the temp. allowed to rise to –15°, then treated with 5 N HCl at –30° → 2-ethyl-1-hexene. Y: 85%. F. e. and geospecific syntheses s. J.-F. Normant et al., Bl. *1974*, 1656; α,β-**ethylenecarboxylic acids** by carboxylation of the vinylcopper compds. cf. J. Organometal. Chem. 77, 281 (1974); **1,3-enynes** s. Tetrah. Let. *1975*, 1465; syntheses via vinylcopper compds. cf. E. J. Corey and R. H. Wollenberg, J. Org. Chem. 40, 2265 (1975).

Magnesium/cuprous iodide $\qquad Mg/CuI$
1,4-Addition to α,β-ethyleneketones ←
s. 30, 453

Magnesium/dichlorobis(triphenylphosphine)nickel(II) $\qquad Mg/NiCl_2(Ph_3P)_2$
Synthesis of ethylene from acetylene derivs. $\qquad C \equiv C \rightarrow CH : CR$
s. 29, 672

Zinc $\qquad Zn$
1,4-Addition $\qquad C : C \rightarrow CHCR$
s. 26, 672

Mercuric acetate s. under HBF_4 $\qquad (CH_3COO)_2Hg$

Magnesium bromide $\qquad MgBr_2$
3-Amino-2-cyclohexenones from ynamines ←

A soln. of the startg. enollactone in acetonitrile added to 2 equivalents $MgBr_2$ (prepared from *Mg*-turnings and *ethylene bromide* in ether) in acetonitrile, then a soln. of N,N-diethylaminopropyne in the same solvent added dropwise at 15–20° with stirring, which is continued 2–3 hrs. at room temp. → 4-benzoyl-3-diethylamino-2,4-dimethyl-2-cyclohexen-1-one. Y: 70%. F. e. s. J. Ficini, J.-P. Genêt, and J.-C. Depezay, Bl. *1973*, 3369.

Zinc chloride/lithium dialkylcuprates $\quad ZnCl_2/Li[CuR_2]$
1,4-Addition with aldol condensation
β-**Hydroxy- and** α,β-**ethylene-ketones**
s. *30*, 489

Aluminum $\quad Al$
5,8-Dialkyl-5,6,7,8-tetrahydro-1-naphthols
from phenols and 1,5-dienes s. *28*, 651

Borane (s. a. under BrCN) $\quad BH_3$
Synthesis of alcohols
from 1,3-dienes and ethylene derivatives via cyclic boranes s. *26*, 716

Borane/lithium triethylcarboxide $\quad BH_3/Et_3C \cdot OLi$
Sym. ketones $\quad 2\,C:C \rightarrow CH \cdot C \cdot CO \cdot C \cdot CH$
from 2 ethylene deriv. molecules

495.

Cyclohexene added with ice-cooling under N_2 to borane in tetrahydrofuran, after 1 hr. at 0° methanol added slowly to form methyl dicyclohexylborinate, then ca. 10% excess α,α-dichloromethyl ether added followed by a soln. of Li-triethylcarboxide (prepared from *n*-butyllithium and triethylcarbinol) in hexane over 5 min., allowed to come to room temp. and maintained there for 0.5 hr., 95%-ethanol added followed by *NaOH* and, slowly, by 30%-H_2O_2, cautiously warmed, and kept 1 hr. at 50–60° → dicyclohexyl ketone. Y: 85%. F. e., also with bulkier ester groups, s. B. A. Carlson and H. C. Brown, Am. Soc. *95*, 6876 (1973); bicyclo-[3.3.1]nonan-9-one from 1,5-cyclooctadiene s. Synthesis *1973*, 776.

Borane/iodine $\quad BH_3/I$
Terminal acetylene from ethylene derivs. $\quad C:C \rightarrow CHC \cdot C \equiv CH$
Stereospecific synthesis with addition of 2 C-atoms s. *29*, 673

Diborane s. under $(CF_3CO)_2O$ $\quad B_2H_6$

Sodium tetrahydridoborate $\quad Na[BH_4]$
Benzenes from cyclopropenium salts
s. *26*, 771

Thexylborane/sodium hydroxide/hydrogen peroxide
β-**Hydroxycarboxylic acids** $\quad C:C \rightarrow CH \cdot C \cdot C(OH)CH_2COOH$
from ethylene derivs.

496/7.

Thexylborane in tetrahydrofuran treated under N_2 in sequence with 1-methylcyclopentene (1 hr. at –30 to –25°), ethyl propiolate (1 hr. at the same temp.), 3 *N* NaOH (5 min. at –25°, then 1 hr. at room temp.), and 30%-H_2O_2, finally warmed 2 hrs. at 40–50° → product. Y: 62–70%. E. Negishi and T. Yoshida, Am. Soc. *95*, 6837 (1973); **review of thexylborane** s. Synthesis *1974*, 77.

Bis-(3,5-dimethyl)borinane
1,4-Addition of branched alkyl groups $\quad COC:C \rightarrow COCHCR$
by bisboracyclanes – Reactivity of α,β-ethyleneoxo compds. s. *27*, 713

Diisobutylaluminum hydride $\qquad$ *i-Bu$_2$AlH*
2-Ethylenealcohols $\qquad$ CO → C(OH)C : C
   from acetylene derivs. and oxo compds. s. *28*, 652

Synthesis of 1,3-diol esters from enolesters   C(OAc) : C → C(OAc)·C·C(OAc)
s. *29*, 674

Diisobutylaluminum hydride/methyllithium $\qquad$ *i-Bu$_2$AlH/MeLi*
Syntheses via lithium alanates $\qquad$ ←
s. *23*, 705

Trialkylborane/hydrogen peroxide-irradiation $\qquad$ R_3B/H_2O_2-*☼*
Synthesis of alcohols from ethylene derivs. $\qquad$ C : C → C(OH)CR
  Stereospecific addition s. *27*, 714

Zeolite $\qquad$ ←
Tricyclo[3.1.0.0^{2,4}]hexanes $\qquad$ ○
  from 2 cyclopropene molecules s. *27*, 715

Thallic sulfate $\qquad$ $Tl_2(SO_4)_3$
3-Chromanols
s. *28*, 653

Boron fluoride (s. a. under CoBr$_2$·2Ph$_3$P) $\qquad$ BF_3
Acetal-enolether condensation $\qquad$ ←
s. *11*, 768

β-Alkoxyketones $\qquad$ C : C(OR) → CH·C(OR)·C·CO·
  from enolethers and ketones s. *28*, 654

Inverse-oriented diene synthesis $\qquad$ ○
s. *28*, 655; s. a. Can. J. Chem. *53*, 616, 619 (1975)

Reactions of p-quinones with enamines
  5-Hydroxyindoles – 3-Amino-6-hydroxycoumarins s. *26*, 717

Fluoroborate $\qquad$ BF_4^-
Cyclopent-3-enones from 1,3-dienes
  via allylsulfonium ylid-type rearrangement and ring expansion of 2-spirocyclopropano-1,3-dithianes s. *28*, 656

Cupric fluoroborate $\qquad$ $Cu[BF_4]_2$
Diene syntheses of cyclic ketones
  from α,β-ethylene-α-chloronitriles with preceding C-alkylation of cyclopentadienes – Electrophilic substitution of cyclopentadienide anions, review, s. *26*, 718

Silver fluoroborate $\qquad$ $AgBF_4$
β-Diketones $\qquad$ C ≡ C → CO·CH·CO
  from acetylene derivs. and carboxylic acid chlorides s. *28*, 104

Tetrahydro-1,2-oxazine ring $\qquad$ ○
  from ethylene derivs. and α-chloronitrones – Enophiles – Cycloaddition with non-activated ethylene derivs. – Prepn. of γ-lactones – Extension of the C-chain s. *28*, 657

Fluoroboric acid/mercuric acetate $\qquad$ $HBF_4/(CH_3COO)_2Hg$
Ar. β-acetoxyethylation $\qquad$ ArH → ArCH$_2$CH$_2$OAc
s. *29*, 675

Diethylaluminum chloride/n-butyllithium $\qquad$ Et_2AlCl/n-$BuLi$
Synthesis of γ,δ-acetylene- from α,β-ethylene-ketones   C : C → CH·C·C ≡ C
s. *28*, 658

Aluminum chloride $\qquad AlCl_3$
Ene synthesis ←
also with organosilicon compds. s. 29, 676
β-Halogenoketones from carboxylic acid chlorides $\quad C:C \rightarrow CHal \cdot C \cdot COR$
s. 27, 213
Diene synthesis with 1,4-dienes ○
via 1,3-dienes conjugated in situ
s. 30, 500
Carbostyrils ⊙
from isocyanates and acetylene derivs. s. 26, 694 suppl. 29
Anilines from pyrroles
via 7-azabicyclo[2.2.1]-2,5-heptadienes s. 26, 719

Aluminum chloride/hydrogen chloride $\qquad AlCl_3/HCl$
Synthesis of indan-1-one-3-carboxylic acids ←
with C-alkyl migration s. 28, 659

Acetone/irradiation $\qquad Me_2CO/⚡$
Cyclobutane ring □
s. 29, 659
1,4-(2π+6π)-Cycloaddition ○
s. 26, 720
Azetidine ring from ethylene derivs.
s. 29, 659

Benzophenone/irradiation $\qquad Ph_2CO/⚡$
Photochemical 1,4-addition $\quad CO \cdot C:C \rightarrow CO \cdot CH \cdot CR$
Carbohydrate derivs. s. 29, 677; s. a. Tetrah. Let. *1975*, 297; prostaglandin derivs. s. G. L. Bundy, Tetrah. Let. *1975*, 1957

Trifluoroacetic anhydride/diborane $\qquad (CF_3CO)_2O/B_2H_6$
Syntheses with cyanoboranes $\quad 3\,C:C \rightarrow (CHC)_3C \cdot OH$
Trialkylcarbinols from ethylene derivs. – Synthesis of ketones with thexylborane s. 27, 716; s. a. Soc. Perkin I *1975*, 129–145

Formic acid $\qquad HCOOH$
[4+2]-Cyclodimerization of 1H-1,2-diazepines ○

498.

A soln. of 1-ethoxycarbonyl-5-methyl-1H-1,2-diazepine in formic acid allowed to stand a few hrs. at room temp. → product. Y: 87%. F. e. and acid catalysts s. B. Willig and J. Streith, Tetrah. Let. *1973*, 4167.

Acetic acid $\qquad CH_3COOH$
5-Subst. 2-pyrrolidones from Δ³-pyrrol-2-ones ←
Addition of indoles to Δ³-pyrrol-2-ones s. 27, 823

Acetic acid/irradiation $\qquad CH_3COOH/⚡$
Cyclobutane ring $\quad 2\,C:C \rightarrow □$
by dimerization of ethylene derivs. s. 27, 704

Acetic acid/cyclohexylamine $\qquad CH_3COOH/C_6H_{11}NH_2$
Monocyanoethylation of ketones $\quad CH_2COCH_2 \rightarrow CH_2COCHCH_2CH_2CN$
s. 27, 717

Thiazolium salt ←
Ketones CHO → CO·C·OH
from aldehydes and ethylene derivs. s. 28, 648 suppl. 30

Thiourea s. under PdCl₂ $(H_2N)_2CS$

Carbon tetrabromide CBr_4
Oxidative dimerization ←
of electron-rich ethylene derivs. s. 26, 721

Cyanogen bromide/borane $BrCN/BH_3$
Synthesis of *trans*-ethylene derivs. C≡C → CH:CR
from acetylene derivs. – Also prepn. of cyanohydrins s. 28, 660

Silica gel SiO_2
Capillary techniques – Preparative chromatography
Diene synthesis with enolesters ○

499.

Reactions such as nucleophilic substitutions and additions may be accelerated by allowing then to take place inside the pores of a high surface area material such as silica gel or alumina. – E: A soln. of tetrafluoro-p-benzoquinone and 1-acetoxy-1,3-butadiene in benzene placed in a column of silica gel at room temp. for 60 hrs. → 5-acetoxy-2,3,4a,8a-tetrafluoro-4a,5,8,8a-tetrahydro-1,4-naphthoquinone. Y: 67–78%; conversion 49.5%. – Reaction in the absence of silica gel requires heating at 98° for several hrs. F. e. s. M. Hudlicky, J. Org. Chem. **39**, 3460 (1974).

Titanium trichloride/hydroxylamine $TiCl_3/H_2NOH$
Ring closure with radical amination
s. 27, 332

Tributylphosphine Bu_3P
Michael addition C:C → CH·C·C(NO₂)
of aliphatic nitro compds. s. 23, 700 suppl. 28

Tricyclohexylphosphine R_3P
Addition of ethylene derivatives ←
to α,β-ethylene-β-dicarboxylic acid esters
s. 26, 722

Triphenylphosphine (s. a. under Ni(0)-complex and [(CH₃CO)₂CH]₂Pd) Ph_3P
Cyclic α-ketoazomethines ⊙
from cyclic ketones and isonitriles – Ring expansion s. 27, 718

Tris(dimethylamino)phosphine/1-pentynylcopper $(Me_2N)_3P/n$-C_3H_7-C≡C-Cu
Syntheses by selective group transfer from cuprates CO·C:C → CO·CHCR
1,4-Addition s. 28, 661; f. reagents, prostaglandin synthesis, s. C. J. Sih et al., Am. Soc. **97**, 865, 857 (1975).

Hydrogen peroxide s. under R₃B H_2O_2

p-Toluenesulfonic acid TsOH
Aminocyclobutanes from enamines □
s. 27, 719

**Diene synthesis with 1,4-dienes
via 1,3-dienes conjugated in situ** ○

500.

1-Methoxy-4-methylcyclohexa-1,4-diene refluxed 44 hrs. with methyl acrylate and p-toluenesulfonic acid → methyl 1-methoxy-4-methylbicyclo[2.2.2]oct-2-ene-6-carboxylate. Y: 82% based on startg. m. consumed. F. e. with $AlCl_3$ or *dichloromaleic anhydride* as catalysts, also isolation of crude 1,3-dienes, s. A. J. Birch and K. P. Dastur, Soc. Perkin I *1973*, 1650; with dichloromaleic anhydride s. a. Am. Soc. *96*, 2605 (1974).

Sulfuric acid (s. a. under $FeSO_4$) $\quad\quad\quad\quad\quad\quad\quad\quad\quad\quad\quad\quad\quad H_2SO_4$
2-Aminoalcohols from ketones $\quad\quad\quad\quad\quad CO \rightarrow C(OH)CH_2NH_2$
via enoxysilanes and α-siloxynitriles – Synthesis with addition of 1 C-atom s. *27*, 720

2-Decalones ○
Bicyclic ketones from dienynes – Solvent effect on stereospecificity s. *29*, 678

Prins reaction with rearrangement
s. *29*, 679

Iodine/cuprous cyanide $\quad\quad\quad\quad\quad\quad\quad\quad\quad\quad\quad\quad\quad\quad I/CuCN$
2-Ethylene-1,4-dinitriles from 1,3-dienes $\quad C:C\cdot C:C \rightarrow C(CN)\cdot C:C\cdot C(CN)$
s. *28*, 662

Hydrogen chloride $\quad\quad\quad\quad\quad\quad\quad\quad\quad\quad\quad\quad\quad\quad\quad\quad HCl$
Dimerization of indoles ←
Oxidative dimerization s. *28*, 663

cis-**Octahydroindol-6-ones** ○
from Δ²-pyrrolines and α,β-ethyleneketones s. *28*, 664

Hydrogen chloride/irradiation $\quad\quad\quad\quad\quad\quad\quad\quad\quad\quad\quad\quad HCl/⚡$
Photodimerization of pyridines
s. *27*, 721

Manganese(III) acetate $\quad\quad\quad\quad\quad\quad\quad\quad\quad\quad\quad\quad (CH_3COO)_3Mn$
C-Carboxymethylation of ethylene derivs. $\quad\quad C:C \rightarrow CHC\cdot CH_2COOH$
s. *28*, 713; s. a. W. J. de Klein, R. *94*, 48 (1975)

Iron carbonyl s. under Rh_2O_3 $\quad\quad\quad\quad\quad\quad\quad\quad\quad\quad Fe(CO)_5$

Ferrous sulfate/sulfuric acid $\quad\quad\quad\quad\quad\quad\quad\quad\quad\quad FeSO_4/H_2SO_4$
Free-radical C-alkylation C
with spirooxaziridines s. *27*, 722

Hydrogen tetracarbonylcobaltate $\quad\quad\quad\quad\quad\quad\quad\quad\quad H[Co(CO)_4]$
Quinolines ○
from azomethines and enolethers s. *26*, 723

Cobaltous acetate $\quad\quad\quad\quad\quad\quad\quad\quad\quad\quad\quad\quad (CH_3COO)_2Co$
Ketones $\quad\quad\quad\quad\quad\quad\quad\quad\quad\quad\quad\quad CHO \rightarrow CO\cdot C\cdot CH$
from aldehydes and ethylene derivs. s. *28*, 648 suppl. *29*

Cobaltous bromide·bis(triphenylphosphine)/boron fluoride $\quad CoBr_2\cdot Ph_3P/BF_3$
Cyclodimerization ○
Highly-condensed polyisocyclics s. *29*, 680

Nickel(0)-complex/triphenylphosphine ←
Cyclocooligomerization
cis,cis,trans-**1,4,7-Cyclodecatrienes**
from acetylene derivs.

5-Decyne and 5 moles butadiene kept in an autoclave at 20° in the presence of 1:1 bis-(*cis,cis*-1,5-cyclooctadiene)nickel(0) catalyst and triphenylphosphine as additional ligand until after 25.4 hrs. the pressure indicates transition from a fast to a slower reaction → product. Y: 93.7%. F. e. s. W. Brenner et al., A. *1973*, 1882; *1975*, 660; cf. ibid. *1975*, 743.

Bis(acrylonitrile)nickel(0)   [*Ni(CH₂ : CHCN)₂*]
Nickel(0)-catalyzed cycloadditions
   Methylenecyclopentane from methylenecyclopropane ring s. 26, 724

Dichlorobis(triphenylphosphine)nickel(II) s. under Mg $NiCl_2(Ph_3P)_2$

Ruthenium triphenylphosphine complex ←
Synthesis of polyhalides   C : C → CHalCR
   from ethylene derivs. s. 19, 776 suppl. 29

Rhodium oxide Rh_2O_3
Sym. 2,4,5-trialkylimidazoles   O
   from ethylene derivs. s. 28, 665

Rhodium oxide/iron carbonyl $Rh_2O_3/Fe(CO)_5$
Amines from ethylene derivs.   C : C → CHC·CH₂N<
   Synthesis with addition of 1 C-atom s. 27, 723

Chlorotris(triphenylphosphine)rhodium(I) $RhCl(Ph_3P)_3$
Ring closures with diynes   O
   Double ring closure s. 26, 725; s. a. A. *1974*, 1876; B. *108*, 237 (1975); via transition metal complexes, review, s. E. Müller, Synthesis *1974*, 761; 1,4-naphthoquinone ring with a Ni-complex cf. Chem. Ztg. *99*, 155 (1975); heterocyclics by triple ring closure cf. P. J. Garratt and S. B. Neoh, Am. Soc. *97*, 3255 (1975)

Bis(triphenylphosphine)(maleic anhydride)palladium ←
2-Piperidone from isocyanates
s. 26, 726

Palladous acetoacetonate/triphenylphosphine   [(CH₃CO)₂CH]₂Pd/Ph₃P
Palladium-catalyzed telomerization ←
   of 1,3-dienes – Palladium compd.-catalyzed reactions, review, s. 27, 724; telomerized silanes cf. J. Langová and J. Hetflejš, Coll. *40*, 420, 432 (1975)

Palladous chloride/sodium acetate $PdCl_2/CH_3COONa$
Stereospecific oxidative cyclodimerization of phenols   O
   O-Heterocyclics by double ring closure s. 27, 725

Palladous chloride/thiourea $PdCl_2/(H_2N)_2CS$
α-Methylene-γ-lactones from 3-acetylenealcohols

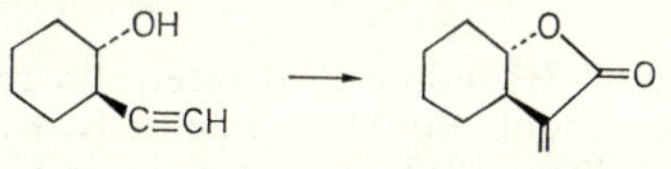

Startg. m. stirred overnight at 50° under 50 psi CO with PdCl₂ and thiourea in acetone → product. Y: 94%. F. e. s. J. R. Norton, K. E. Shenton, and J. Schwartz, Tetrah. Let. *1975*, 51.

Dichlorobis(benzonitrile)palladium(II) $[PdCl_2(PhCN)_2]$
1,3-Dienes $C \equiv C \rightarrow RC:C \cdot C:CR$
   from acetylene derivatives and 2 ethylene deriv. molecules s. *26*, 727

Via intermediates   *v.i.*
Synthesis via di(ene)chloroboranes $2C \equiv C \rightarrow CH:C \cdot C \cdot C:CH$
   1,3-Dienes from 2 acetylene deriv. molecules s. *28*, 666

Dihydroisoquinoline rearrangement   ←
s. *26*, 764

Prolines from α,β-ethyleneoxo compds.   ○
s. *29*, 664

Insertion of carbonyl via C-dichloromethylation   ⊙
   Ring expansion of adamantanes s. *27*, 726

Rearrangement

Hydrogen/Oxygen Type   CC∩HO

Without additional reagents   *w.a.r.*
Cyclopent-2-enecarboxaldehydes   C
   by skeletal rearrangement s. *26*, 728

4,5-Dihydro-3(2H)-pyridazones   ⊙
   from 4-ene-5-hydroxy-Δ^2-pyrazolines s. *27*, 727

Sodium/alcohol   NaOR
α-Diketones from glycols   ←
   with skeletal rearrangement s. *29*, 681

Sulfuric acid/formic acid $H_2SO_4/HCOOH$
O-Heterocyclics by double ring closure   ○
   8-Oxabicyclo[3.2.1]octanes s. *29*, 682

Hydrogen/Carbon Type   CC∩HC

Without additional reagents   *w.a.r.*
Side chain rearrangement of cyclopropanes   ←
   1,3-Dienes from allenes s. *27*, 728

Cope rearrangement   ←
s. *27*, 739

Indoles from indolenines $N:C(CH)CR \rightarrow NHC(CR):C$
by [3.3]-sigmatropic lateral rearrangement

503.

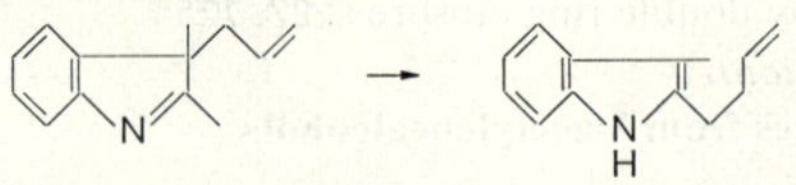

A soln. of 3-allyl-2,3-dimethyl-3H-indole in dry tetralin refluxed 6.25 hrs. under $N_2 \rightarrow$ 2-(but-3-enyl)-3-methylindole. Y: 88%. F. e. s. R. K. Bramely, J. Caldwell, and R. Grigg, Soc. Perkin I *1973*, 1913.

Isoxazoles from 5-ene-Δ^2-isoxazolines   ←
   Alleneisoxazoles by isomerization s. *26*, 729

Cycloisomerization with 1,5-hydrogen shift ○
s. 26, 759; 1,2-dihydronaphthalenes s. suppl. 29; s. a. P. B. Valkovich et al., Am. Soc. 97, 901 (1975)

Intramolecular ene reaction
Cycloisomerization of dienes s. 28, 667

Chromenes by o-Claisen rearrangement
gem-Dimethyl effect s. 27, 729; s. a. H. Omokawa and K. Yamashita, Agr. Biol. Chem. 38, 1731 (1974)

3-Vinylthiophene 1,1-dioxides from diallenesulfones
s. 29, 683

Ring expansion by ring opening C ⊙
by cyclopropane ring opening s. 26, 730

Irradiation (s. a. under i-Bu₃Al, BF₃, PhCOCH₃, ⇝
2-Acetonaphthone and Fe(CO)₅)

Position shift of cyano groups ←
s. 27, 730; with a catalytic amount of NaCN, 5- from 6-cyanouracils, cf. S. Senda, K. Hirota, and T. Asao, J. Org. Chem. 40, 353 (1975)

[3.3]-Sigmatropic rearrangement ←
of allyl dithiocarbamates s. 30, 585

Cycloisomerization ○
of alkaloids and stilbenes – N-Arylenamines s. 26, 731; tetrahydropyrenes s. A. Padwa and A. Mazzu, Tetrah. Let. 1974, 4471

Sodium   Na
C-Phenyl migration ←
s. 26, 732

Sodium/o-chlorotoluene $Na/o\text{-}ClC_6H_4Me$
Isocycles from ethylene derivs. ○
s. 27, 731

Sodium hydride   NaH
N-Propenyl from N-allyl derivs.   N·CH·C:C → N·C:C·CH
s. 26, 739

Sodium hydride/dimethyl sulfoxide NaH/Me_2SO
Mono- and bi-cyclic isocyclics ⊙
by cyclopropane ring expansion s. 29, 982

Potassium hydroxide   KOH
Quaternary benzisoindolinium from diynammonium salts ○

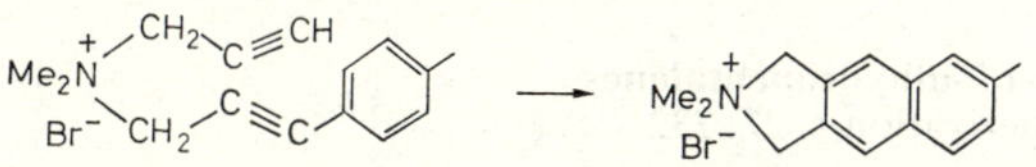

A 2 N KOH-soln. added to an aq. soln. of the startg. m., and the product isolated after 1 hr. → 2,2,6-trimethyl-benz[f]isoindolinium bromide. Y: 92%. F. e. s. El. O. and E. O. Chukhadzhyan, and A. T. Babayan, Zh. Org. Khim. 10, 46 (1974); C. A. 80, 108293.

Potassium tert-butoxide/dimethyl sulfoxide $KOCMe_3/Me_2SO$
Exo- from endo-cyclic carbon-carbon double bonds ←
s. 26, 733

Lithium diisopropylamide/methyl alcohol   *i-Pr$_2$NLi/MeOH*
α,β-**Ethylenemercaptals**   CH·C:C(S·)$_2$ → C:C·CH(S·)$_2$
from ketene mercaptals

505.

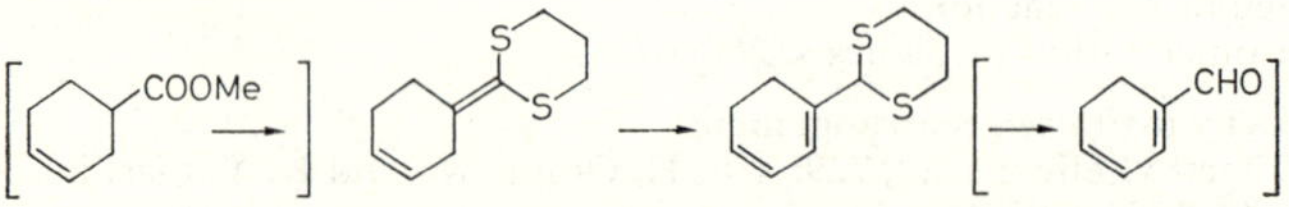

A soln. of 2-(3-cyclohexenylidene)-1,3-dithiane (prepared from methyl 3-cyclohexenecarboxylate according to Synth. Meth. *29*, 552) in tetrahydrofuran added at −78° to Li-diisopropylamide in hexamethylphosphoramide-tetrahydrofuran, allowed to warm to −40° during 1.5 hrs., kept 0.5 hr. at this temp., recooled to −78°, excess methanol added, after several min. warmed to room temp. and poured into water → 2-(1,3-cyclohexadienyl)-1,3-dithiane. Y: 73%. − α,β-**Ethylenealdehydes** can be prepared **from carboxylic acid esters** via the above reaction. F. e., **also** prepn. of α,β-**ethyleneketones**, s. E. J. Corey and A. P. Kozikowski, Tetrah. Let. *1975*, 925.

Lithium diisopropylamide/hexamethylphosphoramide   *i-Pr$_2$NLi/(Me$_2$N)$_3$PO*
β,γ- **from** α,β-**Ethylenecarboxylic acid esters** CH·C:C·COOR → C:C·CH·COOR
s. *29*, 810

Potassium 1,3-propylenediamide   *H$_2$N[CH$_2$]$_3$NHK*
Terminal acetylene derivs.   C≡CH
by contrathermodynamic migration of carbon-carbon triple bonds

506.   CH$_3$CH$_2$[CH$_2$]$_4$−C≡C−[CH$_2$]$_4$CH$_2$CH$_3$ → CH$_3$CH$_2$[CH$_2$]$_4$CH$_2$CH$_2$[CH$_2$]$_4$−C≡CH

7-Tetradecyne injected rapidly under N$_2$ or argon with cooling at 15–20° into a vigorously stirred soln. of the superbase K-1,3-propylenediamide prepared from KH in 1,3-propanediamine, and the mixture quenched with water after 1–2 min. → 1-tetradecyne. Y: 89%. F. e. s. C. A. Brown and A. Yamashita, Am. Soc. *97*, 891 (1975); f. reactions with K-1,3-propylidenediamide s. Chem. Commun. *1975*, 222.

Sodium nitrite s. under HNO$_3$ *NaNO$_2$*

Triethylamine   *Et$_3$N*
Ring contraction by 2 C-atoms
of cyclic α,β-ethyleneketones

507.

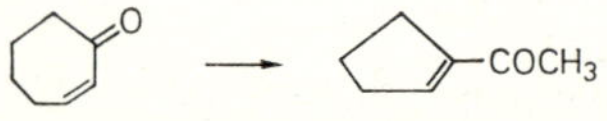

A soln. of cyclohept-2-en-1-one in aq. 60%-ethanol heated 1 week at 140° with triethylamine → 1-acetylcyclopent-1-ene. Y: ca. 100%. F. e. s. E. Lamparter and M. Hanack, Tetrah. Let. *1974*, 1623.

Piperidine   *C$_5$H$_{11}$N*
3-Dialkylamino-1,2-dihydropentalenes
    by cycloisomerization s. *27*, 732

N,N-Diethylaniline   *PhNEt$_2$*
Chromenes by o-Claisen rearrangement
s. *27*, 729 suppl. *28*; s. a. Chem. & Ind. *1975*, 611
6a,11a-Dihydrobenzofuro[3,2-c][1]benzopyrans
    from 1,4-diaryloxy-2-butynes s. *29*, 684

Pyridine s. under Me$_3$SiCl   *C$_5$H$_5$N*

Methylmagnesium bromide s. under FeCl$_3$ *MeMgBr*

Mercuric oxide/sulfuric acid $\qquad HgO/H_2SO_4$
Ring closure with acetylene derivs.
  O-Heterocyclics – Δ^3-Chromenes s. 28, 668

Sodium tetrahydridoborate s. under $NiBr_2 (n-Bu_3P)_2$ $\qquad Na[BH_4]$

Diethylaluminum hydride/ethylene $\qquad Et_2AlH/CH_2:CH_2$
Methylenecyclopentane ring from 1,5-dienes
s. 26, 734

Triisobutylaluminum/irradiation $\qquad i\text{-}Bu_3Al/\text{\textit{hv}}$
cis→trans-**Rearrangement** $\qquad \leftarrow$
s. 28, 669

Alumina $\qquad Al_2O_3$
Isomerization of carbon-carbon multiple bonds $\qquad \leftarrow$
s. 26, 729

Boron fluoride/irradiation $\qquad BF_3/\text{\textit{hv}}$
Cycloisomerization of α,β-ethyleneketones $\qquad \bigcirc$

508.

Startg. m. in benzene containing BF_3-etherate irradiated with Pyrex-filtered light → product. Y: 90%. F. e., reaction conditions, and limitation s. M. Tada et al., Chem. Commun. 1975, 55.

Aluminum chloride $\qquad AlCl_3$
1,3- from 1,4-Dienes
  in situ-Conjugation s. 30, 500

Aluminum chloride/hydrogen chloride $\qquad AlCl_3/HCl$
Gattermann formylation $\qquad ArH \rightarrow ArCHO$
  with C-alkyl migration s. 28, 708

Methylalcohol s. under i-Bu_2NLi $\qquad MeOH$

Acetophenone/irradiation $\qquad PhCOCH_3/\text{\textit{hv}}$
Partial *cis-trans*-rearrangement $\qquad \leftarrow$
s. 28, 669

2-Acetonaphthone/irradiation $\qquad \leftarrow$
trans-cis-**Rearrangement**
s. 28, 669

Cation exchanger $\qquad \leftarrow$
1,3-Dienamines from 1,4-dien-2-amines $\qquad \leftarrow$
s. 30, 23

Formic acid $\qquad HCOOH$
Migration of side chains $\qquad \leftarrow$
s. 28, 670

Acetic acid $\qquad CH_3COOH$
Cyclic alcohols from ketones $\qquad \bigcirc$
s. 28, 768

Benzoic acid   *PhCOOH*
Out-of-ring migration of carbon-carbon double bonds   ←
s. 27, 733

m-Chloroperoxybenzoic acid   *m-ClC$_6$H$_4$COO$_2$H*
Migration of carbon-carbon double bonds   ←
via formation of S-oxides s. 27, 734

Xanthione   ←
1,3-Dienes from allenes   CH·C : C : C → C : C·CH : C
s. 28, 521

Dimethyl sulfoxide s. under NaH   *Me$_2$SO*

Silica   *SiO$_2$*
Thermal 1,5-prototropic shift   ←
s. 27, 735

Trimethylchlorosilane/pyridine   *Me$_3$SiCl/C$_5$H$_5$N*
Migration of carbon-carbon double bonds   ←
s. 28, 46

Stannic chloride   *SnCl$_4$*
Isocyclics by cyclopropane ring expansion   ◉
s. 29, 685

Nitric acid/sodium nitrite   *HNO$_3$/NaNO$_2$*
***cis-trans*-Rearrangement**   ←
s. 28, 671

Hexamethylphosphoramide s. under i-Pr$_2$NLi   *(Me$_2$N)$_3$PO*

Polyphosphoric acid   *H(PO$_3$H)$_x$OH*
Cycloisomerization   ○
Isocyclic ketones s. 26, 735; cyclic phosphonium salts, with 115%-polyphosphoric acid, cf. G. A. Dilbeck, D. L. Morris, and K. D. Berlin, J. Org. Chem. 40, 1150 (1975).

2-Aminoquinolines
from chloroformamidines via α,β-acetyleneamidines s. 28, 672

ang-lin-Rearrangement   ←

509.

7,8-Dihydro-15H-cyclohepta[2,1-a : 4,5-a′]dinaphthalen-15-one added at 115° under anhydrous conditions to polyphosphoric acid, and vigorously stirred 10 hrs. at the above temp. → 13,14-dihydro-6H-cyclohepta[1,2-b : 4,5-b′]dinaphthalen-6-one. Y: 52%. F. e. s. I. Agranat and Y.-S. Shih, Synth. Commun. 4, 119 (1974).

Antimony pentafluoride s. under HF   *SbF$_5$*

p-Toluenesulfonic acid   *TsOH*
Preferential migration of carbon-carbon double bonds   ←
s. 28, 673

1,3- from 1,4-Dienes   ←
in situ-Conjugation s. 30, 500

High dilution cycloisomerization
Indenes from allenes   O

(Ph₂)₂C=C=CHPh → [1,3-diphenylindene structure with Ph groups]

Triphenylallene in a large amount of benzene added dropwise during 1.5 hrs. to a refluxing soln. of p-toluenesulfonic acid in benzene so that the concentration of the product remains below 0.02 M → 1,3-diphenylindene. Y: 73%. T. Greibrokk, Acta Chem. Scand. 27, 2252 (1973).

Double Wagner-Meerwein rearrangement   ←
s. 29, 686; cage compds. s. K. Hirao et al., Am. Soc. 97, 3249 (1975)

Sulfuric acid H_2SO_4
Epimerization of ketosteroids   ←
s. 26, 736

α,β-Ethyleneketones from cyclobutanone ring   C
s. 29, 687

Ring expansion of isocyclics   ☉
s. 17, 780 suppl. 29; s. a. J. Org. Chem. 40, 276 (1975)

Iodine   I
Cycloisomerization of isocyclics   O
s. 27, 736

Silver perchlorate $AgClO_4$
1,3-Dienes from cyclopropenes   C
s. 28, 674

1,3-Dienes from bicyclobutanes
s. 26, 741

Perchloric acid $HClO_4$
Cycloisomerization of thioenolethers   O

Startg. m. stirred 2–3 hrs. at room temp. with 70%-$HClO_4$ → product. Y: ca. 100%. F. e. s. E. R. de Waard, H. R. Reus, and H. O. Huisman, Tetrah. Let. 1973, 4315.

Hydrogen fluoride/antimony pentafluoride HF/SbF_5
Migration of carbon-carbon double bonds   ←
s. 27, 159

Hydrogen chloride (s. a. under $AlCl_3$)   HCl
Cyclic ketones from ethylenealdehydes   O
  Cyclopent-2-enones s. 26, 737

Cyclopent-2-enones by cycloisomerization
s. 29, 688

Hydrogn bromide   HBr
Epimerization of aryl groups   ←
s. 26, 149

1,2,3,4,5,6-Hexahydro-2,6-methano-3-benzazocines   O
  from 1,2,5,6-tetrahydropyridines s. 26, 738

Iron carbonyl	$Fe(CO)_5$
N-Propenyl from N-allyl derivatives	$N \cdot CH \cdot C : C \rightarrow N \cdot C : C \cdot CH$
s. *26*, 739	
Iron carbonyl/irradiation	$Fe(CO)_5/\mathcal{W}$
s. *26*, 739 suppl. 29	
Ferric chloride/methylmagnesium bromide	$FeCl_3/MeMgBr$
Deconjugation	←
β,γ- from α,β-Ethyleneketones s. *27*, 737	
Cobaltous bromide	$CoBr_2$
Pyrazoline and pyrrole ring	○
from azines – Δ^2-Pyrazolines from hydrazones s. *26*, 740	
Nickel phosphite complex/hydrogen chloride	←
Rearrangements with nickel complexes	←
Migration of carbon-carbon double bonds	

512.

A soln. of *cis*-1,4-hexadiene, ethylenebis(tri-o-tolylphosphite)nickel(0), and HCl as co-catalyst in argon-purged toluene stirred 2 hrs. at 24–26° under N_2 → 2,4-hexadiene. Y: 85%. F. e. s. R. G. Miller et al., Am. Soc. *96*, 4211–4235 (1974).

Dibromobis(tri-n-butylphosphine)nickel/	
sodium tetrahydridoborate	$(n\text{-}Bu_3P)_2NiBr_2/NaBH_4$
Redoxdimerization of cyclic dienes	←

513.

Norbornadiene heated 18 hrs. at 80° under N_2 in a sealed tube with dibromobis-(tri-*n*-butylphosphine)nickel and $NaBH_4$ in isopropylamine → *exo*-5-(o-tolyl)-2-norbornene. Y: 69.2%. S. Yoshikawa et al., Tetrahedron *30*, 405 (1974).

Dichlorotris(triphenylphosphine)ruthenium(II)	←
Migration of exocyclic carbon-carbon double bonds	←
s. *28*, 675	
Tetracarbonyldichlorodirhodium	$[Rh(CO)_2Cl]_2$
Double ring opening by isomerization	C
1,3-Dienes from bicyclobutanes s. *26*, 741	
Chlorotris(triphenylphosphine)rhodium(I)	$RhCl(Ph_3P)_3$
O-Propenyl from O-allyl derivs.	$OCH_2CH : CH_2 \rightarrow OCH : CHCH_3$
Allyl as protective group – Carbohydrate derivs. s. *20*, 537 suppl. 29	
Hexacarbonyldichlorodiiridium	$[Ir(CO)_3Cl]_2$
Double ring opening	C
by isomerization s. *26*, 741	
Via intermediates	v.i.
Migration of carbon-carbon double bonds	←
via chlorides s. *23*, 729	
via formation of S-oxides s. *27*, 734	

Oxygen/Nitrogen Type   CC↷ON

Without additional reagents   w.a.r.
4,5-Dihydro-1,2-oxazinium ring opening   C
s. 28, 676

Potassium phenoxide   PhOK
Rearrangement of N-aryloxypyridinium salts   ←
s. 27, 910

Triethylamine Et_3N
s. 27, 910

Oxygen/Carbon Type   CC↷OC

Without additional reagents   w.a.r.
Claisen rearrangement of enolethers   ←
  γ,δ-Ethylenoxo compds. from 2-ethyleneenolethers – Rearrangement of ethylene-carbohydrates s. 27, 738; γ,δ-ethylenealdehydes, steroid derivs., s. R. E. Ireland and D. J. Dawson, Org. Synth. 54, 71 (1974)
  Stereospecific rearrangement s. 26, 742; γ,δ-ethylenealdehydes s. P. A. Grieco, Y. Masaki, and D. Boxler, Am. Soc. 97, 1597 (1975)

Rearrangements of cyclobutane-1,2-diols   R·C(OH)C(OH)·R → CO·C(R)$_2$
  Cyclobutanones and cyclopropyloxo compds. s. 28, 677

Cope rearrangement   ←
  of bicyclo[3.2.0]hepta-3,6-dien-2-ones s. 27, 739

2-Alkylidene-1,3-diones   ←
  by pyrolysis of acoxy-2-acetylenes s. 28, 968

o-Hydroxycarboxylic acid amides   O·CON< → C·CO·N<
  from aryl carbamates s. 28, 678

Pyrolysis of α-alkoxyenolesters   O
  Phthalides from o-oxocarboxylic acid esters, alternate Reformatskii route
  s. 28, 679

α-Aminoketones from 1,2-oxidoamines   ⊙
  Ring expansion s. 29, 689

Spiro-β-diketones from cyclic α,β-oxidoketones   ←

514.

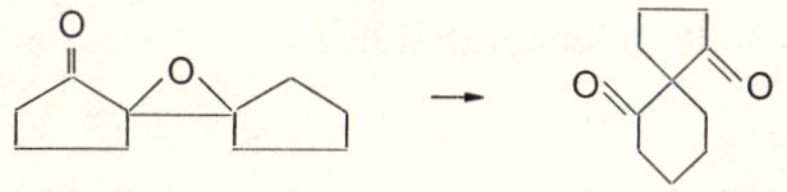

2-Cyclopentylidenecyclopentan-1-one oxide (s. a. 652) heated 15 min. at 225° in a sealed tube → spiro[4.5]decane-1,6-dione. Y: 89%. F. e., also isomers under different conditions, s. J. R. Williams et al., J. Org. Chem. 39, 1028 (1974).

3-Acylindoles from 3,1-benzoxazepines   ⊙
s. 26, 743

Irradiation (s. a. under NaHCO$_3$, CH$_2$:CHCH:CHCH$_3$, PhCOCH$_3$, and Iodine)   ⚡
Migration of hydroxyl   ←
s. 29, 690

2- from 4-Pyrones   ←
s. 28, 680

Cyclic alcohols from ketones   O
  Benzocyclobutenols – Cyclopropanols – Stereospecific ring closure – γ-Hydrogen transfer – 3-Azetidinols s. 26, 744

δ-Hydrogen transfer
2-Hydroxypyrrolidine ring s. *29*, 691

3-Hydroxy-1,2-indandiones from 3-hydroxyflavones

515.

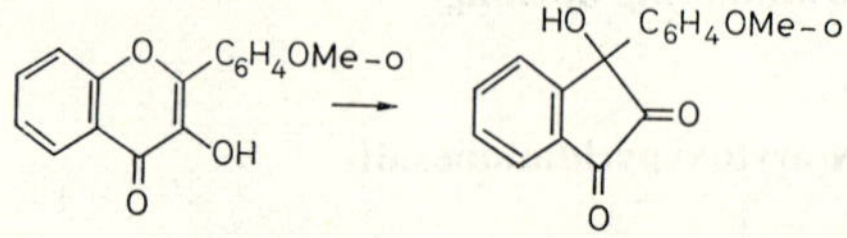

A soln. of 3-hydroxy-2'-methoxyflavone in isopropanol-benzene irradiated 8 hrs. under N_2 with the Pyrex-filtered light (> 290 nm) of a 450 w. high-pressure Hg-lamp → product. Y: ca. 100%. F. e. s. T. Matsuura, T. Takemoto, and R. Nakashima, Tetrahedron *29*, 3337 (1973).

Skeletal rearrangement of quinone epoxides ←
s. *29*, 692

Sodium hydride   NaH

Wittig rearrangement ←
α-Hydroxy- from α-alkoxy-carboxylic acid amides s. *29*, 693

Sodium hydride/dimethyl sulfoxide NaH/Me_2SO

Alcohols from oxido compds. O
with isocyclic ring closure s. *27*, 740

Sodium hydroxide   NaOH

Stereospecific intramolecular aldol condensation
s. *28*, 681

Cyclopropylalcohols from oxido compds. ←
s. *29*, 694

Potassium hydroxide   KOH

Intramolecular aldol condensation O
Cyclic diazoketols s. *26*, 745

Potassium hydroxide/alcohol

4-Hydroxy-Δ^2-pyrazol-5-ones   C(OH)ArCO → COC(OH)Ar
from 5-hydroxy-Δ^2-pyrazol-4-ones by acyloin rearrangement s. *28*, 682; acyloin rearrangement cf. T. Ishiguro, Y. Kondo, and T. Takemoto, Tetrah. Let. *1975*, 315

Sodium, alcohol   NaOR

Indan-1,3-diones from 3-alkylidenephthalides ←
s. *26*, 746

Potassium/tert-butanol K/Me_3COH

Bicyclic alcohols from ethyleneketones O
s. *26*, 747

Potassium tert-butoxide $Me_2C \cdot OK$

3-Hydroxy-1-nitrosoindolines
from o-(N-nitrosamino)ketones s. *29*, 695

2-Alkylthio-Δ^2-oxazoline ring opening   C
α-(N-Alkylthiocarbonylamino)-α,β-ethylenecarboxylic acid esters s. *29*, 696

Butyllithium/diisopropylamine $BuLi/i-Pr_2NH$

3-Hydroxyindolines from o-aminooxo compds. O
s. *30*, 663

Lithium diethylamide Et_2NLi

Alcohols from oxido compds. ←
with isocyclic ring closure s. *28*, 4

Lithium N-isopropylcyclohexylamide/
trimethyl chlorosilane $i\text{-}Pr(C_6H_{11})NLi/Me_3SiCl$
Claisen rearrangement of allyl esters   ←
γ,δ-Ethylenecarboxylic acids s. *28*, 683; γ,δ-ethylene-α-phenylthiocarboxylic acids s. B. Lythgoe, J. R. Milner, and J. Tideswell, Tetrah. Let. *1975*, 2593; preferably with *tert*-butyldimethylchlorosilane cf. J. A. Katzenellenbogen and K. J. Christy, J. Org. Chem. *39*, 3315 (1974).

Sodium amide $NaNH_2$
β-Hydroxy-γ-lactones from α-acetoxyketones   O

516.

5-Acetoxy-5α-cholestan-6-one treated with $NaNH_2$ in liq. NH_3-ether → product. Y: ca. 100%. J. R. Bull and A. Tuinman, Tetrah. Let. *1973*, 4349.

3H-Indol-3-ols from 4H-3,1-benzoxazines
s. *26*, 748

Potassium amide KNH_2
Ring closure with oxidonitriles   ←
Functionally subst. cyclobutanes

517.

Startg. m. stirred overnight with KNH_2 in liq. NH_3-glyme → spirocyclononane deriv. Y: ca. 80%. – With equal substitution at both ends of the oxirane ring, cyclobutanes always form in preference to cyclopentanes, thus leading to a general, non-photochemical synthesis of functionally subst. cyclobutanes. The **strong nucleophilicity of the nitrile anion** permits ring formation even at a quaternary center. F. e. s. G. Stork et al., Am. Soc. *96*, 5270, 5268 (1974); with $NaNH_2$, also cyclopentanes, cf. J. Y. Lallemand and M. Onanga, Tetrah. Let. *1975*, 585; pyrrolidines s. R. Achini and W. Oppolzer, Tetrah. Let. *1975*, 369.

Sodium hydrogen carbonate $NaHCO_3$
Azulene ring   O
by stereospecific transannular ring closure s. *26*, 749

Sodium hydrogen carbonate/irradiation $NaHCO_3/ⵜⵜⵜ$
Cyclobutanol ring from ketones

518.

3,20-Di(ethylenedioxy)-11-oxo-5α,14β,17α-pregnane in satd. ethanolic $NaHCO_3$-soln. irradiated 6.5 hrs. in quartz tubes with a high-pressure Hg-lamp → 3,20-di(ethylenedioxy)-11α-hydroxy-11,19-cyclo-5α,14β,17α-pregnane. Y: ca. 100%. P. Gull et al., Helv. *57*, 863 (1974).

Potassium carbonate $\qquad K_2CO_3$
Ring closures of propargyl ethers
　by o-Claisen rearrangement – Benzofurans – Chromenes s. *29*, 697

Sodium acetate s. under $(CH_3COO)_2Mg$ $\qquad CH_3COONa$

Potassium cyanate $\qquad KOCN$
Isomerization of oxido compds. ←
s. *26*, 750

Triethylamine s. under C_5H_5N $\qquad Et_3N$

Diethylaniline $\qquad PhNEt_2$
Acoxy compds. by Claisen rearrangement ←
s. *27*, 741

Pyridine $\qquad C_5H_5N$
O→C-Carbamyl group migration $\quad C(OCON<) : CH \rightarrow C(OH) : C(CON<)$
s. *28*, 684

Pyridine/triethylamine $\qquad C_5H_5N/Et_3N$
Skeletal rearrangement ←
　of cyclic glycol monosulfonates s. *27*, 974

Lithium dialkylcuprate $\qquad Li[CuR_2]$
Intramolecular aldol condensation O
　with 1,4-addition s. *29*, 669

tert-Butylmagnesium chloride $\qquad Me_3C \cdot MgCl$
　Hindered Grignard compds. s. *28*, 685

Magnesium acetate/sodium acetate $\qquad (CH_3COO)_2Mg/CH_3COONa$
Ring contraction of isocyclics Ö
s. *29*, 698

Alumina $\qquad Al_2O_3$
3β-Acoxy-Δ⁵-steroids C
　from 6β-hydroxy-3α,5α-cyclosteroids s. *29*, 185

Sand ←
Ketones from oxido compds. ←
　with skeletal rearrangement – Hydroazulenes – Synthesis of hydroazulenes, review, s. *28*, 686

Boron fluoride $\qquad BF_3$
α,β-Ethylene-β-hydroxy- from α,β-oxido-thiolic acid esters C
　with migration of the thiolic ester group s. *27*, 742

Isocyclic morphinane ring opening
s. *27*, 915

Hydroxylactones from oxidocarboxylic acids ←
with positional interchange

519.

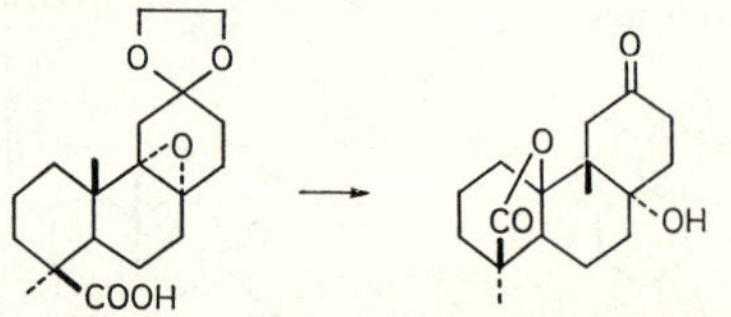

A dil. soln. of the startg. m. in nitromethane treated 0.5 hr. with 2 molar equivalents BF_3-etherate → hydroxy lactone. Y: 55%. – In ether, the acetal group is retained though the yield is lower. W. S. Hancock, L. N. Mander, and R. A. Massy-Westropp, J. Org. Chem. *38*, 4090 (1973).

Silver fluoroborate $\qquad$ $AgBF_4$
Ring closures of propargyl ethers $\qquad$ O
   by Claisen rearrangement s. 29, 788

Fluoroboric acid $\qquad$ HBF_4
Cyclobutanone from 1-oxaspiro[2.2]pentane ring $\qquad$ ←
s. 28, 837

Aluminum chloride $\qquad$ $AlCl_3$
Fries rearrangement $\qquad$ OAc → CAc
s. 28, 127

1,3-Pentadiene/irradiation $\qquad$ $CH_2 : CHCH : CHCH_3$/⫲
O-Heterocyclics
   by double cycloisomerization s. 28, 687

Acetophenone/irradiation $\qquad$ $PhCOCH_3$/⫲
Cycloisomerization of isocyclics
s. 27, 743

S-(−)-Proline $\qquad$ ←
Asym. intramolecular aldol condensation

520.

2-Methyl-2-(3-oxobutyl)-1,3-cyclopentanedione stirred 20 hrs. at 20–22° under argon with 3 mole-% (S)-(−)-proline in anhydrous dimethylformamide → crude (+)-(3aS,7aS)-3a,4,7,7a-tetrahydro-3a-hydroxy-7a-methyl-1,5(6H)-indandione. Y: ca. 100%; optical purity 93.4%. F. e. s. Z. G. Hajos and D. R. Parrish, J. Org. Chem. **39**, 1615 (1974); **40**, 675 (1975); **asym. synthesis with proline derivs.** cf. K. Hiroi and S. Yamada, Chem. Pharm. Bull. **23**, 1103 (1975).

Trifluoroacetic acid $\qquad$ CF_3COOH
o-Claisen rearrangement of allyl ethers $\qquad$ $O \cdot CH_2CH : CH_2$ → $C \cdot CH_2CH : CH_2$
s. 28, 688

Oxalic acid $\qquad$ $(COOH)_2$
Cyclobutanone from 1-oxaspiro[2.2]pentane ring $\qquad$ ←
s. 28, 837

Trimethylsilyl chloride s. under i-Pr(C_6H_{11})NLi $\qquad$ Me_3SiCl

Ethoxytitanium trichloride $\qquad$ $ROTiCl_3$
1,3-Diol monoethers from acetals $\qquad$ ←

521. $\qquad$ $CH_3CH(OEt)_2$ → $CH_3CH(OH)CH_2CH_2OEt$

1,1-Diethoxyethane added slowly with stirring at ca. 0° to a soln. of ethoxytitanium trichloride in methylene chloride, and allowed to stand several hrs. → 4-ethoxy-2-butanol. Y: ca. 85%. F. e. s. P. Mastagli and D. Gibert, C. r. **277** (C), 347 (1973).

Phosphoric acid $\qquad$ H_3PO_4
Cyclic alcohols from ketones $\qquad$ O
   with Wagner-Meerwein rearrangement s. 27, 744

Phosphorus oxide chloride $POCl_3$

Azacyclols
  by intramolecular Vilsmeier reaction s. 28, 689

Antimony pentachloride $SbCl_5$

β-Diketones from α,β-oxidoketones
  Spiro compds. by ring expansion s. 26, 751

Methanesulfonic acid $MeSO_3H$

4H-Pyran-4-carboxaldehydes via *sym*-oxepin oxides
Azo bridge elimination

522.

Startg. m. dissolved in cooled chloroform and allowed to warm to 25° → *sym*-oxepin oxide (Y: ca. 100%) treated a few sec. with a little methanesulfonic acid in chloroform → 4H-pyran-4-carboxaldehyde (Y: ca. 100%). W. H. Rastetter, Am. Soc. 75, 210 (1975).

p-Toluenesulfonic acid TsOH

Acetoxy-2-ethylenes by allyl rearrangement   C(OH)C : C → C : C·C(OAc)
s. 29, 191

Rearrangement of cyclobutane-1,2-diols
s. 28, 677

Skeletal rearrangement
  with O-tosylation s. 29, 699

Fluorosulfonic acid FSO_3H

Ketones from glycols
  with C-rearrangement – β- from α-Ketocarboxylic acid esters s. 29, 700

Sulfuric acid H_2SO_4

γ-Lactones from β-hydroxycarboxylic acids
s. 27, 262

**Cyclic β-hydroxyketones or lactones
from ethylenecarboxylic acids**

523.

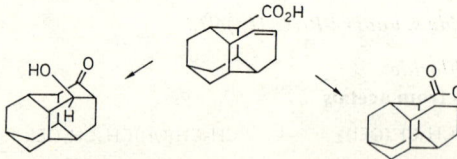

Tetracyclo[7.3.1.0²,⁷.0⁶,¹¹]tridec-3-ene-12-carboxylic acid (prepn. s. 88) heated 5 hrs. at 90–100° in 50%-H_2SO_4 → 5ax-hydroxydiamantan-3-one. Y: 82%. | stirred vigorously 6 min. at room temp. in 96%-H_2SO_4 → lactone. Y: 90%.

F. Blaney, D. Faulkner, and M. A. McKervey, Synth. Commun. 3, 435 (1973).

Dihydropyran ring
s. 28, 690

Sulfuric acid/acetic acid $\quad H_2SO_4/CH_3COOH$
**α,β-Ethylene-α-hydroxy- from α,β-oxido-ketones
with rearrangement
Cyclopent-2-ene-2-ol-1-ones**

524.

A soln. of the startg. m. in acetic acid containing 2% concd. H_2SO_4 warmed 1 hr. at 55° → product. Y: 65%. F. e. s. A. Barco et al., Synthesis *1975*, 104.

Iodine/irradiation $\quad I/\text{hv}$
6-Hydroxyfulvens from 7-oxanorbornadienes
s. *30*, 525

Silver perchlorate $\quad AgClO_4$
Migration of alkoxy groups $\quad$ C
with skeletal rearrangement s. *28*, 691

Perchloric acid $\quad HClO_4$
3β-Alkoxy-Δ⁵-steroids $\quad$ C
from 6β-hydroxy-3α,5α-cyclosteroids s. *22*, 194

Hydrogen chloride $\quad HCl$
Transannular ring closure $\quad$ O
Bicyclic ketones from cyclic acetylenealcohols s. *27*, 745; hydrindan ring cf. P. T. Lansbury et al., Am. Soc. *97*, 394 (1975).
Adamantanes from protoadamantanes
1,2-, 1,4-, and 2,4-Disubst. adamantanes s. *27*, 746
1,2-Dihydronaphthalene-2,3-dicarboxylic acids
from dilactones s. *28*, 692

Hydrogen bromide $\quad HBr$
Cyclic alcohols from ketones $\quad$ O
5-Hydroxy-6,7-benzomorphans s. *27*, 747

Tetracarbonyldichlorodirhodium $\quad [Rh(CO)_2Cl]_2$
6-Hydroxyfulvenes from 7-oxanorbornadienes $\quad$ C

525.

A soln. of the startg. m. and tetracarbonyldichlorodirhodium in chloroform allowed to stand 1 hr. at room temp. → 3,6-dimethyl-1,2-dicarbomethoxy-6-hydroxyfulvene. Y: 80%. F. e. s. H. Hogeveen and T. B. Middelkoop, Tetrah. Let. *1973*, 4325; by irradiation in the presence of iodine, also from oxaquadricyclanes, s. R. K. Bansal et al., Can. J. Chem. *53*, 138 (1975).

Iridium phosphine complex
**Phenols from cyclic dienones
Homogeneous isoaromatization**

526.

A soln. of $IrCl(CO)(Ph_3P)_2$ in *trans,trans*-2,6-dibenzylidenecyclohexanone heated 2 hrs. at 250° → 2,6-dibenzylphenol. Y: 76%. F. e. s. Y. Pickholtz, Y. Sasson, and J. Blum, Tetrah. Let. *1974*, 1263.

Platinum(IV)chloride   *PtCl$_4$*
1,3-Diyn-5-ones   C≡C·CO·C≡C· → C≡C·C≡C·CO·
from 1,4-diyn-3-ones
s. *26,* 752

Nitrogen/Carbon Type   CC∩NC

Without additional reagents   *w.a.r.*
Stevens rearrangement   ←
of carbonyl-stabilized ammonium ylids s. *27,* 748
β-Acoxy-α,β-ethylenecarboxylic acid amides   ←
from acylynamines s. *29,* 134
Imine-enamine tautomers   CH·C(·SSR) : NAc → C : C(·SSR)NHAc
1-Aminoenedisulfides from iminodisulfides s. *29,* 701
C,N-Positional interchange of acyl groups   ←

527.

A soln. of S-acetonyl-N-benzoyl-2-mercaptobenzimidazole in chloroform or benzene stored 4 days at room temp. → S-phenacyl-N-acetyl-2-mercaptobenzimidazole. Y: ca. 100%. Y. Akasaki and A. Ohno, Am. Soc. *96,* 1957 (1974).

N-Condensed 3-amino-2-azetidinone ring
s. *30,* 530

1,2,3,4-Tetrahydroisoquinolines from azomethines
s. *29,* 702

2,3-Dihydro-1H-1,4-diazepines from *cis*-enediazomethines
s. *29,* 703

6-Hydroxyanthyridines
from pyrimido[1,2-a][1,8]naphthyrid-10-ones s. *26,* 465

2-(γ,δ-Ethylene)-5,6-dihydro-4H-1,3-oxazines
via Claisen-type rearrangement of 2-ene-1-(β,γ-ethylene)tetrahydro-1,3-oxazines
s. *29,* 704

Irradiation (s. a. under Me$_2$CO)   ✻
Photochemical N→C-benzyl group migration   N·CH$_2$Ar → C·CH$_2$Ar
s. *29,* 705
Double ring opening by isomerization   C
s. *27,* 749
Cyclopropane from Δ^1-pyrrol-5-one ring   Ö:
s. *26,* 753

Sodium hydride   *NaH*
Stereospecific N→C-alkyl migration   NR → CR
s. *27,* 750

Potassium hydroxide   *KOH*
Stilbenes from azomethines   N(R)CH$_2$Ar → C(Ar) : NR
with and without formation of rearranged azomethines from tert. amines s. *27,* 751
3,4-Dihydroisoquinolinium ring opening   C
with and without naphthalene ring closure s. *28,* 693

Sodium methoxide NaOR
4-Hydroxy-2H-1,2-benzothiazine 1,1-dioxides
 from saccharins – Extended Gabriel-Colman rearrangement s. 28, 694

Potassium/triethylcarbinol $K/Et_3C \cdot OH$
Directed Hofmann degradation
 of cyclimmonium salts s. 28, 695

Butyllithium BuLi
6-Ox-2-azabicyclo[3.1.0]hex-3-enes from 4H-1,3-oxazines

528.

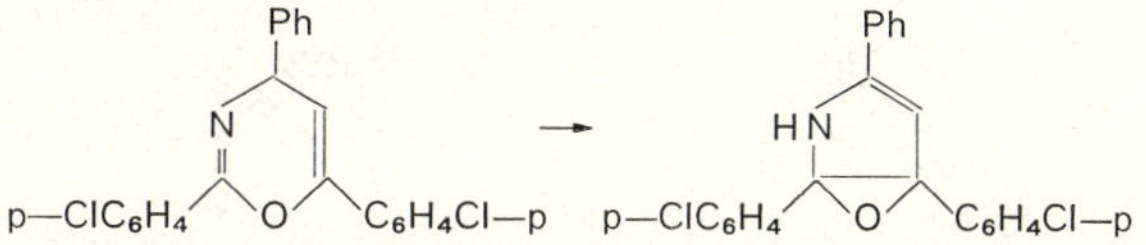

A 20%-soln. of n- or tert-butyllithium in hexane added dropwise at −78° under N_2 during 5 min. to a soln. of the startg. m. in anhydrous tetrahydrofuran, after 20 min. treated with water-chloroform at 0° → 1,5-bis-(4-chlorophenyl)-3-phenyl-6-oxa-2-azabicyclo[3.1.0]hex-3-ene. Y: 75%. F. e. and reactions of the products s. R. R. Schmidt, W. J. W. Mayer, and H.-U. Wagner, A. 1973, 2010.

n-Butyllithium n-BuLi
Quinol-2-ylcarbinols
 from 1-acyl-1,2-dihydroquinolines s. 28, 696

Phenyllithium PhLi
Carbanion ring closure
 Δ^2-Imidazolines from N,N'-dialkylidene-1,1-diamines s. 28, 697

Diethyl sodiomalonate $NaCH(COOEt)_2$
Thiocarbostyrils from isothiocyanates

529.

o-Isothiocyanato-trans-cinnamaldehyde in dry benzene added dropwise during 1 hr. to a suspension of diethyl sodiomalonate in dry benzene, and stirred overnight → 3-formylquinoline-2(1H)-thione. Y: 92%. R. Hull, Soc. Perkin I 1973, 2911.

Lithium diisopropylamide $i\text{-}Pr_2NLi$
2-Azetidinone ring expansion
 by 3 C-atoms s. 29, 706

Sodium amide $NaNH_2$
Stevens aryl migration
 s. 26, 754

Sodium acetate CH_3COONa
N-Condensed 3-amino-2-azetidinone ring

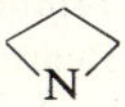

530.

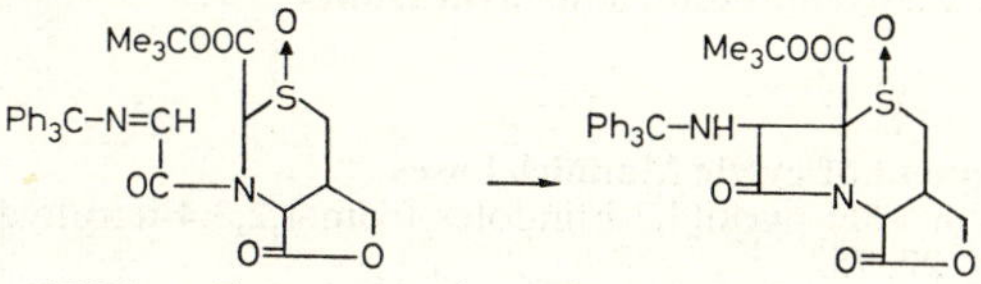

Startg. m. (m.p. 220°) refluxed 3 min. with Na-acetate in ethanol → 6-tert-butoxycarbonyl-3-hydroxymethyl-7-tritylaminocepham-2-carboxylic acid 5-oxide lactone. Y: 75%. – Cyclization of an isomer (m.p. 140°) occurs without Na-acetate. F. e. s. R. Heymès, G. Amiard, and G. Nominé, Bl. 1973, 2343.

Silver oxide Ag_2O
Hofmann degradation of cyclimmonium salts   C
    with dehydrogenation – 2-Amino-6,7-benzotropones s. *29,* 707

Lithium tetrahydridoaluminate $LiAlH_4$
Spirodiamines
by Stevens rearrangement of ammonium ylids
with reduction of lactams

531.

Startg. m. added to a stirred suspension of 10 molar equivalents $LiAlH_4$ in tetrahydrofuran, stirred and refluxed 17 hrs. → 1'-methylindan-2-spiro-2'-piperazine. Y: 64%. F. e. s. H. Kato et al., Chem. Pharm. Bull. *21,* 2661 (1973).

Aluminum chloride $AlCl_3$
Ring closure with nitriles
4-Aminocinnolines from α-cyanohydrazones

532.

A soln. of phenylglyoxylonitrile p-chlorophenylhydrazone in toluene refluxed 1 hr. with 4 moles $AlCl_3$ → 4-amino-6-chloro-3-phenylcinnoline. Y: ca. 100%. F. e. s. M. Lamant, C. r. *277* (C), 319 (1973).

Anilines from 7-azabicyclo[2.2.1]-2,5-heptadienes   C
s. *26,* 719

Acetone/irradiation $Me_2CO/\text{\textit{hv}}$
Skeletal rearrangement of N-heterocyclics   ←

533.

A soln. of 7-*tert*-butoxycarbonyl-2,3-benzo-7-azabicyclo[2.2.1]hepta-2,5-diene in acetone irradiated 0.5 hr. with the Corex-filtered light of a Hanovia 450 w. medium-pressure lamp → 5-*tert*-butoxycarbonyl-5-azatetracyclo[5.4.0.0^{2,4}.0^{3,6}]- undeca-1(7),8,10-triene. Y: 94%. J. S. Swenton, J. Oberdier, and P. D. Rosso, J. Org. Chem. *39,* 1038 (1974).

Trifluoroacetic anhydride $(CF_3CO)_2O$
Rearrangement of 2,5-cyclohexadienone hydrazones   ←
s. *28,* 698

Acetic acid CH_3COOH
Skeletal rearrangement of cyclic Mannich bases   ←
    1,2,3,4-Tetrahydro-5H-pyrid[4,3-b]indoles from 1,2,3,4-tetrahydropyrimido-
    [1,6-a]indoles s. *27,* 752

p-Toluenesulfonic acid $TsOH$
Enacylamines from Δ^2-oxazolines   C
s. *26,* 755

Sulfuric acid H_2SO_4
Pyrazolo[3,4-b]pyridine from pyrazolo[1,5-a]pyrimidine ring
with replacement of cyano groups by hydrogen s. 26, 756

Iodine I
Isocarbostyrils from isocyanates
s. 27, 753

Hydrogen chloride HCl
Rearrangement of 2,5-cyclohexadienone tosylhydrazones
s. 27, 754

*syn-anti-*Rearrangement of amidines

534.

A soln. of *syn*-N,N,N′-trifluorobutyramidine in abs. ether satd. at 20° with HCl-gas, and kept 18 hrs. at this temp. → *anti*-isomer. Y: ca. 90%. F. e. s. A. V. Fokin et al., Izvest. *1973*, 2159; C. A. *80*, 14523.

Iminocyclobutanes from 1-azaspiro[2.2]pentanes

535.

1,2-Diphenyl-1-azaspiro[2.2]pentane in benzene stirred at 0° with a little HCl in benzene → product. Y: 90%. F. e. and methods s. J. K. Crandall and W. W. Conover, J. Org. Chem. *39*, 63 (1974).

Halogen/Carbon Type   CC∩HalC

Without additional reagents w.a.r.
Ring expansion of cobaltocenes
s. 26, 850

Trifluoroacetic acid s. under TsOH CF_3COOH

Titanium trichloride $TiCl_3$
N-Condensed heterocyclics
from N-halogenamines s. 30, 678

p-Toluenesulfonic acid/trifluoroacetic acid $TsOH/CF_3COOH$
Skeletal rearrangement
of halogenometacyclophane rings s. 29, 708

Sulfuric acid H_2SO_4
Double ring closure
with double position shift of substituents of isocycles s. 29, 971

Sulfur/Carbon Type   CC∩SC

Without additional reagents w.a.r.
Rearrangement of ketene mercaptals
s. 26, 582

Ring closures via o-thio-Claisen rearrangement
s. 27, 755; s. a. Y. Makisumi and S. Takada, Chem. Commun. *1974*, 848, 850
3-Acyl-2,3-dihydrothianaphthenes from aryl acetylenesulfoxides via o-thio-Claisen rearrangement and intramolecular Michael addition s. 27, 755

Potassium tert-butoxide $KOCMe_3$
Intramolecular methylene transfer   ▽
    Cyclopropane ring from sulfonium salts s. 28, 699

n-Butyllithium   *n-BuLi*
2-(β,γ-Ethylene)-1,3-dithianes   ←
    s. 28, 883

Lithium diisopropylamide $i\text{-}Pr_2NLi$
[2.3]-Sigmatropic rearrangement   ←
    of ethylenemercaptals – β,γ-Ethylenemercaptals s. 29, 709; thio-Wittig rearrangement s. a. R. H. Mitchell, T. Otsubo, and V. Boekelheide, Tetrah. Let. *1975*, 219; [2,3]-sigmatropic rearrangement of unsatd. thioethers cf. W. Kreiser and H. Wurziger, Tetrah. Let. *1975*, 1669

Lithium diisopropylamide/hexamethylphosphoramide $i\text{-}Pr_2NLi/(Me_2N)_3PO$
1-Alkylation and thio-Wittig rearrangement $CH \cdot S \cdot CS(N<) \rightarrow C(S \cdot) \cdot CS(N<)$
of dithiocarbamic acid esters
α-Mercapto- and α-alkylthio-carboxylic acid thioamides

536.

$$PhCH_2S-C\underset{NMe_2}{\overset{S}{\diagup}}$$

+ MeI

$PhCHS-C\underset{NMe_2}{\overset{S}{\diagup}}$ $PhCHSMe$ $PhCHSH$
$|$ $C\underset{NMe_2}{\overset{S}{\diagup}}$ $C\underset{NMe_2}{\overset{S}{\diagup}}$
CH_3

Benzyl N,N-dimethyldithiocarbamate treated at –60° with Li-diisopropylamide in tetrahydrofuran, and after 5 min. | in n-hexane-tetrahydrofuran-hexamethylphosphoramide, and after 1 hr.
quenched with methyl iodide | or acetic acid
→ α-methylbenzyl N,N-dimethyldithiocarbamate. Y: 99%. | → N,N-dimethyl-(2-methylthio)phenylthioacetamide. Y: 96%. | → N,N-dimethyl-(2-mercapto)phenylthioacetamide. Y: 96%.
F. e. s. T. Hayashi and H. Baba, Am. Soc. 97, 1608 (1975).

Potassium acetate/acetic anhydride CH_3COOK/Ac_2O
Ring expansion of S,S-heterocycles   ☉
s. 30, 649

Triethylamine Et_3N
Alkylthioketene aminals $S \cdot C : C(N<)_2$
    by rearrangement of diaminomethylene sulfonium salts s. 26, 581

Ethylmagnesium bromide   *EtMgBr*
Cyclopropanesulfinic acids   ☉
    from thietane dioxides s. 28, 700

Acetic anhydride s. under CH_3COONa Ac_2O

Remaining Elements/Carbon Type   CC∩RemC

Without additional reagents   *w.a.r.*
B→C-Alkyl migration   BR → CR
s. 27, 867

Phosphorines by rearrangement   ←
    Ar. P-heterocyclics, review s. 29, 710

Carbon/Carbon Type   CC∩CC

Without additional reagents   w.a.r.
Cycloisomerization with 1,5-hydrogen shift   ○
   Ring closure by valence isomerization s. 26, 757

Cyclopolyenes as dienophiles
   Intramolecular diene synthesis – Highly condensed polyisocyclics s. 26, 758; hydro-5,7a-epithioisoindolinium salts by intramolecular diene synthesis s. K. T. Tagmazyan, R. S. Mkrtchyan, and A. T. Babayan, Arm. Khim. Zh. 27, 587 (1974) (Russ); C. A. 81, 169451

1,3-Dienes from cyclobutene ring   C
s. 16, 773; butadiene-2,3-dicarboxylic acid s. P. Dowd and K. Kang, Synth. Commun. 4, 151 (1975); 5,7-cyclooctadiene-1,4-dione s. M. Oda et al., Ang. Ch. 87, 414 (1975)

Cope rearrangement   ←
   of bridged *endo*-vinylcycloalkenes s. 29, 711

Ring expansion of 2 C-atoms   ☉
   of 2-spirocyclopropano-1,3-dithianes s. 28, 656

Silyloxycyclopentene annelation
s. 28, 701

Cycloisomerization of benzocyclobutenes   ○ ☉
s. 27, 756

Irradiation (s. a. under Me$_2$CO, PhCOCH$_3$, and Ph$_2$CO)   ⊬⊬⊬
Cyclopropanes   ▽
   by cycloisomerization of ethylene derivs. s. 26, 759

Skeletal rearrangement of bicyclic dienones   ←
s. 26, 126

1,1a,2,6b-Tetrahydrocycloprop[b]indoles   ←
from 1,2-dihydroquinolines
Stereospecific skeletal rearrangement

537.

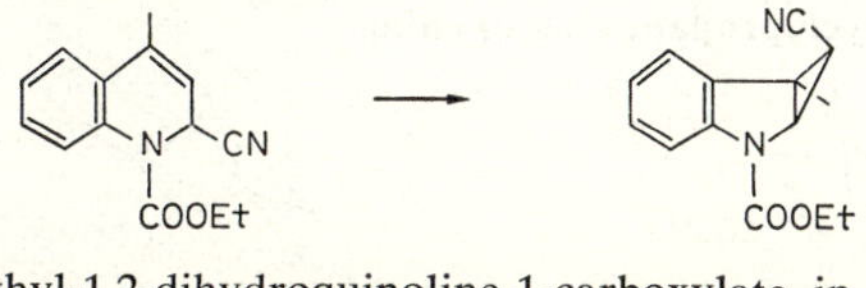

Ethyl 2-cyano-4-methyl-1,2-dihydroquinoline-1-carboxylate in ethanol irradiated with a 350 w. high-pressure Hg-lamp in a Pyrex vessel → ethyl *endo*-1-cyano-6b-methyl-1,1a,2,6b-tetrahydrocycloprop[b]indole-2-carboxylate. Y: 61%. F. e. s. M. Ikeda et al., Chem. Commun. 1974, 433.

Sodium hydroxide   NaOH
Anion-induced skeletal rearrangement   ←
   with formation of ethers from alcohols s. 27, 189

Potassium hydroxide   KOH
3,3'-Bis-(1H-isoindoles)   ←
   from 6,12-dihydrodibenzo[c,h][1,5]naphthyridines s. 29, 712

Sodium acetate/acetic acid   CH$_3$COONa/CH$_3$COOH
Skeletal rearrangement   ←
   of *exo*-tricyclo[4.2.1.0^{2,5}]non-3-enes – Acetolysis of brosylates – O-Deacetylation s. 26, 760

Sodium cyanide s. under Na[BH$_4$]   NaCN

*Sodium tetrahydridoborate/sodium cyanide-
hydrogen chloride* $\quad Na[BH_4]/NaCN\text{-}HCl$
Reductive ring closure ○
of cyclimmonium salts s. 26, 761

Boron fluoride $\quad BF_3$
Transannular ring closure of acetylene derivs.
s. 27, 757
Double ring closure
of vinylogous dicarboxylic acid imides s. 29, 713

Silver fluoroborate $\quad Ag[BF_4]$
Skeletal rearrangement ←
of *seco*-cubane derivs. s. 26, 762

Acetone/irradiation $\quad Me_2CO/\text{\textit{HH}}$
s. 28, 702; 8,9-dehydroadamantane derivs. s. R. K. Murray, Jr., T. K. Morgan, Jr.,
and K. A. Babiak, J. Org. Chem. 40, 1079 (1975).

Acetophenone/irradiation $\quad PhCOCH_3/\text{\textit{HH}}$
Double ring closure by cycloisomerization ○
s. 26, 763

Benzophenone/irradiation $\quad Ph_2CO/\text{\textit{HH}}$
Skeletal rearrangement ←
s. 28, 702

Acetic acid s. under CH_3COONa $\quad CH_3COOH$

Fluorosulfonic acid $\quad FSO_3H$
β- from α-Ketocarboxylic acid esters $\quad CO\cdot C\cdot COOR$
s. 29, 700

Hydrogen chloride (s. a. under $Na[BH_4]$) $\quad HCl$
Dihydroisoquinoline rearrangement ←
Pseudocyanides s. 26, 764

Bis-(1,5-cyclooctadiene)nickel(0) ←
Dienes by double cyclopropane ring opening C

538.

Dicarbomethoxyquadricyclane in toluene added under argon to bis-(1,5-cyclo-
octadiene)nickel(0), and kept 48 hrs. at −15° in a sealed tube → product. Y: 88%.
F. e. and Ni(0)-complexes s. R. Noyori et al., Am. Soc. 97, 812 (1975).

Chloroplatinic acid-alumina $\quad H_2PtCl_6\text{-}Al_2O_3$
Adamantanes by skeletal rearrangement ←
Review of the chemistry of adamantanes s. 27, 758

Exchange ↕

Hydrogen ↑ $\quad$ CC ↕ H

Without additional reagents $\quad$ w.a.r.
Replacement of hydrogen by chloroformyl groups $\quad$ H → COCl
β-Amino-α,β-ethylenecarboxylic acid chlorides s. 29, 714

Quinolines from o-aminostyrenes　　　　　　　　　　　　　　　　　　○
s. 26, 765

Irradiation s. a. under Peroxides, Me₃C·OO·CMe₃ and HCl　　　　ttt

γ-Irradiation s. under Me₂CO

Electrolysis　　　　　　　　　　　　　　　　　　　　　　　　　　　↯
Replacement of hydrogen by cyano groups　　　　　　　H → CN
　Organic electrochemistry, review s. 26, 766

Electrolysis/lithium diisopropylamide　　　　　　　　　　　↯/i-Pr₂NLi
Succinic acid esters　　　　　　　　　　　　　　　　　　2 RH → R·R
　by oxidative dimerization s. 28, 703 suppl. 30

Electrolysis/trifluoroacetic acid　　　　　　　　　　　　　　↯/CF₃COOH
Anodic coupling　　　　　　　　　　　　　　　　　　　2 ArH → Ar·Ar
　Oxidative dimerization of phenolethers s. 28, 703; succinic acid esters by electrochemical oxidation of lithium ester enolates prepared with Li-diisopropylamide cf. M. Tokuda, T. Shigei, and M. Hoh, Chem. Lett. 1975, 621

Sodium hydride　　　　　　　　　　　　　　　　　　　　　　　NaH
Stable sulfonium ylids　　　　　　　　　　　　　　　　　　　C:S<
　from 2-acetylenesulfonium salts s. 26, 767

Potassium hydroxide s. under O₂　　　　　　　　　　　　　　　KOH

Sodium ethoxide　　　　　　　　　　　　　　　　　　　　　　NaOEt
C-Carbalkoxylation with subsequent C-cyanation　　CH₂CN → C(CN)₂COOR
α,α-Dicyanocarboxylic acid esters from nitriles

539.

Diethyl carbonate, toluene, and 2,6-naphthalenediacetonitrile added to Na-ethoxide, stirred and distilled until the b. p. reaches 111°, more toluene added, cooled to 0°, cyanogen chloride distilled at 0–5° into the mixture during ca. 6 hrs., and warmed 2 hrs. at 55–60° → diethyl 2,6-naphthalene-α,α,α',α'-tetracyanodiacetate. Y: 60%. D. J. Sandman and A. F. Garito, J. Org. Chem. 39, 1165 (1974).

Potassium tert-butoxide　　　　　　　　　　　　　　　　　　KOCMe₃
1-Subst. cyclic tert. amines
　from cyclic azomethinium salts s. 29, 716

Potassium tert-butoxide/water-air　　　　　　　　　　　KOCMe₃/H₂O-O₂
1,2-Diquinonylethylenes　　　　　　　　　　　　　　　　2 CH₂ → C:C
　by oxidative dimerization s. 29, 715

Potassium tert-butoxide/phenyl cyanate　　　　　　　　　KOCMe₃/PhOCN
Replacement of hydrogen by cyano groups　　　　　　　　H → CN
s. 28, 704

Lithium diisopropylamide s. a. under ↯　　　　　　　　　　　i-Pr₂NLi

Lithium diisopropylamide/iodine   *i-Pr$_2$NLi/I*
Syntheses with tricarbonylchromium(0) complexes
Nucleophilic ar. substitution by carbanions   ArH → ArR

540.

Coordination of a chromium tricarbonyl unit to an ar. ring via π-bonding increases the reactivity of the ring toward attack by nucleophiles. The unit can be easily attached and removed. – E: Isobutyronitrile added dropwise during 5 min. at –78° under argon to a soln. of Li-diisopropylamide in tetrahydrofuran, maintained 15 min. at 0°, recooled to –78°, a soln. of π-benzenechromium tricarbonyl in tetrahydrofuran added dropwise during 5 min., kept 0.5 hr. at 0°, cooled again to –78°, a soln. of iodine in the same solvent added during 15 min., and stirred 3 hrs. at 24° → phenylisobutyronitrile. Y: 94–98%. F. e. s. M. F. Semmelhack et al., Am. Soc. *97*, 1247 (1975); α-alkylation, also stereospecific, of phenylacetic acid esters cf. M. A. Boudeville and H. des Abbayes, Tetrah. Let. *1975*, 2727; **prepn. of optically active compds.** cf. G. Jaouen and A. Meyer, Am. Soc. *97*, 4667 (1975).

Lithium N-isopropylcyclohexylamide s. under CuBr$_2$ *RR'NLi*

Potassium cyanide (s. a. under K$_3$[Fe(CN)$_6$])   *KCN*
Benzils from 2 aldehyde molecules   2 ArCHO → ArCOCOAr
s. *26*, 768

Triton B   ←
1-Subst. cyclic tert. amines
    from cyclic azomethinium salts s. *29*, 716

Triethylamine   *Et$_3$N*
Reaction of enamines with phosgene   >N·C:CH → >N·C:C·COOR
    β-Amino-α,β-ethylenecarboxylic acid esters s. *26*, 769

Tetrathiofulvalenes from 1,3-dithiolium salts   ←
s. *29*, 717

Pyridine   *C$_5$H$_5$N*
Enediolesters from 2 aldehydes molecules   2 CHO → C(OAc):C(OAc)
    Acylals from aldehydes s. *27*, 759

Pyridine/quinol   ←
Benzene ring   ○
    by trimerization of p-quinones s. *29*, 718

Silver oxide   *Ag$_2$O*
Chromans from phenols and ethylene derivs.
    via o-quinone methids – Trapping of unstable dienes – Reactive electron-rich dienophile s. *27*, 760

Silver carbonate-Celite   ←
Oxidative dimerization of hindered phenols
    Dipheno- and stilbene-quinones s. *27*, 761

Cupric acetate s. under (CH$_3$COO)$_3$Mn   *(CH$_3$COO)$_2$Cu*

Cuprous chloride/N,N,N',N'-tetramethylethylenediamine   *CuCl/TMEDA*
Oxidative dimerization   2 RH → R·R
    Sym. 1,2-dinitriles – Heterocyclic 1,3-diynes s. *28*, 705

Cupric bromide/lithium N-isopropylcyclohexylamide $CuBr_2/RR'NLi$
  Sym. succinic esters – Stereospecific oxidative dimerization s. 28, 706

Methoxymagnesium methyl carbonate $MeOMgOCOOMe$
β-Keto-γ-dicarboxylic acid esters from ketones   ←
s. 26, 770

Fluoroborates BF_4^-
Substitution with carbonium ions   ←
  β,γ-Ethyleneketones from ethylene derivs. and oxocarbonium fluoroborates s. 27, 762; aryl ketones from oxocarbonium hexafluoroantimonates cf. G. A. Olah, H. C. Lin, and A. Germain, Synthesis *1974*, 895

Benzenes from cyclopropenium salts   ☉
s. 26, 771

Aluminum chloride $AlCl_3$
Xanthones and phenoxaphosphine 10-oxides   ○
from diaryl ethers

541.

A mixture of 3-bromophenyl 4-fluorophenyl ether, oxalyl chloride, AlCl₃, and CS₂ refluxed 4 hrs., cooled, and treated with ice-water → 6-bromo-2-fluoroxanthene-9-one. Y: 83%. – Best yields are obtained in dil. soln. | A mixture of 3-bromophenyl 4-chlorophenyl ether, PCl₃, and AlCl₃ refluxed 24 hrs., then poured on ice → 7-bromo-2-chlorophenoxaphosphine 10-oxide. Y: 60%.

F. e. s. I. Granoth, Y. Segall, and A. Kalir, Soc. Perkin I *1973*, 1972.

N-Condensed isatin ring
s. 26, 772

Aluminum chloride/acetic acid $AlCl_3/CH_3COOH$
Replacement of hydrogen by carbamyl   H → $CONH_2$
s. 28, 707

Aluminum chloride/hydrogen chloride $AlCl_3/HCl$
Gattermann formylation   H → CHO
  with and without C-alkyl migration s. 28, 708

Water s. under KOCMe₃ H_2O

Quinol s. under C₅H₅N   ←

Acetone/γ-irradiation Me_2CO/⧘
Radiation-induced reactions   ←
  of purines with alcohols s. 28, 709

α-Diketone s. under Me₃C·OO·CMe₃   RCOCOR

Phenyl cyanate s. under KOCMe₃   PhOCN

tert-Butyl hydroperoxide s. under FeSO₄ $Me_3C·OOH$

Peroxides/irradiation ←
C-α-Hydroxyalkylation with alcohols   H → C(OH)<
 of N-heterocyclics s. 29, 719; without peroxides in the presence of oxygen, lower yields, cf. N. Hata, I. Ono, and H. Suzuki, Bull. Chem. Soc. Japan 47, 2609 (1974)
 8-Substitution of purines s. 29, 719

Di-tert-butyl peroxide/irradiation/α-diketone $Me_3C \cdot OO \cdot CMe_3/\text{\textit{HH}}/RCOCOR$
Photoalkylation of peptides   H → R
s. 29, 720

Picric acid   ←
Oxidative 1,4-thiazine ring dimerization   2 RH → R·R
s. 27, 763

Trifluoroacetic acid s. under ⚡ and VOF_3 CF_3COOH

Dimethyl sulfoxide/perchlorates Me_2SO/ClO_4^-
4-Subst. pyran from pyrylium ring   ←
s. 27, 764

2,3-Dichloro-5,6-dicyanoquinone s. under $HClO_4$ ←

Benzoyl chloride/oxygen $PhCOCl/O_2$
Pyridylation of N-heterocyclics   ←
s. 27, 765

Lead dioxide/acetic acid PbO_2/CH_3COOH
Diphenoquinones from phenols   ←
s. 26, 773

Vanadium oxide fluoride/trifluoroacetic acid VOF_3/CF_3COOH
Nonphenolic oxidative coupling   2 ArH → Ar·Ar
s. 29, 910

Air s. under $KOCMe_3$ O_2

Oxygen s. a. under PhCOCl

Oxygen/potassium hydroxide O_2/KOH
Diphenoquinones from phenols   ←
s. 26, 773

Sulfur   S
Oxidative C-substitution of N-heterocyclics   H → Ar
s. 26, 774

Chlorosulfonyl isocyanate $O:C:N \cdot SO_2Cl$
Replacement of hydrogen by cyano groups   H → CN
s. 23, 521

Molybdenum hexacarbonyl $Mo(CO)_6$
4-Quinazolones from amines   ○
s. 28, 710

Iodine s. under $i\text{-}Pr_2NLi$ I

tert-Butyl hypochlorite $Me_3C \cdot OCl$
o-Substitution of ar. amines   ArH → ArR
s. 28, 762

Perchlorates s. under Me_2SO ClO_4^-

Perchloric acid/2,3-dichloro-5,6-dicyanoquinone   ←
Radical cations   ←
s. 29, 721

Sodium periodate $NaIO_4$
Oxidative cyclodimerization
of o-(ω-hydroxyalkyl)phenols s. 28, 711

Periodic acid HIO_4
Diphenoquinones from phenols
s. 26, 773 suppl. 28

Hydrogen chloride/irradiation   HCl/⫘
Cycloaddition of alcohols
to 2,3-dihydropyridinium rings s. 28, 712

Manganese triacetate $(CH_3COO)_3Mn$
C-Carboxymethylation of ethylene derivs.
with and without migratory retention of the double bond s. 28, 713

Manganese triacetate/cupric acetate $(CH_3COO)_3Mn/(CH_3COO)_2Cu$
γ,δ-Ethylene-α-dicarboxylic acid esters $C : C \cdot C \cdot C(COOR)_2$
from α-dicarboxylic acid esters and ethylene derivs. s. 29, 722

Ferrous sulfate/sulfuric acid/tert-butyl
hydroperoxide $FeSO_4/H_2SO_4/Me_3C \cdot OOH$
Radical C-acylation with aldehydes   H → COR
s. 26, 775

Ferrous sulfate/sulfuric acid/hydrogen peroxide $FeSO_4/H_2SO_4/H_2O_2$
Homolytic C-carbamylation of heterocyclics   H → CON<
s. 25, 480

Potassium hexacyanoferrate(III)/potassium cyanide $K_3[Fe(CN)_6]/KCN$
o-Cyano-N-oxides from cyclic N-oxides   H → CN
s. 26, 776

Ferric chloride $FeCl_3$
Oxidative α-arylation of ketones   CO·CH → CO·CAr

542. PhCOCH₂—⌬ + ⌬ → PhCOCH⌬⌬

A soln. of benzyl phenyl ketone in benzene containing dispersed anhydrous ferric chloride refluxed 7 hrs. → 2,2-diphenylacetophenone. Y: 83%. F. e. s. H. Inoue, Y. Kimura, and E. Imoto, Bull. Chem. Soc. Japan 46, 3303 (1973).

Via intermediates   v.i.
Arylation of quinones via arylquinols   H → Ar
s. 28, 556

Diaryls via telluride dichlorides   2 ArH → Ar·Ar
s. 28, 967

Oxygen ↑   CC ⇈ O

Without additional reagents   w.a.r.
Enamines from amide acetals CH_2 → $C : C \cdot NR_2$
s. 29, 469; C-dimethylaminomethylenation, phosphorus compds., s. M. A. Grassberger, A. 1974, 1872

C-Dimethylaminomethylenation CH_2 → $C : CHNMe_2$
s. 26, 247

C-Formylation with formic acid H → CHO
s. 26, 777

Nucleophilic substitution on 1,3-N,N-heterocycles ←
s. 27, 811

Acyl trifluoromethanesulfonates H → COR
as powerful acylating agents s. 28, 714

Replacement of alkoxy by fluorodinitromethyl groups OR → $CF(NO_2)_2$
s. 27, 766

α-Amino-γ-ketocarboxylic acids from ketones CO·CH → CO·C·C(N<)COOH
s. 28, 715

Di-(7-cycloheptatrienyl)malonic acids 2 ROR' → $R_2C(COOH)_2$
s. 28, 716

**Reactions with
N-dichloromethyleneammonium salts** CO·CH → CHal : C(CON<)
α,β-Ethylene-β-halogenocarboxylic acid amides from ketones via immonium salts s. 28, 717

Polymethine cyanines from triformylmethane ←
Tetramethine cyanines s. 28, 718

Polyfunctional 1-(1-imidazolyl)-1,3-dienes ←
by 3-component reaction s. 26, 778

Reaction of 3,5-dihydroxyisoxazoles ←
with oxo compds. s. 29, 723; s. a. Tetrahedron 31, 1861 (1975)

Cyclic 1,3-dienamines from 1,5-diketones O
s. 27, 950

Ring closure with malonic acid aryl esters
Resorcinol ring s. 27, 383

2-Aminonaphthalenes
s. 29, 724

Furo[3,2-b]quinolines from hydroxyquinolines
s. 26, 939

Pyran ring from carboxylic acid anhydrides

543.

2,3-Dihydro-4-(2-hydroxyphenyl)-1H-1,5-benzodiazepine heated 10 min. with acetic anhydride on a water bath → 8-acetyl-7,8-dihydro-6-methyl[1]benzopyrano-[4,3-b][1,5]benzodiazepine. Y: 45%. F. e. and limitations s. J. Curtze and K. Thomas, A. 1974, 328.

3-Aminocoumarins
from aroxylamines and α,β-acetylenecarboxylic acid esters s. 29, 725

3-Hydroxypyrroles
from α-aminocarboxylic acid esters and allenes s. 27, 784

Indoles from 2-acetyleneamines

[Structure: N-methyl-N-phenyl-CH₂-C≡C-CH₂O-C₆H₄Cl-p + HOO-CO-C₆H₄Cl-m → indole with Me on N, 2-CH₂O-C₆H₄Cl-p and 3-CH₂OCO-C₆H₄Cl-m substituents]

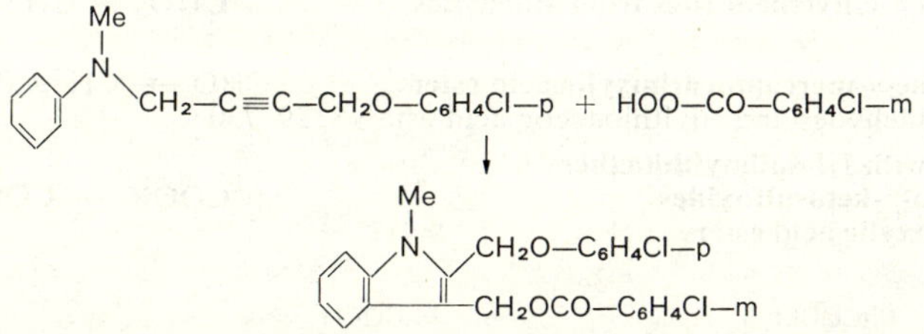

Startg. m. treated 12 hrs. at room temp. with m-chloroperoxybenzoic acid in methylene chloride → product. Y: 81%. B. S. Thyagarajan et al., Tetrah. Let. *1974*, 1999; J. Heterocyclic Chem. *12*, 43 (1975).

3,4-Dihydro-2-pyridone ring
s. *27*, 380

2(1H)-Pyridone-5-carboxylic acid esters
s. *30*, 250

4-Hydroxyquinolines
from o-aminoketones and orthocarboxylic acid esters s. *29*, 726

N-Condensed heterocyclics
by Mannich reaction s. *26*, 779

Pyrazolidines from hydrazines
ethylene derivs., and aldehydes – Double ring closure with unsatd. aldehydes s. *26*, 780

N-Condensed heterocyclics
from 5-alkoxyhydantoins and 1,3-dienes s. *28*, 758

Hexahydropyrimidine ring
by double Mannich reaction – Spiroheterocyclics s. *26*, 781

1,2-Dihydropyrazines
from 2 α-aminoketone molecules s. *29*, 727

2,5-Dihydropyrazine ring
by double Mannich reaction s. *29*, 728

Ring closure with 1-arylpyrroles
4,5-Dihydro-6H-pyrrolo[1,2-a][1,4]benzodiazepines s. *28*, 719

6,7-Dihydroisoquinoline-1(2H),3(5H)-diones
by double ring closure s. *28*, 332

Alkylidene-S-heterocyclics
from α,β-ethylene-α-mercaptocarboxylic acids – 2-Alkylidene-4,5-dihydrothiophen-3-ones s. *29*, 548

Cyclic 1-alkoxy-1-sulfones
from enolethers and sulfur dioxide s. *29*, 729

Irradiation s. under Methylene blue   ☷

Electrolysis   ⚡

γ-Lactones   O
from α,β-ethylenecarboxylic acid esters and oxo compds. s. *28*, 720

Lithium s. under CuI   Li

Sodium   Na

α,β-Ethylenenitriles from oxo compds.   CO → C:C(CN)
s. *26*, 782

Sodium hydride (s. a. under i-Pr₂NLi) *NaH*
α-Alkoxy-α,β-ethylenenitriles from aldehydes CHO → CH : C(OR)CN
s. *29*, 443
α,β-Ethylene-α-mercaptocarboxylic acid esters CHO → CH : C(SH)COOR
from aldehydes and silylthioacetic acid esters s. *29*, 730
Syntheses with 1,1-sulfinylthioethers
α-Alkylthio-β-ketosulfoxides COOR → COC(SR)SOR
from carboxylic acid esters

545.

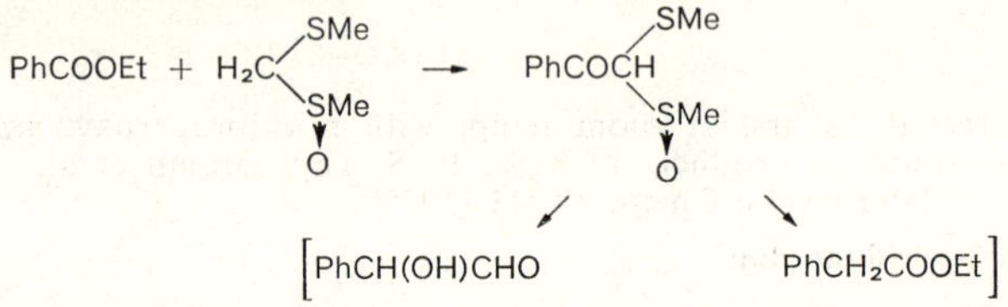

$$\text{PhCOOEt} + \text{H}_2\text{C}\begin{matrix}\text{SMe}\\ \text{SMe}\\ \downarrow\\ \text{O}\end{matrix} \longrightarrow \text{PhCOCH}\begin{matrix}\text{SMe}\\ \text{SMe}\\ \downarrow\\ \text{O}\end{matrix}$$

[PhCH(OH)CHO]   PhCH₂COOEt]

Methyl methylthiomethyl sulfoxide treated 1.5 hrs. at room temp. with NaH in tetrahydrofuran, ethyl benzoate added, and stirred 15 hrs. at room temp. – product. Y: 83%. – **Homologous esters and α-hydroxyaldehydes** can be prepared via the above reaction. F. e. s. K. Ogura, S. Furukawa, and G. Tsuchihashi, Chem. Lett. *1974*, 659.

Cyclopropanes ▽
from active methylene compds. and enol phosphates s. *26*, 783
Indan-1,3-diones ◯
from phthalic acid esters and ketones s. *27*, 767
2-Cyclohexenone ring from α,β-ethyleneketones
Stereospecific Robinson annelation – Effect of Solvent s. *26*, 784
Benzene ring
from α-unsatd. ketones and 2 acetylene deriv. molecules s. *26*, 785
N-Condensed 3,4-dihydro-2-pyridone ring
from O-alkyllactims via enamines and C-decarbobenzoxylation s. *27*, 768
Phospholan-3-one 1-oxide ring
from phosphinous acids and α,β-ethylenecarboxylic acid esters s. *29*, 731
Thiabenzene oxides from α,β-acetyleneketones
s. *27*, 769; f. method cf. M. Watanabe, T. Kinoshita, and S. Furukawa, Chem. Pharm. Bull. *23*, 258 (1975).

α-Carbalkoxy-γ-lactones from oxido compds. ←
Synthesis of carbohydrate derivs. s. *27*, 770

Sodium hydride/dimethyl sulfoxide *NaH/Me₂SO*
Double annelation ◯
with 2,4-dienecarboxylic acid esters s. *29*, 732

Sodium hydride/diethyl phosphorochloridate *NaH/(EtO)₂POCl*
α,β-Ethylenesulfoxides from aldehydes CHO → CH : C·SOR
s. *29*, 733

Sodium hydroxide *NaOH*
Stilbenes from aldehydes ArCHO → ArCH : C·Ar'
via 1,2-diarylethanols s. *26*, 786
2-Amino-1,6-naphthyridines ◯
s. *28*, 739

Potassium hydroxide (s. a. under Fe(CO)₅) *KOH*
α-Methylenealdehydes from aldehydes CH₂CHO → C(: CH₂)CHO
s. *26*, 787

2-Eneindolin-3-ones ←
s. *29*, 890

Phenol ring ○
from α,β-ethylene- and α-hydroxy-ketones
via cyclohex-2-en-4-olone ring

546.

Aq. KOH added dropwise under N_2 at 0° to a soln. of benzoin and methyl vinyl ketone in dioxane, stirred 0.5 hr. with ice-cooling, and allowed to stand 2 days at 20° → crude 4-hydroxy-3,4-diphenylcyclohex-2-enone (Y: 93%) heated 4 hrs. at 150° under N_2 with 85%-phosphoric acid → 3,4-diphenylphenol (Y: 70%). F. e. s. C. Egli, S. E. Helali, and E. Hardegger, Helv. *58*, 104 (1975).

Quinones from α,β-ethyleneoxo compds. and ketones
via 2-cyclohexenones and phenols s. *28*, 721

Sym. condensed pyridine ring synthesis
s. *27*, 771

Pyrazoles from diazo compds. and ketones
s. *26*, 788

Potassium hydroxide-clay ←
Aldol condensation CO ⇸ C:C
in the vapor phase s. *28*, 722

Potassium hydroxide/ethanol
4-Spiro-2-cyclohexenones ○
s. *29*, 734

Potassium hydroxide/dimethyl sulfoxide KOH/Me₂SO
Cyclopent-2-enones
via cyclopent-2-en-4-ol-1-ones s. *27*, 772

Sodium/alcohol NaOR
C-Alkylation of alcohols with alcohols R·OH ⇸ R·R'
s. *27*, 773

2-Styrylazulenes CH_2 ⇸ C:C

547.

Methyl 3-cyano-2-methylazulene-1-carboxylate allowed to react at room temp. with benzaldehyde in anhydrous methanol in the presence of Na-methoxide ⇸ product. Y: 98%. – 2-Methylazulene fails to react. F. e. s. M. Saito, T. Morita, and K. Takase, Chem. Lett. *1974*, 289.

C-Ethoxalylation CH ⇸ C·CO·COOR
s. *30*, 247

α-Azido-α,β-ethylenecarboxylic acid esters CHO ⇸ CH:C(N₃)COOR
from aldehydes – Prepn. of enazides, review s. *26*, 789

Ring closure with o-dialdehydes ○
Condensed p-quinones s. *29*, 735

Annelation
  of pyridine-o-dicarboxylic acid esters s. 28, 723

2-Oximino-6,7-dihydro-4(5H)-benzofuranones
  from 1,3-cyclohexadiones and 1,1-nitroethylene derivs. s. 27, 774

Pyridines from aldehydes
s. 27, 775

α-Diketones from sulfones
  via 3,4-diketotetrahydro- and 3-hydroxy-4-ketotetrahydro-thiophene 1,1-dioxides
  s. 28, 724

6-(4-Amino-1,3-dien-1-yl)fulvenes C
  from 6-amino-1-hydroxy-2,3-dihydropyrans s. 29, 736

Sodium/alcohol-tosylmethylisocyanide $NaOR$-$TsCH_2NC$
Nitriles from ketones   CO → CHCN
  s. 28, 726; cf. J. R. Bull and A. Tuinman, Tetrahedron 31, 2151, 2157 (1975)

Sodium compd.   ←
Cyclic alkylidenemalonic acid esters
  from aldehydes s. 26, 790   CHO → CH : C⟩

Synthesis of thioethers OSO_2R → R'
  from 1-sulfonyloxy-2-thioethers – Branched carbohydrates s. 29, 737

Sodium methoxide $NaOMe$
C-Carbalkoxylation $COCH_2$ → COC(R)COOR'
  with C-alkylation s. 28, 727

Cyclic 2-acyl-β-diketones   O
  from dicarboxylic acid esters s. 28, 728

Potassium/alcohol-pyridine KOR/C_5H_5N
α-Bromocarboxylic acid esters   CO → CBr · COOR
  from oxo compds. via Darzens-Claisen reaction – Synthesis with addition of
  1 C-atom s. 29, 738

Potassium tert-butoxide $KOCMe_3$
Carboxylic acids from oxo compds.   CO → CHCOOH
  via 1-(sulfonyl)eneformylamines – Synthesis with addition of 1 C-atom s. 28,
  729; f. method via α,β-oxidonitriles cf. D. W. White and D. K. Wu, Chem.
  Commun. 1974, 988

α-Aminocarboxylic acid esters   CHO → $CH_2CH(NH_2)$COOR
  from aldehydes and isocyanoacetic acid esters via α-amino-α,β-ethylene-
  carboxylic acid esters, with O-debenzylation s. 29, 739

β-Alkylthio-α-formaminocarboxylic acids   C(SR) · C(NHCHO) · COOH
  from aldehydes and ethyl isocyanatoacetate s. 26, 791

Phenols from α,β-acetyleneketones   O
s. 28, 730

3-Aminothiete 1,1-dioxides   ⟨S⟩
  from ynamines and sulfonic acid esters – Sulfenes as intermediates
  s. 29, 740; review of sulfenes s. J. F. King, Acc. Chem. Res. 8, 10 (1975)

6,7,8,9-Tetrahydro-7a,11a-ethanooxyindolo-   O
[1,2-h][1,7]naphthyridin-10(11H)-ones
  from 6,7-dihydropyrido[1,2-a]indol-9(8H)-ones s. 30, 300

n-Butyllithium (s. a. under WCl_6)
Synthesis of acetylene derivs. from tosylates   *n-BuLi*

$C \equiv CH \rightarrow C \equiv CR$

1.5 M *n*-butyllithium in hexane added slowly at 5° to a soln. of 1,1-dimethyl-2-propynyl tetrahydropyranyl ether in dioxane, stirred 2 hrs. at ca. 5° and 2 hrs. at 25°, (20 S)-6β-methoxy-20-tosyloxymethyl-3α,5-cyclo-5α-pregnane added, and refluxed 72 hrs. → 6β-methoxy-25-(2-tetrahydropyranyloxy)-3α,5-cyclo-5α-cholest-23-yne. Y: 92%. J. J. Partridge, S. Faber, and M. R. Uskoković, Helv. 57, 764 (1974).

α,β-**Acetyleneketones** $C \equiv CH \rightarrow C \equiv C \cdot COR$
from carboxylic acid anhydrides s. 27, 776

Dichloromethyl ketones $COOR \rightarrow COCHCl_2$
from carboxylic acid esters

2.5 moles methylene chloride in tetrahydrofuran added at –80° under N_2 to 2 moles *n*-butyllithium in tetrahydrofuran-ether, after 10 min. cooled to –90°, the startg. ester in ether added, stirring continued 3 hrs. at –90°, then allowed to warm slowly to –60° → product. Y: 88%. F. e. s. C. Bacquet, J. Villieras, and J.-F. Normant, C. r. 278 (C), 929 (1974); **metalation of α-chloro- and α,α-dichlorocarboxylic acid esters** cf. J. Organometal. Chem. 81, 295 (1974).

Δ^2-**Isoxazol-5-ones from oximes**   O
s. 27, 777

Heterocyclics via 1,4-dianions
Pyrazoles from hydrazones – Isoxazoles from oximes s. 26, 792

Quinazoline ring
from phenolethers and 2 nitrile molecules s. 29, 741

n-Butyllithium/dimethylphosphonodiazomethane   *n-BuLi/(MeO)$_2$P(O)CHN$_2$*
Acetylene derivs. from oxo compds. $CO \cdot \rightarrow C \equiv C \cdot$
Insertion of 1 C-atom s. 28, 731

Sodium acetylide/liq. ammonia $HC \equiv CNa/NH_3$
1,1-Diethynylalcohols $COOPh \rightarrow C(OH)(C \equiv CH)_2$
from carboxylic acid phenyl esters s. 26, 793

Lithium diisopropylamide   *i-Pr$_2$NLi*
α-**Lateral metalation of N-heterocyclics** $H \rightarrow COOR$
Preferential lateral α-carbalkoxylation s. 28, 732

Syntheses via carboxylic acid dianions $CHCOOH \rightarrow R \cdot CO \cdot C \cdot COOH$
and trimethylsilyl esters of β-ketocarboxylic acids and ketones – Synthesis of glutaric acids by Michael addition – Direct synthesis of β-ketocarboxylic acids s. 28, 733

γ-**Lactones**   O
from ethylene derivs. and mixed carbonic acid esters s. 27, 778

Metalation of N-nitrosamines
1,2,3-Triazole ring from nitriles s. 29, 742; metalation of N-nitrosamines s. a. B. *108*, 1293 (1975)

Lithium diisopropylamide/sodium hydride   *i-Pr₂NLi/NaH*
Anion-anion condensation   ←
    Poly-β-carbonyl compds. s. 28, 734

Potassium carbonate K_2CO_3
One-step pyrrole ring synthesis
    with hydroxylamine-O-sulfonic acid s. 27, 779

Sodium acetate/acetic anhydride CH_3COONa/Ac_2O
N-Condensed 2-pyridone ring   O
s. 26, 794

Sodium cyanide (s. a. under NH₄Cl)   *NaCN*
1,1,1-Aminodinitriles $\overset{+}{C}(N<)OR \rightarrow C(N<)(CN)_2$
    from alkoxyaminomethinium salts s. 28, 624

N'-Acyl-1,1-cyanohydrazines   CO → C(CN)·N·N·Ac
    from carboxylic acid hydrazides and ketones s. 27, 780

Sodium cyanide/acetic acid $NaCN/CH_3COOH$
α-Aminonitriles   CO → C(N<)CN
    from oxo compds. – Synthesis with addition of 1 C-atom – from azomethines
s. 27, 781

Sodium cyanide/hydrogen chloride   *NaCN/HCl*
Prim. amines from alcohols via nitriles   OH → CH_2NH_2
    Synthesis with addition of 1 C-atom s. 27, 782

Potassium cyanide (s. a. under MeOH and PhCOCl)   *KCN*
α-Cyanohydrazines   CHO → CH(CN)N·N
    from hydrazines and aldehydes s. 26, 828

Potassium fluoride   *KF*
Syntheses with α,β-ethyleneketones   C:C·CO → C:C·C:C
    catalyzed by potassium fluoride – Knoevenagel condensation s. 27, 783

Sodium chloride s. under AlCl₃   *NaCl*

Potassium iodide s. under H₃PO₄   *KI*

Lithium salt   *Li⁺*
1,2-Acoxynitro compds. from acylals $C(OAc)_2 \rightarrow C(OAc) \cdot C(NO_2)$

550.
$$p{-}O_2NC_6H_4{-}CH(OAc)_2 + Li^+ \left[\begin{matrix} CH_3 \\ C{-}NO_2 \\ CH_2 \end{matrix} \right]^- \longrightarrow p{-}O_2NC_6H_4{-}CH(OAc){-}\underset{CH_3}{\overset{CH_3}{C}}{-}NO_2$$

p-Nitrobenzylidene diacetate allowed to react 3 hrs. at room temp. with lithium isopropane nitronate in dimethyl sulfoxide ↠ product. Y: 80%. F. e. s. D. J. Girdler and R. K. Norris, Tetrah. Let. *1975*, 431.

Sodium salt   *Na⁺*
Heterocyclics from allenes   O
    3-Hydroxythiophenes from α-mercapto-, 3-hydroxypyrroles from α-aminocarboxylic acid esters s. 27, 784

Triton B   ←
Arylacetic acid esters   ArCHO → $ArCH_2COOR$
    from aldehydes via 1-sulfinylthioenolethers – Synthesis with addition of 1 C-atom s. 28, 735

n-Butylamine s. under CH_3COOH *n*-$BuNH_2$

tert-Butylamine $Me_3C \cdot NH_2$

Azulenes from 2H-cyclohepta[b]furan-2-ones   ←
s. *28*, 725

Triethylamine (s. a. under (Me$_3$C·OCl) Et_3N
Enamines from iminoesters   ←
Cyclic compds. s. *26*, 795

Enamines from O-alkyllactims
s. *27*, 768
$\underset{C-OR}{\overset{N}{\|}} \rightarrow \underset{C\,:\,C}{\overset{NH}{|}}$

Aliphatic nitro compds. $Ar \cdot C \cdot OH \rightarrow Ar \cdot C \cdot C(NO_2)$
from benzyl alcohols s. *29*, 743

N-Condensed 2-pyridone ring   O
from malonic acid esters

551.

Startg. m. allowed to react with dimethyl malonate in the presence of triethylamine → 5-methoxycarbonyl-6H-indolo[3,2,1-de][1,5]naphthyridin-6-one. Y: 83%.
L. A. Mitscher et al., Heterocycles *3*, 7 (1975).

Ring closure with iminoesters
2-Pyridone ring s. *29*, 744

Base-catalyzed reactions of active methylene compds.
with isocyanates – 5-Carbalkoxybarbituric acids from malonic acid esters and 2 isocyanate molecules s. *28*, 736

Isoxazoles
from aliphatic nitro compds. and acetylene derivs. – Synthesis of semicorrins s. *26*, 796

Pyrrolidine s. under CH_3COOH C_4H_9N

Piperidine (s. a. under CH_3COOH) $C_5H_{11}N$
Knoevenagel condensation $CHO \rightarrow CH:C(SCF_3)CN$
α-(Trifluoromethylthio)-α,β-ethylenenitriles from aldehydes s. *28*, 737

Cyclic ethers from lactols
s. *29*, 745
$\underset{C \cdot OH}{\overset{O}{|}} \rightarrow \underset{C \cdot R}{\overset{O}{|}}$

Ring closures with o-hydroxyaldehydes   O
Coumarins s. *27*, 198; also sultone analogs s. B. E. Hoogenboom et al., J. Org. Chem. *40*, 880 (1975).

Spiro-2H-1-benzopyran-2,2'-benzothiazolines
from benzothiazolium salts and o-hydroxyaldehydes s. *28*, 738

2-Amino-1,6-naphthyridines
s. *28*, 739

Pyridine (s. a. under KOR, TiCl$_4$, and Rh$_6$(CO)$_{16}$) C_5H_5N
Chromenes
s. *26*, 797
from α,β-ethylenealdehydes s. *27*, 785

2-Pyridone from 4-pyrone ring

552.

6-Ethylchromone-3-carboxaldehyde heated 8 hrs. at 120° with an equimolar amount of malonodiamide in pyridine → 3-carbamoyl-5-(5-ethyl-2-hydroxybenzoyl)-2(1H)-pyridone. Y: 71%. F. e. s. A. Nohara, T. Ishiguro, and Y. Sanno, Tetrah. Let. *1974*, 1183.

Lithium dialkylcuprate — $Li[CuR_2]$
Synthesis of ethylene derivs. — $C:C(OAc) \rightarrow C:C \cdot R$
from enol acetates – Ketones from β-acoxy-α,β-ethyleneketones s. *29*, 746

Cupric chloride s. under ZnCl₂ — $CuCl_2$

Cuprous iodide/lithium — CuI/Li
Synthesis of ethylene derivs. ←
from acoxy-2-ethylenes s. *28*, 740

Methylcalcium iodide — $MeCaI$
α,β-Ethyleneoxo compds. — $2\ CH_2CO \rightarrow CH_2 \cdot C:C \cdot CO$
from 2 oxo compd. molecules s. *27*, 786

Magnesium s. under TiCl₃ — Mg

Zinc (s. a. under TiCl₄) — Zn
1,4-Diacyl-1,4-dihydropyridines from pyridines ←
s. *26*, 798

Zinc amalgam s. under BF₃ — Zn,Hg

Ethoxymagnesium bromide — $EtOMgBr$
Annelation of quinones ○
s. *28*, 741

Aryloxymagnesium bromide — $ArOMgBr$
Reactions of aryloxymagnesium halides — $3\ ArH \rightarrow Ar_3C \cdot$
with orthoformic acid esters – Triarylmethanes s. *28*, 742

Mesitoxymagnesium bromide ←
Self-condensation of aldehydes — $2\ CH_2CHO \rightarrow CH_2CH:C \cdot CHO$

553.

$2\ C_2H_5CH_2CHO \rightarrow C_2H_5CH_2CH=C(C_2H_5)-CHO$

Butyraldehyde heated 0.5 hr. at 80° with 2,4,6-trimethylphenoxymagnesium bromide in benzene → product. Y: 90%. F. e., also formation of 1,3-diol monoesters, effect of solvent and reagent, s. G. Casnati et al., Tetrah. Let. *1974*, 959.

Phenyltrichloromethylmercury — $PhHgCCl_3$
Arylcarboxylic acids ←
by dienone-benzene rearrangement s. *26*, 799

Zinc chloride ZnCl$_2$
α,β-**Ethyleneketones** C(OH)CH·COOH → C : C·COR
from β-hydroxycarboxylic acids and carboxylic acid anhydrides

554.

$$\underset{S}{\boxed{}}\!\!-\!\!\underset{\underset{OH}{|}}{\overset{\overset{Ph}{|}}{C}}\!\!-\!\!\overset{\overset{Ph}{|}}{C}H\!\!-\!\!COOH + O(COCH_3)_2 \rightarrow \underset{S}{\boxed{}}\!\!-\!\!\overset{\overset{Ph}{|}}{C}\!\!=\!\!\overset{\overset{Ph}{|}}{C}\!\!-\!\!COCH_3$$

A mixture of 2,3-diphenyl-3-(thien-2-yl)-3-hydroxypropionic acid and acetic anhydride warmed 5 hrs. at 65°, ZnCl$_2$ added, and refluxed 4–5 min. with vigorous CO$_2$-evolution → 3,4-diphenyl-4-(thien-2-yl)but-3-en-2-one. Y: 33%. F. e. s. T. P. Ivanov and D. M. Mondeshka, J. pr. *315*, 993 (1973).

2-Pyridones ○
from 2 β-ketocarboxylic acid amide molecules s. *26*, 800

Isoquinolines
Gattermann-Koch-type synthesis s. *28*, 743

2-Pyridone ring closure ←
by 2H-pyrido[1,2-a]pyrimid-2-one ring opening s. *27*, 787

Zinc chloride-cupric chloride ZnCl$_2$/CuCl$_2$
α,β-**Ethyleneketones from aldehydes** CHO → CH : C·COR
s. *29*, 752

Sodium tetrahydridoborate Na[BH$_4$]
Reductive C-alkylation ←
of isoquinolinium salts s. *26*, 801

Lithium tetrahydridoaluminate s. a. under TiCl$_3$ Li[AlH$_4$]

Lithium tetrahydridoaluminate/aluminum chloride Li[AlH$_4$]/AlCl$_3$
Cyclobutenes ☉
from cyclopropenecarboxylic acid esters s. *27*, 788

Diethylaluminum cyanide Et$_2$AlCN
α,β-**Ethylenenitriles from enolethers** C : C·OR → C : C·CN
s. *27*, 789

Sodium dihydridobis-(2-methoxyethoxo)aluminate NaAlH$_2$(OCH$_2$CH$_2$OMe)$_2$
Reductive C-methylation of ketones >CO → >CMe$_2$
s. *27*, 790

Basic alumina Al$_2$O$_3$
Replacement of acoxy by m-xylyl groups ←
s. *26*, 468

Diethylboryl pivalate Me$_3$C·COOBEt$_2$
Self-condensation of aryl ketones ←
s. *10*, 587 suppl. 29

Boron fluoride BF$_3$
C-Allylation H → CH$_2$CH : CH$_2$
with phosphoric acid esters s. *29*, 747

α-**Subst.** α,β-**ethylenealdehydes** ←
from aldehydes and enolethers s. *26*, 802

β-Acyl-γ-lactones
from β,γ-ethylene-γ-lactones and aldehydes

555.   PhCH$_2$CH$_2$CHO + [α-angelicalactone] → [β-acetyl-γ-phenethyl-γ-butyrolactone]

BF$_3$-etherate in methylene chloride added slowly at 0° to a mixture of α-angelicalactone and β-phenylpropionaldehyde in the same solvent, and stirred 8 hrs. at 0° → β-acetyl-γ-phenethyl-γ-butyrolactone. Y: 94%. F. e. s. T. Mukaiyama et al., Chem. Lett. *1974*, 381, 1181.

Chromans O
   from phenols and 2-ethylenealcohols s. 27, 790a

5-Hydroxyindoles from p-quinones
s. 26, 717

1,3-Dithiins
   from acetylene derivs. and 2 aldehyde molecules – Review of dithiins s. 28, 744

Boron fluoride/zinc amalgam $BF_3/Zn,Hg$
Cyclopropane ring   ▽
   from ethylene derivatives and aldehydes s. 26, 803

Aluminum chloride (s. a. under Li[AlH$_4$]) $AlCl_3$
C-Alkylation of arenes   ArH → ArR
   with phosphorus esters s. 27, 791

Reactions of enolesters with aluminum chloride   H → COR
   C-Acylation s. 26, 804

Chromans   O
   from phenols and 2-ethylene alcohols s. 27, 790a

3,4-Dihydro-2-pyridones
from azomethines and α,β-ethylenecarboxylic acid esters

556.   [PhC(CH$_3$)=NPh] + [H$_2$C=CH-COOMe] → [1,6-diphenyl-3,4-dihydro-2-pyridone]

α-(Phenylimino)ethylbenzene and methyl acrylate allowed to react 1 hr. at room temp. in the presence of AlCl$_3$ → 1,6-diphenyl-3,4-dihydro-2-pyridone. Y: 85%. F. e. s. V. G. Aranda, I. Barluenga, and V. Gotor, Tetrah. Let. *1974*, 977; **1-amino-3,4-dihydro-2-pyridones from hydrazones** s. Synthesis *1974*, 717.

Aluminum chloride/sodium chloride $AlCl_3/NaCl$
Diketones from dicarboxylic acid anhydrides
   Double ring closure s. 28, 745

Methanol/potassium cyanide $MeOH/KCN$
Cyano-N-heterocyclics from cyclic N-oxides   ←
s. 26, 813

Ammonium acetate CH_3COONH_4
Pyridine ring   O
   from β-amino-α,β-ethylenealdehydes s. 26, 805

Triethylammonium acetate (s. a. under P$_2$O$_5$) $Et_3\overset{+}{N}H\ CH_3COO^-$
Knoevenagel-Cope condensation   CO + H$_2$C → C : C
s. 1, 586 suppl. 28

Piperidinium acetate $\quad\quad\quad\quad\quad\quad\quad\quad\quad\quad\quad\quad\quad\quad\quad$ $C_5H_{12}N^+ \ CH_3COO^-$
1,2- and 2,3-Ethylene-1,1-disulfones $\quad\quad\quad\quad\quad\quad\quad\quad\quad\quad\quad\quad\quad$ ←
from aldehydes s. 29, 748

Diethyl azodicarboxylate s. under Ph_3P $\quad\quad\quad\quad\quad\quad\quad\quad$ $ROOC \cdot N : N \cdot COOR$

Dimethylformamide $\quad\quad\quad\quad\quad\quad\quad\quad\quad\quad\quad\quad\quad\quad\quad\quad$ Me_2NCHO
ang-**Subst. *cis*-fused tetrahydroisatins** $\quad\quad\quad\quad\quad\quad\quad\quad\quad\quad\quad\quad$ O
from 1,3-dienes s. 28, 751

Dicyclohexylcarbodiimide $\quad\quad\quad\quad\quad\quad\quad\quad\quad\quad\quad\quad\quad\quad$ $RN : C : NR$
α-**Diazoketones from carboxylic acids** $\quad\quad\quad\quad\quad\quad$ COOH → $COC : N_2$
also via carboxylic alkoxyformic anhydrides, prepn. of β-amino acid peptides
s. 26, 806

Isopropenyl acetate s. under TsOH $\quad\quad\quad\quad\quad\quad\quad\quad\quad$ $CH_2 : C(CH_3)OAc$

Acetic anhydride (s. a. under CH_3COONa, CH_3COOH, *and* $HClO_4$) $\quad\quad$ Ac_2O
Modified aminomethylation of ketones $\quad\quad\quad\quad\quad\quad\quad$ CH → $C \cdot CH_2N<$
s. 26, 807

β-**Amino- and** β-**amino-**α-**methylene-ketones** $\quad COCH_3$ → $CO \cdot C(: CH_2)CH_2NR_2$
from methyl ketones s. 26, 808

o-Acyl-N-heterocyclics $\quad\quad\quad\quad\quad\quad\quad\quad\quad\quad$ $N(O) : CH$ → $N : C \cdot COR$
from cyclic N-oxides and α-acoxynitriles

557.

Quinoline N-oxide heated 5 hrs. with O-benzoyl p-nitrobenzaldehyde cyanohydrin in 5 equivalents acetic anhydride on a water bath → 2-quinolyl p-nitrophenyl ketone O-benzoylcyanohydrin (Y: 75.7%) heated with methanolic 10%-NaOH → 2-quinolyl p-nitrophenyl ketone (Y: 76.5%). – Derivs. of benzaldehyde and p-chlorobenzaldehyde do not react. F. e. s. T. Endo, S. Saeki, and M. Hamana, Heterocycles *3*, 19 (1975).

4-Subst. pyridines from 4(1H)-pyridones $\quad\quad\quad\quad\quad\quad\quad\quad\quad\quad$ ←
4-Alkylpyridines via 5-(4-pyridyl)barbituric acids – 2-(4-Pyridyl)alkanoic acids and anilides s. 28, 746

Quinolizinones $\quad\quad\quad\quad\quad\quad\quad\quad\quad\quad\quad\quad\quad\quad\quad\quad\quad\quad\quad$ O
s. 28, 747

Anion exchanger $\quad\quad\quad\quad\quad\quad\quad\quad\quad\quad\quad\quad\quad\quad\quad\quad\quad\quad$ ←
α,β-**Ethylenealdehydes** $\quad\quad\quad\quad\quad\quad\quad$ $2\ CH_2CHO$ → $CH_2CH : C \cdot CHO$
from 2 aldehyde molecules s. 23, 789 suppl. 28

2,4-Dinitrophenol $\quad\quad\quad\quad\quad\quad\quad\quad\quad\quad\quad\quad\quad\quad\quad\quad\quad\quad$ ←
in-situ-**Claisen rearrangement** $\quad\quad\quad\quad\quad\quad\quad\quad\quad\quad\quad\quad\quad\quad$ ←
with α,β-ethyleneketals – Geospecific synthesis of *trans*-ethylene derivs. – Subsequent reduction of α,β-ethyleneketones to 2-ethylenealcohols – Geospecific synthesis of ethylene derivs., review s. 26, 809

Formic acid $\quad\quad\quad\quad\quad\quad\quad\quad\quad\quad\quad\quad\quad\quad\quad\quad\quad\quad\quad\quad$ HCOOH
Methylene bridge with formaldehyde $\quad\quad\quad\quad\quad\quad\quad\quad\quad\quad$ $R \cdot CH_2 \cdot R$
s. 11, 841 suppl. 29

Acetic acid — CH_3COOH

Aminomethylation – Mannich reaction   H → CH_2NR_2
    β-Aminoketones s. 26, 808   CO·C·CH_2NR_2

Sym. ureas by double Mannich reaction   CO(NH·C·R)$_2$
s. 26, 810

Heterocyclics   O
    from ene-1,2-di(oxysilanes) as acyloin substitutes – Pyrrole ring s. 26, 811

Pyran ring by 3-component reaction
    from β-dioxo compds., aldehydes and acetylene derivs. – also dihydropyran ring s. 26, 812

1,2,3,5,6,6a-Hexahydrocyclopenta[f]chromen-7(8H)-ones
    by stereospecific double ring closure s. 27, 792

1,5-Dinitro-3-azabicyclo[3.3.1]non-6-enes
    from 3,5-bis-*aci*-nitrocyclohexenes s. 26, 48

N-Heterocyclics by Mannich reaction   O
s. 29, 751

2,3-Dihydro-1H-1,4-benzodiazepines and 2-cyanoindoles
    from o-aminoketones via Dimroth-modified Strecker reaction s. 29, 749

Asym. Strecker synthesis   CO → $C(NH_2)COOH$
    of α-aminocarboxylic acids via α-aminonitriles and tetrahydro-1,4-oxazin-2-ones
    – Steric C-chain effect s. 29, 750

1H-Pyrazolo[3,4-b]pyridines   O
s. 24, 182

1-(Alkylideneamino)indolizines
    from 2-acylpyridines and 2 aldehyde molecules s. 27, 793

Double ring closure to N-condensed heterocyclics
    Pictet-Spengler ring closure – Lactam rearrangement with aromatization –
    Benzo[4,5]canthin-6-ones s. 28, 748

Acetic acid/n-butylamine — $CH_3COOH/n\text{-}BuNH_2$

1,1-Nitroethylene derivs. from aldehydes   CHO → $CH:C(NO_2)$
s. 6, 755 suppl. 29

Acetic acid/pyrrolidine — CH_3COOH/C_4H_9N

Knoevenagel-Cope condensation   CHO → CH:C·COR
    α,β-Ethyleneketones from aldehydes – Didehydroparadols s. 29, 752

Acetic acid/piperidine/acetic anhydride   ←

Stepwise double Knoevenagel condensation   ←

558.

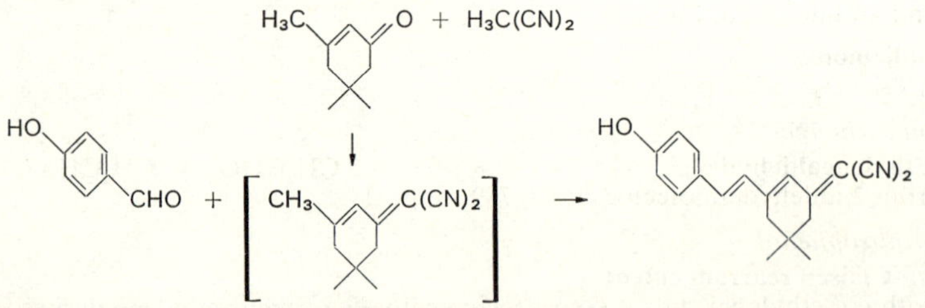

$H_3C\text{-cyclohexenone} + H_3C(CN)_2$

$HO\text{-C}_6H_4\text{-CHO} + [CH_3\text{-cyclohexene=}C(CN)_2] \rightarrow HO\text{-C}_6H_4\text{-CH=CH-cyclohexene=}C(CN)_2$

Piperidine, acetic acid, and acetic anhydride added to a soln. of 3-oxo-1,5,5-trimethylcyclohexene and malononitrile in *dimethylformamide as polar aprotic solvent*, stirred 1 hr. at room temp. and 1 hr. at 80°, p-hydroxybenzaldehyde added, and stirred 1 hr. at 80° → 1-(2-p-hydroxyphenylethenyl)-3-dicyanomethylene-5,5-dimethylcyclohexene. Y: 87%. F. e. and Knoevenagel-catalysts s. R. Lemke, Synthesis *1974*, 359.

Acetic acid/acetic anhydride $\qquad CH_3COOH/Ac_2O$
α,β-**Ethylenecarboxylic acids** $\qquad CH(OR)_2 \rightarrow CH:C\cdot COOH$
   from acetals via dioxolanium salts s. *29*, 753
Propionic acid $\qquad C_2H_5COOH$
γ,δ-**Ethylenecarboxylic acid esters** $\qquad C:C\cdot C\cdot C\cdot COOR$
   from 2-ethylenealcohols – Orthoester Claisen rearrangement s. *27*, 795; steroids cf. R. E. Ireland and D. J. Dawson, Org. Synth. *54*, 74 (1974).
β-**Allenecarboxylic acid esters** ←
   from 2-acetylenealcohols and orthocarboxylic acid esters s. *27*, 796; s. a. C. A. Henrick et al., J. Org. Chem. *40*, 8 (1975)
Trifluoroacetic acid $\qquad CF_3COOH$
Syntheses with acetoxy compds. $\qquad OAc \rightarrow Ar$

559.

A soln. of 3-acetoxymethyl-7β-phenylacetamidoceph-2-em-4α-carboxylic acid and phenol in trifluoroacetic acid allowed to stand 15 min. at room temp. → 3-(4-hydroxybenzyl)-7β-phenylacetamidoceph-2-em-4α-carboxylic acid. Y: 93%. F. e. s. H. Peter et al., Helv. *57*, 2024 (1974).

5-Acyl-1,3-dioxanes from ketones $\qquad$ O
   s. *26*, 659

o-Nitrobenzoic acid ←
Syntheses via Claisen rearrangement of enolethers ←

560.

A soln. of hexa-1,5-dien-3-ol, 1,1,3-triethoxy-2-methylbutane, and o-nitrobenzoic acid in toluene refluxed a total of 48 hrs. with intermittent removal of the resulting ethanol by distillation and repeated addition of more catalyst → 2-methyl-2-vinyl-octa-4,7-dienal. Y: 76%. F. e. s. R. C. Cookson and N. R. Rogers, Soc. Perkin I *1973*, 2741.

Perchlorohomocubancarboxylic acid ←
γ,δ-**Ethyleneketones by Claisen rearrangement** $\qquad C:C\cdot C\cdot C\cdot COR$
   Trisubst. *trans*-ethylene derivs. – Geospecific rearrangement s. *28*, 749
Methylene blue/irradiation ←
9H-Pyrid[3,4-b]indoles from 3-β-aminoindoles $\qquad$ O
   s. *28*, 750
Dimethyl sulfoxide $\qquad Me_2SO$
ang-**Subst.** *cis*-**fused tetrahydroisatins**
   from 1,3-dienes s. *28*, 751
Cyanuric chloride ←
Modified 2-step Vilsmeier aldehyde synthesis $\qquad H \rightarrow CHO$
   s. *27*, 797
Benzoyl chloride/potassium cyanide $\qquad BzCl/KCN$
Cyano-N-heterocyclics from cyclic N-oxides ←
   Direction of substitution s. *26*, 813

Titanium trichloride-magnesium complex $TiCl_3\text{-}Mg$
Sym. ethylene derivs. $2\,R\cdot CO \rightarrow R\cdot C:C\cdot R$
   from 2 oxo compd. molecules s. 28, 761

Titanium trichloride-lithium tetrahydridoaluminate $TiCl_3\text{-}Li[AlH_4]$
s. 30, 561

Titanium tetrachloride $TiCl_4$
Synthesis of mercaptals $H \rightarrow C(S\text{-}]_2$
s. 27, 798

Titanium tetrachloride/pyridine $TiCl_4/C_5H_5N$
Knoevenagel condensation $CO + H_2C \rightarrow C:C$
   Improved procedure s. 26, 814; with malonic acid, porphyrin derivs., s. L. Witte and J.-H. Fuhrhop, Ang. Ch. 87, 387 (1975)

Titanium tetrachloride/zinc $TiCl_4/Zn$
Sym. ethylene derivs. or pinacols $2\,R\cdot CO \rightarrow R\cdot C:C\cdot R$
from 2 oxo compd. molecules

561.
$$Ph\text{-}CH=CH\text{-}Ph \xleftarrow{} 2\,PhCHO \xrightarrow{} PhCH(OH)CH(OH)Ph$$

        Benzaldehyde allowed to react with $TiCl_4$ and Zn-powder

in dioxane at reflux temp. for ca. 4 hrs. → *trans*-stilbene. Y: 98%.	in tetrahydrofuran at 0° for ca. 2 hrs. → 1,2-diphenylethane-1,2-diol. Y: 98%.

F. e. s. T. Mukaiyama, T. Sato, and J. Hanna, Chem. Lett. *1973*, 1041; sym. ethylene derivs. with $LiAlH_4/TiCl_3$ s. J. E. McMurry and M. P. Fleming, Am. Soc. *96*, 4708 (1974).

Stannic chloride (s. a. under $AgClO_4$) $SnCl_4$
C-Glycosides $OAc \rightarrow R$
   from enoxysilanes s. 29, 754; by Wittig synthesis cf. H. Ohrui et al., Am. Soc. *97*, 4603 (1975)

Ar. C-glycosides
s. 29, 755

Phenylhydrazine $PhNHNH_2$
Pyrazolo[2,3-a]quinolines from pyrylium salts   ←
   N-Aminocarbostyrils s. 26, 815

Phosphorus s. under H_3PO_4   P

Triphenylphosphine Ph_3P
α-(Dihalogenomethylene)nitriles $COCN \rightarrow C(:CHal_2)CN$
   from acylcyanides s. 28, 752

Triphenylphosphine/diethyl azodicarboxylate $Ph_3P/ROOC\cdot N:N\cdot COOR$
Alkylation with alcohols $CH + HO\cdot R \rightarrow CR$
   C- and N-Alkylation s. 28, 753

Dimethylphosphonodiazomethane s. under n-BuLi $(MeO)_2P(O)CHN_2$

Phosphorus pentoxide s. a. under H_3PO_4 P_2O_5

Phosphorus pentoxide/triethylammonium acetate $P_2O_5/Et_3N^+H\,CH_3COO^-$
Knoevenagel-Cope condensation $CO + H_2C \rightarrow C:C$
s. 1, 586 suppl. 28

Polyphosphoric acid $\qquad H(PO_3H)_xOH$
C-Alkylation of phenols with alcohols $\qquad$ ArH → ArR
s. 29, 756

Ketones from carboxylic acids $\qquad$ COOH → COR
Deactivation of nitro groups – Prevention of intramolecular ring closure s. 27, 799

2-(o-Acylanilino)quinolines $\qquad$ O
from o-aminoketones and carboxylic acids s. 26, 816

Phosphoric acid/phosphorus/potassium iodide $\qquad H_3PO_4/P/KI$
Reductive dimerization of ketones $\qquad$ ←
s. 27, 800

Phosphoric acid/phosphorus pentoxide/acetic acid $\qquad H_3PO_4/P_2O_5/CH_3COOH$
Lactams from carboxylic acid amides and aldehydes $\qquad$ O
1,2,4,5-Tetrahydro-3H-2-benzazepin-3-ones s. 27, 801

Diethyl phosphorochloridate s. under NaH $\qquad (EtO)_2POCl$

Phosphonitrile chloride $\qquad$ ←
Condensations by phosphonitrile chloride $\qquad$ CHO → CHAr$_2$
Double condensation with aldehydes s. 27, 802

α-Alkylidene-β-ketocarboxylic acid anilides $\qquad$ CO → C : C(COR)CONHAr
from β-ketocarboxylic acid anilides s. 27, 803

Phosphorus tribromide $\qquad PBr_3$
Quinolines and quinazolines $\qquad$ O
from o-aminoketones and nitriles, also prepn. of indoles s. 27, 804

Phosphorus oxide chloride $\qquad POCl_3$
2,4-Diamino-6-chloropyridines
from α-cyanocarboxylic acid amides s. 29, 757

4-Aminopyrimidines
from carboxylic acid amides and 2 formamide molecules s. 26, 817

Phosphorus oxide bromide $\qquad POBr_3$
5-Halogeno-2-pyrrolecarboxaldehydes $\qquad$ ←
from 2-pyrrolones via 2-aminomethylene-5-halogeno-2H-pyrroles – Review of 2-pyrrolones s. 28, 754

Tosylmethylisocyanide s. under NaOR $\qquad TsCH_2NC$

Sulfur trioxide $\qquad SO_3$
Phthalides from carboxylic acids $\qquad$ O
s. 26, 818

p-Toluenesulfonic acid $\qquad TsOH$
Dipyrrylmethanes $\qquad$ ←
s. 28, 755

Enolethers from ketals $\qquad$ CH·C(OR)$_2$ → C : C·OR'
with alkoxyl interchange s. 28, 756

Friedländer quinoline ring synthesis $\qquad$ O
2,3-Dihydro-1H-pyrrolo[3,2-b]quinolines s. 28, 757

One-step synthesis of 4-azaazulenes $\qquad$ ←
s. 26, 819

p-Toluenesulfonic acid/isopropenyl acetate $\qquad TsOH/CH_2 : C(CH_3)OAc$
Benzene ring $\qquad$ O
from α,β-ethyleneoxo compds. via diene synthesis with 1-acoxy-1,3-dienes prepared *in situ* s. 27, 805

β-Naphthalenesulfonic acid ←
N-Condensed heterocyclics
    from 5-alkoxyhydantoins and 1,3-dienes – Stereospecific ring closure s. 28, 758
Sulfuric acid H_2SO_4
Synthesis of carboxylic acids with formic acid ←
    Koch-Haaf synthesis s. 14, 773
Phenanthrenes from α-diketones ○
    s. 27, 806
o-Ring closure with bis(benzylalcohols)
    s. 28, 759
Tropone ring from o-dialdehydes
    s. 29, 847
Benzofuran ring from arenes
    s. 28, 760
Coumarans from phenols and aldehydes
    s. 27, 807
3,4-Dihydroisoquinolines
    from nitriles and alcohols, also from nitriles and cyclopropanes – Ritter-type reaction s. 26, 820
Sulfuric acid/acetic acid H_2SO_4/CH_3COOH
1,2,3,4-Tetrahydroquinolines from ar. amines
    s. 27, 808
Tungsten hexachloride/n-butyllithium $WCl_6/n\text{-}BuLi$
Sym. ethylene derivs. $2\,R\cdot CO \rightarrow R\cdot C\!:\!C\cdot R$
    from 2 oxo compd. molecules s. 28, 761
tert-Butyl hypochlorite/triethylamine $Me_3C\cdot OCl/Et_3N$
Indoles from anilines and α-(alkylthio)oxo compds. ○
    o-Alkylation of ar. amines – 3-Alkylation of 2-aminopyridines – Oxindoles from ar. amines – via halogenosulfenium salts s. 28, 762
Lithium perchlorate (s. a. under HCl) $LiClO_4$
Pyrylium salts from ethyleneketones
    s. 17, 842 suppl. 29
Silver perchlorate-stannic chloride $AgClO_4\text{-}SnCl_4$
Diarylmethanes $Ar\cdot C\cdot OH \rightarrow Ar\cdot C\cdot Ar'$
    s. 28, 763
Perchloric acid $HClO_4$
Condensation via carbonium ions $OH \rightarrow R$
    s. 28, 764
C-Acylation of enollactones $H \rightarrow COR$
    s. 26, 821
Ethyleneketals from 2,3-oxidoalcohols ←
    with ketal cleavage s. 29, 10
3-Acylamino-2-benzopyrylium salts ○
    from arylacetonitriles s. 27, 809
Friedländer pyridine ring synthesis
    s. 28, 757 suppl. 29
Perchloric acid/acetic anhydride $HClO_4/Ac_2O$
Condensation by protonated acetic acid $CHO \rightarrow CH\!:\!C$
    s. 26, 822

Ammonium chloride/sodium cyanide $NH_4Cl/NaCN$
2,6-Dicyanopiperidines   ○
s. *29*, 758

Hydrochlorides   ←
α-**Aminomethyl-β-keto-** and γ-**methylene-β-keto-**   ←
from β-keto-carboxylic acid esters
s. *29*, 759

Heterocyclics by Mannich reaction   ○

562.

Methylenecyclimmonium salts. 2-Methylchromone, piperidine hydrochloride, and paraformaldehyde in benzene-nitrobenzene refluxed 8 hrs. with azeotropic water removal and addition of 3 additional portions of paraformaldehyde → product. Y: 40%. F. e. s. F. Eiden and U. Rehse, B. *107*, 1057 (1974).

9-Amino-9H-pyrrolo[1,2-a]indoles s. *28*, 765

Indole ring from satd. 1,2-N,N-heterocycles   ←

563.

from pyrazolidines. 1-Phenylpyrazolidine hydrochloride and 1.5 moles cycloheptanone refluxed 1 hr. under N_2 in glacial acetic acid → 5-(3-aminopropyl)-5,6,7,8,9,10-hexahydrocyclohept[b]indole hydrochloride. Y: 82%. F. e. s. M. K. Eberle, G. G. Kahle, and S. M. Talati, Tetrahedron *29*, 4045, 4049 (1973).

Hydrogen fluoride   HF
Ketones from carboxylic acids   COOH → COAr
s. *27*, 810

Hydrogen chloride   HCl
Nucleophilic substitution on 1,3-N,N-heterocycles   OR → R'
s. *27*, 811

Xanthylium salts   ○
by dimerization of 1,4-naphthoquinones s. *29*, 760

Quinoline ring from α,β-ethyleneoxo compds.
3H-Imidazo[4,5-f]quinolines s. *26*, 823

1,3-Dichloroisoquinoline ring
from arylacetonitriles s. *26*, 824

Heterocyclics by Mannich reaction
6-Alkoxytetrahydro-2H-1,3-oxazines from aldehydes

564.

A stirred mixture of formaldehyde (trioxane), methylamine hydrochloride, some concd. HCl, and ethanol refluxed 1 hr. under N_2, isobutyraldehyde added dropwise, and refluxing continued 3 hrs. ⇢ 6-ethoxytetrahydro-3,5,5-trimethyl-2H-1,3-oxazine. Y: 50–84%. F. e. and reactions of the product s. P. Y. Johnson, R. B. Silver, and M. M. Davis, J. Org. Chem. *38*, 3753 (1973); **6-hydroxytetrahydro-2H-1,3-oxazines** cf. H. Möhrle and D. Schnädelbach, Arch. Pharm. *308*, 352 (1975).

Thiopyrano[2,2-b]indolium salts from β-diketones
s. *26*, 825

Bicyclic lactolides ←
s. *26*, 826

Hydrogen chloride/lithium perchlorate   HCl/LiClO$_4$
(poly-O)-Macroheterocyclics ←

565.

$$4 \;\square_O + 4\;(CH_3)_2CO \longrightarrow$$

Furan allowed to react at 50–60° with acetone in ethanol in the presence of concd. HCl and Li-perchlorate ⇢ product. Y: 40–45%. M. and F. Chastrette, Chem. Commun. *1973*, 534.

Iron carbonyl/potassium hydroxide   Fe(CO)$_5$/KOH
Reductive C-alkylation with aldehydes   CH ⇢ CR
 Methylation with formaldehyde s. *29*, 761; alkylation with f. aldehydes s. Soc. Perkin I *1975*, 1273

Ferric chloride s. under Pd-C   FeCl$_3$

Hexadecacarbonylhexarhodium/pyridine   Rh$_6$(CO)$_{16}$/C$_5$H$_5$N
N-Subst. maleimides   O
 from acetylene derivs. and nitro compds. s. *28*, 766

Palladium-carbon/ferric chloride/carbon monoxide   Pd-C/FeCl$_3$/CO
Pyridines
 from 3 aliphatic nitro compd. molecules s. *27*, 812

Palladium acetoacetonate/triphenylphosphine   [(CH$_3$CO)$_2$CH]$_2$Pd/Ph$_3$P
Palladium-catalyzed transfer of allylic groups ←
 C-2,7-Octadien-1-ylation – N- and C-Allylation s. *26*, 827

Via intermediates   v.i.
Ethylene derivs.   CHO ⇢ CH : C
 from aldehydes via sec. alcohols s. *29*, 762

Sym. ethylene derivs. from 2 ketone molecules   2 CO ⇢ C : C
via azines, 1,3,4-thiadiazolidines,
and Δ³-1,3,4-thiadiazolines

566.

H$_2$S bubbled 12 hrs. at room temp. through a soln. of adamantanone azine prepared from adamantanone and hydrazine hydrate in *tert*-butanol (cf. Synth. Meth. *1*, 615; Y: 98%) and p-toluenesulfonic acid in benzene-acetone ⇢ adamantanespiro-2'-(1',3',4'-thiadiazolidine)-5'-spiroadamantane (Y: 95%) mixed with benzene, added dropwise with stirring during 1.5 hrs. to a mixture of Pb-tetra-acetate, Ca-carbonate, and benzene, stirring continued 8 hrs. at room temp. → Δ^3-1',3',4'-analog (Y: 94%) heated 12 hrs. at 125–130° with triphenylphosphine under N_2 → adamantylideneadamantane (Y: 74%). Overall Y: 65%. A. P. Schaap and G. R. Faler, J. Org. Chem. *38*, 3061 (1973); s. a. L. K. Bee et al., ibid. *40*, 2212 (1975); **hindered unsym. ethylene derivs.** via Δ^3-1,3,4-thiadiazolines cf. D. H. R. Barton, F. S. Guziec, and I. Shahak, Soc. Perkin I *1974*, 1794.

Prim. amines from alcohols via nitriles OH → CH_2NH_2
   Synthesis with addition of 1 C-atom s. *27*, 782

Nitriles from oxo compds. CO → CHCN
   Synthesis with addition of 1 C-atom s. *29*, 480

α-Alkoxynitriles $C(OR)_2$ → C(CN)OR
from acetals via 1,1-alkoxychlorides

567.
$$PhCH(OEt)_2 \longrightarrow PhCH\begin{matrix}Cl\\OEt\end{matrix} \longrightarrow PhCH\begin{matrix}CN\\OEt\end{matrix}$$

Benzaldehyde diethyl acetal added dropwise below 25° with occasional cooling to a stirred soln. of acetyl chloride and $SOCl_2$, allowed to stand overnight, concd. at 40° water bath temp. under reduced pressure, the residual oil dropped immediately during 1 hr. into a stirred suspension of NaCN in dimethylformamide, and stirring continued several hrs. → 2-ethoxy-2-phenylacetonitrile. Y: 76%. F. e. s. D. M. Bailey et al., J. Med. Chem. *17*, 702 (1974).

α-Aminocarboxylic acids CO → $C(NH_2)COOH$
   from ketones via hydantoins s. *1*, 568

α-Aminocarboxylic acids-¹⁴C from oxo compds.
via cyanohydrins and hydantoins
Synthesis with addition of 1 C-atom

568.
$$\begin{matrix}CHO\\|\\[CH_2]_3\\|\\CHO\end{matrix} + 2\,K\overset{×}{C}N \longrightarrow \begin{matrix}\overset{×}{C}N\\|\\CHOH\\|\\[CH_2]_3\\|\\CHOH\\|\\\overset{×}{C}N\end{matrix} + 2\,NH_4HCO_3 \longrightarrow \begin{matrix}H\\O=\!\!\overset{×}{=}\!\!N\!\!>\!\!=\!\!O\\N\\|\\H\\[CH_2]_3\\|\\H\\N\\O=\!\!\overset{×}{=}\!\!N\!\!>\!\!=\!\!O\\H\end{matrix} \longrightarrow \begin{matrix}\overset{×}{C}OOH\\|\\CHNH_2\\|\\[CH_2]_3\\|\\CHNH_2\\|\\\overset{×}{C}OOH\end{matrix}$$

$K^{14}CN$ followed after 5 min. by non-radioactive KCN added slowly with stirring to a soln. of glutaraldehyde in aq. HCl, neutralized with pyridine to pH 6, stirred 45 min., NH_4HCO_3 and aq. 25%-NH_3 added, heated 3 hrs. at 95° in a stoppered flask, and the resulting hydantoin intermediate heated 4 hrs. at 150° with 65%-H_2SO_4 → 2,6-diaminopimelic acid-1,7-¹⁴C. Overall Y: 74.8%; radiochemical Y: 66%. A. Arendt and A. Kolodziejczyk, J. Labelled Compds. *9*, 457 (1973).

α-Aminocarboxylic acid esters CHO → $CH_2CH(NH_2)COOR$
   from aldehydes and isocyanoacetic acid esters via α-amino-α,β-ethylene-carboxylic acid ester s. *29*, 739

3-Aminosydnone imine salts O
   from hydrazines and aldehydes via α-cyanohydrazines s. *26*, 828

Nitrogen ↑ CC ⇈ N

Without additional reagents w.a.r.

Replacement of nitrosamino by phenyl groups NHNO → Ar
Gomberg-Bachman reaction s. 27, 814

Synthesis of ethylene derivs.
from ketimines C : NH → C : C
s. 27, 815

from azomethines C : N → C : C
s. 28, 767

1,3-Dienes from α,β-ethyleneazomethines
s. 26, 829

Ketones from Mannich bases and enamines C·NR$_2$ → C·C·CO
Cyclic alcohols from ketones s. 28, 768

Enamines from active methylene compds. CH$_2$ → C : CH·N<
s. 30, 281

γ-Hydroxyketones C(N<) : C → CO·C·C·C(OH)
from enamines and oxido compounds s. 26, 830

Syntheses with 2-aminotetrahydropyrans C·NR$_2$ → C·C
s. 28, 769; synthesis with addition of 5 C-atoms of hydroxyketones by subsequent reductive tetrahydropyran ring opening (cf. Synth. Meth. 18, 63) s. C. r. 280 (C), 1471 (1975)

Reactions of 1,1,1-alkoxydiamines ←
N'-(3-Aminoacryloyl)formamidines from carboxylic acid amides – β-Amino-α,β-ethylenecarboxylic acid thioamides from carboxylic acid thioamides s. 28, 770

Reactions with (1,3-dithiolan-2-yl)ammonium salts
C-Formylation via 2-subst. 1,3-dithiolanes H → CHO

569.

Indole added to a stirred soln. of (1,3-dithiolan-2-yl)trimethylammonium iodide in dimethylformamide, and the intermediate isolated after 2 hrs. at 40° → 3-(1,3-dithiolan-2-yl)indole (Y: 78%), dissolved in tetrahydrofuran, added dropwise during 0.5 hr. to red mercuric oxide and BF$_3$-etherate in 1 : 9 tetrahydrofuran-water, and stirred 1 hr. under N$_2$ at room temp. → 3-formylindole (Y: 80%). F. e. s. K. Hiratani, T. Nakai, and M. Okawara, Bull. Chem. Soc. Japan 46, 3510 (1973).

Thionocarboxylic acid esters from diazo compds. CN$_2$ → CHCS·OR
Synthesis with addition of 1 C-atom

570.

Diphenyldiazomethane allowed to react below 20° with ethyl thionoformate in petroleum ether → ethyl diphenylthionoacetate. Y: 67%. F. e. s. R. Mayer and H. Kröber, Z. Chem. 13, 426 (1973).

Homologization with diazomethane Si·C : C → Si·CH$_2$·C : C
of enesilanes s. 29, 763

Stereospecific radical addition CH → CR
to cobaltocenes s. 27, 813

Reactions with 2-diazo-4,5-dicyanoimidazole ←
    4,5-Dicyanoimidazoles s. 28, 771
Cyclic acoxy from ethylene diazo compds. ○

571.   CH$_3$COOH + [structure: cyclopentene-N$_2$HC-CO] → [structure: tricyclic ketone with OAc]

6-*endo*-Diazoacetylbicyclo[3.1.0]hex-2-ene allowed to react with glacial acetic acid at 40° → 6-acetoxytricyclo[3.3.0.0^{2,8}]octan-3-one (55% *endo*, 45% *exo*). Y: 70–80%. R. Malherbe, Helv. 56, 2845 (1973).

peri-**Annelation**
    with cyclic 1,3-dienamines and α,β-ethyleneoxo compds. s. 29, 764
Coumarins
    from o-hydroxyaldehydes and 1-alkoxyenamines s. 26, 831
Spirodilactones
    via C-dimethylaminomethylenation s. 28, 772
Succinimides
    from azides and 2 ketene molecules s. 29, 765
Δ^2-Pyrrol-4-ones from enazides
    and α-diazoketones s. 28, 773
Pyridines ←
    from 2 2-methoxypyrimidine molecules s. 26, 832
N-Heterocycles from 1,2,4-triazines ←
    4H-Azepines – Pyridine ring s. 26, 833
Ring closures with α,α'-bisdiazoketones ○
    via 4-oxopyrazoles s. 29, 766
Δ^2-Isoxazoline oxides from diazo compds.
    s. 28, 774
Ethylene derivs.   C : S : O → C : C
    from sulfines and diazo compds. via thiirane 1-oxides s. 29, 767
2-Chlorothiophene S-oxide ring from diazo compds. ○
    s. 29, 768
1,3-Dithiepane ring from diazo compds.
    s. 29, 769
1,2,4-Triazepines ⊙
    from Δ^1-azirines and 1,2,4,5-tetrazines s. 29, 770
Benzenes from pyridazines ←
    via diene synthesis with inverse electronic demand s. 29, 886
o-Carbalkoxystyrene oxides from 3-hydroxyphthalides
    s. 29, 771

Irradiation
8-Alkylation of purines with amines   CH + H$_2$NR → CR
    s. 29, 719
Ketones   CON< → COR
    from carboxylic acid amides and ethylene derivs. s. 29, 772

Electrolysis
Pyrroles from 2 enamine molecules
    s. 29, 773

Lithium   Li
Synthesis of hydrocarbons   C : NNHTs → CHR
from tosylhydrazones s. 28, 775; also carboxylation s. R. H. Shapiro and T. Gadek, J. Org. Chem. 39, 3418 (1974)

Sodium   Na
Self-condensation   O
of β-amino- and β-cyano-ketones s. 29, 774

Sodium hydride   NaH
C-Acylation with diacylamines   H → COR
s. 26, 679

Syntheses with 2-nitroenamines   ←
Oxo from *aci*-nitro compds.

572.
$$\text{HCO—CH}_2\text{—Ph} + \text{Me}_2\text{N—CH=C(CH}_3\text{)(NO}_2\text{)} \longrightarrow \text{HCO—C(Ph)=CH—C(CH}_3\text{)(NO}_2^-\text{Na}^+\text{)}$$
$$\downarrow$$
$$\text{HCO—C(Ph)=CHCOCH}_3$$

A soln. of 1-dimethylamino-2-nitro-1-propene and phenylacetaldehyde in dimethylformamide treated with NaH, and stirred at room temp. until dissolved → Na-4-*aci*-nitro-2-phenyl-2-pentenal (Y: 83%) dissolved in methanol, treated with silica gel (containing 20% water), the solvent evaporated at 40°, allowed to stand 4–10 hrs. at room temp., then eluted with methylene chloride → 4-oxo-2-phenyl-2-pentenal (Y: 46%). F. e. s. T. Severin, D. König, and H. Lerche, B. 107, 1499, 1509 (1974).

Carboxylic acid amides from ureas   H → CON<
s. 28, 776

Sodium hydride/dimethyl sulfoxide   NaH/Me₂SO
Benzene from pyridine N-oxide ring   ←
s. 29, 775

Lithium hydroxide   LiOH
3-Hydroxythiophene ring   O
from o-nitrocarboxylic acid esters

573.
[structure: methyl 2,6-dinitrobenzoate + HS—CH₂—COOMe → methyl 3-hydroxy-4-nitrobenzo[b]thiophene-2-carboxylate]

LiOH added slowly to a well-stirred ice-cooled soln. of methyl 2,6-dinitrobenzoate and methyl thioglycolate in dimethylformamide, and stirred 0.5 hr. at 0° → methyl 3-hydroxy-4-nitrobenzo[b]thiophene-2-carboxylate. Y: 85%. F. e. s. J. R. Beck, J. Org. Chem. 38, 4086 (1973).

Sodium hydroxide   NaOH
2-Cyclohexenones   ←
from pyridines via 1,4-dihydropyridines s. 27, 816

Potassium hydroxide   KOH
Stilbenes from azomethines   Ar·C:N → Ar·C:C·Ar'
s. 27, 751

Heterocyclics
by displacement of ar. nitro groups – Thianaphthene-2-carboxylic acid esters
s. 29, 776

Lithium methoxide   LiOMe
α-Alkylated nitriles and α-cyanocarboxylic acid esters
from 1,1-cyano(azocarboxylic acid esters) s. 28, 55

Sodium ethoxide   NaOEt
5,6-Dihydro-2H-naphtho[1,2-b]pyran-2-ones
from α-tetralones s. 29, 884

Potassium tert-butoxide $Me_3C \cdot OK$
Nucleophilic substitution of nitrosamino groups

574.

A mixture of 7-chloro-2-(N-nitrosomethylamino)-5-phenyl-3H-1,4-benzodiazepine 4-oxide, malononitrile, K-*tert*-butoxide, and dimethylformamide stirred 0.5 hr. at room temp. → 7-chloro-1,3-dihydro-2-(dicyanomethylene)-5-phenyl-2H-1,4-benzodiazepine 4-oxide. Y: 79%. F. e. s. A. Walser and R. I. Fryer, J. Org. Chem. 40, 153 (1975).

n-Butyllithium (s. a. under BF₃)   n-BuLi
Synthesis of oxo compds.   CON< → COR
from carboxylic acid amides s. 12, 830; o-hydroxyaldehydes cf. H. Christensen, Synth. Commun. 5, 65 (1975).

Sodium amide/diethyl sulfite $NaNH_2/(EtO)_2SO$
3-Hydroxyisothiazoles from 2 nitrile molecules
s. 29, 777

Potassium acetate CH_3COOK
Sym. ethylene derivs. $2 \, CHNO_2$ → C : C
from aliphatic nitro compds.

575. $2 \, Me_2NSO_2\text{—}\langle\rangle\text{—}CH_2NO_2 \rightarrow Me_2NSO_2\text{—}\langle\rangle\text{—}CH=CH\text{—}\langle\rangle\text{—}SO_2NMe_2$

Stilbenes. A mixture of p-(N,N-dimethylsulfamoyl)-α-nitrotoluene, K-acetate, and 95%-ethanol refluxed 21 hrs. → p,p′-bis-(N,N-dimethylsulfamoyl)stilbene. Y: 67.5%. F. e. s. H. Feuer and H. Friedman, J. Org. Chem. 40, 187 (1975).

Sodium cyanide   NaCN
Sugar dilactones from cyanosugars
s. 26, 834

Potassium cyanide   KCN
o-Cyanophenols from nitro compds.
s. 28, 777

o-Alkoxynitriles from nitro compds.
s. 26, 835

Selective replacement of acylamino groups by cyano and organothio groups

576.

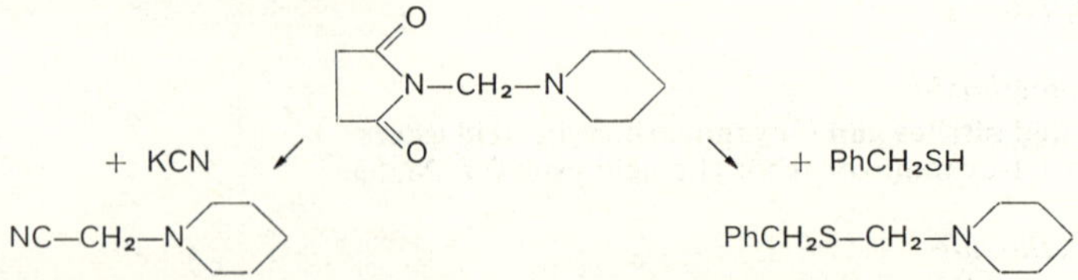

α-Aminonitriles | **1,1-Aminothioethers**

A methanolic soln. of N-(piperidinomethyl)succinimide refluxed, treated with a methanolic soln. of KCN, and refluxing continued 4 hrs. → piperidinoacetonitrile. Y: 82%. | treated with a methanolic soln. of benzyl mercaptan containing NaOH, and refluxed 4 hrs. → N-(benzylthiomethyl)piperidine. Y: 89%.

F. e., also with monoacyl- and sulfonyl-amino compds., s. H. Sakai, K. Ito, and M. Sekiya, Chem. Pharm. Bull. *21*, 2257 (1973).

Sodium nitrite $NaNO_2$
Dipyrrylmethanes
s. *27*, 817

Sodium nitrite/perchloric acid $NaNO_2/HClO_4$
Δ²-Isoxazolines from diazo compds.
s. *28*, 778

Triethylamine (s. a. under POCl₃) Et_3N
Pyrimidines from 1,3,5-oxadiazinium O-salts

577.

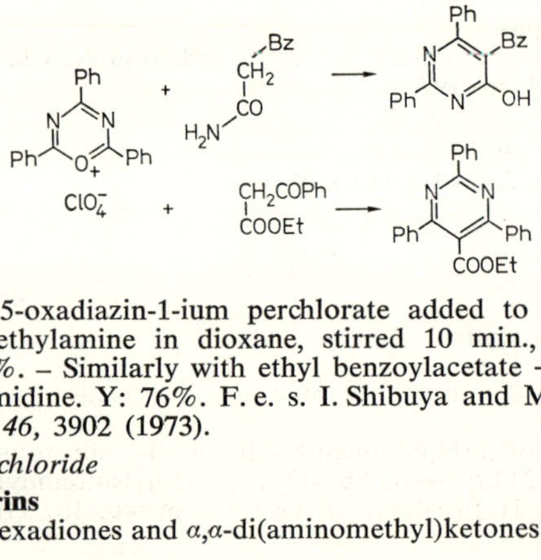

2,4,6-Triphenyl-1,3,5-oxadiazin-1-ium perchlorate added to a soln. of benzoylacetamide and triethylamine in dioxane, stirred 10 min., and refluxed 1 hr. → product. Y: 88%. – Similarly with ethyl benzoylacetate → 5-ethoxycarbonyl-2,4,6-triphenylpyrimidine. Y: 76%. F. e. s. I. Shibuya and M. Kurabayashi, Bull. Chem. Soc. Japan *46*, 3902 (1973).

Pyridine-hydrogen chloride $C_5H_5N\text{-}HCl$
3,4-Dihydrocoumarins
    from 1,3-cyclohexadiones and α,α-di(aminomethyl)ketones s. *26*, 836

Copper Cu
Penicillin from cephalosporin ring skeleton
s. *29*, 778

Copper/magnesium sulfate hydrate $Cu/MgSO_4 \cdot H_2O$
Diaryls from diazonium salts $ArN_2^+ \rightarrow Ar \cdot Ar'$
s. *29*, 279

Cupric acetoacetonate $[(CH_3CO)_2CH]_2Cu$
Expansion and fragmentation of N-heterocycles
    with diazo compds. s. *29*, 779

Cupric acetate $(CH_3COO)_2Cu$
Glycol esters $2\,C:N_2 \rightarrow C(OAc)\cdot C(OAc)$
from 2 diazo compd. molecules s. *29*, 780

Cupric sulfate $CuSO_4$
[2.3]-Sigmatropic rearrangement $C:C:C\cdot SR$
of unsatd. sulfonium ylids – 2-Allenthioethers from diazo compds. s. *29*, 781
Furan ring ◯
from α-alkoxymethyleneketones s. *28*, 779

Cuprous chloride $CuCl$
C-Insertion of diazo compds. ←
into sulfur-carbon bonds with allyl rearrangement s. *27*, 818

Trimethyl phosphite copper(I) iodide/benzoyl peroxide $(MeO)_3P\cdot CuI/(PhCOO)_2$
Cyclopropane ring from ethylene derivs. and diazo compds. ▽
Prepn. and reactions of diazomalonic esters, review s. *28*, 780; tetraphenylethylene as diazoalkane decomposition catalyst cf. C. Ho, R. T. Conlin, and P. P. Gaspar, Am. Soc. *96*, 8109 (1974)

Calcium hydride CaH_2
Δ^3-5-Pyrazolones ◯
from carboxylic acid hydrazides s. *28*, 781

Magnesium sulfate hydrate s. under Cu $MgSO_4\cdot H_2O$

Zinc chloride $ZnCl_2$
Coumarins from phenols via 3,4-dihydrocoumarins
Ring closure with nitriles s. *27*, 819

Trialkylboranes R_3B
β-Aminoketones from α-diazoketones $CO\cdot C\cdot C\cdot N<$
s. *28*, 782
Synthesis of urethans from 1,1-diurethans $C(NCOOR)_2 \rightarrow CR(NCOOR)$
s. *27*, 820

Boron fluoride BF_3
Homologization of ketones $CO\cdot C \rightarrow CO\cdot C\cdot C$
with diazo compds. s. *18*, 858 suppl. *29;* ring expansion, regioselectivity, s. H. J. Liu and S. P. Majumdar, Synth. Commun. *5*, 125 (1975)

Boron fluoride/n-butyllithium $BF_3/n\text{-}BuLi$
Reaction of boranes with diazo compds. $C\equiv CH \rightarrow COCH_2\cdot C\cdot COOR$
γ-Ketocarboxylic acid esters from acetylene derivs. via β,γ-acetylenecarboxylic acid esters – Use of dialkylchloroboranes s. *28*, 783

Fluoroborates BF_4^-
Reaction of quaternary 1,1,1-dialkoxyammonium salts $C\cdot N^+ \leqslant \rightarrow C\cdot C$
with CH-acidic compds. s. *27*, 821
Syntheses with tropenium salts and enamines ←
1,3,5-Cycloheptatrien-7-yl compds. – 4-Subst. α,β-ethyleneketones from 1,3-dienamines s. *27*, 822

Oxonium fluoroborate $R_3O^+\,BF_4^-$
Branched β-ketocarboxylic acid esters ⊙
from ketones – Ring expansion of cyclic ketones with diazo compds., also with ring closure to bicyclic β-diketones s. *26*, 837

Aluminum chloride $AlCl_3$
Reaction of acid azides with arenes $CON_3 \rightarrow NHCOAr$
Acylamines from carboxylic acid azides – Sulfones from sulfonic acid azides s. *29*, 782

Tetraphenylethylene	$Ph_2C : CPh_2$
Cyclopropane ring	▽
from ethylene derivs. and diazo compds.	
s. *28*, 780 suppl. *30*	
Methyl alcohol	MeOH
Ring expansion of α-oxo-N-heterocyclics with diazo compds.	⊙
2-Hydroxy-1,4-dihydroquinolines s. *29*, 783	
Chloroformic acid ester	ClCOOR
Vilsmeier-Haack-type reaction	$CH_2 \rightarrow C : CH(OR)$
Ethoxymethylene compds. s. *27*, 838	
Benzoyl peroxide s. under (MeO)₃P · CuI	$(PhCOO)_2$
Acetic acid	CH_3COOH
Reaction of indoles with Δ³-pyrrol-2-ones	○
5-Subst. 2-pyrrolidinones from 3-pyrrolin-2-ones s. *27*, 823	
Ring closures with 1-alkoxylactams	
s. *29*, 784	
Pyridine ring from β-amino-α,β-ethyleneketones	
Pyrido[2,3-d]pyrimi-2,4-dione ring s. *27*, 794	
2-Aminopyridine ring	
from β-amino-α,β-ethylenenitriles	

578.

β-Dimethylaminoacrylonitrile added dropwise to a stirred refluxing soln. of 6-amino-1,3-dimethyluracil in glacial acetic acid, and refluxing continued 3 hrs. → 7-amino-1,3-dimethylpyrido[2,3-d]pyrimidine-2,4-dione. Y: 50%. E. E. Garcia, Synth. Commun. *3*, 397 (1973).

Diethyl sulfite s. under NaNH₂	$(EtO)_2SO$
Chloranil	←
Diaryls from N'-aryl-n-stannylhydrazines	$ArNHNHSnR_3 \rightarrow Ar \cdot Ar'$
s. *28*, 784	
Phosgene	$COCl_2$
Vilsmeier aldehyde synthesis	$C : CH \rightarrow C : C \cdot CHO$
α,β-Ethylenealdehydes by electrophilic substitution s. *26*, 839	
Lead tetraacetate	$(CH_3COO)_4Pb$
Cycloaddition with hetarynes	←
s. *27*, 824	
Dinitrogen trioxide	N_2O_3
Furoxans	○
from 2 diazo compd. molecules s. *27*, 825	
Phosphorus oxide chloride	$POCl_3$
Vilsmeier aldehyde synthesis	$H \rightarrow CHO$
3-Formyl-4H-flavene s. *26*, 242	
1,3-Dialdehydes	
by double Vismeier synthesis s. *27*, 826	
Chromone-3-carboxaldehydes from o-hydroxyketones	○
s. *29*, 785	

Pyrazole-4-carboxaldehydes from semicarbazones
s. 26, 840
Isothiazoles
from enamines and carboxylic acid amides s. 26, 841
Phosphorus oxide chloride/triethylamine $POCl_3/Et_3N$
Isoxazoles
from aliphatic nitro compds. and enamines s. 29, 786
p-Toluenesulfonic acid $TsOH$
Pyrroles from ketones and hydrazines
via hydrazones s. 26, 842
Sulfuric acid H_2SO_4
C-α-Acylaminoalkylation H → C·NAc
s. 16, 843 suppl. 28
Perchloric acid (s. a. under $NaNO_2$) $HClO_4$
Cyclotribenzylenes from tosylamines ○
Trimerization s. 26, 843
Hydrochlorides ←
Acyl- and aryl-thiomethylation H → $CH_2S·$
s. 26, 844
Hydrobromides ←
2-Arylindoles from α-arylaminoketones ○
Möhlau-Bischler indole ring closure s. 28, 785
Hydrogen chloride HCl
Aminomethylation with hexahydro-1,3,5-triazines H → CH_2NH
p-Aminomethylation of phenols – 3,4-Dihydro-2H-1,3-benzoxazines s. 28, 786
Rhodium(II) acetate $(CH_3COO)_2Rh$
Transition metal-catalyzed reactions of diazo compds. ○
s. 29, 787
Palladous acetate $(CH_3COO)_2Pd$
Cyclopropane ring ▽
from diazo compds. and ethylene derivs. s. 23, 819 suppl. 27; *cis*-addition, limitations s. U. Mende et al., Tetrah. Let. 1975, 629
Via intermediates v.i.
9-Subst. anthracenes ←
s. 28, 644
Homologization of cyclic ketones ◉
via cyclic α-hydroxydiazo compds. s. 26, 661

Halogen ↑   CC ↕ Hal

Without additional reagents w.a.r.
Nucleophilic ar. substitution ArCl → ArAr'
Biphenyls s. 27, 827
Dinitromethylene from diazo compds. $C:N_2$ → $C:C(NO_2)_2$
s. 26, 845
Reaction of polyhalogenocarboxylic acid halides ←
with tert. amines
β-Amino-α,β-ethyleneketones from carboxylic acid halides – C-Acylation of ar. amines s. 26, 846

Diazoacetic from chloroformic acid esters OCOCl → OCOCHN$_2$
s. 27, 828

Heterocumulenes
Keteniminylidenephosphoranes $\geqslant$P : CH$_2$ → $\geqslant$P : C : C : NR
from iminocarbonylchlorides and methylenephosphoranes

579.   Ph$_3$P=CH$_2$ + Cl$_2$C=N—Ph → Ph$_3$P=C=C=N—Ph

N-Phenylcarbonimidoyl chloride added dropwise at room temp. to a stirred salt-free soln. of methylenetriphenylphosphorane in 1 : 1 benzene-tetrahydrofuran, and the product isolated after 1 hr. → N,P,P,P-tetraphenylketeniminylidenephosphorane. Y: 80%. F. e., also thioketenylidenephosphoranes, R$_3$P:C:C:S, s. H. J. Bestmann and G. Schmid, Ang. Ch. 86, 274 (1974); f. heterocumulenes cf. ibid. 87, 34 (1975).

N-Heterocyclic 1-alkylthioazomethinium salts H → C(SR) : $\overset{+}{N}$<
from chlorothioformamidinium chlorides s. 27, 829

Cyclopropanes from ethylene derivs. and halides ▽
1,1,2,2-Tetracyanocyclopropanes s. 26, 847

Anthraquinones ○
from 2-halogeno-1,4-naphthoquinones and ketenacetals s. 29, 788

Reactions with perfluoroalkenes
N-Heterocyclics s. 29, 789

Modified and anomalous Hantzch pyrrole ring synthesis
s. 28, 787

12-Acylindolizino[1,2-c]quinolinium salts
s. 26, 848

4(1H)-Quinolones
s. 28, 788

2-Aminothiophenes
s. 28, 789

Pyrazole ring from o-nitrohalides and hydrazones
Pyrazolo[3,4-d]pyrimidine-4,6-diones – Reactions of hydrazones, review s. 28, 790

6-Hydroxy-5,6-dihydro-4H-1,2-oxazines ⊙
from enamines and 1,2-halogenoximes s. 26, 849

Ring expansion of cobaltocenes
s. 26, 850

Irradiation s. under Li, NaOH, KOH, and Na[BH$_4$] ⋕

Electrolysis ↯
Sec. amines from azomethines and halides C : NR → CR'NHR
s. 22, 833 suppl. 29

Lithium (s. a. under CuI) Li
1,2-Dialkylation of ethylene derivs. C : C → CR·CR
s. 28, 791

Δ^1-Pyrrolines ←
from 2-pyrrolidones via 2-alkylidenepyrrolidines s. 26, 851

Oxido from oxo compds. and 1,1-dihalides CO → ▽$_O$
s. 26, 852

Lithium/irradiation Li/⋕
Sym. diaryls from aryl halides 2 ArHal → Ar·Ar
s. 28, 792

Sodium   *Na*
3(2H)-Furanones   O
from α-bromocarboxylic acid bromides s. 29, 790

Lithium hydride   *LiH*
α-Alkylation of carboxylic acid esters   CHCOOR → CR'COOR
s. 27, 843 suppl. 29

Sodium hydride (s. a. under n-BuLi and CuBr)   *NaH*
Alkylation of ketones   COCH → COCR
s. 26, 853

Oxo compds. from halides   Hal → CO
via 1,1-sulfinylthioethers – Synthesis with addition of 1 C-atom, also via s-trithianes and via thiazoles s. 27, 830

Aldehydes from halides   Hal → C·CHO
via cyclic 1-alkoxyenamines and 1,1-aminoethers – synthesis with addition of 2 C-atoms s. 28, 793

β-Ketocarboxylic acid esters
from halides s. 29, 499   Hal → C(COR)COOR'
from ketones s. 27, 831   COCH → CO·C·COOR

Asym. synthesis of α-aminocarboxylic acids   Hal → C(NH$_2$)COOH
from halides via α-isocyanocarboxylic acid esters s. 29, 791; synthesis of α-aminocarboxylic acids s. a. U. Schöllkopf, D. Hoppe, and R. Jentsch, B. *108*, 1580 (1975)

3,4-Dihydro-2-pyrone ring   O
from α,β-ethylenecarboxylic acid chlorides s. 26, 854

N-Condensed 2-pyridone ring
from carboxylic acid chlorides s. 26, 855

Sodium hydride/tert-butanol   *NaH/Me$_3$C·OH*
α-Alkyl-β,γ-ethylene- from α,β-ethylene-ketones   ←
Stereospecific monoalkylation at low temp.

580.

50%-NaH-mineral oil dispersion added under N$_2$ to a soln. of (+)-17β-*tert*-butoxy-19-(3,5-dimethyl-4-isoxazolyl)-deA-androst-9-en-5-one in dry tetrahydrofuran containing a catalytic amount of *tert*-butanol to speed enolate formation, gently refluxed 18 hrs., cooled in a Dry Ice-acetone bath to –80 to –73°, methyl iodide added via syringe over 2 min., and stirred 8 hrs. at –75° → 17β-*tert*-butoxy-10-[(3,5-dimethyl-4-isoxazolyl)methyl]deA-androst-9(11)-en-5-one. Y: 65.2–78%. F. e. s. J. W. Scott et al., Helv. *57*, 1217 (1974).

Sodium hydride/hexamethylphosphoramide   *NaH/(Me$_2$N)$_3$PO*
α-Monoalkylation of α,β-ethyleneketones   C:CHCO → C:CRCO
via α,β-ethylenehydrazones and via 1,3-dienamines s. 27, 832

Potassium hydride   *KH*
Syntheses with 1,1-sulfinylthioethers   O
Cyclic ketones – Cyclobutanones s. 27, 830 suppl. 29; f. cyclic ketones s. Tetrah. Let. *1975*, 2767

Lithium amalgam   *Li,Hg*
Oxido compds.   CO → ▽O
from oxo compds. and 1,1-dibromides s. 26, 852

Potassium amalgam   *K,Hg*
Sym. succinic acid esters   2 CHalCOOR → ROOC·C·C·COOR
from 2 α-halogenocarboxylic acid ester molecules s. 27, 860

Sodium hydroxide (s. a. under Na[BH$_4$])   *NaOH*
Oxido compds. from aldehydes and halides   CO → ▽O
α'-Diazo-α,β-oxidoketones s. 28, 795 suppl. 29

2-Acylthianaphthene-3-carboxylic acids   ←
from thianaphthenequinones and α-halogenooxo compds. s. 29, 813

Sodium hydroxide/irradiation   *NaOH/ℎѵ*
Halogenocyclopropane ring from ethylene derivs.   C : C → ▽
s. 28, 806

Sodium hydroxide/dimethyl sulfoxide   *NaOH/Me$_2$SO*
Selective C-alkylation of benzoins   CH → CR
also C,O-dialkylation s. 26, 856

Sodium hydroxide/tetrabutylammonium bromide   *NaOH/Bu$_4$N$^+$ Br$^-$*
γ-Diketones   CO·CH(SO$_2$R) → CO·CH·C·COR'
from β-ketosulfones and α-bromoketones via α-sulfonyl-γ-diketones – Ion pair extraction s. 29, 792

α-Alkylation of phosphonic acid esters   CHPO(OR)$_2$ → C(R)PO(OR)$_2$
with n-butyllithium cf. 30, 43; phase transfer catalysis s. E. D'Incan and J. Seyden-Penne, Synthesis 1975, 516

Sodium hydroxide/tetrabutylammonium iodide   *NaOH/Bu$_4$N$^+$ I$^-$*
α-Alkylation of aldehydes   CHCHO → CRCHO
s. 28, 794

Sodium hydroxide/benzyltriethylammonium chloride   *NaOH/PhCH$_2$N$^+$Et$_3$ Cl$^-$*
C-Dichloromethylation   H → CHCl$_2$
s. 27, 726

Cyclopropane ring from ethylene derivs.   C : C → ▽
1,1-Dihalogenocyclopropanes – 1-Chloro-1-arylthiocyclopropane ring s. 27, 833; 1,1-dihalogenospiropentanes s. E. Dunkelblum and B. Singer, Synthesis 1975, 323; 1,1-dihalogenocyclopropanes with hexadecyltrimethylammonium bromide, silanes, cf. R. B. Miller, Synth. Commun. 4, 341 (1974); (gem-dichlorocyclopropyl)amines from enamines s. J. Graefe, M. Adler, and M. Mühlstädt, Z. Chem. 15, 14 (1975); azamacrobicyclic polyethers as phase transfer catalysts cf. M. Cinquini, F. Montanari, and P. Tundo, Chem. Commun. 1975, 393; tert. amines as phase transfer catalysts cf. M. Makosza, A. Kaprowicz, and M. Fedoryński, Tetrah. Let. 1975, 2119; partial 1,1-dihalogeno-cyclopropane ring formation and asym. induction with modified quaternary ammonium phase transfer catalysts cf. T. Hiyama et al., Tetrah. Let. 1975, 3013

Oxido compds. from oxo compds. and halides   CO → ▽O
Darzens-Claisen reaction
s. 28, 795; α,β-oxido- from α-halogeno-sulfones, also α-alkylation of α-halogeno-sulfones, s. J. Org. Chem. 40, 266 (1975)

2,2-Dichloroaziridines from azomethines
Phase transfer catalysis

CHCl$_3$ + Ph—CH=N—Ph → Ph—C(Cl)(Cl)—N—Ph

A soln. of benzylideneaniline in chloroform stirred at 0–20° with aq. 50%-NaOH in the presence of a catalytic amount of benzyltriethylammonium chloride → product. Y: 88%. F. e. s. J. Graefe, Z. Chem. *14*, 469 (1974); **evaluation of phase transfer catalysts** cf. A. W. Herriott and D. Picker, Am. Soc. *97*, 2345 (1975).

1,2-Ethylene-1,3-dihalides from ethylene derivs.
Ring expansion with carbenes s. *29*, 793

Potassium hydroxide   KOH

Quinone methids from polyhalogenomethanes
s. *28*, 796

α,β-**Ethylenesulfonic acids**   CHSO$_2$H → C:C·SO$_3$H
from sulfinic acids and haloforms – Ramberg-Bäcklund rearrangement s. *27*, 627

Potassium hydroxide/irradiation   KOH/∰
Unsym. dihydroxydiaryls   ArHal → Ar·Ar'
s. *26*, 857

Sodium/alcohol   NaOR
Stereospecific C-alkylation   H → R
of 2,5-pyrrolidiones s. *26*, 858

1,2-Nitroethylene derivs.   C:CHal → C:C·C(NO$_2$)
from *α,β*-ethylenehalides s. *28*, 797

C-Alkylation with C-carbalkoxylation   CH$_2$ → CRCOOR'
s. *28*, 727

Phosphonic from *α*-halogenophosphinic acid esters
s. *28*, 798

Cyclopropanes from *β,γ*-ethylenehalides   ▽
with nucleophilic substitution s. *28*, 808

Sodium-tert-butoxide   NaOCMe$_3$
Fragmentation-type ring opening   CH → CR
with C-alkylation s. *27*, 926

Sodium compd.
Fulvenes from 1,1-acoxyhalides
s. *27*, 834

Potassium/tert-butanol   K/Me$_3$C·OH
Modified Darzens-Claisen reaction   CO → (epoxide)—COOR
Aldehydes from ketones via glycidic acid *tert*-butyl esters – Synthesis with addition of 1 C-atom s. *26*, 859

Cyclopropyl ketones
from 2-bromocyclobutanones s. *26*, 860

Potassium tert-butoxide   KOCMe$_3$
Alkenylidenecyclopropanes   ▽
s. *28*, 798a

Methyllithium s. a. under Mg and 3,6-Dimethylborepane   MeLi

Methyllithium/lithium iodide   MeLi/LiI
Tetrahydrofuran ring   O
from ethylene derivs. and 1,1'-dichloroethers s. *29*, 794

Butyllithium *BuLi*

1-Alkylation of indenes CH → CR
Inverse addition s. *29*, 795

Synthesis of α,β-ethyleneoxo compds. C : C·CO
from alkoxyacetylenes via dicarbanions and allenyl ethers s. *27*, 835

α-Metalation and alkylation of sulfones SO$_2$CH → SO$_2$CR
s. *29*, 796; with subsequent reductive removal of the sulfonyl groups s. a. J. Org. Chem. *40*, 150 (1975); synthesis of δ-lactols cf. K. Kondo, E. Saito, and D. Tunemoto, Tetrah. Let. *1975*, 2275

Syntheses with 3-chloro-4,5-dihydrofurans C
3-Acetylenealcohols from halides – Synthesis with addition of 4 C-atoms s. *29*, 797

n-Butyllithium (s. a. under CuI) *n-BuLi*

Acetylene derivs. from halides C≡CH → C≡CR
s. *27*, 851; long-chain terminal acetylene derivs. in tetrahydrofuranhexamethylphosphoramide s. W. Beckmann et al., Synthesis *1975*, 423

C-Alkylation of phenolethers ArH → ArR
of phenolpolyethers s. *27*, 836; 3-alkylation of 2-methoxynaphthalene s. A. Rosowsky et al., J. Med. Chem. *17*, 1217 (1974)

Allenes from oxo compds. and 1-bromoenesilanes CO → C : C : C
s. *29*, 798

Reactions via metalated enamines N·C : C·CH → N·C : C·CR
Masked 3-oxocarbanions s. *29*, 799; s. a. Synthesis *1975*, 512

Alkylation of enamines CO·C : C(N<)CH → CO·C : C(N<)CR
γ-Substitution of β-amino-α,β-ethyleneketones s. *28*, 799; alkylation of enamines with ethylmagnesium bromide instead of *n*-butyllithium cf. G. Stork and S. R. Dowd, Org. Synth. *54*, 46 (1974)

3-Alkylation of quinolines HetH → HetR
s. *28*, 800

Synthesis of ketones from halides CHCOC : P⩽ → C(R)COCH$_2$
via metalation and alkylation
of (α-acylalkylidene)phosphoranes

582.
$$CH_3COCH=PPh_3$$
$$\downarrow$$
$$n-C_9H_{19}Br + [LiCH_2COCH=PPh_3] \rightarrow n-C_9H_{19}CH_2COCH=PPh_3$$
$$\swarrow$$
$$n-C_9H_{19}CH_2COCH_3$$

n-Butyllithium in hexane added under N$_2$ with Dry Ice-acetone cooling to acetylmethylenetriphenylphosphorane in dry tetrahydrofuran until the red ylid anion color persists, stirred 15 min. at –78°, 1-bromononane added, stirred 1 hr. at 0° until the color is almost discharged, the solvent removed under reduced pressure, the residue dissolved in ethanol, water added, and heated 22 hrs. on a steam bath → 2-dodecanone. Y: 93%. – This method is an **alternative to** the classical **acetoacetic ester synthesis** (s. Synth. Meth. *22*, 838). F. e., also isolation of the intermediate alkylated phosphorane, s. M. P. Cooke, Jr., J. Org. Chem. *38*, 4082 (1973).

α,β-Ethylene-(α- or β-)silylketones Hal → COR
from carboxylic acid anhydrides s. *29*, 800

α,β-Ethylenecarboxylic acids from α,β-ethylenebromides C : CBr → C : C·COOH
Geospecific synthesis via 1,1-ethyleneorganolithium compds. s. *27*, 837

α,β-**Acetylenecarboxylic acids**   CHO → C≡C·COOH
from aldehydes via dibromomethylene compds. s. 28, 819

α-**Ketocarboxylic acid amides**   Hal → COCON<
from halides and oxamides s. 29, 801

α-**Alkylation of phosphonic acid esters** $CHPO(OR)_2$ → $C(R)PO(OR)_2$
s. 30, 43

Metalation of phosphoric acid amides   P(O)N·CH → P(O)N·CR

$$Me_2N\text{-}P(=O)(NMe_2)\text{-}N(Me)\text{-}CH_2Ph + ClCH_2OMe \rightarrow Me_2N\text{-}P(=O)(NMe_2)\text{-}N(Me)\text{-}CHPh\text{-}CH_2OMe$$

C-Alkylation. Benzylpentamethylphosphoramide treated at –70° with n-butyllithium in tetrahydrofuran, then monochloromethyl ether added → product. Y: ca. 100%. F. e. s. P. Savignac and Y. Leroux, J. Organometal. Chem. 57, C 47 (1973).

Syntheses with 1,1-sulfinylthioethers   ←
Aldehydes from halides s. 27, 830; also ketones s. 27, 830 suppl. 28

Syntheses with 2-alkylthio-Δ²-thiazolines   S·CH → S·CR
α-Alkylation of thioethers – Synthesis with 2-alkylthiopyridines s. 28, 801

Homologization of halides – β,γ-Ethylenehalides from halides; synthesis with addition of 3 C-atoms s. 28, 802

n-Butyllithium/sodium hydride   *n-BuLi/NaH*
Reactions via stepwise prepared dicarbanions   H → R
γ-Alkylation of β-ketocarboxylic acid esters s. 26, 861

n-Butyllithium/diethylamine   *n-BuLi/Et₂NH*
β-Ketocarboxylic acid esters from halides   C≡C·CH(OR)₂ → CR:C:C(OR)₂
via 1,1-dialkoxy-3-alkylthioallenes
and β-alkylthio-α,β-ethylenecarboxylic acid esters

$$MeS-C\equiv C-CH(OEt)_2 \;+\; BrC_2H_5 \;\rightarrow\; MeS-C(C_2H_5)=C=C(OEt)_2$$
$$\downarrow$$
$$C_2H_5\text{-}C(COCH_2COOMe) \;\xleftarrow{HOMe}\; MeS-C(C_2H_5)=CHCOOEt$$

Diethylamine followed by n-butyllithium in hexane added at –30° under N₂ to dry tetrahydrofuran, cooled to –78°, 3,3-diethoxy-1-methylthiopropyne added during 3 min., stirring continued 2.5 hrs. at –78°, bromoethane added, and stirred 1 hr. at –78° → 1,1-diethoxy-3-methylthio-1,2-pentadiene (Y: 89%) treated with aq. tetrahydrofuran containing a trace of p-toluenesulfonic acid → ethyl 3-methylthio-2-pentenoate (Y: ca. 100% 1 : 1 mixture of E,Z-isomers) dissolved in methanol, HgCl₂ in methanol added, and stirred 1 hr. at 25° → methyl 3-ketopentanoate (Y: ca. 100%). F. e. s. R. M. Carlson and J. L. Isidor, Tetrah. Let. 1973, 4819; **syntheses via allenecarbanions** s. a. ibid. 1975, 1741.

n-Butyllithium/diisopropylamine   *n-BuLi/i-Pr₂NH*
Synthesis of α,β-ethylenealdehydes   C : C·CHO
via 1,3-bis(alkylthio)-1-ethylenes with addition of 3 C-atoms from halides and and 1,3-bis(methylthio)-2-methoxypropane – 1,3-Bis(methylthio)allyl anion – γ-Acetoxy-α,β-ethylenealdehydes from oxido compds. – α,β-Ethylene-γ-hydroxyaldehydes from oxo compds. s. 26, 862; s. a. Org. Synth. 54, 19 (1974)

Geospecific syntheses with allyl dithiocarbamates $\quad$ C:CH$_2$ → C:CHR
via their [3.3]-sigmatropic rearrangement

585.

A soln. of *n*-butyllithium in *n*-hexane added below –55° under argon to a soln. of allyl dimethyldithiocarbamate and diisopropylamine in tetrahydrofuran, then a soln. of ethyl iodide in tetrahydrofuran added below –55°, stirred 15 min., and the resulting crude intermediate refluxed 3 hrs. in chloroform → *trans*-2-pentenyl dimethyldithiocarbamate. Y: 93%. F. e. and reactions of the products s. T. Hayashi and H. Midorikawa, Synthesis *1975*, 100; cf. T. Nakai, H. Shiono, and M. Okawara, Chem. Lett. *1975*, 249.

n-Butyllithium/N,N,N',N'-tetramethylethylenediamine $\qquad$ *n-BuLi/TMEDA*
Metalation of allylic methyl groups $\qquad$ ←
Synthesis of ethylene derivs. from halides

586.

3-Methyl-3-buten-1-ol treated at –78° with *n*-butyllithium-TMEDA-complex in hexane, 1-bromo-3-methyl-2-butene in hexane added, and stirred 3 hrs. at room temp. → product. Y: 70%. G. Cardillo, M. Contento, and S. Sandri, Tetrah. Let. *1974*, 2215; **allylic metalation with trimethylsilylmethylpotassium** s. J. Hartmann and M. Schlosser, Synthesis *1975*, 328.

sec-Butyllithium $\qquad$ $C_2H_5CH(CH_3)Li$
α-Alkylation of thioethers $\qquad$ S·CH → S·CR
s. 28, 842

Synthesis of ketones from thioenolethers $\qquad$ Hal → COCH
s. 29, 802

Phenyllithium $\qquad$ *PhLi*
Ar. C-alkylation $\qquad$ ArH → ArR
s. 29, 803

Trityllithium $\qquad$ $Ph_3C·Li$
α-Monoalkylation of β,γ-ethyleneketones $\quad$ C:C·CH$_2$·CO → C:C·CHR·CO
s. 29, 804

n-Amylsodium/tetramethylethylenediamine $\qquad$ *n-C$_5$H$_{11}$Na/TMEDA*
α-Lateral metalation of isocyclics $\qquad$ Hal → R
$\quad$ Synthesis of hydrocarbons from halides s. 27, 838

Trimethylsilylmethylpotassium $\qquad$ Me_3SiCH_2K
Allylic metalation $\qquad$ ←
s. 30, 586

Lithium/liq. ammonia $\qquad$ *Li/NH$_3$*
Reductive C-alkylation of arenes $\qquad$ ArH → ArR
$\quad$ with subsequent oxidative rearrangement of intermediate 3-alkyl-1,4-cyclohexadienes – Specific alkylation of ar. ring systems s. 27, 839; reductive C-alkylation of furans, also reduction without alkylation, 2,5-dihydrofurans, s. A. J. Birch and J. Slobbe, Tetrah. Let. *1975*, 627

Stereospecific deoxygenative C-alkylation C(OR):C·COOR → CH·C(R')·COOR
1-(Alkoxy)enolethers from ketones s. 28, 803
Lithium amide $LiNH_2$
Directed C-alkylation CH → CR
of subst. succinic acid monoesters – Synthesis via dianions; also directed condensation s. 27, 840
Cyclopropanes from 1,2-dihalides ▽
s. 27, 841
*Lithium diisopropylamide (s. a. under
n-BuLi/i-Pr₂NH and CuI)* i-Pr_2NLi
Cyclic γ-alkyl-α,β-ethyleneketones CH → CR
via 4-alkylation of cyclic 3-keto-1,2-en-1-olethers – 4-Alkyl-2-cyclohexenones s. 29, 805
α-Alkylation of carboxylic acid esters CR'COOR
and lactones s. 27, 843 suppl. 28
α-Lateral metalation of isocyclics Hal → R
of N,N-dialkyltoluamides s. 27, 838 suppl. 28
3-Substitution of 2-azetidinone rings H → COOR
Azomethine-stabilized anions – C-Alkylation of lactams s. 29, 806; 3-alkylation of 1-alkyl-2-piperidones, with Li-dimethylamide, cf. P. Deslongchamps, U. O. Cheriyan, and D. R. Patterson, Can. J. Chem. 53, 1682 (1975)
α-Lateral metalation and alkylation of Δ²-oxazolines CH → CR
Asym. synthesis of carboxylic acids – Syntheses with Δ²-oxazolines s. 29, 807; asym. synthesis of γ-lactones and 1,4-diols s. J. Org. Chem. 40, 1186 (1975)
Syntheses with masked acyl carbanions CHO → COR
Metalation and alkylation of α-siloxynitriles – Ketones from aldehydes via α-siloxynitriles s. 29, 808
Syntheses with γ-alkylthio-α,β-ethylenecarboxylic acid esters
α-Alkylated γ-alkylthio-β,γ-ethylenecarboxylic acid esters C(SR):C·C(R)·COOR

587. MeS⏜COOMe + IMe ⟶ MeS⏜⧵COOMe

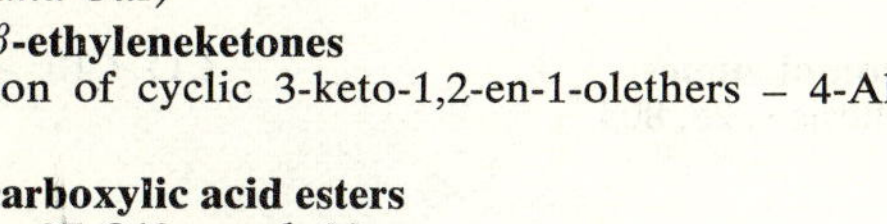

Methyl γ-methylthiocrotonate (prepn. s. Synth. Meth. *10*, 489) allowed to react 0.5 hr. at −78° with Li-diisopropylamide in tetrahydrofuran, then at the same temp. methyl iodide in the presence of *hexamethylphosphoramide*, and the product isolated after 12 hrs. → thioenolether deriv. Y: 92%. F. e. and reactions of the products s. A. S. Kende et al., Tetrah. Let. *1975*, 405.
Syntheses with ketene mercaptal S-monoxides ←
s. 29, 667
α-Alkylation of β,γ-ethylenesulfoxides C:C·CH·SOR → C:C·CR·SOR
s. 29, 217
1-Alkylation of dithiocarbamic acid esters CH·SCS(N<) → CR·SCS(N<)
s. 30, 536
1-Alkylation and rearrangement of allyl dithiocarbamates – Synthesis of ethylene derivs. s. 29, 809
Endocyclic enamines ○
from azomethines and dihalides s. 27, 842; s. a. Org. Synth. *54*, 93 (1974)
Lithium diisopropylamide/hexamethylphosphoramide i-$Pr_2NLi/(Me_2N)_3PO$
Deconjugative alkylation CHC:C·COOR → C:C·CR'COOR
β,γ- from α,β-Ethylenecarboxylic acid esters with and without α-alkylation s. 29, 810

Lithium N-isopropylcyclohexylamide   *i-Pr(C_6H_{11})NLi*
Alkylation of carboxylic acid esters   CH → CR
    α-Alkylation − α-Alkylation and α-deuteriation of lactones − Remote alkylation of polyenecarboxylic acid esters s. 27, 843; stereospecific α-alkylation of esters cf. P. A. Grieco and Y. Masaki, J. Org. Chem. 40, 150 (1975); with Li-diethylamide, α,α-dichlorocarboxylic acid esters via 1,1-halogenoorganolithium compds. (cf. Synth. Meth. 30, 549) cf. J. Villieras et al., Synthesis 1975, 524

Lithium 2,2,6,6-tetramethylpiperidide   ←
Cyclopropanes   C:C → ▽
    from ethylene derivs. and halides s. 26, 847 suppl. 28

Sodium amide $NaNH_2$
γ-Diketones from α-bromoketones   CO·CBr → CO·C·C·CO
    Stereospecific synthesis s. 28, 804

Piperidines   ○
s. 28, 805

Homophthalimides from ar. halides via benzynes
    C-Cleavage of α-dicarboxylic acid esters s. 29, 811

Annelation with ring expansion of cyclic ketones   ○ ⊙
s. 26, 863

Sodium bis(trimethylsilyl)amide $(Me_3Si)_2NNa$
Halogenocyclopropane ring from ethylene derivs.   C:C → ▽
    s. 28, 806; f. method, also ring expansion, cf. S. Miyano, Y. Matsumoto, and H. Hashimoto, Chem. Commun. 1975, 364

Potassium amide KNH_2
N,C-Dianions as intermediates   CH → CR
    C-Alkylation s. 27, 844

C-Alkylation of acylureas via trianions $CHCONHCONH_2$ → $CRCONHCONH_2$
s. 29, 812

Sodium carbonate Na_2CO_3
2-Acylthianaphthene-3-carboxylic acids   ←
    from thianaphthenequinones and α-halogenooxo compds. s. 29, 813

Sodium carbonate Na_2CO_3
Thieno[2,3-d]pyrimidines   ○
s. 27, 845

Sodium hydrogen carbonate s. under $ZnCO_3$ $NaHCO_3$

Potassium carbonate K_2CO_3
α,β-Oxidoketones or α,β-ethylene-β'-hydroxyketones   ←
    from α-halogenoketones and aldehydes s. 26, 864

Cyclopropanes from dihalides   ▽
s. 27, 841

γ-Lactones from α-chloroaldehydes and malonic acid esters   ○

588. $(CH_3)_2\underset{Cl}{C}$—CHO + 2 $H_2C(COOMe)_2$ → [lactone with CH(COOMe)₂ and COOMe substituents]

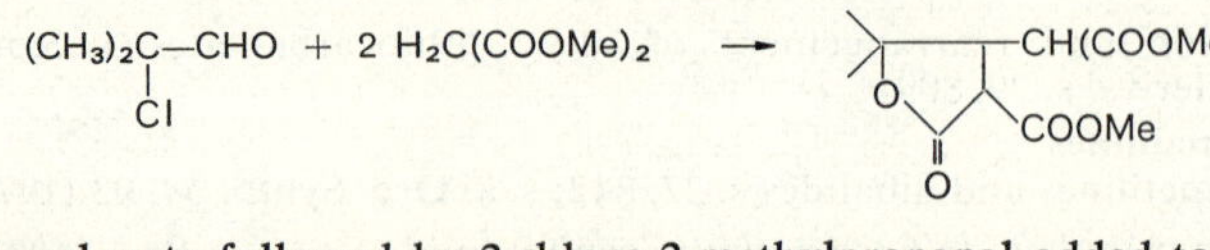

Dimethyl malonate followed by 2-chloro-2-methylpropanal added to an aq. soln. of K-carbonate, and stirred 30 hrs. at room temp. → α-methoxycarbonyl-β-dimethoxycarbonylmethyl-γ,γ-dimethyl-γ-butyrolactone. Y: 82%. F. e. s. A. Takeda, S. Tsuboi, and Y. Oota, J. Org. Chem. 38, 4148 (1973).

Benzofurans from o-hydroxyoxo compds.
s. 28, 807

Sodium cyanide	NaCN
α-Alkoxynitriles from 1,1-alkoxychlorides s. 30, 567	C(OR)Cl → C(OR)CN

Sodium cyanide/tetrabutylammonium bromide	$NaCN/Bu_4N^+ Br^-$
Acylcyanides from carboxylic acid chlorides s. 6, 898 suppl. 29	COCl → COCN

Potassium cyanide	KCN
Cyclopropanes from β,γ-ethylenehalides with nucleophilic substitution s. 28, 808	▽
Fragmentation-type ring opening s. 26, 865	C

Potassium thiocyanate	KSCN
2-Amino-4(3H)-quinazolinethiones s. 29, 814	O

Potassium fluoride s. under Ph_3P	KF
Lithium iodide s. under MeLi	LiI
Ammonia s. under CF_3SO_3Cu	NH_3
Diethylamine s. under n-BuLi	Et_2NH

Triethylamine (s. a. under $AgBF_4$)	Et_3N
Synthesis of enamines from 1-chlorenamines s. 27, 846	C : C(N<)Cl → C : C(N<)R

Ketones   CH : C·N< → C(COR) : C·N<
from carboxylic acid chlorides and enamines
via α-aminomethyleneketones

589.

2-Fluorobenzoyl chloride added dropwise to a stirred soln. of trans-β-dimethyl-amino-2-nitrostyrene and triethylamine in benzene, stirred and refluxed 15 hrs., the resulting intermediate dissolved in dioxane-water, and refluxed 18 hrs. → 2'-fluoro-2-(2-nitrophenyl)acetophenone. Y: 45%. F. e. s. E. E. Garcia and R. I. Fryer, J. Heterocyclic Chem. *11*, 219 (1974).

Synthesis of enolesters CH_2 → C : C(R)OCOR
from 2 carboxylic acid chloride molecules

590.

1-Benzyl-2-methylimidazole in acetonitrile containing triethylamine treated dropwise at 5–25° during 2–4 hrs. with benzoyl chloride → product. Y: ca. 90%. F. e. s. L. A. M. Bastiaansen, A. A. Macco, and E. F. Godefroi, Chem. Commun. *1974*, 36; as first step of a sequence for the conversion of methyl to cyano groups s. J. D. Albright and R. G. Shepherd, J. Heterocyclic Chem. *10*, 899 (1973).

α-**Ketocarboxylic acid amides**   COCl → COCON<
from carboxylic acid chlorides and formamides – Synthesis with addition of
1 C-atom s. 27, 847

β,γ-**Ethylene-β-siloxycarboxylic acid esters**   C : C(OSi ⪕)·C·COOR
from carboxylic acid chlorides and O-silyl O-alkyl keteneacetals s. 28, 809

2-Pyrones   O
from carboxylic and α,β-ethylenecarboxylic acid chlorides s. 27, 848; from
α,β-ethyleneoxo compds. s. a. Can. J. Chem. 53, 195, 201 (1975)

Double ring closure with dicarboxylic acid chlorides
4,5,6,7-Tetrahydrocyclopenta-1,3-dioxin-4-ones s. 27, 849

Heterocyclics from iminochlorides
via nitrile ylids – Pyrroles s. 28, 810

α-**Aminoketones**
from carboxylic acid chlorides and isonitriles via oxazoles – α-Amino-β-ketocarboxylic acid esters s. 28, 811; oxazoles cf. R. Schröder et al., A. 1975, 533

Thiophenes from monothio-β-diketones
via 3-hydroxy-2,3-dihydrothiophenes s. 26, 866

Triethylamine/dimethyl sulfoxide Et_3N/Me_2SO

Condensation with elimination of hydrogen halides   C·Cl + H·C → C·C
s. 29, 815

Tetramethylethylenediamine s. under n-BuLi and n-C_5H_{11}Na   *TMEDA*

Pyridine C_5H_5N

C-Carbalkoxylation   H → COOR
via C-trichloroacetylation s. 29, 816

Lactams by stereospecific double ring closure   O
via N-acylation and intramolecular diene synthesis s. 29, 817

Copper (s. a. under Zn)   *Cu*

(Polyfluoroalkyl)arenes from ar. iodides   ArI → Ar·R
s. 26, 867

2-Iminocyclobutanones from isonitriles   □

591.
$(CH_3)_2$C—CO—C$(CH_3)_2$ + C=N—◯ → [cyclobutanone with =O and =N—◯]
   | |
  Br   Br

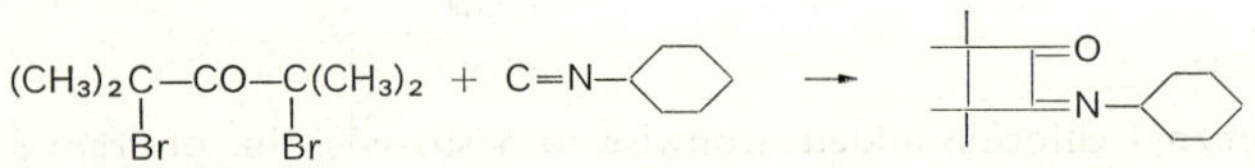

and α,α′-**dibromoketones**. A soln. of 2,4-dibromo-2,4-dimethyl-3-pentanone in
benzene added dropwise at room temp. to a mixture of Cu, excess cyclohexyl
isocyanide, and benzene, stirring continued 12 hrs. → 3,3,4,4-tetramethyl-2-cyclo-
heximiniocyclobutanone. Y: 99%. F. e. s. Y. Ito et al., Synth. Commun. 4, 87
(1974).

Biphenylenes from o-diiodides
s. 27, 850

Copper/potassium carbonate Cu/K_2CO_3
Acetylene derivs. from halides   C≡CH → C≡CR
s. 27, 851

Silver s. under Zn   *Ag*

Cuprous oxide (s. a. under RN : C) Cu_2O
2-Arylation of benzothiazoles   H → Ar
s. 29, 818

Cuprous tert-butoxide $\qquad$ $CuOCMe_3$
s. *29*, 818

Lithium dialkylcuprate $\qquad$ $Li[CuR_2]$
Branched ketones ←
from α-alkylthiomethyleneketones – Incorporation of 3 alkyl groups – 1,4-Addition-alkylation s. *29*, 819; α,β-disubst. ketones from α,β-ethyleneketones, metalation of mercaptals by Li-diisopropylamide, cf. D. Seebach and R. Bürstinghaus, Ang. Ch. *87*, 37 (1975).

Cuprous acetate/ethylenediamine $\qquad$ $CH_3COOCu/H_2NCH_2CH_2NH_2$
Reactions of cuprous acetate with halides $\qquad$ 2 RHal → R·R
Dimerization of halides – 1,5-Dienes s. *29*, 820

Cuprous cyanide (s. a. under EtMgBr) $\qquad$ $CuCN$
Replacement of bromine $\qquad$ Br → CN
by cyano groups and chlorine – Retention of configuration s. *29*, 821

Cuprous trifluoromethanesulfonate/ammonia $\qquad$ CF_3SO_3Cu/NH_3
Reactions of ar. halides $\qquad$ 2 ArHal → Ar·Ar
in homogeneous soln.

592.

Sym. diaryls. A mixture of o-iodonitrobenzene, cuprous trifluoromethanesulfonate, aq. 5%-NH₃, and acetone stirred 5 min. at room temp. under N₂ → 2,2'-dinitrobiphenyl. Y: 92%.

Ar. amines. Methyl o-iodobenzoate treated 24 hrs. at room temp. with cuprous trifluoromethanesulfonate and aq. 5%-NH₃ in acetone → methyl anthranilate. Y: 80%.

F. e., modifications, and limitations s. T. Cohen and J. G. Tirpak, Tetrah. Let. *1975*, 143.

Dilithium tetrachlorocuprate s. under Mg $\qquad$ Li_2CuCl_4

Cuprous chloride s. under Mg and Zn $\qquad$ $CuCl$

Cuprous bromide/sodium hydride $\qquad$ $CuBr/NaH$
C-Arylation of β-dicarbonyl compds. $\qquad$ ArHal → ArC(COR')COOR
with subsequent C-deacylation

593.

Homophthalic acids. A mixture of 2-bromobenzoic acid and a 5–10-fold excess of ethyl acetoacetate heated 0.5–2 hrs. at 60–80° with NaH and 6 mole-% CuBr → intermediate (Y: 96%) dissolved in 2 N NaOH and allowed to stand 20–30 min. at room temp. → homophthalic acid. Overall Y: 92%. F. e. s. A. Bruggink and A. McKillop, Ang. Ch. *86*, 349 (1974); cf. R. G. R. Bacon and J. C. F. Murray, Soc. Perkin I *1975*, 1267.

Cuprous iodide s. a. under Mg	CuI
Cuprous iodide/lithium	CuI/Li
Ketones from carboxylic acid chlorides Hindered ketones s. 27, 852	COCl → COR
2-Allenealcohols **from acetyleneoxido compds.**	←

594.

$$CH_3Br + n-C_4H_9-C\equiv C-\underset{CH_3}{\overset{O}{C}}-CH_2 \longrightarrow \underset{n-C_4H_9}{\overset{CH_3}{>}}C=C=C-CH_2OH \;\; \underset{CH_3}{|}$$

Startg. m. in ether added at –30 to –20° to a stirred ethereal soln. of 6 molar equivalents of lithium dimethylcopper prepared according to Synth. Meth. **23**, 831, and the product isolated after ca. 15 min. → product. Y: 75%. F. e. s. P. R. Ortiz de Montellano, Chem. Commun. **1973**, 709; with Grignard compds. cf. P. Vermeer et al., R. **93**, 46 (1974); f. method, **2-allene-*tert*-alcohols**, cf. G. Léandri, H. Monti, and M. Bertrand, Bl. **1974**, 1919.

Cuprous iodide/n-butyllithium	CuI/n-BuLi
γ,δ-**Ethylenenitriles** from β,γ-ethylenehalides s. 27, 853	C : C·CHal → C : C·C·C·CN
Cuprous iodide/lithium diisopropylamide	CuI/i-Pr₂NLi
γ,δ-**Ethylenecarboxylic acid esters** from β,γ-ethylenehalides – Preferential replacement of bromine s. 27, 854	C : C·Hal → C : C·C·C·COOR
Cupric chloride s. under Zn	CuCl₂
Silver nitrate s. under Mg	AgNO₃
Silver ammonium nitrate s. under Thexylborane	←
Magnesium (s. a. under EtMgBr)	Mg
Grignard syntheses under pressure Bridgehead alkyl derivs. s. 26, 868	RHal + HalR′ → R·R′
Reaction of conjugated enolethers **with Grignard compds.** Synthesis of ethylene derivs. s. 26, 869	C : C(OR) → C : C·R
Synthesis of ethylene derivs. from enamines s. 26, 870	C : C(N<) → C : C·R
Synthesis of 1,5-dienes s. 24, 840	RCl + ClR′ → R·R′
Synthesis of 1,2,4-trienes from 5-chlor-3-en-1-ynes s. 28, 812; s. a. Bl. **1974**, 1119	←
Tert. from sec. amines via N-aminomethylation s. 26, 871	←
Cyclic enamines from lactams	>NCO → >NC(OH)R

595.

Ethereal methylmagnesium bromide added slowly to a soln. of 1,5,5-trimethyl-2-oxo-1,2,3,4,4a,10b-hexahydro-8-*n*-amyl-10-hydroxy-*trans*-5H[1]benzopyrano[4,3-b]-

pyridine in tetrahydrofuran, refluxed 24 hrs., the resulting mixture of enamine and some carbinolamine dissolved in methylene chloride, and stirred 4 hrs. with *molecular sieve 3 A* → 1,2,5,5-tetramethyl-1,4,4a,10b-tetrahydro-8-*n*-amyl-10-hydroxy-*trans*-5H-[1]benzopyrano[4,3-b]pyridine. Y: 81%. M. Cushman and N. Castagnoli, Jr., J. Org. Chem. *39*, 1546 (1974).

Reactions with 4,4-dimethyl-Δ^2-oxazolines Hal → CHO
Aldehydes from halides – Synthesis with addition of 1 C-atom – Deuterioaldehydes – Synthesis of carboxylic acids and esters – Protection of carboxyl groups s. *26*, 872; s. a. Org. Synth. *54*, 42 (1974)

Ketones from carboxylic acid esters COOR → COR′
α-Chloro-α,β-oxidoketones s. *27*, 855

3-Subst. cyclopent-2-ene-1,4-diones ←
also with retention of ketal groups – Prepn. of highly reactive magnesium s. *27*, 856; highly reactive Mg-dispersion s. a. K. J. Klabunde et al., J. Organometal. Chem. *71*, 309 (1974)

γ,δ-Ethyleneketones from carboxylic acids COOH → CO·C·C·C : C
s. *26*, 873

α,β-Acetylenecarboxylic acid esters C≡CH → C≡C·COOR
from acetylene derivs. s. *27*, 857

Grignard synthesis with 1,1′-di(alkoxy) amines N·C(OR) → N·C·R′
s. *28*, 813

2,5-Cyclohexadienone oximes from nitroso compds. ←

596.

1,6-Addition. 2,4,6-Tri-*tert*-butylnitrosobenzene allowed to react 1 hr. at 0° with a large excess of isopropylmagnesium bromide in ether → product. Y: 54%. F. e., also 1,4- and 1,2-addition, s. R. Okazaki, Y. Inagaki, and N. Inamoto, Chem. Commun. *1974*, 414.

Hemiketals from ketoaldehydes O
s. *26*, 874

Lactols from dialdehydes
s. *27*, 858

γ-Hydroxyketones and γ-diketones C
from 2(3H)-furanones s. *29*, 882

Dioxo compd. monoketals
from cyclic 2,2-dialkoxyethers s. *27*, 859

Ring opening of cyclimmonium salts
Mono- from bi-cyclic N-heterocyclics – Abnormal displacement of Grignard compds. s. *29*, 823

Magnesium/dilithium tetrachlorocuprate Mg/Li_2CuCl_4
Cross-coupling with Grignard compds. Hal → R
and tosylates or acetates
s. *29*, 824

Magnesium/cuprous chloride Mg/CuCl
1,6-Diketones CO·C·C·C·C·CO
from 2 carboxylic acid and α,β-ethylenechloride molecules s. *29*, 825

Magnesium/methyllithium/cuprous iodide $\quad Mg/MeLi/Cu$
Cross coupling $\quad\quad\quad\quad\quad\quad\quad\quad\quad\quad\quad\quad\quad$ RHal + Hal'R' → RR'
via mixed lithium dialkylcuprates

597.
$$CH_2=CH[CH_2]_8CH_2Cl \quad\quad MeLi + CuI$$
$$\downarrow Mg \quad\quad\quad\quad\quad\quad\downarrow$$
$$\left[\begin{array}{c} CH_2=CH[CH_2]_8CH_2MgCl \; + \; CuMe \\ \downarrow \substack{Me \\ \diagdown} \\ CH_2=CH[CH_2]_8CH_2 \diagup Cu^- \; +MgCl \end{array} \right] + \; I[CH_2]_{10}COOEt$$
$$\downarrow$$
$$CH_2=CH[CH_2]_8CH_2[CH_2]_{10}COOEt$$

A soln. of 10-undecenylmagnesium chloride prepared from 11-chloroundec-1-ene and Mg-turnings in tetrahydrofuran added at –78° with a syringe to methyl-copper prepared from methyllithium and CuI in tetrahydrofuran, stirring continued 1 hr. at –78°, warmed until a soln. results, immediately recooled to –78°, ethyl 11-iodoundecanate added with a syringe, the resulting suspension stirred 1 hr. at –78°, allowed to come to room temp., and stirred 2 hrs. → ethyl 21-docosenoate. Y: 79%. – This method, which tolerates a variety of functional groups, is particularly well suited for the synthesis of α,ω-disubst., branched, or straight-chain fatty acid esters. F. e. s. D. E. Bergbreiter and G. M. Whitesides, J. Org. Chem. *40*, 779 (1975).

Magnesium/borane/silver nitrate $\quad\quad\quad\quad\quad\quad\quad\quad\quad Mg/BH_3/AgNO_3$
Dimerization of halides $\quad\quad\quad\quad\quad\quad\quad\quad\quad\quad\quad\quad$ 2 RHal → R·R
s. *28,* 814

Magnesium/aluminum isopropoxide $\quad\quad\quad\quad\quad\quad\quad\quad\quad Mg/Al(OR)_3$
α-Vinylketones from α-chloroketones $\quad\quad\quad\quad\quad\quad\quad\quad\quad$ ←

598.

Vinylmagnesium chloride in tetrahydrofuran added dropwise at –30° under N_2 during 3 hrs. to 2-chlorocyclododecanone in the same solvent, after 2 hrs. stirring at –30° tetrahydrofuran removed by distillation under reduced pressure at 30°, the residue dissolved in methylene chloride, Al-isopropoxide added, and refluxed 2 hrs. at 50° → 2-vinylcyclododecanone. Y: 59%. – Without Al-isopropoxide treatment, 80% 2-chloro-1-vinylcyclododecan-1-ol can be obtained. S. Watanabe et al., J. Appl. Chem. Biotech. *23,* 501 (1973).

Magnesium/ferric chloride $\quad\quad\quad\quad\quad\quad\quad\quad\quad\quad\quad\quad Mg/FeCl_3$
Cross-coupling with Grignard compds. $\quad\quad\quad\quad\quad\quad$ RHal + HalR' → R·R'
s. *26,* 875; with dichlorobis(triphenylphosphine)nickel(II) instead of $FeCl_3$, from halogenopyridines, cf. L. N. Pridgen, J. Heterocyclic Chem. *12,* 443 (1975)

Synthesis of ethylene derivs. $\quad\quad\quad\quad\quad\quad\quad\quad\quad\quad$ C : CHal → C : CR
from α,β-ethylenehalides s. *26,* 875; with iron complexes, preferably tris-(dibenzoylmethido)iron(III), $Fe(PhCOCHCOPh)_3$, also stereospecific cross coupling, cf. J. Org. Chem. *40,* 599 (1975); with a Ni-phosphine complex as catalyst, geospecificity, cf. M. Zembayashi, K. Tamao, and M. Kumada, Tetrah. Let. *1975,* 1719; geospecific synthesis with tetrakis(triphenylphosphine)palladium(0) as catalyst, mostly with organolithium compds., cf. M. Yamamura, I. Moritani, and S. Murahashi, J. Organometal. Chem. *91,* C39 (1975)

Magnesium/dichlorobis(triphenylphosphine)nickel(II)   Mg/NiCl$_2$(Ph$_3$P)$_2$
s. *26*, 875 suppl. *30*

Zinc   Zn

Synthesis of unsatd. amines   C : C → CH·CR
s. *29*, 826

3-Ketothioethers   CO·CBr → CO·C·C·SR
from α-bromoketones and 1,1-chlorothioethers

599. 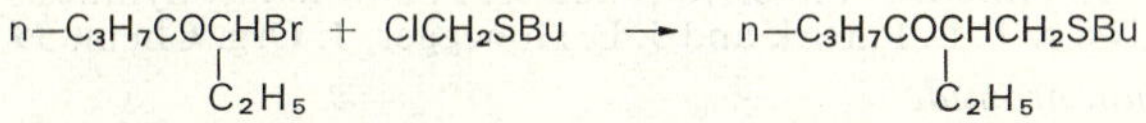   n—C$_3$H$_7$COCHBr + ClCH$_2$SBu   →   n—C$_3$H$_7$COCHCH$_2$SBu
                 | |
                 C$_2$H$_5$ C$_2$H$_5$

A mixture of 3-bromoheptan-4-one and butyl chloromethyl sulfide added gradually with agitation to Zn-filings in anhydrous ethyl acetate, and the product isolated after 1 hr. → 3-butylthiomethyl-4-heptanone. Y: 77%. F. e. s. F. G. Saitkulova, G. G. Abashev, and I. I. Lapkin, Zh. Org. Khim. *9*, 1405 (1973); C. A. *79*, 91529.

Cyclopropanes   ▽
     from ethylene derivs. and 1,1-dihalides s. *28*, 815

2-Azetidinones 
     from azomethines and α-bromocarboxylic acid esters s. *28*, 816

5-Isoxazolidones
     from nitrones and α-bromocarboxylic acid esters s. *28*, 817

Zinc/copper   Zn/Cu

Condensation with elimination of hydrogen bromide   CO·CHal → CO·C·C : C
β,γ-Ethyleneketones
from ethylene derivs. and α-halogenoketones

600. 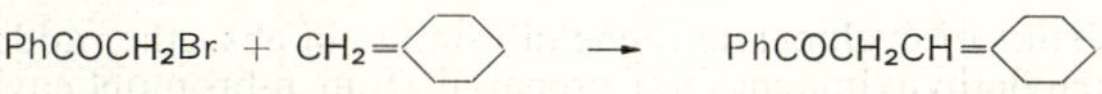   PhCOCH$_2$Br + CH$_2$=⬡   →   PhCOCH$_2$CH=⬡

A stirred mixture of methylenecyclohexane, Zn-Cu-couple, *NaHCO$_3$*, and *NaI* in dimethyl sulfoxide heated 1 hr. at 80° under N$_2$ with excess phenacyl bromide added in 2 portions → product. Y: 54%. F. e. s. E. Ghera, D. H. Perry, and S. Shoua, Chem. Commun. *1973*, 858.

Cyclopropane ring or methylene compds.   ←
from ethylene derivs. and 1,1-dihalides
     Silyloxycyclopropane ring or α-siloxymethylene compds. from enoxysilanes s. *29*, 827

Reductive dimerization   CO·C·C·CO
     γ-Diketones from 2 α,α'-dibromoketone molecules s. *29*, 828

Zinc/copper-iodine   Zn,Cu-I

Simmons-Smith cyclopropane ring synthesis   ▽
s. *26*, 876

Zinc/silver   Zn/Ag

Cyclopropanes   C : C → ▽
     from 1,1-dihalides and ethylene derivs. s. *14*, 858 suppl. *28*

Cyclopropanol ring formation
     and α-methylation of oxo compds. via cyclopropanation of enoxysilanes s. *29*, 829

Zinc/cuprous chloride   Zn/CuCl

Modified Simmons-Smith cyclopropane ring synthesis
s. *26*, 876

Aziridines from azomethines   N : C → △N
s. *28*, 818

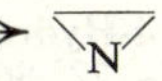

Zinc/cupric chloride $Zn/CuCl_2$
Sym. succinic acid esters   2 CHalCOOR → ROOC·C·C·COOR
    from 2 α-halogenocarboxylic acid ester molecules
    s. 27, 860

Zinc/triphenylphosphine Zn/Ph_3P
Terminal acetylene derivs.   CHO → CH : CBr_2
    from aldehydes via dibromomethylene compds.
    s. 28, 819; cf. J. Villieras, P. Perriot, and J. F. Normant, Synthesis *1975*, 458; thiophene derivs. s. T. B. Patrick and J. L. Honegger, J. Org. Chem. *39*, 3791 (1974)

Ethylmagnesium bromide   *EtMgBr*
α,β-Acetyleneoximes   C(:NOH)Cl → C(:NOH)C≡CR
    from hydroximinochlorides s. 26, 877

α,β-Acetyleneamidines   N : C(N<)Cl → N : C(N<)·C≡C
    from chloroformamidines s. 28, 672

Synthesis with ring opening of isocyclics   C

601.

Polyenynes. A soln. of cyclooctatetraene dibromide in abs. ether added with stirring to p-bromophenylethynylmagnesium prepared from p-bromophenylacetylene and ethylmagnesium bromide in abs. ether, and refluxed 2 hrs. → (3E,5Z,7Z,9E)-1,12-di-p-bromophenyl-3,5,7,9-dodecatetraene-1,11-diyne. Y: 57%. F. e. s. H. Straub, J. M. Rao, and E. Müller, A. *1973*, 1339.

Ethylmagnesium bromide/cuprous cyanide   *EtMgBr/CuCN*
Synthesis of 1,4-enynes   C≡C·C·C : C
    from acetylene derivs. s. 29, 830

Ethylmagnesium bromide/magnesium   *EtMgBr/Mg*
Tetrahydrofuran ring   O
    from ketones and 1,3-chlorohydrins – Spiro-O-heterocyclics s. 28, 609

Isopropylmagnesium chloride   *i-PrMgCl*
2,3-Dihydropyrrole ring
    from enamines and dihalides s. 27, 861

Isopropylmagnesium bromide   *i-PrMgBr*
C-Alkylation of cyclic azomethines   CH → CR
    Thermodynamic and kinetic control s. 29, 831

Zinc carbonate/sodium hydrogen carbonate $ZnCO_3/NaHCO_3$
Coumarin ring   O
    from β-bromo-α,β-ethylenecarboxylic acid esters s. 26, 878

Borane s. under Mg BH_3

Sodium tetrahydridoborate s. a. under $AgBF_4$ $Na[BH_4]$

Sodium tetrahydridoborate/sodium hyroxide/⧖ $Na[BH_4]/NaOH/⧖$
4-Hydroxyspiro(cyclohexa-2,5-dien-1,1′-isoindolin)-3′-ones   O
s. 27, 862   O

Lithium tetrahydridoaluminate — $Li[AlH_4]$
3-Substitution of pyridine H → R
s. 26, 879 H → R

Thexylborane-sodium/alcohol-silver ammonium nitrate ←
Synthesis of *trans*-ethylene derivs. $R·C≡C·Hal → CH(R):CH(R')$
from acetylenehalides s. 28, 820

3,6-Dimethylborepane/methyllithium ←
Sec. alcohols $C≡CH → CH_2CH(OH)R$
from acetylene derivs. and halides s. 28, 821

Lithium hydridotri-sec-butylborate — $Li[BHsec-Bu_3]$
Carboxylic
from α,β-**ethylenecarboxylic acid esters** $C:C·COOR → CH·CR·COOR$
by reduction and reductive alkylation s. 30, 602

Potassium hydridotri-sec-butylborate — $K[BHsec-Bu_3]$
Preferential reductive alkylation $C:C·CO → CH·CR·CO$
of α,β-ethyleneketones

602.

1 equivalent K-tri-*sec*-butylborohydride added at –78° under N_2 to a dry soln. of carvone in tetrahydrofuran, stirred 1 hr. at –78°, methyl iodide injected, and brought to 0° for 10 min. → 1-methyl-1,6-dihydrocarvone. Y: 98%. Also reductions without alkylation s. B. Ganem, J. Org. Chem. **40**, 146 (1975); **carboxylic from α,β-ethylenecarboxylic acid esters** by reduction and reductive alkylation with Li-tri-*sec*-butylborohydride cf. ibid. **40**, 2846.

Diisobutylaluminum hydride/zinc, copper — $i-Bu_2AlH/Zn,Cu$
Cyclopropanes from acetylene derivs. ▽
trans-2-Alkylcyclopropylhalides s. 26, 880

Triethylborane/air — Et_3B/O_2
Coupling of halides $RHal + HalR' → R·R'$
Mixed coupling – Prepn. of iodides from boranes s. 26, 881

Tert. alanes — R_3Al
Tert. azo compds. $C:N·N:C → CR·N:N·CR$
from ketazines via 1,1'-dichloroazo comdps. s. 29, 832

Aluminum isopropoxide s. under Mg — $Al(OR)_3$

Thallous sulfate — Tl_2SO_4
C-Alkylation of cyclopentadienes H → R
s. 26, 718

Boron fluoride s. under H_3PO_4 — BF_3

Copper fluoroborate — $Cu[BF_4]$
Diene synthesis of cyclic ketones ○
from α,β-ethylene-α-chloronitriles with preceding C-alkylation of cyclopentadienes s. 26, 718

Silver fluoroborate — $AgBF_4$
Immoniocyclobutane ring □
from ethylene derivs. and 1-chlorenamines s. 28, 822

30 CC ⇊ Hal

Silver fluoroborate/triethylamine $AgBF_4/Et_3N$
Reaction of ethylene derivs.   CH·C:C → C(OR)CH·C·Ac
    with carboxylic acid chlorides – γ-Alkoxyketones from ethylene derivs. s. *28*, 823

Silver fluoroborate/sodium tetrahydridoborate $AgBF_4/NaBH_4$
6-Oxabicyclo[3.2.1]octanes   O
s. *28*, 823

Aluminum chloride $AlCl_3$
Friedel-Crafts alkylation   ArH → ArR
    with halogenonitriles, also with 5-bromoadamantan-2-one s. *28*, 824; with retention of configuration s. B. S. Masuda, T. Nakajima, and S. Suga, Chem. Commun. *1974*, 954

Polychlorides   H → CCl$_3$
s. *27*, 863

2-Ethyleneamines from 1,1-halogenamines   C:CH → C:C·C·N<
s. *29*, 833

Diaryl ketones   ArCOCl → ArCOAr'
s. *28*, 600

o-Hydroxyacetophenones from phenols   COCl → COR
s. *30*, 67

p-Phenolketones
    from phenols via Fries rearrangement s. *29*, 834

Thiolic acid esters   ArH → Ar·COSR
    from arenes and thiolchloroformic acid esters s. *29*, 835

2,4-Diamino-6-halogenopyridines   O
s. *26*, 882

Aluminum chloride/tert-butyl bromide $AlCl_3/Me_3C·Br$
Polyaryladamantanes   ←
s. *29*, 836

Aluminum chloride/hydrogen chloride $AlCl_3/HCl$
1-Acylation of cycloheptatrienes   COCl → COR
s. *26*, 883

Dimethylformamide Me_2NCHO
Indoles   O
    from o-aminooxo compds. s. *29*, 837

Isonitriles/cuprous oxide $RN:C/Cu_2O$
Cyclopropanes   C:C → ▽
    from ethylene derivs. and halides – Stereospecific ring closure s. *27*, 864

Acetic acid CH_3COOH
2-Benzoyl-1,2,3,4-tetrahydroisoquinolines   O
    from 1,1-alkoxyhalides s. *26*, 884

Dimethyl sulfoxide s. under Et_3N Me_2SO

tert-Butyl bromide s. under $AlCl_3$ $Me_3C·Br$

Zirconocene chloride $(\eta^5\text{-}C_5H_5)_2Zr(Cl)H$
Reactions via hydrozirconation   COCl → COR
**Ketones from ethylene derivs. and carboxylic acid chlorides
with and without rearrangement**

603.

Either 1-octene, or *cis*-, or *trans*-4-octene shaken a few hrs. at room temp. with zirconocene chloride in benzene, then allowed to react with acetyl chloride → 2-decanone. Y: 80%. − In contrast to their air-sensitive and pyrophoric B- or Al-analogs, the alkylzirconium compds. are fairly stable in dry air. They undergo a wide range of transformations under mild conditions and in high yield. The method is particularly attractive when facile migration of the functionalization unit is desired. F. e. s. D. W. Hart and J. Schwartz, Am. Soc. 96, 8115 (1974); reactions with CO (cf. Synth. Meth. 23, 702; 24, 704) s. ibid. 97, 228 (1975); α,β-ethylenehalides by regio- and geo-specific addition to acetylene derivs. cf. ibid. 97, 679 (1975); alcohols from ethylene derivs. (cf. Synth. Meth. 27, 142), oxidation preferably with O_2, s. Tetrah. Let. 1975, 3041.

Stannic chloride $SnCl_4$
Benzyl ketones from carboxylic acid chlorides   COCl → $COCH_2Ph$
s. 27, 865

Direct Bradsher ring closure   O
    Benzene ring − Pyrylium and thiopyrylium ring closure − Se-Heterocyclics
    s. 29, 838

Benzene ring annelation
    with 1,1,1-alkoxydichlorides and allyl derivs. s. 28, 825

7-Halogeno-2,4,9-trioxaadamantanes
    from 3 carboxylic acid chloride molecules and ethylene derivs. s. 29, 839

5,6-Dihydro-4H-1,3-oxazines from ethylene derivs.
    Stereospecific 5,6-dihydro-4H-1,3-oxazine ring synthesis s. 26, 885

Triphenylphosphine s. a. under Zn and $Na_2[Fe(CO)_4]$ Ph_3P

Triphenylphosphine/potassium fluoride Ph_3P/KF
Cyclopropanes   ▽
from ethylene derivs. and 1,1-dihalides

604.

1,1-Dihalogenocyclopropanes. CF_2Br_2 added to a soln. of triphenylphosphine in dry triglyme, stirred 0.5 hr. under dry N_2 followed by addition of trimethylethylene and anhydrous KF, then stirred vigorously for 24 hrs. → 1,1-difluorotrimethylcyclopropane. Y: 67%. F. e., also with *CsF* instead of KF, s. D. J. Burton and D. G. Naae, Am. Soc. 95, 8467 (1973).

Tris(dimethylamino)phosphine $(Me_2N)_3P$
Dichloromethylene compds. from aldehydes   CHO → $CH:CCl_2$
s. 27, 866

α-Halogenoglycidic and β-halogeno-α-ketocarboxylic acid esters   ←
    from oxo compds. s. 26, 886

Hexamethylphosphoramide s. under *i-Pr₂NLi* $(Me_2N)_3PO$

Polyphosphoric acid $H(PO_3H)_xOH$

Condensation with elimination of hydrogen chloride $RCl + HR' \rightarrow R \cdot R'$
Diphenylmethanes s. *8*, 864 suppl. 29

Indenes via indanones and indanols ○
s. *29*, 840

Phosphoric acid · boron fluoride $H_3PO_4 \cdot BF_3$

1,1-Diarylalkanes from 1,1-alkoxyhalides $CHal(OR) \rightarrow CAr_2$
s. *26*, 887

Antimony pentachloride $SbCl_5$

Carboxylic acid esters from halides $Hal \rightarrow COOR$
at low temp. – Synthesis with addition of 1 C-atom – Reactions in liq. SO₂, review, s. *28*, 826

Air s. under *Et₃B* O_2

Trifluoromethanesulfonic acid CF_3SO_3H

Ketones from carboxylic acid chlorides $COCl \rightarrow COR$
s. *28*, 827

Sulfuric acid H_2SO_4

Tscherniac-Einhorn-type reaction ←
C-Imidomethylation via carbonium ions – β-Imido-α-chlorocarboxylic acids – α-Branched β-imidocarboxylic acids from 1,2-ethylene-1,1-dichlorides s. *26*, 888; s. a. B. *106*, 2513 (1973).

Molybdenum hexacarbonyl $Mo(CO)_6$

C-Alkylation of aryl derivs. $ArH \rightarrow ArR$
s. *27*, 867

605. ⌬ + ClC(CH₃)₃ → ⌬-⁺

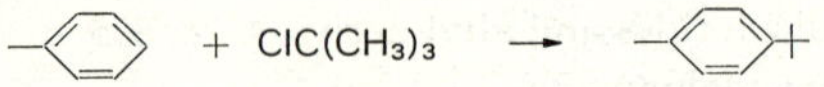

of arenes. A soln. of toluene, *tert*-butyl chloride, and molybdenum hexacarbonyl refluxed 8 hrs. until HCl-gas evolution ceases → *p-tert*-butyltoluene. Y: 72.5–88%. F. e. and molybdenum carbonyl-catalyzed reactions s. J. F. White and M. F. Farona, J. Organometal. Chem. *63*, 329 (1973).

Iodine (s. a. under *Zn,Cu*) I

Diaryl ketones $ArCOCl \rightarrow ArCOAr'$
s. *28*, 600

Perchloric acid/acetic acid $HClO_4/CH_3COOH$

Pyrylium salts ○
from α,β-ethylene-β-halogenaldehydes and ketones s. *29*, 841; review of β-chlorovinylaldehydes cf. H. Teufel, Chem. Ztg. *96*, 606 (1974)

Hydrohalides ←

Reactions with 1,3,3-trichloro-2-azapropenes ←
Quinazolines s. *29*, 842

Chlorides Cl^-

5-Aminopyrazoles ○
from hydrazones and N-dichloromethyleneammonium salts s. *29*, 843

Tetrabutylammonium bromide s. under *NaOH* and *NaCN* $Bu_4N^+ Br^-$

Tetrabutylammonium iodide s. under *NaOH* and *Ni(CO)₄* $Bu_4N^+ I^-$

Benzyltriethylammonium chloride s. under *NaOH* $(PhCH_2)Et_3N^+ Cl^-$

Iron carbonyl/lithium $\qquad Fe(CO)_5/Li$
Synthesis of unsym. ketones from halides $\qquad$ RHal + HalR' → RCOR'
s. *26*, 889

Diiron nonacarbonyl $\qquad Fe_2(CO)_9$
Cyclopent-2-enone ring $\qquad$ O
from α,α'-dibromoketones and enamines s. *27*, 868

4-Cycloheptenones
from α,α'-dibromoketones and 1,3-dienes – Prepn. of troponoids s. *26*, 890; mostly 8-oxabicyclo[3.2.1]oct-6-en-3-ones s. J. Org. Chem. *40*, 806 (1975); troponoids s. a. Chem. Lett. *1975*, 509

3(2H)-Furanones
from α,α'-dibromoketones and carboxylic acid amides s. *29*, 844

Sodium tetracarbonylferrate(-II) $\qquad Na_2Fe(CO)_4$
Syntheses with sodium tetracarbonylferrate(-II) $\qquad$ RHal + HalR' → RCOR'
Ketones from 2 halide molecules s. *27*, 869

Sodium tetracarbonylferrate(-II)/triphenylphosphine $\qquad Na_2[Fe(CO)_4]/Ph_3P$
Aldehydes from halides $\qquad$ Hal → CHO
Synthesis with addition of 1 C-atom s. *26*, 891

Ferric chloride s. a. under Mg $\qquad FeCl_3$

Ferric chloride/hydrogen chloride $\qquad FeCl_3/HCl$
Pyrylium ring $\qquad$ O
from α,β-ethylene-β-halogenoketones s. *28*, 828

Bis-(1,5-cyclooctadiene)nickel(0) $\qquad$ ←
Sym. diaryls from aryl halides $\qquad$ 2 ArHal → Ar·Ar
s. *27*, 870

Nickel carbonyl $\qquad Ni(CO)_4$
Indol-3-ones from azomethines $\qquad$ O
s. *28*, 829

Nickel carbonyl/tetrabutylammonium iodide $\qquad Ni(CO)_4/Bu_4N^+I^-$
γ-Enollactones from carboxylic acid chlorides $\qquad$ O

606.
$$HC \equiv CH + PhCOCl \longrightarrow \underset{Ph}{}\!\!\!\!\diagdown\!\!O\!\!\diagup\!\!\!\!\underset{}{=O}$$

A 1:1 mixture of CO and acetylene passed during 1.5 hrs. into a well stirred mixture of Ni-carbonyl and tetrabutylammonium iodide in acetone with simultaneous dropwise addition of solns. of benzoyl chloride and more Ni-carbonyl in acetone, and stirring continued 0.5 hr. → product. Y: 75%. – Halide ions have a strong activating effect on this reaction. F. e. s. M. Foà and L. Cassar, G. *103*, 805 (1973).

Dichlorobis(triphenylphosphine)nickel(II) s. under Mg $\qquad NiCl_2(Ph_3P)_2$

Palladous cyanide $\qquad Pd(CN)_2$
Arylcyanides from arylhalides $\qquad$ ArHal → ArCN
s. *29*, 845

Palladium triarylphosphine complexes $\qquad$ ←
Acetylene derivs. from halides $\qquad$ C≡CH → C≡CR
with Cu/K$_2$CO$_3$ cf. *27*, 851; with triphenylphosphinepalladium complexes in the presence of a base s. L. Cassar, J. Organometal. Chem. *93*, 253 (1975); H. A. Dieck and F. R. Heck, ibid. *93*, 259

Palladous chloride/potassium acetate $\qquad PdCl_2/CH_3COOK$
Substitution of ethylene derivs. $\qquad C:CH \rightarrow C:CR$
  Arylation s. 27, 871
Via intermediates $\qquad$ v.i.
Synthesis of hydrocarbons $\qquad CO \rightarrow CHR$
  from oxo compds. and halides via alcohols s. 27, 872
Synthesis of ethylene derivs. from halides $\qquad Hal \rightarrow C:C$

607.

A soln. of 9-bromothioxanthene and 1,1-bis(p-ethoxyphenyl)ethylene in dry ether refluxed 2 hrs., allowed to stand overnight, the solvent removed, and the residue digested 15 min. with ethanolic KOH → product. Y: ca. 75%. F. e. s. W. Tadros et al., Helv. 56, 1829 (1973).

α,β-Ethylenealdehydes $\qquad COCl \rightarrow CH:CHCHO$
  from carboxylic acid chlorides via silylethynyl ketones, β-keto- and β-hydroxy-acetals – Synthesis with addition of 2 C-atoms – α,β-Ethyleneketones from enesilanes s. 29, 846
Tropone ring via o-dialdehydes $\qquad$ ○
s. 29, 847
2,4-Diaminothiophenes
  from isothiocyanates, nitriles, and halides s. 26, 892

Sulfur ↑ $\qquad$ CC ⇊ S

Without additional reagents $\qquad$ w.a.r.
Enetriazenes and enamines $\qquad$ ←
  from azides and 2 sulfonium ylid molecules s. 26, 893
Reactions of sulfonium salts with sulfenamides $\qquad$ ←
s. 28, 830
β-Amino-α,β-ethylenesulfones $\qquad C(N<):C \cdot SO_2 \cdot$
  Synthesis with rearrangement s. 28, 831
Cycloheptatrienes $\qquad$ ⊙
  from thiophene 1,1-dioxides and cyclopropenes s. 29, 848; s. a. R. 94, 85 (1975)
2-Pyrones
  from cyclopropenones and carbonyl-stabilized sulfonium ylids – Reactions of cyclopropenones with pyridinium ylids s. 28, 832
Spiroannelation $\qquad$ ←
  with cyclic sulfonium ylids and diketene s. 28, 833
2-Pyridones $\qquad$ ←
  from 4-hydroxythiazolium betaines s. 26, 919
Irradiation (s. a. under Lithium compds.) $\qquad$ ⋎
γ-Diketones from 2 β-ketosulfoxide molecules $\qquad CO \cdot C \cdot C \cdot CO$
s. 26, 894

Electrolysis
1,5-Diketones from 2 β-ketosulfoxide molecules   CO·C·C·C·CO
s. 28, 834

Sodium hydride   NaH
Tetraketonenehydrazones   ←
  from disulfonyldiazomethanes and β-diketones s. 28, 835
Thiocarbamylation with C-sulfonylthioformamides   H → C(S)N<
  C-Thiocarbamylation – Cyanothioformamides s. 28, 836
Ring closures with thetin anions   O
  Glycidic acid esters from ketones via glycidic acids – Stable S-ylid s. 26, 895
Reactions with S-aminosulfoxonium salts
  via their ylids – Ring closures s. 26, 896
Spiroannelation with S-cyclopropylids   ▽
  Spirocyclopentane ring s. 27, 873

Sodium hydride/dimethyl sulfoxide   NaH/Me₂SO
Cyclopropanes
  from halides and 2 sulfoxonium methylid molecules s. 27, 874

Sodium hydroxide   NaOH
Spirocyclobutanones   □
  from α-halogenoketones and sulfonium salts s. 26, 897
Synthesis of oxido from oxo compds.   CO → ▽O
  s. 17, 889 suppl. 28

Potassium hydroxide   KOH
Cyclobutanone ring from ketones
  via 1-oxaspiro[2.2]pentane ring – Annelation complementary   CO → ⌐⌐O
  to the normal Robinson method – Lactone annelation – Replacement of
  carbonyl oxygen by 2 functionalized C-substituents s. 28, 837; s. a. Am. Soc.
  97, 2218, 2224 (1975)

Lithium compd./irradiation   ←
Syntheses with 1,1-nitrosulfones   C(NO₂)SO₂R → C(NO₂)R
Aliphatic 1,2-dinitro compds.

608.
$$\text{Me}_2\text{C}-\text{SO}_2\text{Ph} + \text{HCMe}_2 \longrightarrow \text{Me}_2\text{C}-\text{CMe}_2$$
$$\quad\;\;|\qquad\qquad\;\;\;|\qquad\qquad\qquad\;\;|\quad\;\;|$$
$$\;\;\text{NO}_2\qquad\quad\;\text{NO}_2\qquad\qquad\quad\text{NO}_2\;\text{NO}_2$$

Dimethyl sulfoxide added under N₂ to a mixture of lithio-2-nitropropane and α-nitroisopropyl phenyl sulfone, then stirred 1 hr. at room temp. with irradiation by fluorescent lights → 2,3-dimethyl-2,3-dinitrobutane. Y: 83%. F. e. s. N. Kornblum, S. D. Boyd, and N. Ono, Am. Soc. 96, 2580 (1974).

Sodium/alcohol   NaOR
Ring closure of dialdehydes with sulfur compds.   O

609.

A soln. of bis-(p-nitrobenzyl) sulfide in dimethylformamide added dropwise at 5° to a soln. of 1-benzyl-2-isopropylimidazole-4,5-dicarboxaldehyde in methanol containing dissolved Na, and the product isolated after 6 hrs. → 1-benzyl-2-

isopropyl-5,6-di-(p-nitrophenyl)benzimidazole. Y: 80%. F. e., **also with sulfoxides and sulfones,** and with retention of the sulfur group, s. A. Corvers, A. de Groot, and E. F. Godefroi, R. *92,* 1368 (1973).

2,3-Dihydrofuran ring
   from enesulfonium salts and ketones − Furans from 2-acetylenesulfonium salts s. *28,* 838

Potassium/tert-butanol $K/Me_3C \cdot OH$
Benzene ring from thiopyrylium salts ←
   s. *27,* 875

Potassium tert-butoxide $KOCMe_3$
Cyclopropanes and oxido compds. ▽ ◯
   from sulfonium salts − Synthesis of oxido from oxo compds. s. *17,* 889 suppl. *28*

Ethylenecarboxylic acids C
   from cyclic ketotosylates − Fragmentation-type stereospecific ring opening s. *28,* 839

Potassium tert-butoxide/dimethyl sulfoxide $KOCMe_3/Me_2SO$
Cyclopropanes ▽
   from ethylene derivatives and sulfonium salts − Sulfonylcyclopropanes s. *26,* 898

n-Butyllithium *n-BuLi*
Synthesis of hydrocarbons Hal → R
   from sulfuric acid esters − Polyalkyl compds. from polyhalides − Polymethylthiophenes s. *29,* 849

Biphenyl-2-sulfinic acids ←
   from 2 phenyl sulfone molecules s. *29,* 850

Cyclopropanes ▽
   from ethylene derivs. and sulfonium salts s. *26,* 896 suppl. *28*

Ring closure with sulfonium salts ◯
   Furan ring from α-*n*-butylthiomethyleneketones s. *28,* 840

Lithium/liq. ammonia Li/NH_3
Reactions with *tert*-butyl thioethers $S \cdot CMe_3$ → R
   Synthesis of hydrocarbons s. *29,* 851

Reductive alkylation $C:CH(SR)$ → $C(R')CH_3$
   of α-alkylthiomethyleneketones s. *26,* 899

Potassium carbonate K_2CO_3
Synthesis of heterocyclics ◯
   with 1-sulfonyl-1-isonitriles − Oxazoles from oxo compds. s. *28,* 665

Potassium cyanide *KCN*
Cyanothioformamides $NC \cdot C(S)N<$
   s. *28,* 836

Sodium salt Na^+
Dichloromethylene compds. from sulfines $C:S:O$ → $C:CCl_2$
   s. *29,* 852

Ethyldiisopropylamine *i-Pr₂NEt*
Benzene ring from thiopyrylium salts ←
   s. *27,* 875

Triethylenediamine ←
2-Enethiazolidines ←
   from 2-alkylthio-Δ^2-thiazolines s. *28,* 844

2,6-Lutidine s. under $CuCl_2$ ←

Lithium dialkylcuprate $Li[CuR_2]$
Synthesis of ethylene derivs. from thioenolethers $C:C·SR \rightarrow C:C·R'$
s. 29, 854

Ketones from thiolic acid esters $COS· \rightarrow COR$
s. 29, 853

Branched ketones from α-alkylthiomethyleneketones – ←
Incorporation of 3 alkyl groups s. 29, 819

Cuprous bromide $CuBr$
Ethylene derivs. $CHS^+ < +N_2:C \rightarrow C:C$
from sulfonium salts and diazo compds. s. 27, 876

Cuprous iodide s. under Mg CuI

Cupric chloride $CuCl_2$
Syntheses with sulfur compds. ←
Sulfur compds. in synthetic organic chemistry, review s. 27, 877; chemistry of organic sulfur compds., review, s. P. Neumann and F. Vögtle, Chem. Ztg. 99, 308 (1975)

Cupric chloride/2,6-lutidine ←
s. 27, 877

Calcium carbonate $CaCO_3$
trans-γ,δ-**Ethylenealdehydes from halides** $Hal \rightarrow C:C·C·C·CHO$
via α-alkylation of thioethers and thio-Claisen rearrangement s. 28, 842

Magnesium Mg
Oxo compds. $Hal \rightarrow CO$
from halides and 1-alkylthioazomethinium salts s. 28, 843; f. reactions of 1-alkylthioazomethinium salts cf. R. L. N. Harris, Australian J. Chem. 27, 2635 (1974)

Ketones from thiolic acid esters $COS· \rightarrow COR$
from pyrid-2-yl thiolates s. 29, 853

Magnesium/cuprous iodide Mg/CuI
Synthesis of ethylene derivs. $C:C·SR \rightarrow C:C·R'$
from thioenolethers – Retention of geometry s. 29, 854; s. a. Chem. Lett. 1975, 535

Mercuric oxide HgO
Thio-Claisen rearrangement ←
Replacement of carbonyl oxygen by two functionalized carbon substituents – Synthesis of γ,δ-ethylenealdehydes from ketones s. 26, 900

Zinc chloride $ZnCl_2$
2-Enethiazolidines ←
from 2-alkylthio-Δ^2-thiazolines s. 28, 844

Mercuric chloride s. under $Na[PdCl_4]$ $HgCl_2$

Triphenyl phosphite $(PhO)_3P$
Tetrathiofulvalenes from 1,3-dithiole-2-thiones $2C:S \rightarrow C:C$

A soln. of 4,5-dicyano-1,3-dithiole-2-thione and excess triphenyl phosphite in toluene refluxed 18 hrs. under $N_2 \rightarrow$ tetracyanotetrathiofulvalene. Y: > 70%. Z. Yoshida, T. Kawase, and S. Yoneda, Tetrah. Let. 1975, 331.

Hydrogen peroxide s. under FeSO₄ $\qquad H_2O_2$

Sodium dithionite $\qquad Na_2S_2O_4$

Replacement of sulfonic acid by aminomethyl groups $\qquad SO_3^- \rightarrow CH_2N{<}$
2-Aminomethyl- and 2-hydroxymethyl-anthraquinones s. 27, 878; s. a. A. 1975, 972

Ferrous sulfate/hydrogen peroxide $\qquad FeSO_4/H_2O_2$
Radical C-alkylation with sulfoxides $\qquad H \rightarrow R$
s. 27, 879

Sodium tetrachloropalladate(II)/mercuric chloride $\qquad Na[PdCl_4]/HgCl_2$
Reactions of sulfinic acids with elimination of sulfur dioxide $\qquad \leftarrow$
Diaryls – Styrenes – Sulfinic acids and derivs., review s. 27, 880

Via intermediates $\qquad v.i.$
Ethylene derivs. $\qquad CO \rightarrow C:C$
from oxo compds. and sulfoximines via β-hydroxysulfoximines – Syntheses with sulfoximines, review, s. 29, 855; review of sulfoximines s. P. D. Kennewell and J. B. Taylor, Chem. Soc. Rev. 4, 189 (1975)

Remaining Elements ↑   CC ↓↑ Rem

Without additional reagents $\qquad w.a.r.$
Triarylphosphinerhodium complexes as reactants $\qquad Hal \rightarrow CH_3$
Replacement of halogen by methyl s. 29, 857

Branched hydrocarbons $\qquad \leftarrow$
from tert. alcohols, ketones, or carboxylic acids s. 28, 845

Synthesis of hydrocarbons from halides $\qquad \leftarrow$
with ring contraction s. 27, 881

Wittig synthesis with phosphonium fluorides $\qquad CO \rightarrow C:C$
s. 26, 901

1,3-Dienes $\qquad C:C\cdot C:C$
from nickel(I) π-allyl complexes and diazo compds. s. 26, 915

1,4-Dienes $\qquad C:C\cdot C\cdot C:C$
from enolethers and 2-ethyleneboranes s. 27, 882

Synthesis of 1,1-nitroethylene derivatives $\qquad C(NO_2):C\cdot NR_2 \rightarrow C(NO_2):CR'$
from 2-nitroenamines s. 26, 902

Stereospecific synthesis of tert. alcohols $\qquad CO \rightarrow C(OH)R$
from ketones s. 27, 671

Syntheses with lithium acyl carbonyl metalates $\qquad Hal \rightarrow COR$
Ketones from halides s. 27, 885

Syntheses with organotin(IV) compds. $\qquad C:C(OSnR_3) \rightarrow CR\cdot CO$
Alkylation of organotin(IV) enolates s. 29, 856

Reactions of organotitanium compds. $\qquad \leftarrow$
Carbonylation – Metallocyclics s. 29, 858

Reactions with metal-carbene complexes $\qquad C(OR):W(CO)_5 \rightarrow CO\cdot R'$
Ketones from alkylidenephosphoranes via enolethers – Enolethers from diazo compds. – Transition metal-carbene complexes, review, s. 28, 846

2-Ene-1,4-diones $\qquad CO \rightarrow C:C\cdot CO\cdot R$
from α-diketones and (α-acylalkylidene)phosphoranes s. 28, 847

Enamines from formamides $\qquad N\cdot CHO \rightarrow N\cdot C:C$
s. 29, 859

Ketenimines $\quad$ N:C:O → N:C:C
from alkylidenephosphoranes and isocyanates

611.
$$\text{EtN=C=O} + \text{Ph}_3\text{P=C}\underset{\text{COOEt}}{\overset{\text{CH}_3}{\diagup}} \longrightarrow \text{EtN=C=C}\underset{\text{COOEt}}{\overset{\text{CH}_3}{\diagup}}$$

Ethyl isocyanate added slowly with vigorous stirring under N_2 to the startg. alkylidenephosphorane and dry benzene, then refluxed several hrs. in a slow current of dry N_2 → product. Y: 90%. F. e. s. P. Frøyen, Acta Chem. Scand. *B 28*, 586 (1974); f. method, optically active compds., cf. K. Lee and L. A. Singer, J. Org. Chem. *39*, 3780 (1974).

Synthesis of enesilanes $\quad$ Si·C:C
from 1,1-silyllithium compds. *29*, 860

Reactive 2-silyl groups in benzothiazoles $\quad$ ←
Replacement of silyl groups – Insertion of C-atoms into carbon-silicon bonds s. *27*, 883

β-Hydroxycarboxylic acids and esters from aldehydes CHO → CH(OH)·C·COOR
and ketene silyl hemiacetals or O-silyl O-alkyl keteneacetals respectively s. *27*, 884

Wittig synthesis with lactones $\quad$ ←
with simultaneous migration of the carbon-carbon double bond – with benzofuranones s. *29*, 861

2,5-Disubst. pyrroles from succinimides $\quad$ ←
s. *29*, 862

α,β-Ethylenecarboxylic acid thioamides $\quad$ CO → C:C·CSNHR
from alkylidenephosphoranes, isothiocyanates, and oxo compds. – Wittig synthesis s. *26*, 903

Enollactones from carboxylic acid anhydrides $\quad$ ←
Wittig synthesis – Maleimide-carbohydrate derivs. s. *26*, 238

Cyclopropanes $\quad$ ▽
from ethylene derivs. and selenonium ylids – Stereospecific ring closure s. *29*, 863

Cyclopropylcarbinyl compds. $\quad$ SnC·C·C:C → ▽—C·Hal
from 3-ethylenestannanes s. *28*, 848

Heterocyclics from o-subst. ω-sulfinylacetophenones $\quad$ ○
Coumarins s. *27*, 886

Pyrimidine ring from 2 nitrile molecules
s. *26*, 905

α-Hydroxyazomethines from aziridinones $\quad$ C

612.
$$\text{Me}_3\text{C·Li} + \underset{\text{Me}_3\text{C}}{\overset{\text{O}}{\triangle}}\text{NR} \longrightarrow \underset{\text{Me}_3\text{C}}{\overset{\text{Me}_3\text{C}}{\diagdown}}\text{C}\underset{\text{CH=NR}}{\overset{\text{OH}}{\diagup}}$$
R = 1-adamantyl

Startg. m. treated 4 hrs. at 25–30° under N_2 with 1.1 equivalents *tert*-butyllithium in pentane → product. Y: 77%. F. e. s. E. R. Talaty and C. M. Utermoehlen, Chem. Commun. *1974*, 204.

Ring expansion $\quad$ ☉
of cyclobutadiene triarylphosphine rhodium complexes
Anthraquinones s. *26*, 906

**Reactions of 1,3,5-oxaphosph(V)ol-2-ines
with multiple heterobonds**
   3-Iminoazetines by [3+1]-cycloaddition s. *27,* 887; s. a. Z. Naturf. *29b,* 399 (1974); [3+1]-cycloaddition s. a. Synthesis *1975,* 250

2H-Pyrrolenines
   from 1,3,5-oxazaphosph(V)ol-2-ines s. *28,* 849

Irradiation (s. a. under Br)

Unsym. biphenyls $\quad\quad\quad\quad\quad\quad\quad\quad\quad\quad\quad\quad$ ArTl(OCOCF$_3$)$_2$ → Ar·Ar'
   from arylthallium bis(trifluoroacetates) − Ar. thiocyanates s. *27,* 888

Tert. alcohols from ketones and boranes $\quad$ CO·C·COOR → CR'(OH)·C·COOR
   β-Hydroxycarbonyl compds. s. *27,* 889

Lithium $\quad\quad\quad\quad\quad\quad\quad\quad\quad\quad\quad\quad\quad\quad\quad\quad\quad\quad\quad\quad\quad\quad\quad\quad$ Li

β,γ-Ethyleneketones $\quad\quad\quad\quad\quad\quad\quad\quad\quad\quad\quad\quad\quad\quad\quad$ C:C·CHCO
   from (α-acylalkylidene)phosphoranes s. *28,* 850

Potassium/hexamethylphosphoramide $\quad\quad\quad\quad\quad\quad\quad\quad$ K/(Me$_2$N)$_3$PO

Wittig synthesis $\quad\quad\quad\quad\quad\quad\quad\quad\quad\quad\quad\quad\quad\quad\quad$ CO → C:C
s. *17,* 895 suppl. 29

Sodium hydride $\quad\quad\quad\quad\quad\quad\quad\quad\quad\quad\quad\quad\quad\quad\quad\quad\quad$ NaH

α,β-Ethylene-α-mercaptocarboxylic acid esters $\quad\quad$ CHO → CH:(SH)COOR
   from aldehydes and silylthioacetic acid esters s. *29,* 730

Enecyclopropanes
s. *29,* 864

**Cyclic ethers
from lactols and phosphonic acid esters
Selective hydrolysis of carboxylic acid esters**

613.

Trimethyl phosphonoacetate (cf. Synth. Meth. *23,* 879) added during 2.25 hrs. to a stirred 56%-NaH-mineral oil dispersion in tetrahydrofuran, after 15 min. stirring a soln. of 6-acetoxy-2-hydroxy-2,5,7,8-tetramethylchroman in tetrahydrofuran during 0.5 hr., stirred 18 hrs. at 25°, refluxed 4 hrs., and the resulting crude methyl 6-acetoxy-2,5,7,8-tetramethylchroman-2-acetate stirred 4 hrs. at 25° under N$_2$ in aq.-alc. NaOH → 6-hydroxy-2,5,7,8-tetramethylchroman-2-acetic acid. Y: 80%. J. W. Scott et al., J. Am. Oil Chemists' Soc. *51,* 200 (1974).

Syntheses with enaminophosphine oxides $\quad\quad\quad\quad$ CO → C:C·CO
   α,β-Ethyleneketones − Cyclopropylketones s. *27,* 890

β-Diketones from β-ketocarboxylic acid esters $\quad\quad$ CHOOR → CHCOR'

614.

A soln. of methyl 2-oxocyclohexanecarboxylate in dry tetrahydrofuran treated at 0° with one equivalent NaH, the resulting monoanion treated at 0° with

2 equivalents methyllithium in hexane, and the product isolated after 0.5 hr.
→ 2-acetylcyclohexanone. Y: 83%. F. e. s. S. N. Huckin and L. Weiler, Can. J.
Chem. *52*, 1379 (1974).

Horner synthesis C : C : O → C : C : C·COOR
α-Allenecarboxylic acid esters from ketenes s. *28*, 851

Furans O
from α-hydroxyketones and 2-alkoxyvinylphosphonium salts via 2-alkoxy-2,5-dihydrofurans s. *29*, 865

Cycloalkenylation of oxo compds. ⊙
with cyclopropylphosphonium fluoroborates – Cyclopentene ring – Regiospecific ring closure s. *29*, 866

Sodium hydride/dimethyl sulfoxide NaH/Me_2SO
Ethylenealcohols C
from lactols and phosphonium salts s. *28*, 853; ethylenehydroxysulfonic acids, prostaglandin derivs., s. Y. Iguchi, S. Kori, and M. Hayashi, J. Org. Chem. *40*, 521 (1975)

Sodium hydroxide $NaOH$
Heterogeneous Wittig synthesis CHO → CH : C
with aldehydes s. *29*, 868

Ethylene derivs. CHO → C : C
from aldehydes and 1,1-bi(phosphonium salts) – α,β-Ethylenenitriles s. *27*, 891

Sodium hydroxide/tetrabutylammonium iodide $NaOH/Bu_4N^+ I^-$
Horner synthesis CO → C : C
s. *30*, 645

Lithium/alcohol $LiOR$
α,β-Ethylenealdehydes from oxo compds. CO → C : C·CHO
Wittig synthesis with addition of 2 C-atoms s. *29*, 867

Lithium triethylcarboxide $Et_3C·OLi$
Trialkylcarbinols from boranes $R_3C·OH$
s. *27*, 892

Sodium/alcohol $NaOR$
Horner synthesis CO → C : C
s. *23*, 879 suppl. *29*

Sodium methoxide $NaOMe$
Ethylene derivs.
from 1,1-disilyl compds. and ketones

615. PhCH(SiMe₃)₂ → [PhC̄HSiMe₃ Na⁺] + OCPh₂
                                    ↓
                              PhCH=CPh₂

α,α-Bis(trimethylsilyl)toluene allowed to react under argon with Na-methoxide in *hexamethylphosphoramide*, then treated with benzophenone → triphenylethylene. Y: 79%. F. e. with lower yields s. H. Sakurai, K. Nishiwaki, and M. Kira, Tetrah. Let. *1973*, 4193.

Wittig synthesis
s. *27*, 893

Potassium/alcohol KOR
Double and heterogeneous Wittig synthesis – K-*tert*-butoxide in organic synthesis, review, s. *29*, 868

Potassium/tert-butanol $K/Me_3C \cdot OH$

1,3-Cyclohexadiene ring annelation

616.

K-*tert*-butoxide added at –20° to a slurry of the startg. bi(phosphonium salt) in dry ether, stirred 10 min., a soln. of dihydrocarvone in *tert*-butanol-ether added dropwise, kept 1 hr. at –20°, allowed to warm to room temp., stirred 3 hrs., finally refluxed 4 hrs. → product. Y: 85%. F. e. s. G. Büchi and M. Pawlak, J. Org. Chem. **40**, 100 (1975).

Potassium tert-butoxide $KOCMe_3$

Ethylene derivs. $C(OH)SO_3Na$ → $C:C$

from phosphonium salts and Na-1,1-hydroxysulfonates – Wittig synthesis – 1,3-Dienes s. **28**, 853

Horner synthesis CO → $C:C$

s. **30**, 645

Methyllithium   *MeLi*

Syntheses via enoxyboranes $COC:N_2$ → $COCRR'$

Synthesis of ketones from α-diazoketones s. **27**, 894

α-Hydroxyketones $O(OSi≤):C(OSi≤)$ → $CO \cdot C(OH)R$

from 1,2-di(siloxy)alkenes and halides

617.

An ethereal soln. of 2.2 moles methyllithium added dropwise with vigorous stirring to a soln. of the startg. enediol bis(trimethylsilyl) ether in dry monoglyme, stirred 0.5 hr. at room temp., the solvent replaced by tetrahydrofuran, a soln. of 3-bromo-1-propanol in tetrahydrofuran added, and stirred ca. 15 hrs. → product. Y: ca. 100%. F. e. s. T. Wakamatsu, K. Akasaka, and Y. Ban, Tetrah. Let. **1974**, 3879.

Ring closures by Wittig synthesis   ○

Isocyclics with bridgehead double bonds – Bredt's rule, review, s. **28**, 854

Butyllithium   *BuLi*

Geospecific synthesis of 2-ethylenealcohols CHO → $C:C \cdot C(OH)$

from aldehydes and phosphonium salts s. **28**, 855

γ-Hydroxyketones and 3-ethylenealcohols   ←

from acetylene derivs. and oxido compds. s. **29**, 869

Enisonitriles CO → $C:C \cdot N:C$

from oxo compds. and α-isocyanophosphonic acid esters s. **29**, 870

n-Butyllithium (s. a. under MeSOCl and Iodine)   *n-BuLi*
Insertion of chloromethylene groups   CO → CO·CCl
   α-Chloroketones from oxo compds. s. *29*, 871

Synthesis of ethylene derivs.   CO → C:C·PO(OR)$_2$
   from silanes and oxo compds. s. *28*, 856

Ketene mercaptals   CO → C:C(S)(S)
   from oxo compds. and silylthioformals s. *28*, 857

Syntheses with enaminophosphine oxides   CO → C:C·CO
   α,β-Ethyleneketones s. *27*, 890

Cyclopropanes   C:C → ▽
   from ethylene derivs. and phosphonium salts s. *29*, 872

n-Butyllithium/hydrogen peroxide   *n-BuLi/H$_2$O$_2$*
Oxo compds.   CO
   from mercaptals and boranes s. *29*, 873

Vinyllithium/hydrogen peroxide   *CH$_2$:CHLi/H$_2$O$_2$*
1,4-Diols   ←
from boranes and oxido compds.
via lithium 1-alkenyltriorganoborates

618.   (C$_4$H$_9$)$_3$B + LiCH=CH$_2$ → (C$_4$H$_9$)$_3$B̄—CH=CH$_2$ Li⁺ + (oxirane)

         HO–CH(C$_4$H$_9$)–CH$_2$–CH$_2$–CH(OH) ← (C$_4$H$_9$)$_2$B̄—O—CH(C$_4$H$_9$)... Li⁺

Tributylborane added at 0° under argon to a stirred ethereal soln. of vinyllithium, stirring continued 0.5 hr. at 25°, treated at 0° with methyloxirane, stirred 10 hrs. at room temp., finally treated with alkaline H$_2$O$_2$ → 2,5-nonanediol. Y: 93%. F. e. s. K. Utimoto, K. Uchida, and H. Nozaki, Tetrah. Let. *193*, 4527.

Lithium diisopropylamide   *i-Pr$_2$NLi*
α,β-Ethylenecarboxylic acids from oxo compds.   CO → C:C·COOH
   Carboxylic acid dianions as intermediates s. *29*, 874

Sodium carbonate   *Na$_2$CO$_3$*
Wittig synthesis in aq. medium   CO → C:C
   Styrenes s. *29*, 875

Potassium cyanide   *KCN*
Cyanolysis   ←
   of onium salts – Mixed onium salts s. *27*, 895

Sodium iodide   *NaI*
1,1-Difluorocyclopropane ring   C:C → ▽
   from ethylene derivs. and trihalogenomethylmercury compds. s. *17*, 894 suppl. *26*, *28*

Triethylamine   *Et$_3$N*
Cumulenes by Wittig synthesis   C:C:C:C
   s. *26*, 908

5,6-Dihydro-4H-1,3,4-thiadiazin-5-ones   ⊙
   from 1,3,4-thiadiazolium salts and acylphosphonic acid esters s. *26*, 909

Copper (s. a. under Zn)   *Cu*
γ-Ketocarboxylic acid esters   C : C·OSi≼ → C(R)·CO
    from α-diazocarboxylic acid esters and enoxysilanes s. *28*, 858

Silver oxide Ag_2O
γ-Diketones from 2 enoxysilane molecules   CO·C·C·CO

619.   2 $(CH_3)_2CH-\underset{\underset{OSiMe_3}{|}}{C}=CH_2$ → $(CH_3)_2CHCOCH_2CH_2COCH(CH_3)_2$

Regiospecific coupling. A heterogeneous mixture of the startg. m., Ag_2O, and *dimethyl sulfoxide* stirred and heated 2 hrs. at 65° → 2,7-dimethyl-3,6-octanedione. Y: 81%. F. e., **also** unsym. γ-diketones by **cross coupling** of 2 different enoxysilanes, s. Y. Ito, T. Konoike, and T. Saegusa, Am. Soc. *97*, 649 (1975).

Tetrakis[iodo(tri-n-butylphosphine)copper(I)] $(CuIn$-$Bu_3P)_4$
β-Vinyl- from α,β-ethylene-ketones   COC : C → COCHC·CH : CH_2
s. *26*, 910

Cuprous iodide   *CuI*
Syntheses with tosylates and lithium diorganocuprates(I)   OTs → R
s. *29*, 876

Synthesis by homoconjugate addition to cyclopropanes   C
s. *28*, 859; syntheses by addition to vinylcyclopropanes cf. N. Miyaura et al., Synthesis *1975*, 317

Zinc-copper   *Zn-Cu*
Ethylene derivs.   CO → C : C
from oxo compds. and α-iodophosphonium salts

620. $\underset{Ph}{\overset{CF_3}{\diagdown}}$CO + $Ph_3\overset{+}{P}$—CHFI   I⁻   → $\underset{Ph}{\overset{CF_3}{\diagdown}}$C=CHF

Fluoromethylene compds. α,α,α-Trifluoroacetophenone and triphenylfluoroiodomethylphosphonium iodide allowed to react at 0° with Zn-Cu in *dimethylformamide* → product. Y: 80%. F. e. s. D. J. Burton and P. E. Greenlimb, J. Fluorine Chem. *3*, 447 (1973).

Aluminum chloride $AlCl_3$
α,β-Ethyleneketones from enesilanes   C : C·Si≼ → C : C·COR
s. *29*, 846; s. a. I. Fleming and A. Pearce, Chem. Commun. *1975*, 633

Silylethynyl ketones   COCl → CO·C≡C·Si≼
    from carboxylic acid chlorides s. *29*, 846; cf. M. W. Logue and G. L. Moore, J. Org. Chem. *40*, 131 (1975).

Water s. under Br H_2O

Alcohol   *ROH*
Syntheses with boranes and acetylene derivs.   ←
    Allyl-type rearrangement – α-Alkoxystyrenes from alkoxyacetylenes, also synthesis of 1,4-dienes s. *26*, 911

Methyl alcohol (s. a. under O_2)   *MeOH*
Ethylene derivs.   C
    from cyclopropenes and trialkylboranes s. *29*, 877

Isonitrile   *RN : C*
Syntheses via metallo aldimines   ←
    α-Ketocarboxylic acids – Deuterioaldehydes s. *26*, 279

Trifluoroacetic acid $\qquad CF_3COOH$
Reactions with aryllead tricarboxylates $\qquad ArPb(OCOR)_3 \rightarrow Ar \cdot Ar'$
   Diaryls s. *29*, 878

N-Bromosuccinimide $\qquad \begin{matrix} CH_2CO \\ | \\ CH_2CO \end{matrix} NBr$

α-Halogenoketones $\qquad COC:N_2 \rightarrow COC(Hal)R$
   from α-diazoketones and trialkylboranes – Synthesis of α-halogenocarboxylic acid esters – Review of α-diazoketones s. *28*, 860

Titanium trichloride $\qquad TiCl_3$
Synthesis of glycols $\qquad \leftarrow$
   from 2 carboxylic acid molecules s. *26*, 913

Titanium tetrachloride $\qquad TiCl_4$
Syntheses with enoxysilanes $\qquad CO \rightarrow C(OH) \cdot C \cdot CO$
Cross-aldol reaction

621.
$$\text{PhCH}_2\text{CHO} + \text{CH}_2=\overset{\overset{\text{Me}_3\text{SiO}}{|}}{\text{C}}-\text{Ph} \longrightarrow \text{PhCH}_2\text{CH(OH)CH}_2\text{COPh}$$

β-Hydroxyketones. A soln. of acetophenone trimethylsilyl enolether in methylene chloride added at room temp. under argon to a soln. of phenylacetaldehyde and TiCl$_4$ in the same solvent, stirred 2 hrs., and treated with water → 1,4-diphenyl-3-hydroxybutan-1-one. Y: 78%. F. e. s. T. Mukaiyama, K. Narasaka, and K. Banno, Chem. Lett. *1973*, 1011; Am. Soc. *96*, 7503 (1974); **β-alkoxyketones** and **β-ketoacetals** s. ibid. *1974*, 15.

Triphenylphosphine $\qquad Ph_3P$
Reactions of α-halogenocarboxylic acid silyl esters $\qquad \square$
   2,2-Dibromocyclobutanone ring s. *28*, 861

Air-sodium hydroxide/hydrogen peroxide $\qquad O_2\text{-}NaOH/H_2O_2$
2-Ethylenealcohols from boranes $\qquad \leftarrow$
   Synthesis with addition of 4 C-atoms s. *27*, 896

Oxygen/methanol $\qquad O_2/MeOH$
Synthesis of azomethines $\qquad N:C \cdot C:C \rightarrow N:C \cdot CHCR$
from α,β-ethyleneazomethines and boranes

622.
$$(n\text{-}C_4H_9)_3B + \underset{\underset{CH_3}{|}}{CH}=CHCH=N-CHMe_2 \longrightarrow \left[n\text{-}C_4H_9-\underset{\underset{CH_3}{|}}{CH}CH=CH-\overset{\overset{B(C_4H_9-n)_2}{|}}{N}-CHMe_2 \right]$$

$$\overset{CH_3OH}{\swarrow}$$

$$n\text{-}C_4H_9-\underset{\underset{CH_3}{|}}{CH}CH_2CH=N-CHMe_2$$

Air introduced at 25° just above a stirred mixture of excess N-isopropylcrotonaldimine and tri-*n*-butylborane in tetrahydrofuran, after 0.5 hr. methanol added, and stirred 1 hr. at room temp. → N-isopropyl-β-methylenanthaldimine. Y: 76%. – Excess oxygen decreases the yield. F. e. s. N. Miyaura et al., Chem. Lett. *1974*, 395.

Hydrogen peroxide (s. a. under n-BuLi and CH$_2$: CHLi) $\qquad H_2O_2$
Synthesis of ethylene derivs. $\qquad C:C \cdot SO_2R \rightarrow C:C \cdot R'$
from α,β-ethylenesulfones and boranes

623.
$$\text{Ph}\diagdown\!\!\!\diagup\text{SO}_2\text{Me} + n\text{-Bu}_3B \longrightarrow \text{Ph}\diagdown\!\!\!\diagup\text{Bu-}n$$

Tri-*n*-butylborane added under N$_2$ to a soln. of methyl *cis*-styryl sulfone in tetrahydrofuran, refluxed 48 hrs. with stirring, then treated with alkaline H$_2$O$_2$ → *trans*-

β-*n*-butylstyrene. Y: 76%. F. e., **also with** α,β-**ethylenesulfoxides**, s. N. Miyamoto et al., Bull. Chem. Soc. Japan *47*, 503 (1974); **syntheses with organoboron compds.**, **review**, s. E. Negishi, J. Chem. Educ. *52*, 159 (1975); **synthesis of trisubst. ethylene derivs. from boranes and 1-alkynes** cf. G. Zweifel and R. P. Fisher, Synthesis *1975*, 376.

β-Hydroxythiolic acid esters $\qquad$ CO → C(OH)C·COSR
    from oxo compds., ketenes, and thioborinic acid esters s. *28*, 862

Thionyl chloride $\qquad$ $SOCl_2$
Dimerization of unsatd. Grignard compds. $\qquad$ 2·R·MgHal → R·R
    1,3-Dienes and 1,3-diynes s. *29*, 879

Methanesulfinyl chloride/n-butyllithium $\qquad$ *MeSOCl/n-BuLi*
Acetylene derivs. from boranes $\qquad$ C≡CH → C≡CR
s. *28*, 864 suppl. *29*

Selenous acid $\qquad$ H_2SeO_3
Sym. ethylene derivs. $\qquad$ 2 R·C:P → R·C:C·R
    from 2 alkylidenephosphorane molecules s. *28*, 863

Bromine/water/irradiation $\qquad$ *Br/H_2O/⇝*
Synthesis of branched alcohols from boranes $\qquad$ ←
s. *26*, 914

Iodine/n-butyllithium $\qquad$ *I/n-BuLi*
Syntheses via lithium 1-alkynyltriorganoborates $\qquad$ C≡CH → C≡CR
    Acetylene derivs. from boranes s. *28*, 864

Tetrabutylammonium iodide s. under NaOH $\qquad$ $Bu_4N^+ I^-$

Lithium tetrachloropalladate(II) $\qquad$ $Li_2[PdCl_4]$
***trans*-2-Arylalcohols** $\qquad$ C:C → C(OH)C·Ar
    from ethylene derivs. and arylmercuric chlorides s. *28*, 865

Via intermediates $\qquad$ *v.i.*
Ethylene derivs. $\qquad$ C:NAc → C:C
    from acylimines and phosphonic acid esters via 2-acylaminophosphonic acid esters s. *29*, 880

Synthesis of alcohols $\qquad$ CO → C(OH)R
from oxo compds. and silanes

624.

$$\underset{ClCH_2}{\overset{CH_3}{>}}CO + Me_3SiCH_2CH=CH_2 \longrightarrow \underset{ClCH_2}{\overset{CH_3}{>}}\underset{CH_2CH=CH_2}{\overset{OSiMe_3}{C<}}$$

$$\swarrow$$

$$\underset{ClCH_2}{\overset{CH_3}{>}}\underset{CH_2CH=CH_2}{\overset{OH}{C<}}$$

GaCl$_3$ as Lewis acid added cautiously to an equimolar mixture of chloracetone and allyltrimethylsilane, heated 24 hrs. at 100–110°, treated with ice-water, and the resulting mixture of alcohol and alkoxysilane refluxed 0.5 hr. in methanol containing some water and a little HCl → allyl(chloromethyl)methylcarbinol. Y: 60%. F. e. s. R. Calas et al., J. Organometal. Chem. *69*, C 15 (1974); **hard and soft Lewis acids and bases, review** s. Tse-Lok Ho, Chem. Rev. *75*, 1 (1975).

Syntheses with o-silylheterocyclics $\qquad$ CO → C(OH)Het
    Heterocyclic alcohols from oxo compds. s. *29*, 881

Synthesis of ketones ←
  from acetylene derivs. and boranes s. *29,* 882; Markownikoff-type addition
  cf. H. C. Brown, A. B. Levy, and M. M. Midland, Am. Soc. *97,* 5017 (1975)

Replacement of halogen by acetonyl groups   ArHal → ArCH$_2$COCH$_3$
with nickel(I) π-allyl complexes

625.

A soln. of 6'-bromopapaverine in dimethylformamide added under argon to π-(2-methoxyallyl)nickel bromide in the same solvent, stirred 16 hrs. at 25°, then poured into aq. 3 N HCl to hydrolyze the crude enolether → 6'-acetonylpapaverine. Y: 73%. F. e. s. L. S. Hegedus and R. K. Stiverson, Am. Soc. *96,* 3250 (1974).

α,β-Ethyleneketones   C≡H → C(COR) : CR'$_2$
  from acetylene derivs., boranes, and carboxylic acid chlorides via 2-oxa-3-borolenes s. *28,* 866

β-Ketocarboxylic acid amides   C(OSi⊆) : C → COC·CONHR
from enoxysilanes and isocyanates
s. *30,* 462

Carbon ↑   CC⇅C

Without additional reagents   w.a.r.
Decarboxylative condensation with enamines   C : C·N< → C : C·R

626.

1-(Cyclohex-1-enyl)pyrrolidine refluxed 5.5 hrs. with cyanoacetic acid in dioxane → cyclohex-1-enylacetonitrile. Y: 92%. F. e. s. N. Kumagaya, K. Suzuki, and M. Sekiya, Chem. Pharm. Bull. *21,* 1601 (1973).

Aliphatic nitro compds.   C : C → CHC·C(NO$_2$)
  from ethylene derivs. s. *27,* 897

Mannich reaction with C-deacylation   C·COR → C·CH$_2$N<
s. *28,* 867

2-Ethyleneamines from 1,1-halogenamines   C(Hal)N< → C : C·C·N<
s. *29,* 833

β-Amino-α-methylenenitriles   NH → NCH$_2$C(CN) : CH$_2$
  by decarboxylative Mannich reaction s. *26,* 916; α-methyleneketones from β-ketocarboxylic acids in the presence of piperidine cf. R. A. Micheli et al., J. Org. Chem. *40,* 675 (1975)

Homophthalic acid esters ←
  from 1,3-cyclohexadienes and allene-1,3-dicarboxylic acid esters s. *28,* 868

Enolbetaines ←
  from electron-rich ethylene derivs. and α-diazoketones s. *26,* 917

**Enecyclopropenes
from cyclopropenones and ketenes**

627.

$$R\text{-cyclopropenone} + O=C=C(CF_3)_2 \longrightarrow R\text{-cyclopropene}=C(CF_3)_2$$

$p-MeC_6H_4-=R$

A soln. of di-p-tolylcyclopropenone and bis(trifluoromethyl)ketene in toluene stored 5 days at room temp. → 1,2-di-p-tolyl-4,4-bis(trifluoromethyl)triafulvene. Y: 59%. I. Agranat and M. R. Pick, Tetrah. Let. *1973*, 4079; Tetrahedron *31*, 1163 (1975); f. method cf. T. Eicher, T. Pfister, and N. Krüger, Org. Prep. Proced. Int. *6*, 63 (1974).

Ar. annelation
  Naphthalenes from benzoyl peroxides and 2 acetylene deriv. molecules s. *29*, 883

**3,4,9,10-Tetrahydrophenanthrenes
and 2,3,4,4a,9,10-hexahydro-2,4a-ethanophenanthrenes**
  from α-tetralones via 5,6-dihydro-2H-naphtho[1,2-b]pyran-2-ones s. *29*, 884

Pyrrole ring from α-aminocarboxylic acids
s. *29*, 885

Pyronobenzo[ij]quinolizines
  from 1,2,3,4-tetrahydroquinolines – Double ring closure with malonic acid esters s. *27*, 898

Ring expansion by 2 C-atoms with cyclopropenes
  7,10-Dihydrobenz[a]azulenes s. *27*, 899

Benzenes and pyridines
  from pyridazines and acetylene derivs. via diene synthesis with inverse electronic demand s. *29*, 886

Furans from oxazoles
s. *28*, 869

4-Pyrones
  from 2 2,3-dihydrofuran-2,3-dione molecules s. *28*, 197

2-Pyridones from pyrimidonium betaines
s. *28*, 870

4-Aminoquinolines
  from 4H-3,1-benzoxazin-4-ones and ynamines s. *28*, 871

3H-Azepines from Δ^1-azirines and cyclones
s. *28*, 872

(4+2)π-Cycloaddition-fragmentation
  Azocines from 1,2,4-triazines s. *26*, 918

Heterocyclics from hydroxythiazolium betaines
  Thiophenes or 2-pyridones from 4-hydroxythiazolium betaines s. *26*, 919; thiophenes s. a. J. Org. Chem. *39*, 3627 (1974); pyrroles from 5-hydroxythiazolium betaines cf. ibid. *39*, 3619

1,3,6,9-Tetraazaspiro[4,4]nonanes
  from bis-2-imidazolidinylidenes and 2 isocyanate or isothiocyanate molecules s. *26*, 920

2-Acylbenzo[b]selenophenes
s. *28*, 873

Irradiation (s. a. under SiO₂ and O₂) †††
Photodimerization 2 RCOF → R·R
of perfluorodicarboxylic acid fluorides s. 26, 921

Photochemical C-acylation H → Ac
1-Acetyladamantanes s. 29, 887

Δ^1-**Pyrrolines** ←
from Δ^3-5-oxazolones and ethylene derivs.

628.

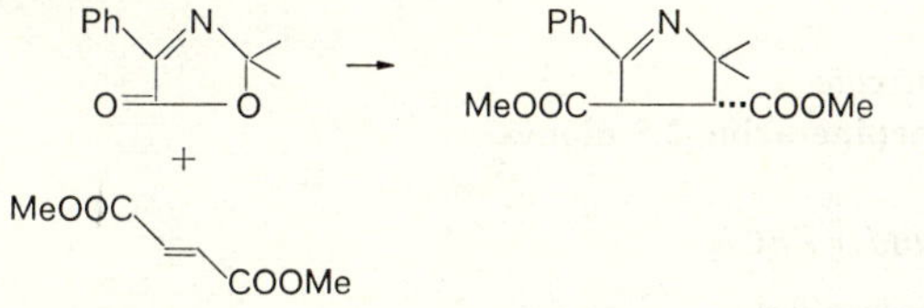

A soln. of 2,2-dimethyl-4-phenyl-Δ^3-oxazolin-5-one and dimethyl fumarate in pentane irradiated 1 hr. under N₂ with 2537 Å light → 2-phenyl-3,4-*trans*-dicarbomethoxy-5,5-dimethyl-Δ^1-pyrroline. Y: 78%. F. e. s. A. Padwa and S. I. Wetmore, Jr., Am. Soc. *96*, 2414 (1974).

Spiro-N-heterocyclics ←
from 2 oxindole molecules with decarbonylation s. 26, 922

Pyrazoles from sydnones ←
s. 27, 900

Sodium hydride NaH
Sym. ethylene derivs. from 2 nitrile molecules 2 CHCN → C : C
Stilbenes s. 27, 901

Aryl ketones CH(OR)CN → COR'
from halides and O-(tetrahydropyran-2-yl)mandelonitriles s. 29, 888

1-Subst. isoquinolines ←
from 2-acyl-1,2-dihydroisoquinaldo-1-nitriles – Synthesis of alkaloids, review s. 29, 889

Sodium hydroxide/alcohol ←
2,1-Benzisoxazoles O
from nitro compds. and nitriles

629.

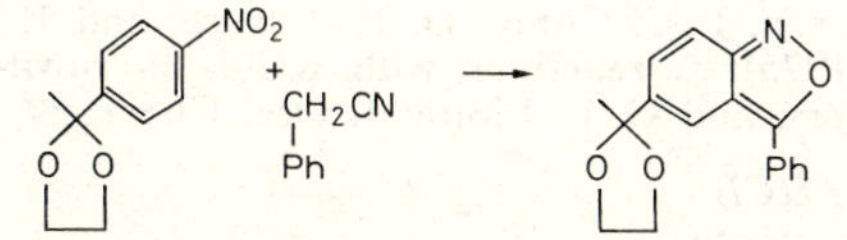

Phenylacetonitrile followed by 2-methyl-2-(4-nitrophenyl)-1,3-dioxolane added to a soln. of NaOH in methanol, and vigorously stirred for 16 hrs. whereby the temp. rises to ca. 55° during the first 0.5 hr. → 5-(2-methyl-1,3-dioxolan-2-yl)-3-phenyl-2,1-benzisoxazole. Y: 87%. R. Y. Ning, P. B. Madan, and L. H. Sternbach, J. Heterocyclic Chem. *11*, 107 (1974).

Potassium hydroxide KOH
2-Eneindolin-3-ones ←
s. 29, 890

Potassium hydroxide/alcohol
Nitrones from ar. nitro compds. ←
s. 27, 902

Sodium/alcohol   *NaOR*
C-Alkylation of alkoxalyl compds.   COCOOR → R
    with elimination of the alkoxalyl group – Annelations s. 27, 903

Azulenes from 2H-cyclohepta[b]furan-2-ones
s. 28, 725

Potassium/tert-butanol $K/Me_3C \cdot OH$
Benzene ring from thiopyrylium salts   ←
s. 27, 875

Potassium tert-butoxide $KOCMe_3$
1-Acyl-3-alkylidenepiperazine-2,5-diones   ←
s. 29, 891

Methyllithium s. under $ZnCl_2$ *MeLi*

Lithium diisopropylamide $i\text{-}Pr_2NLi$
Decarboxylative reactions   COOH → CHO
via carboxylic acid dianions
    Aldehydes and aliphatic nitro compds. from carboxylic acids s. 26, 923

Sodium acetate CH_3COONa
Cyclic enolesters from dicarboxylic acids   ○
s. 29, 892

3-Alkylidenephthalides from phthalic acids
    Labeled compds. s. 26, 924

Potassium cyanide · 18-crown-6 polyether   ←
α-Siloxynitriles from oxo compds.   CO → C(OSi≦)CN
by transcyano-O-silylation

630.
     n—C₅H₁₁CHO + (CH₃)₂C(CN)(OSiMe₃) → n—C₅H₁₁CH(CN)(OSiMe₃)
                                           [+ (CH₃)₂CO]

n-Hexanal stirred 16 hrs. at 25° with an equimolar amount of α-trimethylsilyl-oxyisobutyronitrile in the presence of a catalytic amount of KCN·18-crown-6 polyether complex → product. Y: 76%. F. e. s. D. A. Evans and L. K. Truesdale, Tetrah. Let. *1973*, 4929; s. a. E. J. Corey, D. N. Crouse, and J. E. Anderson, J. Org. Chem. *40*, 2140 (1975); **f. reactions with KCN·18-crown-6 polyether** cf. F. L. Cook, C. W. Bowers, and C. L. Liotta, J. Org. Chem. *39*, 3416 (1974).

Triethylamine (s. a. under HCl) Et_3N
2-Acylthiazoles from thiaminoids   ←
s. 26, 925

2-Azetidinone ring exchange   ←
s. 29, 893

3-Aminopyrazoles   ←
    from 1,3,4-oxadiazolium salts s. 27, 904

Pyrimidines from 1,3,5-oxadiazinium O-salts   ←
s. 30, 577

Ethyldiisopropylamine $i\text{-}Pr_2NEt$
Benzene ring from thiopyrylium salts   ←
s. 27, 875

Piperidine $C_5H_{11}N$
Decarboxylative Mannich reaction $CO \cdot CH \cdot COOH \rightarrow CO \cdot C : CH_2$
s. 26, 916 suppl. 30
2,6-Dicyanoanilines ○
from 2 alkylidenemalononitrile molecules s. 28, 874

Silver nitrate s. under $(NH_4)_2S_2O_8$ $AgNO_3$

Magnesium Mg
Benzylamines $N \cdot C \cdot C \cdot COOR \rightarrow N \cdot C \cdot Ar$
from β-aminocarboxylic acid esters s. 28, 875

Organomagnesium halides $RMgHal$
Reaction of acylcyanides with Grignard compds. $COCN \rightarrow COR$
Ketons s. 27, 905

Zinc chloride/methyllithium $ZnCl_2/MeLi$
Directed aldol condensation via enolate anions $C(OAc) : C \rightarrow CO \cdot C \cdot C(OH)$
β-Hydroxyketones from enolesters s. 29, 894; s. a. Org. Synth. 54, 49 (1974)

Aluminum chloride $AlCl_3$
Benzofuran-3(2H)-ones ○
from phenolethers and α-halogenocarboxylic acid halides s. 28, 876

18-Crown-6 polyether s. under KCN ←

Nitrobenzene $PhNO_2$
Benzene ring from cyclones ←
s. 29, 895

Acetic anhydride Ac_2O
Pyrrolizidines from acetylene derivs. ○

631.

via **dihydropyrrolizines**. N-Formyl-L-proline and ethyl propiolate refluxed 2 hrs. in acetic anhydride → intermediate (Y: 90%) hydrogenated 24 hrs. with 10%-Pd-on-carbon in ethanol under 3 atm. H₂ → ethyl (±)-isoretronecanolate (Y: 93%). M. T. Pizzorno and S. M. Albonico, J. Org. Chem. 39, 731 (1974).

Formic acid $HCOOH$
Stereospecific skeletal rearrangement ←

632.

2-Vinyl-5-(1-hydroxy-1-methylethyl)-4-methoxy-1,7-dimethylbicyclo[2.2.2]oct-2-ene stirred 1 hr. at room temp. in excess formic acid → 11-formyloxy-11,12-dihydronootkatone. Y: 50%. K. P. Dastur, Am. Soc. 96, 2605 (1974).

Chloranil/sulfuric acid ←
Dibenzo[fg,op]naphthacenequinones ○
from phenolethers s. 27, 906

Silica/irradiation $SiO_2/⇝$
Ketones $C : CH_2 \rightarrow CHCOR$
from ethylene derivs. and thionocarboxylic acid esters s. 29, 896

Titanium tetrachloride $\qquad TiCl_4$

Syntheses with enolesters
β-Alkoxyketones from acetals $\qquad C(OR_2) \twoheadrightarrow C(OR) \cdot C \cdot COR$

633. CH₂=C(MeCOO)CH₃ + (MeO)₂CHCH₂CH₂Ph ⟶ MeC(O)CH₂CH(OMe)CH₂CH₂Ph

A soln. of TiCl₄ in methylene chloride added during 10 min. at –10° under argon to a mixture of isopropenyl acetate and β-phenylpropionaldehyde dimethyl acetal in methylene chloride, and stirred an additional 2 hrs. at –10° ⤳ 4-methoxy-6-phenylhexan-2-one. Y: 89%. F. e. s. T. Mukaiyama, T. Izawa, and K. Saigo, Chem. Lett. *1974*, 323.

Lead tetraacetate $\qquad (CH_3COO)_4Pb$

Pyrazoles from hydrazones $\qquad \bigcirc$
   Nitrilimines as intermediates s. *26*, 926

p-Chlorobenzoyl nitrite $\qquad p\text{-}ClC_6H_4CO \cdot ONO$

Reactions with acetanilides as benzyne precursors $\qquad \leftarrow$
   Naphthalenes from cyclones s. *29*, 897

Phosphorus oxide chloride $\qquad POCl_3$

β-Amino-α,β-ethylenealdehydes $\qquad CH_2COOH \twoheadrightarrow C(CHO):CH \cdot NR_2$
from carboxylic acids

634. p—CH₃C₆H₄—CH₂COOH + 2 Me₂NCHO ⟶ p—CH₃C₆H₄—C(CHO)=CHNMe₂

Dimethylformamide added dropwise with vigorous stirring and intermittent cooling at 30° to POCl₃, stirring continued 5 min., then a soln. of p-tolylacetic acid in dimethylformamide added during 5 min., and stirred 18 hrs. at 70° ⤳ 2-p-tolyl-3-(dimethylamino)acrolein. Y: 80%. F. e. s. G. M. Coppola, G. E. Hardtmann, and B. S. Huegi, J. Heterocyclic Chem. *11*, 51 (1974); method s. Z. Arnold, Coll. *26*, 3051 (1961).

Tantalum pentafluoride-hydrogen fluoride $\qquad TaF_5\text{-}HF$

Reactions in superacidic medium $\qquad \leftarrow$
Side chains by disproportionation

635. C₆H₆ ⟶ C₆H₅—CH₃ + C₆H₅—CH₂CH₃

Benzene heated 16 hrs. at 200° in hydrogen fluoride-tantalum pentafluoride in the presence of hydrogen ⤳ ethylbenzene and toluene. Y: 40% and 36% respectively based on startg. m. consumed. F. e. s. M. Siskin and J. Porcelli, Am. Soc. *96*, 3640 (1974).

Oxygen/irradiation $\qquad O_2/ⁿⁿⁿ$

Replacement of cyano by α-hydroxyalkyl groups $\qquad CN \twoheadrightarrow C(OH)R$
s. *28*, 877

Hydrogen peroxide s. under FeSO₄ $\qquad H_2O_2$

Sulfuric acid s. under Chloranil and FeSO₄ $\qquad H_2SO_4$

Ammonium persulfate/silver nitrate $\qquad (NH_4)_2S_2O_8/AgNO_3$
Homolytic C-alkylation $\qquad H \twoheadrightarrow R$
   of N-heterocyclics s. *27*, 907

Pentacarbonylchromium(0)-carbene complex
Transmethylenation via carbenes
Cleavage of carbon-carbon double bonds

636.

$$\text{MeO} \atop \text{Ph} \!\!\diagdown\!\! C\!=\!Cr(CO)_5 + CH_2\!=\!CHN\!\!\diagup\!\!\!\overset{O}{\diagdown} \longrightarrow {\text{MeO} \atop \text{Ph}}\!\!\diagdown\!\! C\!=\!CH_2$$

A soln. of pentacarbonyl(methoxyphenylcarbene)chromium(0) and 1-vinyl-2-pyrrolidone in benzene stirred and heated slowly during several hrs. to 80°, and the product isolated after 7 hrs. → α-methoxystyrene. Y: 48%. Also transfer of subst. methylene groups s. E. O. Fischer and B. Dorrer, B. *107*, 1156 (1974).

Tricarbonylcyclopentadienylmolybdenum dimer	$[Mo(CO)_3Cp]_2$
γ-**Lactones**	O
from α-halogenocarboxylic acid esters and ethylene derivs. s. *27*, 908	
Hydrogen fluoride s. under TaF_5	HF
Hydrogen chloride/triethylamine	HCl/Et_3N
Sym. allenes from ketenes	$2\,R \cdot C:C:O \rightarrow R \cdot C:C:C \cdot R$
s. *28*, 878	
Ferrous sulfate/hydrogen peroxide	$FeSO_4/H_2O_2$
Radical C-carbalkoxylation	H → COOR
with α-ketocarboxylic acid esters s. *28*, 879	
Ferrous sulfate/sulfuric acid	$FeSO_4/H_2SO_4$
Radical C-methylation	H → CH_3
s. *28*, 880	
Nickel acetate	$(CH_3COO)_2Ni$
Pyrroles	O
from acetylene or its derivs. and 3 isonitrile molecules s. *27*, 909	
Via intermediates	v.i.
Interchange of ethylene derivs.	←
via cyclobutanes s. *30*, 467	
α-**Aminocarboxylic acids from isonitriles**	$CH \cdot N:C \rightarrow C(NH_2)COOH$
s. *29*, 898	

Elimination ⇑

Hydrogen ↑   CC⇑H

Without additional reagents	w.a.r.
Pyrazolo[3,4-d]pyrimidine-4,6-diones	O
s. *29*, 899	
Irradiation s. under $(CF_3COO)_3Tl$, N-Bromosuccinimide, O_2, and I)	⋙
Electrolysis	↯
Ring closure by anodic oxidation	
s. *29*, 900	

Sodium hydroxide s. under H_2O_2 and I	NaOH
Sodium hydroxide/tetrabutylammonium iodide	$NaOH/Bu_4N^+ I^-$

Benzene ring annelation ○
via Horner synthesis and cyclodehydrogenation s. *30*, 645

Potassium hydroxide KOH
Azomethines from tert. amines ←
with rearrangement s. *27*, 751

Sodium/alcohol NaOR
Skeletal Stevens-type rearrangement ←
s. *28*, 881

Potassium tert-butoxide (s. a. under $p\text{-}O_2NC_6H_4COOH$ and I) $Me_3C \cdot OK$
β,γ-Ethylenealdehydes from β,γ-ethylenehalides $C:C \cdot CHal \rightarrow C(CHO)C:C$
via rearrangement of N-cyanomethylpyrrolidinium salts s. *29*, 901

Pyrimidines ←
from 1,3-diazabicyclo[3.1.0]hex-3-enes s. *28*, 882

Potassium phenoxide PhOK
Rearrangement of N-aryloxypyridinium salts ←
s. *27*, 910

Butyllithium BuLi
Allenetetramines $(>N)_2C:C:C(N<)_2$
s. *29*, 902

n-Butyllithium (s. a. under $CdCl_2$) n-BuLi
2-(β,γ-Ethylene)-1,3-dithianes ←
s. *28*, 883

Allylsulfonium ylid-type rearrangement ←
s. *28*, 656

Lithium diisopropylamide $i\text{-}Pr_2NLi$
2-(γ,δ-Ethylene)-5,6-dihydro-4H-1,3-oxazines ←
via Claisen-type rearrangement of 2-ene-(β,γ-ethylene)tetrahydro-1,3-oxazines
s. *29*, 704

Lithium diisopropylamide/benzeneselenyl bromide $i\text{-}Pr_2NLi/PhSeBr$
Aromatization of 5-membered heterocyclics ←
Pyrroles from 4-acetylenechlorides via cyclic enamines

637.

Methyl 6-chloro-2-hexynoate, methyl 6-aminohexanoate, NaI, and Na-carbonate added to tetrahydrofuran, and refluxed 20 hrs. under N_2 → methyl N-(5-carbomethoxypentyl)-α-pyrrolidinylideneacetate (Y: 85–100%) in tetrahydrofuran added at −78° under N_2 to a soln. of Li-diisopropylamide in anhydrous tetrahydrofuran, warmed to 0°, then recooled to −78°, treated rapidly with benzeneselenyl bromide, prepared from diphenyl diselenide (s. Synth. Meth. *7*, 546) in tetrahydrofuran, warmed to 0°, water, *acetic acid*, and *30%-H_2O_2* added rapidly, and the product isolated when gas evolution has ceased after 0.5–1 hr. below 25° → N-(5-carbomethoxypentyl)-2-carbomethoxymethylpyrrole (Y: 71%). F. e., **also aromatization to furans and thiophenes**, s. C. H. Wilson and T. A. Bryson, J. Org. Chem. *40*, 800 (1975).

Potassium ethylenediamide $\qquad H_2NCH_2CH_2NHK$
Potassium hydride as superbase source ←
 Aromatization s. *28*, 884; KH as highly reactive base and hydriding agent cf. J. Org. Chem. *39*, 3913 (1974)

Triethylamine $\qquad Et_3N$
2-Aminofurans $\qquad CHC:N^+ \Leftrightarrow C:C \cdot N <$
 from 2,3-dihydrofurfurylideneammonium salts s. *30,* 164

Intramolecular nucleophilic ring closure ○
of cyclimmonium salts
 s. *25,* 645; s. a. T. L. Wimmer, F. H. Day, and C. K. Bradsher, J. Org. Chem. *40,* 1198 (1975)

Rearrangement of N-aryloxypyridinium salts ←
 s. *27,* 910

N-Ethyldiisopropylamine $\qquad i\text{-}Pr_2NEt$
Deprotonation ←
Indeno[2,1-c]-1,2-dithioles
from 8H-indeno[2,1-c]-1,2-dithiolium salts

638.

 A stirred suspension of 3-methyl-8H-indeno[2,1-c]-1,2-dithiolium perchlorate in dry petroleum ether treated with N-ethyldiisopropylamine, and the product isolated after 0.5 hr. → 3-methylindeno[2,1-c]-1,2-dithiole. Y: ca. 100%. F. e. s. K. Hartke and D. Krampitz, B. *107,* 739 (1974).

N,N,N′,N′-Tetramethylethylenediamine s. under CdCl₂ $\qquad TMEDA$

Cuprous bromide s. under PhCOO₂CMe₃ $\qquad CuBr$

Cupric chloride s. under O₂ $\qquad CuCl_2$

Cadmium chloride/n-butyllithium/ $\qquad CdCl_2/n\text{-}BuLi/TMEDA$
N,N,N′,N′-Tetramethylethylenediamine
Dehydrogenation ←
 Aromatization s. *28,* 885

Thallium trifluoroacetate/irradiation $\qquad (CF_3COO)_3Tl/\hslash\nu$
Pheoporphyrins by cyclodehydrogenation ○
 s. *29,* 903

Triphenylcarbonium fluoroborate $\qquad Ph_3C^+ BF_4^-$
Oxidative rearrangement ←
 of 3-alkyl-1,4-cyclohexadienes s. *27,* 866

Aluminum chloride s. under SnCl₄ $\qquad AlCl_3$

Ammonium ceric nitrate $\qquad (NH_4)_2Ce(NO_3)_6$
β,γ-Ethyleneoxo compds. C
 by cyclobutane ring opening – Oxidative fragmentation – Heterolytic fragmentation s. *26,* 927

Water $\qquad H_2O$
2,3-Dihydrofurans ←
 from tetrahydrofurylium salts s. *28,* 105

p-Cymene s. under Pd-C ←

Triphenylcarbinol s. under CF₃COOH $Ph_3C \cdot OH$

Peroxidase/hydrogen peroxide ←

Enzymatic oxidative ring closures ◯
Retention of absolute configuration – Oxidative coupling of phenols – N-Condensed cyclimmonium salts from cyclic tert. amines – Alkaloids by intramolecular coupling, review, s. 29, 904

Acetic anhydride s. under H₂SO₄ Ac_2O

Di-tert-butyl peroxide $Me_3C \cdot OO \cdot CMe_3$

Rearrangement of spirocyclohexadienes ☉
s. 27, 911

tert-Butyl peroxybenzoate/cuprous bromide $PhCOO_2CMe_3/CuBr$

Skeletal rearrangement ←
with dehydrogenation s. 29, 905

Benzoyl peroxide s. under N-Bromosuccinimide $(PhCOO)_2$

p-Nitrobenzoic acid/potassium tert-butoxide $p\text{-}O_2NC_6H_4COOH/Me_3C \cdot OK$

Dehydrogenation of p-nitrophenylhydrazones CHCH → C : C

639.

5α-Cholestan-3-one p-nitrophenylhydrazone in 1,2-dimethoxyethane containing p-nitrobenzoic acid stirred 1 hr. at room temp. with K-*tert*-butoxide under argon → cholesta-1,4-dien-3-one p-nitrophenylhydrazone. Y: 90%. F. e., also with iodine/K-*tert*-butoxide, s. D. H. R. Barton et al., Soc. Perkin I *1973*, 1565.

Trifluoroacetic acid (s. a. under VOF₃ and FSO₃H) CF_3COOH

Skeletal rearrangement ←
with aromatization s. 29, 906

Trifluoroacetic acid/triphenylcarbinol $CF_3COOH/Ph_3C \cdot OH$

Aromatization by dehydrogenation ←

640.

A soln. of 9,10-dihydroanthracene and triphenylcarbinol in trifluoroacetic acid refluxed 6 hrs. → anthracene. Y: ca. 100%. F. e. s. P. P. Fu and R. G. Harvey, Tetrah. Let. *1974*, 3217.

m-Chloroperoxybenzoic acid $m\text{-}ClC_6H_4COO_2H$

1,2,3,4-Tetrahydropyridines from piperidines CHCH → C : C
s. 28, 886

2,3-Dichloro-5,6-dicyanoquinone ←

α,β-Ethylenealdehydes from enols CHC : CH(OH) → C : C·CHO
s. 29, 907; s. a. G. Bozzato, J.-P. Bachmann, and M. Pesaro, Chem. Commun. *1974*, 1005

2-Pyridone from 3,4-dihydro-2-pyridone ring CHCH → C : C
s. 29, 927

Dehydrogenation with skeletal rearrangement ←
s. 28, 887

2,3-Dichloro-5,6-dicyanoquinone/p-toluenesulfonic acid ←
Dehydrogenation ←
s. 28, 921

N-Bromosuccinimide/irradiation/benzoyl peroxide $\begin{matrix}CH_2CO\\|\\CH_2CO\end{matrix}\!\!>\!\!NBr/⇈/(PhCOO)_2$

Aromatization CHCH → C:C
Naphtho[1,2-b]thiophenes

641.

N-Bromosuccinimide added portionwise to a stirred, refluxing, and irradiated soln. of ethyl 4,5-dihydronaphtho[1,2-b]thiophene-2-carboxylate and dibenzoyl peroxide in dry CCl$_4$ with removal of the resulting HBr by a stream of N$_2$, and refluxing continued 15 min. → ethyl naphtho[1,2-b]thiophene-2-carboxylate. Y: 88%. Also with Pd-C s. K. Clarke, D. N. Gregory, and R. M. Scrowston, Soc. Perkin I *1973*, 2956.

N-Bromosuccinimide/benzoyl peroxide $\begin{matrix}CH_2CO\\|\\CH_2CO\end{matrix}\!\!>\!\!NBr/(PhCOO)_2$

Cyclic ketones from cyclic alcohols ☉
   with ring expansion s. 27, 912

1-Chlorobenzotriazole ←
4-Pyrones from 2,3-dihydro-4-pyrones CHCH → C:C
s. 27, 913

Benzeneselenyl bromide s. under i-Pr$_2$NLi PhSeBr

Stannic chloride/aluminum chloride $SnCl_4/AlCl_3$
Dibenzo[a,l]pyrenes by Scholl ring closure ○
s. 27, 914

Lead tetraacetate/perchloric acid $(CH_3COO)_4Pb/HClO_4$
Naphthalenes by cyclodehydrogenation
s. 29, 908

Polyphosphoric acid $H(PO_3H)_xOH$
Benzofuro[2,3-b]quinoxalines
s. 29, 909

Vanadium oxide fluoride/trifluoroacetic acid/ $VOF_3/CF_3COOH/FSO_3H$
fluorosulfonic acid
Nonphenolic oxidative coupling ←
s. 29, 910

Vanadium oxide chloride $VOCl_3$
Cyclodehydrogenation ○
   Oxidative coupling of phenols s. 28, 888

Oxidative coupling of phenols
   Morphinane ring closure with subsequent isocyclic ring opening – Extended quinones s. 27, 915

Air/irradiation $O_2/⇈$
Pyrazoles and β-aminonitriles CHCH → C:C
   from Δ^2-pyrazolines s. 28, 889

Oxygen/irradiation
Cyclodehydrogenation
s. *28*, 890

O

Oxygen/cupric chloride
5-Hydroxypyrazoles from 3-pyrazolidones
also from 1-acyl-3-pyrazolidones s. *28*, 891

$O_2/CuCl_2$
CHCH → C:C

Hydrogen peroxide s. a. under Peroxidase

←

Hydrogen peroxide/sodium hydroxide
Aurones from flavanones
s. *26*, 928

$H_2O_2/NaOH$

Hydrogen peroxide/acetic acid
1,3-Dienes from ethylene derivs.
s. *29*, 911

H_2O_2/CH_3COOH
CH·C:C·CH → C:C·C:C

Sulfur
Aromatization
Benzo[g]pteridines s. *26*, 392

S
CHCH → C:C

p-Toluenesulfonic acid s. under 2,3-Dichloro-5,6-dicyanoquinone

TsOH

Potassium nitrosodisulfonate
3,4-Dihydroisoquinolines and isoquinolines from 1,2,3,4-tetrahydroisoquinolines

$ON(SO_3K)_2$

642.

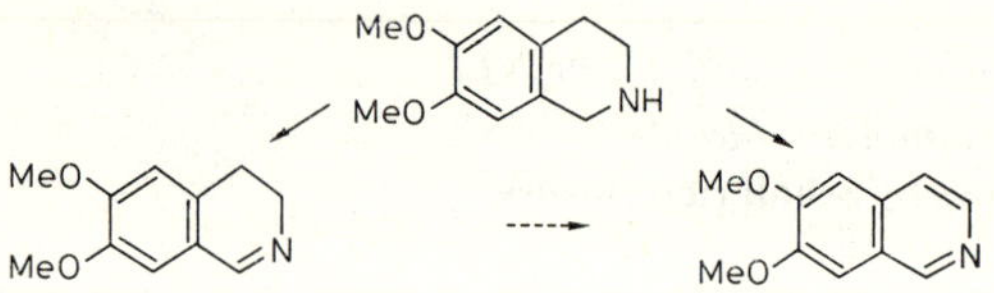

6,7-Dimethoxy-1,2,3,4-tetrahydroisoquinoline hydrochloride and K-nitrosodisulfonate in a 4%-Na-carbonate soln. stirred at room temp.
for 1 hr. → 6,7-dimethoxy-3,4-dihydroisoquinoline. Y: 97%. | for 2 days → 6,7-dimethoxyisoquinoline. Y: 73%.
F. e. s. P. A. Wehrli and B. Schaer, Synthesis *1974*, 288.

Sulfuric acid/acetic anhydride
Ring expansion by cyclopropane ring opening

H_2SO_4/Ac_2O

643.

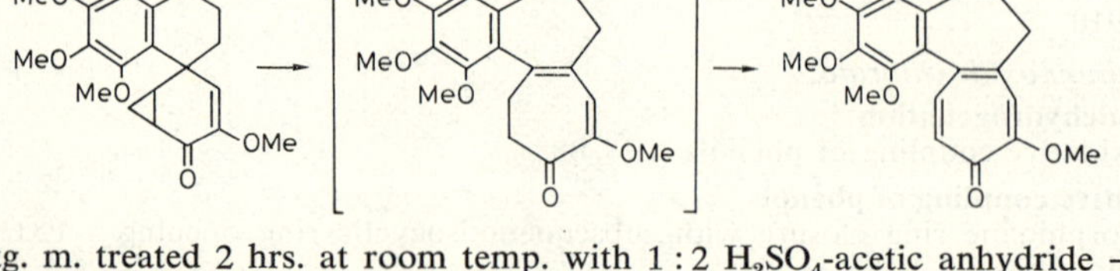

Startg. m. treated 2 hrs. at room temp. with 1:2 H_2SO_4-acetic anhydride → deacetamidoisocolchicine. Y: 90%. E. Kotani, F. Miyazaki, and S. Tobinaga, Chem. Commun. *1974*, 300.

Fluorosulfonic acid s. a. under VOF_3

FSO_3H

Fluorosulfonic acid/trifluoroacetic acid FSO_3H/CF_3COOH
5,6-Dihydronaphtho[2,1-g]pteridin-8(9H)-ones

644.

A stirred soln. of 2-amino-6-phenethylpteridin-4(3H)-one in trifluoroacetic acid treated dropwise with fluorosulfonic acid, allowed to stand 0.5 hr. at room temp., then poured into 95%-ethanol → 10-amino-5,6-dihydronaphtho[2,1-g]pteridin-8(9H)-one. Y: 74%. A. Rosowsky and K. K. N. Chen, J. Org. Chem. *39*, 1248 (1974).

Chlorine   Cl
Dinitrosoarenes from quinone dioximes   ←
s. *26*, 930

Bromine-sodium methoxide   Br-NaOMe
2-Alkoxyazocine ring   ←
s. *26*, 929

Iodine/irradiation   I/∰
Cyclodehydrogenation
and cyclodehydrohalogenation s. *28*, 892; synthesis of helicenes s. C. Jutz and H.-G. Löbering, Ang. Ch. *87*, 415 (1975); R. H. Martin and M. Baes, Tetrahedron *31*, 2135, 2139 (1975); double ring closure by cyclodehydrogenation, hexa[7]circulene, s. P. J. Jessup and J. A. Reiss, Tetrah. Let. *1975*, 1453; cyclodehydrogenation with C-methylene bridge elimination s. W. H. Laarhoven and N. P. J. Kuin, R. *94*, 105 (1975)

Benzene ring annelation
via Horner synthesis and cyclodehydrogenation

645.

A soln. of K-*tert*-butoxide in methanol added dropwise to a stirred and cooled soln. of diethyl 2-thienylmethylphosphonate and 2-methylbenzo[1,2-b : 4,3-b']dithiophene-7-carbaldehyde in dry dimethylformamide, and the product isolated after 2 hrs. → (E)-1-(7-methyl-2-benzo[1,2-b : 4,3-b']dithienyl)-2-(2-thienyl)ethene (Y: 74.5%) and iodine dissolved in benzene, and irradiated 4 hrs. with the quartz-filtered light of a medium-pressure Hg-lamp → product (Y: 73%). F. e. s. P. G. Lehman and H. Wynberg, Australian J. Chem. *27*, 315 (1974); Horner synthesis by phase transfer catalysis with NaOH and tetrabutylammonium iodide s. C. Piechucki, Synthesis *1974*, 869; f. phase transfer catalysts and their effect on geospecificity s. M. Mikołajczyk et al., ibid. *1975*, 278.

Iodine/sodium hydroxide   I/NaOH
Cyclopropanes by dehydrogenation   ▽
s. *28*, 893

Iodine/potassium tert-butoxide $\qquad I/Me_3C \cdot OK$
Dehydrogenation of p-nitrophenylhydrazones $\qquad CHCH \rightarrow C:C$
s. *30*, 639

Perchloric acid s. under $(CH_3COO)_4Pb$ $\qquad HClO_4$

Phenyltrimethylammonium tribromide $\qquad PhMe_3N^+ Br_3^-$
**N-Sulfonylenazo compds.
from sulfonylhydrazones** $\qquad CH \cdot C:N \cdot NHR \rightarrow C:C \cdot N:NR$

$Ph_2CHCH=N-NHTs \longrightarrow Ph_2C=CH-N=NTs$

Phenyltrimethylammonium tribromide added slowly at room temp. under N_2 to a stirred soln. of diphenylacetaldehyde tosylhydrazone in anhydrous tetrahydrofuran, and the product isolated after 10 min. → 2,2-diphenyl-1-tosylazoethylene. Y: 72%. F. e. s. G. Rosini and G. Baccolini, J. Org. Chem. *39*, 826 (1974).

Aromatization $\qquad CHCH \rightarrow C:C$
s. *27*, 916

Pyridine hydrobromide perbromide $\qquad C_5H_5N, HBr \cdot Br_2$
4-Alkylidene-2-ethylene-5-oxazolones $\qquad \bigcirc$
s. *30*, 163

Ferric chloride $\qquad FeCl_3$
Dinitrosoarenes from quinone dioximes $\qquad \leftarrow$
s. *26*, 930

Palladium-carbon $\qquad Pd-C$
Phenols from 2-cyclohexenones $\qquad \leftarrow$
s. *28*, 721; with K_2CO_3-modified Pd-C s. L. Cerveny et al., Chem. Prum. *24*, 503 (1974) (Czech); C. A. *82*, 16371

Coumarins from 3,4-dihydrocoumarins $\qquad CHCH \rightarrow C:C$
s. *27*, 819

Skeletal rearrangement with aromatization $\qquad \leftarrow$
Indans from spiro[4.4]nona-1,3-dienes s. *28*, 894

Palladium-carbon/p-cymene $\qquad \leftarrow$
Dehydrogenative oxidation $\qquad CHCH \rightarrow C:C$
s. *29*, 179

Via intermediates $\qquad v.i.$
Dehydrogenation via alcohols
by autoxidation s. *27*, 917

Dehydrogenation via selenides and thioethers $\qquad CHCHCO \rightarrow C:C \cdot CO$
α,β-Ethyleneoxo compds. – α,β-Ethylene-carboxylic acid esters and -nitriles – α-Methylenelactones s. *29*, 912; s. a. Am. Sc. *97*, 5434 (1975); α,β-ethylene-γ-lactones s. J. Org. Chem. *40*, 542 (1975); dehydrogenation via thioethers s. a. ibid. *40*, 148; reactions via selenides s. a. ibid. *40*, 2570

α,β-Ethylenecarboxylic from carboxylic acids $\qquad CHCHCOOH \rightarrow C:C \cdot COOH$
s. *27*, 918

α-Methylene-γ-lactones $\qquad CHCH \rightarrow C:C$
by dehydrogenation via α-bromination s. *28*, 895

2-Cyclohexenone from dihydropyran ring $\qquad \leftarrow$
via hydroxy- and keto-alkoximes s. *28*, 896

Oxygen   CC⇑O

Without additional reagents   *w.a.r.*
Ethylene derivatives   CHC(OCSN<) → C:C
  from thionocarbamic acid esters s. 26, 931
Enolethers from ketals   CHC(OR)$_2$ → C:C(OR)
  s. 26, 130
Isoxazoles from 5-acetoxy-Δ^2-isoxazolines   ←
  s. 26, 932
Enollactones from ketocarboxylic acids   O
  Isochroman-3-ones s. 27, 919
Ring closure via iminoester fluoroborates and amide acetals
  s. 26, 933
4(1H)-Quinolones
  with hetero-Cope-rearrangement s. 28, 897
o-Hydroxycarboxylic acid amides
  from aryl N-carbalkoxycarbamates via 1,3-benzoxazine-2,4-diones s. 28, 898
6-Hydroxyanthyridines
  s. 26, 465
Ethylene derivs. from tosylates   ←
  with skeletal rearrangement – Stereospecific conversion s. 28, 899
Pyrroles   ←
  from isoxazolium salts via Δ^4-isoxazolines s. 26, 934

Irradiation (s. a. under n-Bu$_3$SnH and HCl)   ⋕
Sym. diaryls from triaryl phosphates   (ArO)$_3$PO → Ar·Ar
  s. 28, 900
Ethylene derivs. from thionobenzoates   CHC(O·CS·R) → C:C
  s. 29, 947
Photocyclization with elimination of alkoxyl   O
  Dibenzofurans – Phenanthrenes – Phenanthridones s. 29, 913; review of preparative photochemistry cf. N. J. Turro and G. Schuster, Science 187, 303 (1975)
Ring closure of cyclic enacylamines
  with elimination of substituents s. 29, 914
Photocyclization of o-chloranil adducts   C
  s. 27, 920
Photolysis of cyclopentadioxinones   ←
  s. 27, 921

Sodium/anthracene   ←
Ethylene derivs. from 1,2-dimesylates   C(OSO$_2$Me)C(OSO$_2$Me) → C:C
  s. 29, 915

Sodium hydride   *NaH*
Alcohols from ethers   CHC(OR) → C:C + HOR
  s. 26, 14
Cyclopentenes from ketones   O
  s. 28, 901
3,4-Dihydro-1H-2,1-benzothiazin-4-one 2,2-dioxides
  s. 28, 902

Sodium hydride/dimethyl sulfoxide $\qquad$ *NaH/Me₂SO*
Heterocycles by intramolecular aldol-type condensation
  Thiophene ring s. *26*, 935

Sodium hydroxide $\qquad$ *NaOH*
β-Elimination of phosphoryl $\qquad$ CHC·OPO(OH)₂ → C : C
s. *29*, 155
1H-Indeno[1,2-b]quinolines $\qquad$ ○
  from 5,6-dihydro-11H-benzo[a]carbazoles via oxidative ring opening – Pyridine from 4-pyridone via 4-chloropyridine ring s. *28*, 903
Ring expansion with tosylates $\qquad$ ⊙
s. *29*, 918

Sodium hydroxide/methanol
Syntheses via cyclopropanol derivs.
  Ring expansion of isocyclics s. *27*, 922
Cyclopent-2-enones from cyclopropylketones
  from acoxycyclopropylketones – also by pyrolysis of δ-cyclopropyl-β-ketocarboxylic acid esters s. *27*, 923

Potassium hydroxide $\qquad$ *KOH*
Partial 1,4-dealcoholation $\qquad$ ←
s. *26*, 936
Benzene ring from 1,3,5-triketones $\qquad$ ○
  Ring closures with poly-β-carbonyl compds. s. *27*, 924; cf. Am. Soc. *97*, 3270 (1975)
Benzofuran ring by cyclodehydration
s. *27*, 925
5,6-Dihydro-2-pyrones
  from β,δ-dihydroxycarboxylic acid *tert*-butyl esters s. *27*, 255

Potassium hydroxide/alcohol
Ethylene derivs. from mesylates $\qquad$ CHC(OSO₂Me) → C : C
  Carbohydrate derivs. s. *26*, 937

Sodium/alcohol $\qquad$ *NaOR*
2-Ethylenealcohols from ketals $\qquad$ ←
  with transesterification s. *28*, 904
Reactions of Reformatskii products $\qquad$ CHC(OAc) → C : C
  α,β-Ethylenecarboxylic acid esters from ketones via β-acetoxycarboxylic acid esters – Bicyclo[2.2.2]octene ring fragmentation s. *26*, 938
2-Cyanoindoles $\qquad$ ○
s. *29*, 749
5,6-Dihydro-2-pyridone ring
s. *29*, 916
Furo[3,2-b]quinolines
  from o-aminoacetic acid esters via acetylenedicarboxylic ester adducts and 3-hydroxyquinolines – O-Acetylation s. *26*, 939

Sodium methoxide $\qquad$ *NaOMe*
α-Alkoxyenephosphonic $\qquad$ CHC(OR)₂PO(OR)₂ → C : C(OR)PO(OR)₂
from α,α-dialkoxyphosphonic acid esters
s. *29*, 607

Sodium ethoxide $\qquad$ *NaOEt*
4-Hydroxypyridazinium betaines
  from α-hydrazinocarboxylic acid esters via Dieckmann cyclization s. *26*, 940

Sodium tert-butoxide $NaOCMe_3$
1,5-Dienes by fragmentation-type ring opening
    with C-alkylation s. *27, 926*

Potassium/n-butanol $K/n\text{-}BuOH$
α-**Amino**-α,β-**ethylene**   C(NHOR)CHCO → C : C(NH₂)CO
from β-**alkoxylamino-ketones**
s. *29, 917*

Potassium tert-butoxide $KOCMe_3$
Ethylene derivs. from acoxy compds.   CHC(OAc) → C : C
s. *27, 927*

2,3-Pyridocyclobutenes from 3,4-dihydroazocines
s. *26, 941*

Methyllithium s. under $K_2[WCl_6]$ $MeLi$

Sodium bis(trimethylsilyl)amide $(Me_3Si)_2NNa$
Dieckmann cyclization
s. *22, 898* suppl. *28*

Intramolecular C-alkylation with mesylates
s. *23, 904* suppl. *29*

Potassium amide KNH_2
1,4-Dealcoholation
s. *26, 942*

Sodium hydrogen carbonate $NaHCO_3$
Phenanthrols
s. *28, 905*

Sodium acetate/acetic acid CH_3COONa/CH_3COOH
Benzimidazoles from imidazoles
s. *28, 922*

Potassium acetate s. a. under Ac_2O CH_3COOK

Potassium acetate/dimethylformamide CH_3COOH/Me_2NCHO
2-Ethylenealcohols from 1,3-ditosylates   C(OTs)CHC(OTs) → C : C · C(OH)
    with epimerization s. *26, 943*

Sodium benzoate $PhCOONa$
Enol tosylates from 1,2-ditosylates   C(OTs)C(OTs) → C : C(OTs)
    with simultaneous formation of acoxy compds. from tosylates s. *26, 944*

Sodium cyanide $NaCN$
Cyclopropane ring formation and ring expansion
    with tosylates s. *29, 918*

Potassium selenocyanate $KSeCN$
Ethylene derivs. from oxido compds.
Stereospecific deoxygenation
s. *30, 662*

Lithium chloride s. under C_5H_5N and HCl $LiCl$
Sodium chloride s. under $AlCl_3$ $NaCl$
Lithium bromide s. under $(Me_2N)_3PO$ $LiBr$
Sodium iodide s. under Zn NaI

Sodium salt   *Na⁺*
Intramolecular C-acylation
with α-enaminocarboxylic acids

647.

Startg. Na-salt added at 80° to acetic anhydride, and the resulting suspension stirred under dry N₂ at 80–83° for 3.75 hrs., then at 25° for 21 hrs. → 9-acetoxy-6-methyl-8-oxo-1,2,5,6,7,8-hexahydro-3H-pyrrolo[1,2-a]indole. Y: 40%. F. e. s. R. J. Friary et al., J. Org. Chem. *38*, 3487 (1973).

Potassium salt   *K⁺*
Ring closures with iminoesters
    Pyrrolo[2,3-d]pyridazines s. *29*, 482

Ammonium hydroxide   *NH₄OH*
1,2,3,4-Tetrahydroisoquinolines
    by cyclodehydration s. *27*, 928

Liq. ammonia   *NH₃*
Carbo-9-fluorenylmethoxyl as N-protective group   CHC(OCON<) → C:C
    Removal by β-elimination s. *26*, 24

Triethylamine (s. a. under ZnCl₂, Ac₂O, and Ph₃P)   *Et₃N*
Fulvenes from 1,1-acoxyhalides   CHC(OAc) → C:C
s. *27*, 834

Solvolytic ring closure of brosylates
s. *27*, 929

Triethylamine/dimethylformamide   *Et₃N/Me₂NCHO*
Ethylene derivs. from mesylates   CH·C·OSO₂Me → C:C
    Furans s. *28*, 906

Piperidine (s. a. under HgCl₂)   *C₅H₁₁N*
N-Condensed 4-hydroxy-2-pyridone ring
s. *29*, 919

1,5-Diazabicyclo[5.4.0]undecene
Thioenolethers   C(OSO₂R)CH(SR′) → C:C(SR′)
    from 1-sulfonyloxy-2-thioethers – Carbohydrate derivs. s. *26*, 945

Pyridine (s. a. under ClCOOEt, PhCOOH, P₂O₅, POCl₃,   *C₅H₅N*
MeSO₂Cl, and SO₃)
Ethylene derivs. from mesylates   CHC(OSO₂Me) → C:C
s. *29*, 622

Acetylene derivs. from enetriflates   CH:C(OSO₂CF₃) → C≡C
s. *29*, 920

Pyridine/lithium chloride   *C₅H₅N/LiCl*
Ethylene derivs. from mesylates   CH·C·OSO₂Me → C:C
s. *28*, 906

Copper s. under Zn   *Cu*
Cupric acetate s. under $FeSO_4$ $(CH_3COO)_2Cu$
Cuprous chloride s. under RN : C : NR   *CuCl*
Calcium carbonate $CaCO_3$
Bicyclo[3.2.1]octenes
    from bicyclo[2.2.2]oct-2-yl tosylates s. 27, 930

Zinc/sodium iodide   *Zn/NaI*
Ethylene derivs. from 1,2-dimesylates $C(OSO_2Me)C(OSO_2Me) \rightarrow C:C$
s. 11, 924 suppl. 29
Ethylene derivs.
    from oxido compds. or 1,2-acetoxyiodides s. 27, 931 $\triangleright\!O \rightarrow C:C$

Zinc/trimethylchlorosilane Zn/Me_3SiCl
Ethylene derivs. from ketones $CH_2CO \rightarrow CH:CH$
s. 17, 933 suppl. 29

Zinc,copper   *Zn,Cu*
Ethylene derivs. from oxido compds. $\triangleright\!O \rightarrow C:C$
s. 27, 932

Zinc chloride $ZnCl_2$
Anthrones via anthranol esters
s. 29, 921

Zinc chloride/triethylamine $ZnCl_2/Et_3N$
Cyclic ketones from carboxylic acids
    via carboxylic alkoxyformic anhydrides – 6,7-Dihydroindol-4(5H)-ones s. 27, 933

Mercuric chloride/piperidine $HgCl_2/C_5H_{11}N$
Aromatization by acid-base catalysis $CHC(OAc) \rightarrow C:C$
    Thiophenes s. 26, 946

Sodium tetrahydridoborate $Na[BH_4]$
Fragmentation-type reductive ring opening
s. 26, 947

Lithium tetrahydridoaluminate (s. a. under $TiCl_3$*)* $Li[AlH_4]$
Cyclopropylalcohols from enelactones
s. 28, 907

Boron fluoride (s. a. under $(CF_3CO)_2O$*)* BF_3
Indene ring
s. 28, 908
Enisocycles from ketones
    Stereospecific synthesis of alkaloids s. 27, 934
Ring closure with aldehydes
    N-Heterocyclics by cyclodehydration s. 29, 922
Ring closure with enol sulfonates
    Thianaphthenes with position shift or ar. substituents, also without position shift s. 27, 935

Fluoroboric acid/di-tert-butylcarbodiimide $HBF_4/RN:C:NR$
Isoquinolines
    from 3,4-dihydroisoquinoline N-oxides s. 29, 923

Aluminum chloride $AlCl_3$
Indoles from 2-aminoenol sulfonates
s. 28, 178

Aluminum chloride/sodium chloride	$AlCl_3/NaCl$
Ceramidonines s. 26, 948	
Aluminum chloride/phosphorus pentachloride	$AlCl_3/PCl_5$
Cyclic ketones from carboxylic acids 4-Chromanones s. 28, 909	O
Bis(tetra-n-butylammonium)oxalate	$(n\text{-}Bu_4N^+)_2(COO^-)_2$
Ethylene derivs. from tosylates s. 27, 936	CHC(OTs) → C : C
Pyrrolidinium acetate	←
Isomeric 2-cyclohexenones from 1,5-dioxo compds. – Regiospecific ring closure s. 28, 910	←
Dimethylformamide s. under Et$_3$N, SOCl$_2$, and HCl	Me_2NCHO
Di-tert-butylcarbodiimide s. under HBF$_4$	RN : C : NR
Dicyclohexylcarbodiimide/cuprous chloride	RN : C : NR/CuCl
Ethylene derivs. from alcohols s. 27, 937	C(OH)CH → C : C
Trialkyloxonium fluoroborate	$R_3O^+ BF_4^-$
N-Condensed cyclimmonium salts s. 28, 911	O
Ethyl chloroformate/pyridine	$ClCOOEt/C_5H_5N$
Dehydration of alcohols	C(OH)CH → C : C

648.

2.0 g. 3-(1-hydroxy-2-phenylethyl)indole treated with ethyl chloroformate in pyridine at 0° for 2 hrs., then at room temp. overnight → 1.45 g. *trans*-styrylindole. F. e. s. W. Carruthers and N. Evans, Soc. Perkin I *1974*, 421.

Acetic anhydride (s. a. under Me$_2$SO and TsOH)	Ac_2O
Ethylene derivs. from 2-amino-1,3-dioxolanes Preferential formation s. 27, 938	C
Enamines from N-oxides Polonovski reaction s. 27, 939	CHCHN(O) → C : C·N
3-Methylene-2-piperidones from piperidine-3-carboxylic acids s. 28, 912	←
Acetic anhydride/potassium acetate	Ac_2O/CH_3COOK
Cyclopropane ring from mesylates with expansion of S,S-heterocycles	←

649.

19-Mesyloxy-4-androstene-3,17-dione 3-thioketal heated 50 min. at 100° with a mixture of K-acetate, acetic anhydride, and acetic acid → product. Y: 73%. J. R. Williams and G. M. Sarkisian, Tetrah. Let. *1974*, 1109.

Acetic anhydride/triethylamine Ac_2O/Et_3N
Isoindoles from isoindoline N-oxides ←
s. *20*, 648a suppl. 29

Trifluoroacetic anhydride s. a. under CF_3COOH $(CF_3CO)_2O$
Trifluoroacetic anhydride/boron fluoride $(CF_3CO)_2O/BF_3$
Ring closure with acetals ○
    Promeranz-Fritzsch isoquinoline ring closure s. *27*, 940

Succinic anhydride s. under PhCOOH $\begin{array}{c}CH_2CO\\|\\CH_2CO\end{array}\!\!\!\!>\!O$

Formic acid   HCOOH
Aromatization of steroid C rings ←
s. *29*, 924

1,2,3,4-Tetrahydro-6H-pyrido[4,3-b]carbazoles ○

650.

A soln. of the startg. mixture of epimeric ethynylcarbinols in anhydrous formic acid refluxed 15 min. → 1,2,3,4-tetrahydro-2-benzylellipticine. Y: 93%. F. e. s. F. Le Goffic, A. Gouyette, and A. Ahond, Tetrahedron *29*, 3357 (1973); **6H-pyrido-[4,3-b]carbazoles** cf. M. Sainsbury and R. F. Schinazi, Chem. Commun. *1975*, 540.

Acetic acid s. under $H(PO_3H)_xOH$ CH_3COOH

Benzoic acid/pyridine/succinic anhydride $PhCOOH/C_5H_5N/\begin{array}{c}CH_2CO\\|\\CH_2CO\end{array}\!\!\!\!>\!O$
Enolethers from ketals $CHC(OR)_2 \twoheadrightarrow C:C(OR)$
s. *29*, 925

Trifluoroacetic acid (s. a. under Ph_3PSe) CF_3COOH
Δ⁴-Oxazol-2-ones   CHC(OH) ↠ C:C
    from 4-hydroxyoxazolid-2-ones s. *28*, 365

Ring closures with β-ketosulfoxides ○
    Cyclic α-(alkylthio)ketones s. *29*, 926
Furoic from 3-deoxy-2-ketoaldonic acids
s. *29*, 252
N-Condensed 2-pyridone ring
    via N-condensed 3,4-dihydro-2-pyridone ring s. *29*, 927
Aromatization by elimination of O-bridges ←
s. *16*, 943 suppl. 29

Trifluoroacetic acid/trifluoroacetic anhydride $CF_3COOH/(CF_3CO)_2O$
Indoles ○
from hydroxylamines wia enehydroxylamines

651.

p-Chlorophenylhydroxylamine refluxed 5–10 hrs. with 5,5-dimethyl-1,3-cyclohexanedione in benzene in the presence of ascorbic acid ↠ intermediate (Y: 70%)

allowed to react at room temp. with trifluoroacetic acid-trifluoroacetic anhydride
→ indole deriv. (Y: 85%). F. e. s. T. Okamoto and K. Shudo, Tetrah. Let. *1973*,
4533.

Dimethyl sulfoxide (s. a. under SO₃) Me_2SO
Stilbenes from 1,2-diarylethanols CH(OH) → C : C
s. *26*, 786

Dimethyl sulfoxide/acetic anhydride Me_2SO/Ac_2O
3-Ethylenealcohols and γ,δ-ethyleneketones ←
  from 2,3-ethylene-1,5-diol 1-(ethyl carbonates) via 2,4-dienones – Stereospecific reduction s. *29*, 928

Reactions with *dialkoxysulfuranes* $Ph_2S(OR)_2$
Dehydration of alcohols CHC(OH) → C : C
s. *27*, 941

Methyl iodide s. a. under Ph₂PLi MeI

Carbon tetrachloride s. under Ph₃P CCl_4

Chloranil s. under HCl ←

2,3-Dichloro-5,6-dicyanoquinone s. under TsOH ←

Trimethylchlorosilane s. under Zn Me_3SiCl

Titanium trichloride-lithium tetrahydridoaluminate $TiCl_3$-$Li[AlH_4]$
Ethylene derivs. from oxido compds. → C : C
  with Co(CO)₄ cf. *30*, 662; with TiCl₃/LiAlH₄, McMurry's
  reagent, s. J. E. McMurry and M. P. Fleming, J. Org. Chem. *40*, 2555 (1975)

Tri-n-butyltin hydride/irradiation n-Bu_3SnH/⇝

652.

A soln. of 2-cyclopentylidenecyclopentan-1-one oxide (s. a. 514) and tri-*n*-butyltin
hydride in benzene irradiated 2 hrs. with a 450 w. Hanovia lamp in a degassed
Pyrex tube → 2-cyclopentylidenecyclopentan-1-one. Y: 95%. F. e. s. J. R. Williams
et al., J. Org. Chem. *39*, 1028 (1974).

Stannous chloride $SnCl_2$
Monovalent reduction ←
  Carbon-carbon double bonds from diols s. *26*, 949

Stannic chloride $SnCl_4$
Steroid skeleton O
  by stereospecific double ring closure s. *29*, 929

Lithium diphenylphosphide/methyl iodide Ph_2PLi/MeI
Ethylene derivs. from oxido compds. → C : C
  via phosphonium betaines – Stereospecific ring opening –
  Geoisomerization of ethylene derivs. s. *27*, 942; geoisomerization by f. methods
  cf. D. Van Ende and A. Krief, Tetrah. Let. *1975*, 2709

Triphenylphosphine/carbon tetrachloride Ph_3P/CCl_4
Carbodiimides from ureas NHCONH → N : C : N
s. *27*, 978

Triphenylphosphine/carbon tetrachloride/triethylamine $Ph_3P/CCl_4/Et_3N$
Allenes from ketones CHCOCH → C : C : C
s. *28*, 913

Hexamethylphosphoramide s. a. under (PhO)₃P⁺Me I⁻   (Me₂N)₃PO

Hexamethylphosphoramide/lithium bromide   (Me₂N)₃PO/LiBr
Cyclic α,β-ethylenealdehydes   Ö:
   from cyclic 2,3-oxidoalcohols with ring contraction s. 29, 930

Phosphorus pentoxide (s. a. under MeSO₃H)   P₂O₅
Chroman ring
s. 28, 914

4(1H)-Quinolone-3-carboxylic acids
s. 28, 915

Pictet-Gams isoquinoline ring closure   O
s. 29, 931

Triphenyl phosphite methiodide/   (PhO)₃P⁺Me I⁻/(Me₂N)₃PO
hexamethylphosphoramide
Preferential dehydration of sec. alcohols   CHC(OH) → C:C
s. 28, 916

Polyphosphoric acid esters   ←
4-Hydroxyquinolines   O
s. 27, 943

Triphenylphosphine selenide/trifluoroacetic acid   Ph₃PSe/CF₃COOH
Ethylene derivs. from oxido compds.   → C:C
   Retention of configuration s. 27, 942 suppl. 28

Polyphosphoric acid (s. a. under POCl₃ and Se)   H(PO₃H)ₓOH
4-Aminoisoquinolinium salts by Schroeter-type dehydration   ←
s. 26, 950

Indenes   O
s. 29, 932

Cycloalkane ring from β-hydroxycarboxylic acid esters
   via cyclic β,γ-ethyleneketones – Dehydration of alcohols s. 29, 933

Isocyclic ring closure with lactones

653.

2-(2,7-Dimethoxy-1-naphthoyl)-3,3-dimethylpentan-4-olide stirred 20 min. at 60° with polyphosphoric acid → 8,9-dihydro-1,6-dimethoxy-8,8,9-trimethylphenaleno-[1,2-b]furan-7-one. Y: 79%. **Also 8,9-dihydrophenaleno[1,2-b]furan-7-ones via Claisen rearrangement** of enolethers s. D. A. Frost and G. A. Morrison, Soc. Perkin I *1973*, 2159.

Combes-Beyer quinoline ring synthesis
   by *anti*-Marckwald cyclization – Marckwald's rule in Skraup reaction s. 27, 944

Quinoline ring from vinylogous carboxylic acid anilides
s. 28, 917

2-Quinazolones from acylureas
s. 29, 934

Polyphosphoric acid/acetic acid $\qquad H(PO_3H)_xOH/CH_3COOH$
Cyclic β-diketones from ketocarboxylic acids
    Spirocyclic β-diketones from cyclic ketocarboxylic acids s. *29*, 935; cyclic β-diketones s. a. M. Shimagaki and A. Tahara, Tetrah. Let. *1975*, 1715

Phosphoric acid $\qquad H_3PO_4$
Phenol from cyclohex-2-en-4-olone ring $\qquad \leftarrow$
    s. *30*, 546
Ring closures of hydroxyketones $\qquad \bigcirc$
    Vinylcyclopropyl ketones from γ,δ-ethylene-ε-hydroxyketones by stereospecific cyclodehydration s. *27*, 945

Phosphorus oxide chloride $\qquad POCl_3$
Dehydration of alcohols $\qquad CHC(OH) \rightarrow C:C$
    Ethylenenitriles s. *26*, 951
Cyclic azomethinium salts from carboxylic acids $\qquad \bigcirc$
via N,N-disubst. carboxylic acid amides

654.

$$\text{CH}_2\text{-COOH} + \text{OS} \begin{matrix} \text{NMe}_2 \\ \text{NMe}_2 \end{matrix} \longrightarrow \text{CH}_2\text{-CONMe}_2 \longrightarrow [\text{azulene fused product}]=\text{NMe}_2 \text{ Cl}^-$$

4-Azulenylacetic acid warmed 2 hrs. at 60° with sulfinylbis(dimethylamine) in benzene → amide (Y: ca. 100%) warmed 2 hrs. at 60° in $POCl_3$ → product (Y: 95%). F. e. s. K. Hafner, K.-P. Meinhardt, and W. Richarz, Ang. Ch. *86*, 235 (1974).

3-Aminobenzofuran ring
    from α-aryloxycarboxylic acid amides s. *29*, 936
2-Iminoindolines from 1-acyl-2-arylhydrazines
s. *28*, 918
4-Chloropyridine ring
    1H-Pyrazolo[3,4-b]pyridines s. *28*, 919
N-Condensed heterocyclics
    from hydroxycarboxylic acid amides via Bischler-Napieralski ring closure – Double ring closure s. *27*, 946

Phosphorus oxide chloride/pyridine $\qquad POCl_3/C_5H_5N$
Indoles from 3-hydroxyindolines $\qquad CHC(OH) \rightarrow C:C$
s. *30*, 663

Phosphorus oxide chloride/polyphosphoric acid $\qquad POCl_3/H(PO_3H)_xOH$
Pictet-Gams isoquinoline ring closure $\qquad \bigcirc$
    with C-aryl migration s. *28*, 920

Phosphorus pentachloride s. under AlCl_3 $\qquad PCl_5$

Thionyl chloride/dimethylformamide $\qquad SOCl_2/Me_2NCHO$
Spirocyclopropane ring $\qquad \triangledown$
    gem-Dimethyl effect on ring closure s. *29*, 937
Reactions with *carbomethoxysulfamoyltriethylammonium* $\qquad MeOCON\text{-}SO_2N^+Et_3$
hydroxide inner salt
Dehydration of alcohols $\qquad CHC(OH) \rightarrow C:C$
s. *26*, 952

Methanesulfonyl chloride/pyridine $\qquad MeSO_2Cl/C_5H_5N$
Cyclobutenes from cyclopropylcarbinols $\qquad \odot$
s. *29*, 938

Sulfur trioxide-pyridine/dimethyl sulfoxide SO_3-C_5H_5N/Me_2SO

α,β-Ethyleneoxo compds.
from 1,3-diol monoesters   C(OAc)CHCH(OH) → C:C·CO
s. *29*, 242

Reactions with *methanesulfonic acid-phosphorus pentoxide* $MeSO_3H$/P_2O_5
2-Cyclopentenones from γ-lactones   ←
s. *11*, 933 suppl. 28

p-Toluenesulfonic acid   TsOH
Indenes from indanols   CHC(OH) → C:C
s. *29*, 840

α,β-Ethylene- from β-hydroxy-ketones   CO·CH·C(OH) → CO·C:C
s. *30*, 489

Furans from 2-alkoxy-2,5-dihydrofurans   ←
s. *29*, 483

Ring closures with β-ketosulfoxides   O
s. *29*, 926

N-Condensed heterocyclics
    from dicarboxylic acid imides via alkoxylactams s. *27*, 947

Bicyclo[2.2.2]octene ring fragmentation   C
s. *26*, 938

p-Toluenesulfonic acid/acetic anhydride $TsOH$/Ac_2O
Dienollactones from ketolactones   ←
s. *26*, 953

p-Toluenesulfonic acid/acetic acid $TsOH$/CH_3COOH
Naphthalenes   ←
    from 1,4-dihydro-1-naphthols s. *28*, 639

p-Toluenesulfonic acid/2,3-dichloro-5,6-dicyanoquinone   ←
Aromatization with migration of methyl groups   ←
    Phenanthrene ring – Dehydrogenation s. *28*, 921

Sulfuric acid H_2SO_4
Ethylene derivs. from alcohols   CHC(OH) → C:C
s. *27*, 917

Dehydration of alcohols   ←
    with conjugation and methyl group migration s. *29*, 939

2-Pyridone from 5-hydroxy-5,6-dihydro-2-pyridone ring   CHC(OH) → C:C
s. *27*, 508

Ring closure of ketones   O
s. *30*, 491

Benzene ring annelation with β-halogenacetals
    Thianaphthenes from thiophenes s. *28*, 922

Cyclic ketones from carboxylic acid esters
    with formation of isocyclics from halides s. *28*, 953

1,2,3,4-Tetrahydroisoquinolines
    by cyclodehydration s. *27*, 928 suppl. 28

9H-Naphth[3,2,1-kl]acrid-9-ones
s. *28*, 923

Ring closure with enolethers
    Carbostyrils s. *26*, 954

4-Substitution of quinolines
via 1,2,4,5-tetrahydro-2,5-methano-3,1-benzoxazepines s. *28,* 924

Benzene from pyran ring

655.

A mixture of 16 g. 3,4-dihydro-3-vinyl-6-ethyl-2H-pyran and aq. 25%-H_2SO_4 heated 3 hrs. at 100° → 10 g. p-ethyltoluene. F. e. s. T. D. J. D'Silva, W. E. Walker, and R. W. Manyik, Tetrahedron *30,* 1015 (1974).

Sulfuric acid/acetic acid H_2SO_4/CH_3COOH
Ethylene derivs. from alcohols   CHC(OH) → C : C
s. *29,* 762

Selenium/polyphosphoric acid $Se/H(PO_3H)_xOH$
11H-Benzo[5,6]cyclohepta[1,2-c]pyrid-11-ones
s. *29,* 940

Potassium hexachlorotungstate/methyllithium $K_2[WCl_6]/MeLi$
Ethylene derivs. from glycols   C(OH)C(OH) → C : C
s. *28,* 925

Tungsten hexachloride/n-butyllithium $WCl_6/n\text{-}BuLi$
Ethylene derivs. from oxido compds.   → C : C
s. *27,* 932

Hydrochlorides
Thiazoles from 4-hydroxy-Δ^2-thiazolines   CHC(OH) → C : C
s. *29,* 941

Tetramethylammonium fluoride $Me_4N^+ F^-$
α,β-Ethylene- from 1,2-acoxy-chlorides   CHClC(OAc) → CCl : C
s. *29,* 942

Hydrogen fluoride   HF
Double ring closure with acetals
s. *28,* 926

Hydrogen chloride   HCl
Aromatization by dehydration of alcohols   CHC(OH) → C : C
    Thiophenes from 3-hydroxy-2,3-dihydrothiophenes
    s. *26,* 866

**Ethylene derivs. with C-alkyl migration
and alcohols
from sulfuric acid monesters**

656.

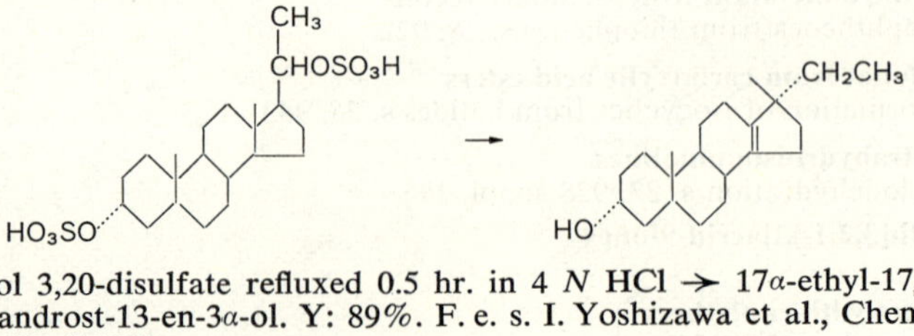

Pregnandiol 3,20-disulfate refluxed 0.5 hr. in 4 N HCl → 17α-ethyl-17β-methyl-18-nor-5β-androst-13-en-3α-ol. Y: 89%. F. e. s. I. Yoshizawa et al., Chem. Pharm. Bull. *21,* 1622 (1973).

Ring closure of ketones
s. *30,* 491

N-Condensed lactams from isocarbostyrils
Ring closure with acetals s. *28,* 329

3-Aminoquinolines
s. *29,* 943

Phenanthrene ring from hydroxyspiroisocyclics

657.

Startg. m. refluxed 24 hrs. under N_2 in a mixture of concd. HCl and ether → product. Y: 74%. I. Monković et al., Am. Soc. *95,* 7910 (1973).

17α-Hydroxy-$\Delta^{8,14}$-steroids
from 9.14β-epoxy-17-keto-8,14-seco-13α-steroids s. *29,* 944

Hydrogen chloride/irradiation   HCl/∰
Dehydrative ring contraction
of O-heterocyclics s. *27,* 948

Hydrogen chloride/lithium chloride/dimethylformamide $HCl/LiCl/Me_2NCHO$
4-Chloro-2,4-dienones
by simultaneous dehydration and dealcoholation s. *29,* 945

Hydrogen chloride/chloranil
Oxidative benzene ring closure

658.

A 2% soln. of HCl in methanol added dropwise to a soln. of the startg. m. in benzene containing 1.5 molar equivalents of chloranil ⇀ product. Y: 84%. J. P. Kutney and D. S. Grierson, Heterocycles *3,* 171 (1975).

Hydrogen bromide   HBr
1,3-Dienes from oxido compds.   CH−C−C−CH → C:C·C:C
                                                 \O/

659.

An ethereal soln. of the startg. m. shaken 0.5 hr. with constant boiling 48%-HBr ⇀ agnosteryl acetate. Y: 93%. R. B. Boar, D. A. Lewis, and J. F. McGhie, Soc. Perkin I *1973,* 1583.

Benzene ring from cyclic oxidoalcohols
Aromatization with migration of methyl groups s. *29,* 946

Hydrogen bromide/acetic acid   *HBr/CH$_3$COOH*
Estra-1,3,5(10)-trienes ←
from steroidal ring B 2,3-oxidoalcohols

660.

17β-Acetoxy-5β,6β-epoxy-7β-hydroxyandrostane refluxed with HBr in glacial acetic acid → 17β-acetoxy-4-methylestra-1,3,5(10)-triene. Y: 30–40%. Also from the 5α,6α,7α- and 5α,6α,7β-isomers and f. e. s. D. Baldwin and J. R. Hanson, Chem. Commun. *1974*, 211.

Ring closure with acetals   O
    4a-Azoniaanthracenes s. *26,* 955

Ferrous sulfate/cupric acetate   *FeSO$_4$/(CH$_3$COO)$_2$Cu*
Ethylenealcohols from hydroperoxides ←
Remote carbon-carbon double bond formation

661.

Startg. hydroperoxide added at room temp. under N$_2$ during 0.5 hr. to a soln. of ferrous sulfate and cupric acetate in acetic acid containing 10% water, the heterogeneous mixture stirred whereupon the color turns from green to brown, and left overnight → product. Y: 60%. F. e. s. Ž. Čeković and M. M. Green, Am. Soc. *96,* 3000 (1974).

Ferric chloride/n-butyllithium   *FeCl$_3$/n-BuLi*
Ethylene derivs. from oxido compds.   → C : C
s. *27,* 932 suppl. 29

Cobalt carbonyl   *Co(CO)$_4$*
Ethylene derivs. from oxido compds.   → C : C
Stereospecific deoxygenation

662.

trans-Dimethyl epoxymethylsuccinate treated 18 hrs. at room temp. with 0.147 equivalent Co$_2$(CO)$_8$ → dimethyl citraconate. Y: 99%. − Larger amounts of reagent give faster but less stereospecific deoxygenation. F. e. s. P. Dowd and K. Kang, Chem. Commun. *1974,* 384; with K-selenocyanate, limitations, s. J. M. Behan, R. A. W. Johnstone, and M. J. Wright, Soc. Perkin I *1975,* 1216.

Bis-(1,5-cyclopentadiene)nickel(0) ←
Ethylene derivs.   C
    from 1,3-dioxolane-2-thiones s. *19,* 962 suppl. 28

Platinum-asbestos ←
Enolethers from acetals   CHCH(OR)$_2$ → C : CH(OR)
s. *26,* 956

Via intermediates   *v.i.*

Indenes   CHCO → C : C
from indanones via indanols s. *29*, 840

Dehydration of alcohols   CHC(OH) → C : C
via thiobenzoylation s. *29*, 947
via mesylates s. *29*, 622

Ethylene derivs. from ketones   CHCO → C : CH
via modified Bamford-Stevens reaction of the tosylhydrazones s. *27*, 949; as part of a 5-step isomerization of α,β-ethyleneketones s. K. M. Patel and W. Reusch, Synth. Commun. *5*, 27 (1975); with LiH instead of MeLi cf. R. M. Moriarty, H. Paaren, and J. Gilmore, Chem. Commun. *1974*, 927

Cyclic ketones   O
from carboxylic acids via carboxylic acid chlorides – 4,5,6,7-Tetrahydrothianaphthen-7-ones s. *26*, 957

2-Cyclohexenone ring
from 1,5-diketones via cyclic 1,3-dienamines s. *27*, 950

Indoles from o-aminooxo compds.   O

663.

via 3-hydroxyindolines. A soln. of butyllithium in hexane added dropwise under argon to a soln. of diisopropylamine in abs. tetrahydrofuran, kept 0.5 hr. at 50°, cooled to –78°, treated dropwise with a soln. of 2-(N-benzylbenzamido)-5-chlorobenzophenone in abs. tetrahydrofuran, stirred 5 hrs. at –78°, the crude intermediate dissolved in pyridine, treated at –20° with POCl₃, and stirred 16 hrs. at room temp. – 1-benzoyl-5-chloro-2,3-diphenylindole. Y: 72%. F. e. s. H. Greuter and H. Schmid, Helv. *57*, 281 (1974).

Nitrogen ↑   CC⇑N

Without additional reagents   *w.a.r.*

Reactions of 1,1,2,2-tetraminoethanes   ←
s. *27*, 951

Tolans from N-tosylenazo compds.   Ar₂C : CH(N : NTs) → ArC≡CAr
s. *28*, 927

Ene-1,2-diamines   C(N<) : C(N<)
s. *24*, 940

Nitriles from enazides   CN
with C-acyl group migration s. *26*, 477

Benzene ring   O
s. *28*, 928

Synchronic 6-electron ring closure of 1,3,5-trienes
s. *28*, 929; benzene ring s. A. *1975*, 874

Δ³-Pyrazol-5-one ring
from enecarbamyl azides s. *27*, 952

Cyclopropane from spirodiazirine ring   ←
s. *29*, 948

2-Cyano-1,3-cyclopentenedione from azidoquinone ring   O
s. *26*, 958

α-**Dioxo compds. from N-imidosulfoximines**
  Ring contraction by fragmentation – Benzocyclobutenediones s. 28, 930
Ring contraction of azides

664.

N-Heterocyclics. 2-Azido-3-methylpyridine 1-oxide in benzene heated at 90° ⇢ 2-cyano-2-methyl-2H-pyrrole 1-oxide. Y: 89%. F. e. s. R. A. Abramovitch and B. W. Cue, Jr., Heterocycles 2, 297 (1974).

Indenones from 1,2,3-triazines
s. 28, 931

Flash vacuum pyrolysis
Isoindoles from 1,2,4-triazoles

665.

1,3,5-Triphenyl-1,2,4-triazole vaporized at 230°/0.01 mm Hg, and the vapor passed at 760° through a quartz tube packed with silicon carbide chips to increase contact time ⇢ 1,3-diphenylisoindole. Y: 95% based on startg. m. consumed; conversion 33%. F. e. s. T. L. Gilchrist, C. W. Rees, and C. Thomas, Soc. Perkin I 1975, 12; flash vacuum pyrolysis s. a. ibid. 1975, 45.

sym-**Oxepin oxides**
  by azo bridge elimination s. 30, 522
Elimination of azoxy bridges
s. 28, 932

Irradiation (s. a. under NaOH, Na₂S, Na⁺, and Cu₂O)
Cyanoketones from cyanohydrins
  with radical-induced nitrile group migration – Nitrite photolysis as radical source s. 28, 933
Double ring closure with ethylenediazo compds.
s. 29, 949

Elimination of azoxy bridges
s. 28, 932

2-Alkoxybenzofurans
  via benzocyclopropene-1,1-dicarboxylic acid esters s. 25, 670
Nitriles from azides
  with ring contractions – 2-Cyano-1,3-cyclopentenedione from azidoquinone ring s. 26, 958
Photofragmentation of spiro-Δ^1-pyrazolines
s. 25, 670
Dibenzo[c,e]-1,2-thiazine 1,1-dioxides
  from 2H-benzo[e]-1,2,3,4-thiatriazine 1,1-dioxides s. 29, 462

Electrolysis
Electrochemical Pschorr ring closure
 Phenanthrenes s. 27, 953

Lithium hydride   LiH
Cyclopropane ring from tosylhydrazones
s. 28, 934

Sodium hydroxide   NaOH
α,β-Ethyleneoximes from 1,3-aminooximes   CHC(N<) → C:C
s. 27, 263

Sodium hydroxide/irradiation   NaOH/⫴
Acetylene derivs.   C(:NNHTs)C(:NNHTs) → C≡C
 from 1,2-bis(tosylhydrazones) – Cycloalkynes s. 27, 954

Sodium hydroxide/methyl iodide   NaOH/MeI
α-H-Labeled α,β-ethylenecarboxylic acids   CH·CH(N<)COOH → C:CDCOOH
 from α-aminocarboxylic acids by Hofmann degradation s. 29, 950

Potassium hydroxide   KOH
2-Cyclohexenone ring from ketoeneurethans

666.

Startg. m. kept 1 hr. at 25° in a 2%-soln. of KOH in 4:1 methanol-water, then heated 2 hrs. at 70° → product. Y: ca. 90%. F. e. s. P. L. Stotter and K. A. Hill, Am. Soc. 96, 6524 (1974).

Sodium/alcohol   NaOR
Ring closure and expansion
 of diazo compds. – Stereospecific ring closure and expansion of acylamino-ketones via N-nitrosoacylamino- and diazo-ketones s. 28, 935

Sodium methoxide   NaOMe
Acetylene derivatives   C(:NNHTs)CH(OAc) → C≡C
 from α-acoxytosylhydrazones s. 26, 959

Δ⁴-Pyrrol-2-ones
 from N-acyl-N'-enehydrazines s. 29, 951

Sodium ethoxide   NaOR
trans-Ethylene derivs. from diazo compds.   CHCN₂ → C:CH
s. 30, 668

Potassium tert-butoxide   Me₃C·OK
Ethylene derivs. from hydrazinium salts   CHC·N⁺(R₂)·NH₂ → C:C

667.

N,N-Dimethyl-N-cyclohexylhydrazinium chloride added to a soln. of K-*tert*-butoxide in dimethyl sulfoxide, and refluxed 1.5 hrs. → cyclohexene. Y: 52%. F. e. s. H. Posvic and D. Rogers, J. Org. Chem. 39, 1588 (1974).

Methyllithium   *MeLi*
Ethylene derivs. from sulfonylhydrazones   CHC(: NNHTs) → C : CH
    Modified Bamford-Stevens reaction s. 27, 949; 1,3-dienes from α,β-ethylene-tosylhydrazones in 1,2-dimethoxyethane, $\Delta^{2,4}$-steroids, s. G. Phillipou and R. F. Seamark, Steroids 25, 673 (1975); with n-butyllithium, 1,1-deuterioethylene derivs., cf. J. E. Stemke and F. T. Bond, Tetrah. Let. *1975*, 1815; review of the Bamford-Stevens reaction s. J. Casanova and B. Waegell, Bl. *1975*, 922; cf. ibid. *1975*, 598

Sodium carbonate Na_2CO_3
Hofmann degradation   CHC(N⁺⇐) → C : C
    of quaternary ammonium salts – α-Methyleneketones s. 26, 960

Benzene ring from N-nitrosoureas   C
s. 27, 955

Sodium sulfide/irradiation Na_2S/⧸⧸⧸
Ethylene derivs.   C(NO₂)C(NO₂) → C : C
    from aliphatic 1,2-dinitro compds. s. 27, 956

Sodium salt/irradiation   Na⁺/⧸⧸⧸
Bicyclic S-heterocyclics   O
    from ethylenetosylhydrazones s. 27, 957

Triethylamine Et_3N
1,1-Nitroethylene derivs.   C(NO)CH(NO₂) → C : C(NO₂)
    from 1,2-nitronitroso compds. s. 29, 318

Copper (s. a. under CuO)   *Cu*
2-Arylpyrroles   O
    and 4,5-dihydropyrrolo[1,2-c]quinazolines via spiro-3H-indoline-3,2′-2′H-pyrroles s. 26, 961

Ring closure and expansion   O   ⊙
    with diazo compds. s. 27, 958

Cuprous oxide/irradiation Cu_2O/⧸⧸⧸
Cyclic ketones from α-diazoketones   O
    Stereospecific functionalization s. 28, 936; regiospecificity s. S. Chakrabarty et al., Synth. Commun. 5, 275 (1975)

Cupric oxide   *CuO*
Enolesters from 2-acoxydiazo compds.   ←
    with acoxy group migration s. 26, 962

Cupric oxide-copper   *CuO-Cu*
Cyclopropane ring by double ring closure   O
    of ethylenediazo compds. s. 22, 935 suppl. 28

Silver oxide Ag_2O
1,3-Dienes from 2-ethylene-tert-amines   CH·C : C·C·NR₂ → C : C·C : C
    via Hofmann degradation of quaternary ammonium salts s. 28, 937

Oxidations with argentic oxide   *AgO*
Acetylene derivs. from osazones   C(: NNH₂)C(: NNH₂) → C≡C
s. 28, 938

Cupric acetoacetonate $(Ac_2CH)_2Cu$
Ring closure with bis-(α-diazoketones)
    with subsequent dehydrochlorination s. 26, 963

Cupric sulfate $CuSO_4$
Cyclic ketones from α-diazoketones   O
s. 28, 936

Cuprous chloride/oxygen   *CuCl/O$_2$*
Acetylene derivs. from α-dihydrazones   C(: NNH$_2$)C(: NNH$_2$) → C≡C
s. *28*, 938

Lithium tetrahydridoaluminate   *Li[AlH$_4$]*
Reductive retrodiene scission   C
with O-deacetylation s. *27*, 959

Boron fluoride (s. a. under RN : C : NR)   *BF$_3$*
Acetylene derivs. from 2-diazoalcohols   CH(OH)C(: N$_2$) → C≡C
Acylacetylenes s. *28*, 939

Anthrasteroid rearrangement   ←
of azodicarboxylic acid ester adducts s. *28*, 940

Alcohol   *ROH*
Ring expansion of cyclic α-hydroxydiazo compds.   ◉
s. *26*, 661

Aldehydes   *RCHO*
Ethylene derivs. and *anti*-aldoximes   CHC(NHOH) → C : C
from hydroxylamines and aldehydes s. *27*, 960

Dicyclohexylcarbodiimide/boron fluoride   *RN : C : NR/BF$_3$*
***cis-* or *trans-*Ethylene derivs.**   CHCN$_2$ → C : CH
from diazo compds.

668.

$$p\text{-MeOC}_6\text{H}_4\text{-CH}_2\text{-C(=N}_2\text{)-COOEt}$$

$$\underset{H}{\overset{p\text{-MeOC}_6\text{H}_4}{>}}C=C\underset{H}{\overset{COOEt}{<}} \qquad \underset{H}{\overset{p\text{-MeOC}_6\text{H}_4}{>}}C=C\underset{COOEt}{\overset{H}{<}}$$

Ethyl α-diazo-p-methoxyphenylpropionate treated
at 0–5° with BF$_3$-etherate and dicyclohexyl- | 1.5 hrs. at 50° with Na-ethoxide
carbodiimide in methylene chloride → ethyl | in tetrahydrofuran → ethyl
cis- | *trans-*
Y: 71%. | p-methoxycinnamate.   Y: 72%.
F. e. s. N. Takamura, T. Mizoguchi, and S. Yamada, Tetrah. Let. *1973*, 4267;
Chem. Pharm. Bull. *23*, 299 (1975).

Trialkyloxonium fluoroborate   *R$_3$O$^+$ BF$^-_4$*
Ring closure with ring expansion   ○ ◉
Bicyclic β-diketones s. *26*, 837

m-Chloroperoxybenzoic acid   *m-ClC$_6$H$_4$COO$_2$H*
Ethylene derivs. from tert. amines   CHCNR$_2$ → C : C
Protection of carbon-carbon double bonds as tert. amines

669.

A soln. of the startg. m. in chloroform treated with a soln. of m-chloroperoxybenzoic acid in the same solvent, allowed to stand overnight at 5°, washed with cold aq. 5%-NaHCO$_3$ and water, then evaporated in vacuo → 2,3-epoxyhelenalin. Y: ca. 100%. The highly reactive α-methylene groups of γ-lactones can be protected as tert. amines. K.-H. Lee et al., J. Med. Chem. *18*, 59 (1975); cf. **Cope elimination; synthesis of α-methylene-γ-lactones, review,** cf. R. B. Gammill, C. A. Wilson, and T. A. Bryson, Synth. Commun. *5*, 245 (1975).

Ethylene derivs. from aziridines C
Stereospecific ring opening s. 26, 964
Methyl iodide (s. a. under NaOH) MeI
Pyrrole and indole ring closure ○
s. 4, 843 suppl. 29
Stannous chloride SnCl₂
Ethylene derivs. C(NO₂)C(NO₂) → C : C
from aliphatic 1,2-dinitro compds. s. 27, 956 suppl. 30
Lead tetraacetate (CH₃COO)₄Pb
Acylcyanides from semicarbazones CH : NNHCONH₂ → COCN
s. 29, 952
Oxygen s. under CuCl O₂
p-Toluenesulfonic acid TsOH
α,β-Ethylene- from β-amino-ketones COCHC(NH₂) → COC : C
s. 28, 204
1,2,3,4-Tetrahydro-2-oxocarbazoles ○
s. 29, 953
Sulfuric acid H₂SO₄
5,11a-Dihydroindolo[3,2-b][1,4]benzoxazines
s. 28, 941
Triphenylcarbonium perchlorate Ph₃C⁺ ClO₄⁻
Ring closure with Mannich bases
Phenalenium ring s. 26, 965
Hydrogen chloride HCl
Anthracenes ←
from 9,10-dihydro-9-nitroanthracenes s. 28, 644
2-Cyclohexenone ring from ketoalkoximes ○
s. 28, 896
3-Hydroxycarbostyrils by ring expansion ⊙
s. 29, 625
Ferrous iodide FeI₂
Ethylene derivs. from aziridines C
s. 22, 938 suppl. 29
Nickel chloride NiCl₂
Pyrrole ring from azines ○
s. 26, 740

Halogen ↑ CC⇑Hal

Without additional reagents w.a.r.
α,β-Ethyleneamidines CHalCHCN → C : C·C(: NR)NR′
from β-halogenonitriles s. 30, 219
1,3-Dienecarbamyl chlorides CHCCl → C : C
from 3-chlor-1-ethylenecarbamyl chlorides s. 27, 545
Rearrangement of 3,4-dihydro-2H-1,4-thiazines
s. 29, 954
Irradiation (s. a. under K/Me₃C·OH) ⚡
Cyclodehydrohalogenation ○
s. 28, 892

Electrolysis

Acetylene from polyhalogenoethylene derivs. C≡C
s. *27*, 961

Sodium hydride NaH
Enecyclopropanes
s. *29*, 864

Intramolecular C-alkylation
of Reissert compds. s. *28*, 942

Sodium hydride/dimethylformamide NaH/Me₂NCHO
Cycloalkenes from dihalides
s. *26*, 966

Sodium hydroxide NaOH
Cyclopropane ring from halides
with retention of α-diazoketo groups s. *26*, 967; cyclopropanes from iodides s. N. O. Brace, J. Org. Chem. *40*, 851 (1975).

4-Phosphabicyclo[3.1.0]hex-2-ene 4-oxides
from 4-iodomethyl-2-pholen 1-oxides s. *27*, 133

Condensed rings
by ar. dehydrochlorination s. *27*, 962

Cyclopropanes from cyclobutanes

670.

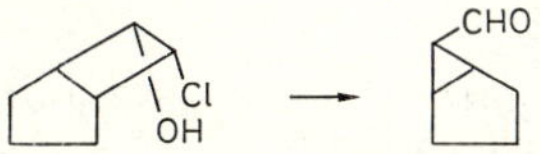

Cyclopropylaldehydes. A mixture of 7-*endo*-chlorobicyclo[3.2.0]heptan-6-*endo*-ol and 1 *N* NaOH stirred 1.5 hrs. at room temp. → bicyclo[3.1.0]hexane-6-*endo*-carbaldehyde (containing ca. 10% *exo*-isomer). Y: 83%. F. e. s. M. Rey and A. S. Dreiding, Helv. *57*, 734 (1974).

Potassium hydroxide KOH
Dehydrofluorination CHal·CH·CHal·CH → C:C·C:C
1,3-Dienes by bis(dehydrohalogenation) s. *27*, 963

Potassium hydroxide/alcohol
Ethylene derivs. from halides CHCHal → C:C
s. *30*, 607

Potassium hydroxide/alcohol-perchloric acid KOH/ROH-HClO₄
Dehydrobromination
with cyclopropane ring opening s. *29*, 955

Sodium/alcohol NaOR
1,2,3-Trienes C:C:C:C
s. *28*, 943

Sodium methoxide NaOMe
Arene oxides CHCBr → C:C
s. *28*, 221

1H-Azepines
from 1,4-dihydrobenzenes s. *26*, 968

Potassium/tert-butanol-irradiation K/Me₃C·OH/⚡
Cyclodehydrohalogenation
Aporphines – Cyclic ketones s. *29*, 956; in liq. NH₃ s. a. Am. Soc. *97*, 2507 (1975).

Potassium tert-butoxide (s. a. under $AgNO_3$) $KOCMe_3$

Potassium tert-butoxide/dimethyl sulfoxide $KOCMe_3/Me_2SO$
Dehydrohalogenation   CHCHal → C:C
s. 28, 944

Dehydrochlorination   CHCCl → C:C
s. 28, 944; s. a. M. F. Grundon and H. M. Okely, Soc. Perkin I 1975, 150

Acetylene derivs. from α,β-ethylenehalides   CH:CHal → C≡C
Cyclic compds. – Prepn. of cyclic acetylene derivs., review s. 27, 964

9-Thiabicyclo[4.2.1]nona-2,4,7-triene 9,9-dioxide ring
by bishomoconjugative Ramberg-Bäcklund rearrangement s. 26, 969

n-Butyllithium   *n-BuLi*
Ethylene derivs.   CBrC(OR) → C:C
from 1,2-alkoxyhalides and 1,2-dihalides – Effect of solvent on stereospecificity
s. 29, 957

Acetylene dervis.   CH:CBr₂ → C≡CH
from dibromomethylene compds. s. 28, 819

α-Chloroketones from 2,2-dichloroalcohols   ←
with rearrangement, also with ring expansion s. 28, 945

Sultams from halogenosulfonic acid amides   ⊙
Intramolecular C-alkylation

671.

A soln. of n-butyllithium in hexane added dropwise below –65° under N₂ to a soln. of N-(2-bromohexyl)-N-methylethanesulfonamide in tetrahydrofuran, and stirring continued 1.5 hrs. → 2-n-butyl-N,1-dimethylpropanesultam. Y: 86%. F. e. s. M. Kobayashi et al., Synthesis 1973, 667.

Cyclic ketones   ⊙
from cyclic 2-hydroxy-1,1-dibromides

672.

Ring expansion. A soln. of 2.1 moles n-butyllithium in hexane added dropwise at –78° under N₂ during 0.5 hr. to a stirred soln. of the startg. m. in dry tetrahydrofuran, stirring continued 0.5 hr. at –78° and 5 min. at 0° → cyclotridecanone. Y: 89%. F. e. s. H. Taguchi, H. Yamamoto, and H. Nozaki, Am. Soc. 96, 6510 (1974).

Pyridines from 2H-pyrrolenines
s. 27, 965

n-Butyllithium/cuprous iodide   *n-BuLi/CuI*
Ethylene derivs. from 1,2-dibromides   CBrCBr → C:C
s. 29, 958

n-Butyllithium/silver iodide   *n-BuLi/AgI*
Cyclobutane ring from dibromides   □
s. 27, 966

Lithium/liq. ammonia   *Li/NH$_3$*
Reductive C-arylation   CO → CHAr
Aryl derivs. from oxo compds.

An ethereal soln. of bromobenzene added to excess Li-foil in anhydrous ether, after 1 hr. an ethereal soln. of 3-heptanone added slowly, stirred 1 hr., then NH$_3$ distilled into the mixture, and when a blue color persists for ca. 10 min. NH$_4$Cl added cautiously during ca. 5 min. → 3-phenylheptane. Y: 96%. – Similarly: (+)-Pulegone → 3-phenyl-4(8)-p-menthene. Y: 85%. F. e., also from aldehydes, s. S. S. Hall and F. J. McEnroe, J. Org. Chem. *40,* 271 (1975).

Lithium diethylamide   *Et$_2$NLi*
α-**Metalation of nitriles**
   Intramolecular C-alkylation – β-Hydroxynitriles from oxo compds. s. *29,* 959

Lithium piperidide   *C$_5$H$_{11}$NLi*
Ketones from 1,2-halogenhydrins
   with rearrangement – Ring expansion – α-Halogenoketones s. *28,* 946

Sodium/liq. ammonia   *Na/NH$_3$*
Ethylene derivs. from 1,2-dihalides   CHalCHal → C:C

7,8-Dichlorobicyclo[4.2.0]oct-2-ene in dry ether added under N$_2$ to a stirred soln. of Na in liq. NH$_3$, stirring continued 1.5 hrs. → bicyclo[4.2.0]octa-2,7-diene. Y: 83%. F. e. s. E. L. Allred, B. R. Beck, and K. J. Voorhees, J. Org. Chem. *39,* 1426 (1974).

Potassium bis(trimethylsilyl)amide   *(Me$_3$Si)$_2$NK*
Stereospecific intramolecular C-alkylation
   Reversal of stereospecificity s. *28,* 947; cyclopropanes (cf. Synth. Meth. *15,* 680) and cyclobutanes by intramolecular alkylation of chlorides with Na-bis(trimethylsilyl)amide cf. G. Stork, J. C. Depezay, and J. d'Angelo, Tetrah. Let. *1975,* 389

Lithium carbonate   *Li$_2$CO$_3$*
1,3-Dienes from 1,1-dibromides
s. *29,* 960

Sodium carbonate   *Na$_2$CO$_3$*
Cyclobutane ring opening
   by heterolytic fragmentation s. *26,* 927; 5-subst. uracils cf. A. Weber, R. J. Balchunis, and J. S. Swenton, Chem. Commun. *1975,* 601

Sodium hydrogen carbonate $NaHCO_3$

$\Delta^{8,14}$-Morphinanes by skeletal rearrangement   ←

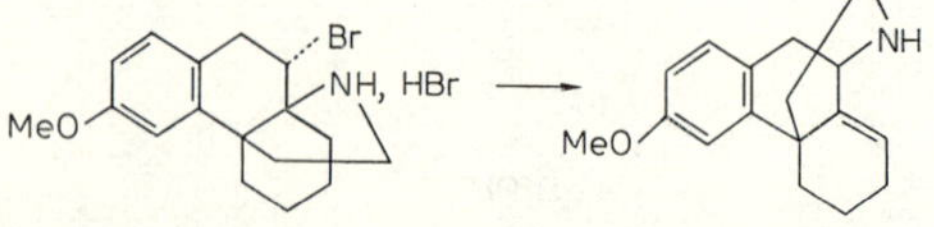

Startg. m. heated 1.5 hrs. at 130–135° with 1 equivalent $NaHCO_3$ in dimethylformamide → 3-methoxy-$\Delta^{8,14}$-morphinane. Y: 70%. I. Monković et al., Am. Soc. 95, 7910 (1973).

Sodium acetate CH_3COONa

2-Alkyltropones   C

by stereospecific ring opening s. 27, 967

Sodium azide NaN_3

α-Azido-α,β-ethyleneketones $COCHICN_3$ → $COC(N_3)$: C

from β-azido-α-iodoketones s. 26, 970

Sodium thiosulfate $Na_2S_2O_3$

Ethylene derivs. from 1,2-dibromides   CBrCBr → C : C

s. 28, 950

Potassium fluoride   KF

Dehydrohalogenation with fluoride ion   CH : CHal·CH ⤨ C≡C·CH / CH : C : C

Acetylene derivs. and allenes from α,β-ethylenehalides s. 29, 985

Lithium chloride/dimethylformamide/ammonia $LiCl/Me_2NCHO/NH_3$

Ethylene- from chloro-lactams   CHCCl → C : C

s. 26, 971

Ammonia s. a. under LiCl NH_3

Triton B   ←

Acetylene derivs. by dehydrohalogenation   C≡C

s. 23, 948 suppl. 29

α,β-Acetylene- from α,β-ethylene-α-halogeno-ketones   CH : CHal → C≡C

$CBr_2=CClCH=ClCOCH_3$ → $CBr_2=CCl-C≡C-COCH_3$

A soln. of 40% excess Triton B in benzene-methylene chloride added all at once at −30° to a soln. of the startg. m., and the product isolated after 3 min. → acetyleneketone. Y: 70%. F. e. s. A. Gorgues and A. Le Coq, C. r. 278 (C), 1153 (1974).

Triethylamine Et_3N

Dichloromethylene from trichloromethyl compds. $CHCl_3$ → $C : CCl_2$

s. 21, 950 suppl. 29

α,β-Ethylene-N-fluoroimines $C(Cl)CHCHNF_2$ → C : C·C : NF

from 3-chloro-N,N-difluoroamines s. 28, 459

N-Sulfonylenazo compds.   CHal·C : N·NH·SO_2R → C : C·N : N·SO_2R

from α-halogenosulfonylhydrazones s. 29, 962

Furo[2,3-b]quinolines   ◯

s. 29, 247

1,5-Diazabicyclo[4.3.0]non-5-ene ←
Dehydrohalogenation CHCHal → C:C
s. 28, 895

Dehydrobromination CHCBr → C:C
with decarbomethoxylation via methyl ester cleavage – Cleavage of hindered carboxylic acid methyl esters – Ketones from β-ketocarboxylic acid esters s. 29, 963

2-Ethylenealcohols from 1,2-iodohydrins C(OH)CICH → C(OH)C:C
with protection of alcohol groups by O-silylation – 1′,2′-Ethylenenucleosides – Unsatd. carbohydrates, review s. 29, 964

1,8-Diazabicyclo[5.4.0]undec-7-ene ←
Cyclopropane ring from halides ▽
with NaOH cf. 26, 967; with 1,8-diazabicyclo[5.4.0]undec-7ene s. T. Kamiya et al., Am. Soc. 97, 5020 (1975)

Diethylaniline PhNEt$_2$
Xanthenes O
from 6-(2-allylphenoxy)-3-chloropyridazines s. 29, 965
Ring closures via o-Claisen rearrangement
Thianaphthenes via thio-Claisen rearrangement

677.

A soln. of 2-chloroallyl phenyl sulfide in N,N-diethylaniline heated ca. 24 hrs. at 225° under N$_2$ → 2-methylbenzo[b]thiophene. Y: 55–80%. Also **benzofurans** by a 2-step process s. W. K. Anderson and E. J. LaVoie, Chem. Commun. *1974*, 174.

Pyridine (s. a. under Cu, Ac$_2$O, and POCl$_3$) C$_5$H$_5$N
Dehydrobromination ←
with skeletal rearrangement of S-heterocyclics s. 29, 966

Copper/pyridine Cu/C$_5$H$_5$N
3-Hydroxypyrazole ring O
3-Hydroxy-1H-pyrazolo[3,4-b]pyridines s. 28, 948

Cuprous iodide s. under n-BuLi CuI

Silver nitrate/potassium tert-butoxide AgNO$_3$/Me$_3$C·OK
Intramolecular C-alkylation of ketones
Tris-bridged cyclophanes s. 29, 967

Silver nitrate/dimethyl sulfoxide AgNO$_3$/Me$_2$SO
Dehydrohalogenation CHCHal → C:C
with rearrangement s. 28, 949

Silver iodide s. under n-BuLi AgI

Zinc Zn
Ethylene derivs. from 1,2-nitrochlorides C(NO$_2$)CCl → C:C
s. 26, 973

Ethylene derivs. from 1,2-acetoxyiodides ClC(OAc) → C:C
s. 27, 931

Ethylene derivs. from 1,2-dihalides CHalCHal → C:C
s. 28, 950

Conjugated carbon-carbon double bonds ←
from ω,ω'-dibromides s. 27, 968

2-Ethylenealcohols from 2-oxidohalides ⟨O⟩–CHal → C(OH)C : C
s. 29, 968

Isopropylmagnesium bromide $Me_2CHMgBr$
Ketones from 1,2-halogenhydrins ←
with rearrangement s. 24, 964

Boron fluoride BF_3
Steroid backbone rearrangement ←
s. 26, 974

Aluminum chloride $AlCl_3$
α,β-Ethyleneketones from 1,2-oxido-1-chlorides ←
with methyl group migration s. 28, 951

Indenones O
from 2-chlorovinyl ketones s. 288, 952

N-Heterocyclics by dehydrobromination
1,2,3,4-Tetrahydroisoquinolines s. 27, 969

Water H_2O
Cyclic ene-α-diones by 1,4-elimination ←
s. 26, 975

Ethylene oxide/tetra-n-butylammonium chloride ⟨O⟩ /n-Bu_4N^+ Cl^-
Ethylene derivs. from halides CHCHal → C : C
Partial dehydrohalogenation – Oxido compds. as bases in the presence of halide ions s. 29, 969

Dimethylformamide s. under LiCl Me_2NCHO

Acetic anhydride/pyridine Ac_2O/C_5H_5N
1,1-Nitroethylene derivs. from 1,2-nitroiodides $CH(NO_2)CI → C(NO_2) : C$
s. 26, 976

Formic acid (s. a. under HClO_4) $HCOOH$
α,β-Ethylenecarboxylic from carboxylic acids CHCHCOOH → C : C·COOH
s. 27, 918

Dimethyl sulfoxide s. under AgNO_3 Me_2SO

Titanium trichloride $TiCl_3$
N-Condensed heterocyclics from N-halogenamines O

678.

Aq. 15%-TiCl_3-soln. added at –10° under N_2 to a soln. of N-chloro-N-allylpent-4-enylamine in 1 : 1 acetic acid-water until a iodometric test is negative → product. Y: 66%. F. e. s. J.-M. Surzur and L. Stella, Tetrah. Let. *1974*, 2191.

Tributyltin hydride Bu_3SnH
Radical ring closure of ethylenehalides
s. 29, 970

Stannic chloride $SnCl_4$
Cyclic ketones from carboxylic acid chlorides
s. 26, 957

Phosphorus oxide chloride/pyridine $POCl_3/C_5H_5N$
Ethylene derivs. from iodohydrins   C(OH)Cl → C:C
s. 27, 970

Sulfuric acid H_2SO_4
Isocyclics from chlorides
    with formation of cyclic ketones from carboxylic acid esters s. 28, 953
Double ring closure
    with double position shift of substituents of isocycles – Indeno[1,2,3-de]quinol-2(3H)-ones s. 29, 971
4,5-Benzotropones with elimination of O-bridges   ←

679.

Startg. m. dissolved in concd. H_2SO_4, and rapidly quenched in ice-water → 4,5-benzotropone. Y: ca. 100%. Also with trifluoroacetic acid s. P. F. Ranken et al., Synth. Commun. **3**, 311 (1973).

Chromous sulfate $CrSO_4$
(Dien)olethers   CHal·C:C·C(OR)$_2$ → C:C·C:C(OR)
    from α,β-ethylene-γ-halogenoketals s. 27, 971

Perchloric acid s. a. under KOH $HClO_4$

Perchloric acid/formic acid $HClO_4/HCOOH$
2-Cyclohexenone ring
from γ,δ-ethylene-δ-chloroketones

680.

2-Methyl-2-(3-chloro-2-butenyl)cyclohexanone refluxed 1.5 hrs. in formic acid containing aq. 60%-$HClO_4$ → product. Y: 98%. F. e. s. M. Kobayashi and T. Matsumoto, Chem. Lett. **1973**, 957.

Tetra-n-butylammonium chloride s. under $n\text{-}Bu_4N^+\,Cl^-$

Hydrogen chloride   HCl
2H-3,4-Dihydropyran-5-carboxaldehydes   ←
    from 7,7-dichloro-2-oxabicyclo[4.1.0]heptanes via 3-chloro-2-*tert*-butoxy-2,5,6,7-tetrahydrooxepins s. 27, 972

Hydrogen bromide   HBr
Cyclopent-2-enones
from dichlorocyclopropylcarbinols

681.

Startg. m. treated 2 hrs. at 100° with 47%-HBr → 3,5-dimethyl-2-cyclopentenone. Y: 83%. F. e. s. T. Hiyama, M. Tsukanaka, and H. Nozaki, Am. Soc. **96**, 3713 (1974).

Nickel carbonyl $Ni(CO)_4$
Ethylene derivs. from 1,2-dihalides   CClCCl → C:C
    cis,cis-Dihydrophosphepine derivs. s. 26, 977

Sulfur ↑   CC⇑S

Without additional reagents   w.a.r.
Ethylene derivs. from sulfonylsulfilimines   C : C
s. 26, 978; s. a. T. Yamamoto and M. Okawara, Chem. Lett. *1975*, 581

Enediazonium hexachloroantimonates   CHal·C : NNHTs → C : C·N_2^+
from 2-halogenotosylhydrazones s. 27, 973; cf. B. *108*, 402 (1975)

3-Alkylidene-1,2-dithioles   ←
from 3-organothio-1,2-dithiolium salts s. 26, 979

Acetylene derivs. from thiirene 1,1-dioxides   C
s. 26, 972

Naphthalenes   ←
from benzo[b]cyclobuta[d]thiophenes s. 28, 954

Irradiation (s. a. under Me$_2$CO)   ⚡
Cyclopentadienones from 4-thiopyrones   Ö
s. 28, 955

Potassium hydroxide   KOH
Aminodiaryls   ←
from ar. sulfonic acid amides s. 28, 956

Potassium hydroxide/tert-butanol s. under CCl$_4$ KOH/$Me_3C·OH$

Sodium/alcohol   NaOR
Benzo[a]pyrenes   O

682.

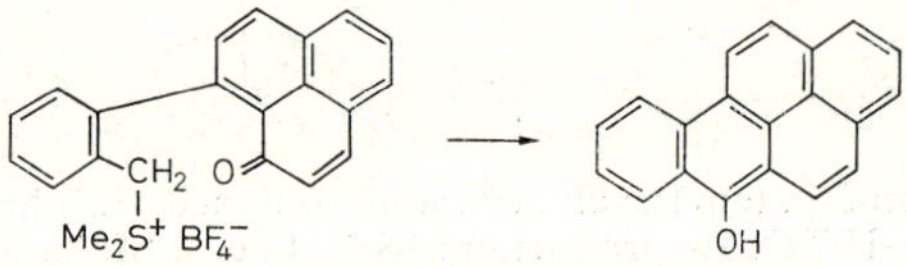

Dimethyl-[o-(9-phenalenonyl)benzyl]sulfonium fluoroborate allowed to react with Na-methoxide in methanol → 6-hydroxybenzo[a]pyrene. Y: 67%. M. S. Newman and L. Lee, J. Org. Chem. *39*, 1446 (1974).

Potassium tert-butoxide s. under Ph$_3$P $Me_3C·OK$

Methyllithium s. under Pyrocatechyl phosphorochloridite   MeLi

n-Butyllithium   n-BuLi
Δ^1-Pyrazoline ring from ethylenetosylhydrazones   O
Intramolecular regiospecific 1,3-dipolar cycloaddition

683.

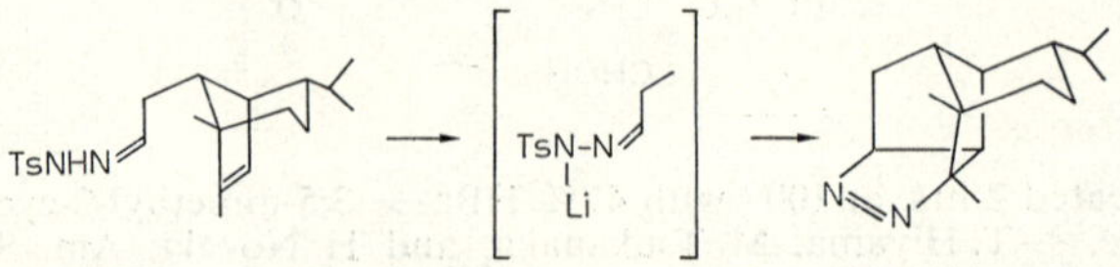

Startg. m. allowed to react with *n*-butyllithium in tetrahydrofuran, and the resulting Li-salt heated at 105–130°/0.1 mm with distillation of the product → pyrazoline deriv. Y: 72%. E. Piers, M. B. Geraghty, and M. Soucy, Synth. Commun. *1973*, 401.

Lithium amide	$LiNH_2$
Thiodienolethers by 1,4-dethiolation	$C(SR)C:C \cdot CHSR \rightarrow C:C \cdot C:C \cdot SR$

684. $Me_3C-S-CH_2CH=CHCH_2-S-CMe_3 \rightarrow CH_2=CHCH=CH-S-CMe_3$

A suspension of $LiNH_2$ in liq. NH_3 added portionwise to a well-stirred mixture of the startg. bisthioether and liq. $NH_3 \rightarrow$ *trans*-1,3-butadienyl *tert*-butyl sulfide. Y: 88%. F. e. s. R. H. Everhardus et al., R. *93*, 90 (1974).

Sodium amide	$NaNH_2$
Ynamines from carboxylic acid thioamides s. *26,* 980	$CH_2CSN \rightarrow C\equiv C \cdot N <$
Potassium amide	KNH_2
Indoles from 4H-3,1-benzothiazines s. *28,* 957	⟵
Sodium hydrogen carbonate s. under CH_3I	$NaHCO_3$
Sodium acetate/acetic acid	CH_3COONa/CH_3COOH
α-Diketones	C
from 3-hydroxy-4-ketotetrahydrothiophene 1,1-dioxides s. *28,* 724	
Lithium bromide s. under $(Me_2NCH_2)_2PPh$	$LiBr$
Pyridine/triethylamine	C_5H_5N/Et_3N
Skeletal rearrangement	⟵
of cyclic glycol monosulfonates s. *27,* 974	
Silver oxide	Ag_2O
Ethylene derivs.	C
from mercaptans and thioethers – Ethyleneacetals from cyclic 1,1-chlorothioethers s. *29,* 972	
Cupric acetoacetonate s. under $N_2CHCOOEt$	$(Ac_2CH)_2Cu$
Silver carbonate	Ag_2CO_3
Ethylene derivatives from mercaptans s. *29,* 972	$CHC(SH) \rightarrow C:C$
Zinc s. under $TiCl_4$	Zn
Aluminum amalgam	Al,Hg
Ethylene derivs. from β-hydroxysulfoximines s. *29,* 855	⟵
Fluoroboric acid	HBF_4
Cyclobutanone ring	□
from 1-phenylthiocyclopropylcarbinols – Spiroannelation s. *28,* 958; cf. J. Org. Chem. *40,* 2013 (1975)	
Aluminum chloride	$AlCl_3$
Δ⁴-Imidazoline-2-thiones	⟵
from dithiohydantoins s. *26,* 981	
Gallium bromide	$GaBr_3$
Imidazoles	⟵
by retro-benzilic acid-type rearrangement s. *28,* 959	
Acetone/irradiation	$Me_2CO/⚡$
Ring opening by elimination of sulfur dioxide s. *17,* 968 suppl. 28	C
Ethyl diazoacetate/cupric acetoacetonate	$N_2CHCOOEt/(Ac_2CH)_2Cu$
Ethylene derivs. from sulfido compds. s. *27,* 976 suppl. 30	$\triangle_S \rightarrow C:C$

Dimethylformamide $\qquad$ Me_2NCHO
Ethylenenitriles $\qquad$ C
from N-chlorosulfonyl-2-azetidinones s. *27*, 975

Succinic acid anhydride $\qquad$ $\begin{matrix} CH_2CO \\ | \\ CH_2CO \end{matrix} \Big\rangle O$

Ethylene derivs. $\qquad$ CHC·SR → C:C
from thioethers via sulfoxides

685.

α-Methylenelactones. Startg. sulfide treated with m-chloroperoxybenzoic acid in methylene chloride, the resulting sulfoxide heated 0.5 hr. at 140° with 3 equivalents succinic anhydride, finally treated with 10%-HCl → dl-avenaciolide. Y: 80%. J. L. Herrmann, M. H. Berger, and R. H. Schlessinger, Am. Soc. *95*, 7923 (1973); **α-methylene-ketones and -carboxylic acids** cf. H. J. Reich and J. M. Renga, Chem. Commun. *1974*, 135; pyrolysis at 110° in toluene and pyridine, **α,β-ethylene-γ-lactones,** cf. K. Iwai et al., Chem. Lett. *1974*, 385, 1237.

Acetic acid s. under CH_3COONa $\qquad$ CH_3COOH

Methyl iodide $\qquad$ CH_3I
Ethylene derivs. and hydrocarbons $\qquad$ ⟨S⟩ → C:C
from sulfido compds. s. *27*, 976; ethylene derivs. from sulfido compds. with ethyl diazoacetate in the presence of Cu(II)-acetoacetonate cf. Y. Hata et al., Am. Soc. *97*, 2553 (1975)

Methyl iodide/sodium hydrogen carbonate $\qquad$ $MeI/NaHCO_3$
Ethylene derivs. from thioethers $\qquad$ CHC(SR) → C:C
s. *28*, 960; with methyl fluorosulfonate and triethylamine, didehydropeptides, cf. D. H. Rich and J. P. Tam, Tetrah. Let. *1975*, 211

Carbon tetrachloride s. a. under Ph_3P $\qquad$ CCl_4

Carbon tetrachloride-potassium hydroxide/tert-butanol $\qquad$ CCl_4-$KOH/Me_3C·OH$
Ethylene derivs. from sulfones $\qquad$ $CHSO_2CH$ → C:C
s. *29*, 973; s. a. P. A. Grieco and D. Boxler, Synth. Commun. *5*, 315 (1975)

Silica $\qquad$ SiO_2
Ethylene derivs. from thiirane 1-oxides $\qquad$ C
s. *29*, 767

Titanium tetrachloride/zinc $\qquad$ $TiCl_4/Zn$
Ethylene derivs. from 2-hydroxythioethers $\qquad$ C(OH)C(S·) → C:C
Thioenolethers from α-hydroxymercaptals s. *29*, 974; ethylene derivs. s. a. Chem. Lett. *1974*, 1161; preferably from 2 benzoyloxythioethers s. ibid. *1974*, 1523

Stannic chloride $\qquad$ $SnCl_4$
Cyclobutanone ring $\qquad$ ⊙
from 1-phenylthiocyclopropylcarbinols s. *28*, 958 suppl. 29

Triphenylphosphine $\qquad$ Ph_3P
Sym. ethylene derivs. from 1,3,4-thiadiazolines $\qquad$ C:C
s. *30*, 566

1,1-Dinitriles from dicyanosulfonium ylids $\qquad$ $>C:SR_2$ → $>CR_2$
s. *27*, 977

Amidines from isothioureas $C \cdot S \cdot C(:NR)NR'_2 \rightarrow C \cdot C(:NR)NR'_2$

686.
$Me_3C \cdot COCH_2—S—C\genfrac{}{}{0pt}{}{NEt_2}{N—CH_2Ph} \rightarrow Me_3C \cdot COCH_2—C\genfrac{}{}{0pt}{}{NEt_2}{N—CH_2Ph}$

β-Ketoamidines. Startg. m. heated 6 hrs. at 120° with 1 equivalent triphenylphosphine → product. Y: 88%. F. e., also in refluxing dioxane, and limitations s. W. Heffe, R. W. Balsiger, and K. Thoma, Helv. 57, 1242 (1974).

Isocyclics from thioozonides ◌
s. 26, 125

Triphenylphosphine/potassium tert-butoxide $Ph_3P/Me_3C \cdot OK$
Ethylene derivs. $CCl \cdot S \cdot CH \rightarrow C:C$
from thioethers via 1,1-chlorothioethers

687. $PhCH_2—S—CH_2Ph \rightarrow PhCHCl—S—CH_2Ph \rightarrow PhCH=CHPh$

Stilbenes. Dil. streams of dibenzyl sulfide in CCl_4 and SO_2Cl_2 in the same solvent run together → mono-α-chlorodibenzyl sulfide (Y: ca. 90%) dissolved in dry tetrahydrofuran under N_2, triphenylphosphine added followed by K-*tert*-butoxide, heated to reflux during 15 min. and refluxing continued 24–36 hrs. → stilbene (Y: 82%; *trans* : *cis* 10 : 1). F. e. s. R. H. Mitchell, Tetrah. Let. *1973*, 4395; **acetylene derivs., tolans, from 1,1,1-dichlorothioethers** cf. Chem. Commun. *1973*, 955.

Triphenylphosphine/carbon tetrachloride Ph_3P/CCl_4
Carbodiimides $NHCSNH \rightarrow N:C:N$
from ureas and thioureas s. 27, 978

Bis-(3-dimethylaminopropyl)phenylphosphine/
lithium bromide $(Me_2NCH_2CH_2CH_2)_2PPh/LiBr$
Ketones from thiolic acid esters $COSR \rightarrow COR$
β-Diketones s. 26, 982

Triethyl phosphite $(EtO)_3P$
Elimination of sulfur ←
Enamines from thioiminoesters – Vinylogous carboxylic acid amides – Ketones from thiolic acid esters – *β*-Diketones s. 26, 982

Pyrocatechyl phosphorochloridite/methyllithium ←
Ethylene derivs. from 2-hydroxythioethers $C(OH)CH_2(SR) \rightarrow C:CH_2$
Terminal ethylene derivs. s. 27, 979

Potassium hydrogen sulfate $KHSO_4$
Thioenolethers from mercaptals $CHC(SR)_2 \rightarrow C:C(SR)$
s. 23, 971 suppl. 28

Thiophenes from enedisulfides ○
s. 27, 980

Sulfuric acid H_2SO_4
α,β-Ethyleneketones $C(SR)CHC(Hal):C \rightarrow C:C \cdot CO \cdot CH$
from 3-ethylene-3-halogenothioethers – Carboxylic acids from their esters s. 28, 961

Iodine I
Ethylene derivs. from sulfido compds. $\triangle_S \rightarrow C:C$
s. 27, 976

Hydrogen chloride   *HCl*
Ring closure with acetals   ◯
    Modified Pomeranz-Fritsch isoquinoline ring closure – Isoquinolines s. 29, 975; 6H-pyrido[4,3-b]carbazoles s. R. W. Guthrie et al., J. Med. Chem. 18, 755 (1975)

Nickel   *Ni*
Aldehydes-1-d and *trans*-stilbenes-α,α'-d_2 from s-trithianes-d_3   CD : CD

688.

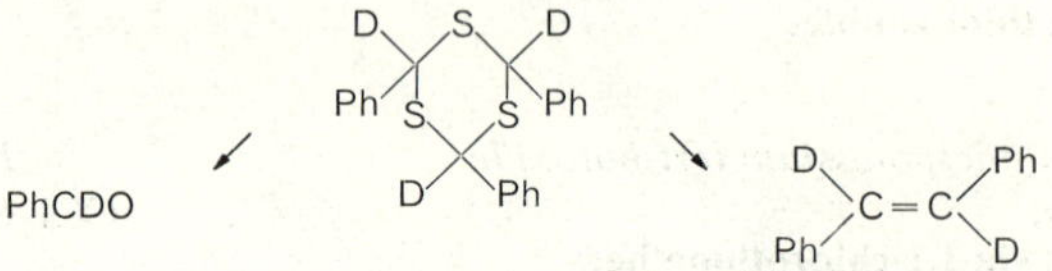

PhCDO   ←   [trithiane-d_3 structure]   →   Ph$_2$C=CD$_2$ (Ph,D / Ph,D)

Triphenyl-s-trithiane-d_3 dissolved
in aq. 90%-methanol, refluxed 2 min. with HgO, a soln. of HgCl$_2$ in the same solvent added, and refluxed 4 hrs. under N$_2$ → α-d-benzaldehyde. Y: 84%. | in dry benzene, and refluxed 5 hrs. with Raney-Ni → α,α'-d_2-*trans*-stilbene. Y: 80%.

F. e. s. J. B. Chattopadhyaya, A. V. R. Rao, and K. Venkataraman, Indian J. Chem. 11, 987 (1973).

Remaining Elements ↑   CC⇑Rem

Without additional reagents   w.a.r.
α-Cyano-α,β-ethylenecarboxylic acid amides   C(P$^+$)⇌CH → C : C
    from 2-carbamyl-2-cyanophosphonium salts s. 28, 588

Irradiation   ⚡
3H-Benzo[c]thiophen-1-ones   Ö:
    from sulfur-donor ligand o-metalated complexes s. 29, 643

Electrolysis   ↯
Hydrocarbons from boranes   2 BR$_3$ → 3 R·R
    Coupling s. 29, 976

Sodium hydride   *NaH*
Allenes   C : C : C

689.

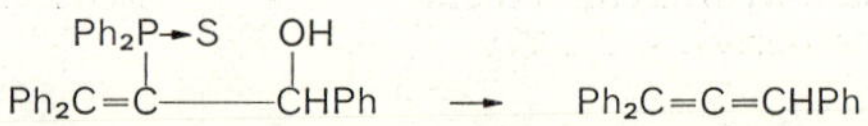

Ph$_2$P→S   OH
Ph$_2$C=C———CHPh   →   Ph$_2$C=C=CHPh

Startg. hydroxyphosphine sulfide deriv. allowed to react at room temp. with NaH in tetrahydrofuran → product. Y: 85%. F. e., also from analogous phosphine oxides, s. M. Baran Marszak, M. Simalty, and A. Seuleiman, Tetrah. Let. 1974, 1905.

Heterocyclics from 2-acetylenephosphonium salts   ◯
    Quinolines by intramolecular Wittig synthesis s. 28, 962

Sodium/alcohol   *NaOR*
2-Cyclohexenone ring
    s. 29, 666

Potassium/tert-butanol   *K/Me$_3$C·OH*
Cyclopropylketones   ▽
    from acoxyphosphonium salts s. 28, 963

Sodium hydroxide s. under I   *NaOH*

Sodium acetate/acetic acid CH_3COONa/CH_3COOH

Ethylene derivs. from 2-hydroxysilanes $C(OH)C(Si\lneq) \rightarrow C:C$
s. 28, 964; cis- or trans-ethylene derivs. by variation of reaction conditions s. Am. Soc. 97, 1464, 2263 (1975)

Sodium iodide   *NaI*

Ethylene derivs. by deoxymercuration $CHC(HgOAc) \rightarrow C:C$
s. 29, 977

Ammonium ceric nitrate $(NH_4)_2Ce(NO_3)_6$

3H-Benzo[c]thiophen-1-ones   Ö:
from sulfur-donor ligand o-metalated complexes s. 29, 643

2-Methyl-2-nitrosopropane $Me_3C\cdot NO$

Ethylene derivs. from boranes   ←
s. 28, 965

Acetic anhydride s. under Me_2SO Ac_2O

Acetic acid s. under CH_3COONa CH_3COOH

Dimethyl sulfoxide/acetic anhydride Me_2SO/Ac_2O

2,3-Dihydro-6H-1,3-thiazine ring   O
Oxidative ring closure – Protection of alcohol groups as β,β,β-trichlorethyl carbonates – Reductive removal of the protective group s. 28, 966

Hydrogen peroxide H_2O_2

Ethylene derivs. from selenides $C(SeR)CH \rightarrow C:C$
s. 29, 912; elimination rate enhancement by electron-withdrawing ar. substituents s. J. Org. Chem. 40, 947 (1975)

Acoxy-2-ethylenes from 2-acoxyselenides $C(OAc)C(SeR)CH \rightarrow C(OAc)C:C$
s. 29, 180

Iodine/sodium hydroxide $I/NaOH$

1,3-Dienes from di(ene)chloroboranes $C:C\cdot B(Cl)\cdot C:C \rightarrow C:C\cdot C:C$
s. 28, 666

Tetraethylammonium fluoride $Et_4N^+F^-$

Allenes $C:C:C$
s. 29, 798

Hydrogen chloride HCl

Ethylene derivs. $C(NAc)C\cdot PO(OR)_2 \rightarrow C:C$
from 2-acylaminophosphonic acid esters s. 29, 880

Nickel   *Ni*

Diaryls via telluride dichlorides $Ar_2TeCl_2 \rightarrow Ar\cdot Ar$
s. 28, 967

Carbon ↑   CC⇑C

Without additional reagents   *w.a.r.*

Cleavage of heterocyclic dimers   ←
s. 30, 480

α,β-Ethyleneketones and 2-alkylidene-1,3-diones   ←
by pyrolysis of acoxy-2-acetylenes s. 28, 968

Ethylene derivs. from β-hydroxycarboxylic acids $C(OH)\cdot C\cdot COOH \rightarrow C:C$
via β-lactones – Retention of geometry s. 27, 981

Allenes from α-chloro-β-lactones
s. *29*, 979

2-Ethylenealcohols
from β,γ-oxidocarboxylic acids s. *27*, 982

$$\text{C·COOH} \rightarrow \text{C(OH)·C:C}$$

Ethylene derivs. and oxo compds.
from oxetanes – Ethylene deriv.-oxo compd. interconversion s. *29*, 978

$$\square\text{O} \rightarrow \text{CO} + \text{C:C}$$

Acetylene derivs. from cyclopropenones
s. *27*, 983

$$\triangledown \rightarrow \text{C}\equiv\text{C}$$

Cyclobutane ring opening
s. *30*, 467

Thermal aromatization
with elimination of an angular methyl group s. *28*, 969

Cyclopropanecarboxylic acid esters
by abnormal Hofmann degradation of quaternary ammonium salts via their hydroxides s. *26*, 983

Fulveneallenes and benzocyclobutenones
from homophthalic acid anhydrides s. *28*, 970

Retrodiene scission
with elimination of sulfenes s. *28*, 971

Transannular ring contraction of azides
9-(1-Naphthyl)anthracenes s. *27*, 984

Oxazoles
by rearrangement of 5-oxazolones s. *27*, 985

Pyrrolo[1,2-b]pyridazin-2-ones
from 9-acyl-1,9-diazabicyclo[4.2.1]nona-4,7-dien-3-ones s. *29*, 983

Irradiation

Photodecarboxylation of acoxy compds.
Replacement of acoxy groups by hydrogen s. *26*, 984

$$\text{OCOR} \rightarrow \text{R}$$

Carbon-carbon double bonds
from dicarboxylic acid anhydrides s. *23*, 976

C:C

2 Ethylene derivs. molecules
from cyclobutane ring s. *27*, 704

Acetylene derivs. from cyclopropenones
s. *27*, 983 suppl. *28*

$$\triangledown \rightarrow \text{C}\equiv\text{C}$$

Photochemical decarbonylation

690.

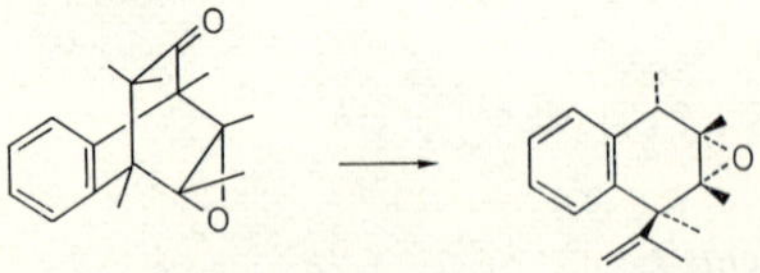

An ethereal soln. of *syn*-7,8-epoxy-1,3,3,4,7,8-hexamethyl-5,6-benzobicyclo[2.2.2]-oct-5-en-2-one irradiated 5 hrs. at 15–20° with the Corex-filtered light of a 450 w. Hg-lamp → product. Y: 95%. R. K. Murray, Jr. et al., J. Org. Chem. *38*, 3805 (1973); Am. Soc. *97*, 938 (1975).

Phenols
by solid-state decarboxylative aromatization s. *28*, 972

Ring contraction by decarbonylation
Cyclopentadienones from o-benzoquinones

691.

A soln. of 3,5-di-*tert*-butyl-*o*-benzoquinone in acetonitrile irradiated 4 hrs. with 2537 Å light in a quartz cylinder → 2,4-di-*tert*-butylcyclopentadienone. Y: ca. 100%. R. C. DeSelms and W. R. Schleigh, Synthesis *1973*, 614.

Indoles from 2,3-dihydro-1H-1,4-benzodiazepines

692.

Startg. m. irradiated 4 hrs. at room temp. under N_2 in methanol with the Pyrex-filtered light of a 450 w. medium-pressure Hg-vapor lamp. → product. Y: 70%. F. e. s. M. Steinman and Y.-S. Wong, Tetrah. Let. *1974*, 2087.

Skeletal rearrangement with decarboxylation
    Azulene from naphthalene ring s. *29*, 980 ←

Electrolysis

Electrochemical oxidative ring closure
    s. *28*, 973; with and without rearrangement s. T. Kametani, K. Shishido, and S. Takano, J. Heterocyclic Chem. *12*, 305 (1975)

Sodium hydride   NaH
Phenols by aromatization
    with elimination of carbalkoxymethyl groups s. *29*, 981 ←

Sodium hydride/dimethyl sulfoxide NaH/Me_2SO
Mono- and bi-cyclic isocyclics
    by cyclopropane ring expansion s. *29*, 982 ←

Sodium hydroxide (s. a. under $K_2S_2O_8$)   NaOH
Cleavage of bisphenols
s. *27*, 986 ←

Cyclic α-hydroxyketones from ethynylglycols
    Ring expansion s. *28*, 974

Potassium hydroxide/alcohol   KOH
2-Subst. benzofurans
    from 3-acylbenzofurans s. *27*, 987 ←

Sodium/alcohol s. a. under $Na[BH_4]$ NaOR
Isoxazolidine ring opening   C

693.

2-Phenyl-3-benzoyl-4,5-dicyanoisoxazolidine allowed to react 12 hrs. with 2 *N* Na-methoxide in methanol → product. Y: 71%. F. e. and products s. M. Joucla, J. Hamelin, and R. Carrie. Tetrahedron *30*, 1121 (1974).

Pyrrolo[1,2-b]pyridazin-2-ones ←
from 9-acyl-1,9-diazabicyclo[4.2.1]nona-4,7-dien-3-ones s. *29*, 983

Potassium tert-butoxide $Me_3C \cdot OK$
**Dieckmann cyclization
with C-decarbalkoxylation**

694.

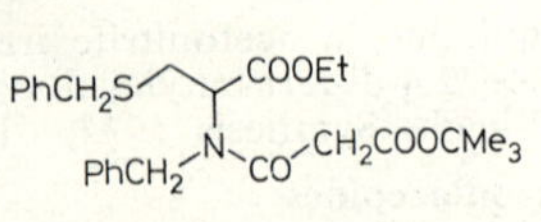

PhCH₂S—C(COOEt)—N(CH₂Ph)—CO—CH₂COOCMe₃ → 1-benzyl-5-benzylthiomethylpyrrolidine-2,4-dione

Startg. m. treated at room temp. with 1 equivalent K-*tert*-butoxide in benzene → *rac*-1-benzyl-5-benzylthiomethylpyrrolidine-2,4-dione. Y: 87%. J. R. Hlubucek and G. Lowe, Chem. Commun. *1974*, 419.

Lithium diisopropylamide $i\text{-}Pr_2NLi$
Oxo compds. from α-alkoxynitriles $CH(OR)CN \to CO \cdot R$
s. *28*, 975

Potassium amide KNH_2
Isoindoles
by benzyne C-C-ring closure s. *28*, 976

Sodium carbonate Na_2CO_3
Ring contraction via chlorination
Ethylene derivs. from α-chloroketones with decarbonylation – Cyclopent-2-enones s. *29*, 984

Potassium carbonate K_2CO_3
Enolethers from α-alkoxy-β-lactones $C:C(OR)$
s. *29*, 985

Aromatization   C
with elimination of a CO-bridge under mild conditions s. *28*, 977

Lithium iodide s. under I LiI

Sodium salt Na^+
Arsabenzenes
s. *28*, 978

Triethylamine Et_3N
Ethylene derivs. by fragmentation $C:C$
s. *28*, 979

Cupric chloride (s. a. under O_2) $CuCl_2$
Oxidative ring contraction of hydroxyquinones
with replacement of hydroxyl by chlorine s. *28*, 980

Calcium oxide CaO
2-Subst. Δ¹-pyrrolines ←
from 1-acyl-2-pyrrolidones s. *28*, 981

Calcium carbonate s. under I $CaCO_3$

Zinc Zn
Ethylene derivs. from 1,3-diol monoesters ←
Fragmentation s. *26*, 985

Zinc acetate $(CH_3COO)_2Zn$
s. *26*, 985

Sodium tetrahydridoborate-sodium/alcohol $\qquad$ $Na[BH_4]$-$NaOR$
Arenes $\qquad$ C
  from N-condensed cyclimmonium salts – Phenols s. 28, 982

Lithium tetrahydridoaluminate $\qquad$ $Li[AlH_4]$
Reductive aromatization of steroid B rings $\qquad$ ←
  in azodicarboxylic acid ester adducts s. 29, 986

Boron fluoride $\qquad$ BF_3
Ring closure with enol acetates $\qquad$ O
  Bicyclo[2.2.1]heptan-2-ones s. 26, 986
Cyclic ketones from oxidonitriles
  Elimination of a 4-methyl group from steroids s. 28, 983

Aluminum chloride $\qquad$ $AlCl_3$
C-Cleavage of polyhalides $\qquad$ ←
  Labeled compds. s. 27, 988
Carbostyrils from cinnamanilides $\qquad$ O
  with alkyl group migration s. 28, 984

Dimethylformamide s. under NH_4Cl $\qquad$ Me_2NCHO
Dimethylformamide dineopentyl acetal $\qquad$ $Me_2NCH(OR)_2$
Decarboxylative elimination $\qquad$ C(OH)C(COOH) → C:C
  Ethylene derivs. from β-hydroxycarboxylic acids s. 28, 985; with dimethylformamide dimethyl acetal cf. S. Hara et al., Tetrah. Let. *1975*, 1545

Formic acid $\qquad$ HCOOH
Skeletal rearrangement $\qquad$ ←
  with ether cleavage – Vinylogous Wagner rearrangement s. 29, 987

Peroxyacetic acid $\qquad$ CH_3COO_2H
Ring contraction by decarbonylation $\qquad$ Ö
s. 28, 986

Phenyl azide $\qquad$ PhN_3
Naphthalenes by elimination of etheno bridges $\qquad$ C
  s. 26, 987; cf. P. C. Buxton et al., Soc. Perkin I *1974*, 2681

Tris(diethylamino)phosphine $\qquad$ $(Et_2N)_3P$
Ethylene derivs. by twofold extrusion
  from 1,3-oxathiolan-5-ones s. 26, 988

Polyphosphoric acid $\qquad$ $H(PO_3H)_xOH$
Cyclic ketones from cyanocarboxylic acid esters $\qquad$ O
  with formation of carboxylic acid amides from nitriles – 4-Piperidone ring s. 26, 989
2H-5,6-Dihydronaphto[1,2-b]pyrones
  by decarboxylative double ring closure s. 27, 989

Oxygen/cupric chloride $\qquad$ $O_2/CuCl_2$
5-Hydroxypyrazoles $\qquad$ ←
  from 1-acyl-3-pyrazolidones s. 28, 891

Sulfuric acid $\qquad$ H_2SO_4
Skeletal rearrangement $\qquad$ ←
  with ether cleavage s. 29, 988

Potassium persulfate/sodium hydroxide $\qquad$ $K_2S_2O_8/NaOH$
Decarboxylative ring closure $\qquad$ O
  Dibenzo[b,d]pyrans s. 28, 987

Iodine/lithium iodide/calcium carbonate   I/LiI/CaCO₃
Ethylene derivs. from carboxylic acid esters   CHC(COOR) → C : C
    with decarbalkoxylation s. 29, 989

Ammonium chloride/dimethylformamide   NH₄Cl/Me₂NCHO
Oxindoles
    from 3,4-oxido-3,4-dihydrocarbostyril-3-carboxylic acid esters s. 29, 988

Hydrogen chloride   HCl
Skeletal rearrangement of isocyclics ←
    by intramolecular retroaldol-aldol condensation – Cleavage of ketals s. 29, 990
Isocyclics from O-heterocyclics ←
    via Birch reduction s. 28, 989

Hydrogen bromide/acetic acid   HBr/CH₃COOH
3,4-Disubst. 2(5H)-furanones ←
    s. 27, 710

Bis(triphenylphosphine)nickel dicarbonyl   Ni(CO)₂(Ph₃P)₂
Ethylene derivs.   C : C
    from *vic*-dicarboxylic acid anhydrides s. 29, 991

Chlorotris(triphenylphosphine)rhodium(I)   RhCl(Ph₃P)₃
Decarbonylation   C≡C·CO·C≡C → C≡C·C≡C
1,3-Diynes from 1,4-diyn-3-ones

695.   Ph—C≡C—CO—C≡C—Ph → Ph—C≡C—C≡C—Ph

Bis(phenylethynyl) ketone refluxed 2 hrs. under N₂ with chlorotris(triphenylphosphine)rhodium(I) in xylene → product. Y: 78%. F. e., also isolation of intermediate Rh-complexes, s. E. Müller and A. Segnitz, A. *1973*, 1583; cf. K. Kaneda et al., Chem. Lett. *1974*, 215.

Palladium-carbon   Pd-C
Aromatization with N-debenzylation ←
    s. 27, 990
4-Pyridones from 4,5'-bisisoxazoles ←
    s. 27, 991

Via intermediates   v.i.
Arenes from N-condensed cyclimmonium salts   C
    Phenols s. 28, 982

Formation of Electron Pair on Nitrogen

Elimination   ⇑

Hydrogen   ElN ⇑ H

Potassium carbonate   K₂CO₃
5,6-Dihydro-1,2-oxazine ring ←
    from 4,5-dihydro-1,2-oxazinium ring s. 28, 676

Propylene oxide ←
α-Aminocarboxylic acids ←
    from their hydrochlorides s. 30, 290

Amberlite IRA-402 ←
Bases from their salts ←
s. *30*, 55
Via intermediates v.i.
Aminocarboxylic acids ←
from their hydrohalides via silylation s. *30*, 290

Oxygen ↑   EIN⇑O

Irradiation s. under KCN and Anthraquinone   ⫲

Sodium hydroxide s. under Na[BH₄]   NaOH

Potassium cyanide/irradiation   KCN/⫲

Nitroso from nitro compds. $NO_2 \twoheadrightarrow NO$
s. *28*, 990

Sodium azide NaN_3

Benzofurazans from benzofurazan oxides   N→O → N
s. *30*, 56

Zinc/acetic acid Zn/CH_3COOH

Deoxygenation of N-oxides   ≧N→O → ≧N
s. *8*, 951; with polymer-bound hydroquinone cf. R. V. Starodubtseva et al., Khim.-Farm. Zh. *8* (2), 33 (1974); C. A. *80*, 121166

Sodium tetrahydridoborate/sodium hydroxide $NaBH_4/NaOH$

Azomethines from nitrones   C:N→O → C:N

$$o\text{-}O_2NC_6H_4\text{—}CH=CH\text{—}CH=\overset{O}{\overset{\uparrow}{N}}\text{—}Ph \longrightarrow o\text{-}O_2NC_6H_4\text{—}CH=CH\text{—}CH=N\text{—}Ph$$

Selective reduction. An aq. soln. of NaBH₄ and NaOH added slowly to a stirred soln. of C-o-nitrostyryl-N-phenylnitrone in methanol, stirred 30 min. at room temp., and allowed to stand overnight → product. Y: 70%. F. e. s. K. Krishan and N. Singh, Indian J. Chem. *12*, 222 (1974).

Polymer-bound hydroquinone ←
s. *8*, 951 suppl. *30*

Anthraquinone/irradiation ←

Photolysis of benzimidazole 3-oxides   N→O → N
N-Deoxygenation – Benzimidazol-2-ones s. *26*, 990

Tri-n-butyltin hydride/azodiisobutyronitrile $n\text{-}Bu_3SnH/(Me_2C(CN)N:)_2$

Cleavage of semipolar bonds by a free radical
Deoxygenation of N-oxides
Cleavage of cyclic N-ylids

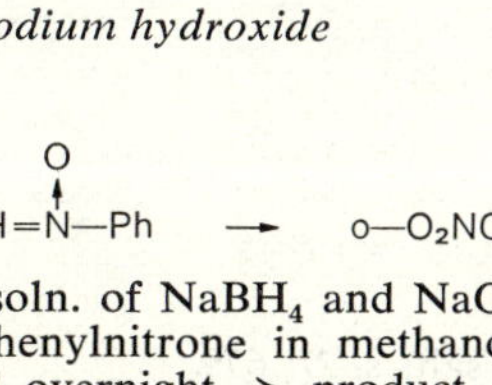

Pyridine N-oxide heated 10 hrs. at 80° with 3 moles tri-*n*-butyltin hydride in benzene in the presence of azodiisobutyronitrile → pyridine. Y: 92%. – Similarly:

N-Benzoyliminopyridinium betaine → pyridine (Y: 80%) and benzamide (Y: ca. 100%). F. e. s. S. Kozuka et al., Chem. & Ind. *1974*, 452; cleavage of semipolar sulfur bonds s. ibid. *1974*, 496.

Triethyl phosphite $(EtO)_3P$
N-Deoxygenation
s. *26*, 991 suppl. *30*

Iron carbonyl $Fe(CO)_5$
s. *26*, 991; with triethyl phosphite, partial and total N-deoxygenation of pyrimidine-1,3-dioxides cf. A. Y. Tikhonov and L. B. Volodarsky, Tetrah. Let. *1975*, 2721

Nitrogen ↑ EIN⇑N

Irradiation ↯
Cleavage of cyclic N-ylids ←
s. *26*, 992

Potassium hydroxide KOH
Isoindoles ←
from cyclic quaternary hydrazinium salts s. *27*, 449

Tri-n-butyltin hydride/azodiisobutyronitrile $n\text{-}Bu_3SnH/(Me_2C(CN)N:)_2$
Cleavage of cyclic N-ylids ←
s. *30*, 697

Carbon ↑ EIN⇑C

Without additional reagents w.a.r.
4-Cyanoquinolines ←
from 4-cyanoquinolinium salts s. *27*, 679

Lithium hydride/n-propylmercaptan LiH/n-PrSH
Tert. amines ≥N⁺–R → ≥N
from quaternary ammonium salts – N-Demethylation
s. *29*, 992; with Li-hydridotriethylborate cf. M. P. Cooke, Jr., and R. M. Parlman, J. Org. Chem. *40*, 531 (1975)

Liq. ammonia NH_3
Dequaternization of cyclimmonium salts

698.

1,2,4,6-Tetramethylpyrimidinium iodide dissolved at –33° in liq. NH_3, and the product isolated after 1 hr. → 2,4,6-trimethylpyrimidine. Y: 65–70%. F. e. s. E. A. Oostveen, H. C. van der Plas, and H. Jongejan, R. *93*, 114 (1974); **demethylation of quaternary ammonium salts** with cuprous phenylmercaptide in pyridine s. G. H. Posner and J.-S. Ting, Synth. Commun. *4*, 355 (1974); **dequaternization of pyridinium salts** with triphenylphosphine s. J. P. Kutney and R. Greenhouse, Synth. Commun. *5*, 119 (1975).

Pyridine C_5H_5N
N-De(methoxymethylation) ←
s. *30*, 701

Cuprous phenylmercaptide   *PhSCu*
Dequaternization of cyclimmonium salts
s. *30*, 698

Lithium hydridotriethylborate   *Li[BHEt₃]*
Tert. amines
from quaternary ammonium salts s. *29*, 992 suppl. *30*

Triphenylphosphine   *Ph₃P*
Dequaternization of cyclimmonium salts
s. *30*, 698

Formation of Electron Pair on Sulfur

Elimination ⇑

Oxygen ↑   EIS⇑O

Irradiation
Thioethers from sulfoxides   >SO → >S
s. *28*, 991

Sodium tetrahydridoborate/cobaltous chloride   *Na[BH₄]/CoCl₂*
s. *26*, 993

Dichloroborane   *BHCl₂*
s. *28*, 991

Reduction with *dithiothreitol*   ←
s. *27*, 992

Azodiisobutyronitrile s. under n-Bu₃SnH   *(Me₂C(CN)N :)₂*

Dithioacetic acid   *CH₃CSSH*
Reductions of sulfur compds.
with dithioacetic acid – Thioethers from sulfoxides s. *29*, 993

Acetyl chloride (s. a. under SnCl₂)   *CH₃COCl*
Thioethers from sulfoxides
s. *29*, 993

Titanium trichloride   *TiCl₃*
s. *28*, 991; titanium(III) compds. in organic chemistry, review, s. J. E. McMurry, Acc. Chem. Res. 7, 281 (1974)

Tri-n-butyltin hydride/azodiisobutyronitrile   *n-Bu₃SnH/(Me₂C(CN)N :)₂*
Cleavage of semipolar sulfur bonds   ←
s. *30*, 697

Stannous chloride $\qquad SnCl_2$
Thioethers from sulfoxides $\qquad >SO \rightarrow >S$
s. 28, 991

Stannous chloride/acetyl chloride $\qquad SnCl_2/CH_3COCl$
Reduction of S-oxides
s. 27, 734

Pyrocatechyl phenyl phosphite $\qquad \leftarrow$
Thioethers from sulfoxides
s. 29, 993

O,O-Dialkyl dithiophosphoric acids $\qquad (RO)_2PSSH$
Reduction of semipolar functional groups
　　Thioethers from sulfoxides s. 26, 993

Thionyl chloride/cyclohexene $\qquad \leftarrow$
Thioethers from sulfoxides
s. 29, 993

Bromine $\qquad Br$
s. 29, 993; with iodine/pyridine-sulfur dioxide complex (cf. Synth. Meth. 29, 287) cf. M. Nojima, T. Nagata, and N. Tokura, Bull. Chem. Soc. Japan 48, 1343 (1975)

Iron $\qquad Fe$
Deoxygenation of sulfur compds. $\qquad S(O_2)S \rightarrow SS$
Disulfides from thiosulfonic acid esters
Thioethers from sulfoxides

699.
$$Ph-\underset{\underset{O}{\|}}{\overset{\overset{O}{\|}}{S}}-S-Ph \rightarrow Ph-S-S-Ph$$

$$Ph-\underset{\underset{O}{\|}}{S}-Ph \rightarrow Ph-S-Ph$$

Phenyl benzenethiolsulfonate heated 10 hrs. at 200°/1 mm with reduced Fe → diphenyl disulfide. Y: 91%. – Similarly: Diphenyl sulfoxide → diphenyl sulfide. Y: ca. 100%. F. e. s. T. Fujisawa, K. Sugimoto, and H. Ohta, Chem. Lett. *1973*, 1241; thioethers from sulfoxides with 5%-Pd-on-charcoal s. K. Ogura, M. Yamashita, and G. Tsuchihashi, Synthesis *1975*, 385.

Cobaltous chloride s. under Na[BH₄] $\qquad CoCl_2$

Nickel $\qquad Ni$
Thioethers from sulfoxides $\qquad >SO \rightarrow >S$
s. 28, 991

Palladium-carbon $\qquad Pd-C$
s. 30, 699

Nitrogen ↑ $\qquad$ EIS⇑N

Sodium nitrite/sulfuric acid $\qquad NaNO_2/H_2SO_4$
Sulfoxides from sulfoximes $\qquad >SO(NH) \rightarrow >SO$
　　Stereochemistry of sulfur compds. s. 28, 992

Nitrous acid s. under NaNO₂/H₂SO₄ $\qquad HNO_2$

Triphenylphosphine Ph_3P
Thioethers and N-sulfonylphosphine imines $>S:NTs \rightarrow\ >S$
from sulfonylsulfilimines and phosphines

700.
$$Ph-S-Me + Ph_3P \rightarrow Ph-S-Me + Ph_3P=NTs$$
$$\underset{NTs}{\|}$$

A soln. of methyl phenyl N-p-tosylsulfilimine in dimethylformamide containing triphenylphosphine heated 6 hrs. at 120° → methyl phenyl sulfide (Y: ca. 100%) and triphenyl N-p-tosylphosphinimine (Y: ca. 100%). F. e. s. T. Aida, N. Furukawa, and S. Oae, Chem. Lett. *1973*, 805.

O,O-Dialkyl dithiophosphoric acids $(RO)_2PSSH$
Thioethers from sulfonylsulfilimines $>S:N\cdot SO_2R \rightarrow\ >S$
s. *26*, 993

Formation of Electron Pair on Remaining Elements

Elimination ⇑

Sulfur ↑ ElRem⇑S

Lithium tetrahydridoaluminate $Li[AlH_4]$
Phosphines from phosphine sulfides $\geqq PS \rightarrow\ \geqq P$
s. *29*, 600

Carbon ↑ ElRem⇑C

Potassium cyanide/dimethyl sulfoxide KCN/Me_2SO
Tert. phosphines from phosonium salts $\geqq \underset{R}{P^+} \rightarrow\ \geqq P$
s. *26*, 994
Pyridine hydrochloride C_5H_5N,HCl
s. *27*, 364

Heteropolar Bond

Uptake ⇓

Addition to Nitrogen HetN

Without additional reagents w.a.r.
Partial nucleophilic substitution ←
of polyhalogenocyclobutenes with rearrangement s. *27*, 993

**N-β-Acoxyethylation by dioxolanium salts
with subsequent N-de(methoxymethylation)
Directed N-substitution of imidazoles via imidazolium salts**

701.

Ar= —C₆H₄F—p

2-(4-Fluorophenyl)-1-methoxymethyl-4-nitroimidazole in dry methylene chloride added to 2-methyl-1,3-dioxolanium fluoroborate in the same solvent, refluxed 40 hrs. with vigorous stirring, dry pyridine added, refluxing continued 5 hrs., the resulting 2-(4-fluorophenyl)-1-(2-acetoxyethyl)-5-nitroimidazole dissolved in 2.5 N HCl, and refluxed 50 min. → 2-(4-fluorophenyl)-1-(2-hydroxyethyl)-5-nitroimidazole. Overall Y: 82.4–92%. E. J. J. Grabowski et al., J. Med. Chem. *17*, 547 (1974).

3-Acetoxyquinuclidinium salts
as intermediates s. *29*, 71

N-Cyanomethylpyrrolidinium salts
s. *29*, 901

Quaternary N-cyanoammonium fluoroborates
s. *27*, 994

Quaternary 1,1,1-dialkoxyammonium salts
s. *27*, 995

Partial formation of hexamethylenetetraminium salts
s. *27*, 541

1,1-Di(pyridinium salts) from 1,1-dihalides
s. *29*, 211

Cyclodiimonium salts
s. *29*, 995; s. a. F. Vögtle and D. Brombach, B. *108*, 1682 (1975)

Carbodiimidium salts from carbodiimides   N : C : NR → N : C : N⁺RR′
s. *27*, 996

Diazenium salts from azo compds.   RN : NR → RN : N⁺RR′
s. *29*, 994

Ethylene oxide

Quaternization of hindered tert. amines $\geqslant N \to \geqslant N^+_R$

702.

2-Cyclopropylethyl bromide added to a soln. of (–)-scopolamine in acetonitrile, refluxed 3 days, evaporated to dryness, the residue dissolved in fresh acetonitrile, ethylene oxide added, and stored 3 days in a stoppered vessel → product. Y: 61%. – Ethylene oxide serves as base to facilitate the removal of by-products. F. e. s. A. Donetti and E. Bellora, Tetrah. Let. *1973*, 3573.

Addition to Sulfur   Het↓S

Trialkyloxonium fluoroborates $R_3O^+ BF_4^-$

Sulfonium salts from cyclic thioethers $\subset S \to \subset S^+_R$
also from mercaptals – Chemistry of trialkyloxonium fluoroborates, review, s. *27*, 997

1,1-Sulfonioammonium salts from 1,1-aminothioethers $>S \to >S^+_R$

703.

$Me_2N-CH_2-SMe \longrightarrow Me_3\overset{+}{N}-CH_2-\overset{+}{S}Me_2 \ 2 \ BF_4^-$
$+ \ 2 \ Me_3O^+ \ BF_4^-$

Dimethyl(methylthiomethyl)amine in nitromethane added dropwise at –10° to a soln. of trimethyloxonium tetrafluoroborate in the same solvent, and the product isolated after 24 hrs. → trimethyl(dimethylsulfoniomethyl)ammonium bis(tetrafluoroborate). Y: 72%. F. e. s. H. Böhme, G. Dähler, and W. Krack, A. *1973*, 1686.

Antimony pentachloride $SbCl_5$
Alkylation with 1,1-chlorothioethers
via carbonium ions generated in situ – Methylthiomethylsulfonium salts s. *28*, 993

Silver perchlorate $AgClO_4$
Cyclic sulfonium salts $\subset S \to \subset S^+_R$
s. *28*, 994

Addition to Remaining Elements   Het↓Rem

Without additional reagents   w.a.r.
Phosphonium salts from *tert*-phosphines $\geqslant P \to \geqslant P^+_R$
s. *27*, 364

S,P,N-Heterocyclics   O
from N,N′-bisphosphinosulfone diimines s. *28*, 995

Rearrangement ↶

Oxygen/Nitrogen Type Het↶ON

Without additional reagents w.a.r.
Diazonium carboxylates ←
  from N-acyl-N-nitrosanilines s. *26*, 995

Exchange Het↓↑

Trialkyloxonium fluoroborates $R_3O^+ BF_4^-$
Quaternary N-cyanoammonium fluoroborates $\geqslant N^+\!\!\!-CN\;\; BF_4^-$
s. *27*, 994
Anion exchanger ←
Quaternary ammonium hydroxides $\geqslant N^+\!\!\!-R\;\; OH^-$
s. *26*, 983
Potassium hydrogen sulfate $KHSO_4$
Peptides from peptide salts $COO^- \to COOH$
s. *28*, 21

Elimination ⇑

Hydrogen ↑ Het⇑H

Anion exchanger ←
Acoxybetaines ←
s. *26*, 996
Sulfuric acid H_2SO_4
Diazonium sulfate betaines ←
s. *26*, 997

Nitrogen ↑ Het⇑N

Perchloric acid $HClO_4$
1,3-Dithiolium salts ←
  from 2-amino-1,3-dithioles s. *26*, 998

Electron ↑ Het⇑El

Antimony pentachloride $SbCl_5$
Oxoazonium salts from N-oxide radicals $>N^\bullet\!\!-O \to N^+\!:\!O$
s. *29*, 996

Heteropolar Bond – Nitrogen Radical   HetN

Lead tetracetate $(CH_3COO)_4Pb$

Nitrogen radical cations $\geqslant \overset{\bullet}{N}{}^+$
s. 27, 998

Other Reactions   Oth

Ferric nitrate $Fe(NO_3)_3$
Carbon-carbon triple bonds $C \equiv C$
   from their cobalt carbonyl complexes s. 27, 140

Index Volumes 26–30

The subject index lists the names of the methods, types of compounds, reagents, etc. For specific compounds and for authors (when a method is not named after them) the reader should consult the indexes of abstract journals. Complex compounds, as those with several functional groups, are referred to under the related simpler compounds, under *special s*. E.g. aminocarboxylic acids are found under carboxylic acids. Only compounds beginning with letters other than those of the parent or main compounds are as a rule listed under this subentry, and entry numbers are omitted because they are specified where the compound is listed as a main entry. Derivatives used for characterization, identification, or separation are listed under the subentry *derivs*. Methods of synthesis for a given substance are indexed under the name of this substance and the subentry *from*, e.g. carboxylic acids from alcohols, hydrocarbons. Syntheses carried out with a particular starting material are indexed under the starting material and the subentry *startg. m. f.* (starting material for the preparation of...), e.g. alcohols, startg. m. f. carboxylic acids, ketones.

Classes of compounds may be designated merely by the functional group that is changed during the reaction. Ring compounds may also refer to the corresponding hydrogenated rings, unless the latter are listed specifically. If that is the case, then, as in other cases, *inversion*, customary in the indexes of Chemical Abstracts, has been used. Dihydrofurans, e.g., may be found under Furans, dihydro-. All single letters, which are separated from the name by a hyphen, are being ignored in regard to alphabetic arrangement, e.g., 'O-Acetyl derivs.' are listed under A. Supplementary entries, such as *17,* 892 suppl. *27,* can be located by consulting the Index of Supplementary References.

Register zu Band 26—30

Im alphabetischen Register finden sich als Schlagworte Methoden, Verbindungsklassen, Reagenzien u. dgl.; Einzelverbindungen und Autoren sind, soweit nicht eine Methode nach ihnen benannt ist, in den Registern der Referatenblätter zu suchen. Auf kompliziertere Verbindungen, z. B. mit mehreren funktionellen Gruppen, ist bei den entsprechenden einfacheren Verbindungen unter *special s.* (spezielle siehe) hingewiesen, z. B. bei Carbonsäuren auf Aminocarbonsäuren. In der Regel wird hier jedoch nur auf Verbindungen hingewiesen, die einen anderen Wortanfang haben und deshalb an anderer Stelle im Register stehen. Derivate, die zur Charakterisierung, Identifizierung oder Trennung dienen, sind unter *derivs.* aufgeführt. Methoden zur Synthese eines bestimmten Endprodukts findet man beim Schlagwort dieses Endprodukts unter *from* (aus) registriert, z. B. carboxylic acids from alcohols, hydrocarbons. Synthesen, die sich mit einem bestimmten Ausgangsmaterial ausführen lassen, sind bei dem Schlagwort des Ausgangsmaterials unter *startg. m. f.* (Ausgangsmaterial für die Darstellung von ...) zu suchen, z. B. Alcohols, startg. m. f. carboxylic acids, ketones.

Die Bezeichnung der Verbindungsklassen kann sich auf funktionelle Gruppen beschränken, die bei der Reaktion verändert werden. Ringbezeichnungen können sich auch auf die entsprechenden hydrierten Ringe beziehen. Sind hydrierte Ringe aber besonders aufgeführt, dann ist bei ihnen, wie auch in anderen Fällen, von der Inversion Gebrauch gemacht worden, wie sie in den Sachregistern der «Chemical Abstracts» üblich ist. Z. B. stehen Dihydrofurane unter Furans, dihydro-. Griechische Buchstaben und Einzelbuchstaben, die vom eigentlichen Namen durch Bindestrich getrennt sind, werden bei der alphabetischen Anordnung nicht berücksichtigt, z. B. stehen «O-Acetyl-derivs.» unter A. Ergänzungen, wie z. B. *17, 892* suppl. *27,* können mit Hilfe der Tabelle der Ergänzungszitate aufgefunden werden.

Abe—Ace 546

abeo- s. under Steroids
Abramov reaction 10, 495
Accelerator s. Activator
Acetallactones, polycyclic
– from
  triols, cyclic, and oxocarboxylic acids **30,** 112
Acetals (s. a. Dialkoxy..., Hemiacetals) **26,** 176; **29,** 187
–, cleavage **26,** 7, 176; **27,** 239;
  29, 8, 10, 11, 867, 990
–, –, preferential **29,** 10
– from
  alcohols, sec. **28,** 141
  aldehydes **6,** 220; **22,** 191;
  27, 220; **30,** 113
  1,1,1-dialkoxyammonium salts,
  quaternary **27,** 821
  enoxysilanes and diols **29,** 193
  ethylene derivatives **26,** 143;
  29, 148
  1,1-hydroxysulfonic acids
  26, 215
  nitro compds., aliphatic
  30, 122
  oxo compds. **28,** 141
  – – and orthoformic acid
  esters **30,** 111
–, ozoniazation **28,** 235
–, retention s. under Retention
– special s.
  acetylene-acetals
  amide-
  amino-
  anhydrosugar –
  O,O-benzylidene derivs.
  carboxy-acetals
  chloral –
  di-
  ethylene-
  formals
  halogen-acetals
  hydroxy-
  ketals
  ketoacetals
–, ring closure with – **27,** 940
–, – –, double, with – **28,** 926
– startg. m. f.
  acoxy compds. with
  rearrangement **30,** 94
  aldehydes **27,** 59
  1,1-alkoxyhalides **30,** 567
  β-alkoxyketones **30,** 633
  α-alkoxynitriles **30,** 567
  carboxylic acid esters **14,** 598
  suppl. **29; 28,** 235
  1,3-diol monoethers **30,** 521
  dioxolanium salts **29,** 753
  enolethers **26,** 956
  – with alkoxyl interchange
  28, 756
  ethers **26,** 59
  α,β-ethylenecarboxylic acids
  29, 753
  α-hydroxyketones **27,** 239
  imidazoles **26,** 304

imines **27,** 393
mercaptals, cyclic **26,** 601
–, **cyclic** (s. a. O,O-Alkylidene...,
  Lactolides)
–, formation, preferential **30,** 151
– from
  1,2-acoxymesylates **27,** 81
– special s.
  bromomethylethylene acetals
  m-dioxanes
  1,3-dioxolanes
– startg. m. f.
  acoxyketones **25,** 99
  aldehydes **27,** 81
  carboxylic acid esters **26,** 147
– via acetals, mixed **10,** 170
  suppl. **29**
–, –, **mixed**
– from
  acoxyaldehydes and orthoformic acid esters **28,** 195
–, **mixed**
–, cleavage, selective **26,** 10
–, –, **O-heterocyclic**
  (s. a. Glycosides)
– from
  enolethers **26,** 138
Acetanilides as benzyne
  precursors **19,** 910; **29,** 897
Acetate ion in benzene **1,** 321
  suppl. **27**
Acetates s. Acoxy...,
  O-Acylation
Acetic acid, protonated,
  condensation by – **26,** 822
– – **esters**
– special s.
  aminoacetic acid esters
  cyclopropylacetic – –
  diazoacetic – –
  silylthioacetic – –
– **acids**
– special s.
  o-ketoacetic acids
– **anhydride**
– as reactant **17,** 149 suppl. **29;**
  22, 153 suppl. **26; 29,** 166, 312,
  392, 921; **30,** 108, 647
– as reagent NC↑O; CC↑O;
  CC↑O; **20,** 648a suppl. **29; 26,**
  137, 256, 374, 498, 645; 793,
  807/8, 976; **27,** 124, 186, 505;
  28, 111, 206, 219, 299, 540, 568,
  713, 966; **29,** 250, 365, 608, 753,
  928; **30,** 12, 53, 163, 188, 631,
  643, 649
– **formic anhydride**
– as reactant **30,** 392
–, S-formylation with – **26,** 600
–, prepn. **16,** 268 suppl. **26**
Acetoacetic ester s. β-Ketocarboxylic acid esters
Acetoacetonates s. Metal
  acetoacetonates
Acetolysis of glycosides **28,** 88

Acetone (s. a. under Sensitizers)
– as nucleophile, weak **27,** 764
– as solvent **28,** 119; **29,** 279
– – –, aprotic, polar **26,** 312
– **cyanohydrin**
– as hydrogen cyanide source
  29, 639
– as reactant **28,** 621
Acetonitrile (s. a. under
  Hydrogen peroxide, Silver
  perchlorate)
– as medium **13,** 131 suppl. **29**
– as reactant **27,** 853
– as reagent **26,** 312; **27,** 421
– as solvent **14,** 608 suppl. **26;**
  26, 262, 416, 621, 704, 782; **27,**
  347, 697, 719; **28,** 595, 664, 962,
  973; **29,** 74, 118, 820, 919, 942;
  30, 302
– as solvent, polar, aprotic
  26, 901
Acetonyl groups s. Replacement
  of halogen by –
Acetophenones (s. a. under
  Sensitizers)
– special s.
  o-hydroxyacetophenones
  ω-sulfinylacetophenones
Acetoxy... s. a. Acoxy...
β-(Acetoxymercury)ketones
– from
  cyclopropanols **26,** 643
Acetyl... s. a. Acyl...
C-Acetyl... s. Methyl ketones
Acetylacetonates s. Metal
  acetoacetonates
Acetylation (s. a. Acylation)
– with ketene, review **6,** 366
  suppl. **28**
Acetyl chloride as reagent
  27, 241, 734; **28,** 462; **29,** 993;
  30, 31, 311, 567
Acetyl derivatives s. Acylation
Acetylene
– startg. m. f.
  γ-enollactones **30,** 606
  pyrroles **27,** 909
–, N-vinylation with – **27,** 330
α-Acetyleneacetals
– startg. m. f.
  α-alkoxyketones **29,** 61
  cis-2-ethyleneethers **29,** 61
Acetyleneacyl peroxides 26, 202
Acetylenealcohols
– special s.
  ethynylalcohols
  vinylethynylcarbinols
– **cyclic**
– startg. m. f.
  ketones, bicyclic **27,** 745
2-Acetylenealcohols
–, ring closure with – **17,** 936
– startg. m. f.
  α,β-acetylenealdehydes **19,** 307
  suppl. **29**

β-allenecarboxylic acid esters
27, 796
allenesulfoxides 28, 565
β-amino-α,β-ethylene-
aldehydes 28, 361
β-amino-α,β-ethyleneketones
30, 246
α,β-ethylenealdehydes 20, 538
suppl. 27
ketones, cleavage 28, 230
3-Acetylenealcohols 18, 735
- from
furans, 4,5-dihydro-3-halogeno-
29, 797
halides, synthesis with
addition of 4 C-atoms
29, 797
- startg. m. f.
2-halogeno-1,3-dienes 26, 537
α-methylene-γ-lactones 30, 502
Acetylenealdehydes
- special s.
enynals
α,β-Acetylenealdehydes
- from
2-acetylenealcohols 19, 307
suppl. 29
α,β-Acetylene-γ-alkoxyketones
- startg. m. f.
3(2H)-furanones 30, 182
α,β-Acetyleneamidines
- from
carbodiimides 28, 617
chloroformamidines 28, 672
- special s.
phenylpropiolamidines
- startg. m. f.
quinolines, 2-amino- 28, 672
Acetyleneamines
- special s.
ynamines
2-Acetyleneamines
- from
α-aminonitriles, synthesis
7, 858 suppl. 27
- startg. m. f.
indoles 30, 544
2-Acetyleneammonium salts
- startg. m. f.
allenes 30, 35
Acetyleneazo compds.
- startg. m. f.
1,2,4,5-tetraazapentalenes
(from 2 molecules) 26, 702
Acetylenebenzofurans
- from
cumulenehalides 26, 252
α,β-Acetylenecarboxylic acid
esters
- from
acetylene derivs. 27, 857
β-ketocarboxylic acid esters
27, 203
5-pyrazolones 27, 203
- startg. m. f.
coumarins, 3-amino- 29, 725
β,γ-Acetylenecarboxylic - -
- from
acetylene derivs. 28, 783
- startg. m. f.
γ-ketocarboxylic acid esters
28, 783
Acetylenecarboxylic acids
- special s.
diynoic acids
-, synthesis 27, 851 suppl. 29

α,β-Acetylenecarboxylic -
- from
aldehydes 28, 292
dihalogenomethylene compds.
28, 292
Acetylene derivs. (s. a. Yn...)
- by
dehydrohalogenation 23, 948
suppl. 29
elimination-rearrangement
12, 929
-, cycloaddition, 1,3-dipolar,
with - 24, 900 suppl. 28
- from
α-acoxytosylhydrazones
26, 959
1,2-bis(tosylhydrazones)
27, 954
boranes, synthesis 28, 864
cyclopropenones 27, 983
2-diazoalcohols 28, 939
α,α-dihalogenosulfones
26, 972
1,1,1-dihalogenothioethers
30, 687
enetriflates 29, 920
α,β-ethylenehalides 27, 964;
29, 961
halides 27, 851
osazones 28, 938
oxo compds., insertion of
1 C-atom 28, 731
polyhalogenoethylene derivs.
27, 961
1,2,3-selenadiazoles 26, 655
N-sulfonylenazo compds.
28, 927
thiirene 1,1-dioxides 26, 972
p-toluenesulfonic acid esters,
synthesis 30, 548
-, hydrogenation HC↓↓CC
-, mercuration 7, 201 suppl. 29
-, oxidation with thallic nitrate
27, 141 suppl. 28
-, protection as cobalt carbonyl
complex 27, 140
-, reaction with
disilanes 30, 425
lithium compds., organo-
30, 479
sulfur dichloride 26, 579
-, ring closure with - 28, 668
- special s.
acoxy-acetylenes
alkoxy-
alkynes
cyanoacetylenes
dienynes
difuranylacetylenes
diynes
ethynyl...
propargyl...
tolans
- startg. m. f.
α,β-acetylenecarboxylic acid
esters 27, 857
β,γ-acetylenecarboxylic - -
28, 783
α,β-acetyleneketones 28, 135
1-acetylenephosphonic acid
esters 28, 596
alcohols, synthesis 28, 821
suppl. 29
-, sec. - 28, 821
allenes with N-ring opening
29, 41
2-aminoeneboranes 29, 597
cis-2-azetidinones 28, 650

azomethines 28, 283
benzene ring (from 2 molecules) 26, 785
benzenes 29, 886
cage compds., isocyclic
28, 640
carbostyrils 26, 694
carboxylic acids 27, 245
cyanohydrins 28, 660
cis,cis,trans-1,4,7-cyclo-
decatrienes, cyclocooligomerization 30, 501
cyclopropanes 26, 880
cyclopropenes, imino- 27, 690
1,1-diboro compds. 30, 427
di(ene)halogenoborans
(from 2 molecules) 28, 666
1,3-dienes, synthesis 26, 727
-, - (from 2 molecules)
28, 666
1,1-dimercury compds. 30, 427
β-diketones 28, 104
α-dioxo compds. 11, 209
suppl. 29; 27, 145
1,2-disilyl-1-ethylene derivs.
29, 602
1,3-dithiins 28, 744
enesilanes 12, 700 suppl. 26
1,4-enynes, synthesis 29, 830
ethylenealcohols 29, 139
2-ethylenealcohols 28, 652;
29, 139
3-ethylenealcohols 29, 869
α,β-ethyleneboronic acid
esters 29, 497
α,β-ethylenecarboxylic acid
amides 26, 688
- - esters 24, 710
- acids 30, 492
ethylene derivs., review
7, 95 suppl. 29
- -, synthesis 23, 693
suppl. 28; 29, 672; 30, 479
cis-ethylene derivs. 7, 95
suppl. 26; 27, 51; 29, 48
cis- and trans-ethylene derivs.
11, 87 suppl. 27; 28, 36
cis- and trans-ethylene derivs.,
synthesis 23, 703 suppl. 29
trans-ethylene derivs.,
synthesis 28, 660
ethylene derivs., terminal
30, 492
- -, trisubst., synthesis
30, 623
α,β-ethylenehalides 29, 497;
30, 603
α,β-ethylene-β-halogeno-
sulfones 27, 598
α,β-ethyleneketones 28, 866
γ,δ-ethyleneketones 23, 705
suppl. 28
trans-γ,δ-ethyleneketones
23, 705 suppl. 28
α,β-ethyleneiminoesters 29, 644
2-halogenothioenolethers,
sym. 26, 579
γ-hydroxyketones 29, 869
α-iminosulfonium ylids 30, 387
isoxazoles 26, 796
2-ketoalkoximes 28, 104
γ-ketocarboxylic acid esters
28, 783
ketones 22, 143 suppl. 27;
29, 140
-, synthesis 29, 882
maleimides, N-subst. 28, 766

(Acetylene derivs.
- startg. m. f.)
naphthalenes 28, 639
- (from 2 molecules) 29, 883
1-naphthols, 1,4-dihydro-
28, 639
2-oxa-3-borolenes 28, 866
2,1-oxazabicyclo[4.1.0]-
heptanes 29, 653
pyridines 29, 886
pyrrolizines, dihydro- 30, 631
semicarbazones 25, 228
suppl. 27
sulfenic acid esters 27, 634
thianaphthenes 26, 579
uracil-1,3-disulfofluorides
26, 698
vinylthallic acetates,
2-acetoxy- 29, 598
-, synthesis with - and boranes
26, 911
- -, **cyclic** 27, 964
-, prepn., review 27, 964
-, ring closure, transannular
27, 757
- -, **terminal** (s. a. 1-Alkynes,
Ethynyl derivs.)
- by
migration of carbon-carbon
triple bonds 30, 506
- from
aldehydes 28, 819
dihalogenomethylene compds.
28, 819
ethylene derivs., synthesis
with addition of 2 C-atoms
29, 673
halides 27, 851 suppl. 30
- **deriv. substitutes**
-, enamines as - 22, 877 suppl. 28
**Acetylenedicarboxylic acid
esters**
- from
alcohols 30, 132
-, insertion of azimines into -
28, 281
- startg. m. f.
anilines 26, 719
azepine ring (from 2 molecules) 28, 635
1,3-benzodioxoles 28, 100
4H-1,3-benzoxazin-4-ones,
2,3-dihydro- 26, 307
α,β-ethylenedicarboxylic acid
esters 28, 626
β,γ-ethylene-β-di-
carboxylic - - 26, 711
furan-3-4-dicarboxylic - -
28, 869
1,3-S,N-heterocyclics 23, 592
Δ³-pyrroline ring 29, 649
2-pyridones 26, 919
4(1H)-pyrimidinone ring,
N-condensed 30, 251
quinolines, 3-hydroxy- 26, 939
quinuclidines, Δ²-dehydro-
4-hydroxy- 30, 471
thiophene ring 26, 919
-, syntheses with -, review
28, 635
- - **fluoride**
-, reactions with - 30, 132
2-Acetylene-1,4-diol monoesters
- startg. m. f.
3-ethylenealcohols 29, 567
Acetylenediols
- special s.
ethynyldiols

2-Acetylene-1,4-diols
- startg. m. f.
1,4-dihydroxy-2,3-diketones
30, 167
3(2H)-furanones, 4-hydroxy-
30, 167
**Acetylene electron
acceptor-donors**
-, reactions 26, 694
-, ring closure 26, 694
Acetyleneethers s. Alkoxyacetylenes, Aryloxy-, Pyrans,
tetrahydro-2-acetylene-
Acetyleneglycols
s. Ethynylglycols
Acetylenehalides
- startg. m. f.
alkoxyacetylenes 27, 210
ethylene derivs., synthesis
(trans-ethylene derivs.)
28, 820
α-ketocarboxylic acid esters
30, 133
2-Acetylenehalides
- startg. m. f.
isoxazoles, 3-nitro- 28, 397
4-Acetylenehalides
- startg. m. f.
enamines, cyclic 30, 637
pyrroles 30, 637
Acetylenehalogenhydrins
- startg. m. f.
pyrroles 27, 451
Acetylene-N-heterocyclics
-, addition of methylene groups,
active to - 29, 662
α,β-**Acetylenehydrazones**
- startg. m. f.
pyrazoles 16, 876 suppl. 26
α,β-**Acetylenehydroxamic acids**
- startg. m. f.
Δ⁴-isoxazol-3-ones 28, 125
Acetyleneketones
- from
α,β-oxidoketones, cyclic
22, 913
- special s.
acetylenealkoxyketones
diynones
silylethynyl ketones
α,β-**Acetyleneketones**
- from
acetylene derivs. 28, 135
carboxylic acid anhydrides
27, 776
- - chlorides 12, 855 suppl. 26
α,β-ethylene-α-halogeno-
ketones 30, 676
-, reactions, review 28, 730
- startg. m. f.
phenols 28, 730
thiabenzene oxides 27, 769
γ,δ-**Acetyleneketones**
- from
α,β-ethyleneketones, synthesis
28, 658
Acetylenemercaptans
- startg. m. f.
thioenolethers, cyclic 26, 587
Acetyleneoxido compds.
- startg. m. f.
2-allenealcohols 30, 594
α,β-**Acetyleneoximes**
- from
hydroximinohalides 26, 877
α,β-**Acetyleneoxo compds.**
- startg. m. f.
pyrazoles 26, 386

**Acetyleneperoxycarboxylic acid
esters** 18, 129 suppl. 26
**1-Acetylenephosphonic acid
esters**
- from
acetylene derivs. and
phosphorochloridates
28, 596
2-Acetylenephosphonium salts
- startg. m. f.
heterocyclics 28, 962
Acetylenesulfenamides
- from
acetylenethiocyanates 28, 260
2-Acetylenesulfonium salts
- startg. m. f.
2-allenethioethers 25, 546
furans 28, 838
sulfonium ylids, stable 26, 767
Acetylenesulfoxides s. Aryl
acetylenesulfoxides
Acetylenethiocyanates
- startg. m. f.
acetylenesulfenamides 28, 260
1-Acetylene-1-thioethers
- startg. m. f.
thioenolethers 26, 715
suppl. 29
2-Acetylenethioethers,
metalation 27, 158
**1-Acetylenethiophosphonic acid
esters** 28, 596
Acetyl halides as reactants
29, 488
Acetyl hypobromite
- as reagent 26, 267
Acetylides s. Copper(I)
acetylides, Mercury(II) -,
Silver -, Sodium
acetylide
Acetyl nitrate
- as reactant 29, 312
- as reagent 26, 328
N-Acetylsaccharin as reactant
29, 394
Acetylsalicyloyl chloride
- startg. m. f.
1,2-acoxyhalides 26, 547
1-Acetyl-2-thiourea
- startg. m. f.
mercaptans 28, 569
Acetyl tosylate as reactant
29, 232
Acid anhydrides, mixed
s. Carboxylic acid anhydrides,
mixed
Acid-base catalysis, aromatization by - 26, 946
Acid chlorides
- startg. m. f.
acid fluorides 3, 485 suppl. 28
- **fluorides**
- from
acid chlorides 3, 485 suppl. 28
Acids s. a. Lewis acids,
Proton -, Superacids
-, **organic, weak** 28, 884
-, **strong** s. Trifluoromethanesulfonic acid
Acoxonium ion rearrangements,
review 22, 261 suppl. 26
- **salts**
- startg. m. f.
1,1,1-dialkoxyammonium salts,
quaternary 27, 995
Acoxy-2-acetylenes
- special s.
1-ethynylacoxy compds.

- startg. m. f.
  2-alkylidene-1,3-diones
  28, 968
  allenes **28**, 115
  a,a'-dihydroxyketones **28**, 115
  a,β-ethyleneketones **28**, 968
1-Acoxy-2-acylamines
- startg. m. f.
  Δ^2-oxazoline ring **27**, 257
3-α-Acoxy-3-acyltriazenes
- from
  3-hydroxytriazenes **29**, 346
Acoxyaldehydes
- startg. m. f.
  acetals, cycle, mixed **28**, 195
- special s.
  formoxyaldehydes
Acoxyalkoxy compds.
  s. Alkoxyacoxy compds.
Acoxybetaines 26, 996
α-Acoxycarboxylic acid amides
- from
  α-acoxynitriles **27**, 709
  acylcyanides and oxo compds.
  27, 709
Acoxycarboxylic acid chlorides
- from
  hydroxycarboxylic acids
  29, 514
- special s.
  acetylsalicyloyl chloride
α-Acoxycarboxylic – –
- as reagents **24**, 268 suppl. **29**
- **acid esters**
- from
  α-hydroxycarboxylic acid
  esters **27**, 92
- startg. m. f.
  carboxylic acid esters **27**, 92
β-Acoxycarboxylic – –
- from
  aldehydes and α-alkoxyenolesters **26**, 687
  ketones **26**, 938
- startg. m. f.
  α,β-ethylenecarboxylic acid
  esters **26**, 938
Acoxycarboxylic acids
- from
  hydroxycarboxylic acids
  30, 127
α-Acoxycarboxylic –
- from
  carboxylic acids (2 molecules)
  27, 178
  α-diazocarboxylic acid
  tert-butyl esters **27**, 201
Acoxy compds. (s. a. Acylation,
  Carboxylic acid esters)
- as intermediates **26**, 161
- by Claisen rearrangement
  27, 741
- from
  acetals with rearrangement
  30, 94
  aldehydes (2 molecules)
  27, 668
  carboxylic acid esters **27**, 68
  ethers **26**, 509 suppl. **28**;
  27, 248; **29**, 232
  ethylene derivs. **29**, 158
  – –, preferential formation
  22, 143 suppl. **29**
  hydrazones **26**, 190
  2-pyranyl ethers, tetrahydro-
  27, 241

sulfonic acid esters **26**, 760
– – – with ring expansion
  27, 226
  p-toluenesulfonic acid
  esters **26**, 212, 944;
  27, 227
- special s.
  diacoxy...
  trifluoroacetoxy...
- startg. m. f.
  ethylene derivs. **27**, 927
  halides **28**, 504
  ketones **26**, 161
  thioethers **26**, 599
-, syntheses with – **30**, 559
-, – – – and Grignard compds.
  29, 824
-, **cyclic**
- from
  ethylenediazo compds. **30**, 571
- –, **prim.**
- from
  ethylene derivs. **29**, 158
- –, **tert. 30**, 146
Acoxycyanohydrins
- startg. m. f.
  chroman-2-carboxylic acids
  30, 183
Acoxycyclopropylketones
- startg. m. f.
  cyclopent-2-enones **27**, 923
2-Acoxydiazo compds.
- startg. m. f.
  enolesters **26**, 962
**N-Acoxydicarboxylic acid
  imides, polymeric**
-, peptide synthesis via –
  18, 435 suppl. **27**
1-Acoxy-1,3-dienes
-, diene synthesis with –
  prepared *in situ* **27**, 805
- startg. m. f.
  α,β-ethylene-γ-halogen-
  aldehydes **26**, 553
1-Acoxy-2-en-4-ynes
- startg. m. f.
  3-allenealcohols **30**, 24
β-Acoxyethylation, ar. 29, 675
N-(β-Acoxyethylation)
- with
  dioxolanium salts **30**, 701
**β-Acoxy-α,β-ethylenecarboxylic
  acid amides**
- from
  acylynamines **29**, 134
β-Acoxy-α,β-ethyleneketones
- startg. m. f.
  ketones **29**, 746
Acoxy-2-ethylenes
- from
  2-acoxyselenides **29**, 180
  2-ethylenealcohols with allyl
  rearrangement **29**, 191
  3-ethylenealcohols **27**, 243
  ethylene derivs. **29**, 173
  – – with double bond shift
  29, 180
- startg. m. f.
  ethylene derivs., synthesis
  28, 740
Acoxy group migration 26, 962
Acoxyhalides
- from
  1-carboxyacetals, cyclic
  27, 583
  ethers, cyclic **26**, 509
-, –, and carboxylic acids
  29, 487

orthocarboxylic acid esters,
  cyclic **28**, 500
- special s.
  formoxyhalides
1,1-Acoxyhalides
- startg. m. f.
  fulvenes **27**, 834
1,2-Acoxyhalides
- from
  1,3-dioxolanes, 2-alkoxy-
  28, 500
  glycols **26**, 547; **29**, 509
  – and nitriles **26**, 547
- special s.
  1,2-formoxyhalides **29**, 942
- startg. m. f.
  α,β-ethylenehalides **29**, 942
Acoxy-N-heterocyclics,
  N-acylation with – **20**, 277
Acoxyketones
- from
  acetals, cyclic **25**, 99
- special s.
  acoxycyclopropylketones
α-Acoxyketones
- startg. m. f.
  β-hydroxy-γ-lactones **30**, 516
  ketones **30**, 27
β-Acoxyketones
- from
  ketones **26**, 659
- special s.
  β-acoxy-α,β-ethyleneketones
Acoxylamines (s. a. Sulfonyloxylamines)
- from
  acyloximes **27**, 42
  amines **26**, 88
  carboxylic acids **26**, 172
  N-tritylacoxylamines **26**, 172
- special s.
  N-tritylacoxylamines
- startg. m. f.
  hydroxylamines **26**, 88
Acoxylation s. a. Replacement
  of hydrogen by acoxy groups
-, ar., orientation in – **24**, 176
  suppl. **26**
C-Acoxylation, review **21**, 197
  suppl. **27**
1,2-Acoxymesylates
- from
  carboxylic acid chlorides
  27, 81
- startg. m. f.
  acetals, cyclic **27**, 81
Acoxymethylene groups
- startg. m. f.
  methyl groups **11**, 128
α-Acoxynitriles
- from
  acylcyanides and oxo compds.
  27, 709
- special s.
  mandelonitrile benzoates
- startg. m. f.
  α-acoxycarboxylic acid
  amides **27**, 709
  o-acyl-N-heterocyclics **30**, 557
  ketones **30**, 557
o-Acoxynitriles
- startg. m. f.
  benzoxazoles **28**, 456
1,2-Acoxynitro compds.
- from
  acylals **30**, 550
- startg. m. f.
  1,2-nitramines **28**, 342

1,1-Acoxynitrones
- startg. m. f.
  acyloximes **30,** 175
Acoxyperchlorates
- from
  acyl perchlorates **30,** 68
Acoxyphosphonium salts
- startg. m. f.
  cyclopropylketones **28,** 963
1,1-Acoxyselenides
- from
  diacyloxyselenuranes **29,** 108
  selenides **29,** 108
2-Acoxyselenides
- from
  ethylene derivs. and
  diselenides **29,** 180
- startg. m. f.
  acoxy-2-ethylenes **29,** 180
3β-Acoxy-Δ⁵-steroids
- from
  3α,5α-cyclosteroids,
  6β-alkoxy- **30,** 108
-, 6β-hydroxy- **29,** 185
Acoxysulfonates
- special s.
  acoxymesylates
Acoxysulfones
- from
  hydroxythioethers **26,** 93
**1-Acoxy-1-sulfonyloxy compds.
27,** 111
1,2-Acoxythioethers
- startg. m. f.
  ethylene derivs. **29,** 974
  suppl. **30**
α-Acoxythiolic acid esters
- from
  β-ketosulfoxides **30,** 100
α-Acoxytosylhydrazones
- startg. m. f.
  acetylene derivs. **26,** 959
Acoxytriazenes
- special s.
  acoxyacyltriazenes
N-Acoxyureas
- from
  N-hydroxyureas **27,** 194
N-Acoxyurethans 26, 412
Acridazines 27, 661
-, 5,10-dihydro-5,10-dihalogeno-
30, 344
Acridinium salts
- startg. m. f.
  dibenzo-1,4-diazepines **28,** 287
Acridinium salts,
sym-**octahydro-**
- from
  amines, prim. **27,** 415
Acridizinium salts, diene
synthesis with – **5,** 503
suppl. **27; 27,** 697
Acrylic acids s. α,β-Ethylene-
carboxylic acids
Acrylonitrile s. Cyanoethylation
Activation (s. a. Functionalization, Long-range activation)
- by
  N-quaternization **18,** 842
  suppl. **27**
- of
  glycol groups as dialkoxy-
    stannanes, cyclic **29,** 111
  halides by adsorption **26,** 252
π-Activation 21, 419 suppl. **27**
Activation, double
-, macrolactonization by –
  30, 168

Activator (s. a. Nucleophiles,
reactive)
- in Grignard reactions
  ethylene bromide **22,** 677
  suppl. **27**
Activity, optical s. under
Stereoisomers
Acyl... s. a. Acet..., Acyl group,
α-Keto..., Replacement
N-Acyl as N-protective group
26, 454
**Acyl alkoxycarbonyl sulfides
26,** 618
N-Acylalkoxylamines
s. Hydroxamic acid esters
(α-Acylalkylidene)phosphoranes
-, metalation and alkylation
30, 582
- startg. m. f.
  2-ene-1,4-diones **28,** 847
  α,β-ethyleneketones **28,** 850
  suppl. **29**
  β,γ-ethyleneketones **28,** 850
  ketones **30,** 582
Acylals (s. a. 1,1-Diacoxy
compds.)
- from
  1,3-dithianes **26,** 211
- startg. m. f.
  1,2-acoxynitro compds.
  30, 550
N-Acylamide acetals 27, 430
Acylamidines
- from
  carboxylic acid amides and
  nitriles **29,** 305
- special s.
  ethyleneacylamidines
**4-Acyl-1-amidinothiosemi-
carbazides**
- from
  acylisothiocyanates **27,** 327
Acylamidrazones
- from
  tetrazoles and carboxylic acid
  amides **30,** 288
Acylamination 29, 334
-, remote
- of
  ketones **29,** 334 suppl. **30**
ω-Acylamination 29, 334
Acylamines (s. a. N-Acyl...,
Carboxylic acid amides,
Amido...)
- from
  acylimines, synthesis **26,** 681
  acylurethans **28,** 463
  amine hydrochlorides **10,** 331
  – – and orthocarboxylic acid
    esters **26,** 384
  amines, tert., cyclic, with
    aromatization **28,** 336
  1,1-aminoethers, cyclic **29,** 164
  ammonium salts, quaternary
    28, 463
  carboxylic acid azides
    29, 782; **30,** 460
  – – chlorides **30,** 460
  isocyanates **30,** 460
  nitriles and alcohols **29,** 307
  – and hydrocarbons **29,** 334
  oxo compds. **29,** 316
- special s.
  ethyleneacylamines
  formamides
  halogenacylamines
  monoacyldiamines
  nitro-acylamines

N-nitroso-
N,N,N-tri-
- startg. m. f.
  amines (s. a. N-Deacylation)
  HN↓↑C; **24,** 491
  N-nitrosoacylamines **28,** 935
  oxo compds. **27,** 199
  ketones **29,** 198
  urethans **30,** 107
Acylamines, sec.
(s. a. Dimethylacylamines)
- from
  amines, tert. **29,** 444
  ammonium salts, quaternary
  29, 440
Acylaminoalcohols
- startg. m. f.
  O-heterocyclics **27,** 264
-, **cyclic**, from ethyleneacylamines **29,** 313
1,1-Acylaminoalcohols
- special s.
  α-acylamino-α-hydroxy-
    carboxylic acid esters
  α-acylamino-α-hydroxy-
    β-diketones
  lactamols
- startg. m. f.
  1,1-acylaminohalides **30,** 332
  acylimines **30,** 332
2-Acylaminoalcohols
- from
  aziridines **26,** 296
  nitriles and oxido compds.
  23, 348
  Δ²-oxazolines **28,** 85
α-Acylaminoaldehydes 28, 66
**α-Acylamino-α-alkoxycarboxylic
acid esters 24,** 145 suppl. **28**
- from
  α-acylamino-α,β-ethylene-
    carboxylic acid esters
    29, 156
C-α-Acylaminoalkylation,
review **16,** 843 suppl. **26**
**α-Acylaminocarboxylic acid
amides**
- from
  4-alkylidene-5-oxazolones
  28, 288
- special s.
  α-acylamino-α,β-ethylene-
    carboxylic acid amides
- startg. m. f.
  α-acylaminocarboxylic acids
  28, 888
  α-aminocarboxylic –
  28, 288
  oxazoles **30,** 165
-, synthesis, asym. **28,** 288
  suppl. **30**
o-Acylaminocarboxylic – –
- from
  o-acylaminocarboxylic acids
  29, 391
  o-aminocarboxylic acid
    amides **28,** 437
  4H-3,1-benzoxazin-4-ones
  29, 391
- startg. m. f.
  4(3H)-pyrimidinone ring
  4, 455; **28,** 437
**Acylaminocarboxylic acid
esters**
- from
  acylaminocarboxylic acids
  26, 224

α-**Acylaminocarboxylic** – –
- from
  Δ²-oxazoline-4-carboxylic acid
    esters **26**, 32
- special s.
  α-acylamino-α-alkoxy-
    carboxylic acid esters
  α-acylamino-α-hydroxy-
    carboxylic – –
Acylaminocarboxylic acids
- startg. m. f.
  acylaminocarboxylic acid
    esters **26**, 224
α-**Acylaminocarboxylic** –
- from
  α-acylaminocarboxylic acid
    amides **28**, 288
  aldehydes **28**, 613
  4-alkylidene-5-oxazolones
    28, 288
  α-aminocarboxylic acids
    30, 451
  ethylene derivs. **28**, 613
- special s.
  β-alkylthio-α-formamino-
    carboxylic acids
- startg. m. f.
  5-oxazolones **30**, 451
– –, **H-labeled 26**, 645
– –, **optically active**
- by
  hydrogenation, asym. **27**, 57
o-**Acylaminocarboxylic** –
- startg. m. f.
  o-acylaminocarboxylic acid
    amides **29**, 391
  4H-3,1-benzoxazin-4-ones
    29, 391
α-**Acylaminocarboxylic acid
thioamides**
- startg. m. f.
  thiazoles, 5-amino- **27**, 641
α-**Acylamino-α,β-ethylene-
carboxylic acid amides**
- from
  4-alkylidene-5-oxazolones
    23, 497 suppl. **29**
– – **esters**
- startg. m. f.
  α-acylamino-α-alkoxy-
    carboxylic acid esters
    29, 156
  α-acylamino-α-hydroxy-
    carboxylic – – **29**, 156
  α-ketocarboxylic acids **29**, 200
Acylaminohalides
- startg. m. f.
  N-heterocyclics **22**, 531
1,1-Acylaminohalides
- from
  1,1-acylaminoalcohols **30**, 332
  acylimines **30**, 332
α-**Acylamino-α-hydroxy-
carboxylic acid esters**
- from
  α-acylamino-α,β-ethylene-
    carboxylic acid esters **29**, 156
α-**Acylamino-α-hydroxy-
β-diketones**
- special s.
  indan-1,3-diones, 2-acylamino-
    2-hydroxy-
Acylaminoketones
– , ring closure with ring
    expansion **28**, 935
α-**Acylaminoketones**
– , ring closures with – **26**, 603

– startg. m. f.
  imidazoles **30**, 267
  1,2,4-oxadiazoles, 3-acyl-
    28, 318
  oxazoles **19**, 752
  thiazoles **26**, 603
o-**Acylaminoketones**
- startg. m. f.
  quinazolines, 3,4-dihydro-
    3-amino-4-hydroxy- **27**, 382
Acylaminomethylation
s. Imidomethylation
α-**Acylaminonitriles**
- from
  cyanohydrins **26**, 375
1-Acylamino-1-peroxides
- from
  enacylamines **28**, 112
**2-Acylaminophosphonic acid
esters**
- from
  acylimines and phosphonic
    acid esters **29**, 880
- startg. m. f.
  ethylene derivs. **29**, 880
Acylaminosilanes
- from
  disilazanes **28**, 412
- special s.
  N-trimethylsilyltrifluoro-
    acetamide
Acylaminosulfenium salts
- startg. m. f.
  N-acylsulfilimines **27**, 300
**N-Acylammonium fluoroborates,
quaternary**
– as acylating agents **30**, 286
Acyl aryl disulfides
- from
  acylsulfenylhalides **28**, 536
1-Acyl-2-arylhydrazines
- startg. m. f.
  indolines, 2-imino- **28**, 918
Acylating agents
  N-acylammonium fluoro-
    borates, quaternary **30**, 286
  acyl dithiocarbamates
    28, 402 suppl. **30**
  carbatricobalt decacarbonyl
    cation **27**, 180
  N-(chlorophosphoryl)-
    pyridinium betaines **27**, 419
  α,α-dichloro-β-ketocarboxylic
    acid esters **29**, 441
  diphenyl phosphite-pyridine
    27, 419 suppl. **28**
  N-sulfenylcarbamic carboxylic
    anhydrides, cyclic **28**, 255
  trifluoromethanesulfonamides
    29, 393
– –, **mild, selective**
– , polymer-based carboxylic
    alkoxyformic anhydrides as –
    19, 233 suppl. **29**
– –, **powerful**
    acyl trifluoromethane-
    sulfonates **28**, 714
Acylation
– , reviews **30**, 271
- special s.
    diacylation
    formylation
– , **oxidative**
– of
    nitro compds., prim. **26**, 322
– , **total 29**, 357
C-Acylation (s. a. Ketones)
  25, 602

- of
  amines, ar. **26**, 846
  cycloheptatrienes **26**, 883
  enollactones **26**, 821
  N-heterocyclics **16**, 872
- with
  enolesters **26**, 804
  polyhalogenocarboxylic acid
    chlorides **26**, 846
–, **intramolecular**
- with
  α-enaminocarboxylic acids
    30, 647
–, **α-lateral**
- of
  N-heterocyclics **28**, 732
    suppl. **29**
–, **photochemical 29**, 887
–, **radical**, with aldehydes
    26, 775
N-Acylation 29, 373
– by different methods **15**, 342
    suppl. **26**
- of
  thioureas **27**, 406
- with
  acoxy-N-heterocyclics
    20, 277
  acyl dithiocarbamates
    28, 402
  diene synthesis, intra-
    molecular **29**, 817
  imidazolium chlorides,
    N-acyl-N'-methyl-
    25, 293 suppl. **28**
  p-nitrophenyl esters **21**, 521
–, **preferential 28**, 344; **29**, 357
–, **regiospecific**, of imidazoles
    30, 302
N-Acylation, selective
    29, 357, 394, 415, 421
- of
  nucleosides **25**, 255 suppl. **26**
- with
  acyl dithiocarbamates **28**, 402
  carboxylic acid p-nitrophenyl
    esters **26**, 363
O-Acylation (s. a. Acoxy-...)
    26, 939; **29**, 394
- of
  carboxylic acid amides,
    vinylogous **26**, 112
  hydroxy compds. OC↓↑O;
    OC↓↑Hal
– –, hindered **29**, 184
–, **preferential 21**, 241; **27**, 191;
    28, 172; **29**, 188
- of
  alcohol groups with retention
    of phenol groups **28**, 86
  carbohydrates **26**, 207;
    26, 175 suppl. **27**
–, **selective 1**, 177 suppl. **26**;
    28, 172
- of
  phenol groups **29**, 206
–, **total**, of carbohydrates
    28, 88
S-Acylation, selective 30, 415
Acylation catalysts
–, pyridines, 4-dialkylamino-
    as – **29**, 184
Acylation-deacylation 13, 428;
    29, 357
Acylazo compds.
– as reagents **29**, 585
- special s.
    diacylazo compds.

O-Acylbenzoins
- from
    benzils (2 molecules) **26,** 221
**N-Acyl-N'-carbalkoxyhydrazines
13,** 434 suppl. **26**
Acyl carbamyl peroxides
- from
    peroxycarboxylic acids and
    isocyanates **26,** 122
- - **sulfides**
- from
    thiolic acids and isocyanates
    30, 382
- **carbanions, masked**
-, syntheses with - **22,** 661;
    27, 830; **29,** 808; **30,** 482
N-Acylcarboxylic acid amides,
    formation, selective **27,** 476
- from
    4-alkylidene-5-oxazolones
    29, 228
    N-silylcarboxylic acid amides
    23, 476 suppl. **29**
-, transamidation via - **27,** 476
- - **thioamides**
- special s.
    N-thioformylacylamines
- startg. m. f.
    ketene N-acyl-S,N-acylals
    29, 575
- S,N-acylals **29,** 575
Acyl compds., active
    s. Carboxylic acid esters,
    active
- -, **vinylogous 26,** 184
Acylcyanides 28, 319
-, condensation, double **7,** 805
    suppl. **29**
- from
    semicarbazones **29,** 952
-, reaction with Grignard
    compds. **27,** 905
- startg. m. f.
    α-acoxycarboxylic acid
    amides **27,** 709
    α-acoxynitriles **27,** 709
    α-aminoketones **8,** 59
    suppl. **27**
    α-(dihalogenomethylene)-
    nitriles **28,** 752
    1,1-iminocyanides **26,** 444
    α-ketocarboxylic acid amides
    27, 207
    - acids **27,** 207
    ketones **27,** 905
N'-Acyl-1,1-cyanohydrazines
- from
    carboxylic acid hydrazides
    and ketones **27,** 780
N-Acylcyclimmonium salts
-, chemistry, review **21,** 821
    suppl. **29**
**Acyldiaminoguanidines,
5-carbamyl-**
- from
    1,3,4-oxadiazolines, 2-imino-
    28, 269
**2-Acyl-1,2-dihydroisoquinaldo-
1-nitriles**
-, reactions **29,** 889
- startg. m. f.
    isoquinolines, 1-subst. **29,** 889
-, synthesis of alkaloids, review
    29, 889
Acyldisulfides
- special s.
    acyl aryl disulfides

Acyl dithiocarbamates
-, acylation with - **28,** 402
    suppl. **30**
N-Acyl-N'-enehydrazines
- startg. m. f.
    Δ⁴-pyrrol-2-ones **29,** 951
Acylglycosides (s. a. Acyl
    glycosides)
- from
    acylhalogenosugars **26,** 203
Acyl glycosides 24, 213
- startg. m. f.
    N-glycosyl-N-heterocyclics
    26, 446
- **group interchange,
    C,N-positional 30,** 527
C-Acyl group migration 26, 477
C→N-Acyl - - **28,** 306
O-Acyl - - **26,** 212
-, prevention s. under
    Prevention
1-Acylguanidines, 3-subst.
- from
    aldehydes **29,** 356
o-Acyl-N-heterocyclics
- from
    N-oxides, cyclic, and α-acoxy-
    nitriles **30,** 557
Acylhydrazines (s. a. Carboxylic
    acid hydrazides, Sulfonyl-
    hydrazines)
- from
    aldehydes and azo compds.
    28, 271
    1,1-alkoxyhydrazones **27,** 336
- special s.
    acylaryl-hydrazines
    acylcyano-
    acylene-
    diacyl-
1-Acylhydrazines, 1-aryl-
- from
    hydrazine-2-sulfonic acids,
    1-aryl- **28,** 409
**α-(1-Acylhydrazino)carboxylic
acid amides 17,** 809 suppl. **28**
Acylhydrazones
- from
    carboxylic acid esters and
    oxo compds. **27,** 377
- startg. m. f.
    halogenazines **27,** 577
    1,3,4-oxadiazoles **26,** 232
S-Acylhydrosulfamines 29, 286
N-Acylhydroxylamines
    s. Hydroxamic acids
O-Acylhydroxylamines
    s. Acoxyamines
**Acylimidophosphoriso-
cyanatidates**
- from
    carboxylic acid azides and
    phosphorisocyanatidites
    27, 308
Acylimines
- from
    1,1-acylaminoalcohols **30,** 332
    1,1-acylaminohalides **30,** 332
- special s.
    quinone iminoximes,
    N,O-diacyl-
- startg. m. f.
    acylamines, synthesis **26,** 681
    2-acylaminophosphonic acid
    esters **29,** 880
    3H-1,2,4-dithiazoles **30,** 398
    ethylene derivs. **29,** 880

4H-1,3-oxazines, 5,6-dihydro-
    6-alkoxy- **29,** 654
-, **cyclic 27,** 797; **29,** 447
- special s.
    pseudoisoindol-1-ones
-, **polyfluoro- 21,** 341 suppl. **29**
N-Acyliminoesters
- from
    carboxylic acid azides and
    enolethers **29,** 395
Acylisocyanates
-, cycloadditions with -,
    review **28,** 266 suppl. **29**
- from
    N-(chloroglyoxyloyl)imino-
    esters **26,** 458
    N-(halogenoformyl)imino-
    halides **30,** 137
    iminoesters **26,** 458
-, review **26,** 458
- special s.
    trichloroacetyl isocyanate
- startg. m. f.
    1,3,5-oxadiazine-2,4-diones
    27, 322
    2H-1,3,5-oxadiazin-4(3H)-ones,
    2-imino- **28,** 266
    1,2,4-triazolidine-3,5-diones,
    4-acyl- **30,** 212
Acylisothiocyanates
- review **27,** 327
- startg. m. f.
    4-acyl-1-amidinothiosemi-
    carbazides **27,** 327
N-Acylisothiouronium salts
- startg. m. f.
    N-acylthiocarbamic acid
    esters **28,** 160
S-Acylisothiouronium -
- from
    thioureas **27,** 625
Acylium - s. Oxocarbonium
    salts
N-Acyllactams
- from
    carboxylic acid chlorides
    27, 471
- startg. m. f.
    azomethines, cyclic, 1-subst.
    28, 465
Acyllactone rearrangement
- with
    amines **27,** 389
β-Acyl-γ-lactones
- from
    β,γ-ethylene-γ-lactones and
    oxo compds. **30,** 555
Acylmercaptans s. Thiolic acids
(2-Acyl-1-methylvinyl)amines
-, protection of amino groups
    as - **27,** 402
Acyl migration s. Acyl group
    migration
- **nitrates**
- startg. m. f.
    nitro compds. **26,** 494
- **nitrites**
- special s.
    p-chlorobenzoyl nitrite
Acylochlorosugars
- from
    carbohydrate orthocarboxylic
    acid esters **27,** 572
Acylohalogenosugars
    (s. a. Acylochlorosugars)
    15, 438 suppl. **26**
- startg. m. f.
    nucleosides **16,** 514

Acyloin rearrangement 28, 682
Acyloins (s. a. α-Hydroxyketones)
- from
  aldehydes (2 molecules) **29,** 630
Acyloin substitutes, ene-1,2-di-(oxysilanes) as – 26, 811
Acyloximes
-, diene synthesis with – **27,** 698
- from
  1,1-acoxynitrones **30,** 175
-, peptide synthesis with –
  20, 293 suppl. **26**
-, protection of carboxyl groups as – **27,** 1
- special s.
  quinone iminoximes, N,O-diacyl-
- startg. m. f.
  acoxylamines **27,** 42
-, **cyclic** s. 1,2-Oxazin-6-ones
1-Acyloximino-2-nitrites
- from
  ethylene derivs. **29,** 336
Acyl perchlorates
- startg. m. f.
  acoxyperchlorates **30,** 68
- **peroxides**
- from
  carboxylic acid chlorides **26,** 202
  – acids **27,** 95
- special s.
  acetyleneacyl peroxides
  acyl carbamyl –
  benzoyl –
- startg. m. f.
  catechol monoesters **25,** 126
  diacyloxyselenuranes **29,** 108
  phenolesters **26,** 166
- –, **cyclic** s. Dicarboxylic acid peroxides
N-Acylperoxycarbamic acids
- startg. m. f.
  1,2,4-dioxazol-3-ones **28,** 212
Acylphosphonic acid esters
- startg. m. f.
  heterocyclics, review **26,** 909 suppl. **29**
  1-hydroxy-1,1-diphosphonic acid esters **28,** 584
  sulfonic acid esters **30,** 109
  4H-1,3,4-thiadiazin-5-ones, 5,6-dihydro- **26,** 909
Acylphosphoric – –
- as intermediates **19,** 86 suppl. **28**
- startg. m. f.
  peroxycarboxylic acids **17,** 145 suppl. **26**
– – –, **cyclic**
-, N-acylation, preferential, with – **28,** 344
Acyl polythiocarbonates
- startg. m f.
  thiolic acid esters **28,** 579
- **radicals**
-, chemistry in soln., review **28,** 271
-, reaction via **28,** 271
- **rearrangement** s. Acyl group migration
N-Acylsaccharins
- as intermediates **18,** 127 suppl. **29**
- special s.
  N-acetylsaccharin

1-Acylsemicarbazides
- startg. m. f.
  carboxylic acid amides **26,** 503
N-Acylsulfenamides
- special s.
  N-formylsulfenamides
Acylsulfenylhalides
- startg. m. f.
  acyl aryl disulfides **28,** 536
Acylsulfides s. Thiolic acid esters
N-Acylsulfilimines
- from
  acylaminosulfenium salts **27,** 300
Acyl sulfonates
-, prepn. **24,** 123 suppl. **26**
-, reactions **24,** 123 suppl. **26**
- special s.
  acetyl tosylate
  acyl trifluoromethanesulfonates
N-Acylsulfonic acid amides 26, 451
- from
  enoxysilanes and sulfonic acid azides **29,** 434
- special s.
  N-aroylsulfonic acid amides
o-Acylsulfonic – –
- startg. m. f.
  1,2-benzisothiazole 1,1-dioxides **30,** 330
1-Acyl-3-sulfonyloxyureas
- startg. m. f.
  Δ²-1,3,4-oxadiazol-5-ones **30,** 193
N-Acylsulfoximines, cyclic (s. a. Benz[d]isothia(IV)zol-3-one 1-oxides) **26,** 360
N-Acylthiolcarbamic acid esters
- from
  N-acylisothiouronium salts **28,** 160
Acylthiomethylation 26, 844
- of 1,1-disulfones **26,** 844 suppl. **28**
Acylthiosemicarbazides
- special s.
  1-phosphinyl-4-acylthiosemicarbazides
Acylthioureas
- special s.
  1-acetyl-2-thiourea
- startg. m f.
  thioureas **29,** 26
Acyl trifluoromethanesulfonates as powerful acylating agents **28,** 714
Acylureas (s. a. Thioacylureas)
-, C-alkylation **29,** 812
- special s.
  1-acyl-3-sulfonyloxyureas
  1,4,6-triazahepta-3,5-dien-7-ones
- startg. m. f.
  2-quinazolones **29,** 934
Acylurethans
- startg. m. f.
  acylamines **28,** 463
Acylynamines
-, reactions **29,** 134
- startg. m. f.
  β-acoxy-α,β-ethylenecarboxylic acid amides **29,** 134

Adamantan-1-amines, N-halogeno-
-, rearrangement, skeletal **28,** 168
Adamantanes (s. a. Heteroadamantanes, Proto-)
- by rearrangement, skeletal **27,** 758
-, chemistry, review **27,** 758
- from
  protoadamantanes **27,** 746; **19;** 780 suppl. **29**
-, ring expansion **27,** 726
- special s.
  spiro[adamantane-2,1'-cyclobutane-2',3'-diones]
-, **1-acyl- 29,** 887
-, **disubst. 27,** 746
-, **polyaryl- 29,** 836
2-Adamantanethiols 27, 47
Adams catalyst s. Platinum catalysts
Addition (s. a. Markownikoff...)
-, direction of –
  to multiple bonds s.
    Addition, regiospecific
- of enolesters with carbon-oxygen cleavage **27,** 314
- of sulfinic acids to nitrogen double bonds **30,** 384
- to carbon-carbon double bonds (s. a. Ethylene derivs., startg. m. f.)
  of ammonia **10,** 274; **13,** 373
  of N-halogenocyanamides **26,** 515
  of hydrocarbons, base-catalyzed, review **20,** 519 suppl. **29**
  of iodonium nitrate **26,** 518
  of solvents, protic **26,** 135
  with 2,4,4,6-tetrabromo-2,5-cyclohexadienone **28,** 479
- to carbon-carbon triple bonds s. Acetylene derivs., startg. m. f.
- to carbon-nitrogen double bonds, cyclic
  of Grignard compds. (preferential addition) **15,** 528 suppl. **26**
- to nitrogen-nitrogen double bonds
  of silanes **27,** 301
  with ring closure, intramolecular, review **27,** 561
-, **conjugate** s. 1,4-Addition, 1,6-Addition, Diene syntheses, Michael addition
-, **dipolar** s. Cycloaddition, dipolar
-, **geospecific** (s. a. *cis-trans*-Isomers) **27,** 530, 598; **30,** 479, 492, 603
-, **homoconjugate**
- to cyclopropanes, synthesis by – **28,** 859
-, **inverse 16,** 225 suppl. **29;**
-, **nucleophilic 29,** 795
- to
  azomethines **26,** 676
  fluoroolefins **17,** 188
-, –, **preferential,** to carbon-nitrogen double bonds **29,** 299
-, **radical** s. under Radical reactions
-, **regiospecific** (s. a. Markownikoff...) **29,** 603

Add—Alc 554

(Addition, regiospecific)
- to
 carbon-carbon triple bonds
  30, 492, 603
-, stereospecific
- to
 alcohols, synthesis 27, 714
 carbon-carbon double bonds
  27, 530
  of hydroxyl groups 19, 802
   suppl. 26
  of oxygen 26, 141; 28, 113;
   30, 91
 carbon-nitrogen double bonds
  30, 459
 cobaltocenes 27, 813
-, -, partial
- to carbon-carbon double
 bonds 29, 158
-, substituting s. under Azodicarboxylic acid esters adducts
-, thiophilic 26, 575
1,2-Addition
- to carbon-carbon double
 bonds
  of 2-different C-functions
   30, 487
- to α,β-ethyleneazomethines
 29, 638
1,3-Addition to benzene ring
 30, 476
1,4-Addition (s. a. Addition,
 conjugate) 28, 661; 29, 671
-, ensilanes, synthesis, by –
 30, 490
- of
 alkyl groups, branched 27, 713
 silyl groups 26, 640
 zinc compds., organo- 26, 672
- to
 α,β-acetylenecarboxylic acid
  esters 23, 693 suppl. 28
 1,3-dienes s. Diene syntheses
 α,β-ethyleneazomethines
  29, 638
 α,β-ethyleneketones 30, 453,
  490/1
 α,β-ethylenesulfones 23, 693
  suppl. 28
- with
 boranes 28, 649
 cyanohydrins, protected
  30, 482
- with simultaneous aldol
 condensation, intramole-
 cular 29, 669
-, geospecific 23, 693 suppl. 27
-, photochemical 29, 677
- stereospecific 23, 705
 suppl. 28
1,5-Addition s. Addition,
 homoconjugate
1,6-Addition
- of hydrogen cyanide 18, 772
 suppl. 26
- to nitroso compds. 30, 596
- to 1,3,5-trienes 30, 82
-, conjugate 17, 765
 suppl. 27
-, -, geospecific 29, 670
1,7-Addition 23, 693 suppl. 28
1,4-Addition-alkylation (s. a 1,2-
 Dialkylation of ethylene
 derivs.) 29, 819
Adenines 30, 256
- from
 furazano[3,4-d]pyrimidines,
  7-acylamino- 27, 521

Adsorption (s. a. Chromatography, preparative)
-, activation of halides by –
 26, 252
Air s. Oxygen
Alane etherates as reactants
 26, 654
Alanes (s. a. Aluminum
 compds., organo-, Hydro-
 alumination)
- special s.
 amino-alanes
 halogen-
 trialkyl-
- startg. m. f.
 alcohols, tert., synthesis,
  stereospecific 27, 671
-, tert., as reactants 29, 832
Alcohols
-, C-alkylation with – s. under
 C-Alkylation
- as reagents 26, 911, 661
-, cycloaddition of – 28, 712
-, dehydration CC⇑O
-, derivatives s. O-Acylation
- from
 acetylene derivs. and halides
  (sec. alcohols) 28, 821
 - -, synthesis 28, 821 suppl. 29
 alcohols, prim., degradation
  with loss of 1C-atom 16,
  305 suppl. 26
 amines, tert. 29, 531
 carboxylic acid amides,
  N,N-disubst. 28, 54
 - - esters, hydrolysis
  HO⇓↑C
 - - -, reduction (prim.
  alcohols) 26, 57, 62; 27, 64;
  28, 50, 53; 29, 39
 - - -, synthesis (tert.
  alcohols) 26, 662
 - acids, degradation with loss
  of 1 C-atom 30, 177
 - -, reduction (prim.
  alcohols) 27, 65; 28, 52;
  29, 57, 243; 30, 29
  sodium carboxylates 18, 947
   suppl. 27
 cyclopropanes, cyano- 29, 87
 diazo compds. 30, 124
 1,3-dienes and ethylene
  derivatives 26, 716
 ethers 26, 14
 -, cyclic 27, 37
 2-ethylenealcohols, synthesis
  29, 660
 α,β-ethylenecarboxylic acids
  18, 947 suppl. 29
 ethylene derivs. (s. a. under
  Hydration) 26, 139; 27, 142;
  29, 150
 - -, cleavage 27, 149
 - -, synthesis 26, 914
  suppl. 27; 27, 714
 - -, - with addition of
  1 C-atom 4, 667 suppl. 26
 glycols, reduction 26, 81
 α-halogenoketones 29, 71
 2-hydroxythioethers 28, 524
 γ-lactones 29, 87
 nitriles 26, 195
 1,2-nitrosohalides 27, 202
 oxido compds., reduction
  26, 55; 27, 35; 28, 32;
  29, 32, 59, 78
 -, - with rearrangement,
  skeletal 29, 31

- -, synthesis 26, 669; 27, 666
- -, via 2-hydroxythioethers
 28, 524
- - with ring closure, isocyclic 27, 740; 28, 4
oximes 26, 185
oxo compds. 26, 662
 aldehydes (sec. alcohols) 8,
  713 suppl. 26: 27, 665;
  29, 762; 30, 450
 ketones (sec. alcohols)
  HC⇓OC; 30, 40
 - (tert. alcohols) 26, 671;
  27, 671; 28, 176, 922;
  30, 606
- and boranes 27, 889
- - and silanes 30, 624
- -, reduction 24, 126
 suppl. 29
sulfuric acid monoesters
 27, 3; 30, 656
thiolic acid esters 26, 76
urethans 28, 8; 30, 7
-, C-α-hydroxyalkylation with –
 29, 719
- special s.
 acetylene-alcohols
 acylamino-
 allene-
 amino-
 aryl-
 azo-
 benzyl-
 carbinols
 carbohydrates
 cyclopropyl-alcohols
 deuterio-
 1,2-diarylethanols
 diazoalcohols
 diols
 halogen-alcohols
 isocyano-
 nitro-
 (N-nitrosoacylamino)-
 oxido-
 phenol-
 phenols, o-(ω-hydroxyalkyl)-
 polyalcohols
 2-α-pyridylethanol
 siloxyalcohols
- startg. m. f.
 acetylenedicarboxylic acid
  esters 30, 132
 acylamines 29, 307
 alkoxysilanes 29, 109
 allophanates 30, 282
 amines 30, 255
 -, prim. 27, 401, 429
 -, -, synthesis with addition
  of 1 C-atom 27, 782
 -, tert. 26, 388; 28, 352
 azides 30, 255
 benzimidazoles, 1-subst.
  26, 369
 benzothiazolines 27, 610
 carbonic acid esters, sym.
  28, 322
 carboxylic acids, carbonylation 28, 612
 -, -, carbonylation, tert. acids
  30, 452
 - -, oxidation 27, 187; 28, 137
 cyclopropane ring 27, 937
 1,1-diacoxy compds., ring
  contraction 28, 133
 dienones 14, 895 suppl. 26
 ethers, sym. 26, 181

ethylene derivs. **27**, 917, 937; **29**, 762
fluoroformic acid esters **30**, 302
halides HalC↓↑O
halogenomethyl ethers **26**, 167
hydrocarbons, reduction **26**, 80; **27**, 88; **30**, 47
–, synthesis **28**, 845
nitriles **27**, 782
nitro compds., aliphatic **26**, 361; **29**, 743
oxo compds. OC↑H
 aldehydes (from prim. alcohols) **28**, 53, 203; **29**, 89
 ketones (from sec. alcohols) OC↑H
 – with ring opening **19**, 211
phosphorodichloridates **28**, 72
phthalimides **27**, 429
sulfonic acid esters **29**, 189
 benzenesulfonic acid esters **27**, 401
 p-toluenesulfonic – – **28**, 52, 119
tetrathiophosphoric – – **28**, 545
thioethers **28**, 547
–, sym. **27**, 609
trifluoromethyl ethers **30**, 106
urethans **5**, 125 suppl. 28
–, **axial 26**, 33
–, **branched**
 – from
  boranes **26**, 914
–, **cyclic**
 – from
  ethyleneketones **26**, 747; **28**, 1/2
  ketones **27**, 744, 747; **28**, 768
 – special s.
  azetidinols
  cyclobutanol ring
 – startg. m. f.
  ω-hydroxyketones **16**, 305 suppl. 26
  ketones, cyclic, ring expansion **27**, 912
–, **heterocyclic**
 – from
  oxo compds. **29**, 881
–, **hindered**
–, acylation **22**, 217 suppl. 27
–, dehydration **6**, 841 suppl. 27
–, **isocyclic**
 – from
  ketones **26**, 744
–, **H-labeled 7**, 73 suppl. 26
–, **optically active 26**, 14
–, **prim.** (s. a. Hydroxymethyl groups)
 – from
  ethers, cyclic **26**, 663
  ethylene derivs., terminal **16**, 194
 – startg. m. f.
  alcohols, degradation with loss of 1 C-atom **16**, 305 suppl. 26
–, sec., dehydration, preferential **28**, 916
 – startg. m. f.
  acetals **28**, 141
–, **tert.**
 – from
  mercaptals and boranes **29**, 873
– startg. m. f.
 hydrocarbons, branched, synthesis **28**, 845
–, synthesis by coupling of ethylene derivs. **26**, 914 suppl. 27
Alcohols-α-D 4, 49 suppl. 26; **27**, 39 suppl. 29
Alcohols-OD, prepn., review **24**, 2 suppl. 27
Alcoholysis s. Trans-esterification, -etherification
Aldehyde acetates s. 1,1-Diacoxy compds.
– **groups**
–, introduction s. C-Formylation
Aldehydes (s. a. Formyl..., Hydroxymethylen..., Oxo compds.)
–, C-acylation, radical, with – **26**, 775
–, α-alkylation **28**, 794
– as reagents **27**, 960
– from
 acetals **27**, 59
 –, cyclic **27**, 81
 alcohols, prim. **28**, 53, 203; **29**, 89
 amines **26**, 194
 2-aminoalcohols, C-cleavage **30**, 179
 1,1-aminoethers, cyclic **28**, 793
 α-aminonitriles **29**, 901
 azides **28**, 149
 carboxylic acid amides, N,N-disubst. **26**, 69
 – – anhydrides **29**, 64
 – – chlorides **27**, 79, 81
 – – esters **28**, 53
 – – halides **27**, 76
 – acids via alcohols, prim. **29**, 89
 – – via carboxylic alkoxyformic anhydrides **28**, 66
 – – via dianions **26**, 923
 1,1-dihalides **29**, 201a, 847
 1,3-dithianes **26**, 211
 α,β-ethyleneacetals **27**, 59
 α,β-ethylenealdehydes **6**, 89 suppl. **29**; **27**, 59
 ethylene derivs., cleavage, oxidative **30**, 149, 154
 glycidic acid tert-butyl esters **26**, 859
 glycols, cleavage, oxidative **3**, 238
 halides, synthesis with addition of 1 C-atom **26**, 872, 891; **27**, 830
 –, – – – – 2 C-atoms **28**, 793
 halogenomethyl compds. **26**, 209
 hydrocarbons (methyl groups) **26**, 157
 ketones, synthesis with addition of 1 C-atom **26**, 859
 mercaptals **26**, 214
 nitriles **26**, 183; **29**, 209
 ozonides **28**, 214
 piperidines, 1-acyl-3-hydroxy- **27**, 247
 thiazolidines **28**, 176
 Δ²-thiazolines **28**, 176
 s-trithianes **30**, 688
–, reactions with ammonia **30**, 253
–, – with thioureas **26**, 286
–, self-condensation **30**, 553
–, separation via benzylic hydrogenolysis **30**, 149
– special s.
 acetylene-aldehydes
 acoxy-
 acylamino-
 amino-
 chloral
 cyano-aldehydes
 cyclopentenecarbox-
 deuterio-
 di-
 ethylene-
 halogen-
 hydroxy-
 keto-
 pyrazolecarbox-
 pyrrolecarbox-
 triformylmethane
– startg. m. f.
 acetals **6**, 220; **22**, 191; **27**, 220; **30**, 113
 α,β-acetylenecarboxylic acids **28**, 292
 acetylene derivs., terminal **28**, 819
 β-acoxycarboxylic acid esters **26**, 687
 acoxy compds. (from 2 molecules) **27**, 668
 1-acylguanidines, 3-subst. **29**, 356
 acylhydrazines **28**, 271
 acyloins (from 2 molecules) **29**, 630
 alcohols, sec. **8**, 713 suppl. **26**; **27**, 665; **29**, 762; **30**, 450
 anti-aldoximes **4**, 395 suppl. **28**; **27**, 960
 α-alkoxy-α,β-ethylenenitriles **29**, 443
 alkylidenemalonic acid esters, cyclic **26**, 790
 β-alkylthio-α-formaminocarboxylic acids **26**, 791
 α-aminocarboxylic acid esters, synthesis with addition of 2 C-atoms **29**, 739
 α-amino-α,β-ethylenecarboxylic acid esters **29**, 739
 arylacetic acid esters, synthesis with addition of 1 C-atom **28**, 735
 α-azido-α,β-ethylenecarboxylic acid esters **26**, 789
 benzils (from 2 molecules) **26**, 768
 benzoins **26**, 675
 carboxylic acid amides **27**, 371
 – – –, synthesis with addition of 1 C-atom **29**, 443
 – acids **30**, 66
 coumarans **27**, 807
 1,1-cyanohydrazines **26**, 828
 1,2-diarylethanols **26**, 786
 dihalogenomethylene compds. **27**, 866; **28**, 819
 1,3-dithiins (from 2 molecules) **28**, 744
 dithiohemiacetals **27**, 608
 enediolesters (from 2 molecules) **27**, 759
 1,1-di(thiolurethans) **29**, 308
 enethiolurethans **26**, 382

Ald – Alk

(Aldehydes
- startg. m. f.)
  2-ethylenealcohols, synthesis
  28, 855
  α,β-ethylenealdehydes,
  α-subst. **26,** 802
  ethylene derivs. **27,** 891;
  29, 762
  1,2-ethylene-1,1-disulfones
  29, 748
  2,3-ethylene-1,1-disulfones
  29, 748
  α,β-ethylene-β'-hydroxy-
  ketones **26,** 257
  α,β-ethyleneketones **29,** 752
  α,β-ethylene-α-mercapto-
  carboxylic acid esters
  29, 730
  hydrocarbons (methyl groups)
  26, 68; **30,** 30
  α-hydroxyamidoximes, synthesis with addition of
  1 C-atom **27,** 667
  α-hydroxyketones **29,** 621
  1,1-hydroxythioethers **28,** 522
  1,1-hydroxythioureas **26,** 286
  imidazole 3-oxides **27,** 412
  imidazoles, 1-hydroxy- **26,** 338
  –, 1-(α-siloxy)- **26,** 290
  4-imidazolidones, 3-hydroxy-
  26, 336
  indolizines, 1-(alkylideneamino)- (from 2 molecules)
  27, 793
  ketones, synthesis **22,** 661
  suppl. **28; 28,** 648; **29,** 808
  lactams **27,** 801
  α-methylenealdehydes **26,** 787
  nitriles **28,** 339; **30,** 277
  4H-1,2,5-oxadiazines, 5,6-di-
  hydro- **27,** 386
  1,2,3-oxathiazetidine 2-oxides
  28, 272
  2H-1,3-oxazines, tetrahydro-
  6-alkoxy- **30,** 564
  α,β-oxidoketones **26,** 864
  pyrazolidines **26,** 780
  pyridines **27,** 775
  quinazolines, 1,2,3,4-tetra-
  hydro-4-acyloximino-
  28, 334
  stilbenes **26,** 786
  1-sulfinylthioenolethers **28,** 735
  sydnone imine salts, 3-amino-
  26, 828
  1,3,5-triazines, hexahydro-
  30, 253
  Δ²-1,2,4-triazolines **27,** 387
  α-(trifluoromethylthio)-
  α,β-ethylenenitriles **28,** 737
  s-trioxanes **30,** 71
-, substitutes
  azomethiones **23,** 435
  suppl. **27**
-, **optically active,** by hydroformylation **4,** 667 suppl. **29**
-, **unsatd.,** ring closure, double,
  of – **26,** 780
Aldehydes-1-d
- from
  carboxylic acid chlorides
  11, 135 suppl. **29**
  1,1-dihalides **29,** 211
  1,1-di(pyridinium salts) **29,** 211
  s-trithianes **30,** 688
- via deuterioazomethines
  26, 912 suppl. **27**
Aldehydes-2-d₂ 27, 651

Aldehydocarboxylic acids
- from
  dicarboxylic acid anhydrides
  29, 64
o-Aldehydocarboxylic –
- special s.
  phthalaldehydic acids
Alder s. Diene syntheses
Aldimines (s. a. Metallo
  aldimines)
- startg. m. f.
  nitriles **27,** 537
Aldol condensation (s. a. Cross-
  aldol reaction, Retroaldol –)
- in the vapor phase **28,** 722
- via enolate anions **29,** 894;
  30, 453
- with
  dicarbanions **26,** 861 suppl. **27**
  simultaneous 1,4-addition
  30, 489
- –, **directed 29,** 894
- –, **intramolecular** (s. a. Retroaldol-aldol condensation,
  intramolecular) **28,** 685
- –, diazoketols, cyclic by –
  26, 745
- with 1,4-addition **29,** 669
- –, –, **asym. 30,** 520
- –, –, **stereospecific 28,** 681
- –, **regiospecific 29,** 894
**Aldol-type condensation, intra-
  molecular,** to heterocycles
  26, 935
Aldonic acids
- from
  carbohydrates, review **4,** 151
  suppl. **28**
- special s.
  3-deoxy-2-ketoaldonic acids
Aldonolactones
- from
  carbohydrates **21,** 285 suppl.
  27; 24, 254 suppl. **26**
Aldoximes (s. a. Oximes)
-, reaction with chlorine **27,** 562
- special s.
  benzaldoxime
  α-ketoaldehyde aldoximes
- startg. m. f.
  carboxylic acid amides **30,** 98
  1,1-dihalides **27,** 562
  hydroxyminohalides **27,** 562
  nitriles **26,** 501; **27,** 513;
  28, 440; **30,** 277
  1,1,1-trinitro compds. **27,** 347
*anti-***Aldoximes**
- from
  aldehydes **4,** 395 suppl. **28;**
  27, 960
Aliquat 336 s. under Phase
  transfer catalysts
Alkali alkoxides s. under
  individual alkali metals
- **halides** (s. a. under individual
  alkali metals) **27,** 214
  suppl. **28**
- **metal catalysts,** ring hydrogenation with – **26,** 46
- **metals** s. under individual
  alkali metals
Alkaloids, cycloisomerization
  26, 731
- by coupling, intramolecular,
  review **29,** 904
Alkene oxides s. Oxido
  compds.

Alkenes s. Ethylene derivs.
Alkenylidenecyclopropanes
  28, 798a
Alkenynes s. Enynes
Alkoxalyl compds.
-, C-alkylation **27,** 903
- **groups** (s. a. Ethoxalylation)
-, elimination **27,** 903
Alkoxides s. Alkali alkoxides
Alkoximes
- from
  alkoxylamines **14,** 430; **27,** 425
- special s.
  hydroxy-alkoximes
  keto-
  (organothio)-
- startg. m. f.
  α-alkoximinooxo compds.
  27, 182
  aziridines **21,** 538 suppl. **29**
2-Alkoximinohalides 17, 916
  suppl. **26**
α-Alkoximinooxo compds.
- from
  alkoximes **27,** 182
Alkoxyacetylenes 14, 279
  suppl. **26**
- from
  acetylenehalides **27,** 210
- special s.
  propargyl ethers
- startg. m. f.
  α-alkoxystyrenes **26,** 911
  allenyl ethers, synthesis
  27, 835
  α,β-ethyleneoxo compds.,
  synthesis **27,** 835
o-Alkoxyacetylenes
- startg. m. f.
  benzofurans **28,** 196
1,1-Alkoxyacoxy compds.
- special s.
  α-alkoxyenolesters
1,1-Alkoxyacylamines
- special s.
  α-acylamino-α-alkoxy-
  carboxylic acid esters
-, **cyclic** s. 1-Alkoxylactams
1-Alkoxy-1-alkoximes
- from
  1-halogeno-1-alkoximes **26,** 197
1-Alkoxy-1-alkylthio-1-amines
  28, 528
1-Alkoxyallenylsilanes 27, 835
Alkoxyamines s. Aminoethers
**o-Alkoxy-o'-amino-N-hetero-
  cyclics**
- from
  dinitriles **30,** 74
Alkoxyaminomethinium salts
- startg. m. f.
  α,α-alkoxyaminonitriles
  28, 624
  amidinium salts, vinylogous
  26, 453
  1,1,1-aminodinitriles **28,** 624
α,α-Alkoxyaminonitriles
- from
  alkoxyaminomethinium salts
  28, 624
1-Alkoxyazomethines
- startg. m. f.
  amines, sec. **27,** 62
3-Alkoxybiurets
- from
  alkoxylamines and iso-
  cyanates **30,** 209

Alkoxyboranes s. Borinic acid esters
Alkoxycarbonium salts
- special s.
  alkoxyaminomethinium salts
- startg. m. f.
  aminocarbonium salts **28**, 359
Alkoxycarbonyl...
  s. Carbalkoxy...
N-Alkoxycarbonylimino-pyridinium ylids
- startg. m. f.
  eneurethans **28**, 376
α-Alkoxycarboxylic acid amides
- startg. m. f.
  α-hydroxycarboxylic acid amides, rearrangement **29**, 693
Alkoxycarboxylic acid esters
- special s.
  acylaminoalkoxycarboxylic acid esters
  alkoxyethylenecarboxylic – –
  glycolic acid esters, O-alkyl-
- startg. m. f.
  hydroxycarboxylic acid esters **26**, 9
β-Alkoxycarboxylic – –
- from
  β-halogeniminoesters **30**, 138
Alkoxy compds. s. a. Dialkoxy..., Ethers, Trialkoxy...
N-Alkoxy – (s. a. Alkoxylamines)
–, Dimroth rearrangement **27**, 168
1,1,1-Alkoxydiamines
- from
  α,α-diminonitriles **28**, 187
–, reactions **28**, 770
- startg. m. f.
  N'-(3-aminoacryloxyl)form-amidines **28**, 770
  β-amino-α,β-ethylenecarboxylic acid thioamides **28**, 770
Alkoxydiazenium salts
  s. O-Alkylnitrosimmonium salts **20**
1-Alkoxy-dienes and -polyenes, review **3**, 727 suppl. **26**
1,1,1-Alkoxydihalides
–, benzene ring annelation with **28**, 825
- special s.
  α,α-dibromomethyl ether
  α,α-dichloromethyl –
1-Alkoxy-1,2-dihalides 3, 428 suppl. **27**
1-Alkoxyenacylamines, cyclic
- from
  iminoesters, cyclic **27**, 466
1-Alkoxyenamines
–, Claisen-type rearrangement **29**, 704
- special s.
  N-chlorocarbonyl-1-alkoxyenamines
- startg. m. f.
  coumarins **26**, 831
–, cyclic
- from
  halides **28**, 793
- special s.
  1,3-oxazines, tetrahydro-2-ene-1-(β,γ-ethylene)-
- startg. m. f.
  1,1-aminoethers, cyclic **28**, 793

α-Alkoxyenephosphonic acid esters
- from
  α,α-dialkoxyphosphonic acid esters **29**, 607
α-Alkoxyenolesters
–, pyrolysis **28**, 679
- startg. m. f.
  β-acoxycarboxylic acid esters **26**, 687
1-(Alkoxy)enolethers
- from
  ketones **28**, 803
α-Alkoxy-α,β-ethylenecarboxylic acid esters
- startg. m. f.
  α-ketocarboxylic acids **29**, 200
β-Alkoxy-α,β-ethylene-carboxylic – –
- from
  β-ketocarboxylic acid esters **26**, 54
–, ring closures with – **30**, 250
- startg. m. f.
  carboxylic acid esters **26**, 54
β-Alkoxy-β,γ-ethylene-carboxylic – –
- from
  α-allenecarboxylic acid esters **29**, 159
- startg. m. f.
  β-ketocarboxylic acid esters **29**, 159
2'-Alkoxy-α,β-ethylenehalides, cyclic
- from
  cyclopropane ring, 1,1-dihalogeno- **26**, 205
β-Alkoxy-α,β-ethyleneketones
  s. 3-Keto-1,2-en-1-olethers
α-Alkoxy-α,β-ethylenenitriles
- from
  aldehydes **29**, 443
- startg. m. f.
  carboxylic acid amides **29**, 443
β-Alkoxy-α,β-ethyleneoximes, cyclic
- from
  β-diketones, cyclic **27**, 163
- startg. m. f.
  β-ketolactams **27**, 163
Alkoxyformic acid anhydrides, mixed, s. Carboxylic alkoxyformic acid anhydrides
α-Alkoxyglycidic acid esters
- from
  β-halogeno-α-ketocarboxylic acid esters **30**, 129
Alkoxy group migration
- with
  skeletal rearrangement **28**, 691
N→C-Alkoxy group migration 28, 384
N→N'-Alkoxy – – 30, 55
Alkoxy groups
–, photocyclization with elimination of – **29**, 913
–, position shift **18**, 338 suppl. **26**
Alkoxyhalides, ring closure **27**, 279
1,1-Alkoxyhalides
- from
  acetals **30**, 567
  diazo compds. **27**, 580
- special s.
  1,1,1-alkoxydihalides

1-alkoxy-1,2-dihalides
1,1'-dihalogenethers
monochloromethyl ether
- startg. m. f.
  α-alkoxynitriles **30**, 567
  1,1-diarylalkanes **26**, 887
  2-halogenosulfenic acid alkoxymethyl esters **28**, 475
  isoquinolines, 1,2,3,4-tetrahydro-2-acyl- **26**, 884
  mercaptals **28**, 566
–, **cyclic**
- startg. m. f.
  ethers, cyclic, synthesis **9**, 881 suppl. **26**
1,2-Alkoxyhalides
- from
  ethylene derivs. **28**, 479
– – and hypochlorites **26**, 513
- special s.
  2-alkoxy-α,β-ethylenehalides
  2-(ω-halogenalkoxy)halides
  β-halogenethyl ethers
- startg. m. f.
  ethers **25**, 64 suppl. **28**
  ethylene derivs. **29**, 957
Alkoxy-N-heterocyclics
- special s.
  alkoxyamino-N-heterocyclics
α-Alkoxy-N-heterocyclics
- special s.
  1,1-alkoxyacylamines, cyclic
1,1-Alkoxyhydrazines
- from
  N-nitrosamines **30**, 314
1,1-Alkoxyhydrazones
- startg. m. f.
  acylhydrazines **27**, 336
1,1-Alkoxyhydroperoxides
- startg. m. f.
  carboxylic acid esters **28**, 215
Alkoxyketones
- special s.
  acetylenealkoxyketones
α-Alkoxyketones
- from
  α-acetyleneacetals **29**, 61
  α-diazoketones **27**, 204
  enolesters **26**, 637
β-Alkoxyketones
- from
  enolesters and acetals **30**, 633
  enolethers and ketones **28**, 654
  enoxysilanes **30**, 621
- special s.
  β-alkoxy-α,β-oxidoketones
γ-Alkoxyketones
- from
  ethylene derivs. **28**, 823
1-Alkoxylactams
- from
  dicarboxylic acid imides **27**, 947
–, ring closures with – **29**, 784
- special s.
  1-alkoxy-2-oximinolactams
- startg. m. f.
  N-heterocyclics, N-condensed **27**, 947
α-Alkoxy-β-lactones
- startg. m. f.
  enolethers **29**, 985
Alkoxylamines (s. a. Aroxylamines, N,N-Dialkoxylamines)
- from
  oxoazonium salts **27**, 125

(Alkoxylamines)
- special s.
  ethylenealkoxylamines
  methoxyamine
- startg. m. f.
  alkoximes **14**, 430; **27**, 425
  3-alkoxybiurets **30**, 209
  alkoxylureas **30**, 209
  alkoxyoxindoles **28**, 384
β-Alkoxylaminoketones
- startg. m. f.
  α-amino-α,β-ethyleneketones
  29, 917
Alkoxylation (s. a. Replacement of hydrogen by alkoxy groups)
- by anodic oxidation **26**, 158
-, **electrochemical**, review
  23, 186 suppl. **29**
Alkoxylcyclimmonium salts
- from
  N-oxides, cyclic **22**, 187
- startg. m. f.
  phosphonic acid esters,
  N-heterocyclic **27**, 654
Alkoxylureas
- from
  alkoxylamines and iso-
  cyanates **30**, 209
Alkoxymethylene compds.
- by Vilsmeier-Haack-type
  reaction **26**, 838
- special s.
  EMCE
  EMME
α-Alkoxymethyleneketones
- startg. m. f.
  furan ring **28**, 779
  heterocyclics **24**, 418 suppl. **28**
α-Alkoxynitriles
- from
  acetals **30**, 567
  1,1-alkoxyhalides **30**, 567
- special s.
  α,α-alkoxyaminonitriles
  methoxyacetonitrile
- startg. m. f.
  oxo compds. **28**, 975
o-Alkoxynitriles
- from
  nitro compds. **26**, 835
Alkoxyoxido compds.
- special s.
  alkoxyglycidic acid esters
β-Alkoxy-α,β-oxidoketones
  6, 192 suppl. **26**
Alkoxyoximes
- special s.
  alkoxyethyleneoximes
o-Alkoxyoximes
- startg. m. f.
  benzoxazoles **26**, 505
1-Alkoxy-2-oximinolactams
- from
  enaminolactams **29**, 311
Alkoxyoxindoles
- from
  alkoxylamines **28**, 384
1,2-Alkoxyperoxides
- from
  ethylene derivs. and trioxides
  30, 77
Alkoxyphosphonic acid esters
- special s.
  dialkoxyphosphonic acid
  esters

Alkoxyphosphonium salts
- special s.
  alkoxytriaminophosphonium
  perchlorates
  alkoxyvinylphosphonium salts
Alkoxysilanes
-, cleavage **29**, 881
- from
  alcohols and silanes **29**, 109
  oxo compds. **29**, 107
- special s.
  alkoxyallenylsilanes
  imidazoles, 1-(α-siloxy)-
- startg. m. f.
  fluorides **29**, 512
-, synthesis, asym. **29**, 107
  suppl. **30**
Alkoxystannanes
- special s.
  dialkoxy-stannanes
  halogenalkoxy-
- startg. m. f.
  oxo compds. **28**, 226
α-Alkoxystyrenes
- from
  alkoxyacetylenes **26**, 911
1-Alkoxy-1-sulfones
- special s.
  α,α-disulfonylethers
  α-halogeno-α-sulfonylethers
-, **cyclic**
- from
  enolethers **29**, 729
Alkoxysulfonium salts
- as intermediates **28**, 499
- from
  sulfoxides **27**, 126
- special s.
  hydroxy(alkoxysulfonium)
  salts
- startg. m. f.
  aminosulfeninum salts **28**, 250
  sulfoxides **27**, 126
Alkoxysulfonylcarbamic acid esters
- startg. m. f.
  alkoxysulfonylisocyanates
  29, 448
Alkoxysulfonylisocyanates
- from
  alkoxysulfonylcarbamic acid
  esters **29**, 448
1-Alkoxy-1-sulfonyloxy compds.
  27, 111
Alkoxythiocarbonylhydrazines
  26, 440
1,2-Alkoxythioethers
- from
  enolethers **18**, 645 suppl. **26**
- special s.
  1,3-bis(methylthio)-2-
  methoxypropane
Alkoxytriaminophosphonium perchlorates
-, monoalkylation of amines
  with – **28**, 358
2-Alkoxyvinylphosphonium salts
- startg. m. f.
  furans **29**, 865
tert-Alkyl...
- special s.
  tert-butyl...
Alkylarenes s. Perfluoro-
  alkylarenes
Alkylating agents 29, 882
- special s.
  methyl fluorosulfonate

-, -, **water-soluble**
-, 1-oxidopyridin-2-yldiazo-
  methane as – **5**, 177 suppl. **29**
Alkylation (s. a. Allylation,
  Di-alkylation, Mono-, Photo-)
- of
  anions, C,O-ambident **14**, 284
- with
  alcohols **28**, 753
  1,1-chlorothioethers **28**, 993
  isoureas, N,N'-dicyclohexyl-
  29, 547
C-Alkylation
- of
  acylureas **29**, 812
  alcohols **27**, 773
  aldehydes **28**, 794
  alkoxalyl compds. **27**, 903
  amines, ar., o-alkylation
  28, 762
  arenes **30**, 605
  – via 1,4-cyclohexadienes,
  3-alkyl- **27**, 839
  – with amines, tert. **22**, 795
  suppl. **29**
  – with phosphorus esters
  27, 791
  arylacetic acid esters **30**, 540
  aryl derivs. s. under Aryl
  derivs.
  azomethines, cyclic **29**, 831
  carboxylic acid esters **27**, 843
  cyclopentadienes **26**, 718
  cyclopropenes **13**, 805
  suppl. **29**
  dithiocarbamic acid esters
  29, 809; **30**, 536
  α,β-ethyleneazomethines
  29, 638 suppl. **30**
  ethylene derivs. (β-alkylation)
  16, 848
  β,γ-ethylenesulfoxides **29**, 217
  α-halogenosulfones **28**, 795
  suppl. **30**
  indenes **27**, 795
  β-ketocarboxylic acid esters
  s. under β-Ketocarboxylic
  acid esters
  3-keto-1,2-en-1-olethers
  (4-alkylation) **29**, 805
  ketones **26**, 853
  β-ketophosphonic acid esters
  s. under β-Ketophosphonic
  acid esters
  lactams **29**, 806
  lactones **27**, 843
  nitriles **27**, 843 suppl. **29**
  phenolethers **27**, 836
  phenols s. under Phenols
  pyridines, 2-amino- **28**, 762
  quinolines **28**, 800
  sulfones s. under Sulfones
  thioethers s. under Thioethers
  tin(IV) enolates, organo-
  29, 856
- special s.
  acoxyethylation
  cycloalkylation
  C-methylation
- via
  N,C-dianions **27**, 844
  C-α-hydroxyalkylation **26**, 665
- with
  alcohols **27**, 773
  sulfones **24**, 865
  sulfonium salts **28**, 623

– with simultaneous
C-carbalkoxylation 28, 727
fragmentation-type ring
opening 27, 926
–, allylic 23, 860 suppl. 28
–, ar. (s. a. Dimethylation, ar.)
29, 803
–, decarboxylative 24, 890
–, deconjugative 29, 810
–, deoxygenative 28, 803
–, directed 4, 775
– of
ketones 19, 881 suppl. 28
succinic acid monoesters,
subst. 27, 840
–, –, kinetically controlled 15,
627 suppl. 29
–, intramolecular 21, 949
– of
halogenosulfonic acid amides
30, 671
ketones 29, 967
nitriles 29, 959
Reissert compds. 28, 942
–, –, stereospecific 28, 947
–, α-lateral
– of Δ²-oxazolines 29, 807
–, radical s. under Radical
reactions
–, reductive (s. a. C-Alkylation,
deoxygenative) 29, 761
– of
alkylthiomethyleneketones
26, 899
arenes 27, 839
aryl ketones 27, 839 suppl. 28
carbonyl oxygen 30, 487
α,β-ethylenecarboxylic acid
esters 30, 602
furans 27, 839 suppl. 30
isoquinolinium salts 26, 801
ketones 27, 790
oxido compds. 27, 666
–, –, preferential
– of α,β-ethyleneketones 30, 602
–, remote
– of polyenecarboxylic acid
esters 27, 843
–, selective
– of benzoins 26, 856
–, stereospecific 28, 803
– of
carboxylic acid esters 27, 843
suppl. 30
2,5-pyrrolidiones 26, 858
–, thermal
– of phenols with ethylene
derivs. 27, 686
N-Alkylation 26, 421, 424;
29, 413
– in dimethyl sulfoxide 27, 457
– in the presence of methylsulfinyl carbanion 16, 434
suppl. 26
– of
carboxylic acid amides 27, 457
N-heterocyclics 27, 457
suppl. 28, 29
hydrazines 29, 414
sulfoximines 28, 337
– with
alcohols 26, 388; 29, 389
amide acetals 27, 373
brosylates 24, 620
chloroformic acid esters
28, 419
fluorosulfonic acid esters
29, 344

oxo compds. 30, 276
trichloromethanesulfonic acid
esters 24, 620 suppl. 29
–, preferential 28, 419
– of N-heterocyclics 17, 490
suppl. 31; 26, 454
– via dianions 26, 417
–, reductive 28, 346
– of
carboxylic acid amides
29, 363
N-heterocyclics 30, 214
–, –, preferential 26, 366
–, selective 26, 415
N- and O-Alkylation, total
29, 410
O-Aikylation 27, 223
– of
alcohols 30, 139
heterocyclics, mesoionic
26, 174
phenols 26, 200; 27, 228
–, partial 26, 189
–, selective 26, 198
– of
uracils 26, 206
S-Alkylation (s. a. Thioethers
from mercaptans)
– of
carboxylic acid thioamides,
vinylogous 26, 574
–, selective 28, 549
Alkylene oxides s. Oxido
compds.
Alkyl group migration
s. Alkyl migration
– groups s. a. Methyl groups
– –, branched, 1,4-addition
27, 713
tert-Alkyl groups
– from
1,1,1-trihalides 8, 857 suppl. 28
– halides s. Halides
– hypohalites s. tert-Butyl
hypohalites, Fluoroxytri-
fluoromethane
Alkylidene... s. a. Methylene...
O,O-Alkylidene... s. a. O,O-Iso-
propylidene...
Alkylideneamino...
s. a. Azomethines
α-Alkylideneaminocarboxylic
acid esters
– startg. m. f.
4-imidazolidones 27, 379
4-Alkylidene-2-amino-1,3-
dioxolanes
– from
α,α'-dihalogenoketones and
carboxylic acid amides
28, 167
Alkylidenebisamines
s. 1,1-Diamines
O,O-Alkylidene derivs.
– of
carbohydrates 26, 171
2-Alkylidene-1,3-diones
– from
acoxy-2-acetylenes 28, 968
3-Alkylidene-1,2-dithioles
– from
3-organothio-1,2-dithiolium
salts 26, 979
Alkylideneglutaconic acids
– startg. m. f.
2,4-dienecarboxylic acids
30, 52
1,3-dienes 29, 91

4-Alkylidene-5-imidazolones
– from
α,β-ethylene-β-(1-tetra-
zolyl)carboxylic acid
amides 29, 463
5-Alkylidene-4-imidazolones
– startg. m. f.
α-aminocarboxylic acids
26, 193
α-Alkylidene-β-ketocarboxylic
acid amides
– from
β-ketocarboxylic acid amides
27, 803
Alkylidenemalonic acid esters,
cyclic
– from
aldehydes 26, 790
– acids s. α,β-Ethylene-α-dicar-
boxylic acids
Alkylidenemalonitriles
– startg. m. f.
anilines, 2,6-dicyano- (from
2 molecules) 28, 874
2-carbamyl-2-cyano-
phosphonium salts 28, 588
2-cyano-α,β-ethylenecarboxylic
acid amides 28, 588
4-Alkylidene-5-oxazolones
– startg. m. f.
α-acylaminocarboxylic acid
amides 28, 288
α-acylamino-α,β-ethylene-
carboxylic acid amides 23,
497 suppl. 29
N-acylcarboxylic acid amides
29, 228
α-alkylaminocarboxylic acids
28, 288
α-aminocarboxylic –
28, 288
–, 2-vinyl- 30, 163
3-Alkylideneoxindoles
–, cycloaddition of enamines
to – 27, 692
Alkylidenephosphines
s. Alkylidenephosphoranes
Alkylidenephosphoranes
– as nucleophiles, review 20,
608 suppl. 26
–, reactions, review 20, 608
suppl. 26
–, soln., salt-free 26, 907
– special s.
(α-acylalkylidene)-
phosphoranes
(α-azoalkylidene)-
(α-carbalkoxyalkylidene)-
methylene-
(α-sulfonylalkylidene)-
triphenylphosphonium
phenacylid
vinylidenephosphoranes
– startg. m. f.
eniminophosphoranes 27, 684
enolethers 28, 846
α,β-ethylenecarboxylic acid
thioamides 26, 903
ethylene derivs., sym.
(from 2 molecules) 28, 863
ketenimines 30, 611
ketones 28, 846
1-phosphonio-1-selenonio-
methylid salts 30, 432
3-Alkylidenephthalimidines
– from
o-dicarboxylic acids 26, 924
–, labeled 26, 924

Alk – All   560

3-Alkylidenephthalmidines
– **from**
   o-dicarboxylic acids **26,** 924
4-Alkylidene-2-quinazolones
– from
   quinolines, 1,2-dihydro-1-
     carbamyl- **26,** 506
Alkylidenesulfuranes
  s. Sulfonium ylids
O-Alkyllactims
– from
   lactams **23,** 80 suppl. 27
–, prepn. and reactions,
   review **20,** 132 suppl. 26
–, ring closure via – **29,** 425
–, – – with – **29,** 350
– startg. m. f.
   1,3-diazetidin-2-one ring,
     4-alkoxy- **29,** 301
   enamines **27,** 768
   Δ^2-imidazoline ring,
     5-hydroxy- **29,** 385
   imines, cyclic **23,** 80 suppl. 27
   lactams, N-alkyl- **28,** 296
   lactimimides **29,** 383
   2-pyridone ring 3,4-dihydro-,
     N-condensed **27,** 768
   4(3H)-pyrimidinone ring,
     5,6-dihydro-, N-condensed
     26, 347
Alkylmagnesium halides
  s. under Magnesium
Alkylmercapto... s. Alkylthio...
Alkyl migration
–, C-alkyl from B-alkyl **27,** 118
–, N-alkyl from O-alkyl **27,** 119;
   29, 323
–, S-alkyl from O-alkyl **27,** 376
– –, **stereospecific**
–, C-alkyl from N-alkyl **27,** 750
C-Alkyl – (s. a. Dihydroisoquinoline rearrangement)
   28, 659, 951, 984; **29,** 939;
   30, 656
–, aromatization with – **28,** 921;
   29, 946; **30,** 259
–, cyclodehydration with –
   4, 832 suppl. **28**
–, Gattermann formylation
   with – **28,** 708
C→N-1,2-Alkyl – **27,** 240
N-Alkyl – **22,** 346
N→C-Alkyl, photochemical
   29, 705
N→O-Alkyl – **30,** 227
O-Alkyl – **27,** 172
Alkyl nitrites
– as reagents
   amyl nitrite **29,** 70
   n-butyl – **29,** 315
   tert-butyl – **26,** 327
   isoamyl – **1,** 321 suppl. **27;**
     12, 359 suppl. **27; 27,** 293;
     28, 159; **30,** 388
   isopropyl – **28,** 319
   n-propyl – **28,** 157; **29,** 343
–, reaction with
   pyridinium salts **27,** 348
   sulfonium – **27,** 348
– **peroxides** s. Di-tert-butyl
   peroxide
– **sulfoxides**
– from
   aryl sulfoxides **29,** 582
Alkylthioallenes
– special s.
   1,1-dialkoxy-3-alkylthio-
     allenes

1-Alkylthioazomethinium salts
–, reactions **28,** 843 suppl. **30**
– startg. m. f.
   oxo compds. **28,** 843 suppl. **30**
– –, **N-heterocyclic**
– from
   chlorothioformamidinium
     chlorides **27,** 829
N-(Alkylthiocarbonyl)ureas
– from
   isocyanates and thioboric
     acid esters **27,** 639
**o-Alkylthiocarboxylic acid
amides**
– from
   o-mercaptocarboxylic acid
     esters **27,** 376
**α-Alkylthiocarboxylic acid
esters 28,** 646
Alkylthiocarboxylic acids
– startg. m. f.
   carboxysulfonium salts **29,** 263
   lactones **29,** 263
α-Alkylthiocarboxylic –
– from
   1,3-oxathiolan-5-ones **26,** 670
     suppl. **27**
– **acid thioamides**
– from
   dithiocarbamic acid esters
     30, 536
**o-Alkylthio-ω-diazoaceto-
phenones**
– startg. m. f.
   thioindoxyls **27,** 644
2-Alkylthio-1,3-dienes
– from
   allenes and thioketones
     28, 521
**β-Alkylthio-α,β-ethylene-
carboxylic acid esters**
– from
   1,1-dialkoxy-3-alkylthioallenes
     30, 584
– startg. m. f.
   β-ketocarboxylic acid esters
     30, 584
**γ-Alkylthio-α,β-ethylene-
carboxylic** – –
– startg. m. f.
   γ-alkylthio-β,γ-ethylene-
     carboxylic acid esters,
     α-alkylated **30,** 587
–, syntheses with – **30,** 587
**γ-Alkylthio-β,γ-ethylene-
carboxylic** – –, **α-alkylated**
– from
   γ-alkylthio-α,β-ethylene-
     carboxylic acid esters
     30, 587
α-Alkylthio-α,β-ethylenenitriles
– special s.
   α-(trifluoromethylthio)-
     α,β-ethylenenitriles
**β-Alkylthio-α-formamino-
carboxylic acids**
– from
   aldehydes **26,** 791
Alkylthio groups
–, introduction, relayed, into
   carbohydrates **29,** 555
Alkylthio-N-heterocyclics,
   hydrolysis **26,** 229
Alkylthioketene aminals
– from
   diaminomethylenesulfonium
     salts, rearrangement **26,** 581

α-(Alkylthio)ketones, cyclic
– from
   β-ketosulfoxides **29,** 926
α-Alkylthio-β-ketosulfoxides
– from
   carboxylic acid esters **30,** 545
γ-Alkylthio-β-lactones
– from
   mercaptans **28,** 534
α-Alkylthiomethyleneketones
–, C-alkylation, reductive **26,** 899
– startg. m. f.
   furan ring **28,** 840
   ketones, branched,
     incorporation of 3 alkyl
     groups **29,** 819
α-(Alkylthio)nitriles
– special s.
   α-(trifluoromethylthio)nitriles
α-(Alkylthio)oxo compds.
– startg. m. f.
   indoles **28,** 762
**α-Alkylthiophosphonic acid
esters**
– from
   phosphorous acid esters
     and 1,1-dithiols **30,** 439
1-Alkylthio-1-pyridinium salts
– startg. m. f.
   N-thioglycosyl-N-heterocyclics
     29, 436
**[(Alkylthio)thiocarbonyl]-
hydrazines 26,** 440
S-Alkylthioureas s. Isothioureas
1-Alkynes s. a. Acetylene
   derivs., terminal
1- and 2-Alkynes 12, 930
   suppl. **26**
Allenealcohols
– from
   2-tosylhydrazonoethers,
     cyclic **27,** 12
– special s.
   5,6-dienols
   2,4,5-trienols
2-Allenealcohols
– from
   acetyleneoxido compds.
     30, 594
– startg. m. f.
   1,3-dienes **29,** 88
2-Allene-tert-alcohols 30, 594
3-Allenealcohols
– from
   1-acoxy-2-en-4-ynes **30,** 24
Alleneamines 29, 390
– special s. allenetetramines
Allenecarbanions, syntheses
   via – **30,** 584
**α-Allenecarboxylic acid amides
6,** 358 suppl. **28; 24,** 204
   suppl. **28**
– – **esters**
– from
   ketenes **28,** 851
   β-ketocarboxylic acid esters
     27, 203
– startg. m. f.
   β-alkoxy-β,γ-ethylene-
     carboxylic acid esters
     29, 159
   β-ketocarboxylic – – **29,** 159
β-Allenecarboxylic acid esters
– from
   2-acetylenealcohols and
     orthocarboxylic acid esters
     27, 796

All–Alu

Allenecarboxylic acid fluorides
- from
  cyclopropenonecarbinols
  28, 501
Allenedicarboxylic acid amides, sym.
- from
  ynamines **27**, 687
Allene-1,3-dicarboxylic acid esters
- startg. m. f.
  homophthalic acid esters
  28, 868
β-Allenedithiocarboxylic – –
- startg. m. f.
  thiophenes, 2-alkylthio-
  26, 582
  thiopyrans, – **26**, 582
Alleneketals
- from
  enecyclopropenes **27**, 156
Alleneketones
- from
  enecyclopropenes **27**, 156
Allenes 30, 689
-, addition of thioketones
  to – **27**, 588
- from
  2-acetyleneammonium salts, quaternary **30**, 35
  acetylene derivs. with N-ring opening **29**, 41
  acoxy-2-acetylenes **28**, 115
  α,β-ethylenehalides **29**, 961
  halogenallenes, synthesis
  23, 831 suppl. **27**, 28
  α-halogeno-β-lactones **29**, 979
  ketones **28**, 913
  oxo compds. and 1-halogenoenesilanes **29**, 798
-, Michael addition to –
  18, 765 suppl. 26
-, reactions **28**, 615
-, reaction with thioketones
  28, 521
-, – – trichloroacetyl isocyanate
  28, 615
- special s.
  alkenylidenecyclopropanes
  aryl-allenes
  cyano-
  1,1-dialkoxy-3-alkylthio-
  di-
  fulvene-
  halogen-
  isoxazoles, allene-
  1,2,4-trienes
- startg. m. f.
  2-alkylthio-1,3-dienes **28**, 521
  1,3-dienes **28**, 521
  -, C-rearrangement **27**, 728
  α,α'-dihydroxyketones **28**, 115
  enolethers **29**, 138
  2-ethylenealcohols **27**, 136
  ethylene derivs. **15**, 536
  suppl. 26
  – –, synthesis **23**, 693 suppl. 28
  – – (cis) **20**, 63 suppl. 26
  ethylenehalogenhydrins **27**, 555
  α,β-ethylene-β'-halogensulfones **29**, 537
  indenes **30**, 510
  pyrroles, 3-hydroxy- **27**, 784
  3-pyrrolidones **26**, 693
  thiophenes, 3-hydroxy- **27**, 784
-, telomerization **27**, 724
  suppl. 28

-, **electron-rich** s. Allenetetramines
-, **optically active,** synthesis, review **19**, 970 suppl. 28
-, **sym.**
- from
  ketenes **28**, 878
-, **terminal 13**, 830 suppl. 26
Allenesulfones
- special s.
  diallenesulfones
Allenesulfoxides
- from
  2-acetylenealcohols and
  sulfenylchlorides **28**, 565
Allenetetramines 29, 902
2-Allenethioethers
- from
  diazo compds. **29**, 781
Allenyl ethers
- from
  alkoxyacetylenes, synthesis
  27, 835
- startg. m. f.
  α,β-ethyleneoxo compds.
  27, 835
Allenylsilanes
  s. Alkoxyallenylsilanes
Allophanates
- from
  alcohols **30**, 282
  isocyanates (2 molecules)
  29, 128
Allophanoyl chlorides
  isocyanates (2 molecules)
  26, 488
- from
- startg. m. f.
  1,3-diazetidine-2,4-diones
  26, 488
Alloxazines 28, 316
-, **2-deoxy- 26**, 482
Allylalcohols s. 2-Ethylenealcohols
Allylation (s. a. 2,7-Octadien-1-ylation) **26**, 827
C-Allylation with phosphoric acid esters **29**, 747
N-Allylation
- of sulfonic acid amides
  27, 481
π-Allyl complexes s. Iron
  π-allyl complexes, Nickel – –,
  Palladium – –
Allyl derivs.
-, benzene ring annelation
  with – **28**, 825
- from
  propenyl derivs. **23**, 719
  suppl. 29
N-Allyl –
- startg. m. f.
  N-propenyl derivs. **26**, 739
Allyl dithiocarbamates
-, 1-alkylation and rearrangement **29**, 809
-, rearrangement,
  [3.3]-sigmatropic **30**, 585
-, syntheses with – **30**, 585
- **ethers,** cleavage **20**, 537
  suppl. **28**; **30**, 8
- **halides** s. β,γ-Ethylenehalides
Allylic C-alkylation
  s. C-Alkylation, allylic
- **functionalization**
  s. Functionalization, allylic

- groups, transfer, palladium-catalyzed **26**, 827
- **metalation** s. Metalation, allylic
- **oxidation** s. Oxidation, allylic
- **synthesis** s. Synthesis, allylic
Allylophilic reagents
  nickel carbonyl **29**, 28
Allyl rearrangement 28, 503;
  29, 217; **28**, 499 suppl. 29
- of
  dithiocarbamic acid allyl
  esters **29**, 809
  2-ethylenealcohols **2**, 626;
  4, 677; **24**, 741; **29**, 191
-, prevention of – **27**, 578
- with C-insertion **27**, 818
Allylsulfonium ylid-type rearrangement 28, 656
Allyl-type rearrangement
  26, 911
Allyl vinyl sulfide as reactant
  28, 842
van Alphen rearrangement
  27, 346
Aluminum 28, 651
Aluminum/mercury(II) chloride
  18, 735 suppl. 27
- alkoxides
  - isopropoxide **26**, 161;
    28, 154; **30**, 598
  - methoxide s. Dialkyl-borane-aluminum methoxide complex
  - 2-methylbutoxide **23**, 49
- amalgam **28**, 22, 176; **29**, 855
- **bromide 28**, 510
- **chloride** (s. a. under Lithium tetrahydridoaluminate)
  CC↓O; CC↓↑Hal; **13**, 699
  suppl. **27**; **18**, 373 suppl. **26**;
  18, 649 suppl. **29**; **20**, 175
  suppl. **29**; **26**, 20, 283, 396, 556,
  606, 635, 674, 685, 719, 772,
  981; **27**, 213, 240, 638, 914,
  969, 988; **28**, 127, 178, 587,
  610, 659, 707, 876, 909, 951/2,
  984; **29**, 402 suppl. **30**; **29**, 420,
  581, 591, 676, 782, 822, 846;
  30, 268, 354, 409, 441, 500,
  532, 541
- –/**hydrogen chloride 26**, 883;
  28, 659, 708
- –/-**polymer 29**, 186
- –/**sodium chloride 26**, 948;
  28, 745
- compds., organo- (s. a. Alanes, Dialkylaluminum..., Diethyl-aluminum chloride, – cyanide, Triisobutyl-aluminum, Tri-methyl-)
  alkylaluminum halides
  as reagents **14**, 756 suppl. 26
- –, –, **optically active,** reduction, asym., by – **23**, 49
  suppl. 27
- **halide 28**, 168; **29**, 498
- **hydride** (s. a. Dialkyl-aluminum hydride, Lithium tetrahydridoaluminate/aluminum chloride **26**, 62, 64
- **nitrate 30**, 239
- **oxide** (s. a. Boron aluminum oxide) **10**, 587 suppl. **28**; **11**,
  224 suppl. **29**; **24**, 923 suppl. **27**;
  26, 252, 368/9, 467, 729; **27**,
  609; **28**, 5; **29**, 1, 185, 452, 495;
  30, 171

(Aluminum oxide)
- –, **basic 19**, 705 suppl. **26; 21**, 33 suppl. **29, 26**, 90, 468
- **silicate 25**, 131 suppl. **27**

Amberlite
- CG-50(H) **27**, 7
- IR 4 B as weak anion exchanger **23**, 789 suppl. **28**
- IR 120(H) **26**, 834; **29**, 252
- IRA 400(F) **26**, 901
- IRA 400(OH) as strong anion exchanger **26**, 237 suppl. **28**
- IRA-402 **30**, 55

Amberlyst 15 29, 30, 251; **30**, 71, 111/2

Ambident s. Electrophiles and under Metalation, allylic

Amidals s. 1,1-Di(acylamines)

Amide... s. a. Carboxylic acid amide...

Amide acetals
- –, N-alkylation with – **27**, 373
- from
  carboxylic acid amides **26**, 933
  1,1,1-dialkoxyammonium salts, quaternary **27**, 430
  α,α-diaminonitriles **28**, 187
  iminoester fluoroborates **26**, 933
- –, reactions with – **18**, 222 suppl. **29**
- –, ring closure via – **26**, 933
- special s.
  formamide acetals
- startg. m. f.
  amide mercaptals **26**, 595
  amidines **26**, 332
  diamidides **26**, 333
  enamines **26**, 247; **29**, 469
- –, **bicyclic**, prepn. and reactions, review **23**, 135 suppl. **26**
- –, **cyclic** s. 1,3-Dioxolanes, 2-amino-
- **halides**
  special s.
  dimethylhalogenoformiminium halides
- **mercaptals**
- from
  amide acetals **26**, 595
- –, **cyclic** s. 1,3-Dithioles, 2-amino-

Amides (s. a. Carboxylic acid amides, Diacylamines, Sulfonic acid amides)
- –, N-halogenation **26**, 267

Amide-N-sulfenyl chlorides
- from
  N-silylcarboxylic acid amides **28**, 259

Amidines
- from
  amide acetals **26**, 332
  aziridines, 2,2-dihalogeno- **28**, 388
  carbodiimides **28**, 617
  carboxylic acid amides **10**, 329; **27**, 416
  iminoesters **30**, 220
  isothioureas **30**, 686
  nitriles **30**, 220
- and amines **26**, 299
- –, syn→anti-rearrangement **30**, 534

- –, ring closures with –, review **25**, 257 suppl. **27**
- special s.
  acetyleneamidines
  amidrazones
  amino-amidines
  carbamyldialkoxy-
  en-
  ethylenehydroxy-
  keto-
  silyl-
  sulfonyl-
  thiocarbamyl-
- startg. m. f.
  amidrazones **28**, 380
  diamidides **26**, 333
  4(3H)-pyrimidinones **25**, 257
- –, transamidation **30**, 289
- –, **bicyclic**
- organic syntheses, review **22**, 959 suppl. **28**
- special s.
  1,5-diazabicyclo[4.3.0]non-5-ene
  1,5-diazabicyclo[5.4.0]undec-5-ene
- –, **cyclic** (s. a. Lactimimides)
- special s.
  indolines, 2-imino-
- startg. m. f.
  oxadiazabicycloalkanes **27**, 316
- –, –, **N-condensed**, ring opening, reductive **26**, 37
- –, **N,N′-disubst.**, from nitriles **30**, 219
- –, **subst.**
- from
  nitriles, halides, and amines **29**, 423

Amidinium salts
- from
  halogenimidium salts **26**, 404
- special s.
  azomethinio-amidinium salts
  carbamyl-
- –, **vinylogous**
- from
  carboxylic acid amides, vinylogous, and alkoxyaminomethinium salts **26**, 453

Amidinothiosemicarbazides
- special s.
  4-acyl-1-amidinothiosemicarbazides

Amidinothioureas
- from
  cyanamides and thioureas **26**, 302

Amido... s. a. Acylamines, Carboxylic acid amides

α-Amidoalkylation s. C-α-Acylaminoalkylation

Amidoximes
- from
  Δ²-1,2,4-oxadiazol-5-ones **26**, 22
- special s.
  hydroxy-amidoximes
  sulfonylform-
- startg. m. f.
  carbodiimides **27**, 516
  1,4,2,5-dioxadiazines **27**, 208
- –, **semicyclic** s. o-Oximino-N-heterocyclics

Amidrazones
- from
  amidines **28**, 380
- –, ring closures with – **26**, 270
- special s.
  acylamidrazones
- startg. m. f.
  1,2,3,5-thiatriazole S-oxides **26**, 270
  1,2,4-triazole ring **26**, 270
  Δ²-1,2,4-triazolines **27**, 387
- –, **unsubst. 12**, 451 suppl. **26**

Amidrazones, N¹-acyl-N³-hydroxy-
- startg. m. f.
  1,2,4-triazoles, 4-hydroxy- **27**, 506

Aminals s. 1,1-Diamines, Ketene aminals

Amination (s. a. under Radical reactions)
- of cycloheptatrienones **11**, 404 suppl. **29**
- special s.
  acyl-amination
  halogen-
  sulfonyl-
- with mesitylenesulfonyloxylamine **28**, 243
- –, **reductive 28**, 346

N-Amination 6, 328

Amine anions
- startg. m. f.
  azides **27**, 438
- **boranes** s. Borazanes
- **hydrochlorides**
- by hydrogenation **27**, 17
- from
  oximes **27**, 17
- startg. m. f.
  acylamines **10**, 331
- **N-nitrimides 30**, 188
- **oxides** (s. a. N-Oxides)
- special s.
  ethyleneamine oxides
  trimethylamine oxide
- –, **O-labeled 9**, 243 suppl. **26**

Amines (s. a. Bases)
- –, alkylation s. N-Alkylation
- as catalyst components **28**, 130
- as reagents **28**, 215, 325
- –, condensation, deaminative **29**, 405
- –, derivs. s. Benzylisothiouronium dithiocarbamates
- from
  acylamines (s. a. N-Deacylation) HN↓↑C
  alcohols **30**, 255
  α-aminocarboxylic acids **28**, 68
  1,1-aminoethers **28**, 57
  o-aminothioureas **30**, 333
  azides **27**, 401; **29**, 324, 424
  aziridines **26**, 55
  azo compds. **29**, 17
  azoles **29**, 561
  carboxylic acid amides, reduction **26**, 64, 389
- acids **26**, 389
  cyanamides **26**, 23
  1,1-diamines **28**, 57
  enamines **28**, 38; **29**, 45
  ethylene derivs., cleavage **27**, 149

– –, synthesis with addition
  of 1 C-atom **27**, 723
formamides s. N-Deformylation
halides NC↓↑Hal
– and acylamines **24**, 491
2-halogenamines with
  rearrangement **29**, 76
hydrazines (s. a. Deamination)
  29, 21
imines **26**, 34
isonitriles **29**, 791
monoacyl-1,1-diamines **28**, 57
nitro compds. **27**, 14/15;
  28, 16, 18; **30**, 10a, 330
N-nitrosamines **28**, 19, 355
nitroso compds. **29**, 16
phosphoramidates **27**, 548
sulfonic acid amides
  s. N-Detosylation
1,2,4-triazolidine-3,5-diones
  29, 21
ureas **26**, 23
urethans (s. a. N-Decarbalkoxylation) **30**, 273
–, methylation, reductive **28**, 346
–, monoalkylation with alcohols
  28, 358
–, reaction with lithium
  compds., organo- **27**, 712
– special s.
  acetylene-amines
  allene-
  benzyl-
  di- (diamino...)
  diazo-
  1,3-dioxane, 5-amino-2,2-
    dimethyl-4-phenyl-
  halogen-amines
  methyl-
  monoacyldi-
  nitr-
  nitros-
  oxido-
  poly-
  tosyloxy-
– startg. m. f.
  acoxylamines **26**, 88
  aldehydes **26**, 194
  amidines **26**, 299
  –, subst **29**, 423
  aminoacetamides
    (from 2 molecules) **28**, 313
  arsines **29**, 609
  azides **27**, 438
  azo compds., sym. **28**, 237
  azomethines **26**, 372; **28**, 283,
    356; **29**, 371, 390, 404, 504;
    30, 468
  diazo compds. **30**, 169
  formamides **26**, 282; **27**, 312
  formamidines **11**, 485; **30**, 281
  hydrazines **26**, 259
  hydroxylamines **7**, 10
    suppl. **29**; **26**, 88
  isothiocyanates **27**, 366
  nitriles **26**, 463
  nitro compds. **28**, 70
  nitroso – **26**, 91
  N-oxides **26**, 90
  phosphines **29**, 609
  4-quinazolones **28**, 710
  quinone monoimines
    (from 2 molecules) **28**, 237
  sulfamides **29**, 287
  sulfamoyl azides **28**, 245
  sulfinodiimines **26**, 273
  thiolurethans **27**, 378; **29**, 531

thiophosphinylthioureas
  28, 396; **30**, 210
thioureas, sym. **26**, 323;
  27, 366
ureas **4**, 413; **26**, 329, 364;
  28, 417; **30**, 229
–, sym. **28**, 322; **30**, 243
urethans **27**, 437; **28**, 322
–, **ar.**
–, C-acylation **26**, 846
–, o-alkylation **28**, 762
–, coupling, anodic **28**, 703
  suppl. **29**
–, o-formylation **28**, 762 suppl. **29**
– from
  ethers **27**, 459
  ketones, cyclic **30**, 259
–, mercuration **30**, 428
–, monohalogenation **28**, 485
– special s.
  p-aminophenyltriphenyl-
    methane
  anilines
  naphthalenes, 2-amino-
– startg. m. f.
  benzothiazoles **30**, 392
  oxindoles **28**, 762
  quinolines, 1,2,3,4-tetra-
    hydro- **27**, 808
–, –, **prim.**, oxidation, review **20**,
  211 suppl. **27**
– from
  halides, ar. **14**, 512; **30**, 592
–, **cyclic** s. Imines, cyclic
–, **hindered** (s. a. Amines, tert.,
  hindered, Amines, sec., –)
– startg. m. f.
  o-aminocarboxylic acid
    amides **11**, 546; **27**, 349
–, α-**labeled**
– from
  N-nitrosamines **28**, 590
–, **prim.**
– from
  alcohols **27**, 401, 429
–, synthesis with addition
  of 1 C-atom **27**, 782
  aldehydes **27**, 414
  dicarboxylic acid imides
    phthalimides **26**, 25;
    27, 429, 480
  halides **27**, 480
  nitriles **26**, 35, 39; **27**, 46;
    28, 34; **30**, 20
  tosylates **26**, 357
–, Mannich reaction with –
  26, 594
– special s.
  (*tert*-carbin)amines, prim.
  isopropylamine
– startg. m. f.
  acridinium salts, *sym*-octa-
    hydro- **27**, 415
  amines, sec. and tert.
    s. a. below
–, – (from 2 molecules)
  28, 378
  ammonium salts, quaternary
    26, 419
  aziridines, 2-cyano- **27**, 468
  1,3,2,4-diazadiphosphetidine
    2,4-dioxides **29**, 294
  N,N-dicyanamides **29**, 342
  dimethylamines **14**, 455
    suppl. **15**, 17, 29
  oxo compds. **24**, 203
  ketones **26**, 187
  phthalimides **29**, 417

– via (disulfen)amines **26**, 357
–, **sec.**
– from
  1-alkoxyazomethines **27**, 62
  amines, prim. s. a. N-Mono-
    alkylation
–, – (2 molecules) **28**, 378
–, –, and halides **23**, 458
  suppl. **26**
–, –, – – (diarylamines)
  29, 418
–, –, and oxo compds.
  29, 371, 390; **30**, 276
–, –, oxo compds., and
  halides **30**, 463
–, tert. **27**, 26, 31
P-aminophosphonium salts
  26, 291
ammonium salts, quaternary
  27, 31
azides and boranes **27**, 485
– and dihalogenoboranes
  27, 485 suppl. **28**
azomethines, reduction **29**, 42,
  371, 390; **30**, 19
–, synthesis **26**, 682; **29**, 634;
  30, 463
carboxylic acids **27**, 392
N-cyclopropylazomethines
  27, 52
α,β-ethyleneazomethines **27**, 43
hydrazones **29**, 399
iminoesters **29**, 369
nitrogen radicals **29**, 12
N-oxide radicals **27**, 13;
  28, 17
N-oxides, tert. **30**, 26
phosphine imines and halides
  26, 291
– special s.
  N-methylamines, ar.
– startg. m. f.
  amines, tert. s. below
  ammonium salts, quaternary
    26, 419
  oxamides, sym. **28**, 322
–, –, **ar.**
– from
  azides **29**, 402
–, –, **cyclic**
– startg. m. f.
  azomethines, cyclic **13**, 383
–, –, **hindered 29**, 371
–, –, **mixed 27**, 31
–, **telomerized 27**, 724
–, **tert.** (s. a. Mannich bases)
–, alkylation of arenes with –
  22, 795 suppl. **29**
– as phase transfer catalysts
  27, 833 suppl. **29**, 30
– as singlet-oxygen quenchers
  29, 147
–, exchange of substituents
  27, 493
– from
  alcohols and phosphoric acid
    amides **28**, 352
  amines, prim., and halides
    26, 423; **28**, 393
–, –, and sulfates **29**, 428
–, sec., and alcohols **26**, 388
–, –, and oxo compds. **28**, 346
–, – via N-aminomethylation
  26, 871
  ammonium salts, quaternary
    29, 992
  halides **28**, 413
  N-halogenamines **30**, 305

Ami

(Amines, tert.)
- special s.
  dimethylamines
  dodecyldimethylamine
- startg. m. f.
  acylamines, sec. **29,** 444
  alcohols **29,** 531
  amines, sec. **27,** 26, 31
  azomethines, rearrangement **27,** 751
  ethylene derivs. **30,** 669
  halides **28,** 508
  ketones, synthesis **28,** 768
  nitriles **22,** 288 suppl. **28**
  sulfones **29,** 559
-, -, **cyclic**
-, Birch reduction **30,** 23
- from
  aminoalcohols **26,** 472
  cyclimmonium salts **21,** 910
  - -, N-condensed **29,** 37
- startg. m. f.
  acylamines with aromatization **28,** 336
  azomethinium salts, cyclic **30,** 152
  cyclimmonium salts, N-condensed **29,** 904
  ethyleneacylamines **26,** 300
  hydroxycyanamides **28,** 277
  lactams **30,** 152
-, -, -, **1-subst.**
- from
  azomethinium salts, cyclic **29,** 716
-, -, **hindered** (s. a. Bases, hindered)
-, N-quaternization **30,** 702
- special s.
  ethyldiisopropylamine
  tri-*n*-butylamine
-, **unsatd.**
-, reaction with Grignard compds. **27,** 712
-, synthesis **29,** 826
Aminimides (s. a. N-Ylids)
-, review **23,** 490 suppl. **28**
- special s.
  aminoaminimides
  pyridinium N-imides
Amin-N-imines, review **27,** 436
2-Aminoacetals
- startg. m. f.
  lactams, N-condensed **28,** 329
  pyrazines, 2,3-dihydroxy- **27,** 511
Aminoacetamides
- from
  amines (2 molecules) **28,** 313
o-Aminoacetic acid esters
- startg. m. f.
  furo[3,2-b]quinolines **26,** 939
  quinolines, 3-hydroxy- **26,** 939
β-Amino acid peptides prepn. **26,** 806
- from
  1,3-oxazin-6-ones, 4,5-dihydro- **30,** 207
-, prepn. **26,** 806
Aminoacids s. Aminocarboxylic acids
N'-(3-Aminoacryloyl)-formamidines
- from
  1,1,1-alkoxydiamines and carboxylic acid amides **28,** 770

O-α-Aminoacylnucleotides 14, 231 suppl. **26**
-, synthesis, review **14,** 231 suppl. **28**
Aminolanes
- special s.
  (dialkylamino)alanes
trans-2-Aminoalcohol groups, diequatorial
-, protection as 2-oxazolidones **26,** 426
Aminoalcohols
- as reducing agents **28,** 16
- from
  dicarboxylic acid anhydrides **28,** 364
- special s.
  (difluoroamino)alcohols
  dihydroxyamines
- startg. m. f.
  amines, tert., cyclic **26,** 472
  aminoethers, rearrangement **28,** 123
  1,3-O,N-heterocyclics **28,** 374
  piperidines **26,** 467
1,1-Aminoalcohols s. Azacyclols, 1,1-Iminoalcohols, cyclic, Pseudobases, N-heterocyclic
2-Aminoalcohols
- from
  α-aminocarboxylic acids **29,** 57
  azomethines **30,** 459
  enamines **19,** 188 suppl. **26**
  2-halogenacylamines **26,** 60
  2-halogenosulfonic acid amides **29,** 213
  β-hydroxycarboxylic acid amides **28,** 309
  ketones, synthesis with addition of 1 C-atom **27,** 720
  1,2,3-oxathiazolidine S-oxide ring **29,** 18
  2-oxazolidone ring **28,** 309
  oxido compds. **13,** 361
  α-siloxynitriles **27,** 720
- special s.
  ethanolamine
- startg. m. f.
  aldehydes, C-cleavage **30,** 179
  2-aminomesylates **30,** 459
  aziridines **26,** 471
  1,3-halogenamines, rearrangement **26,** 534
  oxazolidines **28,** 335
  Δ²-oxazolines, 2-amino- **29,** 335
-, **stereoisomeric,** interconversion **29,** 335
-, -, separation **29,** 18
2-(prim-Amino)alcohols
- from
  oxido compds. **14,** 380
3-Aminoalcohols
- from
  3-isocyanoalcohols **29,** 617
  isonitriles and oxido compds. **29,** 617
  4H-1,3-oxazines, 5,6-dihydro- **29,** 617
- startg. m. f.
  azetidines **28,** 471 suppl. **30**
  azetidinium ring **27,** 509
5-Aminoalcohols
- startg. m. f.
  pyrans, tetrahydro- **29,** 140
Aminoaldehydes
- from
  imines, cyclic **28,** 92

α-Aminoaldehydes 10, 629 suppl. **29**
β-Aminoaldehydes
- startg. m. f.
  1,3-oxazinium salts, tetrahydro-6-hydroxy- **26,** 289
o-Aminoaldehydes
- by pyrimidine ring opening **26,** 229 suppl. **30**
Aminoalkylation
s. Aminomethylation
o-Aminoamidines
- from
  1,2,3-triazine ring **30,** 285
- startg. m. f.
  pyrimidine ring, 4-amino- **30,** 256
β-Aminoaminimides
- startg. m. f.
  2-imidazolidones **28,** 448
Aminoarsines
- startg. m. f.
  As-heterocyclics **26,** 568
Aminoazomethines 29, 404
- special s.
  aminoethyleneazomethines
α-Aminoazomethines 21, 432 suppl. **27**
Aminoboranes
- from
  trialkylboranes **29,** 297
- startg. m. f.
  2-aminoeneboranes **29,** 597
  sulfenamides **29,** 291
Aminocarbonium salts
- from
  alkoxycarbonium salts **28,** 359
- special s.
  alkoxyaminomethinium salts
Aminocarboxylic acid amides
- startg. m. f.
  lactams **26,** 483
α-Aminocarboxylic - -
- by rearrangement **26,** 310
- from
  α-halogenocarboxylic acid amides **30,** 290
- special s.
  aminoacetamides
- startg. m. f.
  α-aminocarboxylic acids **30,** 290
o-Aminocarboxylic - -
- from
  3,1-benzoxazine-2,4-diones and amines **11,** 546; **27,** 349
  pyrimidine ring **16,** 23; **29,** 20
  triazinone - **27,** 15
- startg. m. f.
  o-acylaminocarboxylic acid amides **28,** 437
  N-heterocyclics, N-condensed **26,** 407
  pyrimidine ring, 2-mercapto- **27,** 360
  4(3H)-pyrimidinone ring **4,** 455; **28,** 437
  2,4-quinazolinediones **27,** 349
  uracil ring **27,** 349; **29,** 333
α-Aminocarboxylic acid N-carboxyanhydrides
s. 2,5-Oxazolidinones
Aminocarboxylic acid chlorides
- special s.
  β-amino-α,β-ethylene-carboxylic acid chlorides

Ami

α-**Aminocarboxylic acid derivs.**
- from
  nitriles, synthesis with
    addition of 1 C-atom **29,** 631
β-**Aminocarboxylic** – –
  s. β-Amino acid peptides
Aminocarboxylic acid esters
- special s.
  aminoketocarboxylic acid
    esters
α-**Aminocarboxylic** – –
- from
  aldehydes, synthesis with
    addition of 2 C-atoms
    29, 739
  α-aminocarboxylic acids
    27, 402
  α-amino-α,β-ethylene-
    carboxylic acid esters
    29, 739
  aziridinones **27,** 192
- special s.
  α-alkylideneaminocarboxylic
    acid esters
  O-α-aminoacylnucleotides
- startg. m. f.
  pyrroles, 3-hydroxy- **27,** 784
β-**Aminocarboxylic** – –
- from
  azomethines and α-halogeno-
    carboxylic acid esters
    29, 638
- startg. m. f.
  benzylamines **28,** 875
Aminocarboxylic acids
- from their hydrohalides
    30, 290
- special s.
  aminoketocarboxylic acids
- startg. m. f.
  acylaminocarboxylic acid
    esters **26,** 224
  isocyanatocarboxylic acid silyl
    esters **27,** 512
  lactams **24,** 513; **30,** 331
α-**Aminocarboxylic** –
-, Copper(II) complex salts as
    intermediates **28,** 68 suppl. 29
- from
  α-acylaminocarboxylic acid
    amides **28,** 288
  5-alkylidene-4-imidazolones
    26, 193
  4-alkylidene-5-oxazolones
    28, 288
  α-aminocarboxylic acid
    amides **30,** 290
  carboxylic acids **28,** 312
  halides **29,** 791
  α-halogenocarboxylic acid
    amides **30,** 290
  – acids **26,** 425
  hydantoins **30,** 568
  α-isocyanocarboxylic acid
    esters **29,** 791
  isonitriles **29,** 898
  nitriles, synthesis with
    addition of 1 C-atom
    28, 621
- special s.
  α-amino-β-hydroxycarboxylic
    acids
  carbohydrate-amino acid
    compds.
  α-hydroxymethyl-α-amino-
    carboxylic acids
  isovaline
  tryptophans

- startg. m. f.
  α-acylaminocarboxylic acids
    30, 451
  amines **28,** 68
  2-aminoalcohols **29,** 57
  α-aminocarboxylic acid esters
    27, 402
  α,β-ethylenecarboxylic acids,
    α-H-labeled **29,** 950
  α-hydroxycarboxylic acids
    2, 218 suppl. 26
  α-hydroxymethyl-α-amino-
    carboxylic acids **30,** 451
  pyrrole ring **29,** 885
  P-spiroheterocyclics **28,** 264
-, Strecker synthesis **21,** 829
-, synthesis **29,** 791
-, –, review **30,** 290
-, –, asym. **4,** 643
  suppl. **26; 16,** 731 suppl. 26;
  21, 829 suppl. 27; **28,** 621;
  29, 791
-, –, –, review **17,** 461 suppl. 26
- acids, α-H-labeled **21,** 115
  suppl. 27
- –, β-H-labeled **20,** 106
  suppl. 28
- – and esters, optically
  active, by enzymic resolution
  28, 13
- acids-1-¹⁴C
- from
  oxo compds. **30,** 568
β-**Aminocarboxylic acids**
- from
  azomethines **26,** 682
  Δ²-pyrazolines **27,** 200
γ-**Aminocarboxylic** –
- from
  2-pyrrolidones **26,** 120
- special s.
  γ-aryl-γ-aminocarboxylic
    acids
ο-**Aminocarboxylic** –
- startg. m. f.
  4-quinazolone ring,
    N-condensed **27,** 455
**Aminocarboxylic acid silyl
esters**
- as intermediates **12,** 519
  suppl. 26
α-**Aminocarboxylic borinic
anhydrides 26,** 108
Aminocyanohydrin tosylates
- startg. m. f.
  o-carbalkoxy-N-heterocyclics
    28, 434
N-Aminocyclimmonium salts
- from
  N-heterocyclics **28,** 243
- special s.
  imidazolium ring, N-amino-
Aminocyclopropane ring
- startg. m. f.
  oxo compds. **28,** 161
β-**Amino-α-diazosulfones**
- from
  α-diazosulfones and enamines
    27, 688
**N-Aminodicarboxylic acid
imides**
- from
  dicarboxylic acid anhydrides
    28, 330
- startg. m. f.
  N,N'-dicarboxylic acid
    diimides **29,** 366

1,1,1-Aminodinitriles
- from
  alkoxyaminomethinium salts
    28, 624
  α-aminonitriles **29,** 645
  1,1-iminocyanides **29,** 645
1-Amino-1,3-dinitriles
- from
  α-aminonitriles **29,** 663
- startg. m. f.
  β-cyanoketones **29,** 663
ο-**Aminodisulfides**
- startg. m. f.
  1,4-benzothiazine ring **26,** 624
2-Aminoeneboranes
- from
  aminoboranes and acetylene
    derivs. **29,** 597
1-Aminoenedisulfides
- from
  iminodisulfides **29,** 701
2-Aminoenol sulfonates
- startg. m. f.
  benzils **28,** 178
  indoles **28,** 178
Aminoethers
- from
  aminoalcohols, rearrangement
    28, 123
1,1-Aminoethers
-, cleavage s. N-de(methoxy-
    methylation)
-, protection of amino groups
    as – **28,** 268
- special s.
  1-alkoxy-1-alkylthio-1-amines
  1,1,1-alkoxydiamines
  N-benzyloxymethyl
    1,1'-di(alkoxy)amines
- startg. m. f.
  amines **28,** 57
-, cyclic
- from
  1-alkoxyenamines, cyclic
    28, 793
- special s.
  pyrans, tetrahydro-2-amino-
- startg. m. f.
  acylamines **29,** 164
  aldehydes **28,** 793
**N-Aminoethylation, selective
28,** 273
α-**Amino-α,β-ethylenealdehydes
26,** 870 suppl. 29
β-**Amino-α,β-ethylenealdehydes**
- from
  acetylenealcohols **28,** 361
  carboxylic acids **30,** 634
- startg. m. f.
  1,3-dialdehydes **13,** 786
β-**Amino-α,β-ethyleneazo-
methines**
- from
  azomethines and nitriles
    30, 461
β-**Amino-α,β-ethylenecarboxylic
acid amides**
- startg. m. f.
  4(3H)-pyrimidinones, 5-acyl-
    26, 340
β-**Amino-α,β-ethylenecarboxylic
acid aziridides**
- from
  aziridine ring (2 molecules)
    and cyclopropenones
    26, 306
- – **chlorides 29,** 714

α-**Amino-α,β-ethylenecarboxylic acid esters**
- from
  aldehydes, synthesis with addition of 2 C-atoms **29**, 739
- startg. m. f.
  α-aminocarboxylic acid esters **29**, 739
- – esters
- from
  enamines **26**, 769
- – thioamides
- from
  1,1,1-alkoxydiamines and carboxylic acid thioamides **28**, 770

α-**Amino-α,β-ethyleneketones**
- from
  β-alkoxylaminoketones **29**, 917

β-**Amino-α,β-ethyleneketones**
- from
  2-acetylenealcohols **30**, 246
  carboxylic acid halides **26**, 846
  β-diketones **26**, 331
- special s.
  (2-acyl-1-methylvinyl)amines
  α-aminomethyleneketones
- startg. m. f.
  pyrido[2,3-d]pyrimi-2,4-dione ring **27**, 794
-, γ-substitution **28**, 799

α-**Amino-β,γ-ethylenenitriles**
- startg. m. f.
  heterocyclics **27**, 535

β-**Amino-α,β-ethylenenitriles**
- from
  cyanoallenes **29**, 160
  nitriles (2 molecules) **27**, 677
- startg. m. f.
  α-cyanoketones **29**, 160
  pyridine ring, 2-amino- **30**, 578
-, transamination **18**, 482 suppl. **27**
-, **cyclic**
- startg. m. f.
  dicarboxylic acids **29**, 195

β-**Amino-α,β-ethyleneoxo compds.**
- startg. m. f.
  pyridine ring **26**, 805; **27**, 794
  pyrimidine ring **27**, 408

β-**Amino-α,β-ethylenesulfones**
-, synthesis with rearrangement **28**, 831

Aminogermanes
- special s.
  diaminogermanes

Amino groups s. a. Amines, Replacement

Aminoguanidines
s. Diaminoguanidines

Aminohalides s. Halogenamines

Amino-N-heterocyclics
- special s.
  alkoxyamino-N-heterocyclics

N-**Amino-N-heterocyclics**
-, N-deamination **26**, 480
-, reactions with lead tetraacetate **26**, 480

o-**Amino-N-heterocyclics**
- from
  cyanamide **28**, 604
-, ring closures with – **13**, 535

1,1-Aminohydroperoxides,
review **21**, 161 suppl. **29**

o-**Aminohydroxamic acid esters**
- startg. m. f.
  4-quinazolones, 3-alkoxy- **29**, 406

Aminohydroxamic acids
- from
  lactams **27**, 492
- startg. m. f.
  diisocyanates **27**, 492

α-**Aminohydroxamic** –
- startg. m. f.
  4-imidazolidones, 3-hydroxy- **26**, 336

o-**Aminohydroxamic** –
- startg. m. f.
  hydroxamic acids, cyclic **28**, 385

α-**Amino-β-hydroxycarboxylic** –
- from
  α-isocyanocarboxylic acid esters **22**, 686 suppl. **28**

1,1-Aminohydroxylamines, cyclic
s. Aziridines, C-hydroxamino-

Aminoketals
- startg. m. f.
  isoquinolines **28**, 444

β-**Amino-α-ketocarboxylic acid derivs. 26**, 403

Aminoketocarboxylic acid esters
- special s.
  aminomethylketocarboxylic acid esters

α-**Amino-β-ketocarboxylic** – –
28, 811
- startg. m. f.
  β-ketocarboxylic acid esters **29**, 68

α-**Amino-γ-ketocarboxylic** – –
- from
  ketones **28**, 715

Aminoketones
- special s.
  aminoethylene-ketones
  diamino-
- startg. m. f.
  azomethines, cyclic **28**, 438

α-**Aminoketones**
- from
  acylcyanides **8**, 59 suppl. **27**
  carboxylic acid chlorides and isonitrils **28**, 811
  1,2-di(siloxy)-1-ethylene derivs. **28**, 410
  enamines **28**, 116
  α-halogenoazomethinium salts, rearrangement **28**, 116
  ketones **28**, 323
  oxazoles **28**, 811
  1,2-oxidoamines **29**, 689
  1,2-oxido-1-halides **26**, 403
- special s.
  α-arylaminoketones
- startg. m. f.
  α-dioxo compds. **28**, 153
  Δ²-imidazoline ring, 5-hydroxy- **29**, 385
  pyrazines, 1,2-dihydro- (from 2 molecules) **29**, 727
-, 2,5-dihydro- (from 2 molecules) **28**, 323
-, **cyclic 28**, 116

β-**Aminoketones**
- by Mannich reaction **26**, 807/8
- from
  azomethines **28**, 620
  carboxylic acid amides, vinylogous **29**, 46

α-diazoketones **28**, 782
2-ethylenealcohols **27**, 370
isoxazoles **28**, 204
ketones **26**, 808
-, self-condensation **29**, 774
- special s.
  β-amino-α-methyleneketones
- startg. m. f.
  α,β-ethyleneketones **28**, 204

o-**Aminoketones**
- startg. m. f.
  1H-1,4-benzodiazepines, 2,3-dihydro- **29**, 749
  N-heterocyclics **27**, 804
  indazoles, 2-hydroxy- **26**, 324
  3-indazolones **26**, 324
  indoles **23**, 871 suppl. **26**
  indoles, 2-cyano- **29**, 749
  quinolines, 2-(o-acylanilino)- **26**, 816
-, **4-hydroxy 29**, 726

sec-**Aminoketones**
- startg. m. f.
  azomethinium salts, cyclic **28**, 445

Aminolactols
- from
  lactones **28**, 607

Aminolactones
- from
  azidolactones **29**, 324
- startg. m. f.
  hydroxylactams **29**, 324

Aminolysis 22, 34a
- of
  oxazolidine rings, N-condensed **26**, 332

2-Aminomercaptans
- special s.
  2-mercaptoethylamine
- startg. m. f.
  Δ²-thiazolines **30**, 412

o-**Aminomercaptans**
- startg. m. f.
  1,2,3-benzothiadiazoles **27**, 623
  1,5-benzothiazepine ring **13**, 896
  bis(benzothiazoles), 2,2′-amino- **28**, 538
  o-dithiols **27**, 623
  1,4-thiazin-3-one ring **30**, 393

2-Aminomercuration 29, 595

Aminomercuration, intramolecular 27, 345

2-Aminomesylates
- from
  aminoalcohols **30**, 459
-, **N-heterocyclic**, ring expansion **30**, 459

Aminomethylation with 1,3,5-triazines, hexahydro- **28**, 786

C-**Aminomethylation**
- of
  ethylene derivs. **29**, 833
  N-oxides, cyclic **8**, 822 suppl. **28**
- with
  amines, prim. **3**, 608 suppl. **26**
-, **modified**
- of
  ketones **26**, 807

N-**Aminomethylation 26**, 871
-, amines, tert. from sec, via – **26**, 871
- with
  amines, prim., ar. **11**, 423

O-**Aminomethylation 21**, 404 suppl. **26**, **27**

P-Aminomethylation 29, 605
p-Aminomethylation of phenols
   28, 786
Aminomethylenation
   s. Dimethylaminomethylation
Aminomethylene compds.
- from
   isonitriles 28, 619
α-Aminomethyleneketones
- from
   carboxylic acid chlorides
   30, 589
- startg. m. f.
   ketones 30, 589
β-Amino-α-methyleneketones
- from
   methyl ketones 26, 808
Aminomethylenemalonic acid
   esters
-, transamination 27, 443
β-Amino-α-methylenenitriles
- by
   Mannich reaction,
      decarboxylative 26, 916
α-Aminomethyl-β-ketocarboxylic
   acid esters
- from
   β-ketocarboxylic acid esters
   29, 759
Aminonitriles
-, ring closures with – 27, 334
- special s.
   aminoethylenenitriles
α-Aminonitriles 30, 576
- from
   azomethines 27, 781
   oxo compds., synthesis with
      addition of 1 C-atom
      27, 781; 29, 750
- special s.
   α,α-alkoxyamino-nitriles
   1-amino-1,3-di-
   α,α-diamino-
- startg. m. f.
   2-acetyleneamines, synthesis
      7, 858 suppl. 27
   aldehydes 29, 901
   1,1,1-aminodinitriles 29, 645
   1-amino-1,3-dinitriles 29, 663
   β-cyanoketones 29, 663
   4-imidazolidones 28, 275
   1,1-iminocyanides 19, 553
      suppl. 26; 29, 645
   mercaptans 30, 420
-, transamination 5, 350
   suppl. 29
β-Aminonitriles
- from
   Δ²-pyrazolines 28, 889
- special s.
   β-amino-α-methylenenitriles
o-Aminonitriles
- startg. m. f.
   pyrimidine ring,
      2,4-diamino- 28, 401
trans-β-Amino-2-nitrostyrenes
- startg. m. f.
   indoles 29, 469
o-Amino-N-oxides
- startg. m. f.
   o-azo-N-oxides 27, 283
   1,2,3,5-oxatriazolium salts,
      N-condensed 27, 283
anti-1,2-Aminooximes
- startg. m. f.
   Δ³-imidazoline 3-oxides
      26, 337

syn-1,2-Aminooximes
- startg. m. f.
   4H-1,2,5-oxadiazines,
      5,6-dihydro- 27, 386
Aminooxo compds.
- special s.
   β-amino-α,β-ethyleneoxo
      compds.
o-Aminooxo –
- startg. m. f.
   indoles 29, 837; 30, 663
   indolines, 3-hydroxy- 30, 663
-, 5-subst.
- from
   2,1-benzisoxazoles 28, 470
-, N-subst.
- from
   2,1-benzisoxazoles 26, 18
   2,1-benzisoxazolines 26, 18
Aminoperoxidation
-, 1,2,4-dioxazolidine ring by –
   30, 206
1,1-Aminoperoxides, review
   21, 161 suppl. 29
-, cyclic s. 1,2,4-Dioxazoles
α-Aminophenolethers 9, 887
   suppl. 26
o-Aminophenols
- from
   benzoxazole-2-thiones 26, 21
- startg. m. f.
   benzoxazole-2-aldoximes
      26, 430
   o-hydroxyphosphine imines
      28, 262
   phenoxazines (from 2 mole-
      cules) 28, 378
o-α-Aminophenols
- from
   o-quinone methids 30, 203
p-Aminophenyltriphenyl-
   methane
-, protection of phosphate
   groups with – 25, 88 suppl. 26
Aminophosphines
- special s.
   bis-(3-dimethylamino-
      propyl)phenylphosphine
   phosphorous acid triamides
2-Aminophosphinic acid esters
- from
   azomethines 26, 682
α-Aminophosphonic – –
- from
   azines and dialkyl
      phosphites 29, 610
P-Aminophosphonium salts
- from
   phosphine imines and
      halides 26, 291
-, reactions with – 30, 150
- special s.
   alkoxytriaminophosphonium
      perchlorates
   μ-oxobis(triamino-
      phosphonium salts)
- startg. m. f.
   amines, sec. 26, 291
Aminophosphoryl dichlorides,
   N-halogenosulfonyl- 27, 119
3-Aminophthalides
- from
   3-hydroxyphthalides 28, 324
Aminopyrimidines
   s. Pyrimidines, 4-amino-
      5-nitroso-

4-Aminoquinolines
- startg. m. f.
   quinolines, 1,4-dihydro-
      4-imino-, 1-subst. 26, 401
Amino-p-quinones 11, 404
Aminosilanes
- special s.
   N-silylenamines
   trimethylsilyl-
      N,N-diethylamine
- startg. m. f.
   carbamic acid silyl esters
      29, 438
   isocyanates 26, 443; 29, 438
   isothiocyanates 27, 486
   thioureas, sym. 27, 486
1,1-Aminosilanes
- from
   1,2-silylammonium salts,
      quaternary 30, 444
Aminosiloxanes
- special s.
   silylaminosiloxanes
Aminostibines
- from
   halogenostibines 26, 280
- special s.
   tris(dimethylamino)stibine
Aminostyrenes
- special s.
   aminonitrostyrenes
o-Aminostyrenes
- startg. m. f.
   quinolines 26, 765
Aminosugars
   (s. a. Glycosamines)
-, reaction with isothio-
   cyanates 27, 318
Aminosulfenium salts
   (s. a. Acylaminosulfenium
      salts)
- from
   alkoxysulfonium salts 28, 250
   halogenosulfenium –
      28, 254
   thioethers 28, 243
Aminosulfenyl chlorides
- special s.
   morpholinosulfenyl chloride
- startg. m. f.
   sulfonylthioamines 28, 517
Aminosulfenium ylids
   s. Sulfilimines
o-Aminosulfinic acids
- from
   o-aminosulfones, Smiles
      rearrangement 28, 300
2-Amino-1-sulfinylthioenolethers
- from
   nitriles 29, 631
Aminosulfones
- from
   sulfonic acid amides 27, 602
- special s.
   aminodiazo-sulfones
   β-amino-α,β-ethylene-
α-Aminosulfones 29, 559
o-Aminosulfones
- startg. m. f.
   o-aminosulfinic acids, Smiles
      rearrangement 28, 300
S-Aminosulfoxonium salts
- from
   sulfoxides 28, 243
-, reactions with – 26, 896
-, cyclic, reactions with –
   27, 682

S-Aminosulfoxonium ylids,
  reactions via – 26, 896;
  27, 873
– –, cyclic, reactions via –
  27, 682
o-Aminothiocyanates
– startg. m. f.
  thiazole ring, 2-amino-
  17, 383
Aminothio(dichloroformimido)-
  cyanamides 26, 275
1-Aminothioenolethers 28, 836
  suppl. 29
– from
  dithiocarbamidium salts
  27, 484
1,1-Aminothioethers 30, 576
– from
  Δ³-5-oxazolones 26, 627
– special s.
  1-alkoxy-1-alkylthio-1-amines
– startg. m. f.
  1,1-sulfonioammonium salts,
  quaternary 30, 703
Aminothiolic acids,
  diazotization 27, 293
o-Aminothiolic –
– from
  3,1-benzoxazine-2,4-diones
  27, 293
– startg. m. f.
  3,1,2-benzothiadiazin-4-ones
  27, 293
o-Aminothioureas
– startg. m. f.
  benzimidazoles, 2-amino-
  29, 476
  benzimidazoline-2-thiones
  and amines 30, 333
Aminotropenium salts
  28, 359
Aminoxy compds.
  s. Alkoxylamines
Ammonia 28, 362; 29, 196, 218,
  357, 413, 749
– as reactant 29, 377
– as HCl-scavenger 26, 971
– startg. m. f.
  azines 27, 421
–, liq. (s. a. under Calcium,
  Lithium, Potassium, Sodium,
  Sodium hydrogen sulfide)
  11, 665 suppl. 28; 16, 393
  suppl. 28; 24, 663 suppl. 28;
  26, 24, 87; 29, 384; 30, 146, 698
– as reactant – solvent 23, 423
  suppl. 29
– as solvent 7, 61 suppl. 29
1,3-Ammoniooximes
– startg. m. f.
  α,β-ethyleneoximes
  27, 263
  Δ²-isoxazolines 27, 263
6-Ammoniopurinides, N-alkyl
  migration 22, 346 suppl. 26
Ammoniosulfonyl imides
– special s.
  carbalkoxyammoniosulfonyl
  imides
Ammonium acetate 6, 772
  suppl. 26; 26, 188, 371, 805;
  27, 412; 30, 264
– as reactant 30, 34, 269
– startg. m. f.
  enaminolactams 29, 367
  imidazo[1,2-a]pyrazine ring
  26, 431

quinazolines 30, 223
  1,3,5-triazines 27, 365
– acetates, quaternary
– startg. m. f.
  ethers 26, 182
– azide 26, 428
– carboxylates, quaternary
– startg. m. f.
  carboxylic acid esters 26, 182
– cerium(IV) nitrate 4, 185
  suppl. 26; 9, 347 suppl. 28;
  18, 862 suppl. 26; 24, 196
  suppl. 28; 26, 260, 927; 27, 143;
  28, 70a, 132, 284, 591; 29, 89,
  115, 643
– chloride 24, 354 suppl. 26;
  26, 28; 27, 88; 28, 47, 248,
  362, 988; 29, 4, 32, 75, 148,
  257, 758
– compds., quaternary
– as reagents and catalysts,
  review 28, 795
– halides, – s. under Phase
  transfer catalysts
– hydroxides, –
– as intermediates 7, 811
  suppl. 26
– from
  ammonium salts, quaternary
  26, 983
–, Hofmann degradation,
  abnormal 26, 983
– startg. m. f.
  carboxylic acid esters 26, 182
  ethers 26, 182
– molybdate 26, 3; 29, 99
– nitrate 13, 916 suppl. 28;
  28, 976
– perchlorate 28, 358
– persulfate 27, 907
– salts, cyclic
  s. a. Cyclimmonium salts
– –, quaternary
  (s. a. N-Quaternization
– from
  amines, prim. 26, 419
  –, sec. 26, 419
  nitro compds. 27, 428
– special s.
  acetyleneammonium salts,
  quaternary
  N-acylammonium –, –
  carbalkoxyammonium –, –
  1,1,1-dialkoxyammonium –, –
  (1,3-dithiolan-2-yl)-
  ammonium –, –
  silylammonium –, –
  sulfonioammonium –, –
  tetraethylammonium salts
  ynammonium –, quaternary
– startg. m. f.
  acylamines 28, 463
  –, sec. 29, 440
  amines, tert. 29, 992
  ammonium hydroxides,
  quaternary 26, 983
– –, –, cyclic
–, ring contraction of N-heterocyclics via – 26, 297
– –, –, inner s. Betaines
– sulfate 28, 661
– sulfite 12, 869 suppl. 26
– thiocyanate 26, 323, 413;
  27, 19; 28, 396; 30, 258
– ylids, Stevens rearrangement
  27, 748; 30, 531
Amyl alcohol as antifoam agent
  26, 51

tert-Amyl alcohol as reagent
  27, 50
Amyl nitrite s. Alkyl nitrite
Anchimeric... s. a. Neighboring
  group...
– process 26, 349
Angular s. Annelation, angular,
  Methyl groups, –
Anhydrides, mixed s. Borinic
  carboxylic anhydrides,
  Carboxylic alkoxyformic –
Anhydro-N-carboxy-α-amino-
  carboxylic acids
  s. 2,5-Oxazolidiones
Anhydronucleosides
  s. Cyclonucleosides
Anhydrosugar acetals 26, 216
1,6-Anhydrosugars
– from
  glycosides 26, 241
2,3-Anhydrosugars 29, 101
Aniline as medium 4, 413
  suppl. 26
Anilines (s. a. Amines, ar.)
– from
  7-azabicyclo[2.2.1]-2,5-hepta-
  dienes 26, 719
  pyrroles 26, 719
  stilbenes, diamino- 29, 93
– startg. m. f.
  2,5-cyclohexadienone imines,
  4-alkoxy- 26, 204
  indoles 28, 762
–, 2,6-dicyano-
– from
  alkylidenemalononitriles
  (2 molecules) 28, 874
Anion-anion condensation
  28, 734
Anion exchangers (s. a. Amberlite IRA 400/402, Dowex 1-X2-
  (OH), Dowex 21-K(Cl)) 5, 664
  suppl. 29; 20, 552; 26, 212, 426,
  983, 996; 29, 520
– –, strongly basic 24, 663
  suppl. 28
– –, weakly basic 23, 789
  suppl. 28
– radicals 29, 915
Anions (s. a. Cycloaddition,
  anionic, Metalation)
– special s.
  acetate ion
  azomethine-stabilized anions
  dianions
  dienolate ions
  trianions
Anisole
– as scavenger 29, 17a
– as solvent 29, 326
Annelation 28, 837
– of
  quinones 28, 741
– special s.
  benzene ring annelation
  cycloalkenone – –
  cyclohexadiene – –
  cyclopentane – –
  lactone annelation
  Robinson –
  siloxycyclopentene –
  spiro- –
  steroid D ring –
– via alkoxalyl compds. 27, 903
– with
  dicarboxylic acid anhydrides
  1, 581

o-dicarboxylic acid esters
28, 723
α,β-ethylene-α-silylketones
29, 666
- with simultaneous
ring expansion of ketones,
cyclic **26,** 863
-, **ar. 29,** 883
-, **double,** with 2,4-diene-
carboxylic acid esters **29,** 732
-, **linear 28,** 629
-, - **and angular** s. *lin-ang-*
Rearrangement, *ang-lin-
peri*-**Annelation**
- with 1,3-dienamines, cyclic,
and α,β-ethyleneoxo compds.
29, 764
**Anomerization of carbohydrate
derivs. 29,** 215
- **of**
glycosyl hydrazines **28,** 289
Ansa compds. s. Diansa
compds.
Anthracene s. a. Sodium/-
Anthracenes
- from
anthracenes, 9,10-dihydro-
9-nitro- **28,** 644
-, **9-subst. 28,** 644
-, **9-alkoxy-,** protection of
alcohol groups as - **28,** 194
-, **9,10-dihydro-**
- from
anthraquinones **30,** 32
-, **9,10-dihydro-9-nitro-**
- startg. m. f.
anthracenes **28,** 644
-, **9-nitro-**
-, addition to compds.,
CH-acidic **28,** 644
Anthranils s. 2,1-Benzisoxazoles
Anthranol esters 29, 921
- startg. m. f.
anthrones **29,** 921
Anthraquinone
- as triplet sensitizer **26,** 990
- **ring**
-, cleavage, hydrolytic **26,** 220
Anthraquinones
- from
cyclobutadiene rhodium
triarylphosphine complexes
26, 906
1,4-naphthoquinones, 2-halo-
geno- and keteneacetals
29, 788
- startg. m. f.
anthracenes, 9,10-dihydro-
30, 32
-, **2-aminomethyl- 27,** 878
-, **2-hydroxymethyl- 27,** 878
**Anthrasteroid rearrangement
28,** 906
Anthrasteroids 21, 923 suppl. **26
Anthrones** via anthranol esters
29, 921
9-Anthronyl ethers, 9-phenyl-
-, protection of alcohol groups
as - **26,** 177
**Anthr[2,1-d]oxazole-6,11-diones
26,** 246
9-Anthrylmethyl as protective
group **29,** 3
Anthyridines, 6-hydroxy- **26,** 465
- from
pyrimido[1,2-a][1,8]naph-
thyrid-10-ones **26,** 465

Antifoam agents s. Amyl
alcohol
Antimonanes, pentaalkyl-
- startg. m. f.
antimonanes, tetraalkyl-
28, 79
-, **tetraalkyl-**
- from
antimonanes, pentaalkyl-
28, 79
Antimony chlorides 23, 544
suppl. **29**
- compds. s. a. Hexachloroantimonates
- -, **organo-** s. Antimonanes,
Stib...
- **pentachloride 8,** 748 suppl. **29;
26,** 751; **27,** 604; **28,** 826, 993;
29, 996
- - -**graphite 29,** 521
- - -**liq. sulfur dioxide 28,** 612
- **pentafluoride 27,** 159
- **pentahalides 30,** 411
- **pentasulfide 6,** 571 suppl. **28**
- **trichloride 17,** 148; **29,** 1
Apiezon 27, 335
Aporphine ring opening,
prevention **29,** 206 suppl. **30
Aporphines**
7-oxoaporphines **20,** 672
suppl. **29**
- by cyclodehydrohalo-
genation **29,** 956
- by Pschorr ring closure
23, 213 suppl. **29
Arbuzov-type rearrangements,**
review **12,** 167 suppl. **29
Arbuzov-Michaelis-type
rearrangement 30,** 201
Arenecarboxylic acids
s. Arylcarboxylic acids
Arene oxides 28, 221
- special s.
cis-benzene trioxide
- - -**oxepins,** review
23, 945 suppl. **29
Arenes** (s. a. Aryl derivs.)
- by high-temp. reduction with
lithium tetrahydrido-
aluminate **30,** 48
-, cycloaddition to - **26,** 720
- from
aryl carbamates **15,** 85
suppl. **28**
- sulfonates **29,** 65
cyclimmonium salts,
N-condensed **28,** 982
isoureas, O-aryl- **30,** 50
phenols **29,** 65; **30,** 48, 50
-, hydrogenation, partial **4,** 68
- special s.
alkyl-arenes
nitro-
- startg. m. f.
aryl isopropyl carbonates
26, 170
benzofuran ring **28,** 760
phenolesters **26,** 166
quinones **25,** 389 suppl. **27**
thiolic acid esters **29,** 835
Arenesulfonylimidazolides
as reagent **17,** 169 suppl. **28
Argon 26,** 102
- **laser 30,** 80
Aromatization (s. a. Dehydrogenation, Isoaromatization)
27, 916; **28,** 884

- by
acid-base catalysis **26,** 946
dehydration of alcohols
26, 866
dehydrogenation **30,** 640
dehydrohalogenation
22, 959 suppl. **28**
elimination of C-bridges
21, 725 suppl. **27**
- of
steroid rings s. Steroid
ring aromatization
- to
benzo[g]pteridines **26,** 932
heterocyclics, 5-membered
30, 637
naphtho[1,2-b]thiophenes
30, 641
phenols **30,** 526
- with
C-alkyl migration **28,** 921;
29, 946
C-cleavage **23,** 977
N-deacylation **26,** 498
N-debenzylation **27,** 990
elimination of angular methyl
groups **28,** 969; **29,** 986
- of CO-bridges **28,** 977
- of carbalkoxymethyl groups
29, 981
lactam rearrangement **28,** 748
oxidation of methylene
to keto groups **29,** 179
rearrangement, skeletal
28, 894; **29,** 906
ring opening of N-hetero-
cycles **28,** 336
- - of isocyclics **17,** 777
suppl. **29**
-, **decarboxylative, solid-state
28,** 972
-, **reductive,** of steroid B ring
29, 986
Aroxylamines 15, 26 suppl. **28**
- startg. m. f.
coumarins, 3-amino- **29,** 725
N-Aroylsulfonic acid amides
- from
aryl methyl ketones **26,** 451
Arsabenzenes 28, 978
Arsenanes
- from
arsine dihalides **30,** 440
Arsenious acid esters
- startg. m. f.
arsines, tert. **23,** 657
suppl. **27**
Arsine dihalides
- startg. m. f.
arsenanes **30,** 440
- **imines**
- from
arsines **30,** 199
- startg. m. f.
azo compds. **28,** 240
- **oxides**
- from
arsine sulfides **29,** 119
- startg. m. f.
arsonium ylids **29,** 608
Arsines
- from
amines **29,** 609
- special s.
amino-arsines
halogen-

(Arsines)
- startg. m. f.
  arsine imines **30,** 199
-, tert.
- from
  arsenious acid esters **23,** 657
  suppl. **27**
Arsine sulfides
- startg. m. f.
  arsine oxides **29,** 119
Arsinic acids
- startg. m. f.
  C,As-dihalogenarsines **30,** 344
Arsonium salts
- startg. m. f.
  oxido compds., synthesis **16,** 893 suppl. **26**
- ylids **16,** 893; **29,** 608
- from
  arsine oxides **29,** 608
Arylacetic acid esters
-, α-alkylation **30,** 540
- from
  aldehydes, synthesis with addition of 1 C-atom **28,** 735
  aryl methyl ketones **27,** 162
  1-sulfinylthienolethers **28,** 735
- acids
-, self-acylation **26,** 371
- startg. m. f.
  isoquinol-3(4H)-ones (from 2 molecules) **26,** 371
Arylacetonitriles
- startg. m. f.
  2-benzopyrylium salts, 3-acylamino- **27,** 809
  isoquinoline ring, 1,3-dihalogeno- **26,** 824
Aryl acetylenesulfoxides
- startg. m. f.
  thianaphthenes, 2,3-dihydro-3-acyl- **27,** 755
trans-**2-Arylalcohols**
- from
  ethylene derivs. and arylmercury chlorides **28,** 865
Arylallenes
- startg. m. f.
  naphthalenes, 1,2-dihydro- **26,** 757 suppl. **29**
Arylamines s. Amines, ar.
γ-**Aryl**-γ-**aminocarboxylic acids** **26,** 120
α-**Arylaminoketones**
- startg. m. f.
  indoles, 2-aryl- **28,** 785
C-Arylation (s. a. Heck arylation)
- of
  carbanions **30,** 540
  β-dicarbonyl compds. **30,** 593
  ethylene derivs. **27,** 871
  - -, review **23,** 757 suppl. **28**
  heterocyclics **29,** 7
-, angular, stereospecific **23,** 693 suppl. **29**
-, homolytic
- of
  heterocyclics, review **16,** 851 suppl. **27**
-, oxidative
- of
  ethylene derivs. **23,** 757 suppl. **28**
  N-heterocyclics **26,** 774
  ketones **30,** 542
-, reductive **30,** 678

N-Arylation of azoles,
review **13,** 496 suppl. **26**
- with tricarbonylcyclohexadienoneiron compds. **30,** 266
Arylboronic acids
- from
  arylthallium bis(trifluoroacetates) **27,** 232 suppl. **30**
Aryl N-carbalkoxycarbamates
- startg. m. f.
  1,3-benzoxazine-2,4-diones **28,** 898
  o-hydroxycarboxylic acid amides **28,** 898
- carbamates
- startg. m. f.
  arenes **15,** 85 suppl. **28**
  o-hydroxycarboxylic acid amides **28,** 678
Arylcarboxylic acids
- by dienone-benzene rearrangement **26,** 799
-, ring hydrogenation, partial **26,** 44
- special s.
  arylacetic acids
- startg. m. f.
  phenols **17,** 298 suppl. **26**
Aryl derivs. (s. a. Arenes, Isocyclics, Ring..., Substituents, ar.)
-, C-acylation s. Friedel-Crafts ketone synthesis
-, C-alkylation **27,** 867
-, - with azidoformic acid esters **21,** 815
-, - with diazo compds. **25,** 604
- from
  oxo compds. **30,** 673
- special s.
  adamantanes, polyarylcyclobut-2-enones, 2-hydroxy-4-aryldiaryl...
  stilbenes
  styr...
- glycosides **19,** 201 suppl. **26;** **27,** 218; **19,** 201 suppl. **29**
- halides (s. a. Halides, ar.)
- special s.
  chlorobenzene
  dichlorobenzene
- startg. m. f.
  diaryls, sym. **27,** 870; **28,** 792; **30,** 592
  phenolethers **21,** 237; **30,** 134
- isopropyl carbonates
- from
  arenes **26,** 170
- isothiocyanates as reagents **27,** 360
- ketones
-, alkylation, reductive **27,** 839 suppl. **28**
- from
  halides and mandelonitriles, O-(tetrahydropyran-2-yl)- **29,** 888
  oxocarbonium hexafluoroantimonates **27,** 652 suppl. **30**
- special s.
  acetophenones
  o-aminoketones
  aryl methyl ketones
  o-(N-nitrosamino)ketones
  o-(1,2,4-triazol-4-yl)ketones

- startg. m. f.
  naphthalenes **28,** 639
  1-naphthols, 1-4-dihydro- **28,** 639
Aryllead tricarboxylates
-, reactions **29,** 878
- startg. m. f.
  diaryls **29,** 878
Arylmercaptans 29, 30
- special s.
  p-chlorobenzenethiol
  thiophenol
- startg. m. f.
  imidazoles, 5-arylthio- **29,** 551
Aryl methyl ketones
- startg. m. f.
  N-aroylsulfonic acid amides **26,** 451
  arylacetic acid esters **27,** 162
- migration s. a. Stevens aryl migration
C-Aryl – **26,** 732; **28,** 920
3-Aryloxo compds.
- from
  2-ethylenealcohols **24,** 884 suppl. **27**
Aryloxy... s. a. Phenolethers
Aryloxyacetylenes
- special s.
  1,4-diaryloxy-2-butynes
Aryloxycarbonyl...
  s. a. Carbaryloxy...
Aryloxycarbonyloxy groups,
  substitution, nucleophilic **29,** 224; **30,** 150
α-**Aryloxycarboxylic acid amides**
- startg. m. f.
  benzofuran ring, 3-amino- **29,** 936
Aryloxycarboxylic acids
- startg. m. f.
  hydroxycarboxylic acid aryl esters **29,** 161
Aryloxymagnesium halides
-, reactions with orthoformic acid esters **28,** 742
- special s.
  mesitoxymagnesium bromide
Aryloxysilanes
- from
  phenolethers **29,** 602
- startg. m. f.
  phenolethers **26,** 217
  phenols **29,** 602
Aryloxysulfonylisocyanates **29,** 448
1-Aryloxy-1-sulfonyloxy compds. 27, 111
Arylphosphonic acid esters
- from
  phosphites and halides, ar. **26,** 656
Aryl sulfonates
- startg. m. f.
  arenes **29,** 65
- sulfoxides
- startg. m. f.
  alkyl sulfoxides **29,** 582
Arylthallium bis(trifluoroacetates)
- startg. m. f.
  arylboronic acids **27,** 232 suppl. **30**
  biphenyls, unsym. **27,** 888
  phenols **27,** 232 suppl. **30**
  thiocyanates, ar. **27,** 888

Aryl thioethers
s. Thioethers, ar.
1,1-Arylthiohydrazones
- from
  1,1-halogenohydrazones
  29, 565
Arylthiomethylation 26, 844
Aryl trichloromethyl carbinols
-, reactions with nucleophiles,
  review **22,** 629 suppl. **26**
Arynes s. Benzyne..., Hetarynes
Ascarite 11, 963 suppl. **28**
Ascorbic acid as reagent **30,** 651
Asscher-Vofsi reaction 21, 642
Assistance, intramolecular
s. Neighboring group
participation
Asym. s. Aldol condensation,
intramolecular, asym.,
Hydration, asym., Hydrogenation, -, Reduction, -,
Synthesis, -
Asymmetry s. a. Chirality
Ate complexes, s. Lithium
alanates, - cuprates
Atherton s. Todd
C-Atoms, labeled s. Compds.,
labeled
-, **quaternary** s. Hydrocarbons,
branched
Aurones
- from
  flavanones **26,** 928
Auto... s. a. Self...
Autoxidation 26, 463
- of
  boranes **27,** 138
  carboxylic acid derivs. **28,** 82
  suppl. **30**
  α-cyanocarboxylic acid
  esters **27,** 236
  hydrocarbons **27,** 917
  lactones **30,** 63
  phenols **28,** 130
-, ring opening by - **26,** 131
-, copper salt-amine catalyzed
  28, 130
Avoidance s. Prevention
Axial s. Diaxial
Aza... s. a. N-Heterocyclics
2-Azaadamantanes 29, 313
4-Azaazulenes 26, 819
7-Azabicyclo[2.2.1]-2,5-heptadienes
- from
  pyrroles **26,** 719
- startg. m. f.
  anilines **26,** 719
9-Azabicyclo[3.3.1]nonane-1,5-diol ring 30, 205
3-Azabicyclo[3.3.1]non-6-enes, 1,5-dinitro-
- from
  benzenes, 1,3-dinitro- **26,** 48
  3,5-bis-*aci*-nitrocyclohexenes
  26, 48
Azabullvalenes 28, 702
Azacyanines 27, 446
Azacyclols 26, 119
- by Vilsmeier reaction, intramolecular **28,** 689
1-Aza-4,6-dioxabicyclo[3.3.0]-octanes 29, 407
4-Azadioxindoles
- from
  4-azaoxindoles **27,** 151
2-Azanaphtoquinones 26, 468

4-Azaoxindoles
- startg. m. f.
  4-azadioxindoles **27,** 151
3-Azaphospholes, 4,5-dihydro-26, 639
2-Azapurine nucleosides 17, 341
suppl. **27**
- from
  purine nucleosides **27,** 23
2-Azapyrylium salts, 6-alkoxy-
- from
  1,2-oxazin-6-ones **30,** 110
1-Azaspiro[2.2]pentanes
- startg. m. f.
  cyclobutanes, imino- **30,** 535
Azasteroids 28, 122
- special s.
  diazasteroids
Azeotropic entrainment
(s. a. Water entrainment)
28, 439
Azepine ring
- from
  acetylenedicarboxylic acid
  esters (2 molecules) **28,** 635
- -, **dihydro-** from dialdehydes
  19, 470
1H-Azepines
- from
  benzene ring, 1,4-dihydro-
  26, 968
  benzenes **27,** 420
3H-Azepines
- from
  Δ¹-azirines and cyclones
  28, 872
4H-Azepines
- from
  1,2,4-triazines **26,** 833
3H-Azepines-, 2-amino-
- from
  nitrobenzenes **26,** 379
2,4-Azetidinediones 27, 675
Azetidine ring
- from
  ethylene derivs. **29,** 659
Azetidines
- from
  3-aminoalcohols **26,** 471
  suppl. **30**
-, **N-tosyl- 29,** 470
Azetidinium ring
- from
  3-aminoalcohols **27,** 509
3-Azetidinols 26, 744
2-Azetidinone ring
-, 3-alkoxylation **29,** 177
- by
  Wolff rearrangement **13,** 681
  suppl. **28**
- startg. m. f.
  aziridine ring **26,** 441
-, 3-substitution **29,** 806
-, **3-amino-**, N-condensed
  30, 530
-, **3-carboxy- 30,** 90
- - **exchange 29,** 893
- - **expansion 29,** 706
2-Azetidinones
(s. a. Penicillin ring skeleton)
- by ring opening, desulfurative
  26, 691 suppl. **29**
- from
  acetylene derivs. and nitrones
  (*cis*-2-azetidinones) **28,** 650
  azomethines and α-halogenocarboxylic acid esters
  28, 816

1-halogenenamines **28,** 822
suppl. **29**
thioiminoesters, *cis*-2-azetidinones **7,** 836 suppl. **27**
-, removal of N-substituents
29, 22
- special s.
  N-chlorosulfonyl-2-azetidinones
- startg. m. f.
  4(3H)-pyrimidinone ring,
  5,6-dihydro-, N-condensed
  26, 347
-, synthesis, review **7,** 836
suppl. **28**
-, **4-alkylthio-**
- from
  ketenes and thioiminoesters
  8, 736
-, **3-azido-**
- startg. m. f.
  Δ³-2-imidazolones **28,** 451
-, **N-halogeno-**
- startg. m. f.
  2-halogenisocyanates **28,** 482
-, **3-halogeno-**
- from
  2-aziridinecarboxylic acids
  30, 357
-, **4-imino-**
- from
  ketenes and carbodiimides
  26, 692
  ketenimines and isocyanates
  26, 692 suppl. **28**
Azetines
- from
  cyclopropyl azides **27,** 524
-, **3-imino-**
- from
  1,3,5-oxazaphosph(V)ol-2-ines **27,** 887
Azides
- from
  alcohols **30,** 255
  amines **27,** 438
  benzenesulfonic acid esters
  27, 401
  ethylene derivs. **30,** 222
  halides **26,** 422
  nitric acid esters **27,** 400
-, reactions with mercaptans
  27, 296
-, ring contraction **26,** 958;
  30, 664
-, - -, transannular **27,** 984
- special s.
  carboxylic acid azides
  cyclopropyl -
  di-
  ethylene-
  phenyl azide
  silyl azides
- startg. m. f.
  aldehydes **28,** 149
  amines **27,** 401; **29,** 324, 424
  -, sec. **27,** 485
  -, -, ar. **29,** 402
  aziridine ring **22,** 123 suppl. **28**
  azomethines with rearrangement **29,** 460
  carbalkoxyhyrazones **30,** 191
  carbodiimides, unsym. **27,** 447
  enamines **26,** 983
  enetriazenes **26,** 893
  nitriles with ring contraction
  26, 958

(Azides
- startg. m. f.)
  phosphoramidates **27,** 311
  pyridines, 1,4-dihydro-
  20, 389 suppl. 27
  succinimides **29,** 765
  sulfenamides **27,** 296
  1,2,3,4-thiatriazolines,
  5-sulfonylimino- **30,** 221
  Δ²-1,2,3-triazolines,
  5-amino- **26,** 343
-, **tert. 27,** 400
2-Azidoalcohols
- from
  oxido compds. **30,** 278
- startg. m. f.
  1,2-azidosulfonates **30,** 278
Azidoarylalkoxyl
- as O-protective group,
  photosensitive **27,** 6
Azidoazomethine-tetrazole tautomerism
- s. Tetrazole-azidoazomethine tautomerism
Azidocarboxylic acid esters
- special s.
  azidoethylenecarboxylic acid esters
- startg. m. f.
  α-azido-α,β-ethylenecarboxylic acid esters **26,** 789
α-Azido-α,β-ethylenecarboxylic acid esters
- from
  aldehydes **26,** 789
α-Azido-α,β-ethyleneketones
- from
  β-azido-α-iodoketones **26,** 970
  α,β-dihalogenoketones **26,** 414
Azidoformic acid esters
-, C-alkylation of arenes with – **21,** 815
- startg. m. f.
  Δ²-oxazolines, 2-alkoxy-4-imino- **27,** 432
  urethans **27,** 437
1,2-Azidohalides
- startg. m. f.
  aziridine ring **26,** 405
  – –, N-subst. **29,** 439
  N-(aziridinyl)phosphonates **26,** 405
Azidohalogenoketones
  s. Azidoiodoketones
Azido-N-heterocyclics
- from
  diazo-N-heterocyclics **30,** 185
Azidohydroxynitriles
- startg. m. f.
  O,N-heterocyclics, double ring closure **29,** 464
β-Azido-α-iodoketones
- startg. m. f.
  α-azido-α,β-ethyleneketones **26,** 970
Azidoketones
- special s.
  azidohalogenoketones
α-Azidoketones
- from
  ethylene derivs. **27,** 362
- startg. m. f.
  oxazoles **27,** 477
  pyrazines **26,** 399
Azidolactones
- startg. m. f.
  aminolactones **29,** 324
  hydroxylactams **29,** 324

1,2-Azidonitric acid esters
- from
  ethylene derivs. **28,** 284
Azidonitriles
- special s.
  azidohydroxynitriles
o-Azido-N-oxides
- startg. m. f.
  o-cyano-N-hydroxy-N-heterocyclics, ring contraction **28,** 452
P-Azidophosphonium salts 21, 335 suppl. 29
- special s.
  azidotris(dimethylamino)phosphonium hexafluorophosphate
Azidoquinone ring
- startg. m. f.
  1,3-cyclopentenedione ring, 2-cyano- **26,** 958
Azidoquinones, chemistry, review **30,** 400
-, ring closures with – **30,** 400
Azidosilanes
- special s.
  trimethylsilyl azide
1,2-Azidosulfonates
- from
  2-azidoalcohols **30,** 278
- startg. m. f.
  aziridine ring **21,** 564; **30,** 278
Azidotris(dimethylamino)phosphonium hexafluorophosphate
-, peptide synthesis with – **28,** 144 suppl. 29
Azimines, insertion into carbon-carbon triple bonds **28,** 281
Azine N,N'-dioxides
  (s. a. Benzo[c]cinnoline 5,6-dioxides)
- from
  oximes **26,** 260
Azines
- from
  oxo compds. **27,** 421; **30,** 566
- special s.
  benzalazine
  halogenazines
  ketazines
- startg. m. f.
  α-aminophosphonic acid esters **29,** 610
  azo compds. **26,** 16
  1,5-diazabicyclo[3.3.0]octanes **28,** 642
  pyrazoles **26,** 740
  pyrrole ring **26,** 740
  1,3,4-thiadiazolidines **30,** 566
  -, sym. **27,** 483
Aziridides s. β-Amino-α,β-ethylenecarboxylic acid aziridides
1-Aziridinecarboximidoyl chlorides
- startg. m. f.
  carbodiimides, chloro- **27,** 556
2-Aziridinecarboxylic acids
- startg. m. f.
  2-azetidinones, 3-halogeno- **30,** 357
Aziridine ring
- from
  2-azetidinone ring **26,** 441
  azides and ethylene derivs. **22,** 123 suppl. 28
  1,2-azidohalides **26,** 405

1,2-azidosulfonates **21,** 564; **30,** 278
α,β-ethyleneoximes **29,** 14
oxido compds. **30,** 278
- startg. m. f.
  β-amino-α,β-ethylenecarboxylic acid aziridides (from 2 molecules) **26,** 306
  heterocyclics, ring expansion
  oxazolidines **24,** 692 suppl. 26
- –, **N-acyl-**
- from
  ethylene derivs. and carboxylic acid azides **29,** 398
- –, **N-condensed**
- from
  ethyleneamines **27,** 500
- –, **N-subst.**
- from
  1,2-azidohalides and dihalogenoboranes **29,** 439
- –, **N-unsubst.**
- from
  N-(aziridinyl)phosphonates **26,** 405
- –, **N-cyano-**
- from
  ethylene derivs. **26,** 521
  2-halogenocyanamides **26,** 521
- – **expansion**
-, heterocyclics by –, review **26,** 706
-, review **26,** 706
- – **opening 26,** 118; **28,** 290
- with
  carboxylic acids and derivs. **26,** 296
- – **rearrangement 29,** 319
- special s.
  N-phosphorylaziridines
- startg. m. f.
  2-acylaminoalcohols **26,** 296
  amines **26,** 55
  ethylene derivs. **26,** 964
  2-halogenamines **30,** 346
  Δ²-imidazolines **30,** 217
  Δ⁴-thiazolines, 4-amino- **26,** 706
-, **N-acyl-**
- startg. m. f.
  Δ²-thiazolines **27,** 615
-, **N-amino-** as reactants **22,** 913 suppl. 27
Aziridines (s. a. Ethylenimine)
-, fragmentation **26,** 306
- from
  alkoximes **21,** 538 suppl. 29
  2-aminoalcohols **26,** 471
  Δ¹-azirines **29,** 38
  azomethines **28,** 818
  ethylene derivs. **28,** 458
  2-halogenacylamines **26,** 60
  2-halogenamines **28,** 433
  1,2-halogenurethans **28,** 458
  2-tosyloxyamines **28,** 433
-, **2-cyano-**
- from
  α,β-ethylene-α-halogenonitriles and amines, prim. **27,** 468
-, **N-cyano-**
- from
  ethylene derivs. **28,** 282
-, **2,2-dihalogeno-**

- from
 azomethines **30,** 581
- startg. m. f.
 amidines **28,** 388
-, **C-hydroxyamino-**
- from
 Δ¹-azirines **28,** 276
-, **N-sulfonyl-** 10, 400
-, **2-vinyl-**
- startg. m. f.
 1,3-thiazepines, 4,7-dihydro-
  2-amino- **28,** 526
Aziridinium salts
- special s.
 spiroaziridinium salts
Aziridinones
- startg. m. f.
 α-aminocarboxylic acid esters
  27, 192
 α-halogenocarboxylic acid
  amides **28,** 474
 α-hydroxyazomethines **30,** 612
N-(Aziridinyl)phosphonates
- from
 1,2-azidohalides **26,** 405
- startg. m. f.
 aziridine ring, N-unsubst.
  26, 405
Δ¹-**Azirine-3-carboxamides**
- from
 isoxazoles, 5-amino- **29,** 322
Δ¹-**Azirine-3-carboxylic acid
  esters**
- from
 isoxazoles, 5-alkoxy- **26,** 313
Δ¹-**Azirine ring**
- startg. m. f.
 indole ring **27,** 335
Δ¹-**Azirines**
- as dipolarophiles, strong
 26, 706 suppl. **28**
- from
 ethylene derivs. **18,** 331
  suppl. **27**
-, ring expansion **28,** 643
- startg. m. f.
 3H-azepines **28,** 872
 aziridines **29,** 38
 -, C-hydroxyamino- **28,** 276
 enazides **30,** 457
 Δ³-oxazolines **30,** 458
 Δ³-5-oxazolones **28,** 643
 pyridines **26,** 771 suppl. **29**
 4-pyridones **28,** 634
 Δ¹-pyrrolines **26,** 708
 1,2,4-triazepines **29,** 770
-, **2-amino-**
- from
 1-halogenenamines **27,** 462
-, **2-carbonyl-**
- startg. m. f.
 1,2,4-triazin-6-ones,
  tetrahydro- **28,** 371
Azlactones s. 5-Oxazolones
1-Azoalcohols
- from
 1,1-azohydroperoxides **30,** 10
 hydrazones **30,** 10
(α-**Azoalkylidene)-
 phosphoranes**
- from
 1,1-hydrazonohalides and
  phosphines **29,** 611
α,α'-**Azobis-(N-methyl-
 formamide)** as reagent **29,** 585
Azo bridges, elimination
 s. under Elimination

Azocarboxylic acid esters
- special s.
 N-carbamylazoformic acid
  esters
 cyanoazocarboxylic – –
Azocine ring, 2-alkoxy- 26, 929
Azocines
- from
 1,2,4-triazines **26,** 918
-, **3,4-dihydro-**
- startg. m. f.
 2,3-pyridocyclobutenes **26,** 941
Azo compds.
-, fragmentation **27,** 390
- from
 azines **26,** 16
 diazo-N-sulfides **30,** 198
 N-halogenamines **27,** 284
 hydrazo compds. **26,** 265
 nitroso compds. and arsine
  imines **28,** 240
 ureas **27,** 288
-, hydrogen abstraction by –
 26, 461
-, photochemistry, review
 22, 984 suppl. **28**
-, retrocoupling, oxidative **29,** 284
- special s.
 acetyleneazo compds.
 1H-1,4-diazepinium salts,
  6-azo-
 hydroxyazo compds.
 nitroazo –
 oxyazo –
 sulfonyloxyazo –
- startg. m. f.
 acylhydrazines **28,** 271
 amines **29,** 17
 azoxy compds. **27,** 97
 diazenium salts **29,** 994
 hydrazines, trisubst. **27,** 319
 phosphoric acid diamide
  hydrazides **30,** 200
 sulfones **26,** 604
 sulfonic acid hydrazides
  30, 384
– –, **alicyclic**
- from
 hydrazo compds., alicyclic
  29, 283
– –, **aliphatic**
- from
 azines **4,** 286
- special s.
 acylazo compds.
 azo –, tert.
 enazo –
 1,1-halogenazo –
 phthalans, 1-(azomethyl-
  imino)-
 ynazo compds.
– –, **aliphatic-ar. 29,** 283
- by coupling **9,** 422
- from
 diazo compds. **26,** 284
- startg. m. f.
 hydrazones, subst. **27,** 352
– –, **cyclic**
- by coupling **27,** 354
– –, **sym.**
- from
 amines **28,** 237
 diazonium salts **29,** 279
– –, **tert.**
- from
 1,1'-dihalogenazo compds.
  29, 832
 ketazines **29,** 832

N-Azocyanoboron betaines
 27, 282
Azodicarboxylic acid
 s. Potassium azodicarboxylate
– – **ester adducts**
- of
 steroids, anthrasteroid
  rearrangement **28,** 940
 -, B ring aromatization
  29, 986
-, substituting addition **30,** 208
 addition to enamines
  20, 502 suppl. **26**
– – **esters**
-, addition to ethylene derivs.
 21, 349 suppl. **27**
- special s.
 diethyl azodicarboxylate
Azodienophiles
-, bispyrazolones as –
 22, 314 suppl. **5**
α-**Azo-α,α-dihalogenoketones**
- from
 1,2,3-triketone 2-mono-
  hydrazones **28,** 422
- startg. m. f.
 dihalogenomethylene-
  hydrazones **28,** 422
Azodiisobutyronitrile
- as reactant **27,** 813
- as reagent **18,** 585 suppl. **29;**
 25, 64; **28,** 229, 534;
 30, 426, 697
Azodinitriles s. Azodiisobutyronitrile
Azo N,N'-dioxides
- from
 hydrocarbons **26,** 327
 hydroxylamines, aliphatic
  27, 100
- special s.
 2-nitrosamine dimers
-, **cyclic 27,** 100 suppl. **30**
1-Azoethers
- from
 hydrazones **27,** 185
Azoformamidines
- special s.
 N'-cyanoazoformamidines
Azo groups, N,N-cleavage
 27, 442
-, substitution, electrophilic
 30, 287
1,1-Azohydroperoxides
- startg. m. f.
 1-azoalcohols **30,** 10
Azoketones
- special s.
 azodihalogenoketones
Azoles (s. a. 1,3-Heterocyclics)
-, N-Arylation, review **13,** 496
 suppl. **26**
- special s.
 benzazoles
 di(azoles)
 imidazoles
 oxazoles
 thiazoles
- startg. m. f.
 amines **29,** 561
 thioethers **29,** 561
Azolide analogs, reactions
 with – **30,** 282
Azolides
- special s.
 1,2,4-triazolides

Azomethine imides, cyclic
– from
　diazo compds. **30**, 283
Azomethines (s. a. Imines)
–, addition of phenols to –
　26, 471
– as substitutes of aldehydes
　23, 435 suppl. **27**
– from
　amines and acetylene derivs.
　　28, 283
　– and alcohols **26**, 372
　– and imines **29**, 404
　– and oxo compds. **28**, 356,
　　371, 390, 504; **30**, 468
–, tert., rearrangement
　27, 751
　azides with rearrangement
　　29, 460
　1,N-dihalogenamines **23**, 534
　α,β-ethyleneazomethines
　　and boranes **30**, 622
　isocyanates **28**, 425
　nitro compds. and oxo
　　compds. **28**, 363
　nitrones **30**, 696
　N-nitrosamines **28**, 453;
　　30, 334
　oxaziridine ring **30**, 328
　5-oxazolidones **27**, 536
　urethans and oxo compds.
　　26, 456
–, Michael addition with –
　29, 668
–, reaction with phosgene
　27, 545
– special s.
　aldimines
　1-alkoxyazomethines
　alkylideneamino...
　amino-azomethines
　cyclopropyl-
　deuterio-
　di-
　ethylene-
　halogeno-
　hydroxy-
　imino-
　indolizines, 1-(alkylidene-
　　amino)-
　ketoazomethines
　oximinoazomethines
– startg. m. f.
　amines, sec., reduction
　　29, 42, 371, 390; **30**, 19
–, –, synthesis **26**, 682; **29**,
　634; **30**, 463
　β-aminocarboxylic acid esters
　　29, 638
　– acids **26**, 682
　β-amino-α,β-ethyleneazo-
　　methines, synthesis
　　30, 461
　β-aminoketones **28**, 620
　α-aminonitriles **27**, 781
　2-aminophosphinic acid
　　esters **26**, 682
　2-azetidinones **28**, 816
　aziridines, 2,2-dihalogeno-
　　30, 581
　carboxylic acid amides
　　26, 124
　1,1-diamines **29**, 299
　1,2-di(azomethines) (from
　　2 molecules) **26**, 683
　diazonium salts **26**, 262

　1,N-dihalogenamines **23**, 534
　β-diketones **26**, 685
　γ-diketones, synthesis **30**, 461
　enamines, endocyclic **27**, 842
　ene-1,2-diamines (from
　　2 molecules) **26**, 683
　ethylene derivs., synthesis
　　28, 767; **13**, 784
　α-halogeno-N-carbonylureas
　　28, 473
　β-iminoazomethines **26**, 685
　indol-3-ones **28**, 829
　isoquinolines, 1,2,3,4-tetra-
　　hydro- **29**, 702
　ketones **27**, 890; **29**, 504, 647
　nitriles **28**, 466
　2-pyridones, 3,4-dihydro-
　　30, 556
　2-pyrrolidones, 4-carboxy-
　　28, 616
　Δ²-pyrrol-4-ones **30**, 473
　quinolines **26**, 723
　spiro[pyrrolidin-3,1′-(pyrro-
　　lidino[3,4-c]pyrrolidines)],
　　2,4′,5,6′-tetraoxo- **27**, 701
　stilbenes **27**, 751
　4H-1,3,5-thiadiazine-4-thiones,
　　5,6-dihydro-2-amino- **27**, 591
　4H-1,3,5-thiadiazin-4-ones,
　　5,6-dihydro-2-amino- **26**, 572
　1,1,N-trifluoroamines **29**, 503
–, **cyclic**
–, C-alkylation **29**, 831
– from
　amines, sec., cyclic **13**, 383
　aminoketones **28**, 438
– special s.
　indolenines
　Δ¹-pyrrolines
– –, **1-subst.**
– from
　N-acyllactams **28**, 465
Azomethine-stabilized anions
　29, 806
Azomethine-type cobalt complex
　s. under Cobalt complex
– **complex salts 19**, 744 suppl. **26**
α-Azomethinioamidinium salts
– from
　1,1,2-triaminoethylenes **28**, 279
Azomethinium salts
　s. a. Immonium salts,
　Mannich reagent)
– special s.
　1-alkylthioazomethinium salts
　amidinium –, vinylogous
　azomethinioamidinium –
　N-dihalogenomethylene-
　　ammonium –
　N-(4-en-Δ²-oxazolin-5-
　　yliden)ammonium –
　halogenomethinium –
–, syntheses with – and sulfur
　compds. **25**, 551 suppl. **29**
–, **cyclic**
– from
　amines, tert., cyclic **30**, 152
　sec-aminoketones **28**, 445
　carboxylic acid amides,
　　N,N-disubst. **30**, 654
　– acids **30**, 654
– startg. m. f.
　amines, tert., cyclic, 1-subst.
　　29, 716
4a-Azoniaanthracenes 26, 955
o-Azonitriles
– from
　isatin-3-hydrazones **29**, 282

Azonium salts s. Oxazonium
　salts
o-Azo-N-oxides
– from
　o-amino-N-oxides **27**, 283
　1,2,3,5-oxatriazolium salts,
　　N-condensed **27**, 283
Azosilanes
– from
　hydrazinosilanes **27**, 285
o-Azosulfenylhalides
– startg. m. f.
　benzothiazoles, 2-acyl- **29**, 567
4-Azo-Δ²-5-thiazolones
– startg. m. f.
　1,2,4-triazoles **28**, 406
Azoxy bridges, elimination
　28, 932
– **compds.**
– by coupling **29**, 277
– from
　azo compds. **27**, 97
　nitro compds. (2 molecules)
　　26, 35, 261
– startg. m. f.
　hydroxyazo compds. **30**, 96
　oxyazo – **4**, 183
　sulfonyloxyazo – **4**, 183
　　suppl. **27**
– –, **cyclic**
– from
　hydrazodicarboxylic acid
　　esters, cyclic **26**, 92
　1,2,4-triazolidine-3,5-dione
　　ring **26**, 92
–, **unsym.**
– from
　nitroso compds. and
　　N,N-dichloramines **29**, 280
Azulene ring 10, 667 suppl. **26**
– by ring closure and expansion
　27, 958 suppl. **29**
– – –, transannular, stereo-
　specific **26**, 749
– from
　indan ring **27**, 226
　naphthalene ring **24**, 920;
　　28, 108; **29**, 980
Azulenes
– from
　2H-cyclohepta[b]furan-2-ones
　　28, 725
– special s.
　styrylazulenes
–, **hydro- 28**, 686
–, synthesis, review **28**, 686
–, –, stereospecific **30**, 476

Bachman s. Gomberg
Backbone rearrangement
　s. Steroid backbone
　rearrangement
Bacteria as reagents **21**, 990
　suppl. **26**
Bäcklund s. Ramberg
Baeyer-Villiger oxidation
　26, 137; **26**, 584 suppl. **28**
Ball mill 28, 198
Bamford-Stevens reaction,
　modified **23**, 935; **27**, 949
–, review **27**, 949 suppl. **30**
Barbituric acid...
　s. a. Pyrimidines
– **acids**
– special s.
　hydroxybarbituric acids
　5-(4-pyridyl)barbituric –
–, **5-carbalkoxy-**

– from
  malonic acid esters and
  2 isocyanate molecules
  28, 736
– –, **5,5-dihalogeno-** **28**, 509
– –, **N-monosubst.** **29**, 408
– –, **2-thio-**, N,S-dialkylation
  29, 408
Barium carbonate 26, 553
– **hydroxide 27**, 228; **29**, 499
– **nitrite 29**, 152
– **oxide 27**, 228
– **peroxide 9**, 243 suppl. **26**
Base (s. a. Acid-base...)
–, fluoride ion as – **26**, 901
–, isonitriles as – **27**, 219
Bases (s. a. Lewis bases,
  Superbases, Handle, basic)
–, effect of – on geospecificity
  of β-elimination **28**, 944
– from their salts **30**, 55
– oxido compds. as –, review
  23, 520 suppl. **28**
–, **anionic, bulky** s. Lithium
  alkoxides, bulky, Potassium –, –
–, **heteroar.** s. N-Heterocyclics
–, **hindered**
– special s.
  amines, tert., hindered
  2,6-di-*tert*-butylpyridine
  lithium 2,2,6,6-tetramethyl-
    piperidide
  2,6-lutidine
– **highly reactive**
  potassium hydride **28**, 884
  suppl. **30**
–, **strong, non-nucleophilic**
– special s.
  1,8-bis(dimethylamino)-
  naphthalene
–, –, **proton-specific**
  lithium 2,2,6,6-tetramethyl-
  piperide **26**, 847 suppl. **28**
ψ-**Bases** s. Pseudobases
Bates reagent 30, 260
Beckmann fragmentation,
  ring opening **19**, 562; **27**, 517;
  29, 515
– **rearrangement 26**, 308
– – of
  nitrones **4**, 180; **28**, 122
  pyrrole ring, 3-nitroso-
  30, 99
–, ring opening **26**, 544
– to benzoxazoles **26**, 505
– –, **neutral 18**, 205 suppl. **29**
– –, **photochemical 26**, 150
– –, **reductive**, ring expansion
  26, 61
Bender's salt 26, 618
Benkeser reduction 11, 83;
  22, 51
–, comparison with Birch
  reduction, review **11**, 83
  suppl. **28**
Benzalacetone as reagent **28**, 209
Benzalazine
– startg. m. f.
  hydrazines, monosubst.
  28, 351
Benzaldehyde as reactant **4**, 408
Benzaldoxime – – **30**, 118
**4,5-Benz-2-azaspiro[5,5]-
  undecane-1,3,9-triones 29**, 665
3-Benzazepine ring
– from
  isoquinoline ring **28**, 345

Benzazepine –, dihydro- 27, 505
– startg. m. f.
  rheadan ring skeleton **27**, 505
1(1H)-Benzazepines
– from
  indoles **26**, 705
**3H-2-Benzazepin-3-ones, 1,2,4,5-
  tetrahydro- 27**, 801
**3H-3-Benzazepin-2-ones, 1,2,4,5-
  tetrahydro- 22**, 944 suppl. **27**
**1-Benzazocines, 1,6-dihydro-
  16**, 773 suppl. **28**
Benzazoles
– by ring closure, oxidative
  24, 242 suppl. **29**
–, **2-aryloxy-**, rearrangements
  14, 401 suppl. **29**
**Benz[a]azulenes, 7,10-dihydro-
  27**, 899
Benzene
– as solvent **1**, 321 suppl. **27**;
  27, 671
m-**Benzenedisulfonic acid**
  as catalyst **12**, 205 suppl. **29**
**Benzenephosphonic acid esters,
  2,3-dihydroxy-**
– from
  o-quinones and dialkyl
  phosphites **29**, 592
Benzene ring (s. a. Arenes,
  Cyclohex...) **28**, 928
–, 1,3-addition to – **30**, 476
– from
  acetylene derivs. (2 molecules) and ketones
  26, 785
  cyclones **29**, 895
  enamines **22**, 877 suppl. **28**
  α,β-ethyleneoxo compds.
  27, 805
  N-nitrosoureas **27**, 955
  oxidoalcohols, cyclic
  29, 946
  pyran ring **30**, 655
  pyridine N-oxide ring **29**, 775
  p-quinones, trimerization
  29, 718
  thiopyrrylium salts **27**, 875
  1,3,5-trienes **28**, 892 suppl. **29**;
  28, 929 suppl. **30**
  1,3,5-triketones **27**, 924
– special s.
  estra-1,3,5(10)-trienes
– startg. m. f.
  carboxylic acids **24**, 245
  2,4-cyclohexadienones **28**, 111
– with α,α-dichloromethyl
  ether **29**, 838
– –, **1,4-dihydro-**
– startg. m. f.
  1H-azepines **26**, 968
– – **annelation**
– – with
  allyl derivs. and 1,1,1-alkoxy-
  dihalides **28**, 825
  formamidinium salts, vinylogous **25**, 581 suppl. **29**
  β-halogenoacetals **28**, 922
– via Horner synthesis **30**, 645
– – **closure**, oxidative **30**, 658
Benzenes (s. a. Biphenyls)
– from
  cyclopropenium salts **26**, 771
  pyridazines and acetylene
  derivs. **29**, 886
– startg. m. f.
  1H-azepines **27**, 420

Benzenes, 1,3-dinitro-
– startg. m. f.
  3-azabicyclo[3.3.1]non-6-enes,
  1,5-dinitro- **26**, 48
  3,5-bis-*aci*-nitrocyclohexenes
  26, 48
–, synthesis **26**, 902
Benzeneselenyl bromide
  as reactant **30**, 637
Benzenesulfonic acid esters
– from
  alcohols **27**, 401
– startg. m. f.
  azides **27**, 401
Benzenesulfonyl azide
  as reagent **28**, 321
– **chloride** as reagent **27**, 891
cis-**Benzene trioxide**,
  reactions **30**, 278
Benzhydroximic chloride
  as reagent **28**, 181
Benzhydryl... s. Diphenyl-
  methyl...
**Benzilic acid-type rearrange-
  ment**
–, Δ³-oxazoline-2-thiones,
  4-organothio- **28**, 395
Benzils
– from
  aldehydes (2 molecules)
  26, 768
  2-aminoenol sulfonates **28**, 178
  benzoins **12**, 335 suppl. **28**
  chalcones **27**, 238
– startg. m. f.
  O-acylbenzoins (from
  2 molecules) **26**, 221
  deuterioaldehydes **28**, 64
  mandelonitrile benzoates
  30, 486
–, **unsym.**, photoreduction
  7, 679 suppl. **28**
**2H-Benzimidazole 1,3-dioxides
  28**, 420
Benzimidazole 3-oxides
– startg. m. f.
  benzimidazol-2-ones **26**, 990
– –, **1-hydroxy-**
– from
  benzofurazan oxides **28**, 420
– – and β-ketosulfones
  28, 420
– **ring** from quinoxaline 1-oxide
  ring **30**, 341
Benzimidazoles
(s. a. Imidazoles)
– from
  α-amino-β,γ-ethylene-
  nitriles **27**, 535
  o-diamines and carboxylic
  acids **27**, 410; **30**, 265
  imidazoles **28**, 922
–, review **29**, 476
–, **1-subst.**
– from
  o-nitramines and alcohols
  26, 369
–, **2-alkylthio-**
– from
  o-diamines **28**, 538
–, **2-amino-**
– from
  o-aminothioureas **29**, 476
  o-diamines **30**, 245
Benzimidazoline-2-thiones
– from
  o-aminothioureas **30**, 333

Benzimidazol-2-ones
- from
  benzimidazole 3-oxides
  26, 990
1-Benzimidazolylsuccinimides
- startg. m. f.
  1H-pyrrolo[2,3-b]quinoxalines,
  2,3,3a,4,9,9a-hexahydro-
  29, 329
Benz[b]indeno[2,1-d]thiophenes
28, 531
6H-Benz[b]indolo[3,2,1-d,e]-1,5-
naphthyridines
- from
  2H-1,4-ethanobenzo[b]-1,5-
    naphthyridines, 3,4-dihydro-
    10-aryl- 30, 301
Benzisoindolinium salts,
quaternary
- from
  diynammonium salts,
  quaternary 30, 504
1,2-Benzisothiazole 1,1-dioxides
- from
  o-acylsulfonic acid amides
  30, 330
1,2-Benzisothiazole ring opening
25, 416
1,2-Benzisothiazoles 26, 609
- startg. m. f.
  o,o'-dicyanodisulfides 27, 584
-, 7-amino-
- startg. m. f.
  1,2,3-benzothiadiazoles,
  7-acyl- 28, 239
2,1-Benzisothiazoles
- from
  2,1-benzisoxazoles 29, 554
2,1-Benzisothiazoline
2,2-dioxides 16, 957 suppl. 27
1,2-Benzisothiazolium salts,
3-halogeno-
- startg. m. f.
  thianaphthenes, 2-acyl-
  3-amino- 29, 576
Benz[d]isothiazol-3-one
1,1-dioxides s. Saccharins
Benz[d]isothia(IV)zol-3-one
1-oxides
- from
  o-carbalkoyxsulfoxides
  26, 360
1,2-Benzisoxazoles
- startg. m. f.
  benzoxazoles 22, 95
2,1-Benzisoxazoles
- from
  1,2,3-benzotriazine 3-oxides
  29, 254
  nitro compds. 27, 256
  - - and nitriles 30, 629
- startg. m. f.
  o-aminooxo compds., 5-subst.
  28, 470
  - -, N-subst. 26, 18
  2,1-benzisothiazoles 29, 554
  2,1-benzisoxazolium salts
  26, 18
  tricycloquinazolines
    (from 3 molecules) 30, 269
-, 7-nitro-
- startg. m. f.
  benzofurazans, 4-acyl- 27, 101
-, 3-(2-pyridyl)-
- startg. m. f.
  pyridol[1,2-b]cinnolin-6-ium
  ylids, 5,11-dihydro-11-oxo-
  30, 187

2,1-Benzisoxazolines
- from
  2,1-benzisoxazolium salts
  26, 18
- startg. m. f.
  o-aminooxo compds., N-subst.
  26, 18
2,1-Benzisoxazolium salts
- from
  2,1-benzisoxazoles 26, 18
- startg. m. f.
  2,1-benzisoxazolines 26, 18
Benzo[4,5]canthin-6-ones
28, 748
11H-Benzo[a]carbazoles,
5,6-dihydro-
- startg. m. f.
  11H-indeno[1,2-b]quinolines
  28, 903
Benzo[c]cinnoline 5,6-dioxides
26, 266
3,4-Benzocoumarins
- from
  diphenic acids 26, 254
Benzo[b]cyclobuta[d]thiophenes
- startg. m. f.
  naphthalenes 28, 954
Benzocyclobutenediones
28, 930
Benzocyclobutene ring opening
- to cyclimmonium salts,
  N-condensed 28, 310
Benzocyclobutenes 16, 957
suppl. 27
-, cycloisomerization 27, 756
- from
  isochroman-3-ones 28, 970
- startg. m. f.
  N-heterocyclics, N-condensed
  29, 651
-, 3-subst. 28, 511
-, 1-acyl-
- startg. m. f.
  isochromenes, 3-subst.
  27, 169
Benzocyclobutenols 26, 744
Benzocyclobutenones
- from
  homophthalic acid anhydrides
  28, 970
11H-Benzo[5,6]cyclohepta-
[1,2-c]pyrid-11-ones 29, 940
Benzocyclopropene-1,1-
dicarboxylic acid esters
25, 670
- startg. m. f.
  benzofurans, 2-alkoxy-
  25, 670
Benzocyclopropenes, review
25, 670 suppl. 28
1,2-Benzodiazepine ring
27, 531 suppl. 30
3H-1,3-Benzodiazepines 29, 453
2,3-Benzodiazepines 27, 531
1,4-Benzodiazepines
-, rearrangements. review
  22, 753 suppl. 27
-, review 26, 476
3H-1,4-Benzodiazepines
- by dehydrogenation 25, 354
  suppl. 29
1H-1,5-Benzodiazepines,
4-amino- 29, 320
5H-2,3-Benzodiazepines,
3,4-dihydro- 29, 278

1H-1,4-Benzodiazepines,
2,3-dihydro-
- from
  o-aminoketones 29, 749
  2H-1,4-benzodiazepin-2-ones,
  1,3-dihydro- 29, 58
  2,3-piperazinediones 29, 479
- startg. m. f.
  indoles 30, 692
-, 3,4,5-tetrahydro-
- from
  2H-1,4-benzodiazepin-2-ones,
  1,3-dihydro- 29, 58
-, 2,3,4,5-tetrahydro-4-hydroxy-
- startg. m. f.
  quinoxalines, 1,2,3,4-tetra-
  hydro- 27, 539
2H-1,4-Benzodiazepin-2-one-
3-carboxamides, 1,3-dihydro-
- startg. m. f.
  1H-imidazo[5,1-c][1,4]benzo-
  diazepin-1,3,11-triones,
  2,3,5,10,11,11a-hexahydro-
  30, 241
2H-1,4-Benzodiazepin-2-ones,
1,3-dihydro- 27, 474; 30, 306
- from
  carbostyrils, 3,4-dihydro-
  3-amino-3-hydroxy- 28, 437a
- startg. m. f.
  1H-1,4-benzodiazepines,
  2,3-dihydro- 29, 58
  -, 2,3,4,5-tetrahydro- 29, 58
2H-1,4-Benzodiazepin-2-ones,
1,3-dihydro-3-acoxy-
- startg. m. f.
  3H-1,4-benzodiazepin-3-ones,
  1,2,4,5-tetrahydro-2,5-
  epoxy- 27, 22
5H-1,4-Benzodiazepin-5-ones,
3,4-dihydro- 26, 476
-, -, 2-amino- 27, 333
1H-1,5-Benzodiazepin-2(3H)-
ones, 4(5H)-imino- 29, 344
1H-1,4-Benzodiazepin-2-ones,
2,3,4,5-tetrahydro-
- by Schmidt reaction 4, 326
  suppl. 26
- from
  2H-1,4-benzodiazepin-2-ones,
  1,3-dihydro- 29, 58
3H-1,4-Benzodiazepin-3-ones,
1,2,4,5-tetrahydro-2,5-epoxy-
- from
  2H-1,4-benzodiazepin-2-ones,
  1,3-dihydro-3-acoxy- 27, 22
2H-1,4-Benzodiazepin-2-ones,
1,3,4,5-tetrahydro-4,5-epoxy-
- startg. m. f.
  2(1H)-quinazolones, 3,4-
  dihydro-4-hydroxy-3-alkoxy-
  methyl- 26, 113
2,5-Benzodiazocines,
1,2,3,4,5,6-hexahydro-
- by nucleophilic ring opening
  27, 44
1,3,2-Benzodioxaboroles 26, 638
- from
  ethylene derivs. 26, 638
1,3,2-Benzodioxaphospholenes
s. Pyrocatechyl phosphoro...
Benzodioxepins, 4,5-dihydro-
7-hydroxy-
- from
  p-quinones and oxo compds.
  28, 84

1,3-Benzodioxoles
– from
 acetylenedicarboxylic acid
 esters **28**, 100
 o-hydroxycarbinols **28**, 207
–, protection of keto groups
 as – **16**, 234 suppl. **27**
**1,3-Benzodioxole-2-spirocyclo-
hexadien-4′-ones 26,** 231
1,3,2-Benzodithiaboroles
– from
 1,3,2-benzodithiaboroles,
 2-halogeno- **30,** 379
 o-dithiols **30,** 379
–, 2-halogeno-
– from
 o-dithiols **30,** 379
– startg. m. f.
 1,3,2-benzodithiaboroles
 30, 379
2,3-Benzodithianes
– startg. m. f.
 benzo[c]thiophenes **29,** 586
1,2,4-Benzodithiazines 26, 566
– startg. m. f.
 benzothiazoles **26,** 566
1,3-Benzodithiole-2-thiones
– from
 1,2,3-benzothiadiazoles **27,** 623
– startg. m. f.
 o-dithiols **27,** 623
Benzofuran-2-carboxylic acids
 s. Coumarilic acids
Benzofuranones, Wittig synthesis with – **29,** 861
Benzofuran-3(2H)-ones
– from
 phenolethers and α-halogenocarboxylic acid halides
 28, 876
Benzofuran ring
– by cyclodehydration **27,** 925
– from
 arenes **28,** 760
– –, 3-amino-
– from
 α-aryloxycarboxylic acid
 amides **29,** 936
Benzofurans 23, 927; **28,** 220
–, 2-arylation **24,** 884 suppl. **28**
– by cyclodehydrogenation
 24, 260
– from
 o-alkoxyacetylenes **28,** 196
 o-hydroxyoxo compds. **28,** 807
 propargyl ethers **29,** 697
–, removal of protective benzoin
 ester groups as – **27,** 5
– special s.
 acetylenebenzofurans
 2-oximino-4(5H)-benzo-
 furanones, 6,7-dihydro-
– via o-Claisen rearrangement
 30, 677
–, 3-acyl-
– startg. m. f.
 benzofurans, 2-subst. **27,** 987
–, 2-alkoxy-
– via benzocyclopropene-
 1,1-dicarboxylic acid esters
 25, 670
–, 4-amino-
– startg. m. f.
 indoles, 4-hydroxy- **28,** 297
–, 2,3-dihydro- s. Coumarans
–, 2-subst.
– from
 benzofurans, 3-acyl- **27,** 987

Benzofurazan oxides
– startg. m. f.
 benzimidazole 3-oxides,
 1-hydroxy- **28,** 420
 benzofurazans **30,** 56
 N-heterocyclics **28,** 420
– ring
– from
 o-nitroazides **7,** 172 suppl. **26**
Benzofurazans
– from
 benzofurazan oxides **30,** 56
 o-dinitro compds. **30,** 56
–, 4-acyl-
– from
 2,1-benzisoxazoles, 7-nitro-
 27, 101
**Benzofuro[3,2-c][1]benzopyrans,
6a,11a-dihydro-**
– from
 1,4-diaryloxy-2-butynes **29,** 684
Benzofuro[2,3-c]pyrrole ring
– from
 phenolnitrones, cyclic **30,** 159a
**Benzofuro[2,3-b]quinoxalines
29,** 909
Benzofuroxans
 s. Benzofurazan oxides
Benzoic acid as reagent
 27, 733; **29,** 925
Benzoins (s. a. O-Acylbenzoins)
–, C-alkylation, selective **26,** 856
–, C,O-dialkylation **26,** 856
– from
 aldehydes **26,** 675
– startg. m. f.
 benzils **12,** 335 suppl. **28**
**6,7-Benzomorphans, 5-hydroxy-
27,** 747
Benzonitrile (s. a. Dichlorobis-
 (benzonitrile)palladium(II))
– as solvent **27,** 444, 576; **29,** 326
– oxide as oxidant **28,** 181
Benzophenone
 (s. a. under Sensitizers)
– as reactant **10,** 703
– as reagent **7,** 331
Benzophenone-4-carboxylic acid
– as directive agent **27,** 188
Benzopinacol as reducing agent
 28, 31
Benzo[g]pteridines 26, 392
**1H-Benzopyran-1-imines,
3,4-dihydro-**
– from
 hydroxycarboxylic acid
 amides **24,** 521 suppl. **26**
**[1]Benzopyrano[4,3-b][1,5]-
benzodiazepines 29,** 255
Benzopyrans s. a. Chromans
Benzo[a]pyrenes 30, 682
2-Benzopyr-3-ones
– from
 indans, 2,2-dialkoxy-1,3-
 dihydroxy- **26,** 243
 2-indanones, 1,3-dihydroxy-
 26, 243
 o-ketoacetic acids **26,** 243
**2-Benzopyrylium salts,
3-acylamino-**
– from
 arylacetonitriles **27,** 809
**Benzopyrylium salts, 4-alkoxy-
29,** 192
Benzo[h]quinoline as medium
 27, 962

Benzo[c]quinolizinium salts
– from
 pyrylium salts **27,** 411
Benzoquinone as reagent **26,** 263
o-Benzoquinones
– startg. m. f.
 cyclopentadienones **30,** 691
**Benzo[b]selenophenes, 2-acyl-
28,** 873
–, 2-formyl-
– from
 selenochromenes **26,** 163
**Benzoselenopheno[2,3-b]-
benzoselenophenes**
– from
 1,1-diarylethenes **28,** 592
**2,1,5-Benzothiadiazepin-4-one
2-oxides 26,** 590
**3H-1,2,5-Benzothia(IV)-
diazepines, 4,5-dihydro-**
– from
 1,2,4-benzothia(IV)diazines,
 3,4-dihydro- **30,** 130
**Benzo[e]-1,2,4-thiadiazine
S-oxides 28,** 340
**1,2,4-Benzothia(IV)diazines,
3,4-dihydro-**
– startg. m. f.
 3H-1,2,5-benzothia(IV)-
 diazepines, 4,5-dihydro-
 30, 130
3,1,2-Benzothiadiazin-4-ones
– from
 o-aminothiolic acids **27,** 293
 3,1-benzoxazine-2,4-diones
 27, 293
1,2,3-Benzothiadiazoles
– from
 o-aminomercaptans **27,** 623
– startg. m. f.
 1,3-benzodithiole-2-thiones
 27, 623
–, 7-acyl-
– from
 1,2-benzisothiazoles,
 7-amino- **28,** 239
**2H-Benzo[e]-1,2,3,4-thiatriazine
1,1-dioxides**
– startg. m. f.
 2H-benzo[c]-1,2-thiazete
 1,1-dioxides **29,** 462
 dibenzo[c,e]-1,2-thiazine
 1,1-dioxides **29,** 462
1,5-Benzothiazepine ring
– from
 o-aminomercaptans **13,** 896
**1,5-Benzothiazepin-4(5H)-ones,
2,3-dihydro- 14,** 546 suppl. **26**
**2H-Benzo[c]-1,2-thiazete
1,1-dioxides**
– from
 2H-benzo[e]-1,2,3,4-thia-
 triazine 1,1-dioxides **29,** 462
**2H-1,2-Benzothiazine
1,1-dioxides, 4-hydroxy-**
– from
 saccharins **28,** 694
1,4-Benzothiazine ring
– from
 o-aminodisulfides **26,** 624
 – and ketones **26,** 624
4H-1,3-Benzothiazines 28, 567
4H-3,1-Benzothiazines
– startg. m. f.
 indoles **28,** 957
**1,2-Benzothiazin-3(2H)-one
1,1-dioxides 28,** 443

Ben

**1H-2,1-Benzothiazin-4-one
2,2-dioxides, 3,4-dihydro-
28,** 902
**4H-1,2-Benzothiazin-4-one
1,1-dioxides, 2,3-dihydro-
29,** 465
**1,3-Benzothiazin-4(2H)-ones,
3,4-dihydro- 27,** 344
**2H-1,3-Benzothiazin-2-ones,
3,4-dihydro-**
– from
 o-hydroxyaldehydes and
 isothiocyanates **29,** 550
**1,4-Benzothiazin-3-ones,
3,4-dihydro- 14,** 658
Benzothiazole N-oxides 28, 564
– startg. m. f.
 benzothiazol-2-ones **29,** 269
Benzothiazole ring
– from
 quinones **27,** 617
–, 2-silyl groups, active, in –
 27, 883
Benzothiazoles
– from
 amines, ar. **30,** 392
– special s.
 bis(benzothiazoles),
 2,2′-imino-
Benzothiazoles, 2-acyl-
– from
 o-azosulfenylhalides and
 methyl ketones **29,** 567
Benzothiazolines
– from
 alcohols **27,** 610
Benzothiazolium salts
– startg. m. f.
 spiro[2H-1-benzopyran-
 2,2′-benzothiazolines]
 28, 738
Benzothiazol-2-ones
– from
 benzothiazole N-oxides
 29, 269
Benzo[b]thiepins 18, 759
 suppl. 29
Benzo[b]thiophenes
 s. Thianaphthenes
Benzo[c]thiophenes
– from
 2,3-benzodithianes **29,** 586
 benzothiopyr-3-ones **26,** 634
3H-Benzo[c]thiophen-1-ones
– via sulfur-donor ligand
 o-metalated complexes
 29, 643
2-Benzothiopyr-3-ones
– startg. m. f.
 benzo[c]thiophenes **26,** 634
1H-1,2,4-Benzotriazepines
– startg. m. f.
 1H-indazoles **29,** 483
1,2,3-Benzotriazine 3-oxides
– startg. m. f.
 2,1-benzisoxazoles **29,** 254
**1,2,3-Benzotriazine ring,
3,4-dihydro-**
– startg. m. f.
 indazole ring **28,** 242
1,2,3-Benzotriazines
– from
 indazoles, 1-amino- **28,** 431
1,2,4-Benzotriazines
– from
 2(1H)-quinoxalones,
 1-amino- **24,** 547 suppl. 26

1,2,3-Benzotriazin-4(3H)-ones
– from
 o-carbalkoxytriazines **29,** 452
–, 3-amino-
– startg. m. f.
 3-indazolones **29,** 461
Benzotriazoles
– special s.
 1-chloro-benzotriazole
 1-hydroxy-
**Benzotriazolyl-N-oxytrisdi-
methylaminophosphonium
hexafluorophosphate**
–, peptide synthesis with –
 28, 144 suppl. 30
o-Benzotrithiol-2-ones
– from
 o-dithiols **28,** 515
4,5-Benzotropones
– with
 elimination of O-bridges
 30, 679
Benz[f][1,3,5]oxadiazepines
– from
 quinazoline 3-oxides **24,** 159
**6,7-Benzotropones, 2-amino-
29,** 707
**1H,3H-4,2,1,5-Benzoxathia-
diazocine 2,2-dioxides 26,** 249
Benzoxathianones
– from
 thioketones **29,** 566
3,1-Benzoxathian-4-ones
– from
 o-carboxysulfoxides **29,** 250
3,1-Benzoxazepines
– from
 indoles, 3-acyl- **26,** 743
– special s.
 2,5-methano-3,1-benzoxa-
 zepines
1,3-Benzoxazine-2,4-diones
– from
 aryl N-carbalkoxycarbamates
 28, 898
– startg. m. f.
 o-hydroxycarboxylic acid
 amides **28,** 898
3,1-Benzoxazine-2,4-diones
– startg. m. f.
 o-aminocarboxylic acid
 amides **11,** 546; **27,** 349
 o-aminothiolic acids **27,** 293
 3,1,2-benzothiadiazin-4-ones
 27, 293
 2,4-quinazolinediones **27,** 349
 4-quinazolones **28,** 418
**Benz[d]-1,2-oxazine-1,4-diones,
3,4-dihydro- 27,** 27
2H-1,4-Benzoxazine ring
– startg. m. f.
 benzoxazole ring
 26, 199; **28,** 408
4H-3,1-Benzoxazines
– from
 diazonium salts **28,** 369
– startg. m. f.
 3H-indol-3-ols **26,** 748
1,4-Benzoxazines, condensed
 s. Pyrazolo[4,3-b]-1,4-
 benzoxazines
**2H-1,4-Benzoxazines, 3-amino-
27,** 334
**2H-1,3-Benzoxazines,
3,4-dihydro- 28,** 786

**1H-2,3-Benzoxazines,
4-hydrazino-**
– startg. m. f.
 phthalans, 1-(azomethylimino)-
 28, 294
 1H-1,2,4-triazoles, 3-(o-α-
 hydroxyphenyl)- **28,** 295
**2H-1,3-Benzoxazine-2-thiones,
3,4-dihydro-**
– from
 o-hydroxyaldehydes and
 isothiocyanates **29,** 550
4H-1,3-Benzoxazin-4-one ring
– startg. m. f.
 1,3,5-triazine ring **28,** 421
4H-3,1-Benzoxazin-4-ones
– from
 o-acylaminocarboxylic acids
 29, 391
– startg. m. f.
 o-acylaminocarboxylic acid
 amides **29,** 391
 quinolines, 4-amino- **28,** 871
**2,3-Benzoxazin-1-ones,
 reactions 30,** 268
– startg. m. f.
 phthalazones **30,** 268
 phthalimides **30,** 268
**4H-1,3-Benzoxazin-4-ones,
2,3-dihydro-**
– from
 o-hydroxycarboxylic acid
 amides **26,** 307
– – – and oxo compds. **29,** 374
–, **2,3-dihydro-2-alkoxy**
–, ring closure with 1,3-dienes
 28, 348
**2H-1,3-Benzoxazocines,
2,6-methano-4,3,4,5,6-tetra-
hydro- 29,** 458
Benzoxazole-2-aldoximes
– from
 o-aminophenols **26,** 430
Benzoxazole ring
– from
 2H-1,4-benzoxazine ring
 26, 199; **28,** 408
Benzoxazoles
– from
 o-acoxynitriles **28,** 456
 o-alkoxyoximes **26,** 505
 α-amino-β,γ-ethylenenitriles
 27, 535
 1,2-benzisoxazoles **22,** 95
–, **5-hydroxy-**
– from
 p-quinones, o-amino- **29,** 238
 29, 238
Benzoxazole-2-thiones
– startg. m. f.
 o-aminophenols **26,** 21
2-Benzoxazolone ring
– startg. m. f.
 hydantoin ring, 1-(o-hydroxyaryl)- **26,** 348
– – opening **26,** 348
**6-Benzoxazolones, 2,6-dihydro-
29,** 378
1-Benzoxepins 18, 759 suppl. 29
Benzoyl chloride
– as reagent **26,** 813; **27,** 765;
 28, 462
Benzoyl peroxide
– as reactant **26,** 88
– as reagent **26,** 157, 159, 554,
 588; **27,** 912; **28,** 780; **30,** 641
– **peroxides**

– startg. m. f.
  naphthalenes **29**, 883
Benzvalene ring opening
  27, 549
Benzvalenes, cycloadditions
  to – **29**, 652
Benzyl alcohol as H-donor
  11, 81; **29**, 53
Benzylalcohols
  s. Bis(benzylalcohols)
Benzylamines
– from
  β-aminocarboxylic acid
  esters **28**, 875
O-Benzylation, selective
– of
  nucleosides **20**, 150 suppl. **26**
Benzyl chloride as reactant
  28, 644
– cyanoacetate
– as reactant **27**, 265
– ethers
–, cleavage **27**, 9
– halides s. Benzyl chloride
Benzylic s. under Oxygen,
  Oxidation
O,O-Benzylidene derivs.
– of
  carbohydrates **28**, 166
**Benzylisothiouronium
  dithiocarbamates**
–, characterization of prim. and
  sec. amines as –
  11, 576 suppl. **26**
Benzyl ketones
– from
  carboxylic acid chlorides
  27, 865
N-Benzyloxymethyl
– as N-protective group,
  composite **27**, 456
**Benzyltriethylammonium
  chloride**
– as phase transfer catalyst
  4, 432 suppl. **27**; **11**, 202
  suppl. **28**; **29**, 456; **30**, 417, 581
– as reagent **26**, 535; **27**, 726,
  833; **28**, 795; **29**, 793; **30**, 417
**Benzyltrimethylammonium
  hydroxide** s. Triton B
– salt as reactant
  28, 375
Benzyne intermediates
  23, 458 suppl. **26**
–, ring closure via – **28**, 976;
  29, 811
– precursors
  acetanilides **19**, 910 suppl. **26**;
  29, 897
  diazonium salts **19**, 910
  suppl. **26**
Benzynes s. a. Arynes
**Beryllium compds., organo-,
  optically active**
–, reduction, asym., by –
  23, 49 suppl. **27**
Betaines (s. a. Dipoles, Rings,
  mesoionic, Ylids)
– special s.
  acoxy-betaines
  diazonium sulfate –
  diphosphoric acid diester
    mono-
    enol-
    phenol-
    phosphonium –
    phosphoryl-
  3-pyrazolidone –

pyridazinium –
pyridinium –
pyrimidonium –
sulfinovinyl-
sulfovinyl-
  thetins
1,3-thiazinium –
thiazolium –
thiazolo[3,2-a]pyridinium –
–, tritiation **9**, 510 suppl. **26**
N,N-Betaines s. Aminimides,
  Ammoniosulfonyl imides
Beyer s. Combes
Bi... s. a. **Di...**
Biaryl synthesis s. Ullmann
Bicycloalkenes s. a. Cycloalkenes, bridged
Bicyclo[1.1.0]butane derivs.
  s. Benzvalenes
– ring **23**, 952
–, startg. m. f.
  cyclobutane ring **26**, 45
– – opening, double **26**, 53
Bicyclo[1.1.0]butanes 5, 643
  suppl. **26**; **25**, 587 suppl. **28**
Bicyclo[5.3.0]decane ring
  s. Azulenes
**Bicyclo[3.2.0]hepta-3,6-dien-
  2-ones**
–, Cope rearrangement
  27, 739; **23**, 731
Bicyclo[2.2.1]heptan-2-ones
  26, 986
Bicyclo[3.3.1]nonanes
– with bridgehead functions
  27, 566
Bicyclo[3.2.1]octane ring
– special s.
  bridge bicyclo[3.2.1]octane
    ring
Bicyclo[4.4.0]octa-1,3,5-trienes
  s. Benzocyclobutenes
**Bicyclo[2.2.2]octene ring
  fragmentation 26**, 938
Bicyclo[3.2.1]octenes
– from
  bicyclo[2.2.2]oct-2-yl tosy-
    lates **27**, 930
Bicyclo[2.2.2]oct-2-yl tosylates
–, startg. m. f.
  bicyclo[2.2.2]octenes **27**, 930
Biguanides
–, startg. m. f.
  1,3,5-triazines **30**, 294
Biimidazoles 27, 372
Binuclear catalysis 22, 717
  suppl. **26**
Biochemical reactions
  s. Enzymes, Microorganisms
Biopak 16, 424 suppl. **26**
Biphenyl-2-carboxamides
– startg. m. f.
  phenanthridones **29**, 445
Biphenylenes
– from
  o-dihalides **27**, 850
Biphenyls
– by substitution, nucleophilic,
  ar. **27**, 827
–, unsym.
– from
  arylthallium bis(trifluoro-
    acetate) **27**, 888
Biphenyl-2-sulfinic acids
– from
  phenyl sulfones (2 molecules)
  29, 850

p-Biphenylylurethan group
–, steric control, remote by –
  28, 8
Bi(phosphonium salts)
– special s.
  μ-oxobis(triamino-
    phosphonium salts)
1,1-Bi(phosphonium salts)
– startg. m. f.
  ethylene derivs. **27**, 891
Birch reduction 28, 989
–, comparison with Benkeser
  reduction, review **5**, 47
  suppl. **28**
– of
  amines, tert., cyclic **30**, 23
  pyridines **27**, 816
Bis(acrylonitrile)nickel(0)
  as catalyst **26**, 724
1,3-Bis(alkylthio)-1-ethylenes
– from
  halides **26**, 862
– startg. m. f.
  α,β-ethylenealdehydes **26**, 862
Bisamides s. 1,1-Di(acylamines)
**Bis(benzonitrile)palladium
  dichloride**
– s. Dichlorobis(benzonitrile)-
  palladium(II)
Bis(benzothiazoles), 2,2'-imino-
– from
  o-aminomercaptans **28**, 538
Bis(benzylalcohols)
–, o-ring closure with – **28**, 759
Bisboracyclanes as reactants
  27, 713
– special s.
  bis-(3,5-dimethyl)borinane
**N,O-Bis(carbamyl)hydroxyl-
  amines**
– special s.
  N-carbalkoxy-N,O-bis-
    (carbamyl)hydroxylamines
  – via N-carbo-*tert*-butoxy-
    N,O-bis(carbamyl)hydroxyl-
    amines **26**, 292
Bischler s. a. Möhlau
**Bischler-Napieralski ring
  closure 10**, 665; **23**, 914
  suppl. **26**; **27**, 946
**Bis-(1,5-cyclooctadiene)nickel(0)
  19**, 962 suppl. **28**; **27**, 870;
  30, 538
Bis-(α-diazoketones), ring
  closure with – **26**, 963
α,α'-**Bisdiazoketones**, – – – –
  29, 766
Bis(dibutylacetoxytin)oxide
  as reagent **13**, 70 suppl. **28**
**1,8-Bis(dimethylamino)-
  naphthalene**
– as base, strong, non-nucleophilic **30**, 336
**Bis-(3-dimethylaminopropyl)-
  phenylphosphine**
– as thiophile, basic **26**, 982
Bis(dimethylamino)sulfane
– startg. m. f.
  sulfinodiimines **26**, 273
Bis-(3,5-dimethyl)borinane
– as reactant **27**, 713
**Bis(dimethylglyoximato)-
  cobalt(II)**
–, hydrogenation with – **27**, 55;
  30, 18

1,1-Bis(dithiocarbarmic acid esters)
- startg. m. f.
  1,3-dithietan-2-ylium salts,
  2-amino- **30,** 422
Bis(halogenocarbonyl)amines
- from
  1,2,4-dithiazoli-3,5-diones
  27, 582
Bis(halogenosulfonyl)imides
- from
  ethers **27,** 409
Bishomologization s. Synthesis
  with addition of 2 C-atoms
Bis-2-imidazolidinylidenes
- startg. m. f.
  enolbetaines **26,** 917
  1,3,6,9-tetraazaspiro-
  [4.4]nonanes **26,** 920
3,3′-Bis-(1H-isoindoles)
- from
  dibenzo[c,h][1,5]naph-
  thyridines, 6,12-dihydro-
  29, 712
4,5′-Bisisoxazoles
- startg. m. f.
  4-pyridones **27,** 991
Bis-(3-methyl-2-butyl)borane
  as reactant **29,** 61, 139
Bismethylene transfer 27, 664
1,3-Bis(methylthio)allyl anion
  as intermediate **26,** 862
1,3-Bis(methylthio)-2-methoxy-propane
- startg. m. f.
  α,β-ethylenealdehydes **26,** 862
Bismuth triacetate, reactions
  with - **28,** 424
- **trioxide 26,** 233
3,5-Bis-*aci*-nitrocyclohexenes
- from
  benzenes, 1,3-dinitro- **26,** 48
- startg. m. f.
  3-azabicyclo[3.3.1]non-6-enes
  1,5-dinitro- **26,** 48
Bis(organothio)methylene-hydrazones
- from
  hydrazones **26,** 615
Bisphenols 6, 83
-, cleavage **27,** 986
Bis(phenylsulfonyl) sulfide
  as reactant **27,** 629
N,N′-Bisphosphinosulfones diimines
- startg. m. f.
  S,P,N-heterocyclics **28,** 995
Bispyrazolones as azodienophiles **22,** 314 suppl. **26**
**Bis(pyridine)dimethylform-
  amidedichlororhodium tetrahydridoborate
  22,** 59 suppl. **26**
Bis(tetra-*n*-butylammonium)
  oxalate as reagent **27,** 936
Bis(thiophosphinyl)trisulfides, sym.
- from
  dithiophosphinic acids **26,** 558
Bisthiuram disulfides
- startg. m. f.
  dithiocarbamic acid esters
  26, 620
1,2-Bis(tosylhydrazones)
- startg. m. f.
  acetylene derivs. **27,** 954
p-Bis(trichloromethyl)benzene
  as reagent **29,** 517

Bis(trifluoromethyl)ketene,
  ring closures with - **27,** 691
**Bi(trifluoromethyl)thio-
  ketene, - - - - 27,** 691
**N,O-Bis(trimethylsilyl)-
  acetamide**
- as water scavenger **9,** 951
  suppl. **27**
-, protection of alcohol groups
  with - **29,** 964
-, -, preferential, of phenol
  groups with - **28,** 74
Bis(trimethylsilyl)acetylene
  as reactant **29,** 846
**Bis(triphenylphosphine)(maleic
  anhydride)palladium 26,** 726
Bisulfonyl sulfide
- special s.
  bis(phenylsulfonyl) sulfide
Bi-N-sulfonylsulfinodiimides, sym.
- from
  N,N-dihalogenosulfonamides
  28, 557
- special s.
  bitosylsulfinodiimide
Bitosylsulfinodiimide
-, dehydrogenations with -
  30, 322
- startg. m. f.
  sulfinodiimines, cyclic **26,** 394
Biurets (s. a. Dithiobiurets)
- from
  isocyanates **29,** 300
- special s.
  alkoxybiurets
-, **3-hydroxy- 26,** 330
- from
  N-hydroxyureas **26,** 330
  isocyanates **26,** 293
Bleach 24, 251 suppl. **29**
Blocking (s. a. Protection)
-, **steric 28,** 44
- **groups** s. Protective groups
Bonds s. Carbon-carbon bonds,
  Double -, Triple -, Hetero-,
  Nitrogen-nitrogen -
-, **semipolar,** cleavage **30,** 697
9-Borabicyclo[3.3.1]nonane
-, *anti*-Markownikoff hydration
  with - **16,** 194 suppl. **29**
-, reactions with - **26,** 519
- **ate complexes, lithium -
  27,** 72 suppl. **30**
Boracyclanes s. Bisboracyclanes
Borane (s. a. Diborane) **9,** 938
  suppl. **27; 13,** 106 suppl. **29;
  26,** 30, 56, 641, 716; **27,** 137/8,
  637, 659; **28,** 32 suppl. **29;** 28,
  102/3, 621, 660, 814; **29,** 57,
  89, 673; **30,** 29, 427, 495
Borane-dimethyl sulfide
- as reducing agent **30,** 29
- as storable hydroboration
  agent **21,** 174 suppl. **29;
  26,** 134
Boranes
-, 1,4-additions with **28,** 649
-, applications, synthetic, review
  27, 138
- as intermediates **27,** 232
-, autoxidation **27,** 138
- from
  ethylene derivs., synthesis
  27, 689
  halides **27,** 232
-, α-monohalogenation **28,** 484

-, prepn., review **27,** 689
  suppl. **30**
-, radical reactions with -,
  review **27,** 138
-, reaction with diazo compds.
  28, 782/83, 860
- special s.
  amino-boranes
  cyano-
  enoxy-
  ethylene-
  halogeno-
  oxido-
  trialkyl -
- startg. m. f.
  acetylene derivs., synthesis
  28, 864
  alcohols, branched **26,** 914
  -, tert. **27,** 889; **29,** 873
  amines, sec. **27,** 485
  azomethines, synthesis **30,** 622
  1,4-diols **30,** 618
  2-ene-1,4-diols **28,** 583
  2-ethylenealcohols, synthesis
   with addition of 4 C-atoms
   27, 896
  ethylene derivs. **28,** 965
  - -, synthesis **30,** 623
  α,β-ethyleneketones **28,** 866
  hydrocarbons, coupling
   29, 976
  iodides **26,** 881
  oxo compds., synthesis **29,** 873
  ketones **29,** 882
  1,2-oxaborins, dihydro- **28,** 583
  2-oxa-3-borolenes **28,** 866
  trialkylcarbinols **27,** 892
-, syntheses with - and
  acetylene derivs. **26,** 911
-, **cyclic,** synthesis of alcohols
  via - **26,** 716
-, **prim.** s. Monoalkylboranes
-, **sec.** (s. a. Dialkylboranes)
- from
  borinic acid esters **14,** 65
-, **tert.** s. a. Trialkylboranes
-, -, **mixed**
- startg. m. f.
  alcohols, tert. **26,** 914 suppl. **27**
Borates s. Lithium borates
Borax as buffer **29,** 177
- **-sodium carbonate** as
  pH 9.94 buffer **29,** 418
**8-Boraxanthines, 7,8-dihydro-
  27,** 306
Borazanes, stabilized 26, 108
Borepane, 3,6-dimethyl-
  as reactant **28,** 821
Boric acid 4, 154 suppl. **28; 18,**
  370 suppl. **26; 19,** 237 suppl. **27;
  27,** 160, 191; **29,** 86, 627
- - **derivs.** s. a. Triacyl borates
- - **esters**
  - as reactants **27,** 158
  - as reagents **26,** 367
  - special s.
    trimethyl borate
  - startg. m. f.
    borinic acid esters **27,** 659
Borines s. Boranes
Borinic acid enolesters
  s. Enoxyboranes
- - **esters** (s. a. Thioborinic
  acid esters)
  - as intermediates **30,** 495
  - from
    ethylene derivs. **27,** 649
    halogenoboranes **27,** 649

trialkylboranes **27**, 659
- and hydroxy compds. **28**, 81 suppl. **30**
- startg. m. f.
  boranes, sec. **14**, 65
- – –, **cyclic** s. 1,2-Oxaborins, dihydro-
- **acids**
- from
  α-halogenoboranes **27**, 118
- **carboxylic anhydrides** (s. a. Boryl carboxylates)
- special s.
  α-aminocarboxylic borinic anhydrides
Boron aluminum oxide 14, 922 suppl. **26**
Boronate fragmentation
–, synthesis of 1,5- and 1,6-dienes via –, review **23**, 907 suppl. **26**
Boron bromide 28, 12; **29**, 8; **30**, 262
- **chloride 26**, 488; **27**, 11; **27**, 485 suppl. **28**; **30**, 345
- as reactant **27**, 649
- **compds.** (s. a. Thioboron compds.)
- as selective reagents, review **27**, 716 suppl. **28**
- –, **allyl-**, review **27**, 882
- –, **organo-** (s. a. Boranes)
- –, rearrangements, review **28**, 866
- special s.
  N-azocyanoboron betaines carboranes
  diboro compds.
  potassium triorganocyanoborates
–, syntheses with –, review **30**, 623
- – –, –, **cyclic** s. B-Heterocyclics
- **fluoride** (s. a. under Peroxytrifluoroacetic acid) **2**, 334 suppl. **26**; **7**, 5 suppl. **28**; **7**, 201 suppl. **29**; **12**, 760 suppl. **26**; **15**, 26 suppl. **27**; **18**, 858 suppl. **29**; **21**, 184 suppl. **29**; **23**, 36 suppl. **26**; **26**, 30, 71, 149, 171, 211, 601, 717, 802/3, 974, 986; **27**, 46, 65, 139, 395, 672/3, 742, 757, 790a, 820, 915, 934/5, 940; **28**, 88/9, 178, 348, 506, 549, 654/5, 744, 783, 908, 939/40, 983; **29**, 58, 144, 680, 713, 747, 922; **30**, 108, 120, 141, 216/7, 508, 519, 555, 569, 668
- – –/**phosphoric acid** s. Phosphoric acid/-
- **formate** as reagent **27**, 312
Boronic acid esters
- as intermediates **27**, 158
- special s.
  ethyleneboronic acid esters
- startg. m. f.
  boronic acids **29**, 497
  α,β-ethylenehalides **29**, 497
  hydrocarbons **27**, 84
- – –, **cyclic**
- special s.
  catecholborane
- **acids**
- from
  boronic acid esters **29**, 497
- –, ring closure with – **27**, 307
- special s.
  arylboronic acids

- startg. m. f.
  B,N,S-heterocyclics **27**, 307
Boron oxide (s. a. Boron aluminum oxide) **13**, 131 suppl. **26**
- as dehydrating agent **7**, 339 suppl. **26**
- **sulfide 28**, 550
- **trifluoroacetate**
-, removal of protective groups from peptides with – **15**, 26 suppl. **28**
Boroselenines
- from
  dihalogenoboranes **28**, 581
Borylation with boryl sulfonates **27**, 309
Boryl carboxylates (s. a. Borinic carboxylic anhydrides) **29**, 121
- special s.
  boryl pivalate
- **pivalate, diethyl-** as reagent **28**, 81; **10**, 587 suppl. **29**
- **pivalates** as reagents **29**, 121
- **sulfonates**, borylation with – **27**, 309
Bradsher ring closure, direct 29, 838
Branched alkyl groups s. Alkyl groups, branched
- **carbohydrates** s. Carbohydrates, branched
- **C-chains** s. Alcohols, branched, Ketones, -
Bredt's rule, review **28**, 854 suppl. **29**
Bridge amidines 26, 297
- **bicyclo[3.2.1]octane ring**
- from
  quinones **27**, 711
Bridgehead alkyl derivs. 26, 868
- **carbon-carbon double bonds**, isocyclics with – **28**, 854
- **functionalization 27**, 124
- **functions**
-, bicyclo[3.3.1]nonanes with – **27**, 566
- **halogen interchange 28**, 510
Bridge lactams 23, 689; **26**, 351
Bridges s. a. Azo bridges, Azoxy –, Compds., doublebridged, Ene bridges, Etheno –, Methano..., Methylene bridge..., Methylenecarbonyl bridges
–, **elimination** s. under Elimination
O-Bridges s. Isocycles, O-bridged
S-Bridges 22, 606
Brine 27, 949; **30**, 366
Bromides s. Halides, Replacement
Bromination s. Halogenation, Replacement of hydrogen by halogen
Bromine (s. a. Bromine chloride, N-Bromosuccinimide/bromine) **16**, 785 suppl. **27**; **19**, 444 suppl. **26**; **23**, 703 suppl. **29**; **26**, 267, 409, 914, 929; **27**, 155, 233, 294, 501; **28**, 75, 247, 519, 544; **28**, 226 suppl. **30**; **28**, 860 suppl. **29**; **29**, 180, 993; **30**, 152, 174, 358
- **graphite 28**, 485
- **-water 11**, 348 suppl. **27**

- **chloride** as bromine alternative, review **20**, 417 suppl. **29**
N-Bromoacetamide
- as reagent **12**, 206 suppl. **26**
Bromobenzene as solvent, high-boiling **29**, 651
p-Bromobenzenesulfonic acid esters s. Brosylates
Bromochlorides
  s. Chlorobromides
2-Bromo-2-cyano-N,N-dimethylacetamide
- as brominating agent **30**, 353
1,2-Bromofluorides
- startg. m. f.
  fluorides **25**, 64 suppl. **26**
Bromoform as reagent **28**, 25
2-Bromomalonitrile
- as brominating agent **30**, 353
Bromomethylethylene acetals
-, protection of carbonyl groups as – **28**, 231
Bromonium salts
- startg. m. f.
  phenolethers **29**, 202
N-Bromosaccharin as oxidant **28**, 202
Bromosilanes
- special s.
  trimethylbromosilane
N-Bromosuccinimide
- as reagent HalC↓↑H; **18**, 669 suppl. **28**; **26**, 161, 398, 520/1; **26**, 914 suppl. **28**; **27**, 145, 230, 912; **28**, 860; **30**, 192, 641
- –/**bromine 27**, 564
Bromotrichloromethane
  as reactant **28**, 226, 484
Brosylates
–, ring closure, solvolytic with – **27**, 929
Brown hydration 28, 34
Brush reactor, continuous 21, 401 suppl. **26**
Büchi s. Paterno
Buffer solns. s. Borax, Phosphate buffer, Sodium succinate –
Bulky bases s. Bases, anionic, bulky
- **reducing agents** s. Reducing agents, bulky
Bunte salts s. Thiosulfuric acid S-monoesters
1,3-Butadiene monoxide
- startg. m. f.
  2-ethylenealcohols, synthesis with addition of 4 C-atoms **27**, 896
Δ¹-Butenolides s. α,β-Ethylene-γ-lactones
β,γ-Butenolides
  s. γ-Enollactones
n-Butyl alcohol as reagent **29**, 829 suppl. **30**
tert-Butyl alcohol
- as medium **29**, 973; **30**, 47
- as reactant **29**, 131
- as reagent **29**, 72, 506; **30**, 580
Butylamine as reagent **8**, 922 suppl. **28**; **26**, 46
n-Butylamine – – **6**, 755 suppl. **29**
tert-Butylamine – – **28**, 725

tert-Butyl bromide as reagent
  29, 836
- carbazate as reactant 28, 330
- chromate
- as reagent 23, 195 suppl. 26
sec-Butylcyclohexane as solvent,
  inert 27, 731
tert-Butyl esters s. Carboxylic
  acid *tert*-butyl esters
- ethers
- startg. m. f.
  carboxylic acid esters 26, 168
- hydroperoxide 26, 775
-, C-methylation, radical, with –
  28, 880
- **N-hydroxycarbamate**
- startg. m. f.
  N-carbalkoxysulfonyl-
  oxylamines 27, 110
- **hypohalite**
- hypochlorite 20, 373 suppl. 28;
  23, 311 suppl. 26; 27, 288, 569,
  580; 28, 494, 762; 29, 153, 177,
  243, 507, 645, 984
- **isocyanide** as reactant 29, 340
- peroxybenzoate as reagent
  29, 905
2-*tert*-**Butylpyridine** – –
  20, 150 suppl. 26
tert-Butyl rearrangement 28, 84
- **thioethers**, reactions with –
  29, 851
- **trifluoroacetate** as reactant
  25, 572 suppl. 28
Butyric anhydride – – 27, 741

Cadmium acetate 28, 374
- **carbonate** 13, 699 suppl. 26;
  27, 218, 595; 29, 435/6
- **chloride** 28, 885
- **oxide** 13, 699 suppl. 27
Cadogan ring closure 18, 338
Cage compds., N-heterocyclic
- startg. m. f.
  cage compds., O-heterocyclic
  27, 167
- –, **O-heterocyclic**
- from
  cage compds., N-heterocyclic
  27, 167
-, **isocyclic** 29, 686 suppl. 30
- by ring closure, multiple
  28, 640
Calcium/ammonia, liq. 17, 57
  suppl. 28, 29; 28, 36
- **carbide** 27, 122; 28, 163
- **carbonate** 26, 226, 573; 27, 243,
  930; 28, 656, 842; 29, 989;
  30, 566
- **chloride** 7, 249 suppl. 26; 26,
  107, 27, 64; 28, 194
- **compds., organo-**
  methylcalcium iodide
  as reagent 27, 786
- – –, syntheses with –
  27, 786 suppl. 28
- special s.
  diphenylcalcium
- **hydride** 28, 781
- **hydroxide** 17, 735 suppl. 27
- **hypochlorite** 26, 204; 27, 555
- **oxide** 27, 407; 28, 165, 313,
  431, 465, 981
- **sulfate** as dehydrating agent
  11, 232 suppl. 26; 11, 343
  suppl. 29; 26, 235 suppl. 27
Capillary techniques
  s. Chromatography

Capric acid as reagent 27, 84
N-Carbalkoxy... s. a. Urethans
(α-**Carbalkoxyalkylidene**)-
  phosphoranes
- startg. m. f.
  coumarins 27, 886
**Carbalkoxyammoniosulfonyl
  imides**
- special s.
  methyl(carboxysulfamoyl)-
  triethylammonium
  hydroxide inner salt
**Carbalkoxyammonium
  fluoroborates**
- as reagents 30, 263
- **salts, quaternary**
- startg. m. f.
  lactones 28, 228
**N-Carbalkoxy-N,O-bis-
  (carbamyl)hydroxylamines**
- special s.
  N-carbo-*tert*-butoxy-N,O-
  bis(carbamyl)hydroxyl-
  amines
**N-Carbalkoxycarbamic acid
  esters**
- special s.
  aryl N-carbalkoxycarbamates
Carbalkoxy-α-diazoketones
- from
  dicarboxylic acid anhydrides
  30, 456
- startg. m. f.
  dicarboxylic acid esters
  30, 456
**Carbalkoxydicarboxylic acid
  anhydrides** 27, 216
β-**Carbalkoxyenephosphonic
  acid esters**
- from
  α,β-ethylene-α-nitrocarboxylic
  acid esters 28, 594
o-**Carbalkoxy-N-heterocyclics**
- from
  aminocyanohydrin tosylates
  28, 434
Carbalkoxyhydrazines
  (s. a. Carbazic acid esters)
- special s.
  N-acyl-N'-carbalkoxy-
  hydrazines
Carbalkoxyhydrazones
- from
  azides and N-carbalkoxysulfonyloxylamines 30, 191
Carbalkoxyl
- special s.
  carbobenzoxyl
  carbo-2-p-biphenyl-
  isopropoxyl
  carbo-*tert*-butoxyl
  carbo-9-fluorenylmethoxyl
  carbo-p-methoxybenzoyl
N-Carbalkoxyl
- startg. m. f.
  N-methyl groups 21, 401
  suppl. 28
α-**Carbalkoxy-γ-lactones**
- from
  oxido compds. 27, 770
C-Carbalkoxylation 25, 602
  suppl. 26
- of ketones 26, 770
- via C-trichloroacetylation
  29, 816
- with dialkylcarbonate 30, 539
-, **lateral** 28, 732

- with
  C-alkylation 28, 727
-, **radical** s. under Radical
  reactions
N-Carbalkoxylation
  (s. a. Urethans from amines)
- with
  monothiocarbonic acid esters
  29, 430
  phenyl carbonates 23, 412
  suppl. 28
- with simultaneous
  glycosidation 30, 297
-, **selective** 30, 303
Carbalkoxyl migration, review
  23, 733 suppl. 26
O→N-Carbalkoxyl migration
  29, 15
Carbalkoxymethyl groups
-, aromatization with
  elimination of – 29, 981
1-Carbalkoxysemicarbazides
- startg. m. f.
  N-carbamylazoformic acid
  esters 30, 192
o-**Carbalkoxystyrene oxides**
- from
  3-hydroxyphthalides 29, 771
**N-Carbalkoxysulfamic acid
  esters**
-, reactions 26, 491
- startg. m. f.
  urethans 26, 491
Carbalkoxysulfamoyl chlorides
- startg. m. f.
  1,4,3,5-oxathiadiazine
  4,4-dioxides 29, 290
**N-Carbalkoxysulfamoylsulfone
  diimines**
- startg. m. f.
  1,3,2,4,6-thia(VI)thiatriazin-
  5-one 3,3-dioxides,
  4,5-dihydro- 27, 504
N-Carbalkoxysulfenamides
- from
  halides 29, 412
- startg. m. f.
  mono(carbalkoxy)amines
  29, 412
**N-Carbalkoxysulfonyloxyl-
  amines**
- from
  sulfonic acid chlorides 27, 110
- startg. m. f.
  carbalkoxyhydrazones 30, 191
  2-sulfonyloxylamines 27, 110
o-**Carbalkoxysulfoxides**
- startg. m. f.
  benz[d]isothia(IV)zol-3-one
  1-oxides 26, 360
S-(Carbalkoxy)thiosulfates
- special s.
  sodium benzyloxycarbonyl
  thiosulfate
Carbalkoxythioureas
- startg. m. f.
  thioureas 29, 26
3-Carbalkoxytriazenes 30, 189
o-**Carbalkoxytriazenes**
- startg. m. f.
  1,2,3-benzotriazin-4(3H)-ones
  29, 452
Carbamic acid anhydrides
- from
  carbamyl chlorides 26, 196
- – **chlorides** s. Carbamyl
  chlorides

– – esters s. a. Urethans
– – halides s. Carbamyl halides
– – silyl esters
– from
 aminosilanes 29, 438
– startg. m. f.
 isocyanates 29, 438
– carboxylic anhydrides
– from
 pyridinium salts,
 N-carbamyl- 28, 453
– special s.
 N-sulfenylcarbamic
 carboxylic anhydrides
– – –, cyclic
– special s.
 2,5-oxazolidiones
Carbamides s. Ureas
Carbamidium salts
 s. Dithiocarbamidium salts
Carbamyl s. a. Thiocarbamyl
N-Carbamylamidines
– from
 N-(halogenoformyl)imino-
  halides 28, 383
 isocyanates 26, 295
N-Carbamylamidinium salts
– from
 ureas and carboxylic acid
  amides 29, 370
C-Carbamylation, homolytic,
 of heterocyclics 25, 480
N-Carbamylation (s. a. Ureas
 from amines and isocyanates)
–, selective 29, 304
Carbamyl azides
– special s.
 enecarbamyl azides
N-Carbamylazoformic acid
 esters
– from
 1-carbalkoxysemicarbazides
  30, 192
Carbamyl chlorides
– special s.
 3-chlor-1-ethylenecarbamyl
  chlorides
 1,3-dienecarbamyl –
– startg. m. f.
 carbamic acid anhydrides
  26, 196
– chlorosulfines 29, 103
– from
 α-carbamyl-α,α-dichloro-
  methylsulfenyl chlorides
  29, 103
2-Carbamyl-2-cyano-
 phosphonium salts
– from
 alkylidenemalononitriles
  28, 588
– startg. m. f.
 α-cyano-α,β-ethylenecarboxylic
  acid amides 28, 588
α-Carbamyl-α,α-dichloromethyl-
 sulfenyl chlorides
– from
 α-mercaptoacetamides 28, 492
– startg. m. f.
 carbamyl chlorosulfines
  29, 103
Carbamyl dithiocarboxylates
– from
 dithiocarboxylic acids
  and isocyanates 29, 532
N-Carbamyl group migration
 27, 338

O→C-Carbamyl – – 28, 684
Carbamyl halides
– special s.
 bis(halogenocarbonyl)amines
 carbamyl chlorides
Carbamylhydroxylamines
 s. a. N,O-Bis(carbamyl)-
 hydroxylamines, N,N,O-
 Tris(carbamyl)-
O-Carbamylhydroxylamines
 (s. a. N-Carbamyloxy
 compds.)
– from
 N-hydroxyureas 29, 95
α-(O-Carbamylhydroxylamino)-
 carboxylic acid amides
– from
 α-hydroxylaminocarboxylic
  acid amides 30, 125
– startg. m. f.
 α-ketocarboxylic acid amides
  30, 125
Carbamylisothiocyanates 26, 572
–, reactions 26, 572
– startg. m. f.
 4H-1,3,5-thiadiazin-4-ones,
 5,6-dihydro-2-amino- 26, 572
Carbamylmercapto compds.
 s.Thiourethans
N-Carbamyloxy compds.
– from
 N-hydroxy compds. 29, 126
Carbamyloxy groups,
 substitution, nucleophilic
  27, 374
Carbamyl peroxides
 s. Acyl carbamyl peroxides
Carbamylphosphonium salts
– special s.
 carbamylcyanophosphonium
  salts
3-Carbamylpyridinium salts
–, cyclotetramerization 30, 233
Carbamylsulfonic acids
– from
 isocyanates 29, 285
Carbamylthiocyanates 26, 572
Carbanion ring closure
– of
 N,N'-dialkylidene-1,1-diamines
  28, 697
Carbanions s. Allenecarbanions,
 Dicarbanions, Oxocarbanions,
 Sulfinyl carbanions
–, masked
– special s.
 acyl carbanions, masked
 3-oxocarbanions, –
Carbaryloxy glycosides
– startg. m. f.
 O- and N-glycosides 30, 150
N-Carbaryloxylation 27, 512
Carbatricobalt decacarbonyl
 derivs. 27, 180
Carbazic... s. a. Dithiocarbazic...
Carbazic acid esters
 (s. a. Carbalkoxyhydrazines,
 Thiocarbazic acid esters)
– special s.
 tert-butyl carbazate
– startg. m. f.
 halides 28, 496
Carbazoles, 2,3,4,4a-tetrahydro-
–, rearrangement, oxidative
  27, 131
–, 1,2,3,4-tetrahydro-2-oxo-
  29, 953

Carbene complexes
 s. Metal-carbene complexes
Carbenes
–, reactions via 24, 861; 27, 726;
  30, 636
–, ring expansion with – 29, 793
– special s.
 dicarbenes
 dichloromethylene
 dihalogenocarbenes
Carbenoids s. 1,1-Halogeno-
 organolithium compds.
(tert-Carbin)amines,
 N,N-dichloro-
– startg. m. f.
 ketones 27, 240
–, prim.
– from
 nitriles 29, 636
Carbinolamines
 s. 1,1-Aminoalcohols
Carbinols (s. a. Alcohols,
 Decarbinolation)
– special s.
 quinolyl-carbinols
 trialkyl-
 triaryl-
 trihalogenomethyl-
N-Carbobenzoxylation 28, 394
– in homogenous aq. soln.
  26, 439
–, selective 26, 439
Carbo-2-p-biphenylisopropoxyl
– as N-protective group, acid-
  labile 23, 404 suppl. 26
N-Carbo-tert-butoxy-N,O-
 bis(carbamyl)hydroxylamine
– startg. m. f.
 N,O-bis(carbamyl)hydroxyl-
  amines 26, 292
N-carbo-tert-butoxylation 11, 425
 suppl. 27; 27, 437; 28, 375
– in aq. medium 25, 293
 suppl. 26
N-Carbo-tert-butoxypyridinium
 salts, 4-dimethylamino-
–, N-carbo-tert-butoxylation
 with – 25, 293 suppl. 26
Carbocyclics s. Isocyclics
Carbodiimides (s. a. Heterocumulenes)
– as reactants 29, 923
–, [2+2]-cycloaddition 28, 265
– from
 amidoximes 27, 516
 thionylimines and hydroximinohalides 26, 442
 thioureas 27, 532, 978
 ureas 27, 978
–, reactions with – 29, 131
–, ring closure with dicarboxylic
 acid chlorides 26, 511
– special s.
 di-tert-butyl carbodiimide
 dicyclohexyl-
 imidoyl-carbodiimides
 sulfonyl-
– startg. m. f.
 α,β-acetyleneamidines 28, 617
 amidines 28, 617
 2-azetidinones, 4-imino-
  26, 692
 carbodiimidium salts 27, 996
 guanylureas 29, 131
 1,3-O,N-heterocyclics,
  2-imino- 29, 361
 isothioureas 30, 383
 isoureas 30, 50

(**Carbodiimides**
– startg. m. f.)
  isoureas, O-aryl **30,** 50
  monosulfonylcarbodiimides
   29, 473
  2H-1,3,5-oxadiazin-4(3H)-ones,
   2-imino- **28,** 266
  1,3-oxazetidines **28,** 265
  sulfonylguanidines **26,** 410
  1,2,4-thiadiazetidine 1-oxides,
   3-imino- **29,** 473
  Δ²-1,2,4-thiadiazoline
   1,1-dioxides, 3-amino-
   26, 410
  1,3-thiazetidines **27,** 590
  thioureas **27,** 592
  1,2,4-triazole ring, 3-amino-
   26, 393
–, **chloro-**
– from
  1-aziridinecarboximidoyl
   chlorides **27,** 556
–, **polymeric,** as reagents
  19, 307 suppl. **27**
–, **unsym.**
– from
  azides and isonitriles **27,** 447
–, **vinylogous**
– from
  ureas, vinylogous **27,** 978
–, **water-soluble 19,** 307 suppl. **29**
Carbodiimidium salts
– from
  carbodiimides **27,** 996
– special s.
  N-methyl-N,N'-dicyclohexyl-
   carbodiimidium iodide
Carbo-9-fluorenylmethoxyl
  as N-protective group **26,** 24
Carbohydrate-amino acid
  compds. 28, 342
Carbohydrate derivs., synthesis
  27, 770
–, synthesis, asym., with –
  9, 741 suppl. **29; 26,** 671
– –, **heterocyclic 8,** 657 suppl. **26**
  Δ⁴-imidazoline-2-thiones
   18, 461 suppl. **28**
  oxazolidines **28,** 335; **30,** 120
  –, retention **30,** 9
  Δ²-oxazolines **27,** 257
  –, glycosylation with – **14,** 174
– **orthocarboxylic acid esters**
  27, 217
– startg. m. f.
  acylochlorosugars **27,** 572
  glycosyl phosphates **30,** 59
– Δ²-**oxazoline ring, 2-amino-
   28,** 341
– startg. m. f.
  pyrimidine nucleosides **28,** 341
Carbohydrates
–, O-acylation, total **28,** 88
–, O,O-alkylidene derivs.
  26, 171; **28,** 166
  formals **23,** 248 suppl. **28**
–, chain extension **3,** 554
  suppl. **26**
–, deacylation HO↓↑C
–, degradation via Hofmann
  degradation of amide **27,** 540
– from thioglycosides **30,** 141
–, introduction, relayed, into –
  of alkylthio groups **29,** 555
–, oxidation in carbohydrate
  chemistry, review **19,** 307
  suppl. **26**

–, 1-phosphorylation **27,** 114
–, removal of protective groups
  s. Protective groups, removal
– s. a.
  acylohalogeno-sugars
  amino-
  anhydro-
  cyano-
  disacccharides
  glycos...
  hexoses
  pentoses
  sugar...
– startg. m. f.
  aldonic acids **24,** 121
– –, review **4,** 151 suppl. **28**
  aldonolactones **21,** 285
   suppl. **27; 24,** 254 suppl. **26**
  carbohydrate Δ²-oxazoline
   ring, 2-amino- **28,** 341
  heterocyclics **28,** 446
  pyrimidine nucleosides
   28, 341
  thiazolidines **9,** 666 suppl. **26**
–, synthesis s. a. C-Glycosides
–, Wittig synthesis of – **10,** 633
  suppl. **23; 17,** 892 suppl. **27;**
  25, 626
P-**Carbohydrates 30,** 445
Carbohydrates, branched
– special s.
  nucleosides, sugar-branched
–, synthesis **22,** 661 suppl. **27; 28;**
  29, 737
–, **unsatd.**
–, review **29,** 964
– special s.
  ethylenecarbohydrates
Carbohydrazones s. Thiocarbohydrazones
β-**Carbolines** s. 9H-Pyrid-
  [3,4-b]indoles
γ-**Carbolines** s. 5H-Pyrid-
  [4,3-b]indoles
Carbo-p-methoxybenzoyl
  as N-protective group **27,** 404
Carbonate anion s. Hydrogen
  carbonate anion
Carbonates s. Carbonic acid
  esters, Oxime carbonates
Carbon atoms s. C-Atoms
Carbon-carbon bonds
  (s. a. under Double bonds)
–, formation under mild
  conditions **27,** 118
–, – with nickel complex
  catalysts, review **20,** 531
  suppl. **27**
– –, **multiple, electron-deficient**
–, addition to carbocycles,
  strained, review **26,** 704
Carbon chain s. C-Chain
– **dioxide**
–, activation of amino groups
  in peptide synthesis with –
  18, 435 suppl. **27**
–, elimination s. Decarboxylation
– startg. m. f.
  allenedicarboxylic acid
   amides, sym. **27,** 687
  carbamic acid silyl esters
   29, 438
  4-pyrones **27,** 693
  ureas, sym. **30,** 243
– **eliminators, cyclic**
– startg. m. f.
  heterocyclics **26,** 449

– **disulfide**
– as reactant **27,** 623; **29,** 563
  as reagent **27,** 613
–, insertion of – s. Dithio-
  carboxylation
–, reaction with active methyl
  and methylene groups **30,** 466
–, ring closures with – **27,** 358
– startg. m. f.
  1,4,2-dithiazine 1,1-dioxides
   28, 533
  1,3-dithiolane-2-thiones **28,** 546
  dithiourethans **27,** 594
  (hydroxythiol)thionocarbonic
   acid esters **27,** 175
  N-hydroxythioureas **26,** 334
  isothiocyanates **27,** 366, 486
  orthocarbonic acid esters
   4, 249 suppl. **28**
  thiolactams **27,** 613
  thioureas, sym. **27,** 366, 486;
   30, 243
Carbonic acid anhydrides
  s. Alkoxyformic acid
  anhydrides
– – **chloride esters**
– special s.
  p-nitrophenoxycarbonyl
   chloride
– – **derivs. 26,** 625
– – **ester fluorides**
– startg. m. f.
  urethans **23,** 404 suppl. **26**
– – **groups, cyclic,** migration
  28, 118
– – **esters**
– as reactants **28,** 55
–, C-carbalkoxylation with –
  28, 732
– special s.
  dialkyl carbonate diphenyl –
– startg. m. f.
  ethers **27,** 273
– – –, **active** (s. a. β,β,β-Tri-
  chlorethyl carbonates)
– startg. m. f.
  carbonic acid esters, mixed
   26, 219
– – –, **cyclic**
– from
  diol monosulfonates **30,** 140
  diols **24,** 170; **29,** 197
– special s.
  1,3-dioxanes, 2-oxo-
  1,3-dioxolan-2-ones
– – –, **mixed 27,** 404
– from
  carbonic acid esters, active
   26, 219
– special s.
  aryl isopropyl carbonates
  aryloxycarbonyloxy groups
  ethyl carbonates
  phenyl –
– startg. m. f.
  γ-lactones **27,** 778
  urethans **27,** 404; **28,** 375
– – –, **sym.**
– from
  alcohols **28,** 322
Carbonimidoyl dihalides
  s. Iminocarbonylhalides
**Carbonium hexachloroanti-
  monates** as reagents **29,** 307
– **ions 26,** 262
–, condensation via – **28,** 764
– generated in situ **28,** 993

–, C-imidomethylation via –
  26, 888
–, substitution with – 27, 762
– salts
– special s.
  alkoxyaminomethinium salts
  alkoxycarbonium –
  aminocarbonium –
  carboxonium salts
  cyclopropenium –
  oxocarbonium –
  triphenylcarbonium –
  tropenium –
Carbon monoxide
  (s. a. Carbonylation)
  30, 276, 606
– as reagent 27, 812; 28, 18, 363;
  30, 326
–, elimination s. Decarbonylation
–, reactions with ethylene
  derivs., review 16, 750
  suppl. 27
– startg. m. f.
  α-acylaminocarboxylic acids
  28, 613
  carbonic acid esters, sym.
  28, 322
  carboxylic – – 28, 826
  formamides 26, 282
  imidazoles, 2,4,5-trialkyl-,
  sym. 28, 665
  ketones, cyclic 28, 632
  maleimides, N-subst. 28, 766
  α-methylene-γ-lactones 30, 502
  4-quinazolones 28, 710
  ureas 26, 329
  urethans 28, 322
**Carbon-nitrogen double bonds,
  photochemistry,** review 28, 643
Carbon oxide sulfide
– as reactant 29, 531
– as reagent 29, 531
– **suboxide**
–, reactions, review 17, 652
  suppl. 29
– startg. m. f.
  anhydro-pyrido[1,2-a]-
  pyrimidium hydroxides,
  2-hydroxy-4-oxo- 26, 283
  1,3-thiazinium betaine ring
  28, 558
– **tetrabromide**
– as reactant 28, 819
– as reagent 26, 721
– **tetrachloride** (s. a. under
  Tris(dimethylamino)-
  phosphine)
– as reactant 27, 418; 28, 752;
  29, 511
– as reagent 21, 302 suppl. 29;
  26, 470; 26, 471 suppl. 29; 27,
  415, 418, 514, 978; 28, 144, 263,
  913; 29, 100, 292, 511, 973
– startg. m. f.
  dichloromethylene compds.
  27, 866
  4-quinazolones 28, 710
– **tetrahalides** (s. a. Tetrahalogenomethanes) 29, 510
Carbonylation
–, carboxylic acids and esters
  by –, review 12, 743 suppl. 28
– of
  alcohols 28, 322, 612; 30, 452
  amines 28, 322

ethylene and acetylene
  derivs., review 3, 588
  suppl. 26
titanium compds., organo-
  29, 858
Carbonyl chloride s. Phosgene
– **compds.** s. a. Carboxylic
  acid..., Dicarbonyl compds.,
  Oxo –, Polycarbonyl –
– special s.
  halogenocarbonyl compds.
N,N′-Carbonyldi(azoles)
– special s.
  N,N′-carbonyldiimidazole
  N,N′-carbonyldi-(1,2,4-
  oxadiazol-5-ones)
N,N′-Carbonyldiimidazole
– as reactant 18, 205 suppl. 29;
  20, 637 suppl. 29
– as reagent 13, 159 suppl. 21,
  26; 27, 611; 28, 440
– prepn. 25, 244
– startg. m. f.
  carbonic acid esters, cyclic
  29, 197
**N,N′-Carbonyldi-(1,2,4-
  oxadiazol-5-ones)**
–, peptide synthesis with –
  14, 452 suppl. 28
Carbonyl fluoride as reactant
  30, 381
– **groups**
– from
  carbon-carbon double bonds
  27, 158
  selenocarbonyl groups 27, 231
  thiocarbonyl – 27, 231
– –, **electrophilic,** reactions
  27, 759
– **halides**
– special s.
  carbonyl fluoride
  phosgene
– **oxygen** (s. a. under
  Replacement)
–, alkylation, reductive 30, 487
N-Carbonylsulfamoyl chloride
  s. Chlorosulfonyl isocyanate
Carbonyl sulfide s. Carbon
  oxide sulfide
N-Carbonylureas
– special s.
  α-halogeno-N-carbonylureas
Carboranes 28, 70
–, chemistry, review 28, 70
–, review 28, 70 suppl. 30
**Carbostyril-3-carboxylic acid
  esters, 3,4-dihydro-3,4-oxido-**
– startg. m. f.
  oxindoles 28, 988
Carbostyrils 26, 954
– from
  cinnamanilides 28, 984
  isocyanates and acetylene
  derivs. 26, 694
  succinimide, 3-o-amino-
  arylidene- 26, 152
–, **N-amino-** 26, 815
– **3,4-dihydro-3-amino-
  4-hydroxy-**
– startg. m. f.
  2H-1,4-benzodiazepin-2-ones,
  1,3-dihydro- 28, 437a
– **3-hydroxy-**
– by ring expansion 29, 625
– from
  isatins and diazo compds.
  29, 625

Carbotrichlorethoxyl
– as N-protective group 27, 24
**Carboxamidium salts,
  vinylogous**
– startg. m. f.
  3-keto-1,2-en-1-olethers,
  sym. 26, 184
Carboxonium salts,
  N-alkylation with – 26, 454
– –, **cyclic 22,** 261
1-Carboxyacetals, cyclic
– startg. m. f.
  acoxyhalides 27, 583
**Carboxydicarboxylic acid
  imides** 29, 442
Carboxyethyleneurethans
– startg. m. f.
  ethylenelactones 30, 176
α-Carboxylactams
– from
  α-ketolactams, ring contraction 30, 90
α-Carboxy-γ-lactones
– startg. m. f.
  α-halogeno-γ-lactones
  16, 620 suppl. 28
Carboxylates s. Mercury(II)
  carboxylates, Sodium –
C-Carboxylation 29, 615
– of hydrocarbons 28, 599
–, **ar.,** via diaryl ketones
  28, 600
Carboxyl groups, angular
–, introduction, stereospecific
  29, 226
Carboxylic acid amid...
  s. a. Amid...
– – **amides** (s. a. Carbam...)
–, N-alkylation, reductive
  29, 363
–, cleavage with neighboring
  group participation 26, 28
– from
  1-acylsemicarbazides 26, 503
  aldehydes 27, 371
  –, synthesis with addition
  of 1 C-atom 29, 443
  aldoximes 30, 98
  α-alkoxy-α,β-ethylenenitriles
  29, 443
  azomethines 26, 124
  carboxylic acid esters 26, 358;
  30, 262
– – hydrazides 30, 14
– – methylthiomethyl esters
  29, 230
– acids 25, 400 suppl. 27; 26,
  367/8, 376/7; 28, 144
– – (unsubst. amides) 26, 389
– – and sulfamic acids
  29, 375
– – and sulfurous acid
  diamides 30, 654
– acid thioamides 28, 180
  β-diketones and
  N-halogenamines 28, 416
  α,β-ethyleneketones 27, 134
  halogenimidium salts 26, 404
  hydroxamic acid benzyl
  esters 29, 20
  hydroxamic acids 27, 90
  nitriles 26, 989; 27, 132; 29, 196
  oxaziridines 26, 17
  oximes (unsubst. amides)
  29, 165
  thiolic acids 28, 403
  ureas 28, 776
–, Michael addition of – 28, 647

(Carboxylic acid amides)
- special s.
  acylaminocarboxylic acid
    amides
  acylhydrazinocarboxylic – –
  alkoxycarboxylic – –
  alkylthiocarboxylic – –
  allenecarboxylic – –
  aminocarboxylic – –
  aryloxycarboxylic – –
  biphenyl-2-carboxamides
  (O-carbamylhydroxy(amino)-
    carboxylic acid amides
  diazocarboxylic – –
  ethylenecarboxylic – –
  halogenocarboxylic – –
  hydroxycarboxylic – –
  hydroxylaminocarboxylic – –
  ketocarboxylic – –
  lactams
  nitrocarboxylic acid amides
  pteridine-6-carboxamides
  silylcarboxylic acid amides
  sulfonyloxycarboxylic – –
- startg. m. f.
  acylamidines **29**, 305
  acylamidrazones **30**, 288
  4-alkylidene-2-amino-
    1,3-dioxolanes **28**, 167
  amide acetals **26**, 933
  amidines **10**, 329; **27**, 416
  amines, reduction **26**, 64, 389
  N-carbamylamidinium salts
    29, 370
  carboxylic acids **29**, 194
  3(2H)-furanones **29**, 844
  hydrocarbons
    s. Decarboxamidation
  iminoester fluoroborates
    26, 933
  iminoesters **28**, 139
  iminohalides **27**, 575
  isothiazoles **26**, 841
  ketones, synthesis **29**, 772
  lactams **27**, 801
  phthalans, 1-hydroxy-
    3-acylamino- **21**, 344
    suppl. **26**
  pyrazolo[3,4-d]pyrimidines,
    4-amino- **30**, 252
  pyrimidines, 4-amino- **26**, 817
  4-quinazolones **28**, 418
  thiolic acids **29**, 563
  N,N,N-tri(acylamines) **26**, 420
- , substitution (s. a. under
    N-Alkylation)
  N-substitution **27**, 472
- , transamidation **28**, 424
- – –, **N,N-disubst.**
  (s. a. Dimethylacetamide,
   Dimethylformamide)
- startg. m. f.
  alcohols **28**, 54
  aldehydes **26**, 69
  azomethinium salts, cyclic
    30, 654
- – –, **isomeric 26**, 308
- – –, **N-subst.** (s. a. Peptides)
- from
  carboxylic acid chlorides
    29, 416
  – – esters **24**, 398;
    27, 394, 398
  – acids **23**, 415; **27**, 417;
    28, 353; **29**, 287
  isocyanates **26**, 677; **27**, 676;
    29, 635

- special s.
  carboxylic acid *tert*-
    butylamides
- startg. m. f.
  nitriles **26**, 501, 507
- – –, **vinylogous** (s. a.
    β-Amino-α,β-ethyleneketones)
- , O-acylation **26**, 112
- by desulfuration **26**, 982
- , reactions with – **26**, 660
- special s.
  carboxylic acid anilides,
    vinylogous
- startg. m. f.
  amidinium salts, vinylogous
    26, 453
  β-aminoketones **29**, 46
- – **anhydrides**
- as reactants **27**, 741
- from
  carboxylic acid chlorides
    (2 molecules) **28**, 162
  – acids **27**, 260; **28**, 143/4
  mercury(II) carboxylates
- special s. **28**, 213
  butyric anhydride
  dicarboxylic acid anhydrides
- startg. m. f.
  α,β-acetyleneketones **27**, 776
  aldehydes **29**, 64
  carboxylic acid esters
    by decarbonylation,
    oxidative **26**, 255
  α,β-ethyleneketones **30**, 554
  α,β-ethylene-(α- or β-)silyl-
    ketones **29**, 800
  isocyanates **28**, 400
  pyran ring **30**, 543
- – –, **mixed** (s. a. Carbamic
    carboxylic anhydrides,
    Carboxylic alkoxyformic –)
- special s.
  acetic formic anhydrides
- startg. m. f.
  ketones **20**, 633 suppl. **29**
- – **anilides**
- special s.
  acetanilides
- – –, **vinylogous**
- startg. m. f.
  quinoline ring **28**, 917
- – **aryl esters** (s. a. Phenol-
    esters)
- from
  carboxylic acids and
    phenols **26**, 180; **27**, 193
- – **azides**
- from
  carboxylic acid chlorides
    30, 460
- startg. m. f.
  acylamines **29**, 782; **30**, 460
  acylimidophosphoriso-
    cyanatidates **27**, 308
  N-acyliminoesters **29**, 395
  aziridine ring, N-acyl- **29**, 398
  carboxylic acid hydrazides
    30, 280
  isocyanates **30**, 460
- – **benzhydryl esters** as lipo-
    philic esters **3**, 177 suppl. **28**
- – **benzyl esters 26**, 182
    suppl. **30**
- – **boryl esters** s. Boryl
    carboxylates
- – *tert*-**butylamides**
- as intermediates **27**, 326
- – *tert*-**butyl esters**

- as intermediates **27**, 235
- – **chlorides**
- from
  carboxylic acids **26**, 957; **29**,
    513, 593; **30**, 359, 361
  – acid silyl esters **29**, 524
  1-chlorosulfines **27**, 270
  hydrocarbons s. Replacement
    of hydrogen
- special s.
  chloroformic acid esters
  chloroformyldicarboxylic
    acid anhydrides
  dicarboxylic acid chloride...
  1,3-dithiane, 2-chloro-
    carbonyl –
  ethylenecarboxylic acid
    chlorides
  polyhalogenocarboxylic – –
- startg. m. f.
  α,β-acetyleneketones **12**, 855
    suppl. **26**
  1,2-acoxymesylates **27**, 81
  acylamines **30**, 460
  acyl peroxides **26**, 202
  aldehydes **27**, 79, 81
  aldehydes-1-d **11**, 135
    suppl. **29**
  α-aminoketones **28**, 811
  benzyl ketones **27**, 865
  carboxylic acid amides,
    N-subst. **29**, 416
  – – anhydrides (from
    2 molecules) **28**, 162
  – – azides **30**, 460
  cyclobutanone ring **30**, 487
  diacylamines **26**, 420
  N,N-diacylhydrazines
    27, 467
  β-diketones **28**, 104
  enolesters (from 2 molecules),
    synthesis **30**, 605
  γ-enollactones **30**, 606
  α,β-ethylenealdehydes,
    synthesis with addition
    of 2 C-atoms **29**, 846
  α,β-ethyleneketones **28**, 866
  β,γ-ethylene-β-siloxy-
    carboxylic acid esters
    28, 809
  β-halogenoketones **2**, 617;
    27, 123
  hydroxamic acids, N-aryl-
    27, 460, 479
  isocyanates **28**, 400; **30**, 460
  α-ketocarboxylic acid amides
    27, 847
  β-ketocarboxylic acid esters
    10, 616
  ketones **27**, 852; **28**, 827; **30**,
    589, 603
  –, cyclic **26**, 957; **29**, 671
  2-oxa-3-borolenes **28**, 866
  oxazoles **27**, 477; **28**, 811
  2-pyridone ring, N-condensed
    26, 855
  2-pyrones **24**, 851; **27**, 848
  4-quinazolones, 3-alkoxy-
    29, 406
  silylethynyl ketones **29**, 846
  tetrazoles **26**, 444
  1,3,4-thiadiazoles, 2-alkylthio-
    29, 568
  2,4,9-trioxaadamantanes,
    7-halogeno- (from 3 mole-
    cules) **29**, 839
- – **derivs.**

-, aziridine ring opening with -
 26, 296
- from
  halides, synthesis with
   addition of 1 C-atom
   27, 869 suppl. 28
- special s.
  dienecarboxylic acid derivs.
- - **dianions**
  - as intermediates 29, 874
-, reactions, decarboxylative,
   via - 26, 923
-, syntheses via - 28, 733
- - **esters** (s. a. Acylation,
   Carbalkoxy..., Orthocarboxylic acid esters)
-, α-alkylation 27, 843
- by carbonylation, review
   12, 743 suppl. 28
- from
  acetals 14, 598 suppl. 29;
   28, 235
  -, cyclic 26, 147
  α-acoxycarboxylic acid esters
   27, 92
  β-alkoxy-α,β-ethylene-
   carboxylic acid esters
   26, 54
  1,1-alkoxyhydroperoxides
   28, 215
  ammonium carboxylates,
   quaternary 26, 182
  carboxylic acid anhydrides
   by decarbonylation,
   oxidative 26, 255
  - - methylthiomethyl esters
   29, 230
  - acids (s. a. Esterification)
   OC↓↑O; 26, 421; 27, 419;
   28, 179, 202
  - - and tert-butyl ethers
   26, 168
  - - and enolethers 28, 189
  - - and halides 12, 296;
   27, 219; 28, 165
  - - and isoureas 29, 547
  - - and oxime carbonates
   30, 260
  - - and phosphates 29, 222
  - - and phosphorodiamidites
   28, 142
  - - and sulfonates 26, 944
  - - via ammonium carboxylates, quaternary 26, 182
   sodium carboxylates
   18, 935 suppl. 28
  cyclopropylketones 28, 96
  4,7-dioxaspiro[2,4]heptanes
   29, 493
  1,3-dithianes 28, 177
  -, 2-alkylthio- 28, 177
  dithiocarboxylic acid esters
   29, 214
  α,β-ethylenealdehydes
   28, 94
  ethylenecarboxylic acid
   esters 26, 35
  α,β-ethylenecarboxylic - -
   30, 602
  α,β-ethyleneketones 27, 134
  1-ethynyl-N-nitrosoureas
   28, 150
  1-ethynyl-N-nitrosourethans
   28, 150
  halides, synthesis with
   addition of 1 C-atom
   28, 826

α-halogenoketones with
  rearrangement, skeletal
  27, 209
hydrocarbons
  s. Carbalkoxylation
α-hydroxycarboxylic acid
  esters 27, 92
α-hydroxyketones 29, 162
β-ketocarboxylic acid esters
  26, 54; 29, 499
malonic acid esters 29, 84
thiolic acid 2-pyridyl esters
  30, 142
thionocarboxylic acid esters
  28, 213
trichloromethyl ketones
  29, 227, 816
-, homologization 30, 545
-, hydrolysis HO↓↑C
-, reaction with trimethylchlorosilane and sodium,
  review 25, 589 suppl. 26
- special s.
  acetylenecarboxylic acid
   esters
  acoxy compds.
  acylaminocarboxylic acid
   esters
  alkoxycarboxylic - -
  alkylthiocarboxylic - -
  aminocarboxylic - -
  arylacetic - -
  azidocarboxylic - -
  carbonic - -
  carboxylic acid aryl -
  - - benzhydryl -
  - - benzyl -
  - - tert-butyl -
  - - methyl -
  - - methylthiomethyl -
  - - isopropenyl -
  chloroformic acid esters
  cyanatocarboxylic - -
  cyclopropanecarboxylic - -
  α-diazocarboxylic - -
  dicarboxylic - -
  glycerides
  hydroxycarboxylic acid esters
  hydrazinocarboxylic - -
  hydroxylaminocarboxylic - -
  hydroxyureidocarboxylic - -
  isocyanatocarboxylic - -
  mercaptocarboxylic - -
  mesitoic - -
  oxocarboxylic - -
  pyridinecarboxylic - -
  siloxycarboxylic - -
  silylcarboxylic - -
- startg. m. f.
  acoxy compds. 27, 68
  acylhydrazones 27, 377
  alcohols, prim., reduction
   26, 57, 62; 27, 64; 28, 50, 53;
   29, 39
  -, tert., synthesis 26, 662
  aldehydes 28, 53
  α-alkylthio-β-ketosulfoxides
   30, 545
  carboxylic acid amides
   26, 358; 30, 262
  - - -, N-subst. 27, 394, 398
  - - halides 29, 517
  α-dicarboxylic acid monoesters 28, 598
  dichloromethyl ketones
   30, 549
  1,1-diethynylalcohols 26, 793

α,β-ethylenecarboxylic acid
  esters 29, 912
ethylene derivs. CC↑O
- -, decarbalkoxylation
  29, 989
α,β-ethyleneoxo compds.
  30, 505
α-halogenocarboxylic acid
  esters 27, 560
hydrocarbons HC↓↑C
-, reduction 30, 30
α-hydroxyaldehydes, synthesis
  with addition of 1 C-atom
  30, 545
ketene mercaptals 29, 552
β-ketocarboxylic acid esters
  (from 2 molecules) 3, 638;
  7, 771
γ-ketocarboxylic - -,
  synthesis 29, 667
ketones, cyclic 28, 953
methyl ketones 30, 414
phenolethers 29, 182
O-silyl O-alkyl ketenacetals
  28, 77
succinic acid esters, sym.
  (from 2 molecules) 28, 703
  suppl. 30; 28, 706
thiolic - - 29, 553
-, synthesis 26, 872
-, α-substitution 27, 843
  suppl. 29
α-sulfenylation 30, 414
- - -, **active** (s. a. under
  Peptides from -) 17, 296
  suppl. 26
- by redoxesterification
  26, 175
-, comparison in Merrifield
  synthesis 16, 424 suppl. 29
- special s.
  carboxylic acid p-nitrophenyl
   esters
  - - p-sulfonylphenyl -
  - - -, **cyclic** s. Lactones
  - - -, **hindered** 12, 304
   suppl. 27, 28; 23, 431
   suppl. 28; 25, 141 suppl. 27;
   28, 165; 29, 222
  - - -, **lipophilic** s. Carboxylic
   acid benzhydryl esters
- - **fluorides**
- special s.
  allenecarboxylic acid
   fluorides
  perfluorocarboxylic - -
- - **halides**
- as reactants 21, 556 suppl. 27
- from
  carboxylic acid esters 29, 517
  - acids 30, 364
- special s.
  acetyl halides
  carboxylic acid chlorides
  - - fluorides
  cyanocarboxylic - -
  - acid halides
  halogenocarboxylic - -
  halogenosulfonyl-
   carboxylic - -
- startg. m. f.
  aldehydes 27, 76
  β-amino-α,β-ethyleneketones
   26, 846
  α-halogenocarboxylic acid
   halides 27, 565
- - -, **vinylogous** s. 5-Halogeno-
   2,4-dienals

(Carboxylic acid)
- - **hydrazides** (s. a. Acylhydrazines)
- - from
   carboxylic acid azides
   30, 280
-, oxidation, peptide synthesis
   via - **26,** 398
- special s.
   ketocarboxylic acid
   hydrazides
- startg. m. f.
   N'-acyl-1,1-cyanohydrazines
   27, 780
   carboxylic acid amides **30,** 14
   - acids **28,** 158
   N,N-diacylhydrazines, sym.
   27, 445
   Δ³-5-pyrazolones **28,** 781
   triazenes, 3-acyl- **30,** 189
   1,2,4-triazole ring **28,** 404
- - -, **labeled 18,** 441 suppl. **29**
- - **imides**
- special s.
   halogenocarboxylic acid
   imides
- - **isopropenyl esters**
- special s.
   isopropenyl acetate
- - **methyl esters,** cleavage,
   preferential **30,** 6
- - - -, **hindered**
-, cleavage **29,** 963
- - **p-methylsulfonylphenyl
   esters**
- startg. m. f.
   peptides **18,** 434 suppl. **26**
- - **methylthiomethyl esters**
- startg. m. f.
   carboxylic acid amides
   29, 230
- - - esters **29,** 230
- - **p-nitrophenyl** -
-, N-acylation, selective, with -
   26, 363
- - **2-(p-nitrophenylthio)ethyl** -
-, protection of carboxyl groups
   as - **26,** 3
- **acids** (s. a. Carboxy...,
   Carboxylates)
- by carbonylation, review
   12, 743 suppl. **28**
- by ring opening **26,** 131
- -, derivatives
   isatin-1-ylmethyl esters **28,** 165
- from
   acetylene derivs. **27,** 245
   alcohols, carbonylation
   28, 612
   -, -, tert. acids **30,** 452
   -, oxidation **27,** 187; **28,** 137
   aldehydes **30,** 66
   carboxylic acid amides
   29, 194
   - - esters **28,** 12
   - - hydrazides **28,** 158
   - - silyl esters **28,** 733
   cyclopropane ring, 1-halo-
   geno-1-trifluoroacetoxy-
   29, 22
   o-dicarboxylic acids **12,** 711
   suppl. **28**
   α,β-ethylenecarboxylic acids
   26, 51, 54; **28,** 39
   ethylene derivs., cleavage,
   oxidative **27,** 246
   - -, synthesis **9,** 770

- -, - with addition of
   1 C-atom, review **16,** 750
   suppl. **26**
   halides, synthesis with
   addition of 1 C-atom
   27, 869 suppl. **28**
   hydrocarbons, synthesis with
   addition of 1 C-atom
   s. C-Carboxylation
   hydroximinohalides **28,** 171
   isocycles **26,** 230
   β-ketocarboxylic acid esters
   26, 54
   ketocarboxylic acids **3,** 56
   ketones, hydrolysis **29,** 236
   lactams **28,** 186
   lactols **30,** 93
   lactones, reduction **27,** 32
   mercury compds., dialkyl-
   26, 218
   nitriles **29,** 196
   oxo compds., synthesis with
   addition of 1 C-atom
   28, 729
   ozonides **28,** 214
   1-(sulfonyl)eneformylamines
   28, 729
   tosylhydrazones **28,** 775
   suppl. **30**
-, metalation **26,** 664
- special s.
   acetic acids
   acylaminocarboxylic -
   alkylthiocarboxylic -
   aminocarboxylic -
   arylcarboxylic -
   aryloxycarboxylic -
   capric acid
   cyclopropylcarboxylic -
   dicarboxylic acids
   enaminocarboxylic -
   ethylenecarboxylic -
   β-(2-furyl)propionic -
   guanidinocarboxylic -
   halogenocarboxylic -
   hydroxycarboxylic -
   imidocarboxylic -
   iminocarboxylic -
   mercaptocarboxylic -
   nitrocarboxylic -
   oxidocarboxylic -
   oxocarboxylic -
   perchlorohomocubane-
   carboxylic acid
   propionic -
- startg. m. f.
   α-acoxycarboxylic acids
   (from 2 molecules) **27,** 178
   acoxyhalides **29,** 487
   acoxylamines **26,** 172
   acyl peroxides **27,** 95
   alcohols, degradation with
   loss of 1 C-atom **30,** 177
   -, prim. **27,** 65; **28,** 52;
   29, 57, 243
   aldehydes via alcohols, prim.
   29, 89
   - - carboxylic alkoxyformic
   anhydrides **28,** 66
   - - dianions **26,** 923
   amines **26,** 389
   -, sec. **27,** 392
   α-aminocarboxylic acids
   28, 312
   β-amino-α,β-ethylene-
   aldehydes **30,** 634
   azomethinium salts, cyclic
   30, 152

benzimidazoles **27,** 410;
   30, 265
carboxylic acid amides
   25, 400 suppl. **27;** **26,** 367/8,
   376/7; **28,** 144
- - -, unsubst. **26,** 389
- - aryl esters **26,** 180; **27,** 193
- - esters (s. a. Esterification)
   OC↓↑O; **26,** 421; **27,** 419; **28,**
   179, 202
- - halides **30,** 364
- - chlorides **26,** 957;
   29, 513, 593
- - silyl ester hydrochlorides
   27, 512
N,N'-diacylhydrazines
   13, 434 suppl. **26**
α-diazoketones **26,** 806
1,6-diketones (from 2 molecules) **29,** 825
α,β-ethylenecarboxylic acids
   27, 918
γ,δ-ethyleneketones **26,** 873
glycols (from 2 molecules)
   26, 913
halides, degradation with loss
   of 1 C-atom **26,** 554
α-halogenosulfonylcarboxylic
   acid halides **28,** 541
hydrocarbons, decarboxylation
   HC↑↓C
- (methyl groups) **26,** 66;
   27, 65; **28,** 52
-, branched **28,** 845
α-hydroxycarboxylic acids
   28, 82
1-hydroxy-1,1-diphosphonic
   acids **29,** 593
Δ²-imidazolines **27,** 410
ketones, cyclic **26,** 957; **28,** 909
-, -, via carboxylic alkoxyformic anhydrides **27,** 933
lactones **28,** 205
nitriles **27,** 422
nitro compds., aliphatic
   26, 923
1,3,4-oxadiazoles (from
   2 molecules) **26,** 380
α-oxo-N-heterocyclics **25,** 669
phthalides **26,** 818
quinolines, 2-(o-acylanilino)-
   26, 816
selenolic acid esters **15,** 470
   suppl. **26**
thiolic acid esters **25,** 440,
   457; **26,** 621; **27,** 611
trichlorosilanes **26,** 66
   suppl. **29**
N-tritylacoxylamines **26,** 172
urethans **28,** 353
vinyl ketones **16,** 853 suppl. **29**
-, synthesis **26,** 872
-, -, asym. **29,** 807
- -, **hindered,** reduction **29,** 57
- -, **protonated** s. Acetic acid,
   protonated
- **acid silyl ester hydrochlorides**
- from
   carboxylic acids **27,** 512
- - - **esters**
- special s.
   aminocarboxylic acid silyl
   esters
   halogenocarboxylic - - -
   isocyanatocarboxylic - - -
   isothiocyanatocarboxylic - - -

- startg. m. f.
  carboxylic acid chlorides
  29, 524
  - acids **28,** 733
- - **p-sulfonylphenyl esters**
- special s.
  carboxylic acid p-methyl-
    sulfonylphenyl esters
- - **thioamides** (s. a. Thio-
    carbam...)
- from
  iminohalides **30,** 410
  thioiminoesters **26,** 317
-, reactions, review **27,** 626
- special s.
  acylaminocarboxylic acid
    thioamides
  N-acylcarboxylic - -
  alkylthiocarboxylic - -
  carboxylic acid thio-
    morpholides
  ethylenecarboxylic acid
    thioamides
  hydrocarboxylic - -
  mercaptocarboxylic - -
- startg. m. f.
  β-amino-α,β-ethylene-
    carboxylic acid thioamides
  28, 770
  carboxylic acid amides **28,** 180
  iminodisulfides, sym. **22,** 590
  1-imino-1-sulfenamides **28,** 248
  mercury(II) mercaptides
  26, 567
  nitriles **29,** 456
  oxathiolium salts **28,** 568
  Δ²-thiazolium salts **27,** 626
  thioacylureas **30,** 215
  1,2,4-triazole ring **28,** 404
  ynamines **26,** 980
- - -, **subst.**
- from
  isothiocyanates **14,** 723
- - -, **vinylogous**
-, S-alkylation **26,** 574
- startg. m. f.
  thiopyrans, dihydro- **28,** 637
- - **thiohydrazides** (s. a. Thio-
    acylhydrazines)
- startg. m. f.
  1,3,4-thiadiazolium salts
  28, 553
- - **trifluoroethyl esters**
  as reactants **26,** 57
- - **thiomorpholides**
- from
  α-thioketocarboxylic acids
  27, 496
- - **ureides** s. Acylureas
- - **alkoxyformic anhydrides**
- startg. m. f.
  N-acoxyureas **27,** 194
  aldehydes **28,** 66
  α-diazoketones **26,** 806
  β-ketocarboxylic acid esters
  3, 703 suppl. 26
  ketones, cyclic **27,** 933
- - -, **polymer-based**
  s. Polymer-based carboxylic
    alkoxyformic anhydrides
  C-Carboxymethylation 22, 853
- of
  ethylene derivs. **28,** 713
- ar., review **18,** 905 suppl. 26
  o-Carboxy-N-oxides
- startg. m. f.
  α-oxo-N-heterocyclics **26,** 256

o-Carboxyphenylacetonitriles
- startg. m. f.
  isocoumarins, 3-acylamino-
  27, 470
Carboxysulfonic... s. Sulfocarboxylic...
Carboxysulfonium salts
- from
  alkylthiocarboxylic acids
  29, 263
- startg. m. f.
  lactones **29,** 260, 263
o-Carboxysulfoxides
- startg. m. f.
  3,1-benzoxathian-4-ones
  29, 250
Carboxythioethers
- from
  lactones **27,** 618
- startg. m. f.
  thiolactones **27,** 618
Catalysis s. a. Metal catalysis,
  Electrocatalysis
- of symmetry-disallowed
  reactions, review **20,** 531
  suppl. 29
-, **binuclear** s. Binuclear
  catalysis
-, **intramolecular** s. Assistance,
  intramolecular
Catalysts
- prepared in situ **27,** 61
- special s.
  chromite catalysts
  nickel
  palladium
  platinum
  Raney catalysts
  rhenium sulfide
  rhodium
  ruthenium
-, **homogeneous, polymer-based**
- special s.
  polystyrenetricarbonyl-
    chromium
-, **nucleophilic** (s. a. N-Hydroxybenzenesulfonamide)
  28, 267
Catechol s. a. Pyrocatechol
Catecholborane as reactant
  29, 497
Catechol monoesters
- from
  phenols **29,** 123
- startg. m. f.
  catechols **29,** 123
Catechols
- from
  catechol monoesters
  29, 123
  o-hydroxyacetophenones
  30, 67
  phenols **29,** 123; **30,** 67
Cathodic dimerization
  s. Dimerization, cathodic
Cathylates s. Ethyl carbonates
Cation exchanger, strong
  29, 484
- exchangers (s. a. Amberlyst,
  Dowex, KU-2) **7,** 730
  suppl. **27; 27,** 410; **29,** 30, 255,
  487; **30,** 23, 71
-, comparison **26,** 224
-, **optically active,** resolution
  of racemates on – **5,** 666
  suppl. 26

Cations s. a. Radical cations
-, **heterocyclic,** review **22,** 261
  suppl. 27
Cation specificity 28, 947
Celite 23, 912 suppl. **28;**
  29, 324, 435/6
Cephalosporin ring skeleton
- from
  penicillin ring skeleton
  29, 542
- startg. m. f.
  penicillin ring skeleton
  29, 542
Cephalosporins
- from
  penicillin S-oxides **29,** 585
Cephams, 4-hydroxy- 30, 395
Ceramidonines s. 9H-Naphth-
  [3,2,1-kl]acrid-9-ones
Cerium(IV) acetate
- startg. m. f.
  arylacetic acids **4,** 692
  suppl. 26
- **nitrate** s. Ammonium
  cerium(IV) nitrate
- **salt 24,** 768 suppl. 26
- **salts,** oxidations with –,
  review **28,** 132
Cesium alloys 28, 35
- **fluoride 14,** 606 suppl. **28;**
  30, 604
- as fluoride ion catalyst **25,** 124
  suppl. 26
C-Chain extension (s. a. Synthesis) CC⇓, CC⇑
- of
  carbohydrates **3,** 554 suppl. 26
- **isomerization,** cross-coupling
  with – **26,** 875 suppl. 26
Chains s. a. Hydrocarbon
  chains
-, **branched** s. Branched chains
Chalcone dihalides,
  o'-hydroxy-
- startg. m. f.
  flavones **30,** 170
Chalcones
- by Wittig synthesis **20,** 651
  suppl. 28
- startg. m. f.
  benzils **27,** 238
  pyrans, 2,3-dihydro-2-imino-
  26, 713
Characterization of compds.
  s. subentry 'derivs.'
Chelates s. Complex salts,
  inner
Chelating agents
  zinc chloride **29,** 894
Chelation
-, effect on stereospecificity
  16, 728 suppl. **26, 27**
Chiral compds. s. Compds.,
  optically active
Chirality S→C-transfer, intramolecular **29,** 250 suppl. **30**
Chloracetates, protection of
  alcohol groups as – **26,** 8
2-Chloracrylyl chloride, diene
  synthesis with – **28,** 222
Chloral as reactant **29,** 142
- **acetals**
- from
  ethylenealcohols **29,** 142
- startg. m. f.
  diols **29,** 142
- **condensation 26,** 660

Chloral hydrate
- startg. m. f.
  benzoxazole-2-aldoximines 26, 430
Chlor(alkoxysilanes)
- startg. m. f.
  chlorofluorides 29, 512
  fluoroamines 29, 512
Chloramine (s. a. Dichloramine) 26, 259; 27, 289, 587; 30, 218
- as reactant 28, 248; 29, 316
-, oxidations with – 26, 560
Chloramine-B
- as reactant 26, 451
Chloranil
- as reagent 24, 905 suppl. 29; 26, 462; 27, 906; 28, 784; 30, 256, 658
o-Chloranil adducts
-, photocyclization 27, 920
3-Chlor-1-ethylenecarbamyl chlorides
- from
  α,β-ethyleneazomethines 27, 545
- startg. m. f.
  1,3-dienecarbamyl chlorides 27, 545
Chloride ion as catalyst 28, 712
- – source
  tetraethylammonium chloride 26, 529
  tetramethylammonium – 27, 173
Chlorides s. a. Halides, Hydrochlorides, Replacement
Chlorination s. a. Halogenation, Replacement of hydrogen by halogen
-, ar., under mild conditions 26, 529
Chlorine (s. a. under Dimethyl sulfoxide and Thioanisole) 7, 617 suppl. 27; 18, 647 suppl. 28; 26, 103, 493, 930; 27, 290; 28, 183; 29, 9, 540
- as oxidant 21, 471 suppl. 27
- oxides s. Dichlorine heptoxide
- scavengers
  resorcinol 30, 66
  sulfamic acid 30, 66
Chlorobenzene
- as solvent 27, 700; 28, 548
p-Chlorobenzenethiol
as reagent 29, 412
1-Chlorobenzotriazole
- as reactant 28, 254
- as oxidant, mild 27, 913
-, chlorination with – 27, 564; 28, 489/90
-, cleavage of plumbanes and stannanes with – 27, 310
p-Chlorobenzoyl nitrite
as reagent 29, 897
Chloroborane s. Dichloroborane, Monochloroborane
1,2-Chlorobromides 6, 571 suppl. 29
Chlorocarboyl...
s. a. Carboxylic acid chlorides
N-Chlorocarbonyl-1-alkoxyenamines, cyclic
- from
  iminoesters, cyclic 27, 466

N-(Chlorocarbonyl)isocyanate
-, reactions with – 28, 473
- startg. m. f.
  α-halogeno-N-carbonylureas 28, 473
  heterocyclics 28, 473
Chlorocarbonylsulfenyl chloride
- startg. m. f.
  Δ⁴-2-thiazolones 29, 543
Chloro(chlorosulfates)
as intermediates 27, 579
Chloro(chlorosulfenyl)oxymethyl disulfides
- startg. m. f.
  chlorothionoformic acid esters 26, 632
Chlorodifluoromethane
- startg. m. f.
  trialkylcarbinols 27, 892
N-Chlorodiisopropylamine
as reagent 26, 527
Chlorofluorides
- from
  chlor(alkoxysilanes) 29, 512
- startg. m. f.
  fluoroamines 29, 512
1,1-Chlorofluorides
- from
  α,β-ethylenechlorides 29, 496
- startg. m. f.
  fluorides 16, 24 suppl. 26
Chloroform
- as reagent 26, 535; 27, 17; 29, 456
Chloroformamidine hydrochloride
- startg. m. f.
  pyrimidine ring, 2,4-diamino- 28, 401
Chloroformamidines
- from
  ureas 29, 511
- startg. m. f.
  α,β-acetyleneamidines 28, 672
  formamidinoylisothiocyanates 26, 413
  o- and p-nitroureas 28, 399
  quinolines, 2-amino- 28, 672
-, cyclic
- special s.
  1-aziridinecarboximidoyl chlorides
Chloroformamidinium chlorides
- startg. m. f.
  formamidinium phosphonic acid anhydrides 28, 595
Chloroformic acid esters (s. a. Chloroformoxy-...)
-, N-alkylation with – 28, 419
- as reagents 26, 838
- special s.
  ethyl chloroformate
  isobutyl –
  trichloromethyl –
- startg. m. f.
  diazoacetic acid esters 27, 828
  sulfonic – – 26, 169
Chloroformoxyimines, cyclic
- startg. m. f.
  chlorourethans, cyclic 27, 557
Chloroformyldicarboxylic acid anhydrides 29, 513
N-(Chloroglyoxyloyl)iminoesters
- from
  iminoesters 26, 458

- startg. m. f.
  acylisocyanates 26, 458
1,2-Chloroiodides
- from
  ethylene derivs. 26, 514
2-Chloromethyl-4-nitrophenyl phosphoric acid esters
- as intermediates 25, 92; 26, 222
2-Chloro-2-nitropropane
as 1-electron acceptor 11, 806 suppl. 26
N-Chloronylon-66
as reagent 28, 70a
m-Chloroperoxybenzoic acid
- as reactant 30, 655
- as reagent 9, 342 suppl. 29; 12, 156 suppl. 29; 26, 90/1, 136, 529, 584, 964; 27, 734; 28, 107, 886; 29, 132, 313, 527; 30, 57, 143/4, 669, 685
-, prepn. 20, 153 suppl. 26
Chlorophosphines (s. a. Replacement of silyl by dichlorophosphino groups)
- startg. m. f.
  phosphinic acid chlorides 26, 99
  – – esters 26, 99
  – acids 30, 159
  phosphinous acid esters 30, 159
N-(Chlorophosphoryl)-pyridinium betaines
- as acylating agents 27, 419
N-Chloropolymaleimide
as reagent 11, 609 suppl. 28
N-Chlorosaccharin, reactions 28, 202 suppl. 29
Chlorosilanes
- special s.
  trichlorosilanes
  trimethylchlorosilane
N-Chlorosuccinimide as reagent 13, 573 suppl. 29; 20, 431 suppl. 28; 26, 592; 28, 201, 480; 28, 512 suppl. 30; 29, 105, 504
N-Chlorosuccinimide-triethylamine 28, 201; 28, 512 suppl. 30
Chlorosulfination 25, 431
1-Chlorosulfines
- special s.
  carbamyl chlorosulfines
- startg. m. f.
  carboxylic acid chlorides 27, 270
Chlorosulfonic acid 10, 473 suppl. 26; 21, 600 suppl. 28; 28, 493
- as medium 30, 376
- – esters s. Sulfuric acid chloride esters
N-Chlorosulfonyl-2-azetidinones
- startg. m. f.
  ethylenenitriles 27, 975
Chlorosulfonyl isocyanate
- as chlorinating agent 29, 505
- as reactant 16, 461 suppl. 28
-, radical reactions with ethylene derivs. 26, 613
- startg. m. f.
  biurets, 3-hydroxy- 26, 330
Chlorothio(dichloroformimido)-cyanamide
-, reactions with – 26, 275

- startg. m. f.
  aminothio(dichloro-
  formimido)cyanamides
  26, 275
1,1-Chlorothioethers
-, **alkylation with** – **28,** 993
- from
  thioethers **30,** 687
- special s.
  trichloromethyl thioethers
- startg. m. f.
  ethylene derivs. **30,** 687
**Chlorothioformamidinium
chlorides**
- startg. m. f.
  1-alkylthioazomethinium salts,
  N-heterocyclic **27,** 829
Chlorothionoformic acid esters
- from
  chloro(chlorosulfenyl)oxy-
  methyl disulfides **26,** 632
- startg. m. f.
  trifluoromethyl ethers **29,** 523
o-Chlorotoluene s. under
  Sodium
**N-(2-Chloro-1,1,2-trifluoroethyl)-
diethylamine** s. 2-Chloro-
1,1,2-trifluorotriethylamine
**2-Chloro-1,1,2-trifluorotriethyl-
amine 28,** 501
**Chlorotris(dimethylamino)-
phosphonium perchlorate**
- as reagent **28,** 144
**Chlorotris(triphenylphosphine)-
rhodium(I)**
- as catalyst **17,** 160 suppl. **27;**
  17, 204 suppl. **29; 20,** 537
  suppl. **28, 29; 23,** 323
  suppl. **29; 26,** 507; **27,** 94;
  29, 42, 107, 109, 530; **30,**
  51, 695
- as reactant **26,** 725
Chlorourea as reagent **7,** 617
  suppl. **27**
Chlorourethans, cyclic
- from
  chloroformoxyimines, cyclic
  27, 557
Chloroxine s. 8-Quinolinol,
  5-chloro-
Chroman-2-carboxylic acids
- from
  acoxycyanohydrins **30,** 183
3-Chromanols 28, 653
4-Chromanols
- special s.
  isoflavan-4-ols
6-Chromanols 30, 180
4-Chromanone ring
- via Fries rearrangement
  28, 127
3-Chromanones 29, 237
4-Chromanones 28, 909
- special s.
  isoflavanones
- startg. m. f.
  chromans **24,** 80 suppl. **29**
Chroman ring 28, 914
Chromans (s. a. Benzopyrylium
  salts)
- from
  4-chromanones **24,** 80
    suppl. **29**
  phenols and 2-ethylene-
    alcohols **27,** 790a
- and ethylene derivs. **27,** 760
- special s.
  chromones

- startg. m. f.
  o-γ-halogenophenols
  30, 345
Chromatography
s. a. Adsorption
-, **preparative 9,** 248 suppl. **29;**
  26, 155; **29,** 495; **30,** 499
- –, **gas-liq. 27,** 335
Chromenes 26, 797
- by o-Claisen rearrangement
  27, 729
- by cyclodehydration **28,** 210
- from
  coumarins **27,** 198
  α,β-ethylenealdehydes **27,** 785
  o-hydroxyaldehydes **27,** 198
  propargyl ethers **29,** 697
- special s.
  flavenes
- startg. m. f.
  isoflavan-4-ols **28,** 865
-, **6-hydroxy-**
- startg. m. f.
  quinones **27,** 160
Chromic acid s. Chromium
  trioxide
Chromite catalysts
  s. under Copper
Chromium s. a. Chromyl...
–(II) acetate **26,** 192
–(II) chloride **27,** 554
- **complexes** s. Tricarbonylchromium complexes,
  Pentacarbonylchromium(0)-
  carbene
- oxide s. a. Chromium trioxide
- – -alumina **6,** 388 suppl. **26**
- – -copper(II) oxide
  s. Copper chromite
- oxide-lanthanum oxide-
  alumina-potassium oxide
  28, 543
–(II) perchlorate **22,** 80
  suppl. **28**
- salts, reductions with –,
  review **23,** 968 suppl. **29**
- sulfate **27,** 971; **29,** 792
–(III) sulfate **26,** 139
- trioxide **9,** 213 suppl. **27; 26,**
  165, 235; **27,** 124, 153/4, 183,
  326; **28,** 70, 866, 896; **29,** 233
- – -3,5-dimethylpyrazole
  complex **26,** 235 suppl. **29**
- – -graphite **28,** 203
- – -pyridine **8,** 208 suppl. **26;**
  26, 235; **28,** 135
Chromone-3-carboxaldehydes
- from
  o-hydroxyketones **29,** 785
Chromones
- from
  o-hydroxyketones **26,** 247
- special s.
  isoflavones
Chromosorb s. Kieselguhr
Chromyl chloride 26, 124;
  28, 481
Chugaev reaction, improved 6,
  841 suppl. **26**
α-Chymotrypsin as reagent
  28, 13
Cinnamanilides
- startg. m. f.
  carbostyrils **28,** 984
Cinnamic acids
- startg. m. f.
  thianaphthene-2-carbonyl
    chlorides, 3-chloro- **28,** 548

Cinnolines
- startg. m. f.
  indoles, 1-formamino- **27,** 324
-, **4-amino-**
- from
  α-cyanohydrazones **30,** 532
cis- s. Geospecificity, Isomers,
  Rearrangement
Claisen s. a. Darzens
- **condensation** (s. a. Ester
  condensation)
- –, **stereospecific, low temp.**
  30, 450
- **rearrangement** (s. a. Orthoester Claisen rearrangement,
  Thio-Claisen-)
- of
  allyl esters **28,** 683
  enolethers **26,** 742; **27,** 738;
    30, 653
- –, syntheses via – **30,** 560
  propargyl ethers **29,** 697
- to
  acoxy compds. **27,** 741
  γ,δ-ethyleneketones **28,** 749
- –, **geospecific 28,** 749
- –, **stereospecific**
- of
  enolethers **26,** 742
in situ-**Claisen rearrangement**
- with
  α,β-ethyleneketals **26,** 809
o-Claisen –
-, **direction 2,** 621 suppl. **27**
- of
  allyl ethers **28,** 688
-, **ring closure by** – **29,** 697
-, – **closures via** – **30,** 677
- to
  chromenes **27,** 729
Claisen-type –
- to
  imidazole-4-carboxylic acid
    esters **27,** 502
  oxazines, 5,6-dihydro-2-
    (γ,δ-ethylene)- **29,** 704
Clarke-Eschweiler reaction
-, ring closure by – **19,** 865
Clay (s. a. under Potassium
  hydroxide) **26,** 238
Cleavage (s. a. Lysis)
- of
  boron-carbon bonds **28,** 81
  carbon-carbon –
    s. C-Cleavage
- special s.
  fragmentation
-, **oxidative**
- of
  α-diketones **28,** 193
  ethers, β,γ-unsatd. **26,** 116
  α,β-ethyleneketones **26,** 228
  glycols **29,** 157
  β- to α-hydroxycarboxylic
    acid esters **26,** 253
-, –, **preferential**
- of
  carbon-carbon double bonds
    21, 269 suppl. **28**
-, **preferential** s. a. under
  Ethers
-, **reductive**
- of
  1,3-diene central bonds
    27, 49
  nitrogen-nitrogen –
    29, 17

Cleavage, selective
– of
  acetals, mixed **26**, 10
1,4-Cleavage s. 1,4-De-
  alcoholation
C-Cleavage
– of
  alcohols **29**, 268
  2-aminoalcohols **30**, 179
  bisphenols **27**, 986
  (*tert*-carbin)amines,
    N,N-dichloro- **27**, 240
  carbon-carbon double bonds
    30, 636
  α-dicarboxylic acid esters
    29, 811
  α-ketoaldehyde aldoxime side
    chains **29**, 267
  polyhalides **27**, 988
– to
  1,1-dihalides **30**, 375
  α-oxo-N-heterocyclics **15**, 242
  pyrylium ring, 4-methyl-
    27, 86
– with substitution, electro-
    philic **29**, 312
–, **hydrolytic**
– of
  anthraquinone rings **26**, 220
–, **reductive**, of xanthene-1,8-
    diones, octahydro- **30**, 45
C,O-Cleavage of enolesters
  with addition **27**, 314
N,N-Cleavage
– of
  azo groups **27**, 442
  hydrazines **28**, 331
Cleland's reagent
  s. Dithiothreitol
Clemmensen reduction
– of
  ketones, difunctional, review
    14, 782 suppl. **26**
Clorox 12, 325 suppl. **27**
Cobalt 28, 47
–, Raney-cobalt **14**, 404
  suppl. **28**
–(II) **acetate 26**, 230; **28**, 648
  suppl. **29**
– –/**hydrogen chloride 20**, 126
  suppl. **26**
– **ammine complex 19**, 444
  suppl. **26**
– **boride 26**, 15
–(II) **bromide 26**, 740
– – **bis(triphenylphosphine)**
  29, 680
– **carbonyl** (s. a. Carbatricobalt
    decacarbonyl derivs., Hydro-
    gen tetracarbonylcobaltate
    14, 530 suppl. **28**; **28**, 425, 613;
    30, 662
– – **complex**
–, protection of carbon-carbon
    triple bonds as – **27**, 140
–(II) **chloride 26**, 35, 993; **27**,
    55; **28**, 426
– **compds., organo-**
– startg. m. f.
  halides **26**, 552
– **complexes 23**, 715 suppl. **29**
– as intermediates **17**, 879
  suppl. **26**
– special s.
  bis(dimethylglyoximato)-
    cobalt(II)
– **2-ethylhexanoate** as reagent
  27, 54

–(II) **nitrate 13**, 224 suppl. **26**;
  28, 426
Cobaltocene-oxygen complex
  as oxidant, mild **28**, 193
Cobaltocenes, radical addition
  to – **27**, 813
–, reactions, review **27**, 813
  suppl. **29**
–, ring expansion **26**, 850
Cobalt oxide 26, 463
Codimerization of ethylene
  derivs. **24**, 711 suppl. **29**
Collidine as reagent
  27, 318, 578
Collins reagent, improved
  26, 235
Colman s. Gabriel
Combes-Beyer quinoline ring
  synthesis 27, 944
Competitive substrate,
  methyl acetate as – **12**, 61
  suppl. **28**
Complex compds. s. a. Metal
  complex compds.
– **formation**
–, direction of functionalization
    by – **27**, 188
Complexing agents
  crown polyethers **28**, 198
  diacetals **5**, 194 suppl. **29**
  tris(dimethylamino)phosphine
    28, 661
–, regiospecific effect of –
  29, 619
Complex salts, inner
  s. Copper(II) complex salts,
    inner, Metal acetoacetonates
3-Component reaction
  s. a. Condensation, double,
    Multicomponent reaction)
–, 1-(1-imidazolyl)-1,3-dienes
    by **26**, 542
–, pyran ring by – **26**, 812
Compounds, CH-acidic
  (s. a. C-Hydrogen, active)
–, addition to anthracenes,
    9-nitro- **28**, 644
–, reaction with 1,1,1-dialkoxy-
    ammonium salts, quaternary
    27, 821
–, reaction with Δ²-5-oxazolones
    26, 666
– startg. m. f.
  hydrazines, trisubst. **27**, 319
–, **ar., non-benzenoid**
  s. Azulenes, Fulvenes
–, –, **non-classical**
  s. Metallocenes
–, **bulky** s. Compds., hindered
–, *vic*-**difunctional**
– from
  ethylene derivs., review
    17, 590 suppl. **28**
–, **double-bridged 13**, 482
  suppl. **26**
–, **heteropolar** s. Heteropolar
  compds.
–, **hindered 29**, 383
– special s.
  alcohols, hindered
  carboxylic acid methyl
    esters, –
  carboxylic acids, –
  dialkylboranes, –
  Grignard compds., –
  ketones, –
  nitriles, –
  nitro compds., –

–, **labeled 28**, 218; **29**, 89
– by
  chloromethylation **8**, 824
    suppl. **28**
  Vilsmeier aldehyde synthesis
    26, 242 suppl. **27**
– special s.
  3-alkylidenephthalides,
    labeled
  amines, –
  α-aminocarboxylic acids-1-¹⁴C
  carboxylic acid hydrazides,
    labeled
  cyanohydrins, labeled
  β-diketones, –
  methanesulfonic acid aryl
    esters, –
  peptides, –
  phosphonic acid esters, –
  pyrimidines, 2-amino-, –
  quinolines, –
  resorcinols, –
–, **H-labeled**
–, labeling, preferential **21**, 35
  suppl. **27**
–, –, stereospecific at benzylic
  carbon **29**, 164
– special s.
  α-acylaminocarboxylic acids,
    H-labeled
  alcohols, –
  α-aminocarboxylic acids,
    α-H-labeled
  deuteri...
  α,β-ethylenecarboxylic acids,
    ά-H-labeled
  triti...
–, **N-labeled 23**, 413 suppl. **26**;
  28, 347
–, **O-labeled 21**, 308 suppl. **29**;
  29, 11
– special s.
  amine oxides, O-labeled
–, **lipophilic** s. Lipophilic
  compds.
–, **multifunctional 26**, 688
–, **optically active**
  s. Stereoisomers
–, **organometallic**
  s. Organometallic compds.
–, **proton-active**
–, reactions with ethylene
  derivs., electron-rich, review
    22, 874 suppl. **28**
–, **protonated**, review **26**, 822
–, **unsaturated** s. Acetylene...,
  Ethylene...
–, **unstable** s. Trapping
–, **vinylogous** s. Vinylogs
Comproportionation
  (s. a. Disproportionation)
–, N-monohalogenourethans
  by – **29**, 485
Condensation
– special s.
  aldol condensation
  chloral-
  redox-
– via carbonium ions **28**, 764
– with elimination of hydrogen
  halides **21**, 948; **29**, 815;
    30, 600
–, **anionic** s. Anion-anion
  condensation
–, **decarboxylative**, with
  enamines **30**, 626

Con — Cou

–, **directed**
– with succinic acid monoesters, subst. **27**, 840
–, **double**, of oxo compds.
2, 677; **4**, 742; **7**, 805; **13**, 727; **27**, 802
3-C-Condensation
s. Condensation, double
Configuration (s. a. Geometry, Stereoisomers)
–, change in – s. Epimerization
–, inversion of – (s. a. Double inversion, Walden inversion)
acylamines from 1,1-aminoethers, cyclic **29**, 164
amines from alcohols **27**, 429
ethylene derivs. from oxido compds. **27**, 942
halides from alcohols **24**, 581 suppl. **29**
iodides **27**, 573
phthalimides from alcohols **27**, 429
–, –, double, of –
sulfido from oxido compds. **8**, 662 suppl. **29**
– retention of –
at phosphorus **28**, 550
at sulfur **8**, 370 suppl. **29**
ethylene derivs. from oxido compds. **27**, 942 suppl. **28**
Friedel-Crafts alkylation **28**, 824 suppl. **30**
halides from alcohols **26**, 535 suppl. **29**
Hofmann-type degradation **28**, 477
replacement of halogen **29**, 821
ring closure, enzymatic, oxidative **29**, 904
Zincke reaction **20**, 369 suppl. **29**
Conjugate addition s. Addition, conjugate
Conjugated double bonds
s. Double bonds, conjugated
in situ-**Conjugation 30**, 500
Conrad-Limpach 4(1H)-quinoline ring closure 15, 653
Contact, organic-inorganic 26, 675
Continuous process s. Process, continuous
– **reactor** s. Reactor, continuous
Control (s. a. Orientation, control, Steric –)
–, kinetic and thermodynamic s. under Reactions
Cooligomerization s. Codimerization, Cyclocooligomerization
Cope s. a. under Kimel, Knoevenagel
– **elimination 30**, 669
– **rearrangement** (s. a. Hetero-Cope-rearrangement) **5**, 528
– of
bicyclo[3.2.0]hepta-3,6-dien-2-ones **23**, 731; **27**, 739
cycloalkenes, *endo*-vinyl-bridged **29**, 711
– –, **double 27**, 739
Copper (s. a. under Copper(II) oxide) **22**, 849 suppl. **27**; **26**, 106, 867, 961; **27**, 201, 622,

850/1, 958; **27**, 864 suppl. **28**; **28**, 858, 948; **29**, 279, 463, 587, 778; **30**, 304, 430, 591
–/zinc s. Zinc/copper
–**(I) acetate 29**, 820; **30**, 326
–**(II) acetate 10**, 171 suppl. **29**; 10, 386 suppl. **28**; **17**, 598 suppl. **29**; **28**, 68 suppl. **29**; **29**, 27, 113, 722, 780; **30**, 39, 661
–, reaction with thioureas **27**, 406
–**(II) acetoacetonate** as catalyst **26**, 963; **28**, 619; **29**, 779
–**(I) acetylides 12**, 855 suppl. **26**, **28**; **27**, 652; **28**, 650
–, 1-pentynylcopper as reagent **28**, 661
– startg. m. f.
α,β-acetyleneketones **12**, 855 suppl. **26**
cis-2-azetidinones **28**, 650
–**(I) alkoxides**, reactions with – **30**, 134
– *tert*-butoxide **27**, 652; **29**, 818
–**(II), bis-(8-quinolinolato)-**, removal of copper(II) ions as – **27**, 211
–**(I) bis(trimethylsilyl)amide 14**, 512 suppl. **29**
Copper *tert*-butoxides
(s. a. under Copper(I) alkoxides)
– as reactants **27**, 652
–**(I) carbonyl 26**, 452
– **chromite 26**, 67
–**(II) complex, azomethine-type,** as catalyst **28**, 780
– –, **hydrocarbon-soluble 22**, 935 suppl. **29**
– – **salts, inner**
– of
α-aminocarboxylic acids as intermediates **28**, 68 suppl. **29**
azomethine-type salts **19**, 744 suppl. **26, 29**
o-carboxy-N-heterocyclics **26**, 5; **29**, 27
– **compds., organo-**
(s. a. Copper acetylides, Cuprates, organo-)
– as reactants **29**, 670
–, coupling, oxidative, via – **24**, 761 suppl. **29**
–**(I) compds., organo-**, synthesis with –, review **26**, 669 suppl. **27**
– **compds., organo-, unsatd.**
– special s. –
copper acetylides
vinyl copper compds.
–**(I) cyanide 27**, 622; **28**, 662; **29**, 821, 830
–**(II) fluoroborate 26**, 718
–**(I) halides**
– bromide **16**, 488 suppl. **29**; **29**, 905; **30**, 492, 593
– chloride **22**, 288 suppl. **26**; **26**, 282, 715, 876; **27**, 296, 330, 471, 818, 937; **27**, 442 suppl. **29**; **27**, 958 suppl. **29**; **28**, 130, 429, 586, 705, 818, 938; **29**, 361, 821, 825, 973; **29**, 280 suppl. **30**; **30**, 491
– iodide (s. a. under Potassium hydridotri-*sec*-

butylborate and Trimethyl phosphite copper(I) iodide **27**, 204, 852-4; **27**, 851 suppl. **28**; **28**, 175, 740, 859; **29**, 78, 444, 619, 669–71, 854, 876, 958; **30**, 134, 453, 490, 594, 597
–**(II) halides 25**, 389 suppl. **28**
– bromide **26**, 549; **28**, 706
– chloride (s. a. under Palladium(II) chloride **20**, 409 suppl. **27**; **26**, 170, 365, 514; **27**, 598, 860, 877; **27**, 103 suppl. **29**; **28**, 63, 174/5, 182, 891, 980; **29**, 113, 174, 214, 500; **30**, 63
– –, **complexed 26**, 514
–**(I) hydride** as reagent **29**, 44
– **ions**, removal as copper(II) 8-hydroxyquinolinate **27**, 211
– **-isonitrile complexes**
–, syntheses of heterocyclics via –, review **28**, 606 suppl. **30**
–**(I) mercaptides**
phenylmercaptide **30**, 698
– **nitrate 26**, 143; **29**, 312
–**(I) oxide 25**, 484 suppl. **26**; **27**, 219; **28**, 936; **29**, 92, 617, 818; **30**, 215, 452
–**(II) oxide** (s. a. under Chromium oxide) **26**, 962; **27**, 251, 864; **28**, 175, 606; **29**, 214, 259
–**(II) – -copper 22**, 935 suppl. **28**
– **-phosphorus complex compds. 17**, 765 suppl. **27**; **23**, 693 suppl. **28**
– special s. –
tetrakis[iodo(tri-*n*-butylphosphine)copper(I)]
trimethyl phosphite·copper(I) iodide
– **salt-amine catalyzed autoxidation 28**, 130
–**(I) stearate 26**, 155
–**(II) sulfate 4**, 240 suppl. **29**; **6**, 251, suppl. **27**; **9**, 898 suppl. **26**; **26**, 5; **27**, 622; **28**, 2, 779, 936; **28**, 182 suppl. **29**; **29**, 418, 781; **30**, 408
–**(II) tartrate 29**, 780
–**(I) trifluoromethanesulfonate 27**, 704 suppl. **28**, **29**; **30**, 592
Corey-Winter elimination 19, 962
Cornforth stereoselective olefin synthesis 14, 717
Corrinoid synthesis 26, 982 suppl. **29**
Corrins s. Semicorrins
Coumarans
– from
phenols and aldehydes **27**, 807
Coumarilic acids
from coumarins **28**, 87
Coumarin ring
– from
α,β-ethylene-β-halogenocarboxylic acid esters **26**, 878
– startg. m. f.
pyrimidine ring **26**, 355
Coumarins
– from
(α-carbalkoxyalkylidene)-phosphoranes **27**, 886

(**Coumarins**
- from)
 coumarins, 3,4-dihydro-
  27, 819
 o-hydroxyaldehydes **16**, 803;
  27, 198
 - and 1-alkoxyenamines
  26, 831
 phenols **27**, 819
 succinimides, 3-o-hydroxy-
  arylidene- **26**, 152
- startg. m. f.
 chromenes **27**, 198
 coumarilic acids **28**, 87
-, **3-amino-**
- from
 aroxylamines and
  α,β-acetylenecarboxylic
  acid esters **29**, 725
-, **3-amino-6-hydroxy-**
- from
 p-quinones **26**, 717
-, **3,4-dihydro-**
- by ring closure, oxidative
 27, 252
- from
 1,3-cyclohexadiones and
  α,α-di(aminomethyl)-
  ketones **26**, 836
 phenols **27**, 819
- startg. m. f.
 coumarins **27**, 819
Coumarones s. Benzofurans
Coupling (s. a. Cross-coupling,
 Dimerization)
- of
 halides **26**, 881
 -, prevention **11**, 731 suppl. **27**
-, **anodic 28**, 703
-, **intramolecular**, review **29**, 904
-, **mixed** (s. a. Cross-coupling)
- of
 halides **26**, 881
-, **oxidative 29**, 910
- of
 phenols **27**, 915; **28**, 888;
 29, 904
 -, review **22**, 895 suppl. **28**
- to
 azo compds. **14**, 415
 β,γ-ethylenesulfones **23**, 757
 suppl. **29**
- via copper compds., organo-,
 review **24**, 761 suppl. **29**
-, **regiospecific**, of enoxysilanes
 30, 619
Course s. Direction
Cross-aldol reaction 30, 621
Cross-conjugated s. Dienones,
 cross-conjugated
Cross-coupling (s. a. Coupling,
 mixed)
- of
 enoxysilanes **30**, 619
- via lithium dialkylcuprates,
 mixed **30**, 597
- with
 Grignard compds. **26**, 875;
  29, 824
- with simultaneous
 C-chain isomerization
 26, 875 suppl. **28**
-, **stereospecific 26**, 875 suppl. **30**
Crown heterocyclics 29, 359
- **polyamines 29**, 359
- **polyethers**
- as complexing agents
 (s. a. Potassium cyanide-18-
 crown-6-polyether) **1**, 218
 suppl. **29**; **12**, 619 suppl. **29**;
 23, 235 suppl. **29**; **23**, 988
 suppl. **29**; **28**, 198
- as phase transfer catalysts
 27, 833 suppl. **29**; **30**, 366
-, prepn. **14**, 271 suppl. **29**;
 28, 198
Cubane derivs. 26, 763
 suppl. **27**
seco-**Cubane derivs.,**
 rearrangement, skeletal
 26, 762
Cumene hydroperoxide 27, 144
Cumulenehalides
- startg. m. f.
 acetylenebenzofurans **26**, 252
Cumulenes (s. a. Heterocumulenes)
- by Wittig synthesis **26**, 908
- special s.
 1,2,3-trienes
 ylids, cumulated
Cuprates, organo-
- special s.
 lithium dialkylcuprates
 - divinylcuprates
-, syntheses with -, review
 26, 669 suppl. **27**
-, -, **mixed**, syntheses with -
 28, 661
Curtius degradation 28, 630;
 30, 460
- -, **modified 28**, 400
- -, **simplified 28**, 353
Cyanamide
- startg. m. f.
 o-amino-N-heterocyclics
 28, 604
 benzimidazoles, 2-amino-
 30, 245
 N-cyaniminoesters **27**, 440
 Δ^2-oxazoline ring, 2-amino-
 28, 341
 sulfonylthioureas **27**, 631
Cyanamides
- from
 ureas **29**, 456
- special s.
 N-cyanourethans
 dichloroformimido-
 cyanamides
 di-
 halogeno-
 hydroxy-
 nitroso-
- startg. m. f.
 amidinothioureas **26**, 302
 hydroxyguanidines **26**, 294
 N-nitrosocyanamides
 29, 276
 parabanic acids **28**, 274
 ureas **26**, 23; **30**, 216
-, **cyclic**
- from
 tetrazole ring, N-condensed
 27, 523
Cyanamines s. Aminonitriles
**N-Cyanammonium fluoro-
 borates 27**, 994
Cyanates s. Cyanic acid esters
o-**Cyanatocarboxylic acid esters**
- startg. m. f.
 1,3-oxazin-4-one ring **26**, 314
Cyanic acid esters
 (s. a. Cyanato...)
-, reactions, review **26**, 301
 suppl. **28**
- special s.
 2,4-dimethylphenyl cyanate
 phenyl -
- startg. m. f.
 iminocarbonic acid esters
 27, 214
 1,3,5-triazines, trisaryloxy-
 26, 301
Cyanides s. Nitriles, Replacement, Tetraalkylammonium
 cyanides
Cyanimines
- from
 ethylene derivs. **28**, 282
 guanylhydrazones **28**, 455
**N-Cyaniminodithiocarbonic
 acid esters**
-, reactions **28**, 538
- startg. m. f.
 bis(benzothiazole), 2,2'-
 imino- **28**, 538
 N-cyanoguanidines, cyclic
 28, 311
α-**Cyaniminoesters**
- from
 1,1-dinitriles **29**, 130
N-Cyaniminoesters
- from
 iminoesters **27**, 440
Cyanines s. Aza-cyanines,
 Polymethine -
Cyano... s. a. Di-cyano...,
 Tetra-
α-**Cyanoacetals**
- startg. m. f.
 pyrimidines **27**, 381
Cyanoacetylenes
- startg. m. f.
 pyrimidine nucleosides
 28, 341
Cyanoaldehydes
- from
 2-oximinoalcohols, cyclic
 27, 517
Cyanoalkylation
 s. Cyanoethylation
Cyanoallenes
- startg. m. f.
 β-amino-α,β-ethylenenitriles
 29, 160
 α-cyanoketones **29**, 160
**1,1-Cyano(azocarboxylic acid
 esters)**
- startg. m. f.
 α-cyanocarboxylic acid esters
 28, 55
 nitriles **28**, 55
 -, α-alkylated **28**, 55
N'-Cyanoazoformamidines
- from
 diazocyanides **27**, 442
Cyanoazomethines
 s. 1,1-Iminocyanides
Cyanoboranes, syntheses via -
 27, 716
Cyanocarbon chemistry
 s. Polynitriles
α-**Cyanocarboxylic acid amides**
- special s.
 α-cyano-α,β-ethylene-
 carboxylic acid amides
- startg. m. f.
 pyridines, 2,4-diamino-
 6-halogeno- **29**, 757
Cyanocarboxylic acid esters
- special s.
 benzyl cyanoacetate

- startg. m. f.
  ketones, cyclic **26,** 989
α-Cyanocarboxylic – –
- from
  1,1-cyano(azocarboxylic acid
    esters) **28,** 55
- special s.
  α,α-dicyanocarboxylic acid
    esters
  ethyl cyanoacetate
- startg. m. f.
  α-ketocarboxylic acid esters
    27, 236
  pyrimidines **27,** 398
Cyanocarboxylic acid halides
-, ring closures with – **22,** 560
  suppl. **28**
- **acids**
- special s.
  o-carboxyphenylacetonitriles
o-α-Cyanocarboxylic acids
- startg. m. f.
  imidazole ring, N-condensed
    26, 339
Cyanodithioformic acid esters
-, diene synthesis with – **28,** 637
1-Cyano-1,2-ene-1,2-diamines
- from
  1,2-di(azomethines) **29,** 639
C-Cyanoethylation
  s. C-Monocyanoethylation
S-Cyanoethylation with and
  without ring closure **30,** 397
**α-Cyano-α,β-ethylenecarboxylic
  acid amides**
- from
  alkylidenemalononitriles
    28, 588
  2-carbamyl-2-cyano-
    phosphonium salts **28,** 588
δ-Cyano-α,β-ethyleneketones
- from
  2,4-dienones **18,** 772 suppl. **26**
Cyanogen
- as reagent **27,** 114
Cyanogenation s. Replacement
  of hydrogen by cyano groups
Cyanogen azide
- as reactant **28,** 282
-, ring expansion **29,** 154
- **bromide**
- as reactant **27,** 176, 214;
    28, 277, 660
- as reagent **4,** 52 suppl. **29;**
    28, 143
-, ring closures with –
    s. under Ring closure
- **chloride**
- as reactant **25,** 128 suppl. **28;**
    28, 282; **29,** 335, 342; **30,** 539
Cyano groups (s. a. under
  Replacement)
-, position shift **27,** 730
- –, **active,** addition of sulfenyl-
    chlorides to – **30,** 348
N-Cyanoguanidines, cyclic
- from
  diamines **28,** 311
Cyano-N-heterocyclics
- from
  N-oxides, cyclic **26,** 813
- special s.
  o-cyano-N-hydroxy-N-hetero-
    cyclics
1,1-Cyanohydrazines
- from
  aldehydes **26,** 828

- special s.
  N'-acyl-1,1-cyanohydrazines
- startg. m. f.
  sydnone imine salts,
    3-amino- **26,** 828
α-Cyanohydrazones
- startg. m. f.
  cinnolines, 4-amino- **30,** 532
Cyanohydrins
- from
  acetylene derivs., synthesis
    28, 660
  α-hydroperoxy nitriles **30,** 155
- special s.
  acetone cyanohydrin
  acoxycyanohydrins
  3-hydroxypropionitrile
- startg. m. f.
  α-acylaminonitriles **26,** 375
  cyanoketones with nitrile
    group migration **28,** 933
  hydantoins **30,** 568
  α-hydroxyaldehydes **23,** 223
    suppl. **26**
  α-hydroxycarboxylic acid
    amides **27,** 326
  α-hydroxythiolic acid esters
    27, 624
  α-ketocarboxylic acids **27,** 326
  oxo compds. **30,** 155
-, **labeled** from oxo compds.
    30, 568
-, **protected,** syntheses with –
    30, 487
Cyanohydrin sulfonates
- special s.
  cyanohydrin tosylates
- **tosylates**
- special s.
  aminocyanohydrin tosylates
**o-Cyano-N-hydroxy-
  N-heterocyclics**
- from
  o-azido-N-oxides, ring
    contraction **28,** 452
o-Cyanoisocyanates
- startg. m. f.
  2-quinazolones, 4-halogeno-
    26, 512
Cyanoketenes
- from
  2,5-diazido-p-quinones
    (to 2 molecules) **26,** 496
Cyanoketones
  (s. a. Acylcyanides)
- from
  cyanohydrins with nitrile
    group migration **28,** 933
  ethylene derivs., cyclic **26,** 325
- special s.
  cyanoethyleneketones
- startg. m. f.
  diamines **28,** 362
  N-heterocyclics **6,** 522
  1-hydroxyamidines, cyclic
    26, 298
  1-hydroxyamidoximines,
    cyclic **26,** 298
  hydroxylactams **26,** 121
α-Cyanoketones
- from
  β-amino-α,β-ethylenenitriles
    29, 160
  cyanoallenes **29,** 160
  ketones **23,** 521 suppl. **29**
- startg. m. f.
  pyrimidine ring, 4-amino-
    22, 411

β-Cyanoketones
- from
  aldehydes **28,** 648 suppl. **29**
  1-amino-1,3-dinitriles **29,** 663
  α-aminonitriles **29,** 663
-, self-condensation **29,** 774
γ-Cyanoketones 20, 601
  suppl. **28**
Cyanolysis 27, 895
- of
  onium salts **27,** 895
N-Cyanomethylation 29, 749
Cyanonitromethylene compds.
- by pyrimidine ring opening
    26, 499
2-Cyanonitrones
- startg. m. f.
  imidazoles, 5-arylthio- **29,** 551
o-Cyano-N-oxides
- from
  N-oxides, cyclic **26,** 776
α-Cyanooxo compds.
- startg. m. f.
  pyrazoles, amino- **19,** 461
    suppl. **26**
o-Cyanophenols
- from
  nitro compds. **28,** 777
o-Cyanophosphine imines
- startg. m. f.
  2-phosphaquinazolinium salts
    26, 277
Cyanophosphonium salts
- special s.
  carbamylcyanophosphonium
    salts
α-Cyanophosphonium –
- from
  1,1,1-ethylenenitrohalides
    and phosphines **29,** 606
Cyanosemicarbazones 25, 348
  suppl. **26**
Cyanosilanes
-, reactions with – **28,** 610
- special s.
  trimethylsilyl cyanide
Cyanosugars
- startg. m. f.
  sugar dilactones **26,** 834
β-Cyanosulfones
- from
  ethylene derivs. and sulfonyl-
    cyanides **28,** 638
α-Cyano-N-sulfonylhydrazines
- from
  sulfonylhydrazones **28,** 67
- startg. m. f.
  nitriles **28,** 67
1,1-Cyanothioethers
-, alkylation **23,** 832 suppl. **27**
- special s.
  α-alkylthionitriles
Cyanothioformamides
- from
  C-sulfonylthioformamides
    28, 836
Cyanourethans
- startg. m. f.
  2-oxazolidones, 4-imino-
    26, 311
N-Cyanourethans
- startg. m. f.
  isothioureas **30,** 418
Cyanuric chloride as reagent
    19, 307 suppl. **29**; **26,** 539; **27,**
    532, 797; **28,** 440
- **esters** s. 1,3,5-Triazines,
  trisaryloxy-

Cyanuric fluoride as reagent
3, 463 suppl. 28
Cycl... s. a. Bicycl..., Ring,
Tricycl...
Cycla... s. a. Cycloalk...
Cyclimmonium salts
–, dequaternization 30, 698
–, diene synthesis with – 27, 697
– from
di-*tert*-amines 26, 479
–, ring closure, nucleophilic,
with – 29, 382
–, – –, –, intramolecular, of –
25, 645 suppl. 30
–, – –, reductive 26, 761
–, – opening 22, 42 suppl. 28;
29, 41, 823
– special s.
acyl-cyclimmonium salts
alkoxylcyclimmonium –
aminocyclimmonium –
azonia...
benzisoxazolium salts
cyclodiimonium –
diazepinium –
dicyclimmonium –
halogenocyclimmonium –
imidazolium –
purinium –
pyridazinium –
pyridinium –
pyrrolidinium –
quinolinium –
thiazolium –
s-triazolo[3,4-f][1,2,4]-
triazinium –
– startg. m. f.
N-heterocyclics, N-condensed
25, 644
– –, **N-condensed 28**, 911
– by
cyclo-N-quaternization 27, 527
benzocyclobutene ring
opening 28, 310
– from
amines, tert., cyclic 29, 904
N-heterocyclics, N-condensed
30, 321
–, ring opening, reductive
26, 36
– special s.
acridizinium salts
benzisoindolinium salts
benzo[c]quinolizinium salts
dicyclimmonium salts,
triple-N-condensed
1,2,3,5-oxatriazolium salts,
N-condensed
quinolizinium salts
– startg. m. f.
amines, tert., cyclic 29, 37
arenes 28, 982
N-heterocyclics, N-condensed
30, 321
phenols 28, 982
– –, **methylene-**
– by Mannich reaction 30, 562
Cyclization s. Ring closure
Cycloaddition
–, direction of – 22, 705; 27, 692
– of alcohols 28, 712
– to
arenes 26, 720
benzvalenes 29, 652
4H-pyrazol-4-one N-oxides
26, 701
– with
diaziridinones 30, 212

hetarynes 27, 824
1,2,3-triazolium betaines,
4-hydroxy- 28, 470
ynamines 27, 693
– with simultaneous rearrangement, skeletal 30, 475
–, **anionic** (s. a. Cycloaddition,
polar) 26, 712
–, – review 26, 712 suppl. 29
–, **cationic** (s. a. Cycloaddition,
polar) 27, 868 suppl. 28
–, **dipolar** s. a. Dipolarophiles
–, **1,3-dipolar** (s. a. 1,3-Dipoles)
–, effect of solvents, review
17, 503 suppl. 28
– to cyclobutadiene rings
29, 652
– with
acetylene derivs., review
24, 900 suppl. 28
diazo compds. 25, 507
suppl. 29
nitrones, review 26, 699
suppl. 30
pyridinium betaines 26, 695
–, –, **intramolecular, regiospecific 30**, 683
–, –, **reverse 15**, 283 suppl. 28
–, –, **stereospecific** 24, 692
suppl. 29; 29, 649
–, **inverse-oriented** 26, 699
–, **nickel(0)-catalyzed** 26, 724
–, **polar** (s. a. Cycloaddition,
anionic, –, cationic)
–, review 26, 712 suppl. 28
–, **stereospecific** 26, 690
1,3-Cycloaddition 26, 305
[2+1]-Cycloaddition 27, 690
[2+2]-Cycloaddition
– of
carbodiimides 28, 265
– to
tetramethoxyethylene 27, 329
–, **thermal** 5, 488; 17, 747
– **versus [2+4]-Cycloaddition**
22, 705; 27, 692
[3+1]-Cycloaddition 27, 887
[3+2]-Cycloaddition, anionic
27, 352
[4+2]-Cycloaddition
(s. a. Diene syntheses)
– to
4H-1,3-oxazines, 5,6-dihydro-
6-alkoxy- 29, 654
[4+4]-Cycloaddition 29, 658
[6+4]-Cycloaddition 26, 690
[8+2]-Cycloaddition 26, 690;
27, 707
–, S-heterocyclics by – 28, 636
[δ2+π2+π2]-Cycloaddition
– with enecyclopropanes
29, 655
1,4-(2π+6π)-Cycloaddition
26, 720
Cycloaddition-fragmentation
26, 918
Cycloalkane ring
– from
β,γ-ethyleneketones, cyclic
29, 933
β-hydroxycarboxylic acid
esters 29, 933
Cycloalkenes
– as reactants 29, 261
– from
dihalides 26, 966
–, *endo-***vinyl-, bridged**
–, Cope rearrangement 29, 711

Cycloalkenone ring
s. a. Cyclohexenone ring
– – **annelation,** review 27, 950
suppl. 28
Cycloalkylation 4, 832 suppl. 28
**Cyclobutadiene rhodium
triarylphosphine complexes**
–, ring expansion 26, 906
– startg. m. f.
anthraquinones 26, 906
– **ring,** cycloaddition, 1,3-
dipolar to – 29, 652
**Cyclobuta[a]naphthalenes,
2a,8b-dihydro-** 28, 640
Cyclobutane-1,2-diols
– startg. m. f.
cyclobutanones, rearrangement 28, 677
cyclopropyloxo compds.
28, 677
Cyclobutane-*cis*-1,2-diols
– from
1,2-di(siloxy)cyclobutanes
27, 60
1,2-di(siloxy)cyclobut-1-enes
27, 60
Cyclobutane-1,2-diones
– special s.
spiro[adamantane-2,1'-cyclo-
butane-2',3'-diones]
Cyclobutane-1,3-diones
– from
ketenes 27, 691
–, ring expansion 12, 828
suppl. 27
– **ring**
– as intermediate 30, 181, 467
– by cycloaddition 30, 181
– from
bicyclo[1.1.0]butane ring
26, 45
dihalides 27, 966
ethylene derivs., dimerization
27, 704
– –, –, review 27, 704 suppl. 29
– special s.
immoniocyclobutane ring
– startg. m. f.
cyclopropane ring, review
24, 977 suppl. 27
ethylene derivs. (2 molecules)
27, 704
– – **opening 30,** 467
–, cyclopent-2-enones, 5-β-keto-
by – **30,** 181
–, β,γ-ethyleneoxo compds. by –
26, 927
– – –, **oxidative** 27, 143
Cyclobutanes
– from
ethylene derivs. (2 molecules)
30, 467
halides 28, 947 suppl. 30
uracil dimers 27, 45
– startg. m. f.
cyclopropanes 30, 670
heterocyclics 30, 474
–, **functionally subst. 30,**
467, 517
–, **amino-**
– from
enamines 27, 719
–, **imino-**
– from
1-azaspiro[2.2]pentanes
30, 535

Cyclobutanol ring
- from
 ketones 30, 518
Cyclobutanone ring 29, 949
- as intermediate 30, 487
- from
 cyclobutanone ring,
  2,2-dihalogeno- **29**, 75
 cyclopropylcarbinols, 1-arylthio- **28**, 958
 ethylene derivs. and
  carboxylic acid chlorides
  30, 487
 ketones **28**, 837
 1-oxaspiro[2.2]pentane ring
  28, 837
-, opening **30**, 487
- startg. m. f.
 α,β-ethyleneketones **29**, 687
 γ-lactones **26**, 140
- -, **2,2-dihalogeno- 28,** 861
- startg. m. f.
 cyclobutanone ring **29**, 75
Cyclobutanones 27, 830
 suppl. **29**
- by ring expansion **25**, 520
- from
 cyclobutane-1,2-diols,
  rearrangement **28**, 677
 cyclopropanols, 1-vinyl-
  25, 520
-, **2-halogeno-**
- startg. m. f.
 cyclopropyl ketones
  26, 860
-, **2-imino-**
- from
 isonitriles **30**, 591
 - and α,α-dihalogenoketones **30**, 591
Cyclobutene, 1,2-dicyano-
-, diene synthesis with -
 30, 469
Cyclobutenediones
-, chemistry of phenylcyclobutenediones, review -
 25, 505 suppl. **28**
- special s.
 squaric acid
- startg. m. f.
 cyclobut-2-enones, 2-hydroxy-
  4-aryl- **26**, 674
**Cyclobut-3-ene-1,2-diones,
 3-hydroxy-**
- startg. m. f.
 pyrrolo[1,2-a]benzimidazol-
  1-ones, 3-hydroxy- **27**, 413
Cyclobutene ring 29, 973
- startg. m. f.
 1,3-dienes **16**, 773
- - **opening 16,** 773
- - -, **1-amino-, stereospecific
 28,** 156
Cyclobutenes
- from
 cyclopropenecarboxylic acid
  esters **27**, 788
 cyclopropylcarbinols
  29, 938
- special s.
 squaric acid...
-, **2-amino-3-halogeno-**
- from
 ynamines **28**, 627
Cyclobutenes, polyhalogeno-,
 substitution, nucleophilic,
 on - **27**, 993

**Cyclobut-2-enones, 2-hydroxy-
 4-aryl-**
- from
 cyclobutenediones **26**, 674
Cyclocooligomerization
- to cis,cis,trans-1,4,7-cyclododecatrienes **30**, 501
cis,cis,trans-**1,4,7-Cyclodecatrienes**
- from
 acetylene derivs., cyclocooligomerization **30**, 501
Cyclodehydration
- to
 heterocycles **28**, 216
  chromenes **28**, 210
 N-heterocyclics **18**, 954;
  29, 922
 O-heterocyclics **29**, 251
- with
 C-alkyl migration **4**, 832
 suppl. **28**
-, **stereospecific 27,** 945
Cyclodehydrogenation 28, 888,
 890, 892
-, benzene ring annelation via -
 30, 645
- to
 naphthalenes **29**, 908
 pheoporphyrins **29**, 903
-, **double 28,** 892
Cyclodehydrohalogenation
 (s. a. C-Alkylation, intramolecular, Photocyclization,
 dehydrohalogenative) **28**, 892
- to aporphines **29**, 956
Cyclodextrins, binding to -
-, direction of ar. substitution
 by - **3**, 445 suppl. **26**
Cyclodiimonium salts 29, 995
Cyclodimerization
- to polyisocyclics **29**, 680
-, **oxidative**, of phenols,
 o-(ω-hydroxyalkyl)- **28**, 711
-, -, **electrochemical 29,** 900
-, -, **stereospecific**
- of
 phenols **27**, 725
**[4+2]-Cyclodimerization
 23,** 687
- of
 1H-1,2-diazepines **30**, 498
**Cycloheptadiene ring, fused
 23,** 716 suppl. **28**
2H-Cyclohepta[b]furan-2-ones
- startg. m. f.
 azulenes **28**, 725
Cycloheptatrien... s. a. Trop...
1,3,5-Cycloheptatriene
- startg. m. f.
 benzyl ketones **27**, 865
1,3,5-Cycloheptatrienes
-, 1-acylation **26**, 883
- from
 thiophene 1,1-dioxides and
  cyclopropenes **29**, 848
Cycloheptatrienones, amination
 11, 404 suppl. **29**
**1,3,5-Cycloheptatrien-7-yl
 compds. 27,** 822
- special s.
 di-(7-cycloheptatrienyl)-
 malonic acids
4-Cycloheptenones
- from
 1,3-dienes and α,α'-dihalogenoketones **26**, 890

1,3-Cyclohexadiene ring
- from
 cyclohexenedione ring **27**, 70
 thiophene ring **29**, 640
- - **annelation 30,** 616
Cyclohexadienes (s. a. Spirocyclohexadienes)
- by Wittig synthesis **28**, 854
1,3-Cyclohexadienes
- startg. m. f.
 homophthalic acid esters
  28, 868
1,4-Cyclohexadienes, 3-alkyl-
- from
 arenes **27**, 839
-, rearrangement, oxidative
 27, 839
**2,5-Cyclohexadienone
 hydrazones**
-, rearrangement **28**, 698
- special s.
 2,5-cyclohexadienone
  tosylhydrazones
- **imines, 4-alkoxy-**
- from
 anilines **26**, 204
- **ring** with O-heteroring
 opening **28**, 188
2,5-Cyclohexadienone oximes
- from
 nitroso compds. **30**, 596
Cyclohexadienones
- special s.
 spirooxiranocyclohexadienones
2,4-Cyclohexadienones
- from
 benzene ring **28**, 111
2,5-Cyclohexadienones
 (s. a. Dienones, crossconjugated)
-, photoreactions **27**, 2
- special s.
 2,4,4,6-tetrabromo-2,5-cyclohexadienone
-, **4-halogeno-**
- startg. m. f.
 phenols **27**, 2
-, **4-hydroxy-** s. Hemiquinols
**2,5-Cyclohexadienone,
 2,4,4,6-tetrabromo-**
- as reagent **28**, 485
- **tosylhydrazones**
-, rearrangement **27**, 754
**1,3-Cyclohexadione ring
 rearrangement, skeletal
 29,** 170
1,3-Cyclohexadiones
- startg. m. f.
 coumarins, 3,4-dihydro-
  26, 836
 2-oximino-4(5H)-benzofuranones, 6,7-dihydro-
  27, 274
Cyclohexane
- as solvent **27**, 691
3-Cyclohexanolones
- from
 α,β-ethyleneketones **30**, 478
Cyclohexanone ring
- from
 dienones, cross-conjugated
  29, 665
Cyclohexene (s. a. Palladiumcarbon/cyclohexene,
 Palladium-barium sulfate/-)
- as chlorine scavenger **29**, 993
- as reagent **27**, 610

Cyclohexenedione ring
- startg. m. f.
  1,3-cyclohexadiene ring **27,** 70
Cyclohexenes
- special s.
  3,5-bis-*aci*-nitrocyclohexenes
Cyclohex-2-en-4-olone ring
- from
  α,β-ethylene- and α-hydroxy-
    ketones **30,** 546
- startg. m. f.
  phenol ring **30,** 546
2-Cyclohexenone ring
- from
  1,3-dienamines, cyclic **27,** 950
  1,5-dioxo compds. **27,** 950
  γ,δ-ethylene-δ-halogeno-
    ketones **30,** 680
  α,β-ethyleneketones **26,** 784
  α,β-ethylene-α-silylketones
    29, 666
  ketoalkoximes **28,** 896
  ketoeneurethans **30,** 666
  pyran ring, dihydro- **28,** 896
2-Cyclohexenones
- by
  Horner synthesis, intra-
    molecular **25,** 693 suppl. 28
- from
  α,β-ethyleneoxo compds.
    and ketones **28,** 721
  pyridines **27,** 816
  -, 1,4-dihydro- **27,** 816
  vinylketenimines **29,** 647
- special s.
  spiro-2-cyclohexenones
- startg. m. f.
  phenols **28,** 721
- , **isomeric**
- from
  1,5-dioxo compds. **13,** 858;
    28, 910
-, **4-alkyl-** **29,** 805
-, **3-amino**
- from
  ynamines **30,** 493
**2-Cyclohexenon-3-yl-stabilized
ylids**
-, syntheses with – **27,** 82
  suppl. 29
Cyclohexylamine
  as reagent **27,** 717
Cycloisomerization
  (s. a. Rearrangement, skeletal)
-, double ring closure by –
  26, 763
- of
  N-arylenamines **26,** 731
  benzocyclobutenes **27,** 756
  dienes **28,** 667
  ethylene derivs. **26,** 759
  α,β-ethyleneketones **30,** 508
  N-heterocyclics **26,** 731
    2-pyridones **19,** 812
    suppl. 27
  isocyclics **27,** 736, 743
  thioenolethers **30,** 511
- to
  cyclopent-2-enones **29,** 688
  isocyclics **26,** 735
  pentalenes, 1,2-dihydro-
    3-amino- **27,** 732
  phosphonium salts, cyclic
    26, 735 suppl. 30
- with
  dearomatization **22,** 761
  suppl. 26
  1,5-hydrogen shift **26,** 757

-, **double 19,** 809; **24,** 751
-, O-heterocyclics by – **28,** 687
-, -, **stereospecific 26,** 758
  suppl. 28
Cycloketones s. Cyclones,
  Ketones, cyclic
Cyclols s. Azacyclols
Cyclones
- startg. m. f.
  3H-azepines **28,** 872
  benzene ring **29,** 895
  naphthalenes **29,** 897
**Cyclonucleoside ring opening
29,** 488
2,2'-Cyclonucleoside – –
- with
  formation of thioethers
    from acoxy compds.
    26, 599
**8,5'-Cyclonucleoside ring
  opening, oxidative 30,** 76
Cyclonucleosides
-, ease of formation **24,** 268
  suppl. 26
2,2'-Cyclonucleosides 28, 341
8,2'-Cyclonucleosides 27, 640
S-Cyclonucleosides 26, 628;
  27, 630
8,2'-S-Cyclonucleosides 27, 640
Cycloocta-1,5-diene ring
- from
  cycloocta-1,3,5-triene ring
    29, 56
Cyclooctatetraene dibromide,
  syntheses with – **30,** 601
Cyclooctatetraenes 18, 796
  suppl. 29
Cycloocta-1,3,5-triene ring
- startg. m. f.
  cycloocta-1,5-diene ring
    29, 56
Cyclooligomerization
  (s. a. Cyclo-trimerization,
    -tetramerization)
- with transition metal
  catalysts, review **20,** 531
  suppl. 29
**Cyclopenta[f]chromen-7(8)-ones,
  1,2,3,5,6,6a-hexahydro-** **27,** 792
Cyclopentadienes
-, C-alkylation **26,** 718
- special s.
  fulvenes
- startg. m. f.
  pyridines, 1,4-dihydro-
    20, 389 suppl. 27
Cyclopentadienide anions,
  substitution, electrophilic,
  review **26,** 718
Cyclopentadienones
- from
  o-benzoquinones **30,** 691
  4-thiopyrones **28,** 955
Cyclopentadienylids
- special s.
  pyridinium cyclopenta-
    dienylids
  sulfonium –
Cyclopentadioxinones,
  photolysis **27,** 921
**Cyclopenta-1,3-dioxin-4-ones,
  4,5,6,7-tetrahydro-** **27,** 849
Cyclopentane annelation
  s. Silyloxycyclopentane
    annelation
Cyclopentane ring
- from
  cyclopropane ring **26,** 704

- special s.
  methylenecyclopentane ring
- – **annelation, stereospecific
  30,** 483
Cyclopentanones
- from
  α,α'-dihalogenoketones
    and ethylene derivs. **27,** 868
    suppl. 28
**Cyclopenta[e][1,3,4]oxadiazines,
  4,4a,5,7-tetrahydro-**
- from
  2,3-diazabicyclo[2.2.1]hept-
    5-enes **27,** 166
Cyclopenta[a]phenanthrenes
  s. Steroids
**Cyclopent-2-enecarbox-
  aldehydes**
- by skeletal rearrangement
  26, 728
**1,3-Cyclopentenedione ring,
  2-cyano-**
- from
  azidoquinone ring **26,** 958
**Cyclopent-2-ene-1,4-diones,
  3-subst.** **27,** 856
Cyclopentene ring 29, 866
Cyclopentenes
- from
  ketones **28,** 901
**Cyclopent-2-en-2-ol-1-ones
  30,** 524
**Cyclopent-2-en-4-ol-1-ones
  27,** 772; **27,** 830 suppl. 29
- startg. m. f.
  cyclopent-2-enones **27,** 772
Cyclopent-2-enone ring
- from
  α,α-dihalogenoketones
    and enamines **27,** 668
  furan ring **27,** 388
Cyclopent-2-enones 26, 737
- by
  cycloisomerization **29,** 688
  ring contraction **29,** 984
- from
  acoxycyclopropylketones
    27, 923
  cyclopropylcarbinols,
    dihalogeno- **30,** 681
  cyclopropylketones **27,** 923
  dienones, cross-conjugated
    24, 726 suppl. 29
  γ-diketones **1,** 554
-, prepn., review **1,** 554
  suppl. 28
- startg. m. f.
  cyclopropylacetic acid esters
    26, 709
- via cyclopent-2-en-4-ol-1-ones
    27, 772
-, **5-β-keto-**
- from
  enolesters **30,** 181
- via cyclobutane ring closure
  and opening **30,** 181
Cyclopent-3-enones
- from
  1,3-dienes **28,** 656
Cyclophanes s. Hetero-cyclo-
  phanes, Meta-, Para-
-, **tris-bridged 29,** 967
[2.2]**Cyclophanes 15,** 684
  suppl. 28
Cyclopolyenes as dienophiles
  26, 758

Cyclopropanecarboxylic acid esters
– by
  Hofmann degradation, abnormal **26,** 983
Cyclopropane ring
– from
  alcohols **27,** 937
  cyclobutane ring, review **24,** 977 suppl. **27**
– –, 1,1-dihalogeno- **7,** 738 suppl. **28**
  ethylene derivs. **27,** 833
– – and diazo compds. **28,** 780
– – and 1,1-dihalides **26,** 876; **29,** 827
– – and oxo compds. **26,** 803
  halides **26,** 967
  Δ¹-pyrrol-5-one ring **26,** 753
  spirodiazirine ring **29,** 948
  sulfonates
    mesylates **30,** 649
    tosylates **23,** 904 suppl. **28; 29,** 918
  sulfonium salts **28,** 699
  thiopyran ring, dihydro- **26,** 576
  tosylhydrazones **28,** 934
– special s.
  aminocyclopropane ring
  3,5-cyclosteroids
  methylene-cyclopropane ring
  spiro- –
– startg. m. f.
  cyclopentane ring **26,** 704
  1,3-dihalides **29,** 492
  2-ethylenealcohols **26,** 132
  furan ring **26,** 248
– –, tetrahydro- **28,** 641
  spiroisocycles **28,** 477
– with formation of dihalides from dienes **26,** 516
–, **fused**
– from
  tropone adducts **28,** 97
–, **1,1-dihalogeno-**
– from
  ethylene derivs. **27,** 833
– startg. m. f.
  2′-alkoxy-α,β-ethylenehalides, cyclic **26,** 205
  cyclopropane ring **7,** 738
– –, **halogeno-** suppl. **28**
– from
  ethylene derivs. **28,** 806
– –, **1-halogeno-1-trifluoro-acetoxy-**
– startg. m. f.
  carboxylic acids **29,** 226
– –, **1-halogeno-1-trihalogenomethyl- 17,** 894 suppl. **28**
– – **expansion**
–, isocyclics, mono- and bi-cyclic, by – **29,** 685, 982
– – **opening** (s. a. Addition, homoconjugate)
–, ring expansion by –
  26, 730; **30,** 643
– with dehydrohalogenation **29,** 955
– – –, **double**, dienes by – **30,** 538
– – –, **oxidative 26,** 156
– – –, **radical 28,** 488
– – –, **reductive 27,** 52
– – –, **regiostereospecific 20,** 66 suppl. **28**
– – –, **1,1-dihalogeno-**

–, ring expansion by –
  26, 205, 408; **27,** 972
Cyclopropanes
– by dehydrogenation **28,** 893
– from
  acetylene derivs. **26,** 880
  cyclobutanes **30,** 670
  1,2-dihalides **27,** 841
  ethylene derivs. and 1,1-dihalides **28,** 815; **30,** 604
– – and halides **26,** 847; **27,** 864
– – and phosphonium salts **29,** 872
– – and selenonium ylids **29,** 863
– – and sulfonium salts **26,** 898
– –, cycloisomerization **26,** 759
  β,γ-ethylenehalides **28,** 808
  halides and 2 sulfoxonium methylid molecules **27,** 874
  methylene compds., active, and enol phosphates **26,** 783
–, rearrangement of side chains **27,** 728
– special s.
  alkenylidene-cyclopropanes
  nitro-
  siloxy-
  spirocyclopropan...
– startg. m. f.
  1,3-dialkoxy compds. **27,** 135
  isoquinolines, 3,4-dihydro- **26,** 820
–, synthesis by homoconjugate addition to – **28,** 859
–, **acyl-**
– from
  1,3,2-dioxaphospholenes **23,** 797 suppl. **26**
–, **alkoxy-**
– from
  β-halogenacetals **11,** 956 suppl. **27**
–, **cyano-**
– startg. m. f.
  alcohols **29,** 87
  γ-lactones **29,** 87
– –, **1,1-dihalogeno- 30,** 604
– startg. m. f.
  cyclopropanone mercaptals **29,** 569
Cyclopropanesulfinic acids
– from
  thietane dioxides **28,** 700
Cyclopropanol derivs.,
  syntheses via – **27,** 922
– **ring**
– from
  etheneketones **26,** 29
  oxo compds. **29,** 829
  siloxycyclopropane ring **29,** 829
– startg. m. f.
  etheneketones **26,** 29
Cyclopropanols
– from
  ketones **26,** 744
– startg. m. f.
  β-(acetoxymercury)ketones **26,** 643
– via enoxysilanes **14,** 858 suppl. **28**
–, **1-vinyl-**
– startg. m. f.
  cyclobutanones **25,** 520

Cyclopropanone ketals
– special s.
  4,7-dioxaspiro[2,4]heptanes
– startg. m. f.
  orthocarboxylic acid esters, mixed **29,** 135
– **mercaptals**
– from
  cyclopropanes, 1,1-dihalogeno- **29,** 569
– startg. m. f.
  ketones **29,** 216
Cyclopropanones, chemistry, review **25,** 535 suppl. **29**
Cyclopropenecarboxylic acid esters
– startg. m. f.
  cyclobutenes **27,** 788
  furans, 2-alkoxy- **26,** 155
– **acids**
– startg. m. f.
  γ-enollactones **26,** 155
Cyclopropenes 25, 679 suppl. **26**
–, alkylation **13,** 805 suppl. **29**
– from
  1,3-dihalides **15,** 680 suppl. **27**
–, reactions, review **27,** 899 suppl. **28**
–, ring expansion by 2 C-atoms with – **27,** 899
– special s.
  enecyclopropenes
– startg. m. f.
  1,3,5-cycloheptatrienes **29,** 848
  1,3-dienes **28,** 674
  ethylene derivs., synthesis **29,** 877
  tricyclo[3.1.0.0²,⁴]hexanes (from 2 molecules) **27,** 715
–, **imino-**
– from
  acetylene derivs. and isonitriles **27,** 690
Cyclopropenethiones
–, reaction with enamines **22,** 710 suppl. **29**
– startg. m. f.
  6H-1,3-oxazine-6-thiones **27,** 436
– **diamino- 29,** 574
Cyclopropenium salts
– startg. m. f.
  benzenes **26,** 771
  pyridines **26,** 771 suppl. **29**
–, **halogeno-**, reactions **29,** 574
Cyclopropenonecarbinols
– startg. m. f.
  allenecarboxylic acid fluorides **28,** 501
Cyclopropenones
–, chemistry, review **28,** 832 suppl. **29**
–, reactions **27,** 706
–, – with N-heterocyclics **27,** 706
–, – with pyridinium ylids **28,** 832
– startg. m. f.
  acetylene derivs. **27,** 983
  β-amino-α,β-ethylenecarboxylic acid aziridides **26,** 306
  2,4-dienecarboxylic acid amides **28,** 614
  enecyclopropenes **30,** 627
  N-heterocyclics, N-condensed **27,** 706

(Cyclopropenones
- startg. m. f.)
  α-phosphoranylideneketones
    28, 589
  4-pyridones 28, 634
  2-pyrones 28, 832
  Δ²-pyrrol-4-ones 30, 473
-, diamino- 29, 574
-, hydroxy- 30, 126
Cyclopropenyl azides
- startg. m. f.
  1,2,3-triazines 29, 330
2-Cyclopropen-1-yl ethers,
  2,3-diphenyl-
- startg. m. f.
  disacccharides 15, 247
    suppl. 29
Cycloprop[b]indoles, 1,1a,2,6b-
  tetrahydro-
- from
  quinolines, 1,2-dihydro-
    30, 537
Cyclopropylacetic acid esters
- from
  cyclopent-2-enones 26, 709
Cyclopropylalcohols
- from
  enelactones 28, 907
  oxido compds. 29, 694
- special s.
  cyclopropyl-carbinols
    (vinylcyclopropyl)-
Cyclopropylamines
- startg. m. f.
  ketones 30, 119
Cyclopropyl azides
- special s.
  cyclopropenyl azides
- startg. m. f.
  azetines 27, 524
  ethylene derivs. and nitriles
    27, 524 suppl. 29
N-Cyclopropylazomethines
- startg. m. f.
  amines, sec. 27, 52
Cyclopropylcarbinols
- startg. m. f.
  cyclobutenes 29, 938
-, 1-arylthio-
- startg. m. f.
  cyclobutanone ring 28, 958
-, dihalogeno-
- startg. m. f.
  cyclopent-2-enones 30, 681
Cyclopropylcarbinyl compds.
- from
  3-ethylenestannanes 28, 848
Cyclopropylcarboxylic acids
- startg. m. f.
  lactones 30, 105
trans-Cyclopropylhalides,
  2-alkyl- 26, 880
S-Cyclopropylid, spiro-
  annelation with – 27, 873
Cyclopropylketones
- from
  acoxyphosphonium salts
    28, 963
  enaminophosphine oxides
    27, 890
- special s.
  acoxycyclopropyl-ketones
  vinylcyclopropyl –
- startg. m. f.
  carboxylic acid esters 28, 96
  cyclopent-2-enones 27, 923
  Δ²-pyrroline ring 27, 407

Cyclopropyl ketones
- from
  cyclobutanones, 2-halogeno-
    26, 860
- startg. m. f.
  furans, 2,3-dihydro- 26, 154
Cyclopropylmercury compds.
  11, 779 suppl. 28
Cyclopropyloxido compds.,
  reduction 29, 31
Cyclopropyloxo compds.
- from
  cyclobutane-1,2-diols 28, 677
Cyclopropylphosphonium
  fluoroborates
-, cycloalkenylation with –
    29, 866
Cyclopropylsulfinic acid amides
    27, 682
Cyclopropylsulfonium salts
- startg. m. f.
  1-oxaspiro[2.2]pentane ring
    28, 837
Cyclo-N-quaternization 27, 527
Cyclostannanes s. Replacement
  of tin, cyclic
3α,5α-Cyclosteroid ring opening
    29, 492
3α,5α-Cyclosteroids, 6β-alkoxy-
- startg. m. f.
  3β-acoxy-Δ⁵-steroids 30, 108
-, 6β-hydroxy-
- startg. m. f.
  3β-acoxy-Δ⁵-steroids 29, 185
Cyclotetramerization
- of
  3-carbamylpyridinium salts
    30, 233
Cyclotribenzylenes
- from
  tosylamines, trimerization
    26, 843
Cyclotriboroxanes
  as intermediates 27, 158
Cyclotrimerization
- of
  aldehydes 30, 71
Cyclotrisilazanes
- special s.
  hexamethylcyclotrisilazane
p-Cymene s. under Palladium-
  carbon
Cysteine peptides, cleavage,
  selective 20, 37 suppl. 27
Cystine peptides, cyclic
  25, 417 suppl. 30

Dabco s. Triethylenediamine
Dakin oxidation 30, 67
Darzens-Claisen reaction 29, 738
- -, modified 26, 859
DBU s. 1,8-diazabicyclo-
  [5.4.0]undec-5-ene
Deactivation by nitro groups
  27, 799
Deacylation s. a. Deformylation,
  Hydrolysis
C-Deacylation 29, 85, 236;
  30, 248, 593
- Mannich reaction with –
    28, 867
-, preferential 17, 129 suppl. 28
N-Deacylation 28, 15, 124
- with aromatization 26, 498
- preferential 28, 20; 29, 357

-, selective HN↓↑C; 29, 24;
  30, 120
-, total 28, 20
O-Deacylation 26, 138; 27, 959;
  28, 11, 325, 483; 29, 601; 30, 17
- with lactone rearrangement
    27, 281
-, biochemical 26, 6
-, preferential 28, 13 suppl. 29
-, -, of tert. alcohol groups
    29, 6
-, reductive 26, 760
-, selective 29, 357
- of carbohydrate derivs.
    26, 138; 27, 7
S-Deacylation s. Mercaptans
  from thiolic acid esters
Dealcoholation with dehydration 29, 945
1,4-Dealcoholation, partial
  26, 936
Dealkylation s. a. Elimination
  of alkyl groups, Replacement
  of alkyl groups by hydrogen
- special s.
  de-tert-butylation,
  demethylation
C-Dealkylation
- with
  oxidation of amines, tert.,
    cyclic, to lactams 30, 152
N-Dealkylation 27, 29; 28, 26
-, preferential 29, 23
O-Dealkylation s. Ethers,
  cleavage
Deamination 29, 70; 30, 36
-, cathodic 29, 68
N-Deamination
- of
  N-amino-N-heterocyclics
    26, 480
Dean-Stark apparatus, – trap
  (s. a. Water entrainment,
  azeotropic) 28, 194; 29, 253,
  548, 927
Dearomatization
-, cycloisomerization
  with – 22, 761 suppl. 26
- of
  phenols 27, 159
C-Debenzylation 26, 12
N-Debenzylation
- with
  aromatization 27, 990
O-Debenzylation 27, 456, 768;
  29, 739
S-Debenzylation by reduction,
  electrochemical 27, 18
  suppl. 28
Deblocking s. Protective
  groups, removal
C-De-tert-butylation 26, 83;
  29, 591
N-De-tert-butylation 29, 25
O-De-tert-butylation 30, 49
Decalin
- as medium 29, 931
- as solvent 27, 919; 29, 450
2-Decalones 29, 678
Decarbalkoxylation s. a. Replacement of carbalkoxy
  groups by hydrogen
C-Decarbalkoxylation
-, Dieckmann cyclization with –
    30, 694
N-Decarbalkoxylation
  (s. a. N-Decarbo-tert-
  butoxylation)

Dec – Deu

- in alkaline medium **29**, 19
- with aziridine ring rearrangement **29**, 319
Decarballyloxylation 29, 28
C-Decarbinolation 26, 85
C-Decarbobenzoxylation 27, 768
O-Decarbobenzoxylation 28, 118
N-Decarbo-*tert*-butoxylation **27**, 27; **28**, 330
–, **selective 15**, 26 suppl. **27;** **27**, 110
N-Decarbohalogen-*tert*- butoxylation, selective 28, 23
Decarbonylation
- of
  aldehydes **26**, 87
  carboxylic acids **30**, 177
  cyclopropenones **27**, 983
  dicarboxylic acid imides **26**, 502
  1,4-diyn-3-ones **30**, 695
  ethylenealcohols **27**, 94
  α-halogenoketones **29**, 984
  thiolactones **26**, 634; **27**, 643
–, ring contraction by – **28**, 986; **30**, 691
- with
  dimerization **26**, 922
–, **oxidative**, of carboxylic acid anhydrides **26**, 255
–, **photochemical 30**, 364
Decarboxamidation 26, 84
Decarboxylation (s. a. Carbon dioxide eliminators, Photodecarboxylation, Reactions, decarboxylative) **27**, 768
- without solvent **1**, 610
- with
  rearrangement, skeletal **29**, 980
  ring contraction **29**, 92
  Wolff-Kishner-Huang Minlon reduction **3**, 57 suppl. **26**
–, **oxidative 30**, 414
–, **stereoselective 28**, 733 suppl. **28**
Deconjugation (s. a. C-Alkylation, deconjugative, β,γ- from α,β-Ethylene...) **20**, 533; **23**, 719; **27**, 737
Dedeuteriation, preferential **29**, 81
N-Deformylation, photochemical 28, 27
Degradation s. a. Hofmann degradation
- **with loss of 1 C-atom**
  (s. a. Curtius degradation, Decarbonylation, Decarboxylation)
- of
  alcohols from carboxylic acids **30**, 155
  carboxylic acids to halides **26**, 554
  α-halogenocarboxylic acid halides to ketones **28**, 630
  α-hydroxylactols to hydroxycarboxylic acids **29**, 225
  nitriles to ketones **30**, 155
  side chains **25**, 695
–, **oxidative**
- of
  isocycles in N-heterocyclics **30**, 147
  side chains **26**, 226

Dehalogenation (s. a. Monodehalogenation, Replacement of halogen by hydrogen)
- via pyridinium salts **29**, 80
Dehydrating agents
  calcium sulfate **11**, 232 suppl. **26; 11**, 233 suppl. **29**
Dehydration (s. a. Ring contraction, dehydrative)
- of
  alcohols **26**, 951/2; **27**, 941; **29**, 622, 933; **30**, 648
–, aromatization **26**, 866
- via thiobenzoylation **29**, 947
- with conjugation **29**, 939
- with dealcoholation **29**, 945
- with methyl group migration **29**, 939
–, **partial**, of alcohols **17**, 788 suppl. **26**
–, **preferential**, of alcohols, sec. **28**, 916
trans-**Dehydration 26**, 952 suppl. **29**
Dehydration-cyclization
  s. Cyclodehydration
Dehydrobenzene s. Benzyne
Dehydrocyclization
  s. Cyclodehydrogenation
Dehydrofluorination 27, 963
Dehydrogenation CC↑H
  (s. a. Aromatization, Cyclodehydrogenation, Transferdehydrogenation)
- of hydrazones **30**, 639
- to carbon-nitrogen double bonds **25**, 354
- via alcohols **27**, 917
- via selenides **29**, 912
- via thioethers **29**, 912
- with
  bitosylsulfinodiimide **30**, 322
- with simultaneous
  Hofmann degradation of cyclimmonium salts **29**, 707
  rearrangement, skeletal **28**, 887; **29**, 905
Dehydrohalogenation
  (s. a. Cyclodehydrohalogenation, Dehydrofluorination, Ethylene derivs. from halides, Photocyclization, dehydrohalogenative) **13**, 893
- to
  acetylene derivs. **23**, 948 suppl. **29**
  N-heterocyclics **27**, 969
- with
  fluoride ions **23**, 957; **29**, 961
  simultaneous cyclopropane ring opening **29**, 955
  skeletal rearrangement **29**, 966
–, **ar.**
–, rings, condensed, by – **27**, 962
–, **partial 29**, 969
–, **preferential 3**, 757
N-Dehydroxymethylation 27, 456
Delépine reaction 6, 478
Demercuration (s. a. Deoxymercuration, Solvomercuration-demercuration) **29**, 172

O-Demesylation 30, 7
N-De(methoxymethylation) **30**, 701
Demethylation s. a. Dealkylation
N-Demethylation
- of ammonium salts, quaternary **29**, 992
- via formamides **28**, 136
Demjanov-Tiffeneau ring expansion 1, 539–41
Deoxygenation via phosphoric acid derivs. **28**, 65
N-Deoxygenation 26, 990/1
3-Deoxy-2-ketoaldonic acids
- startg. m. f.
  fuoric acids **29**, 252
Deoxymercuration
–, ethylene derivs. by – **29**, 977
Depsipeptides 24, 184 suppl. **28**
Deprotonation
- of
  1,2-dithiolium salts **30**, 638
N-Dequaternization 25, 324; **30**, 698
- with nucleophiles **25**, 324 suppl. **28**
Derivatives s. Protective groups; subentry 'derivatives'
Desaurins 30, 405
Design s. under Synthesis
Desulfination s. Hydrodesulfination, Replacement of sulfinic acid groups by hydrogen
Desulfuration, ring opening by – **26**, 691
–, **photochemical 26**, 82
1,4-Dethiolation
–, thiodienolethers by – **30**, 684
N-Detosylation 26, 438; **29**, 359
–, **selective, electrochemical** **27**, 18
O-Detosylation, selective 28, 119
Detritylation, review **26**, 27a
N-Detritylation, selective **26**, 27a
Deuteriation
  (s. a. Dedeuteriation)
- of
  ar. and heteroar. compds. **22**, 664 suppl. **28**
  diazo compds. **17**, 705 suppl. **29**
  ketones **17**, 705 suppl. **29**
- via tricarbonylchromium(0) complexes **28**, 591
- with
  replacement of cyclic sulfur by cyclic nitrogen **26**, 432
–, **ar. 22**, 664; **30**, 429
–, **preferential 27**, 650; **29**, 81
- of
  carbon-carbon double bonds **29**, 47
–, **selective**
- of
  carbon-carbon double bonds **29**, 47
Deuterioalcohols
- special s.
  alcohols-α-D
  alcohols-OD
  deuteriomethanol
Deuterioaldehydes 26, 872, 912
- from
  aldehydes **27**, 653
  benzils **28**, 64

Deuterioazomethines
26, 912 suppl. 27
- startg. m. f.
 aldehydes-1-d 26, 912 suppl. 27
Deuterio compds. (s. a. Compds., H-labeled)
- special s.
 indoles-2-D
 monodeuterio compds.
 stilbenes-α,α'-D_2
1,1-Deuterioethylene derivs.
27, 949 suppl. 30
Deuterioketones
- special s.
 dideuterioketones
β-Deuterioketones
- from
 α,β-dideuterioketones 29, 47
 α,β-ethyleneketones 29, 47
Deuteriomethanol as reactant 27, 20
Deuterium, retention, preferential 27, 564
- chloride 27, 549
- oxide 28, 64, 590
Di... s. a. Bi...
Diacetals as complexing agents 5, 194 suppl. 29
β-Diacetals
- startg. m. f.
 pyrimidines, 2-amino- 28, 357
Diacetyl disulfide as reactant 23, 603 suppl. 26
Diacetylene derivs. s. Diynes
Diacoxy compds. (s. a. Diol esters)
- from
 ethers, cyclic, with rearrangement 28, 89
1,1-Diacoxy compds.
- from
 alcohols, ring contraction 28, 133
 oxo compds. 27, 759
- special s.
 acylals
-, mixed 29, 124
1,2-Diacoxy – s. Glycol esters
Diacylamines
-, C-acylation with – 26, 679
- from
 carboxylic acid chlorides 26, 420
-, mixed s. N-Acylcarboxylic acid amides
1,1-Di(acylamines)
- from
 morpholinals 29, 403
α,ω-Di(acylamines)
- from
 α,ω-di(iminoesters) 30, 225
C-Diacylation, mixed 17, 879 suppl. 26
Diacylazo compds.
- special s.
 α,α'-azobis-(N-methyl-formamide)
- startg. m. f.
 4H-1,3,4-oxadiazine ring, 5,6-dihydro- 28, 280
Diacyl disulfides
- special s.
 diacetyl disulfide
N,N-Diacylhydrazines
- from
 carboxylic acid chlorides 27, 467

N,N'-Diacylhydrazines
- from
 carboxylic acids 13, 434 suppl. 26
-, sym.
- from
 carboxylic acid hydrazides 27, 445
Diacyloxyselenuranes
- from
 selenides and acyl peroxides 29, 108
- startg. m. f.
 1,1-acoxyselenides 29, 108
Diacyl sulfides 26, 618
Dialdehydes
- from
 ethylene derivs., cyclic 29, 147
-, ring closures 27, 384
 with diamines 27, 384
 – sulfur compds. 30, 609
- special s.
 glyoxal
- startg. m. f.
 azepine ring, dihydro- 19, 470
 lactols 27, 858
 2-nitro-1,3-diols, cyclic, review 16, 794 suppl. 26
1,3-Dialdehydes
-, by Vilsmeier synthesis, double 27, 826
- from
 β-amino-α,β-ethylene-aldehydes 13, 786
o-Dialdehydes
(s. a. Phthaldehydes)
- startg. m. f.
 p-quinones, condensed 29, 735
 tropone ring 29, 847
1,1-Dialkoxy-3-alkylthioallenes
- from
 halides 30, 584
- startg. m. f.
 β-alkylthio-α,β-ethylene-carboxylic acid esters 30, 584
1,1,1-Dialkoxyamidines 30, 279
1,1'-Di(alkoxy)amines
-, Grignard synthesis with – 28, 813
1,1,1-Dialkoxyamines, cyclic
 s. 1-Aza-4,6-dioxabicyclo-[3.3.0]octanes
1,1,1-Dialkoxyammonium salts, quaternary
- from
 acoxonium salts 27, 995
-, reaction with compds., CH-acidic 27, 821
- startg. m. f.
 acetals 27, 821
 amide acetals 27, 430
1,1-Dialkoxy compds.
 s. Acetals
1,3-Dialkoxy –
- from
 cyclopropanes 27, 135
2,2-Dialkoxyethers, cyclic
- startg. m. f.
 dioxo compd. monoketals 27, 859

N,N-Dialkoxylamines, bicyclic, N-condensed
- from
 4H-1,2-oxazine 2-oxides, 5,6-dihydro- 27, 702
N-Dialkoxymethylation 30, 279
α,α-Dialkoxyphosphonic acid esters
- from
 phosphorous and ortho-carboxylic acid esters 29, 607
- startg. m. f.
 α-alkoxyenephosphonic acid esters 29, 607
Dialkoxyphosphoranes
- special s.
 triphenyldiethoxy-phosphorane
Dialkoxystannanes, cyclic
-, activation of glycol groups as – 29, 111
Dialkoxysulfuranes as reagents 27, 941
Dialkylaluminum hydrides
 s. Diethyl-aluminum hydride, Diisobutyl-
(Dialkylamino)alanes
 as reagents 23, 48 suppl. 29
Dialkylaminosulfur trifluorides
-, as reactants 19, 635 suppl. 29
- special s.
 diethylaminosulfur trifluoride
1,2-Dialkylation
- of ethylene derivs.
 (s. a. 1,4-Addition-alkylation) 28, 791
1,1-C-Dialkylation s. Replacement of carbonyl oxygen by 2 carbon substituents
C,O-Dialkylation of bezoins 26, 856
N,S-Dialkylation of barbituric acids, 2-thio- 29, 408
Dialkylborane-aluminum methoxide complex as intermediate 23, 702 suppl. 26
Dialkylboranes, hindered 24, 48 suppl. 29
Dialkyl carbonate,
 C-carbalkoxylation with – 30, 539
Dialkylchloroboranes
 as intermediates 26, 912 suppl. 29; 27, 485; 28, 783
-, prepn. 27, 649
N,N-Dialkylformamides
 s. Dimethylformamide
N,N'-Dialkylidene-1,1-diamines
- startg. m. f.
 Δ^2-imidazolines 28, 697
Dialkyl phosphites
(s. a. Diaryl phosphites)
- startg. m. f.
 α-aminophosphonic acid esters 29, 610
 benzenephosphonic acid esters, 2,3-dihydroxy- 29, 592
 1-hydroxy-1,1-diphosphonic acid esters 28, 584
 phosphonic acid esters, N-heterocyclic 27, 654
 phosphoric – – 28, 523
- **phosphonates** s. Dialkyl phosphites

- **sulfates** (s. a. Dimethylformamide-dialkyl sulfate)
  - as reactants **30,** 139
  - as reagents **28,** 296
- **Di(alkylthio)methylene compds.** s. Ketene mercaptals
- **N-Di(alkylthio)methylene** –
  s. Iminodithiocarbonic acid esters
- *meso*-**Diallenes, conjugated 19,** 970 suppl. 28
- **Diallenesulfones**
  - startg. m. f.
    thiophene 1,1-dioxides, 3-vinyl- **29,** 683
- **Diamides** s. Diacylamines
- **Diamidides**
  - from
    amidines and amide acetals **26,** 333
  - special s.
    1,3,5-triazahepta-1,3-dien-7-ones
- **Diamines**
  - from
    cyanoketones **28,** 362
    ethylene derivs., cyclic **27,** 149
    nitrocarboxylic acid amides **26,** 64
  - –, ring closures with dialdehydes **27,** 384
  - special s.
    1,8-bis(dimethylamino)-naphthalene
  - startg. m. f.
    N-cyanoguanidines, cyclic **28,** 311
    1,3-N,N-heterocyclics, 2-amino- **26,** 434, 438
    –, 2-oxo- **21,** 393
    –, 2-tosylamino- **26,** 438
    sulfinodiimines, cyclic **26,** 394
- **Di-*tert*-amines**
  - startg. m. f.
    cyclimmonium salts **26,** 479
- **1,1-Diamines**
  - from
    azomethines **29,** 299
  - special s.
    1,1,1-alkoxydiamines
    diaminomethylene...
    α,α-diaminonitriles
    morpholinals
    1,1,2,2-tetraminoethanes
  - startg. m. f.
    amines **28,** 57
- **1,2-Diamines**
  - from
    2-halogenophosphoramidates **28,** 411
  - startg. m. f.
    2-imidazolidone ring **3,** 375
- **1,3-Diamines** (s. a. Triethylenediamine)
  - startg. m. f.
    1,5-diazocines, hexahydro- **27,** 399
- *o*-**Diamines**
  - startg. m. f.
    benzimidazoles **27,** 410; **30,** 265
    –, 2-alkylthio- **28,** 538
    –, 2-amino- **30,** 245

1,4-dicyano-1,3-dienes **28,** 429
1,1-iminoalcohols, 1,3-N,N-heterocyclic, N-condensed **29,** 364
pyrrolo[1,2-a]benzimidazol-1-ones, 3-hydroxy- **27,** 413
- *p*-**Diamines**
  - startg. m. f.
    quinone diimines **30,** 323
- **Diaminogermanes**
  - startg. m. f.
    2-germa-1,3-dioxaheterocycles **29,** 114
- **Diaminoguanidines** s. Acyldiaminoguanidines
- **Diaminoketones**
  - special s.
    α,α-di(aminomethyl)ketones
- **Diaminomethylenesulfonium salts**
  - startg. m. f.
    alkylthioketene aminals, rearrangement **26,** 581
- α,α-**Di(aminomethyl)ketones**
  - startg. m. f.
    coumarins, 3,4-dihydro- **26,** 836
- α,α-**Diaminonitriles**
  - startg. m. f.
    1,1,1-alkoxydiamines **28,** 187
    amide acetals **28,** 187
- **Diaminosulfanes**
  - special s.
    bis(dimethylamino)sulfane
- **Dianions**
  - as intermediates **26,** 417; **27,** 840
  - special s.
    carboxylic acid dianions
    dicarbanions
- **N,C-Dianions** as intermediates **26,** 792; **27,** 844
- **O,C-Dianions** – – **26,** 792; **28,** 732 suppl. 29
- **S,C-Dianions** – – **29,** 620
- **1,4-Dianions**
  - startg. m. f.
    heterocyclics **26,** 792
- **Diansa compds. 22,** 528 suppl. 26
- **1,2-Diarylacetylenes**
  s. Tolans
- **1,1-Diarylakanes**
  - from
    1,1-alkoxyhalides **26,** 887
- **Diarylamines**
  - from
    azides **29,** 402
- **Diaryl disulfides, sym. 24,** 615a suppl. **28**; **26,** 589
- **1,2-Diarylethanols**
  - from
    aldehydes **26,** 786
  - startg. m. f.
    stilbenes **26,** 786
- **Diaryl ethers**
  - startg. m. f.
    phenoxaphosphine 10-oxides **30,** 541
    xanthones **30,** 541
- **1,1-Diarylethylenes**
  - startg. m. f.
    benzoselenopheno[2,3-b]-benzoselenophenes **28,** 592
- **1,2-Diarylethylenes** s. Stilbenes

- **Diaryl ketones 28,** 600
  - –, carboxylation, ar., via – **26,** 600
  - –, cleavage **28,** 600
- α,α-**Diarylketones**
  - from
    α-diketones **2,** 680
- **Diarylmethanes** (s. a. Dipyrrylmethanes) **28,** 763
  - –, polynitro- **11,** 872 suppl. 27
- **1,4-Diaryloxy-2-butynes**
  - startg. m. f.
    benzofuro[3,2-c][1]benzopyrans, 6a,11a-dihydro- **29,** 684
- **Diaryl phosphites** (s. a. Dialkyl phosphites)
  - startg. m. f.
    phosphoromonoamidates **28,** 263
- **Diaryls**
  - from
    aryllead tricarboxylates **29,** 878
    diazonium salts **29,** 279
    N-stannylhydrazines, N'-aryl- **28,** 784
    sulfinic acids **27,** 880
  - via telluride dihalides **28,** 967
- –, **sym.**
  - from
    aryl halides **27,** 870; **28,** 792; **30,** 592
    triaryl phosphates **28,** 900
- –, **amino-**
  - from
    sulfonic acid amides, ar. **28,** 956
- –, **dihydroxy- 26,** 857
- **Diaxial-diequatorial rearrangement 26,** 583
- **1,4-Diazabicyclo[4.n.0]alkan-5-ones 27,** 464
- **2,3-Diazabicyclo[2.2.1]hept-5-enes**
  - startg. m. f.
    cyclopenta[e][1,3,4]oxadiazines, 4,4a,5,7-tetrahydro- **27,** 166
- **1,2-Diazabicyclo[3.2.0]-3-hepten-6-ols**
  - startg. m. f.
    8-oxa-2,3-diazabicyclo-[3.2.1]oct-6-enes **29,** 169
- **1,3-Diazabicyclo[3.1.0]hex-3-enes**
  - startg. m. f.
    pyrimidines **28,** 882
- **1,9-Diazabicyclo[4.2.1]nona-4,7-dien-3-ones, 9-acyl-**
  - startg. m. f.
    pyrrolo[1,2-b]pyridazin-2-ones **29,** 983
- **1,5-Diazabicyclo[4.3.0]non-5-ene** as reagent **28,** 895; **29,** 963/4
- **1,4-Diazabicyclo[2.2.2]octane** s. Triethylenediamine
- **1,5-Diazabicyclo[3.3.0]octanes**
  - from
    azines **28,** 642
- **1,5-Diazabicyclo[3.3.0]octane-2,4,6,8-tetraones 26,** 406
- **1,5-Diazabicyclo[5.4.0]undec-5-ene** s. 1,8-Diazabicyclo-[5.4.0]undec-7-ene

Dia 604

1,8-Diazabicyclo[5.4.0]undec-7-ene as reagent 22, 678
 suppl. 29; 26, 945; 26, 967
 suppl. 30; 27, 661; 29, 963;
 30, 10a
Diazabutadienes s. Trihalogenodiazabutadienes
8,17-Diazadibenzo[c,j]tetracyclo[7.3.1.0²,⁹.0⁷,¹⁶]tridecanes
– from
 isoquinolinium salts
 (2 molecules) and
 methylene compds., active
 28, 645
2,4-Diaza-1,3-diene groups
– startg. m. f.
 2H-thiazolo[3,2-a]-s-triazin-4-ones, 3,4,6,7-tetrahydro-
 27, 321
1,4-Diaza-1,3-dienes
 s. 1,2-Di(azomethines)
1,3,2,4-Diazadiphosphetidine 2,4-dioxides
– from
 amines, prim. 29, 294
1,3,2,4-Diazadiphosphetidines 29, 294
1,3,2-Diazaphospholi-4,5-dione 2-oxides, 2-amino- 26, 418
Diazasteroids 29, 350
8,13-Diazasteroids 29, 350
Diazen... s. a. Azo...
Diazenium salts (s. a. Alkoxydiazenium salts)
– from
 azo compds. 29, 994
1,4-Diazepine-5,7-dione ring 26, 342
1,4-Diazepine ring, 2,3-dihydro- 26, 373
1H-1,2-Diazepines
–, [4+2]-cyclodimerization 30, 498
4H-1,2-Diazepines
– from
 pyrylium salts 26, 370
1H-1,4-Diazepines, 6,7-dihydro-
– from
 cis-enediazomethines 29, 703
1H-1,2-Diazepines, 4,5,6,7-tetrahydro-
– from
 hydrazines 28, 386
1H-1,4-Diazepinium salts, 6-azo- 27, 424
2H,5H-[1,4]Diazepino[6,5-b]indole-2,5-diones, 1,3,4,6-tetrahydro- 21, 554 suppl. 28
1,3-Diazetidine-2,4-diones
– from
 allophanoyl chlorides 26, 488
 isocyanates (2 molecules) 26, 488
1,3-Diazetidin-2-one ring, 4-alkoxy-
– from
 O-allyllactims and isocyanates 29, 301
1,1-Diazides
– from
 1,1-dihalides 26, 428
2,5-Diazido-p-quinones
– startg. m. f.
 cyanoketenes (2 molecules) 26, 496

Diaziridines
– from
 3-pyrazolidone betaines 26, 320
 Δ²-tetrazolines 30, 194
Diaziridinones
–, cycloadditions with – 30, 212
– startg. m. f.
 1,2,4-triazolidine-3,5-diones, 4-acyl- 30, 212
 ureas 29, 131
Diazirine ring s. Spirodiazirine ring
Diazirines
– from
 diazo compds. 29, 328
– startg. m. f.
 diazo compds. 29, 328
Diazoacetic acid esters
– from
 chloroformic acid esters 27, 828
ω-Diazoacetophenones
– special s.
 o-alkylthio-ω-diazoacetophenones
O-Diazoacetylation
– of
 alcohols 26, 201
2-Diazoalcohols
– from
 oxo compds 26, 668
– startg. m. f.
 acetylene derivs. 28, 939
2-Diazoamines (s. a. 1,2,3-Triazoline-2-diazoamin tautomerism)
– from
 diazo compds. and
 enamines 28, 625
Diazo anhydrides s. Diazo oxides
α-Diazocarboxylic acid amides
– startg. m. f.
 lactams 21, 942
– – tert-butyl esters
– startg. m. f.
 α-acoxycarboxylic acids 27, 201
– – esters
– startg. m. f.
 γ-ketocarboxylic acid esters 28, 858
1,5-Diazocines, hexahydro-
– from
 1,3-diamines and
 α,β-ethyleneketones 27, 399
Diazo compds.
–, deuteriation 17, 705
 suppl. 29
– from
 amines 30, 169
 diazirines 29, 328
 imines 27, 444
 hydrazones 27, 444
 methylene groups, active
 s. Diazo group transfer
–, prepn., review 26, 321
 suppl. 27
–, reaction with boranes 28, 782/3, 860
–, reactions with transition
 metal catalysts 29, 371
–, ring closure and expansion
 with – 26, 837; 27, 958

–, ring expansion of –
 N-heterocyclics 29, 779
 heterocyclics 18, 848
 suppl. 29
 α-oxo-N-heterocyclics 18, 848; 29, 783
–, – fragmentation of
 N-heterocyclics 29, 779
– special s.
 acoxydiazo compds.
 diazomalonic acid esters
 diazomethane
 disulfonyldiazomethanes
 ethylenediazo compds.
 hydroxydiazo –
 mercuridiazo –
 1-oxidopyridin-2-yldiazomethane
 phosphonodiazo compds.
– startg. m. f.
 alcohols 30, 124
 1,1-alkoxyhalides 27, 580
 2-allenethioethers 29, 781
 azo compds., aliphatic-ar. 26, 284
 azomethine imides, cyclic 30, 283
 carbostyrils, 3-hydroxy- 29, 625
 cyclopropane ring 28, 780
 diazirines 29, 328
 2-diazoamines 26, 625
 1,3-dienes 26, 915
 dinitromethylene compds. 26, 845
 1,3-dithiepane ring 29, 769
 enazides 30, 457
 enolethers 28, 846
 ethers, sym. 30, 123
 ethylene derivs. 27, 876
 – – (cis or trans) 30, 668
 – –, sym. (cis) 18, 862
 suppl. 26
 furoxans (from 2 molecules) 27, 825
 glycol esters (from 2 molecules) 29, 780
 α-halogenosulfoxides 28, 554
 Δ²-isoxazoline oxides 28, 774
 Δ²-isoxazolines 28, 778
 mercury compds., diorgano-1-halogeno- 27, 655
 peroxides 27, 197
 pyrazoles 26, 788
 selenoacetals 26, 648
 silanes 30, 430
 sulfonium ylids 27, 622
 sulfoxonium ylids 27, 622
 telluroacetals 26, 648
 Δ⁸-1,3,4-thiadiazoline
 1-oxides 25, 535
 thiirane 1-oxides 29, 767
 thionocarboxylic acid esters,
 synthesis with addition of
 1 C-atom 30, 570
 thiophene S-oxide ring, 2-chloro- 29, 768
Diazocyanides
– startg. m. f.
 N′-cyanoazoformamidines 27, 442
 quinone monoimines 27, 442
Diazocyclopentadienes
– startg. m. f.
 pyridinium cyclopentadienylids 29, 397
α′-Diazo-α,β-ethyleneketones 20, 271 suppl. 26; 30, 247

Diazo group transfer
26, 321; 30, 247
–, review 26, 321 suppl. 27
Diazo-N-heterocyclics
– startg. m. f.
 azido-N-heterocyclics 30, 185
2-Diazoimidazole, 4,5-dicyano-
–, reactions with – 28, 771
Diazoketols, cyclic
– by aldol condensation, intramolecular 26, 745
Diazoketones
–, ring closure with ring expansion 28, 935
– special s.
 diazoethyleneketones
α-**Diazoketones** (s. a. Diazoacetyl...)
– from
 carboxylic acids 26, 806
 ketones 30, 247
–, reactions with sulfur dioxide 29, 562
–, review 28, 860
– special s.
 bis-(α-diazoketones)
 α,α'-bisdiazoketones
 carbalkoxy-α-diazoketones
 ω-diazocetophenones
 α'-diazo-α,β-oxidoketones
– startg. m. f.
 α-alkoxyketones 27, 204
 β-aminoketones 28, 782
 enolbetaines 26, 917
 α-halogenoketones, synthesis 28, 860
 α-ketocarboxylic acid esters 27, 204
 ketones, synthesis 27, 894
 –, cyclic 28, 936
 1,4-oxathiin-2(3H)-one 4,4-dioxides 29, 562
 pyrrole ring, 1,2-diamino- 26, 466
 Δ²-pyrrol-4-ones 28, 773
 pyrrolo[1,2-b][1,2,4]triazine ring 26, 466
–, **cyclic**
– startg. m. f.
 pyrazole ring, N-acyl-, N-condensed 29, 657
Di(azoles) s. N,N'-Carbonyldi(azoles)
Diazomalonic acid esters
–, prepn. and reactions, review 28, 780
Diazomethane
– as reagent 27, 409
–, generation in situ 17, 850
–, ring closures with – 25, 587 suppl. 26
Di(azomethines) s. a. N,N'-Dialkylidene-1,1-diamines
1,2-Di(azomethines)
– from
 azomethines (2 molecules) 26, 683
 ene-1,2-diamines 26, 683
–, rearrangement, [3.3]-sigmatropic 24, 746 suppl. 29
– startg. m. f.
 – 1-cyano-1,2-ene-1,2-diamines 29, 639
–, **cyclic** 9, 462 suppl. 26
N-Diazonitrilium group
27, 282

Diazonium carboxylates
– from
 N-nitrosanilines, N-acyl- 26, 995
– **fluoroborates**
– as reactants 30, 236
– startg. m. f.
 halides 26, 549
– **mercuric salt complexes**
– startg. m. f.
 hydrocarbons 30, 36
– **nitrates**
– from
 nitroso compds. 30, 190
– **salt rearrangement 16**, 110 suppl. 26
– **salts**
– as benzyne precursors 19, 910 suppl. 26
– from
 azomethines 26, 262
–, reactions 27, 439
– special s.
 enediazonium salts
– startg. m. f.
 azo compds., sym. 29, 279
 diaryls 29, 279
 diazo-N-sulfides 30, 198
 fluorides 26, 555
 heterocyclics 28, 369
 phenols 26, 117
 tetrazoles 30, 308
 Δ²-1,3,4-thiadiazolines, 5-imino- 30, 234
 1,2,4-triazine ring 29, 358
 ynazo compds. 30, 238
– **sulfate betaines 26**, 997
–, **stable 26**, 997 suppl. 29
Diazo oxides
–, nitration 10, 292 suppl. 29
– startg. m. f.
 o-halogenophenols 26, 550
 phenols 26, 550
α'-**Diazo-α,β-oxidoketones**
26, 806 suppl. 29; 28, 795 suppl. 29
Diazooxo compds.
– special s.
 diazoketones
α-**Diazophosphorus compds.**
–, ring expansions with – 29, 625
Diazo-N-sulfides
– from
 diazonium salts 30, 198
– startg. m. f.
 azo compds. 30, 198
α-**Diazosulfones**, review
19, 357 suppl. 26
– special s.
 β-amino-α-diazosulfones
– startg. m. f.
 β-amino-α-diazosulfones, synthesis 27, 688
 α-halogenosulfones 30, 367
α-**Diazo(sulfonylamidines)**
– from
 sulfonic acid azides and ynamines 29, 310
–, **cyclic**
– special s.
 indolines, 3-diazo-2-sulfonylimino-
Diazotization (s. a. Sodium nitrite)
– of
 aminothiolic acids 27, 293

Dibenzo-1,4-diazepines
– from
 acridinium salts 28, 287
6H-Dibenzo[d,f][1,3]diazepines, 5,7-dihydroxy-
– from
 oxo compds. 29, 362
12H-Dibenzo[d,g][1,3]dioxocin-6-carboxylic acids 29, 203
Dibenzofurans 29, 913
Dibenzo[fg,op]naphthacenequinones
– from
 phenolethers 27, 906
Dibenzo[c,h][1,5]naphthyridines, 6,12-dihydro-
– startg. m. f.
 3,3'-bis-(1H-isoindoles) 29, 712
Dibenzo[b,d]pyrans 28, 987
Dibenzo[a,l]pyrenes
– by Scholl ring closure 27, 914
7H-Dibenzo[de,h]quinol-7-ones, 2-hydroxy- 29, 377
Dibenzo[c,e][1,2]thiazine 1,1-dioxides
– from
 2H-benzo[e][1,2,3,4]-thiatriazine 1,1-dioxides 29, 462
11H-Dibenzo[c,f][1,2,5]-triazepines 27, 354
(Dibenzoyldioxyiodo)benzenes
 as oxidants 27, 116
Diborane (s. a. Borane) 11, 118 suppl. 28; 13, 104 suppl. 26; 16, 54 suppl. 29; 25, 163a suppl. 29; 26, 134; 27, 43, 232, 716; 28, 103; 29, 58; 30, 81
–, reactions with – in methylene chloride 28, 34
–, reductions with – 28, 34
1,1-Diboro compds.
– from
 acetylene derivs. 30, 427
– startg. m. f.
 1,1-dimercury compds. 30, 427
N,N-Dibromobenzenesulfonamide as reagent 14, 598 suppl. 29; 15, 170 suppl. 28
α,α-**Dibromomethyl ether**
–, reactions with carbohydrate derivs. 27, 572 suppl. 28
Di-n-butylamine as reagent 14, 565 suppl. 27; 28, 584
Di-tert-butylcarbodiimide 29, 923
Di-tert-butylketone
 as reagent 29, 35
Di-tert-butyl peroxide
– as initiator 29, 970
– as photoinitiator 29, 720
– as reagent 11, 712 suppl. 29; 27, 69, 285, 301, 911; 28, 271; 30, 94
Di-tert-butylpyridine as base, hindered 29, 183
Dicarbanions
–, aldol condensation with – 26, 861 suppl. 27
– as intermediates 27, 835
– special s.
 β-ketosulfoxide dicarbanions
–, **stepwise prepared**, as intermediates 26, 861; 29, 618; 30, 447

Dicarbenes 16, 836 suppl. **27**
Dicarbonic acid esters
  s. Dithiodicarbonic acid
  esters
Dicarbonyl compds.
– special s.
  enedicarbonyl compds.
β-Dicarbonyl –
–, C-arylation **30,** 593
γ-Dicarbonyl –
– from
  1-sulfinylthioenolethers
  29, 667
Dicarboxydiselenides
– from
  lactones **28,** 585
α-Dicarboxylic... s. a. Malon...
β-Dicarboxylic... s. a. Succin...
γ-Dicarboxylic... s. Glutar...
Dicarboxylic acid amide esters
– special s.
  ethyl oxamate
– startg. m. f.
  dicarboxylic acid imides
  28, 622
  lactams **26,** 502
– – **amides**
– startg. m. f.
  dicarboxylic acid imides
  28, 449
– – **anhydrides**
–, annelation with – **1,** 581
– from
  dicarboxylic acid chlorides
  26, 411
  – acids **9,** 261 suppl. **26;**
  27, 260; **30,** 364
–, reactions with hydrazine
  and subst. hydrazines
  8, 482 suppl. **26**
– special s.
  carbalkoxydicarboxylic acid
  anhydrides
  chloroformyldicarboxylic – –
  homophthalic – –
  hydroxydicarboxylic – –
  succinic – –
– startg. m. f.
  aldehydocarboxylic acids
  29, 64
  aminoalcohols **28,** 364
  N-aminodicarboxylic acid
  imides **28,** 330
  carbalkoxy-α-diazoketones
  30, 456
  dicarboxylic acid chlorides
  29, 513
  – – esters **27,** 127
  – – –, homologous **30,** 456
  – – imides **1,** 332/3; **29,** 442
  – – monoamides **29,** 298
  – – monoesters **28,** 86
  diketones, double ring
  closure **28,** 745
  double bonds, carbon-carbon
  23, 976
  enollactones **26,** 904
vic-**Dicarboxylic** – –
– startg. m. f.
  ethylene derivs. **29,** 991
Dicarboxylic acid chloride esters
– special s.
  o-methoxycarbonylbenzoyl
  chloride

– – **chlorides**
– from
  dicarboxylic acid
  anhydrides **29,** 513
–, ring closure with
  carbodiimides **26,** 511
–. ring closure, double with –
  27, 849
– startg. m. f.
  dicarboxylic acid anhydrides
  26, 411
  – – imides **26,** 411
N,N'-Dicarboxylic acid diimides
– from
  N-aminodicarboxylic acid
  imides **29,** 366
Dicarboxylic acid esters
–, annelation with – **28,** 723
– from
  carbalkoxy-α-diazoketones
  30, 456
  dicarboxylic acid anhydrides
  27, 127
  α-diketones, cyclic **27,** 146
– special s.
  ketodicarboxylic acid esters
– startg. m. f.
  β-diketones, 2-acyl-, cyclic
  28, 728
  lactols **29,** 62
  pyridazine-3,6-diol ring
  21, 563 suppl. **26**
– – –, **homologous**
– from
  dicarboxylic acid anhydrides
  30, 456
– – –, **mixed 27,** 235
– startg. m. f.
  dicarboxylic acid mono-
  esters **27,** 235
α-Dicarboxylic acid esters
  s. Malonic acid esters
β-Dicarboxylic – –
– special s.
  α,β-ethylene-β-dicarboxylic
  acid esters
  β,γ-ethylene-β- – – –
o-Dicarboxylic – –
– special s.
  phthalic acid esters
Dicarboxylic acid halides
– special s.
  dicarboxylic acid chlorides
  halogenodicarboxylic acid
  halides
– – **imides** (s. a. Imido...)
– from
  alcohols **27,** 429
  dicarboxylic acid amide
  esters **28,** 622
  – – amides **28,** 449
  – – anhydrides **29,** 442
  – – chlorides **26,** 411
  sulfenimides **26,** 492
–, hydrogenation, partial and
  selective **26,** 31
– special s.
  N-acoxydicarboxylic acid
  imides
  N-aminodicarboxylic – –
  carboxydicarboxylic – –
  N-halogenodicarboxylic – –
  homophthalimides
  N-hydroxydicarboxylic acid
  imides
  malonimides
  naphthalimides

N-phosphonyldicarboxylic
  acid imides
2,5-pyrrolidiones
succinimides
sulfenimides
N-sulfinyldicarboxylic acid
  imides
– startg. m. f.
  1-alkoxylactams **27,** 947
  N-heterocyclics,
  N-condensed **27,** 947
  ketocarboxylic acid amides
  26, 679
  lactams **26,** 502
–, Wittig synthesis with –
  26, 904
– – –, **vinylogous,** ring closure,
  double **29,** 713
o-Dicarboxylic acid imides
  s. Phthalimides
Dicarboxylic acid monoamides
– from
  dicarboxylic acid anhydrides
  29, 298
– – **monoesters**
– from
  dicarboxylic acid inhydrides
  28, 86
– special s.
  succinic acid monoesters
  – via dicarboxylic acid esters,
  mixed **27,** 235
α-Dicarboxylic acid monoesters
– from
  carboxylic acid esters **28,** 598
Dicarboxylic acid peroxides
– from
  dicarboxylic acids **30,** 54
– **acids**
– from
  β-amino-α,β-ethylenenitriles,
  cyclic **29,** 195
  ditosylates **29,** 195
– special s.
  diphenic acids
  homophthalic –
– startg. m. f.
  dicarboxylic acid anhydrides
  9, 261 suppl. **26; 27,** 260;
  30, 364
  – – peroxides **30,** 54
  enolesters, cyclic **29,** 892
  α-halogenodicarboxylic acid
  halides **3,** 451 suppl. **29**
  pyridazine-3,6-diol ring
  26, 374
α-Dicarboxylic acids
– special s.
  di-(7-cycloheptatrienyl)-
  malonic acids
  ethylene-α-dicarboxylic –
γ-Dicarboxylic –
– special s.
  1,3-diene-1,3-dicarboxylic
  acids
o-Dicarboxylic –
– startg. m. f.
  3-alkylidenephthalides
  26, 924
  carboxylic acids **12,** 711
  suppl. **28**
Dichloracetic acid
– as reactant **29,** 203
– as reagent **19,** 307 suppl. **28**
Dichloramine as reactant
  29, 270

N,N-Dichloramines
- startg. m. f.
  azoxy compds., unsym.
  29, 280
- special s.
  (*tert*-carbin)amines,
  N,N-dichloro-
1,2-Dichlorethane as solvent
29, 833
Dichlorine heptoxide,
  perchlorination with –
  30, 195
o-Dichlorobenzene
- as medium **27,** 492, 729;
  28, 678; **29,** 689
- as solvent **28,** 597
**N,N-Dichlorobenzene-
sulfonamide** as reagent
27, 181
**Dichlorobis(acetonitrile)-
palladium(II) 28,** 973
**Dichlorobis(benzonitrile)-
palladium(II)** as reagent
26, 727; **27,** 250
**Dichlorobis(triphenyl-
phosphine)palladium(II)**
as reagent **26,** 81; **30,** 425
Dichloroborane 28, 991
Dichloroborane-etherate
as reagent **27,** 485 suppl. **28**
**2,3-Dichloro-5,6-dicyano-
quinone**
- as reactant **29,** 655
- as reagent **26,** 160, 231;
  28, 887, 921; **29,** 721, 907, 927
Dichloroformimidocyanamides
s. Aminothio(dichloro-
formimido)cyanamides
**α,α-Dichloro-β-ketocarboxylic
acid esters**
- as acylating agents **29,** 441
Dichloromaleic anhydride
as reagent **29,** 94; **30,** 500
Dichloromethane s. Methylene
chloride
Dichloromethylene as reagent
29, 456
**Dichloromethylenedimethyl-
ammonium chloride**
- as reagent **21,** 302 suppl. **29;
28,** 440 suppl. **29**
α,α-Dichloromethyl ether
- as reactant **27,** 572; **29,** 838;
  30, 495
-, introduction of aldehyde
  groups with – **20,** 617
- startg. m. f.
  trialkylcarbinols **27,** 892
  suppl. **28**
Dichloromethyl ketones 29, 504
- from
  carboxylic acid esters **30,** 549
**N,N-Dichlorophosphor-
amidates 4,** 295 suppl. **26**
1,3-Dichloropropane as solvent
26, 554
2,6-Dichloroquinonechlorimine
-, reactions with phosphoric
  acid derivs. **27,** 4
1,1,1-Dichlorosulfenyl chlorides
- special s.
  α-carbamyl-α,α-dichloro-
  methylsulfenyl chlorides
Dichlorosulfine s. Thiophosgene S-oxide
1,3-Dichloro-1,2,4-triazole
as chlorinating agent
28, 490

N,N-Dichlorourea 28, 486
Dichromate s. Sodium
dichromate
Dicumyl peroxide as photoinitiator **29,** 719
N,N-Dicyanamides
- from
  amines, prim. **29,** 342
**α,α-Dicyanocarboxylic acid
esters**
- from
  nitriles **30,** 147
1,4-Dicyano-1,3-dienes
- from
  o-diamines **28,** 429
o,o′-Dicyanodisulfides
- from
  1,2-benzisothiazoles
  27, 584
1-Dicyanomethyl-2-halides
- from
  ethylene derivs. and
  1,1,1-halogenodinitriles
  26, 707
- startg. m. f.
  γ-lactones **26,** 707
Dicyanosulfonium ylids
- startg. m. f.
  1,1-dinitriles **27,** 977
**Dicylimmonium salts,
triple-N-condensed**
-, by dimerization **26,** 400
**Di-(7-cycloheptatrienyl)malonic
acids 28,** 716
Dicyclohexylamine as reagent
30, 449
- as base **27,** 402
Dicyclohexylcarbodiimide
as reagent **4,** 661 suppl. **28;
24,** 299 suppl. **27; 26,** 104,
172, 469, 557, 806; **27,** 108,
292, 359, 937; **28,** 440; **29,** 187;
29, 250 suppl. **30; 30,** 668
Dicyclohexylethylamine
as reagent **27,** 536
**Di(cyclopentadienyl)metal
compds.** s. Metallocenes
α,β-Dideuterioketones
- from
  α,β-ethyleneketones
  29, 47
- startg. m. f.
  β-deuterioketones **29,** 47
**Dieckmann cyclization
26,** 940
- with C-decarbalkoxylation
  30, 694
N,N-Diels-Alder adducts
- startg. m. f.
  pyrazolidines **27,** 25
Dienals
- special s.
  halogenodienals
1,1′-Dienamines, sym. 30, 264
1,3-Dienamines
-, 2-alkylation **27,** 832
-, diene synthesis with –
  26, 696
- from
  1,4-dien-2-amines **30,** 23
- startg. m. f.
  α,β-ethyleneketones,
  4-subst. **27,** 822
-, cyclic
-, *peri*-annelation with –
  29, 764

- from
  1,5-dioxo compds. **27,** 950
- startg. m. f.
  2-cyclohexenone ring **27,** 950
1,4-Dien-2-amines
- startg. m. f.
  1,3-dienamines **30,** 23
1,3-Dienecarbamyl chlorides
- from
  3-chlor-1-ethylenecarbamyl
  chlorides **27,** 545
  α,β-ethyleneazomethines
  27, 545
**2,4-Dienecarboxylic acid
amides**
- from
  cyclopropenones and
  enamines **28,** 614
**3,5-Dienecarboxylic acid
derivs.**
- from
  hemiquinolesters **28,** 95
2,4-Dienecarboxylic acid esters
-, annelation, double with –
  29, 732
- startg. m. f.
  β,γ-ethylenecarboxylic acid
  esters, 1,6-addition **29,** 670
-, hydrogenation **30,** 25
- acids
- from
  alkylideneglutaconic acids
  30, 52
1,3-Diene central bonds,
cleavage, reductive **27,** 49
1,3-Diene-1,3-dicarboxylic acids
s. Alkylideneglutaconic
acids
Dienediynedials
- startg. m. f.
  difuranylacetylenes **20,** 102
Di(ene)halogenoboranes
- from
  acetylene derivs. (2 molecules) **28,** 666
- startg. m. f.
  1,3-dienes **28,** 666
-, syntheses via – **28,** 666
2,4-Dieneoximes
- startg. m. f.
  pyridine ring **22,** 507
Dienes
- by cyclopropane ring
  opening, double **30,** 538
-, cycloisomerization **28,** 667
- startg. m. f.
  dihalides with cyclopropane
  ring closure **26,** 516
  dioxido compds. **25,** 233
  suppl. **29**
  halogenhydrins, cyclic
  26, 520
  B-heterocyclics **27,** 648
  ethylene derivs. **29,** 54
-, synthesis **23,** 693 suppl. **27**
-, cyclic
- startg. m. f.
  tetrazolylisocyclics,
  halogeno-, bicyclic
  30, 351
1,2-Dienes s. Allenes
1,3-Dienes
- by Wittig synthesis **28,** 853
- from
  acetylene derivs. (2 molecules) **28,** 666

(1,3-Dienes
- from)
  acetylene derivs. and
    ethylene derivs. (2 mole-
    cules) **26,** 727
  alkylideneglutaconic acids
    29, 91
  2-allenealcohols **29,** 88
  allenes **28,** 521
  -, C-rearrangement **27,** 728
  cyclobutene ring **16,** 773
  cyclopropenes **28,** 674
  di(ene)halogenoboranes
    28, 666
  1,4-dienes **30,** 500
  1,1-dihalides **29,** 960
  2-ethylene-*tert*-amines
    28, 937
  α,β-ethyleneazomethines
    26, 829
  ethylene derivs. **29,** 911
  α,β-ethylenehalides (2 mole-
    cules) **27,** 870 suppl. **28**
  ethylenesultines **29,** 105
  α,β-ethylenetosylhydrazones
    27, 949 suppl. **30**
  nickel(I) π-allyl complexes
    and diazo compds. **26,** 915
  oxido compds. **30,** 659
  thiophenes, 2,5-dihydro-
    17, 968
- special s.
  acoxy-1,3-dienes
  2-alkylthio-1,3-dienes
  1,4-dicyano-1,3-dienes
  halogeno-
    1-(1-imidazolyl)-
    vinylketenimines
- startg. m. f.
  alcohols, synthesis **26,** 716
  4-cycloheptenones **26,** 890
  cyclopent-3-enones **28,** 656
  1,2-dioxins, 3,6-dihydro-
    29, 146
  1,2-dithiin 1-oxides,
    3,6-dihydro- **28,** 535
  2,3-ethylene-1,4-dihalides
    29, 490
  2-ethylene-1,4-dinitriles
    28, 662
  furan ring, 2,5-dihydro-
    30, 87
  N-heterocyclics, N-condensed
    28, 758
  isatins, tetrahydro- **28,** 751
  2H-1,2-oxazine ring, 3,6-di-
    hydro-N-acyl- **30,** 244
  2H-1,2-thiaphosph(V)orin
    2-sulfides, 3,6-dihydro-
    29, 612
  thiophenes **28,** 543
  thiopyrans, dihydro- **28,** 637
-, telomerization, palladium-
  catalyzed **27,** 724
- via Hofmann degradation
  of quaternary ammonium
  salts **28,** 937
cis,trans-**1,3-Dienes 28,** 666
1,4-Dienes
- from
  boranes **26,** 911
  enolethers and 2-ethylene-
    boranes **27,** 882
- startg. m. f.
  1,3-dienes **30,** 500
trans-**1,4-Dienes**, synthesis
  23, 705 suppl. **29**

1,5-Dienes 28, 601
- by
  dimerization of halides
    29, 820
  ring opening, fragmentation-
    type **27,** 926
- startg. m. f.
  methylenecyclopentane ring
    26, 734
  1-naphthols, 5,6,7,8-tetra-
    hydro-5,8-dialkyl- **28,** 651
-, synthesis via sulfones **29,** 796
1,5- and 1,6-Dienes
-, synthesis via boronate
  fragmentation, review
  23, 907 suppl. **26**
Diene syntheses (s. a. [4+2]-
  Cyclodimerization,
  Dienophiles)
- with
  1-acoxy-1,3-dienes prepared
    in situ **27,** 805
  acridizinium salts **5,** 503
    suppl. **27; 27,** 697
  acyloximes **27,** 698
  arenes **22,** 355 suppl. **29**
  anthracenes **27,** 694
  azomethines **27,** 696
  carboxylactones **7,** 711
    suppl. **28**
  2-chloracrylyl chloride
    28, 630
  cyanodithioformic acid
    esters **28,** 637
  cyclimmonium salts **27,** 697
  cyclobutene, 1,2-dicyano-
    30, 469
  dicyanoacetylene **21,** 728
    suppl. **29**
  dienamines **26,** 696
  1,4-dienes **30,** 500
  1,3-dioxol-2-ones, halogeno-
    30, 247
  enolesters **30,** 499
  enolethers **27,** 697
  α,β-ethylene-α-halogeno-
    nitriles **26,** 718
  α,β-ethylene-α-halogenooxo
    compds. **30,** 484
  ethylene-N-heterocyclics
    27, 700
  α,β-ethyleneketones **27,** 700
  germanium compds. **27,** 695
  N-heterocyclics **25,** 506
    suppl. **29**
  intermediates prepared
    in situ **22,** 355
-, unstable **22,** 355 suppl. **26**
  lactones **7,** 711 suppl. **28**
  norbornene derivs. **27,** 697
  phosphonium salts **29,** 647
  porphins **21,** 728 suppl. **28**
  quinones **28,** 655
  silicon compds. **27,** 695
  sulfonylcyanides **29,** 648
  O-sulfonyloximes **27,** 698
  vinylketene mercaptals
    21, 728 suppl. **28**
  vinylketenimines **29,** 647
- **synthesis** (s. a. High-pressure
  diene synthesis)
-, order of reactivity of dienes
  30, 484
-, protection of carbon-carbon
  double bonds by − **28,** 631

- with
  inverse electronic demand
    26, 833 suppl. **27; 29,** 886;
    30, 469
  O-transacylation **28,** 629
- −, **double 26,** 758
- −, **intramolecular 26,** 758
- with
  N-acylation **29,** 817
- −, **inverse-oriented 28,** 655
- −, **stereospecific 26,** 691
- with
  thiophenes, 2,5-dihydro-
    26, 691
Dienolate ions as intermediates
  29, 116
Dienolesters
- startg. m. f.
  3-ethylenealcohols **8,** 66
(Dien)olethers
- from
  α,β-ethylene-γ-halogeno-
    ketals **27,** 971
- special s.
  cyclopenta[f]chromenes
Dienollactones
- from
  ketolactones **26,** 953
2,4-Dienols
- from
  2H-pyrans **27,** 669
- startg. m. f.
  furans, 2,5-dihydro-
    25, 123 suppl. **28**
5,6-Dienols
- from
  pyrans, tetrahydro-2-
    acetylene- **29,** 626
**Dienone-benzene
rearrangement**
-, arylcarboxylic acids by −
  26, 799
Dienone-phenol − **27,** 190
Dienones
- from
  alcohols **14,** 895 suppl. **26**
  phenolethers **27,** 159
    suppl. **29**
  phenols **27,** 159
- startg. m. f.
  phenol acetates **27,** 190
-, **bicyclic**
-, isomerization, skeletal
  28, 106
- startg. m. f.
  ethylenecarboxylic acid
    esters, bicyclic **26,** 126
-, **cross-conjugated**
- special s.
  2,5-cyclohexadienones
- startg. m. f.
  cyclohexanone ring **29,** 665
  cyclopent-2-enones **24,** 726
    suppl. **29**
  4-piperidone ring **27,** 328
-, **cyclic**
- startg. m. f.
  phenols **30,** 649
1,4-Dien-3-ones s. Dienones,
  cross-conjugated
2,4-Dienones
- from
  2,3-ethylene-1,5-diol
    1-(ethyl carbonates)
    29, 928
- special s.
  4-halogeno-2,4-dienones

– startg. m. f.
  δ-cyano-α,β-ethyleneketones
  18, 772 suppl. 26
  3-ethylenealcohols 29, 32, 928
  γ,δ-ethyleneketones 29, 928
Dienophiles (s. a. Azodienophiles)
  cyclopolyenes 26, 758
  furan, 2,5-dihydro- 16, 689
    suppl. 22
  maleic thioanhydride 27, 695
–, reaction of thiophenes,
  2,4-diamino- with – 29, 640
–, **electron-deficient** 29, 647
–, **electron-rich** 29, 647
–, –, **reactive** 27, 760
(1,3-Dien)-1-oxysilanes
– from
  α,β-ethyleneketones 27, 83
– startg. m. f.
  3-ethylenealcohols 27, 83
(1,3-Dien)-2-oxysilanes
– from
  α,β-ethyleneketones 29, 116
Dienynes
– startg. m. f.
  ketones, bicyclic 29, 678
Diequatorial s. Diaxial-
  diequatorial rearrangement
Diethylaluminum chloride
  as reagent 27, 142; 28, 658
– **cyanide**
– as reactant 27, 789
–, prepn. 21, 738 suppl. 28
– **hydride** as reagent 26, 734
Diethylamine
– as reagent 10, 495 suppl. 28;
  24, 816; 26, 441, 713; 27, 80;
  28, 306; 29, 534, 625; 30, 584
Diethylaminosulfur trifluoride
  as reactant 30, 365
Diethylaniline
– as medium 11, 224 suppl. 27
– as reagent 26, 161; 27, 741;
  29, 684, 965; 30, 677
Diethyl azodicarboxylate
– as hydrogen abstracting
  agent 28, 317
– as reactant 30, 8
– as reagent 10, 286 suppl. 29;
  26, 105; 26, 175 suppl. 27; 27,
  429; 28, 142, 316/7, 753; 29,
  188; 30, 8
–, C-functionalization of
  pyrimidines with – 30, 208
– startg. m. f.
  alloxazines 28, 316
  pteridines 26, 448
Diethyl Carbitol s. Diethylene
  glycol diethyl ether
Diethylene glycol as solvent
  27, 401
– – **diethyl ether** – – 26, 959
– – **dimethyl ether** – – 26, 148;
  27, 65, 498, 901; 28, 115, 210,
  401, 967; 29, 76, 264, 957, 991
Diethyl malonate
  (s a. Sodiomalonate, diethyl)
– as nitrous acid scavenger
  26, 361
  proton source 12, 127
    suppl. 28
– **methylphosphonite**
  as reagent 26, 379
– **phosphorochloridate**
  as reactant 29, 733
– **phosphorochloridodithioite**
  as reactant 27, 121

– **phosphorocyanidate**
  as reagent 27, 611 suppl. 30;
  28, 353
– **sulfite**
– startg. m. f.
  isothiazoles, 3-hydroxy-
    29, 777
1,1-Diethynylalcohols
– from
  carboxylic acid esters
    26, 793
1,1-Difluorides
– from
  oxo compds. 28, 506; 30, 365
Difluoroamination 27, 351
Difluoroamine, reactions
  26, 285
–, –, review 26, 285
– startg. m. f.
  1-(difluoroamino)alcohols
    26, 285
1,1,1-Difluoroamines
  s. 2-Chloro-1,1,2-trifluoro-
  triethylamine
N,N-Difluoroamines
– special s.
  1,1-dihalogeno-N,N-difluoro-
    amines
  halogeno-N,N-
1-(Difluoroamino)alcohols
  from
  oxo compds. 26, 285
Difluoromethylene compds.
– startg. m. f.
  ynamines 27, 458
1,1,1-Difluoronitro compds.
– special s.
  difluoronitromethyl ketones
Difluoronitromethyl ketones
  29, 422
Difuranylacetylenes
– from
  dienediynedials 30, 102
1,2-Digermanes 26, 650
Diglyme s. Diethylene glycol
  dimethyl ether
Dihalides
– from
  dienes with cyclopropane
    ring closure 26, 516
  ethers, cyclic 16, 597
  ethylene derivs. with
    rearrangement 26, 517
– special s.
  ethylene dihalides
– startg. m. f.
  cycloalkenes 26, 966
  cyclobutane ring 27, 966
  enamines, endocyclic
    27, 842
  O-heterocyclics 27, 213
  P-heterocyclics 28, 597
  imines, cyclic 27, 448
  ketones, cyclic 22, 661
    suppl. 27
  pyrrole ring, 2,3-dihydro-
    27, 861
1,1-Dihalides
– as reactants 26, 850
– from
  aldoximes 27, 562
  ketones, C-cleavage 30, 375
  oxo compds. 29, 518
  1,1,1-trihalides 27, 77; 29, 73
–, reaction with oxo compds.
  24, 844

– special s.
  1,1,1-alkoxydihalides
  1,1-chlorofluorides
  1,1-difluorides
  2-hydroxy-1,1-dihalides
  methylene chloride
– startg. m. f.
  aldehydes 29, 201a, 847
  aldehydes-1-d 29, 211
  cyclopropane ring 26, 876;
    29, 827
  1,1-diazides 26, 428
  1,3-dienes 29, 960
  halides 26, 70
  hydrocarbons 29, 72
  methylene compds. 29, 827
  oxido compds. 26, 852
  thioketones 28, 557
1,2-Dihalides
– from
  ethylene derivs. (s. a.
    Halogenofluorination)
    HalC↓CC
– – (trans-1,2-dihalides)
    27, 552
  oxido compds. (cis-1,2-
    dihalides) 29, 510
– special s.
  ethylene dibromide
– diiodide
– startg. m. f.
  cyclopropanes 27, 841
  1,3-dihalides 29, 498
  1,1-di(pyridinium salts)
    29, 211
  ethylene derivs. CC↑Hal
  cis- and trans-ethylene
    derivs. 2, 822 suppl. 26;
    6, 882 suppl. 26
  furan ring 16, 336
  hydrocarbons 22, 317
    suppl. 29
–, **mixed**
– special s.
  1,2-bromofluorides
  1,2-chlorobromides
  1,2-chloroiodides
1,3-Dihalides
– from
  cyclopropane ring 29, 492
  1,2-dihalides 29, 498
– startg. m. f.
  cyclopropenes 15, 680
    suppl. 27
1,4-Dihalides
– startg. m. f.
  1,2-oxazines, tetrahydro-
    N-carbalkoxy- 30, 295
o-Dihalides
– ring closure with – 28, 560
– startg. m. f.
  biphenylenes 27, 850
α,ω-Dihalides
– startg. m. f.
  dialkylmercury compds.,
    sym. 30, 433
ω,ω'-Dihalides
– startg. m. f.
  double bonds, carbon-carbon,
    conjugated 27, 968
Dihalogenacetonitrile
– startg. m. f.
  α-halogenocarboxylic acid
    esters 29, 738
Dihalogenalanes
– from
  halides 29, 594

2,2-Dihalogenalcohols
- from
  oxo compds. **30,** 449
- startg. m. f.
  α-halogenoketones **28,** 945

α,α-Dihalogenaldehydes
- from
  2-halogenenamines **30,** 370

α,β-Dihalogenaldehydes
- from
  α,β-ethylenealdehydes **27,** 220

1,N-Dihalogenamines
- from
  azomethines **23,** 534
- startg. m. f.
  azomethines **23,** 534

N,N-Dihalogenamines
- special s.
  N,N-dichloramines
  N,N-difluoroamines
- startg. m. f.
  iminothionyl halides **26,** 269

C,As-Dihalogenarsines
- from
  arsinic acids **30,** 344

1,1'-Dihalogenazo compds.
- from
  ketazines **29,** 832
- startg. m. f.
  azo compds., tert. **29,** 832

1,1'-Dihalogenethers
- startg. m. f.
  furan ring, tetrahydro- **29,** 794

Dihalogenoboranes
- from
  ethylene derivs. **19,** 592
  suppl. **28; 27,** 485 suppl. **28**
  halides **26,** 649
- startg. m. f.
  amines, sec. **27,** 485 suppl. **28**
  aziridine ring, N-subst. **29,** 439
  boroselenines **28,** 581
  hydroperoxides **29,** 221

Dihalogenocarbene precursor
-, phenyltrihalogenomethyl-
  mercury as -, review
  17, 894 suppl. **27**

Dihalogenocarbonylimines
s. Iminocarbonylhalides

α,α-Dihalogenocarboxylic acid esters, synthesis **27,** 843 suppl. **30**
- acids
- from
  nitriles **29,** 519

4,6-Dihalogeno-4,6-dien-3-onesteroids 27, 563

1,1-Dihalogeno-N,N-difluoro-amines
- startg. m. f.
  1-halogeno-N-fluoroimines
  26, 487

α,α-Dihalogeno-β-keto-carboxylic acid esters
- special s.
  α,α-dichloro-β-keto-
  carboxylic acid esters

α,α-Dihalogenoketones
- special s.
  α-azo-α,α-dihalogenoketones
  dichloromethyl ketones
- startg. m. f.
  α-ketocarboxylic acids
  29, 208
  pteridines **30,** 299

α,α'-Dihalogenoketones
- special s.
  methyl α,α'-dihalogeno-
  ketones
- startg. m. f.
  4-alkylidene-2-amino-
  1,3-dioxolanes **28,** 167
  cyclobutanones, 2-imino-
  30, 591
  4-cycloheptenones **26,** 890
  cyclopentanones **27,** 868
  suppl. **28**
  cyclopent-2-enone ring
  27, 868
  γ-diketones (from 2 molecules) **29,** 828
  3(2H)-furanones **29,** 844
  ketones, α-subst. **29,** 671

α,β-Dihalogenoketones
- special s.
  chalcone dihalides
- startg. m. f.
  α-azido-α,β-ethyleneketones
  26, 414

C-Dihalogenomethylation 26, 658
-, insertion of carbonyl via -
  27, 726

1-(Dihalogenomethylene)-acylamines
- startg. m. f.
  oxazoles, 5-amino- **30,** 292

N-Dihalogenomethylene-ammonium salts, reactions **28,** 717
-, -, review **28,** 717
- special s.
  dichloromethylenedimethyl-
  ammonium chloride
- startg. m. f.
  pyrazoles, 5-amino- **29,** 843

Dihalogenomethylene compds.
(s. a. Ketene dihalides)
- from
  aldehydes **27,** 866; **28,** 819
  sulfines **29,** 852
- startg. m. f.
  α,β-acetylenecarboxylic
  acids **28,** 292
  acetylene derivs., terminal
  28, 819
  methylene compds. **7,** 738
  suppl. **28**

Dihalogenomethylenedioxy compds. 3, 485 suppl. **27**

Dihalogenomethylene groups,
insertion **26,** 658

Dihalogenomethylene-hydrazones
- from
  α-azo-α,α'-dihalogeno-
  ketones **28,** 422
  1,2,3-triketone mono-
  hydrazones **28,** 422

α-(Dihalogenomethylene)-nitriles
- from
  acylcyanides **28,** 752

Dihalogenomethylenimines
s. Iminocarbonylhalides

Dihalogenomethyl sulfones
- from
  sulfinic acids and haloforms
  27, 627

α,α-Dihalogenonitriles
- special s.
  dihalogenacetonitrile

α,β-Dihalogenonitriles
- from
  α,β-ethylenenitriles **26,** 409
- startg. m. f.
  isoxazoles, 3-amino- **26,** 409

N,N-Dihalogenophosphor-amidates
- startg. m. f.
  2-halogenamines **27,** 548
  2-halogenophosphoramidates
  27, 548
  N-phosphorylaziridines
  29, 411

N,N-Dihalogenosulfonamides
- as reagents **22,** 565 suppl. **29**
- special s.
  N,N-dichlorobenzene-
  sulfonamide
- startg. m. f.
  bi-N-sulfonylsulfinodiimides,
  sym. **28,** 557
  N-sulfonyliminoseleninyl
  halides **28,** 469
  N-sulfonyliminosulfinyl
  halides **28,** 469
  N-sulfonyl-S-(sulfonylimino)-
  sulfinic acid amides
  27, 298
-, sulfonylamination with -
  30, 304

α,α-Dihalogenosulfones
- special s.
  dihalogenomethyl sulfones
- startg. m. f.
  acetylene derivs. **26,** 972
  thiirene 1,1-dioxides **26,** 972

α,α-Dihalogeno-α-sulfonyl-sulfonic acid amides
- from
  α-sulfonylsulfonic acid
  amides **30,** 355

1,1,1-Dihalogenothioethers
- startg. m. f.
  acetylene derivs. **30,** 687

1,1-Dihalogeno(trihalogeno-phosphazo) compds.
- from
  nitriles **27,** 302

S-1,2-Dihalogenovinyl groups
- startg. m. f.
  S-halogenomethyl groups
  26, 77

α-Dihydrazones s. Osazones

Dihydroisoquinoline rearrangement 26, 764

1,1'-Dihydroxyamines, cyclic
s. 9-Azabicyclo[3.3.1]nonane-1,5-diol ring

β,δ-Dihydroxycarboxylic acid esters
- startg. m. f.
  2-pyrones, 5,6-dihydro-
  27, 255
-, tetrahydro-4-hydroxy-
  27, 255

1,4-Dihydroxy-2,3-diketones
- from
  2-acetylene-1,4-diols **30,** 167
- startg. m. f.
  3(2H)-furanones, 4-hydroxy-
  30, 167

α,α-Dihydroxy-β-diketones
- special s.
  indan-1,3-diones,
  2,2-dihydroxy-

2,2'-dihydroxydisulfides
- from
  oxido compds. **28,** 525

- startg. m. f.
  1,2-hydroxymercaptans
  28, 525
a,a'-Dihydroxyketones
- from
  acoxy-2-acetylenes **28**, 115
  allenes **28**, 115
1,2-Di(hydroxylamines)
- startg. m. f.
  imidazolidines,
  1,3-dihydroxy- **29**, 348
Dihydroxyoxido compds.
- from
  dioxido compds. **26**, 114
1α,3β-Dihydroxy-Δ⁵-steroids
  29, 32
Diimide 28, 2, 39
- **oxides** s. Azoxy compds.
Diimides s. Azo compds.,
  Carbodiimides, N,N'-Dicarboxylic acid diimides,
  Sulfinodiimides
α,ω-Di(iminoesters)
- startg. m. f.
  α,ω-di(acylamines) **30**, 225
Diiron nonacarbonyl 26, 890;
  27, 868; **29**, 844
- as reactant **29**, 643
Diisobutylaluminum hydride
- as reactant **26**, 880
- as reagent 7, 914 suppl. **28**;
  20, 7 suppl. **28**, **29**; **27**, 68;
  28, 53, 211, 652; **29**, 62, 674
Diisocyanates
- from
  aminohydroxamic acids
    27, 492
  lactams **27**, 492
Di(isopinocampheyl)borane
  as reagent **28**, 621
Diisopropylamine – –
  6, 89 suppl. **29**; **26**, 862; **27**,
  319; **28**, 11, 82, 598; **29**, 805;
  30, 461, 585, 663
Diisopropyl peroxydicarbonate – – **26**, 613;
  27, 537
- startg. m. f.
  aryl isopropyl carbonates
    26, 170
γ-Diketals
- from
  furans **26**, 115
Diketene
-, spiroannelation with –
  28, 833
- startg. m. f.
  γ-alkylthio-β-lactones
    28, 534
Diketocarboxylic acids,
  synthesis **20**, 607 suppl. 29
β-Diketonates s. Metal
  β-diketonates
α-Diketone monohydrazones
- startg. m. f.
  pyrazoles, 4-hydroxy-
    26, 486
β-Diketone –
- startg. m. f.
  pyrazole ring **28**, 114
  quinoline ring **4**, 835; **13**, 863
Diketones
- from
  dicarboxylic acid anhydrides,
    double ring closure **28**, 745
  glycols, cyclic **29**, 157

- special s.
  ethylene-diketones
  hydroxy-
- startg. m. f.
  1,1-iminoalcohols, 1,3-N,N-
    heterocyclics, N-condensed
    29, 364
  lactamols **28**, 603
  2-naphthols **16**, 871 suppl. 29
α-Diketones (s. a. Benzils)
- as photoinitiators **29**, 720
-, cleavage, oxidative **28**, 193
-, derivatives s. α-Isonitrosoketones
- from
  1,2-di(siloxy)alkenes
    27, 233
  ethylene derivs. **27**, 186
  β,γ-ethylene-α-hydroxyketones, rearrangement
    29, 171
  glycols **26**, 233; **28**, 209
  – with rearrangement
    29, 681
  α-hydroxyketones **26**, 233
  sulfones **28**, 724
  thiophene-1,1-dioxides, tetrahydro-3-hydroxy-4-keto-
    28, 724
-, reaction with lead tetraacetate **27**, 146
-, reactions, photochemical
  26, 744 suppl. 27
- special s.
  (α-diketo)phenols
- startg. m. f.
  α,α-diarylketones **2**, 680
  2-ene-1,4-diones **28**, 847
  ethylene derivs. **27**, 34
  1-halogeno-2-ketothioethers
    26, 602
  α-hydroxyketones, reduction
    27, 34; **28**, 31
  -, synthesis **27**, 674
  ketones, synthesis and
    cleavage **27**, 674
  phenanthrenes **27**, 806
-, **cyclic**
- from
  1,3-dioxolan-2-one ring,
    4,5-dihalogeno- **30**, 247
- startg. m. f.
  dicarboxylic acid esters
    27, 146
β-Diketones
- by desulfuration **26**, 982
- from
  acetylene derivs. and
    carboxylic acid chlorides
    28, 104
  azomethines and nitriles
    26, 685
  β-ketocarboxylic acid esters
    30, 614
  α,β-oxidoketones **26**, 751
-, protection of amino groups
  with – **27**, 402
- special s.
  α-acylamino-α-hydroxyβ-diketones
  α,α-dihydroxyα-hydroperoxy-
- startg. m. f.
  β-amino-α,β-ethyleneketones
    26, 331
  carboxylic acid amides
    28, 416
  1,3-diols **22**, 37 suppl. 29

  enehydrazones, tetraketo-
    28, 835
  α-hydroxyketones **28**, 608
  β-hydroxyketones
    22, 37 suppl. 29
  4-pyridones **26**, 359
  thiopyrano[2,3-b]indolium
    salts **26**, 825
  v-triazolo[3,4-a]pyrimidines
    27, 396
-, **bicyclic 26**, 837
- special s.
  spiro-β-diketones
-, **cyclic**
-, C-alkylation **14**, 781 suppl. 29
- from
  ketocarboxylic acids **29**, 935
- startg. m. f.
  β-alkoxy-α,β-ethyleneoximes,
    cyclic **27**, 163
  β-ketolactams **27**, 163
-, **labeled 1**, 579 suppl. 28
-, **spirocyclic**
- from
  ketocarboxylic acids, cyclic
    29, 935
-, **2-acyl, cyclic**
- from
  dicarboxylic acid esters
    28, 728
γ-Diketones 15, 620 suppl. 28
- from
  azomethines, synthesis
    30, 461
  α,α'-Dihalogenoketones
    (2 molecules) **29**, 828
  enaminoenollactones **27**, 249
  2-ene-1,4-diones **8**, 87
  enoxysilanes (2 molecules)
    30, 619
  α,β-ethyleneketones and
    mercaptals **28**, 175
  2(3H)-furanones **29**, 822
  α-halogenoketones **28**, 804
  – and β-ketosulfones **29**, 792
  γ-ketomercaptals **28**, 175
  β-ketosulfoxides (2 molecules) **26**, 894
  α-sulfonyl-γ-diketones
    29, 920
- special s.
  α-sulfonyl-γ-diketones
- startg. m. f.
  cyclopent-2-enones **1**, 554
  furans **29**, 251
  pyrrole ring, 3-amino-
    28, 331
-, synthesis, review **27**, 249
-, **sym. 26**, 894
1,5-Diketones 15, 620 suppl. 28
- from
  β-aminoketones **8**, 828
  β-ketosulfoxides (2 molecules) **28**, 834
1,6-Diketones
- from
  carboxylic acids (2 molecules) and α,β-ethylenehalides (2 molecules)
    29, 825
(α-Diketo)phenols 27, 278
Dilactones
- special s.
  pulvinic acid dilactones
  sugar –
- startg. m. f.
  naphthalene-2,3-dicarboxylic
    acids, 1,2-dihydro- **28**, 692

Dilithium tetrachlorocuprate
26, 875 suppl. 28; 29, 824
Dilution s. High dilution
Dimerization (s. a. Codimerization, Coupling, Cyclodimerization, Photo-, Redox-, Self-condensation)
– of
  1,3-dienes 23, 715
  halides 28, 814; 29, 820
  heterocyclics 30, 480
  indoles 28, 663
  3-isothiazolones 26, 585
  naphthoquinones 29, 760
  thioketones 29, 536
– to
  dicyclimmonium salts, triple-N-condensed 26, 400
  peptides 26, 365
– with simultaneous decarbonylation 26, 922
–, **anodic** 28, 703 suppl. 29
–, **cathodic**, review 21, 731 suppl. 27
–, **oxidative** (s. a. Dimerization, anodic) 20, 561; 29, 275
– of
  carboxylic acid esters 28, 706
  ethylene derivs., electron-rich 26, 721
  N-heterocyclics 27, 372
  indoles 28, 663
  nitriles 28, 705
  phenolethers 28, 703
  phenols, hindered 27, 761
  1,4-thiazine ring 27, 763
  uracils, 6-amino-5-nitroso- 28, 377
– to
  1,2-diquinonylethylenes 29, 715
  1,2,4,5-tetrazines, hexahydro- 26, 326
–, –, **stereospecific** 28, 706
–, **reductive**
– of
  α,α'-dihalogenoketones 29, 828
  ketones 27, 800
Dimers, heterocyclic
–, cleavage 30, 480
– special s. uracil dimers
1,2-Dimesylates
– startg. m. f.
  aziridines, N-amino- 24, 341
  ethylene derivs. 29, 915
Dimetalation, α,α'-lateral
– of
  isocyclics 27, 838
1,2-Dimethoxyethane as solvent 5, 479 suppl. 28; 16, 922 suppl. 27; 26, 146, 411, 853; 27, 238, 720, 957; 27, 949 suppl. 30; 28, 157, 618, 840, 945; 29, 241, 471, 715, 819; 30, 135, 517, 617
N,N-Dimethylacetamide
– as solvent 26, 508, 598; 27, 444; 28, 341, 453; 29, 585
Dimethylacylamines
s. Dimethylformamide
Dimethylamine as reagent 30, 63
Dimethylamines
– from
  amines, prim. 14, 455 suppl. 15, 17, 29

2-Dimethylaminoethanol
as medium 27, 435
C-Dimethylaminomethylenation 25, 579; 26, 247; 28, 772; 29, 469
4-Dimethylaminopyridine
– as pyridine substitute 24, 733 suppl. 26
– as reagent 9, 267 suppl. 28
Dimethylaniline
– as reagent 18, 725 suppl. 27; 26, 85, 938; 27, 121; 29, 267
p-Dimethylation, ar. 27, 694
Dimethylchloramine as reagent 27, 587
Dimethylchloroformiminium chloride (s. a. Thionyl chloride-dimethylformamide)
– as reagent 27, 826
Dimethylchloromethyleneammonium chloride s. Dimethylchloroformiminium chloride
gem-Dimethyl compds. 28, 845
Dimethyl disulfide as reactant 28, 177
– **dithiobis(thioformate)**
  as reagent 26, 559 suppl. 29
gem-Dimethyl effect s. under Ring closure
Dimethylene compds., vic. 21, 913 suppl. 27
Dimethylformamide (s. a. under Iron(III) chloride, Thionyl chloride)
– as catalyst 30, 347
– as medium 14, 455 suppl. 29; 26, 221, 943; 27, 504; 28, 906; 30, 38, 236
– as reactant 26, 839; 27, 797; 28, 480, 781, 834; 29, 370; 30, 358
– as reagent 25, 324 suppl. 28; 26, 971; 27, 551, 975; 28, 751, 988; 29, 23, 524, 837, 945; 30, 347
– as solvent 3, 632 suppl. 29; 24, 387 suppl. 25; 26, 2, 421; 27, 349, 893; 28, 25, 73, 532; 30, 248
– – –, aprotic 26, 616
– – –, polar, aprotic 11, 510 suppl. 27; 17, 486 suppl. 26; 26, 786; 30, 558
– **-dialkyl sulfate** as reactant 28, 179, 528; 30, 113
– **acetals**
–, N-alkylation with – 27, 373; 30, 320
– special s. dimethylformamide dimethyl acetal
  – dineopentyl –
  – ethylene –
– **chloride** s. Dimethylchloroformiminium chloride
– **dimethyl acetal**
– as reagent 28, 985 suppl. 30
–, heteroring closures with – 29, 469
–, reaction with acidic compds. 27, 373
– startg. m. f.
  diamidides 26, 333
– **dineopentyl** – as reagent 27, 91; 28, 985
– **ethylene** – – – 30, 320

Dimethylglyoxime – – 27, 55
Dimethylhalogenoformiminium halides
– special s. dimethylchloroformiminium chloride
1,1-Dimethylhydrazine
as reagent 7, 19 suppl. 29
2,4-Dimethylphenyl cyanate – – 28, 498
Dimethylpropynylsulfonium bromide – – 26, 376
3,5-Dimethylpyrazole s. under Chromium trioxide
2,4-Dimethylpyridine
as reagent 30, 52
Dimethyl sulfide (s. a. under Borane)
– as reactant 28, 499
– as reagent 27, 188; 28, 109, 201; 29, 199; 30, 29, 149
Dimethylsulfinyl carbanion
s. Methylsulfinyl carbanion
Dimethyl sulfoxide
(s. a. Methylsulfinyl carbanion)
– as medium 8, 531 suppl. 29; 21, 962 suppl. 26; 23, 948 suppl. 28; 26, 131, 733, 856, 898, 969, 994; 27, 145, 457, 764, 769, 772, 964; 28, 617, 944, 949; 29, 122, 391, 815; 30, 140, 608, 619
– as reactant 27, 129
– as reagent (s. a. under Iodine, Iridium, and Iron(II) chloride) 26, 99, 198, 208, 455, 786; 27, 3, 145, 231; 27, 642 suppl. 29; 28, 165, 171, 199, 214/5, 751, 966; 29, 84, 118/9, 207/8, 219, 242, 815, 928; 30, 136, 358
– as solvent 1, 205 suppl. 26; 9, 934 suppl. 26; 9, 768 suppl. 27; 12, 715 suppl. 28; 14, 904 suppl. 27; 15, 398 suppl. 28; 15, 679 suppl. 29; 21, 911 suppl. 28; 24, 291 suppl. 26; 26, 171, 549, 784, 898, 972; 26, 234 suppl. 30; 26, 616 suppl. 27; 28, 694; 29, 362, 961
– as solvent, aprotic, polar 18, 268 suppl. 26; 28, 644, 777, 950
– startg. m. f.
  N-methylthiomethylnitrones 27, 359
– **-chlorine complex** as reagent 28, 200
– **sulfoxide-D$_6$** as reactant 27, 650; 28, 591
Dimethylsulfoxonium methylid, N-methylation with – 26, 437
– startg. m. f.
  thiabenzene oxides 27, 769
Dimethylthallium compds.
as reactants 29, 578
S,S-Dimethyl-N-tosyliminodithiocarbonimidate
– startg. m. f.
  1,3-N,N-heterocyclics, 2-amino- 26, 438
Dimroth s. a. Strecker
– **rearrangement** (s. a. Retro-Dimroth rearrangement) 30, 228, 252
– of
  N-alkoxy compds. 27, 168

Dinitriles (s. a. Dicyano...)
- from
  furazans **27,** 515
  1,2,5-thiadiazole 1,1-dioxide ring **29,** 472
- special s.
  anilines, 2,6-dicyano-azodinitriles
  cyclobutene, dicyano-
- startg. m. f.
  o-alkoxy-o′-amino-N-heterocyclics **30,** 74

1,1-Dinitriles
- from
  dicyanosulfonium ylids **27,** 977
- special s.
  alkylidenemalononitriles
  aminodinitriles
  ethylenedinitriles
  halogenodinitriles
  malononitriles
- startg. m. f.
  α-cyaniminoesters **29,** 130

1,2-Dinitriles, sym.
- from
  nitriles **28,** 705

1,3-Dinitriles 13, 684 suppl. **27**
- special s.
  1-amino-1,3-dinitriles

o-Dinitriles
- startg. m. f.
  N-macroheterocyclics, conjugated **27,** 435
  phthalocyanines **27,** 435

Dinitroalcohols
- from
  nitronitric acid esters **28,** 292

Dinitro compds. s. a. Benzenes, dinitro-
- startg. m. f.
  nitramines **30,** 330

1,2-Dinitro compds., aliphatic
- from
  1,1-nitrosulfones, synthesis **30,** 608
- startg. m. f.
  ethylene derivs. **27,** 956

o-Dinitro compds.
- startg. m. f.
  benzofurazans **30,** 56

2,4-Dinitrofluorobenzene
- startg. m. f.
  carboxylic acid 2,4-dinitrophenyl esters **19,** 264 suppl. **29**

Dinitrogen pentoxide 26, 361
- tetroxide s. Nitrogen dioxide
- trioxide **27,** 147, 825

Dinitromethylene compds.
- from
  diazo compds. **26,** 845

2,4-Dinitrophenol as catalyst **26,** 809

2,4-Dinitrophenyl ethers
as intermediates **27,** 578 suppl. **30**

2,4-Dinitrophenylhydrazones
-, cleavage via hydration **27,** 205
-, - reductive **26,** 188 suppl. **30**

P¹,P²-Di-p-nitrophenyl pyrophosphate
-, phosphorylation with – **28,** 190

Dinitrosoarenes
- from
  quinone dioximes **26,** 930

Diol esters s. a. Diacoxy compds.
1,3-Diol –
- from
  enolesters, synthesis **29,** 674
  ethylene derivs. **28,** 102

Diol monoesters
- from
  ethylenealcohols **27,** 137
- special s.
  acetylenediol monoesters

1,3-Diol –
- startg. m. f.
  ethylene derivs., fragmentation **26,** 985
  α,β-ethyleneoxo compds. **29,** 242

Diol monoethers
- from
  diols **12,** 287 suppl. **28**
- **mono(ethyl carbonates)**
- special s.
  2,3-ethylene-1,5-diol 1-(ethyl carbonates)

1,3-Diol monoethers
- from
  acetals **30,** 521

Diol mononitrates
- from
  nitrous acid esters **30,** 95
- special s.
  glycol mononitrates
- **monosulfonates** (s. a. Glycol monosulfonates)
- startg. m. f.
  carbonic acid esters, cyclic **30,** 140
  ethers, – **26,** 236

Diols (s. a. Dihydroxy...)
-, derivs. s. Acetals, cyclic
- from
  chloral acetals **29,** 142
  ethylenealcohols **29,** 142
  ethylene derivs., cyclic **27,** 149
  ethyleneketones **28,** 103
  glycols, cyclic **29,** 155
  lactones **30,** 17
  peroxides, cyclic **30,** 1
- special s.
  acetylenediols
  o-hydroxycarbinols
  quinols
- startg. m. f.
  acetals **29,** 193
  carbonic acid esters, cyclic **24,** 170; **29,** 197
  diol monoethers **12,** 287 suppl. **28**
  ethers, cyclic **26,** 238; **27,** 941 suppl. **29**
  ethylene derivs. **26,** 949
  formoxyaldehydes **30,** 358
  formoxyhalides **30,** 358
  2-germa-1,3-dioxaheterocycles **29,** 114
  O-heterocyclics **27,** 261
  1,3-O,N-heterocyclics, 2-imino- **29,** 361
  lactones **27,** 100 suppl. **30**
  methylenedioxy compds. **30,** 128

1,1-Diols
- special s.
  ketone hydrates
- startg. m. f.
  ketones **26,** 239

1,2-Diols s. Glycols
1,3-Diols
- from
  β-diketones **22,** 37 suppl. **29**
  1,3-dioxanes **29,** 9
  ethylene derivs. **28,** 102
- special s.
  2-nitro-1,3-diols

1,4-Diols
- from
  boranes and oxido compds. **30,** 618
  1,3-halogenhydrins and ketones **28,** 609

cis-**1,5-Diols, cyclic 13,** 207 suppl. **26**

o-Diols
-, monodehydroxylation **26,** 550
- startg. m. f.
  1,3-dioxole ring, 2-amino- **29,** 204

2,10-Dioxabicyclo[4.4.0]deca-3,8-dienes, 1-amino-
- from
  α,β-ethyleneketones and ynamines **27,** 703

1,4,2,5-Dioxadiazines
- from
  hydroximinohalides and amidoximes **27,** 208

Dioxane
- as solvent **26,** 784

1,3-Dioxane, 5-amino-2,2-dimethyl-4-phenyl-
-, synthesis, asym., with – **29,** 750

1,3-Dioxane ring opening 24, 127

1,3-Dioxanes
-, reactions, review **16,** 210 suppl. **26**
- startg. m. f.
  1,3-diols **29,** 9
-, **5-acyl-**
- from
  ketones **26,** 659
-, **2-oxo-**
- startg. m. f.
  1,3-dioxolan-2-ones **28,** 118
  oxetanes **29,** 266
-, **4-oxo-**
- from
  β-hydroxycarboxylic acids **27,** 196
-, **4-oxo-5-acylamino-**
- from
  5-oxazolones **30,** 451
- startg. m. f.
  α-hydroxymethyl-α-aminocarboxylic acids **30,** 451

1,4-Dioxanes
- from
  ethylene derivs. **28,** 114
  glycols and 1,2-ditosylates **29,** 183

1,4-Dioxan-2-yl ethers,
protection of hydroxyl groups as – **18,** 875 suppl. **28**

1,3,2-Dioxaphospholane, 2-oxo-2-chloro-
-, phosphorylation with – **29,** 110

1,3,2-Dioxaphospholanes
- from
  1,2-dioxetanes **29,** 106

(1,3,2-Dioxaphospholanes)
–, 2-halogeno-
– startg. m. f.
   P-spiroheterocyclics **28**, 264
–, 2-oxo-2-alkoxy- **29**, 110
– startg. m. f.
   phosphoric acid monoesters
   29, 110
1,3,2-Dioxaphospholenes
– startg. m. f.
   cyclopropanes, acyl-
   23, 797 suppl. **26**
   1-halogeno-2-ketothioethers
   26, 602
4,7-Dioxaspiro[2,4]heptanes
– startg. m. f.
   carboxylic acid esters
   29, 493
1,3,2,4-Dioxathiazole 2-oxides
– from
   nitrile oxides **26**, 489
– startg. m. f.
   isocyanates **26**, 489
2,7-Dioxatwistans 28, 223
4H-1,3,5-Dioxazin-4-one-
   5-sulfonylhalides, 5,6-dihydro-
   30, 2111
1,2,4-Dioxazolidine ring
– by aminoperoxidation **30**, 206
1,2,4-Dioxazol-3-ones
– from
   N-acylperoxycarbamic acids
   28, 212
1,2-Dioxetane ring
– from
   ethylene derivs. **30**, 86
1,2-Dioxetanes
– startg. m. f.
   1,3,2-dioxaphospholanes
   29, 106
S-Dioxides, cyclic
– special s.
   thiadiazine S-dioxides
   thiadiazole –
   thiophene –
Dioxido compds.
– from
   dienes **25**, 233 suppl. **29**
– startg. m. f.
   dihydroxyoxido compds.
   26, 114
Dioximes s. Glyoximes
Dioxindoles s. Azadioxindoles
Dioxins s. a. Dioxanes
1,2-Dioxins, 3,6-dihydro-
– from
   1,3-dienes **29**, 146
1,5-Dioxocin-2,6-diones
– from
   furan-2,3-diones, 2,3-dihydro-
   (2 molecules) **28**, 197
Dioxo compd. monoketals
– from
   2,2-dialkoxyethers, cyclic
   27, 859
o-Dioxo compd. monoximes
– startg. m. f.
   imidazole 3-oxides **27**, 412
   2-imidazolones **27**, 412
   oxazole N-oxides **27**, 412
Dioxo compds.
– startg. m. f.
   imines, cyclic **28**, 347
α-Dioxo –
– from
   acetylene derivs. **27**, 145;
   11, 209 suppl. **29**
   α-aminoketones **28**, 153

N-imidosulfoximines **28**, 930
ketones **29**, 176
– special s.
  glyoxal
β-Dioxo –, cleavage **29**, 170
γ-Dioxo –
– startg. m. f.
  heterocyclics **11**, 668 suppl. **26**
o-Dioxo –
– startg. m. f.
  phthalimidines **27**, 490
1,5-Dioxo – (s. a. 1,5-Diketones)
– startg. m. f.
  2-cyclohexenone ring **27**, 950
  2-cyclohexenones, isomeric
  13, 858; **28**, 910
  1,3-dienamines, cyclic
  27, 950
  N-heterocyclics, N-condensed
  26, 345
  pyridine ring **8**, 480
1,3-Dioxolanes (s. a. Acetals,
  cyclic)
– from
  ethylene derivs. and glycols
  29, 174
  oxido and oxo compds.
  24, 128; **30**, 72
–, 2-alkoxy-
– startg. m. f.
  1,2-acoxyhalides **28**, 500
–, 2-amino-
–, protection of glycol groups
  as – **27**, 234
– special s.
  4-alkylidene-2-amino-
  1,3-dioxolanes
– startg. m. f.
  ethylene derivs. **27**, 938
Dioxolanium salts
–, N-β-acoxyethylation with –
  30, 701
– from
  acetals **29**, 753
– startg. m. f.
  α,β-ethylenecarboxylic acids
  29, 753
1,3-Dioxolan-2-one as reactant
  18, 516; **29**, 235
–, 4,5-diphenyl-
–, protection of amino groups,
  prim., with – **28**, 365
– ring, 4,5-dihalogeno-
– startg. m. f.
  α-diketones, cyclic **30**, 247
– – opening **30**, 274
1,3-Dioxolan-2-ones 29, 258
– from
  1,3-dioxanes, 2-oxo- **28**, 118
–, protection of vic-hydroxyl
  groups as – **29**, 197 suppl. **30**
– startg. m. f.
  1,3-dioxole-2-thiones **29**, 557
  oxazolid-2-ones, 4-hydroxy-
  28, 365
  Δ⁴-2-oxazolones **28**, 365
1,3-Dioxolan-2-yltriphenyl-
  phosphonium bromide
– as reactant **29**, 867
1,3-Dioxole ring, 2-amino-
– from
  o-diols and halogeno-
  formiminium halides
  29, 204
1,3-Dioxole-2-thiones
– from
  1,3-dioxolan-2-ones **29**, 557

1,3-Dioxol-2-ones, halogeno-
–, reactions with – **30**, 247
–, – – –, review **30**, 247
Diphenic acids
– startg. m. f.
  3,4-benzocoumarins **26**, 254
Diphenoquinones 10, 228
– from
  phenols **26**, 773; **27**, 761
Diphenylamine as reagent
  28, 68
Diphenylcalcium, metalation
  with – **28**, 599
Diphenyl carbonate as reactant
  21, 265 suppl. **26**; **27**, 640
– **diselenide** as reagent **28**, 471
– **ether**
– as medium **27**, 102; **28**, 898;
  29, 372
– as solvent **30**, 329
Diphenylmethane as medium
  28, 68; **29**, 978
N-Diphenylmethyl groups,
  removal **27**, 28
Diphenyl phosphite as reagent
  27, 419
– –/pyridine
– as acylating agent **27**, 419
  suppl. **28**
– as reagent **30**, 243
– **phosphorazidate** as reagent
  27, 611 suppl. **30**; **28**, 353
– **selenoxide** – – **7**, 514
  suppl. **28**; **29**, 275
– **sulfite** – – **30**, 270
Diphosphines
– startg. m. f.
  P-heterocyclics **28**, 597
Diphosphonic acid esters
  (s. a. Trithiodiphosphonic
  acid esters)
– special s.
  hydroxydiphosphonic acid
  esters
– **acids**
– from
  phosphonic acid dichlorides
  29, 118
Diphosphoric acid diester
  monobetaines 28, 73
– – **diesters**
– special s.
  P¹,P²-di-n-nitrophenyl
  pyrophosphate
– – P¹,P²-diesters, mixed
– from
  thionophosphoric acid
  monoesters **30**, 62
– – **diesters, sym. 29**, 113
– from
  phosphorodichloridates
  29, 118
– – **monoesters 29**, 112
Dipolar s. Cycloaddition,
  dipolar
Dipolarophiles, strong
  Δ¹-azirines **26**, 706 suppl. **28**
Dipoles s. Betaines
1,3-Dipoles (s. a. Cycloaddition, 1,3-dipolar)
–, **functionalized**
– startg. m. f.
  heterocyclics, fused
  24, 692 suppl. **29**
1,1-Dipyrazolyl compds.
– from
  ketones **28**, 426

Dipyrazol-1-yl ketone
– startg. m. f.
  1,1-dipyrazolyl compds. **28,** 426
1,1-Di(pyridinium salts)
– from
  1,1-dihalides **29,** 211
– startg. m. f.
  aldehydes-1-d **29,** 211
α,α'-**Dipyridyl** as reagent
29, 92, 894
2,2'-Dipyridyl disulfide – –
26, 175; **30,** 168
Dipyrrylmethanes 27, 817;
28, 755
1,2-Diquinonylethylenes
  by dimerization, oxidative
  29, 715
Direction (s. a. Regiospecificity)
– s. 'directed' or
  'direction of –' under
  addition
  C-alkylation
  condensation
  cycloaddition
  C-formylation
  functionalization
  glycosylation
  halogenation
  nitration
  ring closure
  – opening
  Schmidt reaction
  stereospecificity
  substitution, ar.
  –, N-heterocyclic
Disaccharides 5, 201 suppl. **28;**
11, 293 suppl. **29; 26,** 129
– from
  2-cyclopropen-1-yl ethers,
  2,3-diphenyl- **15,** 247
  suppl. **29**
  orthocarboxylic acid derivs.
  of carbohydrates **22,** 185
–, α-linked **5,** 201 suppl. **28**
Discharge, microwave
  s. Microwave discharge
Diselenadiazocines,
  perhydro- 28, 593
Diselenides
– from
  selenols **27,** 587
– special s.
  dicarboxydiselenides
  diphenyl diselenide
– startg. m. f.
  2-acoxyselenides **29,** 180
  selenoacetals **26,** 648
Diselenium dichloride
  as reactant **28,** 592
1,2-Diselenolium salts
– startg. m. f.
  1-oxa-6,6a-diselena-
  2-azapentalenes **28,** 315
Disiamylborane s. Bis-
  (3-methyl-2-butyl)borane
Disic acids s. Isoxazoles,
  3,5-dihydroxy-
1,2-Disilacyclohexanes
– from
  1,2,3-trisilacyloheptanes **28,** 582
Disilanes
–, reaction with acetylene
  deriv. **30,** 425
1,2-Disilanes
– from
  halogenosilanes and ethylene
  derivatives **26,** 650

Disilazanes
– special s.
  hexamethyldisilazane
– startg. m. f.
  acylaminosilanes **28,** 412
  isocyanates **26,** 443
1,2-Di(siloxy)alkenes
– as acyloin substitutes **26,** 811
– startg. m. f.
  α-diketones **27,** 233
  α-hydroxyketones, synthesis
  30, 617
1,2-Di(siloxy)cyclobutanes
– from
  1,2-di(siloxy)cyclobut-1-enes
  27, 60
– startg. m. f.
  cyclobutane-*cis*-1,2-diols
  27, 60
1,2-Di(siloxy)cyclobut-1-enes
– startg. m. f.
  cyclobutane-*cis*-1,2-diols
  27, 60
  1,2-di(siloxy)cyclobutanes
  27, 60
1,2-Di(siloxy)-1-ethylene
  derivs.
– startg. m. f.
  α-aminoketones **28,** 410
Disilthianes
– special s.
  hexamethyldisilthiane
1,1-Disilyl compds.
– startg. m. f.
  ethylene derivs., synthesis
  30, 615
1,2-Disilyl-1-ethylene derivs.
– from
  acetylene derivs. **29,** 602
O,N-Disilyliminoesters
– special s.
  N,O-bis(trimethylsilyl)-
  acetamide
C,N-Disilylketenimines
– from
  nitriles and halogenosilanes
  30, 438
N,N-Disilylsiloxylamines
– startg. m. f.
  silylaminosiloxanes **28,** 301
Dismutation s. Disproportionation
Disodium hydrogen phosphate
27, 96, 440; **28,** 293; **30,** 415
Dispersing agents s. Surfaceactive agents
Displacement s. Migration,
  Rearrangement, Replacement, Substitution
Disproportionation
  (s. a. Comproportionation)
–, Friedel-Crafts synthesis
  with – **17,** 879 suppl. **26**
– of
  1,1-aminoethers, cyclic
  29, 164
  disulfides **29,** 526
  ethylene derivs., review
  30, 467 suppl. **30**
  iodides, ar. **19,** 611
  mercury compds.,
  monoorgano- **26,** 647
–, side chains by – **30,** 635
Distannoxanes
– from
  tin oxides, diorgano- **27,** 645
– special s.
  bis (dibutylacetoxytin) oxide

1,1-C-Disubstitution s. Replacement of carbonyl
  oxygen by 2 carbon
  substituents
(Disulfen)amines
– as intermediates **26,** 357
– startg. m. f.
  amines, prim. **26,** 357
C-Disulfenylation 30, 414
– with sulfenimide **28,** 556
Di(sulfenylchlorides)
–, ring closures with – **26,** 608
– startg. m. f.
  1,4-dithienes, 2-chloro-
  26, 608
Disulfide dioxides s. Thiolsulfonic acid esters
Disulfides
– from
  mercaptals **28,** 519; **29,** 527
  mercaptans and sulfenimides
  26, 562
  sulfenic acid esters and
  thioboron compds. **27,** 586
  sulfenylthiocarbonic acid
  esters **26,** 565
  thiolsulfonic acid esters
  30, 699
– special s.
  acyl-disulfides
  amino-
  chloro(chlorosulfenyl)-
  oxymethyl –
  diaryl –
  dicyano-
  2,2'-dipyridyl disulfide
  ene-disulfides
  hydroxy-
  imino-
  phenyl disulfide
  sulfinyl-disulfides
  sulfonyl-
  thioacyl-
– startg. m. f.
  sulfenamides **28,** 258
  sulfenimines **28,** 407
  thioenolethers **30,** 413
  thioethers **26,** 622; **27,** 635,
  637, 642; **28,** 571
  N-thioiminohalides **30,** 347
  thiosulfinic acid esters
  29, 96
  thiolsulfonic – – **27,** 585
–, **cyclic** s. Cystine peptides,
  cyclic
–, **sym.**
– by reduction **26,** 564
– from
  disulfides, unsym., disproportionation **29,** 526
  mercaptans **26,** 559/60;
  30, 377
  nitro compds. **29,** 564
  sulfenylhalides **28,** 516
  sulfonic acid chlorides
  26, 564
  p-toluenesulfonic acid esters
  11, 666 suppl. **27**
–, **unsym.**
– startg. m. f.
  disulfides, sym., disproportionation **29,** 526
Disulfonates
– special s.
  dimesylates
– startg. m. f.
  ethers, cyclic **7,** 355 suppl. **28**

Dis – Dit

Disulfones
- special s.
  disulfonyldiazomethanes
  ethylenedisulfones
1,1-Disulfones (β-disulfones),
  acylthiomethylation
  26, 844 suppl. **28**
- special s.
  α,α-disulfonylethers
Disulfonyldiazomethanes
20, 271 suppl. **28**
- startg. m. f.
  enehydrazones, tetraketo-
  28, 835
2,2-Disulfonylenamines
- from
  2,2-disulfonylenolethers
  30, 272
2,2-Disulfonylenolethers
- startg. m. f.
  2,2-disulfonylenamines
  30, 272
α,α-Disulfonylethers
- from
  α-halogeno-α-sulfonylethers
  29, 583
1,1-Disulfoxides
- from
  mercaptals **29,** 99
  sulfinic acid esters and
  sulfoxides **30,** 394
Disulfurdicarbothionates
- special s.
  dimethyl dithiobis-
  (thioformate)
Dithiacyclophanes 26, 614
Disulfur monoxide as reactant
  28, 535
Ditellurides
- startg. m. f.
  telluroacetals **26,** 648
1,6a-Dithia-6-azapentalenes
  30, 307
- from
  6a-thiathiophthenes **30,** 307
1,4-Dithiafulvenes
- from
  1,2,3-dithiazoles (2 molecules) **29,** 577
2,7-Dithiaisotwistanes 28, 537
1,2-Dithiane 1,1-dioxide ring opening 28, 513
1,3-Dithiane, 2-chlorocarbonyl-
  as reactant **30,** 487
1,3-Dithianes 26, 601
- as intermediates **28,** 729
- special s.
  ethylene-dithianes
  2-spirocyclopropano-1,3-
- startg. m. f.
  acylals **26,** 211
  carboxylic acid esters
  28, 177
  1,3-dithianes, 2-alkylthio-
  28, 177
  α-ketocarboxylic acid esters
  27, 230
  oxo compds. **27,** 230
  aldehydes **26,** 211
-, 2-alkylthio-
- from
  1,3-dithianes **28,** 177
- startg. m. f.
  carboxylic acid esters **28,** 177
1,4-Dithianes, 2α,5α-dichloro-
  26, 573

1,5,3-Dithiazepine ring, N-condensed
- from
  mercuric acetylides and
  isothiocyanates **28,** 573
1,4,2-Dithiazine 1,1-dioxides
  28, 533
- from
  α,β-ethylenesulfonic acid
  amides **28,** 533
1,2,3-Dithiazoles
- startg. m. f.
  1,4-dithiafulvenes (from
  2 molecules) **28,** 577
3H-1,2,4-Dithiazoles
- from
  acylimines **30,** 398
1,2,4-Dithiazoli-3,5-diones
- startg. m. f.
  bis(halogenocarbonyl)amines
  27, 582
1,2,4-Dithiazolium salts
- startg. m. f.
  2H-1,3,5-thiadiazin-2-ones
  27, 636
1,4-Dithienes, 2-chloro-
- from
  di(sulfenylchlorides) **26,** 608
1,3-Dithienium fluoroborate
  as reactant **28,** 656
1,3-Dithiepane ring
- from
  diazo compds. **29,** 769
1,3-Dithietane ring
- from
  3-isothiazolone ring **26,** 585
1,3-Dithietanes
- from
  thioketones, dimerization
  29, 536
-, 2,4-diene-
- from
  1,2-ethylene-1,1-dihalides
  30, 405
1,3-Dithietan-2-ylium salts, 2-amino-
- from
  1,1-bis(dithiocarbamic acid
  esters) **30,** 422
1,2-Dithiete-α-dithione equilibrium 29, 589
1,2-Dithiin 1-oxides, 3,6-dihydro-
- from
  1,3-dienes **28,** 535
Dithiins, review 28, 744
1,3-Dithiins
- from
  aldehydes (2 molecules) and
  acetylene derivs. **28,** 744
1,4-Dithiins, 2,3-dihydro-
- from
  α-halogenoketones **26,** 619
Dithioacetic acid
-, reduction of sulfur compds.
  with – **29,** 993
**Dithiobis(thioformic acid
  esters)** s. Disulfurdicarbothionates
Dithiobiurets s. Isodithiobiurets
Dithiocarbamates (s. a. Acyl
  dithiocarbamates)
- special s.
  benzylisothiouronium
  dithiocarbamates
Dithiocarbamic acid esters
  (s. a. Dithiourethans)

-, 1-alkylation **29,** 809; **30,** 536
- from
  bisthiuram disulfides **26,** 620
- special s.
  allyl dithiocarbamates
  1,1-bis(dithiocarbamic acid
  esters)
- startg. m. f.
  α-alkylthiocarboxylic acid
  thioamides **30,** 536
  α-mercaptocarboxylic acid
  thioamides **30,** 536
- acids
- startg. m. f.
  isothiocyanates **30,** 340
Dithiocarbamidium salts
-, reactions **27,** 484
- startg. m. f.
  1-aminothioenolethers
  27, 484
  1,3-heterocyclics, 2-amino-
  27, 484
Dithiocarbazic acid esters
- from
  N-isothiocyanatoamines
  26, 571
- startg. m. f.
  1,3,4-thiadiazoles, 2-alkylthio-
  29, 568
  1,3,4-thiadiazol-2(3H)-ones,
  5-alkylthio- **28,** 539
**Dithiocarbonic acid
  O,S-diesters** s. Xanthates
Dithiocarboxylates s. Carbamyl
  dithiocarboxylates
C-Dithiocarboxylation 30, 464
Dithiocarboxylic acid esters
  (s. a. Dithio ester...)
- by transesterification **26,** 626
- from
  dithiocarboxylic acids and
  ethylene derivs. **29,** 538
- special s.
  allenedithiocarboxylic acid
  esters
  cyanodithioformic – –
  dithiocarboxylic acid
  propargyl esters
  (thiobenzoylthio)acetic acid
- startg. m. f.
  carboxylic acid esters
  29, 214
  Δ²-thiazolines **30,** 412
- – propargyl esters
- startg. m. f.
  2-ene-1,3-dithioles **30,** 391
- acids
- special s.
  dithioacetic acid
  iminodicarboxylic acids
- startg. m. f.
  carbamyl dithiocarboxylates
  29, 532
  dithiocarboxylic acid esters
  29, 538
**Dithio ester thio-Claisen
  rearrangement, stereospecific**
  28, 842
Dithiohemiacetals
- from
  aldehydes **27,** 608
Dithiohydantoins
- startg. m. f.
  Δ⁴-imidazoline-2-thiones
  26, 981
1,2-Dithiolane-3,5-diones
  as intermediates **28,** 578

1,2-Dithiolanes
- from
  1,3-dithiols **27,** 642
  2-enethiethanes **28,** 532
- startg. m. f.
  thietanes **27,** 642

1,3-Dithiolanes
- from
  1,2-dithiols **27,** 606
-, **2-subst.,** C-formylation via –
  30, 569

1,3-Dithiolane-Δ²,ᵅ-thioacetyl halides, α-cyano-
- from
  1,3-dithiolane-2-thiones
  26, 689

1,3-Dithiolane-2-thiones
- from
  oxido compds. **28,** 546
- startg. m. f.
  1,3-dithiolane-Δ²,ᵅ-thioacetyl halides, α-cyano- **26,** 689

1,3-Dithiolanium salts, 2-amino-, reactions **26,** 577
(1,3-Dithiolan-2-yl)ammonium salts, quaternary
-, reactions with – **30,** 569
Dithiolcarbonic acid esters, cyclic
- special s.
  1,3-dithiol-2-ones
Dithioldicarbonic acid esters
- startg. m. f.
  thiolurethans **27,** 378
1,2-Dithioles
- special s.
  alkylidene-1,2-dithioles
1,3-Dithioles
- special s.
  1,4-dithiafulvenes
  2-ene-1,3-dithioles
-, **2-alkoxy-**
- from
  1,3-dithioles, 2-amino-
  28, 155
-, **2-amino-**
- startg. m. f.
  1,3-dithioles, 2-alkoxy-
  28, 155
  1,3-dithiolium salts **26,** 998
1,3-Dithiole-2-thiones
- startg. m. f.
  tetrathiofulvalenes **30,** 610
α-Dithiolic acids
- startg. m. f.
  thietane-2,4-diones **28,** 578
1,2-Dithiolium salts
-, deprotonation **30,** 638
- special s.
  organothio-1,2-dithiolium salts
- startg. m. f.
  1-oxa-6,6a-dithia-2-azapentalenes **28,** 315
  tetraazamacroheterocyclics **17,** 516 suppl. **28**
1,3-Dithiolium salts
- from
  1,3-dithioles, 2-amino-
  26, 998
- startg. m. f.
  tetrathiofulvalenes **29,** 717
1,3-Dithiol-2-one ring
- from
  α-halogenoketones **29,** 588
  2-ketoxanthates **29,** 588

1,3-Dithiol-2-ones
- startg. m. f.
  α-dithiones **29,** 589
Dithiolortholactones
- from
  hydroxyketene mercaptals
  29, 552
  lactones **29,** 552
Dithiols s. a. Dithiothreitol
1,1-Dithiols
- startg. m. f.
  α-alkylthiophosphonic acid esters **30,** 439
  α-mercaptophosphonic – –
  30, 439
1,2-Dithiols
- from
  sulfido comdps. **28,** 525
- startg. m. f.
  1,3-dithiolanes **27,** 606
1,3-Dithiols
- startg. m. f.
  1,2-dithiolanes **27,** 642
  thietanes **27,** 642
o-Dithiols
- from
  o-aminomercaptans **27,** 623
  1,3-benzodithiole-2-thiones
  27, 623
- startg. m. f.
  1,3,2-benzodithiaboroles
  30, 379
-, 2-halogeno- **30,** 379
  o-benzotrithiol-2-ones **28,** 515
1,1-Di(thiolurethans)
- from
  thiocyanates and aldehydes
  29, 308
α-Dithione-1,2-dithiete equilibrium 29, 589
α-Dithiones
- from
  1,3-dithiol-2-ones **29,** 589
Dithionite 27, 878; **29,** 80, 718;
  30, 330, 392
Dithiophosphinic acids
- startg. m. f.
  bis(thiophosphinyl) trisulfides, sym. **26,** 558
Dithiophosphoric acids, O,O-dialkyl-, as reagents
  23, 581 suppl. **28; 26,** 993;
  28, 550; **29,** 17
Dithiophosphorochloridites
- special s.
  diethyl phosphorochlorido-dithioite
- startg. m. f.
  tetrathiohypodiphosphorous acid esters **30,** 423
Dithiophosphorous acid esters 27, 121
- startg. m. f.
  phosphoric acid monoesters
  27, 121
Dithiothreitol
- as reagent **16,** 167 suppl. **26**
-, reduction with – **27,** 992
Dithiourethans (s. a. Dithio-carbamic acid esters)
- from
  ethylene derivs. **27,** 594
  xanthates **28,** 328
Ditosylates
- startg. m. f.
  dicarboxylic acids **29,** 195

1,2-Ditosylates
- startg. m. f.
  1,4-dioxanes **29,** 183
  enol tosylates **26,** 944
1,3-Ditosylates
- startg. m. f.
  2-ethylenealcohols **26,** 943
1,2-Ditritio compds. 9, 94 suppl. **26**
- from
  ethylene derivs. **22,** 59 suppl. **26**
1,1-Diurethans
- startg. m. f.
  urethans, synthesis **27,** 820
***gem*-Divinyl compds. 17,** 968 suppl. **26**
Diynammonium salts, quaternary
- startg. m. f.
  benzisoindolinium salts, quaternary **30,** 504
Diynes
-, ring closures, double with –
  26, 725
1,3-Diynes
- from
  1,4-diynes **7,** 732 suppl. **26**
  1,4-diyn-3-ones **30,** 695
-, hydroboration, partial
  14, 82 suppl. **26**
- startg. m. f.
  pyridines **29,** 650
1,4-Diynes
- startg. m. f.
  1,3-diynes **7,** 732 suppl. **26**
  stannins, 1,4-dihydro- **30,** 426
2,4-Diynoic acids
- startg. m. f.
  2-pyrones, 4-hydroxy-
  29, 151
1,3-Diyn-5-ones
- from
  1,4-diyn-3-ones **26,** 752
1,4-Diyn-3-ones
- startg. m. f.
  1,3-diynes **30,** 695
  1,3-diyn-5-ones **26,** 752
Dodecyldimethylamine
  as reagent **27,** 128
H-Donors (s. a. Hydride donor)
- special s.
  benzyl alcohol
  ethylene glycol
  hydroquinone
  isopropyl alcohol
  tetrahydrofuran
H⁺-Donors s. Proton donors
Double bonds s. a. Addition, Hydrogenation, Migration
- –, **carbon-carbon**
  (s. a. Ethylene derivs.)
- from
  dicarboxylic acid anhydrides
  23, 976
- startg. m. f.
  carbonyl groups **27,** 158
  double bonds, carbon-nitrogen **28,** 304
- –, –, **conjugated**
  (s. a. 1,3-Dienes)
- from
  ω,ω'-dihalides **27,** 968
- –, **carbon-nitrogen**
  (s. a. Azomethines, Imin...)

(Double bonds, carbon-nitrogen)
– from
  double bonds, carbon-carbon
    28, 304
–, reaction of lead tetraacetate
  with –, review **26**, 190
  suppl. **28**
– inversion **27**, 257
Dowex 1-X2(OH) 26, 426
– **21-K(Cl) 22**, 61 suppl. **29**
Dowex 50–X1 26, 224
Dowex 50W-X2 27, 204
**Dowex 50W-4 cation exchanger
  29**, 145
Dowtherm A as medium
  20, 375 suppl. **26; 26**, 465
Driving force, heterolysis
  as – **30**, 176
DTT s. Dithiothreitol

EDTA s. Ethylenediaminetetraacetic acid
EEDQ (2-ethoxy-N-ethoxy-
  carbonyl-1,2-dihydro-
  quinoline) **23**, 415; **30**, 265
Effect s. a. under Solvents
–, **regiospecific** s. Regiospecific
  effect
–, **steric** s. Steric effect
Einhorn s. a. Tscherniac
– reaction s. Acylamino-
  methylation
Eistert s. Arndt **(25)**
Electrocatalysis s. Hydrogenation, electrocatalytic
Electrolysis (s. a. Oxidation,
  electrochemical, Reduction,
  –, Synthesis, anodic) **22**, 833
  suppl. **29; 26**, 158, 398, 766;
  27, 18, 41, 135, 160, 605, 953,
  961; **28**, 1, 7, 98, 432, 720/1,
  834, 973; **28**, 229 suppl. **29;
  28**, 703 suppl. **29; 29**, 68/9,
  173, 239, 260, 334, 773, 900,
  976; **29**, 362 suppl. **30; 30**, 22,
  177
– at controlled potential
  28, 7; **30**, 433
–, electrochemistry, organometallic, review **18**, 717
  suppl. **28**
–, electroorganic reactions,
  review **25**, 26 suppl. **27;
  26**, 158, 766
– with reversal of electrode
  polarity **28**, 703
Electrolytes, supporting s.
  potassium nitrate
  – perchlorate
  tetrabutylammonium
    fluoroborate
  tetraethylammonium tosylate
  tetramethylammonium
    chloride
Electron acceptors s. Acetylene
  electron acceptor-donors,
  Stilbene – –
Electron-deficient s. Carbon-
  carbon bonds, multiple,
  electron-deficient, Dienophiles, –, Ethylene derivs., –,
  Isoalloxazines, –
Electron donors (s. a.
  Acetylene electron acceptor-
  donors, Stilbene – –)
  potassium **27**, 459

thiophenes, 2,4-diamino-
  29, 640
Electron-rich s. Allenes,
  electron-rich, Dienophiles, –,
  Ethylene derivs., –
π-**Electron systems**, conjugated,
  bi- and poly-cyclic **26**, 757
  suppl. **28**
Electron transfer 29, 13
Electrophiles 27, 843 suppl. **29**
–, opening of N-heterocycles
  at satd. carbon-nitrogen
  bonds by –, review **28**, 336
– special s.
  heteroelectrophiles
–, **ambident**
– special s.
  1,3-dithiolanium salts,
  2-amino-
–, **strong** s. EMME
Electrophilic carbonyl groups
  s. Carbonyl groups,
  electrophilic
– **substitution** s. Substitution,
  electrophilic
– **syntheses** s. Syntheses,
  electrophilic
Elements s. Group... element...,
  Metals, s. a under individual
  elements
Elimination (s. a. Extrusion)
– of
  alkoxy groups, photo-
    cyclization with – **29**, 913
  alkyl – s. under Aromatiza-
    tion and Replacement
    of alkyl groups by
    hydrogen
  azo bridges **30**, 522
  azoxy – **28**, 932
  C-bridges, aromatization
    21, 725 suppl. **27**
  CO-bridges, – **28**, 977
  O-bridges **30**, 679
  carbon dioxide s. De-
    carboxylation
  – monoxide s. Decarbonyla-
    tion
  carboxylic acids **26**, 468
  carbalkoxymethyl groups
    29, 981
  etheno bridges **26**, 987
  functional groups s. a. Re-
    placement... by hydrogen
  hydrogen halide s. Dehydro-
    halogenation
  C-methylene bridges **28**, 892
    suppl. **30**
  N-methylene – **14**, 364
  nitric acid with ring closure
    23, 936
  nitrogen, ring contraction
    by – **5**, 643
  steroid 4-methyl groups
    28, 983
  substituents, ring closure
    with – **26**, 452
  sulfenes **28**, 971
  sulfur s. under Sulfur
  – dioxide **26**, 433;
    29, 587, 848
  1,2,4-triazole ring,
    N-condensed, by –
    26, 490
  sulfur dioxide from sulfinic
    acids **27**, 880
  – – from sulfonic acid
    amides **29**, 213; **30**, 339

– with formation of unsatd.
  carbon-carbon bonds CC⇑
–, **decarboxylative 28**, 985
–, **oxidative**
– of
  phenyl groups **27**, 280
  sulfur dioxide **26**, 75
–, **reductive**
– of
  sulfur dioxide **26**, 75
β-**Elimination** (s. a. Hofmann
  degradation of ammonium
  salts, quaternary)
–, effect of bases on
  geospecificity **28**, 944
– of phosphoryl **29**, 155
–, removal of N-protective
  groups by – **26**, 24
trans-**Elimination**
– special s.
  trans-dehydration
**Elimination, stereospecific
  26**, 491
1,4-Elimination
–, ene-α-diones, cyclic, by –
  26, 975
1,6-Elimination 29, 19
Elimination-rearrangement,
  acetylene derivs. by –
  12, 929
EMCE (ethyl ethoxymethylene-
  cyanoacetate)
– startg. m. f.
  2H-naphtho[1,2-b]pyran-
    2-ones, 5,6-dihydro- **29**, 884
  pyrimidines **26**, 353
Emde degradation, modified
  22, 42
**Emilewicz-Kostanecki flavone
  ring closure 30**, 170
EMME (ethyl ethoxymethylene-
  malonate)
– as electrophile, strong
  5, 269/355 suppl. **26**
– as reactant **26**, 869
Emmons reaction s. Horner
  synthesis
Enacylamines
– from
  Δ^2-oxazolines **26**, 755
–, photocyclization, review
  25, 522 suppl. **29**
– special s.
  eneformylamines
  halogenenacylamines
– startg. m. f.
  1-acylamino-1-peroxides
    28, 112
  enamines **29**, 60
–, **cyclic**, ring closure **29**, 914
Enamides s. Enacylamines
Enamidines
– special s.
  1-halogenenamidines
**Enaminelactone rearrange-
  ment 27**, 198
Enamines (s. a. Aminomethylene...)
–, alkylation **28**, 799
– as substitutes of acetylene
  derivs. **22**, 877 suppl. **28**
–, condensation, decarboxylative, with – **30**, 626
–, cycloaddition to
  3-alkylideneoxindoles **27**, 692
– from
  O-alkyllactims **27**, 768

amide acetals **26**, 247;
   29, 469
enacylamines **29**, 60
α,β-ethylenehalides **27**, 465
formamides **28**, 415; **29**, 859
1-halogeneamines, synthesis
   27, 846
iminoesters **26**, 795
methylene groups, active
   30, 281
N-oxides **27**, 939
oxo compds. **29**, 437
sulfilimines **30**, 387
sulfonium ylids (2 molecules)
   and azides **26**, 893
thioiminoesters **26**, 982
–, metalation **29**, 799
–, reaction with
   cyclopropenethiones **22**, 710
   suppl. **29**
diethyl azodicarboxylate
   20, 502 suppl. **26**
phosgene **26**, 769; **29**, 714
p-quinones **26**, 717
–, ring closures with –
   28, 633; **29**, 543
– special s.
   alkoxyenamines
   2-aminoeneboranes
   1-aminoenedisulfides
   2-aminoenol sulfonates
   β-amino-α,β-ethylene...
   aminomethylene compds.
   β-amino-2-nitrostyrenes
   1-aminothioenolethers
   halogenenamines
   2-nitroenamines
   (polyen)amines
   N-silylenamines
   sulfonylenamines
   1,1,2-triaminoethylenes
– startg. m. f.
   amines **28**, 38; **29**, 45
   2-aminoalcohols **19**, 188
      suppl. **26**
   β-amino-α-diazosulfones
      27, 688
   β-amino-α,β-ethylene-
      carboxylic acid esters
      26, 769
   α-aminoketones **28**, 116
   benzene ring **22**, 877 suppl. **28**
   cyclobutanes, amino- **27**, 719
   cyclopent-2-enone ring
      27, 868
   2-diazoamines **28**, 625
   2,4-dienecarboxylic acid
      amides **28**, 614
   ethylene derivs., synthesis
      26, 870
   α-halogenazomethinium salts
      12, 244; **28**, 116
   O-heterocyclics **28**, 633
   hydrazones **29**, 392
   γ-hydroxyketones **26**, 830
   isothiazoles **26**, 841
   isoxazoles **29**, 786
   isoxazolidines, 4-amino-
      26, 699
   ketones, synthesis **28**, 768;
      30, 589
   4H-1,2-oxazine 2-oxide ring,
      5,6-dihydro- **26**, 700
   4H-1,2-oxazines, 5,6-dihydro-
      6-hydroxy- **26**, 849
   α-oximinoazomethines
      30, 235
   pteridine ring **26**, 392

pteridines **26**, 448
pyrrole ring, N-unsubst.
   25, 275 suppl. **26**
– –, 2,3-dihydro- **27**, 861
pyrroles **21**, 813
– (from 2 molecules) **29**, 773
Δ⁴-pyrrol-2-ones **26**, 703
quinoline-4-carboxylic acid
   amides **28**, 622
thiazole ring, 2-amino-
   27, 619
Δ⁴-2-thiazolones **29**, 543
1,2,3-triazole ring **26**, 321
–, synthesis with tropenium
   salts and – **27**, 822
–, transamination **22**, 425
–, **bulky 17**, 872 suppl. **29**
–, **cyclic**
– from
   4-acetylenehalides **30**, 637
   lactams **30**, 595
   oxido compds. **30**, 254
– special s.
   enamines, endocyclic
– startg. m. f.
   pyrroles **30**, 637
–, **endocyclic**
–, annelation **28**, 664
– from
   azomethines and dihalides
      27, 842
– special s.
   cyclobutene ring, 1-amino-
   Δ²-pyrrolines
–, **prim. 22**, 401 suppl. **27**
α-**Enaminocarboxylic acids**
– startg. m. f.
   N-heterocyclics **30**, 647
Enaminoenollactones
– startg. m. f.
   γ-diketones **27**, 249
Enaminolactams
– from
   ketocarboxylic acid amides
      27, 518
   – acids **29**, 367
– startg. m. f.
   1-alkoxy-2-oximinolactams
      29, 311
   lactams **27**, 53
Enaminophosphine oxides
– startg. m. f.
   cyclopropylketones **27**, 890
   α,β-ethyleneketones **27**, 890
–, syntheses with – **27**, 890
Enazides
– from
   Δ¹-azirines and diazo
      compds. **30**, 457
–, prepn. review **26**, 789
–, ring closures with – **29**, 401
– special s.
   α-azido-α,β-ethylene-...
– startg. m. f.
   furoxans **27**, 401
   N-heterocyclics **24**, 528
   nitriles **26**, 477
   Δ²-pyrrol-4-ones **28**, 773
–, **terminal**
– from
   2-(N-nitrosoacylamino)-
      alcohols **29**, 400
Enazo compds.
– special s.
   sulfonylenazo compds.
Enazomethines as startg. m.
   18, 750 suppl. **27**

– special s.
   β-amino-α,β-ethyleneazo-
      methines
   enediazomethines
Eneboranes
–, reactions via – **29**, 139
– special s.
   2-aminoeneboranes
Ene bridges s. Bis-2-imidazo-
   lidinylidenes
Enecarbamyl azides
– startg. m. f.
   Δ³-pyrazol-5-one ring **27**, 952
Enecyclopropanes 28, 944
   suppl. **29**; **29**, 864
–, [δ²⁺π²⁺π²]-cycloaddition
   with – **29**, 655
– special s.
   methylenecyclopropanes
–, **2-alkoxy-**
– startg. m. f.
   β,γ-ethyleneoxo compds.
      30, 156
Enecyclopropenes 30, 627
– from
   cyclopropenones and ketenes
      30, 627
– special s.
   quinocyclopropenes
– startg. m. f.
   alleneketals **27**, 156
   alleneketones **27**, 156
   fulvenes, 3-amino- **29**, 656
Enecyclopropyl ethers
– startg. m. f.
   β,γ-ethyleneacetals **29**, 145
   β,γ-ethylenealdehyde derivs.
      29, 145
Ene-1,1-diamines s. Ketene
   aminals
Ene-1,2-diamines
– from
   azomethines (2 molecules)
      26, 683
– special s.
   1-cyano-1,2-ene-1,2-diamines
– startg. m. f.
   1,2-di(azomethines) **26**, 683
cis-**Enediazomethines**
– startg. m. f.
   1H-1,4-diazepines,
      6,7-dihydro- **29**, 703
**Enediazonium hexachloro-
   antimonates**
– from
   2-halogenotosylhydrazones
      27, 973
2-Ene-1,4-dicarbonyl compds.,
   reduction **8**, 87 suppl. **29**
Enediolesters
– from
   aldehydes (2 molecules)
      27, 759
2-Ene-1,4-diols
– from
   boranes and furans **28**, 583
   1,2-oxaborins, dihydro-
      28, 583
1,1,1-Enediol sulfates
– special s.
   isobutenylidene sulfate,
      hexafluoro-
Enediones
– special s.
   2-alkylidene-1,3-diones

2-Ene-1,4-diones
- from
  α-diketones and (α-acyl-
    alkylidene)phosphoranes
    28, 847
  furans **28**, 84
- startg. m. f.
  γ-diketones **8**, 87
  furans, 2-α-halogeno- **27**, 571
Ene-α-diones, cyclic
- by 1,4-elimination **26**, 975
2-Ene-1,4-diones, cyclic
  s. Cyclohexenedione ring
Ene-1,2-di(oxysilanes)
- as acyloin substitutes
    26, 811
- startg. m. f.
  heterocyclics **26**, 811
Enedisulfides
- special s.
  aminoenedisulfides
- startg. m. f.
  thiophenes **27**, 980
-, **isocyclic**
- startg. m. f.
  spiro-1,3-S,S-heterocyclics
    30, 390
2-Ene-1,3-dithioles
- from
  dithiocarboxylic acid
    propargyl esters **30**, 391
Eneformylamines
- from
  isonitriles **26**, 680
- special s.
  1-(sulfonyl)eneformylamines
Enehalides s. α,β-Ethylene-
  halides
Eneheterocyclics
- special s.
  3-alkylideneoxindoles
  1,3-dithietanes, 2,3-diene-
  5-ene-Δ²-isoxazolines
  N-(4-en-Δ²-oxazolin-5-
    yliden)ammonium salts
  oxathiol-2-ylidene-
    immonium –
  Δ²-pyrazolines, 4-ene-
    5-hydroxy-
  pyrrolidines, 2-alkylidene-
Enehydrazines
- special s.
  acylenehydrazines
Enehydrazones, tetraketo-
- from
  disulfonyldiazomethanes
    and β-diketones **28**, 835
Enehydroxylamines
- from
  hydroxylamines **30**, 651
  nitroso compds. **27**, 313
    suppl. **30**
- startg. m. f.
  indoles **30**, 651
2-Eneindolin-3-ones 29, 890
5-Ene-Δ²-isoxazolines
- startg. m. f.
  isoxazoles **26**, 729
Enelactones
- startg. m. f.
  cyclopropylalcohols **28**, 907
Enephosphine sulfides
- from
  ynephosphine sulfides,
    synthesis **26**, 715
- startg. m. f.
  poly-*tert*-phosphines **29**, 600

Enephosphonic acid esters
- special s.
  alkoxyenephosphonic acid
    esters
  carbalkoxyenephosphonic – –
Enephosphonium salts
- special s.
  alkoxyvinylphosphonium
    salts
- startg. m. f.
  heterocyclics **30**, 178
4-Enepyran ring
- startg. m. f.
  Δ²-isoxazoline ring **29**, 360
  pyridine N-oxide –
    29, 360
Enesilanes as intermediates
    29, 497
- from
  acetylene derivs. **12**, 700
    suppl. **26**
  2-oxazolidones, 3-nitroso-
    8, 922 suppl. **26**
  1,1-silyllithium compds.
    synthesis **29**, 860
-, homologization **29**, 763
- special s.
  1,2-disilyl-1-ethylene derivs.
  α,β-ethylene-(α- or β-)silyl-
    ketones
  halogenoenesilanes
- startg. m. f.
  α,β-ethyleneketones **29**, 846
  ketones **30**, 143
-, synthesis by 1,4-addition
    30, 490
Enestannanes as intermediates
    28, 658 suppl. **29**
Enesulfonates 27, 107
- special s.
  enetriflates
Enesulfonium salts
- special s.
  diaminomethylenesulfonium
    salts
- startg. m. f.
  furan ring, 2,3-dihydro-
    28, 838
Ene synthesis 29, 676
-, review **17**, 730 suppl. **24**
- with silicon compds.,
  organo- **29**, 676
- –, **intramolecular 28**, 667
2-Enethiazolidines
- from
  Δ²-thiazolines, 2-alkylthio-
    28, 844
2-Enethiethans
- startg. m. f.
  1,2-dithiolanes **28**, 532
Enethiolurethans
- from
  thiolcarbamic acid esters
    and aldehydes **26**, 382
- special s.
  α-(N-alkylthiocarbonyl-
    amino)-α,β-ethylene-
    carboxylic acid esters
Enetriazenes
- from
  sulfonium ylids (2 molecules)
    and azides **26**, 893
Enetriflates
-, reactions **29**, 920
- startg. m. f.
  acetylene derivs. **29**, 920

Eneureas
- special s.
  halogeneneureas
Eneurethans
- from
  α,β-ethylenecarboxylic acid
    azides **29**, 237
  ethylene derivs. and
    N-alkoxycarbonylimino-
    pyridinium ylids **28**, 376
- special s.
  ketoeneurethans
- startg. m. f.
  ketones **29**, 237
Enimines s. Ketenimines
Eniminophosphoranes
- from
  alkylidenephosphoranes and
    nitriles **27**, 684
Enisocyanates
- special s.
  α,β-ethylen-β-isocyanato-
    carboxylic acid esters
  halogenenisocyanates
Enisocycles
- from
  ketones **27**, 934
- startg. m. f.
  ketones, cyclic, ring
    expansion **29**, 154
Enisocyclics s. a. Enecyclo...
Enisonitriles
- from
  oxo compds. and α-isocyano-
    phosphonic acid esters
    29, 870
- special s.
  vinyl isocyanide
- startg. m. f.
  Δ²-thiazolines **30**, 388
Enisothiocyanates
- from
  imines **30**, 242
Enolate, aldol condensation
  via – **29**, 894
- **anions**, α-alkylation of
  ketones via – **19**, 881
    suppl. **28**; **29**, 804
Enolates (s. a. Metal enolates)
-, C-hydroxylation of – **30**, 65
- special s.
  ester enolates
  indenolates
Enolbetaines
- from
  ethylene derivs., electron-
    rich and α-diazoketones
    26, 917
Enolesters
-, C-acylation with – **26**, 804
-, addition with carbon-oxygen
  cleavage **27**, 314
- from
  2-acoxydiazo compds.
    26, 962
  carboxylic acid chlorides
    (2 molecules), synthesis
    30, 605
  α,β-ethyleneketones,
    oxidation **30**, 89
  ketenes (2 molecules) **26**, 710
  ketones **28**, 140
-, peptide synthesis via
  formation of – **26**, 398
-, reactions with aluminum
  chloride **26**, 804
-, ring closure with – **26**, 986

- special s.
  1-acoxy-1,3-dienes
  β-acoxy-α,β-ethylene...
  α-alkoxyenolesters
  borinic acid enolesters
  carboxylic acid isopropenyl
    esters
  halogenenolesters
  vinylthallic acetates,
    2-acetoxy-
- startg. m. f.
  α-alkoxyketones **26,** 637
  β-alkoxyketones **30,** 633
  cyclopent-2-enones,
    5-β-keto- **30,** 181
  1,3-diol esters, synthesis
    29, 674
  ethylene derivs., synthesis
    29, 746
  α,β-ethyleneketones **28,** 229
  γ-halogenenolesters **28,** 229
  β-hydroxyketones **29,** 894
-, syntheses with − **30,** 633
-, **cyclic**
- from
  dicarboxylic acids **29,** 892
- startg. m. f.
  hydroxycarboxylic acids
    28, 109 suppl. **30**
Enolethers
- by transetherification
  5, 213; **27,** 250
- from
  acetals **26,** 956
    − with alkoxy interchange
      28, 756
    α-alkoxy-β-lactones **29,** 985
  alkylidenephosphoranes
    28, 846
  allenes **29,** 138
  diazo compds. **28,** 846
  fluorosulfonic acid esters
    28, 173
  ketals **26,** 130; **29,** 925
  oxo compds. and ortho-
    formic acid esters
    29, 190; **30,** 111
  − −, rearrangement **30,** 103
  − −, synthesis with addition
    of 1 C-atom **14,** 876
-, metalation **29,** 621
-, reaction with Grignard
    compds. **26,** 869
-, reduction, stereospecific
    26, 54
- special s.
  1-alkoxydienes
  α-alkoxyenolesters
  alkoxyenolethers
  α- and β-alkoxy-
    α,β-ethylene...
  alkoxymethylene...
  (dien)olethers
  ethyleneenolethers
  nitroenolethers
  oxo-1,2-en-1-olethers
  sulfonylenolethers
  vinyl ethers
- startg. m. f.
  acetals, mixed, O-hetero-
    cyclic **26,** 138
  N-acyliminoesters **29,** 395
  β-alkoxyketones, synthesis
    28, 654
  1-alkoxy-1-sulfones, cyclic
    29, 729
  1,2-alkoxythioethers **18,** 645
    suppl. **26**

carboxylic acid esters **28,** 189
1,4-dienes **27,** 882
α,β-ethylenealdehydes,
  α-subst. **26,** 802
ethylene derivs., synthesis
  26, 869
α,β-ethylenenitriles **27,** 789
α-halogenacetals **27,** 547
ketals, mixed **26,** 130
4H-1,3-oxazines, 5,6-dihydro-
  6-alkoxy- **29,** 654
quinolines **26,** 723
N-sulfonyliminoesters
  29, 396
-, **bicyclic**
- startg. m. f.
  oximinolactones **29,** 315
-, **cyclic**
- startg. m. f.
  hydroxycarboxylic acid
    esters **28,** 109
Enollactones
-, C-acylation **26,** 821
- from
  dicarboxylic acid anhydrides
    26, 904
  ketocarboxylic acids **27,** 919
- special s.
  dienollactones
  enaminoenollactones
- startg. m. f.
  β-hydroxyketones, cyclic
    20, 7
γ-Enollactones
- from
  carboxylic acid chlorides
    30, 606
  cyclopropenecarboxylic
    acids **26,** 155
- startg. m. f.
  γ-ketocarboxylic acid
    amides **26,** 287
Enol phosphates
- startg. m. f.
  cyclopropanes **26,** 783
  oxo compds. **26,** 251
Enols
- special s.
  α,β-ethylene-
    (α- or β-)hydroxy...
  hydroxymethylen...
  ketoenols
- startg. m. f.
  α,β-ethylenealdehydes
    29, 907
  α,β-ethylenehalides **11,** 941
    suppl. **28**
Enol sulfonates (s. a. Enol
  tosylates)
-, ring closure with − **27,** 935
- special s.
  2-aminoenol sulfonates
- startg. m. f.
  ketosulfones **26,** 588
  thianaphthenes **27,** 935
-, **tosylates**
- from
  1,2-ditosylates **26,** 944
Enophiles
-, α-halogenonitrones as −
  28, 657
**N-(4-En-Δ²-oxazolin-5-yliden)-
  ammonium salts 28,** 206
Enoxyboranes
- from
  ketones **28,** 81
-, synthesis of ketones via −
  27, 894

-, syntheses via − **27,** 894
Enoxysilanes 14, 753 suppl. **26**
- from
  ketones **27,** 720
  oxo compds. **29,** 829
-, ozonization **28,** 109 suppl. **29**
- special s.
  (dien)oxysilanes
  enedi(oxysilanes)
  β,γ-ethylene-β-siloxy-
    carboxylic acid esters
- startg. m. f.
  acetals **29,** 193
  N-acylsulfonic acid amides
    29, 434
  β-alkoxyketones **30,** 621
  cyclopropanols **14,** 858
    suppl. **28**
  γ-diketones (from 2 mole-
    cules) **30,** 619
  C-glycosides **29,** 754
  α-halogenocarbonyl compds.
    30, 374
  α-hydroxyketones **30,** 144
  β-hydroxyketones **30,** 621
  ketoacetals **30,** 621
  β-ketocarboxylic acid
    amides **30,** 462
  γ-ketocarboxylic acid esters
    28, 858
  siloxycyclopropane ring
    29, 827, 829
  α-siloxymethylene compds.
    29, 827
  α-siloxynitriles **27,** 720
-, syntheses with − **30,** 621
-, **cyclic**
- startg. m. f.
  trans-glycols, cyclic **30,** 81
  lactones **30,** 145
  trans-2-siloxyalcohols,
    cyclic **30,** 81
Entrainers in Grignard
  reactions s. Activators
Entrainment, azeotropic
  s. Azeotropic entrainment
Enynals
- special s.
  dienediynedials
Enynes
- special s.
  halogenenynes
  polyenynes
1,3-Enynes
- special s.
  1-organoseleno-1-organothio-
    1-en-3-ynes
- via vinylcopper compds.
  30, 492
trans-**1,3-Enynes,** synthesis
  29, 139
1,4-Enynes
- from
  acetylene derivs., synthesis
    29, 830
Enzymes
- as reagents **26,** 105; **27,** 259;
  28, 13 suppl. **29**
- special s.
  α-chymotrypsin
  glucosyl transferase
  peroxidase
Epichlorhydrin as hydrogen
  bromide scavenger **26,** 208
Epimerization
- in Wittig syntheses
  17, 895 suppl. **26**

(Epimerization)
- of
  acetals **29**, 163
  aryl groups **26**, 149
  glycols **26**, 148
  N-heterocyclics, N-condensed **30**, 321
  hydroxyl groups, sec.
    (s. a. *trans*-2-Ethylene-
    alcohols, epimerized),
    via tosylates **28**, 119
  ketosteroids **26**, 736
  lactones **8**, 193 suppl. **27**
  nucleoside 2′-hydroxyl
    24, 268 suppl. **27**
  steroids **27**, 227; **29**, 188
    14-epimerization **24**, 120
    suppl. **26**
    17-side chain epimerization
    29, 924
  sulfoxides **27**, 126
-, prevention of – during
  oxidation **16**, 319 suppl. **26**
- with formation of oximes
  28, 343
erythro-threo-**Epimerization**
29, 163
Epimers
- special s.
  N-heterocyclics,
  N-condensed, epimeric
Episulfones
- startg. m. f.
  ethylene derivs., review
  20, 609 suppl. **26**
Episulfoxides
- startg. m. f.
  thiolsulfinic acid esters
  (from 2 molecules)
  26, 561
Epoxidation (s. a. Oxido
  compds. from ethylene
  derivs.)
-, preferential **30**, 83
Epoxides s. Oxido...
Eschweiler s. Clarke
Ester condensation s. Dieckmann cyclization
- –, mixed, on polymer
  support **6**, 738 suppl. **27**
- enolates **27**, 843 suppl. **29**
Esterification s. Carboxylic
  acid esters from carboxylic
  acids, Redoxesterification
Esters, active s. Carboxylic
  acid esters, active
Estra-1,3,5(10)-trienes
- from
  steroidal ring B 2,3-oxido-
  alcohols **30**, 660
**2H-1,4-Ethanobenzo[b]-
[1,5]naphthyridines,
3,4-dihydro-10-aryl-**
- startg. m. f.
  6H-benz[b]indolo[3,2,1-d,e]-
  [1,5]naphthyridines
  30, 301
Ethanolamine as reducing
  agent **28**, 16
**7a,11a-Ethanooxyindolo(1,2-h]-
[1,7]naphthyridin-10(11H)-
ones**
-, 6,7,8,9-tetrahydro-
- from
  pyrido[1,2-a]indol-9(8H)-ones,
  6,7-dihydro- **30**, 300
  3-quinuclidones **30**, 300

**2,4a-Ethanophenanthrenes,
2,3,4,4a,9,10-hexahydro-**
- from
  2H-naphtho[1,2-b]pyran-
  2-ones, 5,6-dihydro-
  29, 884
  α-tetralones **29**, 884
Etheno bridges, elimination
  26, 987
Etherates s. Alane etherates
Etherification of phenols
-, partial via total **18**, 280
  suppl. **29**
Ethers (s. a. Alkoxy...,
  O-Alkyl..., Aryloxy...)
-, cleavage **28**, 9, 15; **29**, 7;
  30, 345
-, –, with rearrangement,
  skeletal **29**, 287/8
-, –, with ring closure **26**, 244
-, –, partial, review **21**, 30
  suppl. **29**; **27**, 10 suppl. **28**, 29
-, –, preferential **9**, 29 suppl. **26**
- from
  acetals **26**, 59
  alcohols **27**, 189
  – and phosphorous acid
    diesters **27**, 195
  1,2-alkoxyhalides **25**, 64
    suppl. **28**
  ammonium acetates,
    quaternary **26**, 182
  – hydroxides, quaternary
    26, 182
  carbonic acid esters **27**, 273
  ethylene derivs. **23**, 144
  – –, preferential addition
    27, 148
  oxo compds. **24**, 126
  sulfenic acid esters **27**, 272
  sulfonium salts **27**, 224
  thioethers **28**, 174
-, 1-monohalogenation **26**, 526
-, rearrangement with ring
  closure **30**, 477
- special s.
  acetylene-ethers
  amino-
  9-anthronyl ethers, 9-phenyl-
    benzyl-ethers
  butyl –
  2-cyclopropen-1-yl ethers,
    2,3-diphenyl-
  2,4-dinitrophenyl ethers
  ethylene-
  glycidyl –
  halogenomethyl –
  hydroxy-
  phenol-
  quinol-
  β,β,β-trichlorethyl –
  trifluoromethyl –
  trityl –
- startg. m. f.
  acoxy compds. **26**, 509
    suppl. **28**; **27**, 248; **29**, 232
  alcohols **26**, 14
  amines, ar. **27**, 459
  bis(halogenosulfonyl)imides
    27, 409
  ethylene derivs.
    s. Dealcoholation
  halides **27**, 576
-, cyclic (s. a. O-Heterocyclics)
  22, 280
- from
  1,1-alkoxyhalides, cyclic,
    synthesis **9**, 881 suppl. **26**

diol monosulfonates **26**, 236
diols **26**, 238; **27**, 941
  suppl. **29**
disulfonates **7**, 355 suppl. **28**
ethylenealcohols **27**, 171;
  29, 172
halogenooxo compds. **30**, 171
lactols **29**, 745
– and phosphonic acid
  esters **30**, 613
lactones **27**, 69
- special s.
  crown polyethers
  2,2-dialkoxyethers, cyclic
  oximinoethers, –
  tosylhydrazonoethers, –
- startg. m. f.
  acoxyhalides **26**, 509
  alcohols **27**, 37
  –, prim. **26**, 663
  diacoxy compds., with
    rearrangement **28**, 89
  2-(ω-halogenalkoxy)halides
    27, 553
-, **sym.**
- from
  alcohols **26**, 181
  diazo compds. **30**, 123
-, β,γ-**unsatd.**, cleavage,
  oxidative **26**, 116
C-Ethoxalylation 30, 247
Ethyl alcohol 26, 40; **27**, 562;
  28, 23; **29**, 98
– as additive **29**, 864
Ethylation s. Cyanoethylation
Ethyl benzoate as medium
  27, 679
Ethyl carbonates
- special s.
  diol mono(ethyl carbonates)
Ethyl chloroformate
  (s. a. Chloroformic acid
  esters)
- as reactant **27**, 933; **28**, 66,
  693; **30**, 303
- as reagent **26**, 497; **27**, 193;
  30, 648
- startg. m. f.
  chlorides **28**, 508
  β-ketocarboxylic acid ethyl
    esters **27**, 831
Ethyl cyanoacetate
- special s.
  EMCE
Ethyldiisopropylamine
  as reagent **25**, 141 suppl. **27**;
  27, 79, 417, 471, 584, 875;
  30, 391, 638
- **hydrobromide**
- as proton acid, weak **29**, 457
Ethylene
-, β-acoxyethylation with –
  29, 675
- as reactant **22**, 705 suppl. **27**,
  28; **29**, 884
- as reagent **26**, 734
- startg. m. f.
  β-halogenoketones **2**, 617;
    27, 213
Ethyleneacetals (s. a. Ethyleneketals)
- from
  1,1-halogenothioethers,
    cyclic **29**, 972
α,β-**Ethyleneacetals**
-, hydroformylation **4**, 667
  suppl. **28**

- startg. m. f.
  aldehydes **27,** 59
  2-ethyleneethers **9,** 902
  suppl. **29**
β,γ-**Ethyleneacetals**
- from
  enecyclopropyl ethers
  29, 145
Ethyleneacylamidines
- special s.
  N'-(3-aminoacryloyl)-
  formamidines
Ethyleneacylamines
- from
  amines, tert., cyclic **26,** 300
- special s.
  enacylamines
- startg. m. f.
  acylaminoalcohols, cyclic
  29, 313
2-Ethyleneacylamines
-, alkylation of arenes with -
  13, 696 suppl. **26**
- from
  2-ethyleniminoesters **29,** 325
Ethylenealcohols
- from
  acetylene derivs. **29,** 139
  ethers, cyclic **28,** 5
  O-heterocyclics **26,** 30
  hydroperoxides **30,** 661
  lactols and phosphonium
  salts **28,** 852
  oxido compds., abnormal
  ring opening **28,** 6
- special s.
  dienols
  enols
  (vinylcyclopropyl)carbinols
- startg. m. f.
  chloral acetals **29,** 142
  diol monoesters **27,** 137
  diols **29,** 142
  ethers, cyclic **27,** 171;
  29, 172
  O,O-heterocyclics (from
  2 molecules) **29,** 144
  hydrocarbons **29,** 933
  hydroxydiketones **27,** 184
  oxo compds. **28,** 128
-, **isomeric**
- from
  3-ethyleneoxido compds.
  28, 32 suppl. **29**
2-Ethylenealcohols
-, allyl rearrangement **2,** 626;
  4, 677; **24,** 741
-, decarbonylation **27,** 94
- from
  acetylene derivs. **29,** 139
  acetylene derivs. and oxo
  compds. **28,** 652
  aldehydes and phosphonium
  salts **28,** 855
  allenes **27,** 136
  boranes, synthesis with
  addition of 4 C-atoms
  27, 896
  cyclopropane ring **26,** 132
  1,3-ditosylates **26,** 943
  α,β-ethylenealdehydes,
  reduction **27,** 36, 40
  ethylene derivs. **2,** 144
  α,β-ethyleneketones,
  reduction **26,** 809
  β,γ-ethylenesulfoxides,
  rearrangement **30,** 406
  1,2-halogenhydrins **29,** 964

  ketals **28,** 904
  β,γ-oxidocarboxylic acids
  27, 982
  oxido compds. **28,** 4, 471;
  29, 1
  2-oxidohalides **29,** 968
  β,γ-oxidonitriles **27,** 8
-, isomerization **30,** 101
- special s.
  cyclopropanols, 1-vinyl-
- startg. m. f.
  acoxy-2-ethylenes with allyl
  rearrangement **29,** 191
  alcohols, synthesis **29,** 660
  β-aminoketones **27,** 370
  3-aryloxo compds. **24,** 884
  suppl. **27**
  chromans **27,** 790a
  α,β-ethylenealdehydes
  26, 194
  γ,δ-ethylenecarboxylic acid
  esters **27,** 795
  ethylene derivs., double
  bond migration **30,** 41
  β,γ-ethylenehalides **24,** 840;
  27, 578; **28,** 505
  α,β-ethyleneketones **28,** 209
  suppl. **30**
  β,γ-ethylenesulfoxides
  29, 572
  2-ethylenisothiouronium
  salts, rearrangement
  30, 41
  cis-glycols, cyclic **29,** 142
  β-halogenooxo compds.,
  rearrangement **21,** 601
  oxetane ring with ring
  contraction **26,** 146
  oxido compds. **29,** 1
-, synthesis via α-alkylation
  of β,γ-ethylenesulfoxides
  29, 217
-, -, stereospecific **29,** 217
-, **cyclic**
- startg. m. f.
  ketocarboxylic acids **30,** 148
trans-**2-Ethylenealcohols,
  epimerized**
- from
  cis-2-ethylenealcohols
  30, 406
3-Ethylenealcohols
- from
  acetylene derivs. and
  oxido compds. **29,** 869
  2-acetylene-1,4-diol
  monoesters **29,** 567
  dienolesters **8,** 66
  2,4-dienones **29,** 32, 928
  (1,3-dien)-1-oxysilanes **27,** 83
  ethylene derivs. and oxo
  compds. **29,** 619
  α,β-ethyleneketones **27,** 83
  ketones **28,** 601
- startg. m. f.
  acoxy-2-ethylenes **27,** 243
  4H-1,3-oxazines, 5,6-dihydro-
  26, 309
-, **linear 18,** 736 suppl. **26**
4-Ethylenealcohols
- startg. m. f.
  furans, tetrahydro-
  2-halogenomethyl- **29,** 495
β,γ-**Ethylenealdehyde derivs.**
- from
  enecyclopropyl ethers
  29, 145

Ethylenealdehydes 28, 883
- special s.
  dienals
  enynals
  ethylenehalogenaldehydes
- startg. m. f.
  ketones, cyclic **26,** 737
  α,β-**Ethylenealdehydes 28,** 657
  suppl. **29**
- by substitution, electrophilic
  26, 839
- from
  2-acetylenealcohols **20,** 538
  suppl. **27**
  1,3-bis(alkylthio)-1-ethylenes
  26, 862
  carboxylic acid chlorides,
  synthesis with addition
  of 2 C-atoms **29,** 846
  enols **29,** 907
  2-ethylenealcohols **26,** 194
  α,β-ethylene-β-halogen-
  aldehydes **28,** 58
  1-ethynylacoxy compds.
  28, 191
  halides, synthesis with
  addition of 3 C-atoms
  26, 862
  β-hydroxyacetals **29,** 846
  ketones, synthesis with
  addition of 2 C-atoms
  5, 523 suppl. **26**
  oxo compds., synthesis with
  addition of 2 C-atoms
  29, 867
  silylethynyl ketones **29,** 946
- special s.
  amino-α,β-ethylenealdehydes
  2-methylenaldehydes
- startg. m. f.
  aldehydes **6,** 89 suppl. **29;**
  27, 59
  carboxylic acid esters **28,** 94
  chromenes **27,** 785
  α,β-dihalogenaldehydes
  27, 220
  2-ethylenealcohols, reduction
  27, 36, 40
-, synthesis with addition
  of 3 C-atoms **26,** 862
- via Claisen rearrangement
  25, 576 suppl. **27**
- from
-, **cyclic**
  2,3-oxidoalcohols, cyclic,
  ring contraction **29,** 930
-, **α-subst.**
- from
  aldehydes and enolethers
  26, 802
β,γ-**Ethylenealdehydes 28,** 657
  suppl. **29**
- from
  β,γ-ethylenehalides **29,** 901
  γ,δ-**Ethylenealdehydes 26,** 742
  suppl. **27, 30; 27,** 738
  suppl. **30**
- from
  ketones, synthesis **26,** 900
trans-γ,δ-**Ethylenealdehydes**
- from
  halides **28,** 842
2-Ethylenealkoxyamines
- from
  2-ethyleneamine N-oxides
  30, 101

Ethyleneamidines
- special s.
  enamidines
α,β-Ethyleneamidines
- from
  β-halogenonitriles **30,** 219
Ethyleneamin... s. a. Aminoethylene...
2-Ethyleneamine N-oxides
- startg. m. f.
  2-ethylenealkoxylamines
  30, 101
Ethyleneamines
-, halogenamination, intramolecular, of – **17,** 984
- special s.
  dienamines
  enamines
- startg. m. f.
  aziridine ring, N-condensed
  27, 500
  N-heterocyclics **27,** 345
-, ar.
- startg. m. f.
  N-heterocyclics **17,** 395
2-Ethyleneamines
- from
  α,β-ethyleneazomethines
  27, 43
  α,β-ethylenenitriles **27,** 41
  1,1-halogenamines **29,** 833
2-Ethylene-*tert*-amines
- startg. m. f.
  1,3-dienes **28,** 937
Ethyleneazides
- special s.
  enazides
2-Ethyleneazides
- from
  ethylene derivs. **26,** 325
- special s.
  azidostyrenes **20**
Ethyleneazomethines
- special s.
  enazomethines
α,β-Ethyleneazomethines
-, 1,2- and 1,4-addition
  to – **29,** 638
-, α- and γ-alkylation **29,** 638
  suppl. **30**
- startg. m. f.
  amines, sec. **27,** 43
  azomethines, synthesis
  30, 622
  3-chlor-1-ethylenecarbamyl
  chlorides **27,** 545
  1,3-dienecarbamyl chlorides
  27, 545
  1,3-dienes **26,** 829
  2-ethyleneamines **27,** 43
Ethyleneboranes
- special s.
  eneboranes
2-Ethyleneboranes
- startg. m. f.
  1,4-dienes **28,** 882
α,β-Ethyleneboronic acid esters
- from
  acetylene derivs. and
  boronic acid esters
  29, 497
- acids
- startg. m. f.
  α,β-ethylenehalides **29,** 497
Ethylene bromide (s. a. under Activator)
- as reagent **28,** 895

- startg. m. f.
  magnesium bromide,
  anhydrous **27,** 161; **30,** 493
Ethylenecarbohydrates
-, rearrangement **27,** 738
Ethylenecarboxylic acid amides
- special s.
  dienecarboxylic acid amides
α,β-Ethylenecarboxylic – –
- from
  isonitriles and acetylene
  derivs. **26,** 688
- special s.
  β-acoxy-α,β-ethylenecarboxylic acid amides
  α-acylamino-α,β-ethylene- – –
  β-amino-α,β-ethylenecinnamanilides
  α-cyano-α,β-ethylenecarboxylic acid amides
  α,β-ethylenehalogeno- – –
  α,β-ethylene-β-mercapto- – –
  α,β-ethylene-β-(1-tetrazolyl)- – –
- startg. m. f.
  β-halogenonitriles **28,** 502
- **azides**
- startg. m. f.
  eneurethans **29,** 237
  ketones **29,** 237
- – **chlorides**
- special s.
  β-amino-α,β-ethylenecarboxylic acid chlorides
- startg. m. f.
  2-pyrone ring, 3,4-dihydro-
  26, 854
  2-pyrones **27,** 848
- – **derivs.**
-, reaction with hydrazines,
  review **14,** 424 suppl. **26**
Ethylenecarboxylic acid esters
- special s.
  alkoxyethylene-carboxylic
  acid esters
  diene- – –
  ethylenehydroxy- – –
  ethyleneketo- – –
  ethylenesiloxy- – –
- startg. m. f.
  carboxylic acid esters **26,** 35
- – –, **bicyclic**
- from
  dienones, bicyclic **26,** 126
α,β-Ethylenecarboxylic – –
3, 91 suppl. **28**
- from
  acetylene derivs. **24,** 710
  β-acoxycarboxylic acid esters
  26, 938
  carboxylic acid esters **29,** 912
  α-halogenocarboxylic acid
  esters and phosphonium
  salts **26,** 907
  α-ketocarboxylic acid esters,
  Wittig synthesis **18,** 912
  suppl. **26**
  ketones **26,** 938
  α-silylcarboxylic acid esters
  28, 856 suppl. **29**
- special s.
  α-acylamino-α,β-ethylenecarboxylic acid esters
  alkoxy-α,β-ethylene- – –
  α-(N-alkylthiocarbonylamino)-α,β-ethylene- – –
  alkylthio-α,β-ethylene- – –

β-amino-α,β-ethylene- – –
α-azido-α,β-ethylene- – –
β-carbalkoxyenephosphonic – –
α,β-ethylenehalogenocarboxylic – –
α,β-ethylene-α-mercapto- – –
α,β-ethylene-α-nitro- – –
- startg. m. f.
  carboxylic acid esters
  30, 602
  β,γ-ethylenecarboxylic acid
  esters, α-alkylated **27,** 843
  suppl. **28; 29,** 810
  glycidic acid esters **28,** 110
  γ-lactones **28,** 720
  phospholan-3-one 1-oxide
  ring **29,** 731
  2-pyridones, 3,4-dihydro-
  30, 556
  1,4-thiazin-3-one ring
  30, 393
β,γ-Ethylenecarboxylic – –
- from
  2,4-dienecarboxylic acid
  esters, 1,6-addition **29,** 670
- – –, hydrogenation **30,** 25
  α,β-ethylenecarboxylic acid
  esters **29,** 810
- special s.
  γ-alkylthio-β,γ-ethylenecarboxylic acid esters
- – –, **α-alkylated**
- from
  α,β-ethylenecarboxylic acid
  esters **27,** 843 suppl. **28;**
  29, 810
γ,δ-Ethylenecarboxylic acid esters
- by dithio ester thio-Claisen
  rearrangement **28,** 842
- from
  2-ethylenealcohols **27,** 795
  β,γ-ethylenehalides **27,** 854
Ethylenecarboxylic acids
- from
  ketones, cyclic **30,** 88
  ketotosylates, cyclic **28,** 839
  spirolactones **27,** 505
  suppl. **28**
- special s.
  carboxyethyleneurethans
- startg. m. f.
  β-hydroxyketones, cyclic
  30, 523
  lactones **27,** 505; **28,** 117;
  30, 573
  spirolactones **27,** 505
α,β-Ethylenecarboxylic –
- from
  acetals **29,** 753
  acetylene derivs. **30,** 492
  carboxylic acids **27,** 918
  dioxolanium salts **29,** 753
  α,β-ethylene-α-dicarboxylic
  acids **30,** 53
  α,β-ethylenehalides **27,** 837
  oxo compds. **29,** 874
- special s.
  cinnamic acids
  α,β-ethylene-γ-ketocarboxylic
  acids
  α,β-ethylene-α-mercaptocarboxylic acids
- startg. m. f.
  alcohols **18,** 947 suppl. **29**
  carboxylic acids **26,** 51, 54;
  28, 39

– –, α-**H-labeled**
– from
α-aminocarboxylic acids
29, 950
cis-α,β-**Ethylenecarboxylic acids**
– from
methyl α,α'-dihalogenoketones **30**, 131
β,γ-**Ethylenecarboxylic –**
– startg. m. f.
γ-halogeno-β-lactones
27, 561
γ,δ-**Ethylenecarboxylic –**
28, 683, 713
α,β-**Ethylenecarboxylic acid thioamides**
– from
alkylidenephosphoranes, isothiocyanates, and oxo compds. **26**, 903
– special s.
β-amino-α,β-ethylenecarboxylic acid thioamides
α,β-**Ethylenechlorides**
– startg. m. f.
1,1-chlorofluorides **29**, 496
Ethylene deriv.-oxo compd. interconversion 29, 978
Ethylene derivs. (s. a. Double bonds)
–, addition to azodicarboxylic acid esters **21**, 349 suppl. **27**
–, addition to α,β-ethyleneβ-dicarboxylic acid esters **26**, 722
–, aminomethylation **29**, 833
–, arylation **27**, 871
–, –, oxidative **23**, 757 suppl. **28**
– by dehydrogenation CC⇑H
deoxymercuration **29**, 977
extrusion, twofold **26**, 988
fragmentation **28**, 979
–, C-carboxymethylation
28, 713
–, codimerization **24**, 711 suppl. **29**
–, 1,2-dialkylation **28**, 791
–, disproportionation, review **30**, 467 suppl. **30**
– from
acetylene derivs., review
7, 95 suppl. **29**
– – (*cis*-ethylene derivs.)
7, 95 suppl. **26**; **27**, 51;
29, 48
– – (*cis*- and *trans*-ethylene derivs.) **11**, 87 suppl. **27**
– –, synthesis **23**, 693
suppl. **28**; **29**, 672; **30**, 479
– –, – (*trans*-ethylene derivs.)
28, 660
– –, – (*cis*- and *trans*-ethylene derivs.) **23**, 703 suppl. **29**
acetylenehalides, synthesis (*trans*-ethylene derivs.)
28, 820
acoxy compds. **27**, 927
acoxy-2-ethylenes, synthesis
28, 740
1,2-acoxyhalides **27**, 931
1,2-acoxythioethers **29**, 974 suppl. **30**
2-acylaminophoshonic acid esters **29**, 880

acylimines and phosphonic acid esters **29**, 880
alcohols **27**, 917, 937; **29**, 762
aldehydes **29**, 762
1,2-alkoxyhalides **29**, 957
allenes **15**, 536 suppl. **26**
– (*cis*-ethylene derivs.)
20, 63 suppl. **26**
–, synthesis **23**, 693 suppl. **28**
amines, tert. **30**, 669
aziridines **26**, 964
azomethines, synthesis
13, 784; **28**, 767
1,1-bi(phosphonium salts) and aldehydes **27**, 891
boranes **28**, 965
carboxylic acid esters
CC⇑O
– – –, decarbalkoxylation
29, 989
1,1-chlorothioethers **30**, 687
cyclobutane ring (2 ethylene deriv. molecules) **27**, 704
cyclopropenes and trialkylboranes **29**, 877
cyclopropyl azides **27**, 524 suppl. **29**
vic-dicarboxylic acid anhydrides **29**, 991
diazo compds. (*cis*- or *trans*-ethylene derivs.) **30**, 668
dienes **29**, 54
1,3-dienes **24**, 66 suppl. **29**
1,2-dihalides CC⇑Hal
α-diketones **27**, 34
1,2-dimesylates **29**, 915
1,2-dinitro compds., aliphatic **27**, 956
1,3-diol monoesters, fragmentation **26**, 985
diols **26**, 949
1,3-dioxolanes, 2-amino-
27, 938
1,1-disilyl compds. and ketones **30**, 615
enamines, synthesis **26**, 870
enolesters, – **29**, 746
enolethers, – **26**, 869
2-ethylenealcohols with double bond migration
30, 41
α,β-ethylenehalides, synthesis **26**, 875
–, – with rearrangement
24, 838 suppl. **28**
–, – (*trans*-ethylene derivs.)
27, 837 suppl. **29**
β,γ-ethylenephosphonic acid esters and halides
(*trans*-ethylene derivs.)
30, 43
α,β-ethylenesulfones, reduction **23**, 784 suppl. **28**
– and boranes **30**, 623
α,β-ethylenesulfoxides and boranes **30**, 623
2-ethylenisothiouronium salts **30**, 41
glycols **28**, 925
halides (s. a. Dehydrohalogenation) CC⇑Hal
–, synthesis **30**, 586, 607
halogenhydrins **27**, 970
α-halogenoketones with decarbonylation **29**, 984
α-halogenophosphonium salts and oxo compds.
30, 620

hydrazinium salts **30**, 667
β-hydroxycarboxylic acids
27, 981; **28**, 985
hydroxylamines **27**, 960
2-hydroxysilanes (*cis*- and *trans*-ethylene derivs.)
28, 964 suppl. **30**
β-hydroxysulfoximines
29, 855
2-hydroxythioethers **27**, 979;
29, 974
ketimines, synthesis **27**, 815
ketones **17**, 208; **27**, 949;
28, 65
β-lactones **27**, 981
mercaptans **29**, 972
methylene groups (2 groups)
19, 815
1,2-nitrosohalides **26**, 973
1,3-oxathiolan-5-ones **26**, 988
oxetanes **29**, 978
oxido compds. **27**, 931/2, 942; **30**, 652, 662
phosphonium salts and 1,1-hydroxysulfonic acids
28, 853
pyrimidines **26**, 499
selenides **29**, 912
silanes and oxo compds.
28, 856
sulfido compds. **27**, 976
sulfones **29**, 973
sulfonic acid esters
methanesulfonic acid esters **16**, 933 suppl. **28**;
26, 937, 945; **28**, 906;
29, 622
toluenesulfonic acid esters
27, 936; **28**, 899
sulfonium salts and diazo compds. **27**, 876
sulfonylsulfilimines **26**, 978
sulfoxides **30**, 685
sulfoximines and oxo compds. **29**, 855
sulfuric acid monoesters with C-alkyl migration
30, 34
Δ³-1,3,4-thiadiazolines
30, 566
thiirane 1-oxides **29**, 767
thioenolethers, synthesis
29, 854
thioethers (s. a. Dethiolation)
28, 960; **29**, 972; **30**, 685, 687
thionocarbamic acid esters
26, 931
thionocarboxylic – –
29, 947
tosylhydrazones **27**, 949
–, geoisomerization **7**, 738;
27, 942
–, –, regiospecific **7**, 3 suppl. **28**
–, hydroallylation **29**, 139
–, hydrogenation HC⇓CC
–, hydroperoxymercuration
26, 637
–, interchange **30**, 467
–, oxidation with mercury(II) salts, review **5**, 162 suppl. **27**
–, peroxymercuration **26**, 637
–, reactions with carbon monoxide, review **16**, 750 suppl. **27**
– special s.
acoxyethylenes
alkylidene...
allyl...

(Ethylene derivs.
– special s.)
cumulenes
cycloalkenes
deuterioethylene derivs.
diarylethylenes
dien...
di(siloxy)ethylene derivs.
en...
isocyclics with bridgehead double bonds
methyl groups, allylic
nitroethylene derivs.
olefins
polyenes
propenyl...
stilbenes
styr...
vinyl...
– startg. m. f.
acetals **26**, 143; **29**, 148
acetylene derivs., terminal, synthesis with addition of 2 C-atoms **29**, 673
acoxy compds. **29**, 158
– –, preferential addition **22**, 143 suppl. **29**
– –, prim. **29**, 158
acoxy-2-ethylenes **29**, 173
– with double bond shift **29**, 180
2-acoxyselenides **29**, 180
1-acyloximino-2-nitrites **29**, 336
alcohols **26**, 139; **27**, 142; **29**, 150
–, cleavage **27**, 149
–, synthesis **26**, 914 suppl. **27**; **27**, 714
–, –, with addition of 1 C-atom **4**, 667 suppl. **26**
–, tert., synthesis by coupling of – **26**, 914 suppl. **27**
aldehydes, cleavage, oxidative **30**, 149, 154
1,2-alkoxyhalides **26**, 513; **28**, 479
γ-alkoxyketones **28**, 823
1,2-alkoxyperoxides **30**, 77
amines, cleavage **27**, 149
–, synthesis with addition of 1 C-atom **27**, 723
trans-2-arylalcohols **28**, 865
azetidine ring **29**, 659
azides **30**, 222
α-azidoketones **27**, 362
1,2-azidonitric acid esters **28**, 284
aziridine ring **22**, 123 suppl. **28**
– –, N-acyl- **29**, 398
aziridine ring, N-cyano- **26**, 521
aziridines **28**, 458
–, N-cyano- **28**, 282
Δ¹-azirines **18**, 331 suppl. **27**
1,3,2-benzodioxaboroles **26**, 638
boranes, synthesis **27**, 689
borinic acid esters **27**, 649
carboxylic acids, cleavage, oxidative **27**, 246
– –, synthesis with addition of 1 C-atom, review **16**, 750 suppl. **26**
1,2-chloroiodides **26**, 514
chromans **27**, 760

compds., vic-difunctional, review **17**, 590 suppl. **28**
cyanimines **28**, 282
β-cyanosulfones **28**, 638
cyclobutane ring, dimerization **27**, 704
cyclobutanone – **30**, 487
cyclopentanones **27**, 868 suppl. **28**
cyclopropane ring **26**, 803, 876; **27**, 833; **28**, 780; **29**, 827
– –, 1,1-dihalogeno- **27**, 833
– –, halogeno- **28**, 806
cyclopropanes **26**, 847, 898; **27**, 864; **28**, 815; **29**, 863, 872; **30**, 604
–, cycloisomerization **26**, 759
1,3-dienes **29**, 911
– (from 2 molecules) **26**, 727
1,2-dihalides HalC⇓CC
dihalides with rearrangement **26**, 517
dihalogenoboranes **19**, 592 suppl. **28**; **27**, 485 suppl. **28**
α-diketones **27**, 186
1,3-diol esters **28**, 102
1,3-diols **28**, 102
1,4-dioxanes **28**, 114
1,2-dioxetane ring **30**, 86
1,3-dioxolanes **29**, 174
1,2-disilanes **26**, 650
dithiocarboxylic acid esters **29**, 538
dithiourethans **27**, 594
eneurethans **28**, 376
ethers **23**, 144
–, preferential addition **27**, 148
2-ethylenealcohols **2**, 144
3-ethylenealcohols, synthesis **29**, 619
2-ethyleneazides **26**, 325
γ,δ-ethylene-α-dicarboxylic acid esters **29**, 722
1,2-ethylene-1,3-dihalides **29**, 793
β,γ-ethylenehalides **27**, 569
α,β-ethylene ketones **28**, 131
β,γ-ethyleneketones **27**, 762; **30**, 600
α,β-ethylenesulfones **21**, 642 suppl. **26**
fluorosulfonic acid esters **30**, 78
1,2-formoxyhalides **28**, 480
furan ring, tetrahydro- **29**, 794
glycol esters **29**, 152
– ethers **28**, 114
glycols **27**, 147, 150
trans-glycols **27**, 152
halides HalC⇓CC
–, synthesis **30**, 481, 607
2-halogenacylamines **28**, 478
2-(ω-halogenalkoxy)halides **27**, 553
2-halogenamines **27**, 548
1,2-halogenhydrins **26**, 144
cis-1,2-halogenhydrins **12**, 587 suppl. **26**
halogenoboranes **27**, 649
2-halogenocyanamides **26**, 521
α-halogenoketones **28**, 481
1,2-halogenoperoxides **30**, 350

α-halogenophosphonic acid dihalides **29**, 599
2-halogenophosphoramidates **27**, 548
2-halogenosulfonyl isocyanates **26**, 613
2-halogenothiocyanates **29**, 540
1,2-halogenurethans **27**, 554
hydrocarbons, degradation **30**, 46
hydroperoxides **27**, 138
β-hydroxycarboxylic acids **30**, 496/7
α-hydroxyketones **27**, 184
2-hydroxyselenides **29**, 601
imidazoles, 2,4,5-trialkyl-, sym. **28**, 665
immoniocyclobutane ring **28**, 822
1,2-iodohydrin nitrates **26**, 518
isocyclics **27**, 731
isothiazolid-3-one 1,1-dioxides **26**, 613
α-ketocarboxylic acid ester mercaptals **29**, 661
ketones **27**, 155, 157; **28**, 282, 481
–, cleavage, oxidative **27**, 246 suppl. **28**
–, synthesis **27**, 716; **28**, 648; **29**, 772, 896; **30**, 488
–, sym. (from 2 molecules) **30**, 495
γ-lactones **26**, 707; **27**, 908
mercury compds., monoorgano- **26**, 641
monodeuterio compds. **23**, 63 suppl. **27**
nitriles **14**, 784
1,2-nitroacylamines **28**, 286
nitro compds., aliphatic **27**, 897
1,1-nitroethylene derivs. **29**, 318
1,2-nitrohalides **27**, 550
α-nitroketones **29**, 318
1,2-nitronitroso compds. **29**, 318
1,2-oxazetidines **27**, 329
1,2-oxazine ring, tetrahydro- **28**, 657
4H-1,3-oxazines, 5,6-dihydro- **26**, 885
oxido compds. **26**, 136, 141, 144, 584; **27**, 153, 184; **28**, 113; **29**, 133
oxo compds. with rearrangement **27**, 141
peroxides, unsym. **26**, 128; **29**, 141
N-phosphorylaziridines **29**, 411
polyhalides, synthesis **19**, 776
polyhalogenocarboxylic acid chlorides **22**, 717 suppl. **26**
pyrazolidines **26**, 780
Δ¹-pyrrolines **28**, 606; **30**, 628
silanes s. Hydrosilylation
sulfones, sym. **27**, 599
β-sultones **30**, 389
tetrazoles, azido- **29**, 317
thioethers **27**, 637
thiolic acid esters **27**, 597

trialkylcarbinols, synthesis
  27, 716
2,4,9-trioxaadamantanes,
  7-halogeno- **29**, 839
–, synthesis **29**, 809
–, –, geospecific, review
  26, 809
–, via sulfenes and episulfones,
  review **20**, 609 suppl. **26**
cis- and trans-**Ethylene derivs.**
– from
  1,2-dihalides **2**, 822 suppl. **26**;
  6, 882 suppl. **26**
Ethylene derivs., cyclic
– startg. m. f.
  cyanoketones **26**, 325
  dialdehydes **29**, 147
  diamines **27**, 149
  diols **27**, 149
– –, **electron deficient**
–, reactions with ethylene
  derivs., electron-rich **30**, 467
– startg. m. f.
  Δ¹-pyrrolines **26**, 708
– –, **electron-rich** (s. a. Bis-
  2-imidazolidinylidenes)
–, dimerization, oxidative
  26, 721
–, reactions with compds.,
  proton-active, review
  22, 874 suppl. **28**
–, – – ethylene derivs.,
  electron-deficient **30**, 467
– startg. m. f.
  enolbetaines **26**, 917
– –, **hydroborated,** reaction
  with sodium cyanide
  27, 716
– –, **non-activated,** syntheses
  with – **28**, 657
– –, **strained** s. Benzvalenes
– –, **sym.**
– from
  alkylidenephosphoranes
  (2 molecules) **28**, 863
  diazo compds. (cis-ethylene
  derivs.) **18**, 862 suppl. **26**
  1,1-dihalides **5**, 594
  nitriles (2 molecules)
  27, 901
  nitro compds., aliphatic
  30, 575
  oxo compds. (2 molecules)
  28, 761; **30**, 561, 566
– special s.
  1,2-diquinonylethylenes
  fulvalenes
– –, **terminal**
  (s. a. Methylene...)
– from
  acetylene derivs. **30**, 492
  halides, prim. **23**, 964
    suppl. **26**
  2-hydroxythioethers **27**, 979
  ketones, C-cleavage **25**, 695
    suppl. **26**
–, hydrogenation, homogeneous
  29, 49
–, hydrosilylation,
  preferential **26**, 642
– special s.
  perfluoroalkenes, terminal
– startg. m. f.
  alcohols, prim. **16**, 194
– –, **trisubst.**
– from
  acetylene derivs. and
  boranes **30**, 623

–, prepn., geospecific, review
  25, 634 suppl. **26**
– –, –, trans- **27**, 795; **28**, 749
– –, –, **functionalized,** synthesis
  geospecific **28**, 499 suppl. **29**
– –, –, **geoisomeric,** synthesis
  27, 938 suppl. **29**
Ethylenediamine
– as complexing agent **22**, 80
– as reagent **5**, 441 suppl. **26**;
  29, 673, 820
Ethylenediamines
s. Enediamines
**Ethylenediaminetetraacetic
acid**
–, disodium salt **27**, 421;
  28, 92, 153
Ethylenediazo compds.
–, ring closure, double **29**, 949
– startg. m. f.
  acoxy compds., cyclic
  30, 571
Ethylene dibromide
s. Ethylene bromide
α,β-**Ethylenedicarboxylic acid
esters**
– from
  acetylenedicarboxylic acid
  esters **28**, 626
α,β-**Ethylene-α-dicarboxylic –**
– special s.
  EMME
α,β-**Ethylene-β-dicarboxylic –**
–, addition of ethylene derivs.
  to – **26**, 722
β,γ-**Ethylene-β-dicarboxylic –**
– from
  acetylenedicarboxylic acid
  esters **26**, 711
γ,δ-**Ethylene-α-dicarboxylic –**
– from
  malonic acid esters and
  ethylene derivs. **29**, 722
α,β-**Ethylene-α-dicarboxylic
acids**
– startg. m. f.
  α,β-ethylenecarboxylic acids
  30, 53
1,2-Ethylene-1,1-dihalides
(s. a. Dihalogeno-
methylene...)
– startg. m. f.
  1,3-dithietanes, 2,4-diene-
  30, 405
  β-imidocarboxylic acids,
  α-branched **26**, 888
1,2-Ethylene-1,2-dihalides
– special s.
  1,2-dihalogenovinyl groups
– startg. m. f.
  hydrocarbons **28**, 62
1,2-Ethylene-1,3-dihalides
– from
  ethylene derivs. **29**, 793
2,3-Ethylene-1,4-dihalides
– from
  1,3-dienes **29**, 490
Ethylene diiodide
– as reagent **15**, 473
  suppl. **26**
Ethylenediketones
s. a. Enediones
β,γ-**Ethylene-β'-diketones**
– startg. m. f.
  β,γ-ethyleneketones **29**, 85
2-Ethylene-1,4-dinitriles
– from
  1,3-dienes **28**, 662

**2,3-Ethylene-1,5-diol
1-(ethyl carbonates)**
– startg. m. f.
  2,4-dienones **29**, 928
1,2-Ethylene-1,1-disulfones
– from
  aldehydes **29**, 748
– special s.
  2,2-disulfonylenamines
  2,2-disulfonylenolethers
2,3-Ethylene-1,1-disulfones
– from
  aldehydes **29**, 748
2-(β,γ-**Ethylene**)-1,3-**dithianes**
28, 883
2-Ethyleneenolethers
– startg. m. f.
  γ,δ-ethyleneoxo compds.
  27, 738
2-Ethyleneethers
– from
  α,β-ethyleneacetals **9**, 902
  suppl. **29**
cis-2-**Ethyleneethers**
– from
  acetyleneacetals **29**, 61
α,β-**Ethylene-N-fluoroimines**
– from
  3-halogeno-N,N-difluoro-
  amines **28**, 459
α,β-**Ethylene-β-(2-furyl)-ketones**
– startg. m. f.
  β-(2-furyl)propionic acids,
  5-subst. **29**, 125
Ethylene glycol
– as H-donor **29**, 50
– as medium **27**, 21
– as reactant **29**, 953
– – **dimethyl ether** s. 1,2-
  Dimethoxyethane
Ethylenehalides
–, radical ring closure
  29, 970
– special s.
  ethylenedihalides
  ethylenehalogenamines
  halogenodienes
  polyhalogenoethylene derivs.
α,β-**Ethylenehalides**
– from
  acetylene derivs. **30**, 603
– – and boronic acid esters
  29, 497
  1,2-acoxyhalides **29**, 942
  enols **11**, 941 suppl. **28**
  α,β-ethyleneboronic acids
  29, 497
  ketones **21**, 607 suppl. **27**;
  26, 541
  2-oxazolidones, 3-nitroso-
  8, 922 suppl. **26**
– special s.
  alkoxy-α,β-ethylenehalides
  α,β-ethylenechlorides
  α,β-ethyleneiodides
  1,1,1-ethylenenitrohalides
  halogenoenesilanes
  halogenomethylene...
– startg. m. f.
  acetylene derivs. **27**, 964;
  29, 961
  allenes **29**, 961
  1,3-dienes (from 2 molecules)
  27, 870 suppl. **28**
  1,6-diketones (from 2 molecules) **29**, 825
  enamines **27**, 465

(α,β-**Ethylenehalides**
- startg. m. f.)
  α,β-ethylenecarboxylic acids
    27, 837
  ethylene derivs., synthesis
    26, 875
  - -, - (*trans*-ethylene
    derivs.) **27**, 837 suppl. **29**
  - -, - with rearrangement
    24, 838 suppl. **28**
  ketones **29**, 210
  1,2-nitroethylene derivs.,
    synthesis **28**, 797
  thiazolo[3,2-a]pyridinium
    salts, dihydro- **26**, 611
  thioenolethers **30**, 413
cis- and *trans-*α,β-**Ethylenehalides 29**, 497
- by replacement, geospecific
  29, 497

β,γ-**Ethylenehalides**
- from
  2-ethylenealcohols **24**, 840;
    27, 578; **28**, 505
  ethylene derivs. **27**, 569
  halides, synthesis with
    addition of 3 C-atoms
    28, 802
- special s.
  3-chlor-1-ethylenecarbamyl
    chlorides
- startg. m. f.
  cyclopropanes **28**, 808
  β,γ-ethylenealdehydes
    29, 901
  γ,δ-ethylenecarboxylic acid
    esters **27**, 854
  γ,δ-ethylenenitriles **27**, 853
α,β-**Ethylene-β-halogenaldehydes**
-, review **29**, 841 suppl. **30**
- startg. m. f.
  α,β-ethylenealdehydes
    28, 58
  pyrylium salts **29**, 841
α,β-**Ethylene-γ-halogenaldehydes**
- from
  1-acoxy-1,3-dienes **26**, 553
2,3-Ethylene-2-halogenamines
- from
  1,1-halogenamines **23**, 650
    suppl. **27**
Ethylene-N-halogenamines
- startg. m. f.
  o-α-halogeno-N-heterocyclics **26**, 524; **27**, 530
Ethylenehalogenhydrins
- from
  allenes **27**, 555
α,β-**Ethylene-β-halogenocarboxylic acid amides**
- from
  ketones **28**, 717
- - **esters**
- startg. m. f.
  coumarin ring **26**, 878
α,β-**Ethylene-γ-halogenoketals**
- startg. m. f.
  (dien)olethers **28**, 971
α,β-**Ethylene-α-halogenoketones**
- special s.
  4,6-dihalogeno-4,6-dien-3-onesteroids
- startg. m. f.
  α,β-acetyleneketones **30**, 676

α,β-**Ethylene-β-halogenoketones**
- startg. m. f.
  α,β-ethyleneketones **30**, 39
  indenones **28**, 952
  pyrylium ring **28**, 828
γ,δ-**Ethylene-δ-halogenoketones**
- startg. m. f.
  2-cyclohexanone ring **30**, 680
Ethylenehalogenonitriles
- special s.
  α-(dihalogenomethylene)-
    nitriles
α,β-**Ethylene-α-halogenonitriles**
-. diene synthesis with -
  26, 718
- startg. m. f.
  aziridines, 2-cyano- **27**, 468
  pyrazoles, 3-amino- **30**, 298
α,β-**Ethylene-α-halogenooxo compds.**
-, diene synthesis with -
  30, 484
α,β-**Ethylene-γ-halogenooxo** -
- from
  α,β-ethyleneoxo compds.
    29, 500
α,β-**Ethylene-α-halogenosulfones**
- from
  acetylene derivs. and
    sulfonic acid halides
    27, 598
α,β-**Ethylene-β'-halogenosulfones**
- from
  allenes and sulfonic acid
    halides **29**, 537
β,γ-**Ethylene-γ-halogenosulfoxides**
- startg. m. f.
  α,β-ethyleneketones **29**, 261
3-Ethylene-3-halogenothioethers
- startg. m. f.
  α,β-ethyleneketones **28**, 961
Ethyleneheterocyclics
  s. a. Eneheterocyclics
Ethylene-N-heterocyclics
-, diene syntheses with -
  27, 700
Ethylenehydrazones
- special s.
  enehydrazones
α,β-**Ethylenehydrazones**
-, α-alkylation **27**, 832
- from
  α,β-ethyleneketones **27**, 832
α,β-**Ethylene-γ-hydroperoxy-γ-lactones**
- from
  phenol ring **26**, 227
- startg. m. f.
  ketones **26**, 227
α,β-**Ethylene-α(or β)-hydroxy...**
  s. a. Keto...
γ,δ-**Ethylene-ε-hydroxycarboxylic acid esters**
- from
  oxidocyclobutanones **28**, 101
α,β-**Ethylene-δ-hydroxycarboxylic acids**
- from
  ketones **27**, 663
α,β-**Ethylene-α-hydroxyketones**
- from
  α-halogenoketones **26**, 208

α,β-oxidoketones, rearrangement **30**, 524
- startg. m. f.
  ketones **28**, 51
α,β-**Ethylene-β'-hydroxyketones**
- from
  α-halogenoketones and
    aldehydes **26**, 257
β,γ-**Ethylene-α-hydroxyketones**
- startg. m. f.
  α-diketones, rearrangement
    29, 171
γ,δ-**Ethylene-ε-hydroxyketones**
  28, 101
- startg. m. f.
  vinylcyclopropyl ketones
    27, 945
α,β-**Ethylene-β-hydroxynitriles**
- from
  isoxazole ring **28**, 293
α,β-**Ethylene-β-hydroxythiolic acid esters**
- from
  α,β-oxidothiolic acid esters
    27, 742
Ethyleneketals
- from
  2,3-oxidoalcohols **29**, 10
α,β-**Ethyleneketals**
-, in *situ*-Claisen rearrangement with - **26**, 809
- special s.
  α,β-ethylene-γ-halogenoketals
α,β- and β,γ-**Ethyleneketals**
- from
  α,β-ethyleneketones **30**, 114
Ethyleneketocarboxylic acid esters
- special s.
  methyleneketocarboxylic
    acid esters
γ,δ-**Ethylene-β-ketocarboxylic**
  - - **19**, 766 suppl. **27**
α,β-**Ethylene-γ-ketocarboxylic acids**
- startg. m. f.
  α,β-ethylene-γ-lactones
    27, 258
Ethyleneketones
- from
  cyclopropanol ring **26**, 29
- special s.
  diazoethylene-ketones
  ethylenehalogeno-
  ethylenehydroxy-
  vinylcyclopropyl -
- startg. m. f.
  alcohols, cyclic **26**, 747;
    28, 1/2
  cyclopropanol ring **26**, 29
  diols **28**, 103
α,β-**Ethyleneketones**
-, alkylation, reductive,
  preferential **30**, 574
- by rearrangement,
  [2.3]-sigmatropic **29**, 261
-, cleavage, oxidative **26**, 228
-, cycloisomerization **30**, 508
- from
  acetylene derivs., boranes,
    and carboxylic acid
    chlorides **28**, 866
  acoxy-2-acetylenes **28**, 968
  (α-acylalkylidene)-
    phosphoranes **28**, 850
    suppl. **29**

aldehydes **29**, 752
β-aminoketones **28**, 204
cyclobutanone ring **29**, 687
enaminophosphine oxides
  27, 890
enesilanes **29**, 846
enolesters **28**, 229
2-ethylenealcohols **28**, 209
  suppl. **30**
ethylene derivs. **28**, 131
α,β-ethylene-β-halogeno-
  ketones **30**, 39
β,γ-ethylene-γ-halogeno-
  sulfoxides **29**, 261
3-ethylene-3-halogeno-
  thioethers **28**, 961
γ-halogenenolesters **28**, 229
α-halogenoketones **22**, 959
  suppl. **27**
β-hydroxycarboxylic acids
  and carboxylic acid
  anhydrides **30**, 554
β-hydroxyketones **30**, 489
γ-ketocarboxylic acids
  23, 982
3-keto-1,2-en-1-olethers
  29, 805
2-oxa-3-borolenes **28**, 866
1,2-oxido-1-halides **28**, 951
oxo compds., synthesis
  28, 857 suppl. **29**
1,2,3-trienes **26**, 133
-, isomerization **27**, 949
  suppl. **30**; **28**, 204
-, α-monoalkylation **27**, 832
-, α'-monobromination **28**, 485
- special s.
  β-acoxy-α,β-ethyleneketones
  amino-α,β-
  α-azido-α,β-
  benzalcetone
  chalcones
  δ-cyano-α,β-ethyleneketones
  α,β-ethylene-β-(2-furyl)-
    ketones
  α,β-ethylene-(α- or β-)silyl-
    ketones
  α-methyleneketones
- startg. m. f.
  γ,δ-acetyleneketones,
    synthesis **28**, 658
  carboxylic acid amides
    27, 134
  - - esters **27**, 134
  3-cyclohexanolones **30**, 478
  cyclohex-2-en-4-olone ring
    30, 546
  2-cyclohexenone ring **26**, 784
  β-deuterioketones **29**, 47
  1,5-diazocines, hexahydro-
    27, 399
  α,β-dideuterioketones **29**, 47
  (1,3-dien)-1-oxysilanes
    27, 83
  (1,3-dien)-2- **29**, 116
  γ-diketones, synthesis
    28, 175
  2,10-dioxabicyclo[4.4.0]deca-
    3,8-dienes, 1-amino- **27**, 703
  enolesters, oxidation **30**, 89
  2-ethylenealcohols, reduction
    26, 809
  3-ethylenealcohols **27**, 83
  α,β-ethylenehydrazones
    27, 832
  α,β- and β,γ-ethyleneketals
    30, 114
  β,γ-ethyleneketones **27**, 737

-, α-alkyl- **30**, 580
furans, 2,3-dihydro- **28**, 105
furylium salts, tetrahydro-
  28, 105
α-halogenoketones, synthesis
  28, 860 suppl. **29**
hydrocarbons **23**, 63 suppl. **29**
α-hydroperoxyketones,
  synthesis **8**, 742 suppl. **27**
cis-indol-6-ones, octahydro-
  28, 664
γ-ketomercaptals, synthesis
  28, 175
β-ketomercaptans **29**, 539
ketones, hydrogenation
  23, 63 suppl. **29**; **27**, 50;
  28, 45; **29**, 50
-, α,β-disubst. **29**, 819
  suppl. **30**
phenol ring **30**, 546
1,3-thiazines, 2-amino-
  28, 552
thiazoles **18**, 669 suppl. **26**
1,3,3-trialkoxy compds.
  26, 178
β-vinylketones **26**, 910
-, syntheses with - catalyzed
  by potassium fluoride **27**, 783
-, -, electrophilic, regiospecific with - **30**, 453
-, cyclic
- from
  ethylenenitriles **30**, 121
- ring contraction
  by 2 C-atoms **30**, 507
-, 4-subst.
- from
  1,3-dienamines **27**, 822
trans-α,β-**Ethyleneketones**
  28, 666
β,γ-**Ethyleneketones**
- from
  (α-acylalkylidene)-
    phosphoranes **28**, 850
  ethylene derivs. and
    oxocarbonium
    fluoroborates **27**, 762
  - - and α-halogenoketones
    30, 600
  β,γ-ethylene-β'-diketones
    29, 85
  α,β-ethyleneketones **27**, 737
-, α-monoalkylation **29**, 804
- special s.
  α-vinylketones
-, synthesis **26**, 872 suppl. **29**
- cyclic
- from
  β-hydroxycarboxylic acid
    esters **29**, 933
- startg. m. f.
  cycloalkane ring **29**, 933
-, α-alkyl-
- from
  α,β-ethyleneketones **30**, 580
γ,δ-**Ethyleneketones**
- by Claisen rearrangement
  28, 749
- from
  acetylene derivs. **23**, 705
    suppl. **28**
  - - (trans-γ,δ-ethylene-
    ketones) **23**, 705 suppl. **28**
  2,4-dienones **29**, 928
  carboxylic acids **26**, 873
  (vinylcyclopropyl)carbinols
    28, 124

- special s.
  β-vinylketones
Ethylenelactams
- from
  halogenolactams **26**, 971
- special s.
  enaminolactams
  methylenelactams
Ethylenelactones
- from
  carboxyethyleneurethans
    30, 17
- special s.
  enelactones
- startg. m. f.
  lactones **27**, 56
α,β-exo-**Ethylenelactones**
  by ring closure, double
  19, 786
α,β-**Ethylene-γ-lactones 30**, 685
- from
  α,β-ethylene-γ-keto-
    carboxylic acids **27**, 258
  γ-ketocarboxylic acids
    29, 253
  - - with rearrangement
    28, 217
  γ-lactones **28**, 895;
    29, 912 suppl. **30**
- special s.
  α,β-ethylene-γ-hydroperoxy-
    γ-lactones
β,γ-**Ethylene-γ-lactones**
- startg. m. f.
  β-acyl-γ-lactones **30**, 555
Ethylenemercaptals, rearrangement, [2,3]-sigmatropic
  29, 709
α,β-**Ethylenemercaptals**
- startg. m. f.
  ketene mercaptals **9**, 781
  suppl. **27**
- from
  ketene mercaptals **30**, 505
β,γ-**Ethylenemercaptals 29**, 709
Ethylenemercaptans
- startg. m. f.
  thioethers, cyclic **26**, 587
  suppl. **27**; **27**, 595
α,β-**Ethylene-β-mercapto-
  carboxylic acid amides**
- startg. m. f.
  4H-1,3-thiazin-4-ones,
  2,3-dihydro- **28**, 551
α,β-**Ethylene-α-mercapto-
  carboxylic acid esters**
- from
  aldehydes and silylthioacetic
  acid esters **29**, 730
α,β-**Ethylene-α-mercapto-
  carboxylic acids**
- startg. m. f.
  S-heterocyclics, alkylidene-
    29, 548
2-**Ethylenemesitoates**
-, syntheses, allylic, with -
  28, 601
Ethylenenitriles 26, 951
- from
  N-chlorosulfonyl-
    2-azetidinones **27**, 975
- special s.
  ethylenedi-nitriles
  ethylenehalogeno-
- startg. m. f.
  α,β-ethyleneketones, cyclic
    30, 121

α,β-**Ethylenenitriles**
- from
   aldehydes **27**, 891
   enolethers **27**, 789
   nitriles **29**, 912
   oxo compds. **26**, 782;
     28, 856 suppl. **29**
   ketones **17**, 892 suppl. **27**
   pyridazine ring **28**, 95
-, photoreactions with
   4-thiouracils **30**, 465
- special s.
   acrylonitrile
   α-alkylthio-α,β-ethylene-
     nitriles
   β-amino-α,β-ethylene-
   α-alkoxy-α,β-ethylene-
   1-cyano-1,2-ene-1,2-diamines
   α-methylenenitriles
   tetracyanoethylene
- startg. m. f.
   α,β-dihalogenonitriles **26**, 409
   2-ethyleneamines **27**, 41
   isoxazoles, 3-amino- **26**, 409
   3-pyrazolidones **28**, 379
   pyrrole ring, 1,2-diamino-
     26, 466
   pyrrolo[1,2-b][1,2,4]triazine
     ring **26**, 466
γ,δ-**Ethylenenitriles**
- from
   β,γ-ethylenehalides **27**, 853
Ethylenenitrites
- startg. m. f.
   oximinoethers, cyclic **30**, 97
-, cyclic
-, rearrangement **27**, 343
α,β-**Ethylene-α-nitrocarboxylic
   acid esters**
- startg. m. f.
   β-carbalkoxyenephosphonic
     acid esters **28**, 594
1,1,1-Ethylenenitrohalides
- from
   1,1-nitroethylene derivs.
     26, 528
- startg. m. f.
   α-cyanophosphonium salts
     29, 606
   1-phosphonioiminoesters
     30, 431
Ethylenenitrones
- startg. m. f.
   oxazolidine ring **20**, 284
1′,2′-**Ethylenenucleosides**
   29, 964
**1,1-Ethyleneorganolithium
   compds.**
-, reactions via - **27**, 837;
   30, 413
- special s.
   vinyllithium
**2-Ethyleneorganolithium
   compds.**
-, syntheses, allylic, via -
   28, 601
Ethylene oxide (s. a. Oxido
   compds.)
- as reagent **29**, 969; **30**, 702
- startg. m. f.
   β-halogenethyl ethers **26**, 508
Ethyleneoxido compds.
- special s.
   1,3-butadiene monoxide
3-Ethyleneoxido compds.
- startg. m. f.
   ethylenealcohols, isomeric
     28, 32 suppl. **29**

α,β-**Ethyleneoximes**
- from
   1,3-ammoniooximes **27**, 263
- startg. m. f.
   aziridine ring **29**, 14
   isoxazoles **28**, 204
Ethyleneoxo compds.
- startg. m. f.
   O-heterocyclics, ring
     closure, double **26**, 153
   isoxazolidine ring **20**, 284
α,β-**Ethyleneoxo** -
-, *peri*-annelation with -
   29, 764
- from
   alkoxyacetylenes, synthesis
     27, 835
   allenyl ethers **27**, 835
   carboxylic acid esters
     30, 505
   1,3-diol monoesters **29**, 242
   oxo compds. **29**, 912
   - - (2 molecules) **27**, 786
-, reactivity **27**, 713
- special s.
   β-amino-α,β-ethyleneoxo
     compds.
   α,β-ethylene-γ-halogenooxo -
- startg. m. f.
   benzene ring **27**, 805
   2-cyclohexenones **28**, 721
   α,β-ethylene-γ-halogenooxo
     compds. **29**, 500
   oxo - **29**, 44
   phenols **28**, 721
   prolines **29**, 664
   2-pyrones **27**, 848
   quinoline ring **26**, 823
   β-silyloxo compds. **26**, 640
β,γ-**Ethyleneoxo** -
- by cyclobutane ring
   opening **26**, 927
- from
   enecyclopropanes, 2-alkoxy-
     30, 156
γ,δ-**Ethyleneoxo** -
- from
   2-ethyleneenolethers **27**, 738
**Ethylenephosphosphonic acid
   esters**
- special s.
   enephosphonic acid esters
β,γ-**Ethylenephosphonic** - -
- startg. m. f.
   trans-ethylene derivs.
     30, 43
**Ethylene, tetra(negatively-
   subst.)** s. Tetracyanoethylene
1,1-Ethylenesilanes
- from
   acetylene derivs. **12**, 700
     suppl. **26**
β,γ-**Ethylene-β-siloxycarboxylic
   acid esters**
- from
   carboxylic acid chlorides
     and O-silyl-O-alkyl-
     keteneacetals **28**, 809
α,β-**Ethylene-α-silylketones**
-, annelation with - **29**, 666
- startg. m. f.
   2-cyclohexenone ring
     29, 666

α,β-**Ethylene-α(or β)-silyl-
   ketones**
- from
   carboxylic acid anhydrides
     29, 800
Ethylenestannanes
- special s.
   enestannanes
3-Ethylenestannanes
- startg. m. f.
   cyclopropylcarbinyl compds.
     28, 848
Ethylene sulfate, reactions
   with - **28**, 602
Ethylenesulfides s. Sulfido
   compds.
α,β-**Ethylenesulfones**,
   1,4-addition to -
     23, 693 suppl. **28**
- by Horner synthesis
   23, 879 suppl. **27**
- from
   ethylene derivs. and sulfonic
     acid chlorides **21**, 642
     suppl. **26**
   β-halogenosulfones **3**, 752
     suppl. **27**
   oxo compds. **23**, 776
- special s.
   β-amino-α,β-ethylenesulfones
   α,β-ethylene-β-halogeno-
     sulfones
- startg. m. f.
   ethylene derivs., reduction
     23, 784 suppl. **28**
   - -, synthesis **30**, 623
   1,2,3-triazoles **29**, 429
β,γ-**Ethylenesulfones**
- by coupling, oxidative
   23, 757 suppl. **29**
α,β-**Ethylenesulfones, cyclic**,
   ring expansion **26**, 584
α,β-**Ethylenesulfonic acid
   amides**
- startg. m. f.
   1,4,2-dithiazine 1,1-dioxides
     28, 533
- **acids**
- from
   sulfinic acids and haloforms
     27, 627
   sulfones **29**, 100
Ethylenesulfoxides
- special s.
   ethylenehalogenosulfoxides
α,β-**Ethylenesulfoxides**
- from
   oxo compds. **28**, 856; **29**, 733
-, Michael addition to -
   18, 965 suppl. **28**
- startg. m. f.
   ethylene derivs., synthesis
     30, 623
   2-halogenothioenolethers
     26, 546
β,γ-**Ethylenesulfoxides**
-, α-alkylation, synthesis of
   2-ethylenealcohols via -
     29, 217
- as intermediates, review
   29, 217
- from
   2-ethylenealcohols and
     sulfenylchlorides **29**, 572
   - - -, rearrangement **30**, 406
- startg. m. f.
   2-ethylenealcohols,
     rearrangement **30**, 406

Ethylenesultines
- startg. m. f.
  1,3-dienes **29,** 105
α,β-**Ethylene-*β*-(1-tetrazolyl)-carboxylic acid amides**
- startg. m. f.
  4-alkylidene-5-imidazolones **29,** 463
Ethylenethioethers
- special s.
  ethylenehalogenothioethers thioenolethers
-, synthesis via allylic metalation **29,** 619
Ethylenetosylamines, ring closure **30,** 313
Ethylenetosylhydrazones
- startg. m. f.
  S-heterocyclics, bicyclic **27,** 957
  Δ¹-pyrazoline ring **30,** 683
α,β-**Ethylenetosylhydrazones**
- startg. m. f.
  1,3-dienes **27,** 949 suppl. **30**
Ethyleneurethans
- special s.
  carboxyethyleneurethans eneurethans
Ethylenimine, N-aminoethylation with – **28,** 273
Ethylenimines s. a. Aziridines
2-Ethyleniminoesters
- startg. m. f.
  2-ethyleneacylamines **29,** 325
α,β-**Ethyleniminoesters**
- from
  acetylene derivs. and isonitriles **29,** 644
α,β-**Ethylen-*β*-isocyanato-carboxylic acid esters**
- from
  1,3-oxazin-6-ones,
  2-alkoxy- **27,** 165
- startg. m. f.
  1,3-oxazin-6-ones, 2-alkoxy- **27,** 165
2-Ethyleneisothiouronium salts
- from
  2-ethylenealcohols, rearrangement **30,** 41
- startg. m. f.
  ethylene derivs. **30,** 41
  4H-1,3-thiazines, 5,6-dihydro-2-amino- **28,** 308
Ethyl isocyanatoacetate
- startg. m. f.
  β-alkylthio-*α*-formamino-carboxylic acids **26,** 791
- **malonate**
- as nitrous acid scavenger **26,** 361
- **mercaptan** as reagent **6,** 783 suppl. **29; 29,** 210
N-Ethylmorpholine – – **28,** 398
Ethyl orthoformate
- startg. m. f.
  quinolizinones **28,** 747
  1,3,4-thiadiazolium salts **28,** 553
- **oxamate**
- startg. m. f.
  pyrazines, 2,3-dihydroxy- **27,** 511
- **thionoformate** as reactant **30,** 570
1-Ethynylacoxy compds.
- startg. m. f.
  α,β-ethylenealdehydes **28,** 191

1-Ethynylalcohols
- special s.
  1,1-diethynylalcohols
Ethynyl derivs. s. a. Acetylene derivs., terminal
Ethynyldiols
- from
  lactols **30,** 454
Ethynyl glycols, cyclic
- startg. m. f.
  α-hydroxyketones, cyclic, ring expansion **28,** 974
1-Ethynyl-N-nitrosoureas
- startg. m. f.
  carboxylic acid esters **28,** 150
1-Ethynyl-N-nitrosourethans
- startg. m. f.
  carboxylic acid esters **28,** 150
Exchange (s. a. Interchange, Replacement, Trans...)
- of
  amino groups of squaric acid diamides **28,** 366
  substituents of tert. amines **27,** 493
- special s.
  2-azetidinone ring exchange
Exchangers s. Ion exchangers
Extractor technique 26, 327
Extrusion s. a. Elimination
-, **twofold**
-, ethylene derivs. by – **26,** 988

Favorskii rearrangement
-, geospecific **30,** 131
-, review **24,** 210 suppl. **26**
Favorskii-type –
- of
  α-hydroxyketones **29,** 162
Fenton's reagent 27, 879
Ferrates s. Lithium acyl tetracarbonylferrate,
  Potassium ferrate(VI),
  – hexacyanoferrate(III),
  Sodium tetracarbonyl-ferrate(II), Tetrachloro-ferrates
Ferri-, Ferro- s. Iron
Ferricyanide s. Potassium hexacyanoferrate(III)
Ferrocenes 26, 186; **28,** 325
Fétizon reagent s. Silver carbonate-Celite
Filters s. Light filters
Fischer esterification 2, 208
Fission s. Cleavage
Flash vacuum pyrolysis 30, 665
- of
  sulfonic acid azides,
  β-aryl- **30,** 338
  1,2,4-triazoles **30,** 665
Flavan-4-ols
- startg. m. f.
  flavanones **21,** 285 suppl. **28**
Flavanones
- from
  flavan-4-ols **21,** 285 suppl. **28**
- startg. m. f.
  aurones **26,** 928
  pyrans, 2,3-dihydro-2-imino-**26,** 713

Flavazoles 26, 452
Flav-2-enes 26, 242
Flavones
- from
  chalcone dihalides,
  o'-hydroxy- **30,** 170
-, **3-hydroxy-**
- startg. m. f.
  indan-1,2-diones, 3-hydroxy-**30,** 515
Flavylium salts, 4-alkoxy-**29,** 192
Flow system s. Continuous
Fluo... s. Fluoro...
Fluoride ion
- as base **26,** 901
- as nucleophile **26,** 901
- as dehydrohalogenation agent **23,** 957; **29,** 961
- **ion catalyst,** cesium fluoride as – **25,** 124 suppl. **26**
Fluorides (s. a. Halides, Replacement) **14,** 620;
  19, 635 suppl. **29**
- from
  alkoxysilanes **29,** 512
  1,2-bromofluorides **25,** 64 suppl. **26**
  1,1-chlorofluorides **16,** 24 suppl. **26**
  diazonium salts **26,** 555
  mesylates **12,** 619 suppl. **29**
- special s.
  di-fluorides
  phosphonium –
  tetrabutylammonium fluoride
  tetramethylammonium –
Fluorination
- in liq. phase, review **11,** 601 suppl. **26**
- special s.
  halogeno-fluorination
  hydro-
  nitro-
- under mild conditions **28,** 483
N-Fluorination 30, 196
Fluorine
- as reagent **27,** 95
Fluoro... s. a. Difluor..., Perfluor..., Trifluor...
Fluoroamines (s. a. Difluoro-amines, Trifluoroamines)
- from
  chlor(alkoxysilanes) **29,** 512
  chlorofluorides **29,** 512
Fluoroborates
- as reactants **26,** 370, 453, 771;
  27, 762, 821/2; **28,** 359, 656;
  30, 286, 312
- special s.
  copper(II) fluoroborate
  1,3-dithienium –
  iminoester fluoroborates
  sulfonium –
  trialkyloxonium –
Fluoroboric acid 27, 140, 283;
  28, 104, 837, 958; **29,** 131, 675, 923
Fluorodinitromethane
  as reactant **27,** 766
Fluoroformic acid esters
- from
  alcohols **30,** 302
- startg. m. f.
  trifluoromethyl ethers **30,** 106

N-Fluoroimines
-, reactions **29**, 127
- special s.
  ethylene-N-fluoroimines
  1-halogeno-N-fluoroimines
Fluoromethylene compds.
30, 620
Fluorophosphoranes
- special s.
  phenyltetrafluoro-
  phosphorane
α-**Fluorosulfinylperfluoro-
carboxylic acid fluorides**
- from
  perfluoroalkenes, terminal
  26, 580
Fluorosulfonic acid 12, 417
suppl. **29**; **16**, 610 suppl. **28**;
24, 576 suppl. **27**; **29**, 700, 910;
30, 644
- startg. m. f.
  fluorosulfonic acid esters
  30, 78
- - esters **27**, 112
-, N-alkylation with – **29**, 344
- from
  ethylene derivs. **30**, 78
- special s.
  methyl fluorosulfonate
- startg. m. f.
  alkoxylcyclimmonium salts
  22, 187 suppl. **29**
  enolethers **28**, 173
  iminoesters **28**, 139
Fluorosulfonyl isocyanate
- startg. m. f.
  uracil-1,3-disulfofluorides
  26, 698
Fluoroxyperfluoro compds.
- special s.
  fluoroxytrifluoromethane
Fluoroxytrifluoromethane
-, fluorination with – **27**, 559;
29, 503; **30**, 196
Foam s. Antifoam agents
Formaldehyde (s. a. Aminomethylation, Cyano-,
  Hydroxy-, Mannich reaction)
- as reactant **26**, 289; **21**, 856
- as reagent **26**, 187, 425, 536;
  27, 89
-, C-methylation with – **27**, 761
-, Prins reaction with – **29**, 679
-, reaction with α-halogenoketones **26**, 864
-, - - ketones **26**, 659
-, ring closures with –
  22, 778; **29**, 353
- startg. m. f.
  3-azabicyclo[3.3.1]non-6-enes,
  1,5-dinitro- **26**, 48
  2-imidazolones **27**, 412
  1,3,6-oxadiazepine ring
  26, 362
  2H-1,3-oxazines, tetrahydro-
  6-alkoxy- **30**, 564
  pyrimidine ring, hexahydro-
  26, 781
-, Wittig synthesis with –
  29, 875
Formaldehyde donor 26, 289
Formals
- special s.
  methylenedioxyarenes
Formamide
- as medium **26**, 48
- as reactant **28**, 382

- as reagent **27**, 324
- startg. m. f.
  pyrimidine ring **27**, 408
  pyrimidines, 4-amino-
  (from 2 molecules) **26**, 817
- **acetals** s. a. Dimethylformamide acetals
- startg. m. f.
  purines, 8-*tert*-amino-
  28, 354 suppl. **30**
Formamides as reagents **27**, 53
-, N-demethylation via –
  28, 136
- from
  amines **27**, 312
  - and carbon monoxide
    26, 282
  ketones **27**, 53
- special s.
  N,N-dialkylformamides
  eneformylamines
- startg. m. f.
  amines s. N-Deformylation
  dicarboxylic acid imides
  29, 442
  enamines **28**, 415; **29**, 859
  α-hydroxycarboxylic acid
  amides **23**, 674 suppl. **29**
  α-ketocarboxylic acid amides
  27, 847
  methylamines **9**, 113
  purines, 8-*tert*-amino- **28**, 354
Formamidine acetate
- startg. m. f.
  1,2,4,5-tetrazines, monosubst.
  28, 381
Formamidines
- from
  amines and imidazoles,
  1-(N-iminoformyl)- **30**, 281
  - and 1,3,5-triazine
    11, 485
- special s.
  azoformamidines
  imidazoles, 1-(N-imino-
  formyl)-
Formamidinesulfinic acid
s. Thiourea dioxide
**Formamidiniumphosphonic
acid anhydrides**
- from
  chloroformamidinium
  chlorides **28**, 595
Formamidinoylisothiocyanates
- from
  chloroformamidines **26**, 413
Formates, s. Formoxy...
Formation, stereospecific
s. Stereospecific formation
Formic acid (s. a. Triethylammonium formate,
  Trimethylammonium –)
-, C-carboxylation with –
  29, 615
- as reactant **28**, 612
- as reagent **11**, 841 suppl. **29**;
  26, 96, 151, 315, 373; **27**, 28,
  53, 150, 229, 246, 599, 612, 674,
  918; **28**, 48, 212, 337, 349, 670;
  29, 49, 236, 263, 455, 465, 682,
  924, 987; **30**, 143, 498, 632,
  650, 680
- - **esters**
  s. Formoxy..., Isocyanoto-
  formic acid esters, Isopropenyl formate

Formiminohalides s. Thiobis-
  (sulfonylformimidoyl
  chlorides)
-, **N-halogeno-** s. Sulfonylformiminohalides,
  N-halogeno-
Formoxyaldehydes
- from
  diols **30**, 358
  formoxyhalides **30**, 358
Formoxy compds.
- startg. m. f.
  oxo compds. **30**, 184
Formoxyhalides
- from
  diols **30**, 358
- startg. m. f.
  formoxyaldehydes **30**, 358
1,2-Formoxyhalides
- from
  ethylene derivs. **28**, 480
Formylamines s. Formamides
C-Formylation (s. a. Vilsmeier
  aldehyde synthesis)
- of
  amines, ar. **28**, 762 suppl. **29**
- via 1,3-dithiolanes, 2-subst.
  30, 569
- with
  formic acid **26**, 777
  hexamethylenetetramine
  1, 621; **2**, 702
-, **directed 12**, 830 suppl. **28**
N-Formylation, selective
- with
  isopropenyl formate **27**, 375
S-Formylation with acetic-
  formic anhydride **26**, 600
Formylolefination
s. α,β-Ethylenealdehydes
from oxo compds. **25**
N-Formylsulfenamides
- startg. m. f.
  thiocyanates **28**, 576
Fragmentation (s. a. Cleavage)
- of
  aziridines **26**, 306
  azo compds. **27**, 390
  1,3-diol monoesters **26**, 985
  N-imidosulfoximines **28**, 930
-, ring contraction by –
  s. under Ring contraction
- special s.
  bicyclo[2.2.2]octene ring
  fragmentation
  boronate –
  cycloaddition –
  photo-
- to
  ethylene derivs. **28**, 979
-, **reductive**
- of
  β,γ-oxidonitriles **27**, 8
- -**acylation**
- of
  azo compds. **27**, 390
**Fragmentation-type ring
opening 26**, 865; **27**, 926;
28, 101
- of cyclopropyl azides
  27, 524 suppl. **29**
  N-heterocycles **29**, 779
  O-heterocycles **27**, 12
  ketones, cyclic **30**, 88
- - -, **heterolytic 26**, 927
- - -, **oxidative 26**, 927
- - -, **reductive 26**, 947

– – –, **stereospecific**
– of
 ketotosylates, cyclic **28**, 839
Fragment coupling s. under
 Peptides, synthesis
Frangomeric process 26, 349
Free radical... s. Radical...
Freon 112 as solvent **26**, 554
Freytag s. Löffler
Friedel-Crafts alkylation
 28, 824
–, retention of configuration
 28, 824 suppl. 30
– with
 halogenonitriles **28**, 824
– **ketone synthesis**
–, orientation **1**, 705; **5**, 604
–, solvent effect **5**, 604
– – –, **aliphatic**, review
 6, 814 suppl. 27
– synthesis
– with simultaneous
 disproportionation **17**, 879
 suppl. 26
 position shift **17**, 879
 suppl. 26
**Friedel-Crafts-type acylation
 with little or no catalyst**,
 review **16**, 883 suppl. 27
– synthesis with azides
 29, 402 suppl. 30
**Friedländer pyridine ring
 synthesis 28**, 757
Fries rearrangement 28, 127;
 29, 834
Fritsch s. Pomeranz
Fulvalenes s. Tetrathiofulvalenes, Tetraseleno-
Fulveneallenes
– from
 homophthalic acid
 anhydrides **28**, 970
– startg. m. f.
 fulvenes, 6-alkoxy- **29**, 138
Fulvenes (s. a. Heterofulvenes,
 Triafulvenes)
– from
 1,1-acoxyhalides **27**, 834
–, **6-alkoxy-**
– from
 fulveneallenes **29**, 138
–, **3-amino-**
– from
 enecyclopropenes and
 ynamines **29**, 656
–, **6-(4-amino-1,3-dien-1-yl)-**
– from
 pyrans, 2,3-dihydro-6-amino-
 2-hydroxy- **29**, 736
–, **6-hydroxy-**
– from
 7-oxanorbornadienes **30**, 525
Fumaric acid as catalyst
 30, 114
Functional groups
 (s. a. Compds., multi-
 functional)
–, *vic*-addition of 2 different
 functions to carbon-carbon
 double bonds **30**, 487
–, direction of stereo-
 specificity by – **28**, 43
–, elimination s. Replacement...
 by hydrogen
–, replacement of **carbonyl**
 oxygen by 2-substituents,
 functionalized **26**, 900;
 28, 837

– –, **semipolar**, reduction
 26, 993
Functionalization (s. a. Bridgehead functionalization)
–, direction of – by complex
 formation **27**, 188
– of
 non-activated groups
 27, 528; **28**, 292
 methylene – **27**, 188
 pyrimidines **30**, 621
–, **allylic 29**, 21
–, **remote 27**, 188
 acylamination, remote
 29, 334 suppl. 30
 carbon-carbon double bond
 formation **30**, 661
– –, **stereospecific 28**, 936
Furan-2,3-diones, 2,3-dihydro-
– startg. m. f.
 1,5-dioxocin-2,6-diones
 (from 2 molecules) **28**, 197
 4-pyrones (from 2 molecules)
 28, 197
Furano... s. a. Furo...
2(5H)-Furanone ring
– from
 2H-pyran ring **30**, 153
2(3H)-Furanones
– startg. m. f.
 γ-diketones **29**, 822
 γ-hydroxyketones **29**, 822
3(2H)-Furanones
– from
 α,β-acetylene-γ-alkoxy-
 ketones **30**, 182
 α,α'-dihalogenoketones and
 carboxylic acid amides
 29, 844
 α-halogenocarboxylic acid
 halides **29**, 790
–, **4-hydroxy-**
– from
 2-acetylene-1,4-diols **30**, 167
 1,4-dihydroxy-2,3-diketones
 30, 167
2(5H)-Furanones
– special s.
 vulpinic acids
–, **3,4-disubst.**
– via Michael addition **27**, 710
Furan ring 16, 315; **19**, 891;
 25, 587 suppl. 26
–, closure, oxidative **30**, 158
– from
 α-alkoxymethyleneketones
 28, 779
 α-alkylthiomethyleneketones
 28, 840
 cyclopropane ring **26**, 248
 oxazole ring **24**, 807 suppl. 26
–, hydrogenation **20**, 66
 suppl. **28**; **23**, 66 suppl. **28**
–, opening, oxidative **26**, 117
–, rearrangement **29**, 125
– startg. m. f.
 cyclopent-2-enone ring
 27, 388
– via 1,2-dihalides **16**, 336
–, **2,3-dihydro-**
– from
 enesulfonium salts and
 ketones **28**, 838
–, opening **26**, 538
– –, **2,5-dihydro-**
– from
 1,3-dienes **30**, 87

–, **tetrahydro-**
– from
 1,1'-dihalogenethers and
 ethylene derivs. **29**, 794
 ketones and cyclopropane
 ring **28**, 641
– and 1,3-halogenhydrins
 28, 609
Furans 28, 906
–, C-alkylation, reductive
 27, 839 suppl. 30
– from
 2-acetylenesulfonium salts
 28, 838
 γ-diketones **29**, 251
 furans, 2,5-dihydro-2-alkoxy-
 29, 865
 α-hydroxyketones and
 2-alkoxyvinylphosphonium
 salts **29**, 865
 oxazoles **28**, 869
– special s.
 benzofurans
 difuranylacetylenes
 α,β-ethylene-β-(2-furyl)-
 ketones
– startg. m. f.
 γ-diketals **26**, 115
 2-ene-1,4-diols **28**, 583
 2-ene-1,4-diones **28**, 84
 monosaccharides **27**, 276
 naphthalenes **19**, 917
 suppl. 28
 1,2-oxaborins, dihydro-
 28, 583
–, **2-alkoxy-**
– from
 cyclopropenecarboxylic acid
 esters **26**, 155
–, **2-amino-**
– from
 furfurylideneammonium salts,
 2,3-dihydro- **30**, 164
 γ-ketocarboxylic acid amides
 30, 164
–, **2,3-dihydro-**
– from
 cyclopropyl ketones **26**, 154
 α,β-ethyleneketones **28**, 105
 furylium salts, tetrahydro-
 28, 105
–, synthesis **4**, 667 suppl. 27
–, **2,5-dihydro-**
– by C-alkylation, reductive
 27, 839 suppl. 30
– from
 2,4-dienols **25**, 123 suppl. 28
– startg. m. f.
 furans, tetrahydro-
 3,3-dialkoxy- **26**, 143
–, **2,5-dihydro-2-alkoxy-**
– from
 α-hydroxyketones and
 2-alkoxyvinylphosphonium
 salts **29**, 865
– startg. m. f.
 furans **29**, 865
–, **4,5-dihydro-3-halogeno-**
–, metalation **29**, 797
– startg. m. f.
 3-acetylenealcohols **29**, 797
–, syntheses with – **29**, 797
–, **2-α-halogeno-**
– from
 2-ene-1,4-diones **27**, 571
–, **tetrahydro-** (s. a. Tetrahydrofuran)

(Furans, tetrahydro-)
-, ring closure, stereospecific 7, 355 suppl. 28
- special s.
 oxidotetrahydrofuranolactones
-, **tetrahydro-3,3-dialkoxy-**
- from
 furans, 2,5-dihydro- **26**, 143
-, **tetrahydro-2,3-diimino-**
- from
 isonitriles (2 molecules) and oxido compds. **27**, 673
-, **tetrahydro-2-halogenomethyl-**
- from
 4-ethylenealcohols **29**, 495
-, **tetrahydro-2-imino-**
- from
 isonitriles and oxetanes **27**, 672
Furazano[3,4-d]pyrimidines, 7-acylamino-
- startg. m. f.
 adenines **27**, 521
Furazan oxides s. Furoxans
- **ring opening, reductive**
-, syntheses of N-heterocycles via – **27**, 521
Furazans
- special s.
 furoxans
- startg. m. f.
 dinitriles **27**, 515
 nitriles **28**, 467
Furfurylideneammonium salts, 2,3-dihydro-
- from
 γ-ketocarboxylic acid amides **30**, 164
- startg. m. f.
 furans, 2-amino- **30**, 164
4H-Furo[3,4-d][1,3]benzodiazepin-3-ones, 1,3-dihydro- **26**, 475
Furoic acids
- from
 3-deoxy-2-ketoaldonic acids **29**, 252
4H-Furo[3,2-b]pyrid-2-ones, 5,6-dihydro-
- from
 isoxazolo[2,3-a]pyridinium salts, 4,5,6,7-tetrahydro-4-oxo- **29**, 166
Furo[3,2-c]pyrid-4-ones, 4,5-dihydro- **27**, 508
Furo[2,3-b]quinolines **29**, 247
Furo[3,2-b]quinolines
- from
 o-aminoacetic acid esters **26**, 939
 quinolines, 3-hydroxy- **26**, 939
Furoxan ring (s. a. [1,2,5]-Oxadiazolo[3,4-b]pyridine 1-oxides)
- from
 hydroxylamines **29**, 338
Furoxans
- from
 diazo compds. (2 molecules) **27**, 825
 enazides **29**, 401
-, isomerization **27**, 99
-, **3-amino-**
- from
 glyoximes **27**, 99

- startg. m. f.
 furoxans, 4-amino- **27**, 99
-, **4-amino-**
- from
 furoxans, 3-amino- **27**, 99
 glyoximes **27**, 99
Furylcarbinols
- startg. m. f.
 3-pyrones, 2,6-dihydro-6-hydroxy- **27**, 276
Furylideneacetic acid esters, tetrahydro-
- from
 ε-hydroxy-β-ketocarboxylic acid esters **30**, 44
 β-ketocarboxylic acid esters and oxido compds. **30**, 448
Furylium salts, tetrahydro-
- from
 α,β-ethyleneketones **28**, 105
- startg. m. f.
 furans, 2,3-dihydro- **28**, 105
δ-**(2-Furyl)ketones** **30**, 157, 162
β-**(2-Furyl)propionic acids, 5-subst.**
- from
 α,β-ethylene-β-(2-furyl)-ketones **29**, 125
Fusion s. Nucleosides fusion synthesis

Gabriel-Colman rearrangement **21**, 777
- -, **extended** **28**, 694
Gabriel synthesis **27**, 480
-, alternative to – **26**, 492
- **-type synthesis** **29**, 288
Gallium bromide **28**, 959
- **chloride** **29**, 602; **30**, 624
Galvinoxyl as radical scavenger **22**, 434 suppl. **26**
Gams s. Pictet
Gas evolution with solides, finely divided **30**, 36
- **phase** s. Vapor phase
Gattermann formylation **28**, 708
- with
 C-alkyl migration **28**, 708
Gattermann-Koch-type isoquinoline synthesis **28**, 743
gem-(geminate) (s. a. 1,1)-
s. *gem*-Dimethyl effect, *gem*-Divinyl compds., *gem*-Hydrogenolysis
Geoisomerization s. under Ethylene derivs.
Geoisomers s. *cis-trans*-Isomers
Geometry, retention **27**, 981; **29**, 854
Geoselectivity of hydrohalogenation **12**, 625 suppl. **29**
Geospecificity s. a. Addition, geospecific, Favorskii rearrangement, – Homoallyl rearrangement, –, Migration, –, Orthoester Claisen rearrangement, –, Replacement, –, Ring opening, –, Synthesis, –, Transetherification, –
-, control **28**, 964 suppl. **29**
-, effect of bases on – of β-elimination **28**, 944

-, - - solvents on – **14**, 91 suppl. **29**
2-Germa-1,3-dioxaheterocycles
- from
 diols and diaminogermanes
Germanes **29**, 114
- special s.
 amino-germanes
 dihalogeno-
Germanium
- startg. m. f.
 trihalogenogermanes **27**, 647
Germansultones **26**, 578
Glass powder **30**, 36
Glow discharge **30**, 442
-, reactions by –, review **21**, 687 suppl. **28**; **30**, 442
Glucosyltransferase as reagent **28**, 185
Glutaconic acids
- special s.
 alkylideneglutaconic acids
Glutaric acids by Michael addition **28**, 733
Glycals
- special s. 1',2'-ethylene nucleosides
Glycerides **29**, 188 suppl. **30**
Glycidic acid esters (s. a. Thiolglycidic acid esters)
- from
 α,β-ethylenecarboxylic acid esters **28**, 110
 ketones **26**, 895
- special s.
 alkoxyglycidic acid esters
 halogenoglycidic – –
- - *tert*-butyl esters
- from
 ketones **26**, 859
- startg. m. f.
 aldehydes **26**, 859
Glycidyl ethers **14**, 271 suppl. **27**
Glycitols s. Thioglycitols
-, peracyl-
-, degradation of carbohydrates to – **27**, 540
N-1-Glycityl-N-heterocyclics **30**, 274
Glycocyamines s. α-Guanidinocarboxylic acids
Glycol esters (s. a. Glycol monoesters)
- from
 diazo compds. (2 molecules) **29**, 780
 ethylene derivs. **29**, 152
- **ethers**
- from
 ethylene derivs. **28**, 114
- **groups**, activation as dialkoxystannanes, cyclic **29**, 111
Glycolic... s. α-Hydroxycarboxylic...
- **acid esters, O-alkyl-**
- by redoxidation, intramolecular **30**, 115
Glycol monoesters
- from
 oxido compds. **27**, 128
- **monoethers** s. 2-Methoxyethanol
- **mononitrates**
- from
 hydrocarbons **28**, 132

- **monosulfonates**
- startg. m. f.
  oxido compds. **26**, 237
- –, **cyclic**
- –, rearrangement, skeletal **27**, 974
- **Glycols** (s. a. Diols)
- –, cleavage, oxidative **29**, 157
- –, epimerization **26**, 148
- – from
  carboxylic acids (2 molecules) **26**, 913
  ethylene derivs. **27**, 147, 150
  – – (trans-glycols) **27**, 152
  – – (cis-glycols) **13**, 205
  α,β-ethyleneketones **26**, 134
  1-hydroxy-2-(alkoxysulfonium) salts **27**, 129
  α-hydroxyketones **29**, 36
  α-isonitrosoketones **26**, 185
  oxido compds. **26**, 2; **27**, 129
- – special s.
  acetyleneglycols
  cyclobutane-1,2-diols
  α,β-dihydroxyketones **25**
  ethylene glycol
- – startg. m. f.
  1,2-acoxyhalides **26**, 547; **29**, 509
  alcohols, reduction **26**, 81
  aldehydes, cleavage, oxidative **3**, 238
  α-diketones **26**, 233; **28**, 209
  – with rearrangement **29**, 681
  1,4-dioxanes **29**, 183
  1,3-dioxolanes **29**, 174
  ethylene derivs. **28**, 925
  α-hydroxyketones **21**, 285 suppl. **28**; **26**, 233; **28**, 201 suppl. **29**
  ketones **27**, 259
  – with C-rearrangement **29**, 700
  oxido compds. **27**, 941 suppl. **29**; **29**, 101
- –, **cyclic**
- – startg. m. f.
  diketones **29**, 157
  diols **29**, 155
- cis-**Glycols**, –
- – from
  2-ethylenealcohols **29**, 142
- trans-**Glycols**, –
- – from
  enoxysilanes, cyclic **30**, 81
  trans-2-siloxyalcohols, cyclic **30**, 81
- **Glycols**, sym. s. Pinacols
- **Glycosamines** (s. a. N-Glyco...)
- – special s.
  halogenoglycosamines
- **Glycosidation**
- – with N-carbalkoxylation **30**, 297
- **Glycosides** (s. a. Carbohydrates, Ethers) **28**, 185
- –, acetolysis **28**, 88
- – from
  carbaryloxy glycosides **30**, 150
  thioglycosides **29**, 215
- –, ozonization, stereospecific **28**, 235
- – special s.
  acyl-glycosides
  aryl –
  carbaryloxy-

glycuronides
thioglycosides
- startg. m. f.
  1,6-anhydrosugars **26**, 241
C-Glycosides **9**, 881 suppl. **26**; **29**, 754
- by Wittig synthesis **29**, 754 suppl. **30**
- from
  enoxysilanes **29**, 754
-, ar. **29**, 755
N-Glycosides
- from
  carbaryloxy glycosides **30**, 150
O-Glycosides **30**, 150
-, synthesis, review **1**, 218-20 suppl. **29**
Glycosylation with carbohydrate Δ²-oxazoline derivs. **14**, 174
N-Glycosylation, directed, of N-heterocyclics **19**, 539 suppl. **28**
Glycosyl compds. **29**, 224
- **halides**
  fluorides from chlorides **26**, 551
N-Glycosyl-N-heterocyclics
- from
  acyl glycosides **26**, 446
- special s.
  nucleosides
  N-thioglycosyl-N-heterocyclics
Glycosylhydrazines,
  anomerization **28**, 289
Glycosyl migration, review **20**, 258 suppl. **26**
N-Glycosyl – **26**, 316
Glycosyl phosphates
- from
  carbohydrate orthocarboxylic acid esters **30**, 59
Glycosylureas **10**, 347 suppl. **28**
Glycuronides **28**, 145
Glyme s. 1,2-Dimethoxyethane; s. a. Diglyme
Glyoxal as startg. m. **27**, 951
- **monohydrate**
- startg. m. f.
  aminoacetamides **28**, 313
Glyoxalines s. Imidazoles
Glyoxalyl chloride tosylhydrazone as reactant **26**, 201
Glyoximes
- special s.
  dimethylglyoxime
- startg. m. f.
  furoxans, 3-amino- **27**, 99
-, 4-amino- **27**, 99
Glyoxylic... s. a. α-Ketocarboxylic...
- **acid** as reducing agent **11**, 308 suppl. **29**
- – **hydrate**
- startg. m. f.
  acetallactones, polycyclic **30**, 112
  9H-pyrid[3,4-b]indoles, 1,2,3,4-tetrahydro- **4**, 388 suppl. **27**
Gold s. Tetrachloroauric(III) acid
Gomberg-Bachman reaction **27**, 814
Gould-Jacobs ring closure **3**, 715; **5**, 355

Graphite s. a. Antimony pentachloride-graphite, Bromine- –, Chromium trioxide- –, Potassium- –
Grignard compds. (s. a. organomagnesium halides under Magnesium)
-, addition to carbon-nitrogen double bonds, cyclic **15**, 528
  preferential addition **15**, 528 suppl. **26**
-, constitution, review **22**, 857 suppl. **27**
-, cross-coupling with – **26**, 875; **29**, 824
-, displacement, abnormal, by – **29**, 823
-, reaction with
  acylcyanides **27**, 905
  amines, unsatd. **27**, 712
  enolethers **26**, 782
  nitro compds. **27**, 473
  thioketones **29**, 641
-, ring opening of 1,3-heterocycles **26**, 670
-, **hindered 28**, 685
-, **unsatd. 19**, 745
-, dimerization **29**, 879
-, review **19**, 746 suppl. **26**
-, **reaction**, solvents
  toluene **29**, 636
- **syntheses** (s. a. Magnesium)
- under pressure **26**, 868
- with
  1,1'-di(alkoxy)amines **28**, 813
  enamines **26**, 870
- **synthesis**
- of
  oxo compds. **22**, 857 suppl. **27**
- with
  retention of ketal groups **27**, 856
-, **stereospecific 16**, 886 suppl. **27**
-, effect of chelation on stereospecificity **16**, 728 suppl. **26**; **16**, 886 suppl. **27**
Grignard-type reaction,
  one-step **26**, 662
Group V element(III) compds.
-, reactions with vinyl isocyanide **26**, 639
Groups, functional
  s. Functional groups
-, **leaving** s. Leaving groups
-, **protective** s. Protective groups
Guanidine carbonate
- startg. m. f.
  1-acylguanidines, 3-subst. **29**, 356
  sulfonylguanidines **28**, 252
- **nitrate**
- startg. m. f.
  pyrimidines **30**, 310
Guanidines
- from
  nitroguanidines **12**, 52
-, prepn. **26**, 438 suppl. **28**
- special s.
  acylguanidines
  amidinothioureas
  amino-guanidines
  cyano-
  nitro-
  sulfonyl-
  tetramethylguanidine

(Guanidines)
-, **bicyclic** s. 1 H-Imidazo-
  [1,2-a]imidazoles
-, **cyclic** (s. a. 1,3-N,N-
  Heterocyclics, 2-amino-)
-, prepn. **26**, 438 suppl. 28
-, **monosubst. 27**, 28
-, **tricyclic, N-condensed**
  29, 475
-, **1-acylamino-3-amino-**
- startg. m. f.
  1,2,4-triazoles, 3-hydrazino-
  25, 340
α-**Guanidinocarboxylic acids**,
  synthesis 2, 324 suppl. 26
Guanines, 8-hydroxy- 26, 356
Guanylhydrazones
- startg. m. f.
  cyanimines **28**, 455
Guanylureas
- from
  carbodiimides **29**, 131
Gumbrim (clay) **26**, 238

Halals
- special s.
  chloral
Halide ions
- as activators **30**, 606
-, oxido compds. as bases
  in the presence of – **29**, 969
  review **23**, 530 suppl. 28
Halides (s. a. Replacement)
-, activation by adsorption
  26, 252
- as reactants **30**, 356
- from
  acoxy compds. **28**, 504
  alcohols HalC↓↑O
  - with inversion of configura-
    tion **24**, 581 suppl. 29
  amines, tert. **28**, 508
  carbazic acid esters **28**, 496
  carboxylic acids, degrada-
    tion with loss of 1 C-atom
    26, 554
  cobalt compds., organo-
    26, 552
  diazonium fluoroborates
    26, 549
  1,1-dihalides **26**, 70
  ethers **27**, 576
  ethylene derivs. HalC↓↓CC
  – –, synthesis **30**, 481, 607
  mercury compds., diorgano-
    30, 373
  phosphorodiamidites **30**, 356
  phosphorous acid derivs.
    30, 356
  salicylic acid esters **30**, 360
  stannanes **28**, 511
  sulfonic acid esters **19**, 620
    suppl. **26**, 29
    methanesulfonic acid
      esters (chlorides) **30**, 366
    p-nitrobenzenesulfonic
      acid esters (iodides)
      17, 621 suppl. 28
-, homologization **28**, 802
-, reaction with mercury salts
  30, 135
- special s.
  acetylene-halides
  acoxy –
  acylamino-
  alkoximino-

alkoxy-
aryl-
benzyl –
cumulene-
di-
ethylene-
fluorides
glycosyl-halides
1,1-hydrazono-
metal compds., halogeno-
methyl-
mono-halides
nitro-
oxido-
oxo-
poly-
stannoxy-
triphenylmethyl chloride
- startg. m. f.
  3-acetylenealcohols, synthesis
    with addition of 4 C-atoms
    29, 797
  acetylene derivs. **27**, 851
  – –, terminal **27**, 851 suppl. 30
  alcohols, sec., synthesis
    28, 821
  aldehydes, synthesis with
    addition of 1 C-atom
    26, 872, 891; **27**, 830
  –,– – – – 2 C-atoms **28**, 793
  1-alkoxyenamines, cyclic
    28, 793
  amidines, subst. **29**, 423
  amines NC↓↑Hal
  –, prim. **27**, 480
  –, sec., synthesis **30**, 463
  –, tert. **26**, 423; **28**, 393, 413
  α-aminocarboxylic acids
    29, 791
  P-aminophosphonium salts
    26, 291
  aryl ketones **29**, 888
  azides **26**, 422
  1,3-bis(alkylthio)-1-ethylenes
    26, 862
  boranes **27**, 232
  N-carbalkoxysulfenamides
    29, 412
  carboxylic acid esters
    27, 219; **28**, 165
  – – –, synthesis with
    addition of 1 C-atom
    28, 826
  - acids and derivs.,
    synthesis with addition
    of 1 C-atom **27**, 869
    suppl. 28
  cyclobutanes **28**, 947 suppl. 30
  cyclopropane ring **26**, 967
  cyclopropanes **27**, 874
  1,1-dialkoxy-3-alkylthio-
    allenes **30**, 584
  dihalogenalanes **29**, 594
  dihalogenoboranes **26**, 649
  α,β-ethylenealdehydes,
    synthesis with addition
    of 3 C-atoms **26**, 862
  trans-γ,δ-ethylenealdehydes
    28, 842
  ethylene derivs. (s. a. De-
    hydrohalogenation)
    CC↑↑Hal; **28**, 895
  – –, synthesis **30**, 586, 607
  trans-ethylene –, – **30**, 43
  β,γ-ethylenehalides,
    synthesis with addition
    of 3 C-atoms **28**, 802
  β-halogenethyl ethers **26**, 508

1,2-halogenhydrins, synthesis
  29, 616
α-halogenocarboxylic acid
  esters, synthesis with
  addition of 2 C-atoms
  29, 499
halogenosilanes **28**, 586
hydrocarbons, synthesis
  26, 875; **27**, 838, 872
-, – with ring contraction
  27, 881
α-hydroxyketones, synthesis
  30, 617
hydroxylamines **26**, 402
-, disubst. **27**, 473
isocyclics **28**, 953
α-ketocarboxylic acid
  amides **29**, 801
β-ketocarboxylic acid esters
  29, 499; **30**, 584
ketones (from 2 molecules)
  27, 869
-, synthesis **27**, 885; **30**, 582
-, unsym., synthesis **26**, 889
mercaptans **28**, 569
mono(carbalkoxy)amines
  29, 412
nitric acid esters **30**, 135
oxido compds. **28**, 795
oxo – **27**, 830 suppl. **28**, **29**;
  28, 843; **30**, 358
phosphinic acid esters
  26, 656
- acids **26**, 656
phosphonic acid dichlorides
  26, 651
- - esters **26**, 656
phosphoric acid diesters,
  mixed **28**, 170 suppl. 30
- - monoesters **28**, 170
phthalimides **27**, 480
sulfinic acid esters **29**, 577
1,1-sulfinylthioethers **27**, 830
sulfonic acid esters **27**, 111
thioethers **26**, 616;
  27, 632, 635
-, sym. **18**, 687 suppl. **29**;
  27, 629
vinyl ethers **28**, 163
-, **active**, as reagents
  21, 294 suppl. 29
-, **ar.** (s. a. Aryl halides)
- startg. m. f.
  amines, ar., prim. **14**, 512;
    30, 592
  arylphosphonic acid esters
    26, 656
  homophthalimides **29**, 811
-, **prim.** (s. a. anti-Markownikoff hydrohalogenation)
-, preferential formation
  24, 581 suppl. 26
- startg. m. f.
  ethylene derivs., terminal
    23, 964 suppl. 26
  quinolines, 1,4-dihydro-
    4-imino-, 1-subst. **26**, 401
-, **sec.**, by anti-Markownikoff
  hydrohalogenation **26**, 519
N-Halides, prepn., and
  reactions, review **22**, 262
  suppl. **26**
-, reaction with phosphites
  19, 367
Haloforms
- special s.
  chloroform

- startg. m. f.
 dihalogenomethyl sulfones
 27, 627
 α,β-ethylenesulfonic acids
 27, 627
Halogen (s. a. individual
 halogens, Pseudohalogens)
-, addition HalC↓
- as reagent **26,** 273
-, elimination ↑Hal
-, interchange (s. a. Bridgehead
 halogen interchange, Re-
 placement of individual
 halogens
-, migration s. under
 Migration
-, replacement s. under
 Replacement
-, retention
 formation of imines with –
 26, 433
 O-methylation with – **27,** 228
 reduction with – **8,** 795
 suppl. **26; 13,** 215;
 16, 54 suppl. **29**
Halogen/alkali hydroxide
 s. Hypohalites
- -metal interconversion **26,** 70
α-Halogenacetals
- from
 enolethers **27,** 547
- startg. m. f.
 imidazo[1,2-a]pyrazine ring
 26, 431
β-Halogenacetals
-, benzene ring annelation
 with – **28,** 922
- startg. m. f.
 cyclopropanes, alkoxy-
 11, 956 suppl. **27**
Halogenacetylenes
 s. Acetylenehalides
2-Halogenacylamines
- from
 ethylene derivs. and
 N-halogenocarboxylic acid
 amides **28,** 478
- startg. m. f.
 2-aminoalcohols **26,** 60
 aziridines **26,** 60
Halogenalanes
- special s.
 dihalogenalanes
Halogenalcohols s. Dihalogen-
 alcohols, Halogenhydrins,
 Trihalogenalcohols
Halogenaldehydes
- special s.
 ethylenehalogen-aldehydes
 dihalogen-
α-Halogenaldehydes
- startg. m. f.
 1,2-halogenhydrins **30,** 16
 α-halogenoazomethines
 21, 432 suppl. **26**
 imidazolium ring,
 N-condensed **29,** 349
 γ-lactones **30,** 331
1-Halogen-1-alkoximes
 (s. a. 1-Halogen-1-aroximes)
- startg. m. f.
 1-alkoxy-1-alkoximes **26,** 197
2-(ω-Halogenalkoxy)halides
- from
 ethers, cyclic, and ethylene
 derivs. **27,** 553

Halogenalkoxysilanes
- special s.
 chloralkoxysilanes
(3-Halogenalkoxy)stannanes
- startg. m. f.
 oxetanes **28,** 225
Halogenallenes
- startg. m. f.
 allenes, synthesis **23,** 831
 suppl. **27, 28**
N-Halogenamides
 s. N-Halogenocarboxylic
 acid amides
**Halogenamination,
 intramolecular**
- of ethyleneamines **17,** 984
Halogenamines
- special s.
 ethylenehalogenamines
- startg. m. f.
 indoles **27,** 526
1,1-Halogenamines
- startg. m. f.
 2-ethyleneamines **29,** 833
 2,3-ethylene-2-halogenamines
 23, 650 suppl. **27**
2-Halogenamines
- from
 aziridines **30,** 346
 ethylene derivs. and N,N-
 dihalogenophosphor-
 amidates **27,** 548
 2-halogenophosphoramidates
 27, 548
- startg. m. f.
 amines, with rearrangement
 29, 76
 aziridines **28,** 433
1,3-Halogenamines
- from
 2-aminoalcohols, rearrangement **26,** 534
N-Halogenamines
 (s. a. N,N-Dihalogenamines)
- as intermediates **26,** 204
-, review **25,** 213 suppl. **26**
- special s.
 adamantanamines,
 N-halogeno-
 carbinamines, N,N-dichloro-
 N-chlorodiisopropylamine
 ethylene-N-halogenamines
 N-halogeno-N-nitramines
- startg. m. f.
 amines, tert. **30,** 305
 azo compds. **27,** 284
 carboxylic acid amides
 28, 416
 N-heterocyclics **27,** 530;
 28, 460
 -, N-condensed **30,** 678
1-Halogen-1-aroximes (s. a.
 1-Halogen-1-alkoximes)
- from
 hydroxamic acid esters
 26, 543
Halogenarsines
- special s.
 dihalogenarsines
Halogenation (s. a. Chlorination, Iodination, Mono-
 halogenation) **27,** 563
- of
 2,1,3-benzothiadiazoles
 26, 533
-, orientation **22,** 153 suppl. **29**
- with rearrangement, skeletal
 27, 566

-, **allylic** s. β,γ-Ethylenehalides
 from ethylene derivs.
-, **ar. 25,** 389
-, **directed** (s. a. ω- and
 (ω–1)-Halogenation)
- of
 ketones **22,** 153 suppl. **26**
 sulfones **6,** 572 suppl. **26**
-, radical relay s. Radical
 relay halogenation
-, **ionic 6,** 572 suppl. **28**
-, **preferential 28,** 494; **29,** 504
-, **terminal** s. ω-Halogenation
N-Halogenation 22, 565
 suppl. **29**
- of
 amides **26,** 267
ω-Halogenation
- of
 carboxylic acids **26,** 531
(ω–1)-Halogenation 26, 527
Halogenazines
- from
 acylhydrazones **27,** 577
- special s.
 polyhalogenodiazabutadienes
Halogenazo compds.
- special s.
 dihalogenazo compds.
1,1-Halogenazo –
- from
 hydrazones **27,** 558
α-Halogenazomethinium salts
- from
 enamines **12,** 244; **28,** 116
- startg. m. f.
 α-aminoketones, rearrangement **28,** 116
Halogenenacylamines
- special s.
 1-(dihalogenomethylene)-
 acylamines
1-Halogenenamidines
- from
 nitriles (2 molecules) **29,** 489
- startg. m. f.
 pyrimidines, 4,6-dihalogeno-
 29, 489
1-Halogenenamines
-, ring closures with –
 28, 822 suppl. **29**
- startg. m. f.
 Δ¹-azirines, 2-amino- **27,** 462
 enamines, synthesis **27,** 846
 immoniocyclobutane ring
 28, 822
2-Halogenenamines 28, 487
- startg. m. f.
 α,α-dihalogenaldehydes
 30, 370
α-Halogeneneureas
- from
 α-halogenenisocyanates
 28, 472
 nitriles **28,** 472
α-Halogenenisocyanates
- from
 nitriles **28,** 472
- startg. m. f.
 α-halogeneneureas **28,** 472
γ-Halogenenolesters
- from
 enolesters **28,** 229
- startg. m. f.
 α,β-ethyleneketones **28,** 229

Hal 638

5-Halogen-3-en-1-ynes
- startg. m. f.
  1,2,4-trienes, synthesis
  28, 812
Halogenethers s. Alkoxyhalides
-, **cyclic**
- special s.
  furans, tetrahydro-
  2-halogenomethyl-
Halogenethylen...
  s. Ethylenehal...
β-Halogenethyl ethers
- from
  halides **26**, 508
1,2-Halogenhydrin halogenosulfites
- from
  oxido compds. **29**, 486
Halogenhydrin nitrates
  s. Iodohydrin nitrates
Halogenhydrins
- special s.
  acetylene-halogenhydrins
  ethylene-
- startg. m. f.
  ethylene derivs. **27**, 970
-, **cyclic**
- from
  dienes **26**, 520
1,2-Halogenhydrins
- from
  ethylene derivs. **26**, 144
  - - (cis-1,2-halogenhydrins)
  12, 587 suppl. **26**
  α-halogenaldehydes **30**, 16
  oxido compds. **28**, 471
  oxo compds. and halides
  29, 616
-, rearrangement **24**, 964
- startg. m. f.
  2-ethylenealcohols **29**, 964
  ketones with rearrangement
  28, 946
  oxido compds. **26**, 144
1,3-Halogenhydrins
- startg. m. f.
  1,4-diols, synthesis **28**, 609
  furan ring, tetrahydro-
  28, 609
N-Halogenimides
  s. N-Halogenodicarboxylic
  acid imides
Halogenimidium salts
-, reaction with nucleophiles
  26, 404
- special s.
  halogenoforminium halides
- startg. m. f.
  amidinium salts **26**, 404
  carboxylic acid amides
  26, 404
N-Halogenimines
- from
  oximes **30**, 197
- special s.
  N-fluoroimines
  phenyl N-bromoketimine
β-Halogeniminoesters
- startg. m. f.
  β-alkoxycarboxylic acid
  esters **30**, 138
N-Halogeniminoesters
- special s.
  α,N-dihalogeniminoesters **25**
Halogenisocyanates
- special s.
  halogenenisocyanates

2-Halogenisocyanates
- from
  2-azetidinones, N-halogeno-
  28, 482
1,1-Halogenisothiocyanates
- from
  imines **30**, 242
α-Halogenoazomethines
- from
  α-halogenaldehydes **21**, 432
  suppl. **26**
o-Halogenoazomethines
- from
  nitrones, ar. **30**, 363
Halogenoboranes
- from
  ethylene derivs. **27**, 649
- special s.
  dialkylchloroboranes
  di(ene)halogenoboranes
  dihalogenoboranes
  monochloroborane
- startg. m. f.
  borinic acid esters **27**, 649
α-Halogenoboranes
-, rearrangement **27**, 118
- startg. m. f.
  borinic acids **27**, 118
α-Halogenocarbonyl compds.
- from
  enoxysilanes **30**, 374
α-Halogeno-N-carbonylureas
- from
  azomethines **28**, 473
α-Halogenocarboxylic acid amides
- from
  aziridinones **28**, 474
- startg. m. f.
  α-aminocarboxylic acid
  amides **30**, 290
  - acids **30**, 290
β-Halogenocarboxylic - -
- special s.
  α,β-ethylene-β-halogeno-
  carboxylic acid amides
N-Halogenocarboxylic - -
  (s. a. N-Halides)
- startg. m. f.
  2-halogenacylamines **28**, 478
Halogenocarboxylic acid esters
- from
  halogenocarboxylic acid
  halides **29**, 516
  lactones **29**, 516
α-Halogenocarboxylic - -
- from
  carboxylic acid esters
  27, 560
  halides, synthesis with
  addition of 2 C-atoms
  29, 499
  oxo compds., - - - -
  1 C-atom **29**, 738
-, metalation **30**, 549
- special s.
  α,α-dihalogenocarboxylic
  acid esters
- startg. m. f.
  β-aminocarboxylic acid
  esters **29**, 638
  2-azetidinones **28**, 816
  α,β-ethylenecarboxylic acid
  esters, synthesis **26**, 907
  5-isoxazolidones **28**, 817
  γ-lactones **27**, 908
  succinic acid esters, sym.
  (from 2 molecules) **27**, 860

-, synthesis **28**, 860
β-Halogenocarboxylic - -
- special s.
  β-halogeno-α-ketocarboxylic
  acid esters
Halogenocarboxylic acid halides
- from
  lactones **29**, 516/7
- special s.
  halogenodicarboxylic acid
  halides
- startg. m. f.
  halogenocarboxylic acid
  esters **29**, 516
α-Halogenocarboxylic - -
- from
  carboxylic acid halides
  27, 565
- startg. m. f.
  benzofuran-3(2H)-ones
  28, 876
  3(2H)-furanones **29**, 790
  ketones, degradation with
  loss of 1 C-atom **28**, 630
  oxindoles, 1-amino- **28**, 398
- - **imides**
- startg. m. f.
  4(5H)-oxazolones **27**, 265
- **acids**
- special s.
  α,α-dihalogenocarboxylic
  acids
  β-imido-α-halogenocarboxylic
  acids
- startg. m. f.
  α-aminocarboxylic acids
  26, 425
- **acid silyl esters**
-, reactions **28**, 861
-, Reformatskii synthesis
  with - **27**, 670
2-Halogenocyanamides
- from
  ethylene derivs. **26**, 521
- startg. m. f.
  aziridine ring, N-cyano-
  26, 521
N-Halogenocyanamides
  (s. a. N-Halides)
-, addition to carbon-carbon
  double bonds **26**, 515
o-Halogenocyclimmonium salts
- startg. m. f.
  lactams **30**, 136
α-Halogenodicarboxylic acid halides
- from
  dicarboxylic acids **3**, 451
  suppl. **29**
N-Halogenodicarboxylic acid imides s. N-Halogeno-
  succinimides
5-Halogeno-2,4-dienals
-, reactions with - **30**, 172
- startg. m. f.
  pyrylium salts **30**, 172
2-Halogeno-1,3-dienes
- from
  3-acetylenealcohols **26**, 537
4-Halogeno-2,4-dienones 29, 945
3-Halogeno-N,N-difluoroamines
- startg. m. f.
  α,β-ethylene-N-fluoroimines
  28, 459
N-Halogenodiimide N'-oxides
- from
  nitroso compds. **29**, 270

- special s.
  N-fluorodiimide N'-oxides 25
Halogenodinitriles
- special s.
  1-dicyanomethyl-2-halides
1,1,1-Halogenodinitriles
- startg. m. f.
  1-dicyanomethyl-2-halides
  26, 707
  γ-lactones 26, 707
1-Halogenoenesilanes
- startg. m. f.
  allenes, synthesis 29, 798
Halogenofluorination 29, 494
1-Halogeno-N-fluoroimines
- from
  1,1-dihalogeno-N,N-difluoroamines 26, 487
Halogenoformic acid esters
  s. Chloroformic acid esters,
  Fluoroformic – –
Halogenoformiminium halides
- special s.
  dimethylhalogenoformiminium halides
- startg. m. f.
  1,3-dioxole ring, 2-amino-
  29, 204
N-(Halogenoformyl)iminohalides, reactions 28, 383
- startg. m. f.
  acylisocyanates 30, 137
  N-carbamylamidines
  28, 383
Halogenogermanes
- special s.
  trihalogenogermanes
α-**Halogenogermanes 26, 70**
α-**Halogenoglycidic acid esters**
- from
  oxo compds. 26, 886
Halogenoglycosamines 29, 424
o-**Halogeno-N-heterocyclics**
  14, 608
- from
  N-oxides, cyclic 26, 536
o-α-**Halogeno-N-heterocyclics**
- from
  ethylene-N-halogenamines
  26, 524; 27, 530
- startg. m. f.
  lactams, ring expansion
  17, 263
p-**Halogeno-N-heterocyclics**
- startg. m. f.
  pyrazoles 28, 390
Halogeno-O-heterocyclics
  s. Halogenethers, cyclic
1,1-Halogenohydrazones
  (s. a. 1,1-Hydrazonohalides)
- from
  hydrazones 29, 502
- startg. m. f.
  1,1-arylthiohydrazones
  29, 565
  hydrazonyl sulfides 30, 407
  pyrazoles 16, 876
  thioacylhydrazones 30, 407
  1H-1,2,4-triazoles 29, 420
Halogenohydrins s. Halogenhydrins
S-**Halogenoisothiocarbamyl halides**
- startg. m. f.
  1,2,4-thiadiazolidine-
  3,5-diones 22, 295
  2-thiazolidones, 4-halogeno-
  26, 610

Halogenoketals
- special s.
  ethylenehalogenoketals
Halogenoketenes as intermediates, review 25, 599
  suppl. 26
Halogenoketocarboxylic acid esters
- special s.
  dihalogenoketocarboxylic
  acid esters
β-**Halogeno-α-ketocarboxylic** – –
- from
  oxo compds. 26, 886
- startg. m. f.
  α-alkoxyglycidic acid esters
  30, 129
Halogenoketones
- special s.
  azidohalogeno-ketones
  dihalogenoethylenehalogenohalogenoxidoperhalogenotrihalogeno-
α-**Halogenoketones**
- from
  ethylene derivs. 28, 481
  ketones 30, 353
  1-nitroxido compds. 30, 368
  oxo compds., insertion
  29, 871
  trialkylboranes and
  α-diazoketones 28, 860
- and α,β-ethyleneketones
  28, 860 suppl. 29
- startg. m. f.
  alcohols 29, 71
  carboxylic acid esters with
  rearrangement, skeletal
  27, 209
  γ-diketones 28, 804; 29, 792
  1,4-dithiins, 2,3-dihydro-
  26, 619
  1,3-dithiol-2-one ring 29, 588
  ethylene derivs. with
  decarbonylation 29, 984
  α,β-ethylene-α-hydroxyketones 26, 208
  α,β-ethylene-β'-hydroxyketones 26, 257
  α,β-ethyleneketones
  22, 959 suppl. 27
  β,γ-ethyleneketones,
  synthesis 30, 600
  ketones 26, 71; 27, 75;
  29, 79, 80; 30, 27
  –, hindered, synthesis
  23, 831 suppl. 26
  3-ketothioethers, synthesis
  30, 599
  2-ketoxanthates 29, 588
  1,4-oxazine ring 26, 429
  oxido compds. 28, 59
  α,β-oxidoketones 26, 864
  spirocyclobutanones 26, 897
  α-sulfonyl-γ-diketones 29, 792
  α-vinylketones 30, 598
β-**Halogenoketones**
- from
  carboxylic acid chlorides
  2, 617; 27, 213
o-**Halogenoketones**, ring
  closures with – 22, 467
  suppl. 29

1-Halogeno-2-ketothioethers
- from
  α-diketones and sulfenylhalides 26, 602
- startg. m. f.
  2-ketothioethers 28, 61
2-Halogenolactamois
- startg. m. f.
  2-ketolactams, ring
  contraction 29, 471
Halogenolactams
- startg. m. f.
  ethylenelactams 26, 971
α-**Halogenolactams**
- startg. m. f.
  α-iminocarboxylic acids,
  cyclic 27, 211
N-Halogenolactams
- special s.
  2-azetidinones, N-halogeno-
α-**Halogeno-β-lactones**
- startg. m. f.
  allenes 29, 979
α-**Halogeno-γ-lactones**
- from
  α-carboxy-γ-lactones
  16, 620 suppl. 28
  γ-lactones 28, 895
γ-**Halogeno-β-lactones**
- from
  β,γ-ethylenecarboxylic acids
  27, 561
Halogenolysis 30, 376
Halogenomethyl compds.
- startg. m. f.
  aldehydes 26, 209
Halogenomethylene compds.
- special s.
  fluoromethylene compds.
(**Halogenomethylene**)-
  dimethyliminium halides
  s. Dimethylhalogenoforminium halides
Halogenomethyl ethers
- from
  alcohols 26, 167
S-**Halogenomethyl groups**
- from
  S-1,2-dihalogenovinyl
  groups 26, 77
o-(**Halogenomethylsulfonyl-**
  amino)ketones
-, reactions with ammonia
  26, 433
- startg. m. f.
  quinazolines, 1,2-dihydro-
  26, 433
N-Halogeno-N-nitramines
- from
  N-nitramines 27, 290
Halogenonitriles
- from
  oximes, cyclic 29, 515
- special s.
  dihalogeno-nitriles
  ethylenehalogenohalogenodi-
- startg. m. f.
  nitriles, Friedel-Crafts
  alkylation 28, 824
α-**Halogenonitriles**
- by halogenation 11, 616
- startg. m. f.
  β-hydroxynitriles,
  Reformatskii synthesis
  26, 672
  ketones 26, 718
-, syntheses with – 26, 672

β-Halogenonitriles
- from
  α,β-ethylenecarboxylic acid
    amides 28, 502
- startg. m. f.
  α,β-ethyleneamidines 30, 219
γ-Halogenonitriles
- startg. m. f.
  Δ¹-pyrrolines, 2-amino-
    27, 450
ω-Halogenonitriles 29, 959
Halogenonitrones,
  rearrangement 27, 528
α-Halogenonitrones
- startg. m. f.
  1,2-oxazine ring, tetrahydro-
    28, 657
-, syntheses with – 28, 657
1,1-Halogenoorganolithium
  compds.
-, reactions via – 26, 70; 30, 549
-, – – –, review 21, 863
  suppl. 27
Halogenooxo compds.
- special s.
  ethylenehalogenooxo compds.
- startg. m. f.
  ethers, cyclic 30, 171
α-Halogenooxo –
- startg. m. f.
  imidazole ring, N-condensed
    13, 535
  thianaphthene-3-carboxylic
    acids, 2-acyl- 29, 813
β-Halogenooxo –
- from
  2-ethylenealcohols,
    rearrangement 21, 601
1,2-Halogenoperoxides
- from
  ethylene derivs. and
    O-halogenoperoxides
    30, 350
- via peroxymercuration
    30, 350
O-Halogenoperoxides
- startg. m. f.
  1,2-halogenoperoxides
    30, 350
o-Halogenophenols
- from
  diazo oxides 26, 550
- startg. m. f.
  phenols 26, 550
o-γ-Halogenophenols
- from
  chromans 30, 345
Halogenophosphines
- special s.
  chlorophosphines
α-Halogenophosphinic acid
  esters
- startg. m. f.
  phosphonic acid esters
    28, 798
α-Halogenophosphonic acid
  dihalides
- from
  ethylene derivs. 29, 599
α-Halogenophosphonium salts
- startg. m. f.
  ethylene derivs. 30, 620
Halogenophosphoramidates
  s. a. Dihalogenophosphor-
    amidates

2-Halogenophosphoramidates
- from
  ethylene derivs. and
    N,N-dihalogeno-
      phosphoramidates 27, 548
- startg. m. f.
  1,2-diamines 28, 411
  2-halogenamines 27, 548
Halogenophosphoric acid
  esters s. Phosphorohalidates
Halogenoquinolines 26, 532
N-halogenoquinone monoimines
- special s.
  2,6-dichloroquinone-
    chlorimine
Halogenosilanes
- from
  halides 28, 586
- special s.
  bromo-silanes
  chloro-
    (polyhalogeno)-
- startg. m. f.
  1,2-disilanes 26, 650
  C,N-disilylketenimines
    30, 438
  N-silylketenimines 29, 295
  α-silylnitriles 29, 295
α-Halogenosilanes 26, 70
N-Halogeno-N-sodiosulfonic
  acid amides 27, 305
- special s.
  chloramine-B
  chloramine-T
Halogenostibines
- startg. m. f.
  aminostibines 26, 280
β-Halogenostyrenes
- by replacement, geospecific
    26, 552
N-Halogenosuccinimide
  (s. a. N-Bromo-succinimide,
  N-Chloro-, N-Iodo-)
- as reagents 23, 248 suppl. 28;
    24, 581 suppl. 28;
    28, 487, 496, 499
2-Halogenosulfenic acid
  alkoxymethyl esters 28, 475
- from
  1,1-alkoxyhalides and
    thiirane 1-oxides 28, 475
Halogenosulfenium salts
-, o-alkylation via – 28, 762
- startg. m. f.
  aminosulfenium salts 28, 254
Halogenosulfines
- special s.
  chlorosulfines
  dichlorosulfine
α-Halogenosulfonates
- startg. m. f.
  hydrocarbons 26, 73
Halogenosulfones
- from
  hydroxysulfoxides 27, 567
α-Halogenosulfones
  (s. a. α,α-Dihalogeno-
    sulfones)
-, α-alkylation 28, 795 suppl. 30
- from
  α-diazosulfones 30, 367
- startg. m. f.
  α,β-oxidosulfones
    28, 795 suppl. 30
  sulfones, synthesis
    24, 879 suppl. 29

β-Halogenosulfones
  6, 572 suppl. 29
- special s.
  α,β-ethylene-β-halogeno-
    sulfones
- startg. m. f.
  α,β-ethylenesulfones
    3, 752 suppl. 27
Halogenosulfonic acid amides
- startg. m. f.
  sultams 30, 671
2-Halogenosulfonic – –
- startg. m. f.
  2-aminoalcohols 29, 213
N-Halogenosulfonic – –
  (s. a. N-Halides)
- special s.
  N,N-dihalogenosulfonamides
  N-halogeno-N-sodiosulfonic
    acid amides
α-Halogenosulfonylcarboxylic
  acid halides
- from
  carboxylic acids 28, 541
α-Halogeno-α-sulfonylethers
- startg. m. f.
  α,α-disulfonylethers 29, 583
α-Halogenosulfonylhydrazones
- startg. m. f.
  N-sulfonylenazo compds.
    29, 962
Halogenosulfonyl isocyanates
  (s. a. Chloro-sulfonyl
  isocyanate, Fluoro-)
-, reactions with oxo compds.
    30, 211
2-Halogenosulfonyl –
- from
  ethylene derivs. 26, 613
2-Halogeno-N-sulfonyl-
  N-nitramines
- startg. m. f.
  Δ²-1,2,3-oxadiazoline 2-oxides
    30, 173
Halogenosulfonylsulfonic
  acid amides
- special s.
  dihalogenosulfonylsulfonic
    acid amides
Halogenosulfoxides
- special s.
  ethylenehalogenosulfoxides
α-Halogenosulfoxides 28, 554
- from
  sulfinic acid halides and
    diazo compds. 28, 554
  sulfoxides 6, 572 suppl. 26;
    27, 564
  thioethers 27, 564
-, review 28, 554 suppl. 29
2-Halogenothiocyanates
- from
  ethylene derivs. 29, 540
2-Halogenothioenolethers
- from
  α,β-ethylenesulfoxides 26, 546
2-Halogenothioenolethers, sym.
- from
  acetylene derivs. 26, 579
Halogenothioethers
- special s.
  ethylenehalogenothioethers
1,1-Halogenothioethers
- from
  sulfoxides 28, 497
- special s.
  1,1-chlorothioethers
  1,1,1-dihalogenothioethers

1-halogeno-2-ketothioethers
- startg. m. f.
  3-ketothioethers, synthesis
  30, 599
  thioethers **18**, 125 suppl. **29**
-, **cyclic**
- startg. m. f.
  ethyleneacetals **29**, 972
Halogenothioformamidinium salts
- special s.
  chlorothioformamidinium chlorides
2-Halogenothiolurethans
- from
  thiocyanates and oxo compds. **28**, 618
Halogenothionoformic acid esters
- special s.
  chlorothionoformic acid esters
2-Halogenotosylhydrazones
- startg. m. f.
  enediazonium hexachloro-antimonates **27**, 973
Halogenoureas
- from
  halogenisocyanates **30**, 291
- special s.
  dichlorourea
  halogeneneureas
- startg. m. f.
  Δ²-oxazolines, 2-amino- **30**, 291
Halogenoxido compds.
  s. Oxidohalides
α-**Halogen**-α,β-**oxidoketones 27**, 855
Halogenoximes s. a. Nitrosohalides
1,1-Halogenoximes
  s. Hydroximinohalides
1,2-Halogenoximes
- from
  1,1-nitroethylene derivs. **27**, 570
- startg. m. f.
  1,2-nitroximes **27**, 463
  4H-1,2-oxazines, 5,6-dihydro-6-hydroxy- **26**, 849
-, substitution, nucleophilic **27**, 463
α-**Halogen**-α-**oximinocarboxylic acid esters 28**, 285
Halogen perchlorates
  as reactants **29**, 491
Halogenurethans
- special s.
  chloro-urethans
  monohalogeno-
1,2-Halogenurethans
  15, 373 suppl. **26**
- from
  ethylene derivs. **27**, 554
- startg. m. f.
  aziridines **27**, 458
Halonium salts
- special s.
  bromonium salts
  iodonium -
Handle, basic, in peptide synthesis **9**, 860 suppl. **24**; **22**, 395 suppl. **29**
Hantzsch pyrrole ring synthesis
-, modified and anomalous **28**, 787

Hassner-Ritter reaction 29, 317
Heat s. Pyrolysis
Heck arylation 24, 884
Helicenes (s. a. Heterohelicenes) **20**, 645 suppl. **26**; **28**, 892 suppl. **30**
-, review **28**, 892 suppl. **29**
Helium 26, 45, 461; **27**, 251; **29**, 401, 948
Hematoporphyrin s. under Sensitizers
Hemiacetal esters s. 1,1-Alkoxy-acoxy compds.
Hemiacetals (s. a. Ketene silyl hemiacetals)
-, **cyclic** s. Lactols
Hemiketals
- from
  ketoaldehydes **26**, 874
-, **cyclic** s. Lactols
Hemiquinolesters
- from
  phenols **28**, 98
- startg. m. f.
  3,5-dienecarboxylic acid derivs. **28**, 95
Hemiquinolethers
- from
  phenols **25**, 174 suppl. **27**; **28**, 98, 130
Hemiquinols
- from
  phenolethers **29**, 233
  phenols **28**, 98
Hemithioacetals s. Dithiohemiacetals
Hetarynes, cycloaddition with - **27**, 824
Heteroadamantanes
-, review **19**, 326 suppl. **26**
- special s.
  oxaza-adamantanes
  trioxa-
Heterobonds, multiple
-, reactions of 1,3,5-oxazaphosph(V)ol-2-ines with - **27**, 887 suppl. **29**
Hetero-Cope-rearrrangement 28, 897
Heterocumulenes 30, 579
- from
  heterocyclics **29**, 477
-, reactions with oxaziridines **30**, 213
- special s.
  carbodiimides
  isocyanates
  isothiocyanates
  ketenimines
  keteniminylidene-phosphoranes
- startg. m. f.
  heterocyclics **27**, 675; **30**, 221
S(IV)-Heterocumulenes 26, 276 suppl. **29**
Heterocyclics (s. a. Ring closure)
-, arylation, homolytic, of -, review **16**, 851 suppl. **27**
- by
  aldol-type condensation, intramolecular **17**, 921; **26**, 935
  anodic oxidation **29**, 239
  aziridine ring expansion, review **26**, 706
  cyclodehydration **28**, 216

displacement of nitro groups, ar. **29**, 776
ether cleavage **26**, 657
Mannich reaction **30**, 562, 564
Pummerer rearrangement **29**, 250; **30**, 166
solvomercuration-demercuration, intramolecular **27**, 345
-, C-carbamylation, homolytic, of - **25**, 480
- from
  2-acetylenephosphonium salts **28**, 962
  acylphosphonic acid esters, review **26**, 909 suppl. **29**
  α-alkoxymethyleneketones **24**, 418 suppl. **28**
  α-amino-β,γ-ethylenenitriles **27**, 535
  carbohydrates **28**, 446
  carbon dioxide eliminators, cyclic **26**, 449
  N-(chlorocarbonyl)isocyanate **28**, 473
  cyclobutanes **30**, 474
  diazonium salts **28**, 369
  - - and nitriles **28**, 369
  γ-dioxo compds. **11**, 668 suppl. **26**
  ene-1,2-di-(oxysilanes) **26**, 811
  enephosphonium salts **30**, 178
  heterocumulenes and ketenes **27**, 675
  - and 1,2,3,4-thiatriazolines, 5-sulfonylimino- **30**, 221
  iminohalides **28**, 810
  isoxazole ring, review **29**, 322 suppl. **30**
  ketenes **27**, 675
  β-ketosulfoxides **30**, 166
  lactams, review **22**, 341 suppl. **27**
  lactones, - **22**, 341 suppl. **27**
  nitriles **26**, 792 suppl. **27**
  phosphorus compds., review **28**, 962 suppl. **29**
  silyl derivs. **27**, 488
  ω-sulfinylacetophenones, o-subst. **27**, 886
  1-sulfonyl-1-isonitriles **28**, 841
  tetraacylethylenes **27**, 571
  thiolactones review **22**, 341 suppl. **27**
-, metalation, α-lateral **30**, 480
-, photochemistry, review **28**, 643
-, photodecarboxylation **27**, 534
-, ring expansion with diazo compds. **18**, 848 suppl. **29**
-, 2-silyl groups, reactive in - **27**, 883
- special s.
  7-azabicyclo[2.2.1]heptanes
  azacyclols
  cage compds., heterocyclic
  carbohydrate derivs., heterocyclic
  ethyleneheterocyclics
  heteroadamantanes
  heterocyclophanes
  heterohelicenes
  N-macroheterocyclics
  metallocyclics
  silylheterocyclics
  spiroheterocyclics
    tetra-Het-spirocyclics

**(Heterocyclics
– special s.)
with 1 O-atom**
benzocoumarins
benzofurans
benzopyrans
benzoxepins
cyclohepta[b]furans
cyclopenta[8]chromans
dibenzofurans
dibenzopyrans
isochromans
isocoumarins
naphthopyrans
naphthothiophenes
naphthoxepines
oxa...
oxepins
phenaleno[1,2-b]furans
oxetanes
oxido compds.
pyrans
xanthenes
with 1 N-atom
acridines
aporphines
azaadamantanes
4-azaazulenes
azabicyclo...
1-azaspiro[2.2]pentanes
azacyclols
azepines
azetines
azirines
azocines
azoniaanthracenes
benzazepines
benzisoindolinium salts
benzocarbazoles
benzo[5,6]cyclohepta[1,2-c]-
 pyridines
benzomorphanes
benzo[a]quinolizines
carbazoles
cycloprop[b]indoles
diazocines
dibenzo[de,h]quinolines
indenoquinolines
indoles
2,6-methano-3-benzazocines
morphinanes
9H-naphth[3,2,1-kl]acridines
pyridines
2,3-pyridocyclobutanes
pyrido[1,2-a]indoles
pyrindines
pyrroles
pyrrolo[1,2-a]indoles
quinolines
quinolizines
quinuclidines
spirocyclohexa-2,5-diene-
 1,1′-isoindolines
**with 1 hetero atom other
 than O or N**
acridarsines
arsabenzenes
benz[b]indeno[2,1-d]-
 thiophenes
benzo[b]selenophenes
benzo[b]thiepins
benzothiophenes
benzothiopyrans
9-borabicyclo[3.3.1]nonane
borepanes
phenoxaphosphines

4-phosphabicyclo[3.1.0]-
 hexanes
phosphepines
phosphorins
selenanaphthenes
selenochromans
selenocoumarins
selenophenes
silacyclohexanes
silacyclopentadienes
silacyclopropanes
stannins
9-thiabicyclo[4.2.1]nonanes
thiepins
thietes
thiiranes
thiochromans
thiocoumarins
thiophenes
thiopyrylium salts
thioxanthenes
with 2 O-atoms
benzodioxepine
benzodioxoles
benzofuro[3,2-c][1]benzo-
 pyrans
cyclopentadioxines
dibenzo[d,g][1,3]dioxocins
2,10-dioxabicyclo[4.4.0]-
 decanes
4,7-dioxaspiro[2,4]heptanes
2,7-dioxatwistans
dioxetanes
dioxins
1,5-dioxocins
1,3-dioxoles
with 2 N-atoms
6H-benz[b]indolo[3,2,1-d,e]-
 [1,5]naphthyridines
benzo[4,5]canthines
benzo[c]cinnolines
benzodiazepines
benzodiazocines
1,3-benzodioxole-2-spiro-
 cyclohexanes
diazabicyclo...
8,17-diazadibenzo[c,j]tetra-
 cyclo[7.3.1.0^{2,9}.0^{7,16}]-
 tridecanes
diazepines
diazetidines
diazirines
diazocines
dibenzo-1,4-diazepines
6H-dibenzo[d,f][1,3]-
 diazepines
dibenzo[c,h][1,5]-
 naphthyridines
2H-1,4-ethanobenzo[b][1,5]-
 naphthyridines
imidazo[2,1-a]isoindoles
imidazo[2,1-a]isoquinolines
imidazoles (glyoxalines)
6H-indeno[1,2-e][1,4]-
 diazepines
indolizino[1,2-c]quinolines
indolo[2,1-b]quinazolines
indolo[2,3-a]quinolizines
naphthyridines
 (diazanaphthalenes)
perimidines
phthalazines
pyrazines
pyrazolo[1,5-a]pyridines
pyrazolo[2,3-a]quinolines
pyrid[3,4-b]indoles
pyrid[4,3-b]indoles

pyrido[4,3-b]carbazoles
pyrido[1,2-b]cinnolin-6-ium
 ylids, 5,11-dihydro-
pyrido[1,2-a]pyrimidines
pyrimidines
pyrimido[1,6-a]indoles
pyrrolo[1,2-a]benzimidazoles
1H-pyrrolo[2,1-b][1,3]-
 benzodiazepines
5H-pyrrolo[2,1-c][1,4]-
 benzodiazepines
6H-pyrrolo[1,2-a][1,4]-
 benzodiazepines
pyrrolo[1,2-b]pyridazines
pyrrolo[2,1-b]quinazolines
pyrrolo[1,2-c]quinazolines
pyrrolo[2,3-b]quinolines
pyrrolo[3,2-b]quinolines
pyrrolo[3,4-b]quinolines
quinazolines (1,3-diaza-
 naphthalenes)
quinoxalines (1,4-diaza-
 naphthalenes)
spiro-3H-indolines-3,2′-
 2′H-pyrroles
**with 2 hetero atoms other
 than O or N**
2,3-benzodithianes
benzodithioles
benzoselenopheno[2,3-b]-
 benzoselenophens
diselenols
2,7-dithiaisotwistanes
dithianes
dithiepanes
dithietanes
dithietes
dithiins
dithioles
indeno[2,1-c]-1,2-dithioles
thieno[3,4-c]thiophenes
with 2 different hetero atoms
anthr[2,1-d]oxazoles
3-azaphospholes
benzisothiazoles
benzofuro[2,3-c]pyrroles
benzothiazetes
benzothiazines
benzothiazoles
benzoxathians
benzoxazepines
benzoxazines
benzoxazocines, methano-
benzoxazoles
cephalosporins
cephams
dibenzothiazines
1,2-disilacyclohexanes
furopyridines
furoquinolines
isothiazoles
isoxazoles
isoxazolo[2,3-a]pyridinium
 salts
naphthoxazoles
oxaborins
oxaborolans
oxaphosphorins
4-oxa-5-silaspiro[2.6]nonanes
oxathiins
oxathioles
2-ox-6-azaadamantanes
2,1-oxazabicyclo[4.1.0]-
 heptanes
2,3-oxazabicyclo[3.2.0]-
 heptanes
6-ox-2-azabicyclo[3.1.0]-
 hexanes

oxazepines
oxazines
oxazirines
oxazoles
penicillins
phenoselenazines
phenothiazines
phenoxazines
pyrano[2,3-b]pyridines
pyrano[3,2-c]quinolines
pyronobenzo[ij]quinolizines
pyrrolo[2,1-b][1,3,4H]-
    oxazines
rheadans
selenazoles
spiro(indoline-2,4'-
    [4'H]pyrans]
thiaborolanes
thiaphosphorins
1,4-thiazepines
thiazetidines
thiazines
thiazoles
thiazolo[4,3-a]isoindoles
thiazolo[3,2-a]pyridinium
    salts
thieno[3,2-b]indoles
thiopyrano[2,3-b]indolium
    salts
with 3 hetero atoms
amide acetals, bicyclic
anthyridines
1-aza-4,6-dioxabicyclo[3.3.0]-
    octanes
2H-1,4-benzodiazepines,
    4,5-epoxy-
1,3,2-benzodioxaboroles
1,3,2-benzodithiaboroles
1,2,4-benzodithiazines
benzofurazans
benzofuro[2,3-b]quinoxalines
[1]benzopyrano[4,3-b][1,5]-
    benzodiazepines
benzothiadiazepines
benzothiadiazines
1,2,3-benzothiadiazoles
benzotriazepines
benzotriazines
benzotriazoles
benzotrithioles
benz[f][1,3,5]oxadiazepines
cyclopenta[e][1,3,4]-
    oxadiazines
diazaphospholidines
[1,4]diazepino[6,5-b]indoles
11H-dibenzo[c,f][1,2,5]-
    triazepines
1,3,2-dioxaphospholenes
dioxazines
dioxazoles
1,6a-dithia-6-azapentalenes
dithiazepines
dithiazines
dithiazoles
7a,11a-ethanooxyindolo-
    [1,2-h][1,7]naphthyridines
4H-furo[3,4-d][1,3]benzo-
    diazepines
2-germa-1,3-dioxahetero-
    cycles
germasultones
1H-imidazo[5,1-c][1,4]-
    benzodiazepines
1H-imidazo[1,2-a]imidazoles
imidazopyrazines
3H-imidazo[4,5-f]quinolines
imidazo[1,5-a]quinoxalines
imidazo[2,1-b]thiazoles

indolo[3,2-b][1,4]benzoxazines
isoxazolo[1,2-b]pyrazoles
oxadiazabicyclo...
1,3,6-oxadiazepines
oxadiazines
oxadiazoles
oxathiazetidines
oxathiazoles
oxazaphospholes
10bH-oxazolo[3,2-c][1,3]-
    benzoxazines
2-phosphaquinazolinium salts
pyrazolo[3,4-b]pyridines
pyrazolo[1,5-a]pyrimidines
pyrazolo[1,5-c]pyrimidines
pyrazolo[1,5-a]quinazolines
pyrido[2,1-c][1,2,4]benzo-
    thiadiazines
pyrido[1',2':3,4]imidazo-
    [2,1-b]quinazolines
pyrido[2,3-d]pyrimidines
pyrido[1,2-a][1,3,5]triazines
pyrido[2,1-c][1,2,4]triazines
9H-pyrimido[4,5-b]indoles
2H-pyrimido[1,2-a]-
    pyrimidines
pyrimido[1,6-a]-
    benzimidazoles
pyrimido[1,2-a][1,8]-
    naphthyridines
1H-pyrrolo[2,1-c][1,2,4]-
    benzothiadiazines
pyrrolo[2,3-d]pyridazines
pyrrolo[2,3-d]pyrimidines
pyrrolo[2,3-b]quinoxalines
pyrrolo[1,2-b][1,2,4]triazines
pyrrolo[1,2-d][1,2,4]triazines
12H,6H-quinazolino[2,3-c]-
    [1,4]benzothiazines
selenadiazoles
silasultones
spiro[2H-1-benzopyran-
    2,2'-benzothiazolines]
spiro[pyrrolidin-3,1'-
    (pyrrolidino[3,4-c]-
    pyrrolidones)]
sydnones
thiadiazepines
thiadiazetidines
thiadiazines
thiadiazoles
thiaoxaborepanes
6a-thiathiophthenes
thiazaphospholanes
thiazolo[3,2-a]benzimidazoles
thiazolo[3,4-a]benzimidazoles
thiazolo[3,2-a][1,3]diazepines
thiazolo[2,3-b]quinazolines
thieno[2,3-e][1,4]diazepines
thieno[2,3-d]pyrimidines
1,4,7-triazatricyclo-
    [5.2.1.0^{4,10}]decanes
triazepines
triazines
triazoles
s-triazolo[3,4-a]isoquinolines
s-triazolo[4,3-a]pyridines
trioxaadamantanes
trioxanes
1,2,3-trisilacycloheptanes
trithianes
with 4 or more hetero atoms
alloxazines
azapurines
benzo[g]pteridines
benzo[e][1,2,3,4]thiatriazines
1H,3H-4,2,1,5-benzoxathia-
    diazocines

boraxanthines
boroselenines
carboranes
Crown polyamines
 – polyethers
cyclotriboroxanes
cyclotrisilazanes
1,3,2,4-diazadiphosphetidines
1,4,2,5-diazines
1,4,2,5-dioxadiazines
1,3,2,4-dioxathiazoles
diselenadiazocines
flavazoles
furazano[3,4-d]pyrimidines
hexaaza[26]annulenes
imidazo[2,1-c][1,2,4]triazines
imidazotriazoles
indole-1,2-dicarboximide
    ozonides
isoalloxazines
isoxazolo[4,5-b]quinoxalines
naphtho-[2,1-g]pteridines
1,3,6-oxadiazepino[3,4-a]-
    benzimidazoles
[1,2,5]oxadiazolo[3,4-b]-
    pyridines
1-oxa-6,6a-diselena-
    2-azapentalenes
1-oxa-6,6a-dithia-
    2-azapentalenes
oxathiadiazines
1,2,3,5-oxatriazoles
phosphazenes
purines
pyrazolidino-v-triazolines
pyrazolo[4,3-b][1,4]-
    benzoxazines
pyrazolo[3,4-d]pyrimidines
pyrazolo[3,4-e][1,2,4]-
    triazines
pyrimido[4,5-b][1,8]-
    naphthyridines
7H-pyrimido[4,5-b][1,4]-
    oxazines
pyrimidopteridines
pyrimido[4,5-d]pyrimidines
pyrimido[5,4-e]as-triazines
1,3,5,7-tetraazacyclooctanes
1,2,4,5-tetraazapentalenes
1,3,6,9-tetraazaspiro[4.4]-
    nonanes
tetrathiofulvalenes
tetrazepines
tetrazines
tetrazoles
1,2,3,4-tetrazolopyridines
thia(VI)thiatriazines
1,2,3,5-thiatriazoles
thiazolo[4,5-d]pyrimidines
thiazolo[4,5-b]quinoxalines
thiazolo[3,2-b]thiadiazolines
thiazolo[3,2-a]-s-triazines
triazadiborolidines
triazaphenothiazines
s-triazolo[4,3-a][1,4]-
    benzodiazepines
s-triazolo[4,3-b]benzo-
    thiazoles
s-triazolo[4,3-d][1,3,4]-
    benzotriazepines
s-triazolo[3,4-b][1,3,4]-
    oxadiazoles
s-triazolo[1,5-a]pyrimidines
v-triazolo[3,4-a]pyrimidines
v-triazolo[4,5-d]pyrimidines
triazoloquinazolines
triazolotriazines
tricycloquinazolines

(Heterocyclics)
- startg. m. f.
  heterocumulenes **29,** 477
- -, tautomerism, prototropic,
  review **18,** 397 suppl. **26**
- via
  copper-isonitrile complexes,
  review **28,** 606 suppl. **30**
  1,4-dianions **26,** 792
  nitrile ylids **28,** 810
  Stobbe condensation
  13, 725 suppl. **28**
- -, **fused**
- from
  1,3-dipoles, functionalized
  24, 692 suppl. **29**
- -, **4-membered**
- from
  carbodiimides **28,** 265
- -, **5-membered,** aromatization
  30, 637
- -, **mesoionic**
- -, O-alkylation **26,** 174
- special s.
  ψ-oxazoles
  anhydro-pyrido[1,2-a]-
    pyrimidinium hydroxides,
    2-hydroxy-
    pyrimidines, mesoionic
- -, **C-acyl-**
- from
  heterocyclics, oxido-,
  α,β-lateral **8,** 191 suppl. **29**
- -, **oxido-, α,β-lateral**
- startg. m. f.
  heterocyclics, C-acyl-
  8, 191 suppl. **29**
1,3-Heterocyclics (s. a. Azoles,
  Lithio salts, 1,3-heterocyclic)
  29, 340
- -, ring opening **26,** 670
- -, **2-amino-**
- from
  dithiocarbamidium salts
  27, 484
As-Heterocyclics
- from
  aminoarsines **26,** 568
- special s.
  arsabenzenes
B-Heterocyclics
- from
  dienes **27,** 648
- special s.
  boracyclanes
B,N-Heterocyclics
- special s.
  triazadiborolidines
B,N,S-Heterocyclics
- from
  boronic acids **27,** 307
B,O-Heterocyclics
- special s.
  benzodioxaboroles
B,O,S-Heterocyclics
- special s.
  thiaoxaborepanes
N-Heterocyclics (s. a. Aza...)
- -, C-alkylation, homolytic
  27, 907
- -, N-alkylation **27,** 457
  suppl. **28, 29**
- -, -, preferential **26,** 454
- by
  cyclodehydration **29,** 922, 927
  dehydrohalogenation **27,** 969
  Mannich reaction **29,** 751

ring closure, reductive
  28, 307
- from
  acylaminohalides **22,** 531
  o-aminoketones and nitriles
  27, 804
  benzofurazan oxides
  28, 420
  cyanoketones **6,** 522
  α-enaminocarboxylic acids
  30, 647
  enazides **24,** 528
  ethyleneamines **27,** 345
- -, ar. **17,** 395
  N-halogenamines **27,** 530;
  28, 460
  isocoumarins **24,** 386
  nitroketones **25,** 354
  perfluoroalkenes **29,** 789
  1,2,4-triazines **26,** 833
- -, reactions with sulfonic acid
  azides **28,** 321
- -, rearrangement, skeletal
  30, 533
- -, -, -, oxidative **28,** 99, 129
- special s. (a. under
  Heterocyclics)
  acetylene-N-heterocyclics
  acoxy-
  acyl-
  alkoxy-
  1-alkylthioazomethinium
    salts, H-heterocyclic
  alkylthio-N-heterocyclics
  amino-
  2-aminomesylates,
    N-heterocyclic
  azido-N-heterocyclics
  carbalkoxy-
  N,N-carbonyldi(azoles)
  carboxylic acids,
    N-heterocyclic
  cyano-N-heterocyclics
  cyclimmonium salts
  diazo-N-heterocyclics
  N-glycityl-
  N-glycosyl-
  halogeno-
  hydroxamic acids, cyclic
  hydroxy-N-heterocyclics
  imines, cyclic
  N-macroheterocyclics
  oxime tosylate,
    N-heterocyclic
  oximino-N-heterocyclics
  oxo-
  Reissert compds.
  siloxy-N-heterocyclics
  N-styryl-
- startg. m. f.
  N-aminocyclimmonium salts
  28, 243
  phosphonic acid esters
  27, 660
  1,2,3-triazole ring,
    N-condensed **28,** 321
- -, substitution, oxidative
  26, 774
- via
  furazan ring opening,
    reductive **27,** 521
  nitrenes, review **18,** 338
    suppl. **27**
- -, **bicyclic**
- startg. m. f.
  N-heterocyclics, monocyclic
  22, 42 suppl. **28; 29,** 41, 823
- -, **N-condensed 28,** 436

- by elimination of nitric
  acid **23,** 936
- by ring closure, double
  s. Ring closure, double
- -, epimerization **30,** 321
- from
  1-alkoxylactams **27,** 947
  o-aminocarboxylic acid
  amides **26,** 407
  benzocyclobutenes **29,** 651
  cyclimmonium salts **25,** 644
- - -, N-condensed **30,** 321
  cyclopropenones **27,** 706
  dicarboxylic acid imides
  27, 947
  1,3-dienes and hydantoins,
    5-alkoxy- **28,** 758
  1,5-dioxo compds. **26,** 345
  N-halogenamines **30,** 678
  hydroxycarboxylic acid
    amides **27,** 946
- special s.
  amidines, cyclic,
    N-condensed
  cyclimmonium salts, -
  N,N-dialkoxylamines,
    bicyclic, -
  guanidines, tricyclic,
    N-condensed
  imidazole ring, -
  imidazolium ring, -
  indole ring, -
  isoxazolidine ring, -
  lactams, -
  mercury-N-heterocyclics
  pyrazole ring, N-acyl,
    N-condensed
  pyridone ring, -
  1,2,4-triazole ring, -
- startg. m. f.
  cyclimmonium salts,
  N-condensed **30,** 321
- -, -, **epimeric**
- from
  benzocyclobutenes **29,** 651
- -, **monocyclic**
- from
  N-heterocyclics, bicyclic
  22, 42 suppl. **28;**
  29, 41, 823
N-Heterocyclics, o-subst.
- from
  N-oxides, cyclic, and
    enolethers **20,** 599 suppl. **26**
- -, **m-subst.**
- from
  N-oxides, cyclic **29,** 646
N,N-Heterocyclics (s. a. Azoles)
- special s.
  hydrazinium salts,
    N-condensed
- -, **N,N-condensed,** ring opening
  reductive **22,** 372
1,2-N,N-Heterocyclics
  (s. a. Azo bridges, Azoxy -)
- by recyclization, review
  23, 440 suppl. **28**
- special s.
  N,N-Diels-Alder adducts
- -, **satd.**
- startg. m. f.
  indole ring **30,** 563
- -, **3-oxo-**
- ring contraction **27,** 522
1,3-N,N-Heterocyclics
- from
  diamines and Δ²-oxazolines
  6, 452 suppl. **27**

ureas and oxo compds.,
  review **20**, 289 suppl. **28**
–, review **20**, 289 suppl. **28**
– special s.
  isouronium salts, cyclic
– startg. m. f.
  1,3-N,N-heterocyclics, 2-oxo-
–, **N-condensed**
– special s.
  1,1-iminoalcohols, 1,3-N,N-
  heterocyclic, N-condensed
–, **2-alkoxy-**
– from
  oxime tosylates,
  N-heterocyclic **28**, 121
–, **2-amino-**
– from
  diamines **26**, 438
  – and isothiocyanates **26**, 434
  1,3-N,N-heterocyclics,
  2-tosylamino- **26**, 438
–, **2-oxo-**
– from
  diamines **21**, 393
  1,3-N,N-heterocyclics
  29, 129
–, **perhydro- 27**, 427
–, **2-tosylamino-**
– from
  diamines **26**, 438
– startg. m. f.
  1,3-N,N-heterocyclics,
  2-amino- **26**, 438
O-Heterocyclics (s. a. Oxa...)
– by
  cyclodehydration **29**, 251
  cycloisomerization, double
  28, 687
  pyrolysis **23**, 286; **27**, 268
  Ullmann reaction,
  intramolecular **29**, 259
– from
  acylaminoalcohols **27**, 264
  dihalides **27**, 213
  diols **27**, 261
  enamines **28**, 633
  ethyleneoxo compds., ring
  closure, double **26**, 153
  isocyclics, ring expansion
  26, 127
  oxido compds. **30**, 104
–, hydroboration **26**, 30
–, ring closure with ether
  rearrangement **30**, 477
–, – expansion **27**, 276
– special s.
  acetals, mixed, O-heterocyclic
  spiro-O-heterocyclics
– startg. m. f.
  ethylenealcohols **26**, 30
  isocyclics **28**, 989
O-Heterocyclics, alicyclic
  s. Ethers, cyclic
–, **2-alkoxy-** s. Lactolides
O,N-Heterocyclics
– from
  azidohydroxynitriles,
  double ring closure
  29, 464
1,3-O,N-Heterocyclics
– from
  nitriles and aminoalcohols
  28, 374
–, **2-imino-**
– from
  carbodiimides and diols
  29, 361

O,O-Heterocyclics 28, 184
– from
  ethylenealcohols
  (2 molecules) **29**, 144
–, **bicyclic** s. Lactolides,
  bicyclic
O,O,N-Heterocyclics,
  rearrangment, skeletal **26**, 111
P-Heterocyclics
– from
  dihalides and diphosphines
  28, 597
–, ring expansion **27**, 954
– special s.
  P-carbohydrates
  phosphines, cyclic
–, **ar.**, review **29**, 710
P,N-Heterocyclics 29, 292
– from
  phosphonic acid
  diisocyanates **29**, 302
P,O,N-Heterocyclics
  s. Phosphoramidates, cyclic
S-Heterocyclics (s. a. S-Bridges,
  Thia...)
– by
  [8+2]-cycloaddition **28**, 636
– from
  thioketones **29**, 541
–, reaction with oxygen,
  singlet **26**, 125
– special s.
  poly-S-heterocyclics
  thioethers, cyclic
– startg. m. f.
  isocyclics, ring contraction
  26, 125
  thioozonides **26**, 125
–, **bicyclic**
– from
  ethylenetosylhydrazones
–, **alkylidene-** **27**, 957
– from
  α,β-ethylene-α-mercapto-
  carboxylic acids **29**, 548
(poly-S)-Heterocyclics
  s. Tetrathiospirocyclics
1,3-S,N-Heterocyclics
– from
  acetylenedicarboxylic acid
  esters **23**, 592
  hydroxycarboxylic acid
  thioamides **30**, 421
2,1,3-S,N,N-Heterocyclics
  28, 244
1,3-S,O-Heterocyclics 9, 676
S,P,N-Heterocyclics
– from
  N,N'-bisphosphinosulfone
  diimines **28**, 995
S,S-Heterocyclics, ring
  expansion **30**, 649
Se-Heterocyclics 29, 838
Se,N-Heterocyclics 28, 593
Si-Heterocyclics 30, 425
1,2-Si-Heterocyclics, ring
  expansion **30**, 425
Si,S,N-Heterocyclics 30, 336
Sn-Heterocyclics s. Cyclostannanes, Stannins
Heterocyclic salts (s. a. Cations,
  heterocyclic)
– special s.
  benzothiazolium salts
  cyclimmonium –
  diselenolium –
  dithietanylium –
  dithiolium –

furylium –
isoxazolium –
1,3,5-oxadiazinium –
oxazinium –
oxazolium –
oxonium –, cyclic
2-phosphaquinazolinium –
phospholium –
thiaarenium –
thiazolinium –
Heterocyclophanes
  s. Dithiacyclophanes
Heteroelectrophiles
–, reaction with N-nitrosamines,
  metalated **30**, 437
– special s.
  diselenides
  disulfides
  halogenosilanes
Heterofulvenes (s. Dithia-
  fulvenes) **30**, 248
Heterohelicenes 20, 645
  suppl. **26**
S-Heterohelicenes 28, 892
Heterophanes s. Heterocyclophanes
Heteropolar compds. Het
Heterospirocyclics
  s. Spiroheterocyclics
Heterotwistanoids 29, 596
Heterotwistans
  s. 2,7-Dioxatwistans
1,5,10,14,18,23-Hexaaza[26]-
annulenes 29, 354
Hexaazamacroheterocyclics
  30, 284
– special s.
  hexaazaannulenes
Hexachloroantimonates 28, 254
– special s.
  carbonium hexachloro-
  antimonates
  enediazonium –
  tris-(p-bromophenyl)-
  ammoniumyl hexachloro-
  antimonate
Hexachlorocyclophosphaza-
triene s. Phosphonitrile
  chloride **25**
Hexachloroplatinic acid
  (s. a. under Sodium tetrahydridoborate) **8**, 748
  suppl. **26**; **29**, 403
– – -**alumina 27**, 758
Hexacyanoferrate(III)
  s. Potassium hexacyano-
  ferrate(III)
Hexadecyltributylphosphonium
bromide
– as phase transfer catalyst
  12, 619 suppl. **29**; **16**, 596
  suppl. **29**; **18**, 687 suppl. **29**
Hexadecyltrimethyl-
ammonium –
– as phase transfer catalyst
  27, 833 suppl. **30**
Hexafluoroacetone
  s. Perfluoroacetone
Hexafluoroantimonates
– special s.
  oxocarbonium hexafluoroantimonates
Hexafluorophosphates 27, 180;
  28, 144 suppl. **30**; **28**, 232;
  29, 717
Hexametapol s. Hexamethylphosphoramide

Hex—Hyd 646

Hexamethylcyclotrisilazane
 as reagent **26**, 501
Hexamethyldisilazane
 (s. a. Potassium bis-
 (trimethylsilyl)amide,
 Sodium –)
– as reactant **27**, 83; **28**, 350;
 30, 380
Hexamethyldisilthiane,
 reactions with – **28**, 557
Hexamethylenetetramine
– as reactant **27**, 474
–, C-formylation with –
 1, 621; **2**, 702
**Hexamethylenetetraminium
 salts 27**, 541
Hexamethylphosphoramidate
 s. Hexamethylphosphoramide
Hexamethylphosphoramide
– as additive **30**, 587
– as medium **12**, 43 suppl. **29**;
 18, 736 suppl. **26**; **21**, 771
 suppl. **26**; **26**, 54, 545; **27**, 74,
 87, 843; **28**, 165, 916; **29**, 266,
 799, 959; **30**, 6, 36, 444, 505,
 536, 615
– as reactant **26**, 378, 447; **28**,
 352; **29**, 294; **30**, 200
– as reagent **22**, 803 suppl. **29**;
 26, 308 suppl. **28**; **26**, 501;
 28, 440 suppl. **29**
– as LiHal-solubilizer **19**, 620
 suppl. **26**; **29**, 930
– as solvent **10**, 360 suppl. **28**;
 10, 452 suppl. **27**; **13**, 573
 suppl. **29**; **17**, 705 suppl. **26**;
 20, 446 suppl. **28**; **21**, 869
 suppl. **29**; **23**, 413 suppl. **26**;
 23, 860 suppl. **27**; **26**, 17, 361,
 607, 640, 650, 923; **27**, 223,
 255, 273, 676, 851; **28**, 77, 173,
 312, 547, 776, 802; **29**, 409
–, review **21**, 857 suppl. **26**
–/**lithium diisopropylamide**
 s. under Lithium amides
–/**molybdenum peroxide**
 s. under Molybdenum
 peroxide
–/**potassium** s. under Potassium
–/**sodium** s. under Sodium
– **-thionyl halide** s. under
 Thionyl halide
**Hexamethylphosphorous
 triamide** s. Trisdimethyl-
 aminophosphine
Hexoses
– startg. m. f.
 pentoses **27**, 277
High dilution, technique
 s. under Ring closure
– – **apparatus 5**, 525
 suppl. **26**; **26**, 614
 suppl. **27**
– **pressure diene synthesis
 27**, 696
– – **reactions**
–, review **14**, 391 suppl. **26**;
 27, 696 suppl. **28**
–, ring closure **25**, 507 suppl. **29**
– **temperature medium**, vaseline
 oil as – **26**, 465
Hindrance, steric s. Compds.,
 hindered
**Hinsberg thiophene ring
 synthesis 7**, 773
Histidine peptides 22, 395
 suppl. **29**

Hoffmann s. Woodward
Hofmann degradation
– of
 ammonium salts, quaternary
 26, 960; **28**, 937; **29**, 950
 carboxylic acid amides,
 degradation of carbo-
 hydrates via – **27**, 540
 cyclimmonium salts with
 dehydrogenation **29**, 707
– – – lactone ring opening
 28, 91
– –, **abnormal**
– of
 ammonium hydroxides,
 quaternary **26**, 983
– –, **directed**
– of
 cyclimmonium salts **28**, 695
Hofmann-type degradation
– with
 retention of configuration
 28, 309
**Homoallyl rearrangement,
 geospecific, ethynyl group-
 controlled 17**, 611 suppl. **27**
Homoconjugate addition
 s. Addition, homoconjugate
**Homoelectrocyclic reactions
 26**, 126
Homogeneous s. under Hydrogenation, Metal catalysis,
 Catalysts
Homologization (s. a. Synthesis
 with addition of 1 C-atom)
– of
 carboxylic acid esters **30**, 545
 dicarboxylic acid derivs.
 30, 456
 halides **28**, 802
 ketones, cyclic **11**, 858
 suppl. **26**; **26**, 661
– with
 diazomethane **29**, 763
Homolytic reactions
 s. Radical reactions
Homophthalic acid anhydrides
– startg. m. f.
 benzocyclobutenones
 28, 970
 fulveneallenes **28**, 970
Homophthalic acid esters
– from
 1,3-cyclohexadienes and
 allene-1,3-dicarboxylic
 acid esters **28**, 868
Homophthalic acids 17, 799
 suppl. **26**
Homophthalimides
– from
 halides, ar., and malonic
 acid esters **29**, 811
Hooz reaction of boranes with
 diazo compds. **24**, 817;
 28, 783
Horner synthesis
–, benzene ring annelation
 via – **30**, 645
– of
 a-allenecarboxylic acid
 esters **28**, 851
– –, **intramolecular 25**, 693
Huang Minlon reduction
 s. Wolff-Kishner
Hünig base s. Ethyldiisopropylamine

Hurtley arylation 30, 593
**Hydantoin ring, 1-(o-hydroxy-
 aryl)-**
– from
 2-benzoxazolone ring **26**, 348
Hydantoins
– from
 cyanohydrins **30**, 568
–, reaction with oxido compds.
 28, 267
– special s.
 thiohydantoins
– startg. m. f.
 a-aminocarboxylic acids
 30, 568
–, **5-subst.**
– from
 hydantoins, 5-alkoxy- **24**, 796
–, **5-alkoxy-**
– startg. m. f.
 N-heterocyclics, N-condensed
 28, 758
 hydantoins, 5-subst. **24**, 796
–, **3-hydroxy-**
– from
 a-N-hydroxyureidocarboxylic
 acid esters **27**, 519
 a-isocyanatocarboxylic acid
 esters **27**, 519
Hydrates as reactants **27**, 205
Hydration (s. a. under
 Markownikoff)
–, cleavage of hydrazones
 via – **27**, 205
– of
 carbon-carbon double bonds
 s. a. Alcohols from
 ethylene derivs.
–, **asym. 16**, 196; **22**, 143
 suppl. **27**
–, **preferential**
– of
 carbon-carbon double bonds
 22, 143 suppl. **28**
–, **selective**
– of
 carbon-carbon double
 bonds **27**, 140
Hydrazides (s. a. Carboxylic
 acid hydrazides)
–, reduction with borane
 26, 56
Hydrazinyl radicals, review
 23, 142 suppl. **28**
Hydrazine (s. a. Wolff-Kishner
 reduction) **26**, 15; **27**, 584;
 28, 2, 24, 339; **29**, 16; **30**, 46
–, reaction with dicarboxylic
 acid anhydrides **8**, 482
– startg. m. f.
 acylhydrazones **27**, 377
 4H-1,2-diazepines **26**, 370
 nitriles **28**, 339
 1,3,4-oxadiazoles **26**, 380
 pyrazoles, 3-amino- **30**, 298
 1,2,4,5-tetrazine, monosubst.
 28, 381
– ring, 1,2,3,4-tetrahydro-
 29, 281
 1,2,4-triazoles **29**, 427
–/**nickel 27**, 15
–, **anhydrous 26**, 349; **27**, 444
Hydrazines
–, N-alkylation **29**, 414
– from
 amines **26**, 259
 oxo compds. **13**, 104 suppl. **26**

1,2,4-triazolidine-3,5-diones
  29, 21
-, reaction with
  o-dicarboxylic acid
    anhydrides 8, 482 suppl. 26
  α,β-ethylenecarboxylic acid
    derivs., review 14, 424
    suppl. 26
- special s.
  acyl-hydrazines
  alkoxy-
  amino-
  carbalkoxy-
  cyano-
  glycosyl-
  nitroso-
  phenylhydrazine
  stannyl-hydrazines
  sulfonyl-
  thioacyl-
- startg. m. f.
  amines 29, 21
  1,1-cyanohydrazines 26, 828
  1H-1,2-diazepines, 4,5,6,7-
    tetrahydro- 28, 386
  N-nitrosamines 28, 238
  N-nitrosohydrazines 26, 828
  pyrazoles, review 26, 386
  3-pyrazolidones 28, 379
  pyrazolidines 26, 780
  3(2H)-pyridazones,
    4,5-dihalogeno- 26, 427
  pyrrole ring, 3-amino-
    28, 331
  pyrroles 26, 842
  sydnone imine salts,
    3-amino- 26, 828
  thioacylhydrazines 26, 397
  triazanium salts 29, 271
  1,2,4-triazolidine-3,5-diones,
    4-amino- 29, 341
  Δ³-1,2,4-triazolines,
    3-mercapto- 29, 388
-, cyclic s. N-Amino-N-heterocyclics
-, 1,1-disubst.
- special s.
  1,1-dimethylhydrazine
- startg. m. f.
  oxindoles, 1-amino- 28, 398
-, monosubst.
- from
  N-nitrosamines, prim.
    1, 255 suppl. 29
  oxonium salts 28, 351
  semicarbazides 29, 274
  ureas 29, 274
-, tetrasubst. 28, 346
-, trisubst.
- from
  azo compds. and compds.,
    CH-acidic 27, 319
  N-nitrosamines 29, 433
-, 1,1,2-trisubst.
- from
  hydrazinium salts,
    1,1,1-trisubst. 26, 459
2-Hydrazinesulfonic acids,
  1-aryl-
- startg. m. f.
  1-acylhydrazines, 1-aryl-
    28, 409
Hydrazine-sulfur compds.
  s. Sulfur-hydrazine compds.
Hydrazinium salts
- startg. m. f.
  ethylene derivs. 30, 667
- -, N-condensed

-, ring opening, reductive
  26, 460
- startg. m. f.
  N-methylene bridges 26, 460
- -, inner s. N-Ylids
- -, quaternary
- special s.
  hydrazinium salts,
    1,1,1-trisubst.
- -, -, cyclic 27, 449
- startg. m. f.
  isoindolines 27, 449
- -, 1,1,1-trisubst.
- startg. m. f.
  hydrazines, 1,1,2-trisubst.
    26, 459
α-Hydrazinocarboxylic acid
  esters
- startg. m. f.
  pyridazinium betaines,
    4-hydroxy- 26, 940
o-Hydrazinocarboxylic acid
  esters
- startg. m. f.
  pyrazolo[1,5-a]quinazolin-
    5(4H)-ones 26, 385
Hydrazinosilanes
- startg. m. f.
  azosilanes 27, 285
Hydrazobenzene
- as reagent 26, 263
Hydrazo compds.
- from
  azo compds. 26, 15; 27, 199
- -, alicyclic
- startg. m. f.
  azo compds., alicyclic
    29, 283
Hydrazodicarboxylic acid
  esters
- startg. m. f.
  urethans 30, 208
- - -, cyclic
- startg. m. f.
  azoxy compds., cyclic 26, 92
Hydrazoic acid 26, 308
- as reactant 3, 378 suppl. 29;
    27, 423
Hydrazomethylene compds.
  s. Diaziridines
Hydrazones
- from
  enamines 29, 392
  imines 27, 444
  N-nitrosamines 29, 433
  oxo compds. 26, 940
-, reactions, review 28, 790
-, reaction with lead
    tetraacetate 26, 190
- special s.
  acetylene-hydrazones
  acyl-
  amidrazones
  arylthio-hydrazones
  carbalkoxy-
  cyano-
  cyclohexadienone -
    diaziridines
  dihalogenomethylene-
    hydrazines
  di-hydrazones
  diketone mono-
  2,4-dinotrophenyl-
    ethylene-
  guanyl- 25
  halogeno-
  hydroxy-
  isatin-

  nitro-
  sulfonyl-
  triketone mono-
- startg. m. f.
  acoxy compds. 26, 190
  amines, sec. 29, 399
  1-azoalcohols 30, 10
  1-azoethers 27, 185
  bis(organothio)methylene-
    hydrazones 26, 615
  1,1-halogenazo compds.
    27, 558
  1,1-halogenohydrazones
    29, 502
  nitriles 26, 481; 27, 525
  1,1-nitroazo compds. 30, 231
  1,1-nitrohydrazones 30, 231
  pyrazole ring 28, 790
  pyrazoles 26, 792, 926
  -, 5-amino- 29, 843
  Δ²-pyrazolines 26, 740
  2-pyridones, 3,4-dihydro-
    1-amino- 30, 556
  pyrroles 26, 842
  1,2,4-triazolidine-3-thiones
    29, 306
-, subst.
- from
  azo compds., aliphatic-ar.
    27, 352
1,1-Hydrazonohalides
  (s. a. 1,1-Halogeno-
    hydrazones)
- special s.
  imidazolinylhydrazono-
    halides
  1-(N-tetrazol-5-ylhydrazono)-
    1-halides
- startg. m. f.
  (α-azoalkylidene)-
    phosphoranes 29, 611
Hydrazonyl sulfides
- from
  1,1-halogenohydrazones
    30, 407
  thioacylhydrazines 30, 407
Hydride donor (s. a. H-Donor)
-, poly(hydrosiloxane) as –
    24, 126 suppl. 29
- shift 27, 384
- transfer 29, 13
-, removal of protective
    groups by – 27, 239
Hydriding agent, highly
  reactive
  potassium hydride 28, 884
    suppl. 30
Hydridotetrakis(triphenylphosphine)rhodium(I)
- as hydrosilylation catalyst
    26, 642 suppl. 30
Hydriodic acid s. Hydrogen
  iodide
Hydroallylation of ethylene
  derivs. 29, 139
Hydroalumination 27, 142
Hydroboration (s. a. Brown
  hydration) 29, 158
- of
  O-heterocyclics 26, 30
- 1,2,3-trienes via – 28, 943
    suppl. 29
-, partial
- of
  1,3-diynes 14, 82 suppl. 26
Hydrocarbon chains, long-
  range activation 20, 533
  suppl. 26

Hydrocarbons, hydrocarbon groups (s. a. Hydrogenolysis, Replacement by hydrogen)
- from
 alcohols and trialkylalanes **28**, 845
 -, reduction **26**, 80; **27**, 88; **30**, 47
 - via phosphorodiamidates **28**, 65
 boranes, coupling **29**, 976
 boronic acid esters **27**, 84
 carboxylic acid amides
  s. Decarboxamidation
  - - esters HC↓↑C
  - - -, reduction **30**, 30
  - acids, decarboxylation HC↑C
  - - (methyl groups) **26**, 66; **27**, 65; **28**, 52
 diazonium mercuric salt complexes **30**, 36
 1,1-dihalides **29**, 72
 1,2-dihalides **22**, 317 suppl. **29**
 ethylenealcohols **29**, 933
 ethylene derivs., degradation **30**, 46
 1,2-ethylene-1,2-dihalides **28**, 62
 ethyleneketones **29**, 933
 α,β-ethyleneketones **23**, 63 suppl. **29**
 halides, synthesis **26**, 875; **27**, 838
 -, with ring contraction **27**, 881
 α-halogenosulfonates **26**, 73
 hydroxy compds. HC↑O
 isoureas **30**, 50
 ketals **26**, 55
 mercaptals **28**, 63
 mercaptans **30**, 51
 mercury compds., monoorgano- **30**, 156
 oxido compds. **30**, 30
 oxo compd. N-dervs. **15**, 94 suppl. **26**
 oxo compds. **28**, 47; **29**, 63, 67
  aldehydes s. a. Decarbonylation of aldehydes
  - (methyl groups) **26**, 68; **30**, 30
  ketones (methylene groups) **28**, 48/9; **30**, 31
  - - and halides **27**, 872
 ozonides **30**, 46
 phosphorodiamidates **28**, 65
 sulfido compds. **27**, 976
 sulfonic acid esters, reduction **25**, 62; **29**, 78
 p-toluenesulfonic acid esters **27**, 74; **28**, 50, 52
 - - -, synthesis
 p-toluenesulfonic acid esters **29**, 824
 sulfonium salts **29**, 260
 sulfoxonium ylids **27**, 82
 sulfuric acid esters, synthesis **29**, 849
 - - monoesters **26**, 58
 thioethers **28**, 63
 -, synthesis **29**, 851
 thiolic acid esters (methyl groups) **26**, 76
 tosylhydrazones, reduction **29**, 67

-, synthesis **28**, 775
-, functionalization
 s. Functionalization
- special s.
 arenes
 methyl...
- startg. m. f.
 acylamines **29**, 334
 azo N,N'-dioxides **26**, 327
 carboxylic acid esters
  s. C-Carbalkoxylation
 - acids, synthesis with addition of 1 C-atom
  s. C-Carboxylation
 glycol mononitrates **28**, 132
 nitriles (from methyl groups) **30**, 237
 nitroso compds. **26**, 327
 oxo compds. **26**, 157; **27**, 181
  aldehydes (from methyl groups) **26**, 157
  ketones, oxidation **26**, 162, 165; **27**, 177, 188; **28**, 134, 136; **29**, 179
 N-sulfonylimines **27**, 181
Hydrocarbons, branched
- from
 alcohols, tert. **28**, 845
 carboxylic acids **28**, 845
 ketones **28**, 845
Hydrochlorides s. Pyridine hydrochloride, Triethylamine –
Hydrodesulfination 29, 90
Hydrodisulfides (s. a. Hydropolysulfides) **30**, 378
- startg. m. f.
 trisulfides **26**, 562
Hydrofluorination 29, 494
Hydroformylation
-, aldehydes, optically active by – **4**, 667 suppl. **29**
- of
 α,β-ethyleneacetals **4**, 667 suppl. **28**
Hydrogen s. a. under Palladium-carbon
-, **acidic** s. under Replacement
-, **active** s. a. Compds., proton-active
C-Hydrogen, active s. a. Compds., CH-acidic, Methyl and methylene groups, active, Nucleophiles
Hydrogen abstraction
 s. Oxidation
- **acceptor**
-, benzalacetone as – **28**, 209
Hydrogenation (s. a. Transferhydrogenation) HO, HN, HC
- of
 carbon-carbon double bonds **27**, 54; **28**, 43
 - multiple bonds, unactivated **26**, 47
 rings s. Ring hydrogenation
-, **asym. 27**, 57; **30**, 18
- of proline derivs. **29**, 55
-, review **27**, 57
-, **electrocatalytic 30**, 22
-, **geospecific 15**, 536 suppl. **26**; **30**, 25
-, **homogeneous**
- of
 carbon-carbon double bonds **27**, 57; **28**, 41; **29**, 54

- - -, terminal **29**, 49
-, review **28**, 41
-, **partial**
- of
 arenes **4**, 68
 carbon-carbon double bonds **28**, 41/2
 - triple bonds s. Ethylene from acetylene derivs.
 dicarboxylic acid imides **26**, 31
 quinol monoethers **22**, 49 suppl. **27**
-, **preferential**
- of
 carbon-carbon double bonds **17**, 94; **25**, 46; **26**, 50; **27**, 32, 58; **28**, 44; **29**, 45, 48, 53
 - - -, aliphatic **26**, 72; **28**, 39
-, **selective**
- of
 carbon-carbon double bonds **25**, 46 suppl. **26**; **25**, 554; **26**, 50; **27**, 58; **28**, 40; **29**, 48 suppl. **30**
 - - -, activated **27**, 55
 - - -, conjugated **28**, 18; **30**, 25
 - triple bonds **27**, 51; **29**, 53
 dicarboxylic acid imides **26**, 31
 N-heterocycles **23**, 70
 isocycles in N-heterocyclics **23**, 66 suppl. **29**
-, O-bridged **30**, 33
-, **stereospecific**
- of
 carbon-carbon double bonds **17**, 82 suppl. **26**; **20**, 58 suppl. **27**; **27**, 89; **28**, 42; **23**, 63 suppl. **29**
 enaminolactams **27**, 53
 - with 1,4-dihydropyridines **17**, 80 suppl. **29**
-, **total**
- of
 oxo-N-heterocyclics **26**, 67
Hydrogen atom transfer (s. a. Hydrogen transfer)
- - - **agent**, *tert*-butyllithium as – **29**, 13
- **bromide scavenger**
-, epichlorhydrin as – **26**, 208
- **carbonate anion**
 - as nucleophile **30**, 140
- **chloride** s. under Cobalt(II) acetate
- - **scavenger**, K-carbonate as – **27**, 353
- **cyanide**
 - startg. m. f.
 isoquinolines **28**, 743
- **donors** s. H-Donors
- **fluoride** (s. a. Poly(hydrogen fluoride)) **3**, 170 suppl. **28**; **19**, 161 suppl. **26**; **26**, 438 suppl. **28**; **27**, 159, 810; **28**, 926; **29**, 496; **30**, 106, 635
- -, **liq. 6**, 44 suppl. **26**
- **iodide 26**, 72, 851; **28**, 51; **30**, 32
- - **acceptor**, K-acetate as – **27**, 871
Hydrogenolysis (s. a. Hydrocarbons, Oxygen, benzylic, removal, Replacement by hydrogen)

- of
  benzyl esters and benzyl
  ethers s. O-Debenzylation
  sulfur compds. **30**, 51
-, prevention **29**, 48
-, **benzylic** s. a. Oxygen,
  benzylic, removal
-, separation of aldehydes
  via – **30**, 149
-, -, **partial 28**, 57
-, -, **selective 28**, 57
gem-**Hydrogenolysis 26**, 53
Hydrogen peroxide OC⇓CC; **20**,
  63 suppl. **26**; **22**, 117 suppl. **27**;
  22, 170 suppl. **29**; **26**, 3, 92/3,
  96, 100, 117, 151, 210, 228, 254,
  504, 928; **27**, 96-8, 142, 150/1,
  158, 175, 232, 246, 366, 378, 421,
  642, 714, 716, 879, 896; **28**, 2,
  34, 70a, 83, 102, 193, 491, 583,
  642, 821, 862, 879; **29**, 94, 97,
  99, 180, 218, 220, 408, 873,
  904, 911/2; **30**, 67, 76, 81, 91,
  495/6, 598, 618, 623
-, 30%-, increase of oxidizing
  effectiveness **28**, 111
- –/**acetonitrile** as reagent
  12, 189 suppl. **27**; **27**, 421
- –/**perfluoroacetone**, oxidations
  with – **26**, 137
- **selenide 26**, 282 suppl. **29**;
  26, 652
1,5-Hydrogen shift (s. a.
  Hydrogen transfer) **26**, 757
Hydrogen sulfide 6, 7 suppl. **29**;
  19, 691 suppl. **26**; **27**, 48;
  29, 558; **30**, 156, 378, 420
- startg. m. f.
  1,3-dithiins **28**, 744
  1,2-dithiolanes **28**, 532
  1,2-dithiolium salts **23**, 603
  suppl. **28**
  Δ²-thiazolines **30**, 388
  thioacylhydrazines **30**, 407
  thioethers, sym. **27**, 609
- **tetracarbonylcobaltate**
  26, 723
- **transfer** s. a. Hydrogen atom
  transfer
γ-**Hydrogen** – (s. a. Hydrogen
  shift) **26**, 744
δ-**Hydrogen** – **29**, 691
Hydrohalogenation s. Halides
  from ethylene derivs., Hydrofluorination, anti-Marknowni-
  koff hydrohalogenation
-, geoselectivity **12**, 625
  suppl. **29**
Hydrolysis (s. a. Cleavage,
  De...) HO⇵C; HN⇵C
-, **enzymic**, of carboxylic acid
  esters **28**, 13
-, **partial**
- of carboxylic acid esters
  18, 20
-, **preferential**
- **of**
  carboxylic acid esters
  2-pyridinecarboxylic acid
  esters **26**, 5
  – – tert-butyl esters **27**, 235
  phenolesters **28**, 86
-, **selective**
- of
  carboxylic acid esters
  30, 613
  phosphoramidates **27**, 117
Hydrometalation

- special s.
  hydro-alumination
  -boration
  -zirconation
Hydroperoxide ions as nucleophiles, strong **26**, 210
Hydroperoxides
- as reagents **20**, 211
  suppl. **27**; **27**, 184
- from
  dihalogenoboranes **29**, 221
  ethylene derivs. **27**, 138
  hydrocarbons **26**, 110
-, review **27**, 138 suppl. **30**
- special s.
  alkoxy-hydroperoxides
  amino-
  azo-
  tert-butyl-hydroperoxide
  cumene –
- startg. m. f.
  ethylenealcohols **30**, 661
  peroxides **27**, 197
α-**Hydroperoxy-β-diketones**
  1, 132 suppl. **26**
α-**Hydroperoxyketones**
- from
  α,β-ethyleneketones,
  synthesis **8**, 742 suppl. **27**
γ-**Hydroperoxy-2-lactones**
- special s.
  α,β-ethylene-γ-hydroperoxy-
  γ-lactones
Hydroperoxymercuration
  of ethylene deriv. **26**, 637
α-**Hydroperoxynitriles**
- from
  nitriles **30**, 155
- startg. m. f.
  cyanohydrins **30**, 155
Hydropolysulfides (s. a. Hydrodisulfides) **30**, 378
-, chemistry, review **16**, 623
  suppl. **28**
Hydroquinone (s. a. under
  Polymerization inhibitors)
- as H-donor **29**, 70
-, **polymer-bound**, as reagent
  8, 951 suppl. **29**
Hydroquinones s. Quinols
Hydrosilylation 26, 642; **29**, 603
- of
  oxo compds. **14**, 150; **17**, 160;
  23, 48 suppl. **29**
Hydroxamic acid benzyl esters
- startg. m. f.
  carboxylic acid amides
  29, 20
- – **esters**
- special s.
  aminohydroxamic acid esters
  hydroxamic acid benzyl –
- startg. m. f.
  1-halogen-1-aroximes **26**, 543
- – **halides** s. Hydroximinohalides
- **acids** (s. a. Sulfhydroxamic
  acids)
-, reactions, review **16**, 557
  suppl. **29**
- special s.
  acetylenehydroxamic acids
  aminohydroxamic –
- startg. m. f.
  carboxylic acid amides
  27, 90
  isocyanates **26**, 489

2H-1,2-oxazine ring,
  3,6-dihydro-N-acyl- **30**, 244
-, **cyclic**
- from
  o-aminohydroxamic acids
  28, 385
  lactones **26**, 351
  nitrocarboxylic acids **28**, 432
- special s.
  oxindoles, 1-hydroxy-
- startg. m. f.
  lactams **26**, 351
- –, **N-aryl-**
- from
  carboxylic acid chlorides
  and hydroxylamines, ar.
  27, 460
  – – – and nitroso compds.
  27, 479
- **acid O-sulfates**
-, reactions with – **26**, 364
- startg. m. f.
  ureas **26**, 364
Hydroximino... s. a. Oximino...
Hydroximinohalides
- from
  aldoximes **27**, 562
- special s.
  benzhydroximic chloride
  α-ketohydroximinohalides
- startg. m. f.
  α,β-acetylenoximes **26**, 877
  carbodiimides **26**, 442
  carboxylic acids **28**, 171
  1,4,2,5-dioxadiazines **27**, 208
  nitriles **30**, 337
  1,3,5-oxathiazolines **30**, 408
β-**Hydroxyacetals**
- from
  silylethynyl ketones **29**, 846
- startg. m. f.
  α,β-ethylenealdehydes
  29, 846
o-**Hydroxyacetophenones**
- from
  phenols **30**, 67
- startg. m. f.
  catechols **30**, 67
Hydroxyacetylenes
  s. Acetylenealcohols
Hydroxyacylamines
  s. Acylaminoalcohols
α-**Hydroxyaldehydes 26**, 214
- from
  carboxylic acid esters,
  synthesis with addition of
  1 C-atom **30**, 545
  cyanohydrins **23**, 223
  suppl. **26**
  2-hydroxy-1,1-dihalides
  28, 164
  oxo compds., synthesis with
  addition of 1 C-atom
  27, 830
  ketones, – – – – – – **22**, 661
  suppl. **27**
β-**Hydroxyaldehydes 28**, 176
o-**Hydroxyaldehydes 12**, 830
  suppl. **30**
- from
  o-hydroxymethylphenols,
  hindered **29**, 246
  6-spirooxirano-2,4-cyclo-
  hexadienones **29**, 246
-, ring closures with – **27**, 198
  suppl. **30**

Hyd 650

(o-Hydroxyaldehydes)
- startg. m. f.
  2H-1,3-benzothiazin-2-ones,
    3,4-dihydro- **29**, 550
  2H-1,3-benzoxazine-2-thiones,
    3,4-dihydro- **29**, 550
  chromenes **27**, 198
  coumarins **26**, 831; **27**, 198
  spiro[2H-1-benzopyran-2,2'-
    benzothiazolines] **28**, 738
Hydroxyalkoximes
- from
  pyran ring, dihydro- **28**, 896
- startg. m. f.
  ketoalkoximes **28**, 896
1-Hydroxy-2-(alkoxysulfonium) salts
- from
  oxido compds. **27**, 129
- startg. m. f.
  glycols **17**, 129
C-α-Hydroxyalkylation,
  C-alkylation via – **26**, 665
- with alcohols **29**, 719
N-Hydroxyamidines
- special s.
  N-hydroxyformamidines
1-Hydroxyamidines, cyclic
- from
  cyanoketones **26**, 298
  1-hydroxyamidoximes, cyclic
    26, 298
α-Hydroxyamidoximes
- from
  aldehydes, synthesis with
    addition of 1 C-atom
    27, 667
1-Hydroxyamidoximes, cyclic
- from
  cyanoketones **26**, 298
- startg. m. f.
  1-hydroxyamidines, cyclic
    26, 298
Hydroxyamines s. Aminoalcohols
Hydroxyamino... s. Amino-
  hydroxy...
Hydroxyazides s. Azidoalcohols
o-Hydroxyazo compds.
- from
  azoxy compds. **30**, 96
α-Hydroxyazomethines
- from
  aziridinones **30**, 612
Hydroxybarbituric acids,
  isomerization, review
  25, 120 suppl. **27**
N-Hydroxybenzenesulfonamide
  as catalyst, nucleophilic
  26, 176
**2'-Hydroxybenzophenone-
2-sulfinic acids**
- from
  thioxanthone 10,10-dioxides
    28, 224
- startg. m. f.
  xanthones **28**, 224
1-Hydroxybenzotriazole
  as reagent **16**, 530 suppl. **28**;
  22, 394/5 suppl. **29**
o-Hydroxycarbinols
- startg. m. f.
  1,3-benzodioxoles **28**, 207
Hydroxycarboxylic acid amides
- from
  lactones (unsubst. amides)
    26, 288

- startg. m. f.
  1H-benzopyran-1-imines,
    3,4-dihydro- **24**, 521
  suppl. **26**
  N-heterocyclics, N-condensed
    27, 946
  phthalanimines **24**, 521
  suppl. **26**
α-Hydroxycarboxylic – –
- from
  α-alkoxycarboxylic acid
    amides, rearrangement
    29, 693
  cyanohydrins **27**, 326
- startg. m. f.
  α-ketocarboxylic acid amides
    27, 326
β-Hydroxycarboxylic – –
- startg. m. f.
  2-aminoalcohols **28**, 309
  2-oxazolidone ring **28**, 309
o-Hydroxycarboxylic – –
- from
  aryl N-carbalkoxycarbamates
    28, 898
  – carbamates **28**, 678
  1,3-benzoxazine-2,4-diones
    28, 898
- startg. m. f.
  4H-1,3-benzoxazin-4-ones,
    2,3-dihydro- **26**, 307
Hydroxycarboxylic acid aryl esters
- from
  aryloxycarboxylic acids
    29, 161
- – esters
- from
  alkoxycarboyxlic acid esters
    26, 9
  enolethers, cyclic **28**, 109
  ketocarboxylic acid esters
    1, 42
  lactones **30**, 69
- special s.
  dihydroxycarboxylic acid
    esters
  ethylenehydroxy-
    carboxylic – –
  hydroxycarboxylic acid
    aryl –
  hydroxyketocarboxylic
    acid –
  salicylic – –
α-Hydroxycarboxylic acid esters
- from
  β-hydroxycarboxylic acid
    esters, cleavage, oxidative
    26, 253
- special s.
  α-acylamino-α-hydroxy-
    carboxylic acid esters
- startg. m. f.
  α-acoxycarboxylic acid esters
    27, 92
  carboxylic acid esters
    27, 92
β-Hydroxycarboxylic – –
-, α-alkylation **27**, 843
  suppl. **28**
- from
  oxo compds. **26**, 673
  – – and ketene **29**, 629
  – – and O-silyl O-alkyl
    keteneacetals **27**, 884

- startg. m. f.
  cycloalkane ring **29**, 933
  β,γ-ethyleneketones, cyclic
    29, 933
  α-hydroxycarboxylic acid
    esters, cleavage, oxidative
    26, 253
Hydroxycarboxylic acids
- from
  enolesters, cyclic **28**, 109
  suppl. **30**
  α-hydroxylactols, degradation
    with loss of 1 C-atom
    29, 22
  ketolactones **27**, 505
  ketones, cyclic **29**, 149
  lactones **27**, 56
  oxo compds. **27**, 670
- special s.
  aldonic acids
  salicyclic acid
- startg. m. f.
  acoxycarboxylic acid
    chlorides **29**, 514
  – acids **30**, 127
  lactones **27**, 56, 505;
    30, 168/9
  – with rearrangement
    28, 218
α-Hydroxycarboxylic –
- from
  α-aminocarboxylic acids
    2, 218 suppl. **26**
  carboxylic acids **28**, 82
  α-ketocarboxylic acids
    17, 65 suppl. **28**
- special s.
  α-amino-β-hydroxycarboxylic
    acids
- startg. m. f.
  ketones **27**, 275
β-Hydroxycarboxylic –
- from
  ethylene derivs. **30**, 496/7
  oxo compds. **26**, 664
  – –, Reformatskii synthesis,
    modified **2**, 827 suppl. **26**;
    27, 670
  – – and ketene silyl
    hemiacetals **27**, 884
- startg. m. f.
  1,3-dioxanes, 4-oxo- **27**, 196
  ethylene derivs. **27**, 981;
    28, 985
  α,β-ethyleneketones **30**, 554
  β-lactones **27**, 981
  γ-lactones **27**, 262
Hydroxycarboxylic acid thioamides
- startg. m. f.
  1,3-S,N-heterocyclics **30**, 421
α-Hydroxycarboxylic – –
- from
  oxo compds. and
    thioformamides **29**, 623
Hydroxy compds.
- special s.
  alcohols
  phenanthrols
  phenols
- startg. m. f.
  borinic acid esters
    28, 81 suppl. **30**
  oxo compds. OC↑H
-, **hindered**, acylation **29**, 184
N-Hydroxy compds.
-, catalysis of peptide synthesis
  with – **16**, 424 suppl. **29**

- special s.
  hydroxamic acids
  hydroxylamines
- startg. m. f.
  N-carbamyloxy compds.
  29, 126
Hydroxycyanamides
- from
  amines, tert., cyclic **28**, 277
Hydroxydiazo compds.
- special s.
  diazoalcohols
**α-Hydroxydiazo compds.,
  cyclic**
-, homologization of ketones,
  cyclic, via – **26**, 661
-, ring expansion via –
  26, 661; **29**, 625
**Hydroxydicarboxylic acid
  anhydrides 21**, 434 suppl. **27**
**N-Hydroxydicarboxylic acid
  imides**
-, reactions, review **16**, 557
  suppl. **29**
- special s.
  N-hydroxysuccinimide **25**
2-Hydroxy-1,1-dihalides
- startg. m. f.
  α-hydroxyaldehydes **28**, 164
-, **cyclic**
- startg. m. f.
  ketones, cyclic, ring
  expansion **30**, 672
Hydroxydiketones
- from
  ethylenealcohols **27**, 184
- special s.
  dihydroxydiketones
**1-Hydroxy-1,1-diphosphonic
  acid esters**
- from
  acylphosphonic acid esters
  and dialkyl phosphites
  28, 584
**1-Hydroxy-1,1-diphosphonic
  acids**
- from
  carboxylic acids **29**, 593
Hydroxydisulfides
  s. Dihydroxydisulfides
Hydroxyethers
- special s.
  glycol monoethers
C-β-Hydroxyethylation 29, 624
N-β-Hydroxyethylation 1, 278
O-β-Hydroxyethylation
- of
  phenols **29**, 235
Hydroxyethyleneketones
  s. Ethylenehydroxyketones
N-Hydroxyformamidines
- from
  hydroxylamines and
  isonitriles **30**, 202
Hydroxyguanidines
- from
  cyanamides **26**, 294
**N''-Hydroxyguanidines,
  N,N',N''-trisubst. 26**, 294
  suppl. **28**
Hydroxyhalides s. Halogenhydrins, Halogenophenols,
  Hydroxydihalides
N-Hydroxy-N-heterocyclics
  (s. a. 1-Hydroxyimidazol
  3-oxides **25**)

- from
  2-nitrosamine dimers,
  isocyclic **27**, 331
- special s.
  o-cyano-N-hydroxy-
  N-heterocyclics
α-Hydroxyhydrazones
- startg. m. f.
  ketones, degradation,
  oxidative **27**, 244
1,2-Hydroxyhydroperoxides
- from
  oxido compds. **28**, 69
Hydroxyketene mercaptals
- from
  lactones **29**, 552
- startg. m. f.
  dithiolortholactones **29**, 552
**δ-Hydroxy-β-ketocarboxylic
  acid esters**
- from
  β-ketocarboxylic acid esters
  30, 447
ε-Hydroxy-β-ketocarboxylic – –
- from
  β-ketocarboxylic acid esters
  and oxido compds. **30**, 448
- startg. m. f.
  furylideneacetic acid esters,
  tetrahydro- **30**, 448
Hydroxyketones
-, ring closure of – **27**, 945
- special s.
  diazoketols
  dihydroxyketones
  ethylenehydroxyketones
  hydroxydiketones
  α-hydroxymethyleneketones
α-Hydroxyketones
- from
  acetals **27**, 239
  aldehydes **29**, 621
  α-diketones, reduction
  27, 34; **28**, 31
  -, synthesis **27**, 674
  β-diketones **28**, 608
  1,2-di(siloxy)alkenes and
  halides **30**, 617
  enoxysilanes **30**, 144
  ethylene derivs. **27**, 184
  glycols **21**, 285 suppl. **28**;
  26, 233; **28**, 201 suppl. **29**
  α-ketoaldehydes, synthesis
  30, 446
- special s.
  benzoins
- startg. m. f.
  carboxylic acid esters
  29, 162
  cyclohex-2-en-4-olone ring
  30, 546
  α-diketones **26**, 233
  furans **29**, 865
  glycols **29**, 36
  imidazoles **29**, 384
  ketones, cleavage **27**, 674
  -, degradation, oxidative
  26, 226
  phenol ring **30**, 546
  pyrrole ring, N-unsubst.
  25, 275 suppl. **26**
-, **cyclic**
- from
  ethynylglycols, cyclic, ring
  expansion **28**, 974
-, **monocyclic**
- from
  mercaptals, bicyclic **27**, 229

β-Hydroxyketones 29, 669;
  30, 489
- from
  β-diketones **22**, 37 suppl. **29**
  enolesters **29**, 894
  enoxysilanes **30**, 621
  α,β-oxidoketones **11**, 62
- startg. m. f.
  α,β-ethyleneketones **30**, 489
  β-hydroxy-δ-lactones **26**, 664
-, **cyclic**
- from
  enollactones **20**, 7
  ethylenecarboxylic acids
  30, 523
γ-Hydroxyketones
- from
  2(3H)-furanones **29**, 822
  oxido compds. and
  acetylene derivs. **29**, 869
  – – – enamines **26**, 830
- startg. m. f.
  γ-lactones **29**, 245
o-Hydroxyketones
- special s.
  o-hydroxyacetophenones
- startg. m. f.
  chromone-3-carboxy-
  aldehydes **29**, 785
  chromones **26**, 247
ω-Hydroxyketones
- from
  alcohols, cyclic **16**, 305
  suppl. **26**
**9α-Hydroxy-6-ketosteroids
  29**, 178
δ-Hydroxy-β-ketosulfoxides
- from
  β-ketosulfoxides **29**, 618
1-Hydroxy-2-ketothioethers
- startg. m. f.
  α-hydroxythiolic acid esters
  26, 145
Hydroxylactams
- from
  aminolactones **29**, 324
  azidolactones **29**, 324
  cyanoketones **26**, 121
N-Hydroxylactams
  s. Hydroxamic acids, cyclic
α-Hydroxylactols
- from
  lactols **30**, 64
- startg. m. f.
  hydroxycarboxylic acids,
  degradation with loss
  of 1 C-atom **29**, 225
Hydroxylactones
- from
  oxidocarboxylic acid esters
  29, 234
  – acids **30**, 519
- startg. m. f.
  ketolactones **28**, 208
α-Hydroxylactones
- from
  lactones **30**, 63
β-Hydroxy-γ-lactones
- from
  α-acoxyketones **30**, 516
β-Hydroxy-δ-lactones
- from
  β-hydroxyketones **26**, 664
Hydroxylamine 27, 332
-, reaction with isocyanates
  26, 293

Hyd 652

(Hydroxylamine)
- startg. m. f.
  isoindolenines **29**, 347
  N-oxides, cyclic **22**, 708
    suppl. **26**
  sulfonylformamidoximes
    29, 303
-, **O-protected 23**, 233 suppl. **27**
Hydroxylaminedisulfonate
  28, 721
Hydroxylamine hydrochloride
- as reactant **26**, 430; **29**, 360;
    30, 277
Hydroxylamines
- as reagents **27**, 960
- from
  acoxylamines **26**, 88
  amines **7**, 10 suppl. **29**; **26**, 88
  halides **26**, 402
  nitro compds. **26**, 28
  - -, aliphatic **27**, 16
  nitroso compds. **27**, 313
-, Mannich-type condensation
  with - **27**, 612
- special s.
  amino-hydroxylamines
  carbamyl-
  6H-dibenzo[d,f][1,3]-
    diazepines, 5,7-dihydroxy-
  di-hydroxylamines
  ene-
  keto-
  oximino-
  phenyl-
- startg. m. f.
  enehydroxylamines **30**, 651
  ethylene derivs. **27**, 960
  furoxan ring **29**, 338
  N-hydroxyformamidines
    30, 202
  N-hydroxythioureas **20**, 235
    suppl. **28**; **26**, 334
  indoles **30**, 651
  sulfhydroxamic acids **28**, 257
  sulfonic acid amides **27**, 297
-, aliphatic
- startg. m. f.
  azo N,N'-dioxides **27**, 100
-, ar.
- startg. m. f.
  hydroxamic acids, N-aryl-
    27, 460
-, cyclic
- special s.
  6H-dibenzo[d,f][1,3]-
    diazepines, 5,7-dihydroxy-
-, disubst.
- from
  nitro compds. and halides
    27, 473
Hydroxylamine-O-sulfonic acid
- startg. m. f.
  S-acylhydrosulfamines
    29, 286
  dibenzo-1,4-diazepines **28**, 287
  nitriles **30**, 277
  pyridazinium salts, 1-amino-
    28, 236
  pyrroles **27**, 779
  triazanium salts **29**, 271
α-**Hydroxylaminocarboxylic**
  acid amides
- startg. m. f.
  α-(O-carbamylhydroxyl-
    amino)carboxylic acid
    amides **30**, 125
  α-ketocarboxylic - -
    30, 125

o-**Hydroxylaminocarboxylic**
  acid esters
-, ring closures with - **26**, 346
Hydroxylation s. a. Replacement of hydrogen by
  hydroxyl, Photohydroxylation
C-Hydroxylation of enolates
  28, 82 suppl. **30**; **30**, 65
cis-**Hydroxylation** of carbon-
  carbon double bonds **27**, 147
Hydroxyl groups s. Hydroxy
  compds.
α-**Hydroxymercaptals**
- startg. m. f.
  thioenolethers **29**, 974
β-**Hydroxymercaptals**
- from
  oxido compds. **29**, 624
1,2-Hydroxymercaptans
- from
  2,2'-dihydroxydisulfides
    28, 525
  oxido compds. **26**, 570; **28**, 525
α-**Hydroxymethyl-α-amino-**
  carboxylic acids
- from
  α-aminocarboxylic acids
    30, 451
  1,3-dioxanes, 4-oxo-5-acyl-
    amino- **30**, 451
C-Hydroxymethylation 29, 622
- under acidic conditions
  4, 800 suppl. **26**; **29**, 627
N-Hydroxymethylation 8, 387
(-)-Hydroxymethylenecamphor
-, resolution of amines with -
  2, 725a suppl. **26**
Hydroxymethylene compds.
  s. a. Aldehydes, Enols
α-**Hydroxymethyleneketones**
- startg. m. f.
  β-ketomercaptals, cyclic
    13, 602 suppl. **29**
  α-methyleneketones **27**, 89
  α-methylketones **27**, 89
Hydroxymethyl groups
- startg. m. f.
  methyl groups **27**, 91
o-**Hydroxymethylphenols**
- startg. m. f.
  o-hydroxyaldehydes **29**, 246
  6-spirooxirano-2,4-cyclo-
    hexadienones **29**, 246
Hydroxynitriles (s. a. Cyanohydrins, Cyanophenols)
- special s.
  azidohydroxynitriles
β-**Hydroxynitriles**
- from
  oxo compds. **29**, 959
  - - and α-halogenonitriles
    26, 672
α-**Hydroxyoximes, cyclic**
- startg. m. f.
  oxonitriles **19**, 562
α-**Hydroxyoxo compds.,**
  synthesis **22**, 661 suppl. **27**, **28**
o-**Hydroxyoxo** -
- startg. m. f.
  benzofurans **28**, 807
1,1-Hydroxyperoxides 28, 138
o-**Hydroxyphosphine imines**
- from
  o-aminophenols **28**, 262
α-**Hydroxyphosphine oxides**
- from
  oxo compds. and phosphinyl-
    formic acid esters **29**, 614

α-**Hydroxyphosphinic acids**
- from
  oxo compds. **22**, 656
α-**Hydroxyphosphonic acid**
  derivs.
- from
  phosphoric acid derivs.
    29, 604
- - esters
- startg. m. f.
  phosphoric acid esters **28**, 71
α-**Hydroxyphosphoric acid**
  esters
- from
  oxido compds. **26**, 98
3-Hydroxyphthalides
-, reactions **28**, 324
-, - with amines **28**, 324
- startg. m. f.
  3-aminophthalides **28**, 324
  o-carbalkoxystyrene oxides
    29, 771
3-Hydroxypropionitrile
-, protection of phosphoryl
  hydroxy groups with -
    17, 169 suppl. **29**
Hydroxyquinones
-, ring contraction, oxidative
    28, 980
2-Hydroxyselenides
- from
  ethylene derivs. and
    selenylhalides **29**, 601
2-Hydroxysilanes
- startg. m. f.
  cis- and trans-ethylene
    derivs. **28**, 964 suppl. **30**
3-Hydroxystannanes
- from
  3-stannoxyhalides **27**, 657
Hydroxysteroids
- special s.
  dihydroxy-steroids
  hydroxyketo-
14β-Hydroxysteroids 27, 171
17α-Hydroxy-Δ^{8,14}**-steroids**
- from
  17-keto-13α-steroids, 9,14β-
    epoxy-8,14-seco- **29**, 944
N-Hydroxysuccinimide esters
- special s.
  polymer-N-hydroxy-
    succinimide esters
Hydroxysulfinic acids
- special s.
  2'-hydroxybenzophenone-
    2-sulfinic acids
β-**Hydroxy-α-sulfinylcarboxylic**
  acid esters
- from
  α-sulfinylcarboxylic acid
    esters and oxo compds.
    30, 455
Hydroxysulfonic acids
- from
  thioethers, cyclic **30**, 76
1,1-Hydroxysulfonic acids
- from
  oxo compds. **8**, 641 suppl. **28**
- startg. m. f.
  acetals **26**, 515
  ethylene derivs. **28**, 853
Hydroxysulfoxides
- special s.
  hydroxyketosulfoxides
  hydroxysulfinyl...

- startg. m. f.
  halogenosulfones **27,** 567
3-Hydroxysulfoxides
- from
  oxido compds. **17,** 733
  suppl. **27**
β-Hydroxysulfoximines
- from
  oxo compds. **29,** 855
- startg. m. f.
  ethylene derivs. **29,** 855
Hydroxythioethers
- startg. m. f.
  acoxysulfones **26,** 93
1,1-Hydroxythioethers
  (s. a. 1-Hydroxy-2-keto-
  thioethers)
- from
  aldehydes **28,** 522
-, **cyclic 28,** 574
2-Hydroxythioethers
- from
  oxido compds. **28,** 524
  oxo - **29,** 262
  suppl. **30**
  ketones **29,** 262
- special s.
  2-(phenylthio)ethanol
- startg. m. f.
  alcohols **28,** 524
  ethylene derivs. **27,** 979;
  29, 974
  oxido compds. **29,** 262
-, **cyclic**
- from
  sulfonyloxyoxido compds.
  29, 549
Hydroxythiolic acid esters
- startg. m. f.
  lactones **30,** 142, 169
α-Hydroxythiolic - -
- from
  cyanohydrins **27,** 624
  1-hydroxy-2-ketothioethers
  26, 145
β-Hydroxythiolic - -
- from
  oxo compds., ketenes, and
  thioborinic acid esters
  28, 862
(Hydroxythiol)thionocarbonic
  - - 27, 175
1,1-Hydroxythioureas
- from
  aldehydes **26,** 286
2-Hydroxythioureas
- startg. m. f.
  Δ²-oxazolines **27,** 406
N-Hydroxythioureas
- from
  hydroxylamines **26,** 334
  - and isothiocyanates
    20, 235 suppl. **28**
3-Hydroxytriazenes
- startg. m. f.
  3-α-acoxy-3-acyltriazenes
  29, 346
Hydroxyurea
- startg. m. f.
  isoxazoles, 3-amino- **26,** 409
N-Hydroxyureas
- from
  isocyanates **26,** 293
- startg. m. f.
  N-acoxyureas **27,** 194
  biurets, 3-hydroxy- **26,** 330
  O-carbamylhydroxylamines
  29, 95

α-N-Hydroxyureidocarboxylic
  acid esters
- from
  α-isocyanatocarboxylic acid
  esters **27,** 519
- startg. m. f.
  hydantoins, 3-hydroxy- **27,** 519
N-Hydroxyurethans 26, 412
- special s.
  tert-butyl N-hydroxy-
  carbamate
- startg. m. f.
  1,2-oxazines, tetrahydro-
  N-carbalkoxy- **30,** 295
Hydrozirconation, reactions
  via - **30,** 603
Hypochlorites
- startg. m. f.
  1,2-alkoxyhalides **26,** 513
Hypodiphosphorous acid esters
  (s. a. Tetrathiohypo-
  diphosphorous acid esters)
- from
  phosphorochloridites **28,** 580
Hypohalites s. Acetyl hypobromite, Alkyl hypohalites,
  Hypochlorites, Hypoiodite,
  Potassium hypobromite,
  Sodium hypohalite
Hypoiodite reaction, review
  18, 301 suppl. **27**
Hyposulfite s. Dithionite

Identification of compds.
  s. subentry: derivs.
**1H-Imidazo[5,1-c][1,4]benzo-
  diazepine-1,3,11-triones,
  2,3,5,10,11,11a-hexahydro-**
- from
  2H-1,4-benzodiazepin-2-one-
  3-carboxamides,
  1,3-dihydro- **30,** 241
**1H-Imidazo[1,2-a]imidazoles,
  2,3,5,6-tetrahydro- 27,** 340
**5H-Imidazo[2,1-a]isoindol-
  5-ones, 1,2,3,9b-tetrahydro-
  30,** 249
**Imidazo[2,1-a]isoquinolines,
  2,3,5,6-tetrahydro-6-hydroxy-
  17,** 889 suppl. **26**
Imidazole
- as catalyst **2,** 506 suppl. **29;**
  30, 380
- as reagent **26,** 207; **29,** 554,
  947; **30,** 61, 243
**Imidazole-4-carboxylic acid
  esters**
- by
  Claisen-type rearrangement
  27, 502
Imidazole 3-oxide ring
- special s.
  benzimidazole 3-oxides
Imidazole 3-oxides
- from
  α-dioxo compd. monoximes
  and aldehydes **27,** 412
- ring (s. a. Purines from
  pyrimidines)
- special s.
  2-benzyloxyimidazole ring **25**
Imidazole ring, N-condensed
- from
  o-α-cyanocarboxylic acids
  26, 339
  α-halogenooxo compds.
  13, 535

- - **rearrangements 26,** 20
Imidazoles
-, N-acylation, regiospecific
  30, 302
- by retro-benzilic acid-type
  rearrangement **28,** 959
- from
  α-acylaminoketones **30,** 267
  4H-imidazoles, 5-alkylthio-
  26, 20
  iminoesters and α-hydroxyketones **29,** 384
  nitriles **26,** 304
  purines **29,** 20
-, 2-silyl groups, active, in -
  27, 883
- special s.
  benzimidazoles
  biimidazoles
  N,N'-carbonyldiimidazole
  methylimidazole
  N-thioacylimidazoles
  N-tosylimidazole
- startg. m. f.
  benzimidazoles **28,** 922
  1,3,5-triazines **27,** 365
-, N-substitution, directed
  30, 701
-, **isomeric** s. Imidazoles,
  nitro-, isomeric
-, **5-arylthio-**
- from
  2-cyanonitrones and arylmercaptans **29,** 551
-, **4-chloro- 27,** 546
-, **1-cyano-,** by thermolysis
  28, 450
-, **4,5-dicyano- 28,** 771
- special s.
  2-diazoimidazole, 4,5-dicyano-
-, **1-hydroxy-**
- from
  α-isonitrosoketones and
  aldehydes **26,** 338
-, **1-(N-iminoformyl)-,** reactions
  with - **30,** 281
-, **nitro-,** isomeric **28,** 391
-, **1-(α-siloxy)-**
- from
  aldehydes **26,** 290
-, **2,4,5-trialkyl-,** sym.
- from
  ethylene derivs. **28,** 665
4H-Imidazoles, 5-alkylthio-
- startg. m. f.
  imidazoles **26,** 20
Imidazolides s. 1-Sulfonyl-
  imidazoles
Imidazolidines, 1,3-dihydroxy-
- from
  1,2-di(hydroxylamines)
  29, 348
**Imidazolidine-2-thiones,
  4-hydroxy-**
- from
  isothiocyanates **27,** 318
Imidazolidinetriones
  s. Parabanic acids
Imidazolidinylidenes
  s. Bisimidazolidinylidenes
2,4-Imidazolidiones
  s. Hydantoins
2-Imidazolidone ring
- from
  1,2-diamines **3,** 375
Imidazolidones s. a. Hydantoins

2-Imidazolidones
- from
  β-aminoaminimides **28,** 448
4-Imidazolidones
- from
  α-alkylideneaminocarboxylic
  acid esters **27,** 379
  α-aminonitriles and oxo
  compds. **28,** 275
-, **3-hydroxy-**
- from
  α-aminohydroxamic acids
  and aldehydes **26,** 336
Δ³-Imidazoline 3-oxides
- from
  anti-1,2-aminooximes **26,** 337
Δ²-Imidazoline ring, 5-hydroxy-
- from
  O-alkyllactims and
  α-aminoketones **29,** 385
Imidazoline ring opening
  15, 313
Δ²-Imidazolines
- from
  aziridines and nitriles **30,** 217
  carboxylic acids **27,** 410
  N,N'-dialkylidene-
  1,1-diamines **28,** 697
Δ³-Imidazoline-5-thione ring
- startg. m. f.
  1,2,4,5-tetrazine ring,
  1,2,3,4-tetrahydro- **29,** 281
Δ⁴-Imidazoline-2-thiones
- from
  dithiohydantoins **26,** 981
Imidazolinylhydrazonohalides
- from
  trihalogenodiazabutadienes
  27, 529
- startg. m. f.
  imidazotriazoles, dihydro-
  27, 529
Imidazolium chlorides,
  N-sulfonyl-N'-methyl-,
  N-sulfonation with – **27,** 295
- – , **N-acyl-N'-methyl-,**
  N-acylation with – **25,** 293
  suppl. 28
- **ring, N-amino-**
- startg. m. f.
  pyrazol ring **30,** 485
- **ring, N-condensed**
- from
  α-halogenaldehydes **29,** 349
- salts as intermediates **30,** 701
2-Imidazolone ring
- from
  p-quinones **27,** 405
-, ring closure, anomalous
  to – **27,** 405
Imidazolones
- special s.
  alkylideneimidazolones
2-Imidazolones
- from
  α-dioxo compd. monoximes
  27, 412
  1,2,4-triazin-3(2H)-ones
  27, 289 suppl. 28
Δ³-2-Imidazolones
- from
  2-azetidinones, 3-azido-
  28, 451
4-Imidazolones
- from
  5-oxazolones 4, 309 suppl.
  27, 28; 30, 257

1-(1-Imidazolyl)-1,3-dienes,
  polyfunctional
- by
  3-component reaction **26,** 542
Imidazo[1,2-a]pyrazine ring
  26, 431
Imidazo[1,5-a]pyrazines
  27, 358 suppl. 28
Imidazo[1,5-a]pyrazine-3-thiols
  27, 358
3H-Imidazo[4,5-f]quinolines
  26, 823
Imidazo[1,5-a]quinoxal-4-ones
  29, 467
Imidazo[2,1-b]thiazoles,
  6-chloro- 26, 542
-, 5,6-dihydro- **18,** 669 suppl. 28
Imidazo[2,1-c][1,2,4]triazines
- from
  oxazolium salts **26,** 344
1H-Imidazo[1,2-b]-s-triazoles
  27, 542
3H-Imidazo[1,5-b]-s-triazoles
- from
  nitriles (3 molecules) **27,** 678
Imidazotriazoles, dihydro-
- from
  imidazolinylhydrazono-
  halides **27,** 529
  trihalogenodiazabutadienes
  27, 529
Imides s. Dicarboxylic acid
  imides, Diacylamines
Imidinophosphoranes
- from
  phosphine imines and nitriles
  27, 684
Imidium salts (s. a. Immonium
  salts)
- special s.
  halogenimidium salts
β-Imidocarboxylic acids,
  α-branched
- from
  1,2-ethylene-1,1-dihalides
  26, 888
β-Imido-α-halogenocarboxylic
  acids 26, 888
C-Imidomethylation via
  carbonium ions **26,** 888
Imidophosphorisocyanatidates
- special s.
  acylimidophosphoriso-
  cyanatidates
N-Imidosulfoximines
- startg. m. f.
  α-dioxo compds. **28,** 930
Imidoylcarbodiimides 27, 532
N-Imidoylsulfilimines 30, 198a
Imine-enamine tautomers
  29, 701
Imines (s. a. Azomethines)
- from
  acetals **27,** 393
  ketones **7,** 454 suppl. **26;**
  26, 433
  thioketones **29,** 426
- special s.
  α,β-ethylenimines 25
  ketimines
- startg. m. f.
  amines **26,** 34
  azomethines **29,** 404
  diazo compds. **27,** 444
  enisothiocyanates **30,** 242
  1,1-halogenisothiocyanates
  30, 242

  hydrazones **27,** 444
  thioketones **26,** 605
-, **cyclic** (s. a. Ethylenimines)
- from
  O-alkyllactims **23,** 80
  suppl. 27
  dihalides **27,** 448
  dioxo compds. **28,** 347
  lactams **23,** 80 suppl. **27;**
  30, 531
- special s.
  chloroformoxyimines, cyclic
- startg. m. f.
  aminoaldehydes **28,** 92
Iminium salts s. Azomethinium
  salts
1,1-Iminoalcohols, cyclic
- from
  lactams **22,** 62
  lactols **29,** 380
-, **1,3-N,N-heterocyclic,**
  N-condensed
- from
  diketones and o-diamines
  29, 364
β-Iminoazomethines
- from
  azomethines and nitriles
  26, 685
- startg. m. f.
  pyrimidines **26,** 396
Iminocarbonic acid esters
- as intermediates **29,** 65
- from
  cyanic acid esters **27,** 214
- special s.
  thioiminocarbonic acid
  esters
- – –, **cyclic 27,** 176
- – –, **sym. 27,** 214
Iminocarbonylhalides
-, replacement of halogen
  in – **29,** 578
- special s.
  trichloromethylisocyanide
  dichloride
- startg. m. f.
  iminothiocarbonic acid
  esters **29,** 578
  isothiocyanates **28,** 557
  keteniminylidene-
  phosphoranes **30,** 519
α-Iminocarboxylic acids, cyclic
- from
  α-halogenolactams **27,** 211
1,1-Iminocyanides
- from
  acylcyanides and phosphine
  imines **26,** 444
  α-aminonitriles **19,** 553
  suppl. **26; 29,** 645
- startg. m. f.
  1,1,1-aminodinitriles **29,** 645
-, **cyclic 25,** 354 suppl. 29
-, reactions **25,** 354 suppl. 29
Iminodisulfides
- startg. m. f.
  1-aminoenedisulfides **29,** 701
Iminodithiocarbonic...
  s. a. Sulfonylimino-
  dithiocarbonic...
Iminodithiocarbonic acid
  esters
- from
  iminocarbonylhalides **29,** 578
- special s.
  N-cyaniminodithiocarbonic
  acid esters

β-Iminodithiocarboxylic acids
- from
  ketones 30, 464
- startg. m. f.
  1,3-thiazine-6-thione ring,
  dihydro- 27, 616
Iminoester fluoroborates
- from
  carboxylic acid amides
  26, 933
-, ring closure via - 26, 933
- startg. m. f.
  amide acetals 26, 933
Iminoesters
- from
  carboxylic acid amides and
    fluorosulfonic acid esters
    28, 139
  nitriles 30, 220
  - (free base), preferential
    conversion 30, 75
  orthocarboxylic acid esters
    29, 369
-, ring closure with -
  s. under Ring closure
- special s.
  N-acyl-iminoesters
  N-(chloroglyoxyloyl)-
    cyan-
    di-
    ethylene-
    halogen-
    phosphonio-
    silyl -
  N-sulfonyl-
- startg. m. f.
  acylisocyanates 26, 458
  amines, sec. 29, 369
  N-(chloroglyoxyloyl)imino-
    esters 26, 458
  N-cyaniminoesters 27, 440
  enamines 26, 795
  imidazoles 29, 384
  oxaziridines, 3-alkoxy-
    29, 132
  2-pyridone ring 29, 744
  1,2,4,5-tetrazines,
    monosubst. 28, 381
-, cyclic (s. a. O-Alkyllactims)
- special s.
  1,3-oxazines, 5,6-dihydro-
  oxazolines
- startg. m. f.
  1-alkoxyenacylamines, cyclic
    27, 466
  N-chlorocarbonyl-1-alkoxy-
    enamines, cyclic 27, 466
Iminoester salts, cyclic
- special s.
  furfurylideneammonium
    salts
Iminohalides
- from
  carboxylic acid amides
    27, 575
- special s.
  N-(halogenoformyl)-
    iminohalides
  1,1-hydrazonohalides
  thioiminohalides
- startg. m. f.
  carboxylic acid thioamides
    30, 410
  heterocyclics 28, 810
Iminohalide salts
- special s.
  α-ketohalogenimidium
    salts 25

Iminonitriles s. Iminocyanides
Iminophosgenes s. Imino-
  carbonylhalides
Iminoseleninyl halides
- special s.
  N-sulfonyliminoseleninyl
    halides
1-Imino-1-sulfenamides
- from
  carboxylic acid thioamides
    28, 248
S-Iminosulfinic acid amides
  s. N-Sulfonyl-S-(sulfonyl-
    imino)sulfinic acid amides
Iminosulfinyl halides
  28, 469 suppl. 29
- special s.
  N-sulfonyliminosulfinyl
    halides
1,1-Iminosulfones, cyclic
- from
  sulfonylcyanides 29, 648
- startg. m. f.
  lactams 29, 648
Iminosulfonic acid chlorides
- from
  sulfinic acid amides 28, 468
- - -, cyclic 28, 468
α-Iminosulfonium ylids
- from
  sulfilimines and acetylene
    derivs. 30, 387
Iminothiolic esters
  s. Thioiminoesters
Iminothionyl halides
- from
  N,N-dihalogenamines
    26, 269
Immoniocyclobutane ring
- from
  ethylene derivs. and
    1-halogeneamines 28, 822
Immonium salts (s. a. Imidium
  salts) s.
  azomethinium salts
  cyclimmonium -
  oxathiol-2-ylidenimmonium -
Indan-1,2-diones, 3-hydroxy-
- from
  flavones, 3-hydroxy- 30, 515
Indan-1,3-diones
- from
  3-alkylidenephthalides
    26, 746
  phthalic acid esters and
    ketones 27, 767
- special s.
  isoquinoline-1-spiro-
    2'-(1',3'-indandiones),
    1,2-dihydro-
-, 2-acyl- 28, 728
- startg. m. f.
  6H-indeno[1,2-e][1,4]diazepin-
    6-ones, 2,3-dihydro- 26, 373
  perimidines 27, 497
-, 2-acylamino-2-hydroxy-
  29, 345
-, 2,2-dihydroxy-
- special s.
  ninhydrin
-, 2-isothioureido-
- startg. m. f.
  3H,5H-thiazolo[4,3-a]iso-
    indol-5-ones 26, 319
Indanols
- from
  indanones 29, 840

- startg. m. f.
  indenes 29, 840
Indan-1-one-3-carboxylic acids
-, synthesis with C-alkyl
    migration 28, 659
Indanones 29, 840
- startg. m. f.
  indanols 29, 840
  indenes 29, 840
1-Indanones, 2-oximino-
- startg. m. f.
  isoquinolines, 1,3-dihalogeno-
    26, 544
2-Indanones, 1,3-dihydroxy-
- startg. m. f.
  2-benzopyr-3-ones 26, 243
Indan ring
- startg. m. f.
  azulene ring 27, 226
Indans
- by
  rearrangement, thermal
    17, 777 suppl. 29
- from
  spiro[4.4]nona-1,3-dienes
    28, 894
-, 2,2-dialkoxy-
    1,3-dihydroxy-
- startg. m. f.
  2-benzopyr-3-ones 26, 243
  o-ketoacetic acids 26, 243
Indazole ring
- from
  1,2,3-benzotriazine ring,
    3,4-dihydro- 28, 242
Indazoles 27, 491
1H-Indazoles
- from
  1H,1,2,4-benzotriazepines
    29, 483
Indazoles, 1-amino-
- startg. m. f.
  1,2,3-benzotriazines 28, 431
-, 2-amino- 18, 338 suppl. 28
-, 3 amino-
- startg. m. f.
  quinazolines, 4-amino- 27, 498
-, 2 hydroxy-
- from
  o-aminoketones 26, 324
- startg. m. f.
  3-indazolones 26, 324
3-Indazolones
- from
  o-aminoketones 26, 324
  1,2,3-benzotriazin-4(3H)-ones,
    3-amino- 29, 461
  indazoles, 2-hydroxy- 26, 324
Indene ring 28, 908
Indenes
-, 1-alkylation 29, 795, 932
- from
  allenes 30, 510
  indanols 29, 840
- special s.
  indone ring
- startg. m. f.
  1,4-methanonaphthalenes,
    1,2,3,4-tetrahydro- 28, 628
- via indanones 29, 840
-, 2-aryl- 29, 932
6H-Indeno[1,2-e][1,4]diazepin-
  6-ones, 2,3-dihydro-
- from
  indan-1,3-diones, 2-acyl-
    26, 373
Indeno[2,1-c]-1,2-dithioles
  30, 638

Indenolates
- special s.
  2-(1-pyridinio)-1-indenolates
Indenones 28, 952
- from
  α,β-ethylene-β-halogeno-
    ketones **28,** 952
  1,2,3-triazines **28,** 931
11H-Indeno[1,2-b]quinolines
  30, 325
- from
  11H-benzo[a]carbazoles,
    5,6-dihydro- **28,** 903
Indeno[1,2,3-de]quinolines
  29, 971
Indium 26, 673 suppl. 30
Indole as additive
  19, 22 suppl. 29
Indole-1,2-dicarboximide ozonides
- startg. m. f.
  2-quinazolones **29,** 95
Indole-4,7-diones 24, 528
  suppl. 28
Indolenine ring, 3-halogeno-
- from
  indole ring **28,** 490
Indolenines
- special s.
  thiophene-2-spiro-
    3-indolenines
- startg. m. f.
  indoles, rearrangement,
    lateral **30,** 503
-, **3-halogeno-**
- from
  indoles **29,** 153
- startg. m. f.
  oxindoles **29,** 153
Indole ring
- from
  Δ¹-azirine ring **27,** 335
  1,2-N,N-heterocycles,
    satd. **30,** 563
    pyrazolidines **30,** 563
- startg. m. f.
  indolenine ring, 3-halogeno-
    28, 490
- -, **N-condensed**
- special s.
  isatin ring, N-condensed
Indoles 29, 837
-, 3-alkylation, reductive
  29, 761
-, dimerization **28,** 663
-, -, oxidative **28,** 663
- from
  2-acetyleneamines **30,** 544
  2-aminoenol sulfonates
    28, 178
  o-aminoketones **23,** 871
    suppl. 26
  o-aminooxo compds.
    29, 837; **30,** 663
  anilines and α-(alkylthio)oxo
    compds. **28,** 762
  1H-1,4-benzodiazepines,
    2,3-dihydro- **30,** 692
  4H-3,1-benzothiazines
    28, 957
  enehydroxylamines **30,** 651
  halogenamines **27,** 526
  hydroxylamines **30,** 651
  indolenines, rearrangement,
    lateral **30,** 503
  indolines **25,** 649 suppl. 28
-, 3-hydroxy- **26,** 866
    suppl. 28; **30,** 663

indol-4(5H)-ones, 6,7-dihydro-
  27, 933
isatins **9,** 938 suppl. 27
-, reaction with Δ³-pyrrol-
  2-ones **27,** 823
- special s.
  indoxyls
  isatins
  2-methylindole
  oxindoles
  tryptophanes
- startg. m. f.
  1(1H)-benzazepines **26,** 705
  indolenines, 3-halogeno-
    29, 153
  indolines, 3-diazo-2-sulfonyl-
    imino- **29,** 339
  indoxyls, 2-hydroxy- **30,** 85
  oxindoles **29,** 153
  quinazolines **30,** 223
-, 4-substitution **2,** 621 suppl. 28
- via *trans*-β-amino-2-nitro-
  styrenes **29,** 469
-, **2- from 3-subst. 14,** 937
  suppl. 26
-, **3-acyl-**
- from
  3,1-benzoxazepines **26,** 743
-, **3-β-amino-**
- startg. m. f.
  9H-pyrid[3,4-b]indoles
    28, 750
-, **2-aryl-**
- from
  α-arylaminoketones **28,** 785
-, **2-cyano-**
- from
  o-aminoketones **29,** 749
-, **1-formamino-**
- from
  cinnolines **27,** 324
-, **4-hydroxy-**
- from
  benzofurans, 4-amino-
    28, 297
-, **5-hydroxy-**
- from
  p-quinones **8,** 782; **26,** 717
-, **2-δ-keto-**
- from
  pyrido[1,2-a]indol-6-ones,
    6,7,8,9-tetrahydro- **29,** 637
-, **nitroso**
- startg. m. f.
  quinazolines **28,** 355
3H-Indoles s. Indolenines
Indolines
- special s.
  1-[2-(1-pyrrolinyl)]indolines
- startg. m. f.
  indoles **25,** 649 suppl. 28
-, 3-diazo-2-sulfonylimino-
- from
  indoles **29,** 339
-, **3-hydroxy-**
- from
  o-aminooxo compds. **30,** 663
- startg. m. f.
  indoles **26,** 866 suppl. **28;**
    30, 663
-, **3-hydroxy-1-nitroso-**
- from
  o-(N-nitrosamino)ketones
    29, 695
-, **2-imino-**
- from
  1-acyl-2-arylhydrazines
    28, 918

Indoline-2-thiones 17, 45
  suppl. 29
Indolin-3-ones
- special s.
  2-eneindolin-3-ones
Indolizine ring opening 29, 446
Indolizines,
  1-(alkylideneamino)-
- from
  aldehydes (2 molecules) and
    pyridines, 2-acyl- **27,** 793
**Indolizino[1,2-c]quinolinium
  salts, 12-acyl- 26,** 848
**Indolo[3,2-b][1,4]benzoxazines,
  5,11a-dihydro- 28,** 941
3H-Indol-3-ols
- from
  4H-3,1-benzoxazines **26,** 748
Indol-3-ones
- from
  azomethines **28,** 829
**Indol-4(5H)-ones, 6,7-dihydro-
  27,** 933
- startg. m. f.
  indoles **27,** 933
3H-Indol-2-ones, 1-hydroxy-
- from
  nitro compds., ar. **27,** 499
cis-**Indol-6-ones, octahydro-**
- from
  Δ²-pyrrolines and
    α,β-ethyleneketones
    28, 664
**Indolo[2,1-b]quinazol-12(5H)-
  ones, 5a,6-dihydro- 17,** 395
  suppl. 26
Indolo[2,3-a]quinolizines
- special s.
  2-β-oxiranylindolo[2,3-a]-
    quinolizines
Indone ring
- startg. m. f.
  Δ²-1,2,4-triazoline ring
    27, 341
Indoxyls, 2-hydroxy-
- from
  indoles **30,** 85
Induction, asym. s. Synthesis,
  asym.
Inhibitors s. Polymerization
  inhibitors
Initiator s. Promoter
Insertion
- into
  hydrogen-carbon bonds
    26, 658
- of
  1 C-atom into oxo compds.
    to acetylene derivs. **28,** 731
  C-atoms into carbon-silicon
    bonds **27,** 883
  3 C-atoms into nitrogen-
    carbon bonds **28,** 614
  azimines into carbon-carbon
    triple bonds **28,** 281
  carbon monoxide **28,** 632
  - oxide sulfide into nitrogen-
    carbon bonds **29,** 531
  carbonyl via C-dihalogeno-
    methylation **27,** 726
  dihalogenomethylene groups
    26, 658
  halogenomethylene – **29,** 871
  ketene **4,** 650
  metals into halogen-carbon
    bonds **27,** 647
  oxygen into nitrogen-carbon
    bonds **26,** 123

phosphines into oxygen-
   oxygen bonds **29**, 106
sulfur dioxide into hydrogen-
   carbon bonds **26**, 569
C-Insertion
- into
  sulfur-carbon bonds **27**, 818
Insertion, nitrenoid
s. Nitrenoid insertion
Inter... s. a. Trans...
Intercalates s. Silicate
   intercalates, Graphite
Interchange s. a. Exchange
- of
  ethylene derivs. via
  cyclobutanes **30**, 467
-, **carbonyl-olefin** -
- of
  ethylene derivs., review **30**, 467
-, **positional 26**, 583; **30**, 519
-, **C,N-positional**, of acyl
  groups **30**, 527
Interconversion s. Oxo compd.-
  ethylene deriv. inter-
  conversion
Interfacial reaction conditons
s. 2-Phase medium
Introduction
- of functional groups
  s. a. Aldehyde group,
  Replacement of hydrogen
-, **relayed**, of alkylthio groups
  into carbohydrates **29**, 55
-, **stereospecific**
- of angular groups s. under
  Carboxyl groups, angular,
  Methyl -, -, Vinyl -, -
Inversion s. a. under
  Configuration
- of
  ethylene derivs.
  s. Geoisomerization
Iodides (s. a. Halides,
  Replacement)
- from
  boranes **26**, 881
- special s.
  azidoiodo...
  methyl iodide
- startg. m. f.
  iododichlorides **27**, 543
  iodosoacetates **27**, 543
-, **ar**.
- startg. m. f.
  (perfluoroalkyl)arenes
  26, 867
Iodination (s. a. Halogenation,
  Replacement of hydrogen
  by halogen) **29**, 158
-, **ar**. **17**, 598 suppl. 26
Iodine 16, 410 suppl. **29**; **21**, 600
  suppl. 28; **23**, 603 suppl. 28;
  25, 389 suppl. **29**; **26**, 9, 166,
  214, 255, 269, 841, 876; **27**,
  121, 185, 247, 286, 369, 550,
  560, 568, 642, 679, 736, 753,
  976; **28**, 114, 204, 444, 600,
  662, 666, 816, 864, 892/3; **29**,
  219, 287, 497, 673, 989; **29**, 993
  suppl. **30**; **30**, 32, 481, 525, 540,
  639, 645
- as halogen carrier **30**, 376
-, reductions with
  -/phosphorus/carboxylic acids,
  review **21**, 55 suppl. 26
-/**dimethyl sulfoxide 29**, 219
-/**pyridine 28**, 205
- **azide 29**, 317; **30**, 351

- **monochloride 26**, 514, 518;
  28, 511; **29**, 317
Iodobenzene dibromide 27, 552
- **dichloride 26**, 101, 526;
  28, 485; **29**, 244
Iododichlorides
- from
  iodides **27**, 543
- special s.
  iodobenzene di...
- startg. m. f.
  iodosoacetates **27**, 543
1,2-Iodohydrin nitrates
- from
  ethylene derivs. **26**, 518
Iodonium nitrate
-, addition to carbon-carbon
  double bonds **26**, 518
- salts
- from
  iodosohydroxy tosylates
  27, 574
Iodosoacetates
- from
  iodides **27**, 543
  iododichlorides **27**, 543
Iodosobenzene diacetate
s. Phenyl iodosoacetate
Iodosocarboxylates
- from
  iodoso compds. **27**, 116
- special s.
  (dibenzoyldioxyiodo)-
  benzenes
Iodoso compds.
- startg. m. f.
  iodosocarboxylates **27**, 116
Iodosohydroxy tosylates
- startg. m. f.
  iodonium salts **27**, 574
N-Iodosuccinimide as reagent
  26, 532; **29**, 268
Ion exchangers
- as catalysts, review **26**, 224
- special s.
  Amberlite
  anion exchangers
  cation -
  Dowex
Ionic s. under Halogenation
Ion pair extraction (s. a. Phase
  transfer catalysis)
  29, 414, 792
Iotsich reaction s. Grignard
  reaction
Iridium-carbon 27, 40
- **dimethyl sulfoxide complex
  23**, 69
- **phosphine complex 29**, 54
  suppl. **30**; **30**, 526
- **tetrachloride 26**, 33
- **tricarbonylchloro-, dimer
  26**, 741
Iron (s. a. Ferrates) **26**, 476;
  28, 36; **29**, 15; **30**, 699
-(**III**) **acetoacetonate** as reagent
  27, 87; **30**, 91
- π-**allyl complexes 28**, 632
- **carbonyl** (s. a. Diiron nonacarbonyl, Triiron dodecacarbonyl and under Iron
  complexes) **18**, 124 suppl. 28;
  19, 962 suppl. 28; **26**, 509, 739,
  889, 891, 991; **26**, 739 suppl.
  29; **27**, 77, 447, 600, 723; **28**,
  40; **29**, 761; **30**, 276

-(**II**) **chloride/dimethyl sulfoxide**
-, reaction with polyhalides
  17, 887 suppl. 26
-(**III**) **chloride 26**, 29, 359, 593,
  875, 930; **27**, 156, 257, 737,
  812; **27**, 932 suppl. **29**; **28**, 804,
  828; **29**, 423, 517/8; **30**, 110,
  146, 219, 372, 542
- - -**dimethylformamide
  complex** as oxidant
  22, 895 suppl. 27
- **complexes 26**, 875 suppl. **30**;
  28, 632
-, protection of carbon-carbon
  double bonds as -
  28, 631 suppl. 30
- special s.
  iron π-allyl complexes
  tricarbonylcyclohexa-
  dienoneiron complexes
  tris(dibenzoylmethido)-
  iron(III)
-(**II**) **iodide 22**, 938 suppl. 29
- **nitrate 27**, 140
- **salts** s. a. Ferrates
-(**II**) **sulfate 26**, 775; **27**, 259,
  722, 879; **28**, 879, 880; **30**, 661
-(**III**) **sulfate 28**, 148; **29**, 140
Irradiation (s. a. Light...,
  Photo..., Wavelength
  specificity)
- special s.
  electron irradiation
  laser -
  x-ray -
  red-light -
- with high luminous flux
  27, 559
γ-**Irradiation 4**, 667 suppl. **28**;
  19, 719 suppl. **29**; **28**, 709,
  720; **30**, 93
Isatin, 1-chloromethyl-
-, characterization of carboxylic
  acids with - **28**, 165
Isatin-3-hydrazones
- startg. m. f.
  o-azonitriles **29**, 282
Isatin ring, N-condensed
  26, 772
Isatins
- startg. m. f.
  carbostyrils, 3-hydroxy-
  29, 625
  indoles **9**, 938 suppl. 27
  quinoline-4-carboxylic acid
  amides **28**, 622
-, N-substitution **17**, 355
  suppl. 28
-, **tetrahydro-**
- from
  1,3-dienes **28**, 751
Isatoic anhydrides
s. 3,1-Benzoxazine-2,4-diones
**Isoalloxazine, 8-cyano-
  3,10-dimethyl-**, as oxidant
  30, 377
- **5-oxides**
- from
  uracils, 6-anilino- **28**, 314
-, 6-halogeno- **28**, 314
Isoalloxazines, electron-deficient
- special s.
  isoalloxazine, 8-cyano-
  3,10-dimethyl-
Isoaromatization, homogeneous
- to phenols **30**, 526
Isobasic s. under
  Isoquinolines

Isobutenylidene sulfate,
hexafluoro-
- as reactant 30, 389
Isobutyl chloroformate
as reactant 27, 194
Isocarbostyrils
- from
isocoumarins 28, 329
isocyanates 25, 669; 27, 753
ketocarboxylic acids 28, 329
- startg. m. f.
lactams, N-condensed 28, 329
Isochroman-3-ones, 26, 137
suppl. 27; 27, 919
- startg. m. f.
benzocyclobutenes 28, 970
Isochromene ring, 1-subst.
- from
isoquinoline ring, 1,2,3,4-
tetrahydro- 27, 267
- - opening 27, 544
Isochromenes
- from
isocoumarins 27, 66
-, 3-subst.
- from
benzocyclobutenes, 1-acyl-
27, 169
Isocoumarins
- startg. m. f.
isocarbostyrils 28, 329
isochromenes 27, 66
-, 3-acylamino-
- from
o-carboxyphenylacetonitriles
27, 470
-, 3,4-dihydro- 19, 738 suppl. 27
Isocyanates (s. a. Heterocumulenes)
-, chemistry, review 27, 317
- from
amines 27, 361
aminosilanes 26, 443;
29, 438
carbamic acid silyl esters
29, 438
carboxylic acid anhydrides
28, 400
- - azides 30, 460
- - chlorides 28, 400; 30, 460
1,3,2,4-dioxathiazole 2-oxides
26, 489
disilazanes 26, 443
hydroxamic acids 26, 489
thiolcarbamic acid esters
26, 493
-, prepn., phosgene-free 29, 438
-, production, review 17, 403
suppl. 29
-, reactions, base-catalyzed,
with methylene compds.,
active 28, 736; 29, 635
- special s.
acylisocyanates
N-(chlorocarbonyl)isocyanate
cyano-isocyanates
di-
en-
halogen-
sulfonyl-
- startg. m. f.
acylamines 30, 460
acyl carbamyl peroxides
26, 122
- - sulfides 30, 382
alkoxybiurets 30, 209
alkoxylureas 30, 209

N-(alkylthiocarbonyl)ureas
27, 639
allophanates (from 2 molecules) 29, 128
allophanoyl chlorides (from
2 molecules) 26, 488
2-azetidinones, 4-imino-
26, 692 suppl. 28
azomethines 28, 425
barbituric acids,
5-carbalkoxy- (from 2 molecules) 28, 736
biurets 29, 300
-, 3 hydroxy- 26, 293
N-carbamylamidines 26, 295
O-carbamylhydroxylamines
30, 125
carbamylsulfonic acids
29, 285
carbostyrils 26, 694
carboxylic acid amides,
N-subst. 26, 677; 27, 676
diacylamines 14, 530
1,3-diazetidine-2,4-diones
(from 2 molecules) 26, 488
1,3-diazetidin-2-one ring,
4-alkoxy- 29, 301
N-hydroxyureas 26, 293
isocarbostyrils 25, 669;
27, 753
ketenimines 30, 611
β-ketocarboxylic acid amides
27, 676 suppl. 29; 30, 462
1,3,5-oxadiazine-2,4-diones
27, 322
1,3,4-oxadiazolid-5-ones
30, 213
1,3-oxazetidines 28, 265
2-oxazolidones 27, 317
2-oxazolone ring 30, 293
phthalimidines 27, 490
2-piperidones 26, 726
pyrimidines 28, 338
1,3,6,9-tetraazaspiro[4.4]-
nonanes (from 2 mole-
cules) 26, 920
2H-thiazolo[3,2-a]-s-triazin-
4-ones, 3,4,6,7-tetrahydro-
27, 321
thioacylureas 30, 215
N,N,O-tris(carbamyl)-
hydroxylamines 26, 293
ureas 30, 229
Isocyanatocarboxylic acid
esters
- special s.
ethyl isocyanatoacetate
α-Isocyanatocarboxylic - -
- startg. m. f.
hydantoins, 3-hydroxy-
27, 519
α-N-hydroxyureidocarboxylic
acid esters 27, 519
Isocyanatocarboxylic acid
silyl esters
- from
aminocarboxylic acids
27, 512
Isocyanatoformic acid esters
- startg. m. f.
uracil ring 26, 678
Isocyanides s. Isonitriles
Isocyanoacetic acid esters
- startg. m. f.
α-aminocarboxylic acid
esters 29, 739

2-Isocyanoalcohols
- from
oxo compds. 26, 667
3-Isocyanoalcohols
- from
isonitriles and oxido compds.
29, 617
- startg. m. f.
3-aminoalcohols 29, 617
4H-1,3-oxazines, 5,6-dihydro-
29, 617
α-Isocyanocarboxylic acid
esters
-, α-alkylation 29, 791
- special s.
isocyanoacetic acid esters
- startg. m. f.
α-aminocarboxylic acids
29, 791
α-amino-β-hydroxycarboxylic
acids 22, 686 suppl. 28
Isocyanomethyl p-tolyl sulfone
as reactant 28, 729
α-Isocyanophosphonic acid
esters
- startg. m. f.
enisonitriles 29, 870
α-Isocyanosulfones
- special s.
isocyanomethyl p-tolyl
sulfone
Isocyanuric acids s. Trichloroisocyanuric acid
Isocycles in N-heterocyclics
-, hydrogenation, selective
23, 66 suppl. 29
-, O-bridged
-, hydrogenation, selective
30, 33
-, oxidation, - 30, 33
Isocyclics (isocycles)
(s. a. Aryl derivs., Ring...)
- from
ethylene derivs. 27, 731
halides 28, 953
O-heterocyclics 28, 989
S-heterocyclics, ring
contraction 26, 125
thioozonides 26, 125
- special s.
alcohols, isocyclic
azulenes
benz[a]azulenes
benzene ring
benzocyclobutenes
benzo[a]pyrenes
benzotropones
benzvalenes
bicyclo...
cyclo...
dibenzo[fg,op]naphthacenes
dibenzo[a,l]pyrenes
ene-α-diones, cyclic
enisocyclics
2,4a-ethanophenanthrenes
ethylene derivs., cyclic
fulvenes
helicenes
β-hydroxyketones, cyclic
methanonaphthalenes
naphthalenes
norcaranes
pentalenes
phenalenes
polyisocyclics
protoadamantanes
pyrenes
rotans

tricyclo[3.1.0.0²,⁴]hexanes
tricyclo[4.2.1.0²,⁵]nonanes
tropones
- startg. m. f.
  carboxylic acids 26, 230
  O-heterocyclics, ring
    expansion 26, 127
- with bridgehead double
    bonds 28, 854
-, bicyclic
- special s.
    alcohols, bicyclic
    tetrazolylisocyclics,
    halogeno-, -
-, -, *trans*-fused with methyl
    groups, angular 26, 691
-, condensed s. Isocyclics,
    bicyclic, Polyisocyclics,
    highly condensed
-, mono- and bi-cyclic
- by
    cyclopropane ring expansion
    29, 685, 982
-, strained, addition of carbon-
    carbon multiple bonds to -,
    review 26, 704
Isodithiobiurets
- startg. m. f.
    1,2,4-triazoles, 5-amino-
    3-mercapto- 29, 427
-, 3,5-diamino- 29, 427
Isoflavan-4-ols
- from
    chromenes 28, 865
    isoflavones 29, 52
Isoflavanones
- by Heck arylation 24, 884
    suppl. 29
- from
    isoflavones 29, 52
Isoflavones
- startg. m. f.
    isoflavan-4-ols 29, 52
    isoflavanones 29, 52
Isohydrazones s. Diaziridines
Isoimides s. Dicarboxylic acid
    isoimides 25
Isoindol... s. a. Pseudoiso-
    indol...
Isoindolenine ring, 3-alkoxy-
- startg. m. f.
    phthalazine ring, 1-amino-
    26, 350
Isoindolenines 29, 347
Isoindolenones 29, 447
Isoindoles 30, 261
- by
    benzyne C-C-ring closure
    28, 976
- from
    1,2,4-triazoles 30, 665
- special s.
    bisisoindoles
Isoindolines (s. a. Phthalimi-
    dines) 27, 448, 541
- from
    hydrazines 27, 449
    hydrazinium salts,
    quaternary, cyclic
    27, 449
-, 1,3-diimino-
- startg. m. f.
    phthalocyanines 27, 435
Isomerization (s. a. Migration,
    Rearrangement)
- of
    carbon-carbon multiple
    bonds 26, 729

2-ethylenealcohols 30, 101
hydroxybarbituric acids and
    related compds., review
    25, 120 suppl. 27
oxido compds. 26, 750
-, ring closure by - s. Cyclo-
    isomerization
-, - opening by - s. under
    Ring opening
- special s.
    anomerization
    epimerization
    geoisomerization
    pseudobase isomerization
    valence -
-, photochemical
    s. Photoisomerization
-, prototropic (s. a. Migration
    of multiple bonds)
-, review 21, 749 suppl. 26
-, skeletal, of dienones,
    bicyclic 28, 106
Isomers s. a. Stereoisomers,
    Imidazoles, nitro-
cis-trans-Isomers s. *cis-trans*-
    Ethylene derivs., Geo-
    specificity, *cis-trans*-
    Rearrangement
Isonitrile dihalides s. Imino-
    carbonylhalides
Isonitriles (s. a. Copper-
    isonitrile complexes)
- as reactants 26, 912
- as reagents 27, 219, 864
- from
    formamides 26, 470
-, metalation 29, 617
    syntheses via metalation of -,
    review 29, 617 suppl. 30
-, silyl adducts 16, 152 suppl. 29
- special s.
    tert-butyl isocyanide
    enisonitriles
    isocyano...
    sulfonylisonitriles
- startg. m. f.
    amines 29, 791
    3-aminoalcohols 29, 617
    α-aminocarboxylic acids
    29, 898
    α-aminoketones 28, 811
    aminomethylene compds.
    28, 619
    carbodiimides, unsym.
    27, 447
    cyclobutanones, 2-imino-
    30, 591
    cyclopropenes, imino-
    27, 690
    eneformylamines 26, 680
    α,β-ethylenecarboxylic acid
    amides 26, 688
    α,β-ethyleniminoesters
    29, 644
    furans, tetrahydro-
    2,3-diimino- (from 2 mole-
    cules) 27, 673
-, tetrahydro-2-imino-
    27, 672
    N-hydroxyformamidines
    30, 202
    3-isocyanoalcohols 29, 617
    isothiocyanates 28, 530
    α-ketoazomethines, cyclic
    27, 718
    4H-1,3-oxazines, 5,6-dihydro-
    29, 617
    oxazoles 28, 811

Δ²-oxazolines 28, 606
pyrroles (from 3 molecules)
    27, 909
pyrroline 25, 484 suppl. 26
Δ¹-pyrrolines 28, 606
N-thioformylacylamines
    30, 381
urethans 28, 93
Isonitroso compds.
    s. a. Oximes 30, 381
α-Isonitrosoketones
- startg. m. f.
    glycols 26, 185
    imidazoles, 1-hydroxy-
    26, 338
    nitriles 30, 342
-, cyclic
- startg. m. f.
    tropolone ring 27, 206
Isopropenyl acetate as reactant
    27, 805
- formate, N-formylation,
    selective, with - 27, 375
Isopropyl alcohol
- as H-donor 30, 45
- as medium 28, 445; 30, 515
- as reagent 7, 679 suppl. 28;
    21, 923 suppl. 26; 27, 77;
    30, 23
- as solvent 27, 83
Isopropylamine
- as medium 30, 513
- as reagent 27, 85
Isopropylbenzene as solvent
    26, 443
O,O-Isopropylidene derivs.
    (s. a. Acetals)
-, cleavage 26, 11
-, -, selective 30, 9
- groups, removal 30, 49
Isopropyl sulfide
- as medium 23, 693 suppl. 27
- as solvent 29, 670
Isopyrazole ring
- from
    pyrazolenine ring 27, 346
    suppl. 29
Isoquinoline-1(2H),3(5H)-
    diones, 6,7-dihydro-
- by
    double ring closure
    28, 332
Isoquinoline N-oxides,
    3,4-dihydro-
- startg. m. f.
    isoquinolines 29, 923
- ring (s. a. Pictet-Gams,
    Pomeranz-Fritzsch)
- startg. m. f.
    3-benzazepine ring 28, 345
-, -, 1,3-dihalogeno- (s. a. Iso-
    quinolines, 1,3-dihalogeno-)
- from
    arylacetonitriles 26, 824
-, 1,2,3,4-tetrahydro-
- startg. m. f.
    isochromene ring, 1-subst.
    27, 267
Isoquinolines 29, 931, 975
- by Gattermann-Koch-type
    synthesis 28, 743
- from
    aminoketals 28, 444
    isoquinoline N-oxides,
    3,4-dihydro- 29, 923
    isoquinolines, 1,2,3,4-tetra-
    hydro- 30, 642

(Isoquinolines)
– special s.
  isocarbostyrils
  1-(1'-phthalide)isoquinolines
– startg. m. f.
  isoquinolines, 1,2-dihydro-
  2-acyl- 29, 409
–, **isobasic 30,** 235
–, **1-subst.**
– from
  2-acyl-12-dihydroisoquinaldo-
  1-nitriles 29, 889
–, **1,3-dihalogeno-** (s. a. Isoquinoline ring, 1,3-di-
  halogeno-)
– from
  1-indanones, 2 oximino-
  26, 544
–, **1,2-dihydro-** s. a. Dihydroisoquinoline rearrangement
– special s.
  2-acyl-1,2-dihydroquinaldo-
  1-nitriles
–, **1,4-dihydro-** s. 2-Aza-
  naphthoquinones
–, **3,4-dihydro-**
– from
  isoquinolines, 1,2,3,4-tetra-
  hydro- 30, 334, 642
–, 1,2,3,4-tetrahydro-
  N-nitroso- 30, 334
  nitriles and alcohols 26, 820
– and cyclopropanes 26, 820
– startg. m. f.
  isoquinolines, 1,2,3,4-tetra-
  hydro- 7, 28
–, **1,2-dihydro-2-acyl-**
– from
  isoquinolines 29, 409
–, **3-hydroxy- 26,** 335
–, **3-hydroxy-1-mercapto-,** ring
  closures with – 29, 570
–, **1,2,3,4-tetrahydro-**
– by cyclodehydration 27, 928
– from
  azomethines 29, 702
  isoquinolines, 3,4-dihydro-
  7, 28
– startg. m. f.
  isoquinolines 30, 642
  –, 3,4-dihydro- 30, 334, 642
  –, 1,2,3,4-tetrahydro-
  N-nitroso- 30, 334
–, **1,2,3,4-tetrahydro-2-acyl-**
– from
  1,1-alkoxyhalides 26, 884
–, **1,2,3,4-tetrahydro-N-nitroso-**
– from
  isoquinolines, 1,2,3,4-tetra-
  hydro- 30, 334
– startg. m. f.
  isoquinolines, 3,4-dihydro-
  30, 334
**Isoquinoline-1-spiro-2'-(1',3'-
  indandiones), 1,2-dihydro-
  2-acyl-**
– startg. m. f.
  1-(1'-phthalide)isoquinolines
  30, 44
**Isoquinolinium methylids
  17,** 759
– startg. m. f.
  Δ^3-pyrroline ring 29, 649
– **ring,** opening **30,** 73
–, –, **3,4-dihydro-**
–, opening **28,** 693
– startg. m. f.
  naphthalene ring **28,** 693

– salts
–, C-alkylation, reductive
  26, 801
– startg. m. f.
  8,17-diazadibenzo[c,j]tetra-
  cyclo[7.3.1.0²,⁹.0⁷,¹⁰]-
  tridecanes 28, 645
–, **4-amino-**
– by Schroeter-type dehydration 26, 950
–, **4-hydroxy-**
– from
  isoquinol-4(1H)-ones,
  2,3-dihydro- 26, 462
Isoquinol-3(2H)-ones 28, 368
Isoquinol-3(4H)-ones
– from
  arylacetic acids (2 molecules) **26,** 371
– via self-acylation of arylacetic acids **26,** 371
**Isoquinol-4(1H)-ones,
  2,3-dihydro- 12,** 924 suppl. **26;
  20,** 665 suppl. **26**
– startg. m. f.
  isoquinolinium salts,
  4-hydroxy- 26, 462
**Isothiazole ring, 3-amino-
  30,** 218
Isothiazoles
– from
  enamines and carboxylic acid
  amides **26,** 841
– startg. m. f.
  thiazoles 27, 603
–, **3-hydroxy-**
– from
  nitriles (2 molecules) 29, 777
Isothiazolid-3-one 1,1-dioxides
– from
  ethylene derivs. 26, 613
– **ring opening,** reductive
  30, 42
3-Isothiazolone ring
– startg. m. f.
  1,3-dithietane ring 26, 585
Isothiocarbamyl halides
  s. S-Halogenoisothiocarbamyl
  halides
Isothiocyanates (s. a. Acylisothiocyanates, Hetero-
  cumulenes)
– from
  amines 27, 366
  aminosilanes 27, 486
  dithiocarbamic acids 30, 340
  iminocarbonylhalides 28, 557
  isonitriles **28,** 530
  nitrile oxides **28,** 527
  1,4,2-oxathiazoles **28,** 527
  silylamidines 27, 495
–, reaction with aminosugars
  27, 318
– special s.
  aryl-isothiocyanates
  carbamyl-
  en-
  formamidinoyl-
  halogen-
  sulfonyl-
  thiocarbamyl-
  thiophosphinyl-
– startg. m. f.
  2H-1,3-benzothiazin-2-ones,
  3,4-dihydro- **29,** 550
  2H-1,3-benzoxazine-2-thiones,
  3,4-dihydro- **29,** 550

carboxylic acid thioamides,
  subst. **14,** 723
1,5,3-dithiazepine ring,
  N-condensed **28,** 573
α,β-ethylenecarboxylic acid
  thioamides **26,** 903
1,3-N,N-heterocyclics,
  2-amino- **26,** 434
N-hydroxythioureas **20,** 235
suppl. **28**
imidazolidine-2-thiones,
  4-hydroxy- **27,** 318
2-oxazolidinethiones **29,** 632
phosphinylthiosemicarbazides
  27, 320
2-pyridinethione ring,
  6-amino- **27,** 680
1,3,6,9-tetraazaspiro[4.4]-
  nonanes (from 2 mole-
  cules) **26,** 920
1,2,3,4-thiatriazoles, 5-amino-
  28, 278
1,3-thiazepines, 4,7-dihydro-
  2-amino- **28,** 526
Δ^4-thiazolines, 4-amino-
  26, 706
N-thiocarbamylamidines
  26, 295
thiocarbamylphosphine
  oxides **26,** 636
thiocarbostyrils **30,** 529
thiophenes, 2,4-diamino-
  26, 892
N-Isothiocyanatoamines
– startg. m. f.
  dithiocarbazic acid esters
  26, 571
**Isothiocyanatocarboxylic acid
  silyl esters 21,** 389 suppl. **26**
Isothioureas
– from
  carbodiimides **30,** 383
  N-cyanourethans **30,** 418
– special s.
  indan-1,3-diones,
  2-isothioureido-
– startg. m. f.
  amidines **30,** 686
  1,3,5-triazine ring, hexa-
  hydro-, trimerization
  26, 435
Isothiouronium salts
– by replacement of hydrogen
  26, 593
– special s.
  N-acylisothiouronium salts
  S-acylisothiouronium –
  benzylisothiouronium
  dithiocarbamates
  ethyleneisothiouronium salts
– startg. m. f.
  mercaptans **26,** 593
  selenothiolcarbamic acid
  esters **26,** 652
  thiolcarbamic acid esters
  28, 159
Isotopes s. Compounds,
  labeled
Isoureas
– from
  carbodiimides **30,** 50
– startg. m. f.
  carboxylic acid esters **29,** 547
  hydrocarbons **30,** 50
  thioethers **29,** 547
–, **bicyclic,** ring opening,
  reductive **29,** 33

–, **N-acyl-**, **cyclic** s. Oxazolidine ring, 2-imino-1-acyl-
–, **O-aryl**
– from
 phenols **30**, 50
– startg. m. f.
 arenes **30**, 50
Isouronium salts 26, 173
– from
 ureas **26**, 173
– –, **cyclic 26**, 173
dl-Isovaline as reactant **27**, 414
Isoxazole ring
– startg. m. f.
 α,β-ethylene-β-hydroxy-
  nitriles **28**, 293
– – **conversion** into other heterocycles, review **29**, 322 suppl. **30**
Isoxazoles (s. a. Bisisoxazoles)
– from
 5-ene-Δ²-isoxazolines **26**, 729
 α,β-ethyleneoximes **28**, 204
 Δ²-isoxazolines, 5-acoxy-
  26, 932
 nitro compds., aliphatic, and acetylene derivs. **26**, 796
 – –, –, and enamines **29**, 786
 oximes **26**, 792
– startg. m. f.
 β-aminoketones **28**, 204
–, **5-alkoxy-**
– startg. m. f.
 Δ¹-azirine-3-carboxylic acid esters **26**, 313
–, **allene-** by isomerization **26**, 729
–, **3-amino-**
– from
 α,β-dihalogenonitriles **26**, 409
 α,β-ethylenenitriles **26**, 409
–, **5-amino-**
– startg. m. f.
 Δ¹-azirine-3-carboxamides **29**, 322
–, **3,5-dihydroxy-**, reaction with oxo compds. **29**, 723
–, **3-nitro-**
– from
 2-acetylenehalides **28**, 397
Isoxazolidine ring
– from
 ethylenenitrones **20**, 284
 ethyleneoxo compds. **20**, 284
– – **opening 30**, 693
Isoxazolines, 4-amino-
– from
 enamines and nitrones **26**, 699
5-Isoxazolidones
– from
 nitrones and α-halogeno-
  carboxylic acid esters
  28, 817
Δ²-Isoxazoline oxides
– from
 diazo compds. **28**, 774
Δ²-Isoxazoline ring
– from
 4-ene pyran ring **29**, 360
 nitrile oxides **29**, 652
Δ²-Isoxazolines
– from
 1,3-ammoniooximes **27**, 263
 diazo compds. **28**, 778
– special s.
 5-ene-Δ²-isoxazolines

–, **optically active 10**, 239 suppl. **26**
Δ⁴-Isoxazolines
– from
 isoxazolium salts **26**, 934
– startg. m. f.
 pyrroles **26**, 934
Δ²-Isoxazolines, 5-acoxy-
– startg. m. f.
 isoxazoles **26**, 932
Isoxazolium salts
– startg. m. f.
 Δ⁴-isoxazolines **26**, 934
 pyrroles **26**, 934
5-Isoxazolones
– startg. m. f.
 5-pyrazolones **27**, 397
Δ²-Isoxazol-5-ones
– from
 oximes **27**, 777
Δ⁴-Isoxazol-3-ones
– from
 α,β-acetylenehydroxamic acids **28**, 125
Isoxazolo[1,2-b]pyrazol-3-ones
– from
 4H-pyrazol-4-one N-oxides **26**, 701
Isoxazolo[2,3-a]pyridinium salts, 4,5,6,7-tetrahydro-4-oxo-
– startg. m. f.
 4H-furo[3,2-b]pyrid-2-ones, 5,6-dihydro- **29**, 166
Isoxazolo[3,5-b]quinoxalines 26, 452
Isoximes s. Nitrones

Jacobs s. Gould
Japp-Klingemann cleavage 28, 42

Kakis reaction 27, 155
Kaufmann s. Reissert
Ketal groups, retention in Grignard synthesis **27**, 856
Ketals (s. a. Acetals)
– special s.
 allene-ketals
 amino-
  cyclopropanone –
 di-
 dioxo compd. mono-
 ethylene-
 halogeno-
 nitro-
 spiro-
 1,3,3-trialkoxy compds.
– startg. m. f.
 enolethers **26**, 130; **29**, 925
 2-ethylenealcohols **28**, 904
 hydrocarbons **26**, 55
–, **cyclic** (s. a. Protection of keto groups)
– special s.
 O,O-alkylidene...
–, **mixed**
– via enolethers **26**, 130
–, –, **O-heterocyclic**
– from
 enolethers **26**, 138
Ketazines
– startg. m. f.
 azo compds., tert. **29**, 832
 1,1'-dihalogenazo compds. **29**, 832

Ketene
–, acetylation with –, review **6**, 366 suppl. **28**
–, insertion **4**, 650
– startg. m. f.
 β-hydroxycarboxylic acid esters **29**, 629
 4(3H)-pyrimidinones, 5-acyl- (from 2 molecules) **26**, 340
–/**sulfur dioxide**, reactions with – **26**, 590
Keteneacetals
– startg. m. f.
 anthraquinones **29**, 788
 piperazines (from 2 molecules) **27**, 431
Ketene O,N-acetals
 s. 1-Alkoxyenamines
– **S,N-acetals** s. 1-Amino-
  thioenolethers
– **N-acyl-S,N-acylals**
– from
 N-acylcarboxylic acid thioamides **29**, 575
– **acylals 30**, 70
– **S,N-acylals**
– from
 N-acylcarboxylic acid thioamides **29**, 575
– special s.
 ketene N-acyl-S,N-acylals
– **aminals**
– special s.
 alkylthioketene aminals
 1,1,2-triaminoethylenes
– **dihalides** s. 1,2-Ethylene-1,1-dihalides
– **dimers**
– special s.
 diketene
– **mercaptal monosulfonium salts**
– as 2-C-Michael acceptors **29**, 667
– – **S-monoxides** s. 1-Sulfinyl-
  thioenolethers
– **mercaptals**
– from
 carboxylic acid esters **29**, 552
 α,β-ethylenemercaptals **9**, 781 suppl. **27**
 ketenimines and mercaptans **30**, 386
 oxo compds. and silylthioformals **28**, 857
–, rearrangement **26**, 582
– special s.
 hydroxyketene mercaptals
 ketoketene –
 vinylketene –
– startg. m. f.
 α,β-ethylenemercaptals **30**, 505
 thiolic acid esters **26**, 213
–, substitution, electrophilic **28**, 544
Ketenes (s. a. Thioketenes)
– from
 O-silyl O-alkyl keteneacetals **28**, 227
–, reactions with sulfino-
 diimines **26**, 305 suppl. **28**
– special s.
 bis(trifluoromethyl)ketene
 cyano-ketenes
 halogeno-
 phosphoranylidene-

Ket 662

(Ketenes)
- startg. m. f.
  α-allenecarboxylic acid esters **28**, 851
  allenes, sym. **28**, 878
  2-azetidinones, 4-alkylthio- **8**, 736
  -, 4-imino- **26**, 692
  cyclobutane-1,3-diones **27**, 691
  enecyclopropenes **30**, 627
  enolesters (from 2 molecules) **26**, 710
  β-hydroxythiolic acid esters **28**, 862
  α-oximinocarboxylic acids and derivs. **28**, 285
  succinimides (from 2 molecules) **29**, 765
  sulfocarboxylic acid anhydrides **30**, 79
  1,2,5-thiadiazolid-4-ones **26**, 305
Ketene selenothioacetals
- special s.
  1-organoseleno-1-organothio-1-en-3-ynes
- **silyl hemiacetals**
- startg. m. f.
  β-hydroxycarboxylic acids **27**, 884
Ketenimines
- from
  alkylidenephosphoranes and isocyanates **30**, 611
-, reactions, review **26**, 495
- special s.
  silylketenimines
  vinylketenimines
- startg. m. f.
  2-azetidinones, 4-imino- **26**, 692 suppl. **28**
  ketene mercaptals **30**, 386
  nitriles **26**, 495
  Δ²-oxazolines, 2-alkoxy-4-imino- **27**, 432
  quinolines, 4-amino- **27**, 699; **29**, 647
  thioiminoesters **30**, 386
-, synthesis, review **26**, 495
Ketimines (s. a. Imines)
- special s.
  oxidoketimines
- startg. m. f.
  ethylene derivs., synthesis **27**, 815
-, **enolizable**
-, reaction with electrophiles **5**, 326 suppl. **26**
Keteniminylidenephosphoranes
- from
  iminocarbonylhalides and methylenephosphoranes **30**, 579
α-**Keto**... s. a. Acyl...
β-**Ketoacetals**
- from
  enoxysilanes **30**, 621
o-**Ketoacetic acids**
- from
  indans, 2,2-dialkoxy-1,3-dihydroxy- **26**, 243
- startg. m. f.
  2-benzopyr-3-ones **26**, 243
Ketoalcohols s. Hydroxyketones

α-**Ketoaldehyde aldoximes**
- from
  β-ketocarboxylic acid esters **30**, 319
- startg. m. f.
  1,2,4-triazine-3-thiones **26**, 387
α-**Ketoaldehyde aldoxime side chains**, C-cleavage **29**, 267
- **mercaptals 30**, 402
Ketoaldehydes
- startg. m. f.
  hemiketals **26**, 874
α-**Ketoaldehydes**
- startg. m. f.
  α-hydroxyketones, synthesis **30**, 446
γ-**Ketoaldehydes** **17**, 872 suppl. **29**
- via thio-Claisen rearrangement **28**, 842
Ketoalkoximes
- from
  hydroxyalkoximes **28**, 896
- startg. m. f.
  2-cyclohexenone ring **28**, 896
2-**Ketoalkoximes**
- from
  acetylene derivs. and nitro compds., aliphatic **28**, 104
β-**Ketoamidines 30**, 686
Ketoamines s. Aminoketones
α-**Ketoazomethines, cyclic**
- from
  ketones, cyclic, and isonitriles **27**, 718
Ketocarboxylic acid amides
- from
  dicarboxylic acid imides **26**, 679
- startg. m. f.
  enaminolactams **27**, 518
α-**Ketocarboxylic** – – **27**, 326
- from
  acylcyanides **27**, 207
  α-(O-carbamylhydroxylamino)carboxylic acid amides **30**, 125
  carboxylic acid chlorides and formamides **27**, 847
  halides and oxamides **29**, 801
  α-hydroxycarboxylic acid amides **27**, 326
  α-hydroxylaminocarboxylic acid amides **30**, 125
- startg. m. f.
  α-ketocarboxylic acids **27**, 207, 326
β-**Ketocarboxylic** – –
- from
  isocyanates and enoxysilanes **30**, 462
- and ketones **27**, 676 suppl. **29**
- special s.
  α-alkylidene-β-ketocarboxylic acid amides
- startg. m. f.
  2-pyridones (from 2 molecules) **26**, 800
  ureas **28**, 417
γ-**Ketocarboxylic** – – **26**, 287
- from
  γ-enollactones **26**, 287
- startg. m. f.
  furans, 2-amino- **30**, 164

furfurylideneammonium salts, 2,3-dihydro- **30**, 164
α-**Ketocarboxylic acid chlorides 29**, 524
- – **ester mercaptals**
- from
  ethylene derivs. **29**, 661
Ketocarboxylic acid esters
- special s.
  aminoketocarboxylic acid esters
  ethyleneketocarboxylic – –
  halogenoketocarboxylic – –
  hydroxyketocarboxylic – –
  ketodicarboxylic – –
- startg. m. f.
  hydroxycarboxylic acid esters **1**, 42
α-**Ketocarboxylic** – –
  (s. a. Alkoxalyl...)
-, C-carbalkoxylation, radical, with – **28**, 879
- from
  acetylenehalides **30**, 133
  carboxylic acid esters (2 molecules) **3**, 638; **7**, 771
  α-cyanocarboxylic – – **27**, 236
  α-diazoketones **27**, 204
  1,3-dithianes **27**, 230
- startg. m. f.
  α,β-ethylenecarboxylic acid esters, Wittig synthesis **18**, 912 suppl. **26**
- via α,α-disulfenylation **30**, 414
β-**Ketocarboxylic** – –
-, γ-alkylation **26**, 861
- from
  β-alkoxy-β,γ-ethylenecarboxylic acid esters **29**, 159
  β-alkylthio-α,β-ethylenecarboxylic – – **30**, 584
  α-allenecarboxylic – – **29**, 159
  α-amino-β-ketocarboxylic – – **29**, 68
  carboxylic acid chlorides **10**, 616
  dicarboxylic acid esters, ring closure, s. Dieckmann cyclization
  halides **29**, 499; **30**, 584
  α-ketocarboxylic acid esters **29**, 700
  α-ketoketene mercaptals **30**, 466
  ketones **27**, 831; **30**, 466
- startg. m. f.
  α,β-acetylenecarboxylic acid esters **27**, 203
  β-alkoxy-α,β-ethylenecarboxylic – – **26**, 54
  α-allenecarboxylic – – **27**, 203
  α-aminomethyl-β-ketocarboxylic – – **29**, 759
  carboxylic – – **26**, 54; **29**, 499
  – acids **26**, 54
  β-diketones **30**, 614
  furylideneacetic acid esters, tetrahydro- **30**, 448
  δ-hydroxy-β-ketocarboxylic – – **30**, 447
  ε-hydroxy-β-ketocarboxylic – – **30**, 448
  α-ketoaldehyde aldoximes **30**, 319
  ketones **29**, 963

γ-methylene-β-ketocarboxylic
  acid esters 29, 759
- via carboxylic alkoxyformic
  anhydrides 3, 703 suppl. 26
-- - -, **branched**
- from
  ketones 26, 837
γ-**Ketocarboxylic** - -
- from
  β,γ-acetylenecarboxylic
    acid esters 28, 783
  acetylene derivs. 28, 783
  carboxylic acid esters,
    synthesis 29, 667
  α-diazocarboxylic acid esters
    and enoxysilanes 28, 858
  γ-keto-β-dicarboxylic acid
    esters 29, 86
β-**Ketocarboxylic acid**
  hydrazides
- startg. m. f.
  5-pyrazolones, 1-subst.,
    rearrangement 26, 464
α-**Ketocarboxylic acid ketals**
  s. 1-Carboxyacetals
Ketocarboxylic acids
- by ozonolysis, abnormal
- from   28, 192
  2-ethylenealcohols, cyclic
    30, 148
  lactones 28, 208
- special s.
  aminoketocarboxylic acids
  benzophenone-4-carboxylic
    acid
  diketocarboxylic acids
  o-ketoacetic -
- startg. m. f.
  carboxylic acids 3, 56
  β-diketones, cyclic 29, 935
  enaminolactams 29, 367
  enollactones 27, 919
  isocarbostyrils 28, 329
  lactams 28, 349
  lactones 28, 211
-, -, **cyclic**
- startg. m. f.
  β-diketones, spirocyclic
    29, 935
α-**Ketocarboxylic acids**
- from
  α-acylamino-α,β-ethylene-
    carboxylic acid esters
    29, 200
  acylcyanides 27, 207
  α-alkoxy-α,β-ethylene-
    carboxylic acid esters
    29, 200
  cyanohydrins 27, 326
  α,α-dihalogenoketones
    29, 208
  α-ketocarboxylic acid amides
    27, 207, 326
- special s.
  3-deoxy-2-ketoaldonic acids
  glyoxylic acid
- startg. m. f.
  α-hydroxycarboxylic acids
    17, 65 suppl. 28
-, synthesis 26, 912
β-**Ketocarboxylic** -
- startg. m. f.
  ketones 28, 733
-, synthesis 28, 733
γ-**Ketocarboxylic** -
- from
  β,γ-dihalogeno-β-carboxy-
    carboxylic acids 21, 263

- special s.
  α,β-ethylene-γ-ketocarboxylic
    acids
  levulinic acid
- startg. m. f.
  α,β-ethyleneketones 23, 982
  α,β-ethylene-γlactones
    29, 253
  α,β-ethylene-γ-lactones with
    rearrangement 28, 217
β-**Keto-γ-dicarboxylic acid**
  esters
- from
  ketones 26, 770
γ-**Keto-β-dicarboxylic acid**
  esters
- startg. m. f.
  γ-ketocarboxylic acid esters
    29, 86
Ketoeneurethans
- startg. m. f.
  2-cyclohexenone ring
    30, 666
3-Keto-1,2-en-1-olethers,
-, 4-alkylation 29, 805
- startg. m. f.
  α,β-ethyleneketones 29, 805
-, **sym.**
- from
  carboxamidium salts,
    vinylogous 26, 184
3-Keto-1,3-en-2-ols
  s. Tropolones
α-**Ketohydroximinohalides**
- from
  2-ketosulfonium salts
    27, 348
  β-ketosulfoxides 27, 348
Ketohydroxylamines
- startg. m. f.
  nitrones, cyclic 24, 523
α-**Ketoketene mercaptals**
- from
  ketones 30, 466
- startg. m. f.
  β-ketocarboxylic acid esters
    30, 466
  pyrimidines 30, 310
2-Ketolactams
- from
  2-halogenolactamols, ring
    contraction 29, 471
α-**Ketolactams**
- startg. m. f.
  α-carboxylactams, ring
    contraction 30, 90
β-**Ketolactams**
- from
  β-alkoxy-α,β-ethyleneoximes,
    cyclic 27, 163
  β-diketones, cyclic 27, 163
Ketolactones
- from
  hydroxylactones 28, 208
  lactones 27, 505
- startg. m. f.
  dienollactones 26, 953
  hydroxycarboxylic acids
    27, 505
Ketols s. Hydroxyketones
β-**Ketomercaptals, cyclic**
- from
  α-hydroxymethyleneketones
    13, 602 suppl. 29
γ-**Ketomercaptals**
- from
  α,β-ethyleneketones and
    mercaptals 28, 175

- startg. m. f.
  γ-diketones 28, 175
β-**Ketomercaptans**
- from
  α,β-ethyleneketones 29, 539
Ketone hydrates
- startg. m. f.
  ketones 20, 526 suppl. 26
Ketones (s. a. Oxo compds.)
-, alkylation 23, 832 suppl. 26;
  26, 853
  α-alkylation via enolate
    anions 19, 881 suppl. 28;
    29, 804
  α-monoalkylation 19, 896
    suppl. 26
-, arylation 11, 862 suppl. 27
  α-arylation, oxidative
    30, 542
-, dimerization, reductive 27, 800
- from
  2-acetylenealcohols,
    cleavage 28, 230
  acetylene derivs. 22, 143
    suppl. 27; 29, 140
  - - and boranes 29, 882
  acoxy compds. 26, 161
  β-acoxy-α,β-ethyleneketones
    29, 746
  α-acoxyketones 30, 27
  α-acoxynitriles 30, 557
  (α-acylalkylidene)-
    phosphoranes 30, 582
  acylamines 29, 198
  acylcyanides 27, 905
  alcohols with ring opening
    19, 211
-, sec. oxidation OC↑H
-, tert. 29, 268
  aldehydes, synthesis 22, 661
    suppl. 28; 29, 808
  - and ethylene derivs. 28, 648
  alkylidenephosphoranes
    28, 846
  amines, prim. 26, 187
  -, tert., and enamines 28, 768
  α-aminomethyleneketones
    30, 589
  azomethines 27, 890;
    29, 504, 647
  (*tert*-carbin)amines,
    N,N-dichloro- 27, 240
  carboxylic acid amides and
    ethylene derivs. 29, 772
  - - chlorides 27, 852; 28, 827
  - - - and enamines 30, 589
  - - - and ethylene derivs.
    30, 603
  - esters 27, 855
  - acids 27, 799, 810
  cyclopropanone mercaptals
    29, 216
  cyclopropylamines 30, 119
  α-diazoketones, synthesis
    27, 894
  α-diketones, synthesis and
    cleavage 27, 674
  1,1-diols 26, 239
  enesilanes 30, 143
  eneurethans 29, 237
  α,β-ethylenecarboxylic acid
    azides 29, 237
  ethylene derivs. 27, 155, 157;
    28, 282, 481
  - -, cleavage, oxidative
    27, 246 suppl. 28
  - -, synthesis 27, 716;
    29, 896; 30, 488

(Ketones)
– from)
α,β-ethylenehalides **29,** 210
α,β-ethylene-γ-hydroperoxy-γ-lactones **26,** 227
α,β-ethylene-α-hydroxyketones **28,** 51
α,β-ethyleneketones, hydrogenation **23,** 63 suppl. **29; 27,** 50; **28,** 45; **29,** 50
glycols **27,** 259
  degradation with C-rearrangement **29,** 700
  halides, synthesis **27,** 885; **30,** 582
  – (2 molecules) **27,** 869
α-halogenocarboxylic acid halides, degradation with loss of 1 C-atom **28,** 630
1,2-halogenhydrins with rearrangement **28,** 946
α-halogenoketones **26,** 71; **27,** 75; **29,** 79, 80; **30,** 27
α-halogenonitriles **26,** 718
hydrocarbons (methylene groups) **26,** 162, 165; **27,** 177, 188; **28,** 134, 136; **29,** 179
α-hydroxycarboxylic acids **27,** 275
α-hydroxyhydrazones, degradation, oxidative **27,** 244
–, cleavage **27,** 674
α-hydroxyketones, cleavage **27,** 674
–, degradation, oxidative **26,** 226
β-ketocarboxylic acid esters **29,** 963
– acids **28,** 733
ketone hydrates **20,** 526 suppl. **26**
ketones, α-subst. **27,** 75
2-ketothioethers **27,** 75
mercury halides, monoorgano- **26,** 218
– –, monoorgano-2-hydroxy- **29,** 264
nitriles, degradation with loss of 1 C-atom **30,** 155
oxazolidine N-oxyls **28,** 234
oxido compds. **26,** 149; **27,** 161
– – with rearrangement, skeletal **28,** 686
α,β-oxidoketones **30,** 28
oxime acetates **26,** 192
phenol ring **26,** 227
phosphinous acid esters **30,** 159
phosphoroperoxidates **29,** 248
α-siloxynitriles **29,** 808
sulfines **27,** 270
N-sulfonylimines **27,** 181
thioenolethers **29,** 802
–, synthesis **29,** 802
thioketones **29,** 220
thiolic acid esters **26,** 982; **29,** 853
thionocarboxylic acid esters, synthesis **29,** 896

–, isomerization **14,** 94 suppl. **26**
–, reaction with formaldehyde **26,** 659
– special s.
acetone
acetonyl groups
β-(acetoxymercury)-ketones
acetylene-
acoxy-
acylamino-
alkoxy-
alkoxylamino-
alkylthiomethylene-
allene-
amino-
aryl –
α-azido-
benzyl-
cyano-
cyclopropyl-
deuterio-
α-diazo-
di-*tert*-butylketone
dienones
difluoronitromethyl ketones
diethylene-
furyl-
halogeno-
heterocyclics, C-acyl-
hydroperoxy-ketones
hydroxy-
indoles, 2-δ-ketomethyl ketones
monothiodi-
nitro-
oxido-
oximino-
phenol-
4-piperidone, 1-methyl-
silyl-
sulfonylamino-
tri-
vinyl-
– startg. m. f.
β-acoxycarboxylic acid esters **26,** 938
β-acoxyketones **26,** 659
N'-acyl-1,1-cyanohydrazines **27,** 780
alcohols, cyclic **27,** 744, 747; **28,** 768
–, isocyclic **26,** 744
–, sec. HC↓OC
aldehydes, synthesis with addition of 1 C-atom **26,** 859
1-(alkoxy)enolethers **28,** 803
β-alkoxyketones, synthesis **28,** 654
allenes **28,** 913
2-aminoalcohols, synthesis with addition of 1 C-atom **27,** 720
α-amino-γ-ketocarboxylic acids **28,** 715
α-aminoketones **28,** 323
β-aminoketones **26,** 808
benzene ring **26,** 785
1,4-benzothiazine ring **26,** 624
carboxylic acids, hydrolysis **29,** 236
α-cyanoketones **23,** 521 suppl. **26**
cyclobutanol ring **30,** 518

cyclobutanone ring **28,** 837
2-cyclohexenones **28,** 721
cyclopentenes **28,** 901
cyclopropanols **26,** 744
α-diazoketones **30,** 247
1,1-dihalides, C-cleavage **30,** 375
1,4-diols, synthesis **28,** 609
1,3-dioxanes, 5-acyl- **26,** 659
α-dioxo compds. **29,** 176
1,1-dipyrazolyl compds. **28,** 426
enisocycles **27,** 934
enolesters **28,** 140
enoxysilanes **27,** 720
3-ethylenealcohols **28,** 601
α,β-ethylenealdehydes, synthesis with addition of 2 C-atoms **5,** 523 suppl. **26**
γ,δ-ethylenealdehydes, synthesis **26,** 900
α,β-ethylenecarboxylic acid esters **26,** 938
ethylene derivs. **17,** 208; **27,** 949; **28,** 65
– –, synthesis **30,** 615
– –, terminal, C-cleavage **25,** 695 suppl. **26**
α,β-ethylenehalides **21,** 607 suppl. **27; 26,** 541
α,β-ethylene-β-halogenocarboxylic acid amides **28,** 717
α,β-ethylene-δ-hydroxycarboxylic acids **27,** 663
α,β-ethylenenitriles **17,** 892 suppl. **27**
formamides **27,** 53
furan ring, 2,3-dihydro- **28,** 838
– –, tetrahydro- **28,** 609, 641
glycidic acid *tert*-butyl esters **26,** 859
– – esters **26,** 895
hydrocarbons, reduction **28,** 48/9; **30,** 31
–, branched **28,** 845
α-hydroxyaldehydes, synthesis with addition of 1 C-atom **22,** 661 suppl. **27**
2-hydroxythioethers **29,** 262
imines **7,** 454 suppl. **26; 26,** 433
β-iminodithiocarbonic acids **30,** 464
indan-1,3-diones **27,** 767
α-isonitrosoketones **1,** 317, 783; **2,** 145; **3,** 297; **6,** 390; **11,** 406; **13,** 394; **16,** 415
β-ketocarboxylic acid amides **27,** 676 suppl. **29**
– – esters **27,** 831; **30,** 466
– – –, branched **26,** 837
β-keto-γ-dicarboxylic acid esters **26,** 770
α-ketoketene mercaptals **30,** 466
β-ketosulfoxides **30,** 403
monothiolphosphoric acid esters **28,** 523
nitriles **28,** 55, 67, 726
1-oxaspiro[2,2]pentane ring **28,** 837
oxido compds. **29,** 262

phosphoric acid esters
  28, 523
pyrazines, 2,5-dihydro-
  (from 2 molecules) 28, 323
pyrroles 26, 842
pyrylium salts 29, 841
quinolines 27, 423
thiazoles, 2-amino- 27, 607
tosylhydrazones 27, 949
-, α-sulfenylation 30, 414
-, synthesis via enoxyboranes
  27, 894
-, - via Δ²-oxazolines,
  4,4-dimethyl- 26, 872 suppl. 27
- via
  acoxy compds. 26, 161
  carboxylic acid
    anhydrides, mixed 20, 633
    suppl. 29
  metallo aldimines 26, 912
    suppl. 27
-, ar. s. Aryl ketones
-, bicyclic
- from
  acetylenealcohols, cyclic
    27, 745
  dienynes 29, 678
- special s.
  2-decalones
-, branched, synthesis
  26, 872 suppl. 29
- from
  α-alkylthiomethyleneketones,
    incorporation of 3 alkyl
    groups 29, 819
-, cyclic (s. a. Cycl...)
- by
  carbon monoxide insertion
    28, 632
  diene synthesis 26, 718
- from
  alcohols, cyclic, ring
    expansion 27, 912
  carboxylic acid chlorides
    26, 957; 29, 671
  - - esters 28, 953
  - - acids 26, 957; 28, 909
  - - via carboxylic alkoxy-
    formic anhydrides 27, 933
  cyanocarboxylic acid esters
    26, 989
  α-diazoketones 28, 936
  dihalides 22, 661 suppl. 27
  enisocycles, ring expansion
    29, 154
  ethylenealdehydes 26, 737
  2-hydroxy-1,1-dihalides,
    cyclic, ring expansion
    30, 672
  oxidonitriles 28, 983
  1,1-sulfenylthioethers 27, 830
    suppl. 29, 30
-, homologization 26, 661
-, ring contraction 27, 275
-, - expansion with annelation
  26, 863
- special s.
  tetralones
- startg. m. f.
  amines, ar. 30, 259
  ethylenecarboxylic acids
    30, 88
  hydroxycarboxylic acids
    29, 149
  α-ketoazomethines, cyclic
    27, 718
  lactolides 26, 127; 29, 137
  pyrazoles 26, 788

-, α,β-disubst.
- from
  α,β-ethyleneketones 29, 819
    suppl. 30
-, heterocyclic
- special s.
  o-acyl-N-heterocyclics
-, hindered
- from
  carboxylic acid chlorides
    27, 852
  α-halogenoketones, synthesis
    23, 831 suppl. 26
-, isomeric, separation 8, 641
  suppl. 28
-, α-subst.
- from
  α,α'-dihalogenoketones
    29, 671
- startg. m. f.
  ketones 27, 75
-, sym.
- from
  ethylene derivs. (2 molecules)
    30, 495
  urethans 12, 830 suppl. 28
-, unsym.
- from
  halides, synthesis 26, 889
Ketonitriles s. Cyanoketones
2-Ketonitrones
- startg. m. f.
  oxazoles 28, 219
Ketophenols s. Phenolketones
β-Ketophosphonic acid esters
  7, 217 suppl. 28
-, γ-alkylation 26, 861
  suppl. 28
Ketosteroids, epimerization
  26, 736
- special s.
  hydroxyketosteroids
17-Keto-13α-steroids,
  9,14β-epoxy-8,14-seco-
- startg. m. f.
  17α-hydroxy-Δ⁸,¹⁴-steroids
    29, 944
Ketosulfones
- from
  enol sulfonates 26, 588
β-Ketosulfones
-, alkylation 25, 591 suppl. 26
- startg. m. f.
  benzimidazole 3-oxides,
    1-hydroxy- 28, 420
  γ-diketones 29, 792
  α-sulfonyl-γ-diketones
    29, 792
Ketosulfonium salts
- startg. m. f.
  pyrans, dihydro- 27, 269
2-Ketosulfonium –
- startg. m. f.
  α-ketohydroximinohalides
    27, 348
β-Ketosulfoxide dicarbanions
-, syntheses via – 29, 618
β-Ketosulfoxides
- from
  ketones and sulfinic acid
    chlorides 30, 403
-, Michael addition with –
  18, 965 suppl. 26
- special s.
  α-alkylthio-β-ketosulfoxides
  hydroxyketosulfoxides
  sulfinylacetophenones

- startg. m. f.
  α-acoxythiolic acid esters
    30, 100
  α-(alkoxylthio)ketones,
    cyclic 29, 926
  γ-diketones (from 2 mole-
    cules) 26, 894
  1,5-diketones (from 2 mole-
    cules) 28, 834
  heterocyclics 30, 166
  δ-hydroxy-β-ketosulfoxides
    29, 618
  α-ketohydroximinohalides
    27, 348
  vinyl ketones, synthesis
    29, 618
2-Ketothioethers
- from
  1-halogeno-2-ketothioethers
    28, 61
- special s.
  α-(alkylthio)ketones, cyclic
  1-halogeno-2-ketothioethers
  1-hydroxy-
- startg. m. f.
  ketones 27, 75
3-Ketothioethers
- from
  α-halogenoketones and
    1,1-halogenothioethers
    30, 599
α-Ketothiolic acid esters
  26, 214
Ketotosylates, cyclic
- startg. m. f.
  ethylenecarboxylic acids
    28, 839
3-Ketourethans 22, 143
  suppl. 29
2-Ketoxanthates
- from
  α-halogenoketones 29, 588
- startg. m. f.
  1,3-dithiol-2-one ring 29, 588
Kieselguhr 26, 489
Kimel-Cope reaction 3, 758
Kinetic control s. under
  Reactions
Kishner s. Wolff
Knoevenagel condensation
  27, 783
-, improved procedure 26, 814
- with (trifluoromethylthio)-
  acetonitrile 28, 737
-, double, stepwise 30, 558
Knoevenagel-Cope condensation
  29, 752
Kornblum rule 14, 284 suppl. 28
Kostanecki s. Emilewicz
Kröhnke s. Zecher
KU-2 cation exchanger 10, 111
  suppl. 26; 18, 834 suppl. 28;
  27, 410; 29, 487
Kwart s. Newman

Labeling s. Compds., labeled
Labile s. Unstable
Lactamols
- from
  diketones 28, 603
- special s.
  halogenolactamols
Lactam rearrangement
  with aromatization 28, 748
- ring opening, reductive
  29, 39

Lactams
—, O-alkyl derivs., review **20**, 132 suppl. **26**
—, C-alkylation **29**, 806
— by ring closure, double, stereospecific **29**, 817
— from
  amines, tert., cyclic **30**, 152
  aminocarboxylic acid amides **26**, 483
  — — esters **28**, 30; **29**, 451
  — acids **24**, 513; **30**, 331
  carboxylic acid amides and aldehydes **27**, 801
  dicarboxylic acid amide esters **26**, 502
  α-diazocarboxylic acid amides **21**, 942
  dicarboxylic acid imides **26**, 502
  enaminolactams **27**, 53
  o-halogenocyclimmonium salts **30**, 136
  o-α-halogeno-N-heterocyclics, ring expansion **17**, 263
  hydroxamic acids, cyclic **26**, 351
  1,1-iminosulfones, cyclic **29**, 648
  ketocarboxylic acids **28**, 349
  lactones **26**, 351; **29**, 387
  nitrocarboxylic acid esters **21**, 554
  nitrones, cyclic, ring expansion **28**, 122
  oximes, —, — — **26**, 150
  oxime tosylates, —, — — **18**, 205 suppl. **26**
  spirooxaziridines **29**, 331
  sulfonylcyanides **29**, 648
— special s.
  acyl-lactams
  alkoxy-
  bridge-
  carboxy-
  ethylene-
  halogeno-
  hydroxy-
  keto-
  α-oxo-N-heterocyclics
  phenanthridones
  sulfinyllactams
— startg. m. f.
  O-alkyllactims **23**, 80 suppl. **27**
  aminohydroxamic acids **27**, 492
  carboxylic acids **28**, 186
  diisocyanates **27**, 492
  enamines, cyclic **30**, 595
  heterocyclics, review **22**, 341 suppl. **27**
  imines, cyclic **23**, 80 suppl. **27**; **30**, 531
  1,1-iminoalcohols, cyclic **22**, 62
—, **N-alkyl-**
— from
  O-alkyllactims **28**, 296
—, **N-condensed**
— from
  2-aminoacetals and isocoumarins **28**, 329
  — and ketocarboxylic acids **28**, 329
  isocarbostyrils **28**, 329
α-**Lactams** s. Aziridinones

β-**Lactams** s. 2-Azetidinones
Lactim ethers s. O-Alkyl-lactims
Lactimimides
— from
  O-alkyllactims **29**, 383
Lactolethers s. Lactolides
Lactolides
— from
  ketones, cyclic **26**, 127; **29**, 137
  lactols **28**, 146
— startg. m. f.
  lactones **28**, 233
—, **bicyclic 26**, 826
Lactols
— from
  dialdehydes **27**, 858
  dicarboxylic acid esters **29**, 62
— special s.
  aminolactols
  hydroxylactols
  phthalans, 1-hydroxy-
— startg. m. f.
  carboxylic acids **30**, 93
  ethers, cyclic **29**, 745
  ethylenealcohols **28**, 852
  ethynyldiols **30**, 454
  α-hydroxylactols **30**, 64
  1,1-iminoalcohols, cyclic **29**, 380
  lactolides **28**, 146
  lactones **19**, 308 suppl. **30**; **26**, 235 suppl. **28**
δ-**Lactols**, synthesis **29**, 796 suppl. **30**
Lactone annelation 28, 837
— rearrangement **27**, 281
— **ring expansion 27**, 505
— — opening with Hofmann degradation of cyclimmonium salts **28**, 91
β-**Lactone ring opening 27**, 981
Lactones
—, α-alkylation **27**, 843
— from
  alkylthiocarboxylic acids **29**, 263
  carbalkoxyammonium salts, quaternary **28**, 228
  carboxylic acids **28**, 205
  carboxysulfonium salts **29**, 260, 263
  cyclopropylcarboxylic acids **30**, 105
  diols **27**, 100 suppl. **30**
  enoxysilanes, cyclic **30**, 145
  ethylenecarboxylic acids **27**, 505; **28**, 117; **30**, 573
  ethylenelactones **27**, 56
  hydroxycarboxylic acids **27**, 56, 505; **30**, 168/9
  — — with rearrangement **28**, 218
  hydroxythiolic acid esters **30**, 142, 169
  ketocarboxylic acids **28**, 211
  lactolides **28**, 233
  lactols **19**, 308 suppl. **30**; **26**, 235 suppl. **28**
  mesyloxycarboxylic acid amides **26**, 245
—, ring closure, isocyclic, with — **30**, 653
— special s.
  acetal-lactones
  acyl-

aldono-
amino-
azido-
carboxy-
di-
diazo-
enol-
oxido-
ethylene-
halogeno-
hydroxy-
keto-
methylene-
oximino-
phthalides
spirolactones
— startg. m. f.
  aminolactols **28**, 607
  carboxylic acids, reduction **27**, 32
  carboxythioethers **27**, 618
  dicarboxydiselenides **28**, 585
  diols **30**, 17
  dithiolortholactones **29**, 552
  ethers, cyclic **27**, 69
  halogenocarboxylic acid esters **29**, 516
  — — halides **29**, 516/7
  heterocyclics, review **22**, 341 suppl. **27**
  hydroxamic acids, cyclic **26**, 351
  hydroxycarboxylic acid amides, unsubst. **26**, 288
  — — esters **30**, 69
  — acids **27**, 56
  hydroxyketene mercaptals **29**, 552
  α-hydroxylactones **30**, 63
  ketocarboxylic acids **28**, 208
  ketolactones **27**, 505
  lactams **26**, 351; **29**, 387
  α-methylenelactones **24**, 894 suppl. **28**; **28**, 346
  orthocarboxylic acid esters, cyclic **26**, 179
  thiolactones **26**, 598; **27**, 618
—. Wittig synthesis with — **29**, 861
—, **macrocyclic 30**, 169
β-**Lactones**
— from
  β-hydroxycarboxylic acids **27**, 981
— special s.
  α-alkoxy-β-lactones
  γ-alkylthio-
  halogeno-
— startg. m. f.
  ethylene derivs. **27**, 981
γ-**Lactones**
— from
  carbonic acid esters, mixed, and ethylene derivs. **27**, 778
  cyclobutanone ring **26**, 140
  cyclopropanes, cyano- **29**, 87
  cyclopropanes, 1,1-dicyano- **26**, 707
  1-dicyanomethyl-2-halides **26**, 707
  α,β-ethylenecarboxylic acid esters and oxo compds. **28**, 720
  ethylene derivs. and α-halogenocarboxylic acid esters **27**, 908

– – and 1,1,1-halogeno-
dinitriles **26,** 707
– –, non-activated **28,** 657
α-halogenaldehydes and
malonic acid esters **30,** 331
β-hydroxycarboxylic acids
27, 262
γ-hydroxyketones **29,** 245
oxo compds. **5,** 479 suppl. **26**
– special s.
carbalkoxy-γ-lactones
hydroperoxy-
– startg. m. f.
alcohols **29,** 87
α,β-ethylene-γ-lactones
28, 895; **29,** 912 suppl. **30**
α-halogeno- **28,** 895
α-methylen- **22,** 780;
24, 894 suppl. **25**
Lanthanum oxide s. under
Chromium oxide
Laser s. a. Argon laser
– **irradiation,** reactions by –
30, 80
Lateral s.
C-carbalkoxylation, lateral
metalation, –
rearrangement, –
side chains
Layer s. Phase
Lead(IV) acetate azides
–, reactions with –, review
27, 362
Lead compds., organo-
(s. a. Plumbanes)
– special s.
aryl laed tricarboxylates
– **dioxide 26,** 773; **27,** 254
– **tetraacetate** (s. a. under Trimethylsilyl azide) **18,** 305
suppl. **26**; **26,** 226, 232, 253,
264, 272, 326, 480, 926; **27,** 141
suppl. **28**; **27,** 146, 242-4,
362/3, 444, 500, 595, 824, 998;
28, 133, 158, 241, 246, 309,
377, 430/1, 455; **29,** 908, 952;
30, 64, 199, 230, 566
–, oxidation of nitrogen
compds., organic, with –,
review **23,** 305 suppl. **27**
–, reaction with
N-amino-N-heterocyclics
26, 480
carbon-nitrogen double
bonds, review **26,** 190
suppl. **28**
α-diketones **27,** 146
hydrazones **26,** 190
oximes, review **23,** 305
suppl. **27**
– **thiocyanate 28,** 544
Leaving groups 26, 452
1,2,4-triazol-1-yl **30,** 282
– –, **active**
trifluoromethanesulfonyloxy
group **28,** 495
Lethargic reactions 11, 59
suppl. **26**; **17,** 458; **29,** 383
Leuchs anhydrides
s. 2,5-Oxazolidiones
Levulinic acid as reagent
26, 478
Lewis acid, gallium
chloride – – **30,** 624
– –, **mild,** magnesium
bromide – – **27,** 161
– **acids** as catalysts
15, 154 suppl. **29**

– – **and bases, hard and soft,**
review **30,** 624
– –, **strong** s.
trimethylsilyl perchlorate
– triflate
– **bases 26,** 420
triphenylphosphine **28,** 41
Light s. a. Irradiation
– **filters 29,** 137
– **specificity** (s. a. Wavelength
specificity) **27,** 704; **29,** 328
Limpach s. Conrad
Lipophilic compds.
– special s.
carboxylic acid esters,
lipophilic
Lithiation s. Metalation
Lithio salts, 1,3-heterocyclics
– as intermediates **23,** 838
suppl. **29**
Lithium 26, 279, 662/3, 851/2,
889; **27,** 49, 392, 712, 852; **28,**
601/2, 740, 775, 791/2, 850, 924;
29, 72, 172, 582, 669; **30,** 401,
452a, 460, 594
–/amines
–/ethylamine **26,** 45
–/–/tert-butanol **28,** 65
–/ethylenediamine **20,** 49
suppl. **26**; **26,** 45
–/methylamine **25,** 595
suppl. **26**
–/TMEDA **29,** 669
– **ammonia complex 14,** 88
suppl. **28**
–/**ammonia, liq. 26,** 29, 42/3,
54, 899; **27,** 13, 32, 88, 666,
839; **28,** 47, 803; **29,** 31/2, 37,
851; **30,** 214, 673
–/–, –/**alcohol 22,** 42 suppl. **28;
29,** 928
–/–, –/tert-butanol **28,** 989
–/–, –/isopropyl alcohol
30, 23
–/–, –/1-methoxy-2-propanol
30, 26
–/–, –/**water 26,** 44
–/**naphthalene 14,** 866 suppl. **29;
26,** 664; **27,** 663
– **acetylide** as reactant **29,** 673
– **acetylides**
as reactants **29,** 882
pentyn-1-yllithium as
reagent **29,** 44
– **acyl carbonyl metalates,**
syntheses with – **27,** 885
– **tetracarbonylferrate**
– startg. m. f.
ketones **27,** 885
– **alanates,** syntheses via –
23, 705; **26,** 669 suppl. **30**
– **alkoxides 29,** 867; **30,** 385
– tert-butoxide **29,** 72
– methoxide **29,** 41; **28,** 55;
29, 243
– –, **bulky**
– triethylcarboxide **27,** 892;
30, 495
– **aluminum hydride** s. Lithium
tetrahydridoaluminate
– **amalgam 21,** 958 suppl. **27;
26,** 852
– **amide 26,** 681; **27,** 840/1;
29, 72, 880; **30,** 684
– **amides**
– N-benzyl-tert-butylamide
27, 33

– bis (trimethylsilyl)amide
19, 881 suppl. **28**; **26,** 859
suppl. **27**; **28,** 947; **29,** 116
– dicyclohexylamide **10,** 615
suppl. **29**; **28,** 856 suppl. **29**;
28, 944; **30,** 449
– diethylamide **27,** 843 suppl.
30; **28,** 4; **29,** 634, 959
– diisobutylamide **28,** 647
– diisopropylamide **23,** 674
suppl. **29**; **25,** 487 suppl. **29;
26,** 712, 923; **27,** 319, 778,
842, 854; **27,** 843 suppl. **28,
29; 28,** 82, 312, 598, 646; **29,**
217, 223, 622/3, 667, 706, 709,
742, 805-9, 874; **29,** 819 suppl.
30; 30, 65, 155, 437/8, 481-3,
505, 536, 540, 587, 637
– as metalating agent,
selective **27,** 319
–/hexamethylphosphor-
amide **29,** 810; **30,** 536
– dipropylamide **28,** 4
– N-isopropylcyclohexyl-
amide **10,** 616 suppl. **26;
27,** 560, 843; **28,** 77, 683,
706; **30,** 414
– piperidide **24,** 964 suppl. **27;
28,** 946
– 2,2,6,6-tetramethyl-
piperidide **26,** 847
suppl. **28**; **29,** 139
–, reductions with – **27,** 33
– **ammonia complex,** reduction
with – **27,** 50
– **aroxides**
– 4-methyl-2,6-di-tert-butyl-
phenoxide **19,** 758a suppl.
28; 22, 678 suppl. **29**
– **azide 27,** 400
– **borates** s. a. 9-Borabicyclo-
[3.3.1]nonane ate complexes,
lithium
– –, **1-alkenyltriorgano-**
–, synthesis via – **30,** 618
– –, **1-alkynyltriorgano-**
–, syntheses via – **28,** 864
– **bromide 5,** 467 suppl. **29; 7,**
589 suppl. **29**; **18,** 601 suppl.
26; 21, 771 suppl. **26; 26,** 644,
982; **27,** 317; **28,** 599; **29,** 930;
30, 368
– **carbonate 9,** 968 suppl. **26;
29,** 174, 960
– **chloride 20,** 409 suppl. **27;
26,** 971; **27,** 311 suppl. **30;
27,** 578; **28,** 267, 712, 906;
29, 266, 945
– **compds.** s. a. Silyllithium
compds.
– –, **organo-**
– as reactants **26,** 575; **26,** 875
suppl. **30**; **30,** 479, 614
– as reagents **29,** 633, 670
butyl-lithium **14,** 571 suppl.
29; 20, 538 suppl. **26; 26,**
148, 280, 576, 667/8, 680,
686; **27,** 628/9, 835; **28,** 102,
260, 583, 855; **29,** 617, 795-7,
869/70, 902; **30,** 404, 528, 663
n-butyl- **5,** 473 suppl. **26; 18,**
970 suppl. **26; 19,** 736 suppl.
29; 20, 538 suppl. **27; 20,**
626 suppl. **29; 22,** 803 suppl.
29; 23, 879 suppl. **27; 26,**
70, 84, 357/8, 420, 702, 861/2;
26, 896 suppl. **28; 27,** 12,

(**Lithium compds., organo-**
– as reagents
 n-butyl- lithium **27,**)
 54, 158, 319, 352, 458, 654, 776/7, 836/7, 851, 853, 890, 932, 965/6; **27,** 830 suppl. **29;** **27,** 932 suppl. **29; 27,** 949 suppl. **30; 28,** 65, 82, 140, 175-7, 598/9, 656, 658, 696, 731, 761, 783, 799, 800-2, 819, 840, 856/7, 864, 883, 885, 945; **29,** 139, 572, 604, 618-20, 741, 798-801, 805, 849/50, 855, 871-3, 898, 957/8; **29,** 262 suppl. **30; 30,** 43, 200, 444, 447/8, 480, 495, 548/9, 582-6, 671/2, 683
 sec-butyl- **24,** 331 suppl. **28; 28,** 842; **29,** 802
 tert-butyl- **11,** 731 suppl. **27, 13,** 831 suppl. **28; 28,** 856; **29,** 621; **30,** 413, 490
– as hydrogen atom transfer agent **29,** 13
–, prepn. **29,** 621 suppl. **30**
 ethyl- **30,** 436
 methyl- **27,** 33, 438, 894, 949, 979; **28,** 547, 821, 854, 925; **29,** 666, 794, 804, 894; **30,** 134, 461, 597, 617
 phenyl- **18,** 812 suppl. **26; 27,** 665; **28,** 392, 697; **29,** 803
–, pure **25,** 679 suppl. **26**
 2-thienyl- **25,** 493
–, reaction with acetylene derivs. **30,** 479
–, – with amines **27,** 712
–, – with N-nitrosamines **30,** 314
– special s.
 1,1-halogenoorganolithium compds.
 triphenylmethyllithium
– –, –, **unsatd.**
– special s.
 ethyleneorganolithium compds.
 lithium acetylides
– **cuprates** (s. a. Dilithium tetrachlorocuprate, Lithium diorganocuprates, – hydridocuprate complex)
–, synthesis with – and tosylates **29,** 876
– –, **mixed,** syntheses with – **29,** 671
– special s.
 lithium phenylthio(alkyl)-cuprates
– **deuteridotri-*tert*-butoxoaluminate 11,** 135 suppl. **29**
– **dialkylcuprates 8,** 757 suppl. **26; 20,** 612 suppl. **26; 26,** 669; **29,** 669, 746, 819, 853/4
– special s.
 lithium dimethylcuprate
– –, **mixed**
–, cross-coupling via – **30,** 597
– **diallylcuprate 26,** 669 suppl. **27**
– **dihydridodimesitylborate 17,** 61 suppl. **29**
– **dimethylcuprate** as reagent **30,** 27
– **diorganocuprates 26,** 684 suppl. **29**
– special s.
 lithium dialkylcuprates
 – diallylcuprates

– **diphenylphosphide**
 as reactant **27,** 942
– **divinylcuprates** as intermediates **23,** 693 suppl. **28; 28,** 859
– **enolates** as intermediates **28,** 703 suppl. **30; 29,** 894
– **hydride 23,** 935 suppl. **26; 27,** 843 suppl. **29; 27,** 949 suppl. **30; 28,** 419, 934; **29,** 408, 660, 992
– **hydridocuprate complex 29,** 44
– **hydridotrialkoxoaluminate**
 – hydridotri-*tert*-butoxoaluminate **19,** 57 suppl. **27; 27,** 37; **27,** 76 suppl. **29**
 – hydridotrimethoxoaluminate **27,** 78
– **hydridotri-*sec*-butylborate 25,** 32 suppl. **28; 30,** 602
– **hydridotriethylborate 27,** 35 suppl. **29; 29,** 992 suppl. **30**
– as reducing agent, nucleophilic **28,** 60
– **hydroxide 9,** 93 suppl. **26; 27,** 7; **28,** 8; **30,** 573
– **iodide 6,** 898 suppl. **27; 26,** 71; **28,** 837 suppl. **29; 29,** 794, 989
– **isopropane nitronate**
– as activator **26,** 607
– as reactant **30,** 255, 608
– **mercaptides**
 phenylmercaptide **29,** 671
 n-propylmercaptide **29,** 992
– **nitride 26,** 411
– **nitrite 6,** 390 suppl. **29**
– **oxide 26,** 411
– **perchlorate 17,** 842 suppl. **29; 22,** 318 suppl. **28; 23,** 599 suppl. **28; 30,** 565
– as electrolyte **21,** 35 suppl. **26; 27,** 605; **28,** 7, 98; **29,** 334
– **phenylthio(alkyl) cuprates 29,** 671
– **phosphate 28,** 4
– – **buffer 30,** 90
– **reagents, organo-,** and their modifications, review **26,** 669 suppl. **27**
– **salt 28,** 547; **29,** 620; **30,** 550
– **tetrachloropalladate(II) 24,** 884 suppl. **28; 27,** 157; **28,** 865; **29,** 174
– **tetrahydridoaluminate**
 HC↓OC; HC↓NC; HC↓CC; HC↓↑O; **13,** 883 suppl. **28; 23,** 43a suppl. **27; 26,** 60, 64, 405, 760, 879; **27,** 25, 73, 167, 331, 401, 659, 788, 959; **28,** 63, 307, 520, 525, 907; **29,** 14, 29, 88, 329, 363, 600, 986; **30,** 7, 24, 43, 48, 278, 531, 561
–, high-temp. reductions with – **30,** 48
–, rearrangements, reductive, with – **21,** 538 suppl. **29**
– –/**alcohols** s. a. Lithium hydridotrialkoxoaluminate
 – –/methanol **23,** 82 suppl. **27**
– –/**aluminum chloride** (s. a. Aluminum hydride) **14,** 36 suppl. **26; 21,** 401 suppl. **29; 26,** 62/3; **29,** 34
– –/**sulfuric acid 26,** 64
– –/**titanium tetrachloride,** reductions with – **29,** 77

– –/– **trichloride** s. McMurry reagent
– **tetrahydridoborate 27,** 505, 649
– –, **soluble 12,** 61 suppl. **28**
– **s-trithiane** as S-stabilized carbanion salt **24,** 665
 Löffler-Freytag reaction **15,** 428
– **ring closure 15,** 421
 Long-range activation of hydrocarbon chains **20,** 533 suppl. **26**
 Lossen rearrangement, review **16,** 557 suppl. **29**
 Lucas reagent (ZnCl$_2$ in concd. HCl) **16,** 596
 2,6-Lutidine
– as base, hindered **8,** 146 suppl. **29**
– as reagent **17,** 549 suppl. **29; 22,** 112 suppl. **27; 23,** 242 suppl. **29; 26,** 419/20, **26,** 846 suppl. **27; 27,** 877
 Lysis s. Aceto-lysis, Cata-, Cyano-, Halogeno-, Hydrogeno-, Photo-, Pyro- Solvo-

Macroheterocyclics
– special s.
 crown heterocyclics
 N-Macroheterocyclics
– special s.
 hexaazamacroheterocyclics
 polyethers, azamacrobicyclic
 tetraazamacroheterocyclics
– **, conjugated**
– from
 o-dinitriles **27,** 435
 (*poly*-**O**)-**Macroheterocyclics 30,** 565
– special s.
 crown polyethers
 (*poly*-**S**)-**Macroheterocyclics 29,** 544
 Macrolactonization by double activation **30,** 168
 Märkl cyclic phosphonium salt synthesis 28, 597
 Magnesium (s. a. Grignard)
 CC↓OC; CC↓NC; CC↓↑ Hal; **26,** 107, 422, 620, 640, 650, 714/5; **27,** 161, 232, 473, 657, 712; **28,** 843, 875; **29,** 577, 602, 641/2, 672, 853/4; **30,** 453, 491-3
– as reducing agent **29,** 362
–, highly active, prepn. **27,** 856
– –/**alcohol**
 –/methanol s. Magnesium methoxide
– **acetate 29,** 698
– **alkoxides** s. Magnesium bromide, ethoxy-
– **aluminum hydride** s. Magnesium bis(tetrahydridoaluminate)
– **amalgam 19,** 940 suppl. **26**
– **aroxides** s. Aryloxymagnesium halides
– **bis(tetrahydridoaluminate),** reduction with – **27,** 36
– **bromide 19,** 940 suppl. **26; 27,** 161; **28,** 474; **30,** 493
– as Lewis acid, mild **27,** 161
– –, **aryloxy- 28,** 742
– –, **ethoxy- 28,** 741

- **chloride** as buffer **30,** 278
- **compds., organo-** s. Magnesium, organo- halides
- **enolates** as intermediates **30,** 453
- **halides** s. Aryloxymagnesium halides
- **iodide 17,** 621 suppl. 29
- **methoxide 26,** 87
- **methyl carbonate, methoxy-26,** 770; **28,** 620
- **nitrate 26,** 145
- **, organo- halides** (s. a. Grignard compds.) **27,** 905
  tert-butyl- chloride **28,** 685
  ethyl- bromide **26,** 877; **28,** 596, 609, 672, 700; **28,** 799 suppl. **30; 29,** 830; **30,** 454, 601
  ethyl- iodide **30,** 455
  isopropyl- bromide **29,** 831
  – chloride **27,** 861
  methyl- bromide **27,** 737
  – chloride **26,** 423
  – iodide **29,** 921
  n-propyl- bromide **28,** 567
  vinyl- chloride as reactant **29,** 825
- **oxide 28,** 277
- **perchlorate 28,** 568
- **sulfate 16,** 620 suppl. **26; 27,** 160; **28,** 84, 611; **30,** 33
- – **hydrate 29,** 279
- **Makosza reagent 4,** 432 suppl. **27; 23,** 832; **27,** 726, 833; **28,** 795; **29,** 456
- **Maleic acid anhydride**
  – as reactant **26,** 703
  – as reagent **29,** 897
  – startg. m. f.
    4-thiazolidones **27,** 596
- **Maleimides**
  – startg. m. f.
    spiro[pyrrolidin-3,1'-(pyrrolidino[3,4-c]pyrrolidines)], 2,4',5,6'-tetraoxo-(from 2 molecules) **27,** 701
  – N- and O-substitution **29,** 419
  –, N-subst.
  – by retrodiene scission **17,** 198 suppl. **26**
  – from
    acetylene derivs. and nitro compds. **28,** 766
- –, **N- from O-subst. 29,** 419
- **Malonic acid aryl esters**
  –, ring closure with – **27,** 383
  – – **esters** (s. a. α-Dicarboxylic acid esters)
  –, ring closure, double, with – – special s. **27,** 898
    aminomethylenemalonic acid esters
    diazomalonic – –
    diethyl malonate
    ethylene-α-dicarboxylic acid esters
    malonic acid aryl –
    – – monosilyl –
    urediomalonic acid –
  – startg. m. f.
    barbituric acids, 5-carbalkoxy-**28,** 736
    α-carbalkoxy-γ-lactones **27,** 770
    carboxylic acid esters **29,** 84
    γ,δ-ethylene-α-dicarboxylic acid esters **29,** 722
- homophthalimides **29,** 811
  γ-lactones **30,** 331
  2-pyridone ring, N-condensed **30,** 551
  3H-1,2,6-thia(IV)diazine-3,5-diones, 4,5-dihydro-**26,** 354
- – –, cyclic
- special s.
  alkylidenemalonic acid esters, cyclic
- startg. m. f.
  2-pyrone ring, 4-hydroxy-**15,** 573 suppl. **26**
- – **monosilyl esters**
- as intermediates **14,** 888 suppl. **28**
- **acids** s. α-Dicarboxylic acids
Malonimides s. 2,4-Azetidinediones
Malononitrile
- as reactant **27,** 783; **30,** 558
- startg. m. f.
  pyridines (from 2 molecules) **26,** 882; **27,** 775
Malononitriles
- special s.
  alkylidenemalononitriles
Mandelonitrile benzoates
- from
  benzils **30,** 486
Mandelonitriles, O-(tetrahydropyran-2-yl)-
- startg. m. f.
  aryl ketones **29,** 888
Manganese(III) acetate 5, 162 suppl. **26; 28,** 713; **29,** 722
Manganese dioxide 26, 231; **26,** 683 suppl. **29; 27,** 370/1; **28,** 53, 361, 447; **29,** 157, 546; **30,** 153, 157, 246
β-**Manganese – 25,** 174 suppl. **27**
Mannich bases
- –, chemistry, review **18,** 482 suppl. **29**
- –, ring closure with – **26,** 965
- –, **cyclic,** rearrangement, skeletal **27,** 752
- **reaction** (s. a. Aminomethylation, Retro-Mannich reaction)
- –, control, regiospecific **24,** 765 suppl. **29**
- –, cyclimmonium salts, methylene- by – **30,** 562
- –, heterocyclics by – **29,** 751; **30,** 564
- –, N-heterocyclics, N-condensed, by – **26,** 779
- –, methylene compds. by – **15,** 590
- –, ring closure by – **28,** 765
- – with
  amines, prim. **26,** 594; **30,** 564
  simultaneous C-deacylation **28,** 867
- –, **decarboxylative**
- –, β-amino-α-methylenenitriles by – **26,** 916
- –, **double**
- –, pyrazine ring, 2,5-dihydro-, by – **29,** 728
- –, pyrimidine ring, hexahydro-, by – **26,** 781
- –, ureas, sym., by – **26,** 810
- – **reagent 28,** 782
- synthesis s. a. Retro-Mannich reaction
Mannich-type condensation
- with
  hydroxylamines **27,** 612
  sulfinic acids **20,** 450; **27,** 612
Marchwald's rule 27, 944
Markownikoff addition 26, 637; **27,** 600
- **hydration, abnormal 21,** 174 suppl. **27**
anti-**Markownikoff addition 29,** 158
- to carbon-carbon double bonds **30,** 488
- **hydration 16,** 194 suppl. **29**
- –, **preferential 27,** 202
- **hydrohalogenation 26,** 519, 641
Markownikoff-type addition 29, 538; **29,** 882 suppl. **30**
Masking s. Carbanions, masked, Protection
Mathematical optimization s. Optimization, mathematical
McMurry's reagent
(TiCl₃/LiAlH₄) **30,** 561; **30,** 662 suppl. **30**
Media s. Silicone oil, Solvents
Medium, superacidic
- –, reactions in – **27,** 159; **30,** 635
Meerwein s. Wagner
Meisenheimer complexes 28, 123 suppl. **28**
- **rearrangement 30,** 101
- –, ring expansion by – **26,** 151
Meldrum's acids s. Malonic acid esters, cyclic
Mercaptals
- –, cleavage **28,** 182
- from
  1,1-alkoxyhalides **28,** 566
  morpholinals **30,** 402
  thioenolethers **27,** 600
- special s.
  cyclopropanone mercaptals
  ethylene-
  hydroxy-
  keto-
  ketoaldehyde –
  α-ketocarboxylic acid ester –
  1,1,1-silyl–
- startg. m. f.
  alcohols, tert. **29,** 873
  γ-diketones, synthesis **28,** 175
  disulfides **28,** 519; **29,** 527
  1,1-disulfoxides **29,** 99
  hydrocarbons **28,** 63
  γ-ketomercaptals, synthesis **28,** 175
  oxo compds. **26,** 219
  – –, synthesis **29,** 873
  1,1-sulfinylthioethers **29,** 97
  sulfonium salts **27,** 997
- –, synthesis **27,** 798
- –, **bicyclic**
- startg. m. f.
  α-hydroxyketones, monocyclic **27,** 229
- –, **cyclic**
- –, formation, preferential **16,** 138 suppl. **29**
- from
  acetals **26,** 601
- –, protection of carbonyl groups as – **27,** 271

(Mercaptals, cyclic)
- special s.
  1,3-dithianes
  ketomercaptals, cyclic
  spiro-1,3-S,S-heterocyclics
- startg. m. f.
  oxo compds. **27**, 230; **28**, 656; **30**, 569
-, syntheses via - **22**, 661; **25**, 483
Mercaptans (s. a. sulfhydryl under Replacement, Sulfhydryl groups)
- as reagents **27**, 75
- from
  α-aminonitriles **30**, 420
  halides **28**, 569
  isothiouronium salts **26**, 593
  methyl thioethers **29**, 30
  thiocyanates **28**, 28
  thioketones **27**, 47
  - (tert. mercaptans) **29**, 641
  thiolic acid esters **28**, 147; **30**, 15
  trichloromethyl thioethers **29**, 30
-, radical addition to carbon-carbon multiple bonds, review **16**, 637 suppl. **26**
-, reactions with azides **27**, 296
- special s.
  acetylene-mercaptans
  amino-
  aryl-
  dithiols
  ethylenemercaptans
  ethyl mercaptan
  hydroxy-mercaptans
  keto-
  methyl mercaptan
  n-propylmercaptan
  thioglycitols
- startg. m. f.
  γ-alkylthio-β-lactones **28**, 534
  disulfides **26**, 562
  -, sym. **26**, 559/60; **30**, 377
  ethylene derivs. **29**, 972
  hydrocarbons **30**, 51
  ketene mercaptals **30**, 386
  silyl sulfides **29**, 530
  sulfonic acid bisiminoamides **28**, 247
  N-sulfonyl-S-(sulfonylimino)-sulfinic acid amides **27**, 298
  thioethers **26**, 592; **29**, 547; **30**, 396
  -, synthesis **29**, 620
  thioiminoesters **30**, 386
  thiosulfenimides **28**, 256
  thiosulfenyl halides **28**, 256
  thiosulfuric acid S-mono-esters **26**, 557
Mercaptides s. Mercury mercaptides
α-Mercaptoacetamides
- startg. m. f.
  α-carbamyl-α,α-dichloro-methylsulfenyl chlorides **28**, 492
Mercaptocarboxylic acid amides
- special s.
  ethylenemercaptocarboxylic acid amides

α-**Mercaptocarboxylic** - -
- from
  α-mercaptocarboxylic acids **28**, 327
- special s.
  α-mercaptoacetamides
- **acid esters**
- special s.
  α,β-ethylene-α-mercapto-carboxylic acid esters
- startg. m. f.
  thiophenes, 3-hydroxy- **27**, 784
o-**Mercaptocarboxylic** - -
- startg. m. f.
  o-alkylthiocarboxylic acid amides **27**, 376
α-**Mercaptocarboxylic acids**
- special s.
  α,β-ethylene-α-mercapto-carboxylic acids
- startg. m. f.
  α-mercaptocarboxylic acid amides **28**, 327
  1,3,2-oxathiazolium betaines, 5-hydroxy- **27**, 292
- **acid thioamides**
- from
  dithiocarbamic acid esters **30**, 536
Mercaptoethanesulfonic acid as reagent **23**, 36 suppl. **27**
Mercaptoethanol - - **27**, 259
2-Mercaptoethylamine - - **26**, 8; **30**, 15
α-**Mercapto-α,β-ethylene-carboxylic acids** s. α-Thio-ketocarboxylic acids
Mercaptoketones
s. Ketomercaptans
α-**Mercaptophosphonic acid esters**
- from
  phosphorous acid esters and 1,1-dithiols **30**, 439
2-Mercaptopyrimidine as reactant **27**, 91
Mercuration (s. a. Solvo-mercuration) **29**, 158
- of acetylene derivs. **7**, 201 suppl. **29**
  amines, ar. **30**, 428
- special s.
  amino-mercuration
  hydroperoxy-
  peroxy-
Mercuration-demercuration **29**, 141
Mercury acetamide as reagent **26**, 589
-**(II) acetate 2**, 648 suppl. **27**; **7**, 201 suppl. **29**; **26**, 132, 211, 445, 940; **27**, 157; **28**, 38, 92, 117, 153, 350, 748; **29**, 141, 172, 174, 215, 675; **30**, 156
-**(II) acetylides**
- startg. m. f.
  1,5,3-dithiazepine ring, N-condensed **28**, 573
-**(II) bromide 16**, 514 suppl. **28**; **19**, 539 suppl. **27**; **26**, 316; **27**, 441, 860; **28**, 131
-**(II) carboxylates**
- as reagents **28**, 213; **30**, 135
- startg. m. f.
  carboxylic acid anhydrides **28**, 213

-**(II) carboxylates, optically active**, as reactants **22**, 143 suppl. **27**
- **chloride** (s. a. under Aluminum) **22**, 107 suppl. **29**; **23**, 123 suppl. **29**; **26**, 862; **27**, 345, 419, 482, 880; **28**, 176/7, 656; **29**, 6, 802, 873; **30**, 427, 584, 688
- as acidic catalyst **26**, 946
- **compds., organo-**
  Arylmercury chlorides
- startg. m. f.
  trans-2-arylalcohols **28**, 865
- compds.
- startg. m. f.
  phenols **27**, 232
  Dialkylmercury compds.
- startg. m. f.
  carboxylic acids **26**, 218
- -, sym.
- from
  α,ω-dihalides **30**, 165
  Diarylmercury compds., polyfluoro- **26**, 644
  1,1-Dimercury compds.
- from
  acetylene derivs. **30**, 427
  1,1-diboro compds. **30**, 427
  Diorganomercury compds.
- from
  monoorganomercury compds. **26**, 647
- special s.
  cyclopropylmercury compds.
- startg. m. f.
  halides **30**, 373
  monoorganomercury sulfonates **27**, 593
  Diorganomercury compds., 1-halogeno-
- from
  monoorganomercury halides and diazo compds. **27**, 655
  Mercurybis(phenylacetylide) as reactant **26**, 567
  1,1-Mercuridiazo compds. **30**, 435
  Monoorganomercury compds.
- from
  ethylene derivs. **26**, 641
- special s.
  β-(acetoxymercury)ketones
- startg. m. f.
  hydrocarbons (s. a. De-mercuration) **30**, 156
  diorganomercury compds. **26**, 647
  Monoorganomercury halides
- startg. m. f.
  diorganomercury compds., 1-halogeno- **27**, 655
  ketones **26**, 218
  Monoorganomercury halides, 2-hydroxy-
- startg. m. f.
  ketones **29**, 264
  oxido compds. **29**, 264
  Monoorganomercury sulfonates
- from
  diorganomercury compds. **27**, 593
  Organotrihalogenomethyl-mercury compds.

Phenyltrihalogenomethyl-
mercury
- as dihalogenocarbene
  precursor, review
  17, 894 suppl. 27
- as reactant **26**, 799
- as reagent **26**, 658, 799
–(II) cyanide **19**, 539 suppl. 27
Mercury-N-heterocyclics
- startg. m. f.
  nucleosides **16**, 514
Mercury(II) mercaptides
- from
  carboxylic acid thioamides
  26, 567
  thioureas **26**, 567
–(I) nitrate **17**, 275 suppl. **29;**
  30, 135
–(II) oxide **19**, 539 suppl. 27; **26,**
  211, 255, 900; **27**, 247, 525; **28,**
  87, 153, 177, 668; **29,** 140, 501;
  30, 141, 569, 688
- salt **28,** 213
– – **complex** s. Diazonium
  mercuric salt complexes
- **salts**
–, oxidation of ethylene derivs.
  with –, review **5,** 162
  suppl. 27
–, reaction with halides
  30, 135
- **sulfate 26,** 133; **27,** 136;
  28, 783, 840
–(II) **trifluoroacetate 22,** 143
  suppl. 27; **29,** 142, 325
Merrifield syntheses (s. a. Solid-
  phase multi-step syntheses)
–, attachment to polymer
  support **19,** 264 suppl. 26
–, comparison of active esters
  in – **16,** 425 suppl. 29
–, peptide amides **1,** 307
  suppl. 27; **13,** 364 suppl. 29
–, polymer reactant-substrates
  22, 395 suppl. 28
Mesitoic acid esters
- special s.
  2-ethylenemesitoates
Mesitoxymagnesium bromide
  as reagent **30,** 553
Mesitylenesulfonyl chloride
  as reactant **22,** 101 suppl. 29
Mesitylenesulfonyloxylamine
–, amination with – **28,** 243
- as reactant **20,** 248 suppl. **29;**
  27, 230 suppl. **28; 29,** 278
Mesoionic rings s. Heterocyclics, mesoionic
Mesyl s. a. Methanesulfon...
Mesylation 27, 109 suppl. 30
O-Mesylation 27, 109; **29,** 622
**Mesyloxycarboxylic acid
  amides**
- startg. m. f.
  lactones **26,** 245
meta s. *m*-Activation
Metaboric acid s. Boric acid
Metacyclophane ring, halogeno-
–, rearrangement **29,** 708
Metal... s. a. Transition
  metal...
Metal acetoacetonates
- special s.
  copper(II) acetoacetonate
  iron(III) –
  palladium(II) –

rhodium(I), dicarbonyl-
  trifluoroacetylacetonato-
sodium acetoacetate
vanadyl –
Metalanes
- special s.
  alanes
  germanes
  plumbanes
  stannanes
Metalates s. Lithium acyl
  carbonyl metalates
Metalation (s. a. Hydrometalation, Ring metalation) **28,** 599
- of
  2-acetylenethioethers **27,** 158
  (α-acylalkylidene)-
    phosphoranes **30,** 582
  acylamines **29,** 742
  azo compds., aliphatic-ar.
    27, 352
  azomethines **27,** 352; **30,** 461
  carboxylic acid esters **27,** 843
  – acids **26,** 664
  diazo compds. **26,** 668
  enamines **29,** 799
  enolethers **29,** 621
  furans, 3-halogeno-
    4,5-dihydro- **29,** 797
  α-halogenocarboxylic acid
    esters **30,** 549
  isonitriles **29,** 617
  mercaptals **29,** 819 suppl. **30;**
    29, 873
  mercaptans **29,** 620
  N-nitrosamines **29,** 742
  phosphoric acid amides
    30, 583
  α-siloxynitriles **29,** 808
- special s.
  dimetalation
  mercuration
–, syntheses via –
  to Grignard compds. with
    additional functions,
    review **15,** 522 suppl. 26
- with diphenylcalcium **28,** 599
–, **allylic 29,** 619; **30,** 586
–, ambident character of –
  29, 619
–, **directed**
  substituent-directed **27,** 158
–, α-lateral
- of
  heterocyclics **30,** 480
  N-heterocyclics **28,** 732
  isocyclics **27,** 838
  2-oxazolines **29,** 807
–, **selective 27,** 319
α-Metalation
- of
  nitriles **29,** 959
  sulfones **29,** 796
  sulfoxides **30,** 455
  thioethers **28,** 801
o-Metalation (s. a. Sulfur-donor
  ligand o-metalated complexes)
- of
  amines, tert. **19,** 736 suppl. 29
Metal-carbene complexes (s. a.
  transition metal-carbene
  complexes)
–, reactions with – **28,** 846
Metal carbonyls (s. a. under
  individual metal carbonyls)
- in organic synthesis, review
  25, 517 suppl. 26

- catalysis, homogeneous
  29, 158
- **complex compds.**
- special s.
  π-allyl complexes
  complex salts
  metal acetoacetonates
  metal-carbene complexes
  metallocenes
  nickel complex compds.
  rhodium – –
  ruthenium triarylphosphine
    complex
  sulfur-donor ligand
    o-metalated complexes
  transition metal –
- **compds.** s. a. Organometallic
  compds.
- –, **zerovalent** s. Nickel
  compds., zerovalent,
  Platinum – –, Tricarbonyl-
  chromium(0) complexes
- β-**diketonates** s. Metal
  acetoacetonates
- **enolates,** protection of keto
  groups as – **18,** 59 suppl. 28
–, α-fluoroketones via –
  16, 610 suppl. 29
- special s.
  lithium enolates
  magnesium –
  tin(IV) –, organo–
- **heterocyclics** s. under
  Heterocyclics
Metallo aldimines
- startg. m. f.
  nitriles, hindered **26,** 912
    suppl. 27
–, syntheses via – **26,** 912
Metallocyclics 29, 858
Metallocenes
- special s.
  cobaltocenes
  ferrocenes
  molybdenocenes
  nickelocenes
  titanocene dichloride
  zirconocene chloride
Metal powders, highly active,
  prepn. **27,** 856
Metals, insertion into halogen-
  carbon bonds **27,** 647
–, **highly reactive,** prepn.
  27, 856
Metal salts (s. a. Transition
  metal salts) **26,** 145
- **specificity 26,** 47; **27,** 841
Metathesis s. Interchange
Methanesulfinic acid chloride
  as reactant **28,** 864 suppl. 29
Methanesulfonamide 27, 422
Methanesulfonic... s. a. Mesyl...
- **acid** as reagent **17,** 381
  suppl. **28; 28,** 145; **30,** 54, 88,
  137, 522
- –/**phosphorus pentoxide**
  as polyphosphoric acid
  substitute **11,** 933 suppl. 28
- – **anhydride** as reagent
  19, 307 suppl. 28
- – **aryl esters, labeled 29,** 102
- – **esters** (s. a. Mesylation)
- as intermediates **27,** 109;
  29, 379, 622; **30,** 366
- special s.
  aminomesylates
  dimesylates

(Methanesulfonic
acid esters)
- startg. m. f.
  cyclopropane ring 30, 649
  ethylene derivs. 16, 933
    suppl. 28; 26, 937, 945;
    28, 906; 29, 622
Methanesulfonyl azide
-, diazo group transfer with -
  20, 271 suppl. 29
- chloride
- as reactant 29, 379
- as reagent 27, 578; 29, 938
2,6-Methano-3-benzazocines,
1,2,3,4,5,6-hexahydro-
- from
  pyridines, 1,2,3,6-tetrahydro-
  26, 738
2,5-Methano-3,1-benzoxazepines,
1,2,4,5-tetrahydro-
- from
  quinolines 28, 924
- startg. m. f.
  quinolines 28, 924
Methanol s. Methyl alcohol
1,4-Methanonaphthalenes,
1,2,3,4-tetrahydro-
- from
  indenes 28, 628
Methids s. Quinone methids
Methoxyacetonitrile
- startg. m. f.
  α-alkoxy-α,β-ethylenenitriles
  29, 443
Methoxyamine as reagent
  26, 27
- startg. m. f.
  α-aminocarboxylic acids
  28, 312
- hydrochloride as reactant
  28, 896
o-Methoxycarbonylbenzoyl
chloride
- startg. m. f.
  phthalimides 29, 417
p-Methoxycarbonylperoxy-
benzoic acid as reagent
  26, 584 suppl. 28
2-Methoxyethanol
- as medium 27, 623
- as solvent 30, 473
1-Methoxy-2-propanol s. under
  Lithium/ammonia, liq.
Methyl acetate as competitive
  ester substrate 12, 61 suppl. 28
Methylal
- as solvent 2, 827 suppl. 26;
  28, 815; 29, 638
Methyl alcohol (s. a. Deuteriomethanol)
- as
  medium 22, 153 suppl. 26;
  26, 160
  reactant 30, 427
    cycloaddition 28, 712
    trap, nucleophilic 26, 158
  reagent 23, 57 suppl. 28; 26,
    80, 160, 813; 27, 14; 29, 829,
    783, 877; 30, 337, 505, 622
Methylamine
- as reagent 26, 4, 25
Methylamines
- from
  formamides 9, 113
N-Methylamines, ar.
- from
  N-(succinimidomethyl)-
    amines, ar. 28, 56

α-Methylation of oxo compds.
  29, 829
C-Methylation
- via
  arylthiomethylation 9, 844
    suppl. 27
  chloromethylation 1, 596
- with
  formaldehyde 29, 761
-, intramolecular 23, 787
    suppl. 26
N-Methylation
- with
  dimethylsulfonium methylid
    26, 437
-, reductive, of amines 28, 346
Methyl(carboxysulfamoyl)-
triethylammonium hydroxide
inner salt
- as reagent 26, 952
Methyl Cellosolve
  s. 2-Methoxyethanol
Methylcyclohexane as solvent
  30, 426
N-Methyl-N,N'-dicyclohexyl-
carbodiimidium iodide
- as reactant 27, 573
Methyl α,α'-dihalogenoketones
- startg. m. f.
  cis-α,β-ethylenecarboxylic
    acids 30, 131
Methylen... s. a. Dimethylen...
C-Methylenation 29, 622
α-Methylenealdehydes
- from
  aldehydes 26, 787
Methylene blue s. under
  Sensitizers
Methylene bridges, reaction
  at - 26, 187
N-Methylene -
-, elimination 14, 364
- from
  hydrazinium salts,
    N-condensed 26, 460
Methylenecarbonyl -
  28, 630
Methylene chloride
- as cosolvent 26, 794
- as solvent 26, 519; 27, 213;
  28, 660
-, reactions with diborane
  in - 28, 34
- startg. m. f.
  methylenedioxy compds.
    25, 152; 30, 128
- compds. (s. a. Ethylene
  derivs., terminal) 30, 685
- by Mannich reaction
  15, 590
- from
  dihalogenomethylene compds.
    7, 738 suppl. 28
  ethylene derivs. and
    1,1-dihalides 29, 827
  oxo compds. 13, 831
    suppl. 29
- special s.
  aminomethylene...
  cycliminonium salts,
    methylene-
  difluoromethylene compds.
  dimethylene compds.
  hydroxymethylene...
  siloxymethylene compds.
-, active s. Methyl and
  methylene groups, active

Methylene-D₂ compds. 13, 831
  suppl. 28
Methylenecyclopentane ring
- from
  1,5-dienes 26, 734
  methylenecyclopropane ring
    26, 724
Methylenecyclopropane
  15, 680 suppl. 29
- ring
- startg. m. f.
  methylenecyclopentane ring
    26, 724
Methylenedioxyarenes
-, cleavage 19, 8 suppl. 27;
  27, 10 suppl. 28; 28, 15
-, -, preferential 21, 21
    suppl. 28
Methylenedioxy compds.
  (s. a. Dihalogenomethylenedioxy compds.)
- from
  diols 30, 128
Methylene groups (s. a. Hydrocarbons)
- startg. m. f.
  ethylene derivs. (from
    2 groups) 19, 815
-, active s. under Methyl
  groups
γ-Methylene-β-ketocarboxylic
acid esters
- from
  β-ketocarboxylic acid esters
    29, 759
α-Methyleneketones 26, 960;
  30, 685
- from
  α-hydroxymethyleneketones
    27, 89
-, 1,4-addition to - 30, 491
- special s.
  β-amino-α-methyleneketones
- startg. m. f.
  α-methylketones 27, 89
α-Methylenelactams
- by rearrangement 30, 327
α-Methylenelactones 23, 982
  suppl. 29; 29, 622, 912; 30, 685
- from
  lactones 24, 894 suppl. 28;
    28, 346
-, synthesis, review 29, 622
    suppl. 30
α-Methylene-γ-lactones 28, 657
  suppl. 29; 30, 669
- by
  dehydrogenation 28, 895
- from
  3-acetylenealcohols 30, 502
  γ-lactones 22, 780; 24, 894
    suppl. 25
  2-methylene-1,4-diols 22, 780
α-Methylenenitriles
- special s.
  β-amino-α-methylenenitriles
  α-(dihalogenomethylene)-
    nitriles
Methylenephosphoranes
- startg. m. f.
  keteniminylidene-
    phosphoranes 30, 579
3-Methylene-2-piperidones
- from
  piperidine-3-carboxylic
    acids 28, 912
4-Methylene-2-quinazolones
  26, 506

Methylene transfer s. a. Bismethylene transfer
– –, **intramolecular 28,** 699
Methylenimines
– special s.
dihalogenomethylenimines
Methyl fluorosulfonate
– as alkylating agent, review
 29, 344
– as reactant 28, 960 suppl. 30;
 29, 230, 344, 994
–, cleavage of mercaptals
 with – 28, 182
N-Methylformamide
– as solvent, polar, weakly
 protic 29, 828
Methyl groups (s. a. Hydrocarbons)
– from
 acoxymethylene groups
  11, 128
 hydroxymethyl groups 27, 91
 1,1,1-trihalides 27, 71
 1,1,1-trifluorides 27, 73
–, introduction 16, 848
– special s.
 steroid 4-methyl groups
–, **allylic**, metalation 30, 586
– –, **angular**
–, elimination s. under
 Aromatization
–, introduction, stereospecific
 26, 691
– **and methylene groups, active**
 (s. a. Compds., CH-acidic,
 C-Hydrogen, active)
–, addition to acetylene-
 N-heterocyclics 29, 662
–, phosphorus compds., with
 methylene groups, active,
 review 23, 879 suppl. 26
–, reaction with
 isocyanates 28, 736; 29, 635
– startg. m. f.
 cyclopropanes 26, 783
 8,17-diazabenzo[c,j]tetra-
 cyclo[7.3.1.0²,⁹.0⁷,¹⁶]-
 tridecans 28, 645
 enamines 30, 281
 thioethers 29, 580
N-Methyl groups
– from
 N-carbalkoxyl groups
 21, 401 suppl. 28
Methylids s. Isoquinolinium
methylids
1-Methylimidazole
– as medium 23, 123 suppl. 29
– as reagent 12, 107 suppl. 29
2-Methylimidazole
– as medium 13, 896 suppl. 28
– as reagent 29, 421
2-Methylindole as reagent
 27, 19
Methyl iodide
– as reactant 27, 942; 29, 230,
 263; 30, 466
– as reagent 27, 230, 976;
 28, 14, 34, 802, 960;
 29, 476, 950
– ketones
– from
 carboxylic acid esters 30, 414
– special s.
 aryl methyl ketones
– startg. m. f.
 β-amino-α-methyleneketones
  26, 808

benzothiazoles, 2-acyl-
 29, 567
α-Methylketones
– from
 α-hydroxymethyleneketones
  27, 89
 α-methyleneketones 27, 89
Methyl mercaptan
– as reagent 6, 572 suppl. 26;
 12, 56 suppl. 29
– **methylthiomethyl sulfoxide**
 as reactant 27, 830; 28, 735;
 29, 631
N-Methylmorpholine as solvent
 8, 88 suppl. 29
– as catalyst 26, 827
– as medium 26, 36
– as reagent 9, 453 suppl. 26;
 14, 446 suppl. 26; 20, 678
 suppl. 26; 21, 426 suppl. 27;
 26, 814 suppl. 29; 27, 194;
 30, 260
1-Methylnaphthalene as solvent
 26, 490
2-Methyl-2-nitrosopropane
 as reagent 28, 965
N-Methylpyrrolidine
– as reagent 26, 145
– as solvent 27, 723; 28, 18
–, 3-alkylation with ethylene
 derivs. 28, 646
N-Methyl-2-pyrrolidone
– as additive 10, 250 suppl. 29
– as solvent 7, 830 suppl. 27;
 9, 898 suppl. 26; 27, 869;
 29, 821
Methylsulfinyl carbanion
 (s. a. Sodium hydride/
 dimethyl sulfoxide) 28, 852;
 29, 733; 30, 118
–, N-alkylation in the presence
 of – 16, 434 suppl. 26
**Methylsulfonylammonium
salts, quaternary**
– as mesylating agents
 27, 109 suppl. 30
2-Methyl-Δ²-thiazoline
 as reactant 28, 176
Methyl thioethers
– startg. m. f.
 mercaptans 29, 30
 trichloromethyl thioethers
 29, 30
Methylthiomethyl esters
 s. Carboxylic acid methylthiomethyl esters
N-Methylthiomethylnitrones
– from
 oximes 27, 359
**N-Methyl-N-tosylpyrrolidinium
perchlorate** as reagent
 28, 253
**Methyltrioctylammonium
chloride** s. Aliquat 336
Meyer-Schuster rearrangement
 5, 508 suppl. 26
2-C-Michael acceptors
–, ketene mercaptal mono-
 sulfonium salts as – 29, 667
Michael addition 9, 768;
 27, 830 suppl. 28; 28, 646
–, 2(5H)-furanones, 3,4-disubst.,
 via – 27, 710
– of
 carboxylic acid amides
 28, 647
 β-ketosulfoxides 18, 965
 suppl. 26

5-oxazolones 18, 765 suppl. 26
– to
 allenes 18, 765 suppl. 26
 α,β-ethylenesulfoxides
 18, 965 suppl. 28
 glutaric acids 28, 733
– under aprotic conditions
 29, 666
– with
 azomethines 29, 668
 bases, electrolytically
 generated 25, 633 suppl. 29
 β-ketosulfoxides 18, 965
 suppl. 26
– –, **double 29,** 665, 668
– –, **intramolecular 27,** 755
 suppl. 28
Michaelis s. Arbuzov
Michael-type addition 29, 538
Microorganisms as reagents
 26, 6; 27, 38
– special s.
 bacteria
Microwave discharge s. Glow
 discharge
Migration
– of
 acoxyl s. Acoxy group
 migration
 acyl s. Acyl – –
 alkoxy groups s. Alkoxy – –
 alkyl s. Alkyl –
 aryl s. Aryl –
 bonds, multiple s. a. Isomerization, prototropic
 carbalkoxyl s. Carbalkoxyl
 migration
 carbamyl groups s. Carbamyl
 group –
 carbon-carbon double bonds
 (s. a. Allyl rearrangment,
 Aromatization, Deconjugation) 27, 159; 28, 46; 30, 512
– – – (preferential migration) 28, 673
– – – into rings
 into O-heterocycles 2, 620
 suppl. 26
– – – out of rings 26, 733;
 27, 733
– – – via formation of
 S-oxides 27, 734
– – – with decarbonylation
 26, 87
– – – with replacement of
 diazo groups by hydrogen
 27, 70
– – –, exocyclic 28, 675
– triple bonds 30, 506
 carbonic acid ester groups,
 cyclic 28, 118
 halogen 28, 627; 29, 492
 hydroxyl 29, 690
 nitrile groups 29, 7
 nitro groups s. under Nitro
 groups
 N-nitroso groups s. under
 N-Nitroso groups
 side chains 28, 670
 sulfonyl groups s. Sulfonyl
 group migration
 N-thiocarbamyl groups
 27, 338
 thiolic acid ester groups
 s. Thiolic acid ester groups
 1,3,5-triazin-2-yl groups
 30, 229
–, contrathermodynamic 30, 506

(Migration)
-, geospecific
- of
  carbon-carbon double bonds
  20, 537 suppl. 28
1,2-S→C-Migration s. Thio-
  Wittig-rearrangement
Möhlau-Bischler indole ring
  closure 27, 785
Moffatt s. Pfitzner
Molecular sieve 20, 526 suppl.
  26; 27, 235; 29, 230, 364; 30,
  463, 595
- as catalyst 29, 453
- as water scavenger 8, 550
  suppl. 29; 28, 141, 497
- -, proton-exchanged,
  as catalyst 28, 141
Molybdate s. Ammonium
  molybdate
Molybdenocene hydride
  as catalyst 29, 54
Molybdenum carbonyl complexes as catalysts 27, 908
- hexacarbonyl 17, 153 suppl.
  28; 26, 564; 27, 867; 28, 710;
  30, 605
- hexafluoride 28, 506;
  29, 523
- pentachloride 17, 153 suppl.
  28; 27, 184
- peroxide complex 28, 113;
  30, 65
Monoacyl-1,1-diamines
- startg. m. f.
  amines 28, 57
α-Monoalkylation
- of
  α,β-ethyleneketones 27, 832
  β,γ-ethyleneketones 29, 804
C-Monoalkylation
- of
  ketones 19, 896 suppl. 26
-, stereospecific 30, 580
N-Monoalkylation (s. a.
  Amines, sec., from amines,
  prim., N-Monomethylation)
- of
  amines 28, 358; 29, 369
  sulfonic acid amides
  2, 459 suppl. 27
Monoalkylboranes 14, 65
  suppl. 26, 27
Monoalkyl phosphites
  s. Phosphorous acid
  monoesters
Mono(carbalkoxy)amines
- from
  N-carbalkoxysulfenamides
  29, 412
  halides 29, 412
Monochloroborane
- as reactant 26, 912 suppl. 29;
  28, 666
- as reagent 14, 205 suppl. 28
Monochloromethyl ether
- as reactant 8, 819 suppl. 26;
  27, 364; 28, 803; 30, 583
- startg. m. f.
  isoquinolines, 1,2,3,4-tetra-
  hydro-2-acyl- 26, 884
C-Monocyanoethylation
- of ketones 27, 717
Mono-α-dehalogenation
- of
  oxo compds. 26, 251
Monodehydroxylation
- of o-diols 26, 550

Monodeuterio compds.
- from
  ethylene derivs. 23, 63
  suppl. 27
Monoglyme s. 1,2-dimethoxy-
  ethane
Monohalogenation
- of
  amines, ar. 28, 485
1-Monohalogenation
- of
  ethers 26, 526
α-Monohalogenation
- of
  boranes 28, 484
  ketones 30, 353
N-Monohalogenourethans
  by comproportionation
  29, 485
N-Monomethylation
- of
  amines 26, 383
Mononitration 12, 957
-, ar. 30, 239
Monoperoxydichloromaleic
  acid
- as reagent 29, 94
Monoperoxyphosphonic acid
  esters
- from
  phosphonochloridates 27, 120
Monosaccharides
- from
  furans 27, 276
Monosulfonylcarbodiimides
- from
  carbodiimides 29, 473
  1,2,4-thiadiazetidine 1-oxides,
  3-imino- 29, 473
Monothioacetals, cyclic,
  protection of carbonyl
  groups as – 27, 271
Monothio-β-diketones
- startg. m. f.
  thiophenes 26, 866
-, 2,3-dihydro-3-hydroxy-
  26, 866
-, synthesis, review 26, 866
  suppl. 27
Monothiolcarbonic acid esters
-, N-carbalkoxylation with –
  29, 430
Monothiolphosphoric acid
  diesters
- as intermediates 20, 94;
  26, 175
- – esters
- from
  thionophosphorous acid
  esters and ketones 28, 523
- startg. m. f.
  thionophosphoric acid
  diesters 28, 10
Montmorillonite 29, 93, 544
Morphinane ring closure,
  oxidative 27, 915
- – opening, isocyclic 27, 915
Δ[8,14]-Morphinanes – by rearrangement, skeletal 30, 675
Morpholinals
- startg. m. f.
  1,1-di(acylamines) 29, 403
  mercaptals 30, 402
Morpholine (s. a. 4-Ethyl-
  morpholine)
- as reactant 27, 496
- as reagent 28, 422

Morpholinosulfenyl chloride
  as reagent 26, 558; 29, 529
Mukaiyama reagent 26, 175
Multicomponent reaction
  s. 3-Component reaction

9H-Naphth[3,2,1-kl]acrid-9-ones
  26, 948; 28, 923
Naphthalene s. a. under
  Lithium
Naphthalene-2,3-dicarboxylic
  acids, 1,2-dihydro-
- from
  dilactones 28, 692
Naphthalene ring
- from
  isoquinolinium ring,
  3,4-dihydro- 28, 693
- startg. m. f.
  azulene ring 24, 920; 28, 108;
  29, 980
Naphthalenes 16, 870
- by
  cyclodehydrogenation 29, 908
  elimination of etheno bridges
  26, 987
- from
  acetylene derivs. (2 molecules) and benzoyl
  peroxides 29, 883
  aryl ketones and acetylene
  derivs. 28, 639
  benzo[b]cyclobuta[d]thio-
  phenes 28, 954
  cyclones 29, 897
  furans 19, 917 suppl. 28
  1-naphthols, 1,4-dihydro-
  28, 639
- special s.
  1,8-bis(dimethylamino)-
  naphthalene
-, 2-amino- 29, 724
-, 1,2-dihydro- 26, 757 suppl. 29
- from
  arylallenes 26, 757 suppl. 29
β-Naphthalenesulfonic acid
  as reagent 28, 758
Naphthalimides
- startg. m. f.
  naphthostyrils 29, 478
1-Naphthols, 1,4-dihydro-
- from
  aryl ketones and acetylene
  derivs. 28, 639
- startg. m. f.
  naphthalenes 28, 639
1-Naphthols, 5,6,7,8-tetra-
  hydro-5,8-dialkyl-
- from
  phenols and 1,5-dienes
  28, 651
2-Naphthols
- from
  diketones 16, 871 suppl. 29
Naphtho[2,1-g]pteridin-
  8(9H)-ones, 5,6-dihydro-
  30, 644
2H-Naphtho[1,2-b]pyran-2-ones,
  5,6-dihydro- 27, 989
- from
  α-tetralones 29, 884
- startg. m. f.
  2,4a-ethanophenanthrenes,
  2,3,4,4a,9,10-hexahydro-
  29, 884

phenanthrenes, 3,4,9,10-
 tetrahydro- **29**, 884
1,4-Naphthoquinone ring
 by ring closure, double
 26, 725 suppl. **30**
1,4-Naphthoquinones
– startg. m. f.
 xanthylium salts, dimerization **29**, 760
–, 2-halogeno-
– startg. m. f.
 anthraquinones **29**, 788
Naphthostyrils
– from
 naphthalimides **29**, 478
Naphtho[1,2-b]thiophenes
 30, 641
Naphthoxazoles 26, 395
2H-Naphth[1,8-bc]oxepin-2-ones
 22, 277 suppl. **26**
Naphthyridines 30, 74
**1,6-Naphthyridines, 2-amino-
 28**, 739
**1,6-Naphthyridin-5(6H)-ones
 28**, 333
Napieralski s. Bischler
Nef reaction, modified 30, 122
Neighboring group participation (s. a. Anchimeric
 process) **27**, 132
–, cleavage of carboxylic acid
 amides **26**, 28
– in removal of N-protective
 groups **26**, 28
– **groups, powerful 26**, 631
Neozone D 26, 936
**Newman-Kwart rearrangement
 22**, 625 suppl. **29**
Nickel (s. a. under Urushibara)
 HN↓↑O; HC↓↑OC; HC↓↑CC;
 HC↓↑O; HC↓↑S; **7**, 45 suppl.
 29; **26**, 35, 78, 266, 297, 351,
 388; **26**, 467 suppl. **28**; **27**, 78,
 90, 149; **28**, 288, 362, 382, 967,
 991; **29**, 20, 29, 83, 749, 750;
 30, 184, 208, 335, 688
–, P-2 catalyst, reductions
 with – **29**, 48
–, P-1 catalyst **29**, 48 suppl. **30**
–, Raney nickel D-8 **29**, 245
–, – – W 2 **7**, 45 suppl. **29**;
 12, 90 suppl. **29**; **26**, 76
 prepn. **26**, 76
–, – – W 4 **27**, 91
–, – – W 5 **26**, 76, 78, 691
 prepn. **26**, 76
–, – – W 6
 prepn. **26**, 76
–, – – W 7 **26**, 388 suppl. **29**
–, – – W 7, aged **26**, 266
–, – –, degassed **28**, 967
– -alumina **20**, 75 suppl. **26**
–/hydrazine s. Hydrazine/nickel
– -silica **26**, 181
– -aluminum **26**, 51, 422
– -platinum s. under Platinum
–(0)-catalyzed cycloadditions
 26, 724
– acetate **27**, 909; **29**, 48
– π-allyl complexes as catalysts
 13, 695 suppl. **26**
–(I) – –
–, replacement of halogen
 by acetonyl groups with –
 30, 625
– startg. m. f.
 1,3-dienes **26**, 915
– **boride 27**, 976

– **carbonyl 26**, 977; **28**, 829;
 29, 603; **30**, 606
– as reagent, allylophilic
 29, 28
– **chloride 26**, 656, 740
– **complex catalysts,** formation
 of carbon-carbon bonds
 with –, review **20**, 531
 suppl. **27**
–, rearrangements with –
 30, 512
– – **compds.** (s. a. Nickel
 π-allyl complexes, – phosphine complexes, – phosphite
 complex
–(0) **complexes** as catalysts
 23, 139 suppl. **28**; **30**, 501, 512
– **compds., zerovalent 26**, 724
– special s.
 bis(acrylonitrile)nickel(0)
 bis-(1,5-cyclooctadiene)-
 nickel(0)
– **halides** as catalysts
 26, 669 suppl. **30**
– **hydroxide anode 4**, 187
 suppl. **27**
Nickelocenes, reactions, review
 27, 813 suppl. **29**
Nickel peroxide 22, 355
 suppl. **26**; **26**, 260; **27**, 187
–, oxidation with –, review
 20, 191 suppl. **26**
– **phosphine complexes**
 as catalysts **17**, 28 suppl. **29**;
 19, 778 suppl. **29**; **20**, 531
 suppl. **29**; **26**, 875 suppl. **27**,
 28, **30**; **27**, 364 suppl. **29**
 bis(triphenylphosphine)nickel
 dicarbonyl **29**, 991
 dibromobis(tri-*n*-butyl-
 phosphine)nickel **30**, 513
 dichlorobis(triphenyl-
 phosphine)nickel(II)
 26, 875 suppl. **30**; **29**, 672
– **phosphite complex 30**, 512
**Niementowski 4-quinazolone
 ring synthesis, modified**
–, review **22**, 392 suppl. **29**
Ninhydrin as reactant
 29, 345
–, ring closures with –
 23, 763; **29**, 351
Nitramines
– from
 dinitro compds. **30**, 330
N-Nitramines
– special s.
 N-halogeno-N-nitramines
 N-sulfonyl-
– startg. m. f.
 N-halogeno-N-nitramines
 27, 290
o-**Nitramines**
– startg. m. f.
 benzimidazoles, 1-subst.
 26, 369
1,2-Nitramines
– from
 1,2-acoxynitro compds.
 28, 342
– special s.
 2-nitroenamines
Nitrates s. Aryl nitrates,
 Nitric acid esters
Nitration 27, 356
 (s. a. Mononitration)
– by transition metal-nitrato
 complexes **30**, 240

–, direction of – **26**, 328; **27**, 367;
 30, 240
– of
 diazo oxides **10**, 292 suppl. **29**
 phosphines **27**, 364
– via radical cations **27**, 369
– with nitryl chloride **28**, 320
–, ar. **30**, 240
–, **decarboxylative 26**, 494
Nitrato complexes s. Transition
 metal-nitrato complexes
Nitrenes (s. a. Sulfonylnitrenes)
–, N-heterocyclics via –, review
 18, 338 suppl. **27**, **29**
–, review **26**, 630
N-Nitrenes, review **23**, 381
 suppl. **27**
Nitrenoid insertion 27, 420
Nitric acid 7, 336 suppl. **27**; **26**,
 209, 559; **28**, 90, 671; **29**, 241,
 615; **30**, 87, 147
– **esters**
– from
 alcohols **26**, 361
 halides **30**, 135
 nitrous acid esters **26**, 89
– special s.
 azidonitric acid esters
 diol mononitrates
 nitronitric acid esters
– startg. m. f.
 azides **27**, 400
 4-nitroalcohols **28**, 292
 nitro compds., aliphatic
 26, 361
– **oxide** s. Nitrogen monoxide
Nitrile anions
– as nucleophiles, strong
 30, 517
Nitrile oxides
–, prepn. and reaction, review
 3, 582 suppl. **26**
– special s.
 benzonitrile oxide
– startg. m. f.
 1,3,2,4-dioxathiazole 2-oxides
 26, 489
 isothiocyanates **28**, 527
 Δ²-isoxazoline ring **29**, 652
 1,4,2-oxathiazoles **28**, 527
Nitriles
– as additives **4**, 667 suppl. **29**
– from
 alcohols **27**, 782
 aldehydes **28**, 339; **30**, 277
 aldimines **27**, 537
 aldoximes **26**, 501; **27**, 513;
 28, 440; **30**, 277
 amines **26**, 463
 –, tert. **22**, 288 suppl. **28**
 azides with ring contraction
 26, 958
 azomethines **28**, 466
 carboxylic acid amides
 26, 501, 507; **27**, 514;
 28, 442; **29**, 456
 – – –, preferential conversion
 25, 348 suppl. **26**
 – – –, N-subst. **26**, 501, 507
– acids **27**, 422
– acid thioamides **29**, 456
 1,1-cyano(azocarboxylic acid
 esters) **28**, 55
 α-cyano-N-sulfonylhydrazines
 28, 67
 cyclopropyl azides **27**, 524
 suppl. **29**

(**Nitriles**
– from)
 enazides **26,** 477
 ethylene derivs. **14,** 784
 furazans **28,** 467
 halogenonitriles, Friedel-
  Crafts synthesis **28,** 824
 hydrazones **26,** 481; **27,** 525
 hydrocarbons (methyl groups)
  30, 237
 hydroximinohalides **30,** 337
 α-isonitrosoketones **30,** 342
 ketenimines **26,** 495
 ketones **28,** 55, 67, 726
 α-nitrosophosphonium salts
  28, 319
 oxo compds., synthesis with
  addition of 1 C-atom
  29, 480
 phosphonium salts **28,** 319
 silylamidines **27,** 495
 N-silylcarboxylic acid
  amides **28,** 462
 1-(sulfonyl)enformylamides
  29, 480
 sulfonylhydrazones **28,** 67
 1,2,5-thiadiazole 1,1-dioxide
  ring **29,** 472
 1,2,4-triazoles, 4-amino-
  (2 nitrile molecules) **26,** 480
–, α-metalation **29,** 959
–, α-monoalkylation **27,** 843
 suppl. 29
– special s.
 acetonitrile
 acoxy-nitriles
 acylamino-
 acylcyanides
 alkoxy-nitriles
 alkylthio-
 amino-
 arylaceto-
 aziridines, 2-cyano-
 azonitriles
 cyano...
 1,3-cyclopentenedione ring,
  2-cyano-
 cyclopropanes, cyano-
 diazocyanides
 di-nitriles
 α,β-ethylene-β-hydroxy-
 halogeno-
 hydroperoxy-
 hydroxy-
 iminocyanides
 mandelonitriles O-(tetra-
  hydroxpyran-2-yl)-
 oxido-nitriles
 oxo-
 poly-
 pseudocyanides
 pyridines, 1,4-dihydro-
  4-cyano-
 pyrrolidinium salts,
  N-cyanomethyl-
 Reissert compds.
 siloxynitriles
 silylnitriles
 sulfonylcyanides
– startg. m. f.
 1,2-acoxyhalides **26,** 547
 acylamidines **29,** 305
 acylamines **29,** 307, 334
 alcohols **26,** 195
 aldehydes **26,** 183; **29,** 209
 amidines **26,** 299; **30,** 220
 –, N,N′-disubst. **30,** 219
 –, subst. **29,** 423

amines, prim. **26,** 35, 39;
 27, 46; **28,** 34; **30,** 20
α-aminocarboxylic acid
 derivs., synthesis with
 addition of 1 C-atom
 29, 631
– acids, synthesis with
 addition of 1 C-atom
 28, 621
β-amino-α,β-ethyleneazo-
 methines, synthesis **30,** 461
β-amino-α,β-ethylenenitriles
 (from 2 molecules) **27,** 677
2-amino-1-sulfinylthioenol-
 ethers **29,** 631
2,1-benzisoxazoles **30,** 629
(*tert*-carbin)amines, prim.
 29, 636
carboxylic acid amides
 26, 989; **27,** 132; **29,** 196
– acids **29,** 196
α,α-dicyanocarboxylic acid
 esters **30,** 147
α,α-dihalogenocarboxylic
 acids **29,** 519
1,1-dihalogeno(trihalogenophosphazo) compds. **27,** 302
β-diketones **26,** 685
1,2-dinitriles, sym. **28,** 705
C,N-disilylketenimines
 30, 438
eniminophosphoranes
 27, 684
ethylene derivs., sym. (from
 2 molecules) **27,** 901
α,β-ethylenenitriles **29,** 912
α-halogeneisocyanates **28,** 472
1-halogenenamidines (from
 2 molecules) **29,** 489
α-halogeneureas **28,** 472
heterocyclics **26,** 792
 suppl. 27; **28,** 369
N-heterocyclics **27,** 804
1,3-O,N-heterocyclics **28,** 374
α-hydroperoxynitriles **30,** 155
imidazoles **26,** 304
Δ²-imidazolines **30,** 217
3H-imidazo[1,5-b]-s-triazoles
 (from 3 molecules) **27,** 678
imidinophosphoranes **27,** 684
β-iminoazomethines **26,** 685
iminoesters **30,** 220
–, preferential conversion
 30, 75
isoquinolines, 3,4-dihydro-
 26, 820
isothiazoles, 3-hydroxy-
 (from 2 molecules) **29,** 777
ketones, degradation with
 loss of 1 C-atom **30,** 155
1,2-nitroacylamines **28,** 286
4H-1,3-oxazines, 5,6-dihydro-
 26, 309
4-pyridones **26,** 359
pyrimidines **26,** 396
quinazoline ring (from
 2 molecules) **29,** 741
N-silylenamines **27,** 646
N-silylketenimines **29,** 295
α-silylnitriles **29,** 295
tetrazoles, azido- **29,** 317
tetrazolylisocyclics,
 halogeno-, bicyclic **30,** 351
1,2,4-thiadiazoles **30,** 416
thioacylimines **29,** 533
thioiminoesters **27,** 624
N-thioiminohalides **30,** 347
thiolic acid esters **27,** 624

thiophenes, 2,4-diamino-
 26, 892
1,3,5-triazines, trimerization
 26, 303
–, halogeno- (from 3 molecules) **26,** 457
1,2,3-triazole ring **29,** 742
1H-1,2,4-triazoles **29,** 420
–, **α-alkylated**
– from
 1,1-cyano(azocarboxylic acid
 esters) **28,** 55
–, **hindered,** via metallo
 aldimines **26,** 912 suppl. 27
Nitrile sulfides as intermediates
 30, 416
Nitrile sulfites s. 1,3,2,4-
 Dioxathiazole 2-oxides
– **ylids** as intermediates **28,** 810
Nitriliimines
– as intermediates **26,** 232
– startg. m. f.
 pyrazoles **26,** 926
Nitrilium s. N-Diazonitrilium
 group
– **salts** as intermediates
 29, 209, 423
N-Nitrimides s. Amine
 N-nitrimides
N-Nitrimines
– from
 oximes **26,** 258
–, reactions with – **26,** 390
– startg. m. f.
 nitrones **26,** 390
Nitrite photolysis as radical
 source **28,** 933
Nitrites s. Alkyl nitrites,
 p-Chlorobenzoyl nitrite
1,2-Nitroacylamines
– from
 ethylene derivs. and nitriles
 28, 286
Nitroalcohols
– special s.
 dinitroalcohols
4-Nitroalcohols
– from
 nitric acid esters **28,** 292
Nitroamines s. Nitramines
Nitroarenes, ring hydrogenation, selective **26,** 48
– as reagents **30,** 84
o-Nitroazides
– startg. m. f.
 benzofurazan ring **7,** 162
 suppl. 26
1,1-Nitroazo compds.
– from
 hydrazones **30,** 231
Nitrobenzene
– as medium **28,** 915; **29,** 496,
 508, 895
– as oxidant **26,** 372; **29,** 175
Nitrobenzenes
– startg. m. f.
 3H-azepines, 2-amino-
 26, 379
**p-Nitrobenzenesulfonic acid
 esters**
– startg. m. f.
 halides (iodides) **17,** 621
 suppl. 28
o-Nitrobenzoic acid
 as catalyst **25,** 576 suppl. **27;**
 30, 560
p-Nitrobenzoic – as reagent
 30, 639

o-Nitrobenzoyl groups,
O-protective
–, removal, reductive 29, 4
Nitrocarboxylic acid amides
– startg. m. f.
  diamines 26, 64
– – esters
– special s.
  ethylenenitrocarboxylic acid
    esters
– startg. m. f.
  lactams 21, 554
o-Nitrocarboxylic – –
– startg. m. f.
  thiophene ring, 3-hydroxy-
    30, 573
Nitrocarboxylic acids
– startg. m. f.
  hydroxamic acids, cyclic
    28, 432
  oxindoles 30, 331
  –, 1-hydroxy- 30, 331
Nitro compds.
– from
  acyl nitrates 26, 494
  amines 28, 70
  1,1-nitroethylene derivs.
    28, 37
  nitroso compds. 27, 356;
    30, 315
  stannanes 30, 315
–, reaction with Grignard
  compds. 27, 473
–, ring closure, oxidative
  27, 286
–, – –, reductive 27, 510
– special s.
  acoxynitro compds.
  dinitro...
– startg. m. f.
  o-alkoxynitriles 26, 835
  amines 27, 14/5; 28, 16, 18;
    30, 10a, 330
  ammonium salts, quaternary
    27, 428
  azomethines 28, 363
  azoxy compds. (from
    2 molecules) 26, 35, 261
  2,1-benzisoxazoles 27, 256
  o-cyanophenols 28, 777
  disulfides, sym. 29, 564
  hydroxylamines, disubst.
    27, 473
  maleimides, N-subst. 28, 766
  nitroso compds. 28, 990
  N-oxides, cyclic 28, 464;
    29, 337
  oximes 28, 644; 30, 326
aci-Nitro – (s. a. Nitronates)
– special s.
  3,5-bis-aci-nitrocyclo-
    hexenes
– startg. m. f.
  oxo compds. 30, 572
o-Nitro –, syntheses with –,
  review 18, 338 suppl. 28
Nitro –, aliphatic
– from
  acyl nitrates 26, 494
  alcohols 26, 361; 29, 743
  carboxylic acids 26, 923
  ethylene derivs. 27, 897
  nitric acid esters 26, 361
  nitronitric acid esters 27, 63
  oximes 27, 96
  oxo compds. 24, 62 suppl. 27
– special s.
  1,1,1-difluoronitro compds.

fluorodinitromethane
polynitro compds., aliphatic
– startg. m. f.
  acetals 30, 122
  ethylene derivs., sym. 30, 575
  hydroxylamines 27, 16
  isoxazoles 26, 796; 29, 786
  2-ketoalkoximes 28, 104
  oxo compds. 28, 157; 29, 199
  pseudonitroles 1, 193; 19, 414
  pyridines (from 3 molecules)
    27, 812
– –, ar., reactions, photochemical, with – 30, 84
– startg. m. f.
  3H-indol-2-ones, 1-hydroxy-
    27, 499
  nitrones 27, 902
  phenols 30, 118
– –, hindered, reduction
  17, 345 suppl. 29
– –, prim., acylation, oxidative
  26, 322
– –, unsatd.
– startg. m. f.
  1,2-oxazine ring, 6-hydroxy-
    28, 120
Nitrocyclopropanes 20, 591
  suppl. 26
2-Nitro-1,3-diols, cyclic
– from
  dialdehydes, review 16, 794
  suppl. 26
2-Nitroenamines
– startg. m. f.
  1,1-nitroethylene derivs.,
    synthesis 26, 902
–, syntheses with – 30, 572
2-Nitroenolethers 26, 130
– startg. m. f.
  α-nitroketals 26, 130
1,1-Nitroethylene derivs.
– from
  ethylene derivs. 29, 318
  2-nitroenamines, synthesis
    26, 902
  1,2-nitrohalides 26, 976
  1,2-nitronitroso compds.
    29, 318
– special s.
  dinitromethylene compds.
  ethylenenitrohalides
– startg. m. f.
  1,1,1-ethylenenitrohalides
    26, 528
  1,2-halogenoximes 27, 570
  nitro compds. 28, 37
  4H-1,2-oxazine 2-oxide ring,
    5,6-dihydro- 26, 700
  oximes 28, 37
  2-oximino-4(5H)-benzo-
    furanones, 6,7-dihydro-
    27, 774
1,2-Nitroethylene derivs.
– from
  α,β-ethylenehalides 28, 797
Nitrofluorination 29, 494
Nitrogen compds., organic
–, oxidation of – with lead
  tetraacetate, review 23, 305
  suppl. 27
–, – – – with silver carbonate-
  Celite 27, 100 suppl. 30
– dioxide 26, 361; 28, 234, 512,
  935; 29, 198; 30, 231, 240
– as reactant 29, 422
– startg. m. f.
  1,1,1-trinitro compds. 27, 347

– monoxide 19, 182 suppl. 26;
  28, 234; 30, 190
Nitrogen-nitrogen bonds
–, cleavage, reductive 29, 17
–, –, review 29, 17
Nitrogen oxides s. Dinitrogen
  pentoxide, Nitrogen dioxide,
  – monoxide, N-Oxides
– radical cations 27, 998
– radicals
– startg. m. f.
  amines, sec. 29, 12
N→C-Nitro group migration
  29, 326
Nitro groups (s. a. under
  Replacement)
–, deactivation by – 27, 799
–, reduction with O→N-
  carbalkoxyl migration 29, 15
–, retention 17, 71 suppl. 28;
  26, 62; 27, 46; 29, 58
– –, ar., displacement, heterocyclics by – 29, 776
Nitroguanidines
– from
  amines 5, 346 suppl. 26
– startg. m. f.
  guanidines 12, 52
Nitrohalides
– special s.
  ethylenenitrohalides
1,1-Nitrohalides
– special s.
  2-chloro-2-nitropropane
1,2-Nitrohalides
– from
  ethylene derivs. 27, 550
  1,2-nitrosohalides 21, 123
  suppl. 26
– startg. m. f.
  1,1-nitroethylene derivs.
    26, 976
o-Nitrohalides
– startg. m. f.
  pyrazole ring 28, 790
1,1-Nitrohydrazones
– from
  hydrazones 30, 231
α-Nitroketals
– via 2-nitroenolethers 26, 130
Nitroketones
– startg. m. f.
  N-heterocyclics 25, 354
α-Nitroketones
– from
  ethylene derivs. 29, 318
  1,2-nitronitroso compds.
    29, 318
Nitromethane
– as solvent 23, 242 suppl. 29;
  26, 292; 27, 762; 28, 104, 690;
  30, 519
Nitronates (s. a. aci-Nitro
  compds.)
– special s.
  lithium isopropane nitronate
Nitrones (s. a. N-Oxides)
–, cycloaddition, 1,3-dipolar,
  with –, review 26, 699
  suppl. 30
– from
  halides 22, 458
  hydroxylamines and
    N-nitrimines 26, 390
  nitro compds., ar. 27, 902
– special s.
  acoxy-nitrones
  cyano-

Nit 678

(Nitrones
- special s.)
  ethylene-nitrones
  halogeno-
  keto-
  methylthiomethyl-
- startg. m. f.
  2-azetidinones 28, 650
  azomethines 30, 696
  isoxazolidines, 4-amino-
    26, 699
  5-isoxazolidones 28, 817
  3-pyrrolidones 26, 693
-, ar.
- reactions 30, 363
- startg. m. f.
  o-halogenoazomethines
    30, 363
-, cyclic
- special s.
  isoquinoline N-oxides,
    3,4-dihydro-
  phenanthridine -
  phenolnitrones, cyclic
- startg. m. f.
  lactams, ring expansion
    28, 122
  thiolactams 27, 613
Nitronic acid esters, cyclic
  s. 4H-1,2-Oxazine 2-oxide
  ring, 5,6-dihydro-
Nitronitric acid esters
- startg. m. f.
  dinitroalcohols 28, 292
  nitro compds., aliphatic
    27, 63
1,2-Nitronitroso compds.
- from
  ethylene derivs. 29, 318
- startg. m. f.
  1,1-nitroethylene derivs.
    29, 318
  α-nitroketones 29, 318
Nitronium s. Nitryl
Nitroolefins s. Nitroethylene
  derivs.
Nitrooximes, ring closure,
  reductive, of 30, 335
o-Nitrophenol as reagent
  30, 150
p-Nitrophenoxycarbonyl
  chloride as reactant 26, 426
o-Nitrophenoxydimethylacetyl
  as N-protective group 26, 28
p-Nitrophenyl phosphate
-, phosphorylation with -
  28, 190
o-Nitrophenylsulfenyl
  as N-protective group 27, 19
2-Nitropropane as reagent
  22, 288 suppl. 26
2-Nitrosamine dimers, isocyclic
- startg. m. f.
  N-hydroxy-N-heterocyclics
    27, 331
Nitrosamines s. a. Nitrosanilines
N-Nitrosamines
  (s. a. N-Nitrosation)
-, 1-alkylation 30, 437
- from
  hydrazines 28, 238
-, α-labeling 28, 590
-, metalation 29, 742; 30, 437
-, reaction with lithium
  compds., organo- 30, 314
- special s.
  indolines, 3-hydroxy-
    1-nitroso-

isoquinolines, 1,2,3,4-tetra-
  hydro-N-nitroso-
N-nitrosanilines
- startg. m. f.
  1,1-alkoxyhydrazines 30, 314
  amines 28, 19, 355
  azomethines 28, 453; 30, 334
  hydrazines, trisubst. 29, 433
  hydrazones 29, 433
  1,2,3-triazole ring 29, 742
-, aliphatic, review 18, 325
  suppl. 26
-, metalated
-, reaction with
  electrophiles, review 30, 437
  heteroelectrophiles 30, 437
-, prim.
- startg. m. f.
  hydrazines, monosubst.
    1, 255 suppl. 29
o-Nitrosamines
- startg. m. f.
  pyrazine ring 26, 341
  1,2,4-triazine ring 27, 434
Nitrosamino groups, substitution, nucleophilic 30, 574
o-(N-Nitrosamino)ketones
- startg. m. f.
  indolines, 3-hydroxy-1-nitroso-
    29, 695
N-Nitrosanilines
- special s.
  o-(N-nitrosamino)ketones
- startg. m. f.
  quinoxaline ring 29, 454
-, N-acyl-
- startg. m. f.
  diazonium carboxylates
    26, 995
Nitrosation
- of
  ketones s. α-Isonitrosoketones
-, ar. 25, 389 suppl. 28
-, hydrolytic 29, 315
N-Nitrosation
  (s. a. N-Nitrosamines)
-, steric control 13, 335
  suppl. 26
N-Nitrosoacylamines
- from
  acylamines 28, 935
- special s.
  N-nitrosanilines, N-acyl-
2-(N-Nitrosoacylamino)alcohols
-, reactions with - 29, 327
- startg. m. f.
  enazides, terminal 29, 400
Nitroso compd. dimers
  s. Azo N,N'-dioxides
- compds.
- from
  amines 26, 91
  hydrocarbons 26, 327
  hydroxylamines 27, 100
  nitro compds. 28, 990
  silanes 27, 489
  stannanes 27, 489; 30, 315
- special s.
  dinitrosoarenes
  indoles, 3-nitroso-
  2-methyl-2-nitrosopropane
  nitronitroso compds.
- startg. m. f.
  amines 29, 16
  azo compds. 28, 240
  azoxy compds., unsym.
    29, 280

2,5-cyclohexadienone oximes
  30, 596
diazonium nitrates 30, 190
enehydroxylamines 27, 313
  suppl. 30
N-halogenodiimide N'-oxides
  29, 270
hydroxamic acids, N-aryl-
  27, 479
hydroxylamines 27, 313
nitro compds. 27, 356; 30, 315
1,2-oxazetidines 27, 329
purines 29, 365
triazenes 30, 189
N-Nitrosocyanamides
- from
  cyanamides 29, 276
N-Nitroso groups, migration
  30, 186
Nitrosohalides
  s. a. Halogenoximes
1,2-Nitrosohalides
- startg. m. f.
  alcohols 27, 202
  ethylene derivs. 26, 973
  1,2-nitrohalides 21, 123
  suppl. 26
N-Nitrosohydrazines
- from
  hydrazines 26, 828
Nitrosonium s. Nitrosyl
o-Nitrosophenols
- startg. m. f.
  oxazole ring 26, 395
α-Nitrosophosphonium salts
- from
  phosphonium salts 28, 319
- startg. m. f.
  nitriles 28, 319
N-Nitrosoureas 13, 335
- from
  ureas 29, 274
- special s.
  1-ethynyl-N-nitrosoureas
- startg. m. f.
  benzene ring 27, 955
  semicarbazides 29, 274
Nitrososulfones
- from
  nitrosulfoxides 27, 106
N-Nitrosourethans
- special s.
  1-ethynyl-N-nitrosourethans
Nitrostyrenes
- special s.
  aminonitrostyrenes
1,1-Nitrosulfones
- startg. m. f.
  1,2-dinitro compds.,
    aliphatic, synthesis
    30, 608
-, syntheses with - 30, 608
Nitrosulfoxides
- startg. m. f.
  nitrososulfones 27, 106
Nitrosyl chloride
- as reactant 26, 258;
  30, 237, 369
- as reagent 27, 489; 30, 315
- startg. m. f.
  α-oximinocarboxylic acids
    and derivs. 28, 285
- fluoroborate 26, 262
- salts, reactions with -
  26, 262
- sulfuric acid 28, 436; 29, 282

o- and p-Nitroureas
- from
  chloroformamidines **28,** 399
Nitrous acid cf. Sodium nitrite
- **– esters** (s. a. Nitrites)
- by transesterification **26,** 225
- special s.
  acyloximinonitrites
  ethylenenitrites
- startg. m. f.
  diol mononitrates **30,** 95
  nitric acid esters **26,** 89
  phosphoric – – **28,** 80
- – – –, **cyclic**
- startg. m. f.
  oximinoketones **27,** 342
- – – **scavenger,** ethyl malonate
  as – **26,** 361
-, urea as – **29,** 282
Nitroxides s. N-Oxide radicals
1-Nitroxide compds. 22, 153
  suppl. 27
-, reactions **30,** 368
- startg. m. f.
  α-halogenoketones **30,** 368
1,2-Nitroximes
- from
  1,2-halogenoximes **27,** 463
- startg. m. f.
  1,2-oximinohydroxylamines
  26, 19
Nitroxyls s. N-Oxide radicals
Nitryl chloride as nitrating
  agent **28,** 320
- **fluoroborate**
- as reactant **28,** 286;
  29, 401, 494
- as reagent **27,** 364
Nonaflates 16, 151 suppl. **28**
Norcaranes, 7-*syn*-subst.
27, 134 suppl. **28**
18-Norsteroids 28, 983 suppl. **29**
Nuclei s. Heterocyclics,
  Isocyclics, Rings
Nucleofugal s. Electrophilic
Nucleophiles (s. a. C-Hydrogen,
  active, Trap, nucleophilic)
-, addition with oxidation
  28, 361; **30,** 246
-, alkylidenephosphoranes as –,
  review **20,** 608 suppl. **26**
-, N-dequaternization with –
  25, 324 suppl. **28**
-, reaction of thiirane
  1,1-dioxides with – **30,** 385
-, reaction with
  aryl trichloromethyl
    carbinols, review **22,** 629
    suppl. **26**
  halogenimidium salts **26,** 404
  oxido compds. **26,** 597
-, rearrangement of α-halogenoboranes with – **27,** 118
- special s.
  fluoride ions
  hydrogen carbonate anions
  phosphorus(III) nucleophiles
-, **reactive,** as activators
  of unreactive nucleophiles
  26, 607
-, **strong**
- special s.
  hydroperoxide ions
  nitrile anions
-, **unreactive,** activation by
  reactive nucleophiles **26,** 607

-, **weak**
- special s.
  acetone
Nucleophilic catalysts
  s. Catalysts, nucleophilic
- **reactions**
- s.
  addition, nucleophilic
  ring closure, –
  substitution, –
- with aminosulfenium ylids
  29, 309
- **reducing agents** s. Reducing
  agents, nucleophilic
- **ring opening,** 1,6-benzo-
  diazocines, 1,2,3,4,5,6-hexa-
  hydro-, by – **27,** 44
- special s.
  oxetane ring opening,
  nucleophilic
Nucleoside fusion synthesis
  26, 446 suppl. **29**
- **hydroxyls**
-, phosphorylation, preferential
  26, 97, 105
- **peptides 16,** 424 suppl. **27;**
  22, 394 suppl. **27**
- **pyrophosphates**
- startg. m. f.
  nucleotides **27,** 4
Nucleosides (s. a. N-Glycosyl-
  N-heterocyclics)
-, N-acylation, selective
  25, 255 suppl. **26**
-, O-benzylation, selective
  20, 150 suppl. **26**
- by transglycosidation **27,** 441
-, cleavage **28,** 25
- from
  acylhalogenosugars and
  mercury-N-heterocyclics
  16, 514
- special s.
  2-azapurine nucleosides
  cyclo-
    ethylene-
    purine –
    thio-
    pyrimidine –
- startg. m. f.
  nucleotides **23,** 123 suppl. **29**
- **acyclic-sugar** – **29,** 435
-, **sugar-branched,** synthesis
  23, 879 suppl. **28**
-, **unprotected**
- startg. m. f.
  5'-nucleotides **26,** 97, 105
-, **halogeno- 26,** 545
C-Nucleosides 19, 839 suppl. **28;**
  21, 607 suppl. **28**
Nucleotides (s. a. Oligonucleotides, Poly-)
- from
  nucleoside pyrophosphates
  27, 4
  nucleosides **23,** 123 suppl. **29**
- special s.
  O-α-aminoacylnucleotides
5'-Nucleotides
- from
  nucleosides, unprotected
  26, 97, 105

C-2,7-Octadien-1-ylation 26, 827
Olefins (s. a. Ethylene derivs.)
- as reagents **28,** 482
- as solvents **10,** 650 suppl. **29**

α-**Olefins** s. Ethylene derivs.,
  terminal
Oleum s. Sulfuric acid, fuming
Oligomerization s. Co-oligo-
  merization, Cyclo-, Di-
  merization, Tri-
-, peptides by – **26,** 365
Oligonucleotides
-, solid-phase synthesis **17,** 169
  suppl. **26**
- by redoxphosphorylation
  26, 175 suppl. **29**
-, synthesis on water- and
  pyridine-soluble polymer
  support **25,** 88 suppl. **28**
-, –, review **17,** 169 suppl. **27**
-, –, triester method **25,** 88
  suppl. **28**
-, – via phosphoric acid
  β,β,β-trichlorethyl esters
  25, 88 suppl. **27, 28**
-, – with N-sulfonyl-1,2,4-
  triazolides **17,** 169 suppl. **29**
Oligosaccharides 23, 242
  suppl. **29**
-, **cyclic** s. Cyclodextrins
Onium salts
-, cyanolysis **27,** 895
- special s.
  ammonium salts
  carbonium –
  phosphonium –
  sulfonium –
-, **mixed 27,** 895
Optimization, mathematical
  13, 696 suppl. **27; 18,** 587
  suppl. **28**
Organometallic compds. (s. a.
  under individual metals,
  Metal complex compds.)
-, electrochemistry, review
  18, 717 suppl. **28**
- –, α-**neutral heteroatom-
  subst.,** review **25,** 483
  suppl. **27**
- –, **unsatd.**
- special s.
  copper compds., unsatd.
  Grignard –, –
  lithium –, –
- –, –, **allylic,** review
  19, 745 suppl. **29**
- –, **halogeno-**
- special s.
  1,1-halogenoorganolithium
    compds.
  trifluormethylmetal –
  trihalogenomethylmercury –
**1-Organoseleno-1-organothio-
  1-en-3-ynes**
- by selenophene ring opening
  30, 436
Organothio... s. a. Alkylthio...,
  Arylthio..., Bis(organothio)...,
  Sulfides
1-(Organothio)alkoximes
-, protection of carbonyl
  groups as – **29,** 386
1-Organothioazomethinium salts
  22, 987
- special s.
  alkylthioazomethium salts
**3-Organothio-1,2-dithiolium
  salts**
- startg. m. f.
  3-alkylidene-1,2-dithioles
  26, 979

Orientation s. a. Direction,
Regiospecificity, and under
Friedel-Crafts ketone
synthesis
-, **inverse** s. Cycloaddition,
1,3-dipolar, inverse-oriented,
Diene synthesis, inverse-
oriented
- **control**
 - in acoxylation, ar. **24,** 176
 suppl. 26
 - of nitration **30,** 240
 - of substitution, electrophilic,
 review **25,** 389 suppl. 26
Orthocarbonic acid esters
s. Spiroorthocarbonic acid
esters
Orthocarboxylic acid amides
- special s.
 tris(dimethylamino)methane
- - **esters** (s. a. Thioorthocarboxylic acid esters)
- as reactants **29,** 352
-, review **29,** 607
- special s.
 orthoformic acid esters
- startg. m. f.
 acetals, cyclic, mixed
 28, 195
 β-allenecarboxylic acid
 esters **27,** 796
 α,α-dialkoxyphosphonic acid
 esters **29,** 607
 iminoesters **29,** 369
 4(3H)-pyrimidinones **30,** 250
 quinolines, 4-hydroxy-
 29, 726
-, synthesis, review **21,** 251
 suppl. 29
- - -, **cyclic**
 - by ring closure, oxidative
 16, 317 suppl. 27
-, Claisen rearrangement with -
 27, 795 suppl. 29
- from
 lactones **26,** 179
- special s.
 carbohydrate orthocarboxylic
 acid esters
 2,2-dialkoxyethers, cyclic
- startg. m. f.
 acoxyhalides **28,** 500
- - -, **mixed**
- from
 cyclopropanone ketals
 29, 135
- - -, α,β-**unsatd. 6,** 840
 suppl. 28
**Orthoester Claisen rearrange-
ment 27,** 795
- - -, **geospecific 27,** 795
Orthoformic acid esters
-, reactions with aryloxymagnesium halides **28,** 742
- special s.
 ethyl orthoformate
- startg. m. f.
 acetals **30,** 111
 Benzopyrylium salts,
 4-alkoxy- **29,** 192
 enolethers **29,** 190; **30,** 111
 1,3,5-trialkoxy compds.
 26, 178
- - -, **mixed 8,** 274 suppl. 26
Osazones
- startg. m. f.
 acetylene derivs. **28, 938**

Osmic acid s. Osmium
tetroxide
Osmium tetroxide OC↓CC
Oxa... s. a. O-Heterocyclics
8-Oxabicyclo[5.3.0]decatrienes
- from
 tropones **27,** 707
**2-Oxabicyclo[4.1.0]heptane,
7,7-dichloro-**
- startg. m. f.
 oxepins, 2,5,6,7-tetrahydro-
 3-chloro-2-*tert*-butoxy-
 27, 972
 2H-pyran-5-carboxaldehydes,
 3,4-dihydro- **27,** 972
**3-Oxabicyclo[3.1.0]hexanes
25,** 680 suppl. 26
**6-Oxabicyclo[3.2.1]octanes
28,** 823
**8-Oxabicyclo[3.2.1]octanes
29,** 682
1,2-Oxaborins 10, 587 suppl. **29**
-, **dihydro-**
- from
 boranes and furans **28,** 583
- startg. m. f.
 2-ene-1,4-diols **28,** 583
2-Oxa-3-borolenes
- from
 acetylene derivs., boranes,
 and carboxylic acid
 chlorides **28,** 866
- startg. m. f.
 α,β-ethyleneketones **28,** 866
Oxadiazabicycloalkanes
- from
 amidines, cyclic, and oxido
 compds. **27,** 316
**8-Oxa-2,3-diazabicyclo[3.2.1]oct-
6-enes**
- from
 1,2-diazabicyclo[3.2.0]-
 3-hepten-6-ols **29,** 169
1,3,6-Oxadiazepine ring
s. 1,3,6-Oxadiazepino[3,4-a]-
benzimidazoles, 1,3,4,5-
tetrahydro-
**1,3,6-Oxadiazepino[3,4-a]-
benzimidazoles, 1,3,4,5-
tetrahydro- 26,** 362
1,3,5-Oxadiazine-2,4-diones
- from
 acylisocyanates and
 isocyanates **27,** 322
**4H-1,3,4-Oxadiazine ring,
5,6-dihydro-**
- from
 diacylazo compds. **28,** 280
**4H-1,2,5-Oxadiazines,
5,6-dihydro-**
- from
 syn-1,2-aminooximes and
 aldehydes **27,** 386
1,3,5-Oxadiazinium O-salts
- startg. m. f.
 pyrimidines **30,** 577
**2H-1,3,5-Oxadiazin-4(3H)-ones,
2-imino-**
- from
 acylisocyanates and
 carbodiimides **28,** 266
1,2,5-Oxadiazole 2-oxides
s. Furoxans
**1,3,4-Oxadiazole ring,
2-(pyrrol-2-yl)-**
- startg. m. f.
 pyrrolo[1,2-d][1,2,4]triazine
 ring **29,** 566

1,2,4-Oxadiazole ring, 3-ureido-
- startg. m. f.
 1,2,4-triazol-5-one ring,
 3-acylamino- **26,** 257
1,2,4-Oxadiazoles 28, 318
-, **3 acyl-**
- from
 α-acylaminoketones **28,** 318
1,3,4-Oxadiazoles
- from
 acylhydrazones **26,** 232
 carboxylic acid (2 molecules)
 26, 380
 N-1,2,4-triazol-4-ylamidines
 26, 497
1,2,5-Oxadiazoles s. Furazans
1,3,4-Oxadiazolid-5-ones
- from
 oxaziridines and isocyanates
 30, 213
Δ²-**1,2,3-Oxadiazoline 2-oxides**
- from
 2-halogeno-N-sulfonyl-
 N-nitramines **30,** 173
1,2,4-Oxadiazoline ring
- startg. m. f.
 oxazolidine ring **29,** 168
**1,3,4-Oxadiazoline ring opening,
2-imino- 28,** 269
1,3,4-Oxadiazolines, 2-imino-
- startg. m. f.
 1-acyldiaminoguanidines,
 5-carbamyl- **28,** 269
1,3,4-Oxadiazolium salts
- startg. m. f.
 pyrazole ring, 3-amino-
 27, 904
Δ²-**1,2,4-Oxadiazol-5-ones**
- special s.
 N,N'-carbonyldi-
 (1,2,4-oxadiazol-5-ones)
- startg. m. f.
 amidoximes **26,** 22
**1,3,4-Oxadiazol-2(3H)-ones
6,** 460; **15,** 403
- from
 thionocarbazic acid esters
 30, 174
-, **3-silyl- 16,** 380 suppl. **29**
1,3,4-Oxadiazol-2(5H)-ones
- from
 semicarbazones **26,** 232
 suppl. 28
Δ²-**1,3,4-Oxadiazol-5-ones**
- from
 1-acyl-3-sulfonyloxyureas
 30, 193
**[1,2,5]Oxadiazolo[3,4-b]pyridine
1-oxides**
- from
 1,2,3,4-tetrazolopyridines,
 4-nitro- **27,** 102
**1-Oxa-6,6a-diselena-
2-azapentalenes**
- from
 1,2-diselenolium salts **28,** 315
1-Oxa-6,6-dithia-2-azapentalenes
- from
 1,2-dithiolium salts **28,** 315
Oxalate s. Bis(tetra-*n*-butyl
ammonium) oxalate
Oxalic acid as reagent **7,** 217
suppl. **28; 26,** 912; **27,** 890;
28, 793, 837; **29,** 187, 471, 901;
30, 167, 448
- - **esters** (s. a. Alkoxalyl
 groups)

−, ring closure with − **28,** 724
− − −, **sym. 18,** 472 suppl. **27**
Oxalyl chloride
− as reagent **14,** 107 suppl. **29;
26,** 540; **27,** 361; **29,** 524;
30, 357
−, ring closure with − **30,** 241
− startg. m. f.
N-(chloroglyoxyloyl)imino-
esters **26,** 458
isocyanates **27,** 361
xanthones **30,** 541
Oxamides
− startg. m. f.
α-ketocarboxylic acid amides
29, 801
−, **sym.**
− from
amines, sec. **28,** 322
7-Oxanorbornadienes
− startg. m. f.
fulvenes, 6-hydroxy- **30,** 525
**4H-1,4-Oxaphosphorin 4-oxides
28,** 126
4-Oxa-5-silaspiro[2,6]nonanes
− from
silacyclohexan-2-ones **27,** 708
1-Oxaspiro[2,2]pentane ring
− from
ketones **28,** 837
− startg. m. f.
cyclobutanone ring **28,** 837
−, syntheses via − **28,** 837
**1,4,3,5-Oxathiadiazine
4,4-dioxides 29,** 290
− from
carbalkoxysulfamoyl
chlorides **29,** 290
**1,4-Oxathiane 4,4-dioxides,
2,6-dialkoxy-**
− from
thiophene 1,1-dioxides,
2,5-dihydro- **29,** 148
− startg. m. f.
4H-1,4-thiazine 1,1-dioxides
29, 148
1,4-Oxathian-2-yl ethers,
protection of hydroxyl
groups as − **18,** 875 suppl. **28**
1,2,3-Oxathiazetidine 2-oxides
− from
thionylimines and aldehydes
28, 272
1,4,2-Oxathiazoles
− from
nitrile oxides **28,** 527
− startg. m. f.
isothiocyanates **28,** 527
**1,2,3-Oxathiazolidine S-oxide
ring**
− startg. m. f.
2-aminoalcohols **29,** 18
1,3,5-Oxathiazolines
− from
thioketones and hydroximinohalides **30,** 408
**1,3,2-Oxathiazolium betaines,
5-hydroxy-**
− from
α-mercaptocarboxylic acids
27, 292
1,3,4-Oxathiazol-2-ones
− startg. m. f.
1,2,4-thiadiazoles **30,** 416
**1,2-Oxathiin 2,2-dioxide ring
28,** 540

**1,4-Oxathiin-2(3H)-one
4,4-dioxides**
− from
α-diazoketones **29,** 562
Oxathiins s. a. Oxathianes
**1,4-Oxathiins, 2,3-dihydro-
26,** 619 suppl. **27**
1,2-Oxathiolane 2-oxides 29, 105
1,3-Oxathiolanes 27, 170
− special s.
2-sulfonylimino-1,3-oxa-
thiolanes
1,3-Oxathiolan-5-ones
− startg. m. f.
α-alkylthiocarboxylic acids
26, 670 suppl. **27**
ethylene derivs. **26,** 988
1,3-Oxathiole 3-oxides 26, 896
suppl. **28**
Oxathiolium salts
− from
carboxylic acid thioamides
28, 568
**1,3-Oxathiol-2-ylidenimmonium
salts**
− startg. m. f.
thiophenes **28,** 605
**1,2,3,5-Oxatriazolium salts,
N-condensed**
− from
o-amino-N-oxides **27,** 283
− startg. m. f.
o-azo-N-oxides, cyclic **27,** 283
2-Ox-6-azaadamantanes 29, 596
**2,3-Oxazabicyclo[3.2.0]hepta-
3,6-dienes**
− startg. m. f.
1,3-oxazepines **29,** 332
2,1-Oxazabicyclo[4.1.0]heptanes
− from
6H-1,2-oxazine 2-oxides,
4,5-dihydro-, and acetylene
derivs. **29,** 653
**6-Ox-2-azabicyclo[3.1.0]hex-
3-enes**
− from
4H-1,3-oxazines **30,** 528
1,3,5-Oxazaphosph(V)ol-2-ines
−, reactions with heterobonds,
multiple **27,** 887 suppl. **29**
− startg. m. f.
azetines, 3-imino- **27,** 887
2H-pyrrolenines **28,** 849
1,3-Oxazepine ring
− from
pyridine N-oxide ring **23,** 179
1,3-Oxazepines
− from
2,3-oxazabicyclo[3.2.0]hepta-
3,6-dienes **29,** 332
− startg. m. f.
pyrroles, 3-acyl- **26,** 743
suppl. **28**
1,2-Oxazetidines
− from
ethylene derivs. and nitroso
compds. **27,** 329
1,3-Oxazetidines
− from
carbodiimides and
isocyanates **28,** 265
**4H-1,2-Oxazine 2-oxide ring,
5,6-dihydro-**
− from
1,1-nitroethylene derivs. and
enamines **26,** 700

**4H-1,2-Oxazine 2-oxides,
5,6-dihydro-**
− startg. m. f.
N,N-dialkoxylamines,
bicyclic, N-condensed
27, 702
**6H-1,2-Oxazine 2-oxides,
4,5-dihydro-**
− startg. m. f.
2,1-oxazabicyclo[4.1.0]-
heptanes **29,** 653
1,2-Oxazine ring, dihydro-
− by Meisenheimer rearrangement **26,** 151
− −, **5,6-dihydro-**
− from
1,2-oxazinium ring,
4,5-dihydro- **28,** 676
**2H-1,2-Oxazine ring, 3,6-
dihydro-N-acyl-**
− from
hydroxamic acids and
1,3-dienes **30,** 244
1,2-Oxazine ring, 6-hydroxy-
− from
nitro compds., unsatd.
28, 120
− −, **tetrahydro-**
− from
ethylene derivs. and
α-halogenonitrones **28,** 657
**4H-1,3-Oxazine ring,
5,6-dihydro-**
−, synthesis, stereospecific
26, 885
− −, **tetrahydro-**
− startg. m. f.
oxo compds. **29,** 257
1,4-Oxazine ring
− from
α-halogenoketones **26,** 429
**4H-1,2-Oxazines, 5,6-dihydro-
6-hydroxy-**
− from
enamines and 1,2-halogen-
oximes **26,** 849
**1,2-Oxazines, tetrahydro-
N-carbalkoxy-**
− from
1,4-dihalides and N-hydroxyurethans **30,** 295
1,3-Oxazines, synthesis via −,
review **25,** 612 suppl. **27**
4H-1,3-Oxazines
− startg. m. f.
6-ox-2-azabicyclo[3.1.0]hex-
3-enes **30,** 528
−, **5,6-dihydro-**
− from
ethylene derivs. **26,** 885
3-isocyanoalcohols **29,** 617
isonitriles and oxido compds.
29, 617
nitriles **26,** 309
− startg. m. f.
3-aminoalcohols **29,** 617
pyrroles **28,** 439
pyrrolo[2,1-b][1,3,4H]-
oxazines, dihydro- **28,** 439
−, syntheses with −, review
25, 612 suppl. **28**
−, synthesis of oxo compds.,
branched, with − **25,** 612
−, −, **6-alkoxy-**
− from
acylimines and enolethers
29, 654

Oxa

(4H-1,3-Oxazines, 5,6-dihydro-)
-, -, 2-(γ,δ-ethylene)-
- via Claisen-type rearrangement 29, 704
2H-1,3-Oxazines, tetrahydro-6-alkoxy-
- from
  aldehydes 30, 564
1,3-Oxazines, tetrahydro-2-ene-1-(β,γ-ethylene)- 29, 704
- startg. m. f.
  4H-1,3-oxazines, 5,6-dihydro-2-(γ,δ-ethylene) 29, 704
2H-1,3-Oxazines, tetrahydro-6-hydroxy- 30, 564
1,4-Oxazines s. Morpholines
6H-1,3-Oxazine-6-thiones
- from
  pyridinium 7-imides and cyclopropenethiones 27, 436
1,3-Oxazinium iodide, 5,6-dihydro-N,2,4,4,6-pentamethyl-
- as reactant 28, 793
1,2-Oxazinium ring, 4,5-dihydro-
-, opening 28, 676
- startg. m. f.
  1,2-oxazine ring, 5,6-dihydro-28, 676
Oxazinium salts, tetrahydro-6-hydroxy-
- from
  β-aminoaldehydes 26, 289
1,3-Oxazin-4-one ring
- from
  o-cyanatocarboxylic acid esters 26, 314
1,3-Oxazin-6-one –
- startg. m. f.
  pyridine ring 28, 871
1,2-Oxazin-6-ones
- startg. m. f.
  2-azapyrylium salts, 6-alkoxy- 30, 110
1,3-Oxazin-4-ones
- from
  Δ³-pyrrol-2-ones 29, 175
1,3-Oxazin-6-ones, 2-alkoxy-
- from
  α,β-ethylen-β-isocyanato-carboxylic acid esters 27, 165
- startg. m. f.
  α,β-ethylen-β-isocyanato-carboxylic acid esters 27, 165
-, 4,5-dihydro-
- startg. m. f.
  β-amino acid peptides 30, 207
1,4-Oxazin-2-ones, tetrahydro-
-, synthesis, asym., of α-amino-carboxylic acid via –
  29, 750
1,4-Oxazin-3-ones 26, 250
Oxaziranes s. Oxaziridines
Oxaziridine ring
- startg. m. f.
  azomethines 30, 328
Oxaziridines
-, reactions with heterocumulenes 30, 213

- special s.
  spirooxaziridines
- startg. m. f.
  carboxylic acid amides 26, 17
  1,2,4-oxadiazolid-5-ones 30, 213
-, 3-alkoxy-
- from
  iminoesters 29, 132
Oxazole N-oxides
- from
  α-dioxo compd. monoximes 27, 412
- ring
- from
  o-nitrosophenols and pyridinium salts 26, 395
- startg. m. f.
  furan ring 24, 807 suppl. 26
- – opening 26, 332
Oxazoles
- from
  α-acylaminocarboxylic acid amides 30, 165
  α-acylaminoketones 19, 752
  carboxylic acid chlorides and α-azidoketones 27, 477
  – – – and isonitriles 28, 811
  2-ketonitrones 28, 219
  5-oxazolones, rearrangement 27, 985
  oxo compds. and 1-sulfonyl-1-isonitriles 28, 841
- startg. m. f.
  α-aminoketones 28, 811
  furans 28, 869
ψ-Oxazoles, 5-acylimino-3-amino- 27, 325
Oxazoles, 5-amino-
- from
  1-(dihalogenomethylene)-acylamines 30, 292
Oxazolidine N-oxyls
- startg. m. f.
  ketones 28, 234
- ring
- from
  1,2,4-oxadiazoline ring 29, 168
- –, N-condensed 18, 373 suppl. 26
- –, opening 26, 332
- –, 2-imino-3-acyl-
- startg. m. f.
  2-oxazolidone ring 30, 165
Oxazolidines (s. a. Carbohydrate derivs., heterocyclic)
- from
  2-aminoalcohols and oxo compds. 28, 335
  aziridines 24, 692 suppl. 26
-, synthesis, asym. 28, 335
2-Oxazolidinethiones
- from
  isothiocyanates and oxo compds. 29, 632
2,5-Oxazolidiones 27, 488
2-Oxazolidone ring
- from
  β-hydroxycarboxylic acid amides 28, 309
  oxazolidine ring, 2-imino-3-acyl- 30, 165
- startg. m. f.
  2-aminoalcohols 28, 309

4-Oxazolidone ring s. Spiro-oxazolid-4-one ring
2-Oxazolidones
- from
  isocyanates and oxido compds. 27, 317
-, protection of trans-2-amino-alcohol groups, diequatorial, as – 26, 426
-, 4-β-halogeno-
- from
  pyrrolidines, 3-hydroxy-30, 352
-, 4-hydroxy-
- from
  1,3-dioxolan-2-ones 28, 365
- startg. m. f.
  Δ⁴-2-oxazolones 28, 365
-, 4-imino-
- from
  cyanourethans 26, 311
-, 3-nitroso-
- startg. m. f.
  enesilanes 8, 922 suppl. 26
  α,β-ethylenehalides 8, 922 suppl. 26
5-Oxazolidones
- startg. m. f.
  azomethines 27, 536
Δ²-Oxazoline-4-carboxylic acid esters
- startg. m. f.
  α-acylaminocarboxylic acid esters 26, 32
3-Oxazoline-N-oxides 26, 90
Δ²-Oxazoline ring
- from
  1-acoxy-2-acylamines 27, 257
-, opening, reductive 26, 32
- –, 2-alkylthio-, opening 29, 696
- –, 2-amino-
- special s.
  carbohydrate Δ²-oxazoline ring, 2-amino-
Oxazolines
-, review 19, 432 suppl. 26
Δ²-Oxazolines
- from
  2-hydroxythioureas 27, 406
  isonitriles and oxo compds. 28, 606
-, metalation, α-lateral 29, 807
-, protection of carboxyl groups as – 29, 3
- startg. m. f.
  2-acylaminoalcohols 28, 85
  enacylamines 26, 755
  1,3-N,N-heterocyclics 6, 452 suppl. 27
-, syntheses with –, review 25, 612 suppl. 28; 29, 807
-, 2-alkoxy-4-imino-
- from
  ketenimines and azidoformic acid esters 27, 432
-, 2-amino-
- from
  2-aminoalcohols 29, 335
  halogenisocyanates 30, 291
  halogenoureas 30, 291
-, 4,4-dimethyl-, reactions with – 26, 872

Δ³-**Oxazolines**
- from
  Δ¹-azirines and oxo compds.
  30, 458
Δ³-**Oxazoline-2-thiones,
4-organothio-**
- by
  benzilic acid-type
  rearrangement **28**, 395
Δ²-**Oxazolinium ring opening,
reductive 26**, 32 suppl. **27**
- salts
-, change in configuration
  via – **5**, 154 suppl. **29**
- special s.
  N,4,4-trimethyl-Δ²-oxazo-
  linium iodide
Oxazolium salts
- startg. m. f.
  imidazol[2,1-c][1,2,4]triazines
  26, 344
**10bH-Oxazolo[3,2-c][1,3]-
benzoxazine-2(3H),5-diones
27**, 353
2-Oxazolone ring
- from
  isocyanates and N-oxides,
  cyclic **30**, 293
- as O,N-protective group **27**, 357
4(5H)-Oxazolones
- from
  α-halogenocarboxylic acid
  imides **27**, 265
5-Oxazolones
- from
  α-acylaminocarboxylic acids
  30, 451
-, Michael addition of –
  18, 765 suppl. **26**
- special s.
  4-alkylidene-5-oxazolones
- startg. m. f.
  1,3-dioxanes, 4-oxo-5-acyl-
  amino- **30**, 451
  4-imidazolones **4**, 309
  suppl. **27**; **28**; **30**, 257
  oxazoles, rearrangement
  27, 985
Δ²-**4-Oxazolones**
- startg. m. f.
  1,2,4-triazoles, 3-hydroxy-
  methyl- **27**, 391
Δ²-**5-Oxazolones**
-, reaction with compds.,
  CH-acidic **26**, 666
- startg. m. f.
  Δ²-pyrrol-4-ones, 1-acyl-
  2-amino- **26**, 666
Δ³-**5-Oxazolones**
- from
  Δ¹-azirines **28**, 643
- startg. m. f.
  1,1-aminothioethers **26**, 627
  Δ¹-pyrrolines **30**, 628
Δ⁴-**2-Oxazolones**
- from
  1,3-dioxolan-2-ones **28**, 365
  2-oxazolidones, 4-hydroxy-
  28, 365
sym-**Oxepin oxides 30**, 522
- startg. m. f.
  4-pyran-4-carboxaldehydes
  30, 522
**Oxepins, 2,5,6,7-tetrahydro-
3-chloro-2-*tert*-butoxy-**
- from
  2-oxabicyclo[4.1.0]heptanes,
  7,7-dichloro- **27**, 972

- startg. m. f.
  2H-pyran-5-carboxaldehydes,
  3,4-dihydro- **27**, 972
Oxepins-arene oxides, review
23, 945 suppl. **29**
Oxetane ring
- from
  2-ethylenealcohols with ring
  contraction **26**, 146
- –, opening, nucleophilic
  12, 626
Oxetanes
- from
  1,3-dioxanes, 2-oxo- **29**, 266
  (3-halogenalkoxy)stannanes
  28, 225
- startg. m. f.
  ethylene derivs. and oxo
  compds. **29**, 978
  furans, tetrahydro-2-imino-
  27, 672
2-Oxetanones s. β-Lactones
Oxidants
  N-bromosaccharin as – **28**, 202
  N-chlorosuccinimide-triethylamine as – **28**, 201;
  28, 512 suppl. **30**
  isoalloxazine, 8-cyano-
  3,10-dimethyl- as – **30**, 377
  nitrobenzene as – **26**, 372
-, comparison **28**, 208; **29**, 241
-, **powerful 26**, 137
Oxidation (s. a. Acylation,
  oxidative, Autoxidation,
  Decarboxylation, oxidative,
  Ring closure, –) ON: OS; OC
- in carbohydrate chemistry,
  review **19**, 307 suppl. **26**
- with addition of nucleophiles
  28, 361; **30**, 246
-, allylic **28**, 131; **29**, 173
-, **benzylic 26**, 165
-, –, **double 29**, 178
-, **electrochemical** (s. a. Rearrangement, electrooxidative)
  28, 229 suppl. **29**; **29**, 173
-, acoxylation by – **29**, 173
-, acylamination by – **29**, 334
-, alkoxylation by – **26**, 158
- of
  phenols, hindered **28**, 98
-, review **26**, 158
-, ring closure by – **29**, 900
- to
  heterocyclics **29**, 239
-, **preferential**
- of
  alcohols to oxo compds.
  8, 326 suppl. **26**
- to lactams **30**, 152
-, **selective**
- of
  alcohols to carboxylic acids
  27, 187
-, prim., to aldehydes **19**, 307
  isocycles, O-bridged **30**, 33
-, **stereospecific**
- of
  alcohols, sec. **15**, 254
  suppl. **28**
Oxidation-reduction
  s. Redox...
N-Oxide radicals
-, dimerization **22**, 136 suppl. **29**
-, reactions, review **22**, 136
  suppl. **27**
-, retention **19**, 448 suppl. **29**

- startg. m. f.
  amines, sec. **27**, 13; **28**, 17
  oxoazonium salts **29**, 996
-, synthesis and reactions,
  review **21**, 124 suppl. **26**
-, **cyclic**
- special s.
  oxazolidine N-oxyls
N-Oxides (s. a. Nitrones)
-, deoxygenation **30**, 6
- from
  amines **26**, 90
-, β-rearrangement **29**, 167
- special s.
  amine oxides
  amino-N-oxides
  azine dioxides
  azoxy compds.
- startg. m. f.
  enamines **27**, 939
-, **cyclic 27**, 98; **29**, 94
-, by ring closure, oxidative
  28, 317
- from
  nitro compds. **28**, 464;
  29, 337
  pyridinium ring **22**, 708
  suppl. **26**
-, Meisenheimer rearrangement
  26, 151
-, redoxidation **30**, 161
- special s.
  alloxazine 5-oxides
  o-amino-N-oxides
  o-azido-
  o-azo-
  benzothiazole N-oxides
  1,2,3-benzotriazine 3-oxides
  o-carboxy-N-oxides
  o-cyano-N-oxides
  imidazole 3-oxide ring
  Δ³-imidazoline 3-oxides
  isoalloxazine 5-oxides
  oxazole N-oxides
  3-oxazoline N-oxides
  picolyl, 1-oxy-
  pyridine N-oxides
  pyrimidopteridine 10-oxides
  quinoline-3,4-dione 1-oxides
  1,2,4-triazine 4-oxides
- startg. m. f.
  o-acyl-N-heterocyclics
  30, 557
  alkoxylcyclimmonium salts
  22, 187
  cyano-N-heterocyclics
  26, 813
  o-cyano-N-oxides **26**, 776
  o-halogeno-N-heterocyclics
  26, 536
  N-heterocyclics, m-subst.
  29, 646
  2-oxazolone ring **30**, 293
- via nitration **29**, 337
-, **tert.**
- startg. m. f.
  amines, sec. **30**, 26
S-Oxides, migration of carbon-
  carbon double bonds via –
  27, 734
-, reduction **27**, 734
- special s.
  sulfoxides
-, **cyclic** s. S-Dioxides, cyclic
Oxidoalcohols, cyclic
- startg. m. f.
  benzene ring **29**, 946

2,3-Oxidoalcohols
- startg. m. f.
  ethyleneketals **29,** 10
-, **cyclic**
- special s.
  steroidal ring **B** 2,3-oxido-
    alcohols
- startg. m. f.
  α,β-ethylenealdehydes, cyclic,
    ring contraction **29,** 930
1,2-Oxidoamines
- startg. m. f.
  α-aminoketones **29,** 689
Oxidoboranes 29, 133
Oxidocarboxylic acid esters
- special s.
  o-carbalkoxystyrene oxides
- startg. m. f.
  hydroxylactones **29,** 234
- **acids**
- startg. m. f.
  hydroxylactones **30,** 519
β,γ-**Oxidocarboxylic** −
- startg. m. f.
  2-ethylenealcohols **27,** 982
Oxido compds.
- as bases **29,** 969; **30,** 702
- − −, review **23,** 520 suppl. **28**
-, cleavage, oxidative **21,** 273
- from
  2-ethylenealcohols **29,** 1
  ethylene derivs. (s. a. Epoxida-
    tion) **26,** 136, 141, 144, 584;
    27, 153, 184; **28,** 113; **29,**
    133; **30,** 91
  glycol monosulfonates **26,** 237
  glycols **27,** 941 suppl. **29;**
    29, 101
  1,2-halogenhydrins **26,** 144
  α-halogenoketones **28,** 59
  2-hydroxythioethers **29,** 262
  mercury halides, mono-
    organo-2-hydroxy- **29,** 264
  oxo compds. and arsonium
    salts, synthesis **16,** 893
    suppl. **26**
  − − and 1,1-dihalides **26,** 852
  − − and halides **28,** 795
  ketones and thioethers
    29, 262
-, isomerization **26,** 750
-, reactions, review **26,** 597
-, reaction with
  hydantoins **28,** 267
  nucleophiles **26,** 597
-, rearrangements, basic,
  review **28,** 4
- special s.
  acetyleneoxide compds.
  alkoxyoxido...
  arene oxides
  cyclopropyloxido compds.
  dihydroxyoxido −
  dioxido −
  ethylene oxide
  ethyleneoxido compds.
  glycid...
  heterocyclics, oxido-
    nitroxido compds. **25**
  oxepin oxides
  oxiranyl...
  propylene oxide
  quinone epoxides
  spirooxiranocyclohexa-
    dienones
- startg. m. f.
  2-acylaminoalcohols **23,** 348

  alcohols, reduction **26,** 55;
    27, 35; **28,** 32; **29,** 32, 59, 78
  −, − with rearrangment,
    skeletal **29,** 31
  −, synthesis **26,** 669; **27,** 666
  − via 2-hydroxythioethers
    28, 524
  − with ring closure, trans-
    annular **27,** 740; **28,** 4
  2-aminoalcohols **13,** 361
  2-(*prim*-amino)alcohols
    14, 380
  3-aminoalcohols **29,** 617
  2-azidoalcohols **30,** 278
  aziridine ring **30,** 278
  α-carbalkoxy-γ-lactones
    27, 770
  cyclopropylalcohols **29,** 694
  1,3-dienes **30,** 659
  cis-1,2-dihalides **29,** 510
  2,2'-dihydroxydisulfides
    28, 525
  1,4-diols **30,** 618
  1,3-dioxolanes **24,** 128;
    30, 72
  1,3-dithiolane-2-thiones
    28, 546
  enamines, cyclic **30,** 254
  ethylenealcohols, abnormal
    ring opening **28,** 6
  2-ethylenealcohols **28,** 4, 471;
    29, 1
  3-ethylenealcohols **29,** 869
  ethylene derivs. **27,** 931/32,
    942; **30,** 652, 662
  furans, tetrahydro-
    2,3-diimino- **27,** 673
  furylideneacetic acid esters,
    tetrahydro- **30,** 448
  glycol monoesters **27,** 128
  glycols **26,** 2; **27,** 129
  1,2-halogenhydrin halogeno-
    sulfites **29,** 486
  1,2-halogenhydrins **28,** 471
  O-heterocyclics **30,** 104
  hydrocarbons **30,** 30
  1-hydroxy-2-(alkoxy-
    sulfonium) salts **27,** 129
  1,2-hydroxyhydroperoxides
    28, 69
  ε-hydroxy-β-ketocarboxylic
    acid esters **30,** 448
  γ-hydroxyketones **26,** 830;
    29, 869
  β-hydroxymercaptals **29,** 624
  1,2-hydroxymercaptans
    26, 570; **28,** 525
  α-hydroxyphosphoric acid
    esters **26,** 98
  3-hydroxysulfoxides **17,** 733
    suppl. **27**
  2-hydroxythioethers **28,** 524
  3-isocyanoalcohols **29,** 617
  ketones **26,** 149; **27,** 161
  − with rearrangement,
    skeletal **28,** 686
  oxadiazabicycloalkanes
    27, 316
  4H-1,3-oxazines, 5,6-dihydro-
    29, 617
  2-oxazolidones **27,** 317
  purinium salts, 7-β-hydroxy-
    30, 204
  thioenolethers **26,** 597
-, **acid-sensitive 26,** 584
  suppl. **28**
-, **optically active 26,** 144

  −, **sym.**, from oxo compds.
    (2 molecules) **22,** 803
  1,3-Oxido compds. s. Oxetanes
  Oxidocyclobutanones
  - startg. m. f.
    γ,δ-ethylene-ε-hydroxy-
      carboxylic acid esters
      28, 101
  1,2-Oxido-1-halides
  -, reactions **26,** 403 suppl. **30**
  - startg. m. f.
    α-aminoketones **26,** 403
    α,β-ethyleneketones **28,** 951
  2-Oxidohalides
  - startg. m. f.
    2-ethylenealcohols **29,** 968
  α,β-**Oxidoketimines 21,** 719
    suppl. **26**
  α,β-**Oxidoketones**
  - from
    α-halogenoketones and
      aldehydes **26,** 864
  - special s.
    β-alkoxy-α,β-oxidoketones
    α'-diazo-α,β-oxidoketones
    α-halogen-α,β-oxidoketones
  - startg. m. f.
    α,β-ethylene-α-hydroxy-
      ketones, rearrangement
      30, 524
    β-hydroxyketones **11,** 62
    ketones **30,** 28
  -, **cyclic**
  - startg. m. f.
    spiro-β-diketones **30,** 514
  Oxidolactones
  - special s.
    oxidotetrahydrofurano-
      lactones
  Oxidonitriles
  -, ring closures with −
    28, 983; **30,** 517
  - startg. m. f.
    ketones, cyclic **28,** 983
  α,β-**Oxidonitriles**
  - as intermediates **28,** 729
    suppl. **30**
  β,γ-**Oxidonitriles**
  - startg. m. f.
    2-ethylenealcohols **27,** 8
  α,β-**Oxidooxo compds.**
  - startg. m. f.
    pyrazole ring **26,** 349
  **1-Oxidopyridin-2-yldiazo-
    methane**
  - as alkylating agent, water-
    soluble **5,** 177 suppl. **29**
  1-Oxidopyridin-2-ylmethyl
    as protective group **30,** 12
  Oxidoreduction s. Redox
  5α,6α-**Oxidosteroids 28,** 113
  5β,6β-**Oxidosteroids 30,** 91
  Oxidosulfinic acid amides
    27, 682
  α,β-**Oxidosulfones**
  - from
    α-halogenosulfones **28,** 795
    suppl. **30**
  -, rearrangement **26,** 584
  α,β-**Oxidosulfonic acid amides
    28,** 222
  α,β-**Oxidosulfoxides 26,** 859
    suppl. **28**
  -, rearrangement **26,** 584
  Oxidotetrahydrofuranolactones
  -, rearrangement, stereospecific
    28, 3

2,3-Oxidothioethers 8, 676
 suppl. 28
α,β-**Oxidothiolic acid esters**
 – startg. m. f.
  α,β-ethylene-β-hydroxythiolic
   acid esters 27, 742
Oxime acetates
 –, Beckmann rearrangement
  4, 184 suppl. 27
 – startg. m. f.
  ketones 26, 192
 – **carbonates** as intermediates
  27, 513
 –, esterification with –
  30, 260
 –, peptide synthesis with –
  30, 260
 – **1,4-dianions**, isoxazoles
  via – 26, 792
Oximes
 – from
  nitro compds. 28, 644; 30, 326
  1,1-nitroethylene derivs.
   28, 37
  oxo compds. 4, 395; 28, 343;
   30, 277
 –, reaction with lead tetraacetate, review 23, 305
  suppl. 27
 – special s.
  acetylene-oximes
  acyl-
  ald-
  alkoxy-
  ammonio-
  cyclohexadienone –
  diene-
  di-
  dioxo compd. mon-
  halogen-
  hydroxy-
  isonitroso...
  nitro-oximes
  oxido- 25
  sulfonyl-
 – startg. m. f.
  alcohols 26, 185
  amine hydrochlorides 27, 17
  azine N,N'-dioxides 26, 260
  carboxylic acid amides
   s. a. Beckmann
   rearrangement
  – – – (unsubst. amides)
   29, 165
  N-halogenimines 30, 199
  isoxazoles 26, 792
  Δ²-isoxazol-5-ones 27, 777
  N-methylthiomethylnitrones
   27, 359
  nitro compds., aliphatic
   27, 96
  oxo compds. 26, 188;
   28, 154; 29, 201
–, **N-alkyl-** s. Nitrones
–, **O-alkyl-** s. Alkoximes
–, **cyclic**
 – startg. m. f.
  halogenonitriles 29, 515
  lactams, ring expansion
   26, 150
  N-nitrimines 26, 258
  O-vinyloximes 27, 122
anti-**Oximes**
 – special s.
  anti-aldoximes
syn-**Oximes**, ring closure
  with – 21, 427

Oxime thiocarbamates
 – startg. m. f.
  thiooxime carbamates
   27, 291
 – **tosylates**, equilibration
  17, 549 suppl. 29
 – –, **cyclic**
 – startg. m. f.
  lactams, ring expansion
   18, 205 suppl. 26;
   20, 116 suppl. 29
 – –, **N-heterocyclic**
 – startg. m. f.
  1,3-N,N-heterocyclics,
   2-alkoxy 28, 121
Oximino... s. a. Isonitroso...,
   Hydroximino...
2-Oximinoalcohols, cyclic
 – startg. m. f.
  cyanoaldehydes 27, 517
α-**Oximinoazomethines**
 – from
  enamines 30, 235
**2-Oximino-4(5H)-benzo-
 furanones, 6,7-dihydro-**
 – from
  1,3-cyclohexadiones and
   1,1-nitroethylene derivs.
    27, 774
α-**Oximinocarboxylic acid
  esters**
 – special s.
  α-halogen-α-oximino-
   carboxylic acid esters
 – **acids and derivs.**
 – from
  ketenes 28, 285
Oximinoethers, cyclic
 – from
  ethylenenitrites 30, 97
Oximinohalides s. Halogenoximes
o-**Oximino-N-heterocyclics**
 – special s.
  quinoxalines, 1,2,3,4-tetra-
   hydro-2-oximino-
1,2-Oximinohydroxylamines
 – from
  1,2-nitroximes 26, 19
Oximinoketones
 – from
  nitrous acid esters, cyclic
   27, 342
 – special s.
  α-isonitrosoketones
Oximinolactams
 – special s.
  1-alkoxy-2-oximinolactams
 –, **N-acyl-** 29, 315
Oximinolactones
 – from
  enolethers, bicyclic 29, 315
Oxindoles (s. a. Azaoxindoles,
  Dioxindoles)
 – from
  amines ar. 28, 762
  carbostyril-3-carboxylic acid
   esters, 3,4-dihydro-
   3,4-oxido- 28, 988
  indolenines, 3-halogeno-
   29, 153
  indoles 29, 153
  nitrocarboxylic acids
   30, 331
  quinolines, 1,4-dihydro-
   3-cyano- 29, 231
 –, 3-hydroxy- 26, 504

 – special s.
  alkoxyoxindoles
  3-alkylideneoxindoles
 – startg. m. f.
  spiro-N-heterocyclics
   (from 2 molecules) 26, 922
–, **1-amino-**
 – from
  α-halogenocarboxylic acid
   halides and hydrazines,
    1,1-disubst. 28, 398
–, **1-hydroxy-**
 – from
  nitrocarboxylic acids 30, 331
Oxiranes s. a. Oxido compds.
**2β-Oxiranylindolo[2,3-a]-
 quinolizines, 1,2,3,4,6,7,12,12b-
 octahydro-**
 – startg. m. f.
  spiro-N-heterocyclics 28, 99
Oxo... s. a. Aldehyd..., Ket...
Oxoazonium salts
 – from
  N-oxide radicals 29, 996
 – special s.
  piperidinium chloride,
   1-oxo-2,2,6,6-tetramethyl-
 – startg. m. f.
  alkoxylamines 27, 125
μ-**Oxobis(triaminophosphonium
  salts)**
 – as reagents 30, 260
3-Oxocarbanions, masked,
  as intermediates 29, 799
Oxocarbonium fluoroborates
 – startg. m. f.
  β,γ-ethyleneketones 27, 762
 – **hexafluoroantimonates**
 – startg. m. f.
  aryl ketones 27, 762 suppl. 30
 – **salts** 21, 995 suppl. 27
o-**Oxocarboxylic acid esters**
 – startg. m. f.
  phthalides 28, 679
Oxocarboxylic acids
 – special s.
  aldehydocarboxylic acids
  ketocarboxylic acids
 – startg. m. f.
  acetallactones, polycyclic
   30, 112
Oxo compd. N-derivs.
 –, ring closures, oxidative,
  review 26, 232
 – startg. m. f.
  hydrocarbons 15, 94 suppl. 26
  oxo compds. 26, 188
**Oxo compd.-ethylene deriv.
  interconversion** 29, 978
 – **compds.**
 –, α-alkylation 28, 161
 –, cycloalkenylation of – 29, 866
 – from
  acylamines 27, 199
  alcohols OC↑H; 28, 226
  α-alkoxynitriles 28, 975
  alkoxystannanes 28, 226
  amines, prim. 24, 203
  aminocyclopropane ring
   28, 161
  cyanhydrins 30, 155
  1,3-dithianes 27, 230
  enol phosphates 26, 251
  ethylenealcohols 28, 128
  ethylene derivs. with
   rearrangement 27, 141
  α,β-ethyleneoxo compds.
   29, 44

Oxo 686

(Oxo compds.
— from)
formoxy compds. **30**, 184
glycols, oxidative cleavage **29**, 229
halides **27**, 830 suppl. **29**; **30**, 358
— and 1-alkylthioazomethinium salts **28**, 843
— and 1,1-sulfinylthioethers **27**, 830 suppl. **28**
hydrocarbons **26**, 157; **27**, 181
β-hydroxycarboxylic acid esters **10**, 244
hydroxy compds. OC⇈H
mercaptals **29**, 219
— and boranes **29**, 873
—, cyclic **27**, 230; **28**, 656; **30**, 569
nitro compds., aliphatic **28**, 157; **29**, 199
aci-nitro compds. **30**, 572
4H-1,3-oxazine ring, tetrahydro- **29**, 257
oxetanes **29**, 978
oximes **26**, 188; **28**, 154; **29**, 201
oxo compd. N-derivs. **26**, 188
siloxycyclopropane ring **29**, 829
1,1-sulfinylthioethers **27**, 830; **29**, 667
s-trithianes **29**, 219
trityl ethers **28**, 232
—, Grignard synthesis of — **22**, 857 suppl. **27**
—, hydrates s. 1,1-Diols
—, hydrosilylation **14**, 150; **17**, 160
—, α-methylation **29**, 829
—, photochemistry, review **19**, 764 suppl. **28**
—, reactions with 1,1-dihalides **24**, 844
halogenosulfonyl isocyanates **30**, 211
isoxazoles, 3,5-dihydroxy- **29**, 723
phosphorus(III) acid derivs., review **23**, 797 suppl. **28**
—, reduction **24**, 126 suppl. **29**
— special s.
acetyleneoxo compds.
aldehydes
alkoximinooxo compds.
(alkylthio)oxo —
aminooxo —
carbohydrates
cyanooxo compds.
diazooxo —
dioxo —
ethyleneoxo —
halogenooxo —
hydroxyoxo —
ketones
oxidooxo compds.
polycarbonyl —
quinones
silyloxo compds.
— startg. m. f.
acetals **28**, 141; **30**, 111
acetylene derivs., insertion of 1 C-atom **28**, 731
α-acoxycarboxylic acid amides **27**, 709
α-acoxynitriles **27**, 709

acylamines **29**, 316
β-acyl-γ-lactones **30**, 555
acylhydrazones **27**, 377
alcohols, reduction **24**, 126 suppl. **29**
—, synthesis **30**, 624
—, heterocyclic **29**, 881
alkoxysilanes **29**, 107
allenes, synthesis **29**, 798
amines, sec., synthesis **30**, 463
—, tert. **28**, 346
α-aminocarboxylic acids-1-¹⁴C **30**, 568
α-aminonitriles, synthesis with addition of 1 C-atom **27**, 781; **29**, 750
aryl derivs. **30**, 683
azines **27**, 421; **30**, 566
azomethines **26**, 456; **28**, 356, 363, 371, 390, 504; **30**, 468
benzodioxepines, 4,5-dihydro-7-hydroxy- **28**, 84
4H-1,3-benzoxazin-4-ones, 2,3-dihydro- **29**, 374
carboxylic acids and derivs., synthesis s. Reformatskii synthesis
—, synthesis with addition of 1 C-atom **28**, 729
cyanohydrins, labeled **30**, 568
cyclopropane ring **26**, 803
1,1-diacoxy compds. **27**, 759
2-diazoalcohols **26**, 668
6H-dibenzo[d,f][1,3]-diazepines, 5,7-dihydroxy- **29**, 362
1,1-difluorides **28**, 506; **30**, 365
1-(difluoroamino)alcohols **26**, 285
1,1-dihalides **29**, 518
2,2-dihalogenalcohols **30**, 449
1,3-dioxolanes **30**, 72
enamines **29**, 437
enisonitriles **29**, 870
enolethers **29**, 190; **30**, 111
—, rearrangement **30**, 103
—, synthesis with addition of 1 C-atom **14**, 876
enoxysilanes **29**, 829
ethers **24**, 126
2-ethylenealcohols **28**, 652
3-ethylenealcohols, synthesis **29**, 619
α,β-ethylenealdehydes, synthesis with addition of 2 C-atoms **29**, 867
α,β-ethylenecarboxylic acids **29**, 874
— acid thioamides **26**, 903
ethylene derivs., synthesis **28**, 856; **29**, 855
— —, sym. (from 2 molecules) **28**, 761; **30**, 561, 566
α,β-ethyleneketones, synthesis **28**, 857 suppl. **29**
α,β-ethylenenitriles **26**, 782; **28**, 856 suppl. **29**
α,β-ethyleneoxo compds. **29**, 912
— — (from 2 molecules) **27**, 786
α,β-ethylenesulfones **23**, 776
α,β-ethylenesulfoxides **29**, 733
1,2-halogenhydrins, synthesis **29**, 616

α-halogenocarboxylic acid esters, synthesis with addition of 1 C-atom **29**, 738
α-halogenoglycidic acid esters **26**, 886
β-halogeno-α-ketocarboxylic acid esters **26**, 886
α-halogenoketones, insertion **29**, 871
2-halogenothiolurethans **28**, 618
1,3-N,N-heterocyclics, review **20**, 289 suppl. **28**
hydrazines **13**, 104 suppl. **26**
hydrazones **26**, 940
hydrocarbons, reduction HC⇊O
—, synthesis **27**, 872
α-hydroxyaldehydes, synthesis with addition af 1 C-atom **27**, 830
β-hydroxycarboxylic acid esters **26**, 673; **27**, 884; **29**, 629
— acids **26**, 664; **27**, 884
— —, Reformatskii synthesis, modified **2**, 827 suppl. **26**; **27**, 670
α-hydroxycarboxylic acid thioamides **26**, 623
β-hydroxynitriles **26**, 672; **29**, 959
α-hydroxyphosphine oxides **29**, 614
α-hydroxyphosphinc acids **22**, 656
β-hydroxy-α-sulfinylcarboxylic acid esters **30**, 455
1,1-hydroxysulfonic acids **8**, 641 suppl. **28**
β-hydroxysulfoximines **29**, 855
2-hydroxythioethers **29**, 262 suppl. **30**
β-hydroxythiolic acid esters **28**, 862
4-imidazolidones **28**, 275
2-isocyanoalcohols **26**, 667
ketene mercaptals **28**, 857
γ-lactones **5**, 479 suppl. **26**; **28**, 720
methylene compds. **13**, 831 suppl. **29**
nitriles, synthesis with addition of 1 C-atom **29**, 480
nitro compds., aliphatic **24**, 62 suppl. **27**
oxazoles **28**, 841
oxazolidines **28**, 335
2-oxazolidinethiones **29**, 632
Δ²-oxazolines **28**, 606
Δ³-oxazolines **30**, 458
oxido compds., sym. (from 2 molecules) **22**, 803
oximes **4**, 395; **28**, 343; **30**, 277
pinacols (from 2 molecules) **30**, 561
α-siloxynitriles **28**, 610; **30**, 630
1-(sulfonyl)eneformylamines **28**, 729
1,3-thiazine-6-thione ring, dihydro- **27**, 616

4H-1,3-thiazin-4-ones,
  2,3-dihydro- **28**, 551
thioenolethers **29**, 556
tosylhydrazones **29**, 67
Δ²-1,2,3-triazolines, 5-amino-
  26, 343
Δ³-1,2,4-triazolines,
  3-mercapto- **29**, 388
2,2,2-trihalogenalcohols
  30, 449
– –, **branched**
–, synthesis with 4H-1,3-
  oxazines, 5,6-dihydro- **25**, 612
– –, **N-heterocyclic**
  s. Oxoheterocyclics
3-Oxo-1,2-en-1-olethers
  s. 3-Keto-1,2-en-1-olethers
Oxohalides s. Halogenooxo
  compds.
Oxo-N-heterocyclics
–, hydrogenation, total **26**, 67
– special s.
  (N-vinyl)oxo-N-hetero-
  cyclics
α-**Oxo-N-heterocyclics 29**, 207
– by C-cleavage **15**, 242
– from
  carboxylic acids **25**, 669
  o-carboxy-N-oxides **26**, 256
–, ring expansion with diazo
  compds. **18**, 848; **29**, 783
– special s.
  1,1-alkoxyacylamines, cyclic
  2-azetidinones
  carbostyrils
  naphthostyrils
  2-piperidones
  pyrazole ring, 2-acyl-,
  N-condensed
  2-pyridones
  2-pyrrolidones
  2-pyrrolones
Oxonitriles
–from
  α-hydroxyoximes, cyclic
  19, 562
– special s.
  cyanoaldehydes
  cyanoketones
Oxonium fluoroborates
  s. Trialkyloxonium
  fluoroborates
– **salts** (s. a. Acoxonium salts,
  Trialkyloxonium –)
  hydrazines, monosubst.
  28, 351
– –, **cyclic**
– special s.
  dioxolanium salts
  pyrylium –
  xanthylium –
Oxophlorins 29, 143
Oxothioethers
– special s.
  ketothioethers
2-Oxothioethers, syntheses
  with – **29**, 851
Oxyazo compds.
– from
  azoxy compds. **4**, 183
Oxygen (Air) (s. a. under
  Cobaltocene) **13**, 314 suppl.
  29; **14**, 858 suppl. **28**; **22**, 288
  suppl. **26**; 23, 131 suppl. **27**;
  26, 109, 131, 143, 463, 481,
  503, 683, 881; **27**, 29, 138, 221,
  236, 270, 365, 637, 765, 896;

917; **27**, 444 suppl. **29**; **28**, 26,
82, 99, 129/30, 188, 322, 429,
705, 877, 889-91, 938; **28**, 860
suppl. **29**; **29**, 96, 122, 133,
147, 201, 220, 225, 228, 283,
320, 377/8, 500, 599, 716, 750;
29, 719 suppl. **30**; **30**, 63, 80,
83, 95, 155, 158/59, 223, 275,
603, 622
–, **benzylic**, removal (s. a.
  Hydrogenolysis, benzylic)
  27, 88; **30**, 49
–, **singlet 27**, 179; **29**, 96, 147,
  220, 240; **30**, 86
–, reactions with S-heterocyclics **26**, 125
–, review **26**, 125; **30**, 86
–, quenchers, tert. amines as –
  29, 147
–, source, 1-phospha-2,8,9-
  trioxaadamantane ozonide
  as – **30**, 86
–, acceptor, triethyl phosphite
  as – **27**, 515
N→P-Oxygen transfer 29, 120
N-Oxyls s. N-Oxide radicals
Oxymercuration-demercuration
  28, 117; **29**, 172
–, regio- and stereo-specific
  29, 142
Ozone 25, 111 suppl. **26**; **26**, 162,
218, 227, 230, 255, 361; **27**, 131,
149, 188, 245/6; **28**, 108/9, 192,
194, 235; **29**, 148, 199, 231; **30**,
46, 133, 145, 148/9, 167
–, review **30**, 149
Ozonides (s. a. Thioozonides)
– special s.
  indole-1,2-dicarboximide
  ozonides
  1-phospha-2,8,9-trioxaada-
  mantane ozonide
– startg. m. f.
  aldehydes and carboxylic
  acids **28**, 214
– hydrocarbons **30**, 46
Ozonization
– of
  acetals **28**, 235
  enoxysilanes **28**, 109
  suppl. **29**
  mercury compds., organo-
  26, 218
–, **stereospecific**
– of
  glycosides **28**, 235
Ozonolysis, abnormal
–, ketocarboxylic acids by –
  28, 192
–, **selective 11**, 308 suppl. **29**

Paddlanes 21, 728 suppl. **29**
Palladate s. Lithium tetra-
  chloropalladate(II), Sodium –
Palladium 26, 388 suppl. **30**; **26**,
  642; **28**, 288 suppl. **30**; **29**, 405;
  30, 49, 149
–, non-pyrophoric catalyst
  27, 61 suppl. **29**
–, prepared in situ **27**, 61
– -**alumina 29**, 51
– -**barium sulfate 26**, 73;
  27, 92
– – -/**cyclohexene 28**, 32

– -**calcium carbonate 27**, 58; **28**,
  42; **30**, 50
– -**carbon** HC⇊OC; HC⇊CC;
  HC⇅Hal; **1**, 596 suppl. **28**; **26**,
  12/3, 19, 38, 68, 195; **27**, 16,
  221, 392, 428, 456, 479, 521,
  768, 812, 819, 872, 990/1; **28**,
  57, 66, 128, 161, 194, 721, 894;
  29, 16, 65, 71, 164, 283, 324,
  446, 449, 739; **30**, 2, 33, 158,
  256
–/**cyclohexene 28**, 430
–/**p-cymene 29**, 179
–/**hydrogen** as catalyst **28**, 128
Palladium-thorium dioxide
  29, 54
Palladium(II) acetate 23, 819
  suppl. **27**
– **acetoacetonate 26**, 827; **27**, 724
– π-**allyl complexes** as intermediates **23**, 860 suppl. **28**
– **chloride** (s. a. Lithium palladium chloride) **20**, 519 suppl.
  28; **21**, 167 suppl. **29**; **22**, 420
  suppl. **28**; **25**, 62 suppl. **29**; **26**,
  143; **27**, 61, 725, 871; **29**, 340
– – -**copper(II) chloride 7**, 730
  suppl. **26**
– **chloride complexes 28**, 60
– special s.
  dichlorobis(acetonitrile)-
  palladium(II)
  dichlorobis(benzonitrile)-
  palladium(II)
  dichlorobis(triphenyl-
  phosphine)palladium(II)
– **complexes 24**, 859 suppl. **29**
– special s.
  bis(triphenylphosphine)-
  (maleic anhydride)-
  palladium
  tetrabis(triphenylphosphine)-
  palladium
– –, **azomethine-type**
– –, triarylphosphine- **27**, 851
  suppl. **30**
– **compds.** as catalysts,
  review **27**, 724
–(**II**) **cyanide 29**, 845
Palladium-lead-calcium
  carbonate 29, 66
Palladium oxide 29, 55
Parabanic acids
– from
  cyanamides **28**, 274
Paracyclophanes (s. a. Dithiaparacyclophanes) **25**, 691
  suppl. **28**
[8]Paracycloph-4-enes 30, 472
Paraffin as medium **27**, 679
Paraformaldehydes
  s. Formaldehyde
Participation s. Neighboring
  group participation
Paterno-Büchi reaction,
  intramolecular **26**, 153
Penicillin S-oxide ring,
  rearrangement **26**, 629
–, opening to sulfenic acid
  silyl esters **29**, 528
– **S-oxides**
– startg. m. f.
  cephalosporins **29**, 585
  ring skeleton
– from
  cephalosporin ring skeleton
  29, 542

(Penicillin ring skeleton)
- startg. m. f.
  cephalosporin ring skeleton
  29, 542
  1,4-thiazepin-7-one ring,
  2,3,4,7-tetrahydro- **27**, 604
  Δ^2-thiazolines **26**, 500
Pentabromocarbonates 26, 721
**Pentacarbonylchromium(0)-
carbene complexes**
- as reactants **30**, 636
1,3-Pentadiene as reagent
28, 687
Pentalenes, 1,2-dihydro-3-amino-
- by cycloisomerization **27**, 732
Pentamethine dyes
- startg. m. f.
  β-(1,2,3-triazol-4-yl)acroleins
  22, 281
N,N-Pentamethylenethiourea
- as reactant **24**, 34 suppl. **26**
Pentoses
- from
  hexoses **27**, 277
Peptide amides, Merrifield
synthesis **1**, 307 suppl. **27**;
13, 364 suppl. **29**
- -mediated polymer support
  14, 157 suppl. **29**
Peptide 4-picolyl esters
- as basic handle in peptide
  synthesis **9**, 860 suppl. **26**
Peptides (s. a. Carboxylic acid
amides, subst., Peptoids)
- by di- and oligo-merization
  26, 365
-, cleavage with neighboring
  group participation
  26, 28 suppl. **29**
- from
  carboxylic acid esters, active
    p-methylsulfonylphenyl
    esters **18**, 434 suppl. **26**
    o-nitrophenyl esters
    14, 450 suppl. **28**
    2-pyridyl esters **23**, 386
    suppl. **26**
    8-quinolyl esters, 5-chloro-
    16, 424 suppl. **27**, **29**
  - - -, comparison of
    active esters **16**, 424
    suppl. **26**
  peptide salts **28**, 21
  thiolic acid esters **23**, 386
  suppl. **26**
-, photo-C-alkylation **29**, 720
-, removal of protective
  groups s. Protective groups,
  removal
- special s.
  β-amino acid peptides
  cysteine –
  histidine –
  phosphonyl-
  poly-
  purin-6-yl –
-, synthesis, activation of
  amino groups by carbon
  dioxide **18**, 435 suppl. **27**
-, -, azide route, review
  17, 477 suppl. **29**
-, -, by fragment coupling
  22, 394 suppl. **27**; **28**, 353
-, -, by in situ activation
  26, 398
-, -, catalysis by N-hydroxy
  compds. **16**, 424 suppl. **29**

-, -, comparison of solid-
  phase and solution methods
  22, 394 suppl. **29**
-, -, coupling reagents,
  review **28**, 353
-, -, handle, basic in – **9**, 860
  suppl. **24**; **22**, 395 suppl. **29**
-, -, in aq. medium **8**, 694
  suppl. **26**
-, -, neighboring group
  participation in – **23**, 386
  suppl. **26**
-, -, retention of optical
  activity s. under Stereoisomers
-, -, review **26**, 398; **28**, 353;
  30, 271
-, synthesis with
  acyl hydroxamates,
    polymeric **18**, 435 suppl. **28**
  azidotris(dimethylamino)-
    phosphonium hexafluoro-
    phosphate **28**, 144 suppl. **26**
  benzhydroximic acid chloride
    19, 540 suppl. **26**
  benzotriazolyl-N-oxytris-
    dimethylaminophospho-
    nium hexafluorophosphate
    28, 144 suppl. **30**
  N-carbalkoxyammonium
    fluoroborates, quaternary
    30, 263
  N,N'-carbonyldi-(1,2,4-
    oxadiazol-5-ones) **14**, 452
    suppl. **28**
  N-(chlorophosphoryl)-
    pyridinium betaines
    27, 419
  diphenyl phosphoryl azide
    28, 353
  – sulfite **30**, 270
  μ-oxobis(triaminophospho-
    nium salts) **30**, 270
  phosphonitrile chloride
    24, 414 suppl. **28**
  phosphorous compds. and
    carbon tetrachloride
    27, 418
  polymer-N-hydroxy-
    succinimide esters
    18, 435 suppl. **28**, **29**
  Δ^2-pyrazolines, 4-acyloximino-
    19, 540 suppl. **29**
  quinolines, 2-alkoxy-N-
    alkoxycarbonyl-1,2-dihydro-
    23, 415
  silicon tetrachloride
    14, 459 suppl. **26**
  solubilizing protective
    groups **21**, 426 suppl. **27**
  o-sulfobenzoic anhydride
    30, 271
  N-sulfonyloxy compds.
    30, 271
-, -, asym. **29**, 55
-, -, multistep, one-pot,
  without polymer support
  18, 434 suppl. **29**
-, -, -, solid-phase s. under
  Solid-phase multi-step
  syntheses
- via
  N-acoxydicarboxylic acid
    imides, polymeric **18**, 435
    suppl. **27**
  2,5-thiazolidiones **8**, 694
    suppl. **26**
  5-thiazolones **16**, 384 suppl. **26**

-, **cyclic**
-, by the azide route **13**, 530
  suppl. **26**, **29**
-, comparison of cyclization
  methods **13**, 530 suppl. **26**
- special s.
  cystine peptides, cyclic
-, synthesis, review **13**, 530
  suppl. **28**
-, – on dual-function polymers
  20, 629 suppl. **26**
-, **labeled 22**, 394 suppl. **29**
Peptide salts
- startg. m. f.
  peptides **28**, 21
Peptoids
- special s.
  depsipeptides
  nucleoside peptides
Peracetic acid s. Peroxyacetic
acid
Perbenzoic – s. Peroxybenzoic
acid
Perbromides s. Phenyltrimethylammonium tribromide,
Pyrrolidone hydrotribromide
Perchlorates (s. a. Acyl
perchlorates, Halogen –)
- as reactants **26**, 540, 577;
  27, 484, 764; **28**, 358/9, 445;
  29, 431
N- and O-Perchloration 30, 195
Perchloric acid 1, 603 suppl. **29**;
4, 535 suppl. **28**; **9**, 296 suppl.
29; **17**, 231 suppl. **28**; **23**, 348
suppl. **27**; **24**, 728 suppl. **28**;
26, 180, 473, 821/2, 843, 998;
27, 162, 238, 248, 424, 809; **28**,
206, 764, 778; **29**, 10, 141, 192,
234, 241, 276, 527, 588, 667,
721, 841, 955; **30**, 105, 124, 511,
680
- as reactant **28**, 105
- **esters**
- special s.
  acoxyperchlorates
  trimethylsilyl perchlorate
- – –, **perhalogeno- 29**, 491
**Perchlorohomocubanecarboxylic
acid 28**, 749
Perfluoroacetone s. Hydrogen
peroxide-perfluoroacetone
Perfluoroalkenes
-, reactions with – **29**, 789
- startg. m. f.
  N-heterocyclics **29**, 789
-, **terminal**
- startg. m. f.
  α-fluorosulfinylperfluoro-
    carboxylic acid fluorides
    26, 580
(Perfluoroalkyl)arenes
- from
  iodides, ar. **26**, 867
**Perfluorocarboxylic acid
fluorides**
- special s.
  fluorosulfinylperfluoro-
    carboxylic acid fluorides
  perfluorodi- – –
Perfluoro compds. s. Fluoroxyperfluoro compds.
**Perfluorodicarboxylic acid
fluorides**
-, photodimerization
  26, 921
Perhalogeno compds.
s. a. Perfluoro...

Perhalogenoketones
– special s.
   perfluoroacetone
   1,1,1-trichloro-3,3,3-trifluoro-
   acetone
Perhydrolysis
– of
   oxido compds. **28,** 69
Perimidines
– from
   indan-1,3-diones, 2-acyl-
   27, 497
–, **2-amino- 26,** 434
Periodate 17, 22 suppl. **27; 26,**
   101; **27,** 277; **28,** 207/8, 514,
   711, 903; **29,** 155, 178, 246;
   30, 90
– in organic chemistry, review
   28, 711 suppl. **29**
Periodic acid 21, 273 suppl. **29;**
   26, 265
– in organic chemistry, review
   28, 711 suppl. **29**
Perkadox s. Diisopropyl
   peroxydicarbonate
Permanganate 26, 94; **27,** 104,
   186; **28,** 136; **28,** 198 suppl. **29;**
   29, 104, 283; **30,** 33, 315
–, oxidations in benzene **28,** 198
Peroxidase as reagent **29,** 904
Peroxidation s. Aminoperoxidation
Peroxide chemistry, reviews
   28, 110 suppl. **30**
Peroxides
– from
   hydroperoxides and diazo
   compds. **27,** 197
– as reagents **29,** 719
–, review **27,** 138 suppl. **30**
– special s.
   1-acylamino-1-peroxides
   acyl peroxides
   alkoxy –
   amino-
   di-*tert*-butyl peroxide
   dicumyl –
   halogeno-peroxides
   hydroxy-
–, **cyclic**
– startg. m. f.
   diols **30,** 1
–, syntheses with –, review
   23, 975 suppl. **29**
–, **unsym.**
– from
   ethylene derivs. **26,** 128;
   29, 141
–, *sec*-**alkyl- 21,** 109 suppl. **26**
Peroxyacetic acid 26, 123;
   28, 986; **29,** 152
Peroxybenzoic acid
   (s. a. m-Chloro-peroxy-
   benzoic acid, o-Sulfo-)
   26, 91
– special s.
   p-methoxycarbonylperoxy-
   benzoic acid
Peroxycarbamic acids
   s. N-Acylperoxycarbamic
   acids
Peroxycarboxylic acid esters
– special s.
   acetyleneperoxycarboxylic
   acid esters
   tert-butyl peroxybenzoate
β-**Peroxycarboxylic** – –
   29, 141

Peroxycarboxylic acids
– as reagents **26,** 529
– startg. m. f.
   acyl carbamyl peroxides
   26, 122
Peroxydicarbonates s. Diisopropyl peroxydicarbonate
Peroxydisulfate s. Persulfate
Peroxymaleic acid 26, 136
Peroxymercuration of ethylene
   derivs. **26,** 637
–, 1,2-halogenoperoxides via –
   30, 350
Peroxyphosphonic acid esters
   s. Monoperoxyphosphonic
   acid esters
Peroxyphthalic acid
   as reagent **30,** 85
Peroxysilanes 18, 160 suppl. **28**
Peroxytrifluoroacetic acid
   as reagent **26, 137,** 529;
   28, 491
– –/**boron fluoride,** oxidations
   with –, review **19,** 161
   suppl. **26**
Persulfate 26, 137; 27, 117, 907;
   28, 987; **30,** 89
–, oxidations with – **29,** 445
– special s.
   ammonium persulfate
**Perthiophosphonic acid
   anhydrides**
–, replacement, partial,
   of sulfur, cyclic **29,** 296
– startg. m. f.
   1,4-thiaphosph(V)orin
   4-sulfides, 2,6-diamino-
   29, 613
   2H-1,2-thiaphosph(V)orin
   2-sulfides, 3,6-dihydro-
   29, 612
Pesci reaction, improved
   12, 711 suppl. **28**
Pfitzner-Moffatt oxidation
   19, 307
pH s. Buffer soln., pH-Stat
Phanes s. Cyclophanes
Phase s. a. Solid-phase...
2-Phase medium
–, reactions in – **20,** 271 suppl.
   29; 26, 584 suppl. **28; 27,** 608;
   30, 117
Phase transfer catalysis
   (s. a. Ion pair extraction)
   21, 235 suppl. **29**
–, review **29,** 456; **29,** 456
   suppl. **30**
– – **catalysts** (s. a. Makosza
   reagent)
–, amines, tert. as –
   27, 833 suppl. **29, 30**
–, ammonium halides,
   quaternary as – **28,** 198
   suppl. **29**
   Aliquat 336 (methyltrioctylammonium chloride) **21,** 433
   suppl. **29; 28,** 798a suppl.
   29; 29, 400; **30,** 33, 366
–, evaluation **30,** 581
– special s.
   benzyltriethylammonium
   chloride
   crown polyethers
   hexadecyltributyl-
   phosphonium bromide
   hexadecyltrimethyl-
   ammonium bromide

   polyethers, azamacrobicyclic
   tetrabutylammonium bromide
– iodide
Phenalenium ring 26, 965
**Phenaleno[1,2-b]furan-7-ones,
   8,9-dihydro- 30,** 653
Phenanthrene ring
   by aromatization **28,** 921
– from
   spiroisocyclics, hydroxy-
   30, 657
Phenanthrenes
– by Pschorr ring closure,
   electrochemical **27,** 953
– from
   α-diketones **27,** 806
–, **3,4,9,10-tetrahydro-**
– from
   2H-naphtho[1,2-b]pyran-
   2-ones, 5,6-dihydro- **29,** 884
   α-tetralones **29,** 884
Phenanthridine N-oxides
– startg. m. f.
   phenanthridones **28,** 305
Phenanthridines 16, 957
   suppl. **28**
Phenanthridones
–, by photocyclization **29,** 913
– from
   biphenyl-2-carboxamides
   29, 445
   phenanthridine N-oxides
   28, 305
1,10-Phenanthroline as indicator
   28, 925
Phenanthrols 28, 905
Phenazines (s. a. Quinoxalines)
– by ring closure, reductive
   27, 510
– startg. m. f.
   phenazines, 1-alkoxy- **26,** 109
–, 1-hydroxy **26,** 109
–. **1-alkoxy-**
– from
   phenazines **26,** 109
–, 1-hydroxy-
– from
   phenazines **26,** 109
Phenol
– as reagent **27,** 475; **28,** 15;
   29, 402 suppl. **30**
– **acetates**
   dienones **27,** 190
– special s.
   phenyl acetate
Phenolalcohols
– special s.
   hydroxymethylphenols
Phenolbetaines 26, 381
Phenolesters (s. a. Carboxylic
   acid aryl esters) **16,** 603
– from
   arenes and acyl peroxides
– special s. **26,** 166
   catechol monoesters
   phenol acetates
– startg. m. f.
   phenolketones **19,** 798
   phenols **26,** 83 suppl. **29**
Phenolethers (s. a. Aryloxy...)
   28, 165 suppl. **29**
–, C-alkylation **27,** 836
–, cleavage **26,** 4; **27,** 10; **29,** 602
–, coupling, anodic **28,** 703
– from
   aryl halides **21,** 237; **30,** 134
   aryloxysilanes **26,** 217
   bromonium salts **29,** 202

(Phenolethers
- from)
phenols and carboxylic acid esters 29, 182
- special s.
α-aminophenolethers
diaryl ethers
- startg. m. f.
aryloxysilanes 29, 602
benzofuran-3(2H)-ones 28, 876
dibenzo[fg.op]naphthacenequinones 27, 906
dienones 27, 159 suppl. 29
hemiquinols 29, 233
quinazoline ring 29, 741
Phenolketones
- from
phenolesters 19, 798
phenols, oxidation 26, 160
p-Phenolketones
- from
phenols 29, 834
Phenolnitrones, cyclic 30, 159a
- startg. m. f.
benzofuro[2,3-c]pyrrole ring 30, 159a
Phenol ring
- from
cyclohex-2-en-4-olone ring 30, 546
α,β-ethylene- and α-hydroxyketones 30, 546
- retention 26, 43
- startg. m. f.
α,β-ethylene-γ-hydroperoxyγ-lactones 26, 227
ketones 26, 227
Phenols
-, addition to azomethines 26, 676
-, C-alkylation with alcohols 29, 756
-, - with ethylene derivs. 27, 686
-, p-aminomethylation 28, 786
- as reagents 27, 442
-, autoxidation 28, 130
- by aromatization, decarboxylative 28, 972
- - - with elimination of carbalkoxymethyl groups 29, 981
- - dehydrogenation 9, 919 suppl. 26
-, coupling, oxidative 27, 915; 28, 888
-, -, -, review 22, 895 suppl. 28
-, cyclodimerization, oxidative, stereospecific 27, 725
-, dearomatization 27, 159
- from
α,β-acetyleneketones 28, 730
arylcarboxylic acids 17, 298 suppl. 26
arylmercury compds. 27, 232
aryloxysilanes 29, 602
arylthallium bis(trifluoroacetates) 27, 232 suppl. 30
cyclimmonium salts, N-condensed 28, 982
2,5-cyclohexadienones, 4-halogeno- 27, 2
2-cyclohexenones 28, 721
diazonium salts 30, 117
diazo oxides 26, 550
dienones, cyclic 30, 649

α,β-ethyleneoxo compds. 28, 721
o-halogenophenols 26, 550
nitro compds., ar. 30, 118
phenolesters 26, 83 suppl. 29
- special s.
amino-phenols
bis-
catechol monoesters
catechols
diaryls, dihydroxy-(a-diketo)phenols
2,4-dinitrophenol
halogenophenols
o-hydroxycarbinols
naphthols
nitrosophenols
quinols
resorcinols
- startg. m. f.
arenes 29, 65; 30, 48, 50
catechol monoesters 29, 123
catechols 29, 123; 30, 67
chromans 27, 760, 790a
coumarins 27, 819
-, 3,4-dihydro- 27, 819
dienones 27, 159
diphenoquinones 26, 773; 27, 761
hemiquinolesters 28, 98
hemiquinolethers 25, 174 suppl. 27; 28, 98, 130
hemiquinols 28, 98
o-hydroxyacetophenones 30, 67
isoureas, O-aryl- 30, 50
naphthol, 5,6,7,8-tetrahydro-5,8-dialkyl- 28, 651
phenolethers 29, 182
phenolketones, oxidation 26, 160
p-phenolketones, synthesis 29, 834
quinone monoimines 27, 442
stilbenequinones 27, 761
sulfonic acids, ar. 26, 96
-, hindered
-, dimerization, oxidative 27, 761
-, oxidation, electrochemical 28, 98
Phenols, o-(ω-hydroxyalkyl)-
-, cyclodimerization, oxidative 28, 711
Phenoselenazines 26, 646
Phenothiazine ring with rearrangement 18, 338 suppl. 26; 26, 630
Phenoxaphosphine 10-oxides
- from
diaryl ethers 30, 541
Phenoxazines
- from
o-aminophenols (2 molecules) 28, 378
Phenyl acetate as reagent 27, 645
- azide as reactant 26, 987
- N-bromoketimine as oxidant 27, 253
- carbonates, N-carbalkoxylation with - 23, 412 suppl. 28
- cyanate as reactant 28, 704
- disulfide - - 30, 414
- formate, formylation of amines with - 8, 434 suppl. 26

- groups, elimination, oxidative 27, 280
Phenylhydrazine
- as reactant 26, 815
- as reagent 26, 815; 28, 408
Phenylhydroxylamine as reagent 26, 16
O-Phenylhydroxylamine, 2,4-dinitro- as reactant 30, 277
Phenyl iodosoacetate as reagent 26, 325; 27, 252
- phosphorodiamidate - - 29, 372
Phenylpropiolamidines as reactants 30, 340
Phenylsilane as reagent 21, 988 suppl. 29
Phenyl sulfones
- startg. m. f.
biphenyl-2-sulfinic acids (from 2 molecules) 29, 850
Phenyltetrafluorophosphorane 29, 512
2-(Phenylthio)ethanol
-, protection of phosphoryl hydroxyl groups with - 26, 104 suppl. 27
Phenyltrichlorosilane as reagent 29, 438
Phenyltrimethylammonium tribromide as reagent 27, 916; 30, 646
Pheoporphyrins by cyclodehydrogenation 29, 903
Phlorins
- special s.
oxophlorins
Phosgene
- as reagent 26, 540, 839; 27, 575; 28, 576; 29, 363, 370, 513; 30, 359
-, chemistry, review 28, 539
-, reaction with azomethines 27, 545
enamines 26, 769; 29, 714
- startg. m. f.
β-amino-α,β-ethylenecarboxylic acid chlorides 29, 714
- - esters 26, 769
carbonic acid esters, mixed 27, 404
N-chlorocarbonyl-1-alkoxyenamines, cyclic 27, 466
α-halogeneisocyanates 28, 472
heterocyclics 27, 488
isoquinoline ring, 1,3-dichloro- 26, 824
2-oxazolidones, 4-β-chloro- 30, 352
10bH-oxazolo[3,2-c][1,3]-benzoxazine-2(3H),5-diones 27, 353
pyrimidines,4,6-dihalogeno- 29, 489
quinolines 26, 765
1,3,4-thiadiazol-2(3H)-ones, 5-alkylthio- 28, 539
1,3,5-triazines, chloro- 26, 457
uracil ring 27, 349; 29, 333
- substitute s. Trichloromethyl chloroformate
Phosgeneimmonium salts
s. N-Dihalogenomethyleneammonium salts
Phosphabenzenes s. Phosphorins

4-Phosphabicyclo[3.1.0]hex-2-ene 4-oxides
- from
  2-phospholene 1-oxides,
  4-iodomethyl- **27**, 133
  phospholium salts, 1-iodomethyl- **27**, 133
2-Phosphaquinazolium salts
- from
  o-cyanophosphine imines
  26, 277
Phosphate buffer (s. a. Lithium phosphate buffer) **29**, 211, 630, 904; **30**, 55
Phosphates s. Phosphoric acid esters
Phosphatides
-, synthesis, review **25**, 88 suppl. 27
1-Phospha-2,8,9-trioxa-adamantane ozonide
- as singlet oxygen source **30**, 86
Phosphazenes, cyclic
-, chemistry, review **17**, 173 suppl. 27
- special s.
  phosphonitrile chloride
Phosphazo compds.
- special s.
  trihalogenophosphazo compds.
Phosphepines, dihydro- 26, 977
Phosphinealkylenes
  s. Alkylidenephosphoranes
Phosphine complex compds.
  s. under Nickel, Palladium, Rhodium, Ruthenium
- **dichlorides, polymer-bound,**
  as reagents **30**, 359
- **dihalides** (s. a. Phenyltetrafluorophosphorane, Triphenylphosphine dibromide) **22**, 567 suppl. 26
- **imines**
- special s.
  o-cyanophosphine imines
  o-hydroxyphosphine –
  N-sulfonylphosphine –
- startg. m. f.
  amines, sec. **26**, 291
  P-aminophosphonium salts
  26, 291
  1,1-iminocyanides **26**, 444
  tetrazoles **26**, 444
- **oxides**
- from
  phosphine sulfides **29**, 119
  phosphinic acid esters **27**, 658
- special s.
  enaminophosphine oxides
  hydroxyphosphine –
  thiocarbamylphosphine –
- startg. m. f.
  phosphine sulfides **28**, 550
-, synthesis **27**, 658
-, **sec.** s. Phosphinous acids
Phosphines (s. a. Diphosphines)
- as reagents s. under Rhodium
- from
  amines **29**, 609
  phosphine sufides **29**, 600
-, insertion into oxygen-oxygen bonds **29**, 106
-, nitration **27**, 364
- special s.
  amino-phosphines
  halogeno-

- startg. m. f.
  α-cyanophosphonium salts
  29, 606
  N-sulfonylphosphine imines
  30, 700
-, **cyclic**, as catalyst component
  20, 519 suppl. 28
-, **tert.**
- from
  phosphonium salts **26**, 994;
  27, 364
  phosphoranes, tetrahalogeno-
  7, 670 suppl. 26
- special s.
  poly-*tert*-phosphines
  tributyl-phosphine
  tricyclohexyl-
  triphenyl-
- startg. m. f.
  (α-azoalkylidene)-
  phosphoranes **29**, 611
  phosphonium salts **26**, 994;
  27, 364
Phosphine selenides
- special s.
  triphenylphosphine selenide
- **sulfides**
- as reactants **17**, 664 suppl. 27
- from
  phosphine oxides **28**, 550
- special s.
  enephosphine sulfides
  ynephosphine –
- startg. m. f.
  phosphine oxides **29**, 119
  phosphines **29**, 600
Phosphinic... s. a. Thiophosphinic...
- **acid chlorides**
- from
  chlorophosphines **26**, 99
- startg. m. f.
  phosphinic acid esters
  26, 99
- – **esters 26**, 102
- from
  chlorophosphines **26**, 99
  phosphinic acid chlorides
  26, 99
  phosphonous acid esters
  and halides **26**, 656
- special s.
  aminophosphinic acid esters
  halogenophosphinic – –
- startg. m. f.
  phosphine oxides **27**, 658
- **acids**
- from
  chlorophosphines **30**, 159
  phosphinous acid esters
  30, 159
  phosphonous acid esters and halides **26**, 656
- special s.
  hydroxyphosphinic acids
Phosphinous acid esters
- from
  chlorophosphines **30**, 159
- startg. m. f.
  ketones **30**, 159
  phosphinic acids **30**, 159
- **acids**
- startg. m. f.
  phospholan-3-one 1-oxide ring **29**, 731
1-Phosphinyl-4-acylthiosemicarbazides 27, 320

Phosphinylformic acid esters
- startg. m. f.
  α-hydroxyphosphine oxides
  29, 614
Phosphinylthiosemicarbazides
- from
  phosphonic acid hydrazides
  and isothiocyanates **27**, 320
- special s.
  1-phosphinyl-4-acylthiosemicarbazides
Phosphites (s. a. Phosphorous acid esters)
-, reaction with N-halides
  19, 367
- special s.
  dialkyl phosphites
  diphenyl phosphite
  pyrocatechyl phenyl –
  triisooctyl –
  trimethyl –
- startg. m. f.
  phosphonic acid esters
  4, 597 suppl. **26**; **27**, 660
  arylphosphonic – – **26**, 656
Phospholan-3-one 1-oxide ring
- from
  phosphinous acids and
  α,β-ethylenecarboxylic
  acid esters **29**, 731
2-Phospholene 1-oxides, 4-iodomethyl-
- startg. m. f.
  4-phosphabicyclo[3.1.0]hex-2-ene 4-oxides **27**, 133
Phospholes s. a. 3-Azaphospholes
Phospholium salts, 1-iodomethyl-
- startg. m. f.
  4-phosphabicyclo[3.1.0]hex-2-ene 4-oxides **27**, 133
Phosphonates s. N-(Aziridinyl)-phosphonates, Phosphonic acid esters
Phosphonic... s. a. Thionophosphonic...
- **acid anhydrides**
- s. Formamidinium-phosphonic acid anhydrides, Perthiophosphonic – –
- **derivs.**
- special s.
  hydroxyphosphonic acid derivs.
- – **dichlorides**
- from
  halides **26**, 651
  thionophosphonic acid
  dichlorides **28**, 78
- startg. m. f.
  diphosphonic acids **29**, 118
- – **diisocyanates**
- startg. m. f.
  P,N-heterocyclics **29**, 302
- – **dihalides**
- special s.
  halogenophosphonic acid
  dihalides
  phosphonic acid dichlorides
- – **esters** (s. a. under Replacement)
-, C-alkylation **16**, 866 suppl. **26**; **30**, 43
- from
  α-halogenophosphinic acid esters **28**, 798

(Phosphonic acid esters
- from)
  phosphites and halides
    26, 656
  - and N-heterocyclics **27,** 660
  phosphinic acid **20,** 129
    suppl. **26**
- special s.
  acetylenephosphonic acid
    esters
  acylaminophosphonic – –
  acylphosphonic – –
  alkylthiophosphonic – –
  aminophosphonic – –
  arylphosphonic – –
  benzenephosphonic – –
  diphosphonic – –
  ethylenephosphonic – –
  hydroxyphosphonic – –
  isocyanophosphonic – –
  ketophosphonic – –
  mercaptophosphonic – –
  thionophosphonic – –
  1,3,5-triazinephosphonic – –
- startg. m. f.
  2-acylaminophosphonic acid
    esters **29,** 880
  ethers, cyclic **30,** 613
  ethylene derivs. **29,** 880
- – –, **N-heterocyclic**
- from
  dialkyl phosphites and
    alkoxylcyclimmonium
    salts **27,** 654
- – –, **labeled 4,** 597 suppl. **29**
- – **hydrazides**
- startg. m. f.
  phosphinylthiosemicarbazides
    27, 320
- – **monoesters 22,** 106
    suppl. **29**
- – **acids**
- startg. m. f.
  phosphonic acid esters
    20, 129 suppl. **26**
1-Phosphonioiminoesters
- from
  1,1,1-ethylenenitrohalides
    30, 431
1-Phosphonio-1-selenonio-
   methylid salts
- from
  alkylidenephosphoranes
    and selenide dihalides
    30, 432
Phosphonitrile chloride
  (s. a. Phosphazenes)
- as reagent **8,** 517 suppl. **28;**
    28, 442
- –, condensation by – **27,** 802/3
- –, peptide synthesis with –
    24, 414 suppl. **28**
Phosphonium betaines as intermediates **27,** 942
- special s.
  sulfinovinylphosphonium
    betaines
- **fluorides,** Wittig synthesis
    with – **26,** 901
- **salts** (s. a. Phosphonio...)
- –, diene syntheses with –
    29, 647
- from
  phosphines, tert. **27,** 364
- special s.
  acetylenephosphonium salts
  acoxyphosphonium –
  alkoxyphosphonium –

  aminophosphonium salts
  azidophosphonium –
  benzotriazolyl-N-oxytris-
    dimethylaminophosphonium
    hexafluorophosphate
  bi(phosphonium salts)
  carbamylphosphonium –
  chlorotris(dimethylamino)-
    phosphonium perchlorate
  cyanophosphonium salts
  cyclopropylphosphonium
    fluoroborates
  1,3-dioxolan-2-yl-triphenyl-
    phosphonium bromide
  enephosphonium salts
  ethylenephosphonium –
  halogenophosphonium
    fluoroborates
  hexadecyltributyl-
    phosphonium bromide
  nitrosophosphonium salts
- startg. m. f.
  cyclopropanes **29,** 872
  ethylenealcohols **28,** 852
  2-ethylenealcohols **28,** 855
  α,β-ethylenecarboxylic acid
    esters **26,** 907
  ethylene derivs. **28,** 853
  nitriles **28,** 319
  (2-sulfonylalkylidene)-
    phosphoranes **30,** 404
- with rearrangement **30,** 404
  α-nitrosophosphonium salts
    28, 319
  phosphines, tert. **26,** 994;
    27, 364
  styrenes **29,** 875
- –, **cyclic**
- by cycloisomerization
    26, 735 suppl. **30**
- –, Märkl synthesis **28,** 597
- –, Wittig synthesis with –
    26, 686
- **ylids** s. Alkylidene-
    phosphoranes
- –, cyclic, review **27,** 656
Phosphonochlorides
- startg. m. f.
  monoperoxyphosphonic acid
    esters **27,** 120
1-Phosphonodiazo compds.
- special s.
  phosphonodiazomethane,
    dimethyl-
Phosphonodiazomethane,
  dimethyl-
- as reactant **28,** 731
Phosphonous acid esters
- special s.
  diethyl methylphosphonite
- startg. m. f.
  phosphinic acid esters
    26, 656
- acids **26,** 656
N-Phosphonyldicarboxylic
   acid imides
- from
  phosphites **30,** 201
(*terminal*-**Phosphonyl)peptides**
    23, 415 suppl. **29**
Phosphoramidates
- from
  azides and phosphorous acid
    esters **27,** 311
- hydrolysis, selective **27,** 117
- special s.
  halogenophosphoramidates
  phosphoric acid amides

- – monoamide monoesters
  phosphorodiamidates
  phosphoromonoamidates
  1-phosphoryl-1,4-dihydro-
    pyridines
- startg. m. f.
  amines **27,** 548
  N-thionylphosphorimidates
    28, 249
- –, **cyclic, stereoisomeric 30,** 160
- –, **mixed 27,** 311 suppl. **30**
Phosphoramidites
  s. Phosphorodiamidites
Phosphoranes
- special s.
  alkylidene-phosphoranes
  dialkoxy-
  enimino-
  imidino-
  keteniminylidene-
  phenyltetrafluorophosphorane
- –, **oxy-,** syntheses via –,
    review **20,** 204a suppl. **29**
- –, **tetrahalogeno-**
- startg. m. f.
  phosphines, tert. **7,** 670
    suppl. **26**
- –, **tributyl-**
- –, Wittig synthesis with –
    17, 892 suppl. **29**
α-Phosphoranylideneketones
- from
  cyclopropenones **28,** 589
Phosphorazides
- special s.
  diphenyl phosphorazidate
Phosphoric acid 2, 677 suppl.
    26; 2, 778 suppl. **29; 24,** 850
    suppl. **27; 26,** 109, 356, 360,
    961; **27,** 41, 359, 726, 744,
    800/1, 945; **28,** 378; **30,** 546
- **–/boron fluoride 26,** 887
- **–/phosphorus pentoxide**
    s. under Phosphorus
    pentoxide
- – **amides** (s. a. Phosphor-
    amidates)
- as reactants **26,** 378
- from
  phosphorous acid triamides
    26, 100
- –, metalation **30,** 583
- special s.
  hexamethylphosphoramide
  phosphoric acid diamide
    hydrazides
- – – monoamides
- startg. m. f.
  amines, tert. **28,** 352
  phosphoric acid diamide
    hydrazides **30,** 200
- – **derivs.**
- –, deoxygenation via – **28,** 65
- –, reactions with 2,6-dichloro-
    quinonechlorimine **27,** 4
- startg. m. f.
  α-hydroxyphosphonic acid
    derivs. **29,** 604
- – –, **cyclic 29,** 120
- – **diamide hydrazides**
- from
  phosphoric acid amides and
    azo compds. **30,** 200
- – **diesters 22,** 106 suppl. **29**
- – –, **mixed 28,** 75; **29,** 113
- from
  phosphoric acid monoamide
    monoesters **30,** 60

Pho

– – – monoesters and halides
  28, 170 suppl. 30
– via pyridinium phosphate
  betaines 26, 222
– – esters (s. a. Polyphosphoric
  acid esters)
–, C-allylation with – 29, 747
– from
  dialkyl phosphites and
    ketones 28, 523
  α-hydroxyphosphonic acid
    esters 28, 71
  phosphorous acid esters and
    nitrous acid esters 28, 80
– special s.
  acylphosphoric acid esters
  enol phosphates
  glycosyl phosphates
  hydroxyphosphoric acid
    esters
  phosphoric acid monoesters
  triaryl phosphates
  trimethyl phosphate
– startg. m. f.
  carboxylic acid esters 29, 222
  thioethers 30, 396
– – –, mixed 26, 107
– from
  phosphorodichloridates
    29, 117
–, one-pot process 29, 117
– – ester salts
– special s.
  tetramethylammonium
    dimethyl phosphate
– – hydrazides
– special s.
  phosphoric acid diamide
    hydrazides
– – monoamide monoesters
– startg. m. f.
  phosphoric acid diesters,
    mixed 30, 60
– – monoamides 29, 110
– – monoesters
  (s. a. Phosphorylation)
– from
  1,3,2-dioxaphospholanes,
    2-oxo-2-alkoxy- 29, 110
  halides 28, 170
– special s.
  p-nitrophenyl phosphate
– startg. m. f.
  phosphoric acid diesters,
    mixed 28, 170 suppl. 30
– via dithiophosphorous acid
  esters 27, 121
Phosphoric acids s. Nucleotides
– acid triamides s. Phosphoric
  acid amides
– – β,β,β-trichlorethyl esters
–, oligonucleotide synthesis
  via – 25, 88 suppl. 27
– – vinyl esters s. Enol
  phosphates
Phosphorimidates (s. a.
  N-Thionyl phosphorimidates)
  17, 498 suppl. 28
– startg. m. f.
  quinolines 28, 461
Phosphorins by rearrangement
  29, 710
Phosphorisocyanatidates
– special s.
  imidophosphoriso-
    cyanatidates

Phosphorisocyanatidites
– startg. m. f.
  acylimidophosphoriso-
    cyanatidates 27, 308
Phosphorochloridates
– special s.
  bis-(β,ββ-trichlorethyl)-
    phosphorochloridate
  diethyl phosphorochloridate
  phosphorodichloridates
– startg. m. f.
  1-acetylenephosphonic acid
    esters 28, 596
  phosphoroperoxidates 28, 76
–, cyclic
– special s.
  1,3,2-dioxaphospholane,
    2-oxo-2-chloro-
Phosphorochloridites
(s. a. Dithiophosphoro-
  chloridites)
– special s.
  pyrocatechyl phosphoro-
    chloridite
– startg. m. f.
  hypodiphosphorous acid
    esters 28, 580
Phosphorocyanidates
– special s.
  diethyl phosphorocyanidate
Phosphorodiamidates
– from
  alcohols 28, 65
– special s.
  phenyl phosphorodiamidate
– startg. m. f.
  hydrocarbons 28, 65
Phosphorodiamidites
– startg. m. f.
  carboxylic acid esters 28, 142
  halides 30, 356
Phosphorodichloridates
– from
  alcohols 28, 72
– special s.
  aminophosphoryl
    dichlorides
– startg. m. f.
  diphosphoric acid diesters,
    sym. 29, 118
  phosphoric acid esters,
    mixed 29, 117
Phosphorodithioates, S-acyl
  O,O-dialkyl
–, N-acylation with – 28, 402
  suppl. 29
Phosphorohalidates
– special s.
  phosphorochloridates
Phosphoromonoamidates
– from
  diaryl phosphites 28, 263
Phosphoroperoxidates
– from
  phosphorochloridates 28, 76
– startg. m. f.
  ketones 29, 248
Phosphorous acid amides
  s. Phosphorous acid
    triamides
– – derivs.
– startg. m. f.
  halides 30, 356
– – diesters (s. a. Dialkyl
    phosphites)
– startg. m. f.
  ethers 27, 195
– – –, mixed 29, 205

Phosphorous acid esters
(s. a. Phosphites, Thio-
  phosphorous acid esters)
–, reaction with quinone
  diimines 27, 303
– special s.
  phosphorous acid diesters
– – monoesters
– startg. m. f.
  α-alkylthiophosphonic acid
    esters 30, 439
  α,α-dialkoxy-
    phosphonic – – 29, 607
  α-mercaptophosphonic – –
    30, 439
  N-phosphonyldicarboxylic
    acid imides 30, 201
  phosphoramidates 27, 311
  phosphoric acid esters 28, 80
– – monoesters 26, 103; 29, 205
– – triamides
– special s.
  trisdiethyl-aminophosphine
  trisdimethyl-
– startg. m. f.
  phosphoric acid amides
    26, 100
Phosphorus 26, 72; 27, 800;
  30, 32
–, reductions with
  –/iodine/carboyxlic acids,
  review 21, 55 suppl 26
– startg. m. f.
  phosphorus compds.,
    organo-, review 23, 643
    suppl. 26
–(III) acid derivs.
–, reactions with oxo compds.,
  review 23, 797 suppl. 28
– acid esters s. Phosphorus
  esters
–-carbon bonds, formation
  27, 658
– compds.
– from
  phosphorus, review
    23, 643 suppl. 26
– special s.
  diazophosphorus compds.
–, stereochemistry, review
  25, 701 suppl. 26
  retention of configuration
  28, 550
– with methylene groups,
  active, review 23, 879
  suppl. 26
–(III) compds.
– as reagents 25, 412 suppl. 26
–, protection as phosphonium
  salts 27, 364
– esters
–, C-alkylation with – 27, 791
–(III) nucleophiles
– startg. m. f.
  aziridine ring,
    N-phosphorylated 26, 405
– oxide chloride HalC↓↑O; 11,
  583 suppl. 26; 26, 97, 242, 817,
  840/1, 951; 27, 164, 516, 826,
  946, 970; 28, 354/5, 442, 541,
  689, 754, 918-20; 29, 269, 376,
  757, 785/6, 936; 30, 99, 288,
  634, 654, 663
– as reactant 26, 97
– pentabromide 6, 587 suppl. 26;
  30, 362

(Phosphorus)
- **pentachloride 17,** 549 suppl. **29; 20,** 579 suppl. **27; 26,** 301, 543/4; **26,** 303 suppl. **28; 27,** 422, 583; **28,** 443, 502, 903, 909; **29,** 519; **30,** 360/1
- **pentasulfide 26,** 603; **27,** 614/5; **29,** 554, 557; **30,** 398, 410
-, reactions, review **29,** 557
- **pentoxide** (s. a. under Methanesulfonic acid) **26,** 356, 360; **27,** 444, 801; **28,** 914/5; **29,** 124, 298, 373, 931
- –/**phosphoric acids** s. Polyphosphoric acid
- **silicon** s. Siliconphosphorus
- **sulfide chloride** as reactant **27,** 113
- **tribromide 26,** 537; **27,** 171/2, 566, 804; **29,** 599
- **trichloride 29,** 593, 607
- as reactant **26,** 103; **30,** 541
Phosphoryl, β-elimination of – **29,** 155
Phosphorylation (s. a. Phosphonic acid monoesters from..., Redoxphosphorylation, Thiophosphorylation)
- of
    amines **28,** 262; **29,** 293
    hydroxy compds. **29,** 110
        alcohols **26,** 103, 105; **28,** 190
        - via dithiophosphorous acid esters **27,** 121
        phenols **8,** 146 suppl. **27**
    - with 8-(phosphoryloxy)-quinolines **29,** 113, 293
    p-nitrophenyl phosphate **28,** 190
-, **enzymatic 26,** 105
-, **oxidative 29,** 115
-, **preferential**
- of
    nucleoside hydroxyls **26,** 97, 105
1-Phosphorylation of carbohydrates 27, 114
N-Phosphorylaziridines
- from
    ethylene derivs. and N,N-dihalogenophosphoramidates **29,** 411
Phosphorylbetaines 29, 110
1-Phosphoryl-1,4-dihydropyridines
-, phosphorylation, oxidative, with – **29,** 115
8-(Phosphoryloxy)quinolines
-, phosphorylation with – **29,** 113, 293
N-Phosphorylpyridinium betaines
- special s.
    N-(chlorophosphoryl)-pyridinium betaines
Photo... s. a. Irradiation
Photo-C-alkylation of peptides **29,** 720
Photochemical C-acylation
s. C-Acylation, photochemical
- addition s. Addition, –
- decarbonylation
s. Decarbonylation, –
- reactions with nitro compds., ar. **30,** 84

- **ring closure** s. Photocyclization
Photochemistry, review **20,** 645 suppl. **26**
- of
    azo compds., review **22,** 894 suppl. **28**
    carbon-nitrogen double bonds, review **28,** 643
    heterocyclics, review **28,** 643
    oxo compds., review **19,** 764 suppl. **28**
-, **preparative,** review **24,** 557 suppl. **26; 29,** 913 suppl. **30**
Photocyclization
- of
    o-chloranil adducts **27,** 920
    enacylamines, review **25,** 522 suppl. **29**
    stilbene analogs, review **7,** 863 suppl. **26**
- with
    elimination of alkoxyl **29,** 913
-, **dehydrohalogenative 20,** 672
Photocycloaddition
- of
    thiocarbonyl compds., review **24,** 697 suppl. **26**
Photodecarboxylation
- of
    acoxy compds. **26,** 984
    carboxylic acids **27,** 93
    heterocyclics **27,** 534
Photodimerization
- of
    perfluorodicarboxylic acid fluorides **26,** 921
    pyridines **27,** 721
-, **stereospecific 27,** 704
Photofragmentation
- of
    spiro-Δ¹-pyrazolines **25,** 670
Photohydroxylation of ar. compds., review **24,** 118 suppl. **29**
Photoisomerization
- of
    heterocyclics, 5-membered, review **26,** 603
    isothiazoles **27,** 603
Photolysis
- of
    cyclopentadioxinones **27,** 921
    nitrites as radical source **28,** 933
Photooxidation 26, 157
Photoreactions
(s. a. Irradiation)
- of
    4-thiouracils **30,** 465
Photoreduction 26, 80
- of
    benzils, unsym. **7,** 679 suppl. **28**
    sulfoxonium ylids **27,** 82
Photo-Reimer-Tiemann synthesis 3, 670 suppl. **29**
Photo-sensitive s. Polymer support, photosensitive, Protective groups, –
Photosensitizers
s. Sensitizers
Phthalaldehydes
- startg. m. f.
    phthalans, 1-hydroxy-3-acylamino- **21,** 344 suppl. **26**

Phthalanimines
- from
    hydroxycarboxylic acid amides **24,** 521 suppl. **26**
Phthalans, 1-(azomethylimino)-
- from
    1H-2,3-benzoxazines, 4-hydrazino- **28,** 294
-, **1-hydroxy-3-acylamino-**
- from
    phthalaldehydes and carboxylic acid amides **21,** 344 suppl. **26**
Phthalazine ring, 1-amino-
- from
    isoindolenine ring, 3-alkoxy- **26,** 350
Phthalazines, 1,2-dihydro- 23, 57 suppl. **28**
Phthalazones
- from
    2,3-benzoxazin-1-ones **30,** 268
    phthalides **12,** 427 suppl. **29**
Phthaldialdehydes
s. Phthalaldehydes
Phthalic acid... s. a. o-Dicarboxylic acid...
- – esters
- startg. m. f.
    indan-1,3-diones **27,** 767
1-(1'-Phthalide)isoquinolines
- from
    isoquinoline-1-spiro-2'-(1',3'-indandiones), 1,2-dihydro-2-acyl- **30,** 44
Phthalides
- from
    carboxylic acids **26,** 818
    o-oxocarboxylic acid esters **28,** 679
- special s.
    3-alkylidene-phthalides
    amino-
    hydroxy-
- startg. m. f.
    phthalazones **12,** 427 suppl. **29**
Phthalimides
- from
    alcohols **27,** 429
    amines, prim. **29,** 417
    2,3-benzoxazin-1-ones **30,** 268
    halides **27,** 480
    phthalimidines **27,** 183
-, protection of amino groups, prim., as – **27,** 403
- special s.
    N-sulfonylphthalimides
Phthalimidines 26, 473
- from
    o-dioxo compds. and isocyanates **27,** 490
- special s.
    3-alkylidenephthalimidines
- startg. m. f.
    phthalimides **27,** 183
Phthalocyanines
- from
    o-dinitriles **27,** 435
    isoindolines, 1,3-diimino- **27,** 435
Picolyl esters s. Peptide picolyl esters
2-Picolyl, 1-oxy- as N-protective group **22,** 203 suppl. **27**
Picolyloxycarbonylhydrazides 22, 395 suppl. **29**

Picric acid as reagent **27**, 763
Pictet-Gams isoquinoline ring closure 28, 920; **29**, 931
Pictet-Spengler ring closure **15**, 404; **28**, 748
– – –, stereospecific **8**, 823 suppl. **26**
Piloty pyrrole ring closure, modified 4, 843 suppl. **29**
Pinacols
– from
  oxo compds. (2 molecules) **30**, 561
– special s.
  benzopinacol
(+)-**2-Pinene** in asym. synthesis **28**, 621
2,3-Piperazinediones
– startg. m. f.
  1H-1,4-benzodiazepines, 2,3-dihydro- **29**, 479
2,5-Piperazinediones 27, 454
–, 1-acyl-3-alkylidene **29**, 891
–, 1,4-diaryl- **28**, 392
–, 3,6-epidithio- **29**, 527
Piperazines
– from
  keteneacetals (2 molecules) and sulfonic acid azides (2 molecules) **27**, 431
Piperazinone ring opening 8, 171 suppl. **26**
Piperidine, 2,2,6,6-tetramethyl- 29, 139
Piperidine-3-carboxylic acids
– startg. m. f.
  3-methylene-2-piperidones **28**, 912
Piperidines 28, 805
– from
  aminoalcohols **26**, 467
– startg. m. f.
  pyridines, 1,2,3,4-tetrahydro- **28**, 886
–, 1-acyl-3-hydroxy-
– startg. m. f.
  aldehydes **27**, 247
–, 2,6-dicyano- **29**, 758
Piperidinium chloride, 1-oxo-2,2,6,6-tetramethyl-
– as reagent **29**, 176
4-Piperidone, 1-methyl-
as hydride acceptor **2**, 286 suppl. **27**
– ring **26**, 989
– from
  dienones, cross-conjugated **27**, 328
2-Piperidones
– from
  isocyanates **26**, 726
– special s.
  3-methylene-2-piperidones
4-Piperidones
– startg. m. f.
  quinuclidines, Δ²-dehydro-4-hydroxy- **30**, 471
Piperylene s. 1,3-Pentadiene
Pivalic acid as reagent **27**, 795
– – derivs. as catalysts **28**, 81
– special s.
  boryl pivalate
Plasma chemistry s. Glow discharge, reactions by –

Platinum (s. a. Hexachlorophatinic acid) **27**, 29; **29**, 218
–, catalysts **13**, 314 suppl. **29**; **26**, 940; **27**, 15, 17, 281; **28**, 118; **29**, 21, 371; **30**, 13
– -asbestos **22**, 664 suppl. **28**; **26**, 956
– -carbon **29**, 424; **30**, 331
– -nickel-alumina **26**, 39
– compds., zerovalent
  tris(triphenylphosphine)-platinum(0) **29**, 158
– complexes, protection of amino groups as – **30**, 361
–(IV) chloride **26**, 752
Plumbanes
–, cleavage with 1-chlorobenzotriazole **27**, 310
Polonovski reaction 27, 939
Polyalcohols
– special s.
  triols
– startg. m. f.
  polyamines **27**, 401
Polyalkyl compds.
– from
  polyhalides **29**, 849
Polyamines (s. a. Crown polyamines)
– from
  polyalcohols **27**, 401
Poly-β-carbonyl compds.
–, ring closures **27**, 924
–, synthesis **28**, 734
Polycyclics 28, 640 suppl. **29**
–, rearrangements, silver ion-catalyzed, review **28**, 691 suppl. **30**
– special s.
  polyisocyclics
(Polyen)amines
– special s.
  fulvenes, 6-(4-amino-1,3-dien-1-yl)-
Polyenes (s. a. Cyclopolyenes, Trienes)
Polyenynes, synthesis with isocyclic ring opening **30**, 601
Polyethers s. a. Crown polyethers
–, azamacrobicyclic
– as phase transfer catalysts **27**, 833 suppl. **30**
Polyethyleneimine as reagent **26**, 647
Polyfluorides
–, diazotization **7**, 408 suppl. **21**
– special s.
  diarylmercury compds., polyfluoro- s. under Mercury compds., organovinyllithium, trifluoro-
Polyhalides
–, C-cleavage **27**, 988
– from
  ethylene derivs., synthesis **19**, 776
–, reaction with iron(II) chloride/dimethyl sulfoxide **17**, 887 suppl. **26**
– special s.
  p-bis(trichloromethyl)-benzene

cyclohexa-2,5-dienone, 2,4,4,6-tetrabromoperfluoro…
polyhalogenoalkanes
pyrazolo[1,5-a]pyridinium salts, polyhalogenotrihalides
– startg. m. f.
  polyalkyl compds. **29**, 849
–, synthesis **27**, 863
–, ar. **30**, 376
Polyhalogenoalkanes s. Polyhalogenomethanes, Tetrahalogenomethanes
Polyhalogenocarboxylic acid chlorides
–, C-acylation with – **26**, 846
– from
  ethylene derivs. **22**, 717 suppl. **26**
Polyhalogenodiazabutadienes
–, reactions with – **27**, 529
– special s.
  trihalogenodiazabutadienes
Polyhalogenoethylene derivs.
– special s.
  perfluoroalkenes
– startg. m. f.
  acetylene derivs. **27**, 961
Polyhalogenomethanes
– special s.
  bromoform
  bromotrichloromethane
  carbon tetrabromide
  – tetrachloride
  chlorodifluoromethane
  chloroform
  trichlorofluoromethane
– startg. m. f.
  quinone methids **28**, 796
(Polyhalogeno)silanes, rearrangement **26**, 523
Poly-S-heterocyclics
–, rearrangement, skeletal **29**, 966
– special s.
  (poly-S)-heterocyclics
Poly(hydrogen fluoride)-pyridine
–, reactions with – **24**, 582 suppl. **29**; **29**, 494
Poly(hydrosiloxane) as hydride donor **24**, 126 suppl. **26**
– special s.
  polymethylhydrosiloxane
Polyisocyclics, highly-condensed (s. a. Cubane derivs.) **26**, 758; **29**, 680
–, ring contraction **29**, 92
Polymer, microspherical, as reactant-substrate **11**, 12 suppl. **27**
Polymer-based carboxylic alkoxyformic anhydrides
– as acylating agents, mild, selective **19**, 233 suppl. **29**
– catalysts s. under Catalysts
– reagents
–, reactions with – **29**, 186
– special s.
  aluminum chloride-polymer carbodiimides, polymer-based
–, polymeric hydroquinone, polymer-bound

(Polymer-based reagents
- special s.)
  ion exchangers
  peroxybenzoic acid,
    polymer-based
  phosphine dichlorides,
    polymer-bound
  poly-(3,5-diethylstyrene)-
    sulfonyl chloride
  thioanisole, polymer-based
  tin hydrides, organo-, –
  titanocene dichloride, –
  triphenylphosphine, –
- **sensitizers** s. Sensitizers,
    polymer-based
- **sulfonic acid azide**
    s. Sulfonic acid azide,
    polymer-based
Polymer-N-hydroxysuccinimide esters
–, peptide synthesis with –
  18, 435 suppl. **28**, 29
Polymerization s. a. Oligomerization
- **inhibitors**
  tert-butylcatechol **24**, 981 suppl. 29
  hydroquinone **27**, 697;
    28, 869; **29**, 182;
    30, 469, 484
  phenothiazine **27**, 599
Polymers
- as reactant-substrate
    s. Merrifield syntheses
- as reagents, review
    28, 70a suppl. 29
- special s.
  N-acoxydicarboxylic acid
    imides, polymeric
  acyl hydroxamates, –
  carbodiimides, –
  N-chloronylon-66
  N-chloropolymaleimide
  poly(hydrosiloxane)
–, **dual function**
–, synthesis of peptides, cyclic, on – **20**, 629 suppl. 26
–, **soluble**
–, as protective groups in oligonucleotide synthesis **17**, 169 suppl. 29
–, as reactant-substrate in peptide synthesis **12**, 455 suppl. **28**; **21**, 426 suppl. 27
Polymer support (s. a. under Merrifield syntheses)
–, diol monoethers by use of – **12**, 287 suppl. 28
–, ester condensation, mixed, on – **6**, 738 suppl. 27
–, protection, partial, of dialdehydes on – **12**, 205 suppl. 29
– special s.
  rhodium complexes,
    polymer-supported
–, synthesis, asym., on –
  9, 741 suppl. **29**;
  26, 671 suppl. 28
–, **peptide-mediated 14**, 157 suppl. 27
–, photo-sensitive, removal **30**, 5
– **supports, insoluble**,
  in organic synthesis, review
  9, 741 suppl. 29

Polymethine cyanines 27, 426
- from
  triformylmethane **28**, 718
- special s.
  tetramethine cyanines
Polymethylhydrosiloxane
–, reductions with – **13**, 70 suppl. **28**; **28**, 45
Polynitriles s. Tetracyano...
Polynitro compds., aliphatic
  s. a. 1,1,1-Trinitro compds.
Polynucleotides (s. a. Oligonucleotides)
–, degradation **29**, 155
–, synthesis **25**, 88 suppl. 26
Polyoxo compds. s. Polycarbonyl compds.
Polypeptides, sequential 27, 193
Poly-*tert*-phosphines
- from
  enephosphine sulfides **29**, 600
Polyphosphoric acid
  (s. a. Phosphorus pentoxide-methanesulfonic acid)
  CC↑O; **3**, 170 suppl. **26**; **5**, 260 suppl. **26**; **26**, 308, 371, 380, 502, 735, 816, 989; **27**, 163, 483, 538, 641, 799, 989; **28**, 217, 441, 672; **29**, 375, 467, 756, 840, 909; **30**, 509
–, –, 115%- **26**, 735 suppl. 30
– – esters as reagents **12**, 465 suppl. **26**; **19**, 847 suppl. **26**; **27**, 943; **29**, 374
Polysaccharides
  s. Di-saccharides, Tri-
Polyspiro compds. 14, 858 suppl. 27
Polystyrenetricarbonyl-chromium as catalyst **30**, 25
Polysulfides s. Trisulfides
Pomeranz-Fritsch isoquinoline ring closure 27, 940
–, **modified 29**, 975
Ponsold aziridine ring closure 21, 564
Porphins (s. a. Phlorins, Porphyrins)
–, diene synthesis with –
  21, 728 suppl. 28
Porphyrins
  special s.
  Pheoporphyrins
Position shift (s. a. Interchange, positional, Migration, Rearrangement)
- of
  alkoxy groups **18**, 338 suppl. 26
  cyano groups **27**, 730 substituents
  - of indoles **29**, 153
  - of isocyclics **17**, 879 suppl. **26**; **23**, 458 suppl. **26**; **27**, 935; **28**, 169
- with replacement of halogen by alkoxy groups **27**, 212
- with N-transalkylation **30**, 320
–, –, **double**, of substituents of isocycles **29**, 971
Potassium
- as electron donor **27**, 459
–/**ammonia, liq. 26**, 43
– -**graphite 24**, 699 suppl. 27
–, reduction with – **28**, 29
–/**hexamethylphosphoramide 17**, 895 suppl. **29**; **27**, 87

–/**hexamethylphosphoramide-*tert*-butanol 26**, 47; **28**, 54
–/**sodium** s. Sodium/potassium
- **acetate** as HI-acceptor **27**, 871
- **alkoxides**
  - *n*-butoxide **29**, 917
  - *tert*-butoxide **10**, 566 suppl. **29**; **15**, 679 suppl. **29**; **17**, 889 suppl. **28**; **18**, 815 suppl. **27**; **21**, 962 suppl. **26**; **22**, 530 suppl. **27**; **26**, 131, 144, 220, 276, 639, 733, 747, 791, 859/60, 898, 941, 969; **27**, 211/2, 235/6, 284, 288, 507, 875, 927, 964, 972; **28**, 173, 222, 591, 599, 600, 646, 699, 704, 729, 730, 798a, 839, 853, 882, 944, 963; **29**, 264, 412, 571, 583, 600, 632, 695/6, 716, 739/40, 891, 901, 956, 967; **30**, 400, 459, 574, 616, 639, 645, 667, 687, 694
- in organic synthesis, review **29**, 257
- *tert*-butoxide/water **29**, 715
- *tert*-heptoxide s. triethylmethoxide
–, **bulky**
  - triethylmethoxide **28**, 695
- **amalgam 27**, 860
- **amide 15**, 680 suppl. **27**; **26**, 459 suppl. **27**; **26**, 288, 359, 942; **27**, 459, 844; **28**, 732, 957, 976; **29**, 624, 812; **30**, 517
- **amides** (s. a. Potassium bis(trimethylsilyl)amide)
  ethylenediamine **28**, 884
  1,3-propylenediamide **30**, 506
- **aroxides, bulky** s. Potassium 2,6-di-*tert*-butylphenoxide
- **azodicarboxylate 28**, 39
- **bis(trimethylsilyl)amide 28**, 947
- **bromide 29**, 414
- as catalyst **5**, 162 suppl. 26
- **carbonate** as HCl-scavenger **27**, 353
- **chloride 28**, 267; **29**, 858; **30**, 366
- **compds., organo-**
- as reagents
  trimethylsilylmethylpotassium **30**, 586
- **cyanate 26**, 750; **27**, 636
- **cyanide 26**, 768, 776, 813, 828, 835, 865, 994; **27**, 667, 679, 895; **28**, 64, 339, 603, 777, 808, 836, 990; **29**, 56, 639, 845; **30**, 576
- as reactant **29**, 749
– – . **18-crown-6-polyether 28**, 610 suppl. **29**; **30**, 630
- **2,6-di-*tert*-butylphenoxide 25**, 628
- **ethylxanthate 27**, 629
- **ferrate (VI) 26**, 194
- **fluoride 27**, 783; **28**, 516; **29**, 961; **30**, 464, 604
- **hexachlorotungstate 28**, 925
- **hexacyanoferrate(III) 11**, 644 suppl. **28**; **26**, 776; **27**, 99, 372; **28**, 237, 323; **30**, 323
- **hydride 27**, 830 suppl. 29
- as reactive base and hydriding agent **28**, 884 suppl. 30
- as superbase source **28**, 884; **30**, 506

- **hydridotri-*sec*-butylborate** 12, 64 suppl. 29; 30, 602
- **-/copper(I) iodide**, reductions with - 30, 40
- **hydrogen fluoride** 28, 507
- **- sulfate** 23, 971 suppl. 28; 27, 980; 28, 21
- **hydroxide-clay** 28, 722
- **hydroxide-pumice** 28, 722
- **hypobromite** 26, 532
- **iodide** 26, 183; 27, 446, 464, 642, 800; 27, 856 suppl. 29; 28, 204; 29, 203; 30, 133, 375
- **metabisulfite** 29, 285
- **methylxanthate** 30, 405
- **nitrate** 2, 341 suppl. 27; 29, 337/8
- as supporting electrolyte 28, 834
- **nitrite** 27, 463
- **nitrosodisulfonate** 20, 696 suppl. 29; 30, 642
- -, reactions with -, review 8, 206 suppl. 26
- **osmate** 21, 858 suppl. 28
- **oxide** s. under Chromium oxide
- **perchromate** as singlet oxygen source 27, 179
- **periodate** s. Periodate
- **permanganate** s. Permanganate
- **persulfate** s. Persulfate
- **phenoxide** 27, 910
- **pyrosulfite** 29, 355
- **salt** 26, 416, 618; 27, 95, 282, 480; 29, 182, 482, 588; 30, 218, 271

N-Potassium salts 15, 379 suppl. 27

Potassium sodium tartrate 28, 816
- **tetrahydridoborate** 27, 540; 29, 390
- **thioacetate** 26, 598
- **thiocyanate** 7, 617 suppl. 27; 27, 597, 888; 28, 395; 29, 306, 573, 814
- **triorganocyanoborates**, reactions 27, 282
- **xanthates** as intermediates 6, 841 suppl. 26

Potential s. Electrolysis
Preservation s. Retention
Pressure s. High pressure reactions
Prevention
- of
  O-acylation in peptide synthesis 18, 435 suppl. 26
  O-acyl group migration 13, 255 suppl. 26; 29, 188 suppl. 30
  allyl rearrangement 27, 578
  aporphine ring opening 29, 206 suppl. 8
  coupling of halides 11, 731 suppl. 27
  epimerization during oxidation 16, 319 suppl. 26
  hydrogenolysis 29, 48
  oxidation, allylic 19, 307 suppl. 28
  N-quaternization s. under N-Quaternization

racemization s. retention of optical activity under Stereoisomers
radical addition 27, 600
reduction of ketones in Grignard syntheses 19, 46 suppl. 26
ring closure, intramolecular 27, 799

Prévost reaction, modified 13, 209 suppl. 28

Prins reaction with rearrangement 29, 679

Process, anchimeric s. Anchimeric process

Process, continuous acids and amides from nitriles 29, 196
isocyanates from aminosilanes 26, 443 suppl. 30
phenols from ar. sulfonic acids 19, 279 suppl. 29
Reformatskii synthesis 26, 938 suppl. 29

Process, frangomeric s. Frangomeric process

Product-catalyzed reactions 24, 425 suppl. 26

Product components as solvent 27, 127

(S)-(-)-Proline as reagent 30, 520

Proline derivs.
-, hydrogenation, asym. 29, 55
-, synthesis, asym., with - 21, 835 suppl. 26; 29, 55; 30, 520

Prolines
- from α,β-ethyleneoxo compds. 29, 664

Promoter s. Activator

1,3-Propanediamine as reagent 30, 326, 506

Propargyl esters s. Dithiocarboxylic acid propargyl esters
- **ethers**
-, Claisen rearrangement 29, 697
-, ring closures with - 29, 697

Propellanes 22, 705 suppl. 26
-, small-ring, review 23, 840 suppl. 27

Propenyl derivs.
- startg. m. f.
  allyl derivs. 23, 719 suppl. 29

N-Propenyl -
- from
  N-allyl derivs. 26, 739

Propionic acid as reagent 26, 375; 27, 563, 795/6; 28, 345
- **anhydride** - - 30, 164

n-Propyl alcohol - - 27, 51; 28, 36; 30, 11

n-Propylbenzene as solvent 27, 790

Propylene carbonate - - 20, 413 suppl. 26
- **oxide** as reagent 30, 290

n-Propylmercaptan - - 29, 992
Propyn... s. a. Propargyl...

Prostaglandin synthesis 17, 201 suppl. 29; 28, 661 suppl. 30

Protection, blocking, masking
- of
  aminoalcohol groups 26, 426

amino groups 26, 426
- as
  (2-acyl-1-methylvinyl)-amines 27, 402
  1,1-aminoethers 28, 268
  anthranilylamines 29, 27
  carballyloxyamines 29, 28
  carbisobornyloxyamines 15, 26 suppl. 28, 29
  carbo-*tert*-butoxyamines 11, 425 suppl. 27
  carbocyano-*tert*-butoxyamines 28, 21
  carbo-9-fluorenylmethoxyamines 26, 24
  carbomethylsulfonylethoxyamines 29, 19 suppl. 30
  o-nitrocinnamoylamides 30, 13
  o-nitrophenoxydimethylacetamides 26, 28
  N-nitrosamines 28, 19
  α-picolinoxyamines 29, 27
  platinum complexes 30, 361
  sulfenamides 13, 352 suppl. 26, 29
  1-oxidopyridin-2-ylmethylamines 30, 12
  thiolurethans 27, 378
  tosylamines 22, 871 suppl. 26
  tosylureas 30, 11
  trifluoromethanesulfonamides 29, 288
- by protonation 28, 172
- of aminocarboxylic acids 12, 519 suppl. 26
- -, review 28, 21
- with β-diketones 27, 402
amino groups, prim.
- as
  Δ⁴-oxazol-2-ones 28, 365
  phthalimides 27, 403
carbalkoxyl groups
- as
  ketene mercaptals 29, 552
carbon-carbon double bonds
- as
  amines, tert. 30, 669
  1,4-benzodioxane ring 8, 172 suppl. 29
  1,2-dichlorides 9, 97 suppl. 29
  selenides 29, 912
- by diene synthesis 28, 631
- triple bonds as cobalt carbonyl complex 27, 140
- - -, terminal, as silylacetylenes 23, 656
- double bonds as iron complex 28, 631 suppl. 30
carbonyl groups
- as
  bromomethylethylene acetals 28, 231
  mercaptals, cyclic 27, 271
  mono- or di-2,2,2-trichloroethyl acetals 28, 231
  monothioacetals, cyclic 27, 271
  o-nitrophenylethylene acetals 16, 234 suppl. 29
  1-(organothio)alkoximes 29, 386
  α-siloxynitriles 28, 610
-, preferential protection 28, 610

(Protection
- of
  carbonyl groups
  - as)
    o-xylylene mercaptals,
      cyclic, preferential
      protection 21, 656
    aldehyde groups as acetals,
      cyclic, partial protection
      12, 205 suppl. 29
    – – as amine-adducts
      29, 150
    – – as azomethines 24, 188
    – – as enamines 29, 150
    – – as imidazolidines
      22, 483
    keto groups as 1,3-benzo-
      dioxoles 16, 234 suppl. 27
    – – as metal enolates
      18, 59 suppl. 28
  carboxyl groups
  - as
    acyloximes 27, 1
    9-anthrylmethyl esters
      29, 3
    benzoin esters 27, 5
    N,N'-diisopropylhydrazides
      28, 158
    methylthiomethyl esters
      28, 14
    p-nitrophenyl esters 14, 451
      suppl. 27
    2-(p-nitrophenylthio)ethyl
      esters 26, 3
    Δ²-oxazolines 29, 3
    p-organothiophenyl esters
      p-methylthiophenyl –
      18, 434 suppl. 26
    Δ²-oxazolines 15, 179
      suppl. 27; 26, 872
    phenacyl esters 29, 2
    piperonyl esters 23, 11
      suppl. 27
  hydroxylamines 23, 233
    suppl. 27
  hydroxyl groups
  - as
    carballyloxy derivs.
      29, 28
    1,4-dioxan-2-yl ethers
      18, 875 suppl. 28
    1,4-oxathian-2-yl ethers
      18, 875 suppl. 28
    alcohol groups
  - as
    9-anthronyl ethers,
      9-phenyl- 26, 177
    9-anthryl ethers 28, 194
    benzyl ethers 26, 13
      p-methylbenzyl ethers
      5, 194 suppl. 29
    chloracetates 26, 8
    ketals, heterocyclic,
      mixed 26, 138
    mesylates 30, 7
    4-methoxytetrahydro-
      pyran-4-yl derivs.
      26, 138
    p-methoxytrityl ethers
      12, 288 suppl. 26
    o-nitrobenzoates 29, 4
    o-nitrobenzyl ethers
      30, 5
    1-oxido-2-picolyl ethers
      30, 12
    p-phenylbenzoates 19, 682
      suppl. 26

silyl ethers (s. a.
  O-Silylation) 29, 415,
  964; 30, 61
  tert-butyldimethylsilyl
    ethers 26, 235 suppl.
    29; 28, 44; 30, 4
  dimethylisopropylsilyl
    ethers 28, 44
  subst.-trityl ethers
    12, 288 suppl. 27
  tetrahydro-2-pyranyl
    ethers 27, 139
  β,β,β-trichlorethyl
    carbonates 28, 966
  β,β,β-trichlorethyl
    ethers 29, 5
  urethans 30, 7
  – –, prim., as carboxylic acid
    esters 26, 57 suppl. 28
  glycol groups
  - as
    1,3-dioxolanes, 2-amino-
      27, 234
    aryloxysilanes 28, 74
  – –, steroidal s. under
    Steroid hydroxyl
  vic-hydroxyl groups
  - as
    1,3-dioxolanes, 2-alkoxy-
      28, 268
    1,3-dioxolan-2-ones
      29, 197 suppl. 30
  lactone groups as dithiol-
    ortholactones 29, 552
  phosphoryl hydroxy groups
  - as
    tert-butyl esters 28, 170
    o-nitrobenzyl – 30, 5
    1-oxidopyridin-2-yl-
      methyl – 30, 12
  – with
    2-(arylthio)ethanol 26, 104
      suppl. 27, 28
    –, polymeric, soluble
      17, 169 suppl. 29
    3-hydroxypropionitrile
      17, 169 suppl. 29
    2-α-pyridylethanol 26, 104
  phosphate groups as
    phosphoroanilidates
    13, 355 suppl. 28; 25, 88
    suppl. 26
  phosphorus(III) compds. as
    phosphonium salts 27, 364
  sulfhydryl groups
  - as
    acetamidomethyl thioethers
      21, 60 suppl. 27
    N-(methoxymethyl)thiol-
      urethans 15, 457 suppl. 28
    1-oxido-2-picolyl thioethers
      30, 12
    tetrahydro-2-pyranyl
      thioethers 18, 645
      suppl. 27
    thiolic acid esters 30, 415
    trityl thioethers 24, 44
      suppl. 26
in situ-**Protection** of alcohol
  groups 29, 415
Protection, simultaneous
- by different protective
  groups 28, 268
Protective group
  –, 9-anthrylmethyl as – 29, 3
N-Protective group
  –, acyl as – 26, 454

–, carbodiisopropylmethoxyl
  as – 13, 43 suppl. 26
–, carbo-2-phenylisopropoxyl
  as – 27, 404 suppl. 29
–, carbotrichlorethoxyl as
  27, 24
–, 1-oxy-2-picolyl as – 22, 203
  suppl. 27
–, 2-pyranyl, tetrahydro- as –
  16, 401
– –, **acid labile**
–, carbo-2-p-biphenyliso-
  propoxyl as – 23, 404
  suppl. 26
– –, **composite**
–, benzyloxymethyl as – 27, 456
O,N-Protective group
  –, 2-oxazolone ring as – 27, 357
Protective groups
  s. a. Derivatives
  – –, **photo-sensitive**, review
    29, 194 suppl. 30
  azidoarylalkoxyl 27, 6
  benzoin in ester groups
    27, 5
  carbo-α,α-dimethyl-3,4-di-
    methoxybenzoxyl 22, 26
    suppl. 28
  carbo-2-nitrobenzoyl 22, 26
    suppl. 29
  carbo-6-nitroveratroxyl
    22, 26 suppl. 29
  carboxyl-protective groups
    29, 2, 194
  N-formyl groups 28, 27
  O-o-nitrobenzyl 30, 5
  o-nitrobenzylethylene acetal
    groups 16, 234 suppl. 29
  – –, **2-polyhalogenethyl-
    containing**, removal 28, 7
  – –, **polymeric soluble** 17, 169
    suppl. 29
  – –, **solubilizing**, in peptide
    synthesis 21, 426 suppl. 27
  – –, **removal**
  – by hydride transfer 27, 239
  – from peptides with boron
    bromide 28, 12 suppl. 29
    peptides with boron
    trifluoroacetate 15, 26
    suppl. 28
  – of 9-anthrylmethyl groups
    29, 3
  – with trifluoromethanesulfonic
    acid/anisole 29, 17a
  – –, – –, **electrochemical** 28, 7
    suppl. 29
  – –, – –, **reductive** 27, 24
  – –, – –, **selective**
  – from
    peptides 26, 3, 27a
N-Protective groups, removal
  – by
    β-elimination 26, 24
    1,6-elimination 29, 19
  – of
    tert-butyl groups 27, 538
    carbisobornyloxy groups
      15, 26 suppl. 28
    carbocyano-tert-butoxy
      groups 28, 21
    carbo-9-fluorenylmethoxy
      groups 26, 24
    carbo-2-halogenethoxy
      groups 23, 41 suppl. 27
    1-oxidopyridin-2-ylmethyl
      groups 30, 12

sulfenyl groups **27**, 19;
29, 17
- under neutral conditions
  30, 11
- with pyridinium salts
  26, 27a suppl. **29**
- –, –, **reductive 26**, 26, 28;
  30, 13
- –, –, **selective**
- of
  carbo-*tert*-butoxyl **15**, 26
    suppl. **27**
  carbo-3,5-dimethoxy-
    (α,α-dimethyl)benzoxyl
    15, 26 suppl. **29**
  carbo-1,1-dimethyl-
    2-propynyloxyl **16**, 32
    suppl. **26**
  carbohalogeno-*tert*-butoxyl
    28, 23
  nitroso groups **28**, 19
- –, –, **stepwise 27**, 456
O-Protective –, –
- by Wolff-Kishner reduction
  26, 177
- from
  carbonyl groups **27**, 271
  carboxyl – **27**, 1; **28**, 14
- of
  azidoarylalkoxyl **27**, 6
  tert-butyl groups **28**, 170
- –, –, **enzymatic 28**, 13
- –, –, **oxidative 26**, 104
  suppl. **27**, 28
- from
  carboxyl groups **28**, 158
- –, –, **oxidoreductive**
- of
  9-anthryl groups **28**, 194
- –, –, **photochemical**
- from
  carboxyl groups **29**, 2, 194
- –, –, **preferential**
- of
  chloracetyl groups **26**, 8
- –, –, **reductive 28**, 231, 966
- of
  o-nitrobenzoyl groups **29**, 4
  β,β,β-trichlorethyl groups
    29, 5
- –, –, **selective**
- from
  alcohol groups **30**, 7
  carbonyl – **27**, 271
- of
  benzoin ester groups **27**, 5
  benzyl – **26**, 13
  tert-butyldimethylsilyl –
    30, 4
S-Protective groups,
  comparison **14**, 443 suppl. **26**
Protoadamantanes
- startg. m. f.
  adamantanes **19**, 780
    suppl. **29**; **27**, 746
Proton abstractor, methylmagnesium chloride as –
  26, 423
- acceptor
- –, sodium acetate as – **27**, 79
- **acids, weak**
  ethyldiisopropylamine
    hydrobromide **29**, 457
Protonating agents s. H-Donors
Protonation (s. a. Acetic acid,
  protonated, Compds., –)
- –, protection of amino groups
  by – **28**, 172

Proton donors 26, 899
  ammonium chloride **28**, 47
  benzyl alcohol **23**, 69
    suppl. **26**
  tert-butyl – **28**, 47
  diethyl malonate **12**, 127
    suppl. **28**
  thiophenol **27**, 93
  thiourea **27**, 644
- exchange s. Metalation
Prototropic (s. a. Sigmatropic)
  s. under Isomerization,
  prototropic, Rearrange-
  ment, –, Shift, –
Proximity s. Neighboring
  group...
Pschorr ring closure,
  electrochemical 27, 953
Pseudoaromatic rings
  s. Dithiolium salts, Metal
  complex compds.
Pseudobase isomerization
- –, ring expansion by – **27**, 130
Pseudobases, N-heterocyclic
  27, 811
Pseudocyanides 26, 764
Pseudohalogens 27, 548
- special s.
  thiocyanogen chloride
Pseudoindoles s. Indolenines
Pseudoisoindol-1-ones 27, 797
- as intermediates **27**, 797
Pseudonitrosites s. 1,2-Nitro-
  nitroso compds.
Pseudothiouronium salts
  s. Isothiouronium salts
Pteridine-6-carboxamides
  24, 408 suppl. **28**
Pteridine ring
- from
  purinium salts **29**, 484
  pyrimidines, 4-amino-
    5-nitroso- and enamines
    26, 392
Pteridines
- from
  α,α-dihalogenoketones
    30, 299
  enamines **26**, 448
- –, **6-organothio-**
- from
  pyridinium salts **27**, 620
Pteridin-4(3H)-one ring opening
  26, 27
Pteridin-4(3H)-ones 21, 411
  suppl. **27**
- startg. m. f.
  4(3H)-pyrimidinones,
    4,5-diamino- **26**, 27
Pteridinones, dihydro- 28, 33
Pulvinic acid dilactones
- from
  p-quinones, 2,5-dihydroxy-
    28, 199
- startg. m. f.
  vulpinic acids **30**, 69
Pummerer rearrangement
  (s. a. Silyl-Pummerer
  rearrangement)
- –, heterocyclics by – **29**, 250;
  30, 166
- of β-ketosulfoxides **30**, 100
Purine nucleosides
- startg. m. f.
  2-azapurine nucleosides
    27, 23

Purines
- –, 8-alkylation s. 8-substitution
- from
  pyrimidines **26**, 356; **30**, 208
- –, 4-amino- and nitroso
  compds. **29**, 365
- –, 4-amino-5-nitroso- and
  hydrazones **27**, 433
  ureidomalonic acid esters
    26, 356
- –, one-step synthesis **28**, 382
- –, reaction, radiation-induced,
  with alcohols **28**, 709
- special s.
  adenines
  guanines, 8-hydroxy-
- startg. m. f.
  imidazoles **29**, 20
- –, 8-substitution **29**, 719
- –, **8-subst.**
- from
  pyrimidines, 4-amino-
    29, 343
- –, 4-amino-5-nitroso- **29**, 343
- –, **8-*tert*-amino-**
- from
  pyrimidines, 4-amino-
    5-nitroso- and formamides
    28, 354
  - and formamide acetals
    28, 354 suppl. **30**
- –, **8-aryl-**
- from
  pyrimidines, 4-amino- **30**, 318
Purinesulfonic acids 23, 379
Purinides s. Ammoniopurinides
Purinium salts
- startg. m. f.
  pteridine ring **29**, 484
- –, **7-β-hydroxy-**
- from
  oxido compds. **30**, 204
Purin-6-ylpeptides
- –, degradation of peptides
  via – **24**, 32 suppl. **29**
Pyran, dihydro-
- as protecting agent
  s. 2-Pyranyl...
- –, reactions **18**, 875 suppl. **26**
4H-Pyran-4-carboxaldehydes
- via *sym*-oxepin oxides **30**, 522
2H-Pyran-5-carboxaldehydes,
  3,4-dihydro-
- from
  2-oxabicyclo[4.1.0]heptanes,
    7,7-dichloro- **27**, 972
  oxepins, 2,5,6,7-tetrahydro-
    3-chloro-2-*tert*-butoxy-
    27, 972
Pyran-4-ols, tetrahydro-
- from
  ethylene derivs. **7**, 730
    suppl. **26**
Pyrano[2,3-b]pyridines 26, 244
2H-Pyrano[3,2-c]quinolines
  29, 249
Pyran ring by 3-component
  reaction **26**, 812
- from
  carboxylic acid anhydrides
    30, 543
- special s.
  4-enepyran ring
- startg. m. f.
  benzene ring **30**, 655
2H-Pyran
- startg. m. f.
  2(5H)-furanone ring **30**, 153

4H-Pyran ring
- startg. m. f.
  pyridine ring, 1,4-dihydro-
  27, 315
- -, **4-subst.**
- from
  pyrylium ring **27**, 764
Pyran ring, dihydro- 28, 690
2H-Pyrans
- startg. m. f.
  2,4-dienols **27**, 669
Pyrans, dihydro-
- from
  ketosulfonium salts **27**, 269
-, **2,3-dihydro-6-amino-
  2-hydroxy-**
- startg. m. f.
  fulvenes, 6-(4-amino-1,3-dien-
  1-yl)- **29**, 736
-, **2,3-dihydro-2-imino-**
- from
  chalcones **26**, 713
  flavonones **26**, 713
**4H-Pyrans, 4-(o-nitro-
benzylidene)-**
- startg. m. f.
  spiro[indoline-2,4'-(4'H)-
  pyrans], 1-hydroxy-3-oxo-
  29, 321
Pyrans, tetrahydro-
- from
  5-aminoalcohols **29**, 140
-, **tetrahydro-2-acetylene-**
- startg. m. f.
  5,6-dienols **29**, 626
-, **tetrahydro-2-amino-**
- syntheses with – **28,** 769
-, transamination **5,** 350
  suppl. **28**
2-Pyranthiol, tetrahydro-
-, introduction of protected
  sulfhydryl with – **18,** 645
  suppl. **29**
2-Pyranyl, tetrahydro-
- as N-protective group **16,** 401
**4-Pyranyl, tetrahydro-
4-methoxy-**
- as O-protective group **26,** 138
2-Pyranyl ethers, tetrahydro-
- special s.
  mandelonitriles, O-(tetra-
  hydropyran-2-yl)-
- startg. m. f.
  acoxy compds. **27,** 241
Pyrazine ring 29, 351
- from
  o-nitrosamines **26,** 341
- -, **2,5-dihydro-,** by Mannich
  reaction, double **29,** 728
Pyrazines
- from
  α-azidoketones **26,** 399
  pyrazines, 2,3-dihydro-
  24, 905 suppl. **29**
- special s.
  piperazines
  quinoxalines
- startg. m. f.
  pyrazines, 1,4-dihydro-
  1,4-disilyl- **26,** 279
-, **1,2-dihydro-**
- from
  α-aminoketones (2 molecules)
  29, 727
-, **2,3-dihydro-**
- startg. m. f.
  pyrazines **24,** 905 suppl. **29**

-, **2,5-dihydro-**
- from
  α-aminoketones (2 molecules)
  28, 323
  ketones (2 molecules)
  28, 323
-, **1,4-dihydro-1,4-disilyl-**
- from
  pyrazines **26,** 279
-, **2,3-dihydroxy-**
- from
  2-aminoacetals **27,** 511
Pyrazole-4-carboxaldehydes
- from
  semicarbazones **26,** 840
Pyrazolenine ring
- startg. m. f.
  isopyrazole ring **27,** 346
  suppl. **29**
  pyrazole ring **27,** 346
Pyrazole ring
- from
  β-diketone monohydrazones
  28, 114
  imidazolium ring, N-amino-
  30, 485
  o-nitrohalides and hydrazones
  28, 790
  α,β-oxidooxo compds. **26,** 349
  pyrazolenine ring **27,** 346
- -, **N-acyl-, N-condensed**
  29, 459
- from
  α-diazoketones, cyclic **29,** 657
- -, **3-amino-**
- from
  1,3,4-oxadiazolium salts
  27, 904
- -, **3-hydroxy- 28,** 948
Pyrazoles (s. a. Diazacyclo-
pentadienes)
- from
  α,β-acetylenehydrazones
  16, 876 suppl. **26**
  α,β-acetyleneoxo compds.
  26, 386
  diazo compds. and ketones
  26, 788
  p-halogeno-N-heterocyclics
  28, 390
  1,1-halogenohydrazones
  16, 876
  hydrazines, review **26,** 386
  hydrazones **26,** 792, 926
  Δ²-pyrazolines **28,** 889
  sydnones **27,** 900
- special s.
  3,5-dimethylpyrazole
  dipyrazolyl...
-, **amino-**
- from
  α-cyanooxo compds. **19,** 461
  suppl. **26**
  α,β-ethylene-α-halogeno-
  nitriles **30,** 298
-, **5-amino-**
- from
  hydrazones and N-di-
  halogenomethylene-
  ammonium salts **29,** 843
-, **4-hydroxy-**
- from
  α-diketone monohydrazones
  26, 486
  sulfonium salts **29,** 474
- startg. m. f.
  pyrazoles, 4-oxo- **27,** 254

-, **5-hydroxy-**
- from
  3-pyrazolidones **28,** 891
  -, 1-acyl- **28,** 891
-, **4-oxo-**
- from
  pyrazoles, 4-hydroxy- **27,** 254
-, ring closures via – **29,** 766
Pyrazolidines
- from
  N,N-Diels-Alder adducts
  27, 25
  hydrazines, ethylene derivs.,
  and aldehydes **26,** 780
- startg. m. f.
  indole ring **30,** 563
**Pyrazolidino-v-triazolines
  20,** 244 suppl. **27**
3-Pyrazolidone betaines
- startg. m. f.
  diaziridines **26,** 320
3-Pyrazolidones
- from
  α,β-ethylenenitriles and
  hydrazines **28,** 379
- startg. m. f.
  pyrazoles, 5-hydroxy- **28,** 891
  Δ¹-3-pyrazolones **26,** 264
-, **1-acyl**
- startg. m. f.
  pyrazoles, 5-hydroxy- **28,** 891
Δ¹-Pyrazoline ring
- from
  ethylenetosylhydrazones
  30, 683
Pyrazolines
- from
  azines **26,** 740
Δ¹-Pyrazolines
- special s.
  spiro-Δ¹-pyrazolines
Δ²-Pyrazolines
- from
  hydrazones **26,** 740
- startg. m. f.
  β-aminocarboxylic acids
  27, 200
  β-aminonitriles **28,** 889
  pyrazoles **28,** 889
-, **4-acyloximino-,** peptide
  synthesis with – **19,** 540
  suppl. **26**
-, **4-ene-5-hydroxy-**
- startg. m. f.
  3(2H)-pyridazones,
  4,5-dihydro- **27,** 727
**Pyrazolo[4,3-b][1,4]-
  benzoxazines 27,** 453
4H-Pyrazol-4-one N-oxides
-, cycloaddition to – **26,** 701
- startg. m. f.
  isoxazolo[1,2-b]pyrazol-3-ones
  26, 701
Δ³-Pyrazol-5-one ring
- from
  enecarbamyl azides **27,** 952
Pyrazolones s. a. Bispyrazolones
3-Pyrazolones 26, 474
Δ¹-3-Pyrazolones
- from
  3-pyrazolidones **26,** 264
Δ²-4-Pyrazolones, 5-hydroxy-
- startg. m. f.
  Δ²-5-pyrazolones, 4-hydroxy-
  28, 682

5-Pyrazolones
- from
  5-isoxazolones **27,** 397
- startg. m. f.
  α,β-acetylenecarboxylic acid
  esters **27,** 203
-, **4-hydrazono-3-alkoxy-
  carbonylamino-** **29,** 450
- startg. m. f.
  pyrazolo[3,4-e][1,2,4]triazine-
  3,7-diones, tetrahydro-
  29, 450
Δ²-5-Pyrazolones, 4-hydroxy-
- from
  Δ²-4-pyrazolones, 5-hydroxy-
  28, 682
Δ³-5-Pyrazolones
- from
  carboxylic acid hydrazides
  28, 781
5-Pyrazolones, 1-subst.
- from
  β-ketocarboxylic acid
  hydrazides, rearrangement
  26, 464
Pyrazolo[3,4-b]pyridine ring
- from
  pyrazolo[1,5-a]pyrimidine
  ring **26,** 756
**1H-Pyrazolo[3,4-b]pyridines
5,** 269/355 suppl. **27; 28,** 919
**Pyrazolo[3,4-b]pyridines,
3-hydroxy-** **28,** 948
**Pyrazolo[1,5-a]pyridinium
salts, polyhalogeno-**
- by double ring closure **26,** 484
**Pyrazolo[3,4-d]pyrimidine-
4,6-diones 28,** 790; **29,** 889
Pyrazolo[1,5-a]pyrimidine ring
- startg. m. f.
  pyrazolo[3,4-b]pyridine ring
  26, 756
**Pyrazolo[1,5-a]pyrimidines
30,** 224
**Pyrazolo[1,5-c]pyrimidines
27,** 385
- from
  1,3,5-triketones **27,** 385
**Pyrazolo[3,4-d]pyrimidines,
4-amino-**
- from
  carboxylic acid amides
  30, 252
**Pyrazolo[1,5-a]quinazolin-
5(4H)-ones**
- from
  o-hydrazinocarboxylic acid
  esters **26,** 385
Pyrenes 28, 890 suppl. **29**
Pyrazolo[2,3-a]quinolines
- from
  pyrylium salts **26,** 815
**Pyrazolo[3,4-e][1,2,4]triazine-
3,7-diones, tetrahydro-**
- via
  5-pyrazolones, 4-hydrazono-
  3-alkoxycarbonylamino-
  29, 450
Pyridazine-3,6-diol ring
- from
  dicarboxylic acid esters
  21, 563 suppl. **26**
  - acids **26,** 374
**Pyridazine-3,6-diones,
hexahydro-**
- startg. m. f.
  pyridazines, hexahydro-
  26, 56

Pyridazine 1-oxides
- from
  pyridazines **29,** 94
Pyridazine ring
- from
  pyrrolidine ring **23,** 440
- startg. m. f.
  α,β-ethylenenitriles **28,** 95
Pyridazines
- startg. m. f.
  benzenes **29,** 886
  pyridazine 1-oxides **29,** 94
  pyridazinium salts, 1-amino-
  28, 236
  pyridines **29,** 886
  pyrimidines **29,** 327
-, **6-(2-allylphenoxy)-
  3-halogeno-**
- startg. m. f.
  xanthenes **29,** 965
-, **1,6-dihydro-6-acyl-**
- from
  pyridinium ylids, 1-imino-
  27, 144
-, **hexahydro-**
- from
  pyridazine-3,6-diones,
  hexahydro- **26,** 56
**Pyridazinium betaines,
4-hydroxy-**
- from
  α-hydrazinocarboxylic acid
  esters **26,** 940
- salts
- startg. m. f.
  pyridazinium salts,
  4,5-dihydro-, 4-subst.
  26, 714
-, -, **1-amino-**
- from
  pyridazines **28,** 236
-, -, **4,5-dihydro-, 4-subst.**
- from
  pyridazinium salts **26,** 714
**3(2H)-Pyridazones,
4,5-dihalogeno-**
- from
  hydrazines **26,** 427
  Δ²-pyrazolines, 4-ene-
  5-hydroxy- **27,** 727
9H-Pyrid[3,4-b]indoles
- from
  indoles, 3-β-amino- **28,** 750
- via Fischer indole ring
  synthesis **16,** 951 suppl. **27**
**5H-Pyrid[4,3-b]indoles,
1,2,3,4-tetrahydro-**
- from
  pyrimido[1,6-a]indoles,
  1,2,3,4-tetrahydro- **27,** 752
**Pyridine-sulfur dioxide
complex 29,** 993 suppl. **30**
**2-Pyridinecarboxylic acid
esters,** hydrolysis, preferential
26, 5
**Pyridine-o-dicarboxylic acid
esters,** annelation **28,** 723
Pyridine hydrochloride
  as reagent **26,** 505; **27,** 364;
  28, 196; **29,** 169, 386
- **hydrobromide perbromide** – –
  30, 163
- **N-oxide ring**
- from
  4-enepyran ring **29,** 360
- startg. m. f.
  benzene ring **29,** 775
  oxazepine ring **23,** 179

- **N-oxides**
- special **s.**
  1-oxidopyridin-2-yldiazo-
  methane
  1-oxidopyridin-2-ylmethyl
- **o-position**
-, ring closure, nucleophilic,
  at – **28,** 427
- **ring** (s. a. Quinoline ring)
- from
  β-amino-α,β-ethyleneoxo
  compds. **26,** 805; **27,** 794
  2,4-dieneoximes **22,** 507
  1,5-dioxo compds. **8,** 480
  1,3-oxazin-6-one ring
  28, 871
  4-pyridone – **28,** 903
  quinoline – **30,** 147
  1,2,4-triazines **26,** 833
- startg. m. f.
  pyridine ring, 1,4-dihydro-,
  4-subst. **29,** 633
-, synthesis s. Friedländer...
-, -, **condensed, sym. 27,** 771
-, -, **2-amino-**
- from
  β-amino-α,β-ethylenenitriles
  30, 578
-, -, **1,4-dihydro-**
- from
  4H-pyran ring **27,** 315
-, -, -, **4-subst.**
- from
  pyridine ring **29,** 633
-, -, **4-halogeno- 28,** 919
-, -, **1,2,3,6-tetrahydro-**
- from
  2-pyridone ring **26,** 63
Pyridines (s. a. Piperidines)
- from
  aldehydes **27,** 775
  cyclopropenium salts and
  Δ¹-azirines **26,** 771 suppl. **29**
  1,3-diynes **29,** 650
  nitro compds., aliphatic
  (3 molecules) **27,** 812
  oxazoles, review **16,** 789
  suppl. **26**
  pyridazines and acetylene
  derivs. **29,** 886
  pyrimidines, 2-alkoxy-
  (2 molecules) **26,** 832
  2H-pyrrolenines **27,** 965
-, photodimerization **27,** 721
- special s.
  2,6-di-*tert*-butylpyridine
  2,4-dimethylpyridine
  α,α'-dipyridyl...
  picol...
- startg. m. f.
  2-cyclohexenones **27,** 816
  pyridines, 1,2-dihydro-
  26, 38
-, 1,2-dihydro-1-carbalkoxy-,
  2-subst. **26,** 684
-, 1,4-dihydro- **27,** 816
-, 1,4-dihydro-1-carbalkoxy-,
  4-subst. **26,** 684 suppl. **29**
-, 1,4-dihydro-1,4-diacyl-
  26, 798
-, 3-substitution **26,** 879
-, **4-subst.**
- from
  4(1H)-pyridones **28,** 746
-, **2-acyl-**
- startg. m. f.
  indolizines, 1-(alkylideneamino)- **27,** 793

(Pyridines)
-, 4-alkyl-
- from
  5-(4-pyridyl)barbituric acids
  28, 746
-, 2-alkylthio-, syntheses with -
  28, 801
-, 2-amino-
-, 3-alkylation **28**, 762
- startg. m. f.
  anhydro-pyrido[1,2-a]-
    pyrimidinium hydroxides,
    2-hydroxy-4-oxo- **26**, 283
  pyrido[1,2-a]pyrimidinium
    salts, 3,4-dihydro-2(1H)-
    oxo- **27**, 478
-, dialkylamino-
- from
  pyridones **28**, 352
-, 4-dialkylamino- as acylation
  catalysts **29**, 184
-, 2,4-diamino-6-halogeno-
  26, 882
- from
  α-cyanocarboxylic acid
    amides **29**, 757
-, dihydro-, review **26**, 38
  suppl. 27
-, 1,2-dihydro-
- from
  pyridines **26**, 38
-, 1,2-dihydro-1-carbalkoxy-,
  2-subst.
- from
  pyridines **26**, 684
-, 1,4-dihydro-
- from
  cyclopentadienes and azides
    20, 389 suppl. 27
  pyridines **27**, 816
- special s.
  1-phosphoryl-1,4-dihydro-
    pyridines
- startg. m. f.
  2-cyclohexenones **27**, 816
-, 1,4-dihydro-1-carbalkoxy-,
  4-subst.
- from
  pyridines **26**, 684 suppl. 29
-, 1,4-dihydro-4-cyano- **23**, 678
  suppl. 29
-, 1,4-dihydro-1,4-diacyl-
- from
  pyridines **26**, 798
-, 1,2,3,4-tetrahydro-
- from
  piperidines **28**, 886
-, 1,2,3,6-tetrahydro-
- startg. m. f.
  2,6-methano-3-benzazocines,
    1,2,3,4,5,6-hexahydro-
    26, 738
2-Pyridinethiol, S-acyl-
  s. Thiolic acid 2-pyridyl
  esters
2-Pyridinethione, 3-nitro-
- as reactant **24**, 34 suppl. 28
2-Pyridinethione ring, 6-amino-
- from
  isothiocyanates **27**, 680
2-Pyridinethiones
- startg. m. f.
  thiazolo[3,2-a]pyridinium
    salts, dihydro- **26**, 611
2-(1-Pyridinio)-1-indenolates,
  3-oxo- **30**, 317

Pyridinium betaines
- cycloaddition, 1,3-dipolar,
  with - **26**, 695
- special s.
  pyridinium phosphate
    betaines
    thiazolo[3,2-a]pyridinium -,
    dihydro-
- startg. m. f.
  tropones **26**, 695
-, -, N-acylamino- **28**, 373
- chloride, 4-dimethylamino-
  1-*tert*-butoxycarbonyl-
  as reactant **25**, 293 suppl. 26
- cyclopentadienylids
- from
  diazocyclopentadienes
    29, 397
  pyrylium salts and
    thionylimines **29**, 431
- N-imides
- startg. m. f.
  6H-1,3-oxazine-6-thiones
    27, 436
- phosphate betaines
- as intermediates **26**, 222
- special s.
  N-phosphorylpyridinium
    betaines
-, ring, 2,3-dihydro-, cycloaddition of alcohols to -
    28, 712
- salts
-, dehalogenation via - **29**, 80
- from
  pyrylium salts and thionyl-
    imines **29**, 431
-, reaction with alkyl nitrites
    27, 348
-, removal of N-protective
  groups by - **26**, 27a suppl. 29
- special s.
  1-alkylthio-1-pyridinium
    salts
  di(pyridinium salts)
  carbamylpyridinium salts
  tritylpyridinium fluoroborate
- startg. m. f.
  oxazole ring **26**, 395
  pteridines, 6-organothio-
    27, 620
  β-(1,2,3-triazol-4-yl)acroleins
    22, 281
- -, N-amino- **28**, 373
- -, N-aryloxy-, rearrangement
    27, 910
- -, N-carbalkoxy- s. N-Carbo-
  tert-butoxypyridinium salts,
  4-dimethylamino-
- -, N-carbamyl-
- startg. m. f.
  carbamic carboxylic
    anhydrides **28**, 453
- -, N-hydrogeno- as reactant
    28, 518
- trifluoroacetate as reagent
    19, 307 suppl. 28; **29**, 881
- ylids, reaction via -
    26, 395
-, reactions with cyclo-
    propenones **28**, 832
- special s.
  pyridinium cyclo-
    pentadienylids
- -, 1-imino- **28**, 376
- special s.
  N-alkoxycarbonylimino-
    pyridinium ylids

pyridinium N-imides
- startg. m. f.
  pyridazines, 1,6-dihydro-
    6-acyl- **27**, 144
**Pyrido[2,1-c][1,2,4]benzo-
  thiadiazine 5,5-dioxides,
  7,8,9,10-tetrahydro- 29**, 449
**6H-Pyrido[4,3-b]carbazoles
  29**, 975 suppl. **30; 30**, 650
-, 1,2,3,4-tetrahydro- **30**, 650
**Pyrido[1,2-b]cinnolin-6-ium
  ylids, 5,11-dihydro-11-oxo-**
- from
  2,1-benzisoxazoles,
    3-(2-pyridyl)- **30**, 187
2,3-Pyridocyclobutenes
- from
  azocines, 3,4-dihydro- **26**, 941
**13H-Pyrido[1′,2′:3,4]imidazo-
  [2,1-b]quinazol-13-ones,
  7,8,9,10,10a,11-hexahydro-
  28**, 405
Pyrido[1,2-a]indoles 29, 457
**Pyrido[1,2-a]indol-9(8H)-ones,
  6,7-dihydro-**
- from
  3-quinuclidones **30**, 300
- startg. m. f.
  7a,11a-ethanooxyindole[1,2-h]-
    [1,7]naphthyridin-10(11H)-
    ones, 6,7,8,9-tetrahydro-
    30, 300
**Pyrido[1,2-a]indol-6-ones,
  6,7,8,9-tetrahydro-**
- startg. m. f.
  indoles, 2-δ-keto- **29**, 637
Pyridone s. a. Sodium
  pyridinolates
2-Pyridone as catalyst,
  bifunctional **29**, 451
**2(1H)-Pyridone-5-carboxylic
  acid esters 30**, 250
2-Pyridone ring (s. a. Carbostyrils) **28**, 367
- by 2H-pyrido[1,2-a]pyrimid-
  2-one ring opening **27**, 787
- from
  iminoesters **29**, 744
  2-pyridone ring, 3,4-dihydro-
    29, 927
  4-pyrone ring **30**, 552
- startg. m. f.
  pyridine ring, 1,2,3,6-tetra-
    hydro- **26**, 63
- via 2-pyridone ring,
  5,6-dihydro-5-hydroxy- **27**, 508
- **N-condensed 26**, 794
- from
  carboxylic acid chlorides
    26, 855
  malonic acid esters **30**, 551
- via 2-pyridone ring,
  3,4-dihydro-, N-condensed
    29, 927
-, -, 3,4-dihydro-
- startg. m. f.
  2-pyridone ring **29**, 927
- -, -, N-condensed **27**, 380
- by cyclodehydration **29**, 927
- from
  O-alkyllactims **27**, 768
  enamines **27**, 768
-, -, 5,6-dihydro- **29**, 916
-, -, 5,6-dihydro-5-hydroxy-
  27, 508
- startg. m. f.
  2-pyridone ring **27**, 508

– –, 4-hydroxy-, N-condensed
   27, 503; 29, 919
4-Pyridone ring
– startg. m. f.
   pyridine ring 28, 903
Pyridones
– startg. m. f.
   pyridines, dialkylamino-
   28, 352
2-Pyridones
– by Michael addition
   18, 765 suppl. 26
–, cyloisomerization 19, 812
   suppl. 27
– from
   β-ketocarboxylic acid amides
   (2 molecules) 26, 800
   pyrimidonium betaines
   28, 870
   thiazolium betaines,
   4-hydroxy- 26, 919
   4-pyridones 28, 680
–, 3,4-dihydro- 24, 771
– from
   azomethines and α,β-ethylene-
   carboxylic acid esters
   30, 556
–, –, 1-amino-
– from
   hydrazones 30, 556
4-Pyridones 29, 468
– from
   Δ¹-azirines and cyclo-
   propenones 28, 634
   4,5′-bisisoxazoles 27, 991
   β-diketones and nitriles
   26, 359
– startg. m. f.
   pyridines, 4-subst. 28, 746
   2-pyridones 28, 680
anhydro-Pyrido[1,2-a]pyrimidi-
nium hydroxides, 2-hydroxy-
4-oxo-
– from
   pyridines, 2-amino- 26, 283
Pyrido[1,2-a]pyrimidinium
salts, 3,4-dihydro-2(1H)-oxo-
– from
   pyridines, 2-amino- 27, 478
Pyrido[2,3-d]pyrimi-2,4-dione
ring
– from
   β-amino-α,β-ethyleneketones
   27, 794
2H-Pyrido[1,2-a]pyrimid-2-one
ring opening
–, 2-pyridone ring closure
   by – 27, 787
Pyrido[1,2-a][1,3,5]triazines
   28, 435
4H-Pyrido[2,1-c][1,2,4]triazines,
   1,9a-dihydro- 29, 382
Pyridylation of N-heterocyclics
   27, 765
5-(4-Pyridyl)barbituric acids
– startg. m. f.
   pyridines, 4-alkyl- 28, 746
2-α-Pyridylethanol
–, protection of phosphoryl
   hydroxyl groups with –
   26, 104
2,4-Pyrimidine diones s. Uracils
Pyrimidine nucleosides
– from
   carbohydrate Δ²-oxazoline
   ring, 2-amino- 28, 341
   carbohydrates and cyanoacetylenes 28, 341

–, syntheses, review 28, 341
   suppl. 29
– ring
– from
   β-amino-α,β-ethyleneoxo
   compds. 27, 408
   coumarin ring 26, 355
   nitriles (2 molecules) 26, 905
   1,2,3-triazine ring 29, 273
–, opening 26, 229; 29, 273
   – to o-aminoaldehydes
   26, 229 suppl. 30
– startg. m. f.
   o-aminocarboxylic acid
   amides 16, 23; 29, 20
– –, 4-amino-
– from
   o-aminoamidines 30, 256
   pyrrole ring, 3-nitroso-
   29, 355
– –, 2,4-diamino-
– from
   o-aminonitriles 28, 401
– –, hexahydro-
– by Mannich reaction, double
   26, 781
– –, 4-hydroxy-
– from
   pyrrole ring, 3-amino-
   27, 363
   – –, 3-nitroso- 30, 99
– –, 2-mercapto-
– from
   o-aminocarboxylic acid
   amides 27, 360
Pyrimidines
– from
   α-cyanoacetals 27, 381
   α-cyanocarboxylic acid esters
   27, 398
   1,3-diazabicyclo[3.1.0]hex-
   3-enes 28, 882
   EMCE 26, 353
   isocyanates 28, 338
   α-ketoketene mercaptals
   30, 310
   nitriles and β-iminoazomethines 26, 396
   1,3,5-oxadiazinium O-salts
   30, 577
   pyridazines 29, 327
   pyrimidines, 1,2-dihydro-
   30, 256
   pyrroles 29, 314
–, C-functionalization with
   diethyl azodicarboxylate
   30, 208
– special s.
   aminopyrimidines
   2-mercaptopyrimidine
   thiamine derivs.
   uracils
– startg. m. f.
   ethylene derivs. 26, 499
   purines 26, 356
–, mesoionic 29, 323
–, 2-alkoxy-
– startg. m. f.
   pyridines (from 2 molecules)
   26, 832
–, 2-amino-
– from
   β-diacetals 28, 357
–, –, labeled 28, 357
–, 4-amino-
– from
   carboxylic acid amides
   26, 817

pyrimidines, 1,6-dihydro-
   6-imino- 30, 228
– startg. m. f.
   purines 29, 365
–, 8-aryl- 30, 318
   pyrimidines, 4-amino-
   5-nitroso- 29, 343
–, 4-amino-5-nitroso-
– from
   pyrimidines, 4-amino-
   29, 343
– startg. m. f.
   pteridine ring 26, 392
   purines 27, 433
–, 8-subst. 29, 343
–, 8-tert-amino- 28, 354
–, 4,6-dihalogeno-
– from
   1-halogenenamidines 29, 489
–, 1,2-dihydro- 26, 396 suppl.30
– startg. m. f.
   pyrimidines 30, 256
–, 1,6-dihydro-6-imino-
– startg. m. f.
   pyrimidines, 4-amino- 30, 228
Pyrimidinium ring, 1-benzoxyl-,
   opening 27, 23
4(3H)-Pyrimidinone ring 25, 257
– from
   o-acylaminocarboxylic acid
   amides 4, 455; 28, 437
   o-aminocarboxylic acid
   amides 28, 437
–, opening, reductive 27, 67
4(1H)-Pyrimidinone ring,
N-condensed
– from
   acetylenedicarboxylic acid
   esters 30, 251
4(3H)-Pyrimidinone ring,
5,6-dihydro-, N-condensed
– from
   O-alkyllactims and
   2-azetidinones 26, 347
4(3H)-Pyrimidinones
– from
   β-alkoxy-α,β-ethylene- and
   ortho-carboxylic acid esters
   30, 250
   amidines 25, 257
–, 5-acyl-
– from
   β-amino-α,β-ethylene-
   carboxylic acid amides
   26, 340
–, 4,5-diamino-
– from
   pteridin-4(3H)-ones 26, 27
2-Pyrimidinyl thioethers
   as intermediates 27, 91
Pyrimido[1,6-a]benzimidazoles,
   1,2,3,4-tetrahydro- 26, 352
9H-Pyrimido[4,5-b]indoles
   18, 548 suppl. 27; 29, 466
Pyrimido[1,6-a]indoles,
   1,2,3,4-tetrahydro-
– startg. m. f.
   5H-pyrid[4,3-b]indoles,
   1,2,3,4-tetrahydro- 27, 752
Pyrimido[1,2-a][1,8]naphthyrid-
10-ones 26, 465
– startg. m. f.
   anthyridines, 6-hydroxy-
   26, 465
Pyrimido[4,5-b][1,8]naphthyrid-
4(3H)-ones 28, 437
Pyrimidonium betaine ring
   26, 283; 27, 383

Pyrimidonium betaines
– startg. m. f.
  2-pyridones **28,** 870
7H-Pyrimido[4,5-b][1,4]oxazines
  26, 429
Pyrimidopteridine 10-oxides
  28, 387
2,4,6,8(1H,3H,7H,9H)-Pyrimido-
  [5,4-g]pteridinetetrone
  5-oxides
  – from
    uracils, 6-amino-5-nitroso-
    (2 molecules) **28,** 377
Pyrimido[4,5-b]pyrazines
  s. Pteridines
Pyrimido[4,5-d]pyrimidine ring
  – from
    pyrrolo[2,3-d]pyrimidine ring
    30, 230
Pyrimido[4,5-d]pyrimidine-
  2,4,5,7-tetrones 26, 678
2H-Pyrimido[1,2-a]pyrimid-
  2-ones 26, 485
Pyrimido[5,4-e]*as***-triazine-**
  5,7(6H,1H)-diones 27, 355
Pyrimido[5,4-e]-*as***-triazines**
  28, 360
Pyrindines, dihydro-
  – from
    sulfonic acid azides, β-aryl-
    30, 343
Pyrindoles s. Indolizines
Pyrocarbonic acid esters
  s. Dicarbonic acid esters
Pyrocatechol (s. a. Catechol)
  –, protection of keto groups
  with – **16,** 234 suppl. 27
Pyrocatechyl phenyl phosphite
  as reducing agent **29,** 993
– **phosphorochloridite**
  as reactant **27,** 979
Pyrolusite s. β-Manganese
  dioxide
Pyrolysis (s. a. Thermolysis,
  Flash vacuum pyrolysis)
  – of
    acoxy-2-acetylenes **28,** 968
    α-alkoxyenolesters **28,** 679
  – to
    O-heterocyclics **27,** 268
–, **gas-phase 27,** 268
2-Pyrone ring, 3,4-dihydro-
  – from
    α,β-ethylenecarboxylic acid
    chlorides **26,** 854
4-Pyrone –
  – startg. m. f.
    2-pyridone ring **30,** 552
2-Pyrones
  – from
    cyclopropenones and
    sulfonium ylids **28,** 832
    α,β-ethylenecarboxylic and
    carboxylic chlorides **27,** 848
    α,β-ethyleneoxo compds.
    27, 848
    4-pyrones **28,** 680
  – , **4-dihydro- 26,** 240
  –, **5,6-dihydro-**
    β,δ-dihydroxycarboxylic
    acid esters **27,** 255
–, **5,6-dihydro-3-acyl-4-hydroxy-**
  – from
    4-pyrones-2,3-dihydro-
    5-carbalkoxy- **29,** 265

–, **4-hydroxy-**
  – from
    2,4-diynoic acids **29,** 151
–, **tetrahydro-4-hydroxy-**
  – from
    β,δ-dihydroxycarboxylic acid
    esters **27,** 255
3-Pyrones, 2,6-dihydro-
  6-hydroxy-
  – from
    furylcarbinols **27,** 276
4-Pyrones
  – from
    furan-2,3-diones, 2,3-dihydro-
    (2 molecules) **28,** 197
    4-pyrones, 2,3-dihydro-
    27, 913
    ynamines (2 molecules)
    27, 693
  – startg. m. f.
    2-pyrones **28,** 680
–, **2,3-dihydro-**
  – startg. m. f.
    4-pyrones **27,** 913
–, **2,3-dihydro-5-carbalkoxy-**
  – startg. m. f.
    2-pyrones, 5,6-dihydro-3-acyl-
    4-hydroxy- **29,** 265
–, **tetrahydro- 27,** 213
Pyronobenzo[ij]quinolizines
  – from
    quinolines, 1,2,3,4-tetrahydro-
    27, 898
Pyrophospho... s. Diphospho...
Pyrosulfuryl fluoride, reactions
  with – **27,** 112
Pyrrocolines s. Indolizines
2-Pyrrolecarboxaldehydes,
  5-halogeno-
  – from
    2H-pyrroles, 2-amino-
    methylene-5-halogeno-
    28, 754
  2-pyrrolones **28,** 754
Pyrrole condensation 20, 582;
  22, 805
2H-Pyrrolenines
  – from
    1,3,5-oxazaphosph(V)ol-2-ines
    28, 849
  – startg. m. f.
    pyridines **27,** 965
3H-Pyrrolenines 27, 533
Pyrrole ring (s. a. under
  Hantzsch)
  – from
    α-aminocarboxylic acids
    29, 885
    azines **26,** 740
    ene-1,2-di(oxysilanes) **26,** 811
–, **N-unsubst.**
  – from
    enamines and α-hydroxy-
    ketones **25,** 275 suppl. 26
–, **3-amino- 22,** 963 suppl. 29
  – from
    β-diketones and hydrazines
    28, 331
  – startg. m. f.
    pyrimidine ring, 4-hydroxy-
    27, 363
–, **1,2-diamino-**
  – from
    α-diazoketones and
    α,β-ethylenenitriles **26,** 466
  – startg. m. f.
    pyrrolo[1,2-b][1,2,4]triazine
    ring **26,** 466

–, **2,3-dihydro-**
  – from
    enamines and dihalides
    27, 861
–, **3-nitroso-**
–, expansion **29,** 355; **30,** 99
  – startg. m. f.
    pyrimidine ring, 4-amino-
    29, 355
  – –, 4-hydroxy- **30,** 99
Pyrroles
  – from
    4-acetylenehalides **30,** 637
    acetylenehalogenhydrins
    27, 451
    enamines **21,** 813
    – (2 molecules) **29,** 773
  –, cyclic **30,** 637
    hydrazones **26,** 842
    isonitriles (3 molecules) and
    acetylene derivs. **27,** 909
    Δ^4-isoxazolines **26,** 934
    isoxazolium salts **26,** 934
    ketones and hydrazines
    26, 842
    4H-1,3-oxazines, 5,6-dihydro-
    28, 439
    thiazolium betaines,
    5-hydroxy- **26,** 919 suppl. 30
  – special s.
    dipyrrylmethanes
    tripyrrenes
  – startg. m. f.
    anilines **26,** 719
    7-azabicyclo[2.2.1]-
    2,5-heptadienes **26,** 719
    pyrimidines **29,** 314
  –, β-substitution **5,** 604 suppl. 26
  –, synthesis with hydroxyl-
    amine-O-sulfonic acid **27,** 779
–, **3-acyl-**
  – from
    1,3-oxazepines **26,** 743
    suppl. 28
–, **1-aryl-**, ring closure with –
  28, 719
–, **2-aryl-**
  – via spiro[3H-indoline
    3,2'-2'H-pyrroles] **26,** 961
–, **2,5-disubst.**
  – from
    succinimides **29,** 862
–, **3-hydroxy-**
  – from
    α-aminocarboxylic acid esters
    27, 784
2H-Pyrroles, 2-aminomethylene-
  5-halogeno-
  – from
    2-pyrrolones **28,** 754
  – startg. m. f.
    2-pyrrolecarboxaldehydes,
    5-halogeno- **28,** 754
Pyrrolidine ring
  – startg. m. f.
    pyridazine ring **23,** 440
Pyrrolidines
  – by cycloaddition, anionic
    26, 712
  – special s.
    prolines
–, **2-alkylidene-**
  – from
    2-pyrrolidones **26,** 851
  – startg. m. f.
    Δ^1-pyrrolines **26,** 851
  –, **N-acyl-** by cycloisomeriza-
    tion **28,** 667

−, 3-hydroxy-
– startg. m. f.
  2-oxazolidones, 4-β-halogeno-
  30, 352
Pyrrolidine-2,3,5-triones 29, 635
Pyrrolidinium acetate
  as reagent **28,** 910
– salts
– special s.
  N-methyl-N-tosyl-
  pyrrolidinium perchlorate
– –, **N-cyanomethyl-**
–, rearrangement **29,** 901
4-Pyrrolidinopyridine 24, 733
  suppl. **26**
2,4-Pyrrolidiones 13, 131
  suppl. **29**
2,5-Pyrrolidiones
  (s. a. Succinimides)
–, C-alkylation, stereospecific
  26, 858
2-Pyrrolidone hydrotribromide
  as reagent **9,** 608 suppl. **29**
2-Pyrrolidones
– startg. m. f.
  γ-aminocarboxylic acids
  26, 120
  pyrrolidines, 2-alkylidene-
  26, 851
  Δ¹-pyrrolines **26,** 851
−, **5-subst.**
– from
  Δ³-pyrrol-2-ones **27,** 823
−, **1-acyl-**
– startg. m. f.
  Δ¹-pyrrolines, 2-subst. **28,** 981
−, **4-carboxy-**
– from
  azomethines and succinic
  acid anhydrides **28,** 616
3-Pyrrolidones
– from
  allenes and nitrones **26,** 693
Δ²-Pyrroline ring
– from
  cyclopropylketones **27,** 407
– –, **5-hydroxy- 29,** 691
Δ³-Pyrroline ring
– from
  isoquinolinium methylids
  29, 649
Pyrrolines
– from
  isonitriles **25,** 484 suppl. **26**
Δ¹-Pyrrolines
– from
  Δ¹-azirines and ethylene
    derivs., electron-deficient
    26, 708
  ethylene derivs. and
    Δ³-oxazolones **30,** 628
  isonitriles and ethylene
    derivs. **28,** 606
  pyrrolidines, 2-alkylidene-
    26, 851
  2-pyrrolidones **26,** 851
−, **2-subst.**
– from
  2-pyrrolidones, 1-acyl- **28,** 981
−, **2-amino-**
– from
  γ-halogenonitriles **27,** 450
Δ²-Pyrrolines
– startg. m. f.
  cis-indol-6-ones,
    octahydro- **28,** 664

1-[2-(1-Pyrrolinyl)]indolines
– startg. m. f.
  1H-pyrrolo[2,1-b][1,3]benzo-
  diazepines, 2,3,5,6-tetra-
  hydro- **27,** 339
Pyrrolizidines
– from
  pyrrolizines, dihydro- **30,** 631
Pyrrolizines, dihydro-
– from
  acetylene derivs. **30,** 631
Pyrrolo[1,2-a]benzimidazol-
1-ones, 3-hydroxy-
– from
  cyclobut-3-ene-1,2-diones,
  3-hydroxy- and o-diamines
  27, 413
5H-Pyrrolo[2,1-c][1,4]benzo-
diazepines 26, 478
6H-Pyrrolo[1,2-a][1,4]benzo-
diazepines, 4,5-dihydro-
28, 719
1H-Pyrrolo[2,1-b][1,3]benzo-
diazepines, 2,3,5,6-tetrahydro-
– from
  1-[2-(1-pyrrolinyl)]indolines
  27, 339
1H-Pyrrolo[2,1-c][1,2,4]benzo-
thiadiazine 5,5-dioxides,
2,3-dihydro- 28, 457a
9H-Pyrrolo[1,2-a]indoles,
9-amino- 28, 765
1H-Pyrrolo[1,2-a]indoles,
2,3,9,9a-tetrahydro-
24, 945 suppl. **26**
Δ¹-Pyrrol-5-one ring
– startg. m. f.
  cyclopropane ring **26,** 753
2-Pyrrolones, review **28,** 754
– startg. m. f.
  2-pyrrolecarboxaldehydes,
    5-halogeno- **28,** 754
  2H-pyrroles, 2-amino-
    methylene-5-halogeno-
    28, 754
Δ²-Pyrrol-4-ones
– from
  cyclopropenones and
  azomethines **30,** 473
  enazides **28,** 773
– and α-diazoketones **28,** 773
−, **1-acyl-2-amino-**
– from
  Δ²-5-oxazolones **26,** 666
Δ³-Pyrrol-2-ones
−, reaction with indoles **27,** 823
– startg. m. f.
  1,3-oxazin-4-ones **29,** 175
  2-pyrrolidones, 5-subst.
  27, 823
Δ⁴-Pyrrol-2-ones
– from
  N-acyl-N'-enehydrazines
  29, 951
  enamines **26,** 703
Pyrrolo[2,1-b][1,3,4H]oxazines,
dihydro-
– from
  4H-1,3-oxazines, 5,6-dihydro-
  28, 439
Pyrrolo[2,3-d]pyridazines
29, 368
Pyrrolo[1,2-b]pyridazin-2-ones
– from
  1,9-diazabicyclo[4.2.1]nona-
  4,7-dien-3-ones, 9-acyl-
  29, 983

Pyrrolo[2,3-d]pyrimidine ring
– startg. m. f.
  pyrimido[4,5-d]pyrimidine
  ring **30,** 230
Pyrrolo[1,2-c]pyrimidine ring
  opening 28, 430
Pyrrolo[1,2-c]quinazolines,
4,5-dihydro-
– via spiro-3H-indoline-
  3,2'-2'H-pyrroles **26,** 961
1H-Pyrrolo[2,1-b]quinazol-
4-ones, 2,3,3a,4,9,9a-hexa-
hydro-9a-hydroxy- 26, 119
1H-Pyrrolo[3,2-b]quinolines,
2,3-dihydro- 28, 757
−, **2,3,3a,4-tetrahydro- 29,** 425
Pyrrolo[3,4-b]quinolines 30, 261
Pyrrolo[2,3-b]quinoxalines,
2-amino- 27, 452
1H-Pyrrolo[2,3-b]quinoxalines,
2,3,3a,4,9,9a-hexahydro-
– from
  1-benzimidazolylsuccinimides
  29, 329
Pydrolo[1,2-b][1,2,4]triazine ring
– from
  α-diazoketones and
    α,β-ethylenenitriles **26,** 466
  pyrrole ring, 1,2-diamino-
    26, 466
Pyrrolo[1,2-d][1,2,4]triazine –
– from
  1,3,4-oxadiazole ring,
  2-(pyrrol-2-yl)- **29,** 566
Pyrylium ring 29, 838
– from
  α,β-ethylene-β-halogeno-
    ketones **28,** 828
– startg. m. f.
  4H-pyran ring, 4-subst.
  27, 764
– –, **4-methyl-,** by C-cleavage
  27, 86
– salts
– from
  α,β-ethylene-β-halogen-
    aldehydes and ketones
    29, 841
  5-halogeno-2,4-dienals **30,** 172
– special s.
  benzopyrylium salts
– startg. m. f.
  benzo[c]quinolizinium salts
  27, 411
  4H-1,2-diazepines **26,** 370
  pyrazolo[2,3-a]quinolines
  26, 815
  pyridinium salts **29,** 431

Quadricyclanes 26, 763 suppl. **27**
Quasithioureas 29, 574
Quasiureas 29, 574
N-Quaternization (s. a.
  Ammonium salts, quaternary)
−, activation by – **18,** 842
  suppl. **27**
– of
  amines, tert., hindered **30,** 702
−, prevention **26,** 198
2,4-Quinazolinediones
– from
  o-aminocarboxylic acid
    amides **27,** 349
  3,1-benzoxazine-2,4-diones
    and amines **27,** 349

Qui

Quinazoline 3-oxides
- startg. m. f.
  benz[f][1,3,5]oxadiazepines
  24, 159
- **ring**
- from
  nitriles (2 molecules) and
  phenolethers 29, 741
Quinazolines
- from
  o-aminoketones and nitriles
  27, 804
  indoles 30, 223
  -, 3-nitroso- 28, 355
  1,3,3-trihalogeno-2-aza-
  propenes 29, 842
-, **4-amino-**
- from
  indazoles, 3-amino- 27, 498
-, **1,2-dihydro-**
- from
  o-(halogenomethylsulfonylamino)ketones 26, 433
-, **3,4-dihydro-3-acylamino-
  4-hydroxy-**
- startg. m. f.
  o-(1,2,4-triazol-4-yl)ketones
  29, 455
-, **3,4-dihydro-3-amino-
  4-hydroxy-**
- from
  o-acylaminoketones 27, 382
-, **1,2,3,4-tetrahydro-
  4-acyloximino-**
- from
  aldehydes 28, 334
**4(3H)-Quinazolinethiones,
  2-amino- 29, 814**
**2(1H)-Quinazolinethiones,
  3,4-dihydro- 30, 258**
**12H,6H-Quinazolino[2,3-c]-
  1,4-benzothiazin-12-ones**
- by ring closure, double 26, 469
**4-Quinazolone ring, opening,
  reductive 29, 40**
- -, **N-condensed**
- from
  o-aminocarboxylic acids
  27, 455
2-Quinazolones
- from
  acylureas 29, 934
  indole-1,2-dicarboximide
  ozonides 29, 95
- special s.
  4-alkylidene-2-quinazolones
-, **4-halogeno-**
- from
  o-cyanoisocyanates 26, 512
4-Quinazolones
- from
  amines 28, 710
  3,1-benzoxazine-2,4-diones
  and carboxylic acid amides
-, **3-alkoxy-** 28, 418
- from
  o-aminohydroxamic acid
  esters and carboxylic acid
  chlorides 29, 406
**4(1H)-Quinazolones, 2,3-dihydro-
  28, 302**
**2(1H)-Quinazolones,
  3,4-dihydro-4-hydroxy-
  3-alkoxymethyl-**
- from
  2H-1,4-benzodiazepin-2-ones,
  1,3,4,5-tetrahydro-4,5-epoxy-
  26, 113

**Quinhydrones, intramolecular
  30, 2**
Quinine as reagent 30, 18
Quinocyclopropenes 24, 912
p-Quinodimethanes
- startg. m. f.
  spirocyclics 29, 655
Quinolethers
- special s.
  quinol monoethers
Quinoline
- as medium 28, 16 suppl. 29
- as reagent 29, 92, 463
-, **8-hydroxy-** s, Copper(II),
  bis-(8-quinolinolato)-
**Quinoline-S as regulator
  27, 79**
**Quinoline-4-carboxylic acid
  amides**
- from
  isatins and enamines 28, 622
**Quinoline-2-carboxylic acids,
  4-hydroxy-**
- by
  aromatization with
  N-deacylation 26, 498
Quinoline-3,4-dione 1-oxides
- via 4(1H)-quinolone,
  1,3-dihydroxy- 28, 447
Quinoline N-oxides
- startg. m. f.
  quinolines, 6-alkoxy- 30, 116
- **ring** (s. a. under Combes-
  Beyer and Skraup)
- from
  carboxylic acid anilides,
  vinylogous, 28, 917
  β-diketone monohydrazones
  4, 835; 13, 863
  α,β-ethyleneoxo compds.
  26, 823
- startg. m. f.
  pyridine ring 30, 147
Quinolines
-, 3-alkylation 28, 800
- by Wittig synthesis, intramolecular 28, 962
- from
  o-aminoketones and nitriles
  27, 804
  o-aminostyrenes 26, 765
  azomethines and enolethers
  26, 723
  ketones 27, 423
  2,5-methano-3,1-benzoxaze-
  pines, 1,2,4,5-tetrahydro-
  28, 924
  phosphormidates 28, 461
  quinolinium salts 27, 679
- special s.
  halogeno-quinolines
  phosphoryloxy-
- startg. m. f.
  2,5-methano-3,1-benzoxaze-
  pines, 1,2,4,5-tetrahydro-
  28, 924
  quinolines, 1,4-dihydro-,
  1-subst. 30, 214
-, 4-substitution 28, 924
-, **labeled** 2, 651 suppl. 29
-, **2-(o-acylanilino)-**
- from
  o-aminoketones and
  carboxylic acids 26, 816
-, **6-alkoxy-**
- from
  quinoline N-oxides 30, 116

-, **2-amino-**
- from
  α,β-acetyleneamidines 28, 672
  chloroformamidines 28, 672
-, **3-amino- 29, 943**
-, **4-amino-**
- from
  ynamines and 4H-3,1-
  benzoxazin-4-ones 28, 871
  - and ketenimines 27, 699;
  29, 647
- startg. m. f.
  quinolines, 1,4-dihydro-
  4-imino-, 1-subst. 26, 401
-, **2-anilino-**
- special s.
  quinolines, 2-(o-acylanilino)-
-, **4-cyano-**
- from
  quinolinium salts 27, 679
-, **1,2-dihydro-**
- startg. m. f.
  cycloprop[b]indoles,
  1,1a,2,6b-tetrahydro- 30, 537
-, **1,4-dihydro-**
- startg. m. f.
  quinolinium salts 27, 679
-, -, **1-subst.**
- from
  quinolines 30, 214
-, **1,2-dihydro-1-acyl-**
- startg. m. f.
  quinol-2-ylcarbinols 28, 696
-, **1,2-dihydro-2-alkoxy-
  N-alkoxycarbonyl-**
- as reagents 23, 415; 30, 265
-, **1,2-dihydro-1-carbamyl-**
- startg. m. f.
  4-alkylidene-2-quinazolones
  26, 506
-, **1,4-dihydro-3-cyano-**
- startg. m. f.
  oxindoles 29, 231
-, **1,4-dihydro-4-cyano-**
- from
  quinolinium salts 27, 679
-, **1,4-dihydro-2-hydroxy- 29, 783**
-, **1,4-dihydro-4-imino, 1-subst.**
- from
  quinolines, 4-amino- and
  halides, prim. 26, 401
-, **3-hydroxy-**
- from
  o-aminoacetic acid esters
  26, 939
- startg. m. f.
  furo[3,2-b]quinolines 26, 939
  oxindoles 26, 504
-, **4-hydroxy- 27, 943**
- from
  o-aminoketones and orthocarboxylic acid esters
  29, 726
-, **1,2,3,4-tetrahydro-**
- from
  amines, ar. 27, 808
- startg. m. f.
  pyronobenzo[ij]quinolizines
  27, 898
2(1H)-Quinolinethiones
s. Thiocarbostyrils
Quinolinium salts
- from
  quinolines, 1,4-dihydro-
  27, 679
- startg. m. f.
  quinolines, 4-cyano- 27, 679
-, 1,4-dihydro-4-cyano- 27, 679

8-Quinolinol, 5-chloro-
–, peptide synthesis with –
  16, 424 suppl. **27**
Quinolizidones
– by rearrangement **26**, 312
Quinolizinium ring opening
  27, 350
Quinolizinones 28, 747
Quinol monoethers
–, hydrogenation, partial **22**, 49
  suppl. **27**
– startg. m. f.
  quinones **27**, 274
4(1H)-Quinolone ring closure
  s. a. under Conrad-Limpach
2-Quinolones s. Carbostyrils
4(1H)-Quinolones 21, 926;
  28, 788, 915
– with hetero-Cope-rearrangement **28**, 897
–, **1,3-dihydroxy- 28**, 447
– startg. m. f.
  quinoline-3,4-dione 1-oxides
  28, 447
Quinols (s. a. Quinhydrones)
– as reactants **29**, 718
– from
  quinones **28**, 31; **26**, 263
– startg. m. f.
  quinones **21**, 282 suppl. **28**;
  25, 174 suppl. **26**; **28**, 512,
  611
p-Quinols
– startg. m. f.
  p-quinones, 2,5-diamino-
  27, 368
Quinolsulfonium salts
– from
  quinones and thioethers
  27, 589
Quinol-2-ylcarbinols
– from
  quinolines, 1,2-dihydro-
  1-acyl- **28**, 696
8-Quinolyl sulfate as reactant
  27, 103 suppl. **29**
Quinone diazides, s. Diazo
  oxides
– **diimines**
– from
  p-diamines **30**, 323
–, reaction with phosphorous
  acid esters **27**, 303
– **dioximes**
– startg. m. f.
  dinitrosoarenes **26**, 930
– **epoxides**, rearrangement,
  skeletal **29**, 692
– **iminoximes**
– startg. m. f.
  quinone iminoximes,
  N,O-diacyl- **27**, 461
– –, **N,O-diacyl-**
– from
  quinone iminoximes
  27, 461
– **methids 27**, 266
– from
  polyhalogenomethanes
  28, 796
o-Quinone methids as unstable
  intermediates **27**, 760
–, reactions **30**, 203
– startg. m. f.
  o-α-aminophenols
  30, 203

Quinone monoimines
– from
  amines (2 molecules) **28**, 237
  phenols and diazocyanides
  27, 442
  quinones **27**, 395
– special s.
  N-halogenoquinone
  monoimines
Quinones (s. a. Quinhydrones)
–, C-alkylation, homolytic
  27, 907 suppl. **28**
–, C-allylation **23**, 860 suppl.**27**
–, annelation **28**, 741
–, arylation **28**, 611
– as reagent **27**, 532
–, o-dihydroxylation
  21, 858 suppl. **27**
– from
  arenes **25**, 389 suppl. **27**
  chromenes, 6-hydroxy- **27**, 160
  phenols **27**, 179; **28**, 721
  quinol monoethers **27**, 274
  quinols **21**, 282 suppl. **28**;
  25, 174 suppl. **26**;
  28, 512, 611
– special s.
  azidoquinones
  benzoquinone
  2,3-dichloro-5,6-dicyanoquinone
  1,2-diquinonylethylenes
  hydroxyquinones
  indole-4,7-diones
  naphthoquinones
  tetrachlorobenzoquinone
– startg. m. f.
  benzothiazole ring **27**, 617
  bridge bicyclo[3.2.1]octane
    ring **27**, 711
  quinols **26**, 263; **28**, 31
  quinolsulfonium salts
    27, 589
  quinone monoimines **27**, 395
–, **activated, negatively-subst.,**
  reactions **27**, 662
–, **extended**
– by
  coupling, oxidative **27**, 915
  dehydrogenation **16**, 915
    suppl. **26**
– special s.
  dipheno-quinones
  stilbene-
–, –, **condensed**
– special s.
  dibenzo[fg,op]naphthacenequinones
o-Quinones (s. a. Quinoline-
  3,4-dione...)
– special s.
  o-benzoquinones
  quinoline-5,6-diones **(25)**
– startg. m. f.
  benzenephosphonic acid
    esters, 2,3-dihydroxy-
    29, 592
  p-quinones **27**, 172
p-Quinones
– from
  o-quinones **27**, 172
–, reaction with enamines
  26, 717
– special s.
  amino-p-quinones
  diazido-

– startg. m. f.
  benzene ring, trimerization
    29, 718
  1,3-benzodioxepines,
    4,5-dihydro-7-hydroxy-
    28, 84
  coumarins, 3-amino-
    6-hydroxy **26**, 717
  2-imidazolone ring **27**, 405
  indoles, 5-hydroxy- **8**, 782;
    26, 717
  p-quinones, o-alkoxy- **27**, 174
–, **condensed**
– from
  o-dialdehydes **29**, 735
–, **o-alkoxy-**
– from
  p-quinones **27**, 174
–, **o-amino-**
– startg. m. f.
  benzoxazoles, 5-hydroxy-
    29, 238
–, **2-amino-3-chloro- 28**, 494
–, **2,5-diamino-**
– from
  p-quinols **27**, 368
–, **2,5-dihydroxy-**
– startg. m. f.
  pulvinic acid dilactones
    28, 199
Quinoxaline, 2,3-diphenyl-
  as reagent **30**, 179
– **1-oxide ring**
– startg. m. f.
  benzimidazole ring **30**, 341
– **ring**
– from
  N-nitrosanilines **29**, 454
Quinoxalines s. a. Pyrazines
–, **2-acylamino-**
– via quinoxalines, 1,2,3,4-
  tetrahydro-2-oximino- **28**, 372
–, **1,2,3,4-tetrahydro-**
– from
  1H-1,4-benzodiazepines,
    2,3,4,5-tetrahydro-
    4-hydroxy- **27**, 539
–, **1,2,3,4-tetrahydro-2-oximino-
  28**, 372
– startg. m. f.
  quinoxalines, 2-acylamino-
    28, 372
2(1H)-Quinoxalones
– from
  2(1H)-quinoxalones,
    3,4-dihydro- **26**, 461
–, **3,4-dihydro-**
– startg. m. f.
  2(1H)-quinoxalones **26**, 461
Quinuclidinecarboxylic
  acid esters **28**, 434
**Quinuclidines, Δ²-dehydro-
  4-hydroxy-**
– from
  4-piperidones **30**, 471
Quinuclidinium salts, 3-acetoxy-,
  as intermediates **29**, 71
3-Quinuclidones
– startg. m. f.
  7a,11a-ethanooxyindolo-
    [1,2-h][1,7]naphthyridin-
    10(11H)-ones, 6,7,8,9-
    tetrahydro- **30**, 300
  pyrido[1,2-a]indol-9(8H)-ones,
    6,7-dihydro- **30**, 300

Racemates s. under Stereoisomers
Racemization
–, prevention of – s. retention
of optical activity under
Stereoisomers
Radiation s. Irradiation,
Photo...
Radical anions, ar. 18, 711
suppl. **27**
– **cations 28,** 703; **29,** 721
–, nitration via – **27,** 369
– special s.
  nitrogen radical cations
– **inhibitor 26,** 584 suppl. **27**
  2,6-di-*tert*-butyl-p-cresol
  29, 540
  1,1-diphenylethylene **29,** 202
– **reactions**
  addition of mercaptans,
  review **16,** 637 suppl. **26**
  – of phosphonous acid
    monoesters **11,** 712
    suppl. **29**
  – to carbon-carbon double
    bonds, review **8,** 745
    suppl. **26**
  –, geo- and stereo-specific
    27, 530
  –, stereospecific,
    to cobaltocenes **27,** 813
  C-alkylation **27,** 722, 879, 907;
  28, 880
    – of N-heterocyclics
      27, 907
    – of quinones **27,** 907 suppl. **28**
    amination with ring closure
    27, 332
  arylation, homolytic, of
    heterocyclics, review
    16, 851 suppl. **27**
  C-carbalkoxylation **28,** 879
  C-carbamylation, homolytic
    25, 480
  rearrangement of enol
    sulfonates **26,** 588
  ring closure of ethylene-
    halides **29,** 970
  – –, review **18,** 788 suppl. **26**
  – – to thioenolethers, cyclic
    26, 587
  –, review **27,** 907 suppl. **28**
  – with
    boranes, review **27,** 138
    chlorosulfonyl isocyanate
    and ethylene derivs. **26,** 613
  – **relay halogenation 27,** 188
    suppl. **29**
Radicals
– special s.
  acyl radicals
  anion –
  hydrazidinyl –
  nitrogen –
  N-oxide –
– **scavenger**
–, galvinoxyl as – **22,** 434
  suppl. **26**
– **source,** nitrite photolysis
  as – **28,** 933
Radioactivity s. under Compds.,
  labeled
Ramberg-Bäcklund
rearrangement 27, 627
– –, **bishomoconjugative**
–, 9-thiabicyclo[4.2.1]nona-
  2,4,7-triene 9,9-dioxide ring
  by – **26,** 963

Ramberg-Bäcklund-type
rearrangement 29, 100
Raney catalysts s. Nickel
Rays s. Irradiation
Reaction, terminal
  s. ω-Halogenation
–, **transannular**
  s. Transannular...
Reactions (s. a. Photoreactions,
  Solid-phase reactions)
–, control, kinetic and thermodynamic of – **15,** 627
  suppl. **29; 29,** 831
–, **biochemical** s. Enzymes,
  Microorganisms
–, **decarboxylative**
– special s.
  C-alkylation, decarboxylative
  aromatization, –
  condensation, –
  elimination, –
  Mannich reaction, –
  nitration, –
  – via carboxylic acid dianions
    26, 923
–, **homoelectrocyclic**
  s. Homoelectrocyclic
  reactions
–, **lethargic** s. Lethargic
  reactions
–, **multicomponent** s. Multicomponent reaction
–, **nucleophilic** s. Nucleophilic
  reactions
–, **photochemical** s. Photo...
–, **stereospecific** s. Stereospecific reactions
–, **symmetry-disallowed**
  s. Symmetry-disallowed
  reactions
Reactor s. a. Column reactor,
  Reflux –
–, **continuous** (s. a. Brush
  reactor, continuous)
  26, 39; **28,** 148
Reagents, allylophilic s. Allylophilic reagents
–, **polymer-based** s. Polymer-
  based reagents
Rearrangement
(s. a. Isomerization)
– of
  aminoalcohols **28,** 123
  benzazoles, 2-aryloxy-
    14, 401 suppl. **29**
  boron compds., organo-,
    review **28,** 866
  cyclobutane-1,2-diols
    28, 677
  2,5-cyclohexadienone tosyl-
    hydrazones **27,** 754
  2,2-dihalogenalcohols **28,** 945
  ethers with ring closure
    30, 477
  ethylenecarbohydrates
    27, 738
  β,γ-ethylene-α-hydroxy-
    ketones **29,** 171
  ethylenenitrites, cyclic
    27, 343
  1,2-halogenhydrins **24,** 964;
    28, 946
  α-halogenoboranes **27,** 118
  halogenonitrones **27,** 528
  hydrazinium salts **27,** 459
  ketene mercaptals **26,** 582
  β-ketocarboxylic acid
    hydrazides **26,** 464

5-oxazolones **27,** 985
oxido compds., review **28,** 4
α,β-oxidosulfones **26,** 584
α,β-oxidosulfoxides **26,** 584
penicillin ring S-oxides
  26, 629
(polyhalogeno)silanes
  26, 523
pyridinium salts, N-aryloxy-
  27, 910
side chains in cyclopropanes
  27, 728
spirocyclohexadienes **27,** 911
sulfido compds. **26,** 586
2H-1,4-thiazines, 3,4-dihydro-
  29, 954
triazolotriazines **26,** 315
N-ylids, cyclic **28,** 298
– special s.
  acyllactone rearrangement
  acyloin –
  allyl –
  allylsulfonium ylid-type –
  van Alphen –
  anthrasteroid –
  aziridine ring –
  tert-butyl –
  Claisen –
  configuration, inversion of –
  Cope rearrangement
  diaxial-diequatorial –
  dienone-phenol –
  dihydroisoquinoline –
  elimination –
  enaminelactone –
  furan ring –
  Gabriel-Colman –
  homoallyl –
  lactams –
  lactone –
  Lossen –
  migration
  position shift
  retro-benzilic acid-type
    rearrangement
  steroid backbone –
  Stevens –
  Süs –
  1,2,4-triazolo ring –
  Wagner-Meerwein –
  Walden inversion
  Wittig rearrangement
– to
  α-aminocarboxylic acid
    amides **26,** 310
  α,β-ethylene-α-hydroxy-
    ketones **30,** 524
  2-ethylenisothiouronium
    salts **30,** 41
  1,3-halogenamines **26,** 534
  β-halogenooxo compds.
    21, 601
  α-methylenelactams **30,** 327
  phenothiazine ring **26,** 630
  phosphorins **29,** 710
  quinolizidones **26,** 312
  (α-sulfonylalkylidene)-
    phosphoranes **30,** 404
– with simultaneous
  formation of azomethines
    from azides **29,** 460
  – of dihalides from ethylene
    derivs. **26,** 517
  – of oxo compds. from
    ethylene derivs. **27,** 141
Prins reaction 29, 679
replacement of acoxy by
  amino groups 28, 326

- of halogen by amino
groups **30,** 296
- of halogen by hydrogen
29, 76
- of halogen by hydroxyl
27, 221
-, **ang-lin- 30,** 509
-, *cis→trans-* **28,** 671
- with formation of semi-
carbazones from ketones
2, 379 suppl. **26**
-, *cis→trans-,* partial **28,** 669
-, *exo→endo-* **26,** 517
-, *lin→ang-*
- of
imidazole ring, N-condensed
26, 339
-, *syn→anti-,* of amidines
30, 534
-, *trans→cis-* **28,** 669
-, **anionotropic**
- of
3-acetylenealcohols **26,** 537
-, **electrooxidative 26,** 158
-, **lateral** s. -, [3.3]-sigma-
tropic, -
-, **nucleophilic**
s. N,N-Rearrangement
-, **oxidative**
- of
carbazoles, 2,3,4,4a-tetra-
hydro- **27,** 131
1,4-cyclohexadienes, 3-alkyl-
27, 839
o-hydroxycarbinols **28,** 207
-, radical **27,** 488
-, **[2.3]-sigmatropic**
- of
ethylenemercaptals **29,** 709
sulfonium ylids, unsatd.
29, 781
thioethers, unsatd. **29,** 709
suppl. **30**
- to
α,β-ethyleneketones **29,** 261
-, **[3.3]-sigmatropic**
- of
allyl dithiocarbamates
30, 585
di(azomethines) **24,** 746
suppl. **29**
2-ethyleneiminoesters **29,** 325
-, -, **lateral,** of indolenines
30, 503
-, -, **thermal 28,** 124
-, **skeletal** (s. a. Cycloisomerization, Isomerization,
skeletal) **24,** 750; **28,** 702
- by retroaldol-aldol
condensation **29,** 990
- of
adamantan-1-amines,
N-halogeno- **28,** 168
seco-cubane derivs. **26,** 762
dienones, bicyclic **26,** 126
α,β-ethyleneketones, cyclic
27, 134 suppl. **29**
glycol monosulfonates,
cyclic **27,** 974
N-heterocyclics **30,** 533
O,O,N-heterocyclics **26,** 111
isocycles **28,** 89, 217
Mannich bases, cyclic
27, 752
metacyclophane ring,
halogeno- **29,** 708
poly-S-heterocyclics **29,** 966

pyrrolidinium salts,
N-cyanomethyl- **29,** 901
quinone epoxides **29,** 692
spirocyclics **27,** 173
spirodienones, review **18,** 937
suppl. **28**
tricyclo[4.2.1.0^{2,5}]non-3-enes
26, 760
- special s.
1,3-cyclohexadione ring
rearrangement, skeletal
Stevens-type rearrange-
ment, -
- to
adamantanes **27,** 746, 758
azulenes, hydro- **28,** 686
cyclopent-2-enecarbox-
aldehydes **26,** 728
α-diketones **29,** 681
α-hydroxyketones **28,** 890
Δ^{8,14}-morphinanes **30,** 675
norcaranes **27,** 134 suppl. **28**
- with
alkoxy group migration
28, 691
aromatization **28,** 894;
29, 906
cycloaddition **30,** 475
decarboxylation **29,** 980
dehydrogenation **28,** 887;
29, 905
dehydrohalogenation **29,** 966
ether cleavage **29,** 987/8
formation of alcohols from
oxido compds. **20,** 49
suppl. **28; 29,** 31
- of hydrocarbons
from tosylhydrazones
30, 34
- of lactones **28,** 218
halogenation **27,** 566
ring expansion of O-heterocyclics **30,** 92
O-tosylation **29,** 699
-, -, **anion-induced 27,** 189
-, -, **light-specific 29,** 328
-, -, **oxidative,** of N-heterocyclics **28,** 99, 129
-, -, **stereospecific 27,** 209;
28, 899; **30,** 632
- of
oxidotetrahydrofurano-
lactones **28,** 3
quinolines, 1,2-dihydro-
30, 537
1,5-Rearrangement 26, 588;
28, 124
N,N-Rearrangement, ar.,
nucleophilic 26, 310
α,ω-**Rearrangement 30,** 225
Rearrangements, simultaneous
-, chiral and geometric **30,** 406
-, reductive, by lithium tetrahydroaluminate, review
21, 538 suppl. **29**
-, silver ion-catalyzed, of polycyclics, review **28,** 691
suppl. **30**
Recyclization, 1,2-N,N-hetero-
cyclics by -, review **23,** 440
suppl. **28**
Red-Al s. Sodium dihydrobis-
(2-methoxyethoxo)aluminate
Redistribution s. Comproportionation, Dis-
Red-light irradiation 12, 151
suppl. **28**

Redox... s. a. Ring closure,
redoxidative
Redoxcondensation
-, 1-acylguanidines, 3-subst.
by - **29,** 356
Redoxdimerization of dienes,
cyclic **30,** 513
Redoxesterification
-, carboxylic acid esters,
active, by - **26,** 175
Redoxidation of N-oxides,
cyclic **30,** 161
-, **intramolecular,** glycolic acid
esters O-alkyl-, by - **30,** 115
Redoxphosphorylation 26, 175
Redox reactions, base-catalyzed
23, 401 suppl. **28**
Reducing agents
pyrocatechyl phenyl phosphite
29, 993
- -, **bulky**
potassium hydridotri-*sec*-
butylborate **12,** 64 suppl. **29**
- -, **nucleophilic**
lithium hydridotriethylborate
28, 60
Reduction (s. a. Clemmensen
reduction, Elimination,
reductive, Hydrogenation,
Photoreduction, Ring
closure, reductive)
-, asym. **27,** 38 suppl. **30; 29,** 36
-, -, reversal of stereo-
specificity **17,** 865 suppl. **28**
-, **partial**
- of
nitriles to amines, prim.
26, 39
-, **preferential**
- of
carboxylic acid esters
11, 114 suppl. **27**
α,β-ethylene-α-hydroxy-
ketones **28,** 51
keto to methylene groups
21, 91 suppl. **26**
oxido compds. **27,** 35
-, **regiospecific**
- of
oxido compds. **27,** 35
suppl. **29**
-, **selective 28,** 50
- of
ammonium salts, quaternary
29, 37
carbon-carbon double bonds
26, 49
of enamines **28,** 38
carboxylic acid esters
26, 57; **29,** 39
- acids **29,** 57; **30,** 29
nitro compds. **27,** 14
nitrones **30,** 696
oxo to hydrocarbon groups
26, 65; **29,** 63
thiocarbonyl groups **26,** 40
- with sodium trihydridocyanoborate **27,** 74
- - - -, review **27,** 74 suppl. **30**
-, **stereospecific**
- of
(1,3-dien)oxysilanes **27,** 83
enolethers **26,** 54
α,β-ethyleneketones **26,** 134
- to ketones **27,** 50
α,β-ethylenesulfones to
ethylene derivs. **23,** 784
suppl. **28**

(Reduction, stereospecific
– of)
  ketocarboxylic acid esters
    12, 64 suppl. 29
  ketolactones 27, 505
  ketones to alcohols, sec.
    24, 48 suppl. 29; 26, 33;
    27, 38; 27, 39 suppl. 30;
    28, 30; 29, 35
  oxido compds. 11, 59
    suppl. 26; 27, 35
  oximes to amines 24, 26
  sulfoximines 28, 22
– to
  1α,3β-dihydroxy-Δ⁵-steroids
    29, 32
  cis- and trans-fused rings
    18, 65 suppl. 27
Reductions with
  boron hydrides, sulfurated,
    review 24, 49 suppl. 28
  hydridoalkoxoaluminates,
    review 27, 37
  phosphorus/iodine/carboxylic
    acids, review 21, 55
    suppl. 26
Reductones, review 16, 973
  suppl. 27
Reflux reactor 26, 921
Reformatskii products, reaction
  26, 938
– route, alternate, to
  phthalides 28, 679
– synthesis CC⇓OC.Zn;
  26, 673
– of
  2-azetidinones 28, 816
  β-hydroxynitriles 26, 672
–, review 27, 670
– with
  α-halogenocarboxylic acid
    silyl esters 27, 670
– –, asym. 2, 827 suppl. 26
Regiospecific effect of
  complexing agents 29, 619
Regiospecificity (s. a. Direction)
  29, 509
– of hydroboration 14, 82
  suppl. 26
– special s.
  N-acylation, regiospecific
    addition, –
  aldol condensation, –
  C-alkylation, –
  coupling, –
  geoisomerization, –
  Marknownikoff, reduction, –
    orientation
  oxymercuration-demercura-
    tion, regiospecific
    reduction, –
  ring expansion, –
  ring opening, –
  syntheses, –
**Reimer-Tiemann aldehyde
synthesis** s. Photo-Reimer-
Tiemann synthesis
Reissert compds.
–, C-alkylation, intramolecular
  28, 942
– special s.
  2-acyl-1,2-dihydroisoquinaldo-
    1-nitriles
– startg. m. f.
  carboxylic acids, N-hetero-
    cyclic 16, 111
Reissert-Kaufmann reaction
  15, 587

Relayed introduction
  s. Introduction, relayed
Remote functionalization
  s. Functionalization, remote
– **steric control** s. Steric
  control, remote
Removal s. Elimination,
  Protective groups, removal
Replacement (s. a. Substitution, Trans...)
– **of acoxy groups by**
  amino groups 28, 325
– – with rearrangement
  28, 326
  hydrogen 18, 124; 26, 984;
    28, 46
  m-xylyl groups 26, 468
– **of C-acyl by**
  hydrogen s. C-Deacylation
– **of acylamino groups by**
  cyano groups, selective
    replacement 30, 576
  organothio –, – – 30, 576
– **of aldehyde groups by**
  halogen 29, 525
  hydrogen s. Decarbonylation
    of aldehydes
– **of alkoxy groups by**
  fluorodinitromethyl groups
    27, 766
  halogen s. Halides from
    ethers
  hydrogen, preferential
    replacement 29, 59; 30, 47
– **of alkyl groups by**
  halogen 28, 512
  hydrogen 26, 83
– **of alkylseleno groups by**
  hydrogen 29, 83
– **of alkylthio groups by**
  alkoxy groups 28, 183
  amino groups, ar. 27, 482
  – –, partial replacement
    30, 309
  hydrazino groups 28, 408
  oxygen 29, 408
– **of amino groups by**
  alkoxy groups 26, 191
  halogen 30, 369
  hydrogen s. Deamination
  oxygen 28, 152
  sulfur 27, 621
– **of N-amino groups by**
  hydrogen s. N-Deamination
– **of P-amino groups by**
  P-halogen 22, 540
– **of ammonium groups by**
  fluorine 28, 507
  nucleophiles 28, 507
– **of azo groups by**
  nitro groups 30, 287
– **of bromine** (s. a. halogen) by
  chlorine 26, 258; 27, 581;
    29, 521, 821
–, preferential replacemnt
  30, 135
– **of carbalkoxy groups by**
  hydrogen (s. a. De-
    carbalkoxylation) 14, 132;
    29, 84, 963
– **of carbonyl oxygen by**
  2 carbon substituents 9, 139
    suppl. 28, 29
  – –, functionalized 26, 900;
    28, 837
  sulfur 22, 799
– **of chlorine** (s. a. halogen) by
  bromine 28, 510

fluorine 28, 516; 30, 371
–, preferential replacement
  26, 551
iodine 30, 372
– **of chlorosulfonyloxy
groups by**
  chlorine, preferential
    replacement 26, 556
– **of cyano groups by**
  hydrogen 26, 756; 27, 87
  α-hydroxyalkyl groups 28, 877
– **of diazo groups by**
  hydrogen 27, 70; 29, 69
– **of fluorine** s. halogen
– **of formyl groups**
  s. aldehydes groups
– **of halogen** (s. a. individual
  halogens)
–, preferential replacement
  25, 592; 27, 854
– – – **by**
  acetonyl groups 30, 625
  acoxy – 30, 318
  – – with position shift
    28, 169
  alkoxy groups 30, 318
  – –, preferential replace-
    ment 27, 220
  – – with position shift
    27, 212
  alkyl groups 29, 857
  amino – 26, 310, 422; 27, 459;
    28, 16
  – – with rearrangement
    30, 296
  – – with ring expansion
    28, 389
  azido groups, preferential
    replacement 29, 424
  cyano groups 29, 821
  deuterium 17, 117 suppl. 27;
    21, 35 suppl. 29; 30, 39
  hydrogen HC↓↑Hal; 26, 73;
    29, 669
  –, partial replacement
    26, 70; 29, 73
  –, preferential – 27, 72, 77/8;
    28, 62; 29, 73
  hydroxyl 27, 232; 29, 207;
    30, 135
  – with rearrangement
    27, 221
  oxygen 26, 210
  sulfhydryl 26, 617
  tritium 19, 91 suppl. 29;
    21, 35 suppl. 26
– **of halogen in** iminocarbonylhalides 29, 578
– **of halogen, ar. or vinylic, by**
  cyano groups 29, 845
  hydrogen 29, 77
– **of group IV-element-
halogen by**
  alkylseleno groups 28, 520
  alkylthio – 28, 520
– **of N-halogen by**
  N-hydrogen 27, 548
– **of tert-halogen by**
  hydrogen 27, 72
– **of hydrogen by**
  acoxy groups (s. a. Acoxyla-
    tion) 26, 159
  aldehyde – in ar. rings
    s. C-Formylation
  alkoxy – (s. a. Alkoxylation)
    29, 177
  alkylthio – 27, 638
  aryl s. Arylation

azido groups **30**, 232
carbamyl **28**, 707
chloroformyl groups **29**, 714
cyano groups **26**, 766;
  28, 704; **30**, 539
deuterium s. Deuteriation
halogen HalC↓↑H; **30**, 354
–, preferential replacement
  14, 595 suppl. **28**; **29**, 506
hydroxyl (s. a. Hydroxylation) **27**, 123/4, 917
organothio groups **26**, 622
sulfhydryl **26**, 593
sulfonic acid groups
  s. Sulfonation
thiocyano groups
  s. Thiocyanation
tritium s. Tritiation
– **of hydrogen, acidic by**
alkyl groups **28**, 753
– **of N-hydrogen by**
N-fluorine **24**, 296
– **of hydroxyl** (s. a. oxygen) **by**
amino groups **29**, 372, 379;
  30, 273
– – –, prim. **12**, 454
aryl **11**, 847
halogen HalC↓↑O; **27**, 573;
  28, 496, 498, 980; **29**, 798;
  30, 227, 362, 366
–, partial replacement **27**, 579
–, preferential replacement
  26, 545; **27**, 18; **28**, 499
– with allyl rearrangement
  28, 503
hydrogen HC↑↑O; **27**, 74, 872;
  30, 50
– **of imino groups by**
oxygen **28**, 159
sulfur **28**, 555
– **of iodine** (s. a. halogen) **by**
chlorine **29**, 521
– **of nitro groups by**
azido groups **26**, 607
halogen **26**, 548
sulfonyl groups **26**, 607
– **of nitrosamino groups by**
aryl **27**, 814
– **of oxygen** (s. a. Carbonyl oxygen, hydroxyl) **by**
amino groups **26**, 378, 447;
  28, 463
– –, preferential replacement
  28, 350
halogen **26**, 538; **28**, 903
fluorine **30**, 365
sulfur **28**, 550
– – –, **cyclic, by**
nitrogen, cyclic **30**, 248
sulfur, – **29**, 558
– o**f Si-oxygen by**
Si-fluorine **18**, 602 suppl. **27**
– **of phosphonic acid ester groups by**
hydrogen **29**, 82
– **of phosphonyl groups by**
amino groups **27**, 487
– **of P-selenium by**
P-oxygen **29**, 119
– **of siloxy groups by**
acoxy groups, preferential replacement **29**, 223
– **of silyl groups 27**, 883
– – – – **by**
deuterium **27**, 20
dichlorophosphino groups
  30, 441

– **of sulfhydryl by**
amino groups **2**, 341 suppl. **28**
oxygen **29**, 218
– **of sulfinic acid groups by**
hydrogen **29**, 90
– **of sulfinyl groups by**
halogen **29**, 522
– **of sulfonic acid groups by**
aminomethyl groups **27**, 878
– **of sulfonyl groups by**
halogen **29**, 522
hydrogen **23**, 784; **29**, 792
– **of sulfur by**
oxygen **28**, 180/1
– – –, **cyclic by**
nitrogen, cyclic, with
  deuteriation **26**, 432
– **of P-sulfur by**
P-oxygen **29**, 119
– – –, **cyclic by**
P-silylnitrogen, cyclic,
  partial replacement **29**, 296
– **of thiolcarbalkoxy groups by**
hydrogen **26**, 78
– **of tin, cyclic by**
carbon, cyclic **26**, 625
– **of trifluoromethyl by**
amino groups **27**, 494
– **of trihalogenomethyl**
  s. trifluoromethyl
–, **geospecific**, α,β-ethylenehalides by – **26**, 552; **29**, 497
Resins s. Ion exchangers, Polymers
Resolution s. under Stereoisomers
Resorcinol as chlorine scavenger **30**, 66
– **monoesters** s. 1,3-Diol monoesters
Resorcinols, synthesis **2**, 648
–, labeled **2**, 648 suppl. **27**
–, 2-subst., by aromatization
  20, 674 suppl. **29**
Retention of
acetal groups **29**, 257
activity, optical s. under
  Stereoisomers
carbalkoxy groups in
  reduction **12**, 61 suppl. **28**;
  26, 66 suppl. **29**;
  27, 69 suppl. **29**
carbon-carbon double bonds
  26, 90 suppl. **28**; **27**, 230
Δ⁸-steroid double bonds
  25, 695 suppl. **29**
– triple bonds during
  ozonolysis **11**, 308
  suppl. **29**
chlorosulfonyl groups in
  Friedel-Crafts ketone
  synthesis **17**, 879 suppl. **29**
configuration s. Configuration
cyano groups in Grignard
  syntheses **28**, 607
α-diazoketo groups **26**, 967
fluorine with removal of
  chlorine **16**, 24 suppl. **26**
geometry s. Geometry
halogen s. Halogen
hydroxyl, benzylic **27**, 88
ketal groups s. Ketal groups
lactone groups in Dieckmann
  cyclization **22**, 898 suppl. **28**
– – in reduction **12**, 61
  suppl. **28**

nitro groups s. Nitro groups
N-oxide radicals **19**, 448
  suppl. **29**
oxido groups in hydrogenations **27**, 58
oxygen in hydrogenations
  25, 46 suppl. **26**
–, allylic **27**, 56
phenolether groups **18**, 78
  suppl. **27**; **27**, 11
phenol groups **6**, 209 suppl. **26**
– rings s. Phenol ring
phosphorus in Wittig
  syntheses **26**, 686
Retention, preferential, of
deuterium **27**, 564
Retroaldol-aldol condensation, intramolecular 29, 990
Retroaldol reaction, ring
opening by – **26**, 227
Retro-benzilic acid-type rearrangement
–, imidazoles by – **28**, 959
Retro-Claisen rearrangement
30, 103
Retrocoupling, oxidative
– of azo compds. **29**, 284
Retrodiene scission
– with elimination of sulfenes
  28, 971
– –, **reductive 27**, 959
Retro-Dimroth rearrangement
10, 282
Retro-Mannich reaction 27, 85
Rheadan ring skeleton
– via benzazepine ring,
  dihydro- **27**, 505
Rhenium sulfide 29, 389
Rhodanamine
– startg. m. f.
thiazoles, 2-amino- **27**, 607
–, thiocyanation with – **28**, 542
Rhodanides s. Thiocyanates
Rhodanine ring opening 3, 662
Rhodium-alumina 21, 116
  suppl. **27**; **26**, 52
Rhodium(II) acetate 29, 787
Rhodium-calcium carbonate
4, 667 suppl. **26**
Rhodium-carbon 16, 106 suppl.
  26; **20**, 55 suppl. **29**
Rhodium carbonyl 4, 667 suppl.
  26; **28**, 363, 766
–, reductions with – **28**, 18
– **catalyst, diphosphine-, optically active 27**, 57
– **catalysts 4**, 667 suppl. **28**
– **complexes** as intermediates
  29, 56
–, hydroformylation with –
  4, 667 suppl. **27**, 28
– special s.
  bis(pyridine)dimethylformamidedichlororhodium tetrahydridoborate
rhodium(I) dicarbonyltrifluoroacetylacetonatorhodium phosphine complex
  catalysts
– –, **chiral 23**, 48 suppl. **29**
– –, **polymer-supported**,
  as catalysts **26**, 642
  suppl. **27**
– –, **triarylphosphine-**
– as reactants **29**, 857
– as reagents **27**, 852
  suppl. **28**

(**Rhodium complexes,
triarylphosphine-**)
– special s.
  chlorotris(triphenylphosphine)-
    rhodium(I)
  cyclobutadiene rhodium
    triarylphosphine complexes
  hydridotetrakis(triphenyl-
    phosphine)rhodium(I)
–(I), **dicarbonyltrifluoro-
  acetylacetonato-**
  – as reactant **29**, 195
  – **oxide 27**, 723; **28**, 665
  – **phosphine complex catalysts**
    (s. a. Rhodium complexes,
    triarylphosphine-) **4**, 667
    suppl. **27**; **9**, 129 suppl. **26**;
    26, 50 suppl. **28**
  –, **tetracarbonyldichlorodi-
    9**, 265 suppl. **29**; **26**, 741;
    30, 525
  – **trichloride 24**, 332 suppl. **26**
**Riebsomer 2,3-piperazinedione
  ring closure 6**, 408
Ring-chain isomers
– of
  oxocarboxylic acids and
    derivs., review **21**, 899
    suppl. **29**
– **tautomerism**
– special s.
  tetrazole-azidoazomethine
    tautomerism
  1,2,3-triazoline-2-diazo-
    amine –
Ring cleavage s. Ring opening
– **closure** (s. a. Annelation,
  Cycloaddition, Cyclodehydra-
    tion, Recyclization) O
– by
  Clarke-Eschweiler reaction
    19, 845
  S-cyanoethylation **30**, 397
  isomeriation s. Cycloiso-
    merization
  Mannich reaction s. under
    Mannich reaction
  nitrosation **28**, 315
  oxidation, electrochemical
    29, 900
  substitution, nucleophilic,
    intramolecular **26**, 186
  o-thio-Claisen rearrangement
    27, 755
  valence isomerization **26**, 757
  Wittig synthesis **28**, 854
– by use of
  polyphosphoric acid, review
    8, 899 suppl. **26**
–, gem-dimethyl effect **27**, 729;
  29, 937
–, direction of –, C- or N-ring
  closure **5**, 355
–, direction of –, to O-heterocycles **28**, 107
–, high dilution technique
  12, 950 suppl. **28**; **13**, 318
  suppl. **29**; **24**, 728 suppl. **28**;
  26, 614, 963; **28**, 515; **30**, 510
–, – – –, review **26**, 614 suppl. **28**
– of
  dialdehydes with diamines
    27, 384
  ethylenenitriles **30**, 121
  ethylenetosylamines **30**, 313
  hydroxyketones **27**, 945
  ketones **27**, 934; **30**, 491
  1,3,5-trienes **28**, 929

– special s.
  Bischler-Napieralski ring
    closure
  carbanion ring –
    cycloalkenylation
  Möhlau-Bischler indole ring
    closure
  Pschorr ring –
– via
  O-alkyllactims **29**, 425
  amide acetals **26**, 933
  o-Claisen rearrangement
    30, 677
  iminoester fluoroborates
    26, 933
  pyrazoles, 4-oxo- **29**, 766
– with
  acetals **26**, 304, 955; **27**, 381,
    511; **28**, 357, 439, 444, 922;
    29, 943, 975
  ketals **22**, 927 suppl. **27**
  2-acetylenealcohols **17**, 936
    suppl. **27**
  acetylene derivs. **28**, 668
  – electron acceptor-
    donors **26**, 694
  α-acylaminoketones **26**, 603
  alcohols **8**, 911; **29**, 449
  aldehydes **27**, 801; **28**, 719;
    29, 922, 927
    formaldehyde **22**, 778;
      29, 353
  alkoxyhalides **27**, 279
  1-alkoxylactams **29**, 784
  O-alkyllactims **29**, 350
  amidines, review **25**, 257
    suppl. **27**
  amidrazones **26**, 270
  o-amino-N-heterocyclics
    13, 535
  aminonitriles **27**, 334; **30**, 252
  S-aminosulfoxonium salts
    26, 896; **27**, 873
  azidoquinones **30**, 400
  azo compds. **26**, 482
  bis(α-diazoketones) **26**, 963
  α,α'-bisdiazoketones **29**, 766
  bis(trifluoromethyl)ketene
    27, 691
  bis(trifluoromethyl)thioketene
    27, 691
  boronic acids **27**, 307
  tert-butyl groups
    s. gem-dimethyl effect
  carbodiimides and
    dicarboxylic acid chlorides
    26, 511
  carbon disulfide s. under
    Carbon disulfide
  cyanocarboxylic acid halides,
    review **22**, 560 suppl. **28**
  cyanogen bromide **26**, 479;
    29, 335
  dialdehydes **27**, 384; **29**, 735;
    30, 609
  diazo compds. **29**, 465
  diazomethane **25**, 587
    suppl. **26**
  dienes **28**, 667
  o-dihalides **28**, 560
  dimethylformamide dimethyl
    acetal **29**, 469
  di(sulfenylchlorides) **26**, 608
  diynes **26**, 725
  – via transition metal
    complexes, review **26**, 725
    suppl. **30**
  enacylamines **25**, 522; **29**, 914

enamines **28**, 633; **29**, 543
enazides **29**, 401
enolesters **26**, 986
enolethers **26**, 474, 954
enol sulfonates **27**, 935
ethers **29**, 913
1-halogenenamines **28**, 822
  suppl. **29**
N-(halogenoformyl)imino-
  halides **28**, 383
o-halogenoketones **22**, 467
  suppl. **29**
hydrazones **29**, 467
  aldehyde hydrazones
    27, 355
o-hydroxyaldehydes **27**, 198
o-hydroxylaminocarboxylic
  acid esters **26**, 346
iminoesters **27**, 478;
  29, 482, 744
isoquinolines, 3-hydroxy-
  1-mercapto- **29**, 570
ketenes **27**, 675
β-ketosulfoxides **29**, 926
lactones **30**, 653
malonic acid aryl esters
  27, 383
Mannich bases **24**, 822; **26**, 965
methyl iodide **29**, 676
ninhydrin **23**, 763; **29**, 351
nitriles **27**, 819; **30**, 532
oxalyl chloride **30**, 241
N-oxides **29**, 457
oxidonitriles **28**, 983; **30**, 517
oximes **26**, 478
  syn-oximes **21**, 427
phosgene s. under Phosgene
phosphorimidates **28**, 461
poly-β-carbonyl compds.
  27, 924
propargyl ethers **29**, 697
pyrroles, 1-aryl- **28**, 719
squaric acid 1,2-dihydrazides
  29, 353
sulfone diimines **26**, 354
sulfonic acid esters **28**, 446
sulfonium salts **17**, 889;
  28, 840
sulfoxides **29**, 250
sulfoxonium salts **17**, 889
sulfur **27**, 619; **29**, 534
– compds. **30**, 609
tetracyanoethylene **21**, 728
  suppl. **26**; **30**, 475
thetin anions **26**, 895
thiophosgene **23**, 690
thiosulfonium salts, cyclic
  21, 639 suppl. **28**
tosylates (N-ring closure)
  16, 926; **27**, 507
Vilsmeier salts **27**, 453
– with simultaneous
  amination, radical **27**, 332
  elimination of substituents
    26, 452; **29**, 914
  ether cleavage **26**, 244, 252
  formation of alcohols from
    oxido compds. **27**, 740
  ring expansion **26**, 837;
    27, 958; **28**, 935
  Walden inversion **26**, 471
–, **cationic 17**, 936 suppl. **29**
–, **decarboxylative** (s. a. Ring
  closure, double, decarboxylative) **28**, 987
–, **double** (s. a. Annelation,
  double)

- by
  cyclodehydrogenation 20, 645
  suppl. 27
  cycloisomerization 26, 763
- of
  aldehydes, unsatd. 26, 780
  dicarboxylic acid anhydrides
  28, 745
  – – imides, vinylogous
  29, 713
  ethylenediazo compds.
  18, 962 suppl. 26
  ethyleneketones 26, 747
  ethylenetosylhydrazones
  27, 957
- to
  benz[b]indeno[2,1-d]thiophenes 28, 531
  benzofuro[3,2-c][1]benzopyrans, 6a,11a-dihydro-
  29, 684
  1,4-diazabicyclo[4.n.0]alkan-5-ones 27, 464
  α,β-exo-ethylenelactones
  19, 786
  N-heterocyclics, N-condensed
  26, 345, 407; 27, 380, 946;
  28, 748
  O-heterocyclics 26, 153;
  27, 725; 29, 682
  5H-imidazo[2,1-a]isoindol-5-ones, 1,2,3,9b-tetrahydro-
  30, 249
  imidazole ring, N-condensed
  26, 339
  1,1-iminoalcohols, 1,3-N,N-heterocyclic, N-condensed
  29, 364
  isocyclics 28, 953; 26, 725
  isoquinoline-1(2H),3(5H)-diones, 6,7-dihydro-
  28, 332
  1,4-naphthoquinone ring
  26, 725 suppl. 30
  10bH-oxazolo[3,2-c][1,3]-benzoxazine-2(3H),5-diones
  27, 353
  purines 28, 382
  pyrazole ring 26, 780
  pyrazolo[1,5-a]pyridinium
  salts 26, 484
  1H-pyrrolo[2,1-c][1,2,4]benzothiadiazine 5,5-dioxides,
  2,3-dihydro- 28, 457a
  12H,6H-quinazolino[2,3-c]-1,4-benzothiazin-12-ones
  26, 469
  3H,5H-thiazolo[4,3-b]isoindole-3,5-diones 27, 520
  1,3,5-triazine ring,
  N-condensed 30, 329
  s-triazolo[3,4-b]-1,3,4-oxadiazoles 27, 469
- with
  acetals 28, 926
  dicarboxylic acid chlorides
  27, 849
  diynes 26, 725
  ethylenediazo compds.
  29, 949
  ethylenetosylhydrazones
  30, 683
  o-hydroxylaminocarboxylic
  acid esters 26, 346
  malonic acid esters 27, 898
- with simultaneous positional
  interchange 29, 971

– –, **decarboxylative**,
  to O-heterocyclics 27, 989
– –, –, **reductive**
– to
  5H-1,4-benzodiazepin-5-ones,
  3,4-dihydro- 26, 476
  O,N-heterocyclics 29, 464
– –, –, **stereospecific**
– to
  cyclopenta[f]chromen-7(8H)-ones, 1,2,3,5,6,6a-hexahydro- 27, 792
  lactams 29, 817
  lactamols 28, 603
  Δ²-oxazolines, 2-amno- 29, 335
  steroid skeleton 29, 929
– –, **electrochemical, oxidative**
  28, 973
– –, –, **reductive** 28, 1
– –, **electrocyclic** s. a. Homoelectrocyclic reactions
– –, **1,7-electrocyclic** 27, 531
– –, **enzymatic, oxidative** 29, 904
– –, **intramolecular**, with
  addition, review 27, 561
– –, **multiple**
– to
  cage compds., isocyclic
  28, 640
– –, **nucleophilic**
– with cyclimmonium salts
  29, 382
– –, –, **intramolecular**
– at pyridine o-position 28, 427
– of
  cyclimmonium salts 25, 645
  suppl. 30
– –, **oxidative**
– of
  nitro compds. 27, 286
  oxo compd. N-derivs., review
  26, 232
  thiosemicarbazones 19, 705
– to
  benzene rings 30, 658
  benzazoles 24, 242 suppl. 29
  coumarins, 3,4-dihydro-
  27, 252
  furan ring 30, 158
  morphinanes 27, 915
  orthocarboxylic acid esters,
  cyclic 16, 317 suppl. 27
  1,3,4-oxadiazoles 26, 232
  N-oxides, cyclic 28, 317
  6H-1,3-thiazine ring,
  2,3-dihydro- 28, 966
– –, **oxidative, stereospecific**
  29, 243
– –, **photochemical** s. Photocyclization
– –, **redoxidative** 27, 530
– –, **reductive**
– of
  cyclimmonium salts 26, 761
  nitro compds. 27, 510
  nitrooximes 30, 335
– to
  N-heterocyclics 28, 307
  isocyclics 28, 35
– –, **regiospecific**
– to
  2-cyclohexenones 28, 910
  cyclopentenes 29, 866
– –, **solvolytic**, of brosylates
  27, 929
– –, **stereospecific**
– of
  diazoketones 28, 935

– special s.
  Pictet-Spengler ring closure,
  stereospecific
  Robinson annelation, –
– to
  2-azetidinone ring 7, 836
  suppl. 29
  aziridine ring 26, 60, 405
  cyclopropane ring 27, 864;
  29, 863
  furans, tetrahydro- 7, 355
  suppl. 28
  N-heterocyclics, N-condensed
  28, 758
  1H-imidazo[5,1-c][1,4]benzodiazepine-1,3,11-triones,
  2,3,5,10,11,11a-hexahydro-
  30, 241
  isatins, tetrahydro- 28, 751
  isocyclics 28, 947
  –, tricyclic 26, 744
  lactams 28, 349
  γ-lactones 26, 707; 28, 211
  pyrrolidines, N-acyl- 28, 667
– –, **transannular**
– of
  acetylene derivs., cyclic
  27, 757
  N-halogenamines 28, 460
– to
  alcohols from oxido compds.
  27, 740; 28, 4
  9-azabicyclo[3.3.1]nonane-1,5-diol ring 30, 205
  azulene ring 28, 108
  ketones, bicyclic 27, 745
  2-pyrrolidones 13, 834
  suppl. 26
  tetrazolylisocyclics,
  halogeno-, bicyclic 30, 351
– –, –, **stereospecific**
– to
  azulene ring 26, 749
o-Ring closure with bis(benzylalcohols) 28, 759
Ring contraction ⟲:
– by
  2 C-atoms, of α,β-ethyleneketones, cyclic 30, 507
– –, of N,N-heterocycles
  28, 303
  decarbonylation 28, 986;
  30, 691
  elimination of nitrogen 5, 643
  – of sulfur 25, 691 suppl. 28;
  27, 642
  fragmentation 28, 930
  Wolff rearrangement 13, 681
  suppl. 28; 22, 196
– of
  azides 26, 958; 30, 664
  o-azido-N-oxides 28, 452
  2-ethylenealcohols, cyclic
  26, 146
  α-halogenoketones, cyclic
  29, 984
  2-halogenolactamols 29, 471
  N-heterocyclics via
  ammonium salts,
  quaternary, cyclic 26, 297
  1,2-N,N-heterocyclics, 3-oxo-
  27, 522
  isocyclics 29, 698
  ketones, cyclic 24, 210
  lactams 30, 90
  2,3-oixdoalcohols, cyclic
  29, 930

(Ring contraction
- of)
 α-sulfinyllactams **30**, 311
 1,4,5-thiadiazepin-3-one
  1,1-dioxides, 2,3,4,7-tetra-
  hydro- **28**, 299
- to
 thianaphthenes **25**, 652
- with
 sulfonic acid azides
  29, 396, 434
- with
 simultaneous
  decarboxylation **29**, 92
 synthesis of hydrocarbons
  from halides **27**, 881
- -, **dehydrative**, of O-heterocyclics **27**, 948
- -, **oxidative**
- of
 N-heterocyclics **28**, 133
 hydroxyquinones **28**, 980
- to
 spiro-N-heterocyclics **29**, 136
- -, **reductive**
- of
 1,4-thiazines, 5,6-dihydro-
  27, 48
- -, **stereospecific 9**, 310
 suppl. 29
- -, **transannular**, of azides
 27, 984

Ring expansion ⊙
- by
 2 C-atoms to 3H-azepines
  28, 872
 2 C-atoms with acetylene
  derivs. **27**, 705
 2 C-atoms with cyclopropenes
  27, 899
 2 C-atoms with vinyllithium
  27, 683
 3 C-atoms **26**, 771; **28**, 634;
  29, 706
 cyclopropane ring opening
  s. under Cyclopropane
  ring opening and - - -,
  1,1-dihalogeno-
 Meisenheimer rearrangement
  26, 151
 pseudobase isomerization
  27, 130
 ring opening **26**, 730
 transamidation, intra-
  molecular **22**, 337
- of
 alcohols, cyclic **27**, 912
 2-aminomesylates, N-heterocyclic **30**, 459
 2-azetidinones **29**, 706
 aziridine ring, review **26**, 706
 azirines **28**, 643
 cobaltocenes **26**, 850
 2,2-dihalogenalcohols, cyclic
  28, 945
 ethylene derivs., -
  27, 229
 α,β-ethylenesulfones, -
  26, 584
 ethynylglycols, - **28**, 974
 o-α-halogeno-N-hetero-
  cyclics **17**, 263
 S,S-heterocyclics **30**, 649
 N-heterocyclics **29**, 779
 O-heterocyclics **27**, 276
 P-heterocyclics **27**, 685
 1,2-Si,Si-heterocyclics **30**, 425

 heterocyclic salts **18**, 849
 α-hydroxydiazo compds.
  26, 661; **29**, 625
 2-hydroxy-1,1-dihalides
  30, 672
 isocyclics **20**, 557 suppl. **26**;
  27, 226, 718, 922; **28**, 946
 - to N-heterocyclics **27**, 331
 - to O-heterocyclics **26**, 127
 ketones, cyclic, with diazo
  compds. **26**, 837
 nitrones, cyclic, to lactams
  28, 122
 oxido compds. to ketones
  29, 689
 oximes, cyclic to lactams
  26, 150
 2-oxo-N-heterocyclics with
  diazo compds. **18**, 848;
  29, 783
 Δ²-pyrazolines, 5-hydroxy-
  27, 727
 pyrrole ring, 3-nitroso-
  29, 355
 2-spirocyclopropano-
  1,3-dithianes **28**, 656
 sulfonium ylids, cyclic
  27, 705
- special s.
 cyanogen azide ring
  expansion
 lactone - -
- to
 cyclobutanone ring **28**, 958
 pyridines **25**, 636 suppl. **26**
 pyrimido[4,5-d]pyrimidines
  30, 230
 spiro compds. **26**, 751
- with
 carbenes **29**, 793
 diazo compds. **18**, 849
 tosylates **29**, 918
- with simultaneous
 annelation of ketones, cyclic
  26, 863
 replacement of halogen
  by amino groups **28**, 389
 ring closure **26**, 837; **27**, 958
 N-heterocyclics **29**, 657
- -, **oxidative**
- of
 O-heterocyclics **27**, 276
 thiazolidines **27**, 48
- to lactones **30**, 145
- with simultaneous
 ring expansion of O-heterocyclics **30**, 92
- -, **regiospecific 27**, 524
- -, **stereospecific 26**, 730
 suppl. 29
- **hydrogenation**
- of
 furans **20**, 66 suppl. **28**;
  23, 66 suppl. 28
 N-heterocycles with
  N-alkylation **29**, 389
 isocycles
  benzene rings **26**, 52
- with
 alkali metal catalysts
  26, 46
- -, **partial**
- of
 arylcarboxylic acids **26**, 44
 isocycles **28**, 989; **29**, 51
 - with retention of phenol
  rings **26**, 43

- -, **preferential 27**, 61
- of
 N-heterocycles **26**, 42, 366
 isocycles **26**, 42
- -, **selective**
- of
 nitroarenes **26**, 48
- -, **stereospecific 7**, 89
 suppl. **28**; **27**, 61
- to pyrrolizidines **30**, 631
- **metalation 19**, 736
 (s. a. o-Metalation)
- of
 thiophene ring **27**, 158
- **opening** ⊂
- by
 autoxidation **26**, 131
 desulfuration **17**, 968; **26**, 691
 Hofmann degradation of
  cyclimmonium salts **15**, 672
- -, direction of - (s. a. Ring
  opening, stereospecific,
  directed)
 oxido ring opening **5**, 476; **10**,
  262 suppl. **28**; **12**, 183; **12**,
  376 suppl. **26**; **19**, 658 suppl.
  27; **28**, 471 suppl. **29**
- of
 cyclimmonium salts **22**, 42
  suppl. **28**
 1,3-heterocyclics **26**, 670
 N-heterocyclics at satd.
  carbon-nitrogen bonds by
  electrophiles, review **28**, 336
 isocyclics, synthesis of
  polyenynes **30**, 601
 2-oximinoalcohols, cyclic
  27, 517
- to
 carboxylic acids **26**, 131
 ketones **19**, 211
- with
 ring expansion **26**, 730
- -, **double**
- by isomerization **18**, 796
 suppl. **29**; **26**, 741; **27**, 749
- of
 bicyclo[1.1.0]butane ring
  26, 53
- to
 gem-divinyl compds. **17**, 968
 suppl. **26**
- -, **fragmentation-type**
 s. under Fragmentation
- -, **geospecific 28**, 95
- -, **nucleophilic** s. Nucleophilic ring opening
- -, **oxidative**
- of
 heterocyclics **24**, 150
 N-heterocycles **28**, 903
 isocycles **26**, 156, 228;
  28, 108, 198; **30**, 84
 ketones, cyclic **27**, 154
- -, -, **preferential**
- of
 isocycles **28**, 109; **29**, 147
- -, **photochemical**, of heterocycles **28**, 470
- -, **reductive**
- by
 cleavage of 1,3-diene central
  bonds **27**, 49
- of
 amidines, cyclic,
  N-condensed **26**, 37

cyclimmonium salts,
  N-condensed 26, 36
heterocycles 26, 55
1,2-N,N-heterocyclics,
  N,N-condensed 22, 372
hydrazinium salts,
  N-condensed 26, 460
isocyclics 29, 43
isoureas, bicyclic 29, 33
N-oxides, cyclic 30, 26
– –, –, fragmentation-type
  s. under Fragmentation
– –, regiospecific 28, 524
– of
  oxido compds. 12, 715
    suppl. 28
– –, stereospecific
– of
  aziridines 26, 964
  cyclobutene ring, 1-amino-
    28, 156
  1,2,3-oxathiazolidine ring,
    2-oxo- 29, 18
  oxido compds. 27, 942
  thiirane 1,1-dioxides 30, 385
– –, tropones, 2-alkyl-, by –
  27, 967
– –, –, directed
– of
  oxido compds. 26, 149
– –, transannular 30, 26
– rearrangement
  s. Rearrangement, skeletal
Rings s. a. Adamantanes,
  Alcohols, cyclic, Heterocyclics, Isocyclics, Ketones,
  cyclic, Macrocyclics, Polycyclics, Propellanes
–, condensed, by dehydro-
  halogenation, ar. 27, 962
– highly condensed s. Cage
  compds.
–, cis- and trans-fused s. under
  Reduction, stereospecific
–, mesoionic s. Heterocyclics,
  mesoionic
–, pseudoaromatic
  s. Pseudoaromatic rings
–, strained, functionalized
  29, 31
Ring-O,P-ylids 27, 115
Ritter s. a. Hassner
– reaction 27, 326
– –, extended 28, 286
–, heterocyclics by – 26, 309
– –, modified 29, 307
Ritter-type reaction
–, isoquinolines, 3,4-dihydro-
  by – 26, 820
–, ureas from cyanamides
  30, 216
Robinson annelation,
  stereospecific 26, 784
Rose Bengal (s. a. under
  Sensitizers) 24, 143 suppl. 29
– –, polymer-based 24,
  143 suppl. 29
Rosenmund-Zetzche reduction,
  modified 27, 79
Rotans 14, 858 suppl. 28
Rubidium s. a. under Sodium
– carbonate 26, 46
Rühlmann acyloin condensation
  19, 822; 23, 902
Rupe rearrangement, review
  5, 508 suppl. 26
Ruthenium-carbon 9, 93
  suppl. 26

Ruthenium, chlorohydridotris-
  (triphenylphosphine)- 29, 49
– compds.
  s. a. Sodium ruthenate
– complex 28, 41
–(II), dichlorotris(triphenylphosphine)- 28, 209, 675;
  29, 50
– dioxide 16, 314 suppl. 26; 19,
  313 suppl. 26; 27, 145, 245;
  28, 208; 29, 178
– tetroxide 28, 208; 29, 178;
  30, 154
– triarylphosphine complex
  as reagent 19, 776 suppl. 29;
  23, 69 suppl. 26
– trichloride-sodium hypochlorite, oxidations with –
  27, 280
Rydon reagent s. Triphenyl-
  phosphite methiodide

Saccharides s. Carbohydrates
Saccharins 26, 164; 27, 538
– special s.
  N-acyl-saccharins
  N-bromo-
  N-chloro-
– startg. m. f.
  2H-1,2-benzothiazine
    1,1-dioxides, 4-hydroxy-
    28, 694
Salicylic acid s. Silver
  salicylate
– – esters
– startg. m. f.
  halides 30, 360
Salt additives
–, effect in Grignard reactions
  19, 46 suppl. 26
Salts
–, inner s. Betaines
–, organic s. Heterocyclic salts,
  Onium –
Sand 28, 686
Sandmeyer reaction
–, in dimethyl sulfoxide 26, 549
Saturation s. Hydrogenation
Scavengers s. Hydrogen
  bromide scavenger, Nitrous
  acid –, Radical –, Water –
Schiff bases s. Azomethines
Schmidt reaction
–, 2H-1,4-benzodiazepin-2-ones,
  1,3,4,5-tetrahydro- by –
  4, 326 suppl. 26
– of
  ketones 26, 308
  –, review 26, 308 suppl. 27
  –, cyclic, direction 13, 379
Scholl ring closure
  to dibenzo[a,l]pyrenes 27, 914
Schuster s. Meyer
Selectrides s. Lithium
  hydridotri-sec-butylborate,
  Potassium –
1,2,3-Selenadiazoles 26, 655
– startg. m. f.
  acetylene derivs. 26, 655
Selenanaphthenes 26, 657
Selenazolidines 24, 692
  suppl. 27
Selenic acid as reagent 30, 123
Selenide dihalides
– startg. m. f.
  selenonium ylids 30, 434

1-phosphonio-1-selenonio-
  methylid salts 30, 432
Selenides s. a. Silylselenides
– from
  selenylhalides 29, 912
–, reactions via – 29, 180, 912
– special s.
  acoxy-selenides
  hydroxy-
– startg. m. f.
  1,1-acoxyselenides 29, 108
  diacyloxyselenuranes 29, 108
  ethylene derivs. 29, 912
  selenoxides 26, 101
Selenium (s. a. Diselenium)
  28, 322; 29, 940; 30, 10a, 107
–, prepn. of selenium compds.,
  organic from – 28, 585
– compds., toxicity 28, 471
– –, organic, prepn. from
  selenium 28, 585
– dioxide 21, 124 suppl. 27;
  26, 116, 163; 27, 182; 29, 256
– as reactant 26, 655
– monochloride as reactant
  26, 646; 28, 592
– tetrafluoride 14, 620 suppl. 29
Selenoacetals
– from
  diazo compds. and
    diselenides 26, 648
Selenocarbonyl groups
– startg. m. f.
  carbonyl groups 27, 231
Selenochromenes
– startg. m. f.
  benzo[b]selenophenes,
    2-formyl- 26, 163
Selenocoumarins 26, 635
Selenolic acid esters
– from
  carboxylic acids 15, 470
    suppl. 26
– startg. m. f.
  thiolic acid esters 30, 415
Selenols
– startg. m. f.
  diselenides 27, 587
  selenosulfides 27, 587
Selenonium ylids 30, 434
– from
  selenide dihalides 30, 434
– startg. m. f.
  cyclopropanes 29, 863
Selenonio... s. 1-Phosphonio-
  1-selenoniomethylid salts
Selenonothiolcarbamic acid
  esters
– from
  isothiouronium salts 26, 652
Selenophene ring opening
– to 1-organoseleno-1-organo-
  thio-1-en-3-ynes 30, 436
Selenosulfides
– from
  selenols 27, 587
Selenothioacetals
– special s.
  ketene selenothioacetals
Selenoureas
– from
  isothiouronium salts 26, 652
Selenous acid 28, 863
Selenoxides
– from
  selenides 26, 101
–, oxidations with – 29, 275

(Selenoxides)
– special s.
 diphenyl selenoxide
Selenuranes
– special s.
 diacyloxyselenuranes
Selenylhalides
– special s.
 benzeneselenyl bromide
– startg. m. f.
 2-hydroxyselenides **29**, 601
 selenides **29**, 912
Self... s. a. Auto...
Self-acylation of arylacetic
 acids **26**, 371
Self-condensation
 (s. a. Dimerization)
– of
 aldehydes **30**, 553
 β-aminoketones **29**, 774
 β-cyanoketones **29**, 774
Self-thioalkylation
– of
 thiolsulfonic acid esters
 27, 633
Semi... s. a. Hemi...
Semicarbazide
– startg. m. f.
 1-acyldiaminoguanidines,
 5-carbamyl- **28**, 269
Semicarbazides
– from
 N-nitrosoureas **29**, 274
– special s.
 acylsemicarbazides
 1-carbalkoxysemicarbazides
– startg. m. f.
 hydrazines, monosubst. **29**, 274
Semicarbazones
– from
 acetylene derivs. **25**, 228
 suppl. 27
– special s.
 cyanosemicarbazones
– startg. m. f.
 acylcyanides **29**, 952
 1,3,4-oxadiazol-2(5H)-ones
 26, 232 suppl. 28
 pyrazole-4-carboxaldehydes
 26, 840
 1,2,4-triazol-3-ones **27**, 501
Semicorrins, synthesis **26**, 796
Semipolar bonds. s. Bonds,
 semipolar
– **functions** s. Functional
 groups, semipolar
Sensitizers (s. a. Triplet
 sensitizers)
 2-acetonaphthone **28**, 669
 acetone **17**, 968 suppl. **28; 19**,
 812 suppl. **27; 26**, 720; **28**,
 763 suppl. **27; 28**, 702, 709;
 29, 659; **30**, 533
 acetophenone **26**, 763; **27**, 743;
 28, 669
 benzophenone **26**, 763 suppl.
 27; 27, 93, 132; **28**, 233, 466,
 702; **29**, 677
 dicyclopropyl ketone **19**, 773
 suppl. 27
 N,N-dimethylaniline **26**, 85
 fluorenone **22**, 761 suppl. 26
 hematoporphyrin **27**, 179;
 30, 162
 methylene blue **5**, 130 suppl.
 27; 26, 125; **27**, 270, 365;
 28, 750; **29**, 96, 146/7; **30**,
 223

 Michler's ketone **26**, 763
 suppl. 27
 quinoxaline, 2,3-diphenyl-
 30, 179
 rose Bengal **23**, 131 suppl. **27;
 28**, 26; **29**, 146, 240; **30**, 86
 xylene **26**, 135, 146; **27**, 714
–, **polymer-based 24**, 143
 suppl. 29
Sequestering agent s. Sodium
 pyrophosphate
**Serini reaction, stereospecific
 5**, 150 suppl. 29
Shift s. a. Migration,
 Position shift
**1,5-Shift, prototropic, thermal
 27**, 735
Side chains (s. a. Lateral)
– by disproportionation
 30, 635
–, degradation s. under
 Degradation
–, introduction, stereospecific
 23, 703 suppl. 26
–, rearrangement in cyclopropanes **27**, 728
– special s.
 α-ketoaldehyde aldoxime side
 chains
Sieve s. Molecular sieve
Sigmatropic (s. a. Prototropic)
 s. Rearrangement, – (25)
Sil... s. a. Disil...
Silacyclohexan-2-ones
– startg. m. f.
 4-oxa-5-silaspiro[2,6]nonanes
 27, 708
**1-Silacyclopentadienes
 10**, 639 suppl. 28
Silacyclopropane ring 21, 957
 suppl. 28
Silanes
–, addition to nitrogen-nitrogen
 double bonds **27**, 301
– from
 diazo compds. **30**, 430
 ethylene derivs.
 s. Hydrosilylation
– special s.
 acylamino-silanes
 alkoxy-
 amino-
 aryloxy-
 azido-
 azo-
 cyano-
 di-
 disilyl compds.
 ene-silanes
 enoxy-
 halogeno-
 hydrazino-
 hydroxy-
 peroxy-
 phenylsilane
 silicon hydrides, organo-
– startg. m. f.
 alcohols, synthesis **30**, 624
 alkoxysilanes **29**, 109
 ethylene derivs., synthesis
 28, 856
 nitroso compds. **27**, 489
 silyl sulfides **29**, 530
–, **telomerized 27**, 724 suppl. 30
Silasultones 26, 578
Silazanes s. Disilazanes
–, **cyclic**

– special s.
 cyclotrisilazanes
Silica 27, 735; **28**, 49; **29**, 767
– **gel 9**, 248 suppl. **29; 26**, 155,
 896; **29**, 896; **30**, 98, 499, 572
– **monoester of 1,4-dihydroxy-
 methylbenzene 16**, 424
 suppl. 26
Silicate intercalates
–, reactions in – **29**, 93
Silicic acid 28, 376
Silicon
– startg. m. f.
 halogenosilanes **28**, 586
– **carbide 30**, 665
– **compds.**
– as reagents **26**, 501
–, **organo-** (s. a. Silanes),
 ene synthesis with – **29**, 676
– **dioxide** s. Silica
Silicone oil as medium, inert,
 liq. **28**, 785
Silicon hydrides, organo-
– as reagents **17**, 108 suppl. **26;
 22**, 230 suppl. **26; 29**, 42
– special s.
 poly(hydrosiloxane)
 triethylsilane
**Siliconphosphorus compds.,
 organo- 28**, 587
Silicon tetrachloride, peptide
 synthesis with – **14**, 459
 suppl. 26
– **tetrafluoride 12**, 238 suppl. 28
Siloxanes
– special s.
 aminosiloxanes
Siloxy... s. a. Silyloxy...
***trans*-2-Siloxyalcohols, cyclic**
– from
 enoxysilanes, cyclic **30**, 81
– startg. m. f.
 trans-glycols, cyclic **30**, 81
Siloxycarboxylic acid esters
– special s.
 ethylenesiloxycarboxylic
 acid esters
**Siloxycyclopentene annelation
 28**, 701
Siloxycyclopropane ring
– from
 enoxysilanes **29**, 827, 829
– startg. m. f.
 cyclopropanol ring **29**, 829
 oxo compds. **29**, 829
Siloxy groups, reactivity **27**, 488
Siloxy-N-heterocyclics
– startg. m. f.
 (N-vinyl)oxo-N-hetero-
 cyclics **26**, 445
Siloxylamines
– special s.
 N,N-disilylsiloxylamines
α-Siloxymethylene compds.
– from
 enoxysilanes **29**, 827
α-Siloxynitriles, alkylation
 29, 808
– from
 enoxysilanes **27**, 720
 oxo compds. **28**, 610; **30**, 630
– startg. m. f.
 2-aminoalcohols **27**, 720
 ketones **29**, 808
Silthianes s. Disilthianes
Silver/zinc s. Zinc/silver
– **acetate 26**, 627; **28**, 322
– **acetylides**

- startg. m. f.
  ynazo compds. 30, 238
- **ammonium nitrate complex**
  28, 820
- **bromide 29**, 419
- **carbonate 28**, 191, 625;
  29, 304, 972
- **carbonate-Celite 11**, 314
  suppl. **27**; **19**, 308 suppl. **28**;
  27, 100, 761; **28**, 230, 611
- -, oxidations with – **27**, 100
  suppl. **30**
- **carboxylates** as reagent
  26, 203; **27**, 761
- **chlorodifluoroacetate 27**, 581
- **cyanide 29**, 340
- **fluoride 30**, 371
- **fluoroborate 26**, 551, 762; **28**,
  104, 657, 822/3; **29**, 697; **30**,
  142, 358
- **γ-hydroxyvalerate 26**, 203
- **iodide 27**, 966
- **ion-catalyzed rearrangements**
  of polycyclics, review **28**, 691
  suppl. **30**
- **ions** as catalyst **28**, 674
- **nitrate** (s. a. Silver ammonium
  nitrate) **1**, 218 suppl. **29**; **19**,
  379 suppl. **28**; **26**, 205, 518; **26**,
  762 suppl. **28**; **27**, 155, 177,
  585, 907; **28**, 258, 399, 407, 814,
  949; **29**, 284, 494, 967
- **nitrite 22**, 288 suppl. **29**;
  27, 550
- **oxide 26**, 421; **26**, 524 suppl.
  30; **27**, 160, 237, 760; **28**, 182,
  460, 611, 625, 937; **29**, 707,
  972; **30**, 456, 488, 619
- -**(II) oxide 30**, 488
- -, oxidation with – **15**, 208
  suppl. **26**; **21**, 310 suppl. **27**;
  28, 938
- **perchlorate 15**, 170 suppl. **26**;
  15, 555 suppl. **29**; **19**, 46 suppl.
  26; **19**, 539 suppl. **26**; **26**, 53
  suppl. **29**; **26**, 205, 741; **27**, 369;
  28, 674, 691, 763; **29**, 154; **30**,
  142, 316
- - -**acetonitrile complex 28**, 994
- **salicylate** as reagent **27**, 217
- **salt 26**, 206; **27**, 111; **29**, 419;
  30, 62
- **sulfate 27**, 472
- **trifluoroacetate 26**, 204;
  29, 601

Silylacetylenes
- special s.
  bis(trimethylsilyl)acetylene
  silylynamines

Silyl adducts of isonitriles
16, 152 suppl. **29**

O-Silyl O-alkyl keteneacetals
- from
  carboxylic acid esters **28**, 77
- startg. m. f.
  β,γ-ethylene-β-siloxy-
  carboxylic acid esters
  28, 809
  β-hydroxycarboxylic acid
  esters **27**, 884
  ketenes **28**, 227

Silylamides
- special s.
  bis(trimethylsilyl)acetamide

Silylamidines
- startg. m. f.
  isothiocyanates and nitriles
  27, 495

Silylaminosiloxanes
- from
  N,N-disilylsiloxylamines
  28, 301

1,2-Silylammonium salts,
quaternary
- startg. m. f.
  1,1-aminosilanes **30**, 444

Silylating agents
-, N-silylketenimines as –
  29, 295

Silylation
(s. a. Hydrosilylation)
-, aminocarboxylic acids from
  their hydrohalides via –
  30, 290
-, review **20**, 95 suppl. **27**

N-Silylation, reductive 26, 279

O-Silylation
- of
  alcohol groups **29**, 109; **30**, 61
  hydroxyl, hindered **20**, 95
  suppl. **26**
-, **preferential 4**, 149 suppl. **26**
- of
  phenol groups **28**, 74

S-Silylation 30, 380

Silyl azides
-, reactions **28**, 370, 400
- special s.
  trimethylsilyl azide

N-Silylcarboxylic acid amides
- special s.
  trimethylsilylacetamide
- startg. m. f.
  N-acylcarboxylic acid amides
  23, 476 suppl. **29**
  amide-N-sulfenyl chlorides
  28, 259
  nitriles **28**, 462

α-Silylcarboxylic acid esters
- startg. m. f.
  α,β-ethylenecarboxylic acid
  esters **28**, 856 suppl. **29**

Silyl derivs.
- startg. m. f.
  heterocyclics **27**, 488

N-Silylenamines
- from
  nitriles **27**, 646

Silyl enolethers s. Enoxysilanes
- esters s. Carboxylic acid
  silyl esters, Sulfenic – – –

Silylethynyl ketones
- from
  carboxylic acid chlorides
  29, 846
- startg. m. f.
  α,β-ethylenealdehydes **29**, 846
  β-hydroxyacetals **29**, 846

2-Silyl groups, reactive,
in heterocyclics 27, 883

Silyl hemiacetals s. Ketene
  silyl hemiacetals
- **hydrides** s. Silicon hydrides,
  organo-

Silyliminoesters
  s. a. Disilyliminoesters

O-Silyl iminoesters, cyclic
20, 96

Silylketenimines
- special s.
  disilylketenimines

N-Silylketenimines
- as silylation agents **29**, 295
- from
  nitriles and halogenosilanes
  29, 295

Silylketones
-, review **26**, 640
- special s.
  silylethynyl ketones

1,1-Silyllithium compds.
- startg. m. f.
  enesilanes, synthesis **29**, 860

1,1,1-Silylmercaptals
- special s.
  silylthioformals

α-Silylnitriles
- from
  nitriles and halogenosilanes
  29, 295

β-Silyloxo compds.
- from
  α,β-ethyleneoxo compds.
  26, 640

Silyloxy... s. Siloxy...

Silyl-Pummerer rearrangement
24, 164

Silylselenides as reactants
28, 581

Silyl selenides
- startg. m. f.
  N-sulfonyliminoseleninyl
  halides **28**, 469

Silyl sulfides
- from
  mercaptans and silanes
  29, 530
- startg. m. f.
  N-sulfonyliminosulfinyl
  halides **28**, 469
  thiolic acid esters **29**, 581

Silylthioacetic acid esters
  α,β-ethylene-α-mercapto-
  carboxylic acid esters
  29, 730

Silylthioformals
- startg. m. f.
  ketene mercaptals **28**, 857

Silylynamines 23, 650 suppl. **26**

Simmons-Smith cyclopropane
ring synthesis, modified
26, 876

Singlet oxygen s. Oxygen,
  singlet

Skeletal rearrangement
  s. Rearrangement,
  skeletal

Skraup quinoline ring synthesis
27, 944

Smiles s. a. Truce
- **rearrangement**
- of
  o-aminosulfones **28**, 300

Smiles-type N,N-rearrangement
26, 310

Smith s. Torgov

Sodiobenzenesulfonamide
  as reagent **27**, 181

N-Sodio compds.
  s. N-Halogeno-N-sodio-
  sulfonic acid amides, Sodiobenzenesulfonamide

Sodiomalonate, diethyl- 30, 529

Sodiosulfonic acid amides
- special s.
  N-halogeno-N-sodiosulfonic
  acid amides
  sodiobenzenesulfonamide

Sodium (s. a. Disodium...) **23**,
  943 suppl. **27**; **26**, 710, 732,
  782; **27**, 87, 393, 709/10; **28**,
  427, 523, 580, 617; **29**, 142, 610,
  774, 790; **30**, 30, 47, 423

Sod 718

(Sodium)
- –/ammonia, liq. 26, 55, 74; 27, 8, 9; 29, 43; 30, 21, 41, 674
- –/–, –/*tert*-butanol 28, 204
- –/–, –/ethanol 27, 816
- –/anthracene 29, 915
- –/o-chlorotoluene 27, 731
- –/hexamethylphosphoramide-*tert*-butanol 26, 47
- –/naphthalene 6, 367 suppl. 28; 18, 711 suppl. 27
- –/potassium 29, 797
- –/rubidium 26, 46
- acetoacetonate 30, 434
- acetylide/liq. ammonia 26, 793
- alkoxides 26, 356
  - butoxide 27, 435
    - *tert*-butoxide 27, 263, 926
  - ethoxide 26, 616, 933, 940; 27, 210; 29, 394, 399, 884; 30, 539
  - methoxide 26, 528, 929, 959, 968; 27, 893; 28, 221, 293, 338, 694, 727/8; 29, 101, 607, 951; 30, 107, 397, 615
  - *n*-octanoxide 26, 782 suppl. 27
  - *n*-propoxide 26, 130
- amalgam 26, 891; 27, 42; 29, 73
- amide 5, 467 suppl. 29; 26, 299, 682, 748, 754, 863, 907, 980; 27, 494; 28, 393, 804/5; 29, 777, 811; 30, 296, 443, 450, 516
- amides s. Sodium bis(trimethylsilyl)amide
- azide 8, 404 suppl. 29; 17, 486 suppl. 26; 26, 360, 414, 444, 970; 27, 401, 439, 462; 28, 278, 282, 340, 630; 29, 317, 400, 424, 429; 30, 56, 88, 255, 259, 278, 460
- benzoate 26, 944
- benzylmercaptide as reagent 29, 443
- benzyloxycarbonyl thiosulfate as reactant 26, 439
- bis(trimethylsilyl)amide 15, 656 suppl. 26; 22, 898 suppl. 28; 23, 904 suppl. 29; 28, 806
- borate s. Borax
- boron hydride s. a. Sodium tetrahydridoborate
- – – –, sulfurated 28, 525; 28, 569 suppl. 30
- –, reduction with –, review 24, 49 suppl. 28
- bromate 26, 914 suppl. 28
- carboxylates (s. a. under Ethylenediaminetetraacetic acid)
  startg. m. f.
  alcohols, reduction 18, 947 suppl. 27
  carboxylic acid esters 18, 935 suppl. 28
- chlorate 3, 447 suppl. 28
- chloride (s. a. under Aluminum chloride) 3, 640 suppl. 28; 27, 61; 29, 84, 715
- chlorite 20, 194 suppl. 27; 29, 244; 30, 66
- citrate buffer 29, 276

- compds., organo-
- as reagents
  n-amylsodium 27, 838
- cyanide 13, 196 suppl. 27; 26, 683, 761, 764, 834; 27, 44, 371, 678, 780-2; 27, 730 suppl. 30; 28, 575, 624, 648; 29, 110, 758, 918; 30, 6, 486, 567
- as reactant 6, 898 suppl. 29; 29, 750
- –, reaction with ethylene derivs., hydroborated 27, 716
- deuterioxide 28, 590
- dichromate 26, 164, 234
- dihydridobis-(2-methoxyethoxo)aluminate 14, 46 suppl. 27; 15, 387 suppl. 29; 21, 538 suppl. 29; 27, 35, 658; 28, 60; 29, 46; 30, 17, 35
- as reactant 27, 790
- dihydrogen phosphate 27, 213
- dithionite s. Dithionite
- ethylmercaptide as reagent 27, 10
- fluoride 29, 503; 30, 106, 132
- fluoroborate 27, 364
- formate 14, 455 suppl. 26
- hydride 5, 440 suppl. 28; 6, 841 suppl. 29; 13, 802 suppl. 29; 17, 895 suppl. 29; 18, 324 suppl. 26; 19, 509 suppl. 27; 26, 14, 17, 43, 54, 196, 219, 236, 437, 679, 711, 767, 783-5, 861, 895/6, 935; 26, 154 suppl. 30; 27, 105, 170, 223, 265, 333, 394, 456, 498, 618, 676, 682, 720, 750, 767-70, 830-2, 873, 890, 901; 27, 438 suppl. 30; 28, 76, 99, 129, 220, 377/8, 453, 618, 644, 734, 776, 793, 803, 835/6, 851, 901/2, 942, 962; 28, 140 suppl. 30; 28, 732 suppl. 29; 29, 101, 162, 198, 290, 295, 428, 443, 470/1, 499, 563, 616, 618, 631/2, 661, 693, 730/1, 733, 791, 864-6, 888/9, 947, 981; 30, 37, 128, 159, 178, 232, 244/5, 394, 447/8, 459, 466, 545, 572, 580, 593, 613/4, 689
- as reducing agent 29, 409
- –/dimethyl sulfoxide (s. a. Methylsulfinyl carbanion) 5, 587 suppl. 26; 16, 434 suppl. 26; 26, 299, 966; 27, 664, 740, 874; 28, 433, 852; 29, 410, 732, 775, 982; 30, 118
- –/– sulfoxide-D$_6$ 27, 650
- hydridotris(dimethylaminoethoxo)aluminate 13, 254 suppl. 28
- hydrogen selenide 12, 707 suppl. 29; 26, 652
  - sulfide 26, 841; 27, 630
  - –/ammonia, liq. 29, 564
  - sulfite 24, 990 suppl. 27; 28, 188
- hydrosulfite s. Dithionite
- hydroxylaminedisulfonate s. Hydroxylaminedisulfonate
- hypochlorite (s. a. under Ruthenium trichloride) 26, 140/1, 532; 27, 145, 245, 540; 28, 248; 29, 220, 525; 30, 355
- hypohalite (s. a. Sodium hypochlorite) 27, 305

- hypophosphite 13, 476 suppl. 28
- iodate 27, 368
- iodide 11, 924 suppl. 29; 17, 894 suppl. 26, 28; 26, 539 suppl. 27; 26, 632; 27, 1, 131, 340, 579, 931; 29, 207, 350, 614, 977; 30, 600, 637
- mercaptides
  - as reactants 28, 528
  - special s.
    sodium benzylmercaptide
    - ethylmercaptide
    - methylmercaptide
    - phenylmercaptide
- metabisulfite 27, 548
- metaperiodate s. Periodate
- methylmercaptide as reagent 29, 3
- molybdate 22, 170 suppl. 29
- nitrite OC↓↑N; 26, 23, 324, 361, 828, 997; 27, 147, 274, 283, 355/6, 369, 817; 28, 238, 314/5, 317/8, 397, 428, 436, 671, 778, 992; 29, 273/4, 276, 282, 311, 318, 336, 338; 30, 235, 319
- oxide 28, 590
- perchlorate 27, 797, 998; 29, 773
- periodate s. Periodate
- peroxide 30, 54
- persulate s. Persulfate
- phenylmercaptide as reactant 30, 418
- polysulfides 22, 18 suppl. 29
- pyridinolates as catalysts 27, 398
- pyrophosphate as sequestering agent 26, 422
- pyrosulfite s. Sodium metabisulfite
- ruthenate, oxidations with – 28, 137
- salt 26, 221, 249, 387, 410, 415, 439/40, 451, 577, 599, 944; 27, 214, 298, 309, 531, 561, 784, 957; 28, 409, 495, 517, 524, 644, 978; 29, 316, 359, 852; 30, 357, 647
- succinate buffer of pH 4 15, 457 suppl. 28
- sulfate 29, 112
- sulfide 18, 687 suppl. 29; 27, 608, 956; 28, 17; 29, 549, 574; 30, 419
- sulfite 19, 702 suppl. 29; 26, 324; 28, 115
- tetracarbonylferrate(II) 26, 891; 27, 76, 869; 29, 64
- tetrachloropalladate(II) 24, 252 suppl. 29; 27, 880
- tetradeuterioborate 25, 62 suppl. 29
- tetrahydroaluminate 26, 69
- tetrahydridoborate HC↓OC; HC↓NC; HC↓↑O; 14, 101 suppl. 27; 15, 37 suppl. 27; 18, 92 suppl. 28; 20, 531 suppl. 29; 21, 174 suppl. 29; 26, 9, 26, 35, 48, 75, 79, 185, 336, 761, 771, 801, 947, 993; 27, 46, 61, 72, 82/3, 202, 258, 345, 510, 862, 946/7; 28, 37/8, 56, 59, 109, 117, 525, 585, 793, 828, 982; 28, 57 suppl. 29; 29, 48, 57, 67, 76, 141/2, 172, 464, 846; 30, 34, 75, 145, 261, 513, 696

- –/**hexachloroplatinic acid-carbon** 29, 933
- –/**pyridine** 26, 49
- –/**sodium hydroxide** 26, 185
- –/**transition metal salts,** reduction with – 26, 35
- **tetraperoxymolybdate** 22, 170 suppl. 29
- **tetraphenylborate** as reactant 26, 108
- **thiocyanate** 18, 647 suppl. 28; 28, 562
- as reactant 29, 388
- **thiosulfate** 26, 566, 617; 27, 631; 28, 950; 29, 521
- **p-toluenesulfonate** as catalyst 29, 287
- **trichloroacetate** as reactant 29, 852
- **trihydridocyanoborate**
- as reducing agent, selective, review 27, 74 suppl. 30
- reduction with – 19, 448 suppl. 29; 26, 65; 27, 74; 28, 346/7; 29, 67

Solid-phase multistep syntheses
- of
  oligonucleotides 14, 157 suppl. 27
  peptides 12, 455 suppl. 26; 14, 450 suppl. 28; 15, 26 suppl. 26; 15, 26 suppl. 29; 17, 477 suppl. 26; 22, 286 suppl. 29; 23, 415 suppl. 28; 24, 413 suppl. 29
    evaluation of coupling methods 22, 394 suppl. 29
    improved coupling procedure 12, 455 suppl. 28
    on Biopak 16, 424 suppl. 26
    with polymeric C-terminal diphenylmethylamide 18, 434 suppl. 29
- reactions (s. a. under Polymers)
- special s.
  Merrifield syntheses
- **Wittig synthesis** 17, 892 suppl. 27; 18, 912 suppl. 28

Solids, finely divided (s. a. Glass powder) 30, 36

Solid-state aromatization, decarboxylative 28, 972

Solubilization s. Protective groups, solubilizing

Solvents
–, effect on
  cycloaddition, 1,3-dipolar, review 17, 503 suppl. 28
  Friedel-Crafts ketone synthesis 5, 604
  oxido ring opening 28, 471 suppl. 29
  reaction course 21, 719 suppl. 26; 27, 691; 29, 852
  Reformatskii synthesis 2, 827 suppl. 26
–, –, geoselective 14, 91 suppl. 29
–, –, steric 26, 47, 784; 27, 671; 29, 678, 957
–, product components as – 27, 127
- with readily abstractable hydrogen atoms 30, 83
–, **aprotic** (s. a. Methylene chloride) 27, 531

–, –, **basic** s. Hexamethylphosphoramide
-, –, **polar** s. Acetone, Acetonitrile, Chloroform, Dimethylformamide, Dimethyl sulfoxide, Hexamethylphosphoramide, N-Methylpyrrolidone, Nitromethane, Tetramethylenesulfone
–, **acidic** s. Trimethyl borate
–, **inert** s. sec-Butylcyclohexane
–, **polar** s. Diethylamine, N,N-Dimethylacetamide, Nitrobenzene
–, –, **weakly protic**
  s. N-Methylformamide

Solvolysis s. Aceto-lysis, Alcoho-, Ring closure, solvolytic

Solvomercuration-demercuration, intramolecular 27, 345

Sommer acid s. Hydroxylamine-O-sulfonic acid

Soxhlet apparatus 26, 327, 780
- extractor technique
  s. Extractor technique

(–)-Sparteine
- in Reformatskii synthesis, asym. 2, 827 suppl. 26

Specificity s. Geo-specificity, Metal –, Regio-, Stereo-, Wavelength –

Spiro[adamantane-2,1'-cyclobutane-2'/3'-diones] 27, 233

Spiroannelation
- with
  cyclopropylcarbinols, 1-arylthio- 28, 958
  S-cyclopropylids 27, 873
  sulfonium ylids, cyclic, and diketene 28, 833

Spiroaziridinium salts
- startg. m. f.
  methylene compds. 13, 831 suppl. 29

Spiro[2H-1-benzopyran-2,2'-benzothiazolines]
- from
  benzothiazolium salts and o-hydroxyaldehydes 28, 738

Spiro compds.
- by ring expansion 26, 751
- special s.
  polyspiro compds.

Spirocyclics
- from
  p-quinodimethanes 29, 655
  p-toluenesulfonic acid esters 18, 937 suppl. 26
–, rearrangement, skeletal 27, 173
- special s.
  β-diketones, spirocyclic
tetra-Het-Spirocyclics
  s. Tetrathiospirocyclics

Spirocyclobutanone ring 25, 599 suppl. 29

Spirocyclobutanones
- from
  α-halogenoketones and sulfoxonium salts 26, 897

Spirocyclohexadienes, rearrangement 27, 911

Spiro(cyclohexa-2,5-dien-1,1'-isoindolin)-3'-ones, 4-hydroxy- 27, 862

4-Spiro-2-cyclohexenones 17, 935; 29, 734

Spirocyclopropane ring 11, 779 suppl. 28; 29, 937
- –, **opening** 28, 476
- – –, **double,** [8]paracycloph-4-enes by – 30, 472

2-Spirocyclopropano-1,3-dithiane ring expansion 28, 656

Spirodiamines 30, 531

Spirodiazirine ring
- startg. m. f.
  cyclopropane ring 29, 948

Spirodienones, synthesis and rearrangement, review 18, 937 suppl. 28

Spiro-β-diketones 29, 935
- from
  α,β-oxidoketones, cyclic 30, 514

Spiroheterocyclics 26, 781; 27, 345 suppl. 28
- special s.
  4-oxa-5-silaspiro[2.6]nonanes
  thiophene-2-spiro-3-indolenines

Spiro-N-heterocyclics
- by Pictet-Spengler ring closure 8, 823 suppl. 26
- by ring contraction, oxidative 29, 136
- from
  oxindoles (2 molecules) 26, 922
  2β-oxiranylindolo[2,3-a]quinolizines, 1,2,3,4,6,7,12,12b-octahydro- 28, 99
- special s.
  1-azaspiro[2.2]pentanes
  4,5-benz-2-azaspiro[5.5]-undecanes
  isoquinoline-1-spiro-2'-(1',3'-indandiones), 1,2-dihydro-spirodiamines
  1,3,6,9-tetraazaspiro[4.4]-nonanes

Spiro-O-heterocyclics 26, 236; 28, 609
- special s.
  1,3-benzodioxole-2-spirocyclohexadien-4'-ones
  1-oxaspiro[2.2]pentanes
  spiroketals

P-Spiroheterocyclics, review 28, 264 suppl. 30
- from
  α-aminocarboxylic acids and 1,3,2-dioxaphospholanes, 2-halogeno- 28, 264

Spiro-S-heterocyclics
  s. Tetrathiospirocyclics

Spiro-1,3-S,S-heterocyclics
- from
  enedisulfides, isocyclic 30, 390

Spiro[indoline-2,4'-[4'H]pyrans], 1-hydroxy-3-oxo-
- from
  4H-pyrans, 4-(o-nitrobenzylidene)- 29, 321

Spiro-3H-indoline-3,2'-2'H-pyrroles 26, 961
- startg. m. f.
  pyrroles, 2-aryl- 26, 961
  pyrrolo[1,2-c]quinazolines, 4,5-dihydro- 26, 961

Spiroisocyclics
– from
 cyclopropane ring **28,** 477
–, isomerization **23,** 717
–, **hydroxy-**
– startg. m. f.
 phenanthrene ring **30,** 657
Spiroketals
–, ring opening, reductive
 29, 34
– special s.
 spirostanes **20**
Spirolactones 27, 505; **28,** 772
– from
 ethylenecarboxylic acids
 27, 505
– startg. m. f.
 ethylenecarboxylic acids
 27, 505 suppl. **28**
Spiro[4,4]nona-1,3-dienes
– startg. m. f.
 indans **28,** 894
**Spiroorthocarbonic acid esters
4,** 249 suppl. **28; 26,** 625
Spirooxaziridines
– startg. m. f.
 lactams **29,** 331
Spirooxazolid-4-one ring 29, 243
6-Spirooxirano-2,4-cyclohexadienones
– from
 o-hydroxymethylphenols,
 hindered **29,** 246
– startg. m. f.
 o-hydroxyaldehydes **29,** 246
Spiropentane ring 27, 873
Spiro-Δ¹-pyrazolines, photofragmentation **25,** 670
**Spiro[pyrrolidin-3,1'-(pyrrolidino[3,4-e]pyrrolidines)],
2,4',5,6'-tetraoxo-**
– from
 maleimides (2 molecules)
 and azomethines **27,** 701
Spiro-1,2,4,5-tetrazines, 1,2,3,4-tetrahydro- 29, 281
Squaric acid diamides
– by exchange of amino groups
 28, 366
– by 1,2-dihydrazides, ring
 closure with – **29,** 353
Stannanes (s, a. Tin compds.,
 organo-)
–, cleavage with 1-chlorobenzotriazole **27,** 310
–, prepn., review **16,** 710
 suppl. **26**
– special s.
 alkoxy-stannanes
 hydroxy-
 tin hydrides, organo-
 – oxides, diorgano-
– startg. m. f.
 halides **28,** 511
 nitro compds. **30,** 315
 nitroso – **27,** 489;
 30, 315
–, **tetraalkyl** – **26,** 654
Stannins, 1,4-dihydro-
– from
 1,4-diynes **30,** 426
Stannoxanes s. Distannoxanes
3-Stannoxyhalides
– startg. m. f.
 3-hydroxystannanes **27,** 657
N-Stannylhydrazines, N'-aryl-
– startg. m. f.
 diaryls **28,** 784

Stannyl hydrides s. Tin
 hydrides, organo-
Stark s. Dean
pH-Stat 27, 437
Statistics s. Optimization,
 mathematical
Stereochemistry
– of
 phosphorus compds., review
 25, 701 suppl. **26**
 reduction, cathodic,
 of halides, review **21,** 35
 suppl. **27**
–, electrochemical, review
 28, 1
 sulfur compds. **28,** 992
Stereoisomers (s. a. Asym...,
 Chiral..., Configuration,
 Racem...)
– from
 racemates (one stereoisomer)
 18, 246 suppl. **26**
 meso- compds. **28,** 8
 suppl. **30**
–, prepn. of optically active
 compds. via tricarbonylchromium(0) complexes
 30, 540
–, resolution of racemates
 by gas-liquid
 chromatography **26,** 11
 of alcohols **5,** 666 suppl. **29;
 26,** 14
– –, review **26,** 14 suppl. **28**
 of amines with L-camphorsulfonyl chloride **16,** 368
 suppl. **26**
– – – (-)-hydroxymethylenecamphor **2,** 725a suppl. **26**
 of α-aminocarboxylic acids
 as sulfonates **5,** 666 suppl.
 28, 29
 of α-aminocarboxylic acids
 by chromatography **2,** 842
 suppl. **28**
 of ketones via ketals **26,** 11
 on cation exchanger, optically
 active **5,** 666 suppl. **26**
–, – – –, enzymic, α-aminocarboxylic acids and esters
 28, 13
–, – – –, methods, review **26,** 11
–, retention of optical activity
 during esterification **26,** 175
–, – – – – during hydroformylation **4,** 667 suppl. **29**
–, – – – – during ether
 cleavage **26,** 9
–, – – – – during synthesis
 of O-α-aminoacylnucleotides
 14, 231 suppl. **26**
–, – – – – in peptide synthesis
 24, 396 suppl. **29; 27,** 193
– special s.
 α-acylaminocarboxylic acids,
 optically active
 alcohols, – –
 aldehydes, – –
 allenes, – –
 2-aminoalcohols, stereoisomeric
 Δ²-isoxazolines, optically
 active
 mercuric carboxylates, – –
 oxido compds., – –
 sulfoxides, – –
–, stability s. retention
 of optical activity

Stereospecific formation
– of
 enol tosylates **26,** 944
 ethynyldiols **30,** 454
 hydroperoxides **26,** 110
Stereospecificity
–, direction of – by functional
 groups **28,** 43
– of metals **30,** 452a
–, reversal in C-alkylation
 28, 947
–, – in asym. reduction **17,** 865
 suppl. **28**
–, –, tert. alcohols from ketones
 30, 452a
Stereospecific labeling s. under
 Compds., H-labeled
– **reactions**
 glycols from oxido compds.
 27, 129
 oxo from – – **21,** 767 suppl. **29**
– s. a.
 addition, stereospecific
 aldol condensation,
 intramolecular, –
 alkylation, –
 Claisen condensation, –
 cross-coupling, –
 cyclodehydration, –
 diene synthesis, –
 dithio ester thio-Claisen
 rearrangement, –
 Grignard synthesis, –
 hydrogenation, –
 introduction, –
 C-monoalkylation, –
 oxymercuration-demercuration, –
 ozonization, –
 reduction, –
 ring closure, –
 – opening, –
 Serini reaction, –
 side chains, introduction, –
 synthesis, –
Steric blocking s. Blocking,
 steric
– **control**
– of N-nitrosation of ureas
 13, 335 suppl. **26**
– of N-acylation **28,** 344
– of Darzens-Claisen reactions
 28, 795 suppl. **29**
–, **remote**
– by p-biphenylylurethan groups
 28, 8
– **effect**
– of
 metals **26,** 47
 reactant ratio **27,** 671
 solvents and media s. under
 Solvents
– **hindrance** s. Compds.,
 hindered
Steroidal ring B 2,3-oxidoalcohols
– startg. m. f.
 estra-1,3,5(10)-trienes **30,** 660
**Steroid backbone rearrangement
26,** 974
– Δ⁸-**double bonds,** retention
 25, 695 suppl. **29**
– **4-methyl groups,** elimination
 28, 983
– **B ring aromatization 29,** 986
– **C ring aromatization 29,** 924
– **D ring annelation 22,** 960
 suppl. **26**

Steroids (s. a. Anthrasteroids)
-, epimerization s. under
 Epimerization
- synthesis, total s. under
 Synthesis, total
- special s.
 acoxy-steroids
 cyclo-
 4,6-dihalogeno-4,6-diene-
  3-one-steroids
 estra-1,3,5(10)-trienes
 hydroxy-steroids
 keto-
 nor-
 oxido-
-, **2-halogeno-3-keto-** 29, 492
-, **heterocyclic**
- special s.
 azasteroids
 abeo-**Steroids** 17, 786
 14β-**Steroids** 14, 88 suppl. 28
 $\Delta^{2,4}$-**Steroids** 27, 949 suppl. 30
 $\Delta^{4,6}$-**Steroids, 4-subst.** 29, 573
 2,3-*seco*-Steroids 19, 562
  suppl. 29
Steroid skeleton
- by ring closure, double,
 stereospecific 29, 929
Stevens aryl migration 26, 754
- rearrangement
- of
 ammonium ylids 27, 748;
  30, 531
Stevens-type rearrangement,
 skeletal 28, 881
Stibines
- special s.
 amino-stibines
 halogeno-
Stieglitz-type rearrangement
 24, 504
Stilbene electron acceptor-
 donors 16, 977 suppl. 26
Stilbenequinones
- from
 phenols 27, 761
Stilbenes
- from
 aldehydes 26, 786
 azomethines 27, 751
 1,2-diarylethanols 26, 786
 nitro compds. 30, 575
 thioethers 30, 687
 s-trithianes 30, 688
- special s.
 tetraphenylethylene
-, **diamino-**
- startg. m. f.
 anilines 29, 93
cis-**Stilbenes**
- from
 tolans 26, 41
cis-**Stilbenes-**α,α'-d₂ 26, 41
trans-**Stilbenes** 26, 951
trans-**Stilbenes-**α,α'-d₂ 30, 21
Stiles reagent s. Magnesium
 methyl carbonate
Stobbe condensation, heterocyclics via – 13, 725 suppl. 28
Strecker reaction, Dimroth
 modification 29, 749
- synthesis of α-amino-
 carboxylic acids 21, 829
- -, asym. 29, 750
Styrenes
- by Wittig synthesis in aq.
 medium 29, 875

- from
 sulfinic acids 27, 880
- special s.
 alkoxy-styrenes
 amino-
 halogeno-
 nitro-
2-Styrylazulenes 30, 547
N-Styryl-N-heterocyclics,
 α-carboxy- 30, 254
Substituents s. a. Position
 shift of –
-, **ar.**
-, replacement, alkali metal-
 promoted 27, 459
Substitution (s. a. Replacement)
- with carbonium ions 27, 762
-, **ar.**
-, direction by binding to
 cyclodextrins 3, 445 suppl. 26
-, **double** s. 1,1-Disubstitution
-, **electrophilic**
- of
 azo groups 30, 287
 cyclopentadienide anions,
  review 26, 718
 ketene mercaptals 28, 544
-, orientation control of –,
 review 25, 389 suppl. 26
- to
 α,β-ethylenealdehydes
  26, 839
 thioethers 27, 638
- with C-cleavage 29, 312
-, **N-heterocyclic,** direction
 26, 813
-, **nucleophilic**
- at C-atoms, acetylenic 27, 210
- - -, unreactive 28, 495
- of
 ammonio groups 28, 507
 aryloxycarbonyloxy groups
  29, 224
 carbamyloxy groups 27, 374
 nitrosamino groups 30, 574
 sulfonium ylid groups 27, 222
 2-thiopyridine oxide groups
  28, 174
 tropenium salts, halogeno-
  26, 540
- on 1,2-halogenoximes 26, 463
- on 1,3-N,N-heterocycles
 27, 811
- with
 formation of cyclopropanes
  28, 808
 rearrangement 27, 993
-, -, **ar.**
-, biphenyls by – 27, 827
- by carbanions 30, 540
-, -, **intramolecular** 26, 186
- of
 amino groups, tert. 26, 246
N-Substitution s. N-Alkylation
o-Substitution s. a. 'o-alkylation' under C-Alkylation
o-Substitution-rearrangement,
 ar., review 14, 865 suppl. 28
Substrate, competitive
 s. Competitive substrate
Succinic acid anhydride
 as reagent 29, 925; 30, 685
- - **anhydrides**
- startg. m. f.
 2-pyrrolidones, 4-**carboxy-**
  28, 616

- - **esters, sym.**
- from
 carboxylic acid esters (2 molecules) 28, 703 suppl. 30;
  28, 706
 α-halogenocarboxylic acid
  esters (2 molecules) 27, 860
- - **monoesters, subst.**
-, C-alkylation, directed
 27, 840
-, condensation, directed
 27, 840
- - -, -, **unsatd.** s. Stobbe
 condensation
Succinimides
 (s. a. 2,5-Pyrrolidiones)
- from
 ketenes (2 molecules) and
  azides 29, 765
- special s.
 1-benzimidazolyl-
  succinimides
 N-hydroxysuccinimide esters
- startg. m. f.
 pyrroles, 2,5-disubst. 29, 862
-, **3-o-aminoarylidene-**
- startg. m. f.
 carbostyrils 26, 152
-, **3-o-hydroxyarylidene-**
- startg. m. f.
 coumarins 26, 152
N-(Succinimidomethyl)-
 amines, ar.
- startg. m. f.
 N-methylamines, ar. 28, 56
Süs rearrangement 2, 235;
 11, 246
Sugar anhydrides s. under
 Anhydrosugar
- **dilactones**
- from
 cyanosugars 26, 834
Sugars s. a. Carbohydrates
Sulfamic acid 18, 935 suppl. 27;
 27, 422
- as chlorine scavenger 30, 66
- special s.
 N-carbalkoxysulfamic acid
  esters
- **acids**
- startg. m. f.
 carboxylic acid amides
  29, 375
 sulfuric acid monoesters
  27, 103
-, sulfonation with – 15, 463
 suppl. 28
Sulfamides
- from
 amines 29, 287
- startg. m. f.
 N-sulfamoylsulfilimines
  28, 246
 sulfonic acid amides,
  N-subst. 29, 560
-, **monosubst.** 29, 25
Sulfamoyl azides
- from
 amines 28, 245
- **chloride** s. N-Carbonylsulfamoyl chloride
- **chlorides**
- special s.
 carbalkoxysulfamoyl
  chlorides
- startg. m. f.
 1,2-thiazetidine 1,1-dioxides
  28, 563

Sul

N-Sulfamoylsulfilimines
- from
   sulfamides and thioethers
   28, 246
Sulfamoylsulfone diimines
- s. N-Carbalkoxysulfamoylsulfone diimines
Sulfanes s. Diaminosulfanes
Sulfates, organic s. Sulfuric
   acid esters
Sulfenamides (s. a. Imino-
   sulfenamides, Tri-)
- from
   amines and sulfenimides
   26, 274
   azides 27, 296
   disulfides 28, 258
   sulfenic acid esters and
      aminoboranes 29, 291
-, reactions with sulfonium
   salts 28, 830
- special s.
   acetylenesulfenamides
   N-acylsulfenamides
   N-carbalkoxysulfenamides
- startg. m. f.
   thiolsulfonic acid esters
   26, 563
Sulfenes
- as intermediates 26, 416;
   29, 740
-, review 29, 740 suppl. 30
- startg. m. f.
   episulfones and ethylene
      derivs. 20, 609 suppl. 26
**Sulfenic acid alkoxymethyl
esters**
- special s.
   2-halogenosulfenic acid
      alkoxymethyl esters
- - esters
- from
   thiolsulfinic acid esters
      27, 634
   - - - and acetylene derivs.
      27, 634
- startg. m. f.
   disulfides 27, 586
   ethers 27, 272
   sulfenamides 29, 291
   sulfinic acid esters 30, 57
   thiolsulfonic - - 26, 563
- - silyl esters
- by penicillin S-oxide ring
   opening 29, 528
Sulfenimides
-, C-disulfenylation with -
   28, 556
- startg. m. f.
   dicarboxylic acid imides
      26, 492
   disulfides 26, 562
   sulfenamides 26, 274
   thioenolethers 28, 572
   thioethers 30, 654
   trisulfides 26, 562
Sulfenimines
- from
   disulfides 28, 407
Sulfenium salts
- special s.
   amino-sulfenium salts
   halogeno-
C-Sulfenylation (s. a. C-Disulfenylation) 29, 580;
30, 414
- with
   N-sulfenyl compds. 28, 556

Sulfenylbromides
- from
   sulfenylchlorides 28, 516
**N-Sulfenylcarbamic carboxylic
anhydrides, cyclic 28,** 255
- as acylating agents 28, 255
Sulfenylchlorides
-, addition to cyano groups,
   active 30, 348
- as intermediates, prepn.
   in situ 26, 622
- special s.
   aminosulfenyl chlorides
   chloro(chlorosulfenyl)-
      oxymethyl disulfides
   dichlorosulfenyl chlorides
   di(sulfenylchlorides)
- startg. m. f.
   allenesulfoxides 28, 565
   β,γ-ethylenesulfoxides 29, 572
   sulfenylbromides 28, 516
   sulfonyldisulfides 28, 518
N-Sulfenyl compds.
-, C-sulfenylation with - 28, 556
Sulfenyl groups
- as N-protective groups 27, 19
Sulfenylhalides (s. a. Thiosulfenyl halides)
-, prepn., review 6, 538
   suppl. 26
-, reactions review 26, 602
   suppl. 26
- special s.
   acylsulfenylhalides
   azosulfenylhalides
   sulfenylbromides
   sulfenylchlorides
- startg. m. f.
   1-halogeno-2-ketothioethers
      26, 602
   disulfides, sym. 28, 516
Sulfenylhydrazines
- special s.
   sulfenylhydrazodicarboxylic
      acid esters (25)
N-Sulfenylsulfoximes 26, 94
**Sulfenylthiocarbonic acid
esters**
- startg. m. f.
   disulfides 26, 565
Sulfhydroxamic acids
- from
   hydroxylamines and sulfonic
      acid chlorides 28, 257
- special s.
   N-hydroxybenzene-
      sulfonamide
- startg. m. f.
   sulfonic acid amides 29, 29
   N-sulfonyloxysulfonamides
      26, 95
Sulfhydryl groups (s. a.
Mercaptans and under
Replacement)
- from
   thiocarbonyl groups 26, 40
Sulfides
- s.
   acyl alkoxycarbonyl sulfides
   bisulfonyl -
   diacyl -
   diazo-N-
   di-
   hydrazonyl -
   poly-
   thioethers
Sulfido compds.
-, rearrangement 26, 586

- startg. m. f.
   1,2-dithiols 28, 525
   ethylene derivs. 27, 976
   hydrocarbons 27, 976
Sulfilimines (s. a. Acyl-
sulfilimines, Aminosulfenium
ylids, Sulfonylsulfilimines)
- from
   sulfoxides 24, 299
   sulfonylsulfilimines 18, 501
      suppl. 27
-, reactions, nucleophilic,
   with - 29, 309
-, reaction with acetylene
   derivs. 30, 137
- special s.
   N-imidoylsulfilimines
   N-sulfamoylsulfilimines
   sulfone diimines, N-mono-
      subst.
- startg. m. f.
   enamines 30, 387
   α-iminosulfonium ylids
      30, 387
Sulfimides, cyclic
- special s.
   benz[d]isothiazol-3-one
      1,1-dioxides
Sulfimines s. Sulfilimines
N- and O-Sulfination with
N-sulfinylphthalimides 29, 289
Sulfines
- special s.
   halogenosulfines
- startg. m. f.
   dihalogenomethylene compds.
      29, 852
   ketones 27, 270
   sulfoxides 26, 575
   Δ³-1,3,4-thiadiazoline 1-oxides
      29, 535
   thiirane 1-oxides 29, 767
Sulfinic acid amides
- from
   sulfoximines 28, 22
- special s.
   cyclopropylsulfinic acid
      amides
   S-iminosulfinic - -
   oxidosulfinic - -
- startg. m. f.
   iminosulfonic acid chlorides
      28, 468
- - chlorides
-, prepn., review 13, 551
   suppl. 26
- special s.
   methanesulfinic acid chloride
- startg. m. f.
   β-ketosulfoxides 30, 403
   sulfinic acid amides 27, 297
   sulfoxides 30, 403, 409
   thioethers 30, 409
- - esters
- from
   alcohols 27, 108
   sulfenic acid esters 30, 57
   sulfonic acid hydrazides
      29, 256
   sulfurous acid esters and
      halides 29, 577
- startg. m. f.
   1,1-disulfoxides 30, 394
   sulfones 22, 609
- - **halides** (s. a. Fluoro-
sulfinyl...)
- special s.
   sulfinic acid chlorides

– startg. m. f.
α-halogenosulfoxides **28,** 554
– **acids**
–, addition to nitrogen double
bonds **30,** 384
– and derivs., review **27,** 880
– from
sulfones **26,** 74
– Mannich-type condensation
with – **20,** 450; **27,** 162
–, reactions with elimination
of sulfur dioxide **27,** 880
– special s.
aminosulfinic acids
biphenylsulfinic –
cyclopropanesulfinic –
hydroxysulfinic –
– startg. m. f.
diaryls **27,** 880
styrenes **27,** 880
sulfones **26,** 607
sulfonylthioamines **28,** 517
thiolsulfonic acid esters
26, 563; **27,** 585
thiophene S-oxide ring **29,** 591
–, **α-subst.**
– by
sulfur dioxide insertion
26, 569
Sulfinodiimides
– special s.
bi-N-sulfonylsulfinodiimides
Sulfinodiimines (s. a. Sulfinodiimides)
– from
amines **26,** 273
thionylimines **26,** 276
–, reactions with ketenes
26, 305 suppl. **28**
– startg. m. f.
1,2,5-thiadiazolid-4-ones
26, 305
–, **cyclic**
– from
diamines **26,** 394
–, **sym. 26,** 589
Sulfinovinylbetaines
– from
thiirene 1,1-dioxides **30,** 424
**Sulfinovinylphosphonium
betaines 30,** 424
Sulfinyl... s. a. Sulfoxides,
Thionyl...
ω-**Sulfinylacetophenones,
o-subst.**
– startg. m. f.
heterocyclics **27,** 886
α-**Sulfinylcarboxylic acid esters**
–, reactions **30,** 455
– special s.
β-hydroxy-α-sulfinylcarboxylic
acid esters
– startg. m. f.
β-hydroxy-α-sulfinylcarboxylic
acid esters **30,** 455
**N-Sulfinyldicarboxylic acid
imides**
– special s.
N-sulfinylphthalimides
1,1-Sulfinyldisulfides
– from
thiolsulfinic acid esters
29, 579
α-**Sulfinyllactams**
–, ring contraction **30,** 311
N-Sulfinylphthalimides
–, N- and O-sulfination with –
29, 289

N-Sulfinylsulfoximines 26, 94
1-Sulfinylthioenolethers
– from
aldehydes **28,** 735
– special s.
2-amino-1-sulfinylthioenol-
ethers
– startg. m. f.
arylacetic acid esters **28,** 735
γ-dicarbonyl compds. **29,** 667
–, syntheses with – **29,** 667
1,1-Sulfinylthioethers
– from
halides **27,** 830
mercaptals **29,** 97
– special s.
α-alkylthio-β-ketosulfoxides
methyl methylthiomethyl
sulfoxide
– startg. m. f.
aldehydes **27,** 830
ketones, cyclic **27,** 830
suppl. **29, 30**
oxo compds. **27,** 830 suppl.
28, 29; 29, 667
–, syntheses with – **27,** 830
suppl. **28; 30,** 545
Sulfites, organic s. Sulfurous
acid esters
o-Sulfobenzoic anhydride
–, peptide synthesis with –
30, 271
Sulfocarboxylic acid amides
– from
sulfocarboxylic acids **4,** 350
suppl. **29**
– – **anhydrides**
– from
ketenes and sulfonic acids
30, 79
– special s.
o-sulfobenzoic anhydride
– **acids**
– startg. m. f.
sulfocarboxylic acid amides
4, 350 suppl. **29**
Sulfochlorides s. Sulfonic acid
chlorides
Sulfolane
– as medium **6,** 572 suppl. **28**
– as solvent **26,** 65; **27,** 72, 364,
656; **28,** 355, 516; **29,** 376, 697;
30, 269
– – –, polar, aprotic **24,** 291
suppl. **26**
Sulfonamides s. Sulfonic acid
amides
Sulfonamido... s. Sulfonyl-
amin...
Sulfonates s. Boryl sulfonates,
Sulfonic acid esters
Sulfonation, oxidative 29, 546
N-Sulfonation in aq. medium
27, 295
O-Sulfonation with 1-sulfonyl-
imidazoles **21,** 223 suppl.**26**
–, **preferential 19,** 802; **22,** 101
Sulfone diimines (s. a. N,N′-
Bisphosphinosulfone
diimines)
– ring closure with – **26,** 354
– special s.
sulfamoylsulfone diimines
– startg. m. f.
3H-1,2,6-thia(IV)diazine-
3,5-diones, 4,5-dihydro-
26, 354

– –, **N-monosubst.**
– from
sulfilimines **27,** 294
Sulfones (s. a. Sulfonyl groups
under Replacement)
–, α-alkylation **29,** 796
–, C-alkylation with – **24,** 865
– from
azo compds. **26,** 604
α-halogenosulfones, synthesis
24, 879 suppl. **29**
sulfinic acid esters **22,** 609
– acids **26,** 607
– – and amines, tert. **29,** 559
sulfonic acid azides **29,** 782
– – halides **30,** 411
thioethers **20,** 629; **26,** 3
–, halogenation, directed **6,** 572
suppl. **26**
–, α-metalation **29,** 796
– special s.
acoxy-sulfones
alkoxy-
allene-
amino-
carboxylic acid p-sulfonylphenyl esters
cyano-sulfones
diazo-
di-
ethylene-
halogeno-
imino-
isocyano-
keto-
nitro-
nitroso-
oxido-
phenyl –
sulfonyl...
sulfonylamino-sulfones
trifluoromethane-
– startg. m. f.
α-diketones **28,** 724
ethylene derivs. **29,** 973
α,β-ethylenesulfonic acids
29, 100
sulfinic acids **26,** 74
sulfonic acids **29,** 104
thiophene 1,1-dioxides, tetra-
hydro-3,4-diketo- **28,** 724
–, **cyclic**
– special s.
episulfones
tetramethylenesulfone
thiirane 1,1-dioxides
thiophene –
–, **α-subst. 28,** 495
–, **sym.**
– from
ethylene derivs. **27,** 599
Sulfonhydroxamic acids
– startg. m. f.
N-sulfonyloxysulfonamides
26, 95
Sulfonic acid amides (s. a.
N-Sulfonyl..., Sulfonyl-
amin...)
–, N-alkylation with diazo
compds. **18,** 468 suppl. **26**
–, N-allylation **27,** 481
–, elimination of sulfur dioxide
29, 213; **30,** 339
– from
sulfhydroxamic acids **29,** 29
sulfinic acid chlorides and
hydroxylamines **27,** 297

Sul 724

(Sulfonic acid amides)
 –, N-monoalkylation **2,** 459
 suppl. **27**
– special s.
 acylsulfonic acid amides
 ethylenesulfonic – –
 halogenosulfonic – –
 methanesulfonamide
 oxidosulfonic acid amides
 sodiosulfonic – –
 N-sulfonyloxysulfonamides
 sulfonylsulfonic acid amides
 trifluoromethanesulfonamides
– startg. m. f.
 amines s. N-Detosylation
 aminosulfones **27,** 602
 sulfonylguanidines **28,** 252
 sulfonylsulfilimines **26,** 272;
  26, 272 suppl. **27**
 sulfonylsulfoximines **26,** 272
 trihalogenophosphazosulfonyl
  compds. **27,** 304
– – –, **ar.**
– startg. m. f.
 diaryls, amino- **28,** 956
– – –, **cyclic** s. Sultams
– – –, **N-subst.**
– from
 sulfamides **29,** 560
 sulfonic acid azides **18,** 473
– – **anhydrides**
– special s.
 methanesulfonic acid
  anhydride
– startg. m. f.
 sulfonic acid esters **27,** 107
– – **azide, polymer-based,**
  as reagent **20,** 271 suppl. **29**
– – **azides**
–, reactions with N-heterocyclics **28,** 321; **29,** 339
–, ring contraction with –
 29, 396, 434
– special s.
 methanesulfonyl azide
 tosyl –
– startg. m. f.
 N-acylsulfonic acid amides
  29, 434
 α-diazo(sulfonylamidines)
  29, 310
 piperazines (from 2 molecules) **27,** 431
 sulfones **29,** 782
 N-sulfonyliminoesters **29,** 396
– – –, **β-aryl-**
– startg. m. f.
 pyrindines, dihydro- **30,** 338
– – **bisiminoamides**
– from
 mercaptans **28,** 247
– – **chlorides** (s. a. Sulfonic
  acid halides)
– special s.
 p-toluenesulfonyl chloride
 2,4,6-triisopropylbenzene-
  sulfonyl –
– startg. m. f.
 N-carbalkoxysulfonyloxylamines **27,** 110
 disulfides, sym. **26,** 564
 α,β-ethylenesulfones **21,** 642
  suppl. **26**
 sulfhydroxamic acids
  28, 257
 sulfonic acid fluorides **29,** 520
 sulfonyloxyamines **27,** 110
 sulfonylthioureas **27,** 631

– – **esters** (s. a. Sulfonyloxy...)
– as alkylating agents **29,** 882
– from
 halides **27,** 111
 sulfonic acid anhydrides
  27, 107
– acids and acylphosphonic
  acid esters **30,** 109
– – and alcohols **29,** 189
– – and chloroformic acid
  esters **26,** 169
– special s.
 aryl sulfonates
 azidosulfonates
 benzenesulfonic acid esters
 p-bromobenzenesulfonic – –
 diol monosulfonates
 disulfonates
 enol sulfonates
 fluorosulfonic acid esters
 halogenosulfonates
 methanesulfonic acid esters
 p-nitrobenzenesulfonic – –
 nonaflates
 p-toluenesulfonic acid esters
 trichloromethanesulfonic – –
– startg. m. f.
 acoxy compds. **26,** 760; **27,** 226
 carboxylic acid esters **26,** 944
 halides **19,** 620 suppl. **26**
 thiete 1,1-dioxides, 3-amino-
  29, 740
– – **fluorides**
– from
 sulfonic acid chlorides
  20, 520
– startg. m. f.
 (α-sulfonylalkylidene)-
  phosphoranes **30,** 404
– with rearrangement **30,** 404
– – **germyl esters 26,** 578
 suppl. **28**
– – **halides**
– special s.
 halogenosulfonylcarboxylic
  acid halides
 sulfonic acid chlorides
– – fluorides
– – iodides
– startg. m. f.
 α,β-ethylene-β-halogeno-
  sulfones **27,** 598
 α,β-ethylene-β'-halogeno-
  sulfones **29,** 537
 sulfones **30,** 411
– – **hydrazides**
– from
 azo compds. **30,** 384
– special s.
 cyanosulfonylhydrazines
 tosylhydrazine
 2,4,6-triisopropylbenzene-
  sulfonylhydrazine
– startg. m. f.
 sulfinic acid esters **29,** 256
– – **iodides** as reactants **29,** 537
– – **acids** (s. a. Sulfonic acid
  groups under Replacement)
– as reagents s. p-Toluenesulfonic acid
– from
 sulfones **29,** 104
 thiolcarbamic acid esters
  26, 96
– special s.
 carbamylsulfonic acids
 ethylenesulfonic –

hydroxysulfonic –
 mercaptoethanesulfonic acid
 purinesulfonic acids
– startg. m. f.
 sulfocarboxylic acid
  anhydrides **30,** 79
 sulfonic acid esters **26,** 169;
  29, 189; **30,** 109
– –, **ar.**
– from
 phenols **26,** 96
– **acid silyl esters 26,** 578
 suppl. **28**
– **carboxylic acid anhydrides**
 s. Acyl sulfonates
– – – **imides** s. N-Acylsulfonic
  acid amides
**1,1-Sulfonioammonium salts,
 quaternary**
– from
 1,1-aminothioethers **30,** 703
Sulfoniothioethers
– from
 thiosulfonium salts **21,** 639
Sulfonium cyclopentadienylids
–, reactions **28,** 626
– **fluoroborates** as reactants
 28, 840
– **salts** (s. a. Sulfonio...)
–, C-alkylation with – **28,** 623
– from
 mercaptals **27,** 997
 thioethers **27,** 997
– (2 molecules) **27,** 605
–, reaction with alkyl nitrites
 27, 348
–, reactions with sulfenamides
 28, 830
–, ring closure with – **28,** 840
– special s.
 acetylenesulfonium salts
 alkoxysulfonium –
 carboxysulfonium –
 cyclopropylsulfonium –
 enesulfonium –
 ketene mercaptal mono-
  sulfonium –
 ketosulfonium –
 methylthiomethylsulfonium –
 quinolsulfonium –
 thiopyrylium –
– startg. m. f.
 cyclopropane ring **28,** 699
 cyclopropanes **26,** 898
 ethers **27,** 224
 ethylene derivs. **27,** 876
 hydrocarbons **29,** 260
 pyrazoles, 4-hydroxy- **29,** 474
 thioenolethers **28,** 572
– –, **cyclic 28,** 994
– –, **triaryl-**
– from
 sulfonylsulfilimines **26,** 606
– **ylid groups,** substitution,
  nucleophilic **27,** 222
– **ylids**
– from
 diazo compds. **27,** 622
– special s.
 allylsulfonium ylids
 dicyanosulfonium –
 iminosulfonium –
 sulfonium cyclopentadien-
– startg. m. f.
 cyclopropanes **26,** 898
 enamines (from 2 molecules)
  26, 893

enetriazenes (from 2 molecules) **26**, 893
2-pyrones **28**, 832
– –, **cyclic**, ring expansion **27**, 705
–, spiroannelation with – **28**, 833
– –, **stable**
– from
2-acetylenesulfonium salts **26**, 767
– –, **unsatd.**
–, rearrangement, [2.3]-sigmatropic **29**, 781
Sulfonyl... s. a. Sulfones
(α-**Sulfonylalkylidene)-phosphoranes 27**, 656
– from
phosphonium salts and sulfonic acid fluorides **30**, 404
– – – – – – with rearrangement **30**, 404
Sulfonylamidines
– special s.
diazosulfonylamidines
Sulfonylamines s. a. Sulfonic acid amides
Sulfonylamination
– with N,N-dihalogenosulfonamides **30**, 304
o-Sulfonylaminoketones
s.o-(Halogenomethylsulfonylamino)ketones
α-(**Sulfonylamino)sulfones**
– from
N-sulfonylimines **30**, 384
Sulfonylammonium salts, quaternary
– special s.
methylsulfonylammonium salts, quaternary
Sulfonylcarbodiimides s. Monosulfonylcarbodiimides
Sulfonylcyanides
–, diene synthesis with –**29**, 648
– startg. m. f.
β-cyanosulfones **28**, 638
1,1-iminosulfones, cyclic **29**, 648
lactams **29**, 648
sulfonylformamidoximes **29**, 303
sulfonylformiminohalides, N-halogeno- **26**, 510
thiobis(sulfonylformimidoyl chlorides) **30**, 349
N-Sulfonyldicarboxylic acid imides
– special s.
N-sulfonylphthalimides
α-**Sulfonyl-γ-diketones**
– from
β-ketosulfones and α-halogenoketones **29**, 792
– startg. m. f.
γ-diketones **29**, 920
Sulfonyldisulfides
– from
thiolsulfonic acids and sulfenylchlorides **28**, 518
Sulfonylenamines
– special s.
disulfonylenamines
N-Sulfonylenazo compds.
– from
α-halogenosulfonylhydrazones **29**, 962
sulfonylhydrazones **30**, 646

– startg. m. f.
acetylene derivs. **28**, 927
1-(**Sulfonyl)eneformylamines**
– from
oxo compds. **28**, 729
– startg. m. f.
carboxylic acids **28**, 729
nitriles **29**, 480
Sulfonylenolethers
– special s.
disulfonylenolethers
Sulfonylformamidoximes
– from
sulfonylcyanides **29**, 303
Sulfonylformiminohalides, N-halogeno-
– from
sulfonylcyanides **26**, 510
–, N-alkylthio- **30**, 348
O-Sulfonyl group migration 27, 105
Sulfonylguanidines
– from
carbodiimides **26**, 410
sulfonic acid amides **28**, 252
1,2,4-thiadiazetidine 1-oxides, 3-imino- **29**, 473
Sulfonylhydrazines s. Sulfonic acid hydrazides
Sulfonylhydrazones
– special s.
halogenosulfonylhydrazones
tosylhydrazones
– startg. m. f.
α-cyano-N-sulfonylhydrazines **28**, 67
nitriles **28**, 67
sulfonylenazo compds. **30**, 646
tetrazoles **30**, 308
Sulfonylhydroxylamines s. Sulfhydroxamic acids
1-**Sulfonylimidazoles**
– special s.
arenesulfonylimidazolides
–, O-sulfonation with – **21**, 223 suppl. 26
N-Sulfonylimines
– from
hydrocarbons **27**, 181
– startg. m. f.
ketones **27**, 181
α-(sulfonylamino)sulfones **30**, 384
Sulfonyliminodithiocarbonic acid esters
– special s.
S,S-dimethyl-N-tosyliminodithiocarbonimidate
N-Sulfonyliminoesters
– from
enolethers and sulfonic acid azides **29**, 396
2-**Sulfonylimino-1,3-oxathiolanes**
– startg. m. f.
sulfonylureas **26**, 436
N-Sulfonyliminoseleninyl halides
– from
N,N-dihalogenosulfonamides and silyl selenides **28**, 469
N-Sulfonyliminosulfinyl –
– from
N,N-dihalogenosulfonamides and silyl sulfides **28**, 469
Sulfonylisocyanates
– special s.
halogenosulfonyl isocyanates

1,1-**Sulfonylisocyanates**
– from
1-sulfonyl-1-thiolcarbamic acid esters **26**, 493
1-**Sulfonyl-1-isonitriles**
– special s.
tosylmethylisocyanide
– startg. m. f.
heterocyclics **28**, 841
oxazoles **28**, 841
Sulfonylisothiocyanates
– startg. m. f.
1,2,3,4-thiatriazolines, 5-sulfonylimino- **30**, 221
Sulfonylketones s. Ketosulfones
N-Sulfonyl-N-nitramines
– special s.
halogeno-N-sulfonyl-N-nitramines
Sulfonylnitrenes
– as intermediates, review **18**, 473 suppl. 26
O-Sulfonyloximes, diene synthesis with – **27**, 698
Sulfonyloxyazo compds.
– from
azoxy compds. **4**, 183 suppl. 27
Sulfonyloxycarboxylic acid amides
– special s.
mesyloxycarboxylic acid amides
Sulfonyloxy compds.
– special s.
1-acoxy-1-sulfonyloxy compds.
1-alkoxy-1-
1-aryloxy-1-
N-Sulfonyloxy –
–, peptide synthesis with – **30**, 271
N→C-Sulfonyloxy group migration 27, 164
Sulfonyloxy groups
– special s.
trifluoromethanesulfonyloxy group
Sulfonyloxylamines
– from
N-carbalkoxysulfonyloxylamines **27**, 110
sulfonic acid chlorides **27**, 110
– special s.
N-carbalkoxysulfonyloxylamines
Sulfonyloxyoxido compds.
– startg. m. f.
2-hydroxythioethers, cyclic **29**, 549
N-Sulfonyloxysulfonamides
– from
sulfhydroxamic acids **26**, 95
1-**Sulfonyloxy-2-thioethers**
– startg. m. f.
thioenolethers **26**, 945
thioethers, synthesis **29**, 737
Sulfonyloxyureas
– special s.
1-acyl-3-sulfonyloxyureas
N-Sulfonylphosphine imines
– from
phosphines and sulfonylsulfilimines **30**, 700
N-Sulfonylphthalimides 26, 416
Sulfonylsulfides s. Thiolsulfonic acid esters

Sulfonylsulfilimines
- from
  sulfonic acid amides and
    sulfoxides 26, 272 suppl. 27
  - - - and thioethers 26, 272
- startg. m. f.
  ethylene derivs. 26, 978
  sulfilimines 18, 501 suppl. 27
  sulfonium salts, triaryl-
    26, 606
  sulfonylsulfoximes 26, 94
  thioethers 26, 993; 28, 992
  - and N-sulfonylphosphine
    imines 30, 700
Sulfonylsulfonic acid amides
- special s.
  halogenosulfonylsulfonic
    acid amides
α-**Sulfonylsulfonic** - -
- startg. m. f.
  α,α-dihalogeno-α-sulfonyl-
    sulfonic acid amides
    30, 355
**N-Sulfonyl-S-(sulfonylimino)-
sulfinic acid amides**
- from
  mercaptans and N,N-
    dihalogenosulfonamides
    27, 298
Sulfonylsulfoximines
- from
  sulfonic acid amides and
    sulfoxides 26, 272
  sulfonylsulfilimines 26, 94
Sulfonylthioamines
- from
  sulfinic acids and amino-
    sulfenyl chlorides 28, 517
C-Sulfonylthioformamides
- startg. m. f.
  cyanothioformamides 28, 836
  S-sulfonylthiourethans 28, 514
-, thiocarbamylation with -
    28, 836
**1-Sulfonyl-1-thiolcarbamic acid
esters**
- startg. m. f.
  1,1-sulfonylisocyanates
    26, 493
Sulfonylthioureas
- from
  sulfonic acid chlorides
    27, 631
S-Sulfonylthiourethans
- from
  C-sulfonylthioformamides
    28, 514
N-Sulfonyl-1,2,4-triazolides
-, oligonucleotide synthesis
    with - 17, 169 suppl. 29
Sulfonylureas
- from
  2-sulfonylimino-1,3-oxa-
    thiolanes 26, 436
-, N'-substitution 26, 391
- special s.
  tosylureas
o-**Sulfoperoxybenzoic acids**,
    oxidations with - 27, 152
Sulfovinylbetaines 30, 424
Sulfoxides
-, C-alkylation, radical, with -
    27, 879
-, epimerization 27, 126
- from
  alkoxysulfonium salts 27, 126
  sulfines 26, 575

sulfinic acid chlorides
    30, 403, 409
sulfoximines 28, 992
thioethers 28, 70a; 29, 98;
    30, 685
-, α-metalation 30, 455
-, ring closures with - 29, 250
- special s.
  acetylene-sulfoxides
  alkyl-
  allene-
  aryl-
  carbalkoxy-
  carboxy-
  dimethyl sulfoxide
  disulfoxides
  ethylene-sulfoxides
  halogeno-
  hydroxy-
  keto-
  nitro-
  oxido-
  penicillin ring S-oxides
  sulfinyl...
- startg. m. f.
  alkoxysulfonium salts 27, 126
  S-aminosulfoxonium salts
    28, 243
  ethylene derivs. 30, 685
  α-halogenosulfoxides 6, 572
    suppl. 26; 27, 564
  1,1-halogenothioethers 28, 497
  sulfilimines 24, 299
  sulfonylsulfilimines 26, 272
    suppl. 27
  sulfonylsulfoximines 24, 302;
    26, 272
  thioethers 23, 988 suppl. 29;
    26, 993; 27, 992; 28, 991;
    29, 993; 30, 699
-, syntheses, asym., with -
    29, 634
-, **cyclic**
- special s.
  penicillin S-oxides
-, **optically active** 29, 582
Sulfoximines
-, N-alkylation 28, 337
-, review 29, 855 suppl. 30
-, syntheses with -, review
    29, 855
- special s.
  acyl-sulfoximines
  hydroxy-
  N-imido-
  sulfonyl-
- startg. m. f.
  ethylene derivs., synthesis
    29, 855
  β-hydroxysulfoximines
    29, 855
  sulfinic acid amides 28, 22
  sulfoxides 28, 992
Sulfoxonium methylids
- startg. m. f.
  cyclopropanes (from 2 mole-
    cules) 27, 874
- special s.
  dimethylsulfoxonium
    methylid
- **salts** (s. a. S-Aminosulfoxonium salts)
- startg. m. f.
  spirocyclobutanones 26, 897
- **ylids** (s. a. S-Aminosulfoxonium ylids, Sulfoxonium
  methylids)

- from
  diazo compds. 27, 622
- startg. m. f.
  hydrocarbons 27, 82
Sulfur 11, 644 suppl. 28; 21, 600
    suppl. 28; 24, 615a suppl. 28;
    26, 329, 392, 774; 27, 496; 28,
    17, 525
-, elimination 26, 982
-, -, ring contraction 27, 642
-, reactions with organic
    compds., review 29, 534
-, ring closures with - 27, 619;
    29, 534
- startg. m. f.
  4H-1,4-thiazines, 2,3-dihydro-
    21, 653 suppl. 26
  thiazole ring, 2-amino- 27, 619
Sulfuranes
- special s.
  alkylidene-sulfuranes
  dialkoxy-
Sulfur compds. (s. a. Thi...)
-, chemistry, review 27, 877
    suppl. 30
-, hydrogenolysis 30, 51
- in synthetic organic
    chemistry, review 27, 877
-, reduction with dithioacetic
    acid 29, 993
-, ring closure with
    dialdehydes 30, 609
-, stereochemistry 28, 992
-, syntheses with - 27, 877
-, - - - and azomethinium
    salts 25, 551 suppl. 29
Sulfur(IV) - s. S(IV)-Hetero-
  cumulenes
Sulfur dichloride
-, addition to compds., unsatd.,
    review 26, 579 suppl. 29
- as reactant 20, 444 suppl. 28,
    29; 28, 259; 30, 349
-, reaction with acetylene
    derivs. 26, 579
- **dioxide** 26, 2 suppl. 29;
    29, 143, 148, 903
- as reactant 26, 604
-, elimination s. under
    Elimination
-, insertion of - s. under
    Insertion
-, reactions with α-diazoketones
    29, 562
- startg. m. f.
  1-alkoxy-1-sulfones, cyclic
    29, 729
  sulfones, sym. 27, 599
- -, **liq.** (s. a. under Antimony
    pentachloride) 8, 748 suppl.
    29; 26, 158, 428, 751;
    29, 148, 562
- as medium 28, 826
-, reactions, org. in -, review
    28, 826
-, - with iodine and pyridine
    in - 29, 287
- -/**ketene** s. under Ketene
- -/**pyridine** s. under Pyridine
**Sulfur-donor ligand
o-metalated complexes**
- startg. m. f.
  3H-benzo[c]thiophen-1-ones
    29, 643
- -**hydrazine compds.** 27, 299
Sulfuric acid, fuming,
    containing SO_3, as solvent
    27, 367, 568

– – **amides** s. Sulfamides
– – **chloride esters** (s. a. Chlorosulfonyloxy groups under Replacement)
–, cleavage **27**, 579
– special s.
  chloro(chlorosulfates)
– – **esters**
– as reactants **27**, 127; **28**, 533
– from
  sulfurous acid esters
  27, 104
– special s.
  dialkyl sulfates
– startg. m. f.
  hydrocarbons, synthesis
  29, 849
– – –, **cyclic**
– special s.
  1,1,1-enediol sulfates
  ethylene sulfate
– – **monoaryl esters** as intermediates **29**, 65 suppl. **30**
– – **monoesters 27**, 103
– special s.
  diazonium sulfate betaines
  8-quinolyl sulfate
  sulfuric acid monoaryl
  esters
– startg. m. f.
  alcohols **27**, 3; **30**, 656
  ethylene derivs. with C-alkyl migration **30**, 34
  hydrocarbons **26**, 58
–, synthesis **28**, 602
Sulfur monochloride 29, 544; **30**, 354
Sulfurous acid s. Sulfur dioxide
– – **diamides** s. carboxylic acid amides
– – **ester halides**
– special s.
  1,2-halogenhydrin halogenosulfites
– – **esters**
– special s.
  diethyl sulfite
  diphenyl –
– startg. m. f.
  sulfinic acid esters **29**, 577
  sulfuric – – **27**, 104
Sulfur monochloride 26, 589 suppl. **29**; **30**, 392
– **oxides** s. Disulfur monoxide, Sulfur dioxide, – trioxide
– **reaction**, review **25**, 447 suppl. **26**
– -**transfer agent**
–, N,N'-thiobisphthalimide as **28**, 244
– **trifluorides, amino-** s. Dialkylaminosulfur trifluorides
– **trioxide** (s. a. Sulfuric acid, fuming) **26**, 818; **27**, 367; **28**, 541; **29**, 242, 847
–, oxidations with – **27**, 123
– startg. m. f.
  monoorganomercury sulfonates **27**, 593
– – -**triethylamine 8**, 140
Sulfuryl chloride 26, 573, 622; **27**, 543, 553, 567, 579, 606; **28**, 112, 492, 509; **29**, 98; **30**, 347, 354, 687
Sultams
– from
  halogenosulfonic acid amides
  30, 671

β-**Sultams** s. 2H-Benzo[c]-1,2-thiazete 1,1-dioxides
Sultines 29, 105
– special s.
  ethylenesultines
  1,2-oxathiolane 2-oxides
Sultones
–, α-alkylation **25**, 594 suppl. **27**
– special s.
  germasultones
  1,2-oxathiin 2,2-dioxides
  silasultones
β-**Sultones**
– from
  ethylene derivs. **30**, 389
Superacids s. Medium, superacidic
Superbases 28, 884; **30**, 506
Supercel s. Kieselguhr
Suppression s. Prevention
Surface-active agents (s. a. Phase-transfer catalysts) **26**, 77
Surface, solid, reactions on –
  s. Catalysts, Chromatography, preparative
Sydnone imine salts, N-acylation **15**, 342 suppl. **26**
– – –, **3-amino-**
– from
  1,1-cyanohydrazines **26**, 828
  hydrazines and aldehydes **26**, 828
Sydnones
– startg. m. f.
  pyrazoles **27**, 900
Symmetry-disallowed reactions, catalysis of –, review **20**, 531 suppl. **29**
Syntheses (s. a. Diene syntheses, Grignard –)
– on polymer support s. Solid-phase multi-step syntheses
– via allenecarbanions **30**, 584
– with
  acoxy compds. **30**, 559
  ethylene derivs., non-activated **28**, 657
  α-halogenonitrones **28**, 657
  Δ²-thiazolines, 2-alkylthio- **28**, 801
  vinylcyclopropanes **28**, 859 suppl. **30**
–, **electrophilic, regiospecific**
– with α,β-ethyleneketones **30**, 453
Synthesis CC⇓, CC⇑ (s. a. C-Chain extension, Condensation, Friedel-Crafts ketone synthesis)
–, design **17**, 201 suppl. **29**
– **with addition of 1 C-atom** (s. a. Homologization)
  alcohols from ethylene derivs. **4**, 667 suppl. **26**
  aldehydes from halides **26**, 872, 891
  – from ketones **26**, 859
  amines from ethylene derivs. **27**, 723
  –, prim., from alcohols **27**, 782
  2-aminoalcohols from ketones **27**, 720
  α-aminocarboxylic acid derivs. from nitriles **29**, 631
  – acids from nitriles **28**, 621

– – from oxo compds. **30**, 568
α-aminonitriles from oxo compds. **27**, 781; **29**, 750
arylacetic acid esters from aldehydes **28**, 735
carboxylic acid amides from aldehydes **29**, 443
– – esters from halides **28**, 826
– acids and derivs. from halides **27**, 869 suppl. **28**
– – from alcohols **28**, 612
– – from ethylene derivs., review **16**, 750 suppl. **26**
– – from oxo compds. **28**, 729
enolethers from oxo compds. **14**, 876
α,β-ethylenealdehydes from ketones **19**, 879
α-halogenocarboxylic acid esters from oxo compds. **29**, 738
α-hydroxyaldehydes from carboxylic acid esters **30**, 545
– from oxo compds. **22**, 661 suppl. **27**; **27**, 830
α-hydroxyamidoximes from aldehydes **27**, 667
α-ketocarboxylic acid amides from carboxylic acid chlorides **27**, 847
nitriles from oxo compds. **29**, 480
1,1-sulfinylthioethers from halides **27**, 830
thionocarboxylic acid esters from diazo compds. **30**, 570
– **with addition of 2 C-atoms**
acetylene derivs., terminal, from ethylene derivs. **29**, 673
acoxy-2-ethylenes from ketones **29**, 191
aldehydes from halides **28**, 793
α-aminocarboxylic acid esters from aldehydes **29**, 739
α,β-ethylenealdehydes from carboxylic acid chlorides **29**, 846
– from oxo compds. **29**, 867
– from ketones **5**, 523 suppl. **26**
α-halogenocarboxylic acid esters from halides **29**, 499
– **with addition of 3 C-atoms**
carboxylic acids **18**, 765 suppl. **26**
α,β-ethylenealdehydes **26**, 862
β,γ-ethylenehalides from halides **28**, 802
– **with addition of 4 C-atoms**
3-acetylenealcohols from halides **29**, 797
2-ethylenealcohols from boranes **27**, 896
– **with addition of 5 C-atoms**
hydroxyketones **28**, 769 suppl. **30**
– **with rearrangement**
β-amino-α,β-ethylenesulfones **28**, 831

Synthesis, allylic
- with
  2-ethylene derivs. **28,** 601
  2-ethylenemesitoates **28,** 601
-, **anodic,** of sulfonium salts
  27, 605
-, **asym.** (s. a. Asym.,
  Reformatskii synthesis,
  asym.)
-, effect of C-chains **29,** 750
- of
  α-acylaminocarboxylic acid
  amides **28,** 288 suppl. **30**
  alcohols from oxo compds.
  26, 671
  alkoxysilanes **29,** 107 suppl. **30**
  α-aminocarboxylic acids
  4, 643 suppl. **26; 16,** 731
  suppl. **26; 21,** 829 suppl. **27;**
  28, 621; **29,** 55
  - -, review **17,** 461 suppl. **26**
  - - from halides **29,** 791
  - - via 1,4-oxazin-2-ones,
    tetrahydro- **29,** 750
  carboxylic acids via
  2-oxazolines **29,** 807
  1,4-diols **29,** 807 suppl. **30**
  β-hydroxyarboxylic acid
  esters **2,** 827 suppl. **26**
  γ-lactones **29,** 807 suppl. **30**
  oxazolidines **28,** 335
  peptides **29,** 55
- on polymer support **9,** 741
  suppl. **29; 26,** 671 suppl. **28**
-, review **29,** 55
- with
  carbohydrate derivs. **9,** 741
  suppl. **29; 26,** 671
  α,β-ethylenesulfoxides
  18, 965 suppl. **28**
  proline derivs. **21,** 835
  suppl. **26; 29,** 55; **30,** 520
  sulfoxides **29,** 634
  transition metal catalysts,
  homogeneous, review
  23, 48 suppl. **29**
-, -, **biochemical 24,** 258
-, **geospecific**
- of
  2-ethylenealcohols **28,** 855
  α,β-ethyelnecarboxylic acids
  27, 837
  ethylene derivs. **26,** 875
  suppl. **30**
  - -, review **26,** 809
  - -, trisubst., review **25,** 634
    suppl. **26**
  trans-ethylene derivs. **26,** 809
- with allyl dithiocarbamates
  30, 585
-, **stereospecific**
- of
  acetylene derivs., terminal
  29, 673
  alcohols, sec. **30,** 450
  -, tert. **27,** 671
  alkaloids **27,** 934
  amines, sec. **27,** 485 suppl. **28**
  γ-diketones **28,** 804
  4H-1,3-oxazine ring,
  5,6-dihydro- **26,** 885
-, **total**
- of
  gibberellins **21,** 738 suppl. **26**
  glucose **22,** 2 suppl. **27**
  insulin **25,** 417 suppl. **30**

prostaglandins **23,** 693
  suppl. **28**
steroids **22,** 927 suppl. **27;**
  23, 924; **30,** 491

Tantalum pentafluoride 30, 635
Tartaric acid as reagent **30,** 321
Tautomerism s. a. Ring-chain
  tautomerism
-, **prototropic**
- of
  heterocyclics, review **18,** 397
  suppl. **26**
Tautomers s. Enamine
  tautomers **25,** Imine-enamine
  tautomers
Telluride dihalides 28, 967
- startg. m. f.
  diaryls **28,** 967
Tellurium tetrachloride 28, 967
Telluroacetals
- from
  diazo compds. and
  ditellurides **26,** 648
**Telomerization, palladium-
  catalyzed**
- of
  allenes **27,** 724 suppl. **28**
  1,3-dienes **23,** 715; **27,** 724
Telomerized compds.
- special s.
  amines, telomerized
  silanes, -
Temperature s. High
  temperature
Tetraacylethylenes
- startg. m. f.
  heterocyclics **27,** 571
Tetraalkoxyethylenes
-, reactions **27,** 329 suppl. **30**
- special s.
  tetramethoxyethylene
**Tetraalkylammonium
  carboxylates**
- startg. m. f.
  carboxylic acid esters
  12, 296
- **cyanides** (s. a. Tetrabutylammonium cyanide)
- as bases and nucleophiles
  29, 628
- **salts**
-, reactions with - **26,** 675
1,3,5,7-Tetraazacyclooctanes
  14, 364 suppl. **28**
Tetraazamacroheterocyclics
  30, 233
- from
  1,2-dithiolium salts **17,** 516
  suppl. **28**
- special s.
  porphins
1,2,4,5-Tetraazapentalenes
- from
  acetyleneazo compds.
  (2 molecules) **26,** 702
**1,3,6,9-Tetraazaspiro[4.4]-
  nonane-2,4-diones 26,** 920
**1,3,6,9-Tetraazaspiro[4.4]-
  nonane-2,4-dithiones 26,** 920
**2,4,4,6-Tetrabromo-2,5-cyclo-
  hexadienone**
-, addition to carbon-carbon
  double bonds with - **28,** 479
- as reagent **26,** 560 suppl. **30**
-, bromination with - **28,** 485,
  491

Tetrabutylammonium azide
  as reagent **30,** 460
- **bromide**
- as phase transfer catalyst
  6, 898 suppl. **29; 29,** 792
- as reagent **19,** 46 suppl. **26**
- **chloride** as catalyst
  30, 349
**Tetra-n-butylammonium
  chloride** as chloride ion
  source **29,** 969
Tetrabutylammonium cyanide
  (s. a. Tetraalkylammonium
  cyanides)
- as reagent **26,** 675
- **fluoride 30,** 4
- as reagent **14,** 606 suppl. **26**
- **fluoroborate**
- as electrolyte **29,** 900
- as supporting electrolyte
  30, 433
- **hydroxide** as reagent
  26, 417; **29,** 414
- **iodide**
- as activator **30,** 606
- as phase transfer catalyst
  28, 794; **30,** 139, 645
- **perchlorate** as electrolyte
  27, 953
- **salts** s. a. Bis(tetra-n-butylammonium) oxalate
- **tetrahydridoborate** as reagent
  28, 34
- **trihydridocyanoborate**
  27, 74 suppl. **28**
Tetrachlorethane as solvent
  28, 779
Tetrachloroaluminic acid
  s. Aluminum chloride/
  hydrogen chloride
Tetrachloroauric acid 18, 141
  suppl. **29**
Tetrachloro-o-benzoquinone
  s. o-Chloranil
Tetrachloro-p-benzoquinone
  s. Chloranil
Tetrachloroferrates
- as products **28,** 828
- as reactants **26,** 864
Tetracyanoalkenes s. Tetracyanoethylene
Tetracyanoethylene
- as reactant **17,** 747 suppl. **26**
- as reducing agent **11,** 209
  suppl. **29**
-, ring closures with - **21,** 728
  suppl. **26; 30,** 475
Tetraethylammonium bromide
  30, 274
- **chloride 26,** 529; **30,** 348
- **fluoride 29,** 798
- **halide 28,** 491; **29,** 235
- **iodide 27,** 160; **29,** 235
- **periodate 30,** 244
- **tosylate** as electrolyte,
  supporting **27,** 135; **28,** 1, 350
Tetrafluorobor... s. Fluorobor...
Tetrafluorohydrazine
  as reactant **27,** 351
Tetrahalogenomethanes
  25, 385 suppl. **26; 27,** 417
- special s.
  carbon tetrabromide
  - tetrachloride
Tetrahydrofuran (s. a. Furans,
  tetrahydro-)
- as H-donor **29,** 70
- as reactant **29,** 487

-, δ-hydroxy-*n*-butylation with –
  26, 663
**Tetrakis[iodo(tri-*n*-butyl-
  phosphine)copper(I)] 26**, 910
**Tetrakis(triphenylphosphine)-
  palladium(0) 26**, 875
  suppl. **30; 29**, 201
α-Tetralones
– startg. m. f.
  2,4a-ethanophenanthrenes,
  2,3,4,4a,9,10-hexahydro-
  29, 884
  2H-naphtho[1,2-b]pyran-
  2-ones, 5,6-dihydro-
  29, 884
  phenanthrenes, 3,4,9,10-
  tetrahydro- **29**, 884
β-Tetralones 16, 957 suppl. **29**
Tetramethine cyanines 28, 718
Tetramethoxyethylene
–, [2+2]-cycloadditions to –
  27, 329
Tetramethylammonium chloride
– as chloride ion source **27**, 173
– as electrolyte, supporting
  27, 18 suppl. **28**
– **dimethyl phosphate**
–, dehydrohalogenation with –
  5, 650 suppl. **27**
– **fluoride** as reagent **29**, 942
– **fluoroborate** as electrolyte
  28, 973
– **salts** as reactants **28**, 170, 365;
  29, 205
Tetramethylene sulfone
  s. Sulfolane
**N,N,N′,N′-Tetramethylethylene-
  diamine** s. TMEDA
Tetramethylguanidine as reagent
  23, 412 suppl. **28; 27**, 437;
  27, 404 suppl. **29; 28**, 375
  suppl. **29**
Tetramethylurea as solvent
  26, 564, 755
1,1,2,2-Tetraminoethanes,
  reactions **27**, 951
Tetraphenylethylene as catalyst
  28, 780 suppl. **19**
Tetrathiofulvalenes 16, 820
– from
  1,3-dithiole-2-thiones **30**, 610
  1,3-dithiolium salts **29**, 717
**Tetrathiohypodiphosphorous
  acid esters**
– from
  dithiophosphorochloridites
  30, 423
**Tetrathioorthocarbonic acid
  esters**
–, transesterification **26**, 623
**Tetrathiophosphoric acid
  esters**
– from
  alcohols **28**, 545
Tetrathiospirocyclics
– by transesterification **26**, 623
1,2,4,5-Tetrazepine ring 29, 353
**1,2,4,5-Tetrazine ring,
  1,2,3,4-tetrahydro-**
– from
  Δ³-imidazoline-5-thione ring
  29, 281
1,2,4,5-Tetrazines
– from
  1,2,4,5-tetrazines, 1,6-dihydro-
  28, 428
– special s.
  spiro-1,2,4,5-tetrazines

– startg. m. f.
  1,2,4-triazepines **29**, 770
–, **monosubst.**
– from
  iminoesters **28**, 381
–, **1,6-dihydro-**
– startg. m. f.
  1,2,4,5-tetrazines **28**, 428
–, **hexahydro-**, by dimerization,
  oxidative **26**, 326
**Tetrazole-azidoazomethine
  tautomerism 29**, 466
– –, reviews **17**, 467 suppl. **28**
**Tetrazole ring, N-condensed
  27**, 439
– startg. m. f.
  cyanamides, cyclic **27**, 523
Tetrazoles
– from
  carboxylic acid chlorides
  and phosphine imines
  26, 444
  sulfonylhydrazones and
  diazonium salts **30**, 308
– special s.
  α,β-ethylene-β-(1-tetrazolyl)-
  carboxylic acid amides
– startg. m. f.
  acylamidrazones **30**, 288
–, 1-substitution **29**, 432
–, **2-** from 1-alkyl- **28**, 414
–, **azido-**
– from
  nitriles and ethylene derivs.
  29, 317
Δ²-Tetrazolines
– startg. m. f.
  diaziridines
**1,2,3,4-Tetrazolopyridines,
  4-nitro-**
– startg. m. f.
  [1,2,5]oxadiazolo[3,4-b]-
  pyridine 1-oxides **27**, 102
**1-(N-Tetrazol-5-ylhydrazono)-
  1-halides**
– startg. m. f.
  1,2,4-triazoles, 3-azido-
  28, 457
**Tetrazolylisocyclics, halogeno-,
  bicyclic**
– from
  dienes, cyclic, and nitriles
  30, 351
Teuber reaction, review **8**, 206
  suppl. **26**
Thallation, ar., reactions via –
  25, 389
Thallium 26, 261
– in organic chemistry, review
  27, 141 suppl. **28**
–**(III) acetate 10**, 280 suppl. **29;
  27**, 178
–**(I) alkoxides**
– special s.
  thallium(I) ethoxide
– startg. m. f.
  orthocarbonic acid esters
  4, 249 suppl. **28**
–**(I) carboxylates** as reactants
  13, 209 suppl. **28; 17**, 269
  suppl. **26**
–**(I) compd. 28**, 623
–, **compds., organo-**
–, organothallium chemistry,
  review **25**, 389 suppl. **26**
– special s.
  arylthallium bis(trifluoroacetates)

dimethylthallium compds.
  vinylthallic acetates,
  2-acetoxy-
–**(I) compds., organo-**
  as reactants **21**, 238 suppl. **27;
  28**, 623
–**(I) ethoxide 2**, 259 suppl. **28;
  26**, 424; **29**, 229
–**(III) nitrate 26**, 188; **27**, 162,
  203, 238; **28**, 93; **29**, 229;
  30, 82
–, oxidations with – **27**, 141
–**(I) sulfate 26**, 718
–**(III) sulfate 28**, 653
–**(III) trifluoroacetate 22**, 895
  suppl. **28; 28**, 182; **29**, 143, 903
Thermodynamic control
  s. under Reactions
Thermolysis s. a. Pyrolysis
–, imidazoles, 1-cyano- by –
  28, 450
Thetin anions, ring closures
  with – **26**, 895
Thexylborane
– as reactant **23**, 703 suppl. **29;
  26**, 914 suppl. **27; 27**, 716;
  28, 820; **30**, 496/7
–, review **30**, 496
Thia… s. a. Thio…,
  S-Heterocyclics
Thiaarenium salts
  s. Thopyrylium salts
Thiabenzene oxides
– from
  α,β-acetyleneketones **27**, 769
**9-Thiabicyclo[4.2.1]nona-2,4,7-
  triene 9,9-dioxide ring**
– by Ramberg-Backlund
  rearrangement, bishomo-
  conjugative **26**, 963
1,2-Thiaborolanes
– startg. m. f.
  1,3,4-thiaoxaborepanes **28**, 529
**1,4,5-Thiadiazepin-3-one
  1,1-dioxide ring, 2,3,4,7-
  tetrahydro-**
– startg. m. f.
  4H-1,4-thiazin-3-one
  1,1-dioxide ring,
  2,3-dihydro- **28**, 299
**1,2,4-Thiadiazetidine 1-oxides,
  3-imino-**
– from
  carbodiimides and N-thionylsulfonamides **29**, 473
– startg. m. f.
  monosulfonylcarbodiimides
  and thionylimines **29**, 473
  sulfonylguanidines **29**, 473
**3H-1,2,6-Thia(IV)diazine-
  3,5-diones, 4,5-dihydro-**
– from
  sulfone diimines and
  malonic acid esters **26**, 354
Thiadiazine S-dioxides, review
  20, 308 suppl. **26**
**4H-1,3,5-Thiadiazine-4-thiones,
  5,6-dihydro-2-amino-**
– from
  azomethines and thio-
  carbamylisothiocyanates
  27, 591
**1,3,5-Thiadiazine-2-thiones,
  tetrahydro- 26**, 596
2H-1,3,5-Thiadiazin-2-ones
– from
  1,2,4-dithiazolium salts
  27, 636

**4H-1,3,4-Thiadiazin-5-ones,
5,6-dihydro-**
- from
  1,3,4-thiadiazolium salts and
  acylphosphonic acid esters
  26, 909
**4H-1,3,5-Thiadiazin-4-ones,
5,6-dihydro-2-amino-**
- from
  azomethines and carbamylisothiocyanates **26,** 572
**1,2,5-Thiadiazole 1,1-dioxide
ring**
- startg. m. f.
  nitriles **29,** 472
  dinitriles **29,** 472
Thiadiazole S-dioxides, review
20, 308 suppl. **26**
1,2,4-Thiadiazoles
- from
  1,3,4-oxathiazol-2-ones and
  nitriles **30,** 416
1,3,4-Thiadiazoles, 2-alkylthio-
- from
  carboxylic acid chlorides and
  dithiocarbazic acid esters
  29, 568
1,2,4-Thiadiazolidine-3,5-diones
- from
  S-(halogenocarbonylamino)-
  isothiocarbamyl halides
  22, 295
  S-halogenoisothiocarbamyl
  halides and isocyanates
  22, 295
**1,2,5-Thiadiazolidine 1-oxides
26,** 271
1,3,4-Thiadiazolidines
- from
  azines **30,** 566
- startg. m. f.
  Δ^3-1,3,4-thiadiazolines **30,** 566
1,2,5-Thiadiazolid-4-ones
- from
  sulfonodiimines and ketenes
  26, 305
Δ^2-**1,2,4-Thiadiazoline
1,1-dioxides, 3-amino-**
- from
  carbodiimides **26,** 410
Δ^3-**1,3,4-Thiadiazoline 1-oxides**
- from
  diazo compds. and sulfines
  29, 535
Δ^4-**1,2,4-Thiadiazoline ring,
3-imino- 17,** 657
Δ^2-**1,3,4-Thiadiazolines, 5-imino-**
- from
  thiocyanates and diazonium
  salts **30,** 234
Δ^3-**1,3,4-Thiadiazolines 21,** 283
suppl. **28**
- from
  1,3,4-thiadiazolidines **30,** 566
- startg. m. f.
  ethylene derivs. **30,** 566
1,3,4-Thiadiazolium salts
- from
  carboxylic acid thiohydrazides
  28, 553
- startg. m. f.
  4H-1,3,4-thiadiazin-5-ones,
  5,6-dihydro- **26,** 909
**1,3,4-Thiadiazol-2(3H)-ones,
5-alkylthio-**
- from
  dithiocarbazic acid esters
  28, 539

**3H-[1,2,4]Thiadiazolo[4,3-a]-
pyridines 24,** 297 suppl. **26**
Thiamine derivs. 28, 532
**Thianaphthene-2-carbonyl
chlorides, 3-chloro-**
- from
  cinnamic acids **28,** 548
**Thianaphthene-2-carboxylic
acid esters 29,** 776
**Thianaphthene-3-carboxylic
acids, 2-acyl-**
- from
  thianaphthenequinones and
  α-halogenooxo compds.
  29, 813
Thianaphthenequinones
- startg. m. f.
  thianaphthene-3-carboxylic
  acids, 2-acyl- **29,** 813
Thianaphthenes 26, 657; **27,** 755
suppl. **28; 29,** 545
- by ring contraction **25,** 652
- from
  acetylene derivs. **26,** 579
  enol sulfonates **27,** 935
  ketones **14,** 937 suppl. **27**
  thiophenes **28,** 922
- special s.
  thioindoxyls
- via thio-Claisen rearrangement **30,** 677
-, **2-acyl-3-amino-**
- from
  1,2-benzisothiazolium salts,
  3-halogeno- **29,** 576
-, **2,3-dihydro-3-acyl-**
- from
  aryl acetylenesulfoxides
  27, 755
-, **2-formyl-**
- from
  thiochromenes **26,** 163
**Thianaphthen-7-ones,
4,5,6,7-tetrahydro- 26,** 957
1,3,4-Thiaoxaborepanes
- from
  1,2-thiaborolanes **28,** 529
**1,4-Thiaphosph(V)orin
4-sulfides, 2,6-diamino-**
- from
  ynamines (2 molecules) and
  perthiophosphonic acid
  anhydrides **29,** 613
**2H-1,2-Thiaphosphos(V)orin
2-sulfides, 3,6-dihydro-**
- from
  1,3-dienes and perthio-
  phosphonic acid anhydrides
  29, 612
**1,3,2,4,6-Thia(VI)thiatriazin-
5-one 3,3-dioxides,
4,5-dihydro-**
- from
  N-carbalkoxysulfamoyl-
  sulfone diimines **27,** 504
6a-Thiathiophthenes
- startg. m. f.
  1,6a-dithia-6-azapentalenes
  30, 307
Thiation s. Replacement of
hydrogen by sulfhydryl
1,2,3,5-Thiatriazole S-oxides
- from
  amidrazones **26,** 270
1,2,3,4-Thiatriazoles, 5-amino-
- from
  isothiocyanates **28,** 278

**1,2,3,4-Thiatriazolines,
5-sulfonylimino-**
- from
  azides and sulfonylisothiocyanates **30,** 221
- startg. m. f.
  heterocyclics **30,** 221
Thiazaphospholanes 26, 633
1,3-Thiazepine ring 28, 561
**1,3-Thiazepines, 4,7-dihydro-
2-amino-**
- from
  aziridines, 2-vinyl- and
  isothiocyanates **28,** 526
**1,4-Thiazepin-7-one ring,
2,3,4,7-tetrahydro-**
- from
  penicillin ring skeleton
  27, 604
1,2-Thiazetidine 1,1-dioxides
- from
  sulfamoyl chlorides **28,** 563
1,3-Thiazetidines
- from
  carbodiimides **27,** 590
**1,2-Thiazetidin-3-ones, 1-imino-
26,** 305 suppl. **28**
4H-1,4-Thiazine 1,1-dioxides
- from
  1,4-oxathiane 4,4-dioxides,
  2,6-dialkoxy- **29,** 148
  thiophene 1,1-dioxides,
  2,5-dihydro- **29,** 148
6H-1,3-Thiazine ring, opening,
reductive **30,** 21
-, -, **2,3-dihydro- 28,** 966;
29, 590
1,4-Thiazine ring, dimerization,
oxidative **27,** 763
- -, **condensed 19,** 702
- -, **dihydro- 28,** 446
1,3-Thiazines, syntheses via -,
review **25,** 612 suppl. **27**
-, **2-amino-**
- from
  α,β-ethyleneketones **28,** 552
2H-1,4-Thiazines, 3,4-dihydro-,
rearrangement **29,** 954
**4H-1,3-Thiazines, 5,6-dihydro-
28,** 308
-, -, **2-amino-**
- from
  2-ethyleneisothiouronium salts
  28, 308
**4H-1,4-Thiazines, 2,3-dihydro-
26,** 318
- startg. m. f.
  thiazolidines **27,** 48
**1,3-Thiazine-6-thione ring,
dihydro-**
- from
  oxo compds. **27,** 616
1,3-Thiazinium betaines 28, 558
**4H-1,4-Thiazin-3-one
1,1-dioxide ring, 2,3-dihydro-**
- from
  1,4,5-thiadiazepin-3-one
  1,1-dioxide ring, 2,3,4,7-
  tetrahydro- **28,** 299
1,4-Thiazin-3-one ring
- from
  o-aminomercaptans and
  α,β-ethylenecarboxylic
  acid esters **30,** 393
**4H-1,3-Thiazin-4-ones,
2,3-dihydro-**
- from
  oxo compds. and α,β-

ethylene-β-mercapto-
carboxylic acid amides
28, 551
Thiazole ring 26, 591
-, **2-amino-**
- from
o-aminothiocyanates **17,** 383
enamines **27,** 619
Thiazoles
- as intermediates **27,** 830
- from
α-acylaminoketones **26,** 603
α,β-ethyleneketones **18,** 669
suppl. **26**
isothiazoles **27,** 603
Δ²-thiazolines, 4-hydroxy-
29, 941
-, **2-acyl-**
- from
thiaminoids **26,** 925
-, **2-amino-**
- from
ketones **27,** 607
-, **5-amino-**
- from
α-acylaminocarboxylic acid
thioamides **27,** 641
-, **2-halogeno-**
- startg. m. f.
5H-thiazolo[2,3-b]quinazol-
5-ones **27,** 475
Thiazolidine ring 28, 423
Thiazolidines
- from
carbohydrates **9,** 666 suppl. **26**
4H-1,4-thiazines, 2,3-dihydro-
27, 48
Δ²-thiazolines **28,** 176
- special s.
enethiazolidines
- startg. m. f.
aldehydes **28,** 176
4-Thiazolidone ring 30, 399
- - opening **27,** 323
2-Thiazolidones, 4-halogeno-
- from
S-halogenoisothiocarbamyl
halides **26,** 610
4-Thiazolidones
- from
thiocarbohydrazones **27,** 596
-, **2-imino- 4,** 561
4-Thiazolidone-2-thiones
s. Rhodanines
Δ²-**Thiazoline ring 26,** 631
- - opening **29,** 571
Δ²-**Thiazolines**
- from
aziridines, N-acyl- **27,** 615
dithiocarboxylic acid esters
and 2-aminomercaptans
30, 412
enisonitriles **30,** 388
penicillin ring skeleton
26, 500
- special s.
2-methyl-Δ²-thiazoline
- startg. m. f.
aldehydes **28,** 176
thiazolidines **28,** 176
-, **2-alkylthio-**
-, cleavage **28,** 802
- startg. m. f.
2-enethiazolidines **28,** 844
-, syntheses with - **28,** 801/2
-, - - -, review **28,** 801/2
suppl. **29**

-, **2-amino-**
- from
thiosulfuric acid S-mono-
esters **28,** 570
-, **4-hydroxy-**
- startg. m. f.
thiazoles **29,** 941
Δ⁴-**Thiazolines, 4-amino-**
- from
aziridines and isothiocyanates
26, 706
Δ²-**Thiazolinium salts**
- from
carboxylic acid thioamides
27, 626
Thiazolium betaines, 4-hydroxy-
- startg. m. f.
2-pyridones **26,** 919
thiophenes **26,** 919
-, **5-hydroxy-**
- startg. m. f.
pyrroles **26,** 919 suppl. **30**
- **bromide, N-lauryl-**
- as catalyst **29,** 630
- **salts**
- as catalysts **28,** 648 suppl. **30;**
29, 630 suppl. **30**
Thiazolo[3,2-a]benzimidazoles
18, 669 suppl. **26**
1H,3H-Thiazolo[3,4-a]-
benzimidazoles, 1-imino-
28, 559
Thiazolo[3,2-a]benzimidazol-
3(2H)-ones, reactions **27,** 323
Thiazolo[3,2-a][1,3]diazepines,
5,6,7,8-tetrahydro- 26, 612
3H,5H-Thiazolo[4,3-a]isoindole-
3,5-diones 27, 520
3H,5H-Thiazolo[4,3-a]isoindol-
5-ones
- from
indan-1,3-diones,
2-isothioureido- **26,** 319
Δ⁴-**2-Thiazolones**
- from
enamines **29,** 543
Δ²-**5-Thiazolones**
- special s.
4-azo-Δ²-5-thiazolones
Thiazolo[3,2-a]pyridinium
betaines, dihydro- 26, 611
- **salts, dihydro- 26,** 611
- from
2-pyridinethiones and
α,β-ethylenehalides **26,** 611
Thiazolo[4,5-d]pyrimidine-
5,7-diones, 4,5,6,7-tetrahydro-
26, 591
5H-Thiazolo[2,3-b]quinazol-
5-ones
- from
thiazoles, 2-halogeno- **27,** 475
Thiazolo[4,5-b]quinoxalines,
2,3-dihydro-2-imino-
- from
thioureas **28,** 560
Thiazolo[3,2-b]thiadiazolines,
2-imino- 17, 657 suppl. **29**
2H-Thiazolo[3,2-a]-s-triazin-
4-ones, 3,4,6,7-tetrahydro-
- from
2,4-diaza-1,3-diene groups
and isocyanates **27,** 321
Thieno[2,3-e][1,4]diazepin-
2-ones, 1,3-dihydro- 28, 438

Thieno[3,2-b]indoles
- from
thiophene-2-spiro-3-
indolenines, 3-imino-
30, 343
Thieno[2,3-d]pyrimidines 27, 845
Thieno[3,4-c]thiophenes-
2-S(IV) 29, 539
Thietane-2,4-diones
- from
α-dithiolic acids **28,** 578
Thietane 1,1-dioxides
- startg. m. f.
cyclopropanesulfinic acids
28, 700
Thietanes
- from
1,2-dithiolanes **27,** 642
1,3-dithiols **27,** 642
- special s.
enethiethanes
Thiete 1,1-dioxides
-, cycloaddition with - **26,** 696
-, -, **3-amino-**
- from
ynamines and sulfonic acid
esters **29,** 740
Thiirane 1,1-dioxides
-, reaction with nucleophiles,
strong **30,** 385
- **oxide** as disulfur monoxide
source **28,** 535
- **1-oxides**
- from
sulfines and diazo compds.
29, 767
- startg. m. f.
ethylene derivs. **29,** 767
2-halogenosulfenic acid
alkoxymethyl esters
28, 475
Thiiranes s. Sulfido compds.
Thiirene 1,1-dioxides
- from
α,α-dihalogenosulfones
26, 972
- startg. m. f.
acetylene derivs. **26,** 972
sulfinovinylbetaines **30,** 424
Thio... s. a. Alkylthio...,
Sulfenyl..., Thia...
Thioacetals s. Mercaptals
Thioacetamide as reagent **27,** 19
Thioacylamines s. Carboxylic
acid thioamides
N-Thioacylation
- with
N-thioacylimidazoles **26,** 397
-, **selective 16,** 468 suppl. **28**
Thioacyl disulfides
- startg. m. f.
thionocarboxylic acid esters
27, 225
Thioacylhydrazines
(s. a. Carboxylic acid
thiohydrazides)
- from
1,1-halogenohydrazones
30, 407
hydrazines **26,** 397
- special s.
alkoxythiocarbonyl-
hydrazines
(alkylthio)thiocarbonyl-
hydrazines
- startg. m. f.
hydrazonyl sulfides **30,** 407

N-Thioacylimidazoles, N-thioacylation with – **26,** 397
Thioacylimines
– from
  thioketones and nitriles **29,** 533
Thioacyl thiocyanates 28, 562
Thioacylureas
– from
  carboxylic acd thioamides and isocyanates **30,** 215
Thioalkylation s. Self-thioalkylation
Thioamides s. Carboxylic acid thioamides
Thioanhydrides s. Dicarboxylic acid thioanhydrides
Thioanisole, polymer-based, as reagent **28,** 200 suppl. **30**
Thioanisole-chlorine complex 28, 200
O-Thiobenzoylation 29, 947
(Thiobenzoylthio)acetic acid as reactant **29,** 947
N,N'-Thiobisimides
– special s.
  N,N'-thiobisphthalimide
Thiobis(formimidoyl halides)
– special s.
  thiobis(sulfonylformimidoyl chlorides)
N,N'-Thiobisphthalimide
– as sulfur-transfer agent **28,** 244
Thiobis(sulfonylformimidoyl chlorides)
– from
  sulfonylcyanides **30,** 349
Thioboric acid esters
– startg. m. f.
  N-(alkylthiocarbonyl)ureas **27,** 639
Thioborinic – –
– startg. m. f.
  β-hydroxythiolic acid esters **28,** 862
Thioboron compds.
– startg. m. f.
  disulfides **27,** 586
Thiocarbamic... s. Dithiocarbamic..., Thiolcarbamic..., Thionocarbamic...
N-Thiocarbamylamidines
– from
  isothiocyanates **26,** 295
Thiocarbamylation
– with
  C-sulfonylthioformamides **28,** 836
C-Thiocarbamylation 28, 836
N-Thiocarbamyl group migration 27, 338
Thiocarbamyl halides
– from
  thioformamides **26,** 525
Thiocarbamylisothiocyanates
–, ring closures with **27,** 591
– startg. m. f.
  4H-1,3,5-thiadiazine-4-thiones, 5,6-dihydro-2-amino- **27,** 591
Thiocarbamylphosphine oxides
– from
  isothiocyanates **26,** 636
Thiocarbazic acid esters
– special s.
  thionocarbazic acid esters

Thiocarbimides
s. Isothiocyanates
Thiocarbohydrazones
– startg. m. f.
  4-thiazolidones **27,** 596
Thiocarbonic acid esters
s. Acyl polythiocarbonates, Dithiolcarbonic acid esters, (Hydroxythiol)thionocarbonic acid esters, Monothiol-, Thiol-, Thiono-, Trithio-
N-Thiocarbonylation
s. Thiolurethans from amines
Thiocarbonyl compds.
(s. a. Thiono...)
–, photocycloaddition, review **24,** 697 suppl. **26**
– special s.
  thioketones
– **dichloride** s. Thiophosgene
– **groups**
– startg. m. f.
  carbonyl groups **27,** 231
  sulfhydryl groups **26,** 40
Thiocarbostyrils
– from
  isothiocyanates **30,** 529
Thiocarboxylic acid esters
s. Dithiocarboxylic acid esters, 3-holic – –, Thionocarboxylic – –
Thiochroman-4-ones, 3-cyano- 30, 397
Thiochromenes
– startg. m. f.
  thianaphthenes, 2-formyl- **26,** 163
Thio-Claisen rearrangement 26, 900; **28,** 842
– special s.
  dithio ester thio-Claisen rearrangement
–, thianaphthenes via – **30,** 677
o-Thio-Claisen rearrangement
–, ring closure by – **27,** 755
Thiocoumarins 26, 635
Thiocyanates
– from
  N-formylsulfenamides **28,** 576
–, prepn., review **21,** 676 suppl. **29**
– special s.
  acetylene-thiocyanates amino-carbamyl-halogeno-thioacyl –
– startg. m. f.
  1,1-di(thiolurethans) **29,** 308
  2-halogenothiolurethans **28,** 618
  mercaptans **28,** 28
  Δ²-1,3,4-thiadiazolines, 5-imino- **30,** 234
  thioethers **28,** 575
–, ar.
– from
  arylthallium bis(trifluoro-acetates) **27,** 888
Thiocyanation with rhodanamine **28,** 542
Thiocyanic acid esters
s. Thiocyanates
Thiocyanogen as reactant **28,** 544
– **chloride** as reactant **29,** 540

2-Thiocyanopyrimidines
– from
  pyrimidine-2-thiones, 1-amino-1,2-dihydro- **22,** 628
Thiodienolethers by 1,4-dethiolation **30,** 684
Thioenolethers (s. a. Alkyl-thiomethylene...)
–, cycloisomerization **30,** 511
– from
  1-acetylene-1-thioethers, synthesis **26,** 715 suppl. **29**
  α,β-ethylenehalides and disulfides **30,** 413
  α-hydroxymercaptals **29,** 974
  oxido compds. **26,** 597
  oxo compds. **29,** 556
  sulfonium salts and sulfenimides **28,** 572
  1-sulfonyloxy-2-thioethers **26,** 945
–, reactions, review **25,** 462 suppl. **26**
– special s.
  2-alkylthio-1,3-dienes
  α- or β-alkylthio-α,β-ethylene...
  γ-alkylthio-β,γ-ethylene...
  allyl vinyl sulfide
  1-aminothioenolethers
  1,3-bis(alkylthio)-1-ethylenes
  halogenothioenolethers
  1-sulfinylthioenolethers
– startg. m. f.
  ethylene derivs., synthesis **29,** 854
  ketones **29,** 802
  –, synthesis **29,** 802
  mercaptals **27,** 600
  thioethers, hydrogenation **14,** 90
–, **cyclic**
– from
  acetylenemercaptans **26,** 587
Thioenols
– special s.
  α,β-ethylene-β-mercapto...
Thioethers (s. a. Organothio...)
–, α-alkylation **22,** 688 suppl. **26; 28,** 801, 842
–, dehydrogenation via – **29,** 912
– from
  acoxy compds. **26,** 599
  alcohols **28,** 547
  azoles **29,** 561
  disulfides **26,** 622; **27,** 637, 642; **28,** 571
  – and halides **27,** 635
  – and methylene groups, active **29,** 580
  ethylene derivs. **27,** 637
  – – and halides **27,** 632
  halides **26,** 616
  1,1-halogenothioethers **18,** 125 suppl. **29**
  mercaptans **26,** 592
  – and isoureas **29,** 547
  – and phosphoric acid esters **30,** 396
  –, synthesis **29,** 620
  sulfenimides and halides **30,** 654
  sulfinic acid chlorides **30,** 409
  1-sulfonyloxy-2-thioethers, synthesis **29,** 737
  sulfonylsulfilimines **26,** 993; **28,** 992; **30,** 700

sulfoxides **23**, 988 suppl. **29**;
26, 993; **27**, 992; **28**, 991;
29, 993; **30**, 699
thiocyanates **30**, 417, 419
– and alcohols **28**, 575
thioenolethers, hydrogenation
14, 90
thioiminocarbonic acid esters
28, 575
thioketones (sec. thioethers)
26, 575 suppl. **27**; **29**, 641
thiolsulfonic acid esters
29, 587
–, α-metalation **28**, 801
– special s.
acetylene-thioethers
alkoxy-
allene-
tert-butyl –
carboxylic acid 2-(p-nitrophenylthio)ethyl esters
carboxythioethers
cyclopropylcarbinols,
1-arylthio-
dimethyl sulfide
ethylene-thioethers
halogeno-
hydroxy-
isopropyl sulfide
methyl thioethers
methylthiomethyl...
oxido-thioethers
oxo-
sulfinyl-
sulfonio-
sulfonyloxy-
Δ²-thiazolines, 2-alkylthio-
– startg. m. f.
aminosulfenium salts **28**, 243
1,1-chlorothioethers **30**, 687
ethers **28**, 174
ethylene derivs. **28**, 960;
29, 972; **30**, 685, 687
α-halogenosulfoxides **27**, 564
hydrocarbons **28**, 63
–, synthesis **29**, 851
quinolsulfonium salts **27**, 589
N-sulfamoylsulfilimines
28, 246
sulfones **20**, 629; **26**, 3
sulfonium salts **27**, 997
– – (from 2 molecules)
27, 605
sulfonylsulfilimines **26**, 272
sulfoxides **28**, 70a; **29**, 98;
30, 685
–, ar.
– from
thiolsulfonic acid esters
27, 638
–, **cyclic** (s. a. Sulfido compds.)
– from
ethylenemercaptans **26**, 587
suppl. **27**; **27**, 595
– special s.
1,1-halogenothioethers,
cyclic
– startg. m. f.
hydroxysulfonic acids **30**, 76
–, **sym.**
– from
alcohols **27**, 609
halides **18**, 687 suppl. **29**;
27, 629
thiolic acid esters **29**, 584
p-toluenesulfonic acid esters
11, 666 suppl. **27**

Thioformamides
– special s.
cyanothioformamides
C-sulfonylthioformamides
– startg. m. f.
α-hydroxycarboxylic acid
thioamides **29**, 623
thiocarbamyl halides **26**, 525
N-Thioformylacylamines
– from
isonitriles and thiolic acids
30, 381
Thioglycitols
– special s.
dithiothreitol
Thioglycosides
– startg. m. f.
carbohydrates **30**, 141
glycosides **29**, 215
N-Thioglycosyl-N-heterocyclics
29, 435
– from
1-alkylthio-1-pyridinium
salts **29**, 436
Thiohydantoins
s. Dithiohydantoins
Thiohydroxamic acids, review
15, 481 suppl. **26**
Thiohydroximic acid esters
19, 694
Thioiminocarbonic acid esters
– startg. m. f.
thioethers **28**, 575
Thioiminoesters
– from
ketenimines and mercaptans
30, 386
nitriles **27**, 624
– startg. m. f.
cis-2-azetidinones **7**, 836
suppl. **27**
2-azetidinones, 4-alkylthio-
8, 736
carboxylic acid thioamides
26, 317
enamines **26**, 982
thiolic acid esters **27**, 624
–, transesterification **29**, 571
N-Thioiminohalides
– from
nitriles and disulfides
30, 347
Thioindoxyls
– from
o-alkylthio-ω-diazoaceto-
phenones **27**, 644
Thioketals s. Mercaptals
Thioketenes
– special s.
bis(trifluoromethyl)thio-
ketene
α-**Thioketocarboxylic acid**
esters s. α,β-Ethylene-
α-mercaptocarboxylic acid
esters
– **acids**
– startg. m. f.
carboxylic acid
thiomorpholides **27**, 496
Thioketone S-oxides s. Sulfines
Thioketones
–, addition to allenes **27**, 588
– from
1,1-dihalides **28**, 557
imines **26**, 605
–, reaction with allenes **28**, 521
–, reactions with Grignard
compds. **28**, 497

– special s.
cyclopropenethiones
dithiones
monothioketones
– startg. m. f.
2-alkylthio-1,3-dienes **28**, 521
benzoxathianones **29**, 566
1,3-dithietanes, dimerization
29, 536
S-heterocyclics **29**, 541
imines **29**, 426
ketones **29**, 220
mercaptans **27**, 47
–, tert. **29**, 641
1,3,5-oxathiazolines **30**, 408
thioacylimines **29**, 533
thioethers, sec. **26**, 575
suppl. **27**; **29**, 641
Thiolactams
– from
nitrones, cyclic **27**, 613
Thiolactones
–, decarbonylation **27**, 643
– from
carboxythioethers **27**, 618
lactones **26**, 598; **27**, 618
– startg. m. f.
heterocyclics, review **22**, 341
suppl. **27**
Thiolcarbamic acid esters
(s. a. Thiolurethans)
– from
isothiouronium salts **28**, 159
– special s.
acyl carbamyl sulfides
N-acylthiolcarbamic acid
esters
N-(alkylthiocarbonyl)ureas
selenothiolcarbamic acid
esters
1-sulfonyl-1-thiolcarbamic – –
– startg. m. f.
enethiolurethans **26**, 382
isocyanates **26**, 493
sulfonic acids **26**, 96
Thiolchloroformic – –
– startg. m. f.
thiolic acid esters **29**, 835
Thiolglycidic – – **8**, 842
suppl. **29**
Thiolic acid ester groups,
migration **27**, 742
– **esters** (s. a. Acylthio...,
Thiolcarbalkoxy groups)
– from
acyl polythiocarbonates
28, 579
arenes and thiolchloroformic
acid esters **29**, 835
carboxylic acid esters **29**, 553
– – – and silyl sulfides
29, 581
– acids and disulfides
26, 621
– – and mercaptans **27**, 611
ethylene derivs. **27**, 597
ketene mercaptals **26**, 213
nitriles **27**, 624
selenolic acid esters **30**, 415
thioiminoesters **27**, 624
thionocarboxylic acid esters
27, 601
trithioorthocarboxylic acid
esters **26**, 214
– special s.
α-acoxythiolic acid esters
α,β-ethylene-β-hydroxy-
thiolic acid esters

(**Thiolic acid esters**
- special s.)
  hydroxythiolic acid esters
  α-ketothiolic – –
  α,β-oxidothiolic – –
- startg. m. f.
  alcohols **26**, 76
  hydrocarbons (methyl groups) **26**, 76
  ketones **26**, 982; **29**, 853
  mercaptans **28**, 147; **30**, 15
  peptides **23**, 386 suppl. **26**
  thiethers, sym. **29**, 584
-, **cyclic 27**, 158
- – **2-pyridyl esters**
- startg. m. f.
  carboxylic acid esters **30**, 142
- **acids**
- from
  carboxylic acid amides **29**, 563
- special s.
  amino-thiolic acids
  di-
- startg. m. f.
  acyl carbamyl sulfides **30**, 382
  carboxylic acid amides **28**, 403
  N-thioformylacylamines **30**, 381
Thiolphosphoric acid esters
  s. Monothiolphosphoric acid diesters, Thionophosphoric – –, – – esters
Thiols s. a. Mercaptans
N-Thiols 26, 268
Thiolsulfinic acid esters
- from
  disulfides **29**, 96
  episulfoxides (2 molecules) **26**, 561
-, review **29**, 96
- startg. m. f.
  sulfenic acid esters **27**, 634
  1,1-sulfinyldisulfides **29**, 579
Thiolsulfonic – –
- from
  sulfinic acids and disulfides **27**, 585
  – – and sulfenamides **26**, 563
  – – and sulfenic acid esters **26**, 563
  thiolsulfonic acids **27**, 585
-, self-thioalkylation **27**, 633
- startg. m. f.
  disulfides **30**, 699
  thioethers **29**, 587
  thioethers, ar. **27**, 638
- **acids**
- startg. m. f.
  sulfonyldisulfides **28**, 518
  thiolsulfonic acid esters **27**, 585
Thiolsultones
- special s.
  1,2-dithiane 1,1-dioxides
Thiolthionophosphoric acid esters s. Dithiophosphoric acid, O,O-dialkyl-
Thiolurethans (s. a. Thiolcarbamic acid esters)
- from
  amines and carbon oxide sulfide **29**, 531
  – and dithioldicarbonic acid esters **27**, 378
- special s.
  di(thiolurethans)

enethiolurethans
halogeno-
Thiomethylation s. Acylthiomethylation, Aryl-
Thionocarbamic acid esters
- startg. m. f.
  ethylene derivs. **26**, 931
Thionocarbazic –
- startg. m. f.
  1,3,4-oxadiazol-2(3H)-ones **30**, 174
Thionocarbonic – –
-, dehydration of alcohols via– **6**, 841 suppl. **27**
- – –, cyclic
- special s.
  1,3-dioxole-2-thiones
Thionocarboxylic acid esters
- as reagents **28**, 213
- from
  diazo compds., synthesis with addition of 1 C-atom **30**, 570
  thioacyl disulfides **27**, 225
- special s.
  ethyl thionoformate
- startg. m. f.
  carboxylic acid esters **28**, 213
  ethylene derivs. **29**, 947
  ketones, synthesis **29**, 896
  thiolic acid esters **27**, 601
Thionophosphonic acid dichlorides
- startg. m. f.
  phosphonic acid dichlorides **28**, 78
  thionophosphonic acid esters, mixed **26**, 106
  thionophosphonochloridate **26**, 106
- – esters
- special s.
  acetylenethionophosphonic acid esters
- – – –, mixed
- from
  thionophosphonic acid dichlorides **26**, 106
  thionophosphonochloridates **26**, 106
Thionophosphonochloridates
- from
  thionophosphonic acid dichlorides **26**, 106
  thionophosphonic acid esters, mixed **26**, 106
Thionophosphoramidates
- from
  thionophosphorohalidates **28**, 263
Thionophosphoric acid diesters
- from
  monothiolphosphoric acid esters **28**, 10
- – monoesters
- startg. m. f.
  diphosphoric acid P¹,P²-diesters, mixed **30**, 62
Thionophosphorohalidates
- startg. m. f.
  thionophosphoramidates **28**, 263
Thionophosphorous acid esters
- startg. m. f.
  monothiolphosphoric acid esters **28**, 523

2-Thionucleosides 30, 316
P-Thionucleosides 28, 18
Thionyl bromide 26, 545
-, reaction with – **30**, 364
- **chloride** HalC↓↑O; **10**, 429 suppl. **29**; **16**, 590 suppl. **28**; **26**, 239, 298, 591, 961; **27**, 275, 517, 539, 577, 918; **28**, 73, 78, 590; **29**, 189, 391, 545, 591, 704, 993; **30**, 342, 421
-, dimerization of Grignard compds. with – **29**, 879
- startg. m. f.
  o-benzotrithiol-2-ones **28**, 515
  1,2,5-thiadiazolidine 1-oxides **26**, 271
  thianaphthenes **29**, 545
  thianaphthene-2-carbonyl chlorides, 3-chloro- **28**, 548
  1,2,3,5-thiatriazole S-oxides **26**, 270
  N-thionylphosphorimidates **28**, 249
- – -dimethylformamide **17**, 546 suppl. **27**; **29**, 937; **30**, 227
- **halide-hexamethylphosphoramide 26**, 545
- **halides** as reactants **30**, 344
Thionylimines
- from
  carbodiimides **29**, 473
- startg. m. f.
  carbodiimides **26**, 442
  1,2,3-oxathiazetidine 2-oxides **28**, 272
  pyridinium salts **29**, 431
  sulfinodiimines **26**, 276
N-Thionylphosphorimidates
- from
  phosphoramidates **28**, 249
N-Thionylsulfonamides
- startg. m. f.
  1,2,4-thiadiazetidine 1-oxides, 3-imino- **29**, 473
Thioorthocarbonic acid esters
  s. Tetrathioorthocarbonic acid esters
Thioorthocarboxylic – – s. Trithioorthocarboxylic acid esters
Thioortholactones s. Dithiolortholactones
Thiooxime carbamates
- from
  oxime thiocarbamates **27**, 291
Thiooxo compds. s. Thiocarbonyl compds.
Thioozonides
- from
  S-heterocyclics **26**, 125
- startg. m. f.
  isocyclics **26**, 125
Thiophene 1,1-dioxides
- startg. m. f.
  1,3,5-cycloheptatrienes **29**, 848
- –, **2,5-dihydro-**
- startg. m. f.
  1,4-oxathiane 4,4-dioxides, 2,6-dialkoxy- **29**, 148
  4H-1,4-thiazine 1,1-dioxides **29**, 148
- –, **tetrahydro-3,4-diketo-**
- from
  sulfones **28**, 724

- startg. m. f.
 thiophene 1,1-dioxides,
   tetrahydro-3-hydroxy-
   4-keto- **28**, 724
- –, **tetrahydro-3-hydroxy-4-keto-**
- from
 thiophene 1,1-dioxides,
   tetrahydro-3,4-diketo-
   28, 724
- startg. m. f.
 α-diketones **28**, 724
- –, **3-vinyl-**
- from
 diallenesulfones **29**, 683
- **S-oxide ring**
- from
 sulfinic acids **29**, 591
- – –, **2-chloro-**
- from
 diazo compds. **29**, 768
- **ring**
- by
 aldol-type condensation,
   intramolecular, **26**, 935
- –, opening **27**, 628; **29**, 642
- startg. m. f.
 1,3-cyclohexadiene ring
   29, 640
- – –, **2-amino- 29**, 534
- – –, **3-hydroxy-**
- from
 o-nitrocarboxylic acid esters
 Thiophenes **30**, 573
- –, aromatization by acid-base
   catalysis **26**, 946
- from
 1,3-dienes **28**, 543
 enedisulfides **27**, 980
 monothio-β-diketones **26**, 866
 1,3-oxathiol-2-yliden-
   immonium salts **28**, 605
 thiazolium betaines,
   4-hydroxy- **26**, 919
 thiophenes, 2,3-dihydro-
   3-hydroxy- **26**, 866
- startg. m. f.
 thianaphthenes **28**, 922
- –, **2-alkylthio-**
- from
 β-allenedithiocarboxylic acid
   esters **26**, 582
- –, **2-amino- 28**, 789
- –, **2-aryl- 23**, 831 suppl. 26
- –, **2,4-diamino-**
- as electron donors **29**, 640
- from
 isothiocyanates and nitriles
   26, 892
- –, reaction with dienophiles
   29, 640
- –, reactions **26**, 677
- –, **2,5-dihydro- 19**, 904
 suppl. **28, 29; 25,** 123
 suppl. **28**
- –, diene synthesis, stereospecific, with – **26**, 691
- startg. m. f.
 1,3-dienes **17**, 968
- –, **2,3-dihydro-3-hydroxy-**
- from
 monothio-β-diketones **26**, 866
- startg. m. f.
 thiophenes **26**, 866
- –, **3-hydroxy-**
- from
 α-mercaptocarboxylic acid
   esters **27**, 784
- –, **tetrahydro- 30**, 448

**Thiophene-2-spiro-3-indolenines,
 3-imino-**
- startg. m. f.
 thieno-[3,2-b]indoles **30**, 343
Thiophenol as reagent **27**, 93
Thiophenols s. Arylmercaptans
Thiophile, basic
- –, bis-(3-dimethylaminopropyl)-
   phenylphosphine as – **26**, 982
Thiophilic addition s. Addition,
   thiophilic
Thiophosgene as reactant
   24, 696 suppl. **28; 27**, 495
- startg. m. f.
 enisothiocyanates **30**, 242
 1,1-halogenisothiocyanates
   30, 242
- **S-oxide**, reactions with –
   26, 697
- startg. m. f.
 thiophene S-oxide ring,
   2-chloro- **29**, 768
Thiophosphinic... s. a. Bis-
   (thiophosphinyl)...,
   Dithiophosphinic...
- **acid chlorides**
- startg. m. f.
 thiophosphinylthioureas
   28, 396
- – **halides**
- from
 diphosphine disulfides
   19, 590
Thiophosphinylisothiocyanates
- –, reactions **30**, 210
- startg. m. f.
 thiophosphinylthioureas
   30, 210
Thiophosphinylthioureas
- from
 thiophosphinic acid chlorides
   and amines **28**, 396
 thiophosphinylisothio-
   cyanates – – **30**, 210
Thiophosphonic... s. a. Thionophosphonic...
- **acids, O-alkyl-**
- startg. m. f.
 trithiodiphosphonic acid
   esters **29**, 529
Thiophosphoric acid esters
- special s.
 tetrathiophosphoric acid
   esters
 thiolphosphoric – –
 thiolthionophosphoric – –
Thiophosphorous – – s. Dithiophosphorous acid esters,
   Thionophosphorous – –
Thiophosphorylation 27, 113
Thiopyrano[2,3-b]indolium salts
- from
 β-diketones **26**, 825
Thiopyran ring 30, 470
- – –, **dihydro-**
- startg. m. f.
 cyclopropane ring **26**, 576
Thiopyrans, 2-alkylthio- 26, 582
- –, **dihydro-**
- from
 carboxylic acid thioamides,
   vinylogous **28**, 637
 1,3-dienes **28**, 637
Δ³-**Thiopyrans, dihydro-,**
 α-alkylation **28**, 842
4-Thiopyrones
- startg. m. f.
 cyclopentadienones **28**, 955

Thiopyrylium ring 29, 838
- – – opening **30**, 312
- **salts**
- startg. m. f.
 benzene ring **27**, 875
Thioselenoacetals s. Selenothioacetals
Thiosemicarbazide
- startg. m. f.
 1,2,4-triazine-3-thiones **26**, 387
Thiosemicarbazides
- special s.
 phosphinylthiosemicarbazides
Thiosemicarbazones
- –, ring closure, oxidative
   19, 705
Thiosulfates s. S-(Carbalkoxy)-
   thiosulfates, Thiolsulfuric
   acid S-monoesters
Thiosulfenamides
 s.Thiosulfenimides
Thiosulfenimides
- from
 mercaptans **28**, 256
 thiosulfenyl halides **28**, 256
Thiosulfenyl halides
- from
 mercaptans **28**, 256
- startg. m. f.
 thiosulfenimides **28**, 256
Thiosulfonic... s. Thiolsulfonic...
Thiosulfonium salts
- startg. m. f.
 sulfoniothioethers **21**, 639
- –, **cyclic**
- –. ring closures with – **21**, 639
 suppl. **28**
Thiosulfuric acid S-monoesters
- from
 mercaptans **26**, 557
- –, **retention** s. under Retention
- startg. m. f.
 Δ²-thiazolines, 2-amino-
   28, 570
Thiosultones s. Thiolsultones
4-Thiouracils
- –, photoreactions with
   α,β-ethylenenitriles **30**, 465
Thiourea
- as H-donor **27**, 644
- as reactant **27**, 632; **30**, 41
- as reagent **16**, 970 suppl. **27;
   27**, 592; **30,** 1, 328, 502
- startg. m. f.
 pyrimidine ring **26**, 355;
   30, 344
- **dioxide** as reducing agent
   27, 39
Thioureas (s. a. Isothiour...,
   Quasithioureas)
- –, acylation **27**, 625
- –, N-acylation **27**, 406, 625
- from
 acylthioureas **29**, 26
 carbalkoxythioureas **29**, 26
 carbodiimides **27**, 592
 ureas **27**, 614
- –, prepn., review **29**, 26
- –, reaction with copper(II)
   acetate **27**, 406
- special s.
 acyl-thioureas
 amidino-
 amino-
 carbalkoxy-
 hydroxy-
 N,N-pentamethylenethiourea

(Thioureas
- special s.)
 sulfonyl-thioureas
 thiophosphinyl-
- startg. m. f.
 S-acylisothiouronium salts
  27, 625
 amidinothioureas **26**, 302
 carbodiimides **27**, 532, 978
 mercury(II) mercaptides
  26, 567
 thiazolo[4,5-b]quinoxalines,
  2,3-dihydro-2-imino- **28**, 560
-, **1,1-disubst. 29**, 26
-, **sym.**
- from
 amines **26**, 323; **27**, 366
 - and carbon disulfide
  30, 243
 aminosilanes **27**, 486
N-Thioureidoamidines
 s. Amidinothioureas
Thiourethans s. Dithiourethans,
 S-Sulfonylthiourethans
Thiouronium salts
 s. Isothiouronium salts
Thio-Wittig rearrangement
 29, 709 suppl. **30**
- of dithiocarbamic acid
 esters **30**, 536
Thioxanthone 10,10-dioxides
- startg. m. f.
 2′-hydroxybenzophenone-
  2-sulfinic acids **28**, 224
 xanthones **28**, 224
Thiuram disulfides
 s. Bisthiuram disulfides
Thorium dioxide s. under
 Palladium
Thorpe-Ziegler ring closure
 s. Ziegler ring closure
Tin(II) chloride 8, 748 suppl. **26**;
 26, 189, 665, 949; **27**, 734; **27**,
 956 suppl. **30**; **28**, 239, 242,
 991; **29**, 12, 418; **30**, 155, 180
 –(IV) – **13**, 454 suppl. **26**; **26**,
 446, 885, 957; **27**, 865, 914; **28**,
 763, 825; **28**, 958 suppl. **29**; **29**,
 187, 685, 754/5, 838/9, 929; **30**,
 72, 172
 - **compds. organo-**
  (s. a. Stannanes)
 - in organic synthesis, review
  26, 625 suppl. **27**
 –(IV) –, -
 -, syntheses with – **29**, 856
 - **enolates, organo-**, alkylation
  29, 856
 - **hydrides, organo-** (s. a. Tributyltin hydride, Triphenyltin –) **17**, 108 suppl. **26**
 -, reductions with –, review
  26, 50
 - **oxide, dibutyl-** as reagent
  26, 456
 - **oxides, diorgano-**
 - startg. m. f.
  distannoxanes **27**, 645
 –(IV) salts as Beckmann
  rearrangements catalysts
  7, 224 suppl. **29**
Titanium alkoxides as reactants
 29, 629
 - **amides** as reactants **27**, 416
 - **aroxides** as reagents
  12, 739 suppl. **28**
 - **compds.**, soluble, as reagents
  20, 211 suppl. **27**

- –, **organo-**, reactions **29**, 858
- **dioxide** s. under Vanadium
 pentoxide
- **tetrachloride** (s. a. under
 Lithium tetrahydrido-
 aluminate) **4**, 543 suppl. **28**; **9**,
 902 suppl. **29**; **11**, 194 suppl.
 28; **24**, 437 suppl. **26**, **29**; **24**,
 872 suppl. **28**; **26**, 377, 529,
 814; **27**, 513/4, 530, 798; **28**,
 63, 491; **29**, 210, 371, 556, 974;
 30, 561, 621, 633
-, halogenation with – **28**, 491
- **trichloride 8**, 87 suppl. **29**; **17**,
 345 suppl. **29**; **26**, 188, 913;
 27, 332, 530, 772; **28**, 157, 991;
 29, 79; **30**, 561, 678
-/**lithium tetrahydrido-
 aluminate** s. McMurry
 reagent
- **trichlorides**, alkoxy-
 as reagents **30**, 521
Titanocene dichloride 30, 30
- –, **polymer-based 30**, 30
**TMEDA (N,N,N′,N′-Tetra-
 methylethylenediamine)**
 as reagent **19**, 736 suppl. **29**;
 26, 875 suppl. **28**; **27**, 838; **28**,
 705, 842, 885, 895; **29**, 28,
 619-21, 660; **30**, 479, 586
-/**lithium 29**, 660
**Todd-Atherton reaction
 22**, 106
Tolans 30, 687
- startg. m. f.
 cis-stilbenes **26**, 41
p-Toluenesulfonamide
 as reagent **30**, 150
p-Toluenesulfonic acid
 as reagent OC↓O; **14**, 174
 suppl. **29**; **16**, 813 suppl. **27**;
 20, 526 suppl. **27**; **26**, 65, 115,
 138, 240-2, 251, 368, 382, 472,
 582, 619, 623/4, 755, 819, 842,
 938, 953; **27**, 195/6, 201, 235,
 497, 518, 719, 805, 832, 947;
 28, 194/5, 204, 356, 443, 639,
 673, 677, 755-7, 921; **29**, 174,
 253, 380/1, 437, 483, 552, 647,
 686, 699, 708, 840, 926, 953;
 30, 114/5, 151, 272, 329, 399,
 466, 489, 500, 510, 566, 584
- **hydrate** as reagent **27**, 205
- – **esters** (s. a. O-Tosylation,
 Tosyloxy...)
-, **cleavage** s. O-Detosylation
- from
 alcohols **28**, 52, 119
-, N-ring closure with –
 16, 926; **27**, 507
-, ring expansion with –
 29, 918
- special s.
 di-tosylates
 enol –
 keto-
- startg. m. f.
 acetylene derivs., synthesis
  30, 548
 acoxy compds. **26**, 212, 944;
  27, 227
 amines, prim. **26**, 357
 cyclopropane ring **23**, 904
  suppl. **28**; **29**, 918
 disulfides, sym. **11**, 666
  suppl. **27**
 ethylene derivs. **27**, 936;
  28, 899

 halides (iodides) **17**, 621
 hydrocarbons **27**, 74;
  28, 50, 52
 spirocyclics **18**, 937 suppl. **26**
 thioethers, sym. **11**, 666
  suppl. **27**
-, syntheses with – and
 Grignard compds. **29**, 824
-, – – – and lithium cuprates
 29, 876
- – **hydrate** as reactant **27**, 205
p-Toluenesulfonyl...
 s. a.Tosyl...
- **chloride 23**, 792 suppl. **26**;
 26, 381, 990; **28**, 73, 122, 505,
 547
- as reactant **28**, 547
**Torgov-Smith steroid synthesis
 23**, 924 suppl. **26**
Tosylamines
- special s.
 ethylenetosylamines
- startg. m. f.
 cyclotribenzylenes,
  trimerization **26**, 843
Tosylates s. Oximes tosylates,
 p-Toluenesulfonic acid esters
N-Tosylation, selective 15, 384a;
 28, 253; **30**, 58
O-Tosylation with rearrangement, skeletal **29**, 699
-, **partial 29**, 101
-, **selective 30**, 58
Tosyl azide
- as reactant **26**, 321;
 30, 232, 247
- startg. m. f.
 amines **28**, 422
 azides **26**, 422; **27**, 438
Tosylhydrazine
- as reactant **6**, 121; **29**, 67
- as reagent **26**, 65
Tosylhydrazones
- from
 ketones **27**, 949
 oxo compds. **29**, 67
- special s.
 α-acoxy-tosylhydrazones
 bis(–)
 2,5-cyclohexadienone –
 ethylene-
 glyoxalyl chloride
  tosylhydrazone
 halogenotosylhydrazones
- startg. m. f.
 carboxylic acids **28**, 775
  suppl. **30**
 cyclopropane ring **28**, 934
 ethylene derivs. **27**, 949
 hydrocarbons, reduction
  29, 67
 -, synthesis **28**, 775
 – with rearrangement,
  skeletal **30**, 34
2-Tosylhydrazonoethers, cyclic
- startg. m. f.
 allenealcohols **27**, 12
N-Tosylimidazole, reactions
 with – **29**, 101
Tosylmethylisocyanide
-, as reactant **28**, 726, 841
2-Tosyloxyamines
- startg. m. f.
 aziridines **28**, 433
Tosyltriazene salts as intermediates **26**, 422
Tosylureas, protection of
 amino groups as – **30**, 11

Toxoflavines s. Pyrimido[5,4-e]-
as-triazine-5,7(6H,1H)-diones
trans- s. Geospecificity,
Isomers, Rearrangement
Trans... s. a. Interchange
Transacetalation 26, 223
–, **preferential** 30, 151
N-Transacylation
s. Transamidation
O-Transacylation, diene
synthesis with – 28, 629
N-Transalkylation with position
shift 30, 320
O-Transalkylation
s. Transetherification
Transamidation
– of
amidines 30, 289
carboxylic acid amides
28, 424
– via
N-acylcarboxylic acid amides
27, 476
–, **intramolecular,** ring
expansion by – 22, 337
Transamination
– of
β-amino-α,β-ethylenenitriles
18, 482 suppl. 27
aminomethylenemalonic acid
esters 27, 443
α-aminonitriles 5, 350
suppl. 29
enamines 22, 425
pyrans, tetrahydro-2-amino-
5, 350 suppl. 28
S-Transamination 28, 251
Transannular s. Ring closure,
transannular, Ring contraction, –
Transcyano-O-silylation
–, α-siloxynitriles by – 30, 630
Transesterification 30, 146, 262
– of
carbonic acid esters 26, 219
dithiocarboxylic – – 26, 626
nitrous – – 26, 225
tetrathioorthocarbonic – –
26, 623
thioiminoesters 29, 571
–, **preferential** 27, 235
Transetherification 27, 237;
28, 194
– of
enolethers 27, 250
–, **geospecific** 27, 250
Transfer s. a. Chirality
transfer, Electron –, Hydride
–, Hydrogen –, Hydrogen
atom –, Methylene –,
Oxygen –
–, **palladium-catalyzed**
– of
allylic groups 26, 827
Transfer-alkylation
s. C-Alkylation, remote
Transfer-dehydrogenation
28, 209
Transfer-hydrogenation,
catalytic, review 28, 32
suppl. 30
–, **preferential and selective**
28, 32
Transglycosidation (s. a.
Glycosyl migration)
– to nucleosides 27, 441

Transition metal-carbene
complexes
–, review 28, 846
– special s.
pentacarbonylchromium(0)-
carbene complexes
– **metal catalysts**
–, effect on valence isomerization 26, 762 suppl. 27
–, cyclooligomerization with –,
review 20, 531 suppl. 29
–, reactions of diazo compds.
with – 29, 787
– – –, **homogeneous,** synthesis,
asym. with –, review
23, 48 suppl. 29
– – **complexes**
– as catalysts 28, 675
– as intermediates 17, 879
suppl. 26
–, ring closures with diynes
via –, review 26, 725
suppl. 30
–, syntheses by – , review
23, 863 suppl. 27
– – –, **low-valence,** reactions
with – 28, 761
– – **-nitrato complexes**
–, nitration by – 30, 240
– – **salts** s. under Sodium
tetrahydridoborate
Transmethylenation 30, 636
Transposition s. Rearrangement
Transsilylation s. a. Transcyano-O-silylation
C-Transsilylation 26, 653
Trap, nucleophilic, methanol,
as – 26, 158
Trapping
– of
dienes, unstable 27, 760
– **agent for silylenes** 28, 582
N,N,N-Triacylamines
– from
carboxylic acid chlorides
and – – amides 26, 420
Triacyl borates
– as catalysts 30, 135
– special s.
triformyl borate
Triafulvenes s. Enecyclopropenes
1,3,3-Trialkoxy compds.
– from
α,β-ethyleneketones 26, 178
Trialkylalanes as reactants
27, 671; 28, 354
– as reagents 29, 7
– special s.
triethyl-aluminum
triisobutyl-
trimethyl-
– startg. m. f.
hydrocarbons, synthesis
28, 845
Trialkylboranes
– as reagents 29, 305
– special s.
triethylborane
– startg. m. f.
alcohols, synthesis 27, 714
aminoboranes 29, 297
β-aminoketones 28, 782
borinic acid esters 27, 659;
28, 81 suppl. 30
3-ethylenealcohols 29, 869
ethylene derivs., synthesis
29, 877

α-halogenoketones, – 28, 860
γ-hydroxyketones 29, 869
–, **mixed** 27, 659 suppl. 30
– via borinic acid esters 27, 659
Trialkylcarbinols
– from
boranes 27, 892
ethylene derivs., synthesis
27, 716
Trialkyloxonium fluoroborates
– as reactants 29, 425;
30, 110, 266, 703
– as reagents 26, 173/4, 454, 837,
933; 27, 126, 192, 601, 994,
997; 28, 180, 337, 911; 29, 24,
209, 262; 30, 121
–, chemistry, review 27, 997
– startg. m. f.
sulfonium salts 27, 997
1,1,1-Triamines s. Orthocarboxylic acid amides
1,1,2-Triaminoethylenes
– startg. m. f.
α-azomethinioamidinium
salts 28, 279
Trianions, syntheses via –
29, 812
Triarylcarbinols 29, 122
Triarylmethanes 28, 742
Triaryl phosphates
– startg. m. f.
diaryls, sym. 28, 900
1,3,4,2,5-Triazadiborolidines
26, 281
1,3,5-Triazahepta-1,3-dien-7-ones
– startg. m. f.
1,4,6-triazahepta-3,5-dien-
7-ones 27, 337
1,4,6-Triazahepta-3,5-dien-
7-ones
– from
1,3,5-triazahepta-1,3-dien-
7-ones 27, 337
Triazanium salts
– from
hydrazines 29, 271
1,3,6-Triazaphenothiazines
19, 702 suppl. 29
1,4,7-Triazatricyclo[5,2,1,0[4,10]**]-**
decanes 26, 450
1,3,5,2,4,6-Triazatriphosphorine,
hexachloro- s. Phosphonitrile
chloride
Triazenes
– as intermediates 30, 236
– from
nitroso compds. 30, 189
– special s.
acoxytriazenes
carbalkoxytriazenes
enetriazenes
hydroxytriazenes
tosyltriazene salts
–, **3-acyl-**
– from
carboxylic acid hydrazides
30, 189
– special s.
acoxy-3-acyltriazenes
1,2,4-Triazepines
– from
Δ[1]-azirines and 1,2,4,5-
tetrazines 29, 770
1H-1,2,5-Triazepines,
6,7-dihydro- 29, 381
1,3,5-Triazine
– startg. m. f.
formamidines 11, 485

1,2,4-Triazine 4-oxides
- startg. m. f.
  1,2,4-triazines **18**, 992
  suppl. **28**
1,3,5-Triazinephosphonic acid esters 4, 597 suppl. **26**
1,2,3-Triazine ring 17, 341
- from
  pyrimidine ring **29**, 273
- startg. m. f.
  o-aminoamidines **30**, 285
- - opening **30**, 285
1,2,4-Triazine ring
- from
  diazonium salts **29**, 358
  o-nitrosamines **27**, 434
1,3,5-Triazine –
- from
  4H-1,3-benzoxazin-4-one ring **28**, 421
- -, **N-condensed 30**, 329
- -, **hexahydro-**
- from
  isothioureas, trimerization **26**, 435
1,2,3-Triazines
- from
  cyclopropenyl azides **29**, 330
- startg. m. f.
  indenones **28**, 931
1,2,4-Triazines
- from
  1,2,4-triazine 4-oxides **18**, 992 suppl. **28**
- startg. m. f.
  4H-azepines **26**, 833
  azocines **26**, 918
  N-heterocyclics **26**, 833
  pyridine ring **26**, 833
  1,2,4-triazin-5-ones **28**, 83
- -, **6-subst. 27**, 21
- -, **5-amino- 27**, 334
- -, **6-amino-**
- from
  s-triazolo[3,4-f][1,2,4]-triazinium salts **27**, 21
1,3,5-Triazines
- from
  biguanides **30**, 294
  imidazoles **27**, 365
  nitriles, trimerization **26**, 303
- special s.
  cyanuric chloride
- -, **2,4-disubst.**, review **19**, 494 suppl. **29**
- -, **halogeno-**
- from
  nitriles (3 molecules) **26**, 457
- -, **hexahydro-** (s. a. Isocyanur...
- -, aminomethylation with – **28**, 786
- from
  aldehydes **30**, 253
- -, **o-hydroxyphenyl- 28**, 421
- -, **trisaryloxy-**
- from
  cyanic acid esters **26**, 301
1,2,4-Triazine-3-thiones
- from
  α-ketoaldehyde aldoximes **26**, 387
Triazinone ring
- startg. m. f.
  o-aminocarboxylic acid amides **27**, 15

1,2,4-Triazin-3(2H)-ones
- startg. m. f.
  2-imidazolones **27**, 289 suppl. **28**
  1,2,3-triazoles **27**, 289
1,2,4-Triazin-5-ones
- from
  1,2,4-triazines **28**, 83
1,2,4-Triazin-6-ones, tetrahydro-
- from
  Δ¹-azirines, 2-carbonyl- **28**, 371
1,3,5-Triazin-2-yl groups, migration **30**, 229
1,2,4-Triazole as reagent **28**, 375
1,2,3-Triazole ring
- from
  enamines **26**, 321
  nitriles and N-nitrosamines **29**, 742
  nitro compds. **24**, 438
- -, **N-condensed**
- from
  N-heterocyclics **28**, 321
1,2,4-Triazole ring
- from
  amidrazones **26**, 270
  carboxylic acid hydrazides and thioamides **28**, 404
- -, **N-condensed**
- by elimination of sulfur dioxide **26**, 490
- -, **3-amino-**
- from
  carbodiimides **26**, 393
- - - rearrangement **22**, 345; **26**, 315
1,2,3-Triazoles (s. a. Benzotriazoles)
- from
  α,β-ethylenesulfones **29**, 429
  1,2,4-triazin-3(2H)-ones **27**, 289
- -, **5-alkoxy- 27**, 287
1,2,4-Triazoles
- from
  4-azo-Δ²-5-thiazolones **28**, 406
  1,2,4-triazoline-3-thiones **26**, 82
- special s.
  1,3-dichloro-1,2,4-triazole
- startg. m. f.
  isoindoles **30**, 665
1H-1,2,4-Triazoles
- from
  1,1-halogenohydrazones and nitriles **29**, 420
1,2,4-Triazoles, 3-subst.
- from
  N-1,2,4-triazol-4-ylamidines **26**, 497
- -, **4-amino-**
- startg. m. f.
  nitriles (2 molecules) **26**, 480
4H-1,2,4-Triazoles, 4-amino-3-(2-aminophenyl)-
- startg. m. f.
  s-triazolo[4,3-d][1,3,4]benzotriazepines **29**, 352
1,2,4-Triazoles, 5-amino-3-mercapto-
- from
  isodithiobiurets **29**, 427
- -, **3-azido-**
- from
  1-(N-tetrazol-5-ylhydrazono)-1-halides **28**, 457

-, **1-carbalkoxy-**
- startg. m. f.
  urethans **28**, 375
-, **3,5-diamino-**
- from
  isodithiobiurets **29**, 427
-, **3-hydrazino-**
- from
  guanidines, 1-acylamino-3-amino- **25**, 340
-, **4-hydroxy-**
- from
  amidrazones, N¹-acyl-N³-hydroxy- **27**, 506
-, reactions **27**, 506
-, **3-hydroxymethyl-**
- from
  Δ²-4-oxazolones **27**, 391
1H-1,2,4-Triazoles, 3-(o-α-hydroxyphenyl)-
- from
  1H-2,3-benzoxazines, 4-hydrazino- **28**, 295
1,2,4-Triazolides
- special s.
  N-sulfonyl-1,2,4-triazolides
1,2,4-Triazolidine-3,5-dione ring
- startg. m. f.
  azoxy compds., cyclic **26**, 92
1,2,4-Triazolidine-3,5-diones
- from
  1,2,4-triazolidine-3,5-dione 1,2-ylids **27**, 30
- startg. m. f.
  amines **29**, 21
  hydrazines **29**, 21
-, **4-acyl-**
- from
  acylisocyanates and diaziridinones **30**, 212
-, **4-amino-**
- from
  hydrazines **29**, 341
1,2,4-Triazolidine-3,5-dione 1,2-ylids
- startg. m. f.
  1,2,4-triazolidine-3,5-diones **27**, 30
1,2,4-Triazolidine-3-thiones
- from
  hydrazones **29**, 306
1,2,3-Triazoline-2-diazoamine tautomerism **29**, 466
1,2,4-Triazoline-3,5-dione, 4-phenyl- as reactant **30**, 283, 318
Δ²-1,2,4-Triazoline ring
- from
  indone ring **27**, 341
Δ²-1,2,3-Triazolines, 5-amino-
- from
  azides and oxo compds. **26**, 343
Δ²-1,2,4-Triazolines
- from
  amidrazones and aldehydes **27**, 387
Δ³-1,2,4-Triazolines, 3-mercapto-
- from
  oxo compds. and hydrazines **29**, 388
1,2,4-Triazoline-3-thiones
- startg. m. f.
  1,2,4-triazoles **26**, 82
1,2,3-Triazolium betaines, 4-hydroxy-
-, cycloadditions with – **28**, 270

**1,2,4-Triazolium salts,
3-alkylthio-**
- as intermediates **27,** 81
  suppl. **29**
**s-Triazolo[4,3-a][1,4]benzo-
diazepines 28,** 404; **30,** 324
**s-Triazolo[4,3-b]benzothiazoles
26,** 232
**s-Triazolo[4,3-d][1,3,4]benzo-
triazepines**
- from
  4H-1,2,4-triazoles, 4-amino-
  3-(2-aminophenyl)- **29,** 352
**s-Triazolo[3,4-a]isoquinolines
26,** 490
**1,2,4-Triazol-5-one ring,
3-acylamino-**
- from
  1,2,4-oxadiazole ring,
  3-ureido- **26,** 257
1,2,4-Triazol-3-ones
- from
  semicarbazones **27,** 501
**s-Triazolo[3,4-b][1,3,4]oxa-
diazoles**
- from
  trihalogenodiazabutadienes
  27, 469
**s-Triazolo[4,3-a]pyridines
26,** 490
v-Triazolo[3,4-a]pyrimidines
- from
  β-diketones **27,** 396
**v-Triazolo[4,5-d]pyrimidines
28,** 241
**s-Triazolo[1,5-a]pyrimidines,
7-hydroxy- 29,** 368
**v-Triazolo[3,4-a]pyrimidin-
5(4H)-ones**
- from
  v-triazolo[3,4-a]pyrimidin-
  7(4H)-ones **29,** 272
- startg. m. f.
  v-triazolo[3,4-a]pyrimidin-
  7(4H)-ones **29,** 272
**v-Triazolo[3,4-a]pyrimidin-
7(4H)-ones**
- from
  v-triazolo[3,4-a]pyrimidin-
  5(4H)-ones **29,** 272
- startg. m. f.
  v-triazolo[3,4-a]pyrimidin-
  5(4H)-ones **29,** 272
s-Triazolo[4,3-c]quinazolines
- startg. m. f.
  s-triazolo[1,5-c]quinazolines
  26, 315
s-Triazolo[1,5-c]quinazolines
- from
  s-triazolo[4,3-c]quinazolines
  26, 315
Triazolotriazines, rearrangement **26,** 315
**v-Triazolo[5,1-c]-as-triazines
29,** 358
**s-Triazolo[3,4-f][1,2,4]triazinium
salts**
- startg. m. f.
  1,2,4-triazines, 6-amino-
  27, 21
β-(1,2,3-Triazol-4-yl)acroleins
- from
  pyridinium salts **22,** 281
- via pentamethine dyes
  22, 281
1,2,4-Triazol-1-yl as leaving
group **30,** 282

N-1,2,4-Triazol-4-ylamidines
- startg. m. f.
  1,3,4-oxadiazoles **26,** 497
  1,2,4-triazoles, 3-subst.
  26, 497
O-(1,2,4-Triazol-4-yl)ketones
- from
  quinazolines, 3,4-dihydro-
  3-acylamino-4-hydroxy-
  29, 455
Tri-n-butylamine as reagent
  27, 48, 871; **28,** 61
Tributylphosphine – – **23,** 700
  suppl. **28; 24,** 855 suppl. **27**
Tri-n-butylphosphine – –
  27, 272
Tributylphosphine oxide
  as solubilizer **27,** 317
Tributyltin hydride as reagent
  27, 76; **29,** 970;
  30, 379, 652, 697
**Tricarbonylchromium(0)
complexes** (s. a. Polystyrenetricarbonylchromium)
-, deuteriation via – **28,** 591
-, prepn. of optically active
  compds. via – **30,** 540
-, syntheses with – **30,** 540
**Tricarbonylcyclohexa-
dienoneiron complexes**
-, N-arylation with **30,** 266
**Tricarbonyl-(trans-π-2,4-dienol)
iron complexes, stereo-
isomeric 24,** 61 suppl. **27**
Trichloramine
- as reagent **26,** 522
β,β,β-Trichlorethyl carbonates
- startg. m. f.
  carbonic acid esters, mixed
  26, 219
- **esters** (s. a. Phosphoric acid
  β,β,β-trichlorethyl esters)
-, cleavage **29,** 5
Trichloroacetic acid (s. a.
  Sodium trichloroacetate)
- as reagent **18,** 836 suppl. **26;
  26,** 7; **27,** 293
Trichloroacetonitrile
- as reactant **11,** 619 suppl. **28**
- as reagent **28,** 440
C-Trichloroacetylation 29, 816
Trichloroacetyl halides
  as reactants **26,** 541; **29,** 816
- **isocyanate,** reactions **28,** 615
-, reaction with allenes **28,** 615
1,2,4-Trichlorobenzene
  as solvent **30,** 187
Trichlorofluoromethane
  as solvent **29,** 503
Trichloroisocyanuric acid
  as reagent **26,** 530
**Trichloromethanesulfonic acid
esters**
-, N-alkylation with –
  24, 620 suppl. **29**
**Trichloromethyl chloroformate
14,** 240 suppl. **29**
**Trichloromethylisocyanide
dichloride**
- as reactant **29,** 518
-, reactions with – **29,** 842
Trichloromethyl ketones
  (s. a. C-Trichloroacetylation)
- startg. m. f.
  carboxylic acid esters
  29, 227, 816

- **thioethers**
- from
  methyl thioethers **29,** 30
- startg. m. f.
  mercaptans **29,** 30
Trichlorosilane 26, 66, 564;
  27, 69; **28,** 61
-, reactions in the presence
  of tert. amines, review **26,** 66
Trichlorosilanes
- from
  carboxylic acids **26,** 66
  suppl. **29**
- special s.
  phenyltrichlorosilane
**1,1,1-Trichloro-3,3,3-trifluoro-
acetone,** N-trifluoroacetylation with – **26,** 455
Triclates s. Trichloromethanesulfonic acid esters
Tricyclo[3.1.0.0²,⁴]hexanes
- from
  cyclopropenes (2 molecules)
  27, 715
Tricyclohexylphosphine
  as reagent **26,** 722
**exo-Tricyclo[4.2.1.0²,⁵]non-
3-enes**
-, rearrangement, skeletal
  26, 760
Tricycloquinazolines
- from
  2,1-benzisoxazoles
  (3 molecules) **30,** 269
1,2,3-Trienes (s. a. Cumulenes)
  28, 943
- startg. m. f.
  α,β-ethyleneketones **26,** 133
1,2,4-Trienes
- from
  5-halogen-3-en-1-ynes,
  synthesis **28,** 812
1,3,5-Trienes
- 1,6-addition to – **30,** 82
-, ring closure **28,** 929
- startg. m. f.
  benzene ring **28,** 892 suppl. **29;
  28,** 929 suppl. **30**
2,4,5-Trienols 14, 91 suppl. **29**
Triethylaluminum
- as cocatalyst **23,** 715 suppl. **29**
- as reagent **29,** 7
Triethylamine hydrochloride
  as reagent **27,** 598
**Triethylammonium acetate
  1,** 586 suppl. **28; 25,** 162;
  28, 169
- **formate** as reagent **21,** 317
  suppl. **26**
-, reactions with – **25,** 554
Triethylborane as reagent
  26, 881; **27,** 37
Triethylcarbinol
- as medium **28,** 695
- as reagent **27,** 892; **30,** 495
Triethylenediamine as reagent
  18, 331 suppl. **29; 25,** 324
  suppl. **28; 26,** 420, 576, 972;
  28, 844; **29,** 262, 963
**Triethylene glycol dimethyl
ether** as solvent **26,** 248;
  30, 604
Triethyl phosphite
- as oxygen acceptor **27,** 515
- as reagent **26,** 982; **27,** 420;
  28, 456

Triethylsilane
- as catalyst component
  26, 50 suppl. **28**
- as reagent **29**, 63, 209

Triflamides s. Trifluoromethanesulfonamides

Triflate s. Trifluoromethanesulfon...

Triflones s. Trifluoromethanesulfones

1,1,1-Trifluorides
- startg. m. f.
  methyl groups **27**, 73

Trifluoroacetamides s. N-Trimethylsilyltrifluoroacetamide

Trifluoroacetic acid (s. a. Pyridinium trifluoroacetate)
- as medium **15**, 428 suppl. **26**; **25**, 572 suppl. **28**
- as reactant **29**, 167
- as reagent **1**, 363 suppl. **28**; **8**, 486 suppl. **26**; **14**, 197 suppl. **29**; **16**, 943 suppl. **28**; **17**, 664 suppl. **27**; **18**, 963 suppl. **26**; **26**, 83 suppl. **29**; **26**, 500, 659; **27**, 26, 110, 325; **27**, 141 suppl. **28**; **27**, 942 suppl. **28**; **28**, 170, 268, 308, 365, 423, 439, 611, 688, 703; **29**, 25, 63, 206, 216, 252, 369, 402, 408, 466, 555, 708, 878, 900, 906, 910, 926/7; **30**, 166, 240, 559, 640, 644, 651
- – esters
  - special s.
    tert-butyl trifluoroacetate

Trifluoroacetic anhydride
- as reactant **29**, 167
- as reagent **9**, 243 suppl. **29**; **13**, 916 suppl. **28**; **19**, 328 suppl. **27**; **20**, 90 suppl. **29**; **24**, 299 suppl. **29**; **24**, 408 suppl. **28**; **26**, 300 suppl. **27**; **27**, 96/7, 260, 618, 716, 940; **28**, 578, 698; **30**, 165, 651

Trifluoroacetoxy compds.
- special s.
  cyclopropane ring, 1-halogeno-1-trifluoroacetoxy-

Trifluoroacetyl... s. a. Rhodium(I), dicarbonyltrifluoroacetylacetonato-
- groups, N-protective
- –, removal, reductive, selective **26**, 26

1,1,N-Trifluoroamines
- from
  azomethines **29**, 503

Trifluoroethanol as solvent **27**, 420

Trifluoroethyl esters
s. Carboxylic acid trifluoroethyl esters

Trifluoromethanesulfonamides
- as acylating agents **29**, 393
- –, protection of amino groups as – **29**, 288

Trifluoromethanesulfones and their reactions 29, 796

Trifluoromethanesulfonic acid
- as catalyst **16**, 279 suppl. **29**; **17**, 559 suppl. **28**; **28**, 827
- as reagent **10**, 292 suppl. **29**; **29**, 17a
- – anhydride
- as reactant **27**, 107 suppl. **30**
- – esters
- special s.
  trimethylsilyl triflate

Trifluoromethanesulfonyloxy group
- as leaving group, active **28**, 495

Trifluoromethyl... s. a. Bis-(trifluoromethyl)...

Trifluoromethylation, ar. 26, 867

Trifluoromethyl ethers
- from
  alcohols **30**, 106
  chlorothionoformic acid esters **29**, 523
  fluoroformic acid esters **30**, 106

Trifluoromethyl hypofluorite
s. Fluoroxytrifluoromethane

Trifluoromethylmetal compds. 30, 442

(Trifluoromethylthio)acetonitrile
–, Knoevenagel condensation with – **28**, 737

α-(Trifluoromethylthio)-α,β-ethylenenitriles
- from
  aldehydes **28**, 737

α-(Trifluoromethylthio)nitriles
- special s.
  (trifluoromethylthio)acetonitrile

Triformyl borate
- startg. m. f.
  formamides **27**, 312

Triformylmethane
- startg. m. f.
  polymethine cyanines **28**, 718

Triglyme s. Triethylene glycol dimethyl ether

Trihalides s. a. Trichlor..., Trifluor...

1,1,1-Trihalides
- special s.
  haloforms
  1,1,1-trifluorides
- startg. m. f.
  tert-alkyl groups **8**, 857 suppl. **28**
  1,1-dihalides **27**, 77; **29**, 73
  methyl groups **27**, 71

Trihalogenoacet... s. a. Trichloroacet..., Trifluoroacet...

2,2,2-Trihalogenoalcohols
- from
  oxo compds. **30**, 449

1,3,3-Trihalogeno-2-azapropenes
–, reactions with – **29**, 842
- special s.
  trichloromethylisocyanide dichloride
- startg. m. f.
  quinazolines **29**, 842

Trihalogenodiazabutadienes
- startg. m. f.
  imidazolinylhydrazonohalides **27**, 529
  imidazotriazoles, dihydro- **27**, 529
  s-triazolo[3,4-b][1,3,4]oxadiazoles **27**, 469

Trihalogenogermanes
- from
  halides **27**, 647

Trihalogenoketones
- special s.
  trichloromethyl ketones

Trihalogenomethylcarbinols
- special s.
  aryl trichloromethyl carbinols

Trihalogenomethyl mercury compds. s. Organotrihalogenomethylmercury compds. under Mercury compds., organo-

Trihalogenophosphazocarbonyl compds. 26, 278

Trihalogenophosphazo compds.
- special s.
  1,1-dihalogeno(trihalogenophosphazo) compds.

Trihalogenophosphazosulfonyl compds.
- from
  sulfonic acid amides **27**, 304

Triiron dodecacarbonyl 23, 57 suppl. **28**; **27**, 14; **30**, 337

Triisobutylaluminum as reagent **28**, 669

Triisooctyl phosphite – – **19**, 962 suppl. **23**

2,4,6-Triisopropylbenzenesulfonyl chloride – – **11**, 235 suppl. **26**
- hydrazine – – **17**, 83 suppl. **28**

1,2,3-Triketone 2-monohydrazones
- startg. m. f.
  α-azo-α,α-dihalogenoketones **28**, 422
  dihalogenomethylenehydrazones **28**, 422

Triketones
- special s.
  β-diketones, 2-acyl-

1,3,5-Triketones
- startg. m. f.
  benzene ring **27**, 924
  pyrazolo[1,5-c]pyrimidines **27**, 385

Trimerization (s. a. Cyclotrimerization)
- of
  p-quinones **29**, 718
- to
  cyclotribenzylenes **26**, 843
  1,3,5-triazine ring, hexahydro- **26**, 435
  1,3,5-triazines **26**, 303

Trimetaphosphate 26, 97

Trimethylaluminum as reactant **29**, 552/3

Trimethylamine oxide 29, 120

Trimethylammonium formate 20, 311 suppl. **27**

Trimethyl borate
- as reagent **30**, 29
- as solvent **16**, 242 suppl. **28**
- borate-tetrahydrofuran
  as mildly acidic medium **26**, 673

Trimethylbromosilane
as catalyst **27**, 83

Trimethylchlorosilane
- as reactant **7**, 697 suppl. **27**; **28**, 500 suppl. **29**; **28**, 733
- as reagent **17**, 933 suppl. **29**; **27**, 486, 512; **28**, 46, 350, 683
- –, reaction with carboxylic acid esters and sodium, review **25**, 589 suppl. **26**

N,4,4-Trimethyl-Δ²-oxazolinium iodide
- as reactant **26**, 872

Trimethyl phosphate
- as catalyst **26**, 103
- as medium **26**, 97
- as reagent **26**, 97

- phosphite as reagent 26, 33, 602, 629; 27, 101; 28, 108; 29, 217; 30, 406
- startg. m. f.
  formamidinium phosphonic acid anhydrides 28, 595
- - -copper(I) iodide 28, 780
Trimethylsil... s. a. Sil...
Trimethylsilylacetamide
  as reactant 29, 528
Trimethylsilyl azide as reactant 28, 400; 29, 296
- - -lead tetraacetate
-, reactions with - 26, 325; 27, 362
-, - - -, review 27, 362
- cyanide as reactant 28, 610
Trimethylsilyl-N,N-diethylamine
  as reactant 28, 249
Trimethylsilyl perchlorate
  as Lewis acid, strong 26, 446 suppl. 30
- triflate - - -, 26, 446 suppl. 30
N-Trimethylsilyltrifluoroacetamide as reagent 27, 476
2,4,6-Trinitrobenzenesulfonic acid as reagent 27, 129
1,1,1-Trinitro compds.
- from
  aldoximes 27, 347
Triols, cyclic
- startg. m. f.
  acetallactones, polycyclic 30, 112
2,4,9-Trioxaadamantanes, 7-halogeno-
- from
  carboxylic acid chlorides (3 molecules) and ethylene derivs. 29, 839
1,2,4-Trioxane ring 30, 80
s-Trioxanes from aldehydes 30, 71
Triphenylcarbinol as reagent 30, 640
Triphenylcarbonium salts
  as reagents 28, 232
  fluoroborate as reagent 15, 154 suppl. 27; 27, 239
  perchlorate - - 26, 965; 27, 86
-, hydride abstraction by - 22, 640
Triphenyldiethoxyphosphorane
-, cyclodehydration with - 28, 216
Triphenylmethane
- as medium 1, 397 suppl. 26
- as reagent 29, 804
Triphenylmethyl... s. a. Triphenylcarbonium..., Trityl...
- -lithium 6, 738 suppl. 27; 18, 59 suppl. 28; 28, 895; 29, 804
Triphenylphosphine
- as base 26, 154
  as Lewis base 28, 41
- as reactant 30, 604
- as reagent 4, 677 suppl. 26; 10, 286 suppl. 29; 13, 778 suppl. 26; 14, 976 suppl. 26; 16, 785 suppl. 27; 17, 45 suppl. 29; 20, 55 suppl. 29; 26, 471 suppl. 29; 26, 105, 125, 175, 399, 470, 550, 621, 642, 827, 891; 27, 642 suppl. 28; 27, 417, 429, 477,

514, 718, 724, 977/8; 28, 25, 752/3, 819, 861, 913; 29, 188, 403, 477, 510/1; 30, 10, 167/8, 358, 501, 566, 604, 686/7, 698, 700
- dibromide
- as reagent 9, 357 suppl. 28; 19, 949 suppl. 29; 26, 471; 27, 576, 653
- selenide as reagent 27, 942 suppl. 28
Triphenyl phosphite
- as reagent 27, 419 suppl. 29; 28, 194, 467; 30, 610
- - methiodide 27, 74; 28, 916
Triphenylphosphonium phenacylid as catalyst 20, 239 suppl. 27
Triphenyltin deuteride
  as reagent 29, 47
- hydride
- as reagent 26, 50; 30, 31
Triple bonds s. Acetylene derivs.
Triplet sensitizer
-, anthraquinone as - 26, 990
Tri-n-propylamine as reagent 26, 66, 564
Tripyrrenes 18, 842 suppl. 29
Trisaccharides 11, 293 suppl. 27
Tris-(p-bromophenyl)-ammoniumyl hexachloroantimonate
- as reagent 27, 143
Tris buffer s. Trishydroxymethylaminomethane buffer
N,N,O-Tris(carbamyl)-hydroxylamines
- from
  isocyanates 26, 293
Tris(dibenzoylmethido)iron(III) 26, 875 suppl. 30
Tris(diethylamino)phosphine
- as reactant 27, 418
- as reagent 26, 988; 28, 442
Tris(dimethylamino)methane
  as reactant 28, 772
Tris(dimethylamino)phosphine
- as complexing agent 28, 661
- as reagent 27, 884 suppl. 26; 26, 492, 886; 27, 642 suppl. 28; 27, 642, 866; 28, 661
-/carbon tetrachloride 28, 144, 358
Tris(dimethylamino)stibine, reactions 22, 399 suppl. 26
Trishydroxymethylaminomethane buffer 26, 366
1,2,3-Trisilacycloheptanes
- startg. m. f.
  1,2-disilacyclohexanes 28, 582
Trisulfenamides 29, 314
-, reactions 29, 314
Trisulfides
- from
  hydrodisulfides and sulfenimides 26, 562
s-Trithianes
- as intermediates 27, 830
- startg. m. f.
  oxo compds. 29, 219; 30, 688
  stilbenes 30, 688
1,6,6aS(IV)-Trithiapentalenes s. Thiathiophthene...
Trithiocarbonic acid esters, cyclic s. 1,3-Dithiolane-2-thiones

Trithiodiphosphonic acid esters
- from
  thiophosphonic acids, O-alkyl- 29, 529
Trithioorthocarboxylic acid esters
- startg. m. f.
  thiolic acid esters 26, 214
- - -, cyclic s. 1,3-Dithianes, 2-alkylthio-
Tritiation
- of
  betaines 9, 510 suppl. 26
Tritiation, catalytic, apparatus 21, 102 suppl. 27
Tritio compounds s. Ditritio compounds
Triton B 23, 948 suppl. 29; 26, 270; 27, 456, 917; 28, 735; 29, 716; 30, 676
Trityl... s. a. Triphenylmethyl...
N-Tritylacoxylamines
- from
  carboxylic acids 26, 172
- startg. m. f.
  acoxylamines 26, 172
Trityl chloride
- as reactant 28, 500
- as reagent 21, 294 suppl. 29
- ethers
- startg. m. f.
  oxo compds. 28, 232
Tritylone ethers s. 9-Anthronyl ethers, 9-phenyl-
Tritylpyridinium fluoroborate
  as reactant 8, 275 suppl. 29
Trop... s. a. Cycloheptatrien...
Tropenium salts
- special s.
  aminotropenium salts
-, syntheses with - and enamines 27, 822
- -, halogeno-
- from
  tropones 26, 540
-, substitution, nucleophilic 26, 540
Tropolone ring
- from
  α-isonitrosoketones, cyclic 27, 206
Tropolones
-, rearrangement, skeletal 26, 142
-, 5-hydroxy- 26, 1
Tropone adducts
- startg. m. f.
  cyclopropane ring, fused 28, 97
- ring via o-dialdehydes 29, 847
Tropones 9, 968 suppl. 26
- from
  pyridinium betaines 26, 695
- startg. m. f.
  8-oxabicyclo[5.3.0]decatrienes 27, 707
  tropenium salts, halogeno- 26, 540
-, 2-alkyl-, by ring opening, stereospecific 27, 967
Troponoids 26, 890
- special s.
  tropenium salts
  tropolones
  tropones
Troponoid systems
-, 1,8-addition to - 17, 764

Tropylium s. Tropenium
Truce-Smiles rearrangement,
review **14,** 764 suppl. **29**
Tryptophans, synthesis **6,** 789
Tscherniac-Einhorn-type
reaction (s. a. Einhorn)
26, 888
Tungstates s. Potassium
hexachlorotungstate
Tungsten-carbene complex
as reactant **28,** 846
Tungsten, bis-(η-cyclopenta-
dienyl)dihydrido- as reagent
25, 64 suppl. **28**
- **hexachloride 23,** 441
suppl. **29; 27,** 932; **28,** 761
Twistanoids
s. Heterotwistanoids
Twistans s. Heterotwistans

Ullmann *sec*-amine synthesis,
modified 29, 418
- **diaryl synthesis,** reviews
8, 854 suppl. **28, 29**
- **coupling, intramolecular**
8, 888 suppl. **26**
- **reaction, – 1,** 668
suppl. **29**
-, o-heterocycles by – **29,** 259
Ultrasonic field 2, 78 suppl. **28**
- **waves** s. Compounds,
unstable
Uracil dimers
- startg. m. f.
cyclobutanes **27,** 45
Uracil-1,3-disulfofluorides
- from
acetylene derivates **26,** 698
Uracil ring
- from
amines **26,** 678
o-aminocarboxylic acid
amides **27,** 349; **29,** 333
Uracils (s. a. Thiouracils)
28, 370
-, O-alkylation, selective **26,** 206
- from
isocyanates **28,** 338
-, **3-alkyl- 17,** 495 suppl. **28**
-, **5-subst. 26,** 927 suppl. **30**
-, **6-amino-5-nitroso-**
- startg. m. f.
2,4,6,8(1H,3H,7H,9H)-
pyrimido[5,4-g]pteridine-
tetrone 5-oxides (from
2 molecules) **28,** 377
-, **6-anilino-**
- from
uracils, 6-halogeno- **28,** 314
- startg. m. f.
isoalloxazine 5-oxides **28,** 314
-, **5-cyano- 28,** 454
-, **6-halogeno-**
- startg. m. f.
isoalloxazine 5-oxides **28,** 314
uracils, 6-anilino- **28,** 314
-, **3-hydroxy- 9,** 447 suppl. **26**
Urazoles s. 1,2,4-Triazolidine-
3,5-diones
Urea
- as nitrous acid scavenger
29, 282
- as reagent **21,** 393 suppl. **29;**
21, 434 suppl. **27; 27,** 96;
28, 590; **30,** 240
- startg. m. f.
biurets **29,** 300

carboxylic acid amides **5,** 607
suppl. **26; 28,** 707
nitriles **27,** 422
1,2,4-triazolidine-3,5-diones,
4-amino- **29,** 341
Ureas (s. a. Isoureas,
Quasiureas)
- from
amines and carbon monoxide
26, 329
- and hydroxamic acid
O-sulfates **26,** 364
- and isocyanates (s. a.
N-Carbamylation) **30,** 229
- and β-ketocarboxylic acid
amides **28,** 417
- and urea **4,** 413
cyanamides **26,** 23; **30,** 216
diaziridinones **29,** 13
- special s.
N-acoxy-ureas
acyl-
N-(alkylthiocarbonyl)-
N-carbonyl-
ene-
glycosyl-
guanyl-
halogeno-
hydroxy-
nitro-
quinolines, 1,2-dihydro-
1-carbamyl-
sulfonyl-ureas
sulfonyloxy-
- startg. m. f.
amines **26,** 23
azo compds. **27,** 288
N-carbamylamidinium salts
29, 370
carbodiimides **27,** 978
carboxylic acid amides
28, 776
chloroformamidines **29,** 511
cyanamides **29,** 456
1,3-N,N-heterocyclics, review
20, 289 suppl. **28**
N-nitrosoureas **29,** 274
semicarbazides **29,** 274
thioureas **27,** 614
-, **cyclic** s. 1,3-N,N-Hetero-
cyclics, 2-oxo-
-, **sym.**
- by Mannich reaction, double
26, 810
- from
amines and carbon monoxide
28, 322
- and carbon dioxide **30,** 243
-, **vinylogous**
- startg. m. f.
carbodiimides, vinylogous
27, 978
Ureidomalonic acid esters
- startg. m. f.
purines **26,** 356
Uretdiones s. 1,3-Diazetidine-
2,4-diones
Urethans (s. a. N-Carbalkoxy...,
Carbamyloxy groups,
Thiourethans)
- from
acylamines **30,** 107
alcohols **5,** 125 suppl. **28;**
30, 273
- and amines **28,** 322
amines and azidoformic
acid esters **27,** 437

N-carbalkoxysulfamic acid
esters **26,** 491
carbonic acid ester fluorides
23, 404 suppl. **26**
- – esters, mixed **27,** 404;
28, 375
- acids **28,** 353
1,1-diurethans, synthesis
27, 820
isonitriles **28,** 93
hydrazodicarboxylic acid
esters **30,** 208
1,2,4-triazoles, 1-carbalkoxy-
28, 375
- special s.
acoxy-urethans
acyl-
aryl carbamates
p-biphenylylurethans
carbalkoxycarbamic acid
esters
cyano-urethans
di-
ethylene-
halogeno-
hydroxy-
keto-
mono(carbalkoxy)amines
nitrosourethans
- startg. m. f.
alcohols **28,** 8; **30,** 7
amines (s. a. N-De-
carbalkoxylation) **30,** 273
azomethines **24,** 456
ketones, sym. **12,** 830
suppl. **28**
-, **cyclic**
- special s.
chlorourethans, cyclic
2-oxazolidones
Uronium salts, quaternary,
cyclic s. Pyridinium salts,
N-carbamyl-
Urushibara nickel catalyst,
activated 19, 55 suppl. **29**

Valence isomerization
s. a. Tautomer...
-, ring closure by – **26,** 757
- – **of isocyclics, condensed**
-, effect of catalyst on –
26, 762 suppl. **27**
Valeric acid
- as reagent **27,** 72
Vanadium carbonyl complex
23, 199 suppl. **29**
-**(IV) chloride 28,** 503
-**(V) oxide chloride 27,** 915;
28, 888
- **oxide fluoride 29,** 910
- **pentoxide 29,** 298; **30,** 321
- – -**titanium dioxide 30,** 321
Vanadyl acetoacetonate
as catalyst **24,** 149 suppl. **28**
Vapor phase aldol condensation
28, 722
Vaseline oil as high temperature
medium **26,** 465
vic – (Vicinal) s. 1,2-
Vilsmeier aldehyde synthesis
26, 242, 839
-, labeled compds. **26,** 242
suppl. **27**
- – –, **double 27,** 826
- – –, **modified 27,** 797
- – –, **2-step 27,** 797
- **complex 28,** 754

Vilsmeier-Haack ring closure 15, 600
Vilsmeier-Haack-type reaction 26, 838
- to alkoxymethylene compds. 26, 838
Vilsmeier reaction, intramolecular 28, 689
- **reagent** (s. a. Dimethylhalogenoformiminium halides) 30, 288
- **salts**
-, ring closure with – 27, 453
N-Vinylation
- with acetylene 27, 330
Vinyl chloride
- startg. m. f.
  γ,δ-ethyleneketones (from 2 molecules) 26, 873
- **compounds** s. a. Divinyl compounds
- **copper compds.**, syntheses via – 30, 492
- **cyanide** s. Cyanoethylation
Vinylcyclopropanes, syntheses by addition to – 28, 859 suppl. 30
(Vinylcyclopropyl)carbinols
- startg. m. f.
  γ,δ-ethyleneketones 28, 124
Vinylcyclopropyl ketones
- from
  γ,δ-ethylene-ε-hydroxyketones 27, 945
Vinylene sulfones
s. Thiirene 1,1-dioxides
Vinyl esters
- special s.
  enol phosphates
- **ethers**
- from
  halides 28, 163
Vinylethynylcarbinols 10, 517
Vinyl groups, angular, introduction, stereospecific 23, 693 suppl. 27
Vinylidene chloride
s. 1,1-Dichlorethylene
- **isocyanide,** reactions with group V element(III) compds. 26, 639
- -**phosphoranes** 30, 443
Vinylketene mercaptals, diene synthesis with – 21, 728 suppl. 28
Vinylketenimines
-, diene synthesis with – 29, 647
- startg. m. f.
  2-cyclohexenones 29, 647
Vinyl ketones
- from
  carboxylic acids 16, 853 suppl. 29
  β-ketosulfoxides, synthesis 29, 618
α-**Vinylketones**
- from
  α-halogenoketones 30, 598
β-**Vinylketones**
- from
  α,β-ethyleneketones 26, 910
Vinyllithium
- as reactant 16, 853 suppl. 29; 30, 618
-, prepn. 27, 683 suppl. 30
- ring expansion with – 27, 683

-, **trifluoro-,** syntheses with – 27, 683 suppl. 30
Vinylmagnesium chloride
as reactant 28, 607; 29, 825; 30, 598
Vinylogous s. Wagner rearrangement, –
Vinylogs s.
  acyl compds., vinylogous
  amidinium salts, –
  carbodiimides, –
  carboxamidium salts, –
  carboxylic acid amides, –
  – – halides, –
  – acid thioamides, –
  dicarboxylic acid imides, –
  ureas, –
O-Vinyloximes
- from
  oximes 27, 122
(N-Vinyl)oxo-N-heterocyclics
- from
  siloxy-N-heterocyclics 26, 445
Vinylthallic acetates, 2-acetoxy-
- from
  acetylene derivs. 29, 598
Volfsi s. Asscher
Vulpinic acids
- from
  pulvinic acid dilactones 30, 69

Wagner-Meerwein rearrangement 27, 744
- -, **double** 29, 686
Wagner rearrangement, vinylogous 29, 987
Walden inversion (s. a. under Configuration)
-, ring closure with – 26, 471
Water (s. a. under Potassium tert-butoxide) 30, 16
- as H-donor 26, 899
- as reactant 28, 18
- as reagent 26, 44, 196, 914, 975; 27, 168, 529; 28, 105; 29, 279, 481; 30, 387
- **entrainment, azeotropic** (s. a. Dean-Stark apparatus) 9, 113 suppl. 27; 26, 102, 177, 179/180, 372, 382, 388, 407; 27, 196, 220, 307, 518, 832, 950; 28, 335, 443, 737, 831; 29, 390
- **scavengers**
  acetic anhydride 29, 239
  bis(trimethylsilyl)acetamide 9, 951 suppl. 27
  molecular sieve s. under Molecular sieve
Wavelength specificity (s. a. Light specificity) 23, 976 suppl. 28; 26, 327
Wichterle ring closure, modified 30, 680
Wilkinson rhodium catalyst
s. Chlorotris(triphenylphosphine)rhodium(I)
Willgerodt-Kindler reaction 27, 162
Winter s. Corey
Wittig rearrangement (s. a. Thio-Wittig rearrangement) 29, 693

- **synthesis** 27, 893; 29, 868
-, cation and anion effects 18, 912 suppl. 26
-, geospecific control 18, 912 suppl. 28
-, geospecifity 22, 871 suppl. 29
- in aq. medium 29, 875
- of
  carbohydrates s. under Carbohydrates
  chalcones 20, 651 suppl. 28
  cumulenes 26, 908
  cyclohexadienes 28, 854
  enollactones 26, 904
  α,β-ethylenecarboxylic acid thioamides 26, 608
  α,β-ethylenealdehydes 29, 867
  C-glycosides 29, 754 suppl. 30
-, protection of amino groups 22, 871 suppl. 26
- with
  benzofuranones 29, 861
  formaldehyde 29, 875
  1,1-hydroxysulfonic acids 29, 620
  lactones 29, 861
  phosphonium fluorides 26, 901
  - salts, cyclic 26, 686
  N-sulfonyllactams 26, 904 suppl. 27
  tributylphosphoranes 17, 892 suppl. 29
- with simultaneous migration of carbon-carbon double bonds 29, 861
- -, **double** 17, 892; 29, 868
- -, **heterogeneous** 29, 868
- -, **intramolecular**
- -, quinolines by – 28, 962
- -, **preferential** 17, 895 suppl. 26
- -, **solid-phase** 17, 892 suppl. 27; 18, 912 suppl. 28
Wolff-Kishner-Huang Minlon reduction
- of ozonides 30, 46
- with decarboxylation 3, 57 suppl. 26
Wolff-Kishner reduction H↓↑O v.i.
-, removal of O-protective groups by – 26, 177
Wolff rearrangement 21, 225
-, review 30, 456
- -, **photochemical** 8, 919; 13, 681
Woodward-Hoffmann rules
s. Symmetry-disallowed reactions

Xanthamic acid esters
s. Thionocarbamic acid esters
Xanthates
- special s.
  ketoxanthates
  potassium methylxanthate
- startg. m. f.
  dithiourethans 28, 328
Xanthene-1,8-diones, octahydro-
-, C-cleavage, reductive 30, 45
Xanthenes
- from
  pyridazines, 6-(2-allylphenoxy)-3-halogeno- 29, 965
Xanthione as reagent 28, 521

Xanthones
- from
  diaryl ethers **30**, 541
  2'-hydroxybenzophenone-
  2-sulfinic acids **28**, 224
  thioxanthone 10,10-dioxides
  28, 224
–, **dihydrooxo- 18**, 307 suppl. **28**
Xanthylium salts
- from
  1,4-naphthoquinones,
  dimerization **29**, 760
Xylene as solvent with readily
  abstractable hydrogen atom
  30, 83

Yeast 29, 36
Ylids, prepn. general **27**, 300
- special s.
  aminosulfenium ylids
  nitrile ylids
–, **cumulated**
- special s.
  vinylidenephosphoranes
–, **2-cyclohexenon-3-yl-
  stabilized**, syntheses with –
  27, 82 suppl. **29**
As-Ylids (s. a. Arsonium ylids)
–, reactions via – **16**, 893
  suppl. **26**
N-Ylids, cyclic
–, cleavage **26**, 992; **30**, 697
–, rearrangement **28**, 298
- special s.
  isoquinolinium methylids
  pyridinium ylids
  1,2,4-triazolidine-3,5-dione
  1,2-ylids
  pyrido[1,2-b]cinnolin-6-ium
  ylids, 5,11-dihydro-
P-Ylids s. Alkylidenephosphoranes, Phosphonium
  ylids, cyclic, Ring-O,P-ylids
S-Ylids
- special s.
  S-cyclopropylids
  sulfonium ylids
  sulfoxonium ylids
–, **stable**
  thetin anions **26**, 895

S-N-Ylids s. N-Acylsulfilimines
Se-Ylids s. Selenonium ylids
Ynamines
–, cycloaddition with – **27**, 693
- from
  carboxylic acid thioamides
  26, 980
  difluoromethylene compds.
  27, 458
- special s.
  acylynamines
  silylynamines
- startg. m. f.
  allenedicarboxylic acid
  amides, sym. **27**, 687
  cyclobutenes, 2-amino-
  3-halogeno- **28**, 627
  2-cyclohexenones, 3-amino-
  30, 493
  α-diazo(sulfonylamidines)
  29, 310
  2,10-dioxabicyclo[4.4.0]deca-
  3,8-dienes, 1-amino- **27**, 703
  fulvenes, 3-amino- **29**, 656
  4-pyrones (from 2 molecules)
  27, 693
  quinolines, 4-amino- **27**, 699;
  28, 871; **29**, 647
  1,4-thiaphosph(V)orin
  4-sulfides, 2,6-diamino-
  29, 613
  thiete 1,1-dioxides, 3-amino-
  29, 740
Ynammonium salts
- special s.
  diynammonium salts
Ynazo compds.
- from
  diazonium salts **30**, 238
Ynephosphine sulfides
- startg. m. f.
  enephosphine sulfides,
  synthesis **26**, 715
**Yttrium vanadate phosphor
  29**, 949

**Zecher-Kröhnke pyridine ring
  synthesis 18**, 859
Zeolites 2, 387 suppl. **29;
  27**, 715
  Y-Zeolite(Zn²⁺) **28**, 283

Zerovalent compds. s. Metal
  compds., zerovalent
Zetzsche s. Rosenmund
Ziegler ring closure 18, 819;
  21, 775
Zinc CC⇑Hal; **11**, 924 suppl. **29;
  16**, 940 suppl. **26; 17**, 933 suppl.
  29; 26, 28, 672/3, 798, 876,
  985; **27**, 24, 34, 51, 71, 90, 476,
  521, 670, 860, 931; **28, 28**, 37,
  48/9, 58, 115, 231, 345, 481, 566,
  608, 724, 815-9, 966; **29**, 4, 75,
  142, 274, 475, 638, 792, 826,
  928, 974; **30**, 28, 38/9, 561, 599
–, **liq. 19**, 555 suppl. **27**
–/**copper 26**, 880; **27**, 932; **28**,167;
  29, 5, 827/8; **30**, 39, 600, 620
–/**silver 14**, 858 suppl. **27, 28;
  29**, 829; **30**, 39
- **acetate 25**, 193 suppl. **27;
  26**, 985; **29**, 85
- **amalgam 26**, 803; **28**, 48;
  29, 45
- **bromide 17**, 611 suppl. **28**
- **carbonate 26**, 878
- **chloride 23**, 739 suppl. **27;
  25**, 131 suppl. **26; 26**, 223, 508,
  800; **26**, 446 suppl. **27, 29; 27**,
  190, 572, 646, 787, 819, 933;
  27, 877 suppl. **29; 28**, 63, 378,
  743, 844; **29**, 5, 516, 607, 894,
  921; **30**, 494, 554
- as chelating agent **29**, 894
- **compds., organo-**
  diethylzinc as reagent
  14, 858 suppl. **28**
- **cyanide** as reactant **28**, 708
- **halides** (s. a. Zinc chloride)
  26, 687
- **iodide 28**, 610
- **nitrate 6**, 386 suppl. **26**
- **salt** s. under Zeolites
Zincke reaction, pyridinium
  salts by – **20**, 369
Zirconium salts as Beckmann
  rearrangement catalysts
  7, 224 suppl. **28**
Zirconocene chloride
  as reactant **30**, 603
Zwitterions s. Betaines

Deutscher Schlüssel zum Register (Index)

Abbau
  Degradation
Abspaltung
  Elimination
Acylessigester
  Acylacetic esters
Acylhydrochinonäther
  Acylquinol ethers
Äther
  Ethers
Äthyl-
  Ethyl-
Alkoholate
  Alkoxides
Alkohole
  Alcohols
Alkylideneacetessigester
  Alkylideneacetoacetic esters
Allylumlagerung
  Allyl rearrangement
Ameisensäure
  Formic acid
Amidosulfonsäure
  Sulfamic acid
Anhydrozucker
  Anhydrosugars
Anlagerung
  Addition
Arylierung
  Arylation
Asparaginsäuren
  Aspartic acids
Aufbau
  Synthesis
Austausch
  Replacement
Benzopersäure
  Peroxybenzoic acid
Bernsteinsäure
  Succinic acid

Bimsstein
  Pumice
Blei
  Lead
Blitzthermolyse
  Flash vacuum pyrolysis
Bor
  Boron
Borsäure
  Boric acid
Borsäureanhydrid
  Boron trioxide
Brenzcatechin
  Pyrocatechol
Brenztraubensäure
  Pyruvic acid
Brom
  Bromine
Brom-
  Bromo-
Bromcyan
  Cyanogen bromide
Bromjod
  Iodine bromide
Carbäthoxylierung
  Carbethoxylation
Carbonsäure-
  Carboxylic acid-
Chlor
  Chlorine
Chlor-
  Chloro-
Chloressigsäure
  Chloroacetic acid
Chlorjod
  Iodine monochloride
Chin-
  Quin-
Cumarin
  Coumarin

Dehydratisierung
  Dehydration
Dehydrierung
  Dehydrogenation
Diazokupplung
  Diazo coupling
Dirhodan
  Thiocyanogen
Dithiocarbonsäuren
  Carbodithioic acids
Einführung von funktionellen Gruppen
  Replacement of hydrogen
Eisen
  Iron
Erdalkalien
  Earths, alkaline
Erden, seltene
  Earths, rare
Fluorwasserstoffsäure
  Hydrogen fluoride
Gallensäurederivate
  Bile acid derivatives
Glycerin
  Glycerol
Halbacetale
  Hemiacetals
Halogenide
  Halides
Halogenwasserstoff
  Hydrogen halide
Harnsäuren
  Uric acids
Harnstoff
  Urea
Holzkohle
  Charcoal
Hydrierung
  Hydrogenation
Hypohalogenite
  Hypohalites

Isothioharnstoffe
   Isothioureas
Jod
   Iodine
Jod-
   Iodo-
Jodwasserstoffsäure
   Hydrogen iodide
Kalium
   Potassium
Kern
   Nucleus
Kettenverlängerung
   Chain lengthening
Kohlehydrate
   Carbohydrates
Kohlendioxyd
   Carbon dioxide
Kohlensäure
   Carbonic acid
Kohlenwasserstoffe
   Hydrocarbons
Kugelmühlenreaktor
   Ball mill reactor
Kunstharz-Base
   Resin base
Kupfer
   Copper
Kupplung
   Coupling
Mangan
   Manganese
Molekülverbindungen
   Molecular compounds
Natrium
   Sodium
Natriumäthylat
   Sodium ethoxide
Natronkalk
   Soda lime
Peressigsäure
   Peroxyacetic acid
Phenyljodidchlorid
   Iodobenzene
     dichloride
Polymerisierung
   Polymerization
Quadratsäure
   Squaric acid
Quecksilber
   Mercury

Razemate
   Racemates
Reduktions-
   Tropfverfahren
   Dropping reduction
Rhodan
   Thiocyanogen
Rhodanid
   Thiocyanate
Ringerweiterung
   Ring expansion
Ringöffnung
   Ring opening
Ringschluß
   Ring closure
Ringverengung
   Ring contraction
Salpetrigsäureester
   Nitrous acid esters
Salpetersäure
   Nitric acid
Schutz
   Protection
Schwefel
   Sulfur
Schwefelchloride
   Sulfur chlorides
Schwefelwasserstoff
   Hydrogen sulfide
Seitenketten
   Side chains
Senföle
   Isothiocyanates
Sensibilisator
   Sensitizer
Silicium
   Silicon
Siliciumdioxid
   Silica
Spaltung
   Cleavage
Stellungswechsel
   Position shift
Stickoxydul
   Nitrous oxide
Stickstoff
   Nitrogen
Stickstoffwasserstoff-
   säure
   Hydrazoic acid

Sulfaminsäuren
   Sulfamic acids
Thiohalogenide
   Sulfenylhalides
Thioharnstoff
   Thiourea
Thiokohlensäure-
   disulfide
   Disulfur-
     dicarbothionates
Tonerde
   Alumina
Tschitschibabin
   Chichibabin
Tschugaeff
   Chugaeff
Umacylierung
   Transacylation
Umesterung
   Transesterification
Umkehrung
   Inversion
Umlagerung
   Rearrangement
Unterphosphorige
   Säure
   Hypophosphorous
     acid
Verbindungen
   Compds.
Veresterung
   Esterification
Verseifung
   Hydrolysis
Wanderung
   Migration
Wasserstoffperoxid
   Hydrogen
     peroxide
Weinsäure
   Tartaric acid
Wismutverbindungen
   Bismuth compounds
Wolfram
   Tungsten
Wolframsäure
   Tungstic acid
Zinn
   Tin
Zuckersäuren
   Saccharic acids

Systematic Survey Vol. 30
Systematische Übersicht Bd. 30

Reaction symbol	Volume 30 Page				
HO⇓OO	1	HC⇓CC	31	ORem⇓OC	70
HO⇓OC	1	HC⇅O	38	ORem⇓Rem	70
HO⇓CC	2	HC⇅N	45	ORem∩ON	71
HO∩OO	2	HC⇅Hal	47	ORem∩RemC	71
HO∩OC	2	HC⇅S	51	ORem⇅H	71
HO⇅N	3	HC⇅Rem	53	ORem⇅O	72
HO⇅Hal	3	HC⇅C	54	ORem⇅N	73
HO⇅S	3	HC⇑O	56	ORem⇅Hal	74
HO⇅Rem	4	HC⇑N	58	ORem⇅S	75
HO⇅C	5	HC⇑S	58	ORem⇅Rem	76
HO⇑O	11	HC⇑C	59	ORem⇅C	76
HO⇑N	11	OO⇓OC	60	OC⇓HO	76
HO⇑C	11	OO⇅H	61	OC⇓HC	77
HN⇓N	12	OO⇑H	61	OC⇓OO	79
HN⇓NN	12	ON⇓N	61	OC⇓ON	79
HN⇓NC	12	ON∩HN	62	OC⇓OS	79
HN∩HO	12	ON∩ON	63	OC⇓OC	80
HN∩HC	13	ON∩NC	63	OC⇓NC	82
HN∩ON	13	ON⇅H	63	OC⇓SC	85
HN∩NN	13	ON⇅O	63	OC⇓Rem	86
HN⇅O	13	ON⇅C	64	OC⇓CC	86
HN⇅N	15	ON⇑H	64	OC∩HO	97
HN⇅Hal	15	ON⇑O	64	OC∩HC	99
HN⇅S	15	ON⇑N	64	OC∩ON	100
HN⇅Rem	16	OHal⇅H	65	OC∩OS	102
HN⇅C	17	OHal⇅Hal	65	OC∩OC	102
HN⇑O	22	OS⇓HO	65	OC∩NC	103
HN⇑N	22	OS⇓HS	65	OC∩SC	103
HN⇑C	22	OS⇓S	65	OC∩CC	103
HS⇓SS	23	OS∩HO	67	OC⇅H	106
HS⇅S	23	OS∩ON	67	OC⇅O	110
HS⇅Rem	23	OS⇅H	67	OC⇅N	117
HS⇅C	23	OS⇅O	67	OC⇅Hal	123
HRem⇅Hal	24	OS⇅N	68	OC⇅S	132
HC⇓ON	24	OS⇅Hal	68	OC⇅Rem	137
HC⇓OC	25	OS⇅C	69	OC⇅C	138
HC⇓NC	28	OS⇑C	69	OC⇑H	146
HC⇓SC	31	ORem⇓HO	69	OC⇑O	151
		ORem⇓HRem	70	OC⇑N	157
		ORem⇓OO	70	OC⇑Hal	158
				OC⇑S	160

Reaction symbol	Volume 30 Page				
OC↑Rem	161	NC⇓NC	188	SS↕O	310
OC↑C	162	NC⇓SC	193	SS↕N	310
NN⇓N	167	NC⇓CC	194	SS↕Hal	311
NN∩HN	167	NC∩HN	197	SS↕S	311
NN∩ON	168	NC∩HC	198	SS↕Rem	312
NN↕H	168	NC∩ON	198	SS↕C	312
NN↕O	170	NC∩OC	199	SS↑H	312
NN↕N	170	NC∩NN	200	SS↑C	312
NN↕Hal	171	NC∩NC	201	SRem⇓SC	312
NN↕S	171	NC∩SC	202	SRem↕H	313
NN↕Rem	171	NC∩RemC	202	SRem↕O	313
NN↕C	171	NC∩CC	202	SRem↕N	313
NN↑H	172	NC∩Het	204	SRem↕Hal	314
NN↑O	173	NC↕H	204	SC⇓HC	314
NN↑N	173	NC↕O	214	SC⇓OC	314
NN↑C	173	NC↕N	236	SC⇓NC	315
NHal↕H	174	NC↕Hal	243	SC⇓SS	316
NHal↕O	175	NC↕S	254	SC⇓SC	317
NS⇓HN	175	NC↕Rem	257	SC⇓RemC	317
NS⇓NN	175	NC↕C	260	SC⇓C	318
NS⇓NHal	175	NC↑H	264	SC⇓CC	318
NS⇓NC	176	NC↑O	267	SC∩HO	320
NS⇓S	176	NC↑N	275	SC∩HC	321
NS∩SC	176	NC↑Hal	277	SC∩OC	321
NS↕H	176	NC↑S	279	SC∩NC	321
NS↕O	178	NC↑Rem	281	SC∩HalC	321
NS↕N	178	NC↑C	281	SC∩SS	322
NS↕Hal	178	HalHal⇓Hal	284	SC∩SC	322
NS↕S	179	HalS↕H	285	SC∩CC	322
NS↕Rem	179	HalS↕Hal	285	SC↕H	322
NS↕C	179	HalRem⇓Rem	285	SC↕O	324
NS↑H	180	HalRem↕O	285	SC↕N	329
NRem⇓NN	180	HalRem↕Rem	286	SC↕Hal	331
NRem⇓NC	180	HalC⇓ON	286	SC↕S	337
NRem⇓Rem	180	HalC⇓OC	286	SC↕Rem	339
NRem↕H	181	HalC⇓NC	287	SC↕C	339
NRem↕O	181	HalC⇓SC	288	SC↑O	341
NRem↕N	181	HalC⇓CC	288	SC↑N	342
NRem↕Hal	182	HalC∩HC	291	SC↑Hal	342
NRem↕S	182	HalC∩NC	291	SC↑S	342
NRem↕C	183	HalC∩HalRem	292	SC↑C	343
NC⇓HO	183	HalC∩CC	292	RemRem↕H	344
NC⇓HN	183	HalC↕H	292	RemRem↕Hal	344
NC⇓HC	184	HalC↕O	297	RemRem↑Rem	344
NC⇓ON	184	HalC↕N	305	RemC⇓ORem	345
NC⇓OC	184	HalC↕Hal	306	RemC⇓OC	345
NC⇓NN	187	HalC↕S	307	RemC⇓NC	345
NC⇓NS	187	HalC↕Rem	307	RemC⇓HalC	345
NC⇓NRem	187	HalC↕C	308	RemC⇓SC	346
		SS⇓SS	309	RemC⇓RemRem	346
		SS⇓SC	309	RemC⇓RemC	346
		SS↕H	310	RemC⇓C	347

Reaction symbol	Volume 30 Page				
RemC⇓CC	347	CC⇓CC	373	CC⇑S	524
RemC∩ORem	349	CC∩HO	394	CC⇑Rem	528
RemC↕H	349	CC∩HC	394	CC⇑C	529
RemC↕O	351	CC∩ON	401	ElN⇑H	534
RemC↕N	351	CC∩OC	401	ElN⇑O	535
RemC↕Hal	352	CC∩NC	408	ElN⇑N	536
RemC↕S	354	CC∩HalC	411	ElN⇑C	536
RemC↕Rem	355	CC∩SC	411	ElS⇑O	537
RemC↕C	355	CC∩RemC	412	ElS⇑N	538
RemC⇑H	356	CC∩CC	413	ElRem⇑S	539
RemC⇑Hal	356	CC↕H	414	ElRem⇑C	539
RemC⇑C	357	CC↕O	419	Het⇓N	539
CC⇓HC	357	CC↕N	440	Het⇓S	541
CC⇓OC	358	CC↕Hal	447	Het⇓Rem	541
CC⇓NC	367	CC↕S	470	Het∩ON	542
CC↕SC	372	CC↕Rem	474	Het↕	542
CC⇓RemC	373	CC↕C	483	Het⇑H	542
CC⇓C	373	CC⇑H	489	Het⇑N	542
		CC⇑O	497	Het⇑El	542
		CC⇑N	511	HetN	543
		CC⇑Hal	516	Oth	543

Supplementary References* in Vol. 26—30
Ergänzungszitate* in Bd. 26—30

No.	Suppl. Ref. Vol., Page
Volume 1	
23	28, 10
42	27, 20
69	26, 37
122	26, 56
132	26, 66
133	29, 76
151	29, 93
177	26, 95
191	28, 88
198	27, 113
205	26, 107
217	26, 110
218–20	29, 117
221	26, 111
255	29, 11
278	27, 155
290	29, 7
307	27, 181
321	27, 191
332/3	26, 196; 28, 182
363	28, 143; 29, 162
376	26, 217
397	26, 441
427	26, 267
429	26, 271
474	28, 275
522	26, 330
535	27, 320
539	26, 366 (2)
540	26, 366
541	26, 366 (2)
554	26, 445;
	28, 434; 29, 76
568	27, 352; 28, 373
579	28, 363; 29, 357
581	28, 363
586	28, 364
587	29, 358
591	26, 391
596	28, 372
603	29, 364 (2)
607	26, 139
610	27, 50; 28, 46
615	26, 176; 30, 438
668	29, 452
681	28, 297
712	26, 419
715	28, 421
749	27, 422
752	29, 458
783	29, 95
785	26, 474
Volume 2	
19	26, 5
20	26, 4
38	26, 12; 27, 9
40	26, 11
54	30, 340
78	28, 29
130	29, 58
144	26, 66; 29, 74
151	26, 69
180a	26, 83
184	26, 88
199	29, 104
202	26, 95
205	26, 99
208	29, 109
235	27, 101
259	28, 96
275	28, 103
277	29, 114
286	27, 126
288	28, 116; 29, 135
293	27, 128
296	26, 134
302	28, 131
304	27, 142
324	26, 156
329	26, 161; 27, 161
330	26, 161
334	26, 164
335	26, 166
341	27, 172; 28, 164
353	26, 175
379	26, 188
387	29, 191
392	26, 195; 29, 193
439	29, 211
459	27, 215
475	27, 246
478	27, 211
506	29, 254
518	26, 98
608	29, 21
620	26, 357
621	26, 358; 27, 333; 28, 332
642	26, 374
648	27, 46, 349 (2)
651	27, 348; 29, 351
677	26, 392
680	26, 391
702	27, 369

* In Vol. 1–5 s. Vol. 5, p. 607; in Vol. 6–10 s. Vol. 10, p. 734; in Vol. 11–15 s. Vol. 15, pp. 664; in Vol. 16–20 s. Vol. 20, p. 723; in Vol. 21–25 s. Vol. 25, p. 685.

No.	Suppl. Ref. Vol., Page
725a	26, 176
741	26, 468
758	26, 440
778	29, 441; 30, 222
782	27, 419
816	29, 447
827	26, 326
842	27, 446; 28, 476
882	26, 467

Volume 3

No.	Suppl. Ref.
56	26, 42
57	26, 42; 27, 37; 28, 35
68	29, 50
91	28, 46
97	26, 54
108	27, 64
134	26, 77
141	26, 84
170	26, 131, 297; 28, 122
177	28, 87
196	26, 105
224	29, 119
226	26, 94; 28, 81; 29, 101
227	27, 113
238	28, 111
254	26, 141
290	26, 172
323	29, 187
324	26, 188
330	28, 181
349	27, 202
368	28, 200
375	27, 172; 28, 168
378	29, 212
397	28, 213
420	28, 234
428	27, 252
430	26, 263
442	27, 256
445	26, 265; 29, 252
447	28, 246
451	29, 252
463	28, 250
482	26, 274
485	27, 265; 28, 253
490	27, 259
499	29, 429
514	26, 297; 27, 280
545	26, 315
549	29, 298
554	26, 322
557	26, 324
571	26, 330; 27, 306; 28, 302
582	26, 343
583	28, 317
588	26, 350
608	26, 393
632	26, 388; 29, 355
640	28, 362
662	26, 119
667	29, 373
670	29, 377
679	26, 410
697	29, 395
703	26, 438
705	28, 400
708	29, 426
714	26, 444
715	27, 410
724	28, 435
727	26, 450
729	29, 432
733	29, 458
745	26, 462
752	27, 432; 29, 450
757	26, 467
758	26, 472; 27, 438

Volume 4

No.	Suppl. Ref.
19	27, 7
25	28, 11
26	26, 12
49	26, 21
68	26, 27
100	26, 41
118	27, 44
144	26, 57, 58
149	26, 64
151	28, 56
154	26, 68; 28, 58
182	26, 85
183/4	27, 86
185	26, 87; 29, 97
187	27, 89
189	29, 100
190	27, 93
208	27, 100
209	26, 100
219	26, 101
240	26, 110; 29, 116
249	28, 99
255	28, 106
255 suppl. 28	29, 126
265	28, 111
267	27, 124
276	26, 135
282/3	26, 137
293	26, 141
295	26, 143
309	27, 156; 28, 144
316	26, 159
326	26, 163; 27, 164
357	26, 180
362	26, 182
363	26, 184
368	26, 188
374	28, 180
388	27, 201
395	28, 187
398	26, 201
404	26, 202
408	27, 204
413	26, 204
432	27, 215
438	26, 217
455	26, 235
492	26, 265
499	26, 269
521	29, 262
535	26, 285; 28, 264
539	26, 286; 29, 268
543	28, 265
552	26, 293
556	26, 297
561	28, 274
566	26, 302
596	26, 315; 29, 297
597	26, 315; 29, 298
615	26, 319
622	26, 322
623	29, 300
625	26, 324
635a	26, 328

No.	Suppl. Ref. Vol., Page
Volume 4	continued
643	26, 332
650	27, 310
661	28, 322
667	26, 349; 27, 326; 28, 325; 29, 328
669	28, 330
674	26, 360; 27, 334; 29, 336
677	27, 336
692	26, 371
699	29, 350
714	26, 380
719	26, 381
733	26, 391
742	29, 364
761	27, 369
775	27, 379
799	26, 433
800	26, 437
810	26, 442
818	26, 444
832	28, 443
835	26, 456
840	29, 440
843	29, 445
Volume 5	
45	29, 30
47	28, 25
95	28, 94
97	26, 53; 28, 48
100	27, 56
125	28, 59
130	27, 73
136	26, 75
150	26, 82; 29, 93
154	26, 84; 29, 94, 140
161	26, 90
162	26, 90; 27, 91
166	26, 94
177	29, 109
194	28, 93; 29, 114
201	28, 97
209	28, 101
214	26, 119; 29, 126
245	29, 163
260	26, 164
269	26, 444
279	26, 178; 28, 171
288	26, 185
299	26, 194
319	26, 204
326	26, 209
346	26, 221
350	26, 203; 27, 204; 28, 190; 29, 203
353	26, 230; 27, 228
355	26, 444; 27, 232, 410; 28, 218
398	26, 266
440	28, 276
441	26, 302
444	26, 303
447	28, 278
467	29, 300
473	26, 325
476	26, 325
478	27, 302; 29, 305
479	26, 327; 28, 298
485	26, 332
487	26, 334
488	28, 307; 29, 314
490	26, 338
503	27, 323
508	26, 352; 29, 330
518	28, 330; 29, 332
522	28, 337
523	26, 361
525 suppl. *10*	26, 362
528	29, 342
533	27, 346
545	26, 382
557	28, 370
577	26, 402
587	26, 408
593	26, 403; 29, 387
594	26, 413
596	26, 413
604	26, 419; 27, 387; 28, 400; 29, 395
607	26, 419
610	26, 440
614	26, 444
623	27, 421
642	26, 456; 27, 424
643	26, 458 (2); 27, 426; 29, 443
644	26, 460
650	27, 432
652	26, 467; 27, 433
664	29, 465
666	26, 478; 27, 443; 28, 475; 29, 466
666 suppl. *28*	29, 466
Volume 6	
7	29, 2
43	29, 17
44	26, 17
51	26, 19
52	26, 20
89	29, 33
105	29, 42
121	26, 45
142	26, 50
156	26, 63
158	26, 64
170	27, 66
180	28, 64
186	26, 77
192	26, 78
193	26, 83
196	29, 140
197	26, 86
209	26, 96; 28, 84; 29, 105
211	28, 84; 29, 105
215	26, 99
220	27, 98; 29, 106
227	29, 110
231	26, 104
241	29, 113
249	26, 109
251	27, 110
317	29, 147
325	28, 132
328	29, 150
330	27, 143
341	26, 152
343	29, 163 (2)
348	28, 145

No.	Suppl. Ref. Vol., Page
358	28, 150
366	26, 161; 28, 151
367	28, 152
370	26, 218
386	26, 173
388	26, 175
390	29, 182
408	26, 180
420	26, 192
452	27, 211
460	27, 213
470	28, 203
478	26, 219
479 suppl. 24	28, 204
510	27, 234
519	26, 171
522	28, 225
524	26, 242
531	26, 252
537/8	26, 254
548	28, 236
571	28, 245; 29, 250
572	26, 264; 28, 245; 29, 251
586/7	26, 271
589	26, 272
591	28, 252
633	26, 295
643	26, 297; 28, 272
656	26, 303
663	29, 286
692	27, 308
696	26, 336
705	28, 317
722	26, 362
731	26, 266; 29, 49, 252
738	27, 346
740	26, 376
755	29, 354
757	26, 68
764	28, 297
765	26, 382
769	29, 357
772	26, 388
783	29, 364
789	26, 397
794	26, 402
813	27, 386
814	27, 387
819	28, 400
820	27, 9
831	26, 438; 27, 405
840	28, 430
841	26, 443; 27, 424; 29, 442
854	27, 360
882	26, 466
883	28, 455
884	26, 198, 466; 27, 200
885	27, 433
889	26, 457; 27, 425
898	27, 382; 29, 390
903	29, 463

Volume 7

No.	Suppl. Ref.
3	28, 1
5	28, 4
6	28, 2
10	29, 9
13	29, 3
19	29, 5
28	26, 26
35	26, 12
38	27, 10
45	29, 12
61	29, 19
73	26, 22
77	28, 37
80	27, 26
89	28, 28
91	28, 29
95	26, 35; 29, 36
107	27, 28
147	27, 46
155	27, 50
158	29, 57
159	27, 51
162	26, 55
174	26, 60
181	28, 56; 29, 74
191	26, 100
201	29, 82
207	26, 77
213	29, 88
214	28, 70
217	28, 71
220	29, 93
224	29, 95
228	26, 87
233	26, 91
241	29, 100
249	26, 100
251	28, 88
265	26, 104
286	26, 108
306	28, 99
327	26, 123
331	28, 113; 29, 133
332	29, 26
335	26, 127
336	27, 128
339	26, 381
355	28, 122
369	26, 146; 28, 136
384	27, 155
420	26, 176; 27, 181
434	29, 187
435	26, 184
454	26, 198
459	27, 203
465	26, 204
508	26, 227
514	26, 233; 28, 217
546	30, 490
562/3	27, 256
584	26, 270
589	29, 259
593	26, 275
617	27, 277; 28, 269; 29, 272
623	26, 295
635	27, 285
659	28, 54
670	26, 316
678	26, 319; 28, 293
679	27, 299; 28, 299
680	26, 321; 29, 299
697	27, 303
711	28, 308
730	26, 349; 27, 325
732	26, 354
737	28, 331
738	28, 38
753	28, 338
767	28, 352
769	26, 376; 27, 348
771	28, 354
773	27, 350
798	29, 358

No.	Suppl. Ref. Vol., Page
Volume 7	continued
805	29, 362
811	26, 461
817	26, 395
830	27, 379
833	29, 386
836	26, 412; 27, 380; 28, 392; 29, 387
846	26, 416
858	27, 404
862	28, 424
863	26, 439
864	27, 406
871	29, 429
883	27, 423
914	28, 473

Volume 8	
9	26, 3
18	27, 258
28	29, 14
54	28, 23
59	27, 26
64	29, 30
65	27, 329
66	28, 26
69	26, 33
85	28, 31
87	28, 25; 29, 31
88	27, 35; 29, 40
94	26, 38
102	28, 36
125	26, 48
132	26, 48; 28, 43
139	26, 55
143	27, 248
146	27, 60; 29, 65
157	29, 74
171	26, 154
172	26, 73; 27, 73; 29, 80
174	26, 76
177	26, 77
191	29, 93
192	26, 83
193	27, 85

206	26, 92; 28, 80
208	26, 93
212	27, 95
244	28, 92
254/5	29, 114
256	28, 94
274	26, 109
275	29, 116
281	26, 111
302	27, 119
303	26, 123
312	28, 111
326	26, 128; 27, 129
370	26, 144; 29, 155
372	26, 147
387	26, 153
394	29, 165
404	26, 161; 29, 168
408	26, 167
413	28, 163
434	26, 176; 28, 170
441	27, 182
460	28, 175
469	26, 194
479	26, 200
480	28, 188
482	26, 200
486	26, 201
489	26, 203
493	28, 193
503	27, 219
515	27, 182
517	28, 205
523	28, 206
531	29, 218
532	28, 208
550	29, 183
562	26, 238
594	27, 266; 28, 287
597	26, 262; 28, 241
609	27, 260
614	29, 255
615	26, 255
616	26, 270; 27, 262
620	27, 264
634	27, 267
641	28, 260
655	26, 291
657	26, 293
662	26, 296; 29, 273
676	28, 275
681	26, 302;

	28, 276; 29, 282
694	26, 309
713	26, 323
733	28, 305
736	28, 310
742	27, 321
743	27, 320
745	26, 348; 29, 327
748	26, 349; 29, 328
753	26, 355
782	26, 373
787	26, 379
792	27, 376
795	26, 33
810	27, 356
814	28, 217
819	26, 391
822	28, 372
823	26, 393
824	26, 372; 28, 350
826	29, 364
828	26, 395, 396
830	26, 397
842	29, 300, 376
854	28, 394; 29, 389
857	28, 395
864	29, 398
888	26, 412
899	26, 454; 27, 421; 28, 441
903	28, 364
911	27, 423
917	26, 457
922	26, 458; 27, 427; 28, 448
926	26, 462
927	26, 463
933	26, 466
942	26, 470
951	29, 463; 30, 535

Volume 9	
29	26, 7; 29, 9
37	26, 12
45	26, 216, 245
51	27, 16
69 suppl.	15 29, 24
72	29, 26
77	27, 24

No.	Suppl. Ref. Vol., Page
86	26, 136
93	26, 34
94	26, 35
97	29, 37
99	29, 37
113	26, 38; 27, 181
114	28, 189
122	26, 42
129	26, 42
135	28, 40
139	28, 41; 29, 50
149	27, 45
169	26, 54; 27, 53; 29, 58
176	26, 57
196	26, 65
209	26, 68
213	27, 70
243	26, 53; 28, 74; 29, 94
244	26, 85
248	29, 96
249	26, 88
256	28, 300
261	26, 96
265	29, 108
267	28, 86
291	27, 103
296	29, 112
306	28, 266; 29, 270
310	27, 107; 28, 94; 29, 115
323	26, 112
329	26, 114
337	26, 118; 29, 124
342	29, 127
344	27, 99; 28, 86; 29, 107
345	26, 122
347	28, 109
352	27, 23
355	27, 125
357	28, 113
405	26, 158
415	28, 152
422	28, 163
428	26, 173
437	26, 190
447	26, 186
453	26, 190
456	26, 191; 28, 179
461	29, 194
462	26, 196
485	26, 204
491	28, 193
510	26, 212
526	26, 217
545	27, 226
581	26, 249
598	28, 236
608	29, 246
615	26, 263
644	26, 277
659	26, 292; 29, 272
660	26, 292
666	26, 294
669	26, 295
676	26, 297
677	26, 297 (2)
710	26, 311
717	26, 315
729	26, 319
741	29, 305
749	27, 305
768	27, 319
770	26, 345
781	27, 330
793	28, 335
809	28, 352; 29, 349
823	28, 295
844	27, 356
848	26, 385
850/1	26, 393
860	26, 7
866	29, 370
871	28, 377, 380; 29, 373
881	26, 406
887	26, 410
893	26, 413
898	26, 413
902	29, 390
917	26, 439
919	26, 439
934	26, 449; 29, 430
935	26, 449
938	27, 416
939	27, 417
951	27, 423
968	26, 465
988	29, 466

Volume 10

No.	Suppl. Ref. Vol., Page
11	29, 7
15	29, 8
19	28, 8
35	26, 17
62	27, 352
93	26, 44
101	26, 47
107	26, 49; 27, 49; 28, 44
111	26, 51
113	28, 48
132	26, 68
135	27, 69
151	26, 81
157	27, 92
158	27, 93
170	29, 107
171	29, 109
174	28, 88
210	27, 119
213	27, 120; 29, 126
227	28, 115
239	26, 132
244	26, 136
250	29, 158
255	29, 153
262	28, 144
274	26, 161
275	26, 26
280	29, 170
282	29, 171
286	29, 177
289	27, 66
292	29, 181
296	27, 181
298	26, 209
302	28, 173
315 suppl. 21	28, 181
318	26, 194
328	29, 197
329	29, 198
347	28, 196
354	27, 171
360	28, 201
363	27, 215
386	28, 217
396	26, 241
399	26, 242

No.	Suppl. Ref. Vol., Page
Volume 10 continued	
400	26, 245
421	26, 61; 27, 61
429	26, 264; 29, 250
451	28, 99
452	27, 265
473	26, 287
480	29, 62
489	30, 455
495	27, 292; 28, 285; 29, 289
517	26, 324
522	26, 325
525	26, 328; 27, 305
526	26, 328
561	29, 308
566	29, 345
581	26, 166
583	27, 114
587	28, 363; 29, 356
588	26, 383
594	26, 389; 27, 359
607	29, 369
615	29, 379
616	26, 410
620	29, 52
629	29, 45
639	26, 440; 28, 427
645	26, 446
650	29, 431
665	27, 421
667	26, 440
680	26, 264
686	28, 456
701	26, 478
Volume 11	
2	28, 2
8	26, 3
12	26, 3; 27, 5
17a	29, 6
19	26, 6; 27, 7
41	26, 57
45	27, 16
59	26, 22 (2)
62	27, 23
68	29, 29
69	26, 141
81	29, 30
83	28, 25
84	27, 28
87	27, 29
95	29, 34
114	27, 35
118	28, 34; 29, 40
119	29, 40
128	26, 42
135	29, 48
144	26, 100
149	29, 43
162	26, 56
173	26, 135
194	28, 61
196	29, 81
202	28, 70
209	29, 86
224	27, 89; 29, 98
232	26, 97; 28, 84; 29, 106
233	29, 106
235	26, 98
267	27, 107
293	26, 113; 27, 112; 29, 117
308	27, 123; 29, 128
314	27, 125
327	28, 120
329	26, 129; 29, 139
333	26, 130; 29, 140
348	27, 142
385	29, 167
402	29, 177
404	26, 170; 29, 177
409	27, 16
422	29, 183
423	26, 176
425	26, 178; 27, 182
456	27, 194
457	27, 131
485	27, 205; 28, 191; 29, 204
510	27, 216
537	26, 225; 28, 208; 29, 219
564	26, 139
576	26, 160
578	29, 238
583	26, 253
601	26, 261
609	27, 257; 28, 243
612	26, 264
616	26, 265; 27, 259
619	28, 247
627	27, 264
628	29, 259
633	29, 57
644	28, 257
661	26, 293
665	28, 270; 29, 50
666	27, 278
668	26, 296
689	28, 276
712	29, 292
728	26, 322
730	29, 300
731	27, 300; 28, 295
750	29, 308
768	26, 449; 29, 325
774	29, 327
778	28, 338
779	26, 397; 28, 376
795	28, 339; 29, 339
799	28, 341
803	27, 432
806	26, 371
807	30, 148
811	26, 374
814	26, 376
817	26, 376
824	26, 475
827	26, 80; 28, 360
834	29, 356
841	29, 358
843	26, 386
847	26, 391; 27, 362
858	26, 398
862	27, 367
863	29, 372
864	26, 399
869	28, 383
872	27, 374
880	26, 408
881	29, 383
901	26, 419
920	29, 426
924	29, 431
927	26, 452
933	28, 441
941	26, 271; 28, 250
945	29, 444
956	27, 433
963	28, 467

No.	Suppl. Ref. Vol., Page
Volume 12	
4	26, 2
7	26, 2
31	26, 10
43	29, 14
52	26, 13
56	29, 20
60	27, 22; 29, 26
61	28, 20
64	26, 22; 29, 23
77	29, 136
90	29, 33
99	28, 30
103	29, 26
110	26, 37
124	26, 42
127	28, 38
151	28, 47
155	27, 54
156	29, 60
167	26, 63, 135; 29, 70
169	27, 110
180	29, 365
183	29, 75
189	26, 71; 27, 72
191	27, 77
191a	27, 74
194	26, 75
205	29, 108
206	26, 81; 27, 253; 28, 71
211	26, 84; 28, 73
212	27, 85
218	29, 100
236	28, 182
238	28, 85
249	26, 101
281	26, 109
287	28, 95
288	26, 110; 27, 110; 28, 94
292	28, 96
296	27, 112 (2)
303	29, 120
304	27, 115; 28, 100
305	26, 115; 29, 121, 125
315	27, 120, 125
319	26, 328
325	27, 124
335	28, 116
345	28, 123
355	26, 138
357	29, 149
359	27, 141
362	28, 130
375	26, 152
376	26, 152; 27, 270
379	27, 158
381	27, 158
382	26, 157
387	26, 158
389	29, 166
393	29, 167
400	26, 166; 28, 157
407	26, 168; 27, 168
417	29, 181
419	26, 176
427	26, 178; 29, 184
451	26, 190
454	29, 191
455	26, 194; 28, 181
463	26, 110; 27, 110
464	27, 198
465	26, 198, 296
483	27, 204
488	27, 206
497	29, 208
514	26, 212
519	26, 216; 27, 218; 28, 203
523	26, 263; 29, 249
546	26, 235
555	27, 237
556	26, 239
565	26, 246
569	27, 243
571	30, 45
585	26, 257; 28, 238; 29, 244
586	27, 253
587	26, 259
590	26, 261
599	26, 263
619	26, 274; 28, 252; 29, 259
625	29, 243
626	28, 236
629	29, 261
631	27, 266
639	28, 258
652	28, 265
653	27, 273
667	26, 298
688	27, 289
689	26, 306; 29, 285
700	26, 312; 28, 288; 29, 294
703	26, 315; 27, 296; 29, 298
707	26, 316; 29, 297
711	26, 318; 28, 292
715	28, 295
739	28, 320
743	26, 349; 28, 325
756	28, 332
760	26, 361
761	28, 336
787	28, 354
789	28, 354
816	28, 119
824	27, 364
828	27, 366
830	28, 376, 377; 30, 443
838	30, 226
848	29, 381
855	26, 413; 28, 394
862	26, 417
869	26, 302, 423
898	26, 454
900	27, 422
924	26, 462; 28, 452
929	26, 464
930	26, 465
945	26, 472; 28, 282
948	26, 475
950	28, 434
951	26, 475
957	29, 180
960	27, 77
962	28, 475
Volume 13	
3	26, 2
35	28, 12
43	26, 17; 29, 18
70	28, 22
88	28, 28
105	29, 41
106	28, 34; 29, 41

No.	Suppl. Ref. Vol., Page
Volume 13 continued	
119	27, 190
131	26, 47; 29, 52
148	26, 54
159	26, 61
175	27, 153
190	28, 57
196	27, 71
205	29, 88
207	26, 80; 27, 83
209	28, 238
215	27, 21
223	28, 77
224	26, 89; 28, 78
254	28, 90
255	26, 102; 27, 103
256	27, 103
263	26, 105
266	26, 106
271	27, 106
314	29, 136
318	27, 131; 29, 139
352	26, 148; 29, 158
355	26, 149; 28, 141
361	28, 144
363	26, 154; 28, 17, 146
364	26, 177; 29, 183
365	26, 156; 27, 159
371	29, 167
379	26, 163
381	27, 164
383	28, 217; 29, 225
387	27, 166
390	26, 168
408	29, 184
416	27, 186; 28, 174
422	26, 186
433	27, 191
434	26, 190
437	26, 191
441	29, 192
454	26, 200
473	29, 206
476	28, 197
482	26, 210
492	26, 215
496	26, 217
514	26, 227
515	26, 227; 29, 223
530	26, 236; 28, 220; 29, 227
535	26, 210; 27, 213; 29, 209
541	28, 226
551	26, 254
557	27, 250
573	29, 252
593	26, 214
595	27, 272
602	27, 280; 29, 276
627	26, 305
630	26, 307; 29, 286
637	28, 284
677	26, 335
681	28, 448
684	27, 320
695	26, 350; 27, 326
696	26, 347; 27, 322; 28, 322; 29, 326
699	26, 438; 27, 330
725	28, 353
727	27, 348
735	26, 378; 27, 350
765	26, 389
778	26, 470
784	29, 373
786	29, 110
789	26, 404
802	29, 379
804	26, 409
805	29, 385
811	28, 394
830	26, 428
831	26, 430; 27, 399; 28, 414; 29, 409
834	26, 436
844	26, 442
851	27, 85
870	26, 454, 456; 27, 421; 28, 441
875	26, 458
879	26, 101
882/3	28, 451
888	26, 465
890	28, 455
893	27, 434
896	28, 276
916	28, 476
Volume 14	
34	27, 9
36	26, 12
39	26, 12
42	26, 13
46	27, 11
65	26, 20; 27, 18 (2)
67	28, 19; 29, 21
79	26, 29
81	27, 28
82	26, 32; 28, 25
88	26, 27; 28, 24
90	29, 35
91	28, 25; 29, 31
94	26, 36
101	27, 36
107	29, 256
112	28, 39
116	27, 39
125	26, 100
132	28, 43
133	27, 46
137	29, 54
139	29, 47
140	27, 50
145	26, 53
146	28, 130
155	26, 61; 29, 68
157	27, 62; 29, 69
174	26, 67; 29, 75
188	28, 64
194	27, 77
197	26, 76; 29, 85
198	26, 77
205	28, 71
221	26, 94
231	26, 96; 29, 105
231 suppl.	26, 28, 84
240	29, 180
248	27, 101
252	27, 103
271	27, 107; 29, 115
279	26, 108; 27, 108
282	26, 99
284	28, 95
293	26, 114
329	29, 139
334	26, 130
345	27, 135

No.	Suppl. Ref. Vol., Page
352	26, 102
361	28, 131
363	27, 143
364	28, 132
380	28, 143(2)
391	26, 159
401	29, 174
404	28, 473
415	27, 180
417	26, 13
424	26, 179
427	27, 184
430	26, 184
443	26, 190
446	26, 191
450	27, 96; 28, 180
451	27, 195
452	26, 194; 28, 181
455	26, 195; 27, 356; 28, 174
455 suppl. 17	29, 193
459	26, 197
478	26, 139
512	29, 214
517	28, 205
530	26, 242; 28, 211
546	26, 236
564	26, 244; 29, 234
565	27, 242
571	29, 237
583	26, 255
592	26, 263
595	28, 241
598	29, 250
606	26, 267; 28, 247
608	26, 267
612	26, 269
614	26, 270
620	26, 273; 27, 264; 29, 258 (2)
623	29, 259
638	29, 265
658	28, 269
676	26, 302
709	26, 369; 29, 299
711/2	26, 324
713	26, 379
717	27, 302
719	26, 326
723	28, 301
728	27, 310
733	26, 338
735	28, 314
745	27, 105
753	26, 357
756	26, 361; 28, 335
764	26, 364; 29, 341
765	29, 342
768	29, 343
770	27, 406
773	26, 371; 28, 370; 29, 347
781	28, 391; 29, 387
782	26, 31, 37, 453
791	26, 381
836	28, 380
839	26, 403
858	26, 416; 28, 397; 29, 392
858 suppl. 15	27, 385
864	27, 384
865	28, 395
866	29, 432
876	28, 415
877	29, 411
877 suppl. 16	26, 446
888	28, 422
894	27, 409
895	26, 440; 29, 424
897	26, 440
900	26, 442; 28, 379
904	27, 432
914	29, 429
917	29, 431
920	26, 451
922	26, 452
936	26, 454
937	26, 355, 454; 27, 421
969	26, 466
976	26, 472

Volume 15

26	26, 16; 27, 15; 28, 15; 29, 17
37	27, 21
38	29, 41
56	26, 31; 27, 29; 28, 26
65	26, 34
66	27, 33
69	26, 312
83	26, 41
85	28, 35
94	26, 43
103	28, 41; 29, 380
103 suppl. 28	29, 380
107	28, 42
146	28, 8
154	27, 67; 29, 75, 85
170	26, 81; 28, 239
177	26, 85; 27, 86; 29, 96
179	26, 88; 27, 87
187	26, 227
190	29, 103
194	27, 98
201	26, 101
204	27, 102
208	26, 102
214	28, 93
219	26, 108
225	27, 109
233	29, 138
242	26, 119
247	29, 130
254	28, 115; 29, 134 (2)
261	27, 95
276	27, 136
281	27, 138; 28, 128
283	28, 129
301	29, 165
306	26, 26
310	26, 162
313	26, 164
315	26, 166
329	27, 183
342	26, 189
351	26, 194
360	28, 186
369	28, 193
373	26, 210
379	27, 215
387	29, 31
398	28, 213

No.	Suppl. Ref. Vol., Page
Volume 15 continued	
404	27, 362; 28, 371
410	28, 220
415	26, 240
421	26, 247
428	26, 260
434	27, 259; 29, 252
438	26, 267
441	29, 254
451	27, 243
457	28, 261
463	26, 292; 27, 277; 28, 268
465	29, 271
469	26, 296; 28, 271
470	26, 297
471	29, 271
473	26, 281
481	26, 302
521	29, 306
522	26, 327
528	26, 332; 27, 308; 29, 305
532	26, 335
534	27, 317
536	26, 34
539	27, 322
555	29, 336
573	26, 373
579	28, 83
582	26, 377; 29, 352
587	26, 380
590	27, 353
593	29, 356
599	26, 390
600	29, 362
607	27, 113
609	29, 367
612	29, 373
620	28, 383
624	28, 385
627	29, 378
634	28, 394
647	28, 421(2)
653	28, 431; 29, 427
656	26, 449
659	28, 437
671	27, 427
672	28, 340
679	29, 447
680	26, 465; 27, 431; 29, 449; 30, 519
684	26, 424; 28, 459
698	26, 480
Volume 16	
3	26, 5
11	29, 6
24	26, 44
52	29, 28
32	26, 17
54	29, 29
65	26, 30
104	29, 43
106	26, 42; 29, 43
110/1	26, 43
138	29, 275
151	28, 50
152	29, 78
167	26, 19
175	29, 72
180	26, 396
185	27, 70
194	29, 83 (2)
196	28, 66
198	27, 132
201	26, 5, 79; 28, 9
210	26, 83
219	29, 100
220	26, 92
225	29, 100
228	27, 94; 28, 82
234	26, 99; 27, 99; 29, 108
236	29, 105
242	28, 7
247	27, 105
258	27, 105
263a	29, 78
268	26, 109
278	27, 112
279	26, 115; 29, 119
282	27, 115
287	28, 105
308	29, 129
314	26, 124
315	28, 121
317	27, 128
319	26, 127
324	28, 327
325	28, 122
326	29, 436
329	29, 141
332	26, 131; 27, 196
336	26, 134
345	26, 137
350	27, 141
368	26, 147
380	29, 161
384	26, 153; 27, 131; 28, 144
393	28, 197
401	28, 202
410	29, 175
414	27, 177; 28, 197
416	28, 167
424	26, 177; 29, 183, 184
424 suppl. 20	27, 182
434	26, 211; 28, 200; 29, 210
436	26, 182
437	26, 221
442	26, 191; 29, 191
448	26, 194; 29, 38, 192
454	27, 196
458	28, 185
461	28, 186
464	27, 201
468	28, 191
471	29, 206
472	26, 206
478	26, 207
479	26, 236; 29, 227
488	29, 211, 214
489	29, 154
492	29, 213
499	27, 222
501	27, 222
503	26, 208
505	26, 221; 27, 223
514	28, 208
526	27, 239
530	27, 235; 28, 221
531	26, 237
543	27, 239
550	26, 257; 27, 251
557	29, 238
577	29, 245

No.	Suppl. Ref. Vol., Page					
586	28, 82	851	27, 372	73	26, 25
590	26, 264; 28, 245	853	29, 375	79	28, 25
596	26, 267; 29, 254	863	28, 386	80	29, 31
597	26, 267	864	26, 410	82	26, 32
601	26, 271	866	26, 410	83/4	28, 27
603	27, 96	867	26, 312; 29, 386	94	28, 30
607	26, 273; 28, 252	870	26, 411; 28, 392	96	29, 437
610	28, 331; 29, 252	871	29, 429	108	26, 43
618	29, 260	872	27, 354	114	29, 114
620	26, 276; 27, 266; 28, 254	876	26, 417	117	27, 41
		881	29, 397	122	27, 38
623	27, 18; 28, 256	883	26, 422; 27, 389	128	26, 48
637	26, 286; 28, 264	886	27, 302	129	28, 42
639	29, 269	888	26, 424 (2); 28, 403	135	27, 48
643	29, 269			145	26, 51
645	27, 275	893	26, 428, 429	148	27, 52
647	26, 289	907	26, 438	149	29, 59
653	26, 294	909	26, 439	153	28, 49
660	29, 278	915	26, 440	155	28, 50
683	28, 288	922	27, 410	160	27, 61
710	26, 317	926	29, 227	161	26, 59
726	26, 72	933	28, 439; 29, 435	169	26, 63; 27, 62; 28, 53; 29, 69
728	26, 325; 27, 302	937	26, 455		
729	26, 325	940	26, 456	173	27, 64
731	26, 329	942	26, 126	176	29, 71
735	28, 311	943	29, 441	180	26, 123
737	28, 314	951	27, 429	197	27, 78
750	26, 349; 27, 325	957	26, 465; 27, 432; 28, 454; 29, 449	198	26, 473; 28, 466; 29, 457
761	28, 330			201	29, 88
770	28, 339; 29, 338	960	28, 457	202	26, 80; 27, 82
773	26, 365; 28, 343; 29, 341; 30, 413	967	28, 459	204	29, 93
		969/70	27, 438	206	27, 84
		973	27, 439	215	26, 86, 158; 28, 75
780	28, 346	975	29, 460		
785	27, 97, 345; 28, 350	977	26, 430; 28, 413	227	28, 82
		994	27, 445	231	28, 85
789	26, 374	996	26, 480	263	27, 107
791	26, 376			266	26, 108
792	27, 247	**Volume 17**		269	26, 109
794	26, 377; 27, 349			271	26, 110
803	26, 380	8	26, 3	275	29, 117
811	26, 382	17	29, 8	277	26, 112
813	26, 451; 27, 355	22	27, 8	296	26, 120
820	29, 361	28	29, 12	298	26, 120
831	29, 446	45	29, 21	307	27, 125
836	27, 365	57	28, 21; 29, 23	339	26, 154
841	29, 373	57 suppl. 28	29, 23	341	26, 139; 27, 141
843	26, 400; 28, 380			345	29, 11
848	26, 338; 28, 373; 29, 374	61	29, 25	353	28, 136
		65	**28, 22**	355	28, 193
		71	28, 23	368	29, 183

Supplementary References – Ergänzungszitate

No.	Suppl. Ref. *Vol.*, Page
Volume 17 continued	
379	*27*, 162
381	*26*, 84, 163; *28*, 154
383	*27*, 165
393	*28*, 184
395	*26*, 170
396	*28*, 89
403	*29*, 209
409	*26*, 176
417	*27*, 36
436	*26*, 192
451	*27*, 200 (2)
458	*28*, 187
461	*26*, 202; *28*, 189
467	*26*, 205; *28*, 192
477	*26*, 208; *29*, 208
486	*26*, 211
495	*26*, 214; *27*, 216; *28*, 202
498	*26*, 149; *28*, 142
499	*28*, 203
500	*26*, 216
502	*26*, 216; *29*, 213
503	*28*, 149
516	*28*, 207
528	*27*, 229; *28*, 215
534	*29*, 56
535	*27*, 156, 232
536	*27*, 233
537	*29*, 60
538	*27*, 234
542	*26*, 237
544	*26*, 237
546	*27*, 236; *28*, 222
547	*26*, 238
549	*28*, 224; *29*, 95, 230, 257
559	*26*, 171; *28*, 45
565	*28*, 231
590	*27*, 252; *28*, 239
593	*28*, 240
598	*26*, 262; *29*, 248
601	*28*, 245
611	*27*, 260; *28*, 247; *29*, 253
621	*27*, 264; *28*, 247; *29*, 253
623	*26*, 274
652	*29*, 268
654	*29*, 268
657	*27*, 150; *29*, 159
664	*26*, 296; *27*, 279
703	*26*, 312
705	*26*, 313; *27*, 294; *29*, 294
715	*26*, 316
733	*27*, 300
735	*27*, 302
747	*26*, 335; *28*, 307
747 suppl.	*26*, 340
751	*28*, 312
765	*26*, 345; *27*, 321
772	*27*, 324; *29*, 327
777	*29*, 330
780	*29*, 332 (2)
780 suppl.	*29*, *30*, 399
786	*26*, 359; *27*, 334
788	*26*, 455
792	*28*, 337
793	*28*, 338
798	*29*, 308
799	*26*, 363
802	*28*, 344
809	*28*, 352; *29*, 348
830	*29*, 357
841	*26*, 392
842	*26*, 392; *29*, 363
843	*26*, 393; *29*, 364
846	*26*, 395; *27*, 364
848	*26*, 396
850	*28*, 377
857	*29*, 47
865	*28*, 386
870	*26*, 412; *29*, 387
871	*28*, 393; *29*, 387
872	*29*, 374
879	*26*, 419; *27*, 387; *28*, 400; *29*, 395
884	*26*, 420; *27*, 388
885	*26*, 420
887	*26*, 423
889	*26*, 425 (2), *26*, 426 (3); *27*, 391; *28*, 406 (2); *29*, 144
891	*26*, 428
892	*27*, 396; *29*, 405 (2)
894	*26*, 429; *27*, 396, 397; *28*, 411; *29*, 406
895	*26*, 430; *29*, 407
898	*27*, 245
901	*27*, 403
916	*26*, 262
917	*26*, 442
925	*28*, 435
933	*29*, 432
936	*27*, 419; *29*, 434
937	*26*, 184, 453
946	*26*, 456
948	*28*, 449
949	*26*, 459
955	*26*, 466
962	*29*, 453
964	*26*, 469
968	*26*, 470; *28*, 459; *29*, 454
971	*26*, 403; *28*, 462
972	*26*, 472
975	*26*, 475
984	*29*, 244
Volume 18	
1	*26*, 2
15	*26*, 4
16	*29*, 8
20	*27*, 5
28	*27*, 10
37	*26*, 245
41	*27*, 15
59	*26*, 22; *28*, 21
63	*30*, 440
65	*27*, 23
71	*26*, 25
77	*26*, 27
78	*27*, 28
80	*26*, 31
92	*28*, 32
99	*26*, 40; *29*, 41
124	*27*, 33; *28*, 33
125	*26*, 271; *27*, 262; *28*, 39
127	*28*, 44; *29*, 55
129	*26*, 110
130	*29*, 56

No.	Suppl. Ref. Vol., Page
141	29, 61
143	26, 57
156	27, 63
160	28, 54
176	26, 150
181	26, 72
185	29, 84
188	26, 77
203	27, 85
205	26, 84; 28, 74; 29, 96
209	26, 87
214	28, 78
220	26, 478
222	29, 102
246	26, 103
256	26, 105
257	26, 106
262	26, 302; 28, 269; 29, 21
264	26, 107
268	26, 109
286	28, 105
297	27, 124
301	26, 126; 27, 128
305	26, 127; 29, 132
306	26, 127
307	27, 129; 28, 118
314	26, 131
318	26, 136
324	26, 138
325	26, 139
331	27, 239; 29, 231
338	26, 238; 27, 237; 28, 133, 223
338 suppl. 27	29, 230
340	26, 12
346	26, 145
365	26, 150
370	26, 153
373	26, 153; 27, 157; 28, 145
374	26, 155
388	26, 85
390	27, 169
392	26, 172
394	26, 173
397	26, 183
400	26, 177
414	26, 181
430	26, 188; 28, 176
431	26, 188
434	26, 190; 29, 189
435	26, 190; 27, 191; 28, 178; 29, 189
441	29, 192
450	26, 197
454	26, 199
459	29, 201
461	28, 173
467	26, 202
468	26, 203
472	27, 100
473	26, 203; 27, 205
475	26, 206
482	27, 210; 29, 207
501	27, 12
506	26, 221
516	29, 222
518	27, 227
523	26, 227
533	30, 144
542	26, 237
543	29, 230
544	26, 240; 29, 203
548	27, 240; 29, 232
557	26, 249
571	26, 255
572	26, 255; 29, 243
579	26, 262
585	29, 250
587	28, 197
596	26, 267
601	26, 269
602	27, 262
604	28, 248
622	26, 276
643	26, 286
645	26, 287; 29, 269
647	28, 266
649	26, 288; 29, 270
669	26, 297; 28, 272
687	29, 282
690	29, 283
711	27, 296
717	28, 292
725	26, 319; 27, 299
734	28, 318
735	27, 303
736	26, 326; 27, 303; 29, 306
740	26, 320
744	29, 308
747	26, 332
750	27, 308
751	26, 72
758	26, 339
759	26, 340; 29, 318
765	26, 344
768	29, 324
770	29, 305
772	26, 346; 28, 320
773	26, 38
775	27, 323
776	26, 348
786	27, 330
788	26, 353
790	27, 334
796	29, 342
803	26, 261
813	26, 373
815	27, 346
820	29, 121
827	29, 357
831	29, 396
833	26, 386; 28, 364
834	28, 323
836	26, 388
842	27, 364; 29, 364
848	29, 367
853	26, 399
858	29, 372
858 suppl. 29	30, 445
859	27, 369
862	26, 401; 29, 371
872	27, 372
875	26, 78; 28, 69
881	26, 411
891	29, 390
895	27, 263
897	26, 419; 27, 387
901	28, 400
905	26, 423
912	26, 430; 28, 414
915	26, 65; 28, 51
923	29, 360
925	29, 426
935	27, 100; 28, 86
937	26, 448; 28, 435

No.	Suppl. Ref. Vol., Page				
Volume 18 continued		153	27, 65	464	29, 194
		161	26, 65; 27, 66	465	28, 173, 182
		175	29, 77	470	28, 186
		179	27, 72	483	26, 202
945	29, 435	182/3	26, 73	494	29, 205
947	27, 35; 29, 40	187	28, 63	495	26, 193
954	28, 444	188	26, 75	508	29, 208
959	28, 446	190	26, 76	509	27, 215
960	26, 458; 27, 426; 28, 311; 29, 443	193	27, 158	510	29, 211
		194	27, 79	515	26, 439
		200	26, 80; 27, 82	526	29, 215
962	26, 460	201	26, 111; 29, 4	539	26, 225; 27, 225; 28, 209
963	26, 461	211	28, 76 (2)		
965	26, 343; 28, 315	214	29, 98	540	26, 225; 29, 220
966	27, 141	233	29, 105	553	26, 233, 241
970	26, 464	237	26, 100; 27, 100	554/5	27, 231
979	26, 470	239	29, 109	562	29, 230
985	26, 473	264	26, 109; 29, 116	563	27, 238
990	26, 475	279	29, 120	577	27, 245
992	28, 472	302	26, 123; 28, 110	583	26, 254
993	26, 457	305	27, 229	592	28, 234
994	26, 150	307	26, 125; 27, 126; 28, 114; 29, 133 (2)	611	27, 258
Volume 19				620	26, 267; 29, 253
				629	26, 270
8	26, 4; 27, 6			635	29, 258
10	29, 7	308	28, 115	653	26, 282
22	29, 14	308 suppl. 28		658	27, 270
26	26, 112		30, 147	665	28, 264
32	18, 17	312	29, 136	677	26, 297
41	26, 4	313	26, 128	682	26, 110
46	26, 21	315	27, 129	691	26, 302
51	18, 21; 29, 25	326	26, 131	694	28, 276
55	29, 26	328	27, 133; 28, 123	699	26, 303
56	26, 23	336	26, 134	702	29, 284
57	27, 26; 29, 29	339	28, 124; 29, 143	705	26, 306
60	26, 31	354	26, 141; 27, 241	719	29, 292
73	26, 22, 37; 29, 39	356	27, 144	722	28, 289
		357	26, 143	724	29, 297
76	26, 38	363	28, 141	734	29, 300
82	26, 13	367	28, 142	736	26, 323; 27, 300; 29, 301, 302
85	29, 44	378	28, 274		
86	28, 36, 447	379	28, 150		
91	26, 44; 27, 39; 29, 46	382	27, 161	737	26, 323
		384	26, 161	738	27, 301
99	29, 49	387	26, 162	740	26, 324
100	26, 45	412	26, 171	743	29, 304
101	27, 43, 346	416	28, 165	744	26, 325; 28, 297; 29, 305
103	26, 46	432	26, 181; 28, 174		
107	29, 52	442	28, 178	745	29, 305 (2)
112	27, 47	444	26, 191	746	26, 325 (2)
114	27, 260	448	29, 191	751	27, 304; 28, 299
117	26, 50; 28, 45	461	26, 179	752	28, 122

No.	Suppl. Ref. Vol., Page
758a	28, 305
761	27, 314
764	27, 318; 28, 313; 29, 318, 457
766	27, 439; 29, 457
768	28, 449
770	26, 344
773	26, 348; 27, 323
776	29, 327
778	29, 329
780 suppl. 22	29, 331
783	26, 354
785	28, 328
786	27, 330
798	26, 361; 27, 336
801	28, 338
802	26, 58, 80; 27, 336
806	27, 390; 28, 342
809	27, 341
811	27, 341; 28, 344
812	27, 341 (2)
815	27, 345
825	28, 356
839	28, 363
841	28, 363
845	26, 386; 27, 356
847	26, 389
856	27, 202; 28, 372
862	26, 396; 28, 375
876	26, 407
879	28, 386
881	28, 389
888	29, 119
891	26, 413; 28, 394
896	26, 417
898	26, 419
904	28, 415; 29, 411
907	28, 419
906	27, 400
910	26, 434
911	29, 416
917	28, 423
932	27, 411; 28, 433
937	27, 413; 29, 430
941	26, 450
942	26, 451
947	26, 452; 27, 419
949	29, 437
952	28, 198
962	26, 457; 27, 425; 28, 440, 445; 29, 437
970	28, 454 (2)
978	26, 469
995	29, 37

Volume 20

7	28, 4; 29, 2
9	26, 2
12	26, 2; 29, 68
15	26, 3
22	27, 6, 382
37	27, 13
41	26, 16; 27, 15
49	26, 21; 28, 19
53	29, 138
54	26, 23
55	29, 329
56	29, 27
58	27, 27
61	26, 30
63	26, 33
66	28, 30 (2)
67	29, 444
75	26, 41
76	27, 20
79	28, 39
80	29, 49
82	27, 46
89	26, 57
90	29, 64
95	26, 63; 27, 62
106	28, 29; 29, 365
109	26, 75
111	27, 78; 28, 69
112	26, 77
116	29, 94
123	26, 90
126	26, 94; 28, 82; 29, 101
129	26, 95
132	26, 96; 27, 97 (2)
150	26, 106
153	26, 108
172	26, 115
175	29, 242
184	27, 121
190	29, 24
191	26, 124
192	26, 125; 28, 113; 29, 135
193	27, 127
194	27, 128
201	27, 129
204a	29, 146
208	27, 138
211	27, 142
215	27, 144
216	26, 143
218	26, 145
225	26, 149
235	28, 148
238	26, 156; 27, 187; 28, 179
239	26, 157; 27, 87
244	27, 162
248	29, 169
258	26, 167; 28, 157; 29, 174
271	26, 230; 27, 177, 229; 28, 168; 29, 181
277	26, 177
281	26, 180
284	26, 181; 27, 186; 28, 174
289	28, 180
293	26, 196
302	26, 199
305	29, 201
308	26, 201
311	27, 158
312/3	29, 203
316	29, 188
324	27, 207
328	27, 211
343	26, 214
345	28, 203
363	28, 211; 29, 223
369	29, 224
373	28, 217
375	26, 234; 27, 231
383	26, 129
388	29, 231
389	26, 242; 27, 240; 28, 227
395	29, 238
409	27, 256, 410
413	26, 264

Supplementary References – Ergänzungszitate 766

No.	Suppl. Ref. Vol., Page
Volume 20 continued	
414	28, 244
417	29, 251
431	28, 255
444	28, 268; 29, 272
446	28, 270
448	26, 295
457	26, 300
472	28, 281
473	29, 286
502	26, 329; 30, 370
515	26, 335; 29, 314
518	26, 343
519	28, 314; 29, 320
525	27, 321
526	26, 336; 27, 322
527	26, 348
531	26, 350; 27, 326; 29, 328 (2)
533	26, 353; 28, 327
534	28, 327
537	28, 328; 29, 331
538	26, 354; 27, 330
539	27, 330
541	26, 355
550	27, 338
551	26, 380; 28, 339; 29, 110
552	26, 363
557	26, 367
561	27, 345
565	26, 373
579	27, 359
580	28, 367
582	26, 390
591	26, 396; 27, 366
593	27, 366
599	26, 400
601	28, 306
607	27, 379
608	26, 410
609	26, 412
612	26, 414
615	26, 442
617	27, 387
626	29, 402
629	26, 56
633	26, 438; 29, 412
634	27, 401
637	29, 419
645	26, 441; 27, 409
648a	29, 427
651	28, 411
657	29, 436
660	26, 454
665	26, 455
668	26, 456
672	26, 463; 27, 429; 29, 446
673	29, 446
674	29, 447
678	26, 467
679/80	28, 458
688	28, 460
692	28, 466
696	29, 466
698	27, 444
Volume 21	
9	28, 4
18	28, 6; 29, 4
21	26, 4; 28, 7; 29, 6
27	29, 7
29	26, 6
30	26, 7; 29, 9
35	26, 11, 43 (2); 27, 9, 38; 28, 38; 29, 11, 46
37	28, 11
38	27, 442
44	29, 11
55	26, 19; 28, 18
60	27, 18
61	28, 20
91	26, 42
94	27, 38
97	26, 44
102	27, 42
103	26, 46
108	26, 47
109	26, 47; 27, 45
115	27, 46
116	27, 47
121	27, 50
124	26, 54; 27, 54; 29, 59
134	27, 57; 28, 50
139	29, 63
140	27, 59
152	28, 54
160	27, 68
161	29, 76
166	27, 72; 29, 79
167	29, 79
174	26, 75; 27, 75; 28, 65; 29, 83
177	26, 94
182	28, 434
184	29, 91
192	26, 86, 87
197	27, 90; 29, 99
198	28, 79
204	26, 93
223	26, 102
231	27, 105
235	29, 114
237	27, 107
238	27, 108
248	26, 112
251	29, 119
256	29, 121
265	26, 120; 27, 120; 28, 107
265a	27, 110
269	28, 109
270	26, 123
273	28, 111; 29, 130
276	28, 112; 29, 130
279	28, 73; 29, 116
282	28, 113
283	28, 132
284	26, 125; 27, 20; 28, 20
285	27, 126; 28, 114
289	27, 128
294	28, 67; 29, 141
302	29, 95
303	26, 135
305	29, 306
308	29, 146
310	27, 138
313	28, 130
317	26, 141
329	26, 148
335	29, 161
340	28, 143
341	29, 162
344	26, 153

No.	Suppl. Ref. Vol., Page				
349	26, 329; 27, 157	601	26, 265; 27, 259; 28, 246	884	28, 410; 29, 405
354	29, 165	607	26, 270; 27, 262; 28, 250; 30, 298	887	27, 398; 28, 412
370	26, 165			899	29, 419
373	26, 165			904	27, 409
389	26, 171	614	28, 253	911	28, 435
393	29, 180	630	27, 24; 29, 27	913	27, 413
400	26, 12	635	29, 266	916	26, 451
401	26, 211; 28, 34; 29, 40	639	28, 264	921	28, 438
		642	26, 286; 27, 274; 29, 269	923	26, 452
404	26, 176; 27, 181	653	26, 294	924	27, 420
411	27, 184	676	29, 287	926	26, 454
421	26, 185	683	28, 287	927	29, 437
426	27, 195	687	28, 289	941	29, 162, 442
430	28, 179	714	26, 327; 27, 304	942	26, 458; 27, 426
432	26, 193; 27, 194	719	26, 332	948	29, 446
433	28, 139; 29, 376	725	27, 439	949	28, 452
434	27, 131, 194	728	26, 335; 27, 314; 28, 308, 310; 29, 314	950	29, 447
436	26, 381			952	29, 448
442	29, 194			955	27, 433
444	28, 182, 185			957	28, 456
445	28, 185			958	27, 433
450	29, 201	731	27, 318; 29, 319	961	26, 467
454	27, 204	736	28, 319	962	26, 468
455	27, 262	738	26, 346; 27, 322; 28, 299; 29, 325	967	26, 470
457	26, 203			988	29, 465
459	28, 191; 29, 204	744	27, 328	990	26, 479
468	26, 205; 27, 207	746	28, 327; 29, 330	995	27, 445
471	27, 143	749	26, 354		
492	26, 211	751	27, 330	**Volume 22**	
495	26, 217	761	27, 332	2	27, 1
511	26, 226; 27, 226; 28, 209	767	29, 93	6	30, 5
		771	26, 361	17	28, 10
514	28, 211; 29, 222	786	26, 365; 28, 344; 29, 342	18	29, 12
521	27, 228; 28, 176; 29, 223			26	28, 13; 29, 15
				30	27, 14
522	29, 206	803	26, 378; 27, 20	35/6	26, 20
524	29, 224	810	26, 395; 29, 183	37	29, 23
527	26, 232; 27, 230	812	26, 382	42	28, 23
536	26, 235	813	29, 356	47	26, 26
538	29, 228	821	29, 360	49	27, 28
542	26, 238	824	26, 390	50	29, 30
544	29, 230	829	27, 351	51	26, 29
554	28, 225	835	26, 396; 27, 365; 29, 365	57	27, 31
556	27, 239			59	26, 34; 27, 32; 28, 28; 29, 35
558	28, 226	857	26, 405		
563	26, 244; 28, 229	858	27, 82, 124; 28, 71	61	29, 37
564	26, 242; 29, 234	863	27, 376	62	28, 21
576	26, 250	867	28, 393	65	28, 33
588	27, 250	869	29, 390	70	27, 36
600	27, 258; 28, 246	875	27, 352	73	27, 37
				80	28, 40

Supplementary References – Ergänzungszitate

No.	Suppl. Ref. Vol., Page
Volume 22 continued	
82	29, 50
87	26, 49; 27, 49; 28, 44; 29, 55
91	28, 46
93	26, 53
95	26, 166; 29, 59, 172
99	27, 55
101	29, 64
106	29, 67
112	26, 63; 27, 64
117	27, 66
122	30, 11
123	27, 70; 28, 59, 191
130	26, 71
136	27, 73; 29, 79
143	26, 75; 27, 74 (2); 28, 64, 72; 29, 82 (3), 83
143 suppl. 24	26, 76
153	26, 109, 261; 27, 79; 29, 247
164	26, 168
170	29, 99
177	26, 91; 27, 92
185	26, 95
187	29, 105
196	28, 88
198	29, 110
203	27, 104
209	27, 106
217	27, 108
221	28, 95
224	29, 117
230	26, 114
233	26, 93
252	29, 131
253	26, 125
257	29, 138
260	27, 201
261	26, 133; 27, 134
262	26, 133
263	26, 134
286	29, 13
288	26, 141; 28, 132; 29, 152
292	28, 133
313	26, 153; 27, 157; 28, 145; 29, 163
314	26, 154; 28, 146
317	29, 49
318	28, 148
331	29, 315
335	29, 172
337	26, 165
341	27, 166
346	26, 167
353	29, 180
355	26, 173; 29, 181
357	26, 175
372	26, 9; 27, 8
387	29, 189
390	27, 194
392	26, 194; 29, 192
393	26, 194
394	27, 195; 28, 181; 29, 192
395	29, 192
395 suppl. 24	28, 181; 29, 192
396	26, 194
399	26, 198; 27, 200
401	27, 198; 29, 195
404	26, 198; 27, 126
418	27, 202
420	28, 189
425	27, 204
430	28, 191
434	26, 205
458	29, 210
467	29, 213
473	29, 217
479	28, 207
483	28, 92
484	28, 209
505	26, 233
507	27, 232; 28, 217 (2)
523	27, 240
528	26, 245
530	27, 242
531/2	26, 247
540	27, 248
546	26, 255
548	28, 237
554	27, 251; 28, 230
560	28, 240
565	29, 250
567	26, 267; 29, 255
568	28, 248
588	29, 262
604	27, 161
618	27, 278
624	28, 272
625	26, 58; 27, 281; 29, 270
629/30	26, 301
656	26, 309
659	26, 311
661	26, 313 (2); 27, 116, 294; 28, 290
664	28, 290
666	29, 159
677	27, 298
678	28, 292; 29, 299
686	28, 295
688	26, 324
695	26, 328
701	28, 305
705	26, 341; 27, 318; 28, 312; 29, 318
708	26, 244
710	27, 317; 29, 318
717	26, 344; 28, 318
729	27, 329
732	29, 331
738	27, 406
741	26, 360
742	27, 335
749	26, 361
752	26, 362
753	27, 338
757	26, 364
761	26, 365; 27, 341 (2); 28, 344; 29, 342 (2)
762	26, 359, 366; 27, 341
763	28, 345
769	26, 370
776	26, 372
778	28, 352
780	28, 353

No.	Suppl. Ref. *Vol.*, Page				
782	26, 377	970	26, 471	186	29, 98
795	29, 357	972	26, 324, 472	191	28, 80
799	26, 296; 28, 271			195	26, 93
		Volume 23		199	29, 85
803	26, 389; 29, 361	11	27, 6	213	29, 444
804	26, 390	12	27, 7; 28, 8	215	27, 295
812	26, 394	15	29, 7	218	27, 105
813	27, 372; 30, 366	17	29, 7	223	26, 106
825	28, 380	19	26, 9	233	26, 15; 27, 15; 28, 17
828	27, 372	25	29, 12	235	29, 115
833	29, 375	27	26, 242; 28, 12	237	29, 116
836	29, 310	30	29, 13	242	29, 119
838	30, 452	36	26, 16; 27, 14; 28, 15; 29, 17	244	26, 115
843	27, 382			248	28, 104
844	26, 413	43a	27, 20	255	27, 123
846	28, 93	48	29, 24	264	28, 113
849	27, 385	49	27, 22	267	27, 128
853	26, 422	56	27, 26; 28, 23, 222	268	28, 80
857	27, 393			285	29, 143
865	26, 429	57	28, 24	305	27, 142
871	26, 430; 29, 411	63	27, 30; 29, 32	311	26, 144
872	29, 411	66	28, 28; 29, 34; 30, 36	323	29, 159
874	28, 90, 419			336	26, 237
877	28, 420	67	29, 35	348	27, 160
887	26, 440 (2); 28, 427	69	26, 35; 29, 37	349	28, 149
		79	26, 37	354	29, 167
894	28, 429	80	27, 35; 28, 83	355	27, 161
895	26, 442; 27, 409; 28, 430, 445	80 suppl.	27, 29, 39	363	26, 166
				371	26, 169; 27, 168
		81	26, 38; 27, 58	379	29, 60
898	28, 433	82	26, 150; 27, 36	381	27, 176; 28, 166; 29, 180
899	26, 447	85	26, 42; 29, 298		
906	26, 452; 27, 416	92	27, 39	386	26, 177
912	27, 419	101	26, 47; 27, 44	397	26, 182; 28, 196
913	27, 420	110	27, 100	401	28, 175
927	27, 424	114	27, 52	403	28, 175
929	26, 400; 29, 442	119	26, 247	404	26, 184
935	26, 461; 28, 450; 29, 445	120	26, 56	405	29, 187
		123	29, 67	407	27, 189; 29, 187
937	26, 462	131	27, 66	408	27, 189
938	29, 446	135	26, 67	412	28, 176
941	27, 429	139	28, 59	413	26, 189
944	26, 463; 27, 429; 29, 446	142	28, 61	415	27, 195; 28, 180; 29, 192
		144	28, 61		
953	29, 452	145	27, 77	418	27, 195
959	26, 466; 27, 432; 28, 455	166	27, 84	423	27, 196; 29, 194
		169	26, 84; 27, 85; 30, 338	428	29, 195
960	26, 468			431	28, 186
963	29, 453	176	26, 86	434	29, 203
968	27, 334; 29, 350, 459	179	28, 76	435	27, 204
		180	27, 88	440	28, 192

No.	Suppl. Ref. Vol., Page
Volume 23 continued	
441	29, 206
458	26, 213; 27, 215
462	29, 212
476	29, 220, 256
479	26, 14, 18; 28, 17; 29, 19
480	27, 226
482	28, 214
486	26, 18
490	28, 175
497	29, 163
516	26, 244; 28, 229
520	28, 231
521	26, 248; 28, 348; 29, 347
539	26, 257
542	29, 81
544	29, 246
567	29, 259
573	26, 279; 28, 256
581	28, 262
592	28, 269
599	28, 271
603	26, 297; 28, 272
611	26, 302
612	29, 282
630	28, 281
643	26, 314
650	26, 315; 27, 296
656	28, 42
657	27, 297
669	29, 305
674	29, 301
678	28, 303; 29, 310
689	26, 300, 336; 27, 312; 28, 309; 29, 315
690	28, 309
693	26, 344; 27, 320 (2); 28, 318; 29, 323, 403
700/1	28, 317
702	26, 346 (2); 27, 321; 30, 467
703	26, 346; 29, 325
705	28, 319; 29, 325
715	28, 326; 29, 329
716	28, 327
717	26, 353
719	29, 331
729	27, 332
733	26, 359
736	28, 335
739	27, 337
740	28, 327
757	28, 350; 29, 348
759	26, 133
776	28, 357
782	29, 356
784	28, 41; 29, 356
786	29, 356
787	26, 382
789	28, 363
792	26, 388
797	26, 389; 28, 368
800	28, 366
802	27, 361
810	27, 365
818	28, 378
819	27, 367; 28, 378
819 suppl.	27, 30, 447
825	26, 402
831	26, 413; 27, 382; 28, 394; 30, 460
832	26, 407, 408; 27, 374
833	28, 385
836	29, 383
837	26, 410; 27, 377
838	27, 302, 378; 29, 304, 385
840	27, 380
852	29, 394
853	29, 290
857	26, 420
859	26, 421
860	26, 423; 27, 389; 28, 403; 29, 399
863	27, 390
871	26, 426; 27, 391
879	26, 429 (2); 27, 398; 28, 412; 29, 407 (3); 30, 476
884	26, 433
885	26, 434; 29, 416
896	29, 426
901	26, 444
902	29, 428
903	27, 411
904	26, 445; 28, 434; 29, 428
907	26, 451
910	27, 420
912	26, 453; 28, 439
914	26, 453
915	26, 454
920	29, 438
922	26, 454
924	26, 380
926	26, 239
930	26, 434
935	26, 459
936	26, 460
939	28, 451
942	26, 463; 28, 452
943	27, 430
945	29, 448
948	28, 453; 29, 448
952	29, 448
960	28, 457
964	26, 469
966	27, 434
968	29, 453
969	27, 435; 29, 454
971	28, 460
975	29, 458
976	28, 466; 29, 458
977	26, 473
982	26, 474; 29, 461
986	26, 22; 27, 20, 441; 29, 462
988	29, 464

Volume 24	
2	27, 3
8	27, 5
9	26, 110; 28, 7
32	29, 17
34	26, 16; 28, 15
35	26, 16
41	26, 17
44	26, 19
46	29, 26

No.	Suppl. Ref. Vol., Page
48	29, 25
49	28, 21
53	27, 26
54	26, 25
59	26, 30
61	27, 20; 29, 23
61 suppl.	29 30, 27
62	27, 29
66	29, 33
68	26, 35
75	26, 136; 27, 137; 29, 38
77	27, 34
80	29, 39
82	26, 40; 28, 34
83	26, 40
85	26, 43; 29, 45
86	27, 38
118	27, 66; 29, 74
120	26, 66
122	27, 67
123	26, 67; 27, 68; 29, 129
126	27, 98; 29, 75
128	27, 70
131	27, 70
133	27, 72
140	28, 62
141	29, 81
143	29, 81
145	27, 74; 28, 65
149	28, 68
150	27, 79
152	27, 81
170	28, 78
174	26, 93
176	26, 94
177	28, 81
182	29, 359
184	27, 99; 28, 86
196	28, 91
203	27, 105
204	28, 61
208	26, 107
210	26, 108
228	26, 116
240	28, 108
242	29, 128
251/2	29, 131
251 suppl.	29 30, 146
254	26, 125, 126; 27, 126; 29, 133
257	27, 127
262	26, 125
264	29, 135
268	26, 129; 27, 132; 29, 140
276	29, 142
284	27, 68
289	28, 130
290	29, 150
291	26, 140
294	28, 133; 29, 154
296	26, 143
297	26, 145
299	27, 148; 29, 157
311	26, 154
317	28, 148
319	29, 165
325	29, 118
330	28, 151
331	28, 152
332	26, 162
338	29, 171
340	29, 172
347	29, 185
354	26, 170; 27, 170
371	26, 177
387	28, 174
391	26, 37; 27, 34
396	29, 189
398	29, 227
401	27, 193
406	26, 197; 27, 195
408	28, 181
409	26, 196
413	26, 197; 27, 198; 28, 183; 29, 196
414	28, 185
418	26, 166; 28, 188
425	26, 7, 12; 29, 8
428	27, 204
437	26, 196; 28, 182; 29, 194
447	26, 210
473	26, 218
476	28, 148
479	27, 223
491	26, 227; 27, 226; 28, 210
499	28, 213
501	28, 214
504	30, 265
508	26, 235
513	26, 237
519	26, 239
521	26, 240
528	27, 240 (2); 28, 226
535	29, 235
547	26, 251; 27, 246
557	26, 257
560	26, 312
565	26, 260
567	26, 254
574	26, 264; 27, 257; 29, 249
576	27, 258
579	29, 254
581	26, 270; 28, 250; 29, 255
582	29, 256
585	28, 252
601	27, 271
613	29, 271
615a	28, 269
620	29, 218
624	26, 297
625	28, 10
633	29, 401
638	28, 279
642	27, 289; 28, 282; 29, 286
652	29, 308
663/4	28, 295
667	26, 324; 29, 304
670	26, 327
676	29, 308
678	28, 303
681	28, 304
683	26, 333
689	29, 314
690	27, 313
692	26, 340, 342; 27, 317 (2); 29, 317
693	28, 312
696	27, 269, 318; 28, 259, 310; 29, 318

Supplementary References – Ergänzungszitate 772

No.	Suppl. Ref. Vol., Page
Volume 24	continued
697	26, 342
699	27, 27
704	27, 321; 28, 319; 30, 467
708	29, 327
711	29, 329
719	26, 354
726	29, 333
728	28, 331
733	26, 361
736	26, 361
745	**26, 362**
746	29, 338
750	28, 344; 29, 342
751	26, 365
755	28, 345
761	29, 345
765	29, 346
768	26, 371
792	29, 362
796	26, 391
798	27, 364
802	25, 33 [1]; 28, 27
804	26, 396
807	26, 397
810	26, 398; 28, 377; 29, 370
813	28, 377
817	26, 399; 29, 372
821	27, 370
822	26, 402
825	29, 375
828	26, 406
831	27, 376; 28, 388
838	27, 381; 28, 394
840	29, 253; 30, 298
842	27, 257
843	28, 396
844	26, 416
850	27, 388
853	28, 401
854	29, 397
855	27, 388
859	29, 399
861	27, 390; 29, 378
865	26, 425; 27, 391
870	29, 404
872	28, 104
879	29, 408
882	27, 400
884	26, 433; 27, 401; 28, 418; 29, 413
894	28, 421; 29, 419
895	26, 437
900	28, 423
905	29, 422
912	26, 443
917	29, 429
919	29, 430
920	26, 449; 28, 432
923	27, 418; 28, 439
924	28, 439
940	27, 425; 29, 442
943	26, 459
945	26, 459
950	29, 447
951	28, 453
954	27, 431; 29, 448
955	28, 201
964	27, 433; 28, 457
977	28, 466
981	27, 440; 29, 459
983	27, 6
985	28, 471
989	26, 477
990	27, 443
Volume 25	
4	27, 2
5	26, 2; 27, 2
16	26, 13
22	28, 17
26	26, 20; 27, 19
31	29, 25
32	26, 22; 27, 21; 28, 21; 29, 25
33	28, 21
43	26, 30; 28, 25
46	26, 34; 28, 28
49	28, 31
50	28, 32
52	28, 34; 29, 40
54	26, 442
57	**27, 37**
62	28, 33; 29, 47
64	26, 45; 28, 40; 29, 49
75	26, 55
77	29, 61
78	27, 56
88	26, 63; 27, 62 (2)
88 suppl. 27	28, 53
88 suppl. 28	29, 70
99	26, 69
104	28, 24
105	29, 84
106	26, 76
109	28, 110
110	27, 80
111	26, 79
114	29, 75
120	27, 87
123	28, 76
124	26, 88
128	28, 79
131	26, 91; 27, 91
141	27, 97
146	26, 102
147	29, 110
151	28, 303
158	28, 96; 29, 116
162	27, 112; 29, 117
163a	29, 120
174	26, 121; 27, 120
188	29, 139
193	**27, 132**
200	26, 135
205	26, 138; 29, 150
213	26, 144
215	28, 140
223	27, 100
228	26, 161; 27, 161; 28, 150
229	26, 162
231	27, 165; 29, 172
233	29, 85
241	27, 241
246	28, 164
248	26, 172; 29, 179
249	28, 167

[1] Correction: not 21, 802

No.	Suppl. Ref. Vol., Page				
		505	28, 307	691	28, 461
		506	29, 315	693	26, 194, 472; 28, 462
		507	29, 316		
255	26, 176; 28, 89	509	28, 309	694	29, 457
257	26, 180; 27, 184	517	26, 350	695	26, 473; 29, 458
258	27, 185	520	27, 328; 28, 326; 29, 330	701	26, 478
268	28, 176				
269	29, 189	522	26, 353; 28, 327; 29, 331	**Volume 26**	
275	26, 195, 386				
277	28, 182	527	29, 332	2	29, 1
293	26, 203; 28, 170, 191	533	27, 334	3	26, 16
		535	29, 333	14	28, 9
302	26, 209	536	26, 358	23	30, 85
309	26, 213	549	27, 341	24	28, 14
324	28, 473	551	29, 343 (2)	27a	29, 18
325	26, 226	554	26, 32; 29, 32	28	29, 20
330	29, 223	565	26, 381; 27, 354	32	27, 23
346	26, 238	572	28, 370	33	28, 22
348	26, 239	576	27, 364	38	27, 26
351	28, 224	581	29, 370	47	29, 22
354	26, 241; 29, 225	584	27, 370	50	27, 30; 28, 27, 31; 29, 32
358	26, 243	587	26, 338; 28, 311		
362	26, 248	589	26, 405; 27, 372		
363	26, 222	591	26, 407; 27, 374	53	29, 35
370	29, 239	594	27, 376	54	28, 24
385	26, 262	595	26, 410	57	28, 33
388	27, 256	599	26, 412; 27, 380; 29, 387	66	29, 42
389	26, 263 (2); 27, 256 (2); 28, 241 (2); 29, 249; 30, 209			68	27, 37
		602	26, 412; 28, 393; 29, 388	73	27, 30
				83	29, 56
		612	26, 416; 27, 385; 28, 395 (2), 396; 29, 391	85/6	29, 56
400	27, 261			88	28, 47; 29, 58
401	28, 251			90	28, 47
402	27, 264			92	28, 59; 30, 64
412	26, 277; 28, 134	619	29, 395	104	27, 62; 28, 52
416	27, 267	621	26, 424; 29, 400	105	27, 62
420	26, 282	622	27, 392	117	30, 82
424	27, 275; 28, 266	626	26, 430	121	27, 70
445	29, 279	633	29, 412	126	26, 366
447	26, 301	634	26, 433	137	27, 80
462	26, 280	636	26, 434	138	27, 80; 28, 69; 29, 87; 30, 7
465	27, 292; 29, 290	640	26, 438		
480	28, 350	644	27, 406	148	28, 93
483	27, 301 (2)	645	28, 426; 30, 491	154	30, 105
484	26, 324	649	28, 430	157	28, 77
486	26, 324	652	29, 427	165	28, 81
487	29, 304	669	26, 458; 28, 447	167	28, 82
488	26, 324	670	28, 448; 29, 443	170	29, 104
490	26, 326; 28, 298	679	26, 464	175	27, 98, 99; 28, 85, 141; 29, 107
493	28, 302; 29, 309	680	26, 464; 28, 454		
499	26, 333	683	28, 457		
500/1	26, 334	690	27, 436	177	27, 99

Supplementary References – Ergänzungszitate

No.	Suppl. Ref. Vol., Page
Volume 26 continued	
182	28, 87; 30, 117
183	27, 102; 28, 88
188	30, 120
189	28, 91; 29, 111
190	28, 91
194	28, 82
198	28, 94
203	27, 110; 28, 96
205	27, 111; 29, 119
206	27, 111
207	30, 129
211	28, 290
216	28, 104
229	30, 145
232	28, 115
234	30, 149
235	27, 128; 28, 116; 29, 135
236	27, 129; 29, 430
237	28, 119
242	27, 370
247	28, 350
263	30, 1
272	27, 147; 29, 156
274	27, 148
276	29, 158
282	29, 162
283	27, 154
294	28, 148
300	27, 159
301	28, 149
303	28, 150
305	28, 151
308	27, 164; 28, 74
310	28, 155; 29, 171
315	27, 166; 28, 159
320	29, 177
321	27, 172
323	29, 178
325	28, 167
332	28, 170
335	30, 215
343	27, 186
354	27, 189
355	30, 220
357	27, 190
360	27, 191
366	30, 224
378	26, 227
379	28, 184
388	29, 202; 30, 235
392	27, 206
396	30, 240
403	30, 243
405	28, 198
413	28, 202
424	27, 220; 28, 204
435	30, 255
438	28, 13, 207
443	27, 222; 30, 258
446	27, 226; 29, 221; 30, 259
447	26, 197
451	27, 228
453	29, 224
459	27, 230; 29, 225
467	28, 220
471	29, 230; 30, 272
479	28, 228
495	30, 281
507	28, 233
509	28, 236
519	30, 290
521	28, 229
522	27, 253
524	30, 292
525	27, 255
527	29, 247
535	29, 253
539	27, 262
547	29, 255
559	29, 262
560	30, 310
562	27, 267; 29, 262
565	30, 311
567	28, 258
575	27, 271
578	28, 263
579	29, 267 (2)
584	26, 77; 27, 78; 28, 67
587	27, 276
589	29, 155
593	27, 18
595	27, 277
597	29, 273
602	27, 279
611	29, 280
614	27, 284; 28, 275; 30, 332
616	27, 284
619	27, 286
621	30, 157
625	27, 288
637	29, 290; 30, 289, 347
641	29, 293
642	27, 293; 30, 348
645	27, 294
653	27, 297
658	29, 299
660	27, 119
662	27, 299; 28, 294
664	29, 300
668	28, 295; 30, 359
669	27, 302; 28, 324; 29, 305; 30, 362
670	27, 303
671	28, 297; 30, 363
672	27, 385
673	30, 364
675	29, 307
681	27, 306
683	27, 306; 29, 225
684	29, 310
690	27, 313
691	29, 50
692	28, 309
694	29, 315
699	30, 377
704	27, 317; 30, 378
706	28, 311
708	28, 313; 29, 319
712	28, 316; 29, 322
715	28, 318; 29, 324
720	27, 323
725	27, 327; 30, 393
730	29, 330
731	27, 329; 30, 395
735	30, 398
739	29, 333
740	29, 445
742	27, 333; 30, 401
743	28, 333
744	27, 334; 29, 334
746	29, 335
749	27, 334
757	28, 343; 29, 330
758	28, 343; 29, 314, 341; 30, 413

No.	Suppl. Ref. Vol., Page
759 suppl.	29
	30, 395
762	27, 342; 28, 344
763	27, 342; 28, 345
766	28, 345
769	28, 346
770/1	29, 345
773	28, 348
774	28, 348; 29, 347
776	27, 345
782	27, 346
784	27, 346
792	27, 350
794	28, 359
796	27, 353
806	29, 358
814	27, 359; 28, 367; 29, 361; 30, 434
827	27, 364
833	27, 366
844	28, 380
846	27, 371
847	28, 381
852	27, 372
859	27, 375; 29, 378
861	27, 376; 28, 388; 29, 381
862	29, 382; 30, 453
863	27, 379; 28, 391; 29, 386
866	27, 381; 28, 445
867	29, 388
869	27, 383
870	29, 390
871	28, 170
872	27, 384; 29, 390; 30, 461
875	27, 385; 28, 396; 30, 462 (2)
876	28, 397
885	28, 401
888	28, 402; 30, 468
890	28, 402; 29, 399; 30, 469
891	29, 399
896	27, 309; 28, 406
898	27, 392; 28, 407
904	27, 396
909	29, 411
912	27, 400; 29, 412
914	27, 401; 28, 417
916	30, 483
918	27, 402
919	30, 484
923	29, 418
927	30, 519
934	29, 427
938	26, 326; 27, 413; 29, 306
939	26, 373
942	29, 430
944	27, 116
945	27, 414
952	28, 442; 29, 439
953	27, 423
957	26, 273
958	28, 447; 29, 443
963	27, 428; 29, 444
966	27, 430
967	27, 430; 30, 517, 521
975	28, 458
978	27, 435; 30, 524
982	29, 455
987	30, 533
988	27, 440
990	30, 204
991	30, 536
993	27, 443
997	29, 467

Volume 27

8	30, 6
10	28, 6; 30, 7
10 suppl.	28
	29, 4
15	30, 14
18	28, 12; 29, 14
19	29, 14
24	29, 19
32	27, 23
35	29, 24
37	28, 21
38	30, 27
39	29, 26; 30, 28
46	29, 28
56	27, 133
57	28, 28; 30, 37
61	30, 38
69	29, 42
72	30, 48
74	29, 48 (2); 30, 49
81/2	29, 50
83	28, 42
87	28, 42; 30, 54
88	28, 43
89	30, 57
93	29, 57
96/7	28, 47
99	28, 48
100	28, 130; 30, 64
103	29, 60
106	29, 62
107	29, 63; 30, 67
109	30, 68
112	29, 64
125	29, 100
126	30, 79
129	29, 76
134	28, 62; 29, 80
138	30, 89
141	28, 65, 66 (2)
142	30, 467
155	28, 458
158	29, 91
159	29, 92
162	28, 73
184	28, 81, 217
186	29, 101
190	28, 83
198	30, 427
212	29, 115
214	28, 60
217	28, 97
219	28, 98
230	28, 103
231	29, 123
232	30, 137
235	28, 107
238	28, 108
241	28, 109
246	28, 110
256	28, 119
273	30, 162
299/300	29, 158
302	29, 259
311	30, 183
313	30, 184
319	28, 146

Supplementary References – Ergänzungszitate

No.	Suppl. Ref. Vol., Page
Volume 27 continued	
329	28, 151; 30, 194
330	28, 153
336	28, 157; 29, 173; 30, 199
345	28, 161 (2)
346	29, 176
355	28, 165; 29, 178
357	28, 165; 30, 207
358	28, 165
364	29, 467; 30, 211
402	28, 177; 29, 189
404	29, 189
419	28, 184, 185; 29, 196
421	29, 198
425	29, 200
429	29, 196
430	30, 237 (2)
437	29, 206; 30, 239, 241
438	30, 240
444	29, 153
455	28, 199
457	28, 200; 29, 210
485	28, 208
490	30, 260
501	29, 132
505	27, 133; 28, 72; 29, 227
507	28, 219
521	28, 226
524	29, 231
529	28, 231
531	29, 236; 30, 279
548	29, 15
561	30, 293
565	30, 295
572	28, 248
575	29, 253
578	30, 303
583	28, 255
591	28, 261
598	30, 319
602	29, 270; 30, 321
611	30, 327
613	29, 276
619	29, 278
623	27, 147
624	28, 262
640	28, 282
642	28, 283; 29, 263
647	28, 286
648	28, 288
654	28, 290
658	29, 297
659	30, 355
663	28, 294
670	29, 306
671	28, 299
676	29, 308
683	30, 373
689	30, 374
694	30, 375
696	28, 308; 29, 314 (2)
697	29, 314
704	28, 318; 29, 318, 323; 30, 379
705	29, 360; 30, 378
706	28, 311
714	28, 320
715	28, 320
716	28, 323; 30, 390
720	29, 307
724	28, 326; 30, 393
727	28, 326
729	28, 327; 30, 395
729 suppl. 28	30, 396
730	30, 395
736	28, 330
738	28, 332; 30, 401
745	30, 407
751	28, 376; 29, 369
755	28, 342; 30, 411
756	28, 344; 29, 342
762	30, 417
764	29, 346
765	28, 348
768	27, 50
769	30, 422
776	29, 354
781	28, 359
785/6	28, 361
790	29, 356
794	28, 365; 29, 373
795	28, 366; 29, 360; 30, 433
796	30, 433
813	29, 365
814	28, 373
819	**28, 379**
830	28, 383; 29, 376 (4)
830 suppl. 29	30, 449
833	28, 384; 29, 378 (2); 30, 450
834	29, 378
836	30, 452
837	29, 381
838	28, 389 (2)
839	28, 389; 29, 383; 30, 454
842	30, 455
843	28, 390; 29, 385 (2); 30, 456
848	30, 458
851	28, 386, 394; 29, 379; 30, 452, 469
852	28, 394
855	28, 395
856	29, 390; 30, 461
864	28, 400
868/9	28, 402
870	28, 403
871	29, 400
872	28, 403
873	28, 406
877	29, 402; 30, 473
878	30, 474
887	29, 406; 30, 476
892	28, 413
894	28, 413; 29, 409
899	28, 420
907	28, 423
924	28, 434; 30, 498
928	28, 436
932	29, 432
938/9	29, 433
941	29, 435
942	28, 439; 29, 436; 30, 504
943/4	28, 441
949	27, 427; 28, 446; 30, 511, 514

No.	Suppl. Ref. Vol., Page
950	28, 446
957	28, 449
958	29, 444
969/70	28, 458
973	29, 450; 30, 524
976	30, 526
981	28, 465; 29, 458
983	28, 465
990	27, 429
995	30, 237

Volume 28

No.	Suppl. Ref. Vol., Page
4	29, 1 (2); 30, 2
7	29, 3
8	29, 3; 30, 5
12/3	29, 6
16	29, 213
32	29, 27; 30, 28
44	30, 37
53	29, 136; 30, 45
57	29, 45
66	30, 58
68	29, 56
70	30, 63
70a	29, 61
81	30, 76
82	30, 77
98	29, 98
109	29, 86
110	30, 93
127	29, 97, 337; 30, 104
134	29, 100
140	30, 111
144	29, 197; 30, 230
146	29, 108
152	29, 111
157	29, 111
165	29, 114
170	30, 129
182	29, 123
198	29, 131
200	30, 147
201/2	29, 134
203	29, 135
208	28, 20; 29, 137; 30, 150
209	30, 150
212	29, 76
221	30, 159
226	30, 161
229	29, 146; 30, 163
234	29, 147
243	29, 148; 30, 167
264	30, 182
266	29, 163
278	28, 405
288	30, 197
312	30, 206
322	30, 109
336	30, 130
341	29, 189
352	29, 219
353	30, 230
354	29, 198; 30, 231
356	29, 208
365	29, 203
369	29, 205
374	29, 207
375	29, 193
381	28, 215
383	29, 209
402	29, 217; 30, 254
414	29, 219
428	28, 196
439/40	29, 229
464	29, 232
465	29, 239
466	30, 282
469	29, 241
471	29, 241; 30, 3
478	29, 244
479	28, 245
483	30, 7
499/500	29, 254
506	30, 304
512	30, 148
543	29, 272
554	29, 278
569	30, 336
601	29, 300
605/6	30, 362
610	29, 306, 307, 384; 30, 365
617	30, 368
620	29, 312
631	30, 375
640	29, 319
643 (2)	29, 319
646	29, 322
648	29, 322; 30, 384
655	30, 389
657	29, 325
658	29, 326
661	30, 391
676	28, 322
678	29, 333
680	29, 334
682	30, 402
683	30, 403
684	29, 336
697	29, 339
702	30, 414
703	29, 344; 30, 415
711	29, 347
713	30, 392
717	29, 95, 229
721	30, 496
725	29, 418
726	30, 424
729	29, 239; 30, 424
732	29, 354
733	28, 45
752	29, 361
755	29, 362
757	29, 362
762	29, 347, 363
769	30, 440
775	30, 442
780	30, 445
795	29, 378; 30, 450
798a	29, 378
799	29, 380; 30, 452
801	29, 380, 381
802	29, 381
806	30, 456
811	30, 458
812	30, 460
819	30, 464
821/2	29, 394
824	30, 466
832	29, 400, 401
835	28, 168
836	29, 401
837	29, 337, 401; 30, 471
838	29, 401
843	30, 473
845	29, 404 (2)
850	29, 406
853	30, 477
854	29, 409 (2)

No.	Suppl. Ref. Vol., Page
Volume 28 continued	
856/7	29, 410
859	30, 480
860	29, 413
862	29, 413
864	29, 413
866	29, 415
871/2	29, 416
884	30, 491
885	29, 423
890	29, 425
892	29, 426; 30, 495
895	28, 455; 29, 427
905	28, 7
935	29, 444
936	30, 514
940	29, 445
943	29, 447
944	29, 448; 30, 518
947	30, 519
958	29, 454; 30, 525
960	30, 526
962	29, 456
964	29, 456; 30, 529
973	29, 459; 30, 531
983	29, 461
985	29, 461; 30, 533
989	28, 25; 29, 30
991	30, 537
992	29, 465

Volume 29	
19	30, 17
48	30, 36
54	30, 38
64	30, 58
65	30, 44
67	30, 45
77	30, 49
84	30, 55
88	30, 57
107	30, 70
112	30, 72
132	30, 84
140	30, 88
172	30, 106
188	30, 114
191	30, 115
194	30, 117
206	30, 130
250	30, 154
262	30, 161, 360
280	30, 171
287	30, 538
295	30, 354
313	30, 195
317	30, 289
322	30, 198
324	30, 15
331	30, 203
334	30, 205
359	30, 222
362	30, 223
371	30, 229
394	29, 55
402	30, 241
419	30, 251
428	30, 278
433	30, 257
456	30, 270
469	30, 419
476	30, 280
490	30, 288
512	30, 301
513	29, 223
516	30, 303
542	30, 321
547	30, 111
552	30, 396
585	30, 341
600	30, 347
617	30, 359
621	30, 360
622	30, 361
630	30, 366
651	30, 377
675	30, 379
677	30, 390
686	30, 399
709	30, 412
719	30, 418
723	30, 420
740	30, 424
742	30, 426
754	30, 434
761	30, 438
791	30, 449
796	30, 452
799	30, 452
806/7	30, 455
819	30, 459
829	29, 451; 30, 4
841	30, 468
846	30, 480 (2)
848	30, 470
854	30, 473
855	30, 474
882	30, 483
894	30, 487
907	30, 492
912	30, 496, 529
913	30, 497
935	30, 506
956	30, 517
973/4	30, 526
975	30, 528
992	30, 536
993	30, 538
995	30, 540

Volume 30	
43	30, 450
456	30, 119
549	30, 456
662	30, 504

International Journal of Methods in Synthetic Organic Chemistry

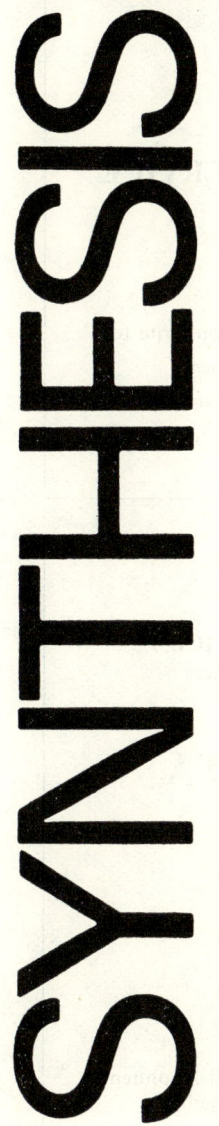

Editors:
G. SCHILL, Freiburg/Br.
G. SOSNOVSKY, Milwaukee/Wis.
H. J. ZIEGLER, Basel

Editorial Office:
R. E. Dunmur and W. Lürken,
D-7000 Stuttgart 1 · P.O. Box 732

SYNTHESIS is published monthly; yearly subscription price: DM 198,– (US $ 75.00) delivery costs not included. Subscription will be considered renewed for one subsequent year unless terminated prior to December 1st.

"Please ask for a sample copy"

Georg Thieme Publishers Stuttgart

Synthetic Methods
of Organic Chemistry

Semimonthly
EXPRESS ABSTRACT SERVICE
for Research Laboratories

For information write to:
Wm. Theilheimer
318 Hillside Avenue
NUTLEY, N.J., USA

SCHWEIZERISCHE
CHEMISCHE GESELLSCHAFT
Verlag Helvetica Chimica Acta
Postfach, CH-4002 Basel

HELVETICA CHIMICA ACTA

Abonnement   Band **59**, 1976 (SFr. 435.–)

Es sind noch auf Lager   **Neudruck-Ausgabe**
Bd. 1–27 (1918–1944)
Bd. 28 vergriffen

Original-Ausgabe
Bd 29–58 (1946–1975)

Wird nur an Mitglieder und Abonnenten
der Original-Ausgabe abgegeben

Bitte verlangen Sie unsere Preisliste